Green Energy and Technology

More information about this series at http://www.springer.com/series/8059

Zhengcheng Zhang · Sheng Shui Zhang
Editors

Rechargeable Batteries

Materials, Technologies and New Trends

Springer

Editors
Zhengcheng Zhang
Chemical Sciences and Engineering Division
Argonne National Laboratory
Lemont, IL
USA

Sheng Shui Zhang
Sensors and Electron Devices Directorate
U.S. Army Research Laboratory
Adelphi, MD
USA

ISSN 1865-3529 ISSN 1865-3537 (electronic)
Green Energy and Technology
ISBN 978-3-319-15457-2 ISBN 978-3-319-15458-9 (eBook)
DOI 10.1007/978-3-319-15458-9

Library of Congress Control Number: 2015939668

Springer Cham Heidelberg New York Dordrecht London

Printed on acid-free paper

Springer International Publishing AG Switzerland is part of Springer Science+Business Media
(www.springer.com)

Preface

Rechargeable batteries are devices that reversibly convert electrical energy into chemical energy and store resulting chemical energy in the unit. In the past two decades, lithium-ion (Li-ion) batteries have been most rapidly developed and widely used in numbers of mobile consumer electronics, such as cellular phones, cameras, laptops, tablets, and power tools, due to their high energy density, high power capacity, and robust performance. Further developments of the Li-ion batteries are aimed at applications in the electric transportations and stationary grids for the effective harvest of renewable solar and wind power, which raise grand challenges in the performance and cost of the batteries. The performances are typically qualified by the energy density, powder capability, cycle life and safety, and the cost by the price of battery materials and engineering. In order to meet the energy and power requirements of these applications, many single cells are electrically connected into modules that are subsequently integrated into a battery pack. Therefore, the challenges are remained not only for the materials and process of single cells but also for the designs and engineering of battery packs. The Li-ion batteries store electrochemical energy through the reversible intercalation and deintercalation of Li^+ ions with a lithiated transitional metal oxide as the cathode material and natural or synthetic graphite as the anode material. The capacities of these Li^+-ion intercalation materials are limited by their crystallographic structure and the present technology has approached to the theoretical values. While acceptable for the applications in mobile consumer electronics, the relatively high cost and limited earth abundance of the transitional metals used in the Li^+-ion cathode materials become grand challenges for the transportation and stationary applications. In order to overcome the challenges of the performance and cost, new materials and concepts are necessary for the development and commercialization of the next generation rechargeable batteries.

The performance enhancement is realized generally by improving the battery materials and design, including the cathode, anode, and electrolyte/separator, and the cost reduction by selecting the low-cost raw materials and process. In response to the increased demands for the transportation and stationary applications, this book is designed to update the latest advancements in the research and development of rechargeable batteries with focus on the materials and technologies for the synthesis and characterization of battery materials, as well as the diagnosis and analysis of single cells and battery packs. According to the battery reactions, this book is composed of Chaps. 1–17 covering the new materials and technologies in lithium-ion battery and Chaps. 18–24 covering the new developments and new trend in the battery systems beyond lithium-ion including lithium-sulfur battery, metal-air battery, magnesium battery, sodium-ion battery and redox flow battery.

This book is contributed by a group of leading scientists, engineers, and professors, who are directly working on the subjected areas. We believe that this book is extremely useful for the researchers who work in the conversion and storage of electrochemical energy, and also serves an excellent textbook or reference for the college/university undergraduates and graduates who are interested in the areas of materials, energy, and electrochemistry in relation to the electrochemical energy storage.

We gratefully acknowledge all chapter authors for their enthusiastic and collaborative contributions. We also wish to thank Ms. Garrett Ziolek, the editorial assistant of Springer, for her guidance and support in preparing this book. Finally, we give our deepest appreciation to our families for their continuous support while we were working on this book.

Zhengcheng Zhang
Sheng Shui Zhang

Contents

Challenges of Key Materials for Rechargeable Batteries

Zhengcheng Zhang and Sheng Shui Zhang

1 Introduction

Energy is the lifeblood of modern society. The concerns over the global warming, finite fossil-fuel supplies and environmental pollution have led to increasing interest in renewable energies such as solar and wind. However, these renewable energy sources are uncontrollable and intermittent, which require reliable electrical energy storage (EES) system for stable and consistent power delivery. On the other hand, the growing number of transportation vehicles has made it urgent to electrify transportation vehicles for reduction of the CO_2 emission and of the dependence on the fossil fuels. Among many EES systems, rechargeable batteries are the most promising to meet these needs because of their high energy density and high energy efficiency [1]. Of particular interest is the lithium-ion battery (LIB) that was commercially launched in 1991 by Sony and since tremendous progress has been made in increasing the energy and power densities and in reducing the material process cost. Now, the LIB has revolutionized the market of portable electronics and power tools, and are penetrating into the market of transportations, such as hybrid electric vehicles (HEVs) and plug-in hybrid electric vehicles (PHEVs), and are being intensively studied for load-leveling stationary EES applications. The success of Li-ion technology in the latter applications largely depends on the component materials used in the batteries.

Z. Zhang
Chemical Sciences and Engineering Division, Argonne National Laboratory,
Lemont, IL 60439, USA
e-mail: zzhang@anl.gov

S.S. Zhang (✉)
Electrochemistry Branch, RDRL-SED-C, Sensors and Electron Devices Directorate,
U.S. Army Research Laboratory, Adelphi, MD 20783, USA
e-mail: shengshui.zhang.civ@mail.mil; shengshui@gmail.com

Z. Zhang and S.S. Zhang (eds.), *Rechargeable Batteries*,
Green Energy and Technology, DOI 10.1007/978-3-319-15458-9_1

In spite of the tremendous success in the portable electronic market, the LIBs still face great challenges for the large scale applications in the transportation vehicles and stationary energy storages. These challenges, varying with the applications, are concentrated on the aspects of energy/power density, cycling life, calendar life span, safety, and overall cost to compete with the internal combustion engine for transportation vehicles and with the fossil fuel energy for stationary EES. This chapter outlines the status and challenges of key materials for the rechargeable batteries, and discusses the strategies to overcome them.

2 Basic Principle of Rechargeable Battery

Cell is the core element of a battery. For many portable electronics, a single cell can satisfy the energy and power requirements. For large scale applications in electric vehicles and stationary EES, however, many cells are required to be electrically integrated into modules that are further packed into a battery (pack) for meeting the energy and power needs. The cell is composed of a positive electrode (cathode), a negative electrode (anode), and an electrolyte, in which the cathode and anode are physically isolated by the electrolyte. When a liquid electrolyte is used, a porous membrane must be placed as the separator between the cathode and anode to avoid their electrical contact. The separator allows the liquid electrolyte to penetrate and mechanically isolates the cathode and anode from each other. The electrolyte is in principle ionically conducting and electronically insulating, which enables the redox reaction on each of electrodes.

Rechargeability of a cell is based on a reversible redox reaction between the cathode and anode materials. The open circuit voltage (U_{ocv}) of a cell equals to the difference of chemical potentials between the cathode (E_+) and anode (E_-) as described by Eq. 1.

$$U_{ocv} = E_+ - E_- \tag{1}$$

where E_+ and E_- vary with the state-of-charge (SOC), more precisely the chemical composition of the electrode materials, as illustrated in Fig. 1a.

Output voltage (U) of the cell depends on the IR polarization, which is described by Eq. 2.

$$U = U_{ocv} - IR \tag{2}$$

where I is the current passing through the cell, and R is the overall internal resistance of the cell, which is affected by many factors, including electrode and electrolyte materials, electrode and cell structure, temperature, and cell's SOC or degree-of-discharge (DOD), the other description form of SOC. In ac impedance spectroscopy (Fig. 1b), the overall resistance of a cell consists of the ohmic bulk resistance (R_b) contributed by the electronic resistance of cell's hardware and the ionic resistance of

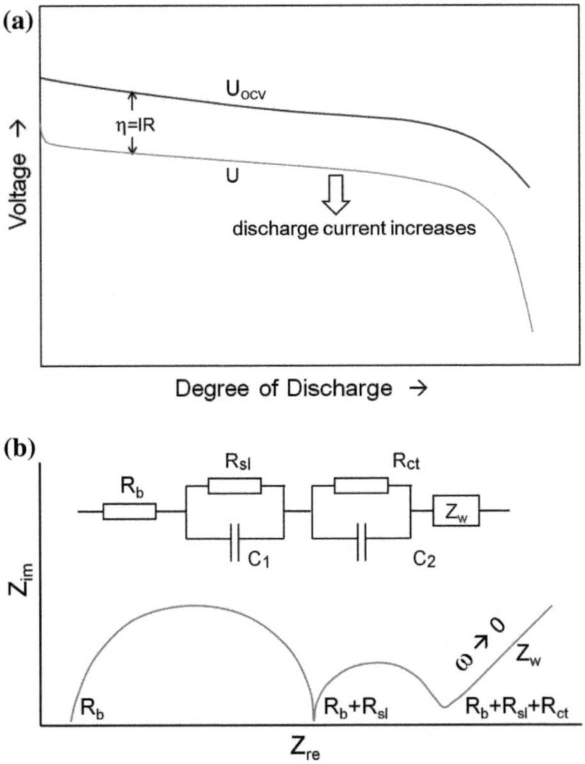

Fig. 1 Typical characteristics of an electrochemical cell. **a** Output voltage versus open-circuit voltage, and **b** ac impedance spectrum

the electrolyte, ohmic surface layer resistance (R_{sl}) on two electrodes, Faradic charge-transfer resistance (R_{ct}) relating to the electrochemical reactions occurring in two electrodes and changing with the SOC, and Warburg impedance (Z_w) corresponding to the ionic diffusion on the electrolyte-electrode interface. In general, the R_{ct} remarkably increases near the end of charge or discharge, which hence results in a steep increase in the cell's polarization. This feature has been used to determine the ending point of charging in many battery chargers.

The energy density, which is generally described by watt-hour per unit weigh or volume, is the product of specific capacity and averaged output voltage, and it is mainly affected by the chemical nature of the cathode and anode materials. However, the power density, which is generally described by watt per unit weigh or volume, is also affected by the design and engineering of the cell in addition to the nature of the electrode materials. For a battery pack, these parameters are further affected by other components such as battery packing materials and battery electrical and thermal managements. The electrolyte affects the performances of a rechargeable cell, including reversibility, coulombic efficiency, rate capability, safety, operating temperature range, and cycle life. With emphasis on the single cell, the status and challenges of key materials for the rechargeable batteries are outlined in the next sections, and accordingly the term "battery" hereafter is referred as to a single cell.

3 Li-Ion Battery

LIBs use the Li^+ ion intercalation materials for both the cathode and anode, between which the Li^+ ions are shuttled across the electrolyte absorbed in the separator. Figure 2 depicts the potential and specific capacity of typical cathode and anode materials suitable for the LIBs [2]. In order to assemble the battery in discharged state, the cathode materials are lithiated transition metal oxides and the anode materials are carbons or compounds capable of intercalating Li^+ ions or alloying with metallic Li. Research focuses have been on the increase of the battery energy/ power density, reduction of material cost, and the improvement of battery safety, which are outlined below.

3.1 *Cathode Material*

3.1.1 Typical Cathode Material

Layered $LiMO_2$ (M=Co, Ni, Mn, and their mixture), spinel LiM_2O_4 (M=Mn, and mixture with Co or Ni), and olivine $LiMPO_4$ (M=Fe, Mn, Ni, Co, and their mixture) are three types of major cathode materials. The layered $LiCoO_2$ was first commercialized in the LIB and dominated the batteries for portable electronics. The $LiCoO_2$, having a two-dimensional Li^+ ion conduction pathway, features good cyclability and moderately high rate capability. In practical cells, $LiCoO_2$ only delivers ~50 % theoretical capacity at ~3.7 V versus Li/Li^+, corresponding to a

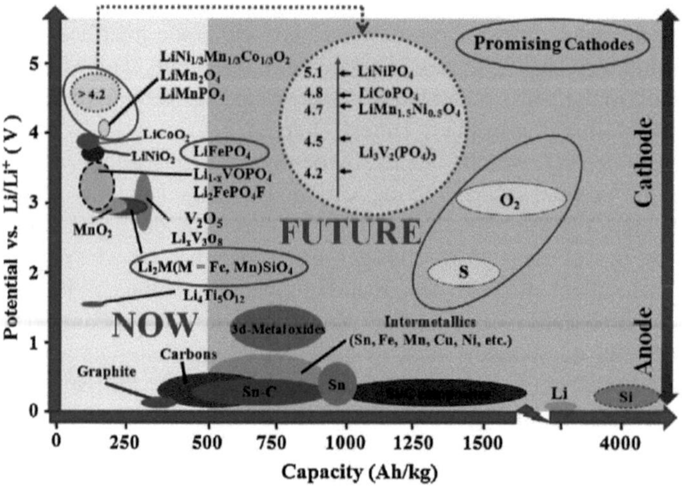

Fig. 2 Potential and specific capacity of selected cathode and anode materials for Li-ion battery. Reproduced with permission of Ref. [2]

140 mAh/g of specific capacity. The limitation is due to the chemical instability at $x < 0.5$ in Li_xCoO_2, arising from a significant overlap of the redox active $Co^{3+/4+}$: t2 g band with the top of the O^{2-}:2p band, which results in oxygen evolution and structure degradation [3]. The generated oxygen not only decomposes electrolyte solvents, but also raises safety hazard. In comparison with $LiCoO_2$, the $LiNiO_2$ and $LiMnO_2$ are more stable and therefore offer higher capacities. However, they both suffer from fast capacity fading due to the irreversible structural transitions, for example the migration of Ni^{3+} ions from the Ni planes to the Li planes for $LiNiO_2$ and the layered-to-spinel transition for $LiMnO_2$ [4]. Partial substitution of Co for Ni stabilizes the layered structure and suppresses cation disorder, which enables $LiNiO_2$ a reversible capacity close to 180 mAh/g. Doping of electrochemically inactive Al, Mg, Ca or Ba suppresses the phase transformation and hence increases cycling stability. Moreover, the doping enhances the thermal stability of $LiNi_{1-x}Co_xO_2$. In general, the substitution and doping improve the structural stability at the cost of small operating voltage and cycling capacity. Major drawbacks of the layered $LiCoO_2$ are high cost, toxicity, and relatively poor rate capability. Therefore, the derivatives of $LiCoO_2$, in which the Co cations are partially substituted by more abundant and environmentally friendly Ni and/or Mn such as $LiNi_xCo_{1\text{-}x\text{-}y}O_2Al_yO_2$ (NCA) and $LiNi_{1/3}Mn_{1/3}Co_{1/3}O_2$ (NMC) have become preferential choice for the portable electronics batteries [5, 6].

Spinel $LiMn_2O_4$, having a three-dimensional Li^+ ion conduction pathway, features high power capability and excellent safety. The capacity of $LiMn_2O_4$ is limited to $\sim$120 mAh/g at $\sim$3.8 V versus Li/Li$^+$, corresponding to a reversible intercalation/deintercalation of $\sim$0.8 Li per $LiMn_2O_4$ formula unit. Further intercalation of Li^+ ions causes a structural transition from cubic to tetragonal symmetry (called Jahn–Teller distortion), resulting in a huge volume change and severe capacity fading [7]. Therefore, over-discharging must be strictly avoided so as to retain the stable cyclability of the $LiMn_2O_4$. Major problems with $LiMn_2O_4$ are the dissolution of Mn^{2+} ion and structural Jahn-Teller distortion at below 3.5 V versus Li/Li$^+$. The dissolution of Mn^{2+} ion is due to the disproportionation of Mn^{3+} ion in the presence of trace amounts of H^+ ions, which not only leaches the active Mn^{2+} ions out of the cathode lattice but also dramatically increases the impedance of graphite anode because the dissolved Mn^{2+} ions migrate to the anode and electrochemically reduce into metal or chemically form much more resistive solid electrolyte interphase (SEI) on the graphite surface. The dissolution of Mn^{2+} ion becomes much severe at elevated temperature (>50 °C), resulting in significant capacity fading. The cationic or anionic substitution is shown to reduce the dissolution of Mn^{2+} ion and suppress dynamic Jahn-Teller distortion, resulting in significantly improved performance. Due to the low cost and environmental friendliness of Mn as well as the high rate capability, spinel $LiMn_2O_4$ is appealing for the transportation and smart grid applications.

Olivine $LiFePO_4$, having a one-dimensional Li^+ ion conduction pathway, features very good cyclability and excellent safety, which are attributed to the facts that the operating potentials of $LiFePO_4$ are far lower than the oxidation potential of electrolyte solvents and the $LiFePO_4$ does not generate oxygen under abuse conditions such as electric shortening and overcharging. Typically, $LiFePO_4$ can achieve a ~170 mAh/g of specific capacity at 3.4 V versus Li/Li^+, corresponding to a reversible intercalation/deintercalation of one Li per $LiFePO_4$ formula unit. The major drawback of $LiFePO_4$ is the intrinsically low electronic and Li^+ ionic conductivities, arising from the one-dimensional Li^+ ion diffusion. Also, the olivine structure is less dense than the layered and spinel structures, resulting in a lower volumetric energy density. Efforts to overcome these drawbacks include the coating of conductive carbon on the $LiFePO_4$ surface and the reduction of $LiFePO_4$ particle size. However, this significantly increases the processing cost of the $LiFePO_4$ material. In spite of these less favorable properties, the low cost, environmental friendliness, and excellent safety make $LiFePO_4$ a viable cathode material for the large size batteries to be used in the electric vehicles and stationary EES.

Two processes that have been frequently employed to improve the structural stability and cyclability of the cathode materials are: (1) substituting or doping of cation or anion, and (2) coating of cathode particles with a stable component or a conductive material. The substitution by the electrochemically active transition metal cation such as Fe, Co, Ni and Mn generally improves both the electronic structure and electrochemical activity of the cathode materials, whereas that by the electrochemically inactive metal cation such as Li, Al, Mg, etc. stabilizes the lattice of the cathode materials. Coating by the electrochemically inactive compounds such as Al_2O_3, MgO, $AlPO_4$, AlF_3 etc. not only enhances the cycling stability of cathode material but also reduces the oxidative decomposition of electrolyte solvents on the cathode surface. The latter considerably suppresses the growth of resistive surface layer on the cathode. Beside the electrochemically inactive compounds, more stable cathode material was also proposed to coat the less stable cathode material. In this strategy, an interesting example is the concentration-gradient cathode material based on a layered NMC cathode, in which the cathode particle has a Ni-rich central core and a Mn-rich outer shell with the Ni concentration decreasing while the Mn and Co concentrations increasing along the radius of spherical particles, as shown in Fig. 3. Thus, the more stable Mn- and Co-rich shell protects the less stable Ni-rich core from contact with the electrolyte. As a result, the Li battery with it as the cathode achieved a 209 mAh/g of initial capacity and remained 96 % retention after 50 cycles at 55 °C between 3.0 and 4.4 V [8].

3.1.2 High Energy Cathode Material

Since energy density of a battery is the product of specific capacity and averaged output voltage, intensive effort has been devoted to the development of the high specific capacity and high voltage cathodes. Among high capacity cathodes, Li-rich layered $Li[Li_{1/3}Mn_{2/3}]O_2$-$LiMO_2$ (M=Ni, Co, Mn) solid solutions, or written as Li_2MnO_3-$LiMO_2$ in the other form of descriptions, are of particular interest. In the solid solution, the electrochemically inactive Li_2MnO_3 phase stabilizes the layered $LiMO_2$ to enable a wide range of operating voltage. These materials typically have a ~250 mAh/g of specific capacity between 2 and 4.8 V versus Li/Li^+ [9].

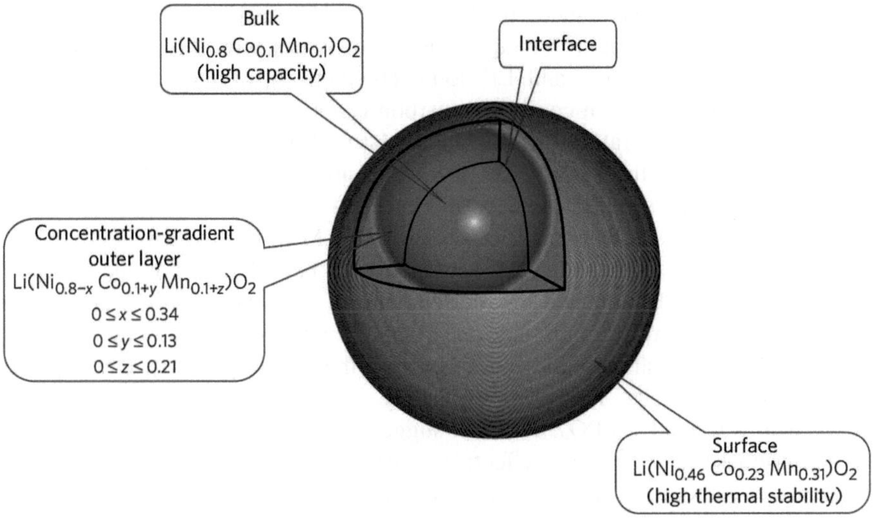

Fig. 3 Schematic diagram of a core-shell structured cathode particle with Ni-rich core surrounded by concentration-gradient outer layer. Reproduced with permission of Ref. [8]

The major issues of these materials are (1) high irreversible capacity (40–100 mAh/g) in the first cycle, (2) increase of voltage hysteresis (delay) with cycling, and (3) poor rate capability. The first issue, which mainly occurs at above 4.5 V in the initial activation step, is due to the irreversible evolution of oxygen and resulting Li^+ ion loss and electrolyte solvent oxidation [10]. The second issue is due to the irreversible structural transition from layered to spinel lattice [11]. The third issue is attributed to the nature of moderately high electronic conductivity of layered $LiMO_2$ and the presence of inactive Li_2MnO_3 phase. Multivalent Li^+ ion compounds, such as polyanionic Li_2MSiO_4 (M=Mn, Fe, Co, Ni) with two Li^+ ions per formula unit corresponding to a 330 mAh/g of theoretical capacity for M=Fe [12] and monoclinic $Li_3V_2(PO_4)_3$ with three Li^+ ions per formula unit equaling to a 197 mAh/g of theoretical capacity [13], have also been investigated as the high capacity cathode material. However, structural instability and capacity's utilization remain challenges for these multiple Li^+ ions materials.

With respect to the high voltage cathode material, spinel $LiNi_{0.5}Mn_{1.5}O_4$ and olivine $LiCoPO_4$ have been intensively studied. Based on the $Ni^{2+/3+}$ and $Ni^{3+/4+}$ couples, $LiNi_{0.5}Mn_{1.5}O_4$ offers a ~130 mAh/g of accessible capacity at ~4.7 V versus Li/Li^+. The high operating voltage and three-dimensional fast Li^+ ion diffusion make $LiNi_{0.5}Mn_{1.5}O_4$ attractive for the large size batteries to be used in the transportation vehicles and stationary EES. The major issues with $LiNi_{0.5}Mn_{1.5}O_4$ are the formation of a $Li_xNi_{1-x}O$ impurity phase and chemical instability in contact with the electrolyte at high operating voltages. Small amount of Fe-substitution stabilizes the structure with cation-disorder in the 16d octahedral sites of the spinel lattice, leading to remarkable improvement in both the capacity retention and rate capability. $LiCoPO_4$ has a theoretical capacity of 167 mAh/g at a ~4.8 V versus Li/Li^+ of averaged operating voltage. The major issues with this material are the intrinsically low electronic and Li^+ ionic conductivity of the olivine structure. Unlike $LiFePO_4$, coating of conductive carbon is not suitable for $LiCoPO_4$ because at the synthesis temperature, the highly oxidative $LiCoPO_4$ will directly oxidize carbon. Similarly to the $LiNi_{0.5}Mn_{1.5}O_4$, small amount of Fe-substitution significantly improves the specific capacity and cycle life [14].

Interestingly, it was observed from both $LiNi_{0.5}Mn_{1.5}O_4$ and $LiCoPO_4$ that the substitution process induced the self-segregation of Fe-rich phase on the surface of cathode particles, leading to the enrichment of the more stable Fe cations. Therefore, the Fe-substitution not only stabilizes the lattice structure but also increases the chemical stability in contact with the electrolyte [15], which greatly enhances the capacity retention, especially at elevated temperatures, of the $LiNi_{0.5}Mn_{1.5}O_4$ and $LiCoPO_4$. This fact suggests that the self-segregation of robust Fe cation phase may be a cost-effective strategy to overcome the chemical instability of the cathode surface in contact with the electrolyte.

Instability of the electrolyte solvents is the major obstacle for the application of high-voltage cathode materials, which is mainly due to the strong oxidation of the high voltage cathode materials and the high catalytic activity of the cathode particles to the electrochemical oxidation of electrolyte solvents. Many problems are associated with the oxidative decomposition of electrolyte solvents on the surface of high voltage cathode particles, such as active material loss (dissolution), low coulombic efficiency, and uncontrollable growth of resistive surface layer. The most studied solution to these problems is the surface coating with more robust compounds like $AlPO_4$, ZnO, Al_2O_3, and Bi_2O_3 [16]. It has been proven that the coating protects the active cathode material surface from direct contact with the electrolyte and thereby suppresses the formation of thick SEI layer. Development of robust cathode surfaces seems to be an attractive strategy to overcome the instability of the high-voltage cathode materials in contact with the liquid electrolyte and thereby improve the energy density, rate capability, and cycle life.

3.1.3 Strategies Toward High Power

Power describes the capability how fast the battery can deliver energy. High power is in particular important for the application of transportation vehicles, such as HEVs and EVs, in which fast discharge and charge are frequently required during the acceleration and braking-up of the vehicles. The power capability is promoted by both the battery design and material optimization. Reducing electrode thickness and increasing the content of conductive carbon in the electrode are two common practices for the improved power by the battery design, however, such improvements are achieved at the cost of energy density due to the introduction of more inactive materials. In the aspect of cathode materials, the high power is achieved by the reduction of particle size and the surface coating of conductive carbon. Nanostructured materials are particularly appealing in the reduction of the particle size. Their advantages include: (1) reducing diffusion length of the Li^+ ion diffusion within the particle for high reaction rate, (2) increasing electrode/electrolyte contact area for more effective and faster electrode reaction, and (3) accommodating volume change caused by the intercalation and deintercalation of Li^+ ions for long cycle life. However, some disadvantages are also accompanied, including (1) parasitic reactions with electrolyte solvents, leading to the formation of thicker (and often more resistive) SEI layer, (2) reduced packing density, leading to lower volumetric energy density, and (3) significantly increased process cost.

Coating of conductive carbon has been widely adopted to improve the rate capability of the low electronically and ionically conductive $LiFePO_4$. However, this technique is only limited to the low voltage cathode materials. For high voltage cathode materials, carbon will be chemically oxidized by the cathode materials in the coating process. Moreover, carbon may be electrochemically oxidized at high potentials that are within the charging potential range of the high voltage cathode materials.

3.2 Anode Material

3.2.1 Carbonaceous Material

Carbon is the standard anode material for the commercial LIBs used in portable electronics. Well-ordered graphite has a 372 mAh/g of theoretical capacity at an operating voltage of ~0.25 V versus Li/Li^+ in accordance with the formation of a LiC_6 graphite intercalation compound (GIC), and ~360 mAh/g can be readily accessed in practical batteries. Disordered hard carbon has higher specific capacity and better rate capability, but suffers larger irreversibility, higher voltage hysteresis, lower packing density, and poorer electrical conductivity, as compared with the ordered graphite [17]. Since the intercalation of Li^+ ions into graphite occurs at lower potentials than the organic solvents electrochemically reduce, organic electrolytes are thermodynamically instable with the graphite anode. Therefore, the electrolyte solvents must be first reduced before the Li^+ ions can be intercalated into graphite. The operation of graphite anode can be enabled only when the reduction products of the electrolyte solvents are stably accumulated on the graphite surface to form a dense and protective film. This film is called SEI, and the formation of the SEI has proven to be an essential step for the manufacture in affecting the cycling performance, cycle life and safety of the LIBs.

The low potential of the GICs is favorable for increasing the battery's operating voltage, however, it meanwhile causes other problems. The potential for the Li^+ ion intercalation is so low (<0.25 V vs. Li/Li^+) that metallic Li can be easily platted on the graphite surface during charging of LIBs, especially at high current rate and/or at cold temperature. The plated Li not only reacts with the electrolyte solvents to degrade the battery's performance, but also forms dendrites, potentially creating a safety hazard. Other drawbacks of the carbonaceous anode materials are the

10–20 % irreversible capacity in the first cycle due to the formation of SEI, low density affecting the battery's volumetric energy density, and unsuitability for fast charging due to the possibility of Li plating.

3.2.2 $Li_4Ti_5O_{12}$ Anode

Spinel $Li_4Ti_5O_{12}$ (LTO) has a 175 mAh/g of theoretical capacity at ~1.5 V versus Li/Li^+ operating voltage, it features very stable cycling stability and excellent safety. Since LTO operates at far above the reduction potentials of carbonate solvents, no SEI is formed in the first cycle. The three-dimensional structure offers LTO excellent reversibility due to the near zero volume strain during the Li^+ ion intercalation and deintercalation cycling. The main drawbacks of the LTO are the destructive gas generation, the relatively low electronic conductivity, and significantly reduced energy density by the high operating potential of the LTO anode. It has been reported that gas generation always occurs during both cycling and storing [18]. The generated gases contain H_2, CO_2 and CO, which are identified to be the products of interfacial reactions between the LTO and carbonate solvents. These reactions result in structural transformation from (111) to (222) plane and formation of (101) plane of anatase TiO_2. Coating of conductive carbon on the LTO surface is shown to be very effective in suppressing the interfacial reactions and resulting gas generation in addition to increasing the electronic conductivity.

3.2.3 Lithium Alloy Based Material

Li alloys with elements such as Sn, Si, Ge etc. are appealing because of their superior capacity and suitable operating potentials [19]. Among many Li alloy forming elements, Si is of particular interest because it offers a specific capacity of as high as 4200 mAh/g at a fairly flat potential plateau of 0.4 V versus Li/Li^+ in accordance with the formation of Si_5Li_{22} alloy. Typically, Si is lithiated at potentials from 0.4 to 0 V versus Li/Li^+, which are lower than the reduction potential of electrolyte solvents. Therefore, the same SEI formation as occurred in the carbonaceous anode materials is necessary for enabling the operation of Si anode.

The major challenge with all Li alloy materials is the poor cyclability due to the dramatic volume change occurring during the alloying and dealloying processes, which leads to the material pulverization, inter-particle contact loss, and severe capacity fading. The basic strategies to overcome the large volume change are to reduce the material's particle size, and/or further embed the active material particles into a conductive (such as carbon and metal) or a non-conductive (such as stable metal oxide) buffer phase. Successful examples are one-dimensional Si nano-wires, Si-C, Si-SiO_2-C nanocomposites, as well as intermetallic alloys with the other metal as the buffer of volume change such as Cu_6Sn_5, InSb and Cu_2Sb. However, these processes greatly increase the material's cost while decreasing the material's specific capacity due to the incorporation of significant amount of inactive materials.

Interestingly, the cellulose based binders significantly enhance the cycling stability of Si anode because of the chemical interactions between the –OH groups in the binder and the Si-O surface groups in the Si particles [20]. This may hint a direction for the advancement of the Li alloy based anode materials. In addition, hybridization of the Li alloy based material with carbon has become the most practiced method to achieve the high capacity and acceptable cycle life. For example, Sony's Nexelion^{TM} battery had a 30 % higher volumetric specific capacity by using a Sn-Co-C amorphous anode as compared with the carbon analogue [21], and Panasonic's 18,650 battery achieved a 4 Ah capacity by using a C-Si composite anode as compared with 3.4 Ah for the carbon analogue [22].

3.2.4 Conversion-Type Anode Material

Convention-type anode materials are based on a displacement reaction between metallic Li and the nanostructured transition metal oxides, sulfides, nitrides, phosphides, or halogates [23]. The metal in these compounds can be capable of alloying with Li or not. SnO_2 is a typical metal oxide (MO), in which the metal is able to alloy with Li. In operation, SnO_2 is first reduced and combined with Li^+ ions from the electrolyte to form nano-sized Li_2O and Sn particles, followed by the Sn particles alloying with Li. Since the Sn particles are finely dispersed in the Li_2O matrix, the Li_2O buffers the mechanical stress caused by the Li alloying and dealloying, and hence results in much improved cycling stability. The drawback of SnO_2 is the large irreversible capacity in the first alloying process and the agglomeration of tin particles during prolonged cycling. The former is due to the irreversible formation of Li_2O, which consumes the limited Li^+ ion source from the cathode. The latter is because the amount of the formed Li_2O matrix is not sufficient to fully isolate Sn-Li alloy particles from contact with each other.

The MO with the metal unable to alloy with Li, such as Mn, Fe, Co, Ni, Cu etc., operates in entirely different mechanism, which can be described by a reversible displacement reaction, for example,

$$MO + 2Li \rightleftharpoons M + Li_2O \tag{3}$$

Such a reaction generally has higher potential and more slopping potential-capacity profile as compared with the Li alloying since the displacement reaction occurs at higher potentials and undergoes larger structural change than the Li alloying reaction. Similarly to the MO, metal halogates (MX_n) also have been intensively investigated as the conversion-type anode material based on the displacement reaction between the MX_n and Li. However, the MX_n are slightly inferior to the oxide analogues in the cycling stability because the MX_n are slightly soluble in the organic electrolytes, resulting in the loss of anode active material. The main challenges of the conversion-type anode materials are the slow reaction kinetics (causing high polarization) and capacity loss with cycling, which are associated with the ineffective solid-solid reactions between M and Li_2O for MO and between M and MX for MX_n in the charging process of a Li half-cell (corresponding to the discharging process of a Li-ion full cell).

3.3 Electrolyte

Electrolyte is an indispensable element of the battery, which enables the electrode reactions and affects the battery's performances. The standard electrolytes used in the current LIBs are a solution of $LiPF_6$ salt dissolved in a mixture of ethylene carbonate (EC) and other linear carbonates with little variation in the salt concentration and solvent ratio. The $LiPF_6$ salt is selected because (1) it offers high ionic conductivity ($\sim$ 10 mS/cm at room temperature), (2) it participates in the formation of stable SEI on the graphite surface, (3) it passivates Al current collector at high potentials, and (4) it is stable over a wide electrochemical window. The EC is the indispensable co-solvent because (1) its reduction products form dense and stable SEI on the graphite surface, and (2) its high polarity favors salt dissociation for high ionic conductivity. The function of the other linear carbonates is to reduce the viscosity and lower the solution's freezing temperature for the high ionic conductivity and wide operating temperature range. The main drawbacks of these electrolytes are that (1) $LiPF_6$ is thermally instable at elevated temperature, especially in the presence of organic solvents, (2) EC phase-separates (crystallizes) at low temperatures, depending on the composition of solvent mixture, (3) the solution is highly flammable, and (4) carbonate solvents are subject to reduction at <1.2 V and oxidation at >4.5 V. In the LIBs, the electrolyte is kinetically stabilized by the *in-situ* formed SEI on the surface of graphite anode during the first charging. The main effort on the electrolyte has been focused on the improvement of battery performance and battery safety by developing better electrolyte components and electrolyte additives.

For the electrolyte component, lithium bis(oxalato)borate (LiBOB) and lithium difluoro(oxalate)borate (LiDFOB) are shown to form favorable SEI with the graphite anode, resulting in significantly improved cycling stability. However, these two salts cannot stand above 4 V versus Li/Li^+ for long time due to the oxidation of oxalate species, and also have inferior ionic conductivity as compared with $LiPF_6$. Lithium bisfluorosulfonyl imide (LiSFI) exhibits higher ionic conductivity than $LiPF_6$ and forms comparable SEI with the graphite anode, however, it fails to passivate the Al current collector at high potentials (3.3 V vs. Li/Li^+) [24]. Therefore, these salts cannot replace $LiPF_6$ to be used as a single salt, but as an additive or a co-salt in combination with $LiPF_6$. With respect to the solvent, non-flammable solvents (mainly containing phosphorus) and room temperature ionic liquids have been developed to reduce the flammability of the electrolytes. Solfonates, aliphatic dinitriles, and fluorinated carbonates have been investigated as the solvent for high voltage electrolytes [25]. In particular, fluoroethylene carbonate (FEC) forms favorable SEI on both surfaces of the graphite anode and high voltage cathode in addition to wide electrochemical window. These unique properties make FEC appealing for the electrolytes of high voltage LIBs.

For the electrolyte additive, much attention has been placed on the formation and stability of the SEI. Beside the intrinsic instability of carbonate solvents at low potentials (<1.2 V vs. Li/Li^+) and high potentials (>4.5 V vs. Li/Li^+), the fresh surfaces of the anode and cathode materials likely catalyze the decomposition of solvents. A desirable additive should be able to deactivate the catalytic sites and form dense SEI to protect the solvents from contact with the electrode particles before the electrode reaches the solvents' decomposing potential. On the graphite anode, vinylene carbonate (VC) seems to be the most effective in reducing the irreversible capacity of the SEI formation and enhancing the stability of the resulting SEI. Most of other additives can be categorized as the derivatives of vinylene compounds [26]. On the cathode, LiBOB and LiDFOB, when used at the additive level, are shown to remarkably enhance the cycling stability of the high voltage $LiNi_{0.5}Mn_{1.5}O_4$ and Li_2CoPO_4 cathodes [27]. The similar improvement has also been achieved by fluorinated phosphates such as tris(hexafluoro-iso-propyl) phosphate [28]. For the safety protection, the compounds that are capable of reversibly oxidizing and reducing in a certain voltage range have been studied as the redox shuttle additive, and the high phosphorus-containing compounds as the fire-retarding additive. However, the improvements by these additives are very limited, and instead, the additive may introduce other adverse effects, for example, an increase in the self-discharge by the redox shuttle additive and an inferior SEI by the fire-retarding additive.

Solid polymer electrolyte and ceramic solid state electrolyte are apparently an effective solution to the safety of LIBs. However, their high process cost and relatively low ionic conductivity make them unsuitable for the large size LIBs to be used in the transportation vehicles and EES. Gel polymer electrolyte offers a compromise to bridge the liquid electrolyte and solid polymer electrolyte. However, it cannot be independently used as the electrolyte/separator due to the substantial loss of the mechanical strength at elevated temperatures. In most cases, the gel polymer electrolyte is used only for modifying the existing separator.

3.4 Separator

Separator is a porous membrane that absorbs liquid electrolyte and physically isolates the cathode and anode from electrical contact. In the operation of batteries, the separator does not participate in any electrochemical reactions, however, it strongly affects the battery's performances, especially the power capability and safety. The separators currently used in the LIBs are typically a microporous polyethylene (PE) or polyprolyene (PP) membrane, which may be in a single layer or a multilayer structure with variations in parameters of porosity, pore size, thickness, and mechanical strength [29]. The PP-PE-PP trilayer separators are of particular interest for the safety of LIBs because of their thermal shutdown feature. The shutdown function is based on the different melting temperature ranges between the PE (120–130 °C) and PP (>165 °C). When the temperature reaches the

PE's melting range, the PE layer melts and closes its pores to shut the current off whereas the PP layers still remain mechanically strong to isolate the electrodes. Since the pore closing is irreversible, the battery will be damaged permanently once the thermal shutdown is initiated. Unfortunately, such a protection is only applicable to the small batteries and under low current densities. For large size batteries and high current densities, the shutdown temperature gap (only 35 °C from the PE's top melting point to the PP's lowest melting point) is so narrow that the temperature can easily rise to or exceed the PP's melting range even if the current is shut off due to the poor heat conduction/radiation of the large size batteries.

Most battery-related accidents are related to the mechanical failure of the separator, such as meltdown and metal penetration. Major energies of an accident are originated from the chemical reactions between the charged cathode and anode materials as a result of the separator failure, which makes two electrodes contact together. Therefore, developing a separator that can retain mechanical strength at elevated temperature seems to be the most effective solution to the safety of the large size LIBs. Towards this, coating a porous ceramic composite layer, such as the ones consisting mainly of nano-sized Al_2O_3 or SiO_2 particles, onto an existing polyolefin separator or onto one or both of the electrodes may be a facile and cost-efficient approach [30].

4 Battery Systems Beyond Li Ion

The current Li-ion technology has approached the capacity limit of the Li^+ ion intercalation compounds, and the cost for raw materials and material processes is relatively high. Therefore, motivation for the high energy density and low cost batteries is driving the research towards the battery systems beyond Li ion. In this subject, sodium-ion batteries and magnesium batteries have been pursued for the low cost, and lithium-air and lithium-sulfur batteries for the high energy density.

4.1 Sodium-Ion Battery

Sodium-ion battery (SIB) is the first choice for the low cost energy storage because of the natural abundance and low cost for sodium raw materials. Due to similarity in the ionic structure and property between Na^+ and Li^+ ions, research on the materials of SIBs has mainly followed the analogues of LIBs. For the cathode materials, the most studied and best performed compounds are sodium transition metal phosphates with small variations in the transition metals, doped cations and anions, such as $NaFePO_4$, $NaFe_2Mn(PO_4)_3$, $Na_3V_2(PO_4)_3$, $Na_3V_2(PO_4)_2F_3$ etc. Other candidates are transition metal oxides $NaMO_2$ (M=Co, Mn etc.) and NaV_6O_{15}, as well as transition metal fluorides ($NaMF_3$, M=Fe, Mn, V, Ni, etc.) [31]. Since Na^+ ion has larger ionic radius than Li^+ ion (0.98 A for Na^+ vs. 0.69 A for Li^+), the intercalation and deintercaltion of Na^+ ions are much slower and suffer from significantly larger structural change, making it difficult to fully utilize the active material. Because of the low material utilization and the higher atomic weight of Na than Li, the Na^+ ion cathode materials have lower specific capacity than the Li^+ ion analogues. Moreover, the Na^+ ion cathode materials are intrinsically poor in the rate capability and cycling stability.

Regarding the anode materials, ordered graphite is unable to accumulate the large size Na^+ ion. Therefore, the research has moved to the disordered carbonaceous materials, such as hard carbon, amorphous carbon black, petroleum-coke carbon, and polymer-pyrolyzed carbon. In particular, a hard carbon obtained from pyrolyzed glucose showed a specific capacity of 300 mAh/g for the Na^+ ion interaction, and the Na^+ ions were indicated to intercalate into both the graphene layers of the graphitic particles and the pores of the hard carbon particles [32]. Beside the carbonaceous materials, $Na_2Ti_3O_7$, a LTO analogue, exhibits reversible intercalation of two Na^+ ions per unit formula, corresponding to a specific capacity of 200 mAh/g at ~0.3 V versus Na/Na^+.

For the electrolyte, the sodium salts are less soluble in the organic solvents than the lithium analogues, which limits the choice of electrolytes. A facile solution is to directly employ a Li^+ ionic electrolyte in the SIB. Fundamentally, this does not affect the performances of the SIB because the electrode materials of the SIB and LIB share the same anonic frameworks. Instead, the substitution of the Li^+ ionic electrolyte for the Na^+ ionic electrolyte results in noticeable improvement in the specific capacity and cyclability of the SIB. For example, using a $LiPF_6$ electrolyte in a graphite/$Na_3V_2(PO_4)_2F_3$ Na-ion cell led to an approximately 300 mAh/g of specific capacity for the graphite anode [33]. This is because the Li^+ ions partially substitute the Na^+ ions in the cathode and anode hosts, and the Li^+ ions are more cyclable than the Na^+ ions. Therefore, hybridization of the Na^+ and Li^+ ions has become the most practiced approach towards improved capacity and cyclability of the SIB, which can be realized either through the electrodes or though the electrolyte. The natural abundance and low cost of the sodium raw materials make the SIB appealing for the large scale stationary energy storage although great challenges remain in aspects of the specific capacity, rate capability, and cyclability.

4.2 Magnesium Battery

Rechargeable magnesium (Mg) batteries are composed of an Mg^{2+} ion interaction cathode, a metallic Mg or Mg alloy anode, and an organic electrolyte. They are appealing for the large scale stationary energy storage due to the high volumetric specific capacity (3832 mAh/cm^3) and atmospheric stability of metallic Mg, the low cost and natural abundance of Mg raw materials, and the excellent safety of Mg batteries.

The most successful cathode material is Chevrel phase $Mg_xMo_6T_8$ (T=S or Se) compounds that typically offer a ~122 mAh/g of specific capacity at an operating potential range of 1.1–1.6 V for a Mg/Mo_6S_8 battery [34]. Other cathode materials are the transition metal oxide such as V_2O_5 and MoO_3 aerogels. The $M_xMo_6T_8$ features high rate capability and good cycling stability but relatively low energy density, whereas the metal oxide aerogels exhibit relatively high voltage and capacity but slow kinetics.

The solid-state diffusion of divalent Mg^{2+} ions in the cathode host is intrinsically slow because of the strong electrostatic interaction between the Mg^{2+} ion and the anionic framework host although the Mg^{2+} ion has similar ionic radius as the Li^+ ion (0.74 A for Mg^{2+} and 0.68 A for Li^+). The slow Mg^{2+} ion diffusion kinetics is reflected by a low intercalation level (low practical capacity) and a large voltage gap between the discharging and charging processes. The strategies to overcome the slow Mg^{2+} ion diffusion include: (1) shielding the divalent Mg^{2+} ion with strong dipoles such as H_2O, (2) reducing the particle size of the cathode material, and (3) combination of the above two strategies [34]. For the first strategy, the V_2O_5 aerogels that contain H_2O in their layered structure showed a fast kinetics because of the shielding effect of the water molecules. However, this strategy cannot be applied to the Mg batteries since H_2O reacts with the Mg anode. For the second strategy, nano-sized V_2O_5 in a 0.1 M Mg(TFSI)$_2$ acetonitrile electrolyte exhibited a 150 mAh/g of specific capacity at 2.3 V versus Mg/Mg^{2+} at 0.5 mA/cm^2.

Beside the Mg^{2+} ion intercalation compounds, elemental sulfur also was explored as the conversion-type cathode. It was reported that a Mg-S battery had a 1200 mAh/g of specific capacity versus S at 0.89 V in the first discharge, but diminished to 394 mAh/g in the second discharge mainly due to the dissolution of long-chain magnesium polysulfide (MgS_n, $n \geq 4$) [35].

Unlike metallic Li that readily forms dendrites and creates safety hazard during the plating and stripping, the cycling of metallic Mg does not produce dendrites. Therefore, metallic Mg can be safely employed as the anode material. Another feature of the Mg anode is that the passivation layer natively formed on the Mg surface does not conduct Mg^{2+} ions. A surface free of the passivation layer is highly desirable for the high coulombic efficiency and good cyclability of the Mg anode. Additionally, Mg alloys with other metals such as Bi, Sn, Sb, and their alloys like $Bi_{1-x}Sb_x$ are suitable for the rechargeable Mg batteries. For the same reason as that of Li alloys in the LIBs, the Mg alloys exhibit better safety but slightly higher reduction potential. The critical requirements for the Mg alloys are the fast magnesium alloying/dealloying kinetics and the high magnesium alloying level with adequate redox potentials over the whole range of magnesiation. A successful example was that a Mg-Bi alloy provided a very stable specific capacity of 222 mAh/g for 100 cycles [36].

Electrolyte is the most difficult challenge for the Mg batteries. Common Mg salts like Mg(TFSI)$_2$, $Mg(SO_3CF_3)_2$, and $Mg(ClO_4)_2$ are almost insoluble in the conventional organic solvents such as ethers, carbonates, and acetonitrile. Therefore, much effort has been centered on the electrolyte system that has acceptable Mg salt solubility for the sufficient Mg^{2+} ionic conductivity. The electrolytes earliest reported to support the reversible cycling of metallic Mg were a series of Grignard ether solutions (RMgX, R=alkyl or aryl group and X=Cl or Br). However, the

Grignard solutions are strongly reductive, poorly conductive (due to low dissociation), and instable upon long time storage. A more recent publication reported that an electrolyte consisting of a 0.1 M $Mg(BH_4)_2$ and a 1.5 M $LiBH_4$ in diglyme showed a close to 100 % coulombic efficiency, free of dendrite formation, and very stable cyclability for Mg plating/stripping [37]. Additionally, Aurbach et al. developed a class of metallorganic based electrolytes by reacting $AlCl_{3-n}R_n$ Lewis acid with R_2Mg Lewis base in ether (THF or glymes). The best electrolyte formed by complexing one equivalent of Bu_2Mg and two equivalents of $EtAlCl_2$ exhibited an ionic conductivity range of 1–1.4 mS/cm, a 100 % Mg coulombic efficiency, and a 2.2 V electrochemical window [38]. However, such electrolytes caused the severe corrosion of stainless steel current collector due to the presence of significant amount of chloride anions, which has been known the source leading to the corrosion of many metals.

In summary, the major challenges of rechargeable Mg batteries are (1) the slow Mg^{2+} ion diffusion in the cathode host, which results in low accessible capacity and high polarization, and (2) the low solubility (and hence low ionic conductivity), and (3) poor chemical stability/compatibility of the electrolyte, which leads to poor rate capability and inferior cyclabilty. These aspects should be addressed in the future research for the practicable Mg batteries.

4.3 Li-Air Battery

Li-air batteries are attractive for the high capacity energy storage due to the highest theoretical capacity of metallic Li (11,248 Wh/kg Li) and the free O_2 from air. Since metallic Li is highly reactive to H_2O and CO_2 in air, the Li anode requires a non-aqueous electrolyte. According to the nature of the electrolytes used, the Li-air batteries can be configured as non-aqueous electrolyte system and aqueous electrolyte system, as shown in Fig. 4, which are different in the electrochemical mechanism and challenge.

The non-aqueous electrolyte Li-air batteries are ideally based on a reversible redox reaction of Eq. 4.

$$O_2 + 2Li \rightleftharpoons Li_2O_2 \quad E = 2.959\ V \tag{4}$$

In practical batteries, however, the formed Li_2O_2 may further reduce into Li_2O or disproportion into O_2 and Li_2O_2, leading to a 4-electron overall reduction, as described by Eq. 5.

$$O_2 + 4Li \rightarrow 2Li_2O \quad E = 2.913\ V \tag{5}$$

Therefore, the discharge products are often a mixture of Li_2O_2 and Li_2O. Since both Li_2O_2 and Li_2O are insoluble in the organic electrolytes, they accumulate on the surface of reaction sites in the cathode and eventually terminate the battery's

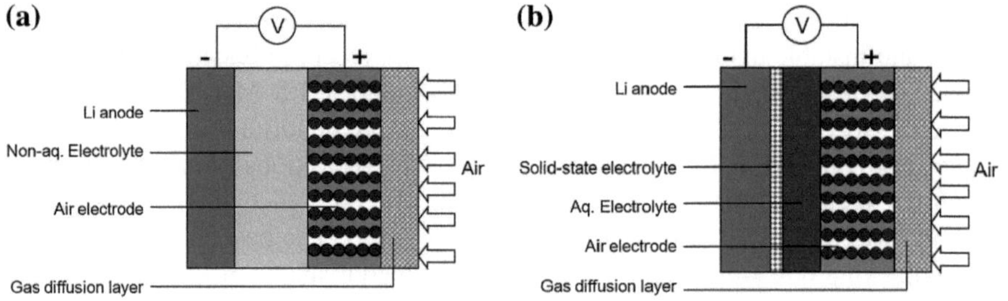

Fig. 4 Schematic structure of Li-air batteries. **a** Non-aqueous electrolyte system, and **b** aqueous electrolyte system

operation by blocking the access of O_2. As a result, the actual capacity of the non-aqueous electrolyte Li-air batteries is determined by the porosity of the air cathode.

The air cathode is generally a porous carbon sheet, which is required to be (1) highly porous for fast O_2 diffusion and high specific capacity, (2) highly catalytic activity to the reversible oxygen reduction reaction (ORR) and oxygen evolution reaction (OER), and (3) capable of blocking CO_2 and H_2O from entering into the battery. A critical challenge for the catalyst is the high selectivity to the ORR. In the ORR products, only Li_2O_2 can be reversibly charged back to O_2 while the Li_2O is produced permanently. In order to make the Li-air battery reversible, the catalyst is required to be highly selective to two-electron ORR (i.e., Eq. 4) while being inactive to four-electron ORR (Eq. 5). Unfortunately, the catalysts catalyze all the 4e-ORR, 2e-ORR, and further reduction or disproportion of the Li_2O_2. Additionally, the Li_2O_2 is a strong oxidative agent, its presence in the discharged state may raise safety hazard. The CO_2 and H_2O are generally isolated by a gas diffusion layer (GDL), which selectively allows O_2 to pass while blocking CO_2 and H_2O from access into the system and electrolyte solvents from evaporation out of the system. The selectivity of GDL is a key challenge for the air cathode. In addition, the anodic stability (namely oxidative corrosion) of the carbon may raise an issue because of the high voltage reached on charging.

Electrolyte is one of the main challenges. In addition to the common challenges for all rechargeable batteries, the Li-air batteries face two particular challenges: (1) the stability to both O_2 and its reduced intermediates/products for good reversibility and cycling stability, and (2) the high solubility and diffusion of O_2 for satisfactory rate capability. Carbonates that are widely used in the LIB electrolytes are chemically incompatible because of the nucleophilic reaction with the ORR intermediates. For example, in a Li-O_2 battery with a propylene carbonate based electrolyte, no Li_2O_2 can be formed in addition to solvent degradation products like $C_3H_6(OCO_2Li)_2$, Li_2CO_3, HCO_2Li, CH_3CO_2Li, CO_2 and H_2O [39]. Turning the electrolyte solvent to ethers or dimethyl sulfoxide (DMSO) led to significantly improved cycliability, however, the long-term stability of these solvents against lithium superoxide intermediate is still a challenge [40].

The challenge for the Li anode has been a long history as faced by all rechargeable Li batteries, which mainly are the formation of Li dendrites and the low coulombic efficiency of Li platting/stripping. In addition, potential reactions with the O_2 dissolved in the liquid electrolyte may be an issue for the Li-air batteries.

Alternatively, the Li anode can be physically protected by a Li^+ ion conductive solid state electrolyte (SSE) protective layer, which leads to an aqueous electrolyte system, as indicated by Fig. 4b. To use an aqueous electrolyte, the Li anode is protected by a thin layer of highly Li^+ ion conductive SSE. As a result, the overall reaction of the Li-air batteries is changed into Eq. 6 and the electrolyte solvent (H_2O) becomes a part of the reactants.

$$O_2 + 4Li + 2H_2O \rightleftharpoons 4LiOH \quad E = 3.446\,V \tag{6}$$

In this case, the ORR products (H_2O_2 and LiOH) can easily dissolve into the aqueous electrolyte, and accordingly the factor to determine the specific capacity of the battery turns to the dissolving capacity of the aqueous electrolyte, which is affected by the amount of the aqueous electrolyte and the solubility of LiOH in water. Meanwhile, the requirement for the 2e-ORR selectivity of the catalyst is no longer necessary because the OER occurs on H_2O, which instead a "solid-liquid-gas" (namely catalyst-electrolyte-oxygen) three-phase reaction is desirable for high power.

Chemical compatibility of the SSE with metallic Li and aqueous electrolyte becomes challenge. LISICON glasses having a general formula of $Li_{(1+x+y)}Al_xTi_{2-}xSi_yP_{(3-y)}O_{12}$ are the most used SSE in the aqueous electrolyte Li-air batteries [41], however, they react with metallic Li. Moreover, cationic exchange between the Li^+ ions in LISCON and H^+ in the aqueous electrolyte occurs unavoidably, which consequently degrades the Li^+ ionic conductivity of the SSE. Therefore, better SSE should be developed to overcome these problems.

In summary, Li-air batteries are attractive for the high capacity energy storage. However, breakthroughs must be made on the challenges stated above before they can become viable for the practical applications. In addition, it must be noted that carbon air electrode, discharge products, and the need of membranes, packaging materials and auxiliary mechanics would significantly reduce the overall energy density.

4.4 Lithium-Sulfur Battery

Lithium-sulfur (Li-S) battery has a theoretical specific capacity of 1675 mAh/g and a theoretical energy density of 2500 Wh/kg (or 2800 Wh/L) according to the complete reduction from elemental sulfur to lithium sulfide (Li_2S) by Eq. 7:

$$2Li + S \rightleftharpoons Li_2S \tag{7}$$

The high theoretical capacity, natural abundance, and low cost of elemental sulfur make Li-S batteries appealing for large scale energy storage. Key challenges of the Li-S batteries are the low coulombic efficiency, fast capacity fading, and high self-discharge rate, all of which are associated with the dissolution of long-chain lithium polysulfide (PS, Li_2S_n, $n \geq 4$) into liquid electrolyte and resulting parasitic reactions with the Li anode. In addition, the insulating nature of sulfur and its reduction intermediates/products limit the power capability of the Li-S batteries. Fundamentally, dissolution in the organic electrolytes is an intrinsic nature of the long-chain PS and is essential for the fast reaction kinetics of the insulating sulfur and PS in the batteries [42]. For the problems described above, considerable effort has been devoted to reducing or eliminating the adverse effects of the PS dissolution by focusing on two directions of (1) confining the dissolved PS within the cathode before reduced into insoluble products (Li_2S_2 or Li_2S), and (2) protecting the Li anode from reaction with the dissolved PS.

On the cathode, much effort has involved in fabrication of the sulfur-carbon composites by incorporating various types of nano-sized carbon materials from micro- through meso- and macro- to hollow-structure. This strategy is based on the concept that the porous carbon particles trap the dissolved PS within pores by physical absorption or chemical adsorption, and meanwhile offers highly conductive channels for the redox reaction of insulating sulfur species [43, 44]. Alternatively, the carbon particles are fabricated as an independent porous carbon sheet, followed by placing the porous carbon sheet as an "interlayer" between the sulfur cathode and separator [45]. All these approaches led to in some degree improvement in both the specific capacity and capacity retention. However, such improvements are achieved at the cost of specific capacity owing to the introduction of large amount of inactive carbon, and are limited by the unavoidable electromigration of the negatively charged PS anions under the condition of battery operation. The complete solutions to the PS dissolution are (1) using inorganic SSE [46], and (2) covalently bonding the short $-S_n-$chains to the carbon surface or polymeric backbones [47]. In the first strategy the Li_2S-P_2S_5 family glasses are the only SSE that is chemically compatible with the Li-S chemistry, and in the second strategy sulfurized carbon has proven to be an excellent option.

While sharing the same strategies as used in the Li-air batteries for protecting the Li anode, considerable effort has involved in the development of liquid electrolytes. In regard to the solvent, ethers are the most suitable solvent whereas carbonates are chemically incompatible with PS because the PS anions nucleophilically react with carbonate molecules [48]. It is unlikely that a single solvent can fulfill all the requirements of the electrolyte. In most cases, the electrolyte solvent is a combination of linear ether and cyclic ether. The combination generally provides a compromise among sulfur utilization, rate capability, operating temperature range, and Li cyclability. For example, linear dimethyl ether (DME) is more reactive with Li, but provides higher PS solubility and kinetics. Whereas cyclic 1,3-dioxolane (DOL) has lower PS solubility and kinetics, but provides a more stable SEI on the Li surface [49]. Combination of these two solvents leads to synergetic effect on the specific capacity and cycle life of the Li-S battery. Interestingly, it was found that fluorinated ether significantly enhances the cycling performance of Li-S batteries in terms of specific capacity, capacity retention, and self-discharge rate probably because the fluorination reduces ether's viscosity and PS's solubility, which facilitates the formation of a denser and more stable SEI on the Li surface [50].

In regard to the salt and additive, quaternary ammoniums such as tetrabutylammonium triflate and quaternary ammonium based room temperature ionic liquids are found to suppress the disproportion of the dissolved PS, and as a result increase the specific capacity and capacity retention [51]. An important finding was that $LiNO_3$ dramatically improves the chemistry and structure of the SEI on the Li surface, which consequently suppresses the redox shuttle of the dissolved PS and increases the coulombic efficiency of the Li-S batteries [52]. Therefore, $LiNO_3$ has been widely used as the co-salt or additive in the Li-S battery electrolytes. As for the Li_2S-P_2S_5 family glasses, they are only SSE that is chemically compatible with the Li-S chemistry. However, they are highly soluble in the organic liquid electrolytes used in the Li-S batteries, cannot be used as the SSE layer to protect the Li anode. Instead, they have proven to be an excellent electrolyte additive, for example, the addition of P_2S_5 into the liquid electrolyte in combination with the PS dissolved in the electrolyte facilitates the formation of a highly Li^+ ion conductive SEI on the Li surface, leading to much improved cycling stability [53].

In summary, Li–S batteries are of particular interest in the low cost and high energy density EES for the transportation and stationary applications. At present, the theoretical capacity of elemental sulfur cannot be fully utilized, and cycling performances (such as coulombic efficiency, capacity retention, cycle life etc.) are far away from the practical applications. Fundamental research and material breakthroughs are needed in future research.

5 Concluding Remarks

Rechargeable batteries are an excellent option for the electrical energy storage in a wide range of energy levels from portable electronics through transportation vehicles to load-leveling stationary storage. At present, the LIBs have satisfied to power most of portable electronics at acceptable price, and their safety may not be an issue for these small batteries. However, the cost must be dramatically reduced and the safety must be remarkably improved for widespread applications in the transportation vehicles and load-leveling stationary storage. Currently, the LIBs have approached the capacity limit of the Li^+ ion intercalation electrode materials, and their cost is high. New materials that are more energetic with lower cost must be developed. Theoretically, the battery systems beyond Li-ion would be an excellent option for these needs.

Among the potential battery systems beyond Li-ion, the Li-air and Li-S batteries best meet the requirements of high energy density, low cost, and environmental friendliness. However, grand challenges remain, especially for the Li-air batteries. In near future, the Li-S batteries may be a better choice than the Li-air batteries although their problems have lasted for many years and have not yet been completely solved. Solid state electrolyte completely avoids the dissolution of lithium polysulfide, however, it does not suit for large-scale production. Improvements made for the liquid electrolyte system, such as those by the carbon-sulfur composites and advanced electrolytes, are very limited and even in exchange of the materials' specific capacity and process cost. Solutions that completely solve the problems caused by the dissolution of lithium polysulfide in the liquid electrolyte and the poor cycling efficiency of the Li anode must be obtained by further understanding the fundamental chemistry of the Li–S batteries. In addition, the cost for the material process and battery manufacture must remain low.

References

1. Chen H, Cong TN, Yang W, Tan C, Li Y, Ding Y (2009) Progress in electrical energy storage system: a critical review. Prog Nat Sci 19:291–312
2. Xu J, Dou S, Liu H, Dai L (2013) Cathode materials for next generation lithium ion batteries. Nano Energy 2:439–442
3. Chebiam RV, Kannan AM, Prado F, Manthiram A (2001) Comparison of the chemical stability of the high energy density cathodes of lithium-ion batteries. Electrochem Commun 3:624–627
4. Capitaine F, Gravereau P, Delmas C (1996) A new variety of $LiMnO_2$ with a layered structure. Solid State Ionics 89:197–202
5. Song SW, Zhuang GV, Ross PN (2004) Surface film formation on $LiNi_{0.8}Co_{0.15}Al_{0.05}O_2$ cathodes using attenuated total reflection IR spectroscopy. J Electrochem Soc 151:A1162–A1167
6. Ohzuku T, Makimura Y (2001) Layered lithium insertion material of $LiCo_{1/3}Ni_{1/3}Mn_{1/3}O_2$ for lithium-ion batteries. Chem Lett 30:642–643
7. Li X, Xu Y, Wang C (2009) Suppression of Jahn-Teller distortion of spinel $LiMn_2O_4$ cathode. J Alloy Compd 479:310–313
8. Sun Y-K, Myung S-T, Park B-C, Prakash J, Belharouak I, Amine K (2009) High-energy cathode material for long-life and safe lithium batteries. Nat Mater 8:320–324
9. Manthiram A (2011) Materials challenges and opportunities of lithium ion batteries. J Phys Chem Lett 2:176–184
10. Armstrong AR, Holzapfel M, Novák P, Johnson CS, Kang S-H, Thackeray MM, Bruce PG (2006) Demonstrating oxygen loss and associated structural reorganization in the lithium battery cathode $Li[Ni_{0.2}Li_{0.2}Mn_{0.6}]O_2$. J Am Chem Soc 128:8694–8698
11. Mohanty D, Kalnaus S, Meisner RA, Rhodes KJ, Li J, Payzant EA, Wood Iii DL, Daniel C (2013) Structural transformation of a lithium-rich $Li_{1.2}Co_{0.1}Mn_{0.55}Ni_{0.15}O_2$ cathode during high voltage cycling resolved by in situ X-ray diffraction. J Power Sources 229:239–248
12. Nyten A, Abouimrane A, Armand M, Gustafsson T, Thomas JO (2005) Electrochemical performance of Li_2FeSiO_4 as a new Li-battery cathode material. Electrochem Commun 7:156–160
13. Saidi MY, Barker J, Huang H, Swoyer JL, Adamson G (2002) Electrochemical properties of lithium vanadium phosphate as a cathode material for lithium-ion batteries. Electrochem Solid-State Lett 5:A149–A151
14. Muraliganth T, Manthiram A (2010) Understanding the shifts in the redox potentials of olivine $LiM_{1-y}M_yPO_4$ (M = Fe, Mn Co, and Mg) solid solution cathodes. J Phys Chem C 114:15530–15540
15. Liu J, Manthiram A (2009) Understanding the improved electrochemical performances of Fe-substituted 5 V spinel cathode $LiMn_{1.5}Ni_{0.5}O_4$. J Phys Chem C 113:15073–15079

16. Liu J, Manthiram A (2009) Understanding the improvement in the electrochemical properties of surface modified 5 V $LiMn_{1.42}Ni_{0.42}Co_{0.16}O_4$ spinel cathodes in lithium-ion cells. Chem Mater 21:1695–1707
17. Buiel E, Dahn JR (1999) Li-insertion in hard carbon anode materials for Li-ion batteries. Electrochim Acta 45:121–130
18. He Y-B, Li B, Liu M, Zhang C, Lv W, Yang C, Li J, Du H, Zhang B, Yang Q-H, Kim J-K, Kang F (2012) Gassing in $Li_4Ti_5O_{12}$-based batteries and its remedy. Sci Rep 2: Article No. 913. doi:10.1038/srep00913
19. Park C-M, Kim J-H, Kim H, Sohn H-J (2010) Li-alloy based anode materials for Li secondary batteries. Chem Soc Rev 39:3115–3141
20. Hochgatterer NS, Schweiger MR, Koller S, Raimann PR, Woehrle T, Wurm C, Winter M (2008) Silicon/graphite composite electrodes for high-capacity anodes: influence of binder chemistry on cycling stability. Electrochem Solid-State Lett 11:A76–A80
21. http://www.sony.net/SonyInfo/News/Press/200502/05-006E/
22. http://www.greencarcongress.com/2009/12/panasonic-20091225.html
23. Cabana J, Monconduit L, Larcher D, Palacin MR (2010) Beyond intercalation-based Li-ion batteries: the state of the art and challenges of electrode materials reacting through conversion reactions. Adv Mater 22:170–192
24. Abouimrane A, Ding J, Davidson IJ (2009) Liquid electrolyte based on lithium bis-fluorosulfonyl imide salt: aluminum corrosion studies and lithium ion battery investigations. J Power Sources 189:693–696
25. Hu M, Pang X, Zhou Z (2013) Recent progress in high-voltage lithium ion batteries. J Power Sources 237:229–242
26. Zhang SS (2006) A review on electrolyte additives for lithium-ion batteries. J Power Sources 162:1379–1394
27. Dalavi S, Xu M, Knight B, Lucht BL (2012) Effect of added LiBOB on high voltage ($LiNi_{0.5}Mn_{1.5}O_4$) spinel cathodes. Electrochem Solid-State Lett 15:A28–A31
28. von Cresce A, Xu K (2011) Electrolyte additive in support of 5 V Li ion chemistry. J Electrochem Soc 158:A337–A342
29. Zhang SS (2007) A review on the separators of liquid electrolyte Li-ion batteries. J Power Sources 164:351–364
30. Zhang SS (2012) Effective approach toward safe Li-ion battery. In: Battery safety 2012, Las Vegas, NV, 6–7 Dec 2012
31. Palomares V, Serras P, Villaluenga I, Hueso KB, Carretero-Gonzalez J, Rojo T (2012) Na-ion batteries, recent advances and present challenges to become low cost energy storage systems. Energy Environ Sci 5:5884–5901
32. Stevens DA, Dahn JR (2000) High capacity anode materials for rechargeable sodium-ion batteries. J Electrochem Soc 147:1271–1273
33. Barker J, Gover RKB, Burns P, Bryan AJ (2006) Hybrid-ion a lithium-ion cell based on a sodium insertion material. Electrochem Solid-State Lett 9:A190–A192
34. Yoo HD, Shterenberg I, Gofer Y, Gershinsky G, Pour N, Aurbach D (2013) Mg rechargeable batteries: an on-going challenge. Energy Environ Sci 6:2265–2279
35. Kim HS, Arthur TS, Allred GD, Zajicek J, Newman JG, Rodnyansky AE, Oliver AG, Boggess WC, Muldoon J (2011) Structure and compatibility of a magnesium electrolyte with a sulphur cathode. Nat Commun 2:427
36. Arthur TS, Singh N, Matsui M (2012) Electrodeposited Bi, Sb and $Bi_{1-x}Sb_x$ alloys as anodes for Mg-ion batteries. Electrochem Commun 16:103–106
37. Shao Y, Liu T, Li G, Gu M, Nie Z, Engelhard M, Xiao J, Lv D, Wang C, Zhang J-G, Liu J (2013) Coordination chemistry in magnesium battery electrolytes: how ligands affect their performance. Sci Rep 3: Article No. 3130. doi:10.1038/srep03130
38. Aurbach D, Lu Z, Schechter A, Gofer Y, Gizbar H, Turgeman R, Cohen Y, Moshkovich M, Levi E (2000) Prototype systems for rechargeable magnesium batteries. Nature 407:724–727

39. Freunberger SA, Chen Y, Peng Z, Griffin JM, Hardwick LJ, Bardé F, Novák P, Bruce PG (2011) Reactions in the rechargeable lithium–O_2 battery with alkyl carbonate electrolytes. J Am Chem Soc 133:8040–8047
40. Mozhzhukhina N, Méndez De Leo LP, Calvo EJ (2013) Infrared spectroscopy studies on stability of dimethyl sulfoxide for application in a Li–air battery. J Phys Chem C 117: 18375–18380
41. Li F, Kitaura H, Zhou H (2013) The pursuit of rechargeable solid-state Li-air batteries. Energy Environ Sci 6:2302–2311
42. Zhang SS (2013) Liquid electrolyte lithium/sulfur battery: fundamental chemistry, problems, and solutions. J Power Sources 231:153–162
43. Yang Y, Zheng G, Cui Y (2013) Nanostructured sulfur cathodes. Chem Soc Rev 42: 3018–3032
44. Wang D-W, Zeng Q, Zhou G, Yin L, Li F, Cheng H-M, Gentle IR, Lu GQM (2013) Carbon-sulfur composites for Li-S batteries: status and prospects. J Mater Chem A 1:9382–9394
45. Manthiram A, Fu Y, Su Y-S (2012) Challenges and prospects of lithium-sulfur batteries. Acc Chem Res 46:1125–1134
46. Nagao M, Hayashi A, Tatsumisago M (2013) Electrochemical performance of all-solid-state Li/S batteries with sulfur-based composite electrodes prepared by mechanical milling at high temperature. Energy Technol (Weinheim, Ger.) 1:186–192
47. Zhang SS (2013) Sulfurized carbon: a class of cathode materials for high performance lithium/sulfur batteries. Front Energy Res 1:10. doi:10.3389/fenrg.2013.00010
48. Gao J, Lowe MA, Kiya Y, Abruna HD (2011) Effects of liquid electrolytes on the charge-discharge performance of rechargeable lithium/sulfur batteries: electrochemical and in-situ X-ray absorption spectroscopic studies. J Phys Chem C 115:25132–25137
49. Mikhaylik Y, Kovalev I, Schock R, Kumaresan K, Xu J, Affinito J (2010) High energy rechargeable Li-S cells for EV application: status, remaining problems and solutions. ECS Trans 25(35):23–34
50. Azimi N, Weng W, Takoudis C, Zhang Z (2013) Improved performance of lithium-sulfur battery with fluorinated electrolyte. Electrochem Commun 37:96–99
51. Zhang SS (2013) New insight into liquid electrolyte of rechargeable lithium/sulfur battery. Electrochim Acta 97:226–230
52. Mikhaylik YV (2008) Electrolytes for lithium sulfur cells. US Patent 7,354,680
53. Lin Z, Liu Z, Fu W, Dudney NJ, Liang C (2013) Phosphorous pentasulfide as a novel additive for high-performance lithium-sulfur batteries. Adv Funct Mater 23:1064–1069

Olivine-Based Cathode Materials

Karim Zaghib, Alain Mauger and Christian M. Julien

1 Introduction

The lithium insertion compounds built with polyanionic groups such as $(SO_4)^{2-}$, $(PO_4)^{3-}$, $(P_2O_7)^{4-}$, $(MoO_4)^{2-}$ or $(WO_4)^{2-}$ are considered as potential positive electrode materials for use in lithium rechargeable batteries [1, 2]. Yet in this family, olivine phosphate and Nasicon-like frameworks are currently the subject of many investigations. In particular, $LiFePO_4$ (LFP) has received a great deal of interest because this cathode material realizes the highest capacity ($\approx$170 mAh g^{-1}) at moderate current densities [3]. In addition, it presents several advantages with regard to low cost, non-toxicity, tolerance on abuse, and high safety, which are determinant with respect to cobalt-oxide-based materials for large-scaled applications such as hybrid electric vehicles (HEV). Nevertheless, the bulk electronic conductivity of olivine is quite low, which may result in losses in the specific capacity during high-rate discharge. To increase the electrochemical performance, it is a common practice in the production of Li-ion battery cathodes to manipulate the active material by (i) adding carbon additives to a olivine matrix [1, 4], (ii) surface coating of particles with thin layers of carbon [5–7] or reducing the particle size [8].

Still, there have been numerous efforts through the years to decrease the size of the particles from a few microns to this "nano" range, for several reasons. One is the increase of the effective contact area of the powder with the electrolyte. A larger effective contact surface with the electrolyte means a greater probability to drain Li^+

K. Zaghib (✉)
IREQ, Varennes, QC, Canada
e-mail: Zaghib.Karim@ireq.ca

A. Mauger
IMPMC, Université Pierre et Marie Curie, Paris, France

C.M. Julien
PHENIX, Université Pierre et Marie Curie, Paris, France
e-mail: christian.julien@upmc.fr

Z. Zhang and S.S. Zhang (eds.), *Rechargeable Batteries*,
Green Energy and Technology, DOI 10.1007/978-3-319-15458-9_2

ions from the electrode, which increases the power density of the cell. A smaller particle size also reduces the length of the path of Li inside the particle, which leads to a greater capacity at higher charge/discharge rates and therefore to a larger power density. Reducing the dimensions of the active particles to nanoscale means, for a given chemical diffusion coefficient of Li^+ ions, D^*, the characteristic time, τ_c, for the intercalation reaction is given by the Fick's law. For non stoichiometric system (one-phase reaction) the characteristic time constant for is expressed by:

$$\tau_c = L^2/4\pi D^*, \tag{1}$$

where L is the diffusion length [9]. In the case of olivine compounds, the insertion operates through a two-phase process, so the characteristic time constant becomes:

$$\tau_c = \frac{F^2}{2V_m}\frac{L^2}{\langle\sigma^i\rangle\Delta\mu_i}, \tag{2}$$

in which F is the Faraday constant, V_m the molar volume, σ^i the ionic conductivity and $\Delta\mu_i$ the chemical potential of ions. Nanoparticles, as well as more tailored nanostructures, are being explored and exploited to enhance the rate capability, even for materials with poor intrinsic electronic conductivity such as olivine frameworks. In addition, the problems raised by the small electronic conductivity that results from the strong ionicity of the bonding have been solved by coating the olivine particles with a thin layer that is conductive for both electrons and Li, usually an amorphous carbon layer [10, 11]. Decreasing the particle size reduces the length of the tunneling barrier for electrons to travel between the surface layer and the core of the particle, while the electrons are driven from the surface of the particle to the current collector via the conductive layer that percolates through the structure. The coat may also decrease the activation energy for Li^+ transfer across the electrode/electrolyte interfaces. Indeed, by preparing the materials at the nanoscale form and by carbon coating, high rates are achievable [10].

The aim of the present chapter is to investigate the physicochemical properties of optimized olivine-like electrode materials. One approach to provide insight into the structural and electronic properties of electrode materials involves a systematic study by a combination of techniques including structural, magnetic and spectroscopic measurements. Furthermore, advantage can be taken of the high sensitivity of some analytical tools for the detection of parasitic impurities that can be grown during synthesis of solid phases. These principles were fully exploited to optimize lithium iron phosphate compounds. This chapter is organized as follows. First, we expose in Sect. 2 the principle of the inductive effect in polyanionic frameworks. Section 3 presents the synthesis route, the structure and morphology of optimized $LiFePO_4$ particles probed by X-ray powder diffractometry (XRD), Scanning Electron Microscopy (SEM), High Resolution Transmission Electron Microscopy (HRTEM), Fourier transform infrared (FTIR) and Raman scattering (RS) spectroscopy. We then complete the analysis (Sect. 4) with magnetic measurements: magnetization curves and electron spin resonance (ESR). In Sect. 5, we examine the electrochemical properties of LFP in various situations including high temperature, high current density, and in humid atmosphere. In the following Sects. 6–8, we explore the other olivine materials, namely $LiMnPO_4$, $LiNiPO_4$ and $LiCoPO_4$.

2 The Inductive Effect

The aspect of tuning the redox potential of an electrode material has been demonstrated by Goodenough et al. [1, 2, 12–14]. They have shown that the use of polyanions $(XO_4)^{n-}$ such as $(SO_4)^{2-}$, $(PO_4)^{3-}$, $(AsO_4)^{3-}$, or even $(WO_4)^{2-}$ lowers the redox energy of the 3*d*-metals to useful levels with respect to the Fermi level of the Li anode. Thus, the most attractive key point of the polyanion frameworks can be seen in the strong *X*-O covalency, which results in a decrease of the Fe–O covalency. This inductive effect is responsible for a decrease of the redox potential in comparison to the oxides [12]. The polyanion PO_4^{3-} unit stabilizes the olivine structure of $LiFePO_4$ and lowers the Fermi level of the $Fe^{2+/3+}$ redox couple through the Fe–O–P inductive effect which results in a higher potential for the olivine material. The discharge voltage 3.45 V is almost 650 mV higher than that of $Li_3Fe_2(PO_4)_3$ [1]. It is also 350 mV higher than that of $Fe_2(SO_4)_3$ [14], which is consistent with the stronger Bronsted acidity of sulphuric versus phosphoric acid. In the case of Li_2FeSiO_4, the lower electronegativity of Si versus P results in a lowering of the $Fe^{2+/3+}$ redox couple [15]. On the other hand, the higher thermal stability of the phospho-olivines and their lower tendency to release oxygen is explained by the strong *X*-O covalency and the rigidity of the $(XO_4)^{n-}$ units decreasing the safety risks. However, $AMXO_4$ compounds and $AM(XO_4)_3$ as well (*A* is an alkali ion) exhibit a very low electronic conductivity because of the separation between MO_6 octahedra and XO_4 tetrahedra that induces a large polarization effect during charge–discharge reaction [16].

Electrochemical extraction of Li from $LiFePO_4$ gives (Fe^{2+}/Fe^{3+}) redox potential at ca. 3.5 V versus Li^0/Li^+. A small but first-order displacive structural change of the framework gives a two-phase separation over most of the solid-solution range $0 < x < 1$ for Li_xFePO_4 and therefore a flat *V*-*x* curve. A reversible capacity of 160 mAh g^{-1} is delivered by the nano-structured cathode particles coated with carbon. Electrochemical characteristics of $LiFePO_4$ are compared with those of other Fe-containing phosphates in Fig. 1. This graph presents the energy of the redox couples against the specific capacity relative to lithium and iron in various phosphate frameworks. Electrochemical tests of optimized $LiFePO_4$ have been conducted under various conditions to assess the influence of the electrolyte on stability and the influence of electrode processing. Post-mortem analysis, i.e. ICP, XRD, SEM, showed that no iron species were detected at the separator-negative electrode interface in cells with anode in lithium metal, graphite, or C–$Li_4Ti_5O_{12}$ [17, 18]. This result is attributed to the high quality of the “optimized” $LiFePO_4$, impurity-free materials used as positive electrodes. This property characterizes the impact of structural fluorine on the inductive effect of the PO_4^{3-} polyanion.

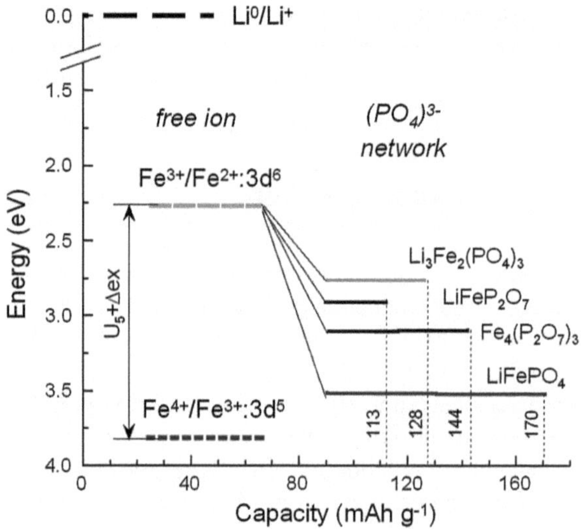

Fig. 1 Energy diagram of the redox couple relative to lithium and iron phosphate frameworks. The graph presents the theoretical capacity for each compound

3 Lithium Iron Phosphate

The electrochemical properties of LFP are known to be sensitive to the mode of preparation and the structural properties [19]. This can be an advantage for potential applications since it allows for an optimization of the material if we can correlate the mode of preparation with the structural and the physical properties. Aiming to this problem, we have first investigated this correlation in $LiFePO_4$ grown by different techniques [20–24]. Different clustering effects have been evidenced [19]. A firing temperature larger than 800 °C increases the fraction of Fe_2P [25], but Fe_2P nano-particles in such a large concentration that they drive superferromagnetism has been detected in samples that have not been heated to such high temperatures [20]. On the one hand, the presence of Fe_2P can increase the electronic conductivity, but on the other hand it also decreases the ionic conductivity so that both the capacity and cycling rates are degraded with respect to the carbon-coated LFP. It is thus desirable to optimize the preparation of the samples so that such clustering effects do not occur. This can be done easily for Fe_2P clusters, but it is more difficult to avoid the presence of a small concentration (1.0×10^{-6} per chemical formula) of γ-Fe_2O_3 nanoparticles [20, 26] in carbon-free LFP. However, we know from the iron industry that hydrogen, carbon monoxide or carbon can reduce Fe_2O_3 through different reduction steps that depend on temperature and other physical parameters such as particle sizes. Nearly all iron produced commercially is made using a blast furnace process covered by most chemistry text books. In essence, at high temperature, Fe_2O_3, is reduced with carbon (as coke) according to the reaction:

$$2Fe_2O_3 + 3C \rightarrow 4Fe + 3CO_2. \quad (3)$$

This is one of the most significant industrial processes in history, and the origins of the modern process are traceable back to a small town called Coalbrookdale in Shropshire (England) around the year 1773. We can then expect that carbon would reduce Fe^{3+} ions directly or through the formation of CO gas thus preventing the formation of γ-Fe_2O_3, if the synthesis temperature might be raised to 1000 °C. However, such a high temperature is prohibited to synthesize nano-LFP. We believe that the carbon deposition process using organic carbon precursors generates reductive gases such as hydrogen that are more active kinetically to reduce Fe^{3+} impurities in the 500–700 °C temperature range used. This is also favored by the fact that the organic precursor is usually mixed with the LFP material or with the LFP chemical precursors by solution processes at a molecular size level. For overview on olivine phosphate material see the reviews recently published [17, 19, 27].

3.1 Synthesis Routes

Many synthesis routes have been used for the preparation of LFP materials [21–48]. Here, we will focus on three methods used by our group: solid-state, hydrothermal and jet milling techniques [21–23]. Then a comparison of the physical and electrochemical properties of a series of carbon-coated $LiFePO_4$ is used to optimize the materials for battery application. Precursors were prepared by mixing various raw materials: iron(II) oxalate [$Fe(C_2O_4)_2 \cdot H_2O$] or [$FePO_4(H_2O)_2$], ammonium dihydrogen phosphate [$NH_4H_2PO_4$] and lithium salt [Li_2CO_3] or [LiOH]. The chemical homogeneity and purity of the powders appear to be sensitive to the conditions of preparation. Thus, the choice of the raw materials, carbon precursor, temperature, and atmosphere is required not only for the reduction of Fe(III) but also for the carbon-coating formation.

A solid-state synthesis method assisted by polymeric carbon additive proved to be efficient to prepare LFP at quite low temperature, 300−400 °C, while the sintering temperature that does not exceed 750 °C makes the existence of any carbothermal effect unlikely [28]. A blend of $Fe^{III}PO_4(H_2O)_2$ and Li_2CO_3 was mixed by ball milling in isopropyl alcohol (IPA) overnight; the blend was dried and mixed with 5 wt% of a polymeric carbon additive, polyethylene-block-poly(ethylene glycol) 50 % ethylene oxide. Overall synthesis was carried out under flowing argon. Note that the carbon source could provide the LFP powder with a carbonaceous deposit after pyrolysis and such a powder pressed at 3750 kg cm^{-2} at room temperature presents an electronic conductivity much higher than 1×10^{-8} S cm^{-1} [11]. Other elements that can be present are hydrogen, oxygen, and nitrogen, as long as they do not interfere with the chemical inertia of the carbon. Preferred polymeric additives include, but are not limited to, hydrocarbons and their derivatives, especially those comprising polycyclic aromatic moieties, e.g. polyolefins,

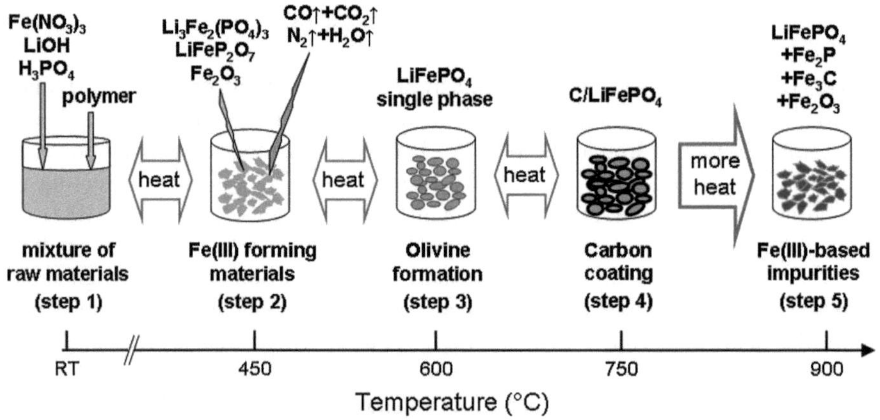

Fig. 2 Scheme of the steps of the polymer-assisted solid-state synthesis of LFP powders. The in situ carbon-coating is realized by polymer precursor at temperature $Ts < 750$ °C

polybutadienes, polyvinyl alcohol, etc. [11]. Figure 2 shows the different steps of the synthesis of carbon-coated LFP particles using the polymer-assisted solid-state method. Note that for $Ts > 750$ °C Fe(III)-containing impurities are formed that poison the electrochemical performance of LFP.

For some of the synthesis routes, the reduction to small size occurs at the expense of crystallinity and the formation of a large concentration of defects [29]. These difficulties have been overcome in the framework of a preparation process in which the mechanical assistance is used only to reduce the size of the particles [11]. In a first step, LFP particles were prepared by melting precursors at 1050 °C for 5 min in a graphite crucible and then cooled under N_2 atmosphere fast enough so that impurities have no time to nucleate [21]. The next step is to decrease the size of the particles down to the desired value, anywhere in the range from the centimeter down to 40 nm (Fig. 3). For this purpose, the ingot is first crushed into centimeter-size particles by using a jaw-crusher with ceramic liner to avoid metal contamination. Then, the roll crusher (ceramic type) is used to obtain millimeter-size particles. The millimeter-size particles are further ground by jet-mill to achieve micrometer-size particles. In the process, the grains enter the grading wheel and are blasted to the collector. The particles obtained at this step are referred as "jet-mill" in the following. There size is the order of 1 μm. To obtain smaller particles, these micrometer-size powders were dispersed in isopropyl alcohol (IPA) solution at 10–15 % of solid concentration and then ground on a bead mill using 0.2-mm zirconia beads to obtain nanometer-sized particles [11]. This final product is referred "wet mill" in the following. One advantage of the process is that we can investigate the properties of the same particles at different stages of the milling, for comparison, so that any difference is a size effect. The particles can be considered as uncoated particles because of the great damage caused by the milling process. In the case of the particles used to obtain the experimental results illustrated in the figures of this

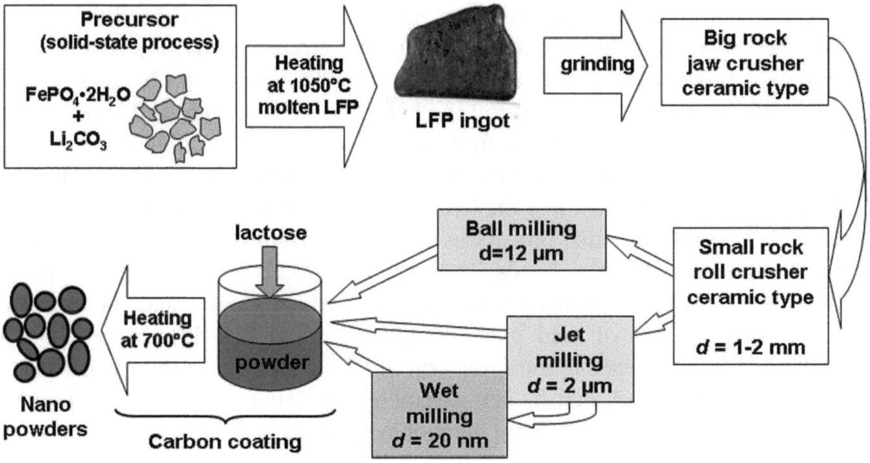

Fig. 3 Milling synthesis from molten ingot. Nanopowders are obtained using crusher ceramic liner and jet milling

review, the carbon-coated particles have been obtained with lactose as the carbon precursor in acetone solution according to the following procedure. The uncoated particles were mixed with the carbon precursor. The dry additive corresponded to 5 wt% carbon in $LiFePO_4$. After drying, the blend was heated at 750 or 700 °C [21, 30] for 4 h under argon atmosphere. This range of temperature is dictated by two considerations. Below 700 °C, the carbon deposit was not conductive enough [21].

The hydrothermal route is particularly successful to control the chemical composition and crystallite size [28, 37–39]. The conventional hydrothermal process involves a reaction time 5–12 h to synthesize LFP [40, 41]. With respect to the previous techniques, the hydrothermal process has the advantage that the synthesis temperature can be as small as 230 °C [42]. Brochu et al. [23] demonstrated the beneficial effect of choosing adapted complexing agent in the hydrothermal solution. Even at this mild temperature the carbon coating of the particles could be achieved by an in situ hydrothermal carbonization of glucose during the synthesis process. The heating temperature to get the particles coated with conductive carbon was still 700 °C, but good results were obtained upon heating during one hour only. Carbon-coated LFP was prepared by hydrothermal synthesis assisted by rotating/stirring tests at different agitation speeds [24]. The process was carried out at 190 °C for 12 h with a solution chemistry of $LiOH \cdot H_2O$, $FeSO_4 \cdot 7H_2O$, H_3PO_4 (85 wt%), and ascorbic acid (as reducing agent) in the stoichiometry 3Li:1Fe:1P:0.2C. Annealing was done at 700 °C under nitrogen atmosphere using lactose as carbon coating source. The $LiFePO_4$–C electrodes prepared by employing agitation during hydrothermal synthesis were found to exhibit higher discharge capacities (~137 mAh g^{-1} at C/12) than those prepared without agitation (~106 mAh g^{-1}). This was equally true for higher current rates, namely C/5 and C/3. Via a series of tests at different speeds (260–1150 rpm) and different concentrations (0.4–0.6 mol.

dm^{-3}) the optimum solution rotating agitation/concentration conditions were determined to be 260–380 rpm and 0.5 mol.dm^{-3}, respectively [24].

A polyol process to synthesize LFP has been developed first by Kim to synthesize particles under the form of rods with average width 20 nm and length 50 nm [37], but different shapes ranging from rods to plates can be obtained [38] with average size of 100–300 nm. Note the polyol process make possible the synthesis of LFP at low temperature just like the hydrothermal process. Bigger particles can be obtained by the solvothermal process in a polyol medium of diethylene glycol, but still under the form of plates or rods. The advantage of this process is that the polyol medium acts not only as a solvent, but also as a reducing agent and stabilizer that limits the particle growth and prevents agglomeration [38–40]. In general, the solvothermal process leads to the formation of LFP under the form of plates, about 50 nm-thick [41], which can be self-assembled by using poly(vinyl pyrrolidone) (PVP) as the surfactant in a benzyl alcohol system [42].

3.2 Structure of Olivine Phosphate

Triphylite is a rather scarce orthophosphate primary mineral found in phosphatic pegmatites and pegmatitic dikes. Its formula is $Li(Mn,Fe)PO_4$ and differs from the other mineral, lithiophilite, by being rich in iron instead of manganese. The structures of the two minerals are the same and form a solid solution, referred as the triphylite series, isomorphous with olivine. Therefore, any differences in physical properties between the two would be related to the iron/manganese percentage. These differences are then best evidenced by comparing the physical properties of the two members at the opposite edges of the triphylite family, namely $LiFePO_4$ and the often associated material $LiMnPO_4$ (LMP), which, in contrast with triphylite and lithiophilite, are artificial ceramics [43]. In addition, triphylite's name in Greek means "family of three" (referring to Fe, Mn, and Li). Any confusion between $LiFePO_4$ and triphylite met recently in the literature should then be avoided. Triphylite alters easily into other phosphate minerals, and geologists show it a lot of respect for making the other phosphate minerals possible. This easy alteration, however, means that this material is not very stable and that it is difficult to make good quality, well-crystallized samples. Despite this drawback, many efforts have been focused on the definite compounds LFP and LMP rather than their solid solutions. They belong to the rich family of olivines of the Mg_2SiO_4-type with the general formula B_2AX_4 [44].

The crystal structure of olivine materials has been studied by several authors [44, 48]. $LiFePO_4$ crystallizes in the orthorhombic system (No. 62) with *Pnma* space group. It consists of a distorted hexagonal-close-packed oxygen framework containing Li and Fe located in half the octahedral sites and P ions in one-eighth of the tetrahedral sites [46]. The FeO_6 octahedra, however, are distorted, lowering their local cubic-octahedral O_h to the C_s symmetry. Corner-shared FeO_6 octahedra are

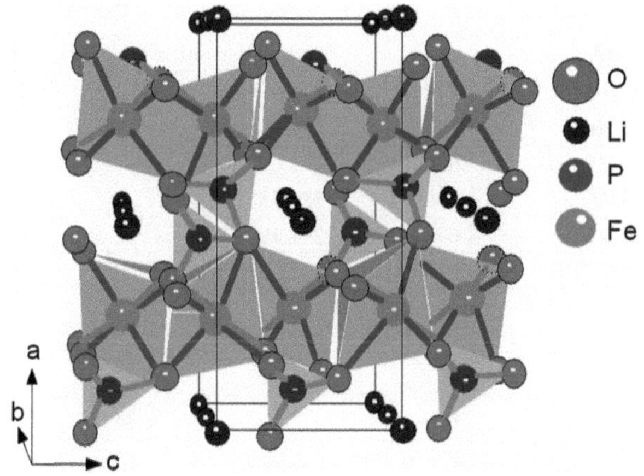

Fig. 4 Crystal structure of $LiFePO_4$ olivine. Corner-shared FeO_6 octahedra are linked together in the *bc*-plane; LiO_6 octahedra form edge-sharing chains along the *b*-axis. The tetrahedral PO_4 groups bridge neighboring layers of FeO_6 octahedra by sharing a common edge with one FeO_6 octahedra and two edges with LiO_6 octahedra

linked together in the *bc*-plane; the LiO_6 octahedra form edge-sharing chains along the *b*-axis. The tetrahedral PO_4 groups bridge neighboring layers of FeO_6 octahedra by sharing a common edge with one FeO_6 octahedron and two edges with LiO_6 octahedra. Remarkably short O–O bonds at the shared PO_4 and FeO_6 edges help to screen the cation charges from each other. This structure is illustrated in Fig. 4 showing the 1-D channels via which the lithium ions can be removed. Corner-shared FeO_6 octahedra are linked together in the *bc*-plane, while LiO_6 octahedra from edge-sharing chains along the *b*-axis. The tetrahedral PO_4 groups bridge neighboring layers of FeO_6 octahedra by sharing a common edge with one FeO_6 octahedra and two edges with LiO_6 octahedra.

The $LiFePO_4$ structure consists in three non-equivalent O sites. Most of the atoms of the olivine structure occupy the 4*c* Wyckoff position except O(3) that lies in the general 8*d* position and Li^+ ions occupying only the 4*a* Wyckoff position (M1 site on an inversion center). The Fe magnetic ions are in the divalent Fe^{2+} state of the FeO_6 units. As a consequence, Fe is distributed so as to form FeO_6 octahedra isolated from each other in $TeOc_2$ layers perpendicular to the (001) hexagonal direction [47]. In addition, the lattice has a strong two-dimensional character, since above a $TeOc_2$ layer comes another one at the vertical of the previous one, to build (100) layers of FeO_6 octahedra sharing corners and mixed layers of LiO_6 octahedra. Nevertheless, these Fe–O planes are strongly linked by the PO_4 bridges, sot that the material is truly three-dimensional, which insures the remarkable thermal stability with respect of the lamellar compounds. The synthetic lithium iron phosphate material differs from the primary mineral triphylite (Mn,Fe)PO_4 by the fact that triphylite is only rich in iron, with some manganese ions also in the M2 site. However, while the triphylite is a naturally occurring mineral, $LiFePO_4$ is an

Table 1 Lattice constants for stoichiometric $LiFePO_4$ materials in the *Pnma* (62) structure

a (Å)	*b* (Å)	*c* (Å)	Unit cell volume (Å^3)	Refs.
10.332(4)	6.010(5)	4.692(2)	291.4(3)	Herle [25]
10.334	6.008	4.693	291.39	Yamada [52]
10.329(0)	6.006(5)	4.690(8)	291.02	Geller [44]
10.31	5.997	4.686	289.73	Santoro [45]
10.3298	6.0079	4.6921	291.19	Andersson [53]
10.334(4)	6.008(3)	4.693(1)	291.392(3)	Padhi [1]

Table 2 Fractional coordinates and site symmetry of atoms in $LiFePO_4$ (*Pnma*)

Atom	*x*	*y*	*z*	Site symmetry
Li	0	0	0	$\bar{1}$ (4*a*)
Fe	0.28222	¼	0.97472	m (4*c*)
P	0.09486	¼	0.41820	m (4*c*)
O(1)	0.09678	¼	0.74279	m (4*c*)
O(2)	0.45710	¼	0.20602	m (4*c*)
O(3)	0.16558	0.04646	0.28478	1 (8*d*)

artificial product. On a fundamental point of view, the main interest lies in the fact that the olivine structure generates magnetic-interaction frustration [49]. However, three olivine structure classes can be distinguished as a function of the site occupation by magnetic ions. In Mn_2SiS_4 and Fe_2SiS_4, the magnetic ion (Mn, Fe) lies in the *M1* and the *M2* site [50], while in $NaCoPO_4$ and $NaFePO_4$, the magnetic ion lies on the *M1* site only [51]. The third class is the phospho-olivine Li*M*PO_4 (*M*=Ni, Co, Mn, Fe) where the magnetic ion lies in the *M2* site with the M1 site occupied by the non-magnetic ion (Li^+).

We used the recent structure determination by Streltsov et al. [46] as a standard reference (Table 1). The orthorhombic unit cell of the olivine structure contains 28 atoms ($Z = 4$). Structural parameters are listed in Table 2. Fe–O distances range from 2.064 to 2.251 Å. The Fe–Fe separation in $LiFePO_4$ is large (3.87 Å). The magnetic interactions between Fe ions are antiferromagnetic superexchange interactions of the form –Fe–O–Fe– and –Fe–O–P–O–Fe–, consistent with the antiferromagnetic ordering observed below $T_N = 52$ K [43]. Figure 5 shows the typical XRD spectra of two $LiFePO_4$ samples with nano- and micro-sized particles. Crystallite sizes 25 nm and 2 μm were estimated using the Scherrer formula.

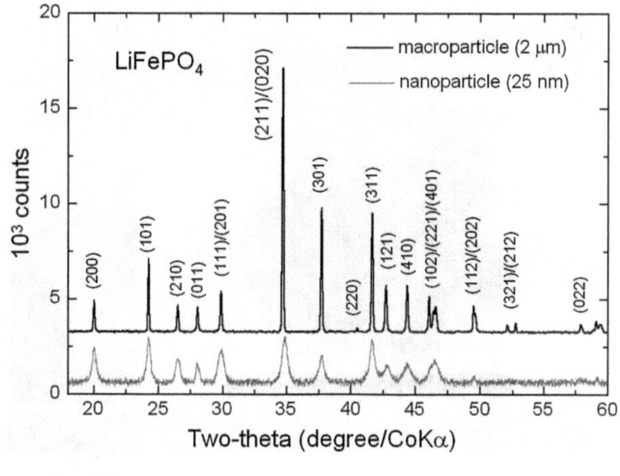

Fig. 5 XRD patterns of LFP samples with nano-sized particles (lower spectrum) obtained after jet milling and micro-sized particles (upper spectrum). Note the large broadening of the XRD lines of the nano-sized particles

3.3 Morphology of LFP

The surface morphologies of the nano-sized LFP powders studied by HRTEM are reported in Fig. 6. The powders are composed of well-dispersed secondary particles that are slightly agglomerated, with a small quantity of fragments. The TEM pictures show similar images at any part of the sample, which is homogeneous at the nanoscale with respect to the area investigated. Images (a and b) are then representative of the free-carbon particles with the tendency of forming agglomerates. Each of the secondary particles is made of a large number of small primary particles. The HRTEM images (c and d) for the carbon-coated sample illustrated polydispersed primary particles with a mean size ≈90 nm, which is larger by a factor 3 with respect to the average size of the monocrystallite grains deduced from the application of the Scherrer law on the XRD pattern. Therefore, the primary particles are polycrystallites of $LiFePO_4$ made of a few (3 on average) monocrystallites of $LiFePO_4$. The amorphous carbon layer is well illustrated in the TEM pictures (Fig. 6c, d). The formation of a carbon network appears in the interstitial grain-boundary region, which could explain the electrical continuity between $LiFePO_4$ crystallites. In the micrographs, the $LiFePO_4$ crystallites appear as the darker regions while the carbon coating is surrounding the primary particle as the greyish region. The average thickness is estimated to be 30 nm. The carbon film is highly porous, which results in an irregular coating of the crystallites well-observed on the HRTEM images but the important point for the electronic conductivity is that it connects the particles. To summarize these results, the HRTEM images clearly depict the beneficial effect of a carbon layer coating the LFP crystallites. XRD and HRTEM data are consistent.

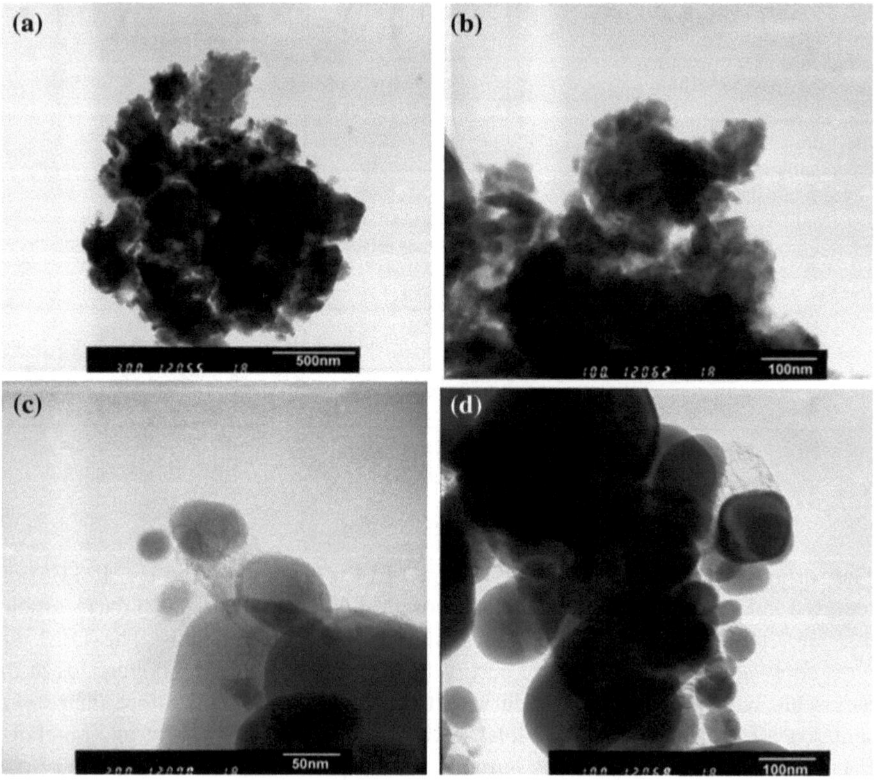

Fig. 6 HRTEM images of LFP nano-sized particles (**a**, **b**) before carbon coating and (**c**, **d**) after carbon deposited by the lactose method

3.4 Local Structure, Lattice Dynamics

Fourier transform infrared (FTIR) spectroscopy probes bulk properties, while Raman scattering (RS) spectroscopy is the tool to perform surface analysis [54–57]. For instance, the amount of carbon on $LiFePO_4$ is too small to be detected by FTIR, but it is well-characterized by RS experiments [54]. The vibrational modes of $LiFePO_4$ are primarily due to motion associated with phosphate and iron the other modes show some lithium contribution [55]. The FTIR spectra of the samples are reported in Fig. 7. We have also reported the position of the peaks intrinsic to this material, already identified in earlier works [52]. Let us recall that the spectra result from absorption measurements, so that they are a probe of the bulk properties, and the amount of carbon in the material is too small to be detected by such experiments. This is the basic reason why the FTIR spectra are characteristics of the $LiFePO_4$ part. The positions of all the IR bands are in agreement with those in Ref. [58]. No extra line is observed with respect to pure LFP. The bands in the range 372–647 cm^{-1} are bending modes (ν_2 and ν_4) involving O–P–O symmetric and

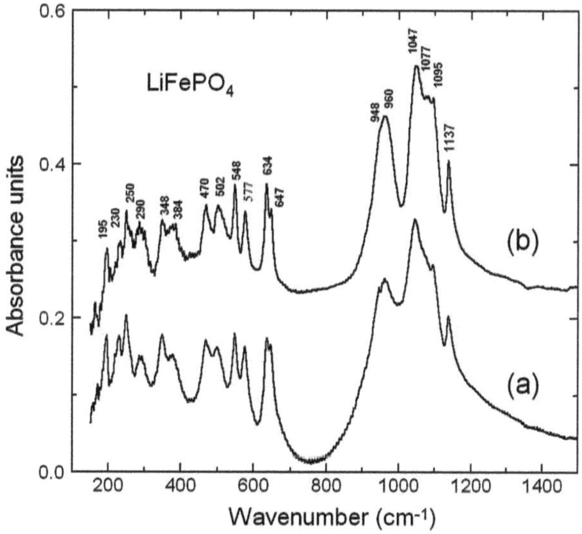

Fig. 7 FTIR absorption spectra of (*a*) carbon-free and (*b*) carbon-coated $LiFePO_4$ samples. Peak positions are marked (in cm^{-1}). Infrared spectra were recorded on pellet of $LiFePO_4$ powders diluted into ICs matrix (1:300)

asymmetric modes and Li vibrations [55]. In particular, the line at 230 cm^{-1} corresponds to the same cage mode of the lithium ions that undergo translation vibrations inside the cage formed by the six nearest-neighbour oxygen atoms [56]. The bands in this range 372–647 cm^{-1} are thus the part of the spectrum that is sensitive to the local lithium environment. This is also the part of the spectrum that is the same in both the carbon-free and carbon-coated samples. We can then infer from this result that the lithium ions do not 'see' the carbon ions, another evidence that the carbon did not penetrate inside the LFP particles. The part of the spectrum in the range 945–1139 cm^{-1} corresponds to the stretching modes of the $(PO_4)^{3-}$ units. They involve symmetric and asymmetric modes of the P–O bonds at frequencies closely related to those of the free molecule, which explains that the frequencies of these modes are the same in both samples. However, the modes in the carbon-free sample are significantly broader than in the carbon-coated sample. This broadening gives evidence of a decrease in the lifetime of the phonons, and thus the existence of defects breaking the periodicity of the lattice sites inside the LFP crystallites of the carbon-free sample. The analysis of magnetic properties in the next section will allow us to identify these defects as γ-Fe_2O_3 nanoparticles.

To explore the surface properties of the LFP particles, Raman spectra have been measured; the penetration depth for carbon with Raman spectroscopy is approximately 30 nm [54]. This is one order of magnitude larger than the thickness of the carbon coat deposited at the surface of the LFP particles in case of a uniform carbon distribution. Therefore, any screening effect of carbon on the $LiFePO_4$ spectra is not expected. The penetration depth inside $LiFePO_4$ is unknown, but it should be small, so that the detector in the Raman experiments collects the signal within the light penetration depth, which basically represents the total amount of carbon and a few per cent of the amount of $LiFePO_4$. Since the total amount of carbon is itself 5 wt%

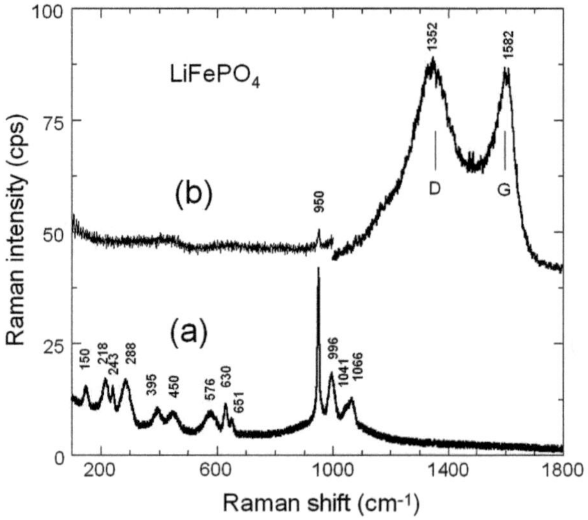

Fig. 8 Raman spectra of the carbon-free and carbon-coated $LiFePO_4$ samples. Spectra were recorded using the 514.5 nm laser line at the spectral resolution 2 cm^{-1}. RS features of the $LiFePO_4$ bulk material are screened by the carbon deposit for which the G- and D-band are observed

of $LiFePO_4$, we can expect that comparable amounts of carbon and $LiFePO_4$ are probed by the sampling depth. This is confirmed by the Raman spectra reported in Fig. 8. The part of the spectrum in the wave number range 100–1100 cm^{-1} is the same in the carbon-free and the carbon-coated sample and only the lines characteristics of $LiFePO_4$ are detected in this range. The peak positions reported in Fig. 8 in this range are within a few cm^{-1} the same as those that have been reported in [57], and we refer to this prior work for their assignment. The largest difference is for the line at 395 cm^{-1}, which is reported at 410 cm^{-1} in Ref. [55]. This line is associated with the PO_4 bending modes v_2, v_4 which are strongly coupled. However, we cannot consider this difference as significant since all the other lines associated with PO_4 have the same position. This is the case in particular for the lines at 620, 940, 986 and 1058 cm^{-1} associated with v_4, v_1, v_3 and v_2 intramolecular stretching modes of PO_4, respectively. The only difference in this range of wave numbers is a shift of the Raman lines by about 10 cm^{-1} towards lower frequencies in the carbon-coated sample. This shift of the Raman lines is in contrast with the absence of any shift of the FTIR lines, which gives evidence that it is a surface effect. This shift of the Raman lines is attributable to the increase of the bonding length in the first layers of LFP_4 particles near the interface with the carbon, taking its origin in the strain induced by the adhesion of the carbon film. For samples with a different mode of preparation, carbon was reported to be responsible for a screening of the signal from bulk, so that only a weak band at 942 cm^{-1} associated with LFP could still be detected [54]. Again, such a screening is not expected for the reasons above mentioned, and it is not observed in the present case.

3.5 Iron(III) Nanoclusters in LFP

Recently, significant effort has been underway to improve $LiFePO_4$ by developing a new synthesis route via carbon coating [59, 60]. The 1-D Li channels make the olivine performance sensitive not only to particle size, but also to impurities and stacking faults that block the channels. Various types of iron-based impurities have been identified in the olivine framework: for examples γ-Fe_2O_3, Fe_3O_4, $Li_3Fe_2(PO_4)_3$, $Fe_2P_2O_7$, Fe_2P, Fe_3P, $Fe_{75}P_{15}C_{10}$, etc. Critical quality control of the product is necessary to obtain a complete understanding of synthesis conditions using combination of experiments such as Raman spectroscopy and magnetic measurements [20–22, 59, 60]. Analysis of the magnetic properties gives an average separation of the magnetic clusters that is too large for interaction between particles (superparamagnetic model). This hypothesis must be released where the number n of magnetic clusters of momentum μ is so large that magnetic interactions between the ferrimagnetic particles become important [20]. At high fields, M_{extrin} saturates to $Nn\mu$ so that this quantity is readily determined as the magnetization at the intersection of the tangent to the magnetization curves at large fields with the ordinate axis at $H = 0$. As a result, we find that $Nn\mu$ does not depend significantly on temperature below 300 K. Therefore, we are in the situation where the cluster magnetization is temperature independent, which amounts to say that the Curie temperature Tc inside the clusters is much larger than 300 K. This is an important information on the nature of the ferromagnetic clusters. In particular, this feature precludes the existence of Fe_2P clusters observed in some LFP samples prepared according to a different procedure [59], since the Curie temperature of these clusters is only 220 K. The nature of the strongly ferromagnetic clusters in the present case is most likely maghemite (γ-Fe_2O_3).

It is remarkable from Fig. 9 that the synthesized LFP sample displays different magnetic features, with a magnetic moment $\mu_{eff} = 5.72$ μ_B due to the existence of Fe (III) containing impurities. The first consequence is an ambiguity in what is called the magnetic susceptibility χ_m since M/H is distinct from dM/dH. The magnetic susceptibility measured with a SQUID at $H = 10$ kOe shows the non-linearity of the magnetic moments attributed to the presence of γ-Fe_2O_3. The best material shows the lowest Curie constant 3.09 emu K mol^{-1}. The effective magnetic moment $\mu_{eff} = 4.98$ μ_B is close to theoretical value 4.90 μ_B calculated from the spin-only value of Fe^{2+} in its high-spin configuration. Departure from the spin-only value is attributable to the presence of Fe^{3+} ions because the orbital-momentum contribution of the Fe^{2+} ions is quenched by the crystal field [26].

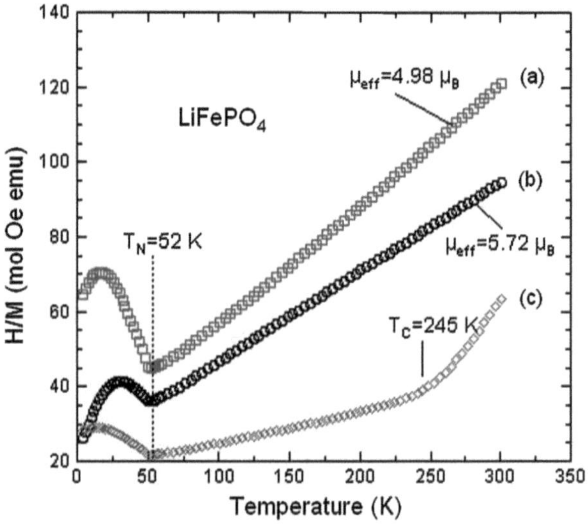

Fig. 9 Temperature dependence of the reciprocal magnetic susceptibility of different $LiFePO_4$ samples. *(a)* optimized pure $LiFePO_4$, *(b)* Fe_2O_3-containing sample, and *(c)* $\chi^{-1}(T)$ of Fe_2P-containing sample. μ_{eff} is deduced from the Curie-Weiss law $\chi = C/(T + \theta)$ above the Néel temperature T_N

3.6 Influence of the Fe_2P Nanoclusters

The electrochemical properties of LFP are known to be sensitive to the mode of preparation and the structural properties [61]. This can be an advantage for potential applications since it allows for an optimization of the material if we can correlate the mode of preparation with the structural and the physical properties. To address this issue, we investigated this relationship in $LiFePO_4$ samples grown under different conditions [62]. Undesirable impurities in the lattice can be introduced during the growth process. For instance, the presence of Fe_2P can increase the electronic conductivity, but on the other hand it also decreases the ionic conductivity so that both the capacity and cycling rates are degraded with respect to C-LFP. In addition, the iron from Fe_2P dissolves into the electrolyte, which reduced the calendar life of the cell. The presence of a small concentration (>0.5 %) of Fe_2P is evidenced in Fig. 9 by the appearance of an abnormal $\chi(T)$ behaviour with the occurrence of a shoulder near $T_C = 265$ K, the Curie temperature of the ferromagnet Fe_2P. Figure 10 shows the Arrhenius plot of the electronic conductivity, σ_{elec}, of three $LiFePO_4$ samples: a pure material, a Fe_2P-containing sample, and a C-LFP. It is obvious that addition of either iron phosphide or carbon enhances greatly σ_{elec} but to the detriment of the capacity for the former compound, as it will be discussed next.

Figure 11 displays the electrochemical charge-discharge profiles of Li//LFP cells cycled at room temperature with pure $LiFePO_4$ and with Fe_2P-containing electrode material. It is obvious that at the rate 2C, the capacity retention decreases significantly for the material containing few % of Fe_2P. A close examination was made for the detection of any iron dissolution that could occur after long-term cycling. The analysis of iron species was investigated at the separator/lithium (SL) interface by SEM cross-section (slice view) as shown in Fig. 12a, b. The micrograph (Fig. 12a)

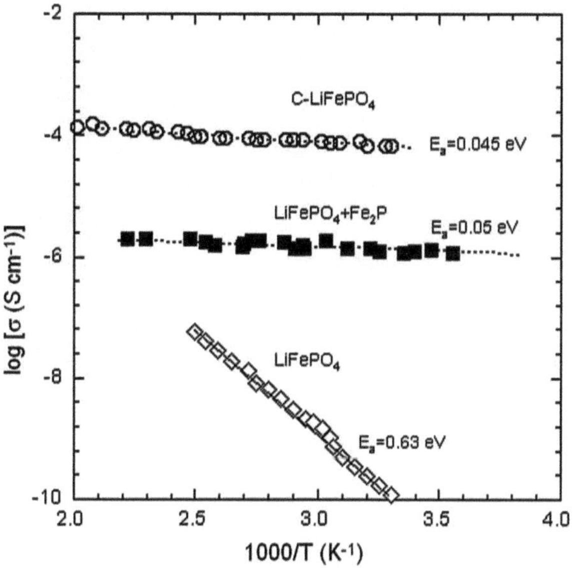

Fig. 10 Electronic conductivity of $LiFePO_4$ samples: pure material, Fe_2P-containing sample, and carbon-coated $LiFePO_4$

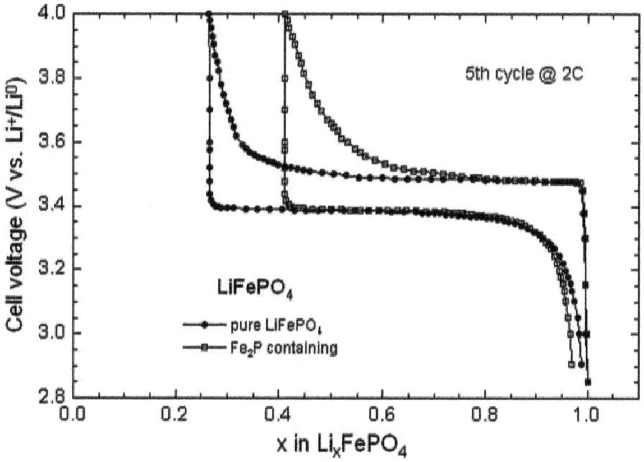

Fig. 11 Electrochemical charge-discharge profiles of Li//$LiFePO_4$ cells cycled at room temperature with pure $LiFePO_4$ and with Fe_2P-containing electrode material

obtained from evaluation of the earlier generation material shows the presence of iron islands at the SL interface. Obviously, some iron particles (or ions) migrate through the electrolyte from the $LiFePO_4$ positive electrode to the lithium negative. The net effect of this migration is a large decrease in capacity retention of the Li// LFP cell. Figure 12b shows the post-mortem micrograph obtained from tests with an optimised electrode in a Li cell with a lithium foil negative. In this case, there is no iron detected at the SL interface, which remained intact after 100 cycles. In fact, this high performance was possible not only because of the optimised synthesis of the LFP powders, but also because of strict control of the structural quality of the materials. Several physical methods were used to analyse the local structure and the electronic properties of the phospho-olivine framework [18].

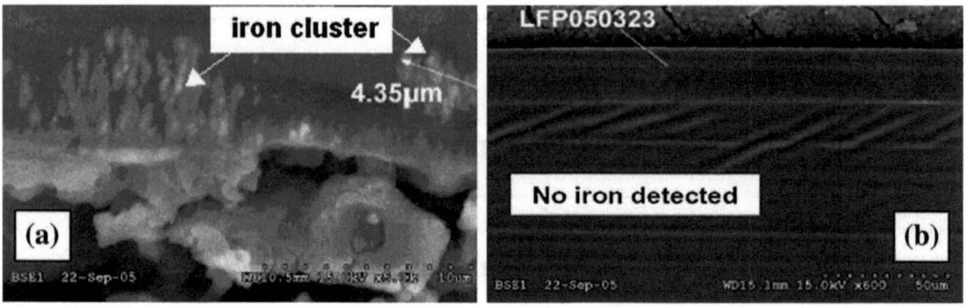

Fig. 12 Post-mortem SEM images of the detection of iron species at the separator/lithium interface. **a** Formation of iron islands at the interface with an earlier generation of $LiFePO_4$. **b** No iron was detected at the surface of Li foil with the optimized $LiFePO_4$

3.7 Carbon Coating

To increase the electronic conductivity, it is a common practice in the production of Li-ion battery electrodes, to add carbon, either by use of carbon additives to the $LiFePO_4$ matrix [3], or by surface coating of LFP particles with thin layers of carbon [5, 6]. A seven-order-of-magnitude increase in the electronic conductivity has been reached by adding sucrose to produce carbon in LFP raw materials by a spray pyrolysis technique [7]. The addition of carbon has then the advantage of combining much better electronic conductivity, and high capacity. In particular, a capacity of about 160 mAh g^{-1} has been found for LFP coated with 1 wt% carbon [61]. Ravet et al. [5, 6] reported two ways to coat carbon: (i) mixing LFP powder with sugar solution and heating the mixture at 700 °C, and (ii) synthesizing LFP with some organic materials added before heating. Although the way to add carbon is not fully optimized yet, the approach that consists in adding a carbon source at the beginning of the synthesis is more promising [6]. Recently, Julien et al. [33] demonstrated that better performance was obtained at high-rate discharge (3C) with 6 % carbon additive in the LFP electrode. This material is suitable for HEV application. This is consistent with the HRTEM images in Fig. 13, showing that the carbon coats the secondary particles with a typical radius of 100 nm and does not penetrate into the LFP particles [23].

3.8 Aging of LFP Particles Exposed to Water

It has been well-known for decades that all the lithium-ion batteries need to be protected against humidity. The main reason is that lithium is very reactive with water according to the chemical reaction $Li + H_2O \rightarrow LiOH + \frac{1}{2}H_2$. Lithium hydroxide (LiOH) is a corrosive alkali hydroxide. When crystallized, it is a white hygroscopic material. It is also soluble in water, a property that has been used to investigate aqueous lithium hydroxide as a potential electrolyte in Li-ion batteries

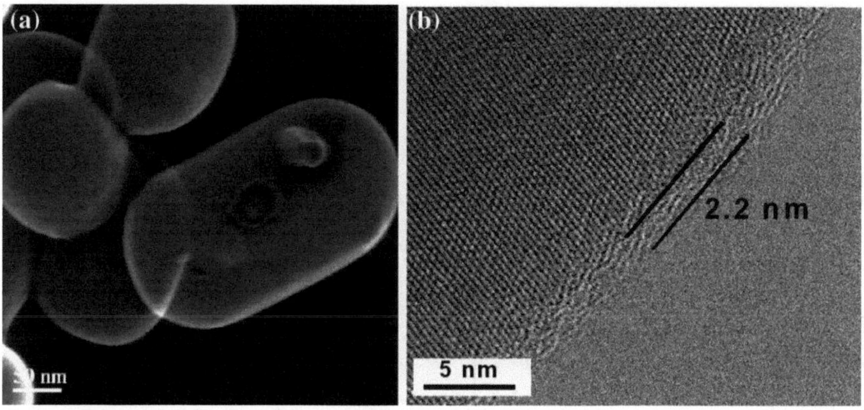

Fig. 13 HRTEM of carbon-coated LFP particles synthesized by hydrothermal method. The carbon coating, ~2-nm thick was realized using the lactose route

with a $LiFePO_4$ cathode. Since the carbon coat is not a barrier for Li^+-ion transport (the reason for the success of C–$LiFePO_4$ as a cathode element of Li-ion batteries), we expect the reaction with water to be effective, implying extraction of Li from LFP grains to interact with water. This delithiation is the only effect that is observed after exposure to H_2O in air, since the surface layer of the particles is also affected [63, 64]. Despite this reaction of lithium with humidity, a drop of $LiFePO_4$ particles into water was used in the laboratory to check the carbon-coating process by separating coated and uncoated particles. This means of characterization is based on the fact that when the C–$LiFePO_4$ powder is dropped into water, part of the carbon that links the particles unties and floats to the surface, retaining with it some of the particles, while the major part sinks. More recently, Porcher et al. [64] have determined that the exposure of C–$LiFePO_4$ particles to water results in the formation of a thin layer of Li_3PO_4 (few nm thick) at the surface of the particles as a result of migration of Fe into the water. In the present work, we investigate the effect of water on carbon-coated $LiFePO_4$ particles and analyze both the particles that have sunk and the floating part. We have shown that the water attacks the particles and that the carbon coat is not a protection because it detaches and is not waterproof [63]. We find that Fe is not the only element that reacts with the water, as the water contains also P and Li species after immersion of the $LiFePO_4$. A strong interaction between $LiFePO_4$ with H_2O molecules was not necessarily expected. Since iron phosphate is hydrophobic, the surface layer protects the iron against oxidation and corrosion. Intuition would then have suggested that, upon immersion of $LiFePO_4$ in water, a delithiation in a thin layer at the surface would lead to the formation of a $FePO_4$ layer that would protect the particles against any other damage. Quantification of magnetic analysis of LFP grains shows that the thickness of the delithiated surface layer is ~3 nm.

The cyclic voltammograms of LFP powders after immersion for 1 h in water are reported in Fig. 14. In these measurements, an initial 3.2 V working potential is

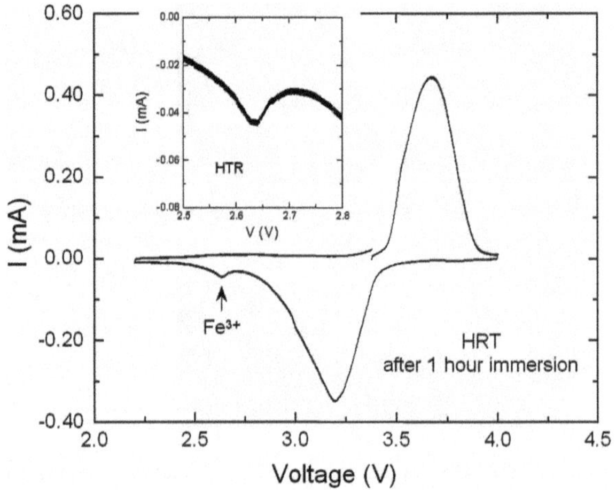

Fig. 14 Electrochemical performance of the C-$LiFePO_4$ synthesized by hydrothermal method in $LiPF_6$-EC-DEC/Li cells at room temperature. Cathode particles were immersed in water for 1 h, then dried for 48 h at 85 °C

applied. Then, the voltage was varied at the rate 1.25 mV per minute, as shown in the figure: a voltage increase up to 4 V is followed by a decrease to 2.2 V and an increase again up to 3.2 V. Besides the peak associated with Fe^{2+}, the part of the curve obtained by decreasing the voltage shows a secondary peak at 2.63 V that is characteristic of the Fe^{3+} in iron oxide (vs. more than 3.5 V in phosphate) [32]. The presence of Fe^{3+} ions in LFP samples confirms the delithiation of the surface layer evident in the previous sections. On the other hand, upon increasing the voltage again, this signal disappeared, which shows that the voltammogram before exposure to H_2O was recovered. Therefore, the surface layer was lithiated again during Li insertion, and the effect of immersion in water was reversed [63]. Note, however, that this recovery could be obtained because we were dealing with a half-cell, i.e. with lithium metal as an anode. In this case, the lithium anode acts as a reservoir of lithium. In a full cell where the negative electrode would be carbon or Si or any metal oxide, however, one would not such a reservoir so that no recovery would have been obtained, so that, in practice, it is important to keep the LFP powder dry.

The same effect holds true for longer immersion times of a few days. In the following experiments, the samples were immersed for 63 h. Then, the samples were dried for 48 h at 85 °C. Moreover, it should be noted that the open-circuit voltage (OCV) decreased by 2.3 % on immersion in water. Since the OCV is directly related to the state of charge of the battery, it can be viewed as an indirect measurement of the delithiation rate of the battery. Indeed, this result is fully consistent with the 4 % delithiation rate deduced from the magnetization measurements, and the 1–3 % loss of Fe and P in the immersion process estimated from the physical and chemical analyses. It thus fully confirms that the delithiation process is located in the surface layer. The effect of H_2O on the electrochemical properties was also evaluated by exposure of the sample to ambient air. This effect is illustrated for a hydrothermal sample in Fig. 15, which shows the change of the capacity as a function of time at different temperatures in dry atmosphere and in ambient air (55 % relative humidity) [63].

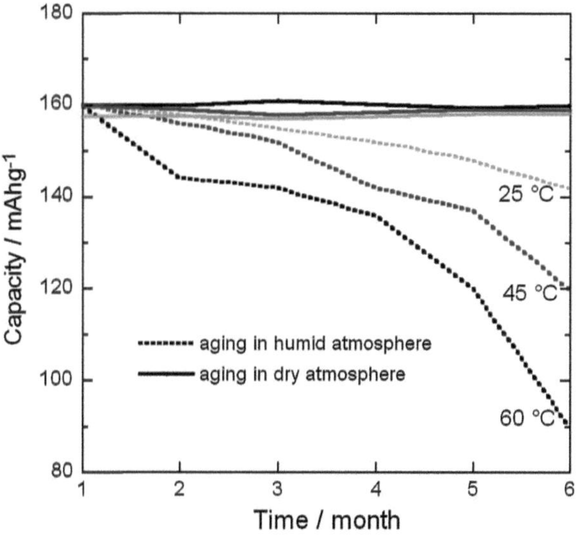

Fig. 15 Capacity of the C-$LiFePO_4$/$LiPF_6$-EC-DEC/Li cells as a function of time spent in dry atmosphere and in ambient atmosphere (55 % relative humidity), at three different temperatures. The temperatures at which the full lines (in dry atmosphere) have been obtained can be distinguished by the fact that they do not overlap, and the higher capacity is obtained at lower temperature

3.9 Non-stoichiometric LFP: Defects and Related Properties

Deviation from ideal stoichiometry of $LiFePO_4$ has been investigated [62]. Any attempt to increase the Li concentration of samples prepared either by the precursor precipitation route or the continuous aqueous precursor synthesis route results in the formation of lithium phosphate impurity, in addition of stoichiometric LFP free of any Li vacancy. On another hand, Li-deficient homogenous solid solutions of composition $Li_{1-2x}Fe_xFePO_4$ could be obtained (Fig. 16). For $x \geq 0.06$, however, a sarcopside impurity phase is formed. Investigations of structural properties allowed us to define the defect responsible for the solid solution as $Fe^{\bullet}_{Li}$+V'_{Li} in the Kroger-Vink notation. Since the chemical formula of the sarcopside is obtained by writing $x = 1/2$ in the chemical formula of the solid solution, this impurity phase can be viewed as a condensation of the $Fe^{\bullet}_{Li}$+V'_{Li} defects. Magnetic measurements showed that isolated lithium vacancies V'_{Li} are also diluted in the $Li_{1-2x}Fe_xFePO_4$ matrix. The negative charge of the isolated V'_{Li} is compensated by the valence change $Fe^{2+} \rightarrow Fe^{3+}$ of an iron ion in its vicinity, forming a small magnetic polaron that is detected by magnetic measurements. The concentration of such polarons, however, remains very small as it saturates to a concentration of 0.2–0.3 mol%, much smaller than the concentration x in V'_{Li} bound to $Fe^{\bullet}_{Li}$. The electrochemical features are significantly damaged by the $Fe^{\bullet}_{Li}$ defects that block the diffusion of lithium along the corresponding channel, while the Li_3PO_4 only acts as an inert mass. Defects

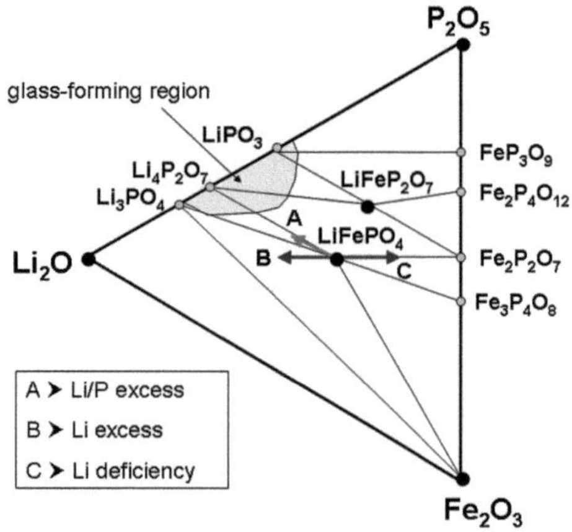

Fig. 16 Ternary phase diagram of off-stoichiometric $LiFePO_4$. The composition moves toward three directions such as *A* for Li/P excess, *B* for Li excess only and *C* for Li deficient

have an important impact on the electrochemical properties. The study of small particles (40 nm) has shown that the presence of defects or structural disorder favors the formation of the solid solution Li_xFePO_4 in the lithiation/delithiation process, instead of the phase separation between a Li-rich and Li-poor phase. This phenomenon has been observed in the surface layer (3 nm thick) alone in case the core of the particles is free of defects and well crystallized. It has also been observed in the whole volume of the particle when not only the surface layer, but the total particle is full of defects and cation vacancies, while the classical separation in two phases is recovered in case the particles (core plus surface layer) are free of defects.

3.10 Electrochemical Performance of LFP

Here, we present an overview of the high-temperature performance for an optimized $LiFePO_4$ sample, i.e. carbon-coated (C-LFP). The coffee-bag cell was charged and discharged at C/8 for the first cycle followed by 12 cycles at C/4 with 1 h rest before each charge and discharge. This high-temperature test was made at 60 °C, which is the appropriate condition to investigate possible iron dissolution in non-aqueous electrolytes [18, 65]. From structural and elemental analysis, no iron, even at the ppm level, was found in the electrolyte solution. Thus, all these data converge to the conclusion that the optimized $LiFePO_4$ is not soluble at 60 °C.

The typical electrochemical profile of the C-LFP/1 mol. L^{-1} $LiPF_6$-EC-DEC/Li 18,650-type cell cycled at 60 °C is shown in Fig. 17a. These experimental conditions (ca. 60 °C) have a severe impact on the kinetics of the Fe^{2+}/Fe^{3+} redox reaction, but the recent report from Hydro-Québec Research Labs showed that this type of C-LFP electrode can be cycled at 60 °C without significant capacity loss for over 200 cycles [35]. Optimized particle size in the range 200–300 nm agrees well with the average diameter of grains *L* that validates Eq. (2) with $D^* \approx 10^{-14}$ cm^2 s^{-1}

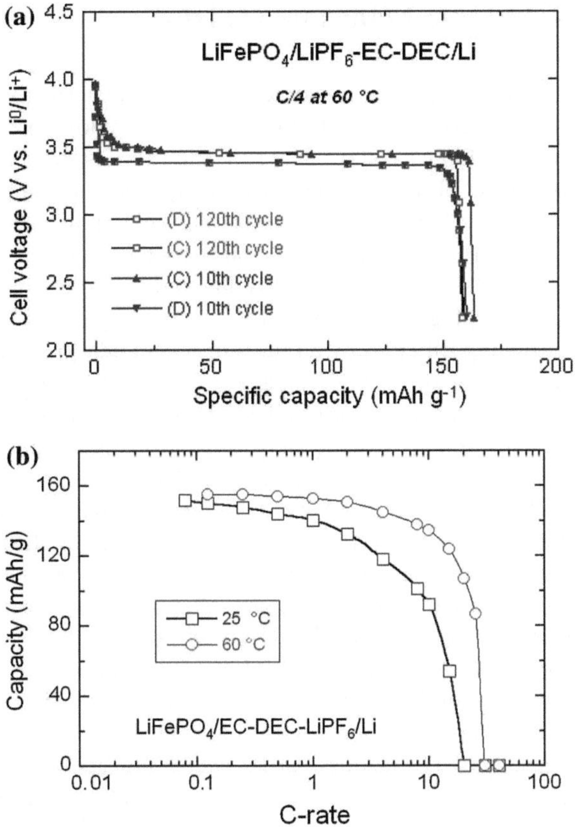

Fig. 17 **a** Voltage-capacity cycle for Li//LFP 18,650-type cell cycled at C/4 rate at 60 °C. **b** Peukert plots of the C-$LiFePO_4$/$LiPF_6$-EC-DEC/Li cells as a function of the working temperature 25 and 60 °C

in the $LiFePO_4$ matrix. The 10th and 120th cycle shows a similar specific capacity of 160 mAh g^{-1}. These results illustrate the excellent electrochemical performance of the carbon-coated olivine material. The electrode can be fully charged up to 4 V, which is its most reactive state. This remarkable performance is attributed to the optimized carbon-coated particles and their structural integrity under a large current in the electrode. Even at such a high cycling rate, C–$LiFePO_4$ exhibits rapid kinetics of lithium extraction and it realizes most of its theoretical capacity (170 mAh g^{-1}). The discharge profile appears with the typical voltage plateau (at ca. 3.45 V vs. Li^0/Li^+) attributed to the two-phase reaction of the $(1 - x)$ $FePO_4$+$x$$LiFePO_4$ system. The modified Peukert plots of cell cycled at 25 and 60 °C are shown in Fig. 17b. The cells were cycled in the potential range 2.5–4.0 V. The discharge capacity and electrochemical utilisation, i.e. the ratio discharge/charge versus cycle number is excellent for the C–$LiFePO_4$//Li cells. At 10C rate, these Li-ion cells provide coulombic efficiencies 85 % at 60 °C.

4 Lithium Manganese Olivine

$LiMnPO_4$ (LMP) is an attractive cathode material that offers several advantages: (i) it is low cost, (ii) due to the Mn^{2+}/Mn^{3+} redox potential of 4.1 V versus Li^0/Li^+ it presents higher voltage than LFP and (iii) it is suitable for the electrolytes used in Li batteries. However, the intrinsically specific capacity is restrained by (i) the inherently low ionic and electrical conductivities of $LiMnPO_4$ that seriously limit Li^+ insertion and extraction and (ii) the high kinetic barrier at the mismatched interface of $MnPO_4/LiMnPO_4$ due to the severe Jahn-Teller (JT) distortion [66]. As shown in Fig. 18, the electrical conductivity of $LiMnPO_4$ is lower than the already insulating $LiFePO_4$ by 5 orders of magnitude [67] making it challenging to achieve high capacity at high rates for $LiMnPO_4$ using methodologies developed for $LiFePO_4$. $LiMnPO_4$ crystallizes in an ordered olivine structure indexed by orthorhombic *Pnmb* space group. Typical crystallographic parameters are a = 6.106 (1) Å, b = 10.452(1) Å, and c = 4.746(1) Å.

LMP was prepared by the same methods described before for other olivine frameworks [68–92]. Figure 19 presents the HRTEM images of LMP samples with different morphologies. Synthesis techniques include solid-state reaction with different carbon sources [91], precipitation method using different precursors such as $NH_4MnPO_4{\cdot}H_2O$, $MnPO_4{\cdot}H_2O$ [68, 95], and modified ionothermal synthesis [69]. Drezen et al. [68] prepared LMP with different crystallite sizes by varying the sintering temperature. Using the polyol route, Wang et al. [69] and Martha et al. [70] obtained carbon-coated LMP particles of 30 nm that delivered a specific capacity of 110 mAh g^{-1} at 1C rate. A flower-like $LiMnPO_4$–C composite prepared by solid-state reaction showed a reversible capacity of 85 mAh g^{-1} at a rate of

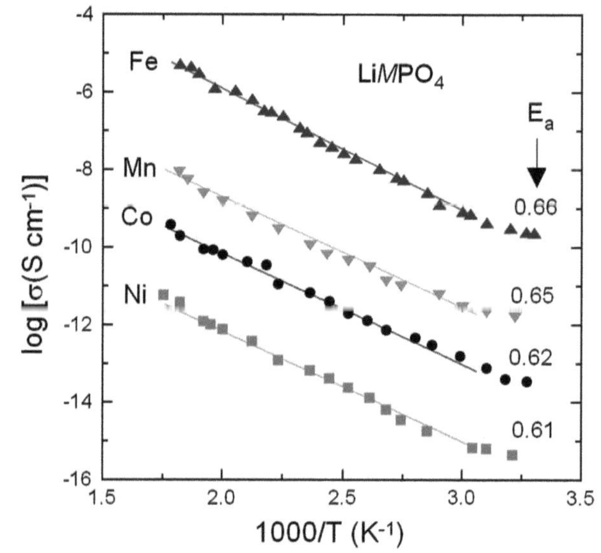

Fig. 18 The electrical conductivity of $LiMPO_4$ (M=Fe, Ni, Co, Mn) olivine materials. Numbers indicate the activation energy in eV. The low conductivity is related to the small free volume and separation of MO_6 octahedra by oxygen atoms of the $(PO_4)^{-3}$ anions

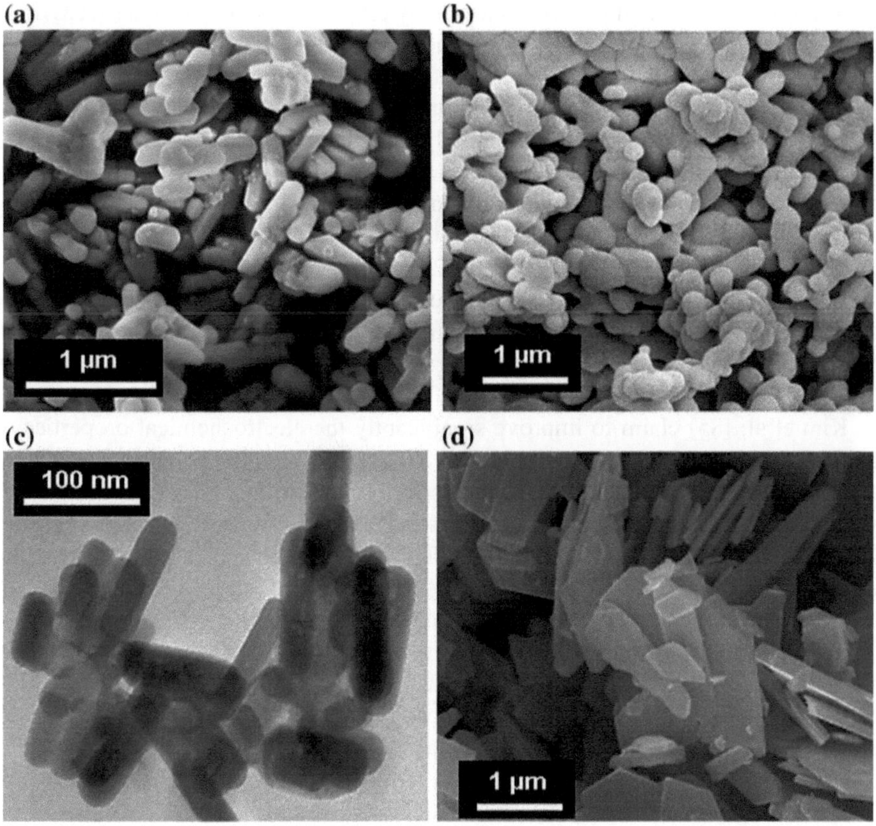

Fig. 19 HRTEM images of LMP samples with different morphologies. LMP nanoparticles were prepared by (**a**) hydro-thermal method with ascorbic acid, **b** polyol reflux method with tetraethylene glycol as solvent, **c** thermal decomposition method using oleic acid as surfactant and benzyl ether as solvent, and **d** precipitation route

0.05C [71]. The solution-based synthetic routes such as spray pyrolysis [72, 73], precipitation, sol-gel [74], hydrothermal method [75], solvothermal route [76, 77] and polyol synthesis [78] provide nanostructured LMP powders with enhanced electrochemical properties, which is mostly attributed to the higher chemical homogeneity and narrow particle size distribution of the material. The best results were obtained for $LiMnPO_4$ prepared via sol–gel, hydrothermal and co-precipitation routes. However, these techniques use the very low concentration of starting materials, which leads to the very low production rate [79]. 5–10 nm thin rod shaped LMP nanoparticles were synthesized by an improved thermal decomposition method using oleic acid as surfactant and benzyl ether as solvent [80]. LMP nanorods were also synthesized by modified polyol and resin coating processes to enhance their conductivity as well as their electrochemical properties with a capacity up to 120 mAh g^{-1} at 1C rate [81]. $LiCu_xMn_{1-x}PO_4/C$ nanorods prepared by a simple solvothermal process followed by calcination show that Cu successfully entered into the lattice of $LiMnPO_4$ and induced a decrease in the lattice parameters; however, a small initial discharge capacity of 87.5 mAh g^{-1} at 0.5C was reported [82]. A microwave irradiated solvothermal method with tetraethylene glycol as solvent was used to prepare dimensionally modulated, single-crystalline $LiMnPO_4$ with nano-thumblike shapes for high-power applications [94]. A two-step procedure for preparation of LMP with small particle size (15–20 nm) and embedded in a carbon matrix was presented by Pivko et al. [83]. Despite the nanosized particles limited capacity of 130 mAh g^{-1} at C/20 rate was observed at 55 °C after 100 cycles. A sequential precipitation method was applied to $LiMnPO_4$ olivine by confining $Mn_3(PO_4)_2$ precipitation on surface of a precursor seed of Li_3PO_4 that results in discharge capacity of 62 mAh g^{-1} at 5C rate for the size of LMP particles limited to less than 100 nm [84].

Kim et al. [85] claim to improve significantly the electrochemical properties of LMP by small amounts of co-doping of Fe and Co. The 4.0 V redox potential is achieved, while the kinetics are enhanced through formation of a local solid-solution like phase in the matrix, which is expected to lower the nucleation barrier of the delithiated phase and reduced the JT distortion. Lee et al. [86] evidenced the intercrystallite ionic transport in 40 nm × 200 nm sized $LiMnPO_4$ nanorods synthesized by the modified polyol method that resolves the problem of particle inhomogeneity occurring in stoichiometric synthesis. LMP nanoplates with thickness of 50 nm grew in the (100) plane via solid-state reaction in molten hydrocarbon such as paraffin and delivered a capacity 130 mAh g^{-1} at C/10 rate [87]. The effect of different carbon sources on the electrochemical properties of rod-like LMP/C nanocomposites was reported by Li et al. [88]. A reversible capacity of ~153 mAh g^{-1} at a rate of C/10 was obtained with beta-cyclodextrin as the carbon source, which is much better than that obtained from ascorbic acid, citric acid, glucose and sucrose. A carbon matrix, for restricting growth of LMP crystallites, was built on the small Li_3PO_4 crystallites precipitated from aqueous solutions, by the pyrolysis of sucrose. Small crystallite-size (8–12 nm) particles were successfully prepared using the carbon coated Li_3PO_4 as one of the reactants (the other reactant is $MnSO_4$) and the nuclei by a solvothermal method [89]. Chen and Richardson [90] demonstrated that the release of oxygen from LMP material occurs at 150 °C followed by the combustion of the electrolyte at 215 °C for a total heat of 884 J g^{-1} that is less thermal stability than LFP. The thermal stability mechanism on charged LMP electrode can be expressed as:

$$\underset{pmna}{MnPO_4} \xrightarrow{150-180\,^{\circ}C} \underset{JT\ distortion}{MnPO_4} \xrightarrow{490\,^{\circ}C} \underset{C2/m}{\tfrac{1}{2}Mn_2P_2O_7} + \tfrac{1}{2}O_2\uparrow \qquad (4)$$

From in-situ XRD, XPS and EDAX results on the charged and discharged LMP cathode, the charged state of $MnPO_4$ undergoes structural changes due to Jahn-Teller effect at above 180 °C and followed by reduction into pyrophosphate $Mn_2P_2O_7$ at ca. 490 °C. TGA-DSC measurements shown that the weight loss up to 450 °C for charged $MnPO_4$ shows similar behavior as that of discharged $LiMnPO_4$

cathode, indicating that the reaction is a decomposition of the SEI layer formed during high-voltage electrochemical cycling process. Note that a sharp exothermic peak is observed in the DSC scan during reduction due to the oxygen and CO_2 gas released after reacting with conductive carbon additive.

5 Lithium Mn–Fe Olivine

The solid solution $LiMn_yFe_{1-y}PO_4$ (LMFP) looks promising because it operates at 3.4–4.1 V that is not so high as to decompose the organic electrolyte but not so low as to sacrifice energy density [95–101]. Gardiner and Islam [99] have discussed the formation of the intrinsic defect-type with the lowest energy in the cation anti-site defect, in which Li and Fe/Mn ions exchange position in 4*c* sites. Migrations energies for Fe and Mn anti-site cations on Li sites suggest that Mn defects would impede bulk Li mobility in $LiMnPO_4$ to a greater extent than Fe anti-site defects in $LiFePO_4$. The electrochemical properties of the LMFP materials prepared by chelating-assisted hydrothermal method have been investigated by a number of research groups (see [100] and references herein). Trottier et al. [100] investigated the electrochemical properties of $LiMn_yFe_{1-y}PO_4$ ($0.5 \le y \le 0.8$) mixed-metal phospho-olivines (LMFP) grown by hydrothermal route assisted by ascorbic acid as chelating agent. Hong et al. [94] synthesized $LiMn_{1-x}Fe_xPO_4$ by a facile solvothermal approach with an excess of 20 wt% sucrose to yield carbon coating. Kosova et al. [95] reported structural studies of nanosized $LiFe_{0.5}Mn_{0.5}PO_4$ prepared by mechanochemically assisted carbothermal reduction route under cycling by in situ synchrotron diffraction. Wang et al. [96] reported the synthesis of $LiMn_{1-x}Fe_xPO_4$ nanorods on reduced graphene oxide sheets.

The rate capability for Li/C–$LiMn_{0.8}Fe_{0.2}PO_4$ cell is shown in Fig. 20. The cell was charged and discharged at the same C-rate starting from C/12 to 20C. For instance, the composite electrode including C–$LiMn_{0.8}Fe_{0.2}PO_4$ electrode 5 wt% Denka and 5 wt% VGCF delivers a capacity 92 mAh g^{-1} at 1C rate. For $LiMn_{0.8}Fe_{0.2}PO_4$, the electrode exhibited two reversible discharge plateaus at 3.95 and 3.52 V versus Li^0/Li^+ related to the Mn^{3+}/Mn^{2+} and Fe^{3+}/Fe^{2+} redox couples, respectively. From these profile, several conclusions can be drawn: (i) the charge-discharge profile exhibits similar shape for all compositions, (ii) the two pairs of peaks in dx/dV curves shift slightly with y as a result of the inductive effect (charge: 3.54 and 4.12 V; discharge 3.95 and 3.47 V for $y = 0.6$), (iii) the potentials of the reduction peaks are more sensitive than those of the oxidation peaks to the Mn^{2+}-substitution, (iv) a small increase of the over-voltage is seen as a function of y, and (v) the highest specific capacity is 130 mAh g^{-1} for $y = 0.5$, when the electrode was kept at 4.5 V until C/50 rate. Note that the polarization potential increases slightly for the Mn-rich compounds, which we attribute to a Jahn-Teller effect increased by the alloying process, rather than the formation of the intrinsic defect-type with the lowest energy in the cation anti-site defect, in which Li and Fe/Mn ions exchange positions.

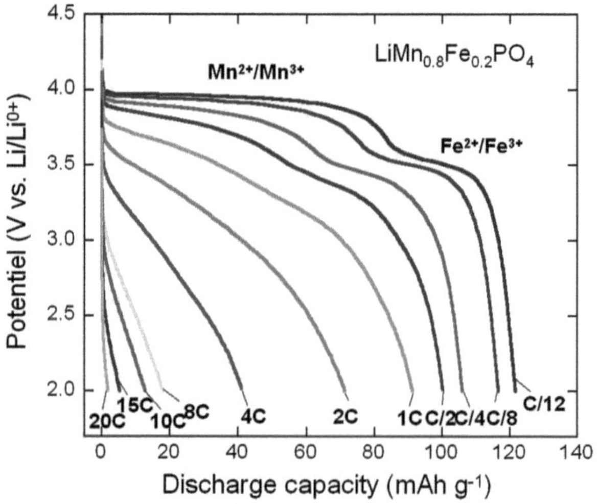

Fig. 20 Typical discharge curves of the Li//$LiMn_yFe_{1-y}PO_4$ ($y = 0.8$) cell

Magnetic properties of the solid solution $LiMn_yFe_{1-y}PO_4$ showed that for $y \leq 0.6$, all the Mn^{3+} ions in delithiated LMFP are in the high-spin state ($S = 2$). At larger manganese concentration, however, the Mn^{3+} ions in excess of the critical concentration $y_c = 0.6$ undergo a transition to the low-spin state ($S = 1$). The spin-transition of Mn^{3+} in concentration ($y - y_c$) to the low-spin state is at the origin of the strain fields at the molecular scale that increase with y for $y > 0.6$, and ultimately prevents the full delithiation for $y > 0.8$. This result sheds light on the reason for the degradation of cathode properties in Mn-rich compounds of the heterosite–purpurite series, while the electrochemical properties are good in the range $y \leq 0.6$ but only at slow rates, due to the very small hopping mobility of the small polaron [97, 98]. It is obvious that LMP exhibits inferior electrochemical performance compared with LFP. Optimizing the synthesis process and carbon coating should lead to promising electrochemical properties for LMP nanoparticles.

6 Lithium Cobalt Phosphate

Lithium cobalt phosphate, $LiCoPO_4$ (LCP) has attracted attention since it offers both flat high potential (at approximately 4.8 V vs. Li^0/Li^+), good theoretical capacity (167 mAg^{-1}) and smaller structure volume change. As other olivine compounds, LCP crystallizes with the orthorhombic symmetry (*Pnma* S.G.) with lattice parameters: $a = 10.2048$ Å, $b = 5.9245$ Å, and $c = 4.7030$ Å. However, like in the case of $LiMnPO_4$, the electrochemical performance of pristine LCP is very poor due to the low intrinsic electronic and ionic conductivity [67, 102–116]. Zhao et al. [116] prepared LCP micro-rods with the diameter of ca. 500 nm and length of ca. 5 μm by a hydrothermal method, which delivered a discharge capacity of only

65 mAh g^{-1} at C/10 rate. Solution-based methods are appealing to control LCP particles size, shape and morphology of nanocrystals. Truong et al. [102] used the supercritical fluid (SCF) processing with different amines having long alkyl chain such as hexamethylene diamine for adjusting physicochemical properties. Hedgehog-like LCP with hierarchical microstructures was first synthesized via a simple solvothermal process in water–benzyl alcohol mixed solvent at 200 °C [103]. The hedgehog-like microstructures in the size of about 5–8 μm were composed of large numbers of nanorods in diameter of ca. 40 nm and length of ca. 1 μm, which are coated with a carbon layer of ca. 8 nm in thickness by in situ carbonization of glucose during the solvothermal reaction. As a 5 V positive electrode material for rechargeable lithium battery, the hedgehog-like $LiCoPO_4$ delivered an initial discharge capacity of 136 mAh g^{-1} at C/10 rate and retained its 91 % after 50 cycles. LCP nanocrystals with (010) orientation were prepared by solvothermal method employing ethanol as the solvent and poly(vinyl pyrrolidone) as the carbon source and template [104].

In the early work by Amine et al. [117], it is demonstrated that Li can be reversibly removed from LCP at an average voltage of 4.8 V versus Li^+/Li with only a small contraction in the unit cell volume of the olivine lattice and the formation of a second olivine-like phase upon Li extraction from Li_xCoPO_4 with limited Δx = 0.42 lithium per formula unit. The electrochemical properties of $LiCoPO_4$ have been studied as a function of synthesis parameters. Effects on the discharge capacity have been investigated and capability improvements include: mixing LNP–LCP to obtained solid solutions as cathodes [105], carbon coating [106, 107], effect of oxygen partial pressure on the discharge capacity [108]. Wolfenstine et al. [109] have studied the structural evolution of $LiCoPO_4$ delithiated by the chemical oxidation. Okada et al. [110] have shown that $LiCoPO_4$ exhibited the highest 4.8 V discharge plateau of 100 mAh g^{-1} after initial charging to 5.1 V giving an energy density of 480 Wh kg^{-1} comparable to that of $LiCoO_2$. Electrical conductivity of LCP and mixed (Co, Ni), (Co, Mn) compounds [111], and of doped LCP [112] have also been explored. LCP having low electron conductivity, its use as the cathodic material is possible only in the case of synthesis of the $LiCoPO_4$/C composite [103]. Such composites can be discharged at the potentials of 4.7–4.8 V. However, cyclability of such composites is very low, as decomposition of liquid electrolyte occurs under charging in the potential range of 4.8–5.1 V simultaneously with oxidation of Co^{2+} to Co^{3+}. The initial discharge capacity of $LiCoPO_4$/C is close to the theoretical one that is about 167 mAh g^{-1}. Figure 21 compares the discharge curves of various lithium cells including $LiCoPO_4$ (LCP), $LiCoO_2$ (LCO), $LiFePO_4$ (LFP) and $Li_3V_2(PO_4)_3$ (LVP).

The phase transitions occurring upon lithium insertion-extraction of $LiCoPO_4$ have been investigated by several groups [111–115]. A two phase mechanism was confirmed by in situ synchrotron diffraction [111]. An amorphization of the phosphate was observed after electrochemical or chemical oxidation [113]. Nagayama et al. [114] suggested from X-ray absorption spectroscopy a hybridization effect between the Co 3*d* and O 2*p* orbitals and the polarization effect introduced by Li ions. Bramnik et al. [115] revealed the appearance of two

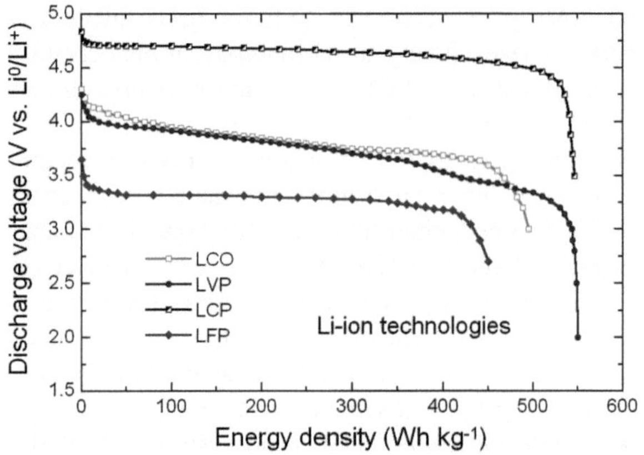

Fig. 21 Comparison of discharge curve of several cathode materials: $LiCoPO_4$ (LCP), $LiCoO_2$ (LCO), $LiFePO_4$ (LFP) and $Li_3V_2(PO_4)_3$ (LVP)

orthorhombic phases upon electrochemical Li extraction. The $LiCoPO_4$ and the Li deficient phases, $Li_{0.7}CoPO_4$ and $CoPO_4$, are responsible for the two voltage plateaus at 4.8 and 4.9 V versus Li^+/Li. The hedgehog-like $LiCoPO_4$ microstructures in the size of about 5–8 μm are composed of large numbers of nanorods in diameter of ca. 40 nm and length of ca. 1 μm, which are coated with a carbon layer of ca. 8 nm in thickness by in situ carbonization of glucose during the solvothermal reaction. As a 5 V positive electrode material for rechargeable lithium battery, the hedgehog-like $LiCoPO_4$ delivers an initial discharge capacity of 136 mAh g^{-1} at C/10 rate and retains its 91 % after 50 cycles [111]. Surface modification of LCP particles results in a satisfactory cyclability for $LiCoPO_4$ to be used as a 5 V cathode material [115]. The capacity retention of Al_2O_3-coated LCP was 105 mAh g^{-1} after 50 cycles at $T = 55$ °C. Jang et al. [118] claimed that $LiFePO_4$ coated LCP particles prepared by SSR method (~100–150 nm) show improved battery performance with an initial discharge capacity of 132 mAh g^{-1}, but did neither mention the C-rate nor the current density. The ability to suppress thermal runaway of $LiMPO_4$ olivine frameworks is attributed to the high covalent feature of the P–O bonds in the tetrahedral (PO_4) units, which stabilizes the olivine structure and prevents oxygen release from the charged (delithiated) olivine materials up to 600 °C. This is still controversial in LCP and LNP olivine lattices. In particular, a thermal instability has been reported in the charged (i.e. delithiated) state of $LiCoPO_4$ [119]. Both olivine-like phases Li_zCoPO_4 (z = 0.6) and $CoPO_4$ appearing during the delithiation of $LiCoPO_4$ are unstable upon heating, and decompose readily in the range 100–200 °C. The decomposition of lithium-poor phases leads to gas evolution and the crystallization of $Co_2P_2O_7$. Incorporation of lithium bis(oxalato)borate (LiBOB) as additive in conventional electrolyte solutions enhances the electrochemical performance of LCP electrode [120].

7 Lithium Nickel Phosphate

Compared with $LiFePO_4$ and $LiMnPO_4$, $LiNiPO_4$ (LNP) displays much higher redox potential [121–143]. Wolfenstine and Allen [121] determined the Ni^{3+}/Ni^{2+} redox potential between 5.1 and 5.3 V for the LNP olivine lattice. To overcome the problems with low electrolyte stability, a 1 mol L^{-1} $LiPF_6$ in tetramethylene sulfone electrolyte was used because of its high oxidative stability, around 5.8 V versus Li^0/Li^+. These experimental values are in excellent agreement with the theoretical predictions [139–141]. Recently, a review has presented the progress in the fabrication of LNP powders and the general synthesis approaches to circumvent the drawbacks of LNP [142]. Among the various synthesis approaches, solution-based methods have been successful for LNP. Standard hydrothermal route starts with mixing Li salt with either $Ni(C_2H_3O_2)_2{\cdot}4H_2O$ or $Ni(NO_3){\cdot}6H_2O$ in different reaction media such as NH_4OH, cotton fibers, 1,2-propanediol and ethylene glycol [36, 143–152]. Tsai [145] produced $LiNiPO_4$ nanocrystals via a microemulsions method and determined that the water-to-surfactant ratio directly affects the particle size and shape. Various attempts were made to prepare nanosized LNP particles using sol–gel method. Prabu and Selvasekarapandian [146] prepared plate-like LNP (surface area 0.25 m^2 g^{-1}) by adding citric acid and polyethylene glycol as chelating agent for gelation. Yang and Xu [144] obtained phase pure $LiNiPO_4$ with desirable particles sub-micron size and free of agglomeration. Nanometric crystallite size were obtained by using both the acrylamide and N,N′-methylenebis(acrylamide) as gelling agents [147]. Lanthanum- and zinc-doped $LiNiPO_4$ was prepared by the polyol method from metal acetates and 1,2 propanediol as polyol medium [121–122, 148]. Nano-particles <50 nm were obtained. Dimesso et al. [123] reported the synthesis of graphitic carbon foam—$LiNiPO_4$ composites prepared by a Pechini-assisted sol–gel method using water as solvent. The difficulties in obtaining a LNP pure phase have been pointed out by several groups [121, 123–125].

Solid-state reaction has been widely used for LNP production. Impurities such as NiO, Ni_3P, $Ni_2P_2O_7$ and $Li_2Ni_3(P_2O_7)_2$ have been identified. The carbothermal-reaction method was employed by Herle et al. [126] using heat-treatment in flowing argon atmosphere. The most commonly used precursors are Li_2CO_3, $LiOH{\cdot}H_2O$, Li_3PO_4, but also LiH_2PO_4 for Li, $Ni(COO)_2{\cdot}2H_2O$, $Ni(C_2H_3O_2)_2{\cdot}4H_2O$, $Ni(C_2H3O_2)_2$, NiO, $Ni(CO_3)$, $Ni(OH)_2$, $Ni_3(PO_4)_2$ for Ni, and $NH_4H_2PO_4$, $(NH4)_2HPO_4$, a mixture of $(NH_4)_2HPO_4$ and P_2O_5 or $(NH_4)_3PO_4$ for P, respectively. Sugiyama et al. [127] prepared phase pure $LiNiPO_4$ by planetary ball milling and calcination under argon at 750 °C for 6 h. Wang et al. [128] only obtain phase pure $LiNiPO_4$ after calcination above 800 °C, while lower temperature processes provide samples containing Li_3PO_4 and NiO_x. The olivine analogue compound containing mixed (Mg, Ni) cations contained Li_3PO_4 as a second phase upon synthesis; however a carbothermal reduction method produced a single-phase compound [129]. Solid-state methods are also of importance in terms of obtaining ordered crystal structure. As an example, Julien et al. [125, 130] reported the growth of $LiNiPO_4$ with high purity using modified solid-state reaction with different sets of reactants such as (i)

Fig. 22 SEM images of $LiNiPO_4$ samples synthesized by **a** the ammonium-hydrogen-phosphate route, and **b** the lithium-dihydrogen-phosphate route

Li_2CO_3, $(NH_4)_2HPO_4$, and NiO (ammonium-hydrogen-phosphate route), and (ii) NiO and LiH_2PO_4 (lithium-dihydrogen-phosphate route) dissolved in acetone solution. Single phases of LNP were obtained after sintering at 800 °C for 36 h. Figure 22 shows the typical SEM images of LNP powders synthesized by solid-state reactions [125]. Structural characteristics of particle synthesized by the ammonium-hydrogen-phosphate route were studied in detail using analytical electron microscopy. Results are shown in Fig. 23. The high-resolution transmission electron microscopy (HRTEM) and selected area electron diffraction (SAED) images indicate that the grown $LiNiPO_4$ is well-crystallized in the olivine structure without any indication of crystallographic defects such as dislocations or misfits.

The absence of redox peaks when LNP was heated under argon suggests that this material has a very low intrinsic electrical conductivity and hence, additional treatments such as carbon coating are required for LNP to exhibit Li insertion/deinsertion. Indeed, Rissouli et al. [67] determined that the electrical conductivity of $LiNiPO_4$ was 2–3 decades lower than that for $LiCoPO_4$ and $LiMnPO_4$ (Fig. 18). Such a poor electronic conductivity has been recently confirmed by ac measurements on nano-crystalline powders [131]. Herle et al. [126] have pointed out that $LiNiPO_4$ exhibits a percolating nano-network of metal-rich phosphides (i.e. Ni_3P, Ni_2P, or NiP_3) responsible for enhanced conductivity. Magnetic anisotropy in Li-phosphates and the origin of their magneto-electric properties have been investigated [125, 132, 133]. Magnetic properties of LCP and LNP show that antiferromagnetic *M*–O–*M* superexchange interactions couple the spins closely in planes parallel to (100) [133] with a Néel temperature $T_N = 21$ K. The magnetically ordered phases and spin dynamics of magnetoelectric $LiNiPO_4$ have been studied in magnetic field as high as 17.3 T along the *c* axis and up to 16 T along the *a*- and *b*-axis [134]. Vaknin et al. reported the commensurate-incommensurate magnetic phase transition in magneto-electric single crystal $LiNiPO_4$ [135]. Local environment and bonding strength of cations were studied by vibrational spectroscopy, i.e. Raman and FTIR [125, 136].

The information available about the electrochemical performance of LNP is very limited. Few reports have shown that LNP is not electrochemically active when it is

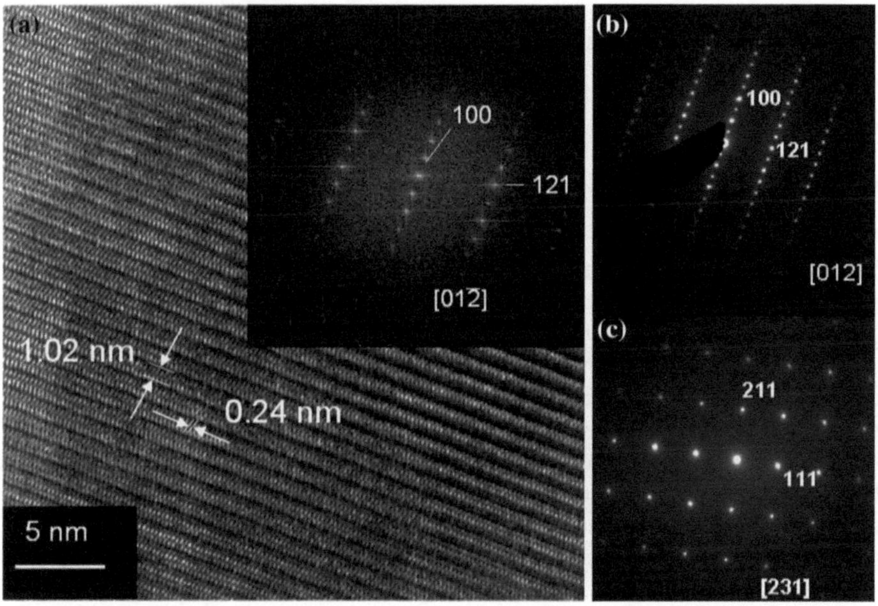

Fig. 23 **a** HRTEM image of $LiNiPO_4$. The well-resolved lattice fringes indicate the crystallinity of the sample. The measured values of the lattice fringe spacing are indicated. Inset shows the FFT of the HRTEM image indicating the direction in which the crystallite was examined and lattice planes contributing to the diffraction. **b**, **c** SAED patterns of $LiNiPO_4$. The assignments of diffraction maxima and the direction of view are indicated

charged over 5.2 V [137–139, 149]. However, Wolfenstine and Allen [121] mentioned the electrochemical activity of LNP powders prepared by SSR method under high purity argon with addition of a thin layer of carbon coat. The voltammetry displayed an oxidation peak at ~5.3 V and a reduction peak at ~5.1 V.

Recently, Dimesso et al. [123, 150] have reported the preparation of LNP and LCP by a Pechini assisted sol–gel process that provides material exhibiting redox peaks at ~5.2 and ~4.9 V versus Li^0/Li^+, respectively. Mg-substituted LNP/graphitic carbon foams composite was also synthesized by the same method that shows a discharge capacity of 126 mAh g^{-1} at C/10 rate by substituting 0.2 Mg for Ni [151]. Modifications of the local structure and lattice parameters in $LiFe_{1-y}Ni_yPO_4$ ($0 < x < 1$) olivine-type solid solutions have been studied by X-ray absorption spectroscopy (XAS) [152, 153]. Results indicate ordering of Li in the M1 site, and of Fe and Ni in the M2 site of the olivine structure. Ni doping is found to induce an anisotropic shrinking of the unit cell with both Fe and Ni six-coordinated with oxygens, occupying distorted octahedral sites. $LiNi_{1-x}Cu_xPO_4$ ($0 < x < 0.99$) samples were successfully synthesized by non-aqueous sol–gel method followed by microwave (MW) annealing were found to be electrocatalysts towards oxygen reduction which is apparently due to nanosized particles as obtained via MW annealing [152]. The solid solution $LiNi_xFe_{1-x}PO_4$ ($0 < x < 1$)

synthesized via a solid-state reaction under Ar ambient shows lattice parameters following the Vegard's law but the formation of agglomerates reduces the specific capacity that cannot exceed 45 mAh g^{-1} [153].

8 Concluding Remarks

The use of lithiated frameworks has revolutionized the concept of lithium rechargeable batteries for high-power sources applied to hybrid electric vehicles (HEVs) and plug-in hybrid electric vehicles (PHEVs). To date, most interest has been focused on the $LiFePO_4$ phase, which is already in commercial use. In this chapter, we have provided the state-of-the-art of olivine (orthophosphate) compounds used as positive electrodes for the new generation of Li-ion batteries. The successful application of lithium iron phosphate has triggered great interest in the R&D of high-power batteries with remarkable safety concerns despite the fact that all these materials are insulators. Other olivine $LiMPO_4$ (M=Mn, Ni, Co) compounds, which have been identified as potential high-voltage cathode, are creating intensive studies both from scientific and technological points of view and some of the more prominent follow-up activities have been briefly discussed. Unfortunately these materials all suffer from a number of drawbacks that must be overcome before they can compete with $LiFePO_4$. The limitation in capacity attributed to the diffusion-limited transfer of lithium across the two-phase interface $LiMPO_4$–MPO_4 as end members without much solubility has been overcome with reducing the particle at the nanometer scale and by painting them with conducting carbon. For example, $LiMnPO_4$ has poorer Li-ion conductivity than $LiFePO_4$, $LiNiPO_4$ also exhibits extremely poor electronic conductivity and extremely low charge/discharge capacity. Nevertheless, the use of nanosized solid-state materials not only improves the power density, but also facilitates Li-ions insertion/extraction in/out of the storage materials, which improves the cycle life of these batteries. $LiCoPO_4$ (4.8 V) and $LiNiPO_4$ (5.2 V) with higher operating voltage are appealing for increasing the energy density. Owing to this, large room for improvements in the performances in Li-cells of LCP and LNP is expected by assessing the synthesis route, the coatings, the doping/substitution and the electrolyte additives in order to boost its ability to reversibly cycle lithium. However, the use of these cathodes requires challenges in the development of more stable and robust electrolytes.

References

1. Padhi K, Nanjundaswamy KS, Goodenough JB (1997) Phospho-olivines as positive-electrode materials for rechargeable lithium batteries. J Electrochem Soc 144:1188–1194
2. Padhi K, Nanjundaswamy KS, Masquelier C, Okada S, Goodenough JB (1997) Effect of structure on the Fe^{3+}/Fe^{2+} redox couple in iron phosphates. J Electrochem Soc 144:1609–1613

3. Huang H, Yin SC, Nazar LF (2001) Approaching theoretical capacity of $LiFePO_4$ at room temperature at high rates. Electrochem Solid State Lett 4:A170–A172
4. Dominko D, Gaberscek M, Drofenik J, Bele M, Jamnik J (2003) Influence of carbon black distribution on performance of oxide cathodes for Li ion batteries. Electrochim Acta 48:3709–3716
5. Ravet N, Goodenough JB, Besner S, Simoneau M, Hovington P, Armand M (1999) Improved iron based cathode material. In: Proceedings of the 196th ECS meeting, Honolulu, Oct 1999, Extended Abstract no 127
6. Ravet N, Chouinard Y, Magnan JF, Besner S, Gauthier M, Armand M (2001) Electroactivity of natural and synthetic triphylite. J Power Sources 97:503–507
7. Bewlay SL, Konstantinov K, Wang GX, Dou SX, Liu HK (2004) Conductivity improvements to spray-produced $LiFePO_4$ by addition of a carbon source. Mater Lett 58:1788–1791
8. Julien CM, Mauger A, Ait-Salah A, Massot M, Gendron F, Zaghib K (2007) Nanoscopic scale studies of $LiFePO_4$ as cathode material in lithium-ion batteries for HEV application. Ionics 13:395–411
9. Weppner W, Huggins R (1977) Determination of the kinetic parameters of mixed-conducting electrodes and applications to the system Li_3Sb. J Electrochem Soc 124:1569–1578
10. Chen Z, Dahn JR (2002) Reducing carbon in $LiFePO_4$/C composite electrodes to maximize specific energy, volumetric energy and tap density. J Electrochem Soc 149:A1184–A1189
11. Ravet N, Besner S, Simoneau M, Vallée A, Armand M, Magnan JF (2005) Electrode materials with high surface conductivity. US Patent 6,962,666, 8 Nov 2005
12. Nanjundaswamy KS, Padhi AK, Goodenough JB, Okada S, Ohtsuka H, Arai H, Yamaki J (1996) Synthesis, redox potential evaluation and electrochemical characteristics of NASICON-related-3D framework compounds. Solid State Ionics 92:1–10
13. Pahdi AK, Manivannan M, Goodenough JB (1998) Tuning the position of the redox couples in materials with NASICON structure by anionic substitution. J Electrochem Soc 145:1518–1520
14. Manthiram A, Goodenough JB (1989) Lithium insertion into $Fe_2(SO_4)_3$ frameworks. J Power Sources 26:403–408
15. Nyten A, Abouimrane A, Armand M, Gustafsson T, Thomas JO (2005) Electrochemical performance of Li_2FeSiO_4 as a new Li-battery cathode material. Electrochem Commun 7:156–160
16. Barker J, Saidi MY, Swoyer JL (2003) Electrochemical insertion properties of the novel lithium vanadium fluorophosphate, $LiVPO_4F$. J Electrochem Soc 150:A1394–A1398
17. Zaghib K, Mauger A, Julien CM (2012) Overwiew of olivines in lithium batteries for green transportation and energy storage. J Solid State Electrochem 16:835–845
18. Julien CM, Zaghib K, Mauger A, Groult H (2012) Enhanced electrochemical properties of $LiFePO_4$ as positive electrode of Li-ion batteries for HEV application. Adv Chem Eng Sci 2:321–329
19. Fergus JW (2010) Recent developments in cathode materials for lithium ion batteries. J Power Sources 195:939–954
20. Ait-Salah A, Mauger A, Julien CM, Gendron F (2006) Nanosized impurity phases in relation to the mode of preparation of $LiFeO_4$ effects. Mater Sci Eng, B 129:232–244
21. Ait-Salah A, Mauger A, Zaghib K, Goodenough JB, Ravet N, Gauthier M, Gendron F, Julien CM (2006) Reduction of Fe^{3+} impurities in $LiFePO_4$ from the pyrolysis of organic precursor used for carbon deposition. J Electrochem Soc 153:A1692–A1701
22. Zaghib K, Dontigny M, Charest P, Labrecque JF, Guerfi A, Kopec M, Mauger A, Gendron F, Julien CM (2010) $LiFePO_4$: from molten ingot to nanoparticles with high-rate performance in Li-ion batteries. J Power Sources 195:8280–8288
23. Brochu F, Guerfi A, Trottier J, Kopeć M, Mauger A, Groult H, Julien CM, Zaghib K (2012) Structure and electrochemistry of scaling nano C-$LiFePO_4$ synthesized by hydrothermal route: complexing agent effect. J Power Sources 214:1–6

24. Vediappan K, Guerfi A, Gariépy V, Demopoulos GP, Hovington P, Trottier J, Mauger A, Julien CM, Zaghib K (2014) Stirring effect in hydrothermal synthesis of C-$LiFePO_4$. J Power Sources 266:99–106
25. Herle PS, Ellis B, Coombs N, Nazar LF (2004) Nano-network electronic conduction in iron and nickel olivine phosphates. Nat Mater 3:147–152
26. Ait-Salah A, Dodd J, Mauger A, Yazami R, Gendron F, Julien CM (2006) Structural and magnetic properties of $LiFePO_4$ and lithium extraction effects. Z Allg Inorg Chem 632:1598–1605
27. Ellis BL, Lee KT, Nazar LF (2010) Positive electrode materials for Li-ion and Li-batteries. Chem Mater 22:691–714
28. Ravet N, Gauthier M, Zaghib K, Goodenough JB, Mauger A, Gendron F, Julien CM (2007) Mechanism of the Fe^{3+} reduction at low temperature for $LiFePO_4$ synthesis from polymeric precursor. Chem Mater 19:2595–2602
29. Laffont L, Delacourt C, Gibot P, Wu MY, Kooyman P, Masquelier C, Tarascon JM (2006) Study of the $LiFePO_4$/$FePO_4$ two-phase system by high-resolution electron energy loss spectroscopy. Chem Mater 18:5520–5529
30. Zaghib K, Mauger A, Gendron F, Julien CM (2008) Surface effects on the physical and electrochemical properties of thin $LiFePO_4$ particles. Chem Mater 20:462–469
31. Yang S, Zavajil PY, Whittingham MS (2001) Hydrothermal synthesis of lithium iron phosphate cathodes. Electrochem Commun 3:505–508
32. Sato M, Tajimi S, Okawa H, Uematsu K, Toda K (2002) Preparation of iron phosphate cathode material of $Li_3Fe_2(PO_4)_3$ by hydrothermal reaction and thermal decomposition processes. Solid State Ionics 152–153:247–251
33. Dokko K, Koizumi S, Kanamura K (2006) Electrochemical reactivity of $LiFePO_4$ prepared by hydrothermal method. Chem Lett 35:338–339
34. Dokko K, Koizumi S, Sharaishi K, Kananura K (2007) Electrochemical properties of $LiFePO_4$ prepared via hydrothermal route. J Power Sources 165:656–659
35. Jin B, Gu HB (2008) Preparation and characterization of $LiFePO_4$ cathode materials by hydrothermal method. Solid State Ionics 178:1907–1914
36. Murugan AV, Muraliganth T, Manthiram A (2009) One-pot microwave-hydrothermal synthesis and characterization of carbon-coated $LiMPO_4$ (M=Mn, Fe, and Co) cathodes. J Electrochem Soc 156:A79–A83
37. Kim DH, Kim J (2006) Synthesis of $LiFePO_4$ nanoparticles in polyol medium and their electrochemical properties. Electrochem Solid State Lett 9:A439–A442
38. Kim DH, Kim J (2007) Synthesis of $LiFePO_4$ nanoparticles and their electrochemical properties. J Phys Chem Solids 68:734–737
39. Kim DH, Lim JS, Kang JW, Kim EJ, Ahn HY, Kim J (2007) A new synthesis route to nanocrystalline olivine phosphates and their electrochemical properties. J Nanosci Nanotechnol 7:3949–3953
40. Azib T, Ammar S, Nowak S, Lau-Truing S, Groult H, Zaghib K, Mauger A, Julien CM (2012) Crystallinity of nano C-$LiFePO_4$ prepared by the polyol process. J Power Sources 217:220–228
41. Saravanan K, Reddy MV, Balaya P, Gong H, Chowdari BVR, Vittal JJ (2009) Storage performance of $LiFePO_4$ nanoplates. J. Mater. Chem 19:605–610
42. Yang H, Wu XL, Cao MH, Guo YG (2009) Solvothermal synthesis of $LiFePO_4$ hierarchically dumbbell-like microstructures by nanoplate self-assembly and their application as a cathode material in lithium-ion batteries. J Phys Chem C 113:3345–3351
43. Arcon D, Zorko A, Dominko R, Jaglicic Z (2004) A comparative studies of magnetic properties of $LiFePO_4$ and $LiMnPO_4$. J Phys C Condens Matter 16:5531–5548
44. Geller S, Durand JL (1960) Refinement of the structure of $LiMnPO_4$. Acta Crystallogr 13:325–329
45. Santorro RP, Newnham RE (1987) Antiferromagnetism in $LiFePO_4$. Acta Crystallogr 22:344–347

46. Streltsov VA, Belokoneva EL, Tsirelson VG, Hansen NK (1993) Multipole analysis of the electron density in triphylite $LiFePO_4$ using X-ray diffraction data. Acta Crystallogr B 49:147–153
47. Rousse G, Rodriguez-Carvajal J, Patoux S, Masquelier C (2003) Magnetic structures of the triphylite $LiFePO_4$ and its delithiated form $FePO_4$. Chem Mater 15:4082–4090
48. Losey A, Rakovan J, Huges J, Francis CA, Dyar MD (2004) Structural variation in the lithiophilite-triphylite series and other olivine-group structures. Canad Mineral 42:1105–1108
49. Andersson AS, Thomas JO (2001) The source of first-cycle capacity loss in $LiFePO_4$. J Power Sources 97–98:498–502
50. Nyten A, Thomas JO (2006) A neutron powder diffraction study of $LiCo_xFe_{1-x}PO_4$ for x = 0, 0.25, 0.40, 0.60 and 0.75. Solid State Ionics 177:1327–1330
51. Beale AM, Sankar G (2002) Following the structural changes in iron phosphate catalysts by in situ combined XRD/QuEXAFS technique. J Mater Chem 12:3064–3072
52. Yamada A, Chung SC, Hinokuma K (2001) Optimized $LiFePO_4$ for lithium battery cathodes. J Electrochem Soc 148:A224–A229
53. Andersson AS, Thomas JO (2001) The source of first-cycle capacity loss in $LiFePO_4$. J Power Sources 97–98:498–502
54. Julien CM, Zaghib K, Mauger A, Massot M, Ait-Salah A, Selmane M, Gendron F (2006) Characterization of the carbon-coating onto $LiFePO_4$ particles used in lithium batteries. J Appl Phys 100:063511
55. Ait-Salah A, Jozwiak P, Zaghib K, Garbarczyk J, Gendron F, Mauger A, Julien CM (2006) FTIR features of lithium-iron phosphates as electrode materials for rechargeable lithium batteries. Spectrochim Acta A 65:1007–1013
56. Julien CM, Ait-Salah A, Gendron F, Morhange JF, Mauger A, Ramana CV (2006) Microstructure of $LiXPO_4$ (X=Ni Co, Mn) prepared by solid-state chemical reaction. Scripta Mater 55:1179–1182
57. Burba CM, Frech R (2006) In situ transmission FTIR spectroelectrochemistry: A new technique for studying lithium batteries. Electrochem Acta 52:780–785
58. Paques-Ledent MT, Tarte P (1974) Vibrational studies of olivine-type compounds—II orthophosphates, -arsenates and -vanadates $A^IB^{II}X^VO_4$. Spectrochim Acta, Part A 30: 673–689
59. Ait-Salah A, Zaghib K, Mauger A, Gendron F, Julien CM (2006) Magnetic studies of the carbothermal effect on $LiFePO_4$. Phys Status Solidi A 203:R1–R3
60. Doeff MM, Hu Y, McLarnon F, Kostecki R (2003) Effect of surface carbon structure on the electrochemical performance of $LiFePO_4$. Electrochem Solid State Lett 6:A207–A209
61. Arnold G, Garche J, Hemmer R, Ströbele S, Vogler C, Wohlfgang-Mehrens M (2003) Fine-particle lithium iron phosphate $LiFePO_4$ synthesized by a new low-cost aqueous precipitation technique. J Power Sources 119–121:247–251
62. Axmann P, Stinner C, Wohlfahrt-Mehrens M, Mauger A, Gendron F, Julien CM (2009) Non-stoichiometric $LiFePO_4$: defects and related properties. Chem Mater 21:1636–1644
63. Zaghib K, Dontigny M, Charest P, Labrecque JF, Guerfi A, Kopec M, Mauger A, Gendron F, Julien CM (2008) Aging of $LiFePO_4$ upon exposure to H_2O. J Power Sources 185:698–710
64. Porcher W, Moreau P, Lestriez B, Jouanneau S, Guyomard D (2008) Electrocvhem Solid State Lett 11:A4–A8
65. Zaghib K, Ravet N, Gauthier M, Gendron F, Mauger A, Goodenough JB, Julien CM (2006) Optimized electrochemical performance of $LiFePO_4$ at 60 °C with purity controlled by SQUID magnetometry. J Power Sources 163:560–566
66. Bramnik NN, Ehrenberg H (2008) Precursor-based synthesis and electrochemical performance of $LiMnPO_4$. J Alloys Compd 464:259–264
67. Rissouli K, Benkhouja K, Ramos-Barrado JR, Julien C (2003) Electrical conductivity in lithium orthophosphates. Mater Sci Eng B 98:185–189
68. Drezen T, Kwon NH, Bowen P, Teerlinck I, Isono M, Exnar I (2007) Effect of partic1e size on $LiMnPO_4$ cathodes. J Power Sources 174:949–953

69. Wang D, Buqa H, Crouzet M, Deghenghi G, Drezen T, Exnar I, Kwon NH, Miners J, Poletto L, Gratzel M (2009) High-performance. nano-structured $LiMnPO_4$ synthesized via a polyol method. J Power Sources 189:624–628
70. Martha SK, Markovsky B, Grinblat J, Gofer Y, Haik O, Zinigrad E, Aurbach D, Drezen T, Wang D, Deghenghi G, Exnar I (2009) $LiMnPO_4$ as an advanced cathode material for rechargeable lithium batteries. J Electrochem Soc 156:A541–A552
71. Liu JL, Hu DG, Huang T, Yu AS (2012) Synthesis of flower-like $LiMnPO_4/C$ with precipitated $NH_4MnPO_4 \cdot H_2O$ as precursor. J Alloys Compd 518:58–62
72. Bakenov Z, Taniguchi I (2010) $LiMg_xMn_{1-x}PO_4/C$ cathodes for lithium batteries prepared by a combination of spray pyrolysis with wet ballmilling. J Electrochem Soc 157:A430–A436
73. Doan TNL, Bakenov Z, Taniguchi I (2010) Preparation of carbon coated $LiMnPO_4$ powders by a combination of spray pyrolysis with dry ball-milling followed by heat treatment. Adv Powder Technol 21:187–196
74. Xiao J, Xu W, Choi D, Zhang J (2010) Synthesis and characterization of lithium manganese phosphate by a precipitation method. J Electrochem Soc 157:A142–A147
75. Fujimoto D, Lei Y, Huang ZH, Kang F, Kawamura J (2014) Synthesis and electrochemical performance of $LiMnPO_4$ by hydrothermal method. Int J Electrochem 2014:768912
76. Devaraju MK, Honma I (2012) Hydrothermal and solvothermal process towards development of $LiMPO_4$ (M=Fe, Mn) nanomaterials for lithium-ion batteries. Adv Ener Mater 2:284–297
77. Zhou F, Zhu P, Fu X, Chen R, Sun R, Wong C (2014) Comparative study of $LiMnPO_4$ cathode materials synthesized by solvothermal methods using different manganese salts. Cryst Eng Comm 16:766–774
78. Wang D, Buqa H, Crouzet M et al (2009) High-performance, nano-structured $LiMnPO_4$ synthesized via a polyol method. J Power Sources 189:624–628
79. Bakenov Z, Taniguchi I (2011) $LiMnPO_4$ olivine as a cathode for lithium batteries. Open Mater Sci J 5:222–227
80. Kwon NH, Fromm KM (2012) Enhanced electrochemical performance of <30 nm thin $LiMnPO_4$ nanorods with reduced amount of carbon as a cathode for lithium ion batteries. Food Chem 133:1435–1440
81. Kumar PR, Venkateswarlu M, Misra M, Mohanty AK, Satyanarayana N (2011) Carbon coated $LiMnPO_4$ nanorods for lithioum batteries. J Electrochem Soc 158:A227–A230
82. Yuan LF, Ge LL, Shen YH, Zhang H, Wang CP, Xie AJ (2013) Synthesis and electrochemical properties of Cu-doped $LiMnPO_4/C$ nanorods as cathode materials of lithium-ion batteries. J Nano Res 25:1–7
83. Pivko M, Bele M, Tchernychova E, Logar NZ, dominko R, Gaberscek M (2012) Synthesis of nanometric $LiMnPO_4$ via a two-step technique. Chem Mater 24:1041–1047
84. Kim TA, Park HS, Lee MH, Lee SY, Song HK (2012) Restricted growth of $LiMnPO_4$ nanoparticles evolved from a precursor seed. J Power Sources 210:1–6
85. Kim J, Seo DH, Kim SW, Park YU, Kang K (2010) Mn based olivine electrode material with high power and energy. Chem Commun 46:1305–1307
86. Lee KT, Kan WH, Nazar LF (2009) Proof of intercrystallite ionic transport in LiMPO4 electrodes (M=Fe, Mn). J Am Chem Soc 131:6044–6045
87. Choi D, Wang D, Bae IT, Xiaot J, Nie Z, Wang W, Yiswanathan Y, Lee Y, Zhang JG, Graff G, Yang Z, Liu J (2010) $LiMnPO_4$ nanoplate grown via solid-state reaction in molten hydrocarbon for Li-ion battery cathode. Nano Lett 10(8):2799–2805
88. Li L, Liu J, Chen L, Xu H, Yang J, Qian Y (2013) Effect of different carbon sources on the electrochemical properties of rod-like LMP/C nanocomposites. RSC Adv 3:6847–6852
89. Zhao M, Fu Y, Xu N, Li G, Wu M, Gao X (2014) High performance $LiMnPO_4/C$ prepared by a crystallite size control method. J Mater Chem A 2:15070–15077
90. Chen G, Richardson TJ (2010) Thermal instability of olivine-type $LiMnPO_4$ cathodes. J Power Sources 195:1221–1224
91. Zhong S, Xu Y, Li Y, Zeng H, Li W, Wang J (2012) Synthesis and electrochemical performance of $LiMnPO_4/C$ composites cathode materials. Rare Met 31:474–478

92. Wang L, Sun W, Li J, Gao J, He X, Jiang C (2012) Synthesis of electrochemically active $LiMnPO_4$ via $MnPO_4 \cdot H_2O$ with different morphology prepared by facile precipitation. Int J Electrochem Sci 7:3591–3600
93. Barpanda P, Djellab K, Recham N, Armand M, Tarascon JM (2011) Direct and modified ionothermal synthesis of $LiMnPO_4$ with tunable morphology for rechargeable Li-ion batteries. J Mater Chem 21:10143–10152
94. Murugan AV, Muraliganth T, Ferreira PJ, Manthiram A (2009) Dimensionally modulated, single-crystalline $LiMPO_4$ (M=Mn, Fe Co, and Ni) with nano-thumblike shapes for high-power energy storage. Inorg Chem 48:946–952
95. Kosova NV, Devyatkina ET, Ancharov AI, Markov AV, Karnaushenko DD, Makukha VK (2012) Structural studies of nanosized $LiFe_{0.5}Mn_{0.5}PO_4$ under cycling by in situ synchrotron diffraction. Solid State Ionics 225:564–569
96. Wang H, Yang Y, Liang Y, Cui LF, Casalongue HS, Li Y, Hong G, Cui Y, Da H (2011) $LiMn_{1-x}Fe_xPO_4$ nanorods grown on graphene sheets for ultrahigh-rate-performance lithium ion batteries. Angew Chem Int Ed 50:7364–7368
97. Kopec M, Yamada A, Kobayashi G, Nishimura S, Kanno R, Mauger A, Gendron F, Julien CM (2009) Structural and magnetic properties of $Li_x(Mn_yFe_{1-y})PO_4$ electrode materials for Li-ion batteries. J Power Sources 189:1154–1163
98. Zaghib K, Mauger A, Gendron F, Massot M, Julien CM (2008) Insertion properties of $LiFe_{0.5}Mn_{0.5}PO_4$ electrode materials for Li-ion batteries. Ionics 14:371–376
99. Gardiner GR, Islam MS (2010) Anti-site defects and ion migration in the $LiFe_{0.5}Mn_{0.5}PO_4$ mixed-metal cathode material. Chem Mater 22:1242–1248
100. Trottier J, Mathieu MC, Guerfi A, Zaghib K, Mauger A, Julien CM (2013) $LiMn_yFe_{1-y}PO_4$ ($0.5 \leq y \leq 0.8$) cathode materials grown by hydrothermal route: electrochemical performance. ECS Trans 50–24:109–114
101. Hong Y, Tang Z, Hong Z, Zhang Z (2014) $LiMn_{1-x}Fe_xPO_4$ (x = 0, 0.1, 0.2) nanorods synthesized by a facile solvothermal approach as high performance cathode materials for lithium-ion batteries. J Power Sources 248:655–659
102. Truong QD, Devaraju MK, Ganbe Y, Tomai T, Honma I (2013) Controlling the shape of $LiCoPO_4$ nanocrystals by supercritical fluid process for enhanced energy storage properties. Sci Rep 4:3975
103. Wang F, Yang J, Nuli Y, Wang J (2011) Novel hedgehog-like 5 V $LiCoPO_4$ positive electrode material for rechargeable lithium battery. J Power Sources 196:4806–4810
104. Su J, Wei BQ, Rong JP, Yin WY, Ye ZX, Tian XQ, Ren L, Cao MH, Hu CW (2011) A general solution-chemistry route to the synthesis $LiMPO_4$ (M=Mn, Fe and Co) nanocrystals with [010] orientation for lithium ion batteries. J Solid State Chem 184:2909–2919
105. Wolfenstine J, Allen J (2004) $LiNiPO_4$–$LiCoPO_4$ solid solutions as cathodes. J Power Sources 136:150–153
106. Ni J, Gao L, Lu L (2013) Carbon coated lithium cobalt phosphate for Li-ion batteries: Comparison of three coating techniques. J Power Sources 221:35–41
107. Wolfenstine J, Read J, Allen J (2007) Effect of carbon on the electronic conductivity and discharge capacity $LiCoPO_4$. J Power Sources 163:1070–1073
108. Wolfenstine J, Lee U, Poese B, Allen J (2005) Effect of oxygen partial pressure on the discharge capacity of $LiCoPO_4$. J Power Sources 144:226–230
109. Wolfenstine J, Poese B, Allen J (2004) Chemical oxidation of $LiCoPO_4$. J Power Sources 138:281–282
110. Okada S, Sawa S, Egashira M, Yamaki JI, Tabuchi M, Kageyama H, Konishi T, Yoshino A (2001) Cathode properties of phospho-olivine $LiMPO_4$ for lithium secondary batteries. J Power Sources 97–98:430–432
111. Bramnik NN, Bramnik KG, Baehtz C, Ehrenberg H (2005) Study of the effect of different synthesis routes on Li extraction–insertion from $LiCoPO_4$. J Power Sources 145:74–81
112. Wolfenstine J (2006) Electrical conductivity of doped $LiCoPO_4$. J Power Sources 158:1431–1435

113. Wang F, Yang J, Li YN, Wang J (2011) Novel hedgehog-like 5 V $LiCoPO_4$ positive electrode material for rechargeable lithium battery. J Power Sources 196:4806–4810
114. Nakayama M, Goto S, Uchimoto Y, Wakihara M, Kitayama Y, Miyanaga T, Watanabe I (2005) X-ray absorption spectroscopic study on the electronic structure of $Li_{1-x}CoPO_4$ electrodes as 4.8 V positive electrodes for rechargeable lithium ion batteries. J Phys Chem B 109:11197–11203
115. Bramnik NN, Nikolowski K, Baehtz C, Bramnik KG, Ehrenberg H (2007) Phase transition occurring upon lithium insertion-extraction of $LiCoPO_4$. Chem Mater 19:908–915
116. Zhao Y, Wang S, Zhao C, Xia D (2009) Synthesis and electrochemical performance of $LiCoPO_4$ micron-rods by dispersant-aided hydrothermal method for lithium ion batteries. Rare Met 28:117–121
117. Bramnik NN, Nikolowski K, Trots DM, Ehrenberg H (2008) Thermal stability of $LiCoPO_4$ cathodes. Electrochem Solid State Lett 11:A89–A93
118. Jang IC, Lim HH, Lee SB, Karthikeyan K, Aravindan V, Kang KS, Yoon WS, Cho WI, Lee YS (2010) Preparation of $LiCoPO_4$ and $LiFePO_4$ coated $LiCoPO_4$ materials with improved battery performance. J Alloys Compd 497:321–324
119. Bramnik NN, Nikolowski K, Trots DM, Ehrenberg H (2008) Thermal stability of $LiCoPO_4$ cathodes. Electrochem Solid State Lett 11:A89–A93
120. Aravindan V, Cheah YL, Chui Ling WC, Madhavi S (2012) Effect of LiBOB additive on the electrochemical performance of $LiCoPO_4$. J Electrochem Soc 159:A1435–A1439
121. Wolfenstine J, Allen J (2005) Ni^{3+}/Ni^{2+} redox potential in $LiNiPO_4$. J Power Sources 142:389–390
122. Karthickprabhu S, Hirankumar G, Maheswaran A, Daries-Bella RS, Sanjeeviraja C (2014) Structural, morphological, vibrational and electrical studies on Zn doped nanocrystalline $LiNiPO_4$. Mater Sci Forum 781:145–153
123. Dimesso L, Becker D, Spanheimer C, Jaegermann W (2012) Investigation of graphitic carbon foams/$LiNiPO_4$ composites. J Solid State Electrochem 16:3791–3798
124. Piana M, Arrabito M, Bodoardo S, D'Epifanio A, Satolli D, Croce F, Scrosati B (2002) Characterization of phospho-olivines as materials for Li-ion celle cathodes. Ionics 8:17–26
125. Julien CM, Mauger A, Zaghib K, Veillette R, Groult H (2012) Structural and electronic properties of the $LiNiPO_4$ orthophosphate. Ionics 18:625–633
126. Herle PS, Ellis B, Coombs N, Nazar LF (2004) Nano-network electronic conduction in iron and nickel olivine phosphates. Nat Mater 3:147–152
127. Sugiyama J, Nozaki H, Harada M, Kamazawa K, Ikedo Y, Miyake Y, Ofer O, Mansson M, Ansaldo EJ, Chow KH (2012) Diffusive behaviour in LiMPO4 with M=Fe Co, Ni probed by muon-spin relaxation. Phys Rev B 85:054111
128. Wang D, Xiao J, Xu W, Zhang JG (2010) Investigation of $LiNiPO_4$ as a cathode material for lithium ion battery. The 15th international meeting on lithium batteries, IMLB2010, abstract no°372
129. Minakshi M, Singh P, Ralph D, Appadoo D, Blackford M, Ionescu M (2012) Structural characteristics of olivine $Li(Mg_{0.5}Ni_{0.5})PO_4$ via TEM analysis. Ionics 18:583–590
130. Ramana CV, Ait-Salah A, Utsunomiya S, Becker U, Mauger A, Gendron F, Julien CM (2006) Structural characteristics of lithium nickel phosphate olivine studied using analytical electron microscopy and raman spectroscopy. Chem Mater 18:3788–3794
131. Cheruku R, Govindaraj G (2014) Structural and electrical conductivity studies of nanocrystalline olivine type $LiNiPO_4$ material. Int J ChemTech Res 6:2017–2201
132. Goñi A, Lezama L, Barberis GE, Pizarro JL, Arriortua MI, Rojo T (1996) Magnetic properties of the $LiMPO_4$ (M=Co, Ni) compounds. J Magn Magn Mater 164:251–255
133. Santoro RP, Segal DJ, Newnham RE (1966) Magnetic properties of $LiCoPO_4$ and $LiNiPO_4$. J Phys Chem Solids 27:1192–1193
134. Toft-Petersen R, Jensen J, Jensen TBS, Andersen NH, Christensen NB, Niedemayer C, Kenzelmann M, Skoulatos M, Le MD, Lefmann K, Hensen SR, Li J, Zarestky JL, Vaknin D (2011) High-field magnetic phase transition and spin excitations in magnetoelectric $LiNiPO_4$. Phys Rev B 84:054408

135. Vaknin D, Zarestky JL, Rivera JP, Schmid H (2004) Commensurate-incommensurate magnetic phase transition in magnetoelectric single crystal $LiNiPO_4$. Phys Rev Lett 92:207201
136. Fomin VI, Gnezdilov VP, Kurnosov VS, Peschanskii AV, Yeremenko AV, Schmid H, Rivera JP, Gentil S (2002) Raman scattering in a $LiNiPO_4$ single crystal. Low Temp Phys 28:203–209
137. Ficher CAJ, Prieto VMH, Islam MS (2008) Lithium battery materials $LiMPO_4$ (M=Mn, Fe, Co and Ni): insights into defect association, transport mechanisms and doping behaviour. Chem Mater 20:5907–5915
138. Garcia-Moreno O, Alvarez-Vega M, Garcia-Alvarado F, Garcia-Jaca J, Garcia-Amores JM, Sanjuan ML, Amador U (2001) Influence of the structure on the electrochemical performance of lithium transition metal phosphates as cathodic materials in rechargeable lithium batteries: a new high-pressure form of $LMPO_4$ (M=Fe and Ni). Chem Mater 13:1570–1576
139. Zhou F, Cococcioni M, Kang K, Ceder G (2004) The Li intercalation potential of $LiMPO_4$ and $LiMSiO_4$ olivines with M=Fe, Mn Co, Ni. Electrochem Commun 6:1144–1148
140. Howard WF, Spotnitz RM (2007) Theoretical evaluation of high-energy lithium metal phosphate cathode materials in Li-ion batteries. J Power Sources 165:887–891
141. Chevrier VL, Ong SP, Armiento R, Chan MKY, Ceder G (2010) Hybrid density functional calculations of redox potentials and formation energies of transition metal compounds. Phys Rev B 82:075122
142. Rommel SM, Schall N, Brünig C, Weihrich R (2014) Challenges in the synthesis of high voltage electrode materials for lithium-ion batteries: a review on $LiNiPO_4$. Monatsh Chem 145:385–404
143. Kausarjanjua N, Mumtaz M, Yabuq A, Sabahat S, Mujtaba A (2014) Electrocatalytic activity of $LiNiPO_4$ and the copper doped analogues towards oxygen reduction. Nucleus 51:109–115
144. Yang J, Xu JJ (2006) Synthesis and characterization of carbon-coated lithium transition metal phosphates $LiMPO_4$ (M=Fe, Mn Co, Ni) prepared via nonaqueous sol-gel route. J Electrochem Soc 153:A716–A723
145. Tsai A (2011) Synthesis and characterization of $LiNiPO_4$ nanocrystals via microemulsion method as a new class of electrocatalyst for oxygen reduction. Master's thesis, New Jersey Graduate School, New Brunswick
146. Prabu M, Selvasekarapandian S (2012) Dielectric and modulus studies of $LiNiPO_4$. Mater Chem Phys 134:366–370
147. Gangulibabu DB, Kalaiselvi N, Jayaprakash N, Periasamy P (2009) CAM sol-gel synthesized $LiMPO_4$ (M=Co, Ni) cathodes for rechargeable lithium batteries. J Sol-Gel Sci Technol 49:137–144
148. Karthickprabhu S, Hirankumar G, Maheswaran A, Daries-Bella RS, Sanjeeviraja C (2013) Structural and conductivity studies on lanthanum doped $LiNiPO_4$ prepared by polyol method. In: Chowdari BVR, Kawamura J, Mizusaki J, Amezawa K (eds) Proceedings of the 13th Asian conference on solid state ionics, Sendai, Japan 17–20 July 2012, World Scientific, Singapore
149. Piana M, Arrabito M, Bodoardo S, D'Epifanio A, Satolli D, Croce F, Scrosati B (2002) Characterization of phospho-olivines as materials for Li-ion celle cathodes. Ionics 8:17–26
150. Dimesso L, Jacke S, Spanheimer C, Jaegermann W (2012) Investigation on LiCoPO4 powders as cathode materials annealed under different atmospheres. J Solid State Electrochem 16:3911–3919
151. Dimesso L, Spanheimer C, Jaegermann W (2013) Effect of the Mg-substitution on the graphitic carbon foams—$LiNi_{1-y}Mg_yPO_4$ composites as possible cathodes materials for 5 V applications. Mater Res Bull 48:559–565
152. Tabassam L, Giuli G, Moretti A, Nobili F, Marassi R, Minicucci M, Gunnella R, Olivi L, DiCicco A (2012) Structural study of $LiFePO_4$-$LiNiPO_4$ solid solutions. J Power Sources 213:287–295
153. Qing R, Yang MC, Meng YS, Sigmund W (2013) Synthesis of $LiNi_xFe_{1-x}PO_4$ solid solution as cathode materials for lithium ion batteries. Electrochim Acta 108:827–832

Layered and Spinel Structural Cathodes

Ying Chun Lyu, Jie Huang and Hong Li

Layered and spinel materials have been used successfully as intercalation-type cathode active materials in commercial Li-ion batteries. The physical and chemical properties, electrochemical reactions, structure evolution mechanisms, stability and safety issues have been widely investigated. Based on comprehensive fundamental researches, since 1980s, their electrochemical performances are improved continuous after various modifications, such as doping, surface coating, forming solid solution and composite, controlling morphology, size and crystallinity. Here, basic features of layered and spinel materials are summarized.

1 Introduction

In order to achieve excellent electrochemical performances in lithium ion batteries, several requirements on cathode materials should be satisfied [1]:

1. The material should contain Li^+ ions and work as Li source, providing both the Li^+ transferred between the cathode and anode, and the expend for SEI on the surface of the anode;
2. It should offer a high electrode potential for a high output potential;
3. The electrode potential should be stable for a stable output potential;
4. High capacity and high energy density;
5. High Li^+ diffusion coefficient, highly stable electrode interface for high rate performance;
6. High structural stability during cycling, for perfect cycle performance;
7. High electronic conductivity and ion conductivity;
8. Low price and environmental-friendly.

Actually, it is difficult for one material system to satisfy all requirements simultaneously. For various applications, the developers need to consider the

Y.C. Lyu · J. Huang · H. Li (✉)
Institute of Physics, Chinese Academy of Sciences, Beijing 100190, China
e-mail: hongli0920hli@iphy.ac.cn

Z. Zhang and S.S. Zhang (eds.), *Rechargeable Batteries*,
Green Energy and Technology, DOI 10.1007/978-3-319-15458-9_3

preferential performance targets and the strategy for meeting the minimum demanding. The exploring of the cathode materials is a balance of art. Up to now, only three types of cathode materials have been used and purposed by a few scientists.

In 1980, John B. Goodenough's research group at Oxford University discovered the usefulness of $LiCoO_2$ as an intercalation electrode for lithium ion batteries [2]. In 1984, Thackeray et al. found that $LiMn_2O_4$ spinel can be used as a cathode material for lithium ion batteries [3]. In 1997, $LiFePO_4$ was developed as a choice for cathode material by Goodenough [4]. Based on these findings, layered compounds $LiMO_2$, $Li_{1-x}M_{1+x}O_2$, spinel compounds LiM_2O_4, $LiNi_{0.5}Mn_{1.5}O_4$, and olivine compounds $LiMPO_4$ (M = transition metal elements such as Fe, Co, Ni and Mn) become the most important three families applied in lithium ion batteries. Figure 1 and Table 1 show the electrochemical properties of the major cathode materials in commercial use [5]. In this chapter we will summary the layered and spinel compounds, while the olivine compounds are discussed in Chap. 4.

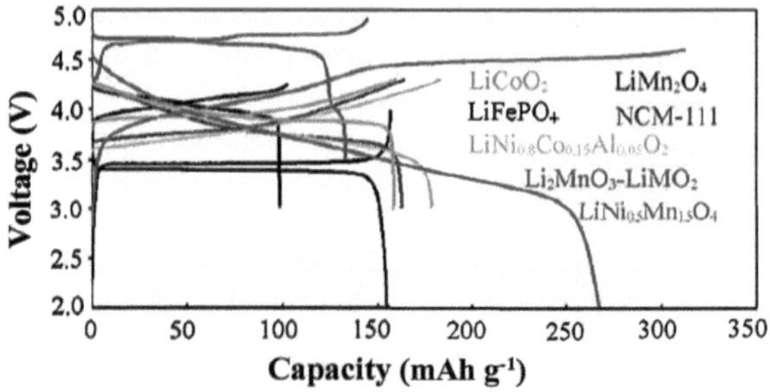

Fig. 1 The major cathode materials for Li-ion batteries and their properties [5]

Table 1 The properties of commercial positive electrode materials [5]

	$LiCoO_2$	NCM	NCA	$LiMn_2O_4$	$LiFePO_4$
Structure	Layered	Layered	Layered	Spinel	Olivine
Voltage (V)	3.7	3.6	3.6	3.8	3.4
Capacity ($mAh\ g^{-1}$)	140–200	160–190	180–200	100–120	150–165
Life (cycles)	500–1000	500–3,000	500–2,000	500–3,000	1000–20,000
Price	High	Medium	High	Low	Low
Safety	Medium	Good	Good	Good	Excellent
Application area	3C	3C, HEV, EV	EV, 3C	HEV, EV, stationary	HEV, EV, stationary…

NCM refers to $LiNi_xCo_yMn_zO_2$, x + y + z = 1, NCA refers to $LiNi_{0.8}Co_{0.15}Al_{0.05}O_2$

2 Layered Structural Cathodes

2.1 Early Studies on Layered Cathodes

Layered cathode materials were developed since the birth of the rechargeable lithium batteries. The first use of layered materials was dichalcogenides. TiS_2 was considered as the most appealing consideration as an energy storage electrode [6, 7]. Titanium disulfide has a hexagonal close-packed sulfur lattice with the titanium ions in octahedral sites between alternating sulfur sheets. The TiS_2 sheets are stacked directly on top of one another, giving the sulfur anion stacking sequence ABAB, as shown in Fig. 2. It forms a single phase with lithium over the entire composition range of Li_xTiS_2 for $0 \leq x \leq 1$ [8]. Figure 3 shows the typical insertion/removal galvanostatic cycle for a Li/TiS_2 cell [6]. Most of other dichalcogenides are also electrochemically active, and show similar single phase behavior with lithium intercalation, such as TaS_2, MoS_2 [9], so do the diselenide. But VSe_2 is an exception, showing a two-phase behavior [10].

Layered oxides with the same structures as the layered dichalcogenides were studied, such as, V_2O_5, V_3O_{16}, CrO_3, Cr_8O_{21}, MoO_3. V_2O_5 has a layered structure with weak vanadium-oxygen bonds between the layers, and it reacts by an intercalation mechanism with lithium: $xLi + V_2O_5 = Li_xV_2O_5$. Another vanadium oxide that has received much attention is LiV_3O_8, which has a layered structure composed of octahedral and trigonal bipyramidal ribbons that can be swelled just like other layered compounds and can be intercalated by lithium.

Although LiV_3O_8 contains a lithium ion in the lattice, it still needs to be discharged first to be lithiated due to the feature of V^{5+}. Alkali-metal compounds of the oxides of manganese, cobalt, chromium, and others had been extensively studied by a number of groups. The structures of these layered oxides and chalcogenides were extensively studied in the 1970s. Whereas lithium only occupies octahedral sites in the strictly layered materials when the Li/transition-metal ratio is unity or less, the larger alkali ions often occupy trigonal prismatic sites.

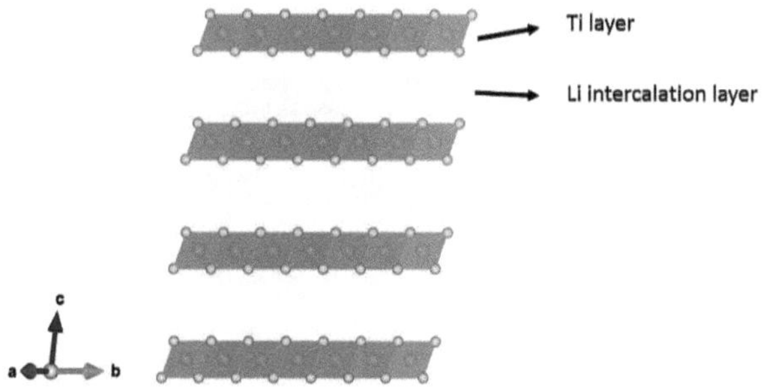

Fig. 2 The structure of layered TiS_2

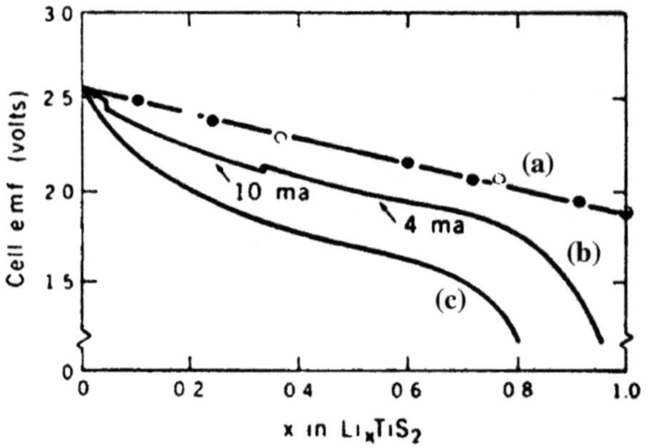

Fig. 3 Discharge/charge curve of Li/TiS_2 cell. *Curvea* open circuit voltage obtained on discharge (*solid dots*) of TiS_2 and recharge of (*hollow dots*) $LiTiS_2$, *curve b* voltages on discharge at 10 and 4 mA, and *curve c* voltage on discharge after 1000 cycles at 4 mA. A mixture of finely divided TiS_2 and Teflon (9 to 1 by weight) was hot-pressed into a stainless steel grid of area 2 cm^2 [6]. Copyright 1977 by AAAS

2.2 Layered $LiCoO_2$ Compounds

Goodenough recognized that $LiCoO_2$ had a structure similar to the layered structures of the dichalcogenides and showed that the lithium could be removed electrochemically, thus making it a viable cathode material [2]. SONY combined the $LiCoO_2$ cathode with a coke anode to make the first successful Li-ion battery [11], which now dominates the lithium battery market. Li-ion batteries are much safer than Li metal batteries. During the development of Li-ion batteries, a series layered metal oxides with a general formula of $LiMO_2$ (M = Transition metal elements such as Co, Ni and Mn) were found. In these materials, the lithium ions and transition metal ions occupy the alternate (111) planes of the rock salt structure, as shown in Fig. 4. The structure with MO_2 layers allows a reversible extraction/insertion of lithium ions from/into the lithium layer.

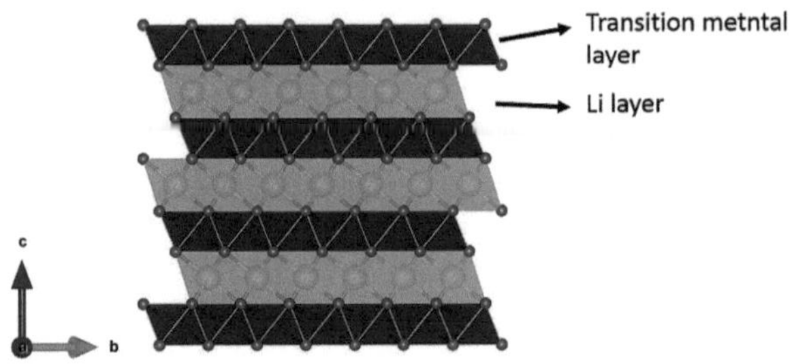

Fig. 4 Schematic diagram of the layered $LiMO_2$ (M = Transition metal elements such as Co, Ni and Mn) structure

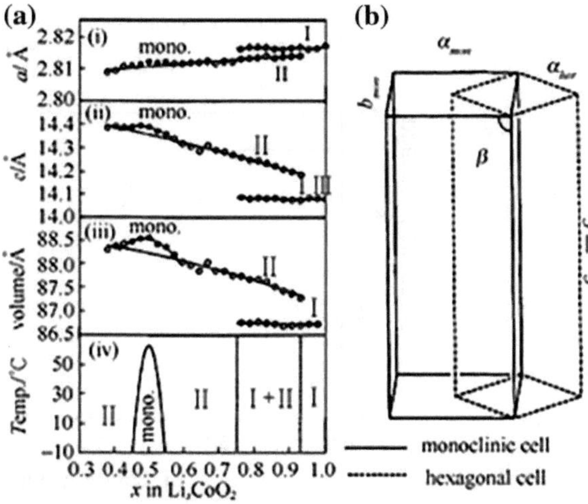

Fig. 5 **a** Unit cell constants a-(i), c-(ii), and cell volume-(iii) as a function of lithium concentration x in Li_xCoO_2. Combined with the electrochemical results, (*i*), (*ii*), and (*iii*) determine a global phase diagram-(*iv*) for Li_xCoO_2; **b** outlines for the monoclinic (*solid*) and hexagonal (*dashed*) unit cells [12]. Copyright 1992 by The Electrochemical Society

The theoretical capacity of $LiCoO_2$ is 274 mAh g^{-1}. In a commercial cell, during the delithiation of Li_xCoO_2 ($0.4 < x < 1$), it undergoes a solid solution reaction with three distinct phase transition processes as shown in Fig. 5 [12]: The first phase transition is shown to be the first order (coexisting phases are observed for $0.75 \le x \le 0.93$), involving a significant expansion of the *c*-lattice parameter of the hexagonal unit cell. The phase has a delocalized-electron type behavior, and a metal–non-metal transition is strictly associated to the energy change because of the Mott transition [13, 14]. The other two transitions are situated slightly above and below $x = 0.5$ and they are caused by an order/disorder transition of the lithium ions. After that it undergoes an irreversible transition to monoclinic phase [15, 16]. Van der Ven et al. [17, 18] calculated the phase transition at low Li content ($0 \le x \le 0.5$). As shown in Fig. 6, with more Li delithiation, O3 phase transforms to (1) a mixture of O3 and a stage two phase (A six-layer unit cell with adjacent CoO_2 sandwiches stacked with O1-CoO_2 and O3-Li_xCoO_2.) called H1-3; (2) H1-3 phase; and (3) a mixture of H1-3 and O1. This has been approved by in situ XRD measurements [19]. This irreversible transition often leads to the reduction of the electrochemical properties, so $LiCoO_2$ is usually cycled with an upper cutoff voltage of about 4.2 V with respect to lithium metal. Charge-discharge cycling with this upper cutoff corresponds to repeatedly extracting and inserting about 0.5 Li per $LiCoO_2$ and gives a specific capacity of about 140 mAh g^{-1}. HAADF-STEM images show that during charging the phase irreversibly changes from O3-type in pristine $LiCoO_2$ to O1-type Li_xCoO_2 ($x \approx 0.50$) after the first electrochemical Li extraction and back to O2-type Li_xCoO_2 ($x \approx 0.93$) rather than to O3-stacking after

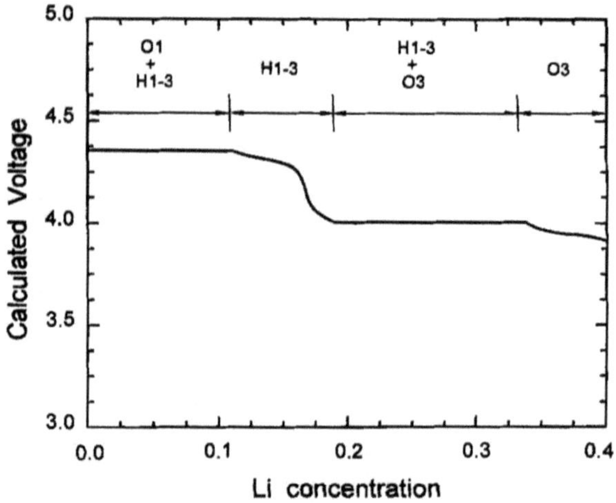

Fig. 6 Calculated intercalation voltage curve of Li_xCoO_2 as a function of Li concentration at T = 30 °C [17]. Copyright 1998 by The Electrochemical Society

the first electrochemical lithiation [20]. This discovery makes a connection between the two separated $LiCoO_2$ phases, i.e., O2 and O3 systems (shown in Fig. 7).

The diffusion coefficient in LixCoO$_2$ was calculated from the first-principles method by Van der Ven et al. [21]. They suggested that lithium ions can hop according to two migration paths depending on the local environment around the hopping ion. Furthermore, it was found that the activation barrier for a particular hopping mechanism depend strongly on the local lithium-vacancy configuration. Kinetic Monte Carlo simulations predict that lithium diffusion in Li_xCoO_2 is mediated by divacancies even at high lithium concentrations where the

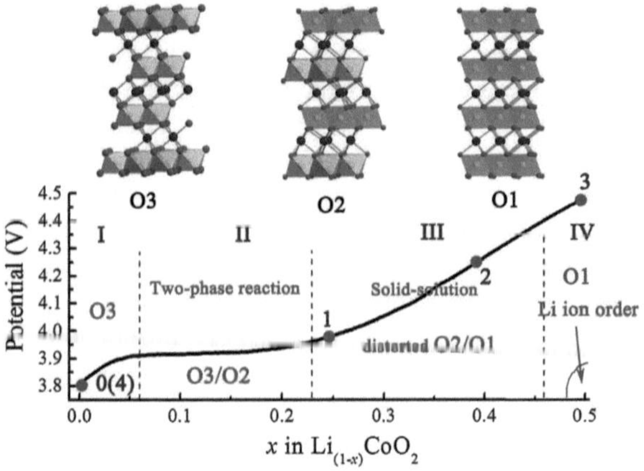

Fig. 7 Phase diagram of $Li_{1-x}CoO_2$ ($0 \le x \le 0.50$) nanoparticle [20]. Copyright 2012 by American Chemical Society

concentration of divacancies is low. The simulations have also shown that the strong concentration dependence of the activation barrier results in a diffusion coefficient that spans several orders of magnitude with lithium concentration as shown in Figs. 8 and 9.

To obtain a higher capacity from $LiCoO_2$, one must charge it to a potential above 4.2 V to use more than 0.5 Li per $LiCoO_2$. However, charge-discharge cycling using a higher upper cutoff voltage often results in rapid capacity fade, thought to be caused by the structural instability of Li_xCoO_2 with $x < 0.5$, side reactions with electrolyte at high potentials [12] and the problem of cobalt dissolution in typical electrolytes [22].

First principles calculations show that the structural stability of $LiCoO_2$ upon deep lithium extraction is strongly associated with the electronic structures of Co-3d, which is very flexible and can be exhibited as different electronic configurations [23]. In $LiCoO_2$, Co^{3+} is non-magnetic and holds the $(t_{2g}\uparrow)^3$ $(t_{2g}\downarrow)^3$ electronic configuration. Upon lithium deintercalation, some Co^{3+} ions lose one electron and become Co^{4+} with $(t_{2g}\uparrow)^3$ $(t_{2g}\downarrow)^2$ configuration. Both structures are stable since the distortion of the CoO_6 octahedron is small, and thus these structures

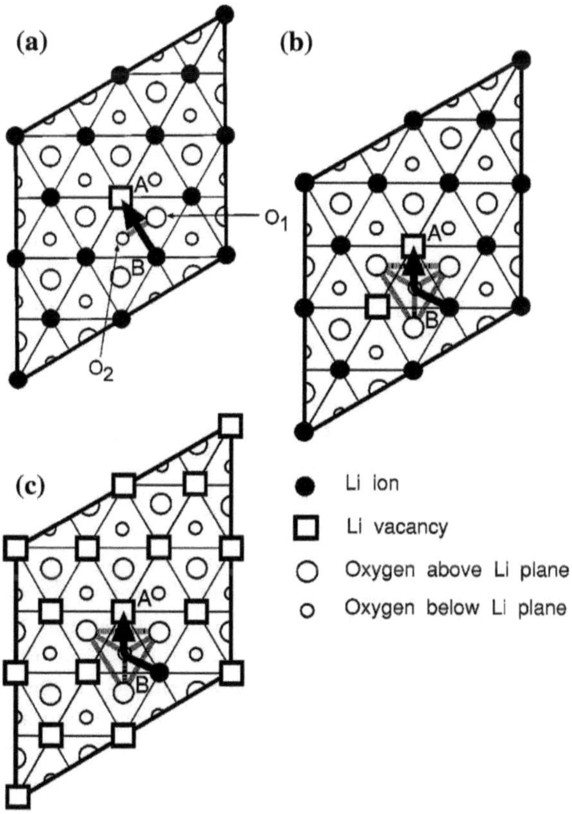

Fig. 8 Lithium migration paths in Li_xCoO_2 determined with the elastic band method. The triangular lattice corresponds to the lithium sites and the *filled circles* are lithium ions. The *largeempty circles* are oxygen ions above the lithium plane and the *smallempty circles* are oxygen ions below the lithium plane [21]. Copyright 2001 by The American Physical Society

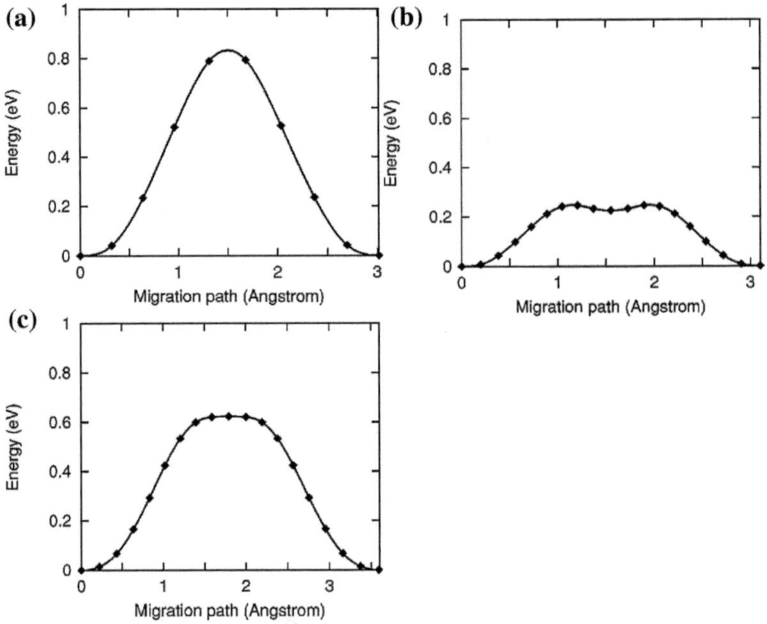

Fig. 9 Energy along the migration path in different lithium vacancy environments determined with the elastic band method. Refer to Fig. 8 for the corresponding lithium-vacancy environments. **a** Migration of an isolated vacancy at x = 11/12. **b** Migration of an isolated lithium according to a tetrahedral site hop (TSH) at x = 1/12. **c** Migration of a lithium ion into a divacancy according to a TSH at x = 10/12 [21]. Copyright 2001 by The American Physical Society

do not contribute much to the instability. On the other hand, oxygen vacancy is one of the important reasons to the structural instability. Upon high voltage charging of $LiCoO_2$, oxygen release is observed from experiments and oxygen vacancies are formed in the lattice. It is found that spin flip occurs to the electronic structure of Co-3d close to oxygen vacancies. Co^{3+} holds the $(\uparrow)^4$ $(\downarrow)^2$ electronic configuration and magnetizs with 2 μ_B magnetic moment. Furthermore, some Co^{3+} obtain one electron and become Co^{2+} near the oxygen vacancy. In these cases, the charge distribution around the Co atom is not symmetric and the local structure is obviously distorted, which can further accelerate the process of the structural degradation (Figs. 10, 11 and 12).

There are three different structures in $LiCoO_2$ materials: the most important is O3 type (O in ABC stacking) as discussed above [24], O2 type (O in ABAC stacking) [25] and spinel [26] $LiCoO_2$ also exist, since synthesis precursors, synthesis temperature, heat-treatment time and cooling rate could be attributed to different cation and oxygen arrangements in the layered materials, showing in Fig. 13. In this chapter, if without special mention, all the discussions below are based on O3-$LiCoO_2$.

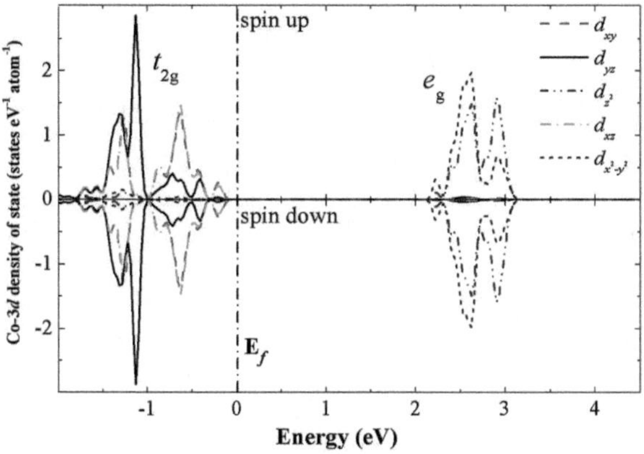

Fig. 10 Co-3d projected density of states of Co^{3+} in $LiCoO_2$. The triplet and duplet of the d-orbit are denoted as t_{2g} and e_g, respectively. The Fermi level is set to 0 eV [23]. Copyright 2012 by ESG

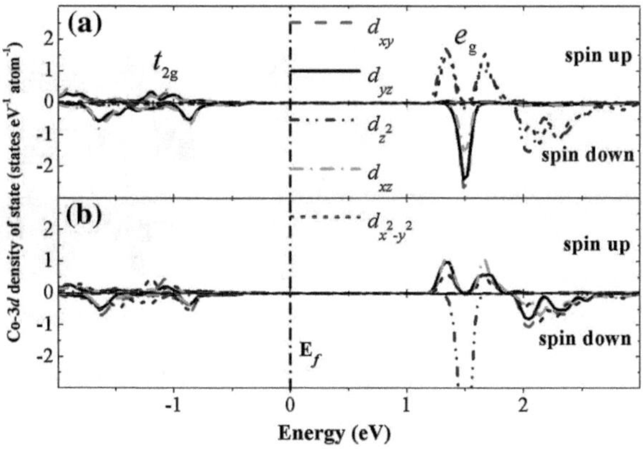

Fig. 11 Co-3d projected density of states of Co^{4+} ions in Li_xCoO_2 along the Co–O bond direction (**a**) and c-vector direction (z-axis) of the lattice (**b**). The five "d-orbitals" in (**b**) is not really corresponding to the split of an octahedral crystal field. The triplet and the duplet of the d-orbital are denoted as t_{2g} and e_g, respectively. The Fermi level is set to 0 eV [23]. Copyright 2012 by ESG

$LiCoO_2$ synthesized at high temperature (>800 °C) adopts the O3 layered structure as shown in Fig. 14. The O3-$LiCoO_2$ obtained by the high temperature procedure exhibits excellent electrochemical properties and has been commercial used in Li-ion batteries since 1991. Nowadays, it is still used in electronic market, for example in wireless communications and laptop computers. In contrast, synthesis at lower temperature (~400 °C) results in a lithiated spinel $LiCoO_2$ phase with cation distribution of $[Li_2]_{16c}[Co_2]_{16d}O_4$ [27]. It is usually designated as LT-$LiCoO_2$, showing poor electrochemical properties. When heating the LT-$LiCoO_2$ sample to high temperature (>800 °C), the cation ions will reorder and convert into O3 phase. Another variety of $LiCoO_2$ is O2 layered structure

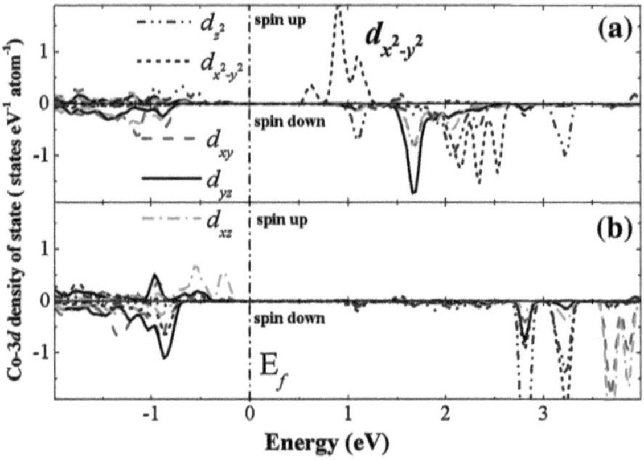

Fig. 12 Co-3d projected density of states of Co^{3+} ions (**a**) and Co^{2+} ions (**b**) next to an O vacancy in $Li_{0.75}CoO_2$. The project is along the Co–O bond direction. The Fermi level is set to 0 eV [23]. Copyright 2012 by ESG

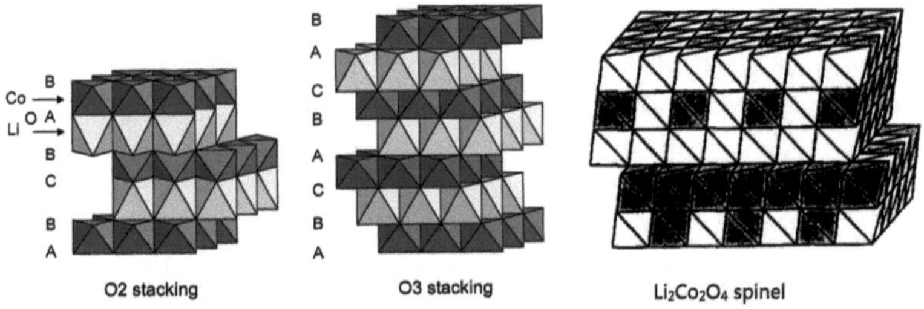

Fig. 13 The three different oxygen packings of $LiCoO_2$ [27]. Copyright 1993, Elsevier B.V.

synthesized by ion-exchange from P2-$Na_{0.70}CoO_2$ [25]. O2-$LiCoO_2$ is a metastable phase [28], Fig. 14 shows the XRD patterns of O2-$LiCoO_2$ during thermal treatments at different temperatures [28]. It transforms to well-crystallized O3-$LiCoO_2$ irreversibly when heated up to 400 °C [29]. Electrochemical properties of the O2-$LiCoO_2$ phase are competitive with those of the conventional O3-$LiCoO_2$ and their DSC measurements indicate that the O2-$LiCoO_2$ system charged to 4.2 V is about as safe as the O3 one at the same potential [30]. But the synthesis is really difficult.

Coating or annealing can improve the properties of Li_xCoO_2 at high voltage by forming a more stable surface or by decreasing the active surface area. Using a ZrO_2 or Al_2O_3 coating, Cho et al. showed that $LiCoO_2$ can deliver a capacity near 170 mAh g^{-1} with little capacity loss during the first 70 cycles between 2.75 and 4.4 V [31, 32]. The understanding of the coating mechanism was argued for a long time. Cho et al. think that the thin solid solution film of $LiCo_{1-x}M_xO_2$ (M = Zr, Al, Ti, B) at the surface effectively suppresses the lattice-constant changes during

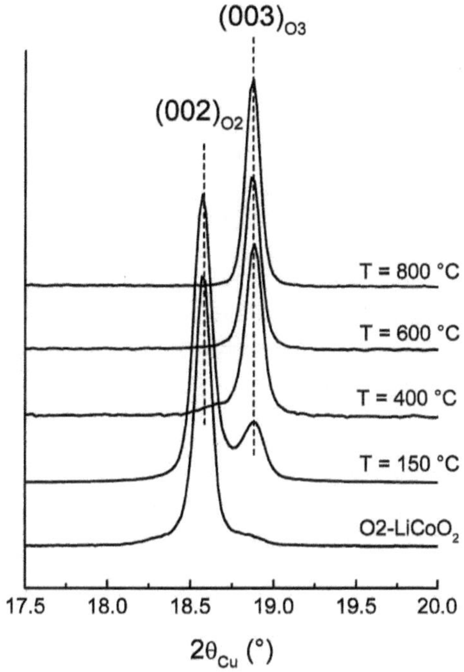

Fig. 14 Expansion of the first Bragg reflection of the XRD patterns observed for the samples obtained after different thermal treatments of O2-$LiCoO_2$ [28]. Copyright 2001, Elsevier B.V.

electrochemical cycling and thereby suppresses phase transitions. Chen and Dahn [33] argued that ZrO_2 coating had no effect on the expansion of Li_xCoO_2 during charge or discharge, but it may inhibit side reactions involving oxygen loss from Li_xCoO_2 to the electrolyte and hence improve the cycling stability. Based on in situ synchrotron XRD studies, Liu et al. [34, 35] announced that the phase transitions of surface coated Li_xCoO_2 from O3 to H1-3, and finally to O1 during charge are still clearly recognizable, but the surface coating make these phase transitions reversible. This agrees well with Chen's results [19]. At the same time, modifying the surface of the $LiCoO_2$ particle may be helpful to suppress the release of oxygen, which always leads to instability of the interior structure and decomposition of the electrolyte [36]. Besides ZrO_2 and Al_2O_3, other coating strategies include $AlPO_4$ [37], AlF_3 [38] and MgO [36] etc. Moreover doping such as Mg [39, 40], Al [40] and Ti/Mg double substitution [41] is also demonstrated as another effective method to improve its cycleability for high voltage application. The addition of a little of the redox-inactive element reduces the capacity fading on cycling, this inert element prevents the complete removal of all the lithium and thus minimizes possibly structural collapse [1].

$LiCoO_2$ has high theoretical density and tap-density. This makes it the highest volumetric energy density cathode materials. Recent efforts on improving its reversible capacity to 220 mAh/g have made by surface coating and doping enhance its value further. Shaking its top one position in the market seems more difficult, even for the high gravimetric capacity Li-rich layered compounds.

2.3 Layered $LiNiO_2$ Compounds

Another well-known $LiMO_2$ family member is $LiNiO_2$. Its crystal structure also belongs to the hexagonal system with cell constant of a = 2.88 Å and c = 14.18 Å. Its working voltage is about 3.7 V and theoretical capacity is 275 mAh g^{-1}. Nickel is more abundant and cheaper than cobalt, so $LiNiO_2$ was planned to be developed to replace expensive $LiCoO_2$ for battery applications. But two disadvantages of $LiNiO_2$ exist: the first drawback of practical application is the anti-sites of Li/Ni in the pristine material. During synthesis, there is a tendency toward loss of lithium and reduction of some Ni to the +2 oxidation state. The Ni^{2+} migrates to Li^+ 3a sites, because of the small size difference between Li^+ (0.9 Å) and Ni^{2+}(0.83 Å) [42]. Another drawback of practical application is its thermal instability with low lithium concentration, which relates to exothermic oxidation of the organic electrolyte with the collapsing delithiated Li_xNiO_2 structure. It should be mentioned that the second lithium can be inserted either chemically or electrochemically into $LiNiO_2$, as in $Li_{1.8}Ni_{1+y}O_2$, which is a mixture as expected of "$LiNiO_2$" and "Li_2NiO_2" [43]. Substitution of Mg for Ni in $LiNi_{1-x}Mg_xO_2$ was adopted to provide a partial solution to the safety concern on $LiNiO_2$. A number of investigations also show that the partly substitution of small amount of Mn, Co and Al for Ni in $LiNiO_2$ can also enhance its thermal stability. The most commonly used electrode material related to $LiNiO_2$ is $LiNi_{0.8}Co_{0.15}Al_{0.05}O_2$, which will be discussed in Sect. 2.7.

2.4 Layered $LiMnO_2$ Compound

Layered $LiMnO_2$ has been expected as a very attractive material from an economical and environmental point of view. $LiMnO_2$ has two structures: orthorhombic structure (space group: Pmmn) and rhombohedral structure. Orthorhombic $LiMnO_2$ is a thermodynamically stable structure. Ion-exchange technique was used to obtain layered $LiMnO_2$ [44]. It exhibits a small monoclinic (C2/m) deformation from the ideal rhombohedral structure due to Jahn-Teller distortion of Mn^{3+}. It is thermodynamically unstable, and unable to synthesized directly because of the low-spin state of Mn^{3+} in $LiMnO_2$ [45].

Both orthorhombic and rhombohedral $LiMnO_2$ will transform into a spinel structure during cycling and its poor crystallinity of this phase leads to poor electrochemical properties and a spinel-like drop in the voltage profile. Owing to the same cubic-close-packed (ccp) sublattice of oxygen ions adopted by layered $LiMnO_2$ and spinel $LiMn_2O_4$ phases. This conversion from layered to spinel-like structure is not difficult. It requires only cation diffusion and this is a common phenomenon in almost all the layered structure cathode materials.

To inhibit the transformation, other layered Li–Mn–O materials with non-ccp close-packed structures, such as, $Li_{2/3}[Li_{1/6}Mn_{5/6}]O_2$ with an O2 type structure, are synthesized by ion exchange from $Na_{2/3}[Li_{1/6}Mn_{5/6}]O_2$ (P2 structure). It exhibits a different oxygen sublattice from orthorhombic, rhombohedral $LiMnO_2$ and spinel $Li_2Mn_2O_4$ or $LiMn_2O_4$ [46]. Unfortunately, the electrochemical performance of $Li_{2/3}[Li_{1/6}\text{-}Mn_{5/6}]O_2$ is compromised because of its poor crystallinity caused by stacking faults. The partial substitution of the manganese ions by cobalt, iron, or nickel is found to significantly increase the electronic conductivity of $LiMnO_2$.

2.5 Layered $LiNi_{0.5}Mn_{0.5}O_2$ Compound

There are more than three transitional metal elements (Ni, Co, Mn) for forming layered transition-metal oxides with the α-$NaFeO_2$ structure, [47]. To improve the performance of $LiNiO_2$ and avoid the presence of toxic and expensive of Co ions, the approach of mixing the $LiNiO_2$ and $LiMnO_2$ with 1:1 ratio, $LiNi_{0.5}Mn_{0.5}O_2$ was reported in 1992 by Dahn et al. [48], but it did not receive much attention because of poor electrochemical behavior. This system was revisited in 2001 by Ohzuku et al. [49]. In $LiNi_{0.5}Mn_{0.5}O_2$, Mn becomes the structure stabilizer and Ni the electrochemically active ion. Ni and Mn ions have valence +2 and +4 oxidation states respectively [50], so in principle, it should be free from either a transition into a spinel-like phase or any Jahn-Teller distortion of Mn^{3+} during charge and discharge. Mn could suppress the migration of Ni to the Li layers in $LiNi_{0.5}Mn_{0.5}O_2$. The precise details of the structure of $LiNi_{0.5}Mn_{0.5}O_2$ are complex. Long-range order has been detected by TEM, NMR etc. [51, 52]. The ordered arrangement of Mn, Ni and Li in the metal layer that satisfy the available experimental evidence is a "flower" pattern illustrated in Fig. 15 [53]. The Li ions in the flower pattern are surrounded by a hexagonal ring of Mn ions. The rings of Mn in turn are surrounded by a larger ring of Ni so that no nearest neighbor Li–Ni contacts exist in the TM layer. During charge and discharge, the nickel ion cycles between Ni^{2+} and Ni^{4+}, Mn remains inactive throughout normal operating cell voltages [52, 54]. It shows

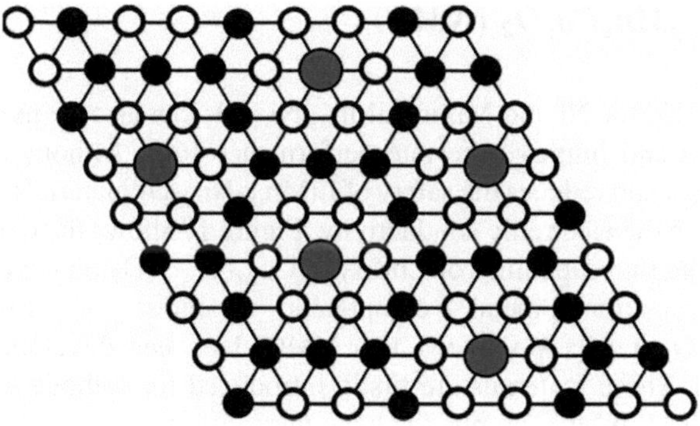

Fig. 15 Transition metal layer in $LiNi_{0.5}Mn_{0.5}O_2$ showing the flower pattern. Li is represented by the *large gray circles*, Mn by *small white circles*, and Ni by *black circles* [53]. Copyright 2004, Elsevier B.V.

200 mAh g^{-1} of rechargeable capacity with relatively small irreversible capacity and low polarization and little capacity fading even after 100 cycles at a low rate [55]. Partial exchange of Li and Ni ions is always observed $LiMn_{0.5}Ni_{0.5}O_2$ made by a conventional high-temperature synthesis. It contracts the space through which Li can move, resulting in a low-rate cathode material. XRD results show an 8–9 % displacement of Li^+ and Ni^{2+} between 3a and 3b sites when a structural model based on R-3 m is assumed [56]. This cation disorder impedes the kinetics of Li diffusion as described above for $LiNiO_2$.

The structural changes accompanying the removal of Li from a number of this material have been investigated. $LiMn_{0.5}Ni_{0.5}O_2$ has a higher thermal stability. Yang et al. show that the charge–discharge process of $LiNi_{0.5}Mn_{0.5}O_2$ above 2 V is accompanied by a reversible phase transition between H1 to H2, similar to the $LiNiO_2$ system. However, the formation of the H3 phase that occurs at voltages above 4.3 V in the $LiNiO_2$ system is suppressed [57]. So it can get a high capacity. Other advantages of $LiNi_{0.5}Mn_{0.5}O_2$ are lower thermal runaway, better structural thermal stability than $LiCoO_2$ or $LiNiO_2$, and greater inhibition to reaction with electrolytes in the charged state [55]. Cycled $Li_xNi_{0.5}Mn_{0.5}O_2$ (x = 0.2, 0.9) also shows a higher onset temperature of oxygen release and reduces oxygen loss compared to $Li_{0.3}NiO_2$ and $Li_{0.3}Ni_{0.7}Co_{0.15}Al_{0.15}O_2$ as a result of increased structural stability [58]. But the anti-site mixing is thought to be a thermodynamically favored feature of the ion ordering and thus intrinsic to the structure. Simple manipulation of conditions during direct synthesis is not likely to result in a near-ideal layered structure. Kang et al. [59] have successfully synthesized $LiNi_{0.5}Mn_{0.5}O_2$ with very little intralayer disordering by ion-exchange from $NaNi_{0.5}Mn_{0.5}O_2$. It displays an exciting combination of high rate and high capacity (183 mAh g^{-1} at 6 C rate), shown in Fig. 16. Yang et al. also reported that excess lithium could be intercalated to $LiNi_{0.5}Mn_{0.5}O_2$, this introduces possibly the phase $Li_2Ni_{0.5}Mn_{0.5}O_2$ [52, 57].

2.6 $LiNi_{1-y-z}Mn_yCo_zO_2$ (NMC)

Substitution of Co for Ni and Mn in $LiMn_{0.5}Ni_{0.5}O_2$ can also be used to reduce the Li/Ni exchange and improve the rate performance [60]. Although the use of Co increases the cost and reduces the safety of $LiMn_{0.5}Ni_{0.5}O_2$ materials, Co substitution can also increase the electronic conductivity. Figure 17 shows the dependence of the charge-discharge rate capability of $LiMn_{0.5-x}Co_{2x}Ni_{0.5-x}O_2$ on Co content.

Ternary transition metal component systems of Li–Ni–Mn–Co–O ($LiNi_xMn_yCo_zO_2$ with x + y + z = 1 or NMC) has been developed as a cathode active material. These materials are firstly introduced for cathode materials by Liu et al. in 1999 [61]. NMC compounds have the same structure as $LiCoO_2$ [62]. Ni, Co and Mn ions are in a random distribution in 3b sites. The Li/Ni anti-site can still be found in the compounds. Commonly, in the XRD patterns, if the value of the intensity ratio of (003)/(104) peaks are larger than 1.2, and the peaks of (006)/(012)

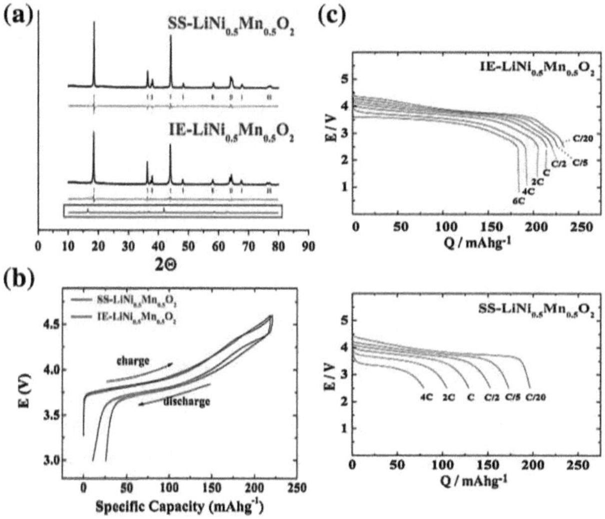

Fig. 16 **a** XRD patterns of solid state reaction (SS)-$LiMn_{0.5}Ni_{0.5}O_2$ (*top*) and ion exchange (IE)-$LiMn_{0.5}Ni_{0.5}O_2$ (*bottom*). **b** First charge/discharge curves of SS-$LiMn_{0.5}Ni_{0.5}O_2$ and IE-$LiMn_{0.5}Ni_{0.5}O_2$ with the voltage window of 3.0–4.6 V at C/20 rate. **c** The discharge curve at various C rates for IE-$LiMn_{0.5}Ni_{0.5}O_2$ (*top*) and SS-$LiMn_{0.5}Ni_{0.5}O_2$ (*bottom*) [59]. Copyright 2006, AAAS

Fig. 17 Dependence of the charge-discharge rate capability of $LiMn_{0.5-x}Co_{2x}Ni_{0.5-x}O_2$ on Co content. Assuming that the discharge capacity at C/7 is 100 %. [60] Copyright 2004 by The Electrochemical Society

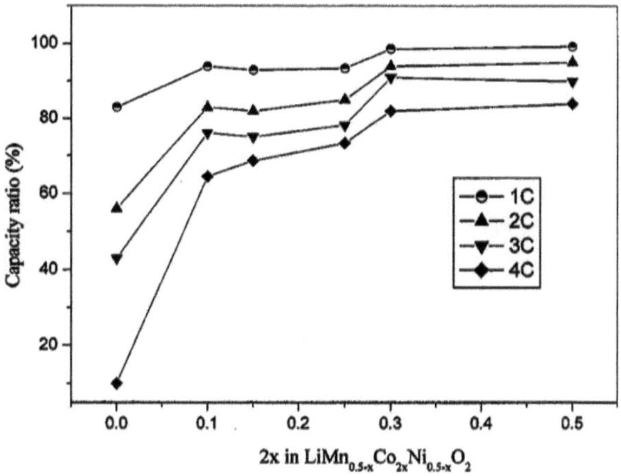

and (018)/(110) are pronounced splitted, they indicate a well-developed layered structure [26]. The most representative composition is $LiNi_{1/3}Mn_{1/3}Co_{1/3}O_2$ reported in 2001 by Ohzuku and Makimura [63]. It provides a somewhat higher capacity than $LiCoO_2$ below 4.3 V versus Li^+/Li (typically about 160 mAh g^{-1} compared to 140 mAh g^{-1} of $LiCoO_2$). In $LiNi_{1/3}Mn_{1/3}Co_{1/3}O_2$, and other compounds having the general formula $LiNi_xCo_{1-2x}Mn_xO_2$, which can be considered as a solid solution of $LiNi_{0.5}Mn_{0.5}O_2$ and $LiCoO_2$, Ni, Co, and Mn adopt valence states of +2, +3 and +4 respectively. Co is reversibly oxidized from Co^{3+} to Co^{4+} at the higher potentials while Ni is oxidized from Ni^{2+} to Ni^{4+} at low potentials during cycling. However, there is no change of oxidation number of Mn at Mn^{4+} during lithiation/delithiation processes [62, 64]. Therefore, manganese provides structural stability. Also, the absence of trivalent manganese suppresses manganese dissolution. Neutron and X-ray power diffraction has shown that there is Ni ions in the lithium metal layers [64]. The cobalt reduces the number of nickel ions in the lithium layer.

The material also shows a small volume change during lithiation/delithiation processes, actually zero-volume change can be seen in x = 0–0.67 in the formula of $Li_{1-x}Ni_{1/3}Mn_{1/3}Co_{1/3}O_2$ [65]. The material shows good rate capability and excellent safety properties at a high state of charge, compared to $LiNiO_2$, $LiCoO_2$ and $LiNi_{0.8}Co_{0.15}Al_{0.05}O_2$ [65]. Capacities in excess of 200 mAh g^{-1} can be achieved when cells are charged to higher voltage limits, although this usually results in diminished cycle life [66].

The synthetic methods have a significant impact on the performance of the layered cathode materials, especially for the binary or ternary compounds. NCMs are prepared from a homogeneous mixed solution of Ni, Mn and Co precursors. Spherical particles of three component hydroxide or carbonate are obtained by co-precipitation, followed by calcination with Li_2CO_3 or $LiOH{\cdot}H_2O$ [67]. Preparing spherical NCM precursors with specific molar ratio has developed as a professional production in industry now.

Besides $LiNi_{1/3}Mn_{1/3}Co_{1/3}O_2$, Ni-rich compositions (x > 1/3), such as $LiNi_{0.5}Mn_{0.3}Co_{0.2}O_2$, $LiNi_{0.4}Co_{0.2}Mn_{0.4}$ etc. also attract much attention for their good electrochemical performance. The electrochemical behavior of a number of different compositions over a range of current densities have been studied, and two of these studies identified the $LiNi_{0.4}Co_{0.2}Mn_{0.4}$ material as having the highest capacity and maintaining its capacity on cycling. All these compounds show higher thermal stabilities compared to the Co-free compounds, and they are gradually replacing $LiCoO_2$ in consumer batteries (in some cases, a mixture of the two are used as the cathode) and is under consideration for some vehicular applications. Up to now, Ni-rich NCM phases, such as $LiNi_{0.8}Co_{0.1}Mn_{0.1}$ and $LiNi_{0.6}Co_{0.2}Mn_{0.2}$, are attractive due to their high energy densities, but still suffers from instability of oxygen. Further efforts are needed.

2.7 $LiNi_{1-y-z}Co_yAl_zO2$ (NCA)

A part of the nickel in $LiNiO_2$ can be replaced by cobalt and aluminum. Cobalt assists in ordering the structure, that is keeping the nickel in the nickel layer, and aluminum, being redox inactive, prevents the complete removal of all the lithium, thus additionally stabilizing the structure and preventing any phase changes that might occur at very low or zero lithium content. NCA with more than 80 % Ni contents delivers higher specific capacity than $LiCoO_2$ does, showing good cycleability and better structural stability due to no cation exchange of Ni with Li in the Li layers.

However, NCA has safety problems related to oxygen evolution at overcharge situations. Figure 18 shows the different phase-distributions and the phase transition path of $Li_{0.15}Ni_{0.8}Co_{0.15}Al_{0.05}O_2$ during heating [68]. The overcharged $Li_{0.15}Ni_{0.8}Co_{0.15}Al_{0.05}O_2$ has a multiple phase of layered, spinel and rock salt from bulk to surface. During heating, the surface structures propagate towards to the core, and the rocksalt phase becomes the dominant structure due to the migration of Ni cations to the original Li sites at high temperature. Fortunately, the structure stability can be enhanced by surface modifying such as coating [69] and doping [70], just like $LiCoO_2$.

NCA is a promising cathode material for next generation lithium ion batteries, and it has attracted considerable attention. In 2005, Sony and Hitachi have applied it in their new lithium ion batteries. NCA has become the cathode materials in Li-ion batteries for Tesla, provided by Panasonic.

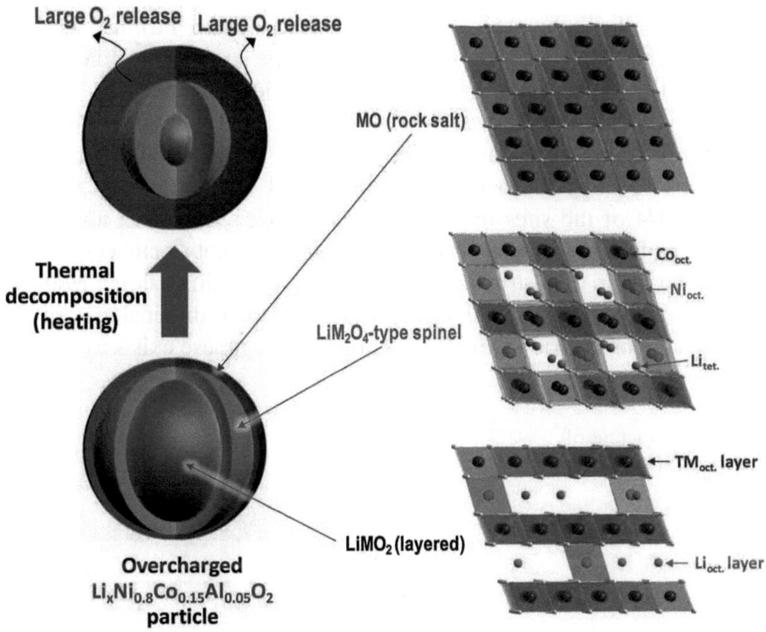

Fig. 18 Schematic of the mechanism of thermal decomposition of the overcharged $Li_{0.15}Ni_{0.8}Co_{0.15}Al_{0.05}O_2$ cathode during heating. The *left side* shows phase propagation from the surface to core for the overcharged particle. The *right side* shows the changes in crystal structure and cation distribution [68]. Copyright 2013 by John Wiley & Sons

2.8 *Other Layered Oxides*

$LiFeO_2$ (R-3 m structure synthetised by ion exchange of $NaFeO_2$) would be an ideal low cost cathode material, but it is not suitable for lithium ion batteries due to low and highly sloping voltage profiles and poor cycling properties [71]. It can be used as additive to improve high-temperature cycling performance of $LiCoO_2$ [72]. $LiCrO_2$ [73, 74] and $LiVO_2$ [75] are also impossible due to irreversible migration of Cr or V into Li layers in surface regions after further charging (oxidation) ($LiVO_2$ can be used as anode for Li-ion batteries). Li_2MnO_3 can be considered a layered structure similar to $LiCoO_2$ with the formula of $Li[Li_{1/3}Mn_{2/3}]O_2$. xLi_2MnO_3-$(1 - x)LiMO_2$ (M = Ni, Co, Mn), referred to a compound between Li_2MnO_3 and $LiMO_2$ (M = Ni, Co, Mn), is another promising oxide cathode material for high capacity lithium ion batteries [76]. This will be discussed in Chap. 20.

3 Spinel Structural Cathodes

Another alternative to $LiCoO_2$ is spinel $LiMn_2O_4$ by Thackeray et al. in 1984 [3].

$LiMn_2O_4$ has the cubic spinel structure (space group Fd-3m) with a lattice parameter of $a = 8.2449$ Å at room temperature. The structure of LiM_2O_4 spinel is shown in Fig. 19. As discussed in 2.3, the anion lattice contains cubic close-packed oxygen ions and is closely related to the α-$NaFeO_2$ layer structure, only differing in the distribution of the cations among the available octahedral and tetrahedral sites.

Mn ions still occupy the octahedral site but 1/4 of them are located in the Li layer, leaving 1/4 of the sites in transition metal layer vacant. Li ions occupy the tetrahedral sites in Li layer that share faces with the empty octahedral sites in the transition metal layer. The structure is based on a three-dimensional MO_2 host and the vacancies in transition metal layer ensure the three-dimensional Li diffusion pathways. In stoichiometric $LiMn_2O_4$, half of the manganese exists as Mn^{3+} and the

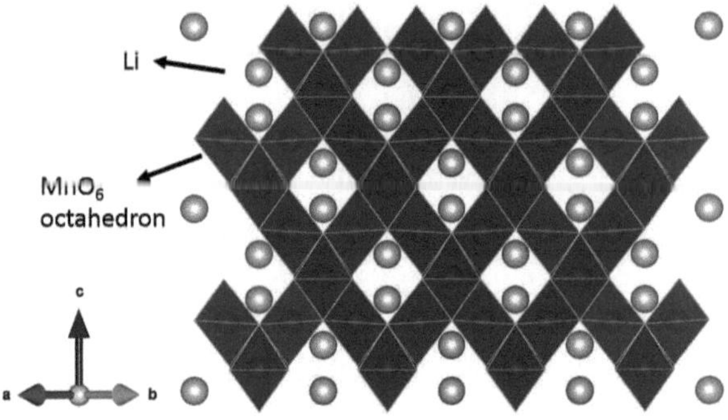

Fig. 19 Crystal structure of spinel LiM_2O_4 [5]

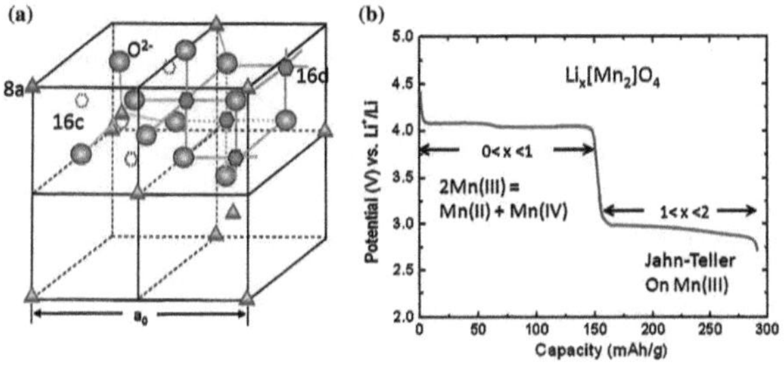

Fig. 20 **a** Two quadrants of the cubic spinel structure, **b** V(x) profile of $Li_xMn_2O_4$ [78]. Copyright 2012, Springer

other half as Mn^{4+}, with the ions distributed randomly throughout the crystal at room or elevated temperature. $LiMn_2O_4$ exhibits an operating voltage of 4.1 V and its theoretical capacity is 148 mAh g^{-1} [3]. The reversible capacity is about 140 mAh g^{-1}. Li ions can also reversible inset in the $LiMn_2O_4$ structure with a ~3.0 V potential despite working on the same Mn^{4+}/Mn^{3+} redox couple caused by a cooperative Jahn-Teller orbital order on Mn^{3+} ions (see Fig. 20) [77, 78].

Although $LiMn_2O_4$ spinel and its derivatives show promising features as the cathodes of high-power lithium batteries for transportation applications, a slow capacity fade has been encountered in the high- voltage range, particularly at elevated temperatures (>50 °C) [79]. The capacity loss can be attributed to several possible factors [80]:

1. an instability of the organic-based electrolyte at the high voltages reached when charging cells,
2. a slow dissolution of the $Li_xMn_2O_4$ electrode into the electrolyte (as Mn^{2+}) according to the disproportionation reaction: $2Mn^{3+} \rightarrow Mn^{4+} + Mn^{2+}$,
3. the onset of the Jahn-Teller effect in deeply discharged $Li_xMn_2O_4$ electrodes (i.e. at $x \approx 1$).

To solve these problems, three approaches were adopted to increase the average manganese-ion valence marginally above 3.5 to decrease the amount of Mn^{3+}:

1. synthesizing stoichiometric spinels of general formula $Li_{1+\delta}Mn_{2-\delta}O_4$, i.e. by replacing some manganese with lithium. For example, when $\delta = 0.05$ the spinel has the formula $Li_{1.05}Mn_{1.95}O_4$ in which manganese ion valence is 3.56,
2. replacing a small amount of manganese with a univalent metal cation such as Mg^{2+} or Zn^{2+}. For example, when $\delta = 0.025$ in $LiMg_{\delta/2}Mn_{2-\delta}O_4$ manganese ion valence is 3.54 (note that this is a non-stoichiometric spinel with a slight cation deficiency),
3. synthesizing defects (cation-deficient) spinels $Li_{1-\delta}Mn_{2-2\delta}O_4$, for example, $Li_{0.975}Mn_{1.95}O_4$ ($\delta = 0.025$) in which manganese ion valence is 3.60.

The instability of the structure is another problem of $LiMn_2O_4$. A phase transition has been observed slightly below room temperature from cubic to orthorhombic phase [81]. This is attributed to charge ordering of Mn^{3+} and Mn^{4+} ions as a result of cooperative interactions between Jahn–Teller-distorted Mn^{3+} ions on the octahedral site. The presence of oxygen vacancies, which can be introduced by altering the synthesis conditions and starting reagents [82], leads to a higher proportion of Mn^{3+} relative to Mn^{4+}, which affects the degree of Jahn–Teller-induced ordering. For example, a tetragonal spinel phase (space group $I4_1/amd$) has been observed in $LiMn_2O_{4-\delta}$ above room temperature [83, 84]. Mn_3O_4 containing 1/3 soluble Mn^{2+} ions and tetragonal $LiMn_2O_{4-\delta}$ have been directly observed on the surface of $Li_xMn_2O_4$ [85]. So it is important to stabilize the surface structure of $LiMn_2O_4$ at both pristine and charged states. This material has been plagued by self-discharge when left under full charge, particularly at elevated temperatures; however, this problem may have been solved by switching from the fluoride-containing $LiPF_6$ salt, which can generate HF in the presence of traces of moisture, to salts such as LiBOB [86].

Like $LiCoO_2$, to improve the electrochemical performances at elevated temperature and the storage performance, different methods of doping and coating have been attempted. Among them, Al doping and $LiAlO_2$ (or Al_2O_3) coating are excellent [87]. Sun et al. [87] developed a sol-gel method for $LiAlO_2$ coating to improve its electrochemical performance at elevated temperatures. The capacity retention of $LiAlO_2$-modified spinel $LiMn_2O_4$ is more than 94 % after 500 cycles at room temperature and more than 90 % after 200 cycles at 55 °C at 1 C charge-discharge rate (shown in Fig. 21). The improved electrochemical performance of surface-modified spinel $LiMn_2O_4$ is attributed to the protection effect of the surface solid solution shell while the improved stability of crystalline structure is assigned to the entrance of Li^+ and Al^{3+} into the spinel crystalline structure. An alternate solution pioneered by the Korean school is to coat the surface of the spinel particles with materials such as ZrO_2 or $AlPO_4$, which are believed to act as getters for any HF. $LiMn_2O_4$ core with $LiM_xMn_{2-x}O_4$ (M = Ni, Co) shell can improve the cycle stability attributed to the suppression of Jahn–Teller distortion on the surface of spinel $LiMn_2O_4$ particles during cycling [88–90].

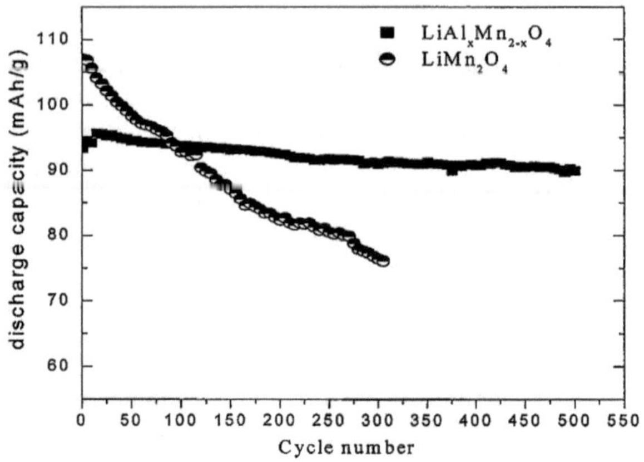

Fig. 21 Cycling behaviors of the samples at room temperature. Comparison of cyclability of $LiAl_xMn_{2-x}O_4$ with $LiMn_2O_4$ at 1 C rate [87]. Copyright 2003 by The Electrochemical Society

When 1/4 Mn is replaced by Ni^{2+}, all the Mn will be displaced as Mn^{4+}. $LiNi_{0.5}Mn_{1.5}O_4$ shows the best overall electrochemical performances among the above. The voltage of these materials can be raised to more than 4.7 V, therefore the energy densities are highly improved [91]. This will be discussed in Chap. 19.

$LiMn_2O_4$ is ideal as a high-capacity Li-ion battery cathode material by virtue of its low toxicity, low cost, the high natural abundance of Mn, high rate performances and excellent cyclic performances. It is suitable for large scale power batteries used for EVs, e-bike, power grid and power tools.

4 The Combination of Layered and Spinel Cathode

Several works have been reported on the mixed spinel $LiMn_2O_4$ and layered $LiMO_2$ composite cathode materials [92–95]. The mixing can decrease the manganese dissolution of $LiMn_2O_4$ and capacity fading [92]. The storage performance of $LiMn_2O_4$ spinel at high temperatures can also be improved [93]. It is suggested that the mixed electrode is considered to be a quite promising positive electrode material to be applied for lithium-ion batteries. The mixture of $LiMn_2O_4$ with NCM is also used widely in EV batteries.

Another approach to combine the two structure is forming core–shell structured particles. The highly stable spinel phase can used to enhance the thermal stability of the layered structure [96–98]. A heterostructured $Li[Ni_{0.54}Co_{0.12}Mn_{0.34}]O_2$ cathode material, with $Li[Ni_{0.54}Co_{0.12}Mn_{0.34}]O_2$ as core and $Li_{1+x}[CoNi_xMn_{2-x}]_2O_4$ as shell, possesses both high energy density and safety. [99] The material demonstrates reversible capacity of 200 mAh g^{-1} and retains 95 % capacity retention under the most severe test condition of 60 °C. In addition, the amount of oxygen evolution from the lattice in the cathode with two heterostructures is reduced by 70 %, compared with the reference sample. Layered $LiCoO_2$ and $LiNi_{1-x}Co_xO_2$ are also used for coating to improve the structure stability of $LiMn_2O_4$ at elevated temperature [100–104].

5 Summary

Significant progress in cathode materials for lithium ion batteries has been made in the past decades. So far, layered and spinel cathode materials still occupy the large ratio in battery market and industry applications. $LiCoO_2$ has been successful in consumer electronic devices, in which volumetric energy density is the preferential factor. But cobalt is nocuous the price of cobalt is expensive. Many efforts have been attempted by researchers to avoid its weakness. Among them, NCM and NCA are the most successful candidates, for high capacity and low price. Nowadays, with the success of improving the stability of $LiCoO_2$ at high voltage, $LiCoO_2$ may get the second success in high capacity and high energy density use in near future.

Surface modified $LiMn_2O_4$ will continue its success in large scale energy storage filed, such as smart grid and uninterrupted power supply. In spinel cathodes, high voltage $LiNi_{0.5}Mn_{1.5}O_4$ has been deeply researched in laboratory and tested in industry. Maybe it will be used in commercial application for high energy density lithium ion batteries after solving challenges in high voltage electrolyte or solid electrolyte.

References

1. Whittingham MS (2004) Lithium batteries and cathode materials. Chem Rev 104:4271–4302
2. Mizushima K, Jones PC, Wiseman PJ, Goodenough JB (1980) Li_xCoO_2: a new cathode material for batteries of high energy density. Mater Res Bull 15:783–789
3. Thackeray MM, Johnson PJ, de Picciotto LA, Bruce PG, Goodenough JB (1984) Electrochemical extraction of lithium from $LiMn_2O_4$. Mater Res Bull 19:179–187
4. Padhi AK, Nanjundaswamy KS, Goodenough JB (1997) Phospho-olivines as positive-electrode materials for rechargeable lithium batteries. J Electrochem Soc 144:1188–1194
5. Ma C, Lyu Y, Li H (2014) Fundamental scientific aspects of lithium batteries (VII)—positive electrode materials. Energy Storage Sci Tech 3:53–65
6. Whittingham MS (1976) Electrical energy storage and intercalation chemistry. Science 192:1126–1127
7. Whittingham MS, Chianelli RR (1980) Layered compounds and intercalation chemistry: an example of chemistry and diffusion in solids. J Chem Edu 57:569
8. Whittingham MS (1976) The role of ternary phases in cathode reactions. J Electrochem Soc 123:315–320
9. Whittingham MS (1974) Electrointercalation in transition-metal disulphides. J Chem Soc Chem Commun 328–329
10. Whittingham MS (1978) The electrochemical characteristics of VSe_2 in lithium cells. Mater Res Bull 13:959–965
11. Ozawa K (1994) Lithium-ion rechargeable batteries with $LiCoO_2$ and carbon electrodes: the $LiCoO_2$/C system. Solid State Ionics 69:212–221
12. Reimers JN, Dahn JR (1992) Electrochemical and in situ X-Ray diffraction studies of lithium intercalation in $LixCoO_2$. J Electrochem Soc 139:2091–2097
13. Menetrier M, Saadoune I, Levasseur S, Delmas C (1999) The insulator-metal transition upon lithium deintercalation from $LiCoO_2$: electronic properties and 7Li NMR study. J Mater Chem 9:1135–1140
14. Marianetti CA, Kotliar G, Ceder G (2004) A first-order Mott transition in Li_xCoO_2. Nat Mater 3:627–631
15. Ohzuku T, Ueda A (1994) Solid-State Redox Reactions of $LiCoO_2$ ($R\bar{3}m$) for 4 Volt secondary lithium cells. J Electrochem Soc 141:2972–2977
16. Shao-Horn Y, Levasseur S, Weill F, Delmas C (2003) Probing lithium and vacancy ordering in O3 layered Li_xCoO_2 ($x \approx 0.5$) an electron diffraction study. J Electrochem Soc 150:A366–A373
17. Van der Ven A, Aydinol MK, Ceder G (1998) First-principles evidence for stage ordering in $LixCoO_2$. J Electrochem Soc 145:2149–2155
18. Van der Ven A, Aydinol MK, Ceder G, Kresse G, Hafner J (1998) First-principles investigation of phase stability in $LiCoO_2$. Phys Rev B 58:2975–2987
19. Chen Z, Lu Z, Dahn JR (2002) Staging phase transitions in Li_xCoO_2. J Electrochem Soc 149: A1604–A1609

20. Lu X et al (2012) New insight into the atomic structure of electrochemically delithiated O3-$Li_{(1-x)}CoO_2$ ($0 \leq x \leq 0.5$) nanoparticles. Nano Lett 12:6192–6197
21. Van der Ven A, Ceder G, Asta M, Tepesch PD (2001) First-principles theory of ionic diffusion with nondilute carriers. Phys Rev B 64:184307
22. Amatucci GG, Tarascon JM, Klein LC (1996) Cobalt dissolution in $LiCoO_2$-based non-aqueous rechargeable batteries. Solid State Ionics 83:167–173
23. Xiong F, Yan HJ, Chen Y, Xu B, Le JX, Ouyang CY (2012) The atomic and electronic structure changes upon delithiation of $LiCoO_2$: from first principles calculations. Int J Electrochem Sci 7:9390–9400
24. Delmas C, Fouassier C, Hagenmuller P (1980) Structural classification and properties of the layered oxides. Phys B+C **99**:81–85
25. Delmas C, Braconnier J-J, Hagenmuller P (1982) A new variety of $LiCoO_2$ with an unusual oxygen packing obtained by exchange reaction. Mater Res Bull 17:117–123
26. Gummow RJ, Thackeray MM, David WIF, Hull S (1992) Structure and electrochemistry of lithium cobalt oxide synthesised at 400 °C. Mater Res Bull 27:327–337
27. Gummow RJ, Liles DC, Thackeray MM (1993) Spinel versus layered structures for lithium cobalt oxide synthesised at 400 °C. Mater Res Bull 28:235–246
28. Carlier D, Saadoune I, Croguennec L, Ménétrier M, Suard E, Delmas C (2001) On the metastable O2-type $LiCoO_2$. Solid State Ionics 144:263–276
29. Mendiboure A, Delmas C, Hagenmuller P (1984) New layered structure obtained by electrochemical deintercalation of the metastable $LiCoO_2$ (O_2) variety. Mater Res Bull 19:1383–1392
30. Paulsen JM, Mueller Neuhaus JR, Dahn JR (2000) Layered $LiCoO_2$ with a different oxygen stacking (O2 structure) as a cathode material for rechargeable lithium batteries. J Electrochem Soc 147:508–516
31. Cho J, Kim YJ, Park B (2001) $LiCoO_2$ cathode material that does not show a phase transition from hexagonal to monoclinic phase. J Electrochem Soc 148:A1110–A1115
32. Cho J, Kim YJ, Kim TJ, Park B (2001) Zero-strain intercalation cathode for rechargeable Li-ion cell. Angew Chem Int Ed Engl 40:3367–3369
33. Chen Z, Dahn JR (2002) Effect of a ZrO_2 coating on the structure and electrochemistry of Li_xCoO_2 when cycled to 4.5 V. Electrochem Solid-State Lett 5:A213–A216
34. Liu L et al (2004) Electrochemical and in situ synchrotron XRD studies on Al_2O_3-coated $LiCoO_2$ cathode material. J Electrochem Soc 151:A1344–A1351
35. Li H, Wang Z, Chen L, Huang X (2009) Research on advanced materials for Li-ion batteries. Adv Mater 21:4593–4607
36. Wang Z, Huang X, Chen L (2003) Performance improvement of surface-modified $LiCoO_2$ cathode materials: an infrared absorption and X-Ray photoelectron spectroscopic investigation. J Electrochem Soc 150:A199–A208
37. Cho J, Lee J-G, Kim B, Park B (2003) Effect of P_2O_5 and $AlPO_4$ coating on $LiCoO_2$ cathode material. Chem Mater 15:3190–3193
38. Sun YK, Han JM, Myung ST, Lee SW, Amine K (2006) Significant improvement of high voltage cycling behavior AlF_3-coated $LiCoO_2$ cathode. Electrochem Commun 8:821–826
39. Luo W, Li X, Dahn JR (2010) Synthesis and characterization of Mg substituted $LiCoO_2$. J Electrochem Soc 157:A782–A790
40. Zhou F, Luo W, Zhao X, Dahn JR (2009) Relative impact of Al or Mg substitution on the thermal stability of $LiCo_{1-z}M_zO_2$ (M = Al or Mg) by accelerating rate calorimetry. J Electrochem Soc 156:A917–A920
41. Gao Y, Yakovleva MV, Ebner WB (1998) Novel $LiNi_{1-x}Ti_{x/2}Mg_{x/2}O_2$ compounds as cathode materials for safer lithium-ion batteries. Electrochem Solid State Lett 1:117–119
42. Rougier A, Gravereau P, Delmas C (1996) Optimization of the composition of the $Li_{1-z}Ni_{1+z}O_2$ electrode materials: structural, magnetic, and electrochemical studies. J Electrochem Soc 143:1168–1175
43. Dahn JR, von Sacken U, Michal CA (1990) Structure and electrochemistry of $Li_{1\pm y}NiO_2$ and a new Li_2NiO_2 phase with the $Ni(OH)_2$ structure. Solid State Ionics 44:87–97

44. Armstrong AR, Bruce PG (1996) Synthesis of layered $LiMnO_2$ as an electrode for rechargeable lithium batteries. Nature 381:499–500
45. Huang Z-F, Du F, Wang C-Z, Wang D-P, Chen G (2007) Low-spin $Mn3^+$ ion in rhombohedral $LiMnO_2$ predicted by first-principles calculations. Phys Rev B 75:054411
46. Paulsen JM, Thomas CL, Dahn JR (1999) Layered Li-Mn-oxide with the O_2 structure: a cathode material for Li-ion cells which does not convert to spinel. J Electrochem Soc 146:3560–3565
47. Miyazaki S, Kikkawa S, Koizumi M (1983) Chemical and electrochemical deintercalations of the layered compunds $LiCrO_2$, $LiCoO_2$ and $NaCrO_2$, $NaFeO_2$, $NaCoO_2$, $NaNiO_2$. Synth Met 6:211–217
48. Rossen E, Jones CDW, Dahn JR (1992) Structure and electrochemistry of $Li_xMn_yNi_{1-y}O_2$. Solid State Ionics 57:311–318
49. Ohzuku T, Makimura Y (2001) Layered lithium insertion material of $LiNi_{1/2}Mn_{1/2}O_2$: a possible alternative to LiCoO2 for advanced lithium-ion batteries. Chem Lett 30:744–745
50. Reed J, Ceder G (2002) Charge, potential, and phase stability of layered $Li(Ni_{0.5}Mn_{0.5})O_2$. Electrochem Solid State Lett 5:A145–A148
51. Meng YS, Ceder G, Grey CP, Yoon W-S, Shao-Horn Y (2004) Understanding the crystal structure of layered $LiNi_{0.5}Mn_{0.5}O_2$ by electron diffraction and powder diffraction simulation. Electrochem Solid State Lett 7:A155–A158
52. Yoon W-S, Paik Y, Yang X-Q, Balasubramanian M, McBreen J, Grey CP (2002) Investigation of the local structure of the $LiNi_{0.5}Mn_{0.5}O_2$ cathode material during electrochemical cycling by X-ray absorption and NMR spectroscopy. Electrochem Solid State Lett 5:A263–A266
53. Van der Ven A, Ceder G (2004) Ordering in $Li_x(Ni_{0.5}Mn_{0.5})O_2$ and its relation to charge capacity and electrochemical behavior in rechargeable lithium batteries. Electrochem Commun 6:1045–1050
54. Yoon W-S, Grey CP, Balasubramanian M, Yang X-Q, McBreen J (2003) In situ X-ray absorption spectroscopic study on $LiNi_{0.5}Mn_{0.5}O_2$ cathode material during electrochemical cycling. Chem Mater 15:3161–3169
55. Makimura Y, Ohzuku T (2003) Lithium insertion material of $LiNi_{1/2}Mn_{1/2}O_2$ for advanced lithium-ion batteries. J Power Sources 119–121:156–160
56. Ohzuku T, Brodd RJ (2007) An overview of positive-electrode materials for advanced lithium-ion batteries. J Power Sources 174:449–456
57. Yang X-Q, McBreen J, Yoon W-S, Grey CP (2002) Crystal structure changes of $LiMn_{0.5}Ni_{0.5}O_2$ cathode materials during charge and discharge studied by synchrotron based in situ XRD. Electrochem Commun 4:649–654
58. Yabuuchi N, Kim Y-T, Li HH, Shao-Horn Y (2008) Thermal instability of cycled $Li_xNi_{0.5}Mn_{0.5}O_2$ electrodes: an in situ synchrotron X-ray powder diffraction study. Chem Mater 20:4936–4951
59. Kang K, Meng YS, Bréger J, Grey CP, Ceder G (2006) Electrodes with high power and high capacity for rechargeable lithium batteries. Science 311:977–980
60. Sun Y, Ouyang C, Wang Z, Huang X, Chen L (2004) Effect of Co content on rate performance of $LiMn_{0.5-x}Co_{2x}Ni_{0.5-x}O_2$ cathode materials for lithium-ion batteries. J Electrochem Soc 151:A504–A508
61. Liu Z, Yu A, Lee JY (1999) Synthesis and characterization of $LiNi_{1-x-y}Co_xMn_yO_2$ as the cathode materials of secondary lithium batteries. J Power Sources 81–82:416–419
62. Koyama Y, Tanaka I, Adachi H, Makimura Y, Ohzuku T (2003) Crystal and electronic structures of superstructural $Li_{1-x}[Co_{1/3}Ni_{1/3}Mn_{1/3}]O_2$ ($0 \leq x \leq 1$). J Power Sources 119–121:644–648
63. Ohzuku T, Makimura Y (2001) Layered lithium insertion material of $LiCo_{1/3}Ni_{1/3}Mn_{1/3}O_2$ for lithium-ion batteries. Chem Lett 30:642–643
64. Kim J-M, Chung H-T (2004) The first cycle characteristics of $Li[Ni_{1/3}Co_{1/3}Mn_{1/3}]O_2$ charged up to 4.7 V. Electrochim Acta 49:937–944

65. Yabuuchi N, Ohzuku T (2003) Novel lithium insertion material of $LiCo_{1/3}Ni_{1/3}Mn_{1/3}O_2$ for advanced lithium-ion batteries. J Power Sources 119–121:171–174
66. Choi J, Manthiram A (2005) Role of chemical and structural stabilities on the electrochemical properties of layered $LiNi_{1/3}Mn_{1/3}Co_{1/3}O_2$ cathodes. J Electrochem Soc 152:A1714–A1718
67. Noh M, Cho J (2013) Optimized synthetic conditions of $LiNi_{0.5}Co_{0.2}Mn_{0.3}O_2$ cathode materials for high rate lithium batteries via co-precipitation method. J Electrochem Soc 160: A105–A111
68. Nam K-W et al (2013) Combining in situ synchrotron X-ray diffraction and absorption techniques with transmission electron microscopy to study the origin of thermal instability in overcharged cathode materials for lithium-ion batteries. Adv Funct Mater 23:1047–1063
69. Cho Y, Lee Y-S, Park S-A, Lee Y, Cho J (2010) $LiNi_{0.8}Co_{0.15}Al_{0.05}O_2$ cathode materials prepared by TiO_2 nanoparticle coatings on $Ni_{0.8}Co_{0.15}Al_{0.05}(OH)_2$ precursors. Electrochim Acta 56:333–339
70. Kondo H et al (2007) Effects of Mg-substitution in Li(Ni, Co, Al)O_2 positive electrode materials on the crystal structure and battery performance. J Power Sources 174:1131–1136
71. Li JG, Li JJ, Luo J, Wang L, He XM (2011) Recent advances in the $LiFeO_2$-based materials for Li-ion batteries. Int J Electrochem Sci 6:1550–1561
72. Park K-S, Im D, Benayad A, Dylla A, Stevenson KJ, Goodenough JB (2012) $LiFeO_2$-incorporated Li_2MoO_3 as a cathode additive for lithium-ion battery safety. Chem Mater 24:2673–2683
73. Jones C, Rossen E, Dahn J (1994) Structure and electrochemistry of $Li_xCr_yCo_{1-y}O_2$. Solid State Ionics 68:65–69
74. Lyu Y, Ben L, Sun Y, Tang D, Xu K, Gu L, Xiao R, Li H, Chen L, Huang X (2015) Atomic insight into electrochemical inactivity of lithium chromate ($LiCrO_2$): Irreversible migration of chromium into lithium layers in surface regions. J Power Sources 73:1218–1225
75. De Picciotto LA, Thackeray MM, David WIF, Bruce PG, Goodenough JB (1984) Structural characterization of delithiated $LiVO_2$. Mater Res Bull 19:1497–1506
76. Johnson CS, Kim JS, Lefief C, Li N, Vaughey JT, Thackeray MM (2004) The significance of the $L_{i2}MnO_3$ component in 'composite' $xLi_2MnO_3(1-x)LiMn_{0.5}Ni_{0.5}O_2$ electrodes. Electrochem Commun 6:1085–1091
77. Thackeray MM, David WIF, Bruce PG, Goodenough JB (1983) Lithium insertion into manganese spinels. Mater Res Bull 18:461–472
78. Goodenough JB (2012) Rechargeable batteries: challenges old and new. J Solid State Electrochem 16:2019–2029
79. Momchilov A, Manev V, Nassalevska A, Kozawa A (1993) Rechargeable lithium battery with spinel-related MnO_2 II. Optimization of the $LiMn_2O_4$ synthesis conditions. J Power Sources 41:305–314
80. Gummow RJ, de Kock A, Thackeray MM (1994) Improved capacity retention in rechargeable 4 V lithium/lithium-manganese oxide (spinel) cells. Solid State Ionics 69:59–67
81. Yamada A, Tanaka M (1995) Jahn-Teller structural phase transition around 280 K in $LiMn_2O_4$. Mater Res Bull 30:715–721
82. Yonemura M et al (2004) Synthesis, structure, and phase relationship in lithium manganese oxide spinel. J Mater Chem 14:1948–1958
83. Huang R et al (2011) Oxygen-vacancy ordering at surfaces of lithium manganese(III, IV) oxide spinel nanoparticles. Angew Chem Int Ed 50:3053–3057
84. Strobel P, Le Cras F, Seguin L, Anne M, Tarascon JM (1998) Oxygen nonstoichiometry in Li–Mn–O spinel oxides: a powder neutron diffraction study. J Solid State Chem 135:132–139
85. Tang D, Sun Y, Yang Z, Ben L, Gu L, Huang X (2014) Surface structure evolution of $LiMn_2O_4$ cathode material upon charge/discharge. Chem Mater 26:3535–3543
86. Amine K et al (2004) Improved lithium manganese oxide spinel/graphite Li-ion cells for high-power applications. J Power Sources 129:14–19

87. Sun Y, Wang Z, Chen L, Huang X (2003) Improved electrochemical performances of surface-modified spinel $LiMn_2O_4$ for long cycle life lithium-ion batteries. J Electrochem Soc 150:A1294–A1298
88. Myung S-T, Lee K-S, Kim D-W, Scrosati B, Sun Y-K (2011) Spherical core-shell Li $[(Li_{0.05}Mn_{0.95})_{0.8}(Ni_{0.25}Mn_{0.75})_{0.2}]_2O_4$ spinels as high performance cathodes for lithium batteries. Energy Environ Sci 4:935–939
89. Hwang BJ, Santhanam R, Huang CP, Tsai YW, Lee JF (2002) $LiMn_2O_4$ core surrounded by $LiCo_xMn_{2-x}O_4$ shell material for rechargeable lithium batteries: synthesis and characterization. J Electrochem Soc 149:A694–A698
90. Li X, Xu Y, Wang C (2009) Novel approach to preparation of $LiMn_2O_4$ core/$LiNi_xMn_{2-x}O_4$ shell composite. Appl Surface Sci 255:5651–5655
91. Wang L, Li H, Huang X, Baudrin E (2011) A comparative study of Fd-3m and P4332 "$LiNi_{0.5}Mn_{1.5}O_4$". Solid State Ionics 193:32–38
92. Myung S-T, Cho MH, Hong HT, Kang TH, Kim C-S (2005) Electrochemical evaluation of mixed oxide electrode for Li-ion secondary batteries: $Li_{1.1}Mn_{1.9}O_4$ and $LiNi_{0.8}Co_{0.15}Al_{0.05}O_2$. J Power Sources 146:222–225
93. Kitao H, Fujihara T, Takeda K, Nakanishi N, Nohma T (2005) High-temperature storage performance of Li-ion batteries using a mixture of Li–Mn spinel and Li–Ni–Co–Mn oxide as a positive electrode material. Electrochem Solid State Lett 8:A87–A90
94. Numata T, Amemiya C, Kumeuchi T, Shirakata M, Yonezawa M (2001) Advantages of blending $LiNi_{0.8}Co_{0.2}O_2$ into $Li_{1+x}Mn_{2-x}O_4$ cathodes. J Power Sources 97–98:358–360
95. Tran HY, Täubert C, Fleischhammer M, Axmann P, Küppers L, Wohlfahrt-Mehrens M (2011) $LiMn_2O_4$ spinel/$LiNi_{0.8}Co_{0.15}Al_{0.05}O_2$ blends as cathode materials for lithium-ion batteries. J Electrochem Soc 158:A556–A561
96. Cho J, Kim G (1999) Enhancement of thermal stability of $LiCoO_2$ by $LiMn_2O_4$ coating. Electrochem Solid State Lett 2:253–255
97. Wu F et al (2013) Spinel/layered heterostructured cathode material for high-capacity and high-rate Li-ion batteries. Adv Mater 25:3722–3726
98. Wu F et al (2014) Ultrathin spinel membrane-encapsulated layered lithium-rich cathode material for advanced Li-ion batteries. Nano Lett 14:3550–3555
99. Cho Y, Lee S, Lee Y, Hong T, Cho J (2011) Spinel-layered core-shell cathode materials for Li-ion batteries. Adv Energy Mater 1:821–828
100. Park S-C, Kim Y-M, Han S-C, Ahn S, Ku C-H, Lee J-Y (2002) The elevated temperature performance of $LiMn_2O_4$ coated with $LiNi_{1-X}Co_XO_2$ (X = 0.2 and 1). J Power Sources 107:42–47
101. Liu Z, Wang H, Fang L, Lee JY, Gan LM (2002) Improving the high-temperature performance of $LiMn_2O_4$ spinel by micro-emulsion coating of $LiCoO_2$. J Power Sources 104:101–107
102. Cho J, Kim GB, Lim HS, Kim CS, Yoo SI (1999) Improvement of structural stability of $LiMn_2O_4$ cathode material on 55 °C cycling by sol-gel coating of $LiCoO_2$. Electrochem Solid State Lett 2:607–609
103. Park S-C et al (2001) Electrochemical properties of $LiCoO_2$-coated $LiMn_2O_4$ prepared by solution-based chemical process. J Electrochem Soc 148:A680–A686
104. Kannan AM, Manthiram A (2002) Surface/chemically modified $LiMn_2O_4$ cathodes for lithium-ion batteries. Electrochem Solid State Lett 5:A167–A169

Polyanion Compounds as Cathode Materials for Li-Ion Batteries

X.B. Wu, X.H. Wu, J.H. Guo, S.D. Li, R. Liu, M.J. McDonald and Y. Yang

The development of high energy density Li-ion batteries depends on finding electrode materials that can meet increasingly stringent demands, in particular cathode materials [1, 2]. Cathodes are not only the primary factor producing the working potential of Li-ion batteries, but also determine the number of Li ions (i.e., the practical capacity) which can be utilized. Due to $LiFePO_4$ having succeeded as a prime example of high powered Li-ion battery material, polyanion-type compounds have attracted wide interests in the field of cathode research for the last two decades. Although polyanion compounds exhibit disadvantages of weight and volume (i.e., they have smaller theoretical gravimetric or volumetric capacities) compared with layered oxide compounds, their inherent advantages are also clear. They have very stable frameworks that provide long-term structural stability, which is essential for extensive cycling and combating safety issues. In addition, the chemical nature of polyanions allows the monitoring of a given $M^{n+}/M^{(n-1)+}$ redox couple through the inductive effect introduced by Goodenough [3, 4], and gives rise to higher voltage values versus Li^+/Li^0 than in oxides. Finally, a large number of atomic arrangements and crystal structures can be adopted by polyanion compounds, which have an extreme versatility with respect to cation and anion substitutions for a given structural type.

In the last two decades, compounds with different polyanion groups such as phosphates (PO_4^{3-}), pyrophosphates ($P_2O_7^{4-}$), silicates(SiO_4^{4-}), sulfates(SO_4^{2-}), borates (BO_3^{3-}) as well as their fluorinated compounds have been widely investigated in the literature. In this chapter, some recent studies of polyanion compounds for use in Li-ion batteries are introduced and summarized. Some review papers in this field can also be found in the literature [5, 6]. Here, we mainly focus on the different polyanion compounds in use as cathode materials, except for olivine-type $LiFePO_4$ and its analogues.

X.B. Wu · X.H. Wu · J.H. Guo · S.D. Li · R. Liu · M.J. McDonald · Y. Yang (✉)
State Key Lab for Physical Chemistry of Solid Surfaces, College of Chemistry and Chemical Engineering, Xiamen University, Xiamen 361005, China
e-mail: yyang@xmu.edu.cn

Z. Zhang and S.S. Zhang (eds.), *Rechargeable Batteries*,
Green Energy and Technology, DOI 10.1007/978-3-319-15458-9_4

1 Silicates

As part of the wider group of new polyanion electrode materials that are lower in cost, safer, and have a higher capacity than older compounds [7], "tetrahedral" silicates with the formula Li_2MSiO_4 (M = Fe, Mn, Co, Ni, V) are an exciting area of study that shows great promise. Not only do these materials display a high structural stability arising from the strong Si–O covalent bond, they are environmental friendly, of low cost to produce and have a theoretical two electron exchange per formula unit during cycling, which leads to a high theoretical capacity of about 330 mAh/g [8]. Due to these and other factors, orthosilicates have prompted significant research efforts since 2000 [9]. The theoretical voltages of the redox processes M^{2+}/M^{3+} and M^{3+}/M^{4+} were predicted in 2006 [10] with some of these predictions later being experimentally confirmed [11]. However, despite intensive research efforts having been devoted to this area, the hunt for a reversible two electron reaction in Li_2MSiO_4 has thus far been met with limited success. Of the different types of orthosilicate cathode materials, Li_2MnSiO_4 attracted the attention of researchers when it was hypothesized that at moderate voltages, both lithium ions in each formula unit could potentially be extracted [12]. This process would utilize the $Mn^{2+/3+}$ and $Mn^{3+/4+}$ electrochemical couples for the extraction of the 1st and 2nd respective lithium ions in currently available electrolytes. A similar reaction is also theoretically possible for the Fe and Co analogues, but in these cases, the $Fe^{3+/4+}$ and $Co^{3+/4+}$ couples are predicted to be outside the voltage stability window of common electrolytes, introducing a major complication [10]. The Ni analogue Li_2NiSiO_4 has not yet been synthesized, but theoretical calculation predicts prohibitively high lithium extraction voltages. With the voltage plateau for the second Li-ion deintercalation at about 5.0 V, this compound's working voltage is also beyond the electrochemical window of standard electrolytes. Further narrowing down of the viable options comes from the fact that Li_2VOSiO_4 is toxic, while Co, Ni and V are scarce and thus expensive resources [13, 14], so research on Li_2MSiO_4 has mainly concentrated on Li_2FeSiO_4 and Li_2MnSiO_4.

The crystal chemistry of the family of Li_2MSiO_4 analogues remains quite ambiguous, due to its rich polymorphism and hence the difficulties encountered in obtaining single phase samples. Nyten et al. [8] performed Rietveld refinements on a powder X-ray diffraction profile of pristine Li_2FeSiO_4 and from this proposed an α-Li_3PO_4-based structure, crystallizing in the orthorhombic space group $Pmn2_1$ with lattice parameters a = 6.27 Å, b = 5.33 Å, and c = 5.01 Å. This structure's tetrahedral sites are generated by a distorted hexagonal close packing of oxygen atoms and are one half occupied by cations, thus face sharing between pairs of tetrahedral sites is avoided. There are 8 different polymorphs that are known for these tetrahedral structures, which adopt a large variety of crystal structures related to Li_3PO_4 polymorphs but show a different connectivity for each of the $[SiO_4]$, $[LiO_4]$, and $[MO_4]$ tetrahedral units [15]. In addition, they exist in two main classes, labeled α and β, which differ in their respective orientations of filled tetrahedra. All the tetrahedra are aligned in the same direction in α phases, specifically

Table 1 Unit cell parameters reported for polymorphs of Li_2MSiO_4 with M = Fe, Mn, Co. Reproduced with permission from Ref. [13]

	Space group	*a* (Å)	*b* (Å)	*c* (Å)	*β* (°)
Li_2FeSiO_4	$Pmn2_1$	6.26	5.32	5.01	90
	Pmnb	6.285	10.659	5.036	
	$P2_1In$	6.2835	10.6572	5.0386	89.941
	$P2_1$	8.229	5.0200	8.2344	99.203
	$P2_1In$	8.2253	5.0220	8.2381	99.230
Li_2MnSiO_4	$Pmn2_1$	6.31	5.38	4.96	90
	Pmnb	6.3126	10.7657	5.0118	90
	$P2_1In$	6.33	10.91	5.07	90.99
Li_2CoSiO_4	$Pmn2_1$	6.253	10.685	4.929	90
	$Pbn2_1$	6.2599	10.6892	4.9287	90
	Pmnb	6.20	10.72	5.03	90
	$P2_1In$	6.284	10.686	5.018	90.60

perpendicular to the close packed planes, and share corners with each other. But for *β* phases, the tetrahedra are arranged in groups of three with the central tetrahedron pointing in the opposite direction to the outer two, with which it shares edges. The difference between the *α* and *β* polymorphs corresponds to the temperature, where low temperatures form *α* polymorphs and high temperatures form *β* polymorphs. Both *α* and *β* structures can be further divided into different sub-variants, marked by distinct distortions, orderings, etc. Due to these sub-variants having similar formation energies, they can co-exist in fabricated Li_2MSiO_4 materials if reaction conditions (temperature, pressure, and so on) are not carefully controlled. Pure sub-variant structures have been obtained and characterized, with data shown in Table 1.

To study the various structures of Li_2MSiO_4 polymorphs, different powerful analytical tools have been put into use. Nishimura et al. utilized high-resolution synchrotron XRD (HR-XRD) experiments to investigate the crystal structure of Li_2FeSiO_4 that was synthesized by a ceramic-type process at 800 °C, with results that the material had monoclinic symmetry with parameters $a = 8.23$ Å, $b = 5.02$ Å, $c = 8.23$ Å, and $\beta = 99.20°$ [16]. The refined crystal structure is shown in Fig. 1. This was later confirmed by Boulineau and Sirisopanaporn who discovered and described the crystal structure of a new metastable polymorph, obtained by rapid quenching at room temperature from 900 °C [17, 18]. Furthermore, Armstrong et al. [19] found that Li_2FeSiO_4 transformed from the $P2_1/n$ structure to the $Pmn2_1$ structure after cycling. Ex situ XRD study was also carried out on charged Li_2FeSiO_4 by Kojima et al. [20] who found that upon the extraction of one Li^+ ion, $LiFeSiO_4$ transformed into a new, ion-disordered (Li^+ and Fe^{3+}) structure, isostructural with pristine Li_2FeSiO_4 in $P2_1/n$ symmetry. The sizeable amount of conflicting reports on structures and mechanisms may have resulted from differences in the sample polymorphs, due to different synthesis temperatures or preparation methods, since as

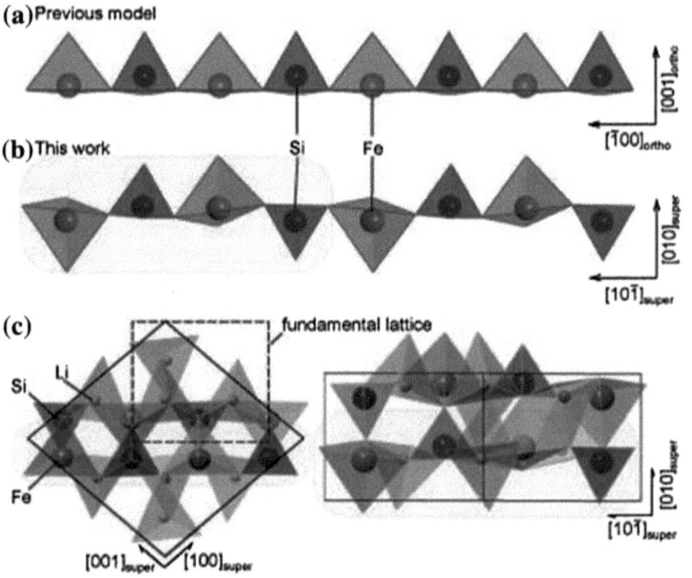

Fig. 1 Corner-shared FeO_4–SiO_4 one-dimensional chains in both **a** $Pmn2_1$ and **b** $P2_1$ polymorphs of Li_2FeSiO_4. **c** Refined crystal structure of Li_2FeSiO_4. Reproduced with permission from Ref. [14]. Copyright 2006, American Chemical Society

previously remarked upon, the presence or coexistence of various polymorphs is sensitive to reaction conditions. Four possible Li_2MnSiO_4 forms have been considered, with the Pmnb, $Pmn2_1$ (β-Li_3PO_4 derivatives), $P2_1/n$ (γ-Li_3PO_4 derivatives) and Pn space groups. All four known ambient temperature polymorphs have Li, Mn and Si cations in tetrahedral co-ordination in a distorted, hexagonally close-packed oxygen array and differ only in the arrangement of the tetrahedra [21]. In addition, the structures are all related to either the α- or β-Li_3PO_4 structures. The pair of orthorhombic forms ($Pmn2_1$ and Pmnb) have two-dimensional pathways for Li-ion diffusion while the two monoclinic forms ($P2_1/n$ and Pn) are framework structures with Li-ion positions interconnected in three dimensions, as shown in Fig. 2. Dominko et al. [22] first synthesized Li_2MnSiO_4 via a modified Pechini sol–gel synthesis route and found that Li_2MnSiO_4 crystallized with a slightly distorted orthorhombic crystal structure. The phase of Li_2MnSiO_4 was quite pure, with a small amount of observable impurities that included MnO and Li_2SiO_3. XRD profiling revealed a $Pmn2_1$ space group, the same structure proposed for Li_2FeSiO_4, although here with cell parameters of a = 6.3109 Å, b = 5.3800 Å, and c = 4.9662 Å [22]. A second orthorhombic form with Pmnb symmetry was calculated to have very similar thermodynamic stability to the $Pmn2_1$ form, suggesting that it would be challenging to prepare as a pure phase [23]. Recently, however, exactly that was achieved [24]. Rietveld refinement of diffraction data from this phase pure Li_2MnSiO_4 showed that this form displayed no Li/Mn site/anti-site disorder,

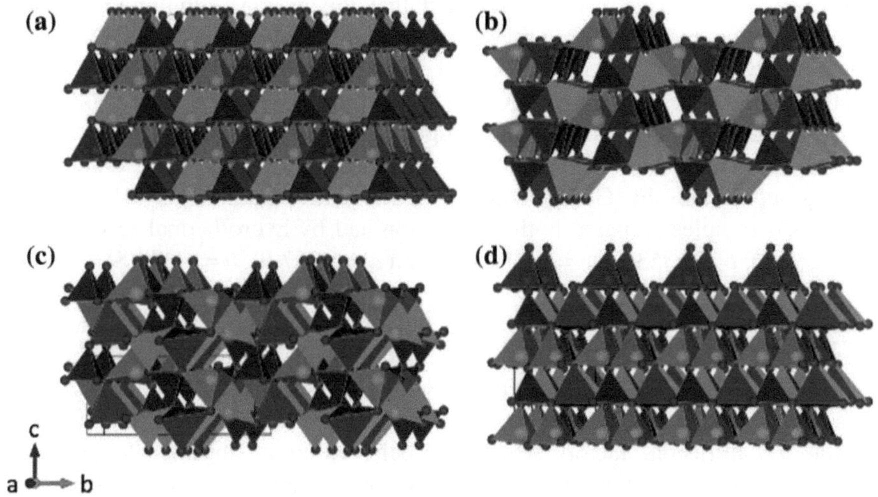

Fig. 2 Crystal structures of the four known ambient pressure polymorphs of Li_2MnSiO_4: **a** $Pmn2_1$, **b** Pmnb, **c** $P2_1/n$ and **d** Pn. Li tetrahedra are shown in *green*, Mn tetrahedra in *purple* and Si tetrahedra in *blue*, with *red spheres* representing oxygen atoms (Color figure online). Reproduced with permission from Ref. [21]. Copyright 2014, Elsevier

potentially leading to good Li-ion diffusion properties supposed that a suitable coating technique is found for this pure material. Politaev et al. [25] first synthesized the high temperature form of Li_2MnSiO_4 with the monoclinic $P2_1/n$ space group (with parameters $a = 6.3368[1]$ Å, $b = 10.9146[2]$ Å, $c = 5.0730[1]$ Å and $\beta = 90.987$ [1]°) at 1150 °C. Subsequently, other synthesis methods producing this structural form were reported by Arroyoy de Dompablo et al. [23] (the sol-gel method) and Mali et al. [26] (the hydrothermal method). Arroyoy de Dompablo et al. [27] pointed out that the denser $Pmn2_1$ polymorph can be obtained via high-pressure and high-temperature treatment of either the $P2_1/n$ or Pmnb polymorphs or their mixtures, but the $Pmn2_1$ form remains stable even at 80 kbar and 900 °C [27]. Due to this dynamic, Li_2MnSiO_4 is usually described as an orthorhombic structure with the space group $Pmn2_1$, with the $MnSiO_4$ layer repeating along the *a* and *c* axes and the LiO_4 tetrahedrons linked together along the *b*-axis. In these layers, the SiO_4 tetrahedrons and MnO_4 tetrahedrons are connected by multiple points. Each Li ion holds two $MnSiO_4$ layers of the tetrahedron between these positions. In addition, the other common orthosilicate cathode material, Li_2FeSiO_4, is isostructural with Li_2MnSiO_4, also crystallizing with the $Pmn2_1$ space group but demonstrating better stability during the charging and discharging processes. Recently ion-exchange from Na_2MnSiO_4 was used to produce a novel metastable polymorph of Li_2MnSiO_4 that adopted the Pn space group. This polymorph was unstable at elevated temperatures, transforming into the stable polymorph above 370 °C. Amidst the proliferation of details on various polymorphs, theoretical calculations were made on Li_2MnSiO_4 in order to investigate the influence of the crystal structure on its electrochemical performance [28]. The results concluded that the average lithium intercalation voltage for the two electron process was only slightly affected by the crystal structure, with GGA + U calculated voltages (according to the reaction: Host-Li_2MnSiO_4 $\longleftrightarrow$ Host-$MnSiO_4$ + 2Li) being 4.18, 4.19, and 4.08 V for Pmnb, $Pmn2_1$ and $P2_1/n$ polymorphs, respectively.

The crystal system of Li_2CoSiO_4 was shown to be orthorhombic with space-group $Pmn2_1$ by Gong et al. [13]. The lattice parameters calculated for Li_2CoSiO_4 samples prepared both by solution and by hydrothermal reaction were $a = 6.287(6)$, $b = 5.353(1)$, $c = 4.939(4)$ Å and $a = 6.267(9)$, $b = 5.370(8)$, $c = 4.939$ (4) Å, respectively.

Although novel cathode materials in the Li_2MSiO_4 family have attracted the attention of researchers since 2000, the first notable cathode performance was reported by Nyten et al. [8] in 2005. As with phosphate-based cathodes, the family's key drawbacks have been an extremely low electronic conductivity and slow lithium ion diffusion, factors which are believed to be intrinsic to polyanion compounds. To counter these limitations, analogous approaches for improving the conductivity of $LiFePO_4$ are also used with lithium orthosilicates. However, this can be difficult as polyanion Li_2MSiO_4 materials possess very low intrinsic conductivity (at room temperature, about 6×10^{-14} S/cm for Li_2FeSiO_4 and about 5×10^{-16} S/cm for Li_2MnSiO_4) [29, 30]. Electronic conductivity can be enhanced to various extents by coating nanoparticles of active material with conductive layers, typically carbon. This can also prevent the agglomeration of particles, facilitating short reaction pathways for the insertion/removal of Li ions. The application of materials with specific morphologies could enhance the structural properties of the base material, improving their electrochemical performance. This can also be modified by changing the source and content of the carbon applied. Meanwhile, ionic conductivity can be improved by decreasing the particle size as well as coating the active material. This leads to the general wisdom that improving the electrochemical performance of Li_2MSiO_4 entails adjusting various preparation processes to control the particle size and introducing surface coating techniques to increase the conductivity.

Li_2FeSiO_4 was the first of the silicates family to be synthesized and characterized, by Nyten et al. [8]. Although the theoretical capacity of Li_2FeSiO_4 is high (333 mAh/g), coming from two Li ions per formula unit, the first work on the compound reported an initial charge capacity of only 165 mAh/g that eventually stabilized at around 140 mAh/g, indicating that less than one Li^+ ion per formula unit was being effectively extracted. Correspondingly, it was hypothesized that with carbon coated and nano-sized particles, the intrinsically low conductivity of silicate materials could be overcome. A voltage profile of this material cycled at 60 °C is plotted in Fig. 3. The observed lowering of the potential plateau from 3.10 to 2.80 V during the first cycle was explained by Nyten et al. as a structural rearrangement to maximize stability, in which some of the Li ions (in the 4b sites) and Fe ions (in the 2a sites) become interchanged [31].

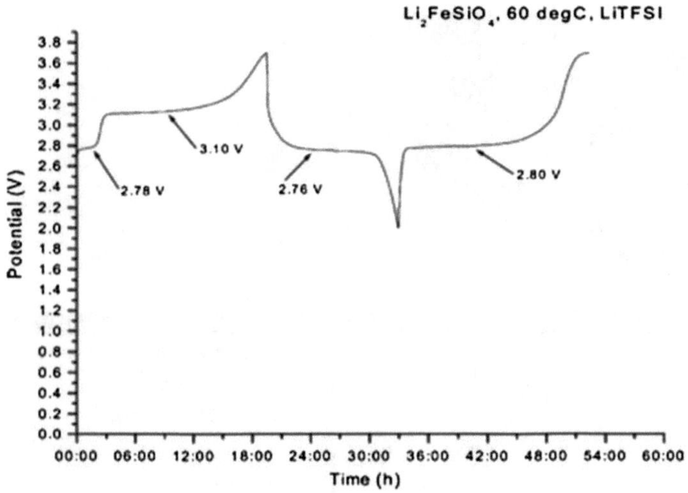

Fig. 3 A voltage profile of pristine Li_2FeSiO_4 cycled at 60 °C at a C/16 rate, using 1 M LiTFSI in EC:DEC 2:1 electrolyte. Reproduced with permission from Ref. [8]. Copyright 2005, Elsevier

Muraliganth et al. [32] synthesized nanostructured and carbon-coated Li_2FeSiO_4 and Li_2MnSiO_4 via a microwave-solvothermal approach. The nanosphere-like morphology of the resulting Li_2MSiO_4/C nanocomposite material is shown in Fig. 4, with Fig. 4a giving a particle diameter of approximately 150 nm. However, upon examination of the magnified SEM image shown in Fig. 4b, it becomes apparent that the larger nanospheres are the result of an agglomeration of smaller nanoparticles of Li_2FeSiO_4, in turn with an average particle size of around just 20 nm. This microwave-solvothermal synthesis method created a structure similar to what Nyten et al. observed, with characteristics such as an oxidation peak shift also in evidence, although the particle sizes were smaller. This decrease in particle size had the effect of lowering the inherent lithium ion diffusion path length, with the reversible capacity greatly benefiting from this improved morphology, reaching 150 mAh/g at 25 °C as shown in Fig. 5.

Dominko et al. [33] prepared Li_2FeSiO_4 samples by three different synthesis techniques (hydrothermal synthesis, modified Pechini synthesis and Pechini synthesis). The obtained samples demonstrated some interesting differences in morphology and particle size, as seen in Fig. 6, but fewer impurities were found in the samples that did not contain any in situ carbon or were prepared via hydrothermal synthesis.

Lv et al. [34] prepared Li_2FeSiO_4 samples by a novel sol-gel synthesis method. SEM images of the resulting Li_2FeSiO_4/C composite are shown in Figs. 7a and 8b. As can be seen in the image, the particles are micron-sized with irregular shapes, which can be attributed to interconnection between the carbon frameworks formed during the heat treatment. However, as with other reports, at a higher magnification (Fig. 7b) the large particles can be seen to consist of smaller nanosized spheres,

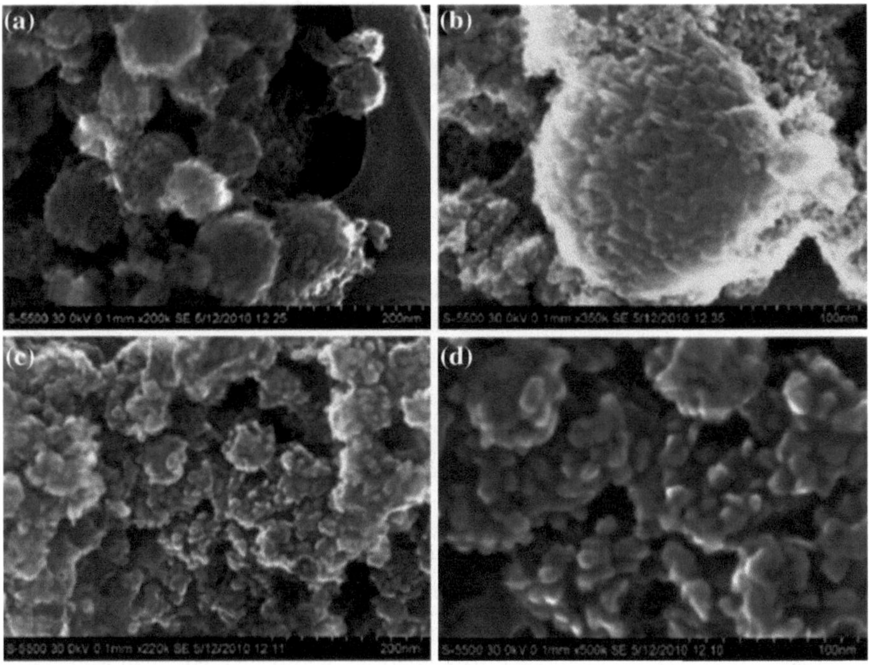

Fig. 4 **a** Low and **b** high magnification FE-SEM images of the Li_2FeSiO_4/C nanocomposite material. **c** Low and **d** high magnification FE-SEM images of the Li_2MnSiO_4/C nanocomposite material. Reproduced with permission from Ref. [32]. Copyright 2010, American Chemical Society

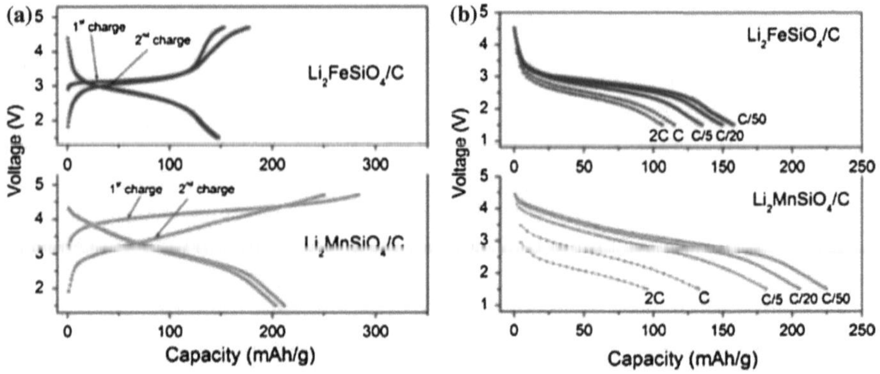

Fig. 5 **a** Charge-discharge profiles recorded at a cycling rate of C/20 and at 55 °C together with voltage-capacity graphs of Li_2FeSiO_4/C and Li_2MnSiO_4/C at different cycling rates (**b**), demonstrating their rate capabilities. Reproduced with permission from Ref. [32]. Copyright 2010, American Chemical Society

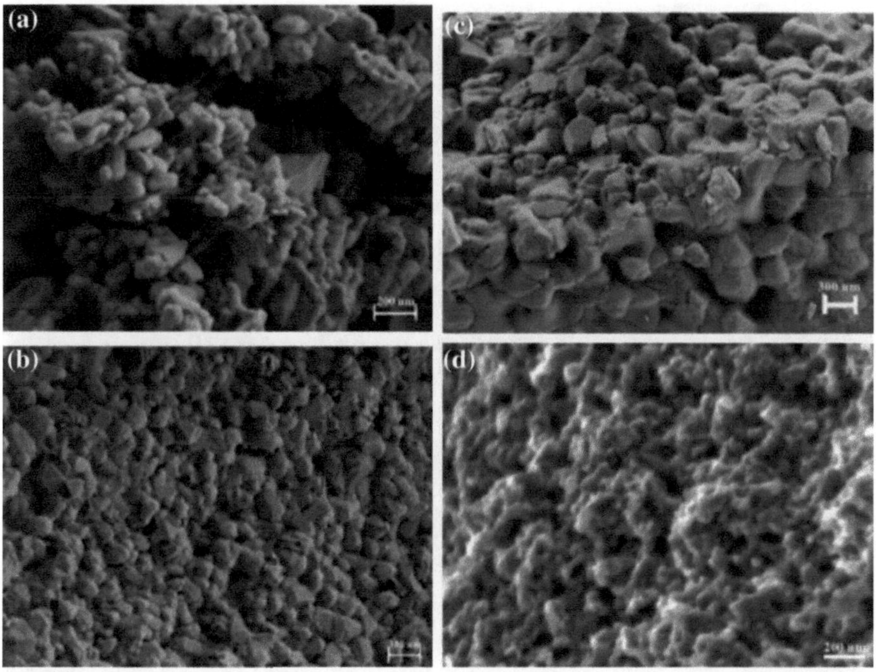

Fig. 6 SEM micrographs of Li_2FeSiO_4 samples corresponding to **a** hydrothermal synthesis (HTS), **b** modified Pechini synthesis involving heating at 700 °C (MPS700), **c** modified Pechini synthesis involving heating at 900 °C (MPS900) and **d** Pechini synthesis (PS-Li2FeSiO4). Reproduced with permission from Ref. [33]. Copyright 2008, Elsevier

these being 40–50 nm in diameter and having a uniform size distribution. TEM imagery (Fig. 7d) confirms that the Li_2FeSiO_4 particles are in the size range of 30–50 nm and are tightly connected by carbon. The HRTEM image in Fig. 7e additionally indicates that the Li_2FeSiO_4 material is highly crystalline and coated with carbon. This type of structure is likely to enhance electrochemical performance, especially at high current densities. It was also found that when cycling at room temperature, Li_2FeSiO_4 could deliver a reversible high capacity that corresponded to more than one lithium ion extracted per formula unit by charging to the relatively high voltage of 4.8 V. In the same work, ex situ Mössbauer was used to find significant oxidation of Fe^{3+} to Fe^{4+} when lithium ions were deintercalated from the novel Li_2FeSiO_4/C composite that was obtained using a solution polymerization approach, with testing performed up to 4.8 V versus Li^+/Li^0.

In order to maximize the charge capacity, a strategy was developed to promote Li^+ diffusion in polyanion cathode materials such as $0.8Li_2FeSiO_4/0.4Li_2SiO_3/C$ with the incorporation of Li_2SiO_3 as a lithium ionic conductive matrix [35]. It was shown that the presence of Li_2SiO_3 separates the Li_2FeSiO_4 particles into small domains of a few nanometers and provides a fast Li^+ diffusion channel, bolstering diffusion in the $0.8Li_2FeSiO_4/0.4Li_2SiO_3/C$ composite. As a result, the composite

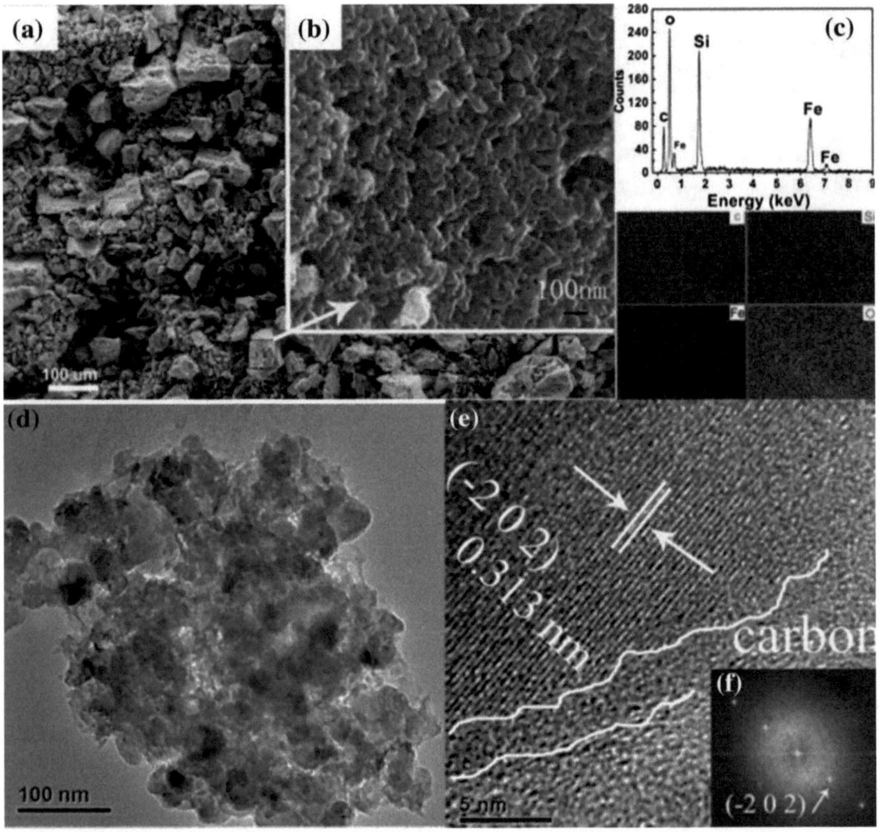

Fig. 7 **a** SEM imagery of Li_2FeSiO_4/C, **b** magnification of a particle from (**a**), **c** the EDX result and elemental C, Si, Fe, and O mappings, **d** TEM imagery, **e** HRTEM and **f** its corresponding FT image. Reproduced with permission from Ref. [34]. Copyright 2011, Royal Society of Chemistry

material showed enhanced electrochemical performance and delivered a capacity as high as 240 mAh/g (corresponding to 1.44 electrons exchanged per active Li_2FeSiO_4 formula unit) together with good cyclic stability at 30 °C, as can be seen in Fig. 8. Nevertheless, as was predicted by theoretical calculations, a slow capacity decay can be observed in the data at a smaller potential window, due to irreversible structural rearrangements. A similar result was achieved by in situ X-ray absorption near edge structure (XANES) research on $Li_2Fe_{0.5}Mn_{0.5}SiO_4$, where the $Fe^{3+/4+}$ redox couple was found to be involved in the electrochemical reaction at high voltages and to also contribute to the observed high capacity [36]. Further exploration of the $Fe^{3+/4+}$ couple would be very interesting and could lead to the substantial increase in capacity of some iron based polyanion cathode materials.

In related work, the electrochemical mechanism of Li_2FeSiO_4 with the reversible extraction/insertion of more than one Li^+ from/into the structure has been studied by techniques such as in situ synchrotron XANES and X-ray diffraction (XRD) by

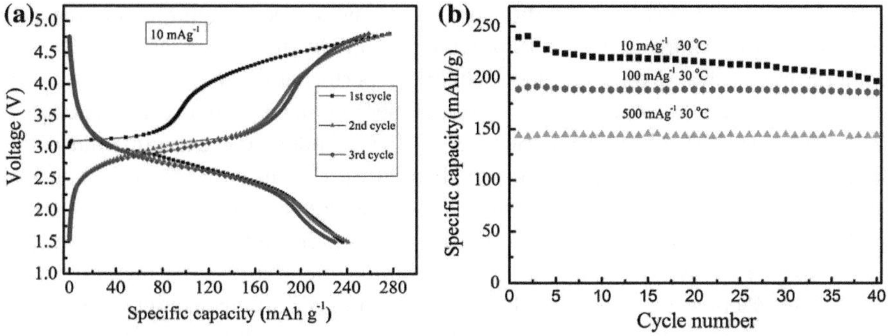

Fig. 8 **a** The charge/discharge profiles of the initial three cycles of $0.8Li_2FeSiO_4/0.4Li_2SiO_3/C$, cycled at a current density of 10 mA/g and at 30 °Cin a voltage range of 1.5–4.8 V versus Li^+/Li^0. **b** Cycling performance of $0.8Li_2FeSiO_4/0.4Li_2SiO_3/C$ at current densities of 10, 100, 500 mA/g and at 30 °C in a voltage range of 1.5–4.8 V versus Li^+/Li^0. Reproduced with permission from Ref. [35]. Copyright 2012, Royal Society of Chemistry

Lv et al. [37], as seen in Fig. 9. The in situ XRD/XANES study together with density functional theory (DFT) calculations demonstrate that the two plateaus observed during charging of Li_2FeSiO_4 can be ascribed to two different two-phase transformation processes, one in the voltage range of open circuit potential (OCP) —4.1 V versus Li^+/Li^0 and the other from 4.1–4.8 V versus Li^+/Li^0. From OCP to 4.1 V versus Li^+/Li^0, the reaction is $Li_2Fe(II)SiO_4$ (α) → $LiFe(III)SiO_4$ (β) + Li^+ + e^-, in which Fe^{2+} ions are oxidized to Fe^{3+} ions and the crystal structure transforms into a charged $LiFeSiO_4(\beta)$. When the electrode is then charged to voltages above 4.1 V versus Li^+/Li^0, more Li^+ will be extracted through the following two-phase reaction: $LiFe(III)SiO_4$ (β) → $Li_yFe(III)_yFe(IV)_{1-y}SiO_4$ (γ) + $(1-y)$ Li^+ + $(1-y)$ e^- $(0 \leq y < 1)$.

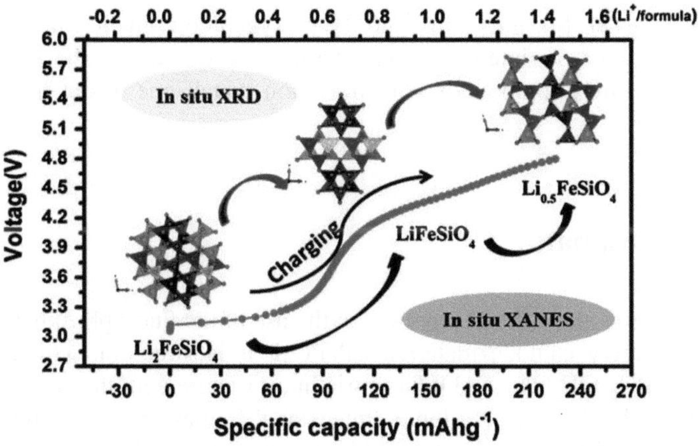

Fig. 9 Schematic of the electrochemical mechanism of Li_2FeSiO_4 during the charging process. Reproduced with permission from Ref. [37]. Copyright 2013, American Chemical Society

In addition to various preparation processes and surface coating techniques, the calcining temperature plays an important role in the formation of particles in a cathode material and thus its electrochemical performance. Li_2MnSiO_4 was synthesized with an all-acetate precursor sol-gel method under reducing atmospheres of 600, 700, and 800 °C by Belharouak et al. [38]. Under their experimental conditions, the Li_2MnSiO_4 samples contained impurities such as MnO at 600 °C and Mn_2SiO_4 and Li_2SiO_3 at 800 °C, while the sample prepared at 700 °C was in a pure phase with a structure analogous to the Li_3PO_4 orthorhombic phase. Li et al. [39] synthesized Li_2FeSiO_4/C cathodes by combining the wet-process method with a solid-state reaction. Of note is the fact that the primary particle size increased as the temperature increased from 600 to 750 °C. The Li_2FeSiO_4/C sample that was synthesized at 650 °C demonstrated good electrochemical performance and an initial discharge capacity of 144.9 mAh/g, with the discharge capacity remaining at 136.5 mAh/g after 10 cycles.

As part of the wave of new polyanion-based cathode materials, the Li_2MSiO_4 family remains attractive due to its low cost, environmentally friendly components and high theoretical capacity. However, it is also faced with many scientific issues that need to be explored and clarified with a combination of experimental and theoretical approaches. In particular, special attention should be paid to the stability of oxygen coordination of polyhedral environments of transition metal ions in Li_2MSiO_4 and delithiated $Li_{2-x}MSiO_4$. As Fe^{2+}, Fe^{3+} and even possibly Fe^{4+} can be found to be stable in tetrahedral coordination, it is unsurprising that Li_2FeSiO_4 possesses both high thermal and cyclic stability. In terms of the lithium intercalation voltage, of the various analogues in the Li_2MSiO_4 family, Li_2MnSiO_4 is the best candidate for taking full advantage of the two electron process. However, joint computational and experimental work has demonstrated that the crystal structure of Li_2MnSiO_4 has a tendency to collapse after the removal of lithium, forming a $MnSiO_4$ structure built by edge-sharing Mn^{4+} octahedra. This suggests that the poor cycle performance of Li_2MnSiO_4 is perhaps caused by the instability of Mn^{4+} coordination. An irreversible phase transformation in Li_2MnSiO_4 during delithiation can arise when Mn ions rearrange their oxygen coordination upon oxidation. Thus, it could be very interesting and worthwhile to explore the synthesis and testing of Li_2MSiO_4 compounds with transition metal ions in octahedral oxygen coordination.

2 Fluorophosphates

In 2003, Barker et al. [40] first reported on the use of the fluorophosphate $LiVPO_4F$ as a Li-ion battery cathode material. $LiVPO_4F$ is isostructural with the mineral tavorite, $LiFePO_4{\cdot}OH$ [41], and it has a triclinic structure with the space group P-1. The structure of $LiVPO_4F$ is a three-dimensional framework that is built up from PO_4 tetrahedra and VO_4F_2 octahedra. Lithium ions are statistically distributed in the two available crystallographic sites. A small inflection is observed in the charge

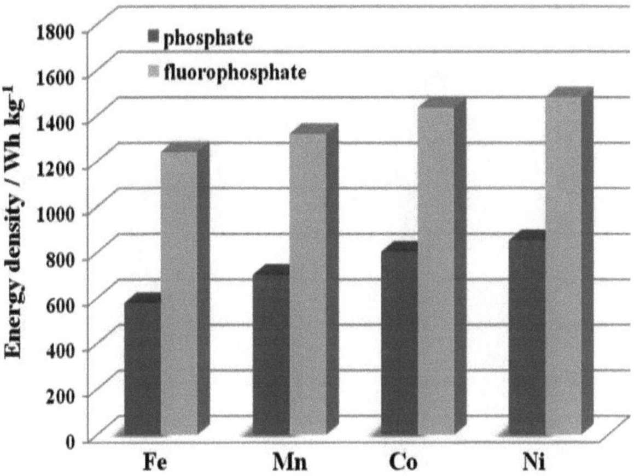

Fig. 10 Comparison of the theoretical energy densities of Li_2MPO_4F and $LiMPO_4$ (M = Fe, Mn, Co, Ni)

curve, which indicates that two energetically nonequal reactions exist. This is due to the formation of an intermediate phase, $Li_{0.67}VPO_4F$ [42]. However, in the discharge data, a single differential capacity peak can be seen, corresponding to a two-phase reaction mechanism. The average discharge voltage of $LiVPO_4F$ is about 4.2 V versus Li^+/Li^0, 0.3 V higher than that of $Li_3V_2(PO_4)_3$ because of the inductive effect of fluorine. By optimizing preparation conditions, the reversible capacity can be as high as 155 mAh/g with a small voltage polarization, which is very close to the theoretical capacity of 156 mAh/g [43]. Additionally, long-term cycling stability was achieved for $LiV_{0.96}Mn_{0.04}PO_4F$, with a capacity retention of 90 % after 1000 cycles [44]. $LiVPO_4F$ exhibits high ionic conductivity but poor electronic conductivity [45]. This can be countered through suitable carbon coating procedures, and with these $LiVPO_4F$ shows excellent rate performance. Furthermore, the thermal stability of $LiVPO_4F$ in a charged state is even better than that of $LiFePO_4$ [46]. The high energy and power densities, long-term cycling stability and high thermal stability of $LiVPO_4F$ make it a competitive cathode material for large-format Li-ion batteries.

Since the demand for higher energy density Li-ion batteries has led to a search for cathode materials with higher capacities and working voltages [47], in addition to $LiVPO_4F$, attention has been directed toward fluorophosphates of the form A_2MPO_4F (A = Li, Na; M = Fe, Mn, Co, Ni) as materials with strong potential [48–54] Introducing the F^- anion allows fluorophosphates to exchange two electrons per formula unit and achieve a higher capacity. Furthermore, the higher electronegativity of fluorine compared to that of oxygen increases the ionicity of M–X bonds and their redox plateau. Figure 10 compares the theoretical energy densities of phosphates and fluorophosphates. It can be seen that the theoretical energy densities of fluorophosphates are about twice that of phosphates, bolstering the case for fluorophosphates as promising high energy density cathode materials for Li-ion batteries.

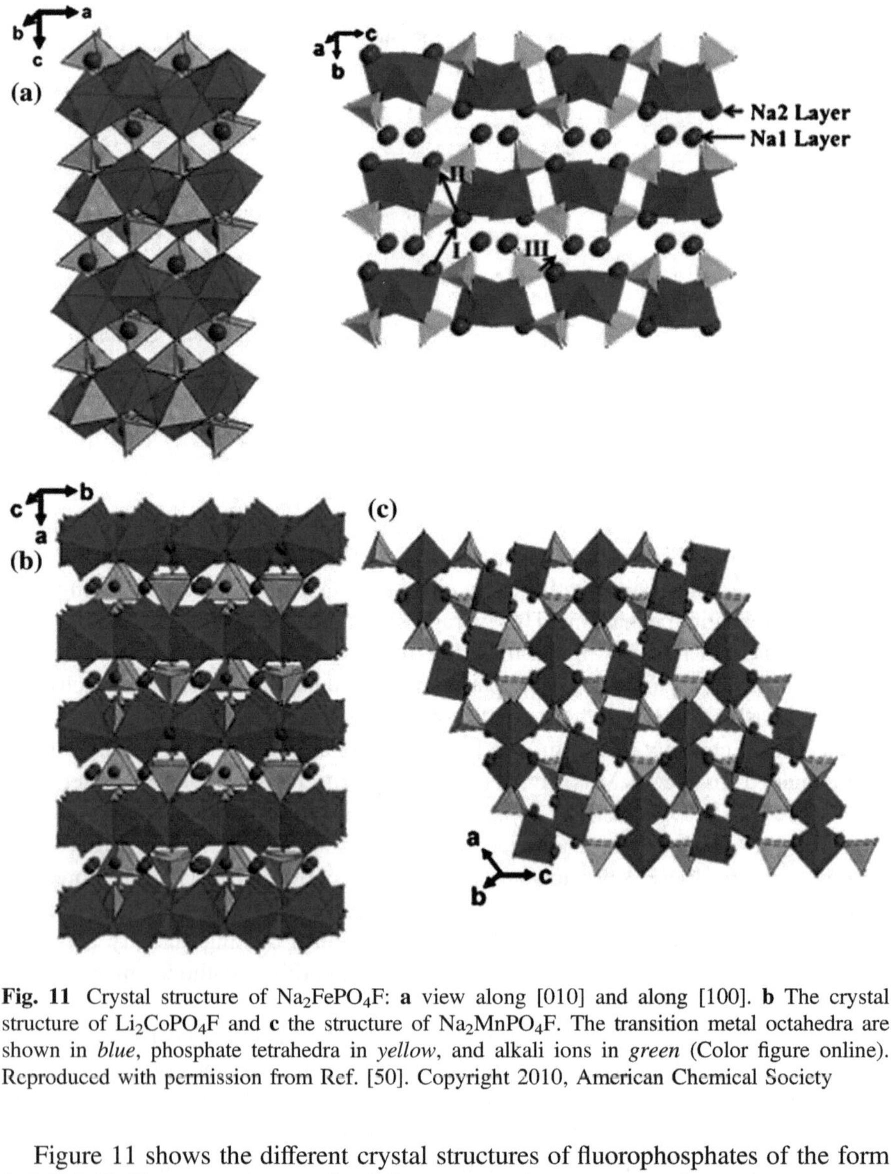

Fig. 11 Crystal structure of Na_2FePO_4F: **a** view along [010] and along [100]. **b** The crystal structure of Li_2CoPO_4F and **c** the structure of Na_2MnPO_4F. The transition metal octahedra are shown in *blue*, phosphate tetrahedra in *yellow*, and alkali ions in *green* (Color figure online). Reproduced with permission from Ref. [50]. Copyright 2010, American Chemical Society

Figure 11 shows the different crystal structures of fluorophosphates of the form A_2MPO_4F. There are three main different crystal structures: a layered structure as seen in Na_2FePO_4F and Na_2CoPO_4F, a "stacked" structure used by Li_2CoPO_4F and Li_2NiPO_4F, and a 3D tunnel structure as displayed by Na_2MnPO_4F. The transition metal ions for all three of these crystal structures are located in the center of the MO_4F_2 octahedra, coordinated with four oxygen ligands and two fluorine ligands. However, the connection type of the octahedra differs from structure to structure, mixing face-shared and corner-shared in the layered type, edge-shared in the "stacked" type and corner-shared in the 3D type [50]. The origin of the different crystal structures found in fluorophosphates of the form A_2MPO_4F lies in the matching between the radii of the alkali metal ions and the interstitial formed by the MO_4F_2 octahedra. Therefore, the crystal structure of these materials should be able to be regulated by cation doping.

In 2007, Ellis et al. [49] first reported on Na_2FePO_4F as a hybrid Li-ion battery cathode material. Na_2FeO_4F has an orthorhombic structure with the space group Pbcn and is isostructural with Na_2FePO_4OH [55], Na_2MgPO_4F [56] and Na_2CoPO_4F [57]. The layered framework of Na_2FePO_4F facilitates the two-dimensional migration of Li/Na ions, thus enabling the material to exhibit excellent rate performance [50]. The sloping charge/discharge profiles for Na_2FePO_4F in hybrid Li-ion batteries indicate solid solution electrochemical behavior. In subsequent cycling, Li ions become more and more predominant in the mixed Li/Na insertion process, due to the excess of lithium in the cell. The processes of Li/Na exchange spontaneously occur at the open circuit voltage in a hybrid Li-ion battery, which results in an open circuit voltage drop from 2.960 to 2.905 V versus Li^+/Li^0. In addition, the oxidation of Na_2FePO_4F to form $NaFePO_4F$ leads to a 3.7 % unit-cell volume contraction. This value is much smaller than that resulting from the oxidation of $LiFePO_4$ to $FePO_4$, and thus represents a lower strain deintercalation process. Unfortunately, only 0.8 Li could be reversibly inserted/removed from this material as originally synthesized. However, in 2011, our group adopted the sol-gel method of preparing the Na_2FeO_4F/C nanocomposite [53]. This produced Na_2FePO_4F that was capable of delivering a reversible capacity as high as 182 mAh/g (corresponding to 1.46 electrons exchanged per formula unit) along with good cycling stability. The more than one electron exchanged per formula unit indicates the feasibility of a second alkaline ion being reversibly deintercalated/intercalated, accompanied with the $Fe^{3+/4+}$ redox couple.

Manganese-based materials often exhibit higher redox potentials than iron-based materials. In the hopes of achieving further increases in energy density, Na_2MnPO_4F has become attractive to researchers. The open anionic framework of Na_2MnPO_4F suggests a possible cationic migration pathway [54]. However, during testing, Ellis et al. [50]. could not achieve any electrochemical activity in Na_2MnPO_4F, nor even charge to 5.0 V versus Li^+/Li^0. In response, they claimed that a reduction in the particle size to reduce the ion transport length may be feasible, in order to detect some activity in Na_2MnPO_4F. Later, our group carefully controlled the particle size and added a uniform carbon coating, being the first to achieve electrochemical activity in Na_2MnPO_4F [53]. Later, Kim et al. [58] also successfully synthesized Na_2MnPO_4F that showed electrochemical activity, by the solid-state method. An ion-exchange method was used to prepare Li_2MnPO_4F, which is isostructural with Na_2MnPO_4F. The discharge capacities of Na_2MnPO_4F and Li_2MnPO_4F are 120 and 140 mAh/g, respectively, corresponding to one electron exchange per formula unit. However, both Na_2MnPO_4F and Li_2MnPO_4F show significant polarization during charge and discharge processes. Recently, our group used spray drying followed by a high-temperature sintering method to synthesize a Na_2MnPO_4F/C nanocomposite that exhibits a more than one electron exchange per formula unit in a Na-ion battery [59]. Zheng et al. [60] used DFT calculations within the GGA + U framework to study the structure and electronic properties of Na_2MnPO_4F. The voltage plateaus of extracting 0.5 and 1.0 Na are at 3.71 and 3.76 V against Na, respectively. However, the voltage of the extraction of the second Na ion is 4.76 V against Na, which is higher than the stability window of commonly used electrolytes. In addition, the energy gap of Na_2MnPO_4F is about 3.68 eV, which leads to poor electronic conductivity. Therefore, carbon coating should be commonly adopted in order to improve the electrochemical performance.

Na_2MnPO_4F exhibits a discharge plateau at about 3.9 V against Na, which is higher than that of Na_2FePO_4F in a hybrid Li-ion battery. However, both Na_2MnPO_4F and Li_2MnPO_4F show significant polarization during charge and discharge processes. In order to take advantage of Fe-based and Mn-based materials, the $Na_2Fe_{1-x}Mn_xPO_4F$ solid solution material is a good choice. Recham et al. [61] used the ionthermal method to synthesize $Na_2Fe_{1-x}Mn_xPO_4F$. When the manganese content is increased to 0.25, the structural changes from a layered structure to a 3D tunnel structure. This is because the ionic radius of Mn^{2+} is larger than that of Fe^{2+}, so in order to minimize the M-M repulsive interactions, the structure morphs from an edge-shared to a corner-shared framework that exhibits a larger M-M distance. However, when manganese content increases, the electrochemical performance of $Na_2Fe_{1-x}Mn_xPO_4F$ rapidly decays. Our group used a sol-gel method to synthesize a $Na_2Fe_{1-x}Mn_xPO_4F/C$ nanocomposite [53]. The resulting $Na_2Fe_{0.3}Mn_{0.7}PO_4F$ material exhibited a distinct plateau at ~4 V against Na along with a high reversible capacity and good cycling performance. Thus, $Na_2Fe_{1-x}Mn_xPO_4F$ was shown to be able to combine the high working voltage of Mn-based materials with the high electrochemical activity of Fe-based materials as $LiFe_{1-x}Mn_xPO_4$ materials have done, and have become a promising Li-ion battery cathode material.

In 2005, Okada et al. [52] reported Li_2CoPO_4F as a high voltage Li-ion battery cathode material. The crystallization process of Li_2CoPO_4F involves two steps, first forming $LiCoPO_4$ and then reacting with LiF to form Li_2CoPO_4F [62, 63]. Thus, the synthesis process often also includes two steps [52, 62, 64]. Wang et al. [63] used LiF as the only lithium source when preparing Li_2CoPO_4F, however, the as-obtained Li_2CoPO_4F was carbon free and could only cycle at low current density. Our group optimized the synthesis conditions and prepared a Li_2CoPO_4F/C nanocomposite with a reversible capacity of 138 mAh/g [65]. Furthermore, this Li_2CoPO_4F exhibits excellent rate capability, with an 86 % retention of the capacity at 1 C achieved at 20 C. However, Li_2CoPO_4F suffers from poor cycling performance, due to a high working voltage and the powerful catalytic effects of cobalt. Techniques such as employing a high voltage electrolyte, surface modification and using film-forming electrolyte additives have been applied to improve the cycling performance [63, 66, 67]. Our group used Li_3PO_4 coating and the LiBOB electrolyte additive to stabilize the electrode/electrolyte interface [67]. With the proper modifications, the Li_2CoPO_4F cathode material can show excellent long-term cycling performance, with an 83.8 % capacity retention after 150 cycles at a 1 C current rate, as shown in Fig. 12.

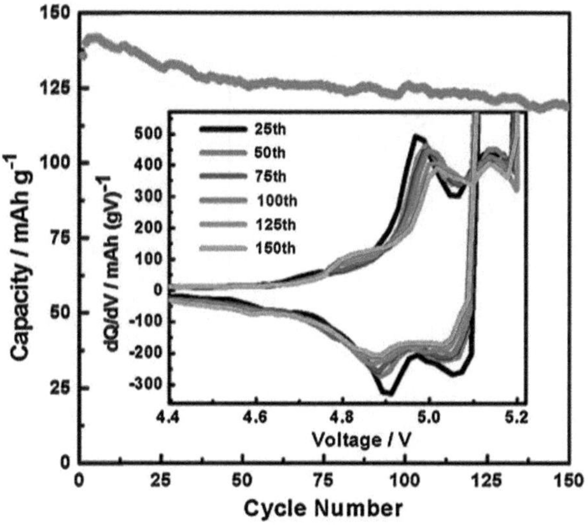

Fig. 12 Long-term cycling stability of the Li_3PO_4-coated Li_2CoPO_4F electrolyte with LiBOB additive. The *inset* gives the corresponding dQ/dV versus voltage plots at different cycles. The voltage range is 2–5.2 V versus Li^+/Li^0. The current density is 143 mA/g. Reproduced with permission from Ref. [67]. Copyright 2011, Royal Society of Chemistry

In the first charge process, Li_2CoPO_4F suffers from an irreversible structural change, which results in unit-cell volume expansion. The "modified" framework is then expected to favor lithium ion transport because of the larger interstitial space [68]. In practice, Li_2CoPO_4F exhibits excellent rate performance, and shows structural stability in subsequent cycling, as demonstrated by ex situ XRD experiments [65]. Because of this, Li_2CoPO_4F has joined the ranks of promising high energy and power density cathode materials for Li-ion batteries that also have considerable cycling stability.

3 Pyrophosphates

The series of $Li_xMP_2O_7$ and $M_2P_2O_7$ (M = V, Cr, Mn, Fe, Co, etc.) compounds, a subset of the pyrophosphate family, has had much of its early interest centered on the investigation of its interesting crystal structure. The $LiM^{III}P_2O_7$ (Fe, Mn, Co, V, etc.) compounds form cage structures, where the framework is composed of PO_4 tetrahedra and MO_6 octahedra that share corners. Li atoms with a tetrahedral coordination are located at voids in the framework [69]. In an early study, $LiFeP_2O_7$ employed as a Li-ion battery cathode showed a relatively limited capacity, but its average discharge voltage of 2.9 V versus Li^+/Li^0 was slightly higher than that of another iron-based polyanion cathode, $Li_3Fe_2(PO_4)_3$ [70]. However, there is no obvious advantage for $LiFeP_2O_7$ compared to $LiFePO_4$, due to the inherent

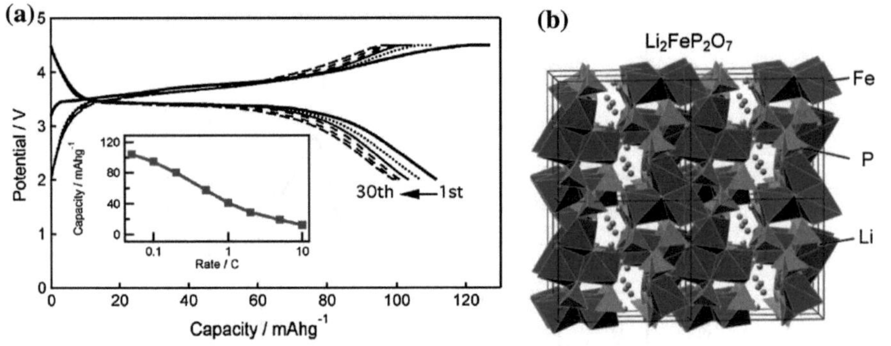

Fig. 13 **a** Galvanostatic charge–discharge characteristics of $Li_2FeP_2O_7$. The *inset* shows the capacity as a function of discharge rate. **b** Crystal structure of $Li_2FeP_2O_7$. Reproduced with permission from Ref. [72]. Copyright 2010, American Chemical Society

limitations of using a single-electron redox reaction. Ramana et al. [71] have reported the structure and electrochemical performance of $LiFe_{1.5}P_2O_7$, demonstrating that the material has a stable structure for the first cycle with the potentials of the redox peaks at 3.33 and 3.22 V versus Li^+/Li^0.

A new pyrophosphate compound with the formula $Li_2FeP_2O_7$ and a three-dimensional $P2_1/c$ structure (Fig. 13a) was first reported by Nishimura et al. [72] as a promising cathode material. It presented reversible voltage plateaus at 3.5 V versus Li^+/Li^0, higher than the 3.4 V exhibited by $LiFePO_4$. Interestingly, its remarkable electrochemical activity can be realized without a complex synthesis process. In addition, its potential theoretical capacity would reach 220 mAh/g if a Fe^{4+}/Fe^{2+} two-electron reaction could occur. Further understanding of the electrochemical reaction mechanisms occurring upon cycling in the $Li_2FeP_2O_7$ electrode for Li-ion batteries is crucial to account for its electrochemical characteristics and a prerequisite to investigating the possibility of more than one Li^+ per formula unit being extracted/inserted. However, there is ongoing debate on these charge/discharge reaction mechanisms. Kim et al. [73] demonstrated that the charging process of a $Li_2FeP_2O_7$ electrode in a Li cell is mainly controlled by a two-phase reaction with some degree of single-phase behavior occurring before the two-phase reaction, while Shimizu et al. [74] proposed that the (de)lithiation process in $Li_{2-x}FeP_2O_7$ consisted of a single-phase solid-solution mechanism with small volume change during battery cycling.

As a useful modification method, carbon coating has been proven to be an effective method for improving the electrochemical performance of $Li_2FeP_2O_7$ [75]. In addition, Barpanda et al. [76] developed an eco-efficient splash combustion synthesis method for $Li_2FeP_2O_7$/C nanocomposite. This method is also applicable to the preparation of the family of $Li_2(Fe_{1-x}Mn_x)P_2O_7$ and $Li(Fe_{1-x}Mn_x)PO_4$ ($x = 0$–1) nanocomposites.

In order to clarify influencing factors of electrochemical behavior, it is necessary to further study the crystal structure of the material, underlying defects, and

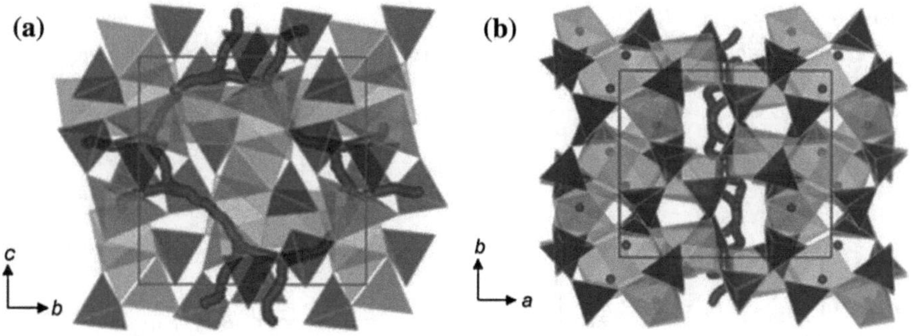

Fig. 14 The calculated paths for long-range Li^+ migration: **a** viewed along the a-axis; **b** viewed along the c-axis, Reproduced with permission from Ref. [77]. Copyright 2012

transport properties on an atomic scale. For detailed structural analysis, XRD and neutron powder diffraction are known as powerful tools. Studies have confirmed that $Li_2FeP_2O_7$ crystallizes in a monoclinic unit cell with the space group P21/c. It was also found that there are three different transition metal sites and five different Li sites in the $Li_2FeP_2O_7$ crystal structure [73]. When using advanced simulation and neutron diffraction techniques to investigate the local structure, defect chemistry and lithium-diffusion pathways in the crystal structure of lithium transition metal pyrophosphates, research confirmed that Li/Fe antisite defects existed along with a quasi-two-dimensional (2D) Li-ion diffusion channel along the bc-plane, as shown in Fig. 14 [77]. However, different occupancies of Li5 sites have led to different results. When there are partial occupancies of Li5 and M3 occurred at Li5 sites in the pyrophosphates, the Li^+ diffusion pathways form a 3D network from two continuous but not interconnected 2D pathways [78]. The magnetic structure of this pyrophosphate $Li_2FeP_2O_7$ was reported by Barpanda et al. [79] from neutron powder diffraction measurements, revealing that the magnetic moments of Fel and Fe3 sites are parallel to [100] while the moment of Fe2 sites is in the ac-plane.

Early in 2008, Adam et al. [80] reported the synthesis of $Li_2MnP_2O_7$ by a two-step solid state method. This compound exhibits a novel tunnel structure, with the presence of Mn_2O_9 units and lithium cations arranged in the form of $[Li_2O_5]^{\infty}$ and $[Li_2O_6]^{\infty}$ layers, as shown in Fig. 15. Unlike with $Li_2FeP_2O_7$, electrochemical activity in $Li_2MnP_2O_7$ was hardly ever observed until 2012. By adding carbon black to form smaller particles, Tamaru et al. [81] found Mn^{3+}/Mn^{2+} redox activity centered at 4.45 V versus $Li^+/Li^0$when cycling at 40 °C. This is the highest achievable voltage reported so far for the Mn^{3+}/Mn^{2+} redox couple among all Mn-based cathode materials. By way of comparison, the redox potentials obtained by ab initio calculations for the Mn^{3+}/Mn^{2+}, Mn^{4+}/Mn^{3+}, Fe^{3+}/Fe^{2+}, and Fe^{4+}/Fe^{3+} redox couples in $Li_2MP_2O_7$ compounds are located at 4.47, 5.3, 3.8, and 5.2 V versus Li^+/Li^0, respectively [82].

$Li_2MnP_2O_7$ attracts much interest because of the presence of environmentally benign manganese and the high potential of the Mn^{3+}/Mn^{2+} redox couple at 4.45 V

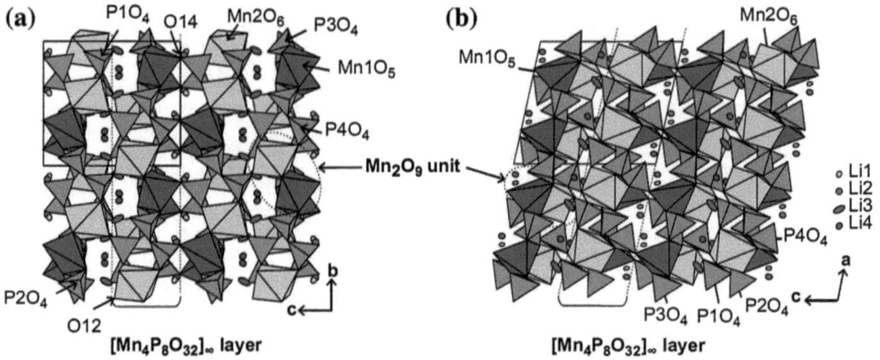

Fig. 15 Projections of the crystal structure of $Li_2MnP_2O_7$ **a** along [100] and **b** along [010]. Reproduced with permission from Ref. [80]. Copyright 2008, Elsevier

versus Li^+/Li^0, which is still compatible with most commonly used electrolytes. However, in practice the actual capacity has only reached around 60 mAh/g at best, due to the large lattice distortions induced by Jahn-Teller active Mn^{3+} ions and the compound's very low electronic conductivity [81]. Mn^{2+} substitution into a parent material $Li_2FeP_2O_7$ was successfully employed by Furuta et al. [83] to form a solid-solution between pyrophosphate phases of Fe and Mn, with lattice parameters increasing along with the manganese content. It was found that the introduction of Mn into the pyrophosphate structure may stabilize the two distinct Fe^{3+}/Fe^{2+} redox reactions by Fe ions in octahedral and trigonal-bipyramidal sites, hence shifting the Fe^{3+}/Fe^{2+} redox potential upward to close to 4 V versus Li^+/Li^0. Regarding other materials, further study on the novel $Li_2M_xFe_{1-x}P_2O_7$ (M = Mn, Co, Mg) family of compounds was carried out, revealing that the higher Fe^{3+}/Fe^{2+} redox potential was caused by stabilized edge sharing local structural arrangement and an associated larger Gibbs free energy in the charged state, as shown in Fig. 16 [84]. Kim et al. [73] reported for the first time that $Li_{2-x}CoP_2O_7$ electrodes are electrochemically active, with a redox potential of 4.9 V versus Li^+/Li^0 and delivering a reversible capacity approaching 85 mAh/g. Although the reversible capacity of the $Li_2CoP_2O_7$ electrode suffers seriously from electrolyte decomposition, the development of a high voltage electrolyte is anticipated to assist in realizing its potential performance. This prospect offers exciting opportunities for the development of next generation cathode materials for Li-ion batteries.

The novel solid solution pyrophosphate $Li_2Fe_{1/3}Mn_{1/3}Co_{1/3}P_2O_7$ shows advantageous properties as compared to single-component analogues and other multi-component polyanion compounds [85]. $Li_2Fe_{1/3}Mn_{1/3}Co_{1/3}P_2O_7$ follows a single-phase process for Li insertion/extraction with an extremely small volume difference of ~0.7 % between the fully charged and discharged states. Generally, a larger strain during the charge/discharge process may lead to electrode materials undergoing large volume changes and destabilizing the contact between electrode and current collector, greatly limiting the cycling life and rate capability. Compared

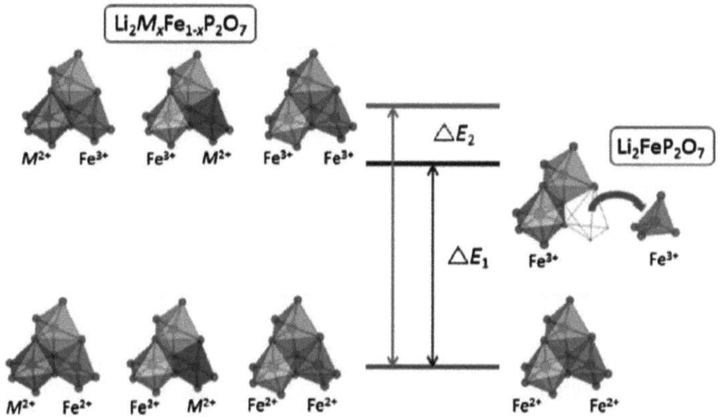

Fig. 16 Schematic description of the free energy difference between starting and delithiated materials. Reproduced with permission from Ref. [84]. Copyright 2012, American Chemical Society

to the $Li_2M_xFe_{1-x}P_2O_7$ (M = Mn, Co, Mg) family of compounds, $Li_2Fe_{1/3}Mn_{1/3}Co_{1/3}P_2O_7$ provides high potential from the mixed redox reactions of active Fe, Mn and Co ions without much specific capacity loss.

Li-ion battery safety is a matter of crucial importance for many applications, and cathode materials with high thermal stabilities will contribute to overall battery safety. Since pyrophosphates are usually generated by the removal of oxygen from phosphates at high temperatures, pyrophosphates tend to be more thermally stable than their phosphate counterparts. Researchers have very closely studied the high temperature behaviors of the delithiated states of $Li_2FeP_2O_7$ and $Li_2MnP_2O_7$ [85]. It was shown that all of the delithiated states could withstand higher temperatures than their olivine counterparts ($FePO_4$ and $MnPO_4$). In particular, both $Li_2FeP_2O_7$ and its partially delithiated compounds were stable up to 600 °C. Similarly, $Li_{1.4}MnP_2O_7$ was stable up to 450 °C [86]. Materials such as these will contribute to future efforts to design high voltage cathode materials with high thermal stabilities.

New iron-based mixed-polyanion compounds with the formula $Li_xNa_{4-x}Fe_3(PO_4)_2(P_2O_7)$ (x = 0–3) were synthesized via a topotactic ion-exchange of the Na cation to the Li cation within the lattice, by refluxing in a solution of LiBr [87]. The new compounds contained three-dimensional (3D)-sodium/lithium paths that were supported by P_2O_7 pillars in the crystal. The combined $(PO_4)^{3-}$ and $(P_2O_7)^{4-}$ polyanion groups, which provided a new crystal framework that could accommodate the naturally abundant Fe redox centers as well as lithium ions, exhibited fast and stable cathode activity in Li-ion batteries. Electrochemical measurements showed that about 92 % of the theoretical capacity was obtained at a C/20 rate, with an average voltage of 3.4 V versus Li^+/Li^0. Battery operation at higher current densities showed that 75 % of the initial capacity could be delivered at a 1 C rate and 60 % even at a 2 C rate.

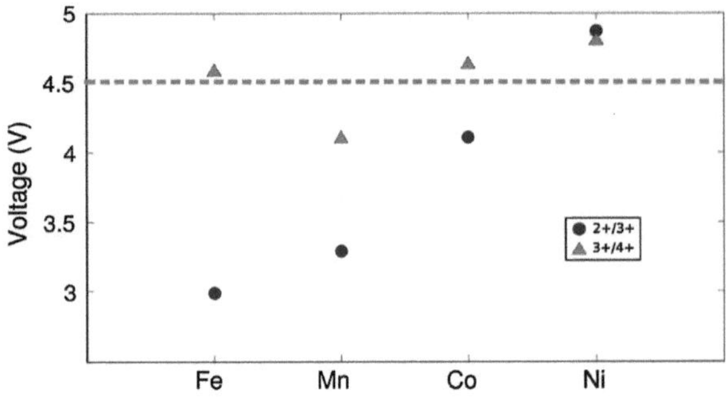

Fig. 17 Voltages for the $2^+/3^+$ (*blue circles*) and $3^+/4^+$ (*red triangles*) redox couples for the delithiation of $Li_3MCO_3PO_4$ in the sidorenkite crystal structure (M = Fe, Mn, Co, Ni) (Color figure online). Reproduced with permission from Ref. [92]. Copyright 2012, American Chemical Society

4 Carbonophosphates

In 2011, Hautier et al. [88] reported on a series of new mixed polyanionic compounds with the formula $Li_3MCO_3PO_4$ (M = Fe, Mn, Co) as possible cathode materials for Li-ion batteries. Ab initio high-throughput computing and screening results suggested several novel carbonophosphates as potential high capacity (>200 mAh/g) and specific energy (>700 Wh/kg) cathode materials [89]. This class of compounds has the potential to retain the safety characteristics of $LiFePO_4$ while having a theoretical specific energy that is almost 50 % greater [90, 91].

Chen et al. [92] reported that the high-throughput computational approach using DFT could obtain details on several material properties including voltage, structural stability and so on. The resulting computed Li-extraction voltages (versus Li^+/Li^0) for the $Li_3MCO_3PO_4$ compounds are shown in Fig. 17. The Fe-(Fe^{2+}/Fe^{3+}: 3 V), Mn-(Mn^{2+}/Mn^{3+}: 3.2 V and Mn^{3+}/Mn^{4+}: 4.1 V), and Co-(Co^{2+}/Co^{3+}: 4.1 V) based compounds are predicted to be electrochemically active in a voltage range suitable for today's electrolytes. In addition, their sidorenkite structures are more likely to remain stable upon the extraction of multiple Li^+ ions, as evidenced by the very low volume changes predicted: 1.1 % for Fe (after one lithium removed) and 2.4 % for Mn (after 2 lithium removed), compared to a volume change of 4.6 % for $LiFePO_4$, 4.9 % for $Li_3V_2(PO_2)_3$, and 7 % for $LiFe(SO_4)F$. As it is accepted that large volume changes are detrimental to electrode cycling, the small values exhibited for these sidorenkite structures favors them as promising electrode materials [93].

Both $Li_3FeCO_3PO_4$ and $Li_3MnCO_3PO_4$ have been synthesized through Li–Na ion-exchange from the stable sodium compounds $Na_3MPO_4CO_3$ (M = Mn, Fe), as the lithium-containing carbonophosphates are not thermodynamic ground states [89]. ICP results show that ion-exchange was almost fully completed for the Fe sample, with the elemental ratio being Li:Na:Fe:P = 2.95:0.08:1.01:0.95. However, the elemental ratio for the Mn sample was Li:Na:Mn:P = 2.67:0.63:0.96:1, indicating that a fairly stable phase exists close to this composition, preventing further

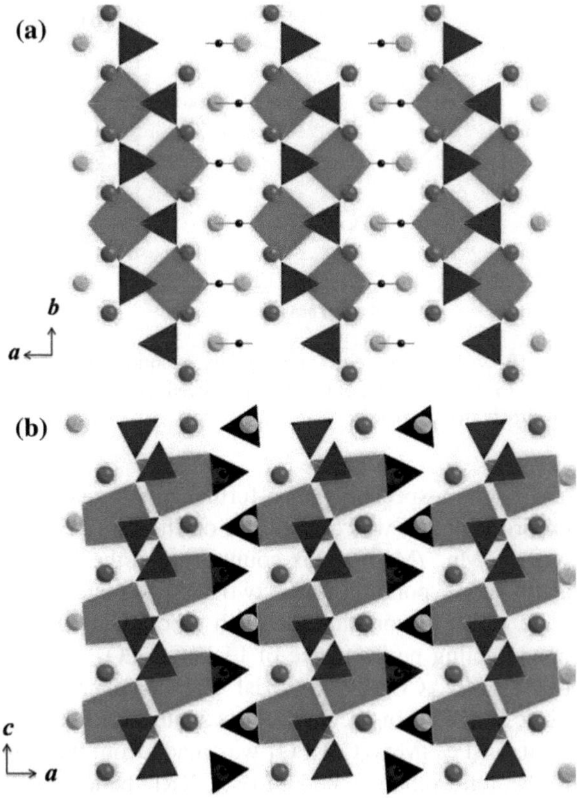

Fig. 18 Structure of $Na_3MPO_4CO_3$ (M = Mn, Fe) viewed from [001] (**a**) and [010] (**b**). Transition metal octahedra are purple, PO_4 tetrahedra are *blue*, CO_3 planar triangles are *black*, Na1 are *green balls* and Na2 are *yellow balls* (Color figure online). Reproduced with permission from Ref. [92]. Copyright 2012, American Chemical Society

Na exchange. Neither a longer exchange time, higher exchange temperature nor repeated exchanging could significantly increase the Li:Na ratio in the final product.

TEM and synchrotron XRD results have revealed that the $Li_3MCO_3PO_4$ structure transforms from monoclinic to triclinic after ion exchange [89, 92]. Ab initio computations have also shown that the $_{lowest}$ energy structure of the Li compound is approximately the same as its Na precursor, as shown in Fig. 18. In the Li compound, the polyhedral are slightly turned and tilted so that the 2-fold axis and mirror plane present in the $P2_1/m$ space group are lost and only the inversion symmetry is preserved, giving the computed cell P1 symmetry.

The first electrochemically active $Li_3FeCO_3PO_4$ was reported by Chen et al. [92] with a reversible capacity of around 110 mAh/g while cycled at a C/5 rate at room temperature, which matches the theoretical capacity of 115 mAh/g. Moreover, capacity of about 100 mAh/g was obtained at between 2 and 4.5 V in subsequent cycles. The plateau at 3.0 V in the voltage curve is consistent with the theoretically computed voltage for the Fe^{2+}/Fe^{3+} redox couple in this structure. As for its Mn counterpart, the discharge capacity of $Li_{2.67}Na_{0.63}Mn_{0.96}CO_3PO_4$ was only ~50 mAh/g, but two voltage "plateaus" were observed, consistent with computationally predicted voltages for the Mn^{2+}/Mn^{3+} and Mn^{3+}/Mn^{4+} redox couples (3.2

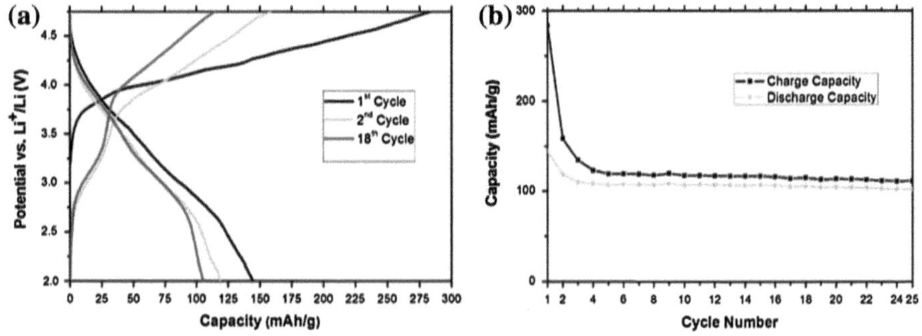

Fig. 19 Voltage profile (**a**) and capacity retention (**b**) of $Li_3Fe_{0.2}Mn_{0.8}CO_3PO_4$ cycled between 2 and 4.75 V and at a rate of C/50. Reproduced with permission from Ref. [94]. Copyright 2013, The Electrochemical Society

and 4.1 V, respectively). To further investigate the redox activity of Mn, XANES experimentation was carried out on the samples. The results showed that not only was the Mn^{2+}/Mn^{3+} redox couple active, the Mn^{3+}/Mn^{4+} couple was also partially active, corresponding to the two voltage plateaus observed on the voltage profile.

In order to combine the high theoretical capacity of $Li_3MnCO_3PO_4$ with the good cycling performance of $Li_3FeCO_3PO_4$, in 2013, Matts et al. [94] reported a $Li_3Fe_{0.2}Mn_{0.8}CO_3PO_4$ material with a rather good electrochemical performance. First, $Na_3Fe_{0.2}Mn_{0.8}CO_3PO_4$ was synthesized via a hydrothermal synthesis method. Next, $Li_3Fe_{0.2}Mn_{0.8}CO_3PO_4$ was made from its sodium-containing precursor via a Li–Na ion-exchange method [92]. XRD patterns and ICP analysis showed that the Li–Na ion exchange was carried out almost (>99 %) to completion [94]. The now ion-exchanged compound $Li_3Fe_{0.2}Mn_{0.8}CO_3PO_4$ was tested as a Li-ion cathode at a rate of C/50 at room temperature in a 2.00–4.75 V window. The voltage versus capacity data for one of these tests is shown in Fig. 19a. On charging following the first cycle, a distinct plateau can be seen near 3.0 V, which is the predicted voltage of the Fe^{2+}/Fe^{3+} couple and near that of the Mn^{2+}/Mn^{3+} couple. However, both the charge capacity and discharge capacity drop significantly over the first 4 cycles, from 284 to 120 mAh/g and from 143 to 108 mAh/g, respectively. After the fourth cycle the charge and discharge capacities remain relatively constant, and drop only 5 mAh/g over the next 20 cycles, as seen in Fig. 19b.

Despite performance that is well under theoretical values, this demonstrates a significant improvement over the performance of $Li_3MCO_3PO_4$ (M = Mn, Fe), withthis doped compound exhibiting higher discharge voltages and better cycling performance. These results lead to new opportunities to improve the performance of novel carbonophosphate cathode materials by doping and structural tuning.

5 Sulfates

Multiple bimetallic sulfates with the formula $Li_2M(SO_4)_2$ (M = Fe, Mn, Co) have been proposed as new polyanionic Li-ion battery cathode compounds since the discovery of $LiFePO_4$ as a promising positive electrode material [5, 95, 96]. The Fe-based $Li_2Fe(SO_4)_2$ exhibits an open circuit voltage of 3.83 V versus Li^+/Li^0, which is the highest potential ever obtained for the Fe^{3+}/Fe^{2+} redox couple in an iron-based, fluorine-free compound, and is only matched by the triplite phase of $LiFe(SO_4)F$ [97, 98]. This finding has not only paved the way for the development of a totally new class of fluorine-free compounds but could also reveal fundamental structure–property relationships in Li-ion cathode materials.

In the early 1990s, Touboul et al. [99, 100] first mentioned the existence of anhydrous lithium metal sulfate compounds, with the formula $Li_2M_y(SO_4)_{1+y}$ ($y = 0.5$ with M = Zn, $y = 1$ with M = Co, Ni and Zn and $y = 2$ with M = Mg and Mn). However, they were unable to produce a phase-pure sample and could not determine the exact structures of the compounds.

In 2012, Tarascon et al. [101] reported on the preparation and electrochemical characterization of $Li_2M(SO_4)_2$ (M = Fe, Co, Mn), which are attractive new electrode materials for Li-ion batteries. As previously mentioned, the iron compound was noted for displaying an open circuit voltage of 3.83 V versus Li^+/Li^0, involving one lithium ion per formula unit and the Fe^{3+}/Fe^{2+} redox couple [98]. Phase pure $Li_2M(SO_4)_2$ (M = Fe, Co, Mn) compounds were prepared by a three-step process. First, stoichiometric amounts of Li_2SO_4 and MSO_4 were thoroughly mixed using a Spex Miller 8000M, with the resulting powder next pressed into a pellet. Finally, the pellet is annealed at 325 °C for 12–48 h, either under air (M = Co, Mn) or in a quartz tube sealed under vacuum (M = Fe) [102].

$Li_2M(SO_4)_2$ (M = Fe, Co) has been fully indexed to the monoclinic unit cell with the space group $P2_1/c$ [98, 102]. The structure, as shown in Fig. 20 for M = Fe and Co, consists of isolated MO_6 (M = Fe, Co) octahedra linked through shared oxygen vertices with surrounding SO_4 tetrahedra. Figure 21a and the inset of Fig. 21b illustrate the Fe compound from a different perspective, with the remainder of Fig. 21b giving its electrochemical results, displaying the unusually high voltage plateau. Each octahedron is linked to six SO_4 tetrahedra which are oriented in a star or pinwheel pattern when viewed along the *a*-axis. Conversely, each SO_4 group is only bound to three MO_6 octahedra, with the fourth corner of the non-shared tetrahedron pointing to an open channel where lithium ions reside. Later, in 2013, Reynaud et al. [102] reported that $Li_2Mn(SO_4)_2$ crystallized into the same structure as the iron and cobalt compounds, which was confirmed by Rietveld refinement of the XRD data.

The plateau in the voltage-composition trace on charge and discharge suggests a two-phase intercalation process, as confirmed by PITT measurements [98]. During the first charge of $Li_2Fe(SO_4)_2$ to 4.5 V versus Li^+/Li^0, nearly 1 Li^+ ion per formula unit is removed from the structure. On discharge only 0.86 Li^+ ions per formula unit are reinserted, giving a reversible capacity of around 88 mAh/g of active material as

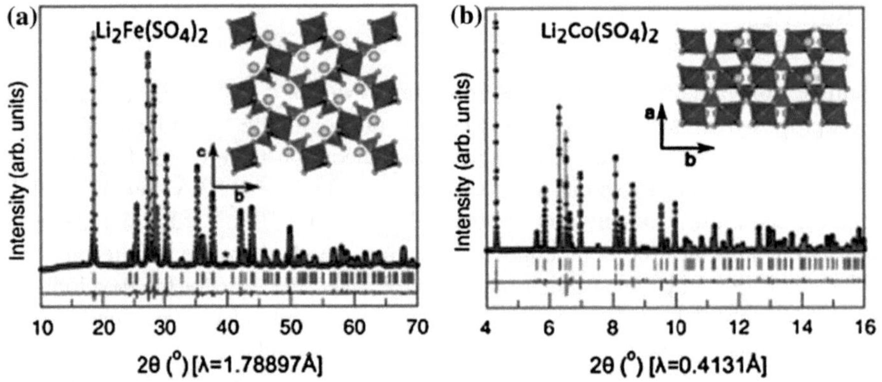

Fig. 20 Rietveld refinements of **a** $Li_2Fe(SO_4)_2$ and **b** $Li_2Co(SO_4)_2$. *Black circles*, *red lines* and *blue lines* represent the observed, calculated and difference patterns, respectively. Reproduced with permission from Ref. [98]. Copyright 2012, Elsevier

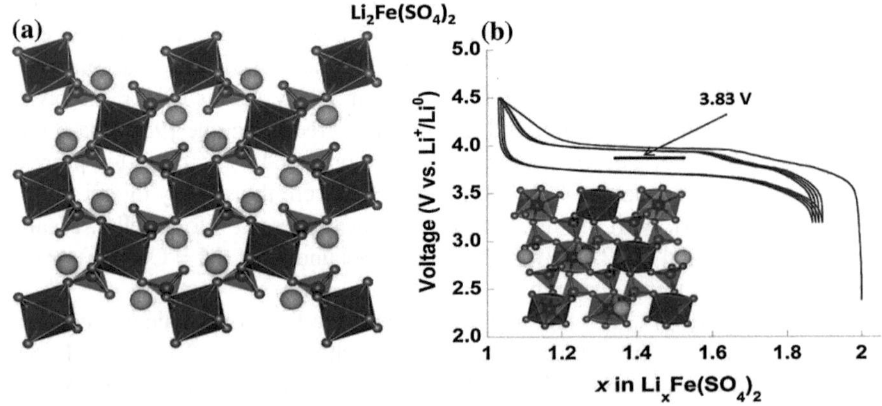

Fig. 21 **a** Structure (view perpendicular to [100]) and **b** electrochemical properties of the marinite $Li_2Fe(SO_4)_2$ phase. Reproduced with permission from Ref. [103]. Copyright 2014, American Chemical Society

compared to the theoretical specific capacity of 102 mAh/g. Subsequent charge/discharge curves neatly superimpose on each other, with steady capacity retention upon cycling. Moreover, the 85 % capacity retention at a 1 C rate also reflects the excellent rate capability of the material.

In contrast, neither the Co nor the Mn phases present any electrochemical activity up to 5 V [103]. As DFT calculations have predicted the redox potential of $Li_2Co(SO_4)_2$ to exceed 5 V in the tavorite and marinite structures, the problem is likely that electrolytes that function in such a high voltage region are not currently available for these compounds [104]. Regarding $Li_2Mn(SO_4)_2$, its theoretical potentials should range from 4 to 4.6 V, a voltage domain fully accessible by

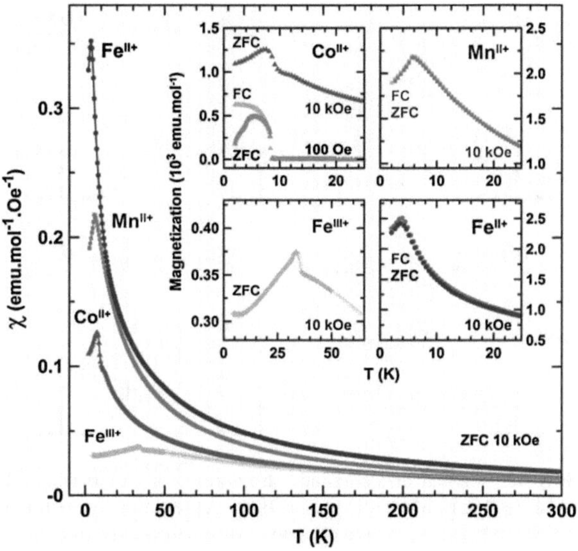

Fig. 22 Temperature dependence of the magnetic susceptibility (χ) for four marinite compounds, measured under zero-field-cooled (ZFC) conditions with a field of 10 kOe and between 300 and 2 K. Reproduced with permission from Ref. [102]. Copyright 2013, American Chemical Society

present electrolytes [104, 105]. However, a rapid decrease in electrochemical activity is observed on the voltage profile, making it an inappropriate candidate for a Li-ion battery electrode materials.

Moreover, magnetic properties of the marinite phases of $Li_2M(SO_4)_2$ (M = Fe, Co, Mn) and $Li_1Fe(SO_4)_2$ were proposed in order to aid the investigation of their magnetic structures [102]. In 2013, Reynaud et al. [102] reported on the temperature-dependent magnetic susceptibilities of the four marinite compounds and used low-temperature neutron powder diffraction to determine their magnetic structures. The $Li_1Fe(SO_4)_2$ was prepared by chemical delithiation from $Li_2Fe(SO_4)_2$. The macroscopic magnetic properties of these four compounds were determined by a SQUID magnetometer in both ZFC and FC conditions and under 10 kOe. As shown in Fig. 22, all compounds show the cusps of a long-range antiferromagnetic ordering. $Li_2Mn^{II}(SO_4)_2$ and $Li_2Fe^{II}(SO_4)_2$ present typical antiferromagnetic behavior, with the ZFC and FC curves superimposed. The ZFC and FC curves of $Li_1Fe^{III}(SO_4)_2$ deviate at temperatures below T_N, which may result from either a ferromagnetic impurity or some ferromagnetic/ferrimagnetic contributions. A similar behavior is observed with $Li_2Co^{II}(SO_4)_2$, having a nonlinearity in the moment versus applied field data, as the curves recorded under a field of 100 Oe lead to magnetization larger than expected from the value obtained at 10 kOe.

Low-temperature NPD measurements were also performed on the four target compounds and Rietveld refinement patterns were obtained, in order to determine the magnetic structure of each compound. A symmetry analysis was then performed using Bertaut's method [106, 107] to determine all possible spin configurations compatible with the crystal symmetry of the nuclear structure, as shown in Fig. 23. The ground-state magnetic structures of the four compounds were finally

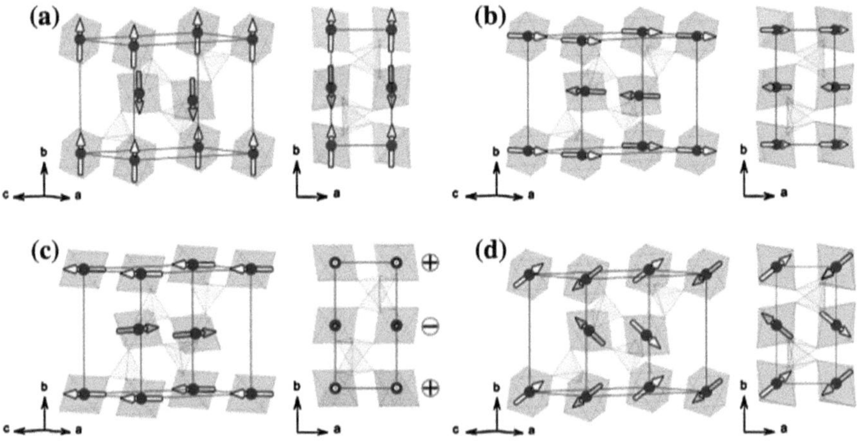

Fig. 23 Nuclear and magnetic structures of **a** $Li_2Co(SO_4)_2$, **b** $Li_2Mn(SO_4)_2$, **c** $Li_1Fe(SO_4)_2$, and **d** $Li_2Fe(SO_4)_2$. Magnetic moments are represented by vectors through the 3D metal atoms. *Yellow vectors* represent positive moments while *orange vectors* represent negative moments. For the sake of clarity, Li atoms are omitted (Color figure online). Reproduced with permission from Ref. [102]. Copyright 2014, American Chemical Society

established, with all magnetic structures able to be explained by three negative super-super-exchange paths, in agreement with Goodenough–Kanamori–Anderson rules.

A careful study of the crystal structures of these compounds would help to determine the cause of the increased potentials observed and confirm that marinite structures of the form $Li_2M(SO_4)_2$ (M = Co, Fe, Mn) are future positive electrode materials for Li-ion batteries with great promise.

6 Fluorosulfates

In 2010, Tarascon et al. [108] proposed the fluorosulphate $LiFeSO_4F$ as a novel 3.6 V Li-ion battery cathode material. Owing to the higher electronegativity of sulphur compared to phosphorus, the replacement of phosphate groups by sulfates can improve the open circuit voltage, as demonstrated by isostructural NASICONs [109]. However, to balance the overall charge, adding another anion such as fluorine anion is required. In addition, due to the inductive effect of the F^- anion, the substitution of F^- can further increase the working voltage. In 2002, Piffard et al. [110] reported on the lithium fluorosulphate $LiMgSO_4F$ with high Li-ion conduction. The authors concluded that "Lithium ion conduction in $LiMgFSO_4$ suggests that isostructural transition metal analogues, $LiMSO_4F$ (M = Mn, Fe, Co), would be important for redox extraction/insertion of lithium involving M^{II}/M^{III} oxidation states". Unfortunately, fluorosulphates suffer from thermal instability at high temperatures as well as the fact that the sulphates are highly soluble in commonly used electrolytes, which has resulted in research stagnation.

Until 2010, Tarascon and co-workers successfully prepared $LiFeSO_4F$ by a non-aqueous solvothermal method. $LiFeSO_4F$ powder was synthesized by a two-step process. First, the dehydration of $FeSO_4{\cdot}7H_2O$ was used to prepare monohydrate $FeSO_4{\cdot}H_2O$ under vacuum. Secondly, as-prepared $FeSO_4{\cdot}H_2O$ was mixed with LiF and placed in a polytetrafluoroethylene liner bomb, using EMI-TFSI as the reacting media. The role of the ionic liquid is to shift the dehydration of $FeSO_4{\cdot}H_2O$ to higher temperatures and to allow $FeSO_4{\cdot}H_2O$ and LiF to react at those higher temperatures. $LiFeSO_4F$ is also insoluble in the ionic liquid. The resultant formation of $LiFeSO_4F$ is based on a topotactic reaction where F^- replaces the OH^- and Li^+ replaces H^+ from the water molecule. The crystal structure of $LiFeSO_4F$ is isostructural with $LiMgSO_4F$ [110]. FeO_4F_2 octahedra are corner-shared with fluorine ligands along the c axis to form chains. The chains share corners with SO_4 tetrahedra to form a three-dimensional structure. Thermal gravimetric analysis indicates that $LiFeSO_4F$ is thermally unstable and starts to decompose at around 450 °C. However, this level of thermal instability might not be an issue for many practical battery applications, supposing that no serious side reactions happen. $LiFeSO_4F$ shows a voltage plateau at 3.6 V and about 0.85 Li per formula unit can be re-inserted during discharge, corresponding to a reversible capacity of about 130 to 140 mAh/g. It also exhibits steady capacity retention in subsequent charge/discharge cycles. Furthermore, increasing the charge/discharge rate does not increase the polarization of the material, indicating that the $LiFeSO_4F$ cathode material possesses excellent rate capability. Conductivity measurements show that the values of the ionic and electronic conductivities are 7.0×10^{-11} and 5.2×10^{-11} S/cm, respectively. Ab initio molecular dynamics studies demonstrate that $LiFeSO_4F$ operates electrochemically with three-dimensional lithium diffusion [111]. The 3D diffusion channel framework for $LiFeSO_4F$ is speculated to have less defects than the 1D structure of $LiFePO_4$, allowing the material to exhibit an acceptable rate capability without having to carefully control the particle size and/or apply a carbon coating. In situ XRD measurements suggest that $LiFeSO_4F$ undergoes a two-phase reaction process during charge/discharge processes, which is in accord with the flat charge/discharge curves.

In order to increase the energy density, using a manganese-based fluorosulphate can be a good choice, due to the higher redox voltage of Mn compared to Fe. However, $LiMnSO_4F$ does not have a tavorite structure but a triplite structure, which does not show any electrochemical activity. Interestingly, substituting in just 5 % manganese in $LiFeSO_4F$ can cause its structure to change from tavorite to triplite [97]. Figure 24 compares these two structures. For both, the MO_4F_2 octahedra are coordinated with four oxygen ligands and two fluorine ligands. However, the MO_4F_2 octahedra are linked together by edge-sharing in the triplite structure, while corner-sharing is used in the tavorite structure. Also, the fluorine atoms in the triplite structure are in cis configuration, but in trans configuration in the tavorite structure. Furthermore, Li and M in the triplite structure are randomly distributed in the metal sites. An increase in manganese content for $LiFe_{1-x}Mn_xSO_4F$ in the

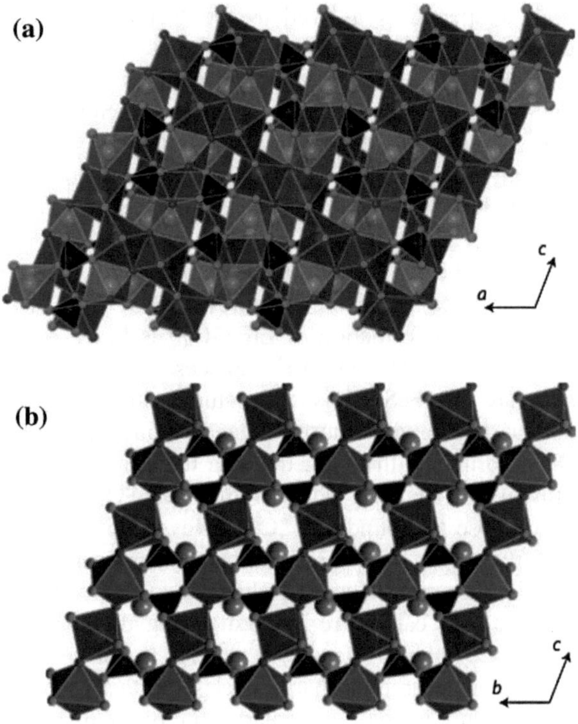

Fig. 24 The crystal structure of the triplite (**a**) and tavorite (**b**) structure type. Reproduced with permission from Ref. [97]. Copyright 2011, Macmillan Publishers Limited

triplite configuration causes the cation disorder of Li and M to become more serious. As shown in Fig. 25, the tavorite phase of $LiFe_{0.9}Mn_{0.1}SO_4F$ shows similar charge/discharge curves to tavorite $LiFeSO_4F$, but with slightly decreasing discharge capacity due to the electrochemical inactivity of manganese. The triplite phase of $LiFe_{0.9}Mn_{0.1}SO_4F$ exhibits a reversible capacity of 0.7–0.8 Li^+ per formula unit during cycling, with a working voltage of 3.9 V. This is the highest $Fe^{2+/3+}$ redox voltage that has thus far been reported for this class of Fe-based inorganic compound. It can be seen that triplite $LiFe_{0.9}Mn_{0.1}SO_4F$ shows a larger polarization than the tavorite phase, indicating slow kinetics during charge/discharge processes. This phenomenon arises from the cation disorder of Li and M in the triplite that hinders the efficient diffusion of lithium. When the manganese content is increased to 40 %, $LiFe_{1-x}Mn_xSO_4F$ does not show any electrochemical activity, due to Mn obstructing the diffusion tunnels. In situ XRD measurements suggest a two-phase Li insertion-extraction mechanism which is in agreement with charge/discharge curves. Also, triplite $LiFe_{0.9}Mn_{0.1}SO_4F$ exhibits excellent cycling performance and high structural stability. The volume change is only 0.6 % between lithiated and delithiated triplite phases, while it is a substantially higher 10.4 % for tavorite phases. The small volume change of the triplite structure makes it much more suitable to industrial application.

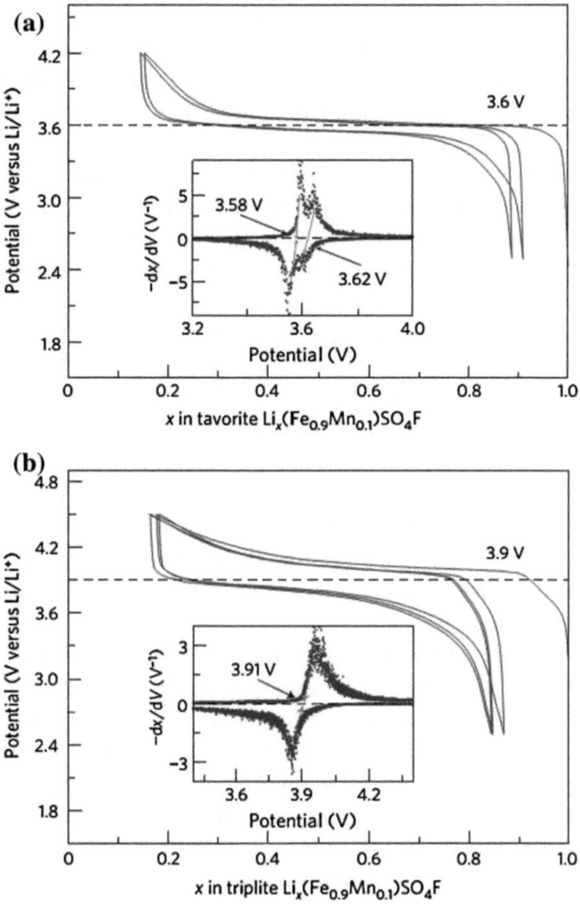

Fig. 25 Charge and discharge curves for tavorite (**a**) and triplite (**b**) phase Li $(Fe_{0.9}Mn_{0.1})SO_4F$, at a C/20 rate. Reproduced with permission from Ref. [97]. Copyright 2011, Macmillan Publishers Limited

Zinc substitution for iron in $LiFe_{1-x}Zn_xSO_4F$can also force the tavorite phase to change to the triplite phase and exhibit a working voltage of 3.9 V. When the content of zinc further increases, the structure of $LiFe_{1-x}Zn_xSO_4F$ changes to the sillimanite phase with a redox voltage of 3.6 V [112, 113]. Tarascon et al. [112] and Huang et al. [112, 113] used a traditional ceramic method to synthesize pure triplite $LiFeSO_4F$ without substitution. The high working voltage of 3.9 V observable for triplite $LiFeSO_4F$ makes it a competitive cathode material compared to $LiFePO_4$, although $LiFeSO_4F$ has a slightly lower theoretical specific capacity. However, triplite $LiFeSO_4F$ also shows poorer electrochemical performance than tavorite $LiFeSO_4F$. Furthermore, the capacity quickly decays with an increase in charge/discharge current. Experiments with chemical oxidation by NO_2BF_4 indicate that $LiFeSO_4F$ cannot form a completely delithiation phase because of atomic disorder. In addition, galvanostatic intermittent titration technique (GITT) measurements show that when more than half of the Li is extracted from triplite $LiFeSO_4F$, larger polarization results [114]. A core-shell model was proposed to explain this mechanism, where if the lithium diffusion length increases through the core, the polarization increases. Therefore, both careful control of the triplite $LiFeSO_4F$ particle morphology and the use of conductive additives are needed in order to improve the material's electrochemical performance.

The tavorite phase is obtained when an ionic liquid is used as the reacting media or if a solid-state method is used with a slow initial heating rate. Conversely, the triplite phase is obtained when using a solid-state method with a fast initial heating rate [112]. Therefore, the polymorphs of $LiFeSO_4F$ that are present after preparation are strongly dependent on synthesis conditions. The EMI-TFSI ionic liquid can restrain the dehydration of $FeSO_4{\cdot}H_2O$ and then promote the reaction between $FeSO_4{\cdot}H_2O$ and LiF by topotactic conversion. Using a fast heating rate makes the dehydration $FeSO_4{\cdot}H_2O$ form α-$FeSO_4$, rather than proceeding by topotactic conversion with LiF. The moisture in Teflon can solubilize LiF and then promote the reaction between $FeSO_4$ and LiF to form triplite $LiFeSO_4F$. However, there is some controversy about the transformation between the tavorite and triplite phases. Ati et al. [112] claimed that $LiFeSO_4F$ is not able to transform from one phase to another simply by extending the heat treatment time. However, Tripathi et al. [115] and Radha et al. [116] have successfully achieved this transformation, from tavorite to triplite, by using a longer reaction time in the experiment. Isothermal acid solution calorimetry measurements indicate that $LiFe_{1-x}Mn_xSO_4F$ with $x < 0.2$ in the tavorite phase is more energetically stable than in the triplite phase. The cation disorder in the triplite phase makes for a higher entropy than in the tavorite phase. The $T\Delta S$ term becomes more significant at high temperatures that act to stabilize the triplite phase, and the entropy drives the formation of the disordered triplite polymorph. The free energy change between tavorite–triplite phases during transformation is almost zero by calculating configurational entropies using measured enthalpies at room temperature, while it becomes exothermic at 500 °C. These results indicate that both tavorite and triplite phases are thermodynamically stable at room temperature, while the triplite phase is thermodynamically favored at high temperatures. The transformation of $LiFeSO_4F$ from the tavorite phase to the triplite phase is thus feasible by extending the heat treatment time, indicating that the tavorite phase is only an intermediate in the reaction process. Figure 26 summarizes the obtained phase as a function of cation radius and temperature.

Several works have been devoted to understanding the different working voltages of the tavorite and triplite phases. The average metal to ligand bond lengths are 2.1034 and 2.1508 Å for the tavorite phase and triplite phase, respectively. A longer bond length corresponds to a greater ionic character, and therefore a higher voltage $Fe^{2+/3+}$ redox couple as observed in the triplite [117, 118]. DFT + U calculations, crystallographic and electrostatic analyses were used to further explain the voltage increase [105]. The working voltage is closely related to the energy difference between the lithiated phase and delithiated phase involved. The stabilization of the lithiated phase and/or the destabilization of the delithiated phase leads to a high voltage. Additionally, a shorter F-F distance means that there is more repulsive F–F interactions in the triplite phase than in the tavorite phase. This difference

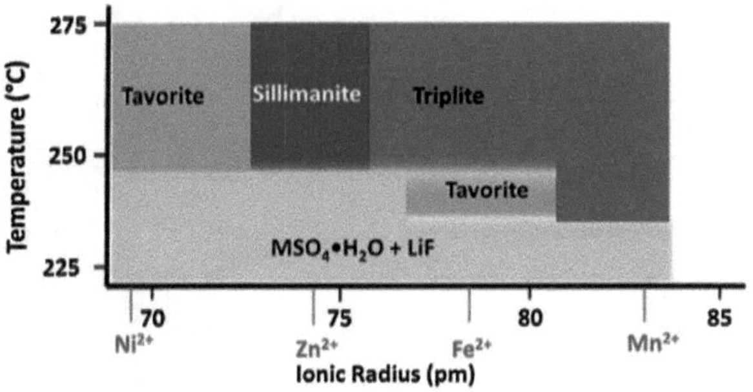

Fig. 26 A scheme which depicts the phases obtained under solvothermal reaction conditions (with reaction time of 3 days) as a function of the average cation (M^{2+}) radius. Cations with a lower radius such as Ni^{2+} (0.69 Å) tend to result in formation of the tavorite structure while cations with larger radii tend to adopt the triplite structure; for example, Mn^{2+} (0.83 Å) triplite can be crystallized at temperatures as low as 230 °C. Cations with an intermediate radius (0.76–0.79 Å) can be crystallized either as tavorite or triplite. Zn^{2+} and similarly sized ions (~0.72–0.76 Å) remain sillimanite at least up to the higher bound of considered temperatures. Reproduced with permission from Ref. [115]. Copyright 2012, Royal Society of Chemistry

corresponds to different configurations that are caused by the fluorine atoms around transition metals in these two structures: the fluorine atoms lie in cis-configuration for the triplite polymorph but in trans-configuration for the tavorite polymorph. The lower number of repulsive F-F interactions in the delithiated tavorite phase lead to it showing a higher stability than the delithiated triplite phase. However, by contrast both the lithiated phases have very similar energies, because there are also more attractive interactions to compensate for the increased amount of repulsive interactions in the triplite phase. Therefore, the different working voltages of the tavorite and triplite phases end up being strongly related to the different F-F distances.

7 Borates

Various structural forms of transition-metal borates have been considered as Li-ion battery electrodes. Generally, transition metal borate families that have attracted interest include MBO_3, $M_2B_2O_5$, $M_3B_2O_6$ and $LiMBO_3$ (M = transition metal) [119–123]. Among these compounds, the first three types are uncompetitive in energy density due to their low operating voltages on discharge, hindering their application to lithium ion batteries. However, a new class of $LiMBO_3$ has recently gathered increasing research interest worldwide. Its crystal structure contains trigonal planar BO_3 building blocks, and since the BO_3^{3-} group has a lower weight than the PO_4^{3-} group, the material has an advantage in terms of theoretical capacity compared to some competitor materials. Furthermore, the B–O bond is weaker than the P–O bond. This gives Li^+ ions higher mobility in a $LiMBO_3$ crystal compared to those in a $LiMPO_4$ crystal, resulting in a higher Li-ion conductivity. A structural

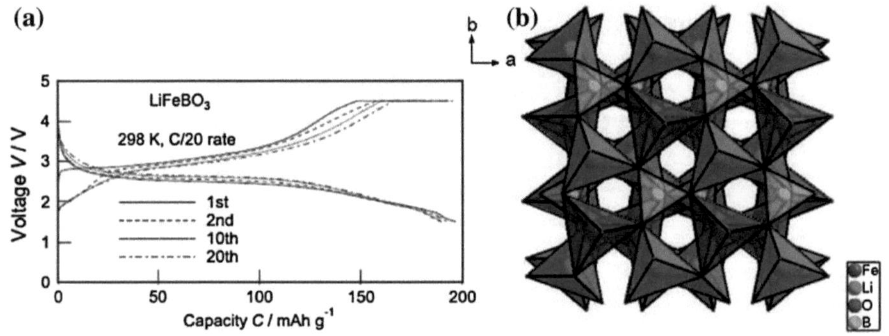

Fig. 27 **a** Charge-discharge curves of Li_xFeBO_3 measured at room temperature at a C/20 rate; **b** crystal structure of monoclinic $LiFeBO_3$. Reproduced with permission from Refs. [124, 126]. Copyright 2010, Wiley and 2012, The Royal Society of Chemistry

analysis of $LiMBO_3$ (M = Mg, Mn, Fe, Co, Zn) compounds was reported by Legagneur et al. [121] demonstrating that $LiMBO_3$ compounds (M = Mg, Mn, Co, Zn) primarily displayed a triclinic structure with the space group C2/c. In addition, $LiMnBO_3$ also stabilized into a hexagonal structure. This series of $LiMBO_3$ compounds have lately emerged as one of the more promising families of cathode materials for next generation Li-ion batteries.

The crystal structure of $LiFeBO_3$ contains the three-dimensional $[FeBO_3]_n^{n-}$ framework that consists of BO_3 planar triangles and FeO_5 trigonal bipyramids. The FeO_5 bipyramids use edge-shared connections to form single chains running along the [−101] direction, as shown in Fig. 27. The planar BO_3 groups link three pseudo one-dimensional FeO_5 chains via corner sharing. The possible Li diffusion pathways are parallel to the [001] direction. Many endeavors have been devoted to realizing the large theoretical reversible capacity of the compound. However, the achieved reversible capacities have been very low in initial reports. In 2010, through avoiding surface exposure to air, Yamada et al. [124] managed a significant performance improvement of $LiFeBO_3$, which reached a capacity of >190 mAh/g at around ca. 3 V versus Li^+/Li^0. The theoretical average voltage plateau that was located at 3 V, as obtained from DFT calculations, was in accordance with the experimental value. Carbon-coated synthesized $LiFeBO_3$ also exhibited excellent electrochemical performance due to the reduction of the crystallite size and the limitation of exposure to air. Bo et al. [125] studied the degradation mechanism of the nano-size $LiFeBO_3$/C composite. It was demonstrated that the degradation of nanoscale $LiFeBO_3$ is caused by moisture/air reacting on the surface of the particles and moving inward along a spherical front. This surface oxidation of Fe(II) changes the original balance of charge, leading to Li loss. The degradation product has lower ionic/electronic conductivity and exists on the surface of $LiFeBO_3$ particles, severely hindering the performance of the material. Tao et al. [126] proposed an average "1Fe 2Li" structural model for $LiFeBO_3$ powder, where Li atoms partially

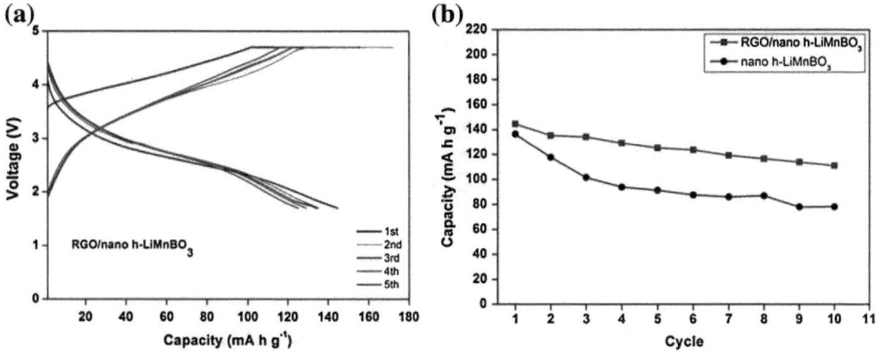

Fig. 28 **a** Charge/discharge curves of RGO/nano h-$LiMnBO_3$ at a C/20 rate. **b** Discharge capacity as a function of cycle number for nano h-$LiMnBO_3$ and RGO/nano h-$LiMnBO_3$. Reproduced with permission from Ref. [129]. Copyright 2013, Elsevier

occupy two split sites and Fe atoms are randomly located within a slightly elongated ellipsoid.

$LiMnBO_3$ crystallizes into two different polymorphs (hexagonal and monoclinic phases) based on the heat-treatment temperature. The hexagonal and monoclinic phases exist in lower and higher temperature regions, respectively. Kim et al. [127] reported on the successful preparation of monoclinic and hexagonal $LiMnBO_3$ by a conventional solid state method. The as-synthesized monoclinic $LiMnBO_3$ exhibited a second cycle discharge capacity of 100 mAh/g with good capacity retention over multiple cycles. By contrast, the hexagonal $LiMnBO_3$ generally exhibited a low discharge capacity above 1.5 V along with a very large polarization, indicative of poor ionic/electronic transport within this phase [128]. However, carbon coating together with nanostructuring have been considered effective methods to improve the performance of the hexagonal $LiMnBO_3$ [129]. The h-$LiMnBO_3$/C nanocomposite prepared by a sol-gel method delivered a first discharge capacity of 136 mAh/g at a C/20 rate [129]. After the nano h-$LiMnBO_3$/C was ball-milled together with graphite oxide, followed by heat treatment at 200–250 °C under a nitrogen atmosphere, the RGO/nano h-$LiMnBO_3$ composite was obtained. This material delivered a higher discharge capacity than that of the nano h-$LiMnBO_3$/C in the first 10 cycles, as shown in Fig. 28a, b. Li et al. [130] used in situ carbothermal solid state synthesis to prepare carbon-coated monoclinic and hexagonal $LiMnBO_3$ composites, and then compared their electrochemical performance. Both monoclinic and hexagonal $LiMnBO_3$ exhibited good cycling stability (with a reversible capacity of 86.7 and 78.5 mAh/g achieved after 40 cycles, respectively).

A complete solid solution of $Li(Mn_xFe_{1-x})BO_3$ in monoclinic form was synthesized by an optimized solid state reaction [131]. All samples exhibited reversible electrode activity, with the reversible capacity decreasing as the Mn content increased. Overall higher potentials were observed for manganese-substituted materials. Nevertheless, the manganese-rich phases suffered from structural

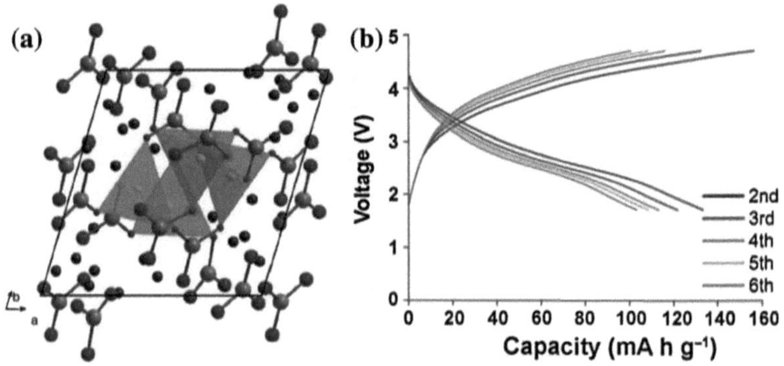

Fig. 29 **a** View of the crystal structure of $Li_7Mn(BO_3)_3$ along the [001] direction, with B in *green*, O in *red* and Li in *black*. **b** Charge/discharge curves of $Li_7Mn(BO_3)_3$ (Color figure online). Reproduced with permission from Ref. [132]. Copyright 2013, Wiley

instability upon charging. The low Mn^{3+}/Mn^{2+} redox activity in oxyanion compounds was attributed to five 3*d* electrons in Mn^{2+} that were all in the up-spin state hybridized with the O 2*p* state below the Fermi level, leading to a much larger contribution from O^{2-} to O^{-} upon charging rather than that from Mn^{2+} to Mn^{3+}.

Afyon et al. [132] found a new lithium-rich compound, $Li_7Mn(BO_3)_3$, with a network of MnO_4 tetrahedra and BO_3^{3-} units as characteristic features of a new structure type. Interestingly, compared to other Mn^{2+} oxide compounds that typically consist of MnO_6 octahedra, MO_5 trigonal bipyramidal or square-pyramidal geometry, this material has Mn^{2+} in tetrahedral coordination, a completely novel observation. This tetrahedral coordination may help to stabilize the boratev-manganate framework up to higher oxidation states Mn^{n+} with $n > 4$. The carbon-coated $Li_7Mn(BO_3)_3$ nanocomposite delivered approximately 154 mAh/g within a potential window of 4.7–1.7 V at a 10 mA/g rate in the first cycle. From the second cycle onwards, the charge/discharge capacities decreased gradually. The irreversible capacity loss might be attributed to amorphization of the electrode material, the formation of electrochemically inactive phases, contact losses owing to volume changes or solid electrolyte interface (SEI) formation (Fig. 29).

That the electrochemical activity of $LiCoBO_3$ is centered at 4 V was first demonstrated by Yamashita et al. [133]. But the accessible capacity of this material was only 30 mAh/g, which is much lower than the 1-electron theoretical capacity (ca. 215 mAh/g). Reasons for the discrepancy included poor Li^+ diffusivity, progressive Co dissolution and amorphization at the electrode-electrolyte interface. The development of $LiCoBO_3$ as a possible high-voltage cathode material for Li-ion batteries should continue, with attention required at further optimization of the synthesis conditions.

References

1. Goodenough JB, Kim Y (2010) Challenges for rechargeable Li batteries. Chem Mater 22:587–603
2. Tarascon JM, Armand M (2001) Issues and challenges facing rechargeable lithium batteries. Nature 414:359–367
3. Manthiram A, Goodenough JB (1987) Lithium insertion into $Fe_2(MO_4)_3$ frameworks—comparison of M = W with M = Mo. J Solid State Chem 71:349–360
4. Manthiram A, Goodenough JB (1989) Lithium insertion into $Fe_2(SO_4)_3$ frameworks. J Power Sources 26:403–408
5. Gong ZL, Yang Y (2011) Recent advances in the research of polyanion-type cathode materials for Li-ion batteries. Energ Environ Sci 4:3223–3242
6. Masquelier C, Croguennec L (2013) Polyanionic (Phosphates, Silicates, Sulfates) frameworks as electrode materials for rechargeable Li (or Na) batteries. Chem Rev 113:6552–6591
7. Eames C, Armstrong AR et al (2012) Insights into changes in voltage and structure of Li_2FeSiO_4 polymorphs for lithium-ion batteries. Chem Mater 24:2155–2161
8. Nyten A, Abouimrane A et al (2005) Electrochemical performance of Li_2FeSiO_4 as a new Li-battery cathode material. Electrochem Commun 7:156–160
9. Armand M, Michot C, Ravet N et al (2000) US Patent, 6,085,015
10. Arroyoy de Dompablo ME, Armand M et al (2006) On-demand design of polyoxianionic cathode materials based on electronegativity correlations: an exploration of the Li_2MSiO_4 system (M = Fe, Mn Co, Ni). Electrochem Commun 8:1292–1298
11. Lyness C, Delobel B et al (2007) The lithium intercalation compound Li_2CoSiO_4 and its behaviour as a positive electrode for lithium batteries. Chem Commun 4890–4892
12. Dominko R, Bele M et al (2007) Li_2MnSiO_4 as a potential Li-battery cathode material. J Power Sources 174:457–461
13. Gong ZL, Li YX et al (2007) Synthesis and electrochemical performance of Li_2CoSiO_4 as cathode material for lithium ion batteries. J Power Sources 174:524–527
14. Prakash AS, Rozier P et al (2006) Electrochemical reactivity of Li_2VOSiO_4 toward Li. Chem Mater 18:407–412
15. Sirisopanaporn C, Masquelier C et al (2011) Dependence of Li2FeSiO4 electrochemistry on structure. J Am Chem Soc 133:1263–1265
16. Nishimura SI, Hayase S et al (2008) Structure of Li_2FeSiO_4. J Am Chem Soc 130:13212–13213
17. Boulineau A, Sirisopanaporn C et al (2010) Polymorphism and structural defects in Li_2FeSiO_4. Dalton T 39:6310–6316
18. Sirisopanaporn C, Boulineau A et al (2010) Crystal structure of a new polymorph of Li_2FeSiO_4. Inorg Chem 49:7446–7451
19. Armstrong AR, Kuganathan N et al (2011) Structure and lithium transport pathways in Li_2FeSiO_4 cathodes for lithium batteries. J Am Chem Soc 133:13031–13035
20. Kojima A, Kojima T et al (2012) Crystal structure and electrochemical performance of a new lithium trivalent iron silicate. J Electrochem Soc 159:A725–A729
21. Gummow RJ, He Y (2014) Recent progress in the development of Li_2MnSiO_4 cathode materials. J Power Sources 253:315–331
22. Dominko R, Bele M et al (2006) Structure and electrochemical performance of Li_2MnSiO_4 and Li_2FeSiO_4 as potential Li-battery cathode materials. Electrochem Commun 8:217–222
23. Arroyoy de Dompablo ME, Dominko R et al (2008) On the energetic stability and electrochemistry of Li_2MnSiO_4 polymorphs. Chem Mater 20:5574–5584
24. Gummow RJ, Sharma N et al (2012) Crystal chemistry of the Pmnb polymorph of Li_2MnSiO_4. J Solid State Chem 188:32–37
25. Politaev VV, Petrenko AA et al (2007) Crystal structure, phase relations and electrochemical properties of monoclinic Li_2MnSiO_4. J Solid State Chem 180:1045–1050

26. Mali G, Meden A et al (2010) Li-6 MAS NMR spectroscopy and first-principles calculations as a combined tool for the investigation of Li_2MnSiO_4 polymorphs. Chem Commun 46:3306–3308
27. Arroyoy de Dompablo ME, Amador U et al (2009) Polymorphs of Li_3PO_4 and Li_2MSiO_4 (M = Mn, Co) the role of pressure. J Power Sources 189:638–642
28. Kobayashi G, Nishimura SI et al (2009) Isolation of solid solution phases in size-controlled Li_xFePO_4 at room temperature. Adv Funct Mater 19:395–403
29. Dominko R (2008) Li_2MSiO_4 (M = Fe and/or Mn) cathode materials. J Power Sources 184:462–468
30. Kokalj A, Dominko R et al (2007) Beyond one-electron reaction in Li cathode materials: designing $Li_2Mn_xFe_{1-x}SiO_4$. Chem Mater 19:3633–3640
31. Nyten A, Kamali S et al (2006) The lithium extraction/insertion mechanism in Li_2FeSiO_4. J Mater Chem 16:2266–2272
32. Muraliganth T, Stroukoff KR et al (2010) Microwave-solvothermal synthesis of nanostructured Li_2MSiO_4/C (M = Mn and Fe) cathodes for lithium-ion batteries. Chem Mater 22:5754–5761
33. Dominko R, Conte DE et al (2008) Impact of synthesis conditions on the structure and performance of Li_2FeSiO_4. J Power Sources 178:842–847
34. Lv DP, Wen W et al (2011) A novel Li2FeSiO4/C composite: synthesis, characterization and high storage capacity. J Mater Chem 21:9506–9512
35. Bai JY, Gong ZL et al (2012) Nanostructured 0.8Li(2)FeSiO(4)/0.4Li(2)SiO(3)/C composite cathode material with enhanced electrochemical performance for lithium-ion batteries. J Mater Chem 22:12128–12132
36. Gong ZL, Li YX et al (2006) Synthesis and characterization of $Li_2Mn_xFe_{1-x}SiO_4$ as a cathode material for lithium-ion batteries. Electrochem Solid State 9:A542–A544
37. Lv DP, Bai JY et al (2013) Understanding the high capacity of Li_2FeSiO_4: in situ XRD/XANES study combined with first-principles calculations. Chem Mater 25:2014–2020
38. Belharouak I, Abouimrane A et al (2009) Structural and electrochemical characterization of Li_2MnSiO_4 cathode material. J Phys Chem C 113:20733–20737
39. Li LM, Guo HJ et al (2009) Effects of roasting temperature and modification on properties of Li_2FeSiO_4/C cathode. J Power Sources 189:45–50
40. Bao LY, Gao W et al (2013) Progression of the silicate cathode materials used in lithium ion batteries. Chinese Sci Bull 58:575–584
41. Pizarrosanz JL, Dance JM et al (1994) The natural and synthetic tavorite minerals—crystal-chemistry and magnetic-properties. Mater Lett 18:327–330
42. Mba JMA, Croguennec L et al (2012) Lithium insertion or extraction from/into tavorite-type $LiVPO_4F$: an in situ X-ray diffraction study. J Electrochem Soc 159:A1171–A1175
43. Barker J, Gover RKB et al (2005) Structural and electrochemical properties of lithium vanadium fluorophosphate, $LiVPO_4F$. J Power Sources 146:516–520
44. Sun XF, Xu YL et al (2013) High performance $LiV_{0.96}Mn_{0.04}PO_4F$/C cathodes for lithium-ion batteries. J Mater Chem A 1:2501–2507
45. Xiao PF, Lai MO et al (2013) Transport and electrochemical properties of high potential tavorite $LiVPO_4F$. Solid State Ionics 242:10–19
46. Zhou F, Zhao XM et al (2009) Reactivity of charged $LiVPO_4F$ with 1 M $LiPF_6$ EC: DEC electrolyte at high temperature as studied by accelerating rate calorimetry. Electrochem Commun 11:589–591
47. Tarascon JM (2010) Key challenges in future Li-battery research. Philos Trans R Soc A 368:3227–3241
48. Dutreilh M, Chevalier C et al (1999) Synthesis and crystal structure of a new lithium nickel fluorophosphate Li-2[NiF(PO_4)] with an ordered mixed anionic framework. J Solid State Chem 142:1–5
49. Ellis BL, Makahnouk WRM et al (2007) A multifunctional 3.5 V iron-based phosphate cathode for rechargeable batteries. Nat Mater 6:749–753

50. Ellis BL, Makahnouk WRM et al (2010) Crystal structure and electrochemical properties of A(2)MPO(4)F fluorophosphates (A = Na, Li; M = Fe, Mn Co, Ni). Chem Mater 22:1059–1070
51. Nagahama M, Hasegawa N et al (2010) High voltage performances of Li_2NiPO_4F cathode with dinitrile-based electrolytes. J Electrochem Soc 157:A748–A752
52. Okada S, Ueno M et al (2005) Fluoride phosphate Li_2COPO_4F as a high-voltage cathode in Li-ion batteries. J Power Sources 146:565–569
53. Wu XB, Zheng JM et al (2011) Sol-gel synthesis and electrochemical properties of fluorophosphates $Na_2Fe_{1-x}Mn_xPO_4F/C$ (x = 0, 0.1, 0.3, 0.7, 1) composite as cathode materials for lithium ion battery. J Mater Chem 21:18630–18637
54. Yakubovich OV, Karimova OV et al (1997) The mixed anionic framework in the structure of $Na_2\{MnF[PO_4]\}$. Acta Crystallogr C 53:395–397
55. Kabalov YK, Simonov MA et al (1974) Crystalline-structure of basic iron ortho-phosphate, Na_2Fe $[PO_4](Oh)$. Dokl Akad Nauk SSSR 215:850–853
56. Swafford SH, Holt EM (2002) New synthetic approaches to monophosphate fluoride ceramics: synthesis and structural characterization of $Na_2Mg(PO_4)F$ and $Sr_5(PO_4)(3)F$. Solid State Sci 4:807–812
57. Sanz F, Parada C et al (2001) Crystal growth, crystal structure and magnetic properties of disodium cobalt fluorophosphate. J Mater Chem 11:208–211
58. Kim SW, Seo DH et al (2012) A comparative study on Na_2MnPO_4F and Li_2MnPO_4F for rechargeable battery cathodes. Phys Chem Chem Phys 14:3299–3303
59. Lin XC, Hou X et al (2014) Exploiting Na_2MnPO_4F as a high-capacity and well-reversible cathode material for Na-ion batteries. RSC Adv 4:40985–40993
60. Zheng Y, Zhang P et al (2013) First-principles investigations on the Na_2MnPO_4F as a cathode material for Na-ion batteries. J Electrochem Soc 160:A927–A932
61. Recham N, Chotard JN et al (2009) Ionothermal synthesis of sodium-based fluorophosphate cathode materials. J Electrochem Soc 156:A993–A999
62. Kosova NV, Devyatkina ET et al (2012) In situ and ex situ X-ray study of formation and decomposition of Li_2CoPO_4F under heating and cooling. Investigation of its local structure and electrochemical properties. Solid State Ionics 225:570–574
63. Wang DY, Xiao J et al (2011) Preparation and electrochemical investigation of Li_2CoPO_4F cathode material for lithium-ion batteries. J Power Sources 196:2241–2245
64. Amaresh S, Kim GJ et al (2012) Synthesis and enhanced electrochemical performance of Li_2CoPO_4F cathodes under high current cycling. Phys Chem Chem Phys 14:11904–11909
65. Wu XB, Gong ZL et al (2012) Sol-gel synthesis of Li_2CoPO_4F/C nanocomposite as a high power cathode material for lithium ion batteries. J Power Sources 220:122–129
66. Amaresh S, Karthikeyan K et al (2013) Facile synthesis of ZrO_2 coated Li_2CoPO_4F cathode materials for lithium secondary batteries with improved electrochemical properties. J Power Sources 244:395–402
67. Wu XB, Wang SH et al (2014) Promoting long-term cycling performance of high-voltage Li_2CoPO_4F by the stabilization of electrode/electrolyte interface. J Mater Chem A 2:1006–1013
68. Khasanova NR, Gavrilov AN et al (2011) Structural transformation of Li_2CoPO_4F upon Li-deintercalation. J Power Sources 196:355–360
69. Vitins G, Kanepe Z et al (2000) Structural and conductivity studies in $LiFeP_2O_7$, $LiScP_2O_7$, and $NaScP_2O_7$. J Solid State Electr 4:146–152
70. Padhi AK, Nanjundaswamy KS et al (1997) Effect of structure on the Fe^{3+}/Fe^{2+} redox couple in iron phosphates. J Electrochem Soc 144:1609–1613
71. Ramana CV, Ait-Salah A et al (2007) Novel lithium iron pyrophosphate ($LiFe_{1.5}P_2O_7$) as a positive electrode for Li-ion batteries. Chem Mater 19:5319–5324
72. Nishimura S, Nakamura M et al (2010) New lithium iron pyrophosphate as 3.5 V class cathode material for lithium ion battery. J Am Chem Soc 132:13596–13597

73. Kim H, Lee S et al (2011) Neutron and X-ray diffraction study of pyrophosphate-based $Li_{2-x}MP_2O_7$ (M = Fe, Co) for lithium rechargeable battery electrodes. Chem Mater 23:3930–3937
74. Shimizu D, Nishimura S et al (2012) Electrochemical redox mechanism in 3.5 V $Li_{2-x}FeP_2O_7$ ($0 \leq x \leq 1$) pyrophosphate cathode. Chem Mater 24:2598–2603
75. Du J, Jiao LF et al (2013) Synthesis and characterization of $Li_2FeP_2O_7$/C nanocomposites as cathode materials for Li-ion batteries. Electrochim Acta 103:219–225
76. Barpanda P, Ye T et al (2012) Eco-efficient splash combustion synthesis of nanoscale pyrophosphate ($Li_2FeP_2O_7$) positive-electrode using Fe(III) precursors. J Mater Chem 22:13455–13459
77. Clark JM, Nishimura S et al (2012) High-voltage pyrophosphate cathode: insights into local structure and lithium-diffusion pathways. Angew Chem Int Edit 51:13149–13153
78. Lee S, Park SS (2012) Structure, defect chemistry, and lithium transport pathway of lithium transition metal pyrophosphates ($Li_2MP_2O_7$, M: Mn, Fe, and Co): atomistic simulation study. Chem Mater 24:3550–3557
79. Barpanda P, Rousse G et al (2013) Neutron diffraction study of the Li-ion battery cathode $Li_2FeP_2O_7$. Inorg Chem 52:3334–3341
80. Adam L, Guesdon A et al (2008) A new lithium manganese phosphate with an original tunnel structure in the A(2)MP(2)O(7) family. J Solid State Chem 181:3110–3115
81. Tamaru M, Barpanda P et al (2012) Observation of the highest Mn^{3+}/Mn^{2+} redox potential of 4.45 V in a $Li_2MnP_2O_7$ pyrophosphate cathode. J Mater Chem 22:24526–24529
82. Zhou H, Upreti S et al (2011) Iron and manganese pyrophosphates as cathodes for lithium-ion batteries. Chem Mater 23:293–300
83. Furuta N, Nishimura S et al (2012) Fe^{3+}/Fe^{2+} redox couple approaching 4 V in $Li_{2-x}(Fe_{1-y}Mn_y)P_2O_7$ pyrophosphate cathodes. Chem Mater 24:1055–1061
84. Ye T, Barpanda P et al (2013) General observation of Fe^{3+}/Fe^{2+} redox couple close to 4 V in partially substituted $Li_2FeP_2O_7$ pyrophosphate solid-solution cathodes. Chem Mater 25:3623–3629
85. Shakoor RA, Kim H et al (2012) Site-specific transition metal occupation in multicomponent pyrophosphate for improved electrochemical and thermal properties in lithium battery cathodes: a combined experimental and theoretical study. J Am Chem Soc 134:11740–11748
86. Tamaru M, Chung SC et al (2013) Pyrophosphate chemistry toward safe rechargeable batteries. Chem Mater 25:2538–2543
87. Kim H, Park I et al (2012) New iron-based mixed-polyanion cathodes for lithium and sodium rechargeable batteries: combined first principles calculations and experimental study. J Am Chem Soc 134:10369–10372
88. Hautier G, Jain A et al (2011) Novel mixed polyanions lithium-ion battery cathode materials predicted by high-throughput ab initio computations. J Mater Chem 21:17147–17153
89. Chen HL, Hautier G et al (2012) Synthesis, computed stability, and crystal structure of a new family of inorganic compounds: carbonophosphates. J Am Chem Soc 134:19619–19627
90. Huang H, Yin SC et al (2001) Approaching theoretical capacity of $LiFePO_4$ at room temperature at high rates. Electrochem Solid State 4:A170–A172
91. Yamada A, Chung SC et al (2001) Optimized $LiFePO_4$ for lithium battery cathodes. J Electrochem Soc 148:A224–A229
92. Chen HL, Hautier G et al (2012) Carbonophosphates: a new family of cathode materials for li-ion batteries identified computationally. Chem Mater 24:2009–2016
93. Arora P, White RE et al (1998) Capacity fade mechanisms and side reactions in lithium-ion batteries. J Electrochem Soc 145:3647–3667
94. Matts I, Chen HL et al (2013) Electrochemical properties of $Li_3Fe_{0.2}Mn_{0.8}CO_3PO_4$ as a Li-ion battery cathode. ECS Electrochem Lett 2:A81–A83
95. Melot BC, Tarascon JM (2013) Design and preparation of materials for advanced electrochemical storage. Acc Chem Res 46:1226–1238
96. Padhi AK, Nanjundaswamy KS et al (1997) Phospho-olivines as positive-electrode materials for rechargeable lithium batteries. J Electrochem Soc 144:1188–1194

97. Barpanda P, Ati M et al (2011) A 3.90 V iron-based fluorosulphate material for lithium-ion batteries crystallizing in the triplite structure. Nat Mater 10:772–779
98. Reynaud M, Ati M et al (2012) $Li_2Fe(SO_4)(2)$ as a 3.83 V positive electrode material. Electrochem Commun 21:77–80
99. Touboul M, Edern P et al (1992) Li_2SO_4-MSO_4 (M = Co, Ni) phase-diagrams and considerations about alpha-$Li_{2-2x}M_xSO_4$ solid-solutions and $Li_2M_y(SO_4)_{1+y}$ compounds (M: divalent ions). Solid State Ionics 50:323–328
100. Touboul M, Lesamedi E et al (1993) Binary-systems with Li_2SO_4 as one of the components. J Therm Anal 40:1151–1156
101. Reynaud M, Ati M et al (2013) Bimetallic sulfates $A(2)M(SO_4)(2)$ center dot nH(2)O (A = Li, Na and M = transition metal) as new attractive electrode materials for Li- and Na-ion batteries. ECS Trans 50:11–19
102. Reynaud M, Rousse G et al (2013) Marinite $Li_2M(SO_4)_2$ (M = Co, Fe, Mn) and $Li_1Fe(SO_4)_2$: model compounds for super-super-exchange magnetic interactions. Inorg Chem 52:10456–10466
103. Rousse G, Tarascon JM (2014) Sulfate-based polyanionic compounds for li-ion batteries: synthesis, crystal chemistry, and electrochemistry aspects. Chem Mater 26:394–406
104. Frayret C, Villesuzanne A et al (2010) $LiMSO_4F$ (M = Fe, Co and Ni): promising new positive electrode materials through the DFT microscope. Phys Chem Chem Phys 12:15512–15522
105. Ben Yahia M, Lemoigno F et al (2012) Origin of the 3.6 V to 3.9 V voltage increase in the $LiFeSO_4F$ cathodes for Li-ion batteries. Energ Environ Sci 5:9584–9594
106. Bertaut EF, Buisson G et al (1967) Structure magnetique et proprietes magnetiques de $BiMn_2O_5$. Solid State Commun 5:25–30
107. El Khayati N, El Moursli RC et al (2001) Crystal and magnetic structures of the oxyphosphates $MFePO_5$ (M = Fe Co, Ni, Cu). Analysis of the magnetic ground state in terms of superexchange interactions. Eur Phys J B 22:429–442
108. Recham N, Chotard JN et al (2010) A 3.6 V lithium-based fluorosulphate insertion positive electrode for lithium-ion batteries. Nat Mater 9:68–74
109. Padhi AK, Manivannan V et al (1998) Tuning the position of the redox couples in materials with NASICON structure by anionic substitution. J Electrochem Soc 145:1518–1520
110. Sebastian L, Gopalakrishnan J et al (2002) Synthesis, crystal structure and lithium ion conductivity of $LiMgFSO_4$. J Mater Chem 12:374–377
111. Ramzan M, Lebegue S et al (2011) Hybrid density functional calculations and molecular dynamics study of lithium fluorosulphate, a cathode material for lithium-ion batteries. J Phys Chem C 115:2600–2603
112. Ati M, Melot BC et al (2011) Synthesis and electrochemical properties of pure $LiFeSO_4F$ in the triplite structure. Electrochem Commun 13:1280–1283
113. Liu L, Zhang B et al (2011) A 3.9 V polyanion-type cathode material for Li-ion batteries. Prog Nat Sci 21:211–215
114. Dong JP, Yu XQ et al (2013) Trip lite $LiFeSO_4F$ as cathode material for Li-ion batteries. J Power Sources 244:716–720
115. Tripathi R, Popov G et al (2012) Lithium metal fluorosulfate polymorphs as positive electrodes for Li-ion batteries: synthetic strategies and effect of cation ordering. Energ Environ Sci 5:6238–6246
116. Radha AV, Furman JD et al (2012) Understanding the stability of fluorosulfate Li-ion battery cathode materials: a thermochemical study of $LiFe_{1-x}Mn_xSO_4F$ ($0 \le x \le 1$) polymorphs. J Mater Chem 22:24446–24452
117. Ati M, Melot BC et al (2011) Structural and electrochemical diversity in $LiFe_1$-delta Zn delta SO_4F solid solution: a Fe-based 3.9 V positive-electrode material. Angew Chem Int Edit 50:10574–10577
118. Tripathi R, Popov G et al (2013) Ultra-rapid microwave synthesis of triplite $LiFeSO_4F$. J Mater Chem A 1:2990–2994

119. Debart A, Revel B et al (2003) Study of the reactivity mechanism of $M_3B_2O_6$ (with M = Co, Ni, and Cu) toward lithium. Chem Mater 15:3683–3691
120. Kawano T, Suehiro T et al (2010) Preparation, crystal structure and photoluminescence of Mn^{2+}-doped magnesium pyroborates solid solutions, $(Mg_{1-x}Mn_x)(2)B_2O_5$. J Lumin 130:2161–2165
121. Legagneur V, An Y et al (2001) $LiMBO_3$ (M = Mn, Fe, Co): synthesis, crystal structure and lithium deinsertion/insertion properties. Solid State Ionics 139:37–46
122. Okada S, Tonuma T et al (2003) Anode properties of calcite-type MBO_3 (M: V, Fe). J Power Sources 119:621–625
123. Rowsell JLC, Gaubicher J et al (2001) A new class of materials for lithium-ion batteries: iron (III) borates. J Power Sources 97–8:254–257
124. Yamada A, Iwane N et al (2010) Lithium iron borates as high-capacity battery electrodes. Adv Mater 22:3583–3587
125. Bo SH, Wang F et al (2012) Degradation and (de)lithiation processes in the high capacity battery material LiFeBO. J Mater Chem 22:8799–8809
126. Tao L, Rousse G et al (2014) Preparation, structure and electrochemistry of $LiFeBO_3$: a cathode material for Li-ion batteries. J Mater Chem A 2:2060–2070
127. Kim JC, Moore CJ et al (2011) Synthesis and electrochemical properties of monoclinic $LiMnBO_3$ as a Li intercalation material. J Electrochem Soc 158:A309–A315
128. Chen L, Zhao YM et al (2010) Structure and electrochemical properties of $LiMnBO_3$ as a new cathode material for lithium-ion batteries. J Alloy Compd 494:415–419
129. Afyon S, Kundu D et al (2013) Nano $LiMnBO_3$, a high-capacity cathode material for Li-ion batteries. J Power Sources 224:145–151
130. Li SL, Xu LQ et al (2013) In-situ controllable synthesis and performance investigation of carbon-coated monoclinic and hexagonal $LiMnBO_3$ composites as cathode materials in lithium-ion batteries. J Power Sources 236:54–60
131. Yamada A, Iwane N et al (2011) Synthesis and electrochemistry of monoclinic $Li(Mn_xFe_{1-x})BO_3$: a combined experimental and computational study. J Mater Chem 21:10690–10696
132. Afyon S, Wörle M et al (2013) A lithium-rich compounds $Li_7Mn(BO_3)_3$ containing Mn^{2+} in tetrahedral coordination: a cathode candidate for Li-ion batteries. Angew Chem Int Ed 125:12773–12776
133. Yamashita Y, Barpanda P et al (2013) Demonstration of Co^{3+}/Co^{2+} electrochemical activity in $LiCoBO_3$ cathode at 4.0 V. ECS Electrochem Lett 2:A75–A77

Carbonaceous Anode Materials

Yoong Ahm Kim, Yong Jung Kim and Morinobu Endo

1 Introduction

In the last decade, lithium-ion batteries (LIBs) have been commercialized as the best energy storage devices due to high energy densities and high operation voltage up to 3.6 V. The advantages of LIBs have supported the recent remarkable development in cellular phones, personal computers and high speed CPU. In addition, LIBs have been expected to become alternative energy sources for electric vehicles. The applicability of the energy storage device strongly depends on the constituent materials of electrodes. The diversity of the electrode materials causes a necessity to examine many parameters in order to enhance the battery performance. Up to now, carbon materials have been selected as anode materials of the commercial LIBs with high specific capacity, good cyclic efficiency and long cycle life [1, 2]. Moreover, the potential use of LIBs in electric vehicles requires superiority in both energy density and power density [3–5]. The performance of Li-ion batteries strongly depends on the thermal history and morphology of carbon materials [6]. Carbon materials have a wide variety of microstructures, texture, crystallinity and morphology. Thus, it is critical to design and choose the anode material [7, 8]. Among various carbon materials, two types of carbon materials (i.e., soft carbon prepared by high-temperature thermal treatment at 2800 °C and hard carbon

Y.A. Kim (✉)
School of Polymer Science and Engineering, Chonnam National University,
77 Yongbong-ro, Buk-gu, Gwangju 500-757, Republic of Korea
e-mail: yak@jnu.ac.kr

Y.J. Kim
Research Institute of Industrial Science and Technology (RIST),
San32, Hyoja-dong, Nam-gu, Pohang 790-600, Republic of Korea

M. Endo
Institute of Carbon Science and Technology, Shinshu University,
4-17-1 Wakasato, Nagano 380-8553, Japan

Z. Zhang and S.S. Zhang (eds.), *Rechargeable Batteries*,
Green Energy and Technology, DOI 10.1007/978-3-319-15458-9_5

prepared at a low temperature around 1100 °C [9, 10]) have been generally used as anode materials in commercial batteries.

This chapter reviews various carbon materials as promising candidates for anode materials of LIBs. Moreover, nanosized carbon materials (e.g., carbon nanotube and graphene) have been considered as alternative electrode materials or as additives to increase the stability on the charge/discharge cycles. Moreover, demand and trend in the market of LIBs are also described.

2 Battery Performances for Conventional Carbon Materials

2.1 Typical Charge/Discharge Profile Dependent on the Crystallinity of Carbon

Three types of charge and discharge profiles at the second cycle for representative carbon and graphite materials are shown in Fig. 1. The shape of the profiles and the capacity strongly depends on the category of materials and the thermal history for preparation. Graphitic carbon shows a reversible capacity of 280–330 mAh/g, and Li discharge/charge plateaus are clearly observed below 0.2 V. In the first cycle, all types of carbon materials show the irreversible capacity. After the second cycle the irreversible capacity is reduced, and the electrode exhibits stable cyclic properties. Although the graphitic materials have some merits on discharge profile, the limited capacity of Li storage by LiC_6 stoichiometry can be said to be a main drawback.

Furthermore, the preparation of graphitic carbon materials requires high temperature heat-treatment, which is disadvantageous in production cost. Non-graphitic carbon heat-treated at around 700 °C shows excellent capacities over 1000 mAh/g

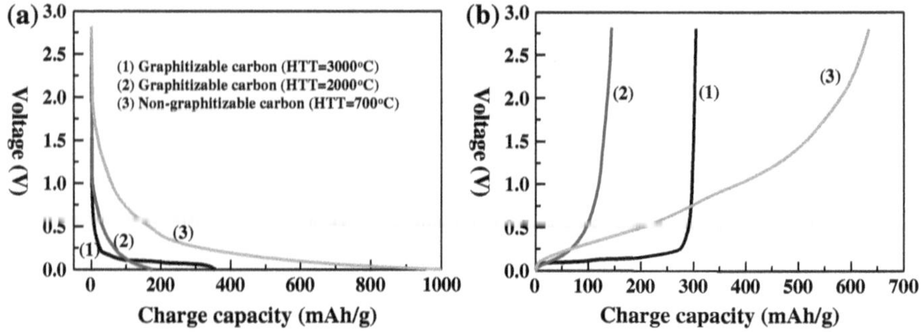

Fig. 1 Typical voltage profiles from carbon materials with different heat-treatment temperatures and different structural variations. The profiles were obtained at the second cycle, and **a** and **b** show discharge and charge cycles by using (*1*) graphitic carbon heat-treated at 3000 °C, (*2*) graphitic carbon heat-treated at 2000 °C and (*3*) non-graphitic carbon heat-treated at 700 °C. Reproduced from Ref. [6]

at charge and 600 mAh/g at discharge. The high capacities of non-graphitic carbons can be explained by (i) formation of Li_2 molecules between layers [10], (ii) cluster formation Li ions in nano-cavity [11], and (iii) hydrogen content of single graphene layers [12]. In addition, Yazami and Munshi [13] suggested that the excess Li capacity (more than LiC_6) originates from Li multi-layers on graphite sheets. On the other hand, Zhou et al. [14] suggested that Li–C–H bonds contribute to the excess capacity. Although the mechanism causing the excess capacity is not thoroughly clarified yet, it would have close relation with the nanostructure of carbon materials. Non-graphitic carbons show different output properties from those of graphitic carbons, i.e., somewhat constant and slightly inclined discharge potentials.

Figure 2 shows the second cycle charge capacity as a function of crystallite thickness, Lc_{002}, on various carbon-fiber and PPP-based-carbon electrodes [6]. Well-ordered graphite ($Lc_{002} > 20$ nm) and low crystalline materials ($Lc_{002} < 3$ nm) have larger capacities. However, the intermediate crystallite sizes lower capacities to the minimum at around 10 nm. Dahn et al. [12] reported a similar kind of dependence of the charge capacity on heat treatment temperatures, and classified carbon materials suited for commercial Li-ion batteries into three groups. In the HTT-dependence of capacity, interestingly, the highly disordered PPP-700 carbon (prepared by the heat-treatment at 700 °C) exhibits a large charge capacity of 680 mAh/g. As Lc_{002} becomes smaller than that of graphite crystal, the charge capacity decreases monotonically until the Lc_{002} value reaches about 10 nm, which is based on the Li^+ intercalation of turbostratic carbon structures. On the other hand, a different process of doping and undoping of Li ions may occur in $Lc_{002} < 10$ nm, and is largely enhanced by decreasing the crystallite thickness. For Lc_{002} around 10 nm, both reaction processes occur incompletely, which might cause a minimum in the capacity.

Fig. 2 Charge capacities (reversible capacity at the second cycle) from various carbon materials as a function of crystallite thickness. Reproduced from Ref. [15]

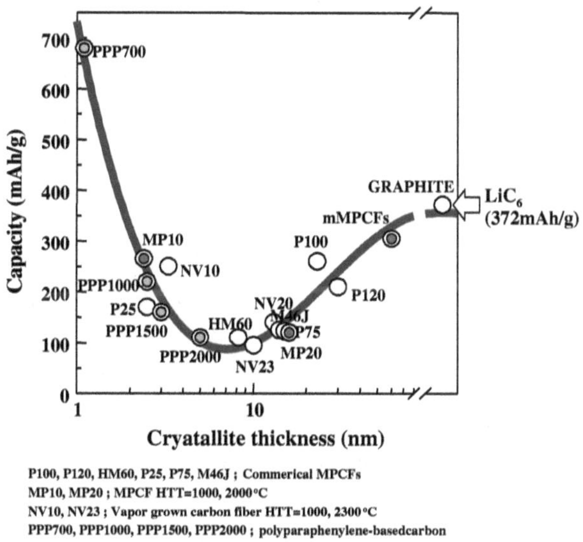

3 Non-graphitic Carbons

The performance of Li ion batteries depends strongly on the crystalline structure of carbon materials used for the anode electrodes [6, 15]. Especially, Kovacic-method-based poly(*p*-phenylene) (PPP) carbons (see below for Kovacic method) heat-treated at the low temperature of 700 °C are proved to possess a superior Li storage capacity (more than 1100 mAh/g), which is three times higher than that of well-ordered graphite-based first stage intercalation compounds. This is also four times higher than the Yamamoto-method-based PPP carbons (see below for Yamamoto method), which was obtained by the heat-treatment at the same temperature (280 mAh/g) [16]. To enhance the cell capacities in Li-ion batteries, different types of low-temperature carbonization and non-graphitizing carbons have been extensively investigated for anode materials such as meso-carbon microbeads [17] and phenolic resin [18]. A high excess of Li storage capacity (corresponding to LiC_3) in the range of 550–700 mAh/g has been reported [19]. However, the Li storage mechanisms in such kinds of disordered carbonaceous systems are not yet understood well, and the mechanisms might be largely affected by the structure of the low HTT forms of carbons. These promising Li storage properties for anode materials of Li-ion batteries have been considered to depend on the size of the defective carbon networks, the amount of hydrogen bonded to the periphery of carbons, and the coexisting nanopores. In such defective structures, Li is considered to be stored as LiC_2, to form nanoclusters in the pores and to be absorbed on a single layer of carbons.

Figure 3 shows the charge/discharge profiles for PPP-based carbon, which was heat-treated from 700 to 3000 °C. Two different PPPs have been used as precursors, and synthesized by the Kovacic- and Yamamoto-methods [20, 21]. PPP from the Kovacic method is prepared through a polymerization of benzene or of a

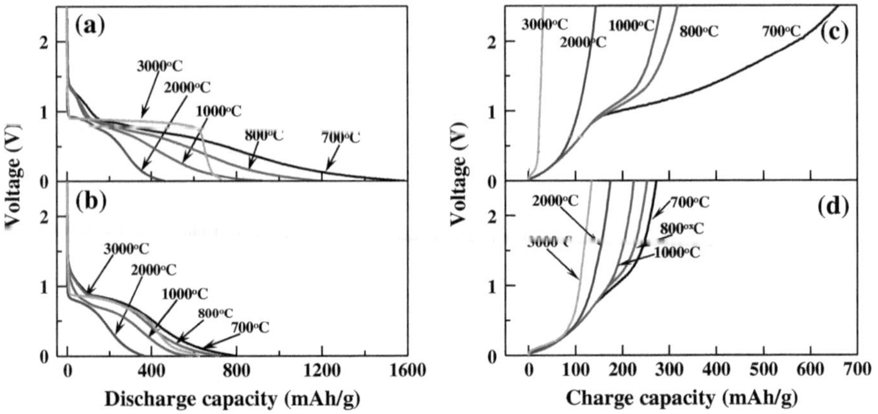

Fig. 3 Discharge (**a**, **b**) and Charge (**c**, **d**) curves of poly(para-phenylene)-based carbon, which was synthesize by Kovacic-(**a**, **c**) and Yamamoto-method (**b**, **d**) as a function of HTT

benzene-derived reagent by using an $AlCl_3$–$CuCl_2$ catalyst. The polymerization is promoted by the formation of benzene radicals. Therefore, the procedure leads to the formation of long polymer chains with some structural defects such as branching in ortho- or meta-positions. The Yamamoto-method PPP is prepared from 1,4-dibromobezene by using a Grignard reagent and a transition metal catalyst, resulting in a shorter polymer chain. In carbonization by heat treatment, Kovacic-method-based PPP has provided higher yields, which is advantageous to reduce the production cost.

Kovacic-method-based PPP carbons at the HTT below 2000 °C show a larger Li insertion capacity than that of Yamamoto-method-based PPP carbons. On the other hand, for the HTT above 2000 °C, Yamamoto-method-based carbons show a larger battery capacity than that of Kovacic-method-based PPP carbons. These results might be related to that Kovacic-method-based PPP carbons at the HTT below 2000 °C have more porous structure than Yamamoto-method-based PPP carbons. As for anode materials in Li-ion batteries, the porous texture of Kovacic-method-based PPP carbons could be useful for penetration of electrolytes and Li ions. Furthermore, Kovacic-method-based PPP carbons heat-treated at 700 °C show the largest insertion capacity. This might be related to the quinoid type of plate-like graphene structure and the homogeneously developed disordered carbon structure. Propylene carbonates (PC) are not likely to be proper for both types of PPP carbons heat-treated over 2000 °C. During the first discharging of both samples heat-treated at 3000 °C, irreversible capacity and long plateaus are observed at about 0.8 V (SEI formation).

It is reported that polyaramid-based carbon, which is obtained from Kevlar (i.e., poly(*p*-phenylene terephthalamide)) [22, 23], can give an additional profit from the viewpoint of manufacturing process. That is, stabilization step to maintain the morphology of polyaramid fibers through carbonization is not required. Advantageously as anode materials, furthermore, the porous texture of polyaramid-based activated carbon fibers (ACFs) is large but the pore sizes are restricted to be very small, i.e., micropores, which are expected to play a special role in intercalation and deintercalation of Li ions. Figure 4 shows the galvanostatic charge/discharge voltage profiles of Kevlar fibers which have treated by single and two-step carbonization processes. Both carbons exhibit a higher charging capacity than graphite (372 mAh/g) and the charge/discharge behavior similar to that usually observed in nongraphitic carbons. The results obtained for the carbon fibers prepared with an intermediate heating step are proved to be better than that of samples without an intermediate heating step. Particularly, their discharge capacity at near 0 V (vs. Li/Li^+) is notably high. Besides, their reversible capacity is higher than that of the fibers prepared by the single-step pyrolysis process. The introduction of an intermediate isothermal step in the pyrolysis process leads to both higher carbon yields and higher amounts of micropores. The latter feature can justify the higher Li insertion capacity of the fibers prepared by the two-step pyrolysis process [24].

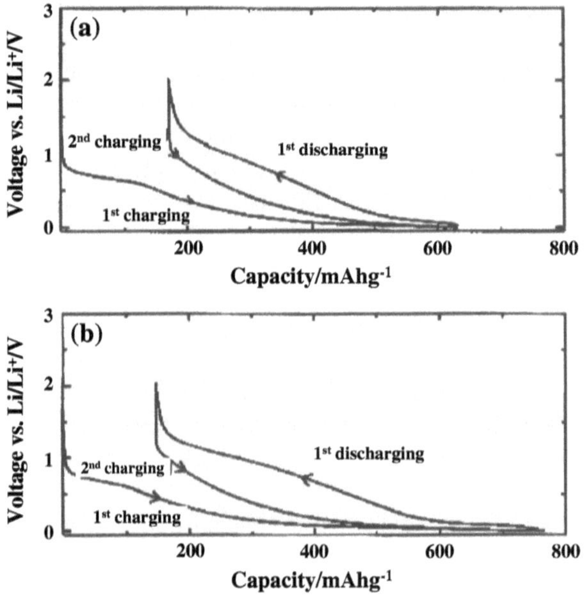

Fig. 4 Galvanostatic charge/discharge voltage profiles of Kevlar-derived carbon fibers prepared **a** in a single step and **b** in two steps with an intermediate heating at 410 °C. Reproduced from Ref. [23]

3.1 Graphitic Carbons

Mesophase pitch-based carbon fibers (MPCFs) have widely been used as a filler for carbon-carbon composite due to their excellent mechanical properties such as high strength, modulus, etc. MPCFs have anisotropic structure, and it contributes to mechanical, electrical, magnetic, thermal as well as chemical properties. These anisotropies are directly related to the layered structure with strong intralayer interactions and very weak van der Waals interplanar interactions between adjacent graphene sheets. As a functional material, MPCFs with anisotropy-originating chemical and physical properties have also been utilized for anode materials for Li-ion batteries [25, 26]. The anode performance of synthetic carbons and graphitic carbons in Li-ion batteries depends strongly on the precursor materials and the synthesis conditions. Graphitic carbons have been attractive due to their good irreversibility on repetitive charge/discharge cycles and the high amount of capacity close to the theoretical value of LiC_6. Figure 5 shows the typical charge-discharge profiles of the graphitized MPCFs (a, b) and the artificial graphite electrode (c, d) at a low rate of 0.25 mA/cm^2. The charge profile for the graphitized MPCFs has one short upper potential plateau at 0.2 V and two long potential plateaus at 0.1 and 0.07 V. The reversible capacity and the coulombic efficiency of the graphitized MPCFs for the first cycle were 303 mAh/g and 94.5 %. The irreversible capacity for the first cycle was 17.6 mAh/g. The reversible capacity and the efficiency during the second cycle were 305 mAh/g and 99.0 %. As for the graphite, plateaus were clearly observed as indicated by arrows. The capacity and the efficiency for the first cycle were 359 mAh/g near $x = 1$ in Li_xC_6 and 91 %. The irreversible capacity was

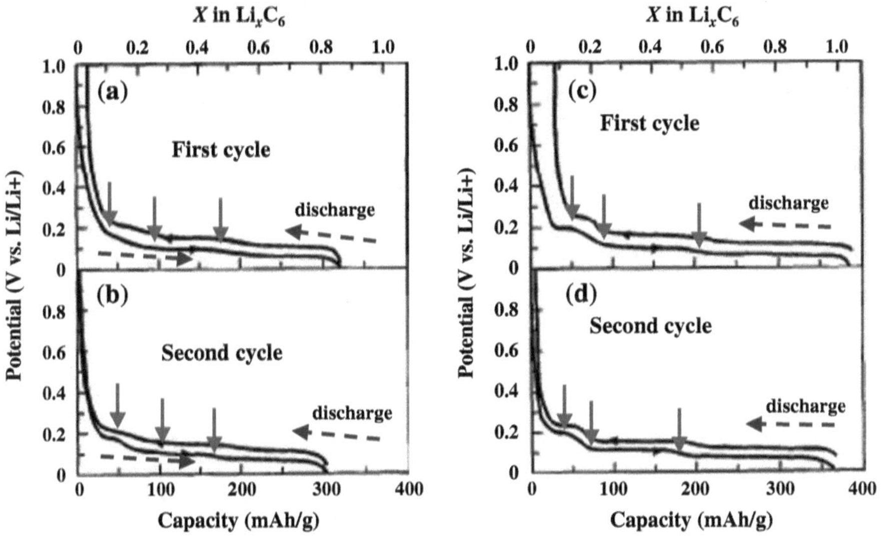

Fig. 5 Charge-discharge profiles of graphitized mesophase pitch-based carbon fibers at 3000 °C and artificial graphite. **a**, **c** were obtained at the first cycle. **b**, **d** were obtained at the second cycle. Reproduced from Ref. [26]

33 mAh/g. The reversible capacity was 360 mAh/g and the efficiency was 97.9 %. Comparing the both electrodes, the maximum reversible capacity of the graphite is larger than that of the graphitized MPCF due to the higher crystallinity. However, the efficiencies for the graphitized MPCF are higher than those of the graphite. Furthermore, MPCFs have excellent cell stability on the repetitive charge/discharge cycles, which maintains 86 % of its initial capacity even after 400 cycles [26].

As another representative graphitic carbon, mesocarbon microbeads (MCMBs) has been intensively investigated [17, 27], due to their comparable performance with natural graphite [28]. Figure 6 shows the discharge curves (a) of the MCMBs heat-treated at the high temperature ranging from 2000 to 2800 °C (200 °C increment for each sample) and (b) of the MCMBs treated at the HTTs of 1000, 2000, 2300 and 2800 °C. The discharge capacity increases with increasing the HTT, and shows longer plateau below 0.25 V. Ohzuku et al. [28] have reported that the electrochemical deintercalation of Li in natural graphite proceeds below *ca.* 0.25 V. This suggests that the charge reaction of MCMBs below 0.25 V in EC + DEC is the deintercalation of Li from the graphitic structure with the *AB* stacking order. It is reasonable to consider that the graphitic stacking order has an effect on the Li intercalation reaction. As for Li intercalation, the stacking of grapheme layers along the c axis becomes *AA* stacking. However, the change to the *AA* stacking will not be allowed in turbostratic structure because of the crystal defects and residual strains. In graphitic carbons, thus, it is considered that the crystallites of the graphitic stacking structure give the charge capacity only in the potential range of 0–0.25 V.

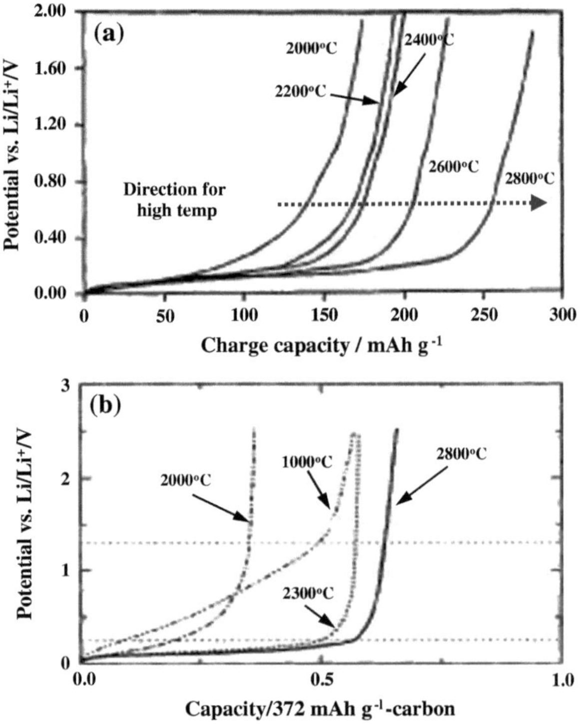

Fig. 6 Charge-discharge profiles of mesocarbon microbeads prepared at 2000–2800 °C (**a**), typical potential change of the mesocarbon microbeads by the heat-treatment temperature (**b**). The capacity is normalized by the theoretical capacity of LiC_6. Reproduced from Refs. [17, 27]

Since Li-ion batteries were commercialized by Sony, their energy density has been improved by ca. 10 % every year to reach 2.5 times higher value than that of the first commercial cell. Among the diverse carbon materials as anode candidates, Hitachi Chemical Co. Ltd. [29–31] has developed a remarkable material, i.e., massive artificial graphite (MAG), for the anode in Li-ion battery, which has realized the high capacity (362 mAh/g) comparable to the theoretical value of LiC_6 (372 mAh/g). Some plateaus have been observed at 190, 95 and 65 mV in the charge and at 105, 140 and 230 mV in the discharge, which imply the formation of graphite intercalation compounds (GICs) by Li-intercalation and -deintercalation. MAGs have spherical shape and isotropic structures with high crystallinity, which can contribute to energy density. Compared to the other graphitic materials such as natural graphite and graphitized mesophase, MAGs show good stability of the discharge capacity with the increase of current density and electrode density. The stability depending on current density and electrode density is attributed to pores existing at the inside of the particles. The spaces inside the particles can be expected to play a role of moderating the displacement of graphite lattice during charge and discharge (insertion/desertion of Li ions).

3.2 Heteroatom Incorporated Carbon Electrodes

Carbon materials, especially with low thermal history, usually include some impurities. Heteroatoms also can be considered as the one of impurities. Importantly, many results have reported that heteroatoms can be introduced into the carbon lattice intentionally, which is defined as "doping" [32, 33]. Doping has been used extensively in the past [34], mainly in order to change the distribution of electrons between energy levels in the carbon materials, as well as to affect the graphitization process and to modify the chemical state of the surface of the carbon materials. The doping with boron (B), nitrogen (N) and phosphorous (P) has mainly been investigated for graphite materials. Particularly, B-doped carbon materials have been experimentally and theoretically investigated not only from fundamental scientific aspects (e.g., electronic properties) but also for potential applications such as high temperature oxidation protectors for carbon/carbon (C/C) composite and anode materials of Li-ion batteries. B-doping generates electron acceptor levels [35, 36], so that the capacity has been expected to be enhanced. There have been many reports about the preparation methods of B-doped carbons by co-deposition of B-containing organic molecules in chemical vapor deposition (CVD).

Figure 7 describes the typical voltage profiles of the second discharge and charge for the B-doped graphite cells. The samples were prepared from a mixture of the

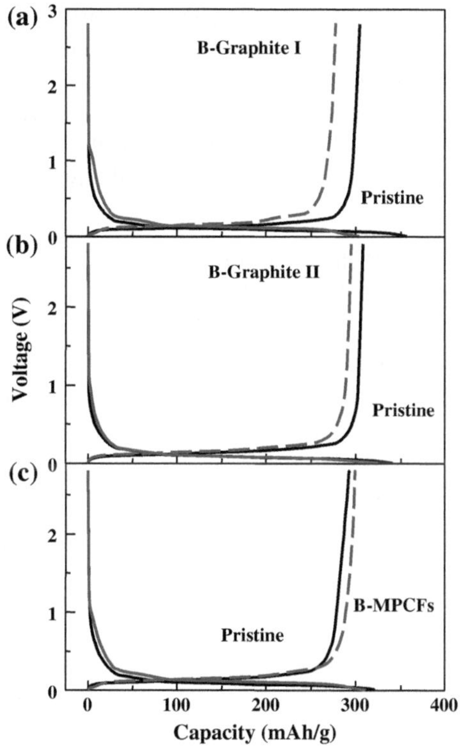

Fig. 7 Change in potential during the second discharge and charge cycle of the graphitized and boron doped samples for various types of graphite host. Reproduced from Ref. [6]

pristine material and boron carbide by heat-treatment at 2800 °C in argon atmosphere. The long plateaus below 0.2 V correspond to the reversible intercalation of Li in the graphitized pristine and the B-doped samples. It should be noted that the second discharge/charge capacities of B-doped graphitizable carbon I (B-Graphite I, Fig. 7a) and graphitizable carbon II (B-Graphite II, Fig. 7b) were slightly lower than that of the pristine graphite. However, in the case of B-doped MPCFs, the second charge capacity is larger than that of the undoped pristine (Fig. 7c). The reduced charge capacity of the B-doped samples might be related to boron atoms occupying the Li insertion active sites, such as edge-type sites in the graphite layers, which would inhibit the Li insertion process. In the discharging cycle for B-doped samples, negligible shoulder plateaus are characteristically observed at about 1.3 V, which may be caused by the induction of an electron acceptor level, so that Li insertion yields a higher voltage compared to undoped samples [36]. It is interesting that the irreversible capacity loss for some B-doped samples is lower than that of the corresponding undoped samples. These results might be related to the redistribution of the Fermi level of the B-doped samples, which is lowered by B-doping, i.e., by the introduction of an electron acceptor to the lattice.

The contents of the B atoms doped have been investigated by X-ray photoelectron spectrometry (XPS). Figure 8 shows boron 1s (B_{1s}) peak at higher resolution. The B_{1s} peak appears in the B-doped samples, although the peak position and shape are different depending on the samples. Particularly, the B_{1s} peak of B-Graphite I was split to three peaks at 185.6, 187.7, and 189.8 eV, which were assigned to the binding energies (BE) originating from boron carbide and the boron clusters substituted in graphite plane and incorporated with N atoms, respectively. From these results, the appearance of the B_{1s} peak of B-Graphite I near 190 eV corresponds to the substitutionally incorporated B atoms into graphite lattice, which preferentially make bonding with N atoms existing in the heat-treatment atmosphere. It is also possible that the residual N atoms in the raw materials react with

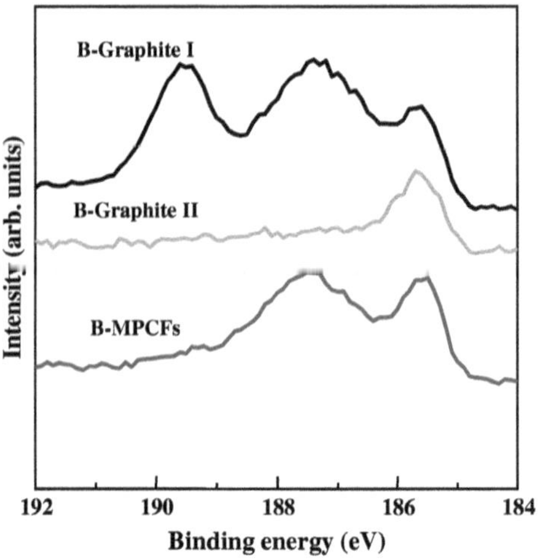

Fig. 8 High resolution B_{1s} spectra by XPS of two different types of boron-doped graphite and mesophase pitch carbon fibers. Reproduced from Ref. [6]

the boron carbide during the carbonization step and then form the boron nitride and/or BC_xN compounds during the graphitization step. These phenomena should be taken into consideration for industrial process using Acheson-type furnace. Consequently, the degradation of the Li insertion capacity observed in some kinds of B-doped graphite might be highly related to the presence of B atoms in the form of boron nitride and boron carbide. Also, the unexpected opposite effects of B-doping could be related to the heterogeneous growth of the crystallites dimension, La, due to the boron acting as graphitization catalyst.

In order to demonstrate topological variation by B-doping, surface analysis by scanning tunneling microscopy (STM) has been carried out [37]. Figure 9 shows the contrast between the STM images of a three-dimensional surface plot with a scan range of 5 nm and its sectional analysis of the pristine HOPG (a) and B-doped HOPG (b) graphene surface. The surface graphene layer of the pristine HOPG with a trigonal lattice shows a perfect superstructure with *ABAB* stacking as observed in graphite. At the B-doped HOPG graphene surface, four substituted B atoms with the highest intensity of the electron groups (Ⓐ–Ⓓ) are found in the image. Each bright area consists of the B atoms with the highest electron density located in the center of the six surrounding medium-intensity site that corresponds to carbon atoms. Also, the electron density distributions of a substituted B atom and the surrounding six carbon atoms appear in the three-dimensional surface plot (the inset of Fig. 9b). The substituted B atom clearly shows the highest electron density in the center. The six carbon atoms closest to the B atom also show a higher electron density than the next neighbor β-site carbon atoms. This indicates that the substituted B atoms affect the electronic structure of the adjacent six β-site carbon atoms. The substitution of B atoms, which are in an electron-deficient state compared to the

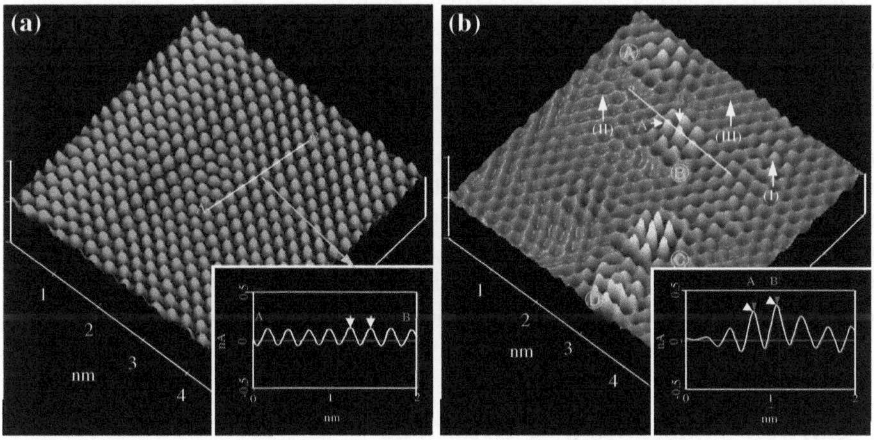

Fig. 9 Scanning tunneling microscopic images of three-dimensional surface plot (**a**) and its section analysis (**b**) of the boron-doped highly oriented pyrolytic graphite surface. Reproduced from Ref. [37]

carbon host, produces vacancies at the top of the valence π-band, resulting in an increase in the density of states near the Fermi level. Due to the enhancement of the density of states (DOS), the boron sites become brighter than the surrounding sites in the STM lattice image. The effect is not restricted to the boron sites themselves but extends to the surrounding carbon sites due to the delocalized nature of the boron-induced defect.

As shown in the inset of Fig. 9a, the closest distance between the β-site carbon atoms observed in the section analysis of the STM image is measured to be 0.246 nm for the pristine HOPG. The value is similar to the a_0 spacing in the graphite lattice. On the other hand, the distance between the boron and carbon atoms located at the β-sites is 0.276 ± 0.005 nm, as is shown in the inset of Fig. 9b. The distance is slightly longer than the corresponding distance of 0.246 nm in the pristine HOPG. Turbostratic structures are found at the surface of B-doped HOPG. As can be seen in Fig. 16b, some regions show a three-dimensional trigonal image (spot image) of the superstructure (Fig. 9b-I) consistent with *ABAB* stacking. Some regions show hexagonal (Fig. 9 b-II) or linear (Fig. 16b-III) images, indicating a turbostratic stacking [38]. Thus, the substitution of B atoms in the hexagonal network affects the stacking nature of the host material presumably because of the lattice defects and strain associated with the B substitution.

Figure 10 shows a schematic model for B-substituted graphite. The average distance between the boron and closest carbon (C_1) atoms at β-sites is 0.276 ± 0.005 nm, as is described above. The bond distance between the boron and adjacent carbon atoms is calculated to be 0.159 nm, and for the β-site the B–C distance is 0.276 nm. The distance of C_1–C_1 measured in the STM is also 0.276 nm. The C–C bond distance at the next β-site is evaluated to be the same as in HOPG (0.246 nm). Therefore, the substituted B atom might be located at a slightly higher position than the surrounding C atoms in the basal plane of HOPG. The substitution of boron should slightly deform the flatness of basal plane. Hach et al. reported [39] the similar results that the bond distance between B and C atoms is 0.154 nm for $B_2C_4H_6$ and 0.152 nm for $B_2C_{52}H_{18}$. By restricting the bond length between the B–C atoms obtained from Fig. 9b, our group has simulated the optimized structure of a graphene sheet. As shown in Fig. 10b, an improper torsion angle (B at the apex) is calculated to be 164°, while the original plane is almost flat with an angle of 179°. It has been reported that the electronic structure is also modified largely by B-doping, as shown by susceptibility measurement [40]. Thus, both atomically and electronically the graphite planes could be tailored by B-doping, which can contribute to modifying and controlling the properties of graphite.

4 Carbon Nanotubes and Graphene

Carbon nanotubes (CNTs), which can be regarded as a representative of nano-carbon architectures, have received much attention since the Iijima's report of 1991 [41]. The interesting new allotropes of carbon, CNTs, are attractive materials

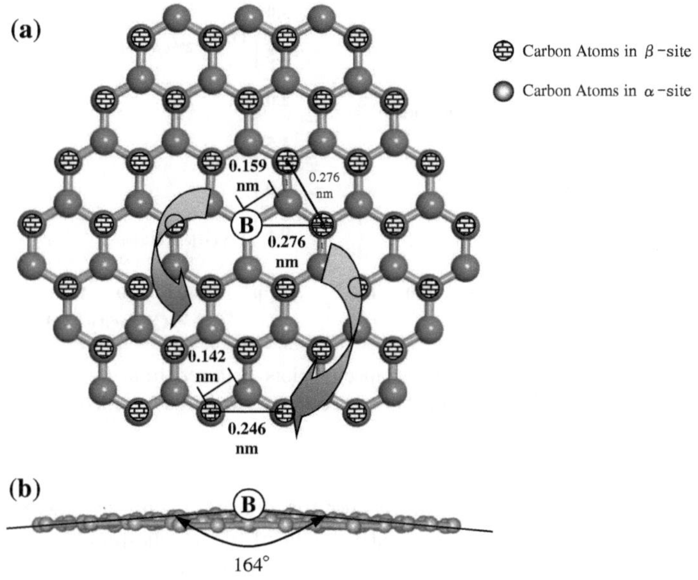

Fig. 10 Schematic models of the *top* (**a**) and *side* (**b**) view for a boron-substituted graphene sheet based on the measured dimensions of B–C_1 and C_1–C_1. Reproduced from Ref. [37]

for Li-ion batteries. In fact, the application widely extends to additive materials enhancing the battery performances in cathode and anode. In this section, the possibility of CNTs to batteries, including the practical application, is discussed.

4.1 Active Material of Electrodes

4.1.1 SWCNTs and MWCNTs

Feasibility of high Li capacity in battery application has been suggested and investigated by many researchers. If all the interstitial sites (inter-shell van der Waals spaces, inter-tube channels, and inner hollows) were accessible for Li intercalation, CNTs could achieve tremendous amount of Li storage. CNTs can be classified into two categories of SWCNTs and MWCNTs. Electrochemical intercalation of MWCNTs [42–56] and SWCNTs [57–66] has been actively investigated. The CNTs can be regarded as promising materials in Li-ion batteries since the large number of nanoscale sites for intercalation exceed those of the commonly used graphite electrodes. CNTs generally have bundle structure due to Van der Waals force existing between their graphene layers. Intertubular vacancies (typically denoted by B) can serve as the space for Li ion storage. This becomes a clue to discuss the charge/discharge curves different from those of graphitic carbons.

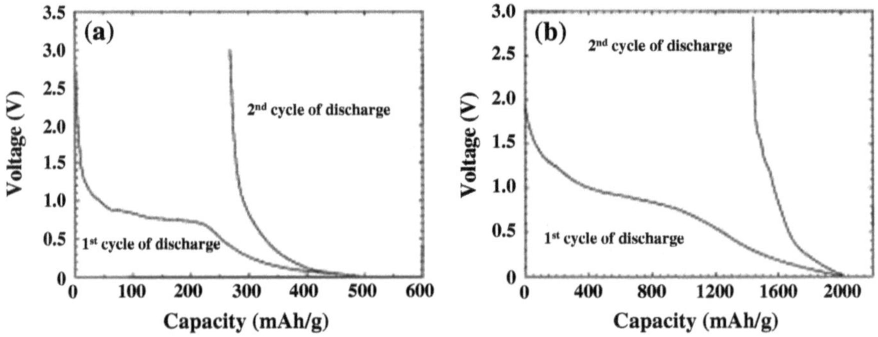

Fig. 11 Voltage versus specific lithium capacity plots obtained from the pristine MWCNTs (**a**) and purified SWCNTs (**b**). The both data were collected under galvanostatic mode at 50 mA/g. Reproduced from Ref. [60]

Figure 11 shows representative curves for the charge/discharge of CNTs with Li ions. Figure 11a shows the representative data of electrochemical intercalation on arc-discharge MWCNTs, which were used without purification [66]. 1M solution of $LiClO_4$ in 1:1 (volume fraction) of EC (ethylene carbonate) and DMC (dimethyl carbonate) was used as an electrolyte. After a long plateau around 0.75 V, the voltage gradually dropped with the further increase of the Li insertion. The voltage profile similar to that of non-graphitic carbon suggests that there is no stage transition as for graphite. Total amount of Li insertion was 500, and 250 mAh/g was obtained by Li deinsertion. There was no drastic decrease of charge amount after the second cycle. In some reports, a reversible capacity (C_{rev}) of 100–640 mAh/g has been observed, which depends on the sample processing and annealing conditions [65, 66]. In general, well-graphitized MWNTs, such as synthesized by arc-discharge methods, have a lower C_{rev} than those prepared by CVD methods. Structural studies [67, 68] have shown that alkali metals can be intercalated into the inter-shell spaces within the individual MWNTs through defect sites. Figure 11b shows typical charge/discharge profiles of SWCNTs, which indicate higher reversible and irreversible capacities than those of MWCNTs. SWCNTs have shown the remarkably high capacity of 2000 mAh/g ($Li_{5.4}C_6$) in fully lithiated state and the C_{rev} of 600 mAh/g ($Li_{1.6}C_6$). Ball-milling process is considered to be an effective method to improve the crucial difference of capacity between the charge and discharge of CNTs. Gao et al. [60] have reported the electrochemical properties of ball-milled SWCNTs for Li ions. Figure 12 shows the charge/discharge curve obtained from the ball-milled and then purified SWCNTs. Note that the C_{rev} has increased by 50 % to reach 1000 mAh/g. No drastic deterioration of capacity has been observed by repetitive charge/discharge cycles. The mechanical process can induce disorder in structure and cut strands of tubes. These structural and morphological variations would be at least partially related to the kinetics of the intercalation reaction, although the exact mechanism is not understood well.

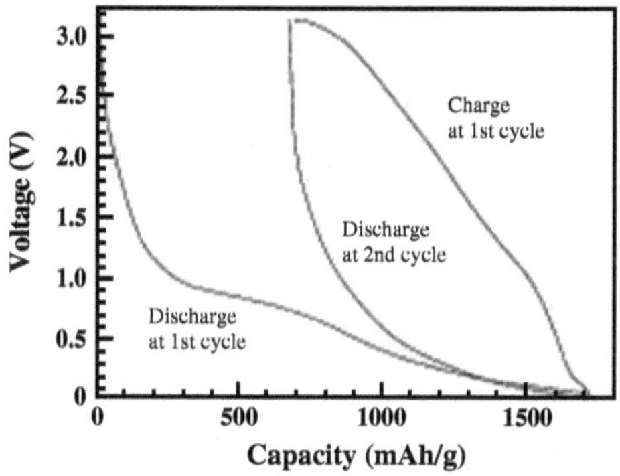

Fig. 12 The charge/discharge characteristics of purified SWCNTs after ball-milling. The reversible capacity increased by 50 % to 1000 mAh/g, and the irreversible part reduced to 600 mAh/g. Reproduced from Ref. [60]

4.1.2 Double-Walled Carbon Nanotube (DWCNTs)

Double-walled carbon nanotubes (DWCNTs) consist of two concentric graphene cylinders. DWCNTs, categorized different from both MWCNTs and SWCNTs, are expected to exhibit mechanical and electronic properties superior to SWCNTs. Endo et al. have suggested that DWCNTs have potential utility in the various application fields [69–71]. DWNTs are expected to be more advantageous than SWNTs due to the structural stability and different pore structure originating from the features on bundling formation [70]. The pore structure characteristic of DWCNTs can offer a much better adsorption field for H_2 than SWNTs [72]. Furthermore, a thin paper comprised of DWCNTs could be an excellent candidate of anode materials, because it offers a short path for Li ion transfer to improve rate capability. Figure 13 shows a photograph of the DWCNT paper used as an electrode (a), scanning electron microscopy (SEM) and transmission electron microscopy (TEM) images of a SWCNT paper (b, d), SEM and TEM images of a DWCNT paper (c, e), and a high resolution TEM image of DWCNTs (f). The fabrication of each nanotube paper is so easy that no binder is used. Figure 14 shows voltage-capacity curves obtained from the SWCNT and DWCNT papers. Note that Li^+ insertion into carbon materials takes place during discharging process, because the electrode of the carbon materials is the cathode, and the counter electrode corresponds to the Li metal. The features of their voltage profiles, which monotonically decreased, are very similar to non-graphitic carbons. Although both samples have very large Li-ion storage capacities (SWCNT; close to 2000 mAh/g, DWCNT; *ca.* 1600 mAh/g) above 1500 mAh/g, the C_{rev} values are 510 and 300 mAh/g for SWCNTs and DWCNTs, respectively. Such a big difference between charge and discharge near 1000 mAh/g is attributed to electrolyte decomposition and the formation of solid-electrolyte interphase (SEI). This is also supported by the long plateau at *ca.* 0.9 V observed for both samples in the first discharge.

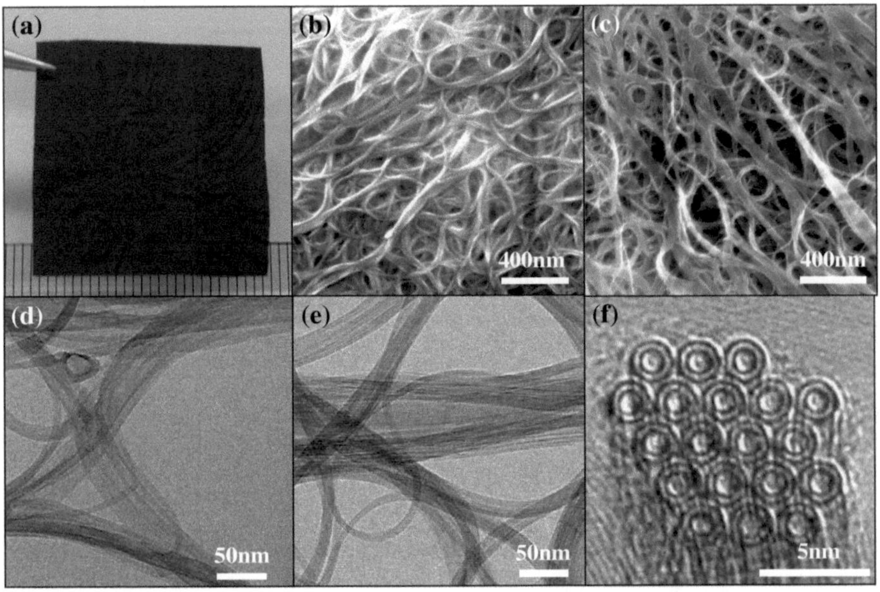

Fig. 13 **a** SEM images of SWCNT and DWCNT papers (**b**, **c**), Low-resolution TEM images of SWCNT and DWCNT papers (**d**, **e**). It is noteworthy that both samples exist as structures with large-sized bundles, cross-sectional images by high-resolution TEM, which are stacked in a hexagonal array (**f**). Reproduced from Ref. [71]

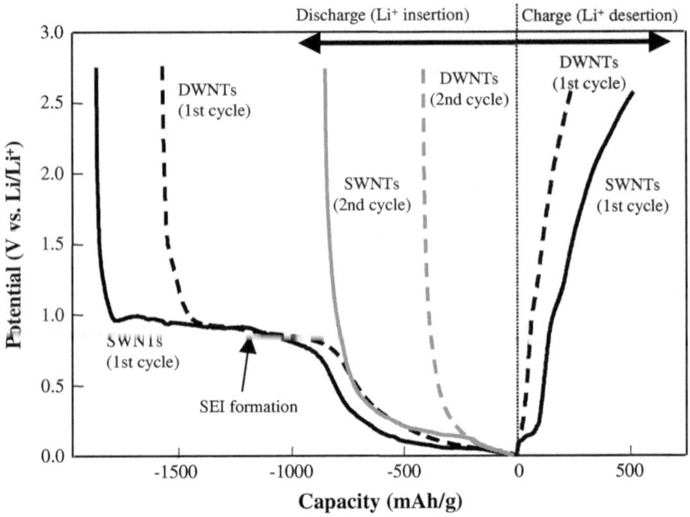

Fig. 14 Voltage profiles of highly pure and bundled single- and double-walled carbon nanotubes for discharging and charging cycles. Reproduced from Ref. [71]

4.2 Additives to Anode and Cathode for Lithium Ion Battery

One-dimensional structure of CNTs makes it easy to form a net-work by the entanglement due to their relatively large aspect ratios. Furthermore, the structural integrity and its excellent electric conductivity make CNTs promising conducting materials. The problematic drawback that CNTs have suffered from is high production cost originating from low production efficiency, which seriously inhibits their industrial application. However, SDK (Showa Denko K.K.) shows one solution by accomplishing the mass production of MWNTs. MWNTs have been synthesized by the decomposition of hydrocarbons, such as benzene and methane, using transition metal particles as a catalyst at a growth temperature of 1000–1300 °C [73–75]. The mass-produced but high-quality MWNTs (i.e., commercial products are graphitized) have one-dimensional morphology with highly preferred orientation of the graphitic basal planes parallel to the fiber axis, which gives rise to excellent mechanical properties and electrical and thermal conductivity. Therefore, MWNTs have high possibility of application to fillers to both anode and cathode material in Li-ion batteries by utilizing their high conductivity and surface-to-volume ratios. In fact, MWNTs have been applied to commercial Li-ion batteries. To achieve maximum battery performance, generally, electrodes used in a battery system are required to possess sufficiently high electrical and thermal conductivity, mechanical strength enough to sustain volume changes during the charge and discharge processes, and favorable penetration of the electrolyte. Therefore, in the case that VGCFs are used as electrode fillers, there are necessities not only to evaluate the basic properties of the nanotube itself but also to characterize the packed state of the nanotubes. For example, Fig. 15a shows the carbon anode sheet made up of synthetic graphite and MWNTs,

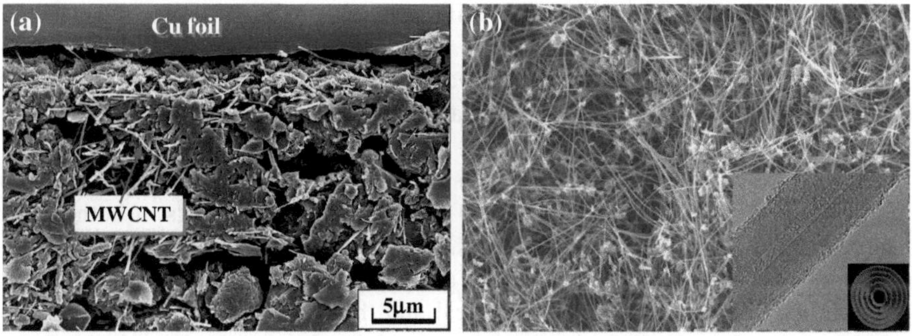

Fig. 15 **a** SEM image for carbon anode sheet in a commercial cell, in which the carbon sheet was made up of synthetic graphite. **b** SEM image of MWCNTs. An inset at the photo on the *right bottom* shows the TEM photograph of single MWCNTs and the model based on computational simulation. Reproduced from Ref. [76]

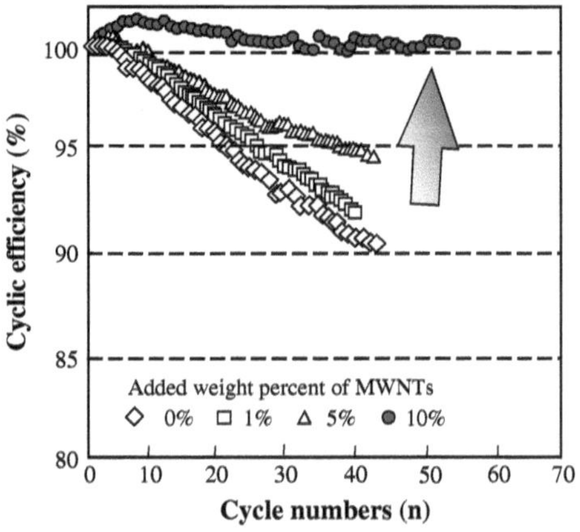

Fig. 16 Cycle efficiency of synthetic graphite (HTT = 2900 °C) as a function of the added weight percent of MWCNTs (cycle characteristic of 0.2 mA/cm^2 at 0–1.5 V). Reproduced from Ref. [76]

which is used in a commercial cell. As seen clearly, the MWNTs interconnect well the synthetic graphite powders with each other, contributing to the improvement of electrical conductance and the reinforcement of the electrode. A typical SEM image of MWNTs is shown in Fig. 15b. The inset in Fig. 15b shows a TEM image of a single MWNTs and a drawing of its computational cross-section model.

The cyclic efficiency of a synthetic graphite (HTT = 2900 °C) anode as a function of added amount of MWNT is shown in Fig. 16 [76]. As the addition of MWNTs increases, the cyclic efficiency of the synthetic graphite anodes increases continuously. Particularly, addition of 10 wt% MWNTs has maintained almost 100 % of cycle efficiency up to 50 cycles. At the high addition percentage, graphite powder particles are interconnected by MWNTs to form a continuous conductive network. Thus, the addition of MWNTs to anode materials improves the conductivity of the anode. In addition, MWNTs contributes to absorb and retain significant amounts of electrolyte and to provide resiliency and compressibility to the electrode structure. Therefore, the use of MWNTs as additives enhances the performance of Li-ion batteries. Furthermore, as compared to conventional whiskers, relatively high capability of VGCFs for Li-ion intercalation would also be beneficial for battery performance.

The effectiveness of carbon nanotubes as conductive filler to cathode of lithium ion batteries (Fig. 17) was demonstrated by adding small amounts of both carbon nanotubes and acetylene blacks to $LiCoO_2$-based active materials [77]. The merits of using carbon nanotubes together with acetylene blacks as cathode fillers include not only the enhancement of the electrical and the thermal properties of the electrode, but also the enhancement of the density of the electrode and the shortening of the electrolyte absorption time. We envisage that the use of carbon nanotubes as multi-functional fillers will increase in both cathode and anode materials for lithium ion secondary batteries.

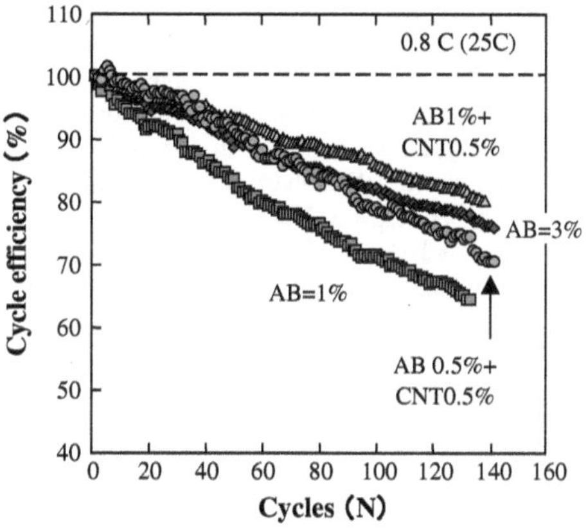

Fig. 17 Cycle performance of the cathode having different amount of carbon blacks and MWCNTs. Reproduced from Ref. [77]

5 Conclusions

Modern society with the drastic development of mobile device and zero emission vehicles has required further thin, light and small energy devices with higher energy and power densities. Such demands will continue to generate many research activities for the development of new cell materials and configurations. As was mentioned in the above sections, many companies have developed the anode and cathode materials with new concepts. This review has focused on carbon host materials for anodes in all configurations of cells. The important thing is that the effort to improve the battery performance should be made by multidisciplinary approaches such as organic and inorganic chemistry, physics, surface science and corrosion. Through the comprehensive study we can expect significant improvements in energy density. The energy density of batteries is generally limited by the low density of host materials and their defects. Fuel cells are expected to overcome such a limited energy problem in the future. For the time being, however, Li-ion batteries will have played an important role in high energy systems.

References

1. Brandt K (1994) Historical development of secondary lithium batteries. Solid State Ionics 69:173–183
2. Delhaes P, Manceau KP, Guerard D (1980) Physical properties of first and second stage lithium graphite intercalation compounds. Synth Met 2:277–284
3. Kang K, Meng YS, Breger J et al (2006) Electrodes with high power and high capacity for rechargeable lithium batteries. Science 311:977–980
4. Takada K, Ohta N, Zhang L et al (2008) Interfacial modification for high-power solid-state lithium batteries. Solid State Ionics 179:1333–1337
5. Wen Z, Huang S, Yang X et al (2008) High rate electrode materials for lithium ion batteries. Solid State Ionics 179:1800–1805

6. Endo M, Kim C, Nishimura K (2000) Recent development of carbon materials for Li ion batteries. Carbon 38:183–197
7. Oberlin A (1989) TEM studies of carbonization and graphitization. In: Thrower PA (ed) Chemistry and physics of carbon, vol 22. Marcel Dekker, New York, pp 1–143
8. Imanishi N, Takeda Y, Yamamoto O (1998) Development of the carbon anode in lithium ion batteries. In: Wakihara M (ed) Lithium ion batteries: fundamentals and performance. Wiley-VCH, New York, pp 1–98
9. Edwards IAS (1997) Structure in carbons and carbon forms. In: Marsh H (ed) Introduction to carbon science. Butterworth-Heinemann, Portland, pp 1–36
10. Sato K, Noguchi M, Demachi A et al (1994) A mechanism of lithium storage in disordered carbons. Science 264:556–558
11. Fujimoto H, Mabuchi A, Tokumitsu K (1995) Irreversible capacity of lithium secondary battery using meso-carbon micro beads as anode material. J Power Sour 54:440–443
12. Dahn JR, Zheng T, Liu Y et al (1995) Mechanisms for lithium insertion in carbonaceous materials. Science 270:590–593
13. Yazami R, Munshi MZA (1995) Novel anodes for solid state batteries. In: Munshi MZA (ed) Handbook of solid state batteries and capacitors. World Scientific, Singapore, pp 425–460
14. Zhou P, Papanek P, Lee R et al (1997) Local structure and vibrational spectroscopy of disordered carbon for Li batteries: neutron scattering studies. J Electrochem Soc 144:1744–1750
15. Endo M, Nishimura Y, Takahashi T et al (1996) Lithium storage behavior for various kinds of carbon anodes in Li ion secondary battery. J Phys Chem Solids 57:725–728
16. Endo M, Kim C, Hiraoka T et al (1998) Li storage behavior in polyparaphenylene (PPP)-based disordered carbon as a negative electrode for Li ion batteries. Mol Cryst Liq Cryst 310:353–358
17. Mabuchi A, Tokumitsu K, Fujimoto H (1995) Charge-discharge characteristics of the mesocarbon miocrobeads heat-treated at different temperatures. J Electrochem Soc 142:1041–1046
18. Zheng T, Zhong Q, Dahn JR (1995) High-capacity carbons prepared from phenolic resin for anodes of lithium-ion batteries. J Electrochem Soc 142:L211–L214
19. Matsumura Y, Wang S, Mondori J (1995) Interactions between disordered carbon and lithium in lithium ion rechargeable batteries. Carbon 33:1457–1462
20. Kovacic P, Kyriakis A (1963) Polymerization of benzene to p-polyphenyl by aluminum chloride-cupric chloride. J Am Chem Soc 85:454–458
21. Yamamoto T, Hayashi Y, Yamamoto A (1978) A novel type of polycondensation utilizing transition metal-catalyzed C–C coupling I: preparation of thermostable polyphenylene type polymers. Bull Chem Soc Japan 51:2091–2097
22. Villar-odil S, Suarez-garcia F, Paredes JI et al (2005) Activated carbon materials of uniform porosity form polyaramid fibers. Chem Mater 17:5893–5908
23. Ko KS, Park CW, Yoon SH et al (2001) Preparation of Kevlar-derived carbon fibers and their anodic performances in Li secondary batteries. Carbon 39:1619–1625
24. Zheng T, Xing W, Dahn JR (1996) Carbons prepared from coals for anodes of lithium-ion cells. Carbon 34:1501–1507
25. Kim WS, Chung KI, Lee CB et al (2002) Studies on heat-treated MPCF anodes in Li ion batteries. Microchem J 72:185–192
26. Takami N, Satoh A, Hara M et al (1995) Rechargeable lithium-ion cells using graphitized mesophase-pitch-based carbon fiber anodes. J Electrochem Soc 142:2564–2571
27. Tatsumi K, Iwashita N, Sakaebe H et al (1995) The influence of the graphitic structure on the electrochemical characteristics for the anode of secondary lithium batteries. J Electrochem Soc 142:716–720
28. Ohzuku T, Iwakoshi Y, Sawai K (1993) Formation of lithium-graphite intercalation compounds in nonaqueous electrolytes and their application as a negative electrode for a lithium ion (shuttlecock) cell. J Electrochem Soc 140:2490–2498

29. Yoshito I (2006) Carbon anode material for lithium-ion rechargeable battery. Tanso 225:382–390
30. Ishii Y, Fujita A, Nishida T et al (2001) High-performance anode material for lithium-ion rechargeable battery. Hitachi Chem Tech Rep 36:27–32
31. Ishii Y, Nishida T, Suda S et al (2006) Anode material for high energy density rechargeable lithium-ion battery. Hitachi Chem Tech Rep 47:29–32
32. Lowell CE (1967) Solid solution of boron in graphite. J Am Ceram Soc 50:142–144
33. Kouvetakis J, Kaner RB, Sattler ML et al (1986) A novel graphite-like material of composition BC_3, and nitrogen–carbon graphites. J Chem Soc Chem Commun 24:1758–1759
34. Marchand A (1971) Electronic properties of doped carbons. In: Walker PL (ed) Chemistry and physics of carbon, vol 7. Marcel Dekker, New York, pp 155–191
35. Nakajima T, Koh K, Takashima M (1998) Electrochemical behavior of carbon alloy C_xN prepared by CVD using a nickel catalyst. Electrochim Acta 43:883–891
36. Nishimura Y, Yakahashi T, Tamaki T et al (1996) Anode performance of B-doped mesophase pitch-based carbon fibers in lithium ion secondary batteries. Tanso 172:89–94
37. Endo M, Hayashi T, Hong SH et al (2001) Scanning tunneling microscope study of boron-doped highly oriented pyrolytic graphite. J Appl Phys 90:5670–5674
38. Endo M, Oshida K, Kobori K et al (1995) Evidence for glide and rotation defects observed in well-ordered graphite fibers. J Mater Res 10:1461–1468
39. Hach CT, Jones LE, Crossland C et al (1999) An investigation of vapor deposited boron rich carbon-a novel graphite-like material–part I: the structure of BC_x (C_6B) thin films. Carbon 37:221–230
40. Matthews MJ, Dresselhaus MS, Dresselhaus G et al (1996) Magnetic alignment of mesophase pitch-based carbon fibers. Appl Phys Lett 69:430–432
41. Iijima S (1991) Helical microtubules of graphitic carbon. Nature 354:56–58
42. Nalimova VA, Sklovsky DE, Bondarenko GN et al (1997) Lithium interaction with carbon nanotubes. Synth Met 88:89–93
43. Lee JH, Kim GS, Choi YM et al (2008) Comparison of multiwalled carbon nanotubes and carbon black as percolative paths in aqueous-based natural graphite negative electrodes with high-rate capability for lithium-ion batteries. J Power Sour 184:308–311
44. Frackowiak E, Gautier S, Gaucher H et al (1999) Electrochemical storage of lithium multiwalled carbon nanotubes. Carbon 37:61–69
45. Maurin G, Bousquet Ch, Henn F et al (1999) Electrochemical intercalation of lithium into multiwall carbon nanotubes. Chem Phys Lett 312:14–18
46. Leroux F, Metenier K, Gautier S et al (1999) Electrochemical insertion of lithium in catalytic multi-walled carbon nanotubes. J Power Sour 81–82:317–322
47. Lu W, Chung DDL (2001) Anodic performance of vapor-derived carbon filaments in lithium-ion secondary battery. Carbon 39:493–496
48. Yang Z, Wu HQ, Simard B (2002) Charge-discharge characteristics of raw acid-oxidized carbon nanotubes. Electrochem Commun 4:574–578
49. Frackowiak E, Beguin F (2002) Electrochemical storage of energy in carbon nanotubes and nanostructured carbons. Carbon 40:1775–1787
50. Shin HC, Liu M, Sadanadan B et al (2002) Electrochemical insertion of lithium into multi-walled carbon nanotubes prepared by catalytic decomposition. J Power Sour 112:216–221
51. Chen WX, Lee JY, Liu Z (2003) The nanocomposites of carbon nanotube with Sb and $SnSb_{0.5}$ as Li-ion battery anodes. Carbon 41:959–966
52. Yoon SH, Park CW, Yang HJ et al (2004) Novel carbon nanofibers of high graphitization as anodic materials for lithium ion secondary batteries. Carbon 42:21–32
53. Wang X, Liu H, Jin Y et al (2006) Polymer-functionalized multiwalled carbon nanotubes as lithium intercalation hosts. J Phys Chem B 110:10236–10240
54. Deng D, Lee JY (2007) One-step synthesis of polycrystalline carbon nanofibers with periodic dome-shaped interiors and their reversible lithium ion storage properties. Chem Mater 19:4198–4204

55. Park MS, Needham SA, Wang GX et al (2007) Nanostructured SnSb/carbon nanotube composites synthesized by reductive precipitation for lithium-ion batteries. Chem Mater 19:2406–2410
56. Chen Liu Y, Minett AI et al (2007) Flexible, aligned carbon nanotube/conducting polymer electrodes for a lithium-ion battery. Chem Mater 19:3593–3597
57. Garau C, Frontera A, Quiñonero D et al (2004) Ab initio investigations of lithium diffusion in single-walled carbon nanotubes. Chem Phys 297:85–91
58. Claye AS, Fischer JE, Huffman CB et al (2000) Solid-state electrochemistry of the Li single wall carbon nanotube system. J Electrochem Soc 147:2845–2852
59. Jouguelet E, Mathis C, Petit P (2000) Controlling the electronic properties of single-wall carbon nanotubes by chemical doping. Chem Phys Lett 318:561–564
60. Gao B, Bower C, Lorentzen JD et al (2000) Enhanced saturation lithium composition in ball-milled single-walled carbon nanotubes. Chem Phys Lett 327:69–75
61. Yang ZH, Wu HQ (2001) The electrochemical impedance measurements of carbon nanotubes. Chem Phys Lett 343:235–240
62. Morris RS, Dixon BG, Gennett T et al (2004) High-energy, rechargeable Li-ion battery based on carbon nanotube technology. J Power Sour 138:277–280
63. Ng SH, Wang J, Guo ZP et al (2005) Single wall carbon nanotube paper as anode for lithium-ion battery. Electrochim Acta 51:23–28
64. Udomvech A, Kerdcharoen T, Osotchan T (2005) First principles study of Li and Li^+ adsorbed on carbon nanotube: variation of tubule diameter and length. Chem Phys Lett 406:161–166
65. Wu GT, Wang CS, Zhang XB et al (1998) Lithium insertion into CuO/carbon nanotubes. J Power Sour 75:175–179
66. Gao B, Kleinhammes A, Tang XP et al (1997) Electrochemical intercalation of single-walled carbon nanotubes with lithium. Chem Phys Lett 307:153–157
67. Zhou O, Fleming RM, Murphy DW et al (1994) Defects in carbon nanostructures. Science 263:1744–1747
68. Suzuki S, Tomita M (1996) Observation of potassium-intercalated carbon nanotubes and their valence-band excitation spectra. J Appl Phys 79:3739–3743
69. Endo M, Muramatsu H, Hayashi T et al (2005) Buckypaper' from coaxial nanotubes. Nature 433:476
70. Muramatsu H, Hayashi T, Kim YA et al (2005) Pore structure and oxidation stability of double-walled carbon nanotube-derived bucky paper. Chem Phys Lett 414:444–448
71. Kim YA, Kojima M, Muramatsu H et al (2006) In situ Raman study on single- and double-walled carbon nanotubes as a function of lithium insertion. Small 2:667–676
72. Miyamoto J, Hattori Y, Noguchi D et al (2006) Efficient H_2 adsorption by nanopores of high-purity double-walled carbon nanotubes. J Am Chem Soc 128:12636–12637
73. Tibbetts GG (1983) Carbon fibers produced by pyrolysis of natural gas in stainless steel tubes. Appl Phys Lett 42:666–668
74. Oberlin A, Endo M, Koyama T (1976) Filamentous growth of carbon through benzene decomposition. J Cryst Growth 32:335–349
75. Tibbetts GG, Gorkiewicz DW, Alig RL (1993) A new reactor for growing carbon fibers from liquid- and vapor-phase hydrocarbons. Carbon 31:809–814
76. Endo M, Kim YA, Hayashi T et al (2001) Vapor-grown carbon fibers (VGCFs): basic properties and their battery applications. Carbon 39:1287–1297
77. Sotowa C, Origi G, Takeuchi M et al (2008) The reinforcing effect of combined carbon nanotubes and acetylene blacks on the cathode electrode of lithium ion batteries. ChemSusChem 1:911–915

Lithium Titanate-Based Anode Materials

Hailei Zhao

Graphitic carbon is the most widely used anode material in commercial Li-ion batteries due to its low lithiation potential, long cycle life, abundant resources and low cost. However, Li-ion batteries using graphite as anode material give rise to rate, safety and life problems. During lithium intercalation/deintercalation process, graphite undergoes a considerable volume change ($\sim 10\ \%$ [1]), which could cause particle cracking and even peeling off of anode film from the current collector, leading to gradual capacity degradation of the electrode [2, 3]. Safety concerns arise when the cells experience fast charging, long-term cycling, or low temperature charging owing to the propensity of formation of lithium dendrites, which is induced by the low lithiation potential of the graphite anode (close to 0 V vs. Li/Li^+) and the low lithium ion diffusivity in the graphite lattice [4, 5]. As an alternative anode material to carbon, $Li_4Ti_5O_{12}$ has been extensively studied for the potential use in large-scale Li-ion batteries. $Li_4Ti_5O_{12}$ shows stable charge/discharge platform at ca. 1.55 V versus Li/Li^+, and possesses excellent cycling stability and unique safety characteristic owing to its negligible volume change and high redox potential upon Li-ion intercalation/deintercalation. However, coarse $Li_4Ti_5O_{12}$ exhibits poor rate performance because of its low electronic conductivity and sluggish lithium ion diffusivity [6, 7]. In some cases, especially when aging at elevated temperature or cycling in a long-term regime, gas generation frequently occurs in $Li_4Ti_5O_{12}$-based batteries [8]. In past decades, many efforts have been devoted to overcoming these problems and significant advancements have been achieved, which make $Li_4Ti_5O_{12}$ viable for practical application in batteries for various electrical energy storage, such as electric/hybrid electric/plug-in hybrid electric vehicles (EV/HEV/PHEV), grid load leveling, integration of renewable energy sources, etc.

H. Zhao (✉)
University of Science and Technology Beijing, 30 Xueyuan Rd,
Haidian District, Beijing 100083, China
e-mail: hlzhao@ustb.edu.cn

Z. Zhang and S.S. Zhang (eds.), *Rechargeable Batteries*,
Green Energy and Technology, DOI 10.1007/978-3-319-15458-9_6

1 Lattice Structure and Electrochemical Characteristics

The $Li_4Ti_5O_{12}$ compound has a defective spinel structure with cubic space group $Fd\bar{3}m$ [9], where the 32e positions are occupied by oxygen atoms, the tetrahedral 8a positions are taken by Li atoms and the octahedral 16d positions are shared by Ti and Li atoms in a ratio of 5:1. Upon lithiation, Li atoms at the 8a sites are moved to the 16c sites and the additional Li atoms fill the remaining 16c vacancies, resulting in phase transformation from spinel structure of $Li_4Ti_5O_{12}$ ($[Li]_{8a}[Li_{1/3}Ti_{5/3}]_{16d}O_4$) to rocksalt structure of $Li_7Ti_5O_{12}$ ($[Li_2]_{16c}[Li_{1/3}Ti_{5/3}]_{16d}O_4$) [9, 10]. The two crystal structures are illustrated in Fig. 1 [11]. The structural transformation generates only a slight lattice contraction, from 8.3595 Å to 8.3538 Å [9] for $Li_4Ti_5O_{12}$ and $Li_7Ti_5O_{12}$, respectively, corresponding to a volume shrinkage of about 0.2 %. Due to the negligible volume change, $Li_4Ti_5O_{12}$ is widely considered to be a “zero strain” material for lithium insertion and removal. This gives active $Li_4Ti_5O_{12}$ superior structural stability and guarantees the mechanical integrity of the electrode by maintaining good $Li_4Ti_5O_{12}$ particle contact with binder and conductive carbon matrix during charge/discharge process, leading to an extremely long cycle stability.

One mole of $Li_4Ti_5O_{12}$ can uptake three moles of Li ions, corresponding to a theoretical specific capacity of 175 mAh g^{-1}. The electrochemical lithiation of $Li_4Ti_5O_{12}$ is commonly regarded as a two-phase process between $Li_4Ti_5O_{12}$ and $Li_7Ti_5O_{12}$, which delivers a long and flat plateau at 1.55 V versus Li/Li^+ [9, 10]. The typical charge/discharge curve of $Li_4Ti_5O_{12}$ is depicted in Fig. 2 [12]. The high redox potential makes lithium dendrite formation impossible, thereby averting the safety hazards of negative electrodes that operate close to the potential of metallic lithium [13]. Furthermore, the high operating potential of $Li_4Ti_5O_{12}$ can effectively avoid the formation of solid-electrolyte interface (SEI) film. Therefore, the

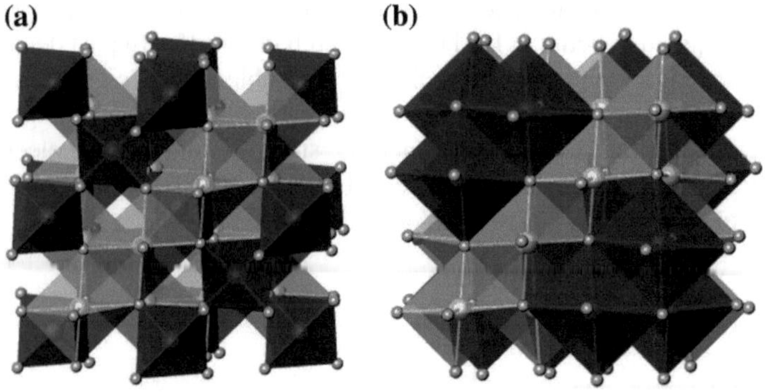

Fig. 1 **a** $Li_4Ti_5O_{12}$ spinel structure. **b** $Li_7Ti_5O_{12}$, rock salt. Blue (*dark*) *octahedra* represent lithium, and green (*light*) *octahedra* represent disordered lithium and titanium (reproduced with permission from [11])

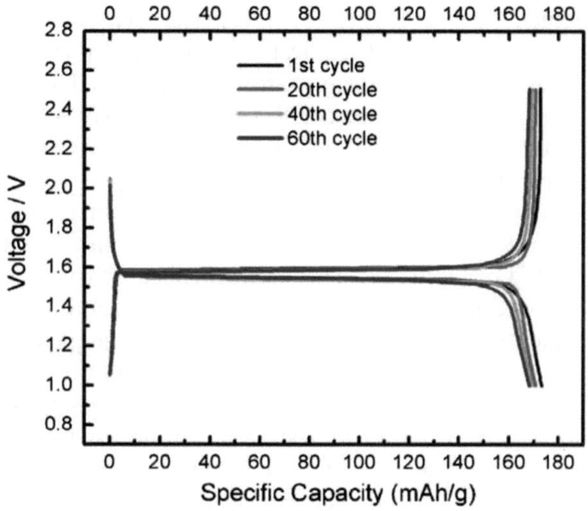

Fig. 2 Typical charge-discharge curves of $Li_4Ti_5O_{12}$ versus Li for 1st, 20th, 40th and 60th cycles at 0.5 C (reproduced with permission from [12])

consumption of lithium from cathode material for the generation of SEI film can be eliminated and the risk of SEI film decomposition at high temperature, which may release heat thereby triggering the reaction of cathode with electrolyte and consequently generating even large quantity of heat, can be decreased, making the electrode possess high coulombic efficiency and good thermal stability [14].

2 Electronic Conductivity, Ionic Diffusivity and Lithiation/Delithiation Mechanism

$Li_4Ti_5O_{12}$ is insulating in character due to the large bandgap between the occupied oxygen p-states and the empty Ti d-states. Literature reports a wide-ranging bandgap width for $Li_4Ti_5O_{12}$ from 1.8 to 3.8 eV determined by experiments [15–18] to 1.7–2.3 eV calculated by ab initio [19–23]. The relatively low calculated value compared to that experimentally determined is not surprising because density functional theory (DFT) calculation tends to underestimate the bandgap of materials [19, 24]. The large bandgap endows $Li_4Ti_5O_{12}$ with a low electronic conductivity $<10^{-13}$ S cm^{-1} [25]. However, after lithium ions are inserted with concomitant incorporation of electrons, the Ti d-states become partly filled, and the electronic structure of $Li_7Ti_5O_{12}$ changes to be metallic [26]. This means that $Li_4Ti_5O_{12}$ is insulating only at the beginning of lithiation process.

Ionic diffusivity of active materials is another factor dominating electrode reaction kinetics. $Li_4Ti_5O_{12}$ has a low Li ion diffusion coefficient on the order of 10^{-8}–10^{-15} cm^2 s^{-1} [27–31]. The huge difference in the reported Li ion diffusivities of $Li_4Ti_5O_{12}$ is attributable to the different testing methods and the different lithiation depths of the employed electrode. The low electronic conductivity and poor ionic diffusivity lead to poor rate performance of $Li_4Ti_5O_{12}$ anode and thus limit the practical and direct application of coarse $Li_4Ti_5O_{12}$ in high-power batteries. To overcome this problem, numerous methods have been devised to enhance the electronic and/or ionic conductivities of $Li_4Ti_5O_{12}$, including lattice doping, surface coating, compositing and nanostructuring (see Sect. 3), with the eventual aim of improving the rate-capability.

It is imperative to understand the lithium insertion and extraction mechanism that affects the structural stability during cycling and the kinetic performance of the $Li_4Ti_5O_{12}$ electrodes. The electrochemical lithiation and delithiation of $Li_4Ti_5O_{12}$ is believed to occur through a two-phase reaction between $Li_4Ti_5O_{12}$ and $Li_7Ti_5O_{12}$, which is responsible for the very flat voltage plateau around 1.55 V versus Li/Li^+ [9]. The electrochemical lithium insertion in the voltage range of 1.0–2.5 V versus Li/Li^+ causes the filling of 16c sites and at the same time drives the original lithium to migrate from 8a to 16c sites, leading to the full occupation of 16c sites in $Li_7Ti_5O_{12}$. The evacuation of lithium ion from 8a sites is most likely due to the Coulombic repulsion between nearest Li ions occupying 8a–16c sites (separated at a distance of 1.81 Å) [32]. The lithiation/delithiation process of $Li_4Ti_5O_{12}$ can be described by Eq. (1). Further lithiation gives rise to the partial re-occupation of 8a site, resulting in the formation of $Li_{8.5}Ti_5O_{12}$ at a low potential (ca. 0.05 V vs. Li/Li^+) accompanied by an approximately 0.4 % lattice expansion, which is confirmed by experiment and first principle calculation [26, 33, 34]. Although the framework of $[Li_1Ti_5]_{16d}O_{12}$ is not changed upon lithium insertion down to low potential, a structural distortion may be induced. The insertion of 4.5 mol lithium per $Li_4Ti_5O_{12}$ provides a theoretical capacity of ca. 262 mAh g^{-1}, 1.5 times higher than that of. $Li_7Ti_5O_{12}$.

$$\begin{aligned}&[\mathrm{Li}]^{8a}[\,]^{16c}[\mathrm{Li}_{1/3}\mathrm{Ti}_{5/3}]^{16d}\mathrm{O}_4^{32e} + xe^- + x\,\mathrm{Li}^+ \overset{\text{ca. 1.55 eV vs. Li/Li}^+}{\leftrightarrows}\\&(1-x)[\mathrm{Li}]^{8a}[\,]^{16c}[\mathrm{Li}_{1/3}\mathrm{Ti}_{5/3}]^{16d}\mathrm{O}_4^{32e} + x\,[\,]^{8a}[\mathrm{Li}_2]^{16c}[\mathrm{Li}_{1/3}\mathrm{Ti}_{5/3}]^{16d}\mathrm{O}_4^{32e}\end{aligned} \quad (1)$$

The observation of co-occupied 8a and 16c sites in the lithiated intermediate product $Li_{4+x}Ti_5O_{12}$ has led to reconsideration of the lithiation/delithiation mechanism of $Li_4Ti_5O_{12}$. Wagemaker et al., based on neutron and X-ray diffraction measurements, suggested that the electrode reaction of $Li_4Ti_5O_{12}$ is not a two-phase reaction but a solid-solution process at room temperature, with 8a and 16c sites being co-occupied. With increasing lithiation depth, the occupancy at 16c sites increases while that at 8a sites decreases gradually. The real two-phase reaction between $Li_4Ti_5O_{12}$ and $Li_7Ti_5O_{12}$ is only stable below 100 K [32]. The mixed occupation at 8a and 16c sites was also reported in a single-crystal study [35]. However, more detailed investigation revealed that the solid solution can be described as well dispersed distinct domains with either 8a or 16c Li occupation. The domain's length scale is less than 10 nm at 373 K [36]. The two-phase character is clearly observed by aberration-corrected scanning transmission electron microscopy (STEM) [37], which shows a sharp, dislocation free coherent heterophase boundary (as illustrated

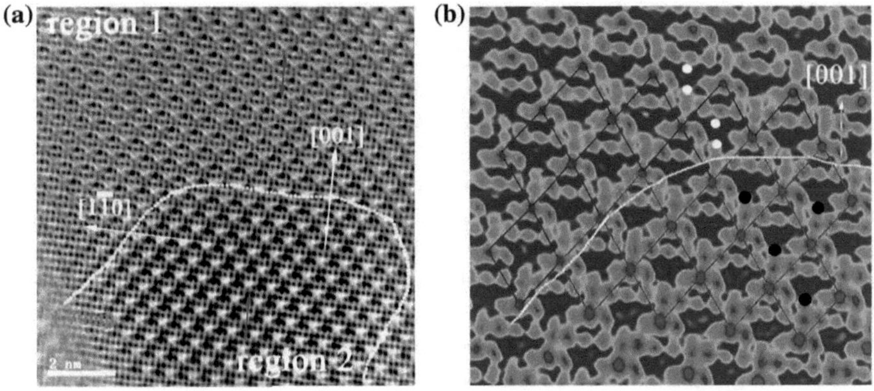

Fig. 3 Interfacial structure in a chemically lithiated $Li_4Ti_5O_{12}$ sample with approximately 0.15 mol Li insertion per formula unit along the [110] direction. **a** ABF image near the interface between $Li_4Ti_5O_{12}$ phase (region 1) and $Li_7Ti_5O_{12}$ phase (region 2). The yellow (*light*) line indicates the boundary of the interface. **b** Colored ABF image of the two phases near the interface, where the 8a sites occupied in $Li_4Ti_5O_{12}$ and the 16c sites occupied in $Li_7Ti_5O_{12}$ are marked as yellow (*light*) and *black dots*, respectively (reproduced with permission from [37])

in Fig. 3). This is beneficial for Li ion migration through the grain-boundaries and thereby contributes much to the good rate-capability of $Li_4Ti_5O_{12}$. The two-phase mechanism for lithiation/delithiation usually corresponds to a poor rate-capability due to the existence of strain and interfacial energy between the two phases.

Curved charge-discharge voltage profiles in high potential range (>1.55 V) and extra specific capacity exceeding its theoretical value are commonly observed for nanosized $Li_4Ti_5O_{12}$ electrodes. This is believed to be associated with the energetically favorable Li occupation at 16c sites near the surface region and high accumulation of Li ions at the surface layer, because the surface environment helps relax the strain caused by the repulsion between Li ions co-occupied at 8a and 16c sites [38]. Deep lithiation can lead to a surface composition above that of $Li_{8.5}Ti_5O_{12}$ and therefore cause large structural distortion, or surface reconstruction or even mechanical failure of a thin surface layer. These changes will passivate the particle surface and deteriorate the electrochemical performance, especially the rate-capability of $Li_4Ti_5O_{12}$ electrodes. In addition, the surface structure variation of nanosized $Li_4Ti_5O_{12}$ can cause irreversible capacity loss, leading to a lower initial coulombic efficiency. The surface structure of nanoparticles contributes more fraction to the whole particle structure as compared to that of the bulk material. Therefore, from the point of view of overall performance, there is an optimum particle size for $Li_4Ti_5O_{12}$ electrodes depending on the voltage windows, although nanoparticles could significantly reduce the diffusion distance of Li ions. As lithium ion diffusion conducts along the 16c-8a-16c pathway [39, 40], therefore, Li occupation at 8a sites in addition to 16c sites will hinder the diffusion process and further restrict the rate performance of $Li_4Ti_5O_{12}$ electrodes.

3 Advances in Performance Improvement

The superior structural stability endowed by the "zero strain" characteristics upon lithiation/delithiation, the excellent safety feature ensured by the high redox potential and the resource abundance in raw materials render the $Li_4Ti_5O_{12}$ a promising anode material of lithium ion batteries used for stationary energy storage and electric vehicles. At the same time, the demand to further improve the unsatisfactory rate performance and resolve the gas generation issue of $Li_4Ti_5O_{12}$ electrodes has motivated the design and preparation of $Li_4Ti_5O_{12}$ materials with novel particle morphology or chemical composition. With these efforts, great progress has been made in optimization of electrochemical properties and better understanding of electrode reaction kinetics of $Li_4Ti_5O_{12}$ electrodes. Recently, $Li_4Ti_5O_{12}$ was found to be a good Na storage material, which make it be another research hot-spot as electrode material making it another source of research activity surrounding its possible use as electrode material for Na-ion batteries.

3.1 *Rate-Capability*

Rate-capability is one of the important electrochemical properties of batteries, which ensures a high and stable delivery of electrochemical capacity under high current density for the batteries. Many factors can affect the rate-capability of electrodes, including active material features (electronic and ionic conductivity, electrochemical activity, particle size and morphology), electrode recipe, and electrode geometric dimension, etc. Here, we place an emphasis on the intrinsic properties of $Li_4Ti_5O_{12}$ active material that affect the electrochemical performance under high current density environments.

3.1.1 Lattice Doping

Lattice doping is a common strategy to improve the electronic and ionic conductivity of materials by producing effectively charged point defects and/or altering lattice parameters. Donor doping with high valence elements substituting for Li (Mg^{2+}, Al^{3+}, La^{3+}, Ca^{2+}, Zn^{2+}, Sn^{4+}) [25, 41–46] or Ti (Nb^{5+}, W^{6+}, V^{5+}, Mo^{6+}, Ta^{5+}) [47–51] or O^{2-} (F^-, Br^-) [43, 52, 53] can yield mixed valence Ti^{3+}/Ti^{4+} as charge compensation and thereby generate n-type electronic conduction in $Li_4Ti_5O_{12}$. The Mg substitution increases the conductivity of $Li_{4-x}Mg_xTi_5O_{12}$ by several orders of magnitude from 10^{-13} S cm^{-1} for x = 0 to 10^{-2} S cm^{-1} for x = 1 [25]. The rate capability of $Li_4Ti_5O_{12}$ can be significantly improved by appropriate Ca substitution for Li sites, delivering a specific capacity of ~120 mAh g^{-1} at 20 C in the cut-off voltages of 2.5–1.0 V [44]. The Nb-doped $Li_4Ti_{4.95}Nb_{0.05}O_{12}$ exhibits an enhanced rate capability with a reversible capacity of 135 mAh g^{-1} at 10 C and 127 mAh g^{-1} at 20 C [47].

The generation of electronic defects by aliovalent ion doping can enhance the electronic conductivity of materials by increasing the delocalized electron concentration, while the increase in lattice parameter by large size ion substitution or oxygen vacancy generation can facilitate the lithium ion diffusion in lattice, both of which are essential for a good rate-performance of electrode. It is reported that some dopants, such as Ru, W, Sr and Zr, increase the lattice parameter of $Li_4Ti_5O_{12}$, and therefore promote lithium ion diffusion [17, 48, 54, 55]. In addition, the isovalent doping of Na for Li [56, 57] and the Li substitution for Ti (Li excessive $Li_{4+x}Ti_{5-x}O_{12}$) [58, 59] have a positive effect in improving the rate performance of $Li_4Ti_5O_{12}$, which can be related to the enlarged lattice parameter induced by large ion substitution or oxygen vacancy generation.

Because the lithium ion diffusion in the lattice of $Li_4Ti_5O_{12}$ during charge/discharge process occurs via 8a-16c-8a route, the occupancy of foreign ions at 8a site may affect the lithium insertion kinetics and even the specific capacity of doped $Li_4Ti_5O_{12}$ [41, 42, 60]. In spinel $Li_4Ti_5O_{12}$, the same Li ions take two different sites (8a and 16d) while the same 16d sites are occupied with two different ions (Li and Ti). To elucidate the exact occupying site of the dopant ions in $Li_4Ti_5O_{12}$, especially for Li substitution, is extremely important for understanding the doping mechanism and the resultant electrochemical performance variation. In this regard, many advanced techniques are employed to probe the structural details of the doped $Li_4Ti_5O_{12}$ that should be responsible for the variation of the electrochemical properties, including Electron energy-loss spectroscopy (EELS), X-ray absorption spectroscopy (XAS), electron paramagnetic resonance (EPR), ^7Li nuclear magnetic resonance magic-angle spinning (NMR-MAS), inductive couple plasma-atomic emission spectrometry (ICP-AES), neutron diffraction (ND) as well as first principle calculation [17, 53, 60–63].

From the viewpoint of defect chemistry, either lower valence ion substitution for Ti site or charge compensation substitution for both Li and Ti sites (co-doping) cannot cause charged electronic point defects in $Li_4Ti_5O_{12}$ for improving the electronic conductivity. However, several reported works demonstrated that the rate-capability of $Li_4Ti_5O_{12}$ can be remarkably enhanced by low valence doping at Ti sites, such as $Li_4Ti_{5-x}M_xO_{12}$ (M = Mn^{2+} [63], Al^{3+} [64], Sc^{3+} [65]), and charge compensated co-doping for Li and Ti sites, such as $Li_{4-x/3}M_xTi_{5-2x/3}O_{12}$ (M = Cr [22], Al [66]), $Li_{3.9}Ni_{0.15}Ti_{4.95}O_{12}$ [54] and $Li_{3.95}M_{0.15}Ti_{4.9}O_{12}$ (M = Al, Ga, Co) [67]. The possible reason for the improvement is that these kinds of doping change the electronic structure and narrow the band-gap energy, leading to a decreased activation energy for electron conduction.

The substitution of elements with fixed valence for Ti in $Li_4Ti_5O_{12}$ can commonly enhance the structural stability against the shock of high current density, therefore, co-doping with two different elements, one with fixed valence for Ti and another acting as donor dopant for Li or O, can be expected to yield a good electrochemical performance under fast charge/discharge condition. The Mg, Zr co-doped system $Li_{3.95}Mg_{0.05}Ti_{4.95}Zr_{0.05}O_{12}$ displays excellent rate-capability and cycling stability [68].

Foreign element doping can sometimes cause the change of particle size because it usually alters the total energy of the lattice, the specific surface energy of crystals as well as the ion diffusion activities. Small particle size reduces the diffusion distance of Li ions and provides more surface area to come in contact with the electrolyte solution for electrode reaction, leading to an improved electrochemical property, especially the rate-capability. Doping with Sr, La and Zr was reported to decrease particle size and produce less particle agglomeration, which are part of the reasons that contribute to the improved rate performance of $Li_4Ti_5O_{12}$ [54, 69, 70].

In fact, the lattice ion doping may have influence on not just one aspect but simultaneously several aspects of the properties of the host material, such as electronic structure, point defect species and concentration, lattice distortion, lattice energy and specific surface energy, which in turn exert effect on the electronic conductivity, Li ion diffusivity, particle size and facet orientation. Therefore, the improvement of electrochemical performance by lattice doping is usually a synergistic effect of several factors. To effectively regulate the properties of $Li_4Ti_5O_{12}$, it is important to distinguish the dominant factors and establish the correlation among the dopant feature, electronic and lattice structure and electrochemical performance.

3.1.2 Surface Modification

Carbon coating is a common approach to improve the electrochemical performance of many electrode materials, including rate-capability and cycling stability. Owing to its high electronic conductivity and good chemical stability, carbon layers can remarkably enhance the electronic conductivity, increase interparticle contact, and help to form a uniform SEI layer on active particle surface and thus diminish the side reaction between the active material and electrolyte. In addition, the uniformly coated carbon layer can extend the effective reaction interface between the active particle and electrolyte, homogenize the current density and structural stress, and thereby improve the rate capability and cycling stability of the electrode. The carbon sources and carbon layer thickness have a strong impact on the physical and electrochemical performance of carbon coated $Li_4Ti_5O_{12}$/C electrodes [71–73]. The commonly used carbon sources are glucose, sucrose, pitch, epoxy, polyacrylate acid (PAA), citric acid (CA), maleic acid (MA), polyvinyl alcohol (PVA) and polyaniline (PANI) [73–78]. If the carbon coating is deposited via a CVD process, toluene vapor carried by an inert gas is usually employed as the carbon source [72, 79]. When the carbon coating process is required to carry out at low temperature to avoid the decomposition or vaporization of substance, acetylene is suggested.

The degree of graphitization of coated carbon has an effect on the electronic and ionic conductivity of the coated $Li_4Ti_5O_{12}$/C particles. Although high crystallinity of the carbon layer corresponds to a high electronic conductivity, it is actually unfavorable for lithium ion diffusion if the graphitic carbon grows with an orientated plane along the particle surface [72, 80], as illustrated in the scheme of Fig. 4. Therefore, there is a compromise between the electronic and lithium ionic

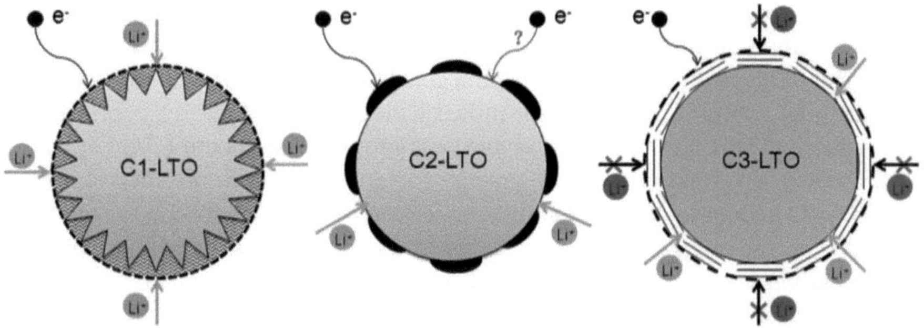

Fig. 4 Schematic illustration showing the effect of carbon coating on the electron and lithium ion conductions on a surface of carbon coated $Li_4Ti_5O_{12}$ (reproduced with permission from Ref. [80])

Fig. 5 Effect of carbon content on the rate performance of $Li_4Ti_5O_{12}$/C electrode (reproduced with permission from Ref. [77])

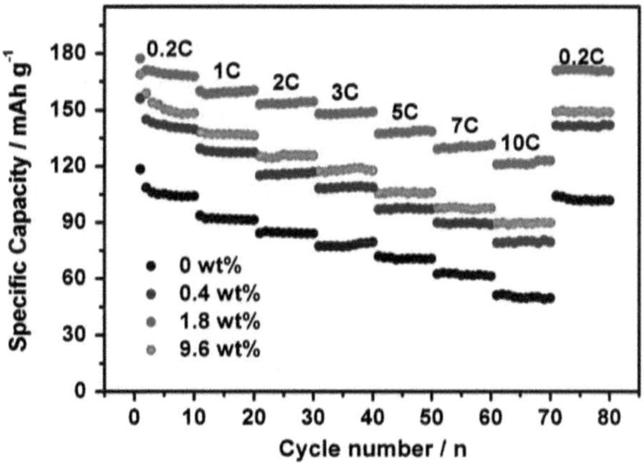

conductivities when optimizing the carbon coating parameters. To avoid excessive graphitization, high calcination temperature for carbon source pyrolysis should be averted. Similarly, a thick carbon layer is not recommended from the point of view of lithium ion diffusion (insertion/extraction). The optimized carbon layer thickness is in the range of 0.7–5 nm [72, 80, 81]. Figure 5 shows the effect of carbon content on the rate performance of $Li_4Ti_5O_{12}$/C electrodes [77].

An in situ Raman study revealed that the defects and vacancies existing in the carbon coating layer provide passages for lithium ion diffusion, and thus can promote the interfacial electrode reaction [77]. The N-doped carbon, when compared with pristine carbon, shows better improvement in the rate capability and cycling stability of $Li_4Ti_5O_{12}$/C electrodes due to the enhanced electronic conductivity [79, 82, 83]. Other advantages of carbon coatings on $Li_4Ti_5O_{12}$ include (1) limiting the particle growth during calcination and thus shortening the lithium ion diffusion distance during charge/discharge process [74, 75]; (2) reducing $Li_4Ti_5O_{12}$ to generate $Ti^{3+/4+}$ mixed valency on the particle surface, enhancing the electronic conductivity [78, 84].

In order to achieve high tap density of $Li_4Ti_5O_{12}$ powders, large secondary particles composed of small primary particles are prepared. As shown in the work reported by Zhu et al., nano-TiO_2 was first coated with carbon by mixing with sugar and calcining at 600 °C and then ball-milled with Li_2CO_3, followed by spray drying and further calcining at 800 °C to obtain nanoprous micro-sphere LTO/C particles [75]. Shen et al. proposed a novel strategy for preparation of core/shell structured $Li_4Ti_5O_{12}$/C nanoparticles via a simple solid-state reaction method by using metal oxyacetyl acetonate as titanium and carbon sources [85]. Nanosized TiO_2 particles with a carbon coating was first formed during heating process and then lithium (Li_2CO_3) diffused through the carbon layer to react with TiO_2 in a limited space, forming nano-sized $Li_4Ti_5O_{12}$ coated with a thin and discrete carbon layer. This specially synthesized $Li_4Ti_5O_{12}$/C structure displays an excellent rate capability, ca. 53 % of the capacity at 0.2 C is delivered at 90 C. With a similar method, Wang et al. prepared nano-sized $Li_4Ti_5O_{12}$ particles with double surface modification layers of Ti^{3+} and carbon with polyaniline (PANI) as a carbon precursor, which prevents the $Li_4Ti_5O_{12}$ particle growth during heat treatment and simultaneously reduces the LTO particle surface to generate Ti^{3+}-containing layer [78].

Besides carbon coating, metal nanoparticles are often employed to modify the surface of $Li_4Ti_5O_{12}$. The metal nanoparticles (e.g. Au [86], Ag [87, 88], Cu [89]) with a size range of 2–10 nm and highly dispersed on the $Li_4Ti_5O_{12}$ particle surface via a wet chemistry route can enhance the electrical contact between $Li_4Ti_5O_{12}$ particles and the current collector, promoting the electrode reaction kinetics and thereby improving the rate performance. As shown in Fig. 6, the Ag modified $Li_4Ti_5O_{12}$ nanocomposite delivers an excellent rate performance with a specific capacity of 131 mAh g^{-1} at 30 C [88].

Surface modification with oxides is another strategy to improve the rate performance of $Li_4Ti_5O_{12}$ anode material. Feng et al. [90] reported a modification of $Li_4Ti_5O_{12}$ with an aqueous CrO_3 solution, which leads to the generation of Li_2CrO_4, Cr_2O_5 and anatase TiO_2 on the $Li_4Ti_5O_{12}$ particle surface. The first two have a positive effect in improving the rate capability of $Li_4Ti_5O_{12}$, resulting in a capacity improvement of ca. 60 % from its original 80 to 130 mAh g^{-1} at 30 C. CeO_2 is also suggested as a suitable coating oxide for improving the rate performance of $Li_4Ti_5O_{12}$ electrodes, which could enhance both electronic and lithium ionic conductivity because partial CeO_2 enters the $Li_4Ti_5O_{12}$ lattice as a dopant [91]. The nominal compositions $Li_4Ti_5Cu_xO_{12+x}$ with two spinel phases $Li_2CuTi_3O_8$ and $Li_4Ti_5O_{12}$ were synthesized by Wang et al. [92]. The component of $Li_2CuTi_3O_8$ is decomposed into Cu, Li_2O and $Li_4Ti_5O_{12}$ during the first lithiation process, and the in situ generated Cu dispersing uniformly with $Li_4Ti_5O_{12}$ promotes the electron transport and improves the rate performance of the $Li_4Ti_5O_{12}$-based dual-phase electrode. TiN with high electronic conductivity is also employed as a surface modification material to facilitate the electron transport and promote the electrode reaction kinetics of $Li_4Ti_5O_{12}$, which can be generated on the particle surface of $Li_4Ti_5O_{12}$ by simply thermal treating in NH_3 atmosphere [93, 94]. Besides the formation of TiN, the surface of $Li_4Ti_5O_{12}$ particles is reduced in NH_3

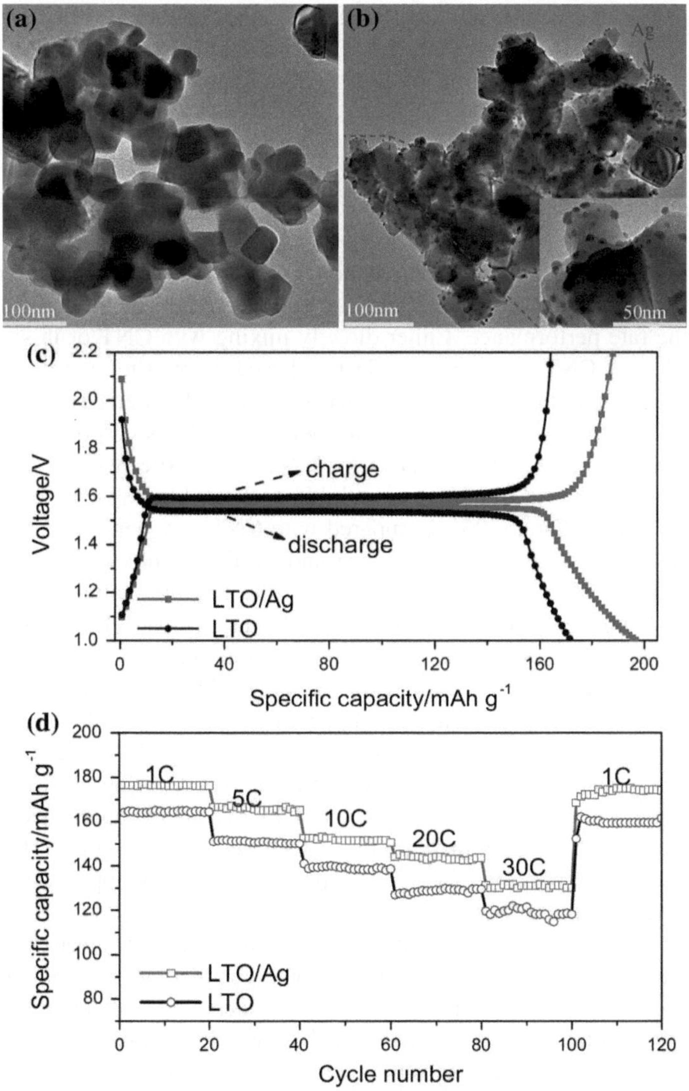

Fig. 6 TEM images and electrochemical performance of $Li_4Ti_5O_{12}$ and $Li_4Ti_5O_{12}$/Ag composite (reproduced with permission from Ref. [88])

atmosphere during heat treatment, leading to the generation of partial Ti^{3+}, which can also contribute to the fast electrode reactions.

Heat-treating the sample in a hydrogen containing atmosphere at high temperature [6] or immersing the sample in formaldehyde aqueous solution at room temperature can also induce the reduction of Ti ions from Ti^{4+} to Ti^{3+}, as confirmed by XPS examination, and thereby increase the electronic conductivity of $Li_4Ti_5O_{12}$ [95]. Wolfenstine's work revealed that the electrical conductivity of $Li_4Ti_5O_{12}$ can be increased from less than 10^{-9} to 10^{-5} S cm^{-1} after heat-treated in 3 vol.% H_2/Ar for 12 h at 800 °C [6]. Therefore, it is expected that high performance $Li_4Ti_5O_{12}$ could be obtained by preparing or post heat-treating the materials in a reducing atmosphere.

3.1.3 Compositing

Due to the extremely high electronic conductivity, graphene and carbon nanotubes (CNT) are frequently incorporated into electrode materials, including $Li_4Ti_5O_{12}$, to enhance the rate performance. Either directly mixing with CNT or in situ growing of $Li_4Ti_5O_{12}$ on CNT can create $Li_4Ti_5O_{12}$-based composites with a remarkably improved rate capability [96–98]. In order to ensure a high electrical contact area, the CNT is commonly employed together with amorphous carbon to enhance the electrode performance. The $Li_4Ti_5O_{12}$/C/CNT composite with a total carbon amount of 6 wt.% was reported to exhibit a reversible capacity of more than 140 mAh g^{-1} at 10 C-rate [98]. Compared with CNT, graphene can deliver a even greater improvement in rate performance and cycling stability of $Li_4Ti_5O_{12}$ electrodes due to its high aspect ratio, especially the flexible feature, which enables good contact between graphene sheets and $Li_4Ti_5O_{12}$ particles, allowing a fast charge transfer process in the electrode reaction. Oh et al. [99] reported a graphene-wrapped $Li_4Ti_5O_{12}$ composite, which exhibits an excellent rate performance with a reversible capacity of 147 mAh g^{-1} at 10 C and 105 mAh g^{-1} at 100 C. For the synthesis process, as illustrated in Fig. 7, the graphene oxide wrapped TiO_2 nanoparticles are preferentially prepared via an electrostatic interaction between graphene oxide (GO) and P25 (TiO_2) nanoparticles in an acid environment, which are then mixed with Li_2CO_3 and calcined at 850 °C in 4 %H_2/Ar, leading to the formation of $Li_4Ti_5O_{12}$ particles tightly wrapped with graphene. In the synthesis step, the graphene acts as a buffer to prevent the $Li_4Ti_5O_{12}$ particle aggregation by entangling the particles within the graphene sheets, while in the electrode reaction step, it provides an electronic conducting network for fast electrode reaction. Another interesting work was conducted by Shen et al. [100], where the $Li_4Ti_5O_{12}$ particles were in situ formed on graphene sheets with a controlled size and a high loading density through a hydrothermal reaction. The $Li_4Ti_5O_{12}$ nanoparticles anchored onto graphene can effectively prevent the restacking of graphene sheets and provide void space for electrolyte well penetration. This kind of structure offers an excellent rate performance of ca. 85 mAh g^{-1} at 60 C for 100 cycles (Fig. 8).

The tight contact between $Li_4Ti_5O_{12}$ particles and reduced graphene oxide (rGO) sheets after solvothermal treatment was shown by micro-Raman and X-ray photoelectron spectroscopy studies, which revealed the formation of chemical bonds and internal electron transfer between $Li_4Ti_5O_{12}$ and graphene [101]. The transfer of a π electron from the C_6 unit of rGO to $Li_4Ti_5O_{12}$ will cause the partial occupation of the conduction band of Ti 3d t_{2g} in the particle surface area, leading to an enhanced charge transfer ability at the interface of $Li_4Ti_5O_{12}$ and graphene.

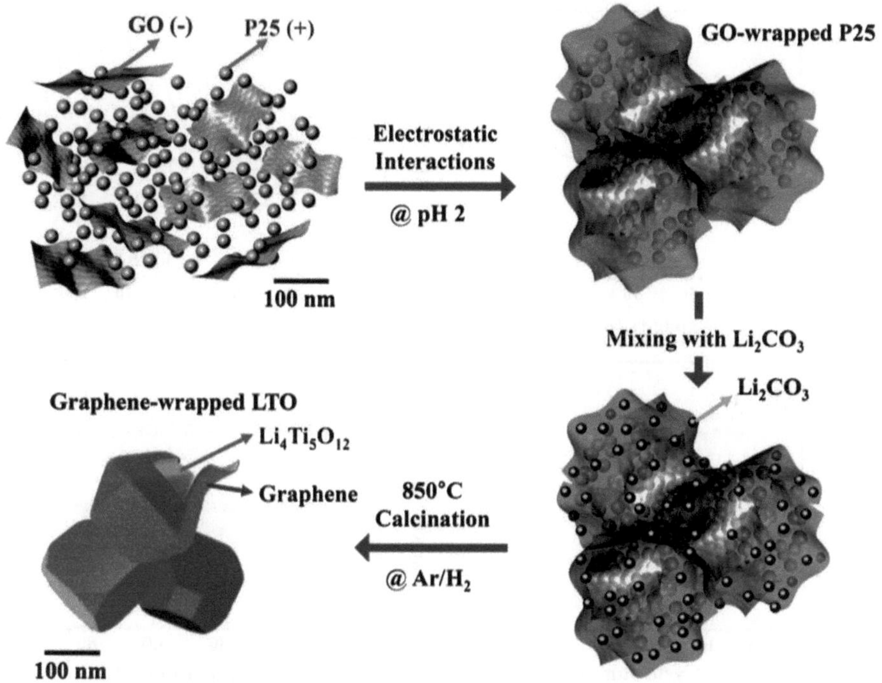

Fig. 7 Schematic for the effective graphene wrapping on individual LTO particles (reproduced with permission from Ref. [99])

This kind of hybridization between $Li_4Ti_5O_{12}$ and graphene was confirmed by first principle calculation based on DFT [79], which revealed the existence of strong bonding between the graphene coating layer and Ti-terminated $Li_4Ti_5O_{12}$ surface.

With the aim of increasing the specific capacity and enhancing the rate capability of $Li_4Ti_5O_{12}$, various metals and metal oxides with lithium storability are applied to composite with $Li_4Ti_5O_{12}$, such as Sn [102], SnO_2 [103], Fe_2O_3 [104] and CuO [105]. Most of those composites delivered a high initial specific capacity in concomitant with an unsatisfactory cycling stability, due to the large volume change of these active materials upon lithium insertion/extraction. A successful example for the compositing of $Li_4Ti_5O_{12}$ is the $Li_4Ti_5O_{12}$-TiO_2 composite, which delivers remarkably enhanced electrochemical properties in terms of specific capacity, cycling stability and rate capability when compared to pristine $Li_4Ti_5O_{12}$. The heterophase boundary between $Li_4Ti_5O_{12}$ and TiO_2 is considered to play an important role in improving the electrochemical performance of $Li_4Ti_5O_{12}$, which may help to store electrolyte and offer more channels for Li^+ ion insertion/extraction reaction [106, 107]. A more rational mechanism of TiO_2 in improving the rate capability and reversible capacity of $Li_4Ti_5O_{12}$ is its lithium storability. TiO_2 in the form of rutile or anatase is an active material towards Li-ion storage with a theoretical capacity of ca. 336 mAh g^{-1} and a pair of redox potential

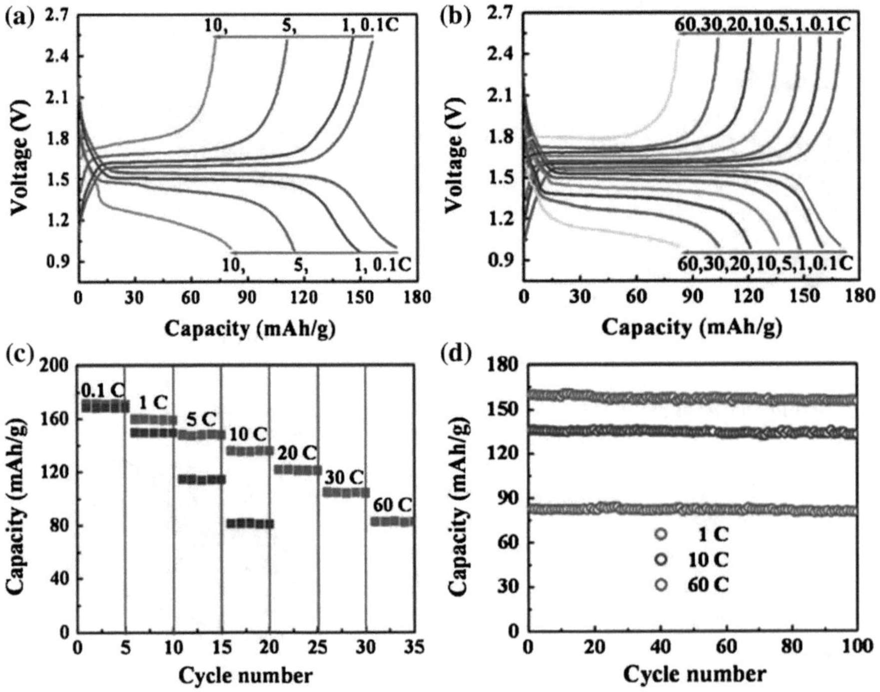

Fig. 8 Capacity-voltage profile of $Li_4Ti_5O_{12}$ (**a**) and $Li_4Ti_5O_{12}$/GNS (**b**). Comparison of rate capabilities of $Li_4Ti_5O_{12}$/GNS with $Li_4Ti_5O_{12}$ (**c**). Cycle performance of $Li_4Ti_5O_{12}$/GNS electrode at different current densities (**d**). GNS means graphene nano sheets (reproduced with permission from Ref. [100])

at ca. 1.7 and 2.0 V, respectively [108, 109]. It is reported that TiO_2 has fast lithium ion diffusivity [110]. During the first lithiation process, Li-ions can preferentially insert into TiO_2 to form Li_xTiO_2, in which the generated Ti^{3+} makes the Li_xTiO_2 phase a highly conductive phase and therefore capable of promoting the electrode reaction of $Li_4Ti_5O_{12}/TiO_2$ composites. Compared to the carbon coating layer, TiO_2 not only provides fast lithium ion and electron transportation, but also offers a high reversible capacity in the voltage range of 1.0–2.5 V, which endows the $Li_4Ti_5O_{12}/TiO_2$ composite with high specific capacity, excellent rate capability and stable cycling performance. Wang et al. prepared $Li_4Ti_5O_{12}$ nanosheets with rutile-TiO_2 at the edges via a facile solution-based route [111]. The rutile-TiO_2 as a coating layer enhances the lithium ion and electron conductivity of $Li_4Ti_5O_{12}$ and thus facilitates the electrode reaction kinetics, leading to an excellent electrochemical performance of high specific capacity and superior rate capability, as shown in Fig. 9.

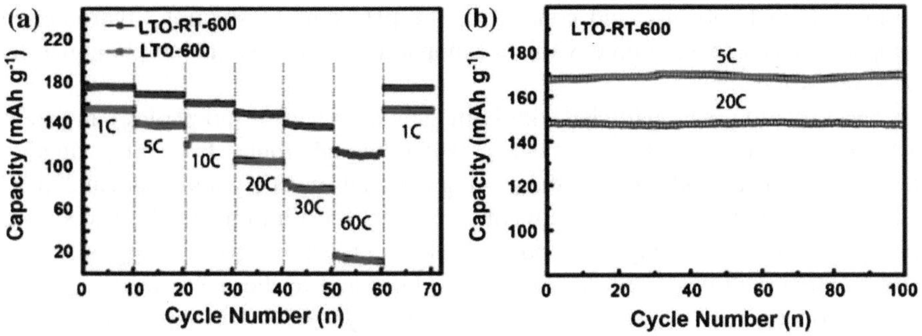

Fig. 9 Electrochemical performance of LTO-RT-600 and LTO-600 NSs: **a** rate performance; **b** cycle performance at 5 and 20 C. LTO-600 NSs and LTO-RT-600 refer to pure LTO nanosheets and LTO nanosheets with a thin rutile-TiO_2 terminated layer at the edges (reproduced with permission from Ref. [111])

3.1.4 Nano-Structuring

The reversible capacity and cycling performance of $Li_4Ti_5O_{12}$ are greatly influenced by particle morphology. Reducing the geometrical size of particles is an important strategy to improve the rate performance of electrode materials. The small particle size provides a short distance for lithium ion diffusion and electron transport, and more surface area for electrode reaction, ensuring fast lithiation/delithiation kinetics and thereby a remarkably improved rate performance of the electrode. Additionally, nanoparticles usually have various defects and unsaturated coordination sites at the surface. As stated above, these defects sometimes induce new lithium storage mechanisms [38] or provide sites for the generation of conductive layers [79, 93, 94], and consequently contribute to some the improvement of electrochemical performance of $Li_4Ti_5O_{12}$. Nanostructured $Li_4Ti_5O_{12}$ with morphologies of nanosheets [112, 113], nanorods [114, 115], nanotubes [116], nanowires [117], nanoflakes [118] and nanoflowers [119–121], are designed and prepared via various routes, including solid-state, hydrothermal, sol-gel, microwave, combustion, molten salt, sonochemical, rheological phase, and spray pyrolysis, and improved electrochemical performance, especially rate-capability, was demonstrated.

The synthesis route and starting materials have strong influence on the crystal structure, particle morphology and therefore electrochemical performance of $Li_4Ti_5O_{12}$. Solvothermal techniques, including hydrothermal, is one of the important methods of preparing nanoparticles. Because no long distance diffusion of ions is required in the liquid phase, as is often encountered in solid-state reactions, the product can be usually finalized by a low-temperature post heat-treatment after the hydrothermal reaction, during which the intermediate product is subjected to the decomposition of remaining organic groups and lattice rearrangement of $Li_4Ti_5O_{12}$ particles. It is believed that the post heat treatment can improve the electrochemical stability of electrode materials [122]. By varying the species and concentration of surfactants and solvents, and controlling the pH value of solution, hydrothermal temperature and soaking time, various nanoparticles with distinct morphologies can be synthesized.

The $Li_4Ti_5O_{12}$ nanoparticles with homogenous spherical morphology in size of 10–20 nm were synthesized via a solvothermal route by using titanium tetra-isopropoxide and LiOH as the starting materials in a polyol medium at 235 °C, combined with a subsequent treatment of 500 °C for 5 h in air [123]. The solvent polyol is believed to play an important role in preventing particle agglomeration and creating well-dispersed nanoparticles. Compared to the microsized $Li_4Ti_5O_{12}$ (1–2 μm) prepared by solid state reaction, the synthesized nanoparticles exhibit high specific capacity and excellent rate capability with a reversible capacity of 159 and 137 mAh g^{-1} at 30 C and 60 C, respectively. Feckl et al. reported a nanoscale porous framework of $Li_4Ti_5O_{12}$ composed of ultrasmall interconnected nanoparticles in the size range of a few nanometers (3–4 nm), which delivers a capacity of ca. 175 mAh g^{-1} at 1–50 C and can maintain 74 % of the maximum capacity at an extremely high rate of up to 800 C (corresponding to 4.5 s charge/discharge time) without any decline for a thousand cycles (Fig. 10) [124]. Such a porously structured $Li_4Ti_5O_{12}$ was synthesized in tert-butanol by a solvothermal route with Li*Ot*Bu and $Ti(OBu)_4$ as the starting materials in the presence of Pluraonic polymer (P123). Their work demonstrates that besides the solvent, the similar reactivity of the precursors is essential for the formation of a stoichiometric compound via a solvothermal reaction.

Hollow structured $Li_4Ti_5O_{12}$ particles were frequently reported to have a remarkably enhanced rate performance and can be prepared by hydrothermal reaction [119] or SiO_2 microsphere-based template routes [125]. Electrospinning is a simple, versatile, fast and inexpensive technology to produce one-dimensional (1D) fibers at micro- or nanoscales. When combined with conventional sol-gel processing, it offers the possibility to produce ceramic or organic/inorganic composite fibers with either a solid, porous or hollow structure [126]. $Li_4Ti_5O_{12}$/C fibers were reported to be produced from the electrospinning technique [127, 128], in which the $Li_4Ti_5O_{12}$ nanopartices are coated with carbon or dispersed in carbon matrix to form $Li_4Ti_5O_{12}$/C 1D composite fibers. The unique structural characteristics of those fibers with well dispersed dual-phase structure, nanoscale diameters and high aspect ratios ensure a shortened distance for Li ion and electron transport and provide more contact area with the electrolyte for electrode reaction, leading to fast electrode reaction kinetics and therefore an excellent electrochemical performance, with a specific capacity of more than 140 mAh g^{-1} achieved at a 10 C rate [128] (Fig. 11).

Self-supported $Li_4Ti_5O_{12}$ nanosheet arrays grown directly on conductive Ti foil was prepared by hydrothermal reaction between Ti foil and LiOH solution [129]. The excellent stability of the well aligned self-supported $Li_4Ti_5O_{12}$ nanosheet enables the electrode to have flexibility while the advantages of good conductivity, high surface area and shortened Li ion diffusion distance endow the electrode with excellent electrochemical performance, a high capacity of 163 and 78 mAh g^{-1} at 20 and 200 C, respectively, and a capacity of 124 mAh g^{-1} after 3000 cycles at

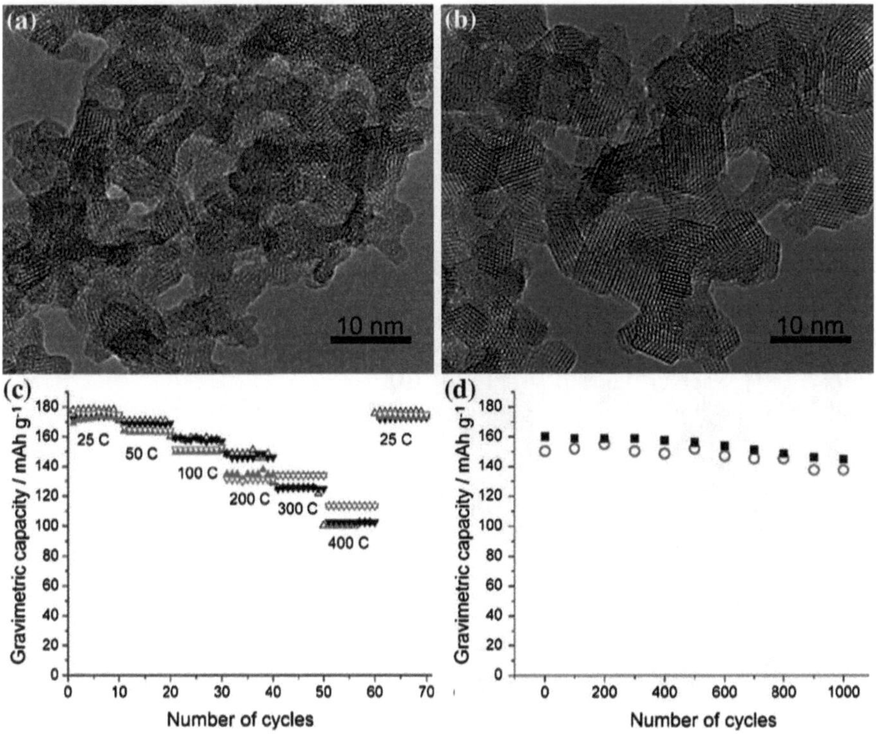

Fig. 10 HR-TEM images of nanosized $Li_4Ti_5O_{12}$ heated at 400 °C (**a**) and 500 °C (**b**), respectively. Multicycling stability at different rates (**c**) and at a rate of 100 C (**d**). The *gray* and *black symbols* correspond to the samples heated at 400 °C and 500 °C, respectively. The open and the filled symbols correspond to charge and discharge cycles, respectively. The cut-off potentials are 1.0 V and 2.4 V versus Li. The thickness of the film is about 0.5 μm, corresponding to a loading of about 0.14 mg cm^{-2} (reproduced with permission from Ref. [124])

50 C. Recently, self-supported $Li_4Ti_5O_{12}$/C nanotube arrays with uniform carbon layers on inner and outer tube surfaces was reported by Liu et al., which were directly grown on stainless steel foil via a template-based solution route [130]. The structure schematic, SEM images and electrochemical performance of such electrodes are shown in Fig. 12. The hollow structure expands the electroactive interface for electrode reaction and reduces the distance for Li-ion diffusion, while the carbon layer on the inner and outer surface of the tube enhances the electronic conductivity. Such electrode exhibits outstanding cycling performance of ca. 7 % capacity loss after 500 cycles at 10 C and excellent rate capability with a reversible capacity of 135, 105, and 80 mAh g^{-1} at 30, 60 and 100 C, respectively.

Although nano-structured $Li_4Ti_5O_{12}$ particles could provide outstanding rate performance, the high specific surface area could induce high irreversible capacity loss and the low tap density reduce the volumetric energy density of batteries. Hollow structured particles especially cannot resist the high pressure rolling for

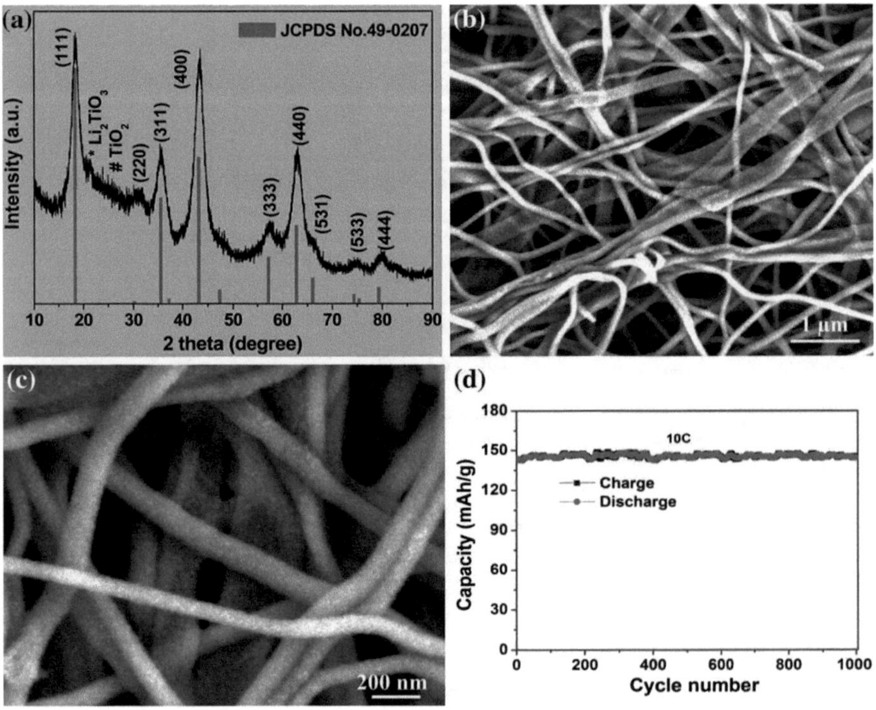

Fig. 11 **a** XRD pattern of 1D $Li_4Ti_5O_{12}$@C nanofibers, all the main diffraction peaks are indexed as spinel $Li_4Ti_5O_{12}$ with *Fd3m* space group; **b** and **c** Low and high magnification SEM images of the final annealed $Li_4Ti_5O_{12}$@C hierarchical nanofibers, these hierarchical nanofibers have a diameter of ca. 100–200 nm and are randomly oriented forming an interconnected fiber network; **d** charge/discharge capacities of $Li_4Ti_5O_{12}$@C hierarchical nanofibers over 1000 cycles at 10 C (reproduced with permission from Ref. [128])

practical electrode preparation. An effective way to overcome these problems while maintaining the excellent electrochemical performance is to fabricate hierarchical structures with microsized secondary particles composed of nanosized primary particles. The microsized particle increases the tap density while the nanosized particle and the connected pores inside the microsized particle provide not only a shortened distance for Li ion diffusion but also a percolative path for electrolyte penetration, thus maintaining all the advantages of nanoparticles. Such $Li_4Ti_5O_{12}$ particles with a nano-/micro-level combined structure were prepared by Amine et al. via a colloidal solution method, which deliver a high specific capacity and excellent rate-capability compared with microsized $Li_4Ti_5O_{12}$ particles, and show lower specific area impedance than microsized $Li_4Ti_5O_{12}$ and carbon electrodes when coupled with $Li_{1+x}Mn_{2-x}O_4$ cathode material [131]. Lin et al. [132] and Shen et al. [133] prepared hierarchically porous $Li_4Ti_5O_{12}$ microspheres with nanosized primary particles and inside rich nanopores by using commercial TiO_2 powders with an average size of 10 nm and LiOH as starting materials via a

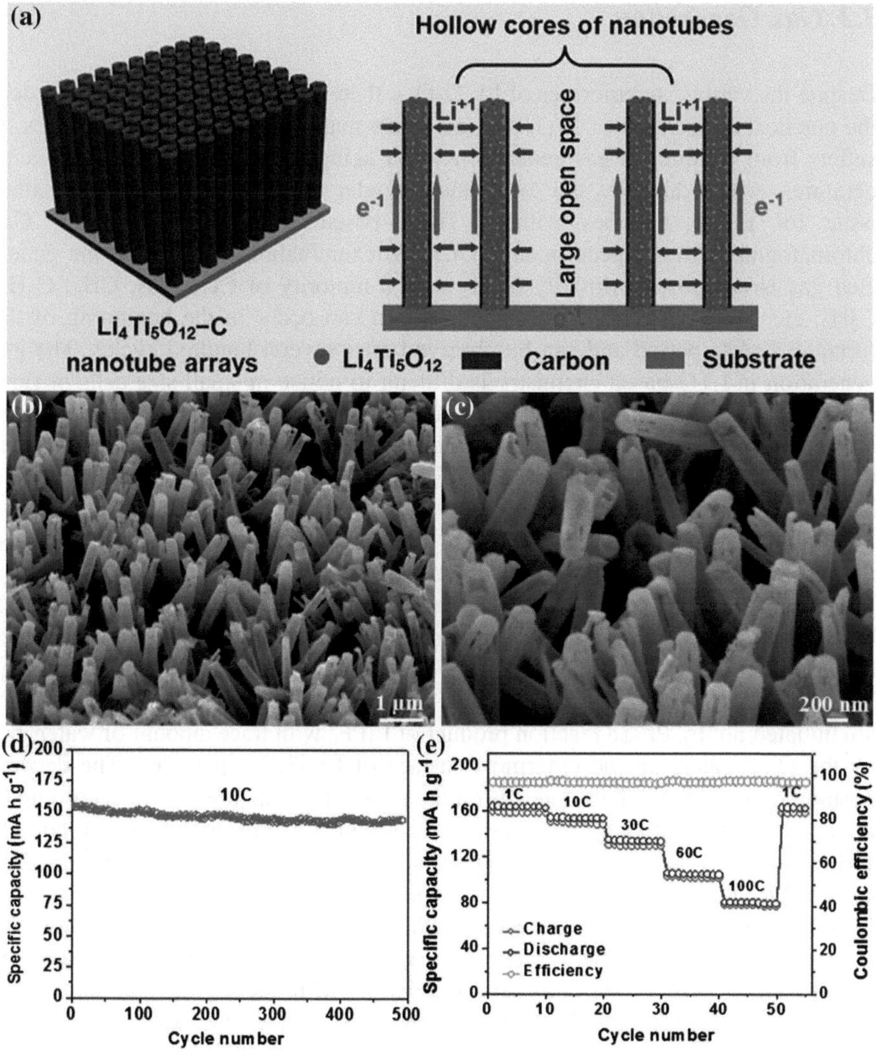

Fig. 12 The structure schematic, SEM images and electrochemical performance of self-supported $Li_4Ti_5O_{12}$/C nanotube arrays electrode (reproduced with permission from Ref. [130])

hydro/solvothermal reaction route. These hierarchical particles exhibit excellent rate performance and high capacity retention over long-term cycles. Especially, a high tap density of 1.62 g cm^{-3} can be achieved for submicrospheres with a 60 mm secondary particle size and 20–100 nm primary particle size [132]. Such micro/nanoscale combined particles with high tap densities will find wide application in practical high energy batteries.

3.2 Gas Generation

Despite the various superiorities of $Li_4Ti_5O_{12}$, there is still one obstacle that hinders the practical application of $Li_4Ti_5O_{12}$ as anode material in lithium ion batteries. It suffers from continuous gas generation when aging or operating at elevated temperatures, which damages the cycle and calendar life and poses a serious safety issue for Li-ion batteries with $Li_4Ti_5O_{12}$-based anodes [8, 134–138]. Gas chromatography/mass spectrometry (GC/MS) examination reveals that the generated gas is composed primarily of H_2 with a minority of CO, CO_2, CH_4, C_2H_4, C_2H_6, etc. [136, 138, 139]. This phenomenon can occur in the beginning of the formation cycle period and can be observed after several hundred cycles. The gas generation in LTO-based chemistry is difficult to notice in small size cells or rigid packaged cells due to the small amount of gas released but becomes more noticeable in large scale soft-package cells. Three factors are considered to contribute to the gas generation of $Li_4Ti_5O_{12}$: lithiated $Li_4Ti_5O_{12}$, lithium salt and carbonate solvent [136, 140].

The lithium salt in electrolyte solvent was found to have an obvious impact on the gas generation. Compared with $LiPF_6$, $LiBF_4$ salt releases less gas in LTO-based batteries [136]. However, He et al. demonstrated that the lithium salt has little impact on the gassing reactions [8], whereas the solvent plays the more important role in the gas generation [8, 138]. They found that gassing reactions include decarboxylation, decaronylation and dehydrogenation of solvents, which are initiated not by PF_5, a reaction product of $LiPF_6$ with trace amount of water, but by the (111) plane on the outermost surface of $Li_4Ti_5O_{12}$ particles. The decomposition product of solvents depends on the molecular structure [138]. According to the results of IR and GC, linear carbonates produce mainly hydrogen and soluble species, while cyclic carbonates generates alkylene gas, Li_2CO_3 and dilithium alkyl carbonates. PC-contained electrolyte can form thicker/denser layer than the EC-contained on $Li_4Ti_5O_{12}$ surface, which can mitigate further decomposition of the solvents and thus generate minimum gases. By comparing different solvents with 1 mol L^{-1} $LiPF_6$, Wu et al. revealed that DMC produces the maximum gases while PC + DMC (1:1 in vol.) the minimum gases. Besides the gases, anatase TiO_2 [8] and α-Li_2TiO_3 [141] are found as by-products on $Li_4Ti_5O_{12}$ particle surface, which may also exert influence on the cycling performance of $Li_4Ti_5O_{12}$ electrodes.

Among the various reasons for gas generation, the main suspect should be the lithiated $Li_4Ti_5O_{12}$ ($Li_{4+x}Ti_5O_{12}$), which directly participates in the gas generation reaction by transferring electrons from LTO to the electrolyte resulting in the reduction of the electrolyte. The in situ XANES measurements on $Li_4Ti_5O_{12}$ electrodes at high temperature reveal the continuous shift of Ti K-edge energy to higher values during aging, implying the progressive increase in the average valence of Ti ions, which is caused by the electron loss due to the self-discharge of $Li_4Ti_5O_{12}$ electrodes. Lower temperature leads to a longer transition time from Ti^{3+} to Ti^{4+}. Because the reduction reaction is carried out at the particle surface of $Li_4Ti_5O_{12}$, the particle surface chemistry [8] and particle morphology [137] have an important influence on the gas generation. Cutting off the transport path of electrons from $Li_4Ti_5O_{12}$ to electrolyte by forming a smooth surface layer on $Li_4Ti_5O_{12}$ should be an effective way to suppress the gas generation in LTO-based batteries. He et al. reported that addition of vinylene carbonate to the electrolyte of 1 M $LiPF_6$/EC + DMC + EMC could facilitate the rapid formation of a protective SEI film on $Li_4Ti_5O_{12}$ electrodes [137]. The chlorosilane additive also has an obvious effect in controlling the gas generation of $Li_4Ti_5O_{12}$ electrodes in electrolyte of $LiPF_6$ in EC/EMC [136]. Another approach to form a protective layer on $Li_4Ti_5O_{12}$ is to directly coat an inert layer onto $Li_4Ti_5O_{12}$ electrode or particles before cycling. The inert coating layer should be an electronic insulator but thin enough not to impede the lithium ion transport. Atomic layer deposition (ALD) [136, 140] and carbon coating techniques [142] are often employed to produce a protective thin layer on $Li_4Ti_5O_{12}$ electrode or particles, which can mitigate the gas generation problem. The carbon coating can help to form successive SEI film on $Li_4Ti_5O_{12}$ particles and thus prevent the gassing reaction between $Li_4Ti_5O_{12}$ and electrolyte. This has the added benefit in that the carbon coating can improve the rate-capability of $Li_4Ti_5O_{12}$ at the same time.

3.3 Performance as Anode Material for Na-Ion Battery

Apart from lithium-ion batteries, $Li_4Ti_5O_{12}$ can also be used as an electrode material for sodium-ion batteries. With the advantages of abundant and low cost of sodium sources, sodium-ion battery is deemed as an alternative of lithium-ion battery for large-scale energy storage applications [143]. Zhao et al. [144] first reported that $Li_4Ti_5O_{12}$ can be a Na-ion storage material, though the radius of Na ion (1.02 Å) is ca. 34 % larger than Li ion (0.76 Å). The sodiation behavior of $Li_4Ti_5O_{12}$ is much more complicated and quite different from the two-phase reaction of $Li_4Ti_5O_{12}$/$Li_7Ti_5O_{12}$ system. It usually presents a three-phase separation mechanism [145], as described in Eq. (2), with a theoretical capacity of 175 mAh g^{-1}.

$$2Li_4Ti_5O_{12} + 6Na^+ + 6e^- \leftrightarrow Li_7Ti_5O_{12} + Na_6LiTi_5O_{12} \quad (2)$$

More detailed investigation reveals that the lithium insertion behavior into $Li_4Ti_5O_{12}$ is strongly particle size dependent [146]. A solid solution mechanism for sodium insertion is found in nano-sized $Li_4Ti_5O_{12}$ particles. The prepared $Li_4Ti_5O_{12}$ can deliver a reversible Na storage capacity about 155 mAh g^{-1} with an average charge/discharge voltage of ~0.9 V versus Na^+/Na, as illustrated in Fig. 13. Based on various in situ techniques, the Na^+ ion apparent diffusion coefficient in $Li_4Ti_5O_{12}$ particles is estimated at around 10^{-16} cm^2 s^{-1} [146, 147], which is consistent with the result of DFT calculation [145]. The sluggish diffusion kinetics of Na^+ ions in $Li_4Ti_5O_{12}$ necessitates the preparation of nanosized $Li_4Ti_5O_{12}$ particles for suitable use as anode material in sodium-ion batteries.

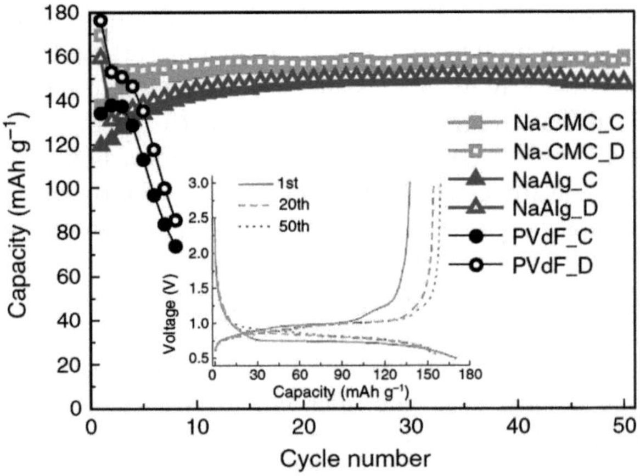

Fig. 13 Elecrochemical performance of $Li_4Ti_5O_{12}$ in sodium-ion batteries. The $Li_4Ti_5O_{12}$ electrode with PVdF, NaAlg and Na-CMC as binder, respectively, was cycled in NaFSI/EC:DEC electrolyte at a current rate of C/10 (reproduced with permission from Ref. [145])

Liu et al. [128] prepared $Li_4Ti_5O_{12}$@C hierarchical nanofibers with tiny $Li_4Ti_5O_{12}$ nanoparticles embedded in carbon by an electrospinning technique, which exhibited high and stable reversible capacity of about 162.5 mA h g^{-1} for 100 cycles at 0.2 C as anode material for Na-ion battery. The nanofiber shape combined with uniformly distributed nanosized $Li_4Ti_5O_{12}$ in carbon matrix offers short transport distance for Na ion and electron transport, high contact area with electrolyte, and good conducting phase percolation, ensuring excellent electrochemical properties.

A free-standing CNT/$Li_4Ti_5O_{12}$/C composite nanofiber with $Li_4Ti_5O_{12}$ nanoparticles and CNTs uniformly dispersing in a 1D carbon nanofiber matrix was also reported to deliver a good rate capability as anode material for Na-ion batteries [148]. Yu et al. [147] revealed the peseudocapacitive behavior of $Li_4Ti_5O_{12}$ upon Na ion insertion, especially for nanoparticles with sufficient surface defects, which contribute much of the specific capacity and rate-capability without trade-off of structural phase transformation.

4 Performance in Full-Cells

Because of the high safety, long cycling life and low cost, batteries with $Li_4Ti_5O_{12}$ as anode materials are ready for practical applications in the fields of electrical vehicles (EV) and large scale energy storage devices. However, the high redox potential of $Li_4Ti_5O_{12}$ results in some reduction of cell working voltage and thus the energy density. This issue can be overcome by coupling $Li_4Ti_5O_{12}$ anode with cathodes having relatively high redox potentials. Figure 14 displays the various voltages of Li-ion cells with $Li_4Ti_5O_{12}$ anode and different common cathode materials [149].

Apparently, $LiMn_{1.5}Ni_{0.5}O_4$ (LMNO) electrode offers the highest voltage of 3.2 V when combined with $Li_4Ti_5O_{12}$ anode. The electrochemical performance of $Li_4Ti_5O_{12}$—containing batteries depends strongly on the match design of negative and positive electrodes. Wu et al. [150] compared the electrochemical performance of three cell designs with different LNMO-to-LTO ($Li_4Ti_5O_{12}$) loadings (positive-electrode limited, negative-electrode limited, and positive/negative capacity ratio = ~1) and demonstrated that the negative-limited LNMO/LTO full cells delivered the best electrochemical performance, with 98 % of the first cycle capacity after 1000 cycles. At the same time, the cells with $Li_4Ti_5O_{12}$ limiting capacity exhibits less limitation of electrolyte choice than the cells with LNMO limiting capacity [151]. Actually, due to the high redox potential of LMNO, the electrode/electrolyte interfacial reactivity at high potential is usually the dominating factor for the electrochemical performance of LMNO/LTO cells [152, 153]. Recently, Kim et al. [154] reported a Ti-substituted cathode material $LiNi_{0.5}Mn_{1.5-x}Ti_xO_4$ (LMNTO), which delivers longer cycle life, higher cell operating voltage, higher coulombic efficiency and lower electrode polarization compared with Ti-free LMNO when paired with LTO negative electrode. Besides, the Ti substitution can improve the capacity retention of LMNTO/LTO cell at high temperature (45 °C). The improvement is mainly ascribed to the retardation of electrolyte oxidation at the cathode side.

Although LMNO provides high working voltage for LMNO/LTO cells, its high redox potential (ca. 4.8 V vs. Li^+/Li) imposes stringent requirements on electrolyte systems. By contrast, $LiCoO_2$, $LiNi_{1/3}Co_{1/3}Mn_{1/3}O_2$ and $LiMn_2O_4$ cathodes coupling with LTO can operate well in most of the conventional electrolytes and thus attracts more attention. In spite of the relatively lower working voltage, these battery systems deliver much stable cycling performance. Toshiba's SCiB™

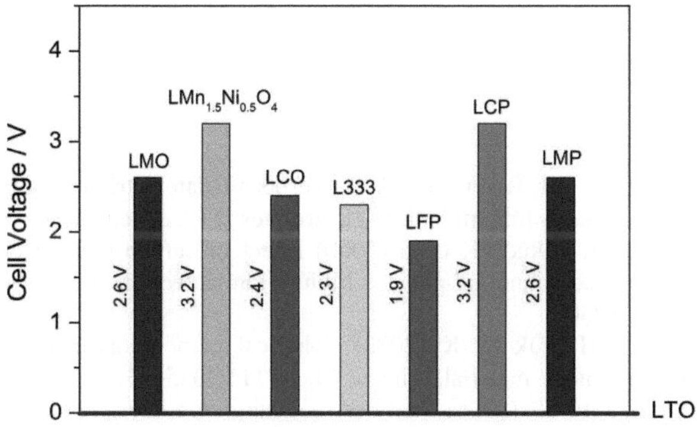

Fig. 14 Voltage of $Li_4Ti_5O_{12}$-based cells with different cathode materials. LMO—$LiMn_2O_4$, LCO—$LiCoO_2$, L333—$LiNi_{0.33}Co_{0.33}Mn_{0.33}O_2$, LFP—$LiFePO_4$, LCP—$LiCoPO_4$, LMP—$LiMnPO_4$ (reproduced with permission from Ref. [149])

rechargeable battery with LTO as anode and $LiNi_{1/3}Co_{1/3}Mn_{1/3}O_2$ as cathode exhibits excellent cycling stability with a capacity retention of 90 % after 10000 cycles and superior rate-capability taking only 6 min. to reach 80 % State of Charge level [155]. With respect to olivine $LiFePO_4$ (LFPO), owing to its intrinsic structural stability and electrochemical safety, the nanoengineered LFPO/LTO cell with a working voltage of 1.9 V presents especially promising performance with capacity loss rates of 0.003 %/cycle at 5 C rate after 200 cycles [156]. The mixed metal compound $LiMn_yFe_{1-y}PO_4$ (LMFP) has high redox potential at 3.6–4.1 V (vs. Li^+/Li), while it is not high enough to decompose the conventional electrolyte solutions based on alkyl carbonate solvents. The full cell LMFP-LTO operates at two potentials of 2.5 V (80 %) and 2 V (20 %) and shows good rate capability, impressive cycling stability and high safety features, making it a competitive power system in load leveling applications [157].

5 Conclusion

$Li_4Ti_5O_{12}$ is a potential Li-ion battery anode material of for use in large-scale energy storage, considering its high safety, excellent cycling stability, environmental friendliness and low cost. Its intrinsic low electronic conductivity and sluggish Li-ion diffusivity has driven research to design and prepare nanosized particles with controlled morphologies with the aim of enhancing the rate-capability. The gassing issue of $Li_4Ti_5O_{12}$-based batteries is the main obstacle that hinders its practical application. Surface coatings on $Li_4Ti_5O_{12}$ electrode or $Li_4Ti_5O_{12}$ particles as well as employing effective additives for electrolyte are potential approaches to form stable film on $Li_4Ti_5O_{12}$ electrode to circumvent the gas generation problem. The cathode materials, electrolyte systems as well as capacity matching of the two electrodes impose important influences on the cycling performance of $Li_4Ti_5O_{12}$-based batteries.

References

1. Winter M, Wrodnigg GH, Besenhard JO et al (2000) Dilatometric investigations of graphite electrodes in nonaqueous lithium battery electrolytes. J Electrochem Soc 147(7):2427–2431
2. Yoshio M, Wang H, Fukuda K et al (2000) Effect of carbon coating on electrochemical performance of treated natural graphite as lithium-ion battery anode material. J Electrochem Soc 147(4):1245–1250
3. Yoshio M, Wang H, Fukuda K (2003) Spherical carbon-coated natural graphite as a lithium-ion battery-anode material. Angew Chem 115:4335–4338
4. Zhang SS, Xu X, Jow TR (2006) Study of the charging process of a $LiCoO_2$-based Li-ion battery. J Power Sources 160(2):1349–1354
5. Zheng SS (2006) The effect of the charging protocol on the cycle life of a Li-ion battery. J Power Sources 161(2):1385–1391

6. Wolfenstine J, Lee U, Allen JL (2006) Electrical conductivity and rate-capability of $Li_4Ti_5O_{12}$ as a function of heat-treatment atmosphere. J Power Sources 154(1):287–289
7. Wunde F, Berkemeier F, Schmitz G (2012) Lithium diffusion in sputter-deposited $Li_4Ti_5O_{12}$ thin films. J Power Sources 215:109–115
8. He Y-B, Li B, Liu M, Zhang C et al (2012) Gassing in $Li_4Ti_5O_{12}$-based batteries and its remedy. Sci Rep 2:913
9. Scharner S, Weppner W, Schmid-Beurmann P (1999) Evidence of two-phase formation upon lithium insertion into the $Li_{1.33}Ti_{1.67}O_4$ spinel. J Electrochem Soc 146(3):857–861
10. Ohzuku T, Ueda A, Yamamoto N (1995) Zero-strain insertion material of $Li[Li_{1/3}Ti_{5/3}]O_4$ for rechargeable lithium cells. J Electrochem Soc 142(5):1431–1435
11. Sorensen EM, Barry SJ, Jung H-K et al (2006) Three-dimensional ordered macroporous $Li_4Ti_5O_{12}$: effect of wall structure on electrochemical properties. Chem Mater 18(2):482–489
12. Sun L, Wang J, Jiang K et al (2014) Mesoporous $Li_4Ti_5O_{12}$ nanoclusters as high performance negative electrodes for lithium ion batteries. J Power Sources 248:265–272
13. Ferg E, Gummow RJ, de Kock A et al (1994) Spinel anodes for lithium-ion batteries. J Electrochem Soc 141(11):L147–L150
14. Chen Z, Belharouak I, Sun Y-K et al (2013) Titanium-based anode materials for safe lithium-ion batteries. Adv Funct Mater 23:959–969
15. Kostlanova T, Dědeček J, Krtil P (2007) The effect of the inner particle structure on the electronic structure of the nano-crystalline Li–Ti–O spinels. Electrochim Acta 52(5):1847–1856
16. Kellerman DG, Mukhina NA, Zhuravlew NA et al (2010) Optical absorption and nuclear magnetic resonance in lithium titanium spinel doped by chromium. Phys Solid State 52 (3):459–464
17. Jhan Y-R, Duh J-G (2012) Electrochemical performance and low discharge cut-off voltage behavior of ruthenium doped $Li_4Ti_5O_{12}$ with improved energy density. Electrochim Acta 63:9–15
18. Kim C, Norberg NS, Alexander CT et al (2012) Mechanism of phase propagation during lithiation in carbon-free $Li_4Ti_5O_{12}$ battery electrodes. Adv Funct Mater 23(9):1214–1222
19. Tsai P-C, Hsu W-D, Lin S-K (2014) Atomistic structure and ab initio electrochemical properties of $Li_4Ti_5O_{12}$ defect spinel for Li ion batteries. J Electrochem Soc 161(3):A439–A444
20. Lippens P-E, Womes M, Kubiak P et al (2004) Electronic structure of the spinel $Li_4Ti_5O_{12}$ studied by ab initio calculations and X-ray absorption spectroscopy. Solid State Sci 6 (2):161–166
21. Ouyang CY, Zhong ZY, Lei MS (2007) *Ab initio* studies of structural and electronic properties of $Li_4Ti_5O_{12}$ spinel. Electrochem Commun 9(5):1107–1112
22. Song H, Yun SW, Chun HH et al (2012) Anomalous decrease in structural disorder due to charge redistribution in Cr-doped $Li_4Ti_5O_{12}$ negative-electrode materials for high-rate Li-ion batteries. Energy Environ Sci 5:9903–9913
23. Ding Z, Zhao L, Suo L et al (2011) Towards understanding the effects of carbon and nitrogen-doped carbon coating on the electrochemical performance of $Li_4Ti_5O_{12}$ in lithium ion batteries: a combined experimental and theoretical study. Phys Chem Chem Phys 13:15127–15133
24. Chan MKY, Ceder G (2010) Efficient band gap prediction for solids. Phys Rev Let 105:196403
25. Chen CH, Vaughey JT, Jansen AN et al (2001) Studies of Mg-substituted $Li_{4-x}Mg_xTi_5O_{12}$ spinel electrodes ($0 \leq x \leq 1$) for lithium batteries. J Electrochem Soc 148(1):A102–A104
26. Zhong Z, Ouyang C, Shi S et al (2008) Ab initio Studies on $Li_{4+x}Ti_5O_{12}$ compounds as anode materials for lithium-ion batteries. ChemPhysChem 9(14):2104–2108
27. Wang Y, Liu H, Wang K et al (2009) Synthesis and electrochemical performance of nano-sized $Li_4Ti_5O_{12}$ with double surface modification of Ti(III) and carbon. J Mater Chem 19:6789–6795

28. Fang W, Zuo P, Ma Y et al (2013) Facile preparation of $Li_4Ti_5O_{12}$/AB/MWCNTs composite with high-rate performance for lithium ion battery. Electrochim Acta 94:294–299
29. Yi F-T, Xie Y, Wu Q et al (2012) High rate cycling performance of lanthanum-modified $Li_4Ti_5O_{12}$ anode materials for lithium-ion batteries. J Power Sources 214:220–226
30. Sun Y-K, Jung D-J, Lee YS et al (2004) Synthesis and electrochemical characterization of spinel $Li[Li_{(1-x)/3}Cr_xTi_{(5-2x)/3}]O_4$ anode materials. J Power Sources 125:242–245
31. Zhu G-N, Wang C-X, Xia Y-Y (2011) A comprehensive study of effects of carbon coating on $Li_4Ti_5O_{12}$ anode material for lithium-ion batteries. J Electrochem Soc 158(2):A102–A109
32. Wagemaker M, Simon DR, Kelder EM et al (2006) A kinetic two-phase and equilibrium solid solution in spinel $Li_{4+x}Ti_5O_{12}$. Adv Mater 18(23):3169–3173
33. Wang F, Wu L, Ma C et al (2013) Excess lithium storage and charge compensation in nanoscale $Li_{4+x}Ti_5O_{12}$. Nanotechnology 24:424006. doi:10.1088/0957-4484/24/42/424006
34. Ganapathy S, Wagemaker M (2012) Nanozize storage properties in spinel $Li_4Ti_5O_{12}$ explained by anisotropic surface lithium insertion. ACS Nano 6(10):8702–8712
35. Kataokaa K, Takahashia Y, Kijima N et al (2009) A single-crystal study of the electrochemically Li-ion intercalated spinel-type $Li_4Ti_5O_{12}$. Solid State Ionics 180(6–8):631–635
36. Wagemaker M, Eck ERH, Kentgens APM et al (2009) Li-ion diffusion in the equilibrium nanomorphology of spinel $Li_{4+x}Ti_5O_{12}$. J Phys Chem B 113(1):224–230
37. Lu X, Zhao L, He X et al (2012) Lithium storage in $Li_4Ti_5O_{12}$ spinel: the full static picture from electron microscopy. Adv Mater 24(24):3233–3238
38. Borghols WJH, Wagemaker M, Lafont U et al (2009) Size effects in the $Li_{4+x}Ti_5O_{12}$ spinel. J Am Chem Soc 131(49):17786–17792
39. Ven AV, Jishnu Bhattacharya J, Belak AA (2013) Understanding Li diffusion in Li-intercalation compounds. Acc Chem Res 46(5):1216–1225
40. Laumann A, Boysen H, Bremholm M (2011) Lithium migration at high temperatures in $Li_4Ti_5O_{12}$ studied by neutron diffraction. Chem Mater 23(11):2753–2759
41. Zhao H, Li Y, Zhu Z (2008) Structural and electrochemical characteristics of $Li_{4-x}Al_xTi_5O_{12}$ as anode material for lithium-ion batteries. Electrochim Acta 53:7079–7083
42. Wang W, Jiang B, Xiong W et al (2013) A nanoparticle Mg-doped $Li_4Ti_5O_{12}$ for high rate lithium-ion batteries. Electrochim Acta 114:198–204
43. Ji M, Xu Y, Zhao Z (2014) Preparation and electrochemical performance of La^{3+} and F^- co-doped $Li_4Ti_5O_{12}$ anode material for lithium-ion batteries. J Power Sources 263:296–303
44. Zhang Q, Zhang C, Li B et al (2013) Preparation and electrochemical properties of Ca-doped $Li_4Ti_5O_{12}$ as anode materials in lithium-ion battery. Electrochim Acta 98:146–152
45. Zhang B, Du H, Li B et al (2010) Structure and electrochemical properties of Zn-doped $Li_4Ti_5O_{12}$ as anode materials in Li-ion battery. Electrochem Solid-State Let 13(4):A36–A38
46. Zhang B, Huang Z-D, Oh SW et al (2011) Improved rate capability of carbon coated $Li_{3.9}Sn_{0.1}Ti_5O_{12}$ porous electrodes for Li-ion batteries. J Power Sources 196:10692–10697
47. Tian B, Xiang H, Zhang L et al (2010) Niobium doped lithium titanate as a high rate anode material for Li-ion batteries. Electrochim Acta 55:5453–5458
48. Zhang Q, Zhang C, Li B et al (2013) Preparation and characterization of W-doped $Li_4Ti_5O_{12}$ anode material for enhancing the high rate performance. Electrochim Acta 107:139–146
49. Yi T-F, Shu J, Zhu Y-R et al (2010) Advanced electrochemical performance of $Li_4Ti_{4.95}V_{0.05}O_{12}$ as a reversible anode material down to 0 V. J Power Sources 195 (1):285–288
50. Yi T-F, Xie Y, Jiang L-J et al (2012) Advanced electrochemical properties of Mo-doped $Li_4Ti_5O_{12}$ anode material for power lithium ion battery. RSC Adv 2:3541–3547
51. Wolfenstine J, Allen JL (2008) Electrical conductivity and charge compensation in Ta doped $Li_4Ti_5O_{12}$. J Power Sources 180(1):582–585
52. Ma Y, Ding B, Ji G et al (2013) Carbon-encapsulated F-doped $Li_4Ti_5O_{12}$ as a high rate anode material for Li^+ batteries. ACS Nano 7(12):10870–10878

53. Du G, Sharma N, Peterson VK et al (2011) Br-doped $Li_4Ti_5O_{12}$ and composite TiO_2 anodes for Li-ion batteries: synchrotron X-ray and in situ neutron diffraction studies. Adv Funct Mater 21(20):3990–3997
54. Wu H, Chang S, Liu X et al (2013) Sr-doped $Li_4Ti_5O_{12}$ as the anode material for lithium-ion batteries. Solid State Ionics 232:13–18
55. Yi T-F, Chen B, Shen H-Y et al (2013) Spine $Li_4Ti_{5-x}Zr_xO_{12}$ ($0 \le x \le 0.25$) materials as high-performance anode materials for lithium-ions batteries. J Alloy Compd 558:11–17
56. Xiao CW, Ding Y, Zhang JT et al (2014) $Li_{4-x}Na_xTi_5O_{12}$ with low operation potential as anode for lithium ion batteries. J Power Sources 248:323–329
57. Yi T-F, Yang S-Y, Li X-Y et al (2014) Sub-micrometric $Li_{4-x}Na_xTi_5O_{12}$ ($0 \le x \le 0.20$) spinel as anode material exhibiting high rate capability. J Power Sources 246:505–511
58. Yoshikawa D, Kadoma Y, Kim J-M et al (2010) Spray-drying synthesized lithium-excess $Li_{4+x}Ti_{5-x}O_{12-\delta}$ and its electrochemical property as negative electrode material for Li-ion batteries. Electrochim Acta 55(6):1872–1879
59. Ge H, Li N, Li D et al (2008) Study on the effect of Li doping in spinel $Li_{4+x}Ti_{5-x}O_{12}$ ($0 \le x \le 0.20$) materials for lithium-ion batteries. Electrochem Commun 10(7):1031–1034
60. Capsoni D, Bini M, Massarotti V et al (2008) Cations distribution and valence states in Mn-substituted $Li_4Ti_5O_{12}$ structure. Chem Mater 20(13):4291–4298
61. Song H, Yun S-W, Chun H-H et al (2012) Anomalous decrease in structural disorder due to charge redistribution in Cr-doped $Li_4Ti_5O_{12}$ negative-electrode materials for high-rate Li-ion batteries. Energy Environ Sci 5:9903–9913
62. Capsoni D, Bini M, Massarotti V et al (2009) Cr and Ni doping of $Li_4Ti_5O_{12}$: cation distribution and functional properties. J Phys Chem C 113(45):19664–19671
63. Kaftelen H, Tuncer M, Tu S et al (2013) Mn-substituted spinel $Li_4Ti_5O_{12}$ materials studied by multifrequency EPR spectroscopy. J Mater Chem A 1:9973–9982
64. Lin J-Y, Hsu C-C, Ho H-P et al (2013) Sol–gel synthesis of aluminum doped lithium titanate anode material for lithium ion batteries. Electrochim Acta 87:126–132
65. Zhang Y, Zhang C, Lin Y et al (2014) Influence of Sc^{3+} doping in B-site on electrochemical performance of $Li_4Ti_5O_{12}$ anode materials for lithium-ion battery. J Power Sources 250:50–57
66. Park JS, Baek S-H, Jeong Y-I et al (2013) Effects of a dopant on the electrochemical properties of $Li_4Ti_5O_{12}$ as a lithium-ion battery anode material. J Power Sources 244:527–531
67. Huang S, Wen Z, Zhu X et al (2007) Effects of dopant on the electrochemical performance of $Li_4Ti_5O_{12}$ as electrode material for lithium ion batteries. J Power Sources 165:408–412
68. Qiu C, Yuan Z, Liu L et al (2013) Sol-gel synthesis and electrochemical performance of $Li_{4-x}Mg_xTi_{5-x}Zr_xO_{12}$ anode material for lithium-ion batteries. Chin J Chem 31:819–825
69. Gao J, Jiang C, Wan C (2010) Synthesis and characterization of spherical La-doped nanocrystalline $Li_4Ti_5O_{12}$/C compound for lithium-ion batteries. J Electrochem Soc 157(2): K39–K42
70. Li X, Qu M, Yu Z (2009) Structural and electrochemical performances of $Li_4Ti_{5-x}Zr_xO_{12}$ as anode material for lithium-ion batteries. J Alloy Compd 487:L12–L17
71. Li B, Han C, He Y-B et al (2012) Facile synthesis of $Li_4Ti_5O_{12}$/C composite with super rate performance. Energy Environ Sci 5:9595–9602
72. Zhu G-N, Wang C-X, Xia Y-Y (2011) A comprehensive study of effects of carbon coating on $Li_4Ti_5O_{12}$ anode material for lithium-ion batteries. J Electrochem Soc 158(2):A102–A109
73. Hu X, Lin Z, Yang K et al (2011) Effects of carbon source and carbon content on electrochemical performances of $Li_4Ti_5O_{12}$/C prepared by one-step solid-state reaction. Electrochim Acta 56:5046–5053
74. Luo H, Shen L, Rui K et al (2013) Carbon coated $Li_4Ti_5O_{12}$ nanorods as superior anode material for high rate lithium ion batteries. J Alloys Compds 572:37–42
75. Zhu G-N, Liu H-J, Zhang J-H et al (2011) Carbon-coated nano-sized $Li_4Ti_5O_{12}$ nanoporous micro-sphere as anode material for high-rate lithium-ion batteries. Energy Environ Sci 4:4016–4022

76. Jung H-G, Kim J, Scrosati B et al (2011) Micron-sized, carbon-coated $Li_4Ti_5O_{12}$ as high power anode material for advanced lithium batteries. J Power Sources 196:7763–7766
77. Zhu Z, Cheng F, Chen J (2013) Investigation of effects of carbon coating on the electrochemical performance of $Li_4Ti_5O_{12}$/C nanocomposites. J Mater Chem A 1:9484–9490
78. Wang Y, Liu H, Wang K et al (2009) Synthesis and electrochemical performance of nano-sized $Li_4Ti_5O_{12}$ with double surface modification of Ti (III) and carbon. J Mater Chem 19:6789–6795
79. Ding Z, Zhao L, Suo L et al (2011) Towards understanding the effects of carbon and nitrogen-doped carbon coating on the electrochemical performance of $Li_4Ti_5O_{12}$ in lithium ion batteries: a combined experimental and theoretical study. Phys Chem Chem Phys 13:15127–15133
80. Guo X, Xiang HF, Zhou TP et al (2014) Morphologies and structures of carbon coated on $Li_4Ti_5O_{12}$ and their effects on lithium storage performance. Electrochim Acta 130:470–476
81. Nugroho A, Chang W, Kim SJ et al (2012) Superior high rate performance of core-shell $Li_4Ti_5O_{12}$/carbon nanocomposite synthesized by a supercritical alcohol approach. RSC Adv 2:10805–10808
82. Li H, Shen L, Yin K et al (2013) Facile synthesis of N-doped carbon-coated $Li_4Ti_5O_{12}$ microspheres using polydopamine as a carbon source for high rate lithium ion batteries. J Mater Chem A 1:7270–7276
83. Zhao L, Hu Y-S, Li H et al (2011) Porous $Li_4Ti_5O_{12}$ coated with N-doped carbon from ionic liquids for Li-ion batteries. Adv Mater 23:1385–1388
84. Jung H-G, Myung S-T, Yoon CS et al (2011) Microscale spherical carbon-coated $Li_4Ti_5O_{12}$ as ultra high power anode material for lithium batteries. Energy Environ Sci 4:1345–1351
85. Shen L, Li H, Uchaker E et al (2012) General strategy for designing core-shell nanostructured materials for high-power lithium ion batteries. Nano Lett 12:5673–5678
86. Li CC, Li QH, Chen LB et al (2012) A facile titanium glycolate precursor route to mesoporous Au/$Li_4Ti_5O_{12}$ spheres for high-rate lithium-ion batteries. ACS Mater Interfaces 4:1233–1238
87. Krajewski M, Michalska M, Hamankiewicz B et al (2014) $Li_4Ti_5O_{12}$ modified with Ag nanoparticles as an advanced anode material in lithium-ion batteries. J Power Sources 245:764–771
88. Liu Z, Zhang N, Wang Z et al (2012) Highly dispersed Ag nanoparticles (<10 nm) deposited on nanocrystalline $Li_4Ti_5O_{12}$ demonstrating high-rate charge/discharge capability for lithium-ion battery. J Power Sources 205:479–782
89. Cheng C, Liu H, Xue X et al (2014) Highly dispersed copper nanoparticle modified nano $Li_4Ti_5O_{12}$ with high rate performance for lithium ion battery. Electrochim Acta 120:226–230
90. Feng X, Ding N, Dong Y et al (2013) A chromium oxide solution modified lithium titanium oxide with much improved rate performance. J Mater Chem A 1:15310–15315
91. Yang X, Huang Y, Wang X et al (2014) High rate capability core-shell lithium titanate@ceria nanosphere anode material synthesized by one-pot co-precipitation for lithium-ion batteries. J Power Sources 257:280–285
92. Wang D, Xu H-Y, Gu M et al (2009) $Li_2CuTi_3O_8$-$Li_4Ti_5O_{12}$ double spinel anode material with improved rate performance for Li-ion batteries. Electrochem Commun 11:50–53
93. Park K-S, Benayad A, Kang D-J et al (2008) Nitridation-driven conductive $Li_4Ti_5O_{12}$ for lithium ion batteries. J Am Chem Soc 130:14930–14931
94. Wan Z, Cai R, Jiang S et al (2012) Nitrogen- and TiN-modified $Li_4Ti_5O_{12}$: one-step synthesis and electrochemical performance optimization. J Mater Chem 22:17773–17781
95. Cai R, Jiang S, Yu X et al (2012) A novel method to enhance rate performance of an Al-doped $Li_4Ti_5O_{12}$ electrode by post-synthesis treatment in liquid formaldehyde at room temperature. J Mater Chem 22:8013–8021
96. Fang W, Zuo P, Ma Y et al (2013) Facile preparation of $Li_4Ti_5O_{12}$/AB/MWCNTs composite with high-rate performance for lithium ion battery. Electrochim Acta 94:294–299

97. Shen L, Yuan C, Luo H et al (2011) In situ growth of $Li_4Ti_5O_{12}$ on multi-walled carbon nanotubes: novel coaxial nanocables for high rate lithium ion batteries. J Mater Chem 21:761–767
98. Li X, Qu M, Huai Y et al (2010) Preparation and electrochemical performance of $Li_4Ti_5O_{12}$/carbon/carbon nano-tubes for lithium ion battery. Electrochim Acta 55:2978–2982
99. Oh Y, Nam S, Wi S et al (2014) Effective wrapping of graphene on individual $Li_4Ti_5O_{12}$ grains for high-rate Li-ion batteries. J Mater Chem A 2:2023–2027
100. Shen L, Yuan C, Luo H et al (2011) In situ synthesis of high-loading $Li_4Ti_5O_{12}$-graphene hybrid nanostructures for high rate lithium ion batteries. Nanoscale 3:572–574
101. Han SY, Kim IY, Jo KY et al (2012) Solvothermal-assisted hybridization between reduced graphene oxide and lithium metal oxides: a facile route to graphene-based composite materials. J Phys Chem C 116:7269–7279
102. Cai R, Yu X, Liu X et al (2010) $Li_4Ti_5O_{12}$/Sn composite anodes for lithium-ion batteries: synthesis and electrochemical performance. J Power Sources 195:8244–8250
103. Han SY, Kim IY, Lee SH et al (2012) Electrochemically active nanocomposites of $Li_4Ti_5O_{12}$ 2D nanosheets and SnO_2 0D nanocrystals with improved electrode performance. Electrochim Acta 74:59–64
104. Chem M, Li W, Shen X et al (2014) Fabrication of core–shell α-Fe_2O_3@$Li_4Ti_5O_{12}$ composite and its application in the lithium ion batteries. ACS Appl Mater Interfaces 6:4514–4523
105. Hu M, Jiang Y, Yan M (2014) High rate $Li_4Ti_5O_{12}$-Fe_2O_3 and $Li_4Ti_5O_{12}$-CuO composite anodes for advanced lithium ion batteries. J Alloys Compds 603:202–206
106. Rahman MM, Wang J-Z, Hassan MF et al (2011) Amorphous carbon coated high grain boundary density dual phase $Li_4Ti_5O_{12}$-TiO_2: a nanocomposite anode material for Li-ion batteries. Adv Energy Mater 1:212–220
107. Rahman MM, Wang JZ, Hassan MF et al (2010) Basic molten salt process-a new route for synthesis of nanocrystalline $Li_4Ti_5O_{12}$-TiO_2 anode material for Li-ion batteries using eutectic mixture of $LiNO_3$-LiOH-Li_2O_2. J Power Sources 195:4297–4303
108. Hu YS, Kienle L, Guo YG et al (2006) High lithium electroactivity of nanometer-sized rutile TiO_2. Adv Mater 18:1421–1426
109. Chen JS, Lou XW (2009) Anatase TiO_2 nanosheet: an ideal host structure for fast and efficient lithium insertion/extraction. Electrochem Commun 11:2332–2335
110. Sushko ML, Rosso KM, Liu J (2010) Mechanism of Li^+/electron conductivity in rutile and anatase TiO_2 nanoparticles. J Phys Chem C 114:20277–20283
111. Wang Y-Q, Gu L, Guo Y-G et al (2012) Rutile-TiO_2 nanocoating for a high-rate $Li_4Ti_5O_{12}$ anode of a lithium-ion battery. J Am Chem Soc 134:7874–7879
112. Chiu H-C, Brodusch N, Gauvin R et al (2013) Aqueous synthesized nanostructured $Li_4Ti_5O_{12}$ for high-performance lithium ion battery anodes. J Electrochem Soc 160(5): A3041–A3047
113. Xiao L, Chen G, Sun J et al (2013) Facile synthesis of $Li_4Ti_5O_{12}$ nanosheets stacked by ultrathin nanoflakes for high performance lithium ion batteries. J Mater Chem A 1:14618–14626
114. Song K, Seo D-H, Jo MR et al (2014) Tailored oxygen framework of $Li_4Ti_5O_{12}$ nanorods for high-power Li ion battery. J Phys Chem Let 5:1368–1373
115. Li Y, Pan L, Liu JW et al (2009) Preparation of $Li_4Ti_5O_{12}$ nanorods as anode materials for lithium-ion batteries. J Electrochem Soc 156(7):A495–A499
116. Lee SC, Lee SM, Lee JW et al (2009) Spinel $Li_4Ti_5O_{12}$ nanotubes for energy storage materials. J Phys Chem C 113:18420–18423
117. Kim J, Cho J (2007) Spinel $Li_4Ti_5O_{12}$ nanowires for high-rate Li-ion intercalation electrode. Electrochem Solid-State Let 10(3):A81–A84
118. Chou S-L, Wang J-Z, Liu H-K et al (2011) Rapid synthesis of $Li_4Ti_5O_{12}$ microspheres as anode materials and its binder effect for lithium-ion battery. J Phys Chem C 115:16220–16227

119. Zhang Z, Li G, Peng H et al (2013) Hierarchical hollow microspheres assembled from N-doped carbon coated $Li_4Ti_5O_{12}$ nanosheets with enhanced lithium storage properties. J Mater Chem A 1:15429–15434
120. Tang YF, Yang L, Qiu Z et al (2008) Preparation and electrochemical lithium storage of flower-like spinel $Li_4Ti_5O_{12}$ consisting of nanosheets. Electrochem Commun 10:1513–1516
121. Lin Y-S, Tsai M-C, Duh J-G (2012) Self-assembled synthesis of nanoflower-like $Li_4Ti_5O_{12}$ for ultrahigh rate lithium-ion batteries. J Power Sources 214:314–318
122. Fattakhova D, Krtil P (2002) Electrochemical activity of hydrothermally synthesized Li-Ti-O cubic oxides toward Li insertion. J Electrochem Soc 149(9):A1224–A1229
123. Lim J, Choi E, Mathew V et al (2011) Enhanced high-rate performance of $Li_4Ti_5O_{12}$ nanoparticles for rechargeable Li-ion batteries. J Electrochem Soc 158(3):A275–A280
124. Feckl JM, Fominykh K, Doblinger M et al (2012) Nanoscale porous framework of lithium titanate for ultrafast lithium insertion. Angew Chem 124:7577–7581
125. Yu L, Wu HB, Lou XW (2013) Mesoporous $Li_4Ti_5O_{12}$ hollow spheres with enhanced lithium storage capability. Adv Mater 25:2296–2300
126. Li D, Mccann JT, Xia Y et al (2006) Electrospinning: a simple and versatile technique for producing ceramic nanofibers and nanotubes. J Am Chem Soc 89(6):1861–1869
127. Xu H, Hu X, Luo W et al (2014) Electrospun conformal $Li_4Ti_5O_{12}$/C fibers for high-rate lithium-ion batteries. ChemElectroChem 1:611–616
128. Liu J, Tang K, Song K et al (2013) Tiny $Li_4Ti_5O_{12}$ nanoparticles embedded in carbon nanofibers as high-capacity and long-life anode materials for both Li-ion and Na-ion batteries. Phys Chem Chem Phys 15:20813–20818
129. Chen S, Xin Y, Zhou Y et al (2014) Self-supported $Li_4Ti_5O_{12}$ nanosheet arrays for lithium ion batteries with excellent rate capability and ultralong cycle life. Energy Environ Sci 7:1924–1930
130. Liu J, Song K, van Aken PA et al (2014) Self-supported $Li_4Ti_5O_{12}$-C nanotube arrays as high-rate and long-life anode materials for flexible Li-ion batteries. Nano Lett 14:2597–2603
131. Amine K, Belharouak I, Chen Z et al (2010) Nanostructured anode material for high-power battery system in electric vehicles. Adv Mater 22:3052–3057
132. Lin C, Fan X, Xin Y et al (2014) Monodispersed mesoporous $Li_4Ti_5O_{12}$ submicrospheres as anode materials for lithium-ion batteries: morphology and electrochemical performances. Nanoscale 6:6651–6660
133. Shen L, Yuan C, Luo H et al (2010) Facile synthesis of hierarchically porous $Li_4Ti_5O_{12}$ microspheres for high rate lithium ion batteries. J Mater Chem 20:6998–7004
134. Pasquier AD, Plitz I, Menocal S et al (2003) A comparative study of Li-ion battery, supercapacitor and nonaqueous asymmetric hybrid devices for automotive applications. J Power Sources 115:171–178
135. Belharouak I, Koenig GM, Tan T et al (2012) Performance degradation and gassing of $Li_4Ti_5O_{12}/LiMn_2O_4$ lithium-ion cells. J Electrochem Soc 159(8):A1165–A1170
136. Qin Y, Chen Z, Amine K (2011) Functionalized surface modification agents to suppress gassing issue of $Li_4Ti_5O_{12}$-based lithium-ion chemistry. In: Fiscal year 2011 annual progress for report energy storge R&D, U.S. Department of Energy, Washington, D.C., p 321
137. He Y-B, Liu M, Huang Z-D et al (2013) Effect of solid electrolyte interface (SEI) film on cyclic performance of $Li_4Ti_5O_{12}$ anodes for Li ion batteries. J Power Sources 239:269–276
138. Wu K, Yang J, Liu Y et al (2013) Investigation on gas generation of $Li_4Ti_5O_{12}$/$LiNi_{1/3}Co_{1/3}Mn_{1/3}O_2$ cells at elevated temperature. J Power Sources 237:285–290
139. Wu K, Yang J, Zhang Y et al (2012) Investigation on $Li_4Ti_5O_{12}$ batteries developed for hybrid electric vehicle. J Appl Electrochem 42:989–995
140. Qin Y, Chen Z, Amine K (2012) Functionalized surface modification agents to suppress gassing issue of $Li_4Ti_5O_{12}$-based lithium-ion chemistry. In: Fiscal year 2012 annual progress for report energy storge R&D, U.S. Department of Energy, Washington, D.C., p 363
141. Kitta M, Akita T, Maeda Y et al (2012) Study of surface reaction of spinel $Li_4Ti_5O_{12}$ during the first lithium insertion and extraction processes using atomic force microscopy and analytical transmission electron microscopy. Langmuir 28(33):12384–12392

142. He Y-B, Ning F, Li B et al (2012) Carbon coating to suppress the reduction decomposition of electrolyte on the $Li_4Ti_5O_{12}$ electrode. J Power Sources 202:253–261
143. Pan H, Hu Y-S, Chen L (2013) Room-temperature stationary sodium-ion batteries for large-scale electric energy storage. Energy Environ Sci 6:2338–2360
144. Zhao L, Pan H-L, Hu Y-S et al (2012) Spinel lithium titanate ($Li_4Ti_5O_{12}$) as novel anode material for room-temperature sodium-ion battery. Chin Phys B 21(2):028201
145. Sun Y, Zhao L, Pan H et al (2013) Direct atomic-scale confirmation of three-phase storage mechanism in $Li_4Ti_5O_{12}$ anodes for room-temperature sodium-ion batteries. Nat Commun 4:1870
146. Yu X, Pan H, Wan W et al (2013) A size-dependent sodium storage mechanism in Li4Ti5O12 investigated by a novel characterization technique combining in situ X-ray diffraction and chemical sodiation. Nano Lett 13:4721–4727
147. Yu P, Li C, Guo X (2014) Sodium storage and pseudocapacitive charge in textured $Li_4Ti_5O_{12}$ thin films. J Phys Chem C 118:10616–10624
148. Wang J, Li W, Yang Z et al (2014) Free-standing and binder-free sodium-ion electrodes based on carbon-nanotube decorated $Li_4Ti_5O_{12}$ nanoparticles embedded in carbon nanofibers. RSC Adv 4:25220–25226
149. Yi T-F, Jiang L-J, Shu J et al (2010) Recent development and application of $Li_4Ti_5O_{12}$ as anode material of lithium ion battery. J Phys Chem Solids 71(9):1236–1242
150. Wu HM, Belharouak I, Deng H et al (2009) Development of $LiNi_{0.5}Mn_{1.5}O_4/Li_4Ti_5O_{12}$ system with long cycle life. J Electrochem Soc 156(12):A1047–A1050
151. Xiang HF, Zhang X, Jin QY et al (2008) Effect of capacity matchup in the $LiNi_{0.5}Mn_{1.5}O_4$/ $Li_4Ti_5O_{12}$ cells. J Power Sources 183:355–360
152. Li SR, Chen CH, Xia X et al (2013) The impact of electrolyte oxidation products in $LiNi_{0.5}Mn_{1.5}O_4/Li_4Ti_5O_{12}$ cells. J Electrochem Soc 160(9):A1524–A1528
153. Dedryvère R, Foix D, Franger S et al (2010) Electrode/electrolyte reactivity in high-voltage spinel $LiMn_{1.6}Ni_{0.4}O_4/Li_4Ti_5O_{12}$ lithium-ion battery. J Phys Chem C 114:10999–11008
154. Kim J-H, Pieczonka NPW, Sun Y-K et al (2014) Improved lithium-ion battery performance of $LiNi_{0.5}Mn_{1.5-x}Ti_xO_4$ high voltage spinel in full-cells paired with graphite and $Li_4Ti_5O_{12}$ negative electrodes. J Power Sources 262:62–71
155. https://www.toshiba.com/tic/industrial/rechargable-battery
156. Jaiswal A, Horne CR, Chang O et al (2009) Nanoscale $LiFePO_4$ and $Li_4Ti_5O_{12}$ for high rate Li-ion batteries. J Electrochem Soc 156(12):A1041–A1046
157. Borgel V, Gershinsky G, Hu T et al (2013) $LiMn_{0.8}Fe_{0.2}PO_4/Li_4Ti_5O_{12}$, a possible Li-Ion battery system for load-leveling application. J Electrochem Soc 160(4):A650–A657

Alloy-Based Anode Materials

D. Pribat

1 Introduction/Context

Without lithium-ion batteries (LIBs), modern mobile devices and systems such as lap top computers, tablets, smart phones and electric vehicles just would not exist. Quite paradoxically, LIBs are also the weakest link of such devices and systems, limiting their autonomy or their driving range. For instance, in order to increase the available energy of portable electronic devices, manufacturers are now distributing several batteries inside the casings, wherever they can find room, so that the owner cannot even change those batteries by her/himself when they are out of order. Actually, after more than 25 years of steady progress, LIBs seem to have reached their asymptotic capacity values with the present combination of graphite at the anode and insertion oxide or phosphate materials at the cathode. Progress is now rather marginal, while new applications, particularly all-electric vehicles, necessitate higher capacity systems in order to be competitive with vehicles propelled by internal combustion engines. Such new applications are pushing the development of novel electrode materials/configurations with (much) higher Li storage capabilities, for both electrodes.

As we shall see in this chapter, Li can react with elements or compounds in different ways, namely by means of insertion, alloying or conversion reactions. When alloying or conversion reactions take place, they are usually accompanied by a strong electrode volume change and restructuration, including bond breaking and phase transformations. Schematically, the more Li can be packed in the host anode material, the larger the volume expansion and the associated mechanical degradation problems upon repeated alloying–dealloying. Moreover, lithium alloys are highly ionic (there are many Zintl phases) and consequently they are brittle and do

D. Pribat (✉)
Department of Energy Science, Sungkyunkwan University, Suwon, South Korea
e-mail: pribat2@yahoo.fr

Z. Zhang and S.S. Zhang (eds.), *Rechargeable Batteries*,
Green Energy and Technology, DOI 10.1007/978-3-319-15458-9_7

not withstand mechanical stresses induced by volume changes. This also contributes to a rapid degradation and pulverization of the electrode as lithiation and delithiation cycles are applied.

Over the past few years, research efforts have focused on the use of nanostructured anode materials (nanoparticles, nanocrystals, nanowires, nanorods etc., including complex core-shell and composite structures) in order to mitigate the effects of volume change upon Li uptake. Nanostructured materials present the add-on advantage of shorter diffusion distances for Li species, thus offering the possibility to increase charging and discharging rates (i.e., battery power). This chapter will essentially concentrate on such recent developments.

The chapter is organized as follows: after this short introduction, the next paragraph briefly shows why Li metal cannot be used as an active anode material. Section 3 then establishes a list of attributes for a good anode. In Sect. 4, the different types of Li reaction with elements and compounds are presented. Section 5 describes Li alloying with single elements, emphasizing column IV elements, then column VI and column III, by order of importance. In Sect. 6, Li alloying with intermetallic compounds and multi-element alloys is briefly presented. Conversion reactions are discussed in Sect. 7, and section height presents a short conclusion.

2 Why Alloy/Conversion Anodes for Li-Ion Batteries?

Metallic lithium (Li, atomic number 3 and atomic mass 6.94 g) crystallizes in the body centered cubic (bcc) system with a lattice parameter of 351 pm. This corresponds to an atomic density of $\sim 4.597 \times 10^{22}$ at/cm^3, yielding a specific weight of ~ 0.53 g/cm^3, which is the lowest of all metals. Li also exhibits the lowest redox potential of the periodic table (−3.045 V against a standard hydrogen electrode). Because of those properties, lithium should obviously be a first choice for anodes of rechargeable lithium-based batteries. Unfortunately, Li tends to grow in the form of dendrites when re-plated on the anode surface during battery charging [1] (even when polymer electrolytes are employed, see Fig. 1), which after a number of charge-discharge cycles can lead to short circuits between anode and cathode, inducing risks of thermal runaway and fire.

Because of the above-mentioned security problems, Li metal has been replaced by an insertion-type anode, resulting in the concept of Li-ion technology, where no Li metal appears anymore [2]. Instead, Li ions are intercalated in host materials exhibiting very different Li chemical potentials at either the cathode or the anode. For most today commercial products, Li ions are shuttled back and forth between a graphite anode and an oxide cathode ($LiMO_2$, with M being a transition metal such as Co, Mn or Ni) or a phosphate cathode (e.g., $LiFePO_4$) as the battery is charged and discharged (the so-called "rocking chair" operation). Graphite is an interesting insertion material because it exhibits only a ~ 10 % increase of the interlayer graphite spacing (c-axis) and less than 1 % expansion in the basal graphite plane at maximum Li insertion [3]. However, graphite can only store 1 Li for 6 carbon

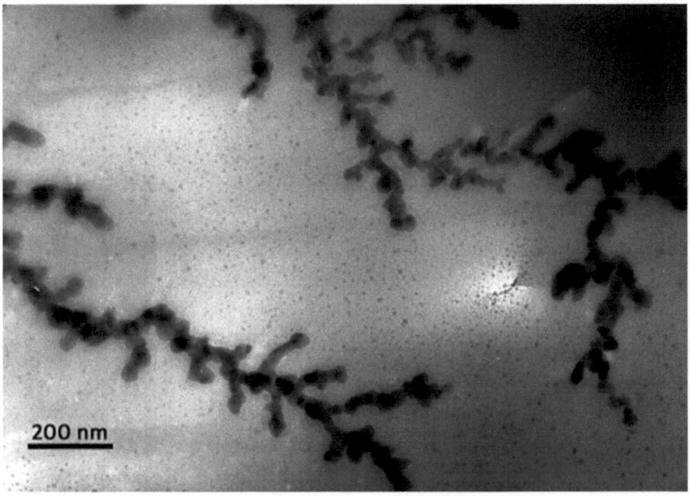

Fig. 1 An example of Li metal dendritic growth through a polymer electrolyte (Image courtesy of G.M. Stone/UC Berkeley and LBNL)

atoms (LiC_6), which corresponds to a modest capacity of 372 mAh/g. This is the main reason why other anode materials are currently being actively studied. Before entering the heart of the subject, let us examine in the next paragraph what should be the attributes of a good anode.

3 What Should Be the Attributes of a Good Anode?

The requirements for battery electrodes can vary according to the type of application. For instance, thousands charge-discharge cycles are needed for an electric vehicle (EV), whereas few hundreds might be satisfactory for mobile electronic devices. On the other hand, smaller battery volumes are more adapted to portable devices. Also, a small charging time and an increased capacity are game changing improvements for EVs, ultimately eliminating battery swapping, contributing to increase the acceptance of EVs by the public at large.

Whatever the application, one of the main parameters characterizing an anode is its capacity, which can be defined as the number of Li-based charges that can be reversibly stored in a unit mass (specific capacity, traditionally expressed in Ah/g; 1 Ah = 3600 C) or in a unit volume (volumetric capacity, expressed in Ah/cm^3) of the active anode material. For instance, the specific capacity of an anode made of pure lithium is just the number of Li atoms/g in the metal multiplied by the elementary charge (since lithium redox reactions involve one electron only), i.e., $\sim$3.860 Ah/g or 3860 mAh/g (corresponding to $\sim$2046 mAh/cm^3). For an alloy such as Li_xM_y, the specific capacity can be expressed as:

$$C = \frac{x \times q \times N_{Av}}{y \times M_{at} \times 3600}\ \text{Ah/g}, \tag{1}$$

where q is the elementary charge (1.602×10^{-19} C), N_{Av} is the Avogadro number and M_{at} is the atomic mass of element M. For example, the capacity of the LiAl compound calculated with expression (1) is 991 mAh/g, while that of $Li_{22}Sn_5$ is 993 mAh/g and that of $Li_{15}Si_4$ is 3579 mAh/g, close to that of pure Li.

At this point, it should be emphasized that the capacity of an electrode is not an absolute parameter, since it varies with the charge/discharge rate of the electrode. Actually, one has to realize that in a battery, electrons are not directly available as they would be in a capacitor. Rather, the availability of electrons results from interface reactions ($Li^+ + e^- \Leftrightarrow Li$), which can be slow. Moreover, concerning alloy or insertion anodes, Li has to diffuse inside the host material to reach the interface with the electrolyte, which can take time ($\tau \approx l^2/D$, where τ is the diffusion time, l the distance to travel and D the diffusion coefficient of Li in the host material). Finally, Li^+ transport in the electrolyte (which must balance the electron transport in the outside circuit) is generally slower than electron transport in the external circuit. For all the above reasons, the capacity of the electrode decreases as its charge/discharge rate increases and capacity values are usually specified for low charge/discharge rates, e.g., C/20 or even C/50. A C/N rate corresponds to a complete charge/discharge of the battery in N hours; for instance, a C/20 rate means charging/discharging the battery in 20 h; a 2 C rate means charging/discharging the battery in ½ h.

Once the capacity of a particular material is known, there are some fabrication constraints for the corresponding anode [4], so that a more relevant parameter is often the electrode capacity per unit area of the current collector, expressed in mAh/cm^2. This number incorporates the mass loading of active material per unit area of the current collector, a quantity which is always limited in practice. For example, as quoted above, graphite exhibits a maximum capacity of 372 mAh/g, which corresponds to the formation of the LiC_6 insertion compound (note that this is ~10 times lower than the capacity of pure Li). For the fabrication of commercial anodes, graphite powders are mixed with some inactive components, including a binder and a conductive carbon additive, so that the graphite loading is only 85–90 wt% of the total mass. Moreover, the thickness of the graphite-based deposit on the current collector is commonly below ~80 μm, because of delamination/adhesion problems at larger thickness values. Taking a density of 2 g/cm^3 for graphite and assuming a ~30 % porosity of the graphite-based deposit (the porosity is necessary for electrolyte permeation), this yields an anode capacity of 3–4 mAh/cm^2 (per side of the current collector), depending on the exact thickness of the graphite-based deposit and also depending on the exact porosity value.

A second important parameter is the potential of the anode (versus Li^+/Li^0), which, for a given cathode material, determines the final voltage delivered by the battery. A higher potential versus Li/Li^+ of the negative electrode induces a lower cell voltage. For instance, the potential of many Li alloys is comprised between ~0.2 and ~1.0 V versus Li^+/Li^0 whereas it is only ~0.1 V versus Li^+/Li^0 for graphite.

The thermodynamics of Li-based batteries are detailed in Chap. 1 of this book (or elsewhere in the literature [5]), and it is not our purpose to review this topic here. From Chap. 1 or Ref. [5], we understand that the open circuit voltage V_{OC} of a lithium cell results from the difference in the chemical potential of lithium between the cathode (μ_{Li-C}) and the anode (μ_{Li-A}) and can be expressed as:

$$V_{OC} = -\frac{\mu_{Li-C} - \mu_{Li-A}}{nF}, \tag{2}$$

where F is the Faraday constant and n = 1 for the $Li^+ + e^- \Leftrightarrow Li$ equilibrium. The chemical potential of Li in a host material is related to its thermodynamic activity, a_{Li} (which can be viewed as an effective concentration), so that:

$$\mu_{Li-A,C} = \mu^0_{Li} + RTLn(a_{Li-A,C}), \tag{3}$$

where R is the gas constant, T the absolute temperature and $a_{Li-A,C}$ is the Li activity either in the anode (A) or the cathode (C); a_{Li} in pure metallic lithium is equal to unity.

The situation is summarized on Fig. 2a, which is reproduced from a recent review by Goodenough and Kim [9]. It follows from expression (2) that for a given cathode material, in order to maximize the cell voltage, one has to use an anode material in which the chemical potential of Li in the host structure (which translates into a voltage against Li^+/Li^0 according to an expression similar to (2), when the host/alloying material is tested as a positive electrode in the so-called half-cell configuration using a pure Li anode as the negative electrode) must be as close to that of pure Li as possible, i.e., zero. Figure 2b shows the voltages versus Li^+/Li^0 of some insertion/alloying compounds, including some popular cathode materials (which are treated in other chapters of this book). Figure 2b also shows the window stability of the mostly used non-aqueous electrolyte, namely $LiFP_6$ in a mixture of ethylene carbonate (EC) and diethyl carbonate (DEC). This window stability corresponds to the energy difference between the lowest unoccupied molecular orbital (LUMO) and the highest occupied molecular orbital (HOMO) of the organic components of the electrolyte. If the chemical potential of the anode (or the cathode) lies outside the stability window of the electrolyte, the latter will be reduced (or oxidized in the case of the cathode). In both cases, the unwanted decomposition of the electrolyte will alter the proper functioning of the battery, leading to quick failure after a few charge-discharge cycles if the electrolyte decomposition is not blocked.

Unfortunately, as shown in Fig. 2b, if we except $Li_4Ti_5O_{12}$ and TiS_2, most alloy or insertion anode materials exhibit a potential versus Li^+/Li^0 which falls outside the stability window of the $LiFP_6$/EC/DEC electrolyte. This means that the electrolyte will be reduced as the battery is charged, i.e., when the voltage is brought below ~1 V versus Li^+/Li^0 (unless otherwise mentioned, all voltage values are expressed with reference to Li^+/Li^0). However, as indicated in Fig. 2a, this situation

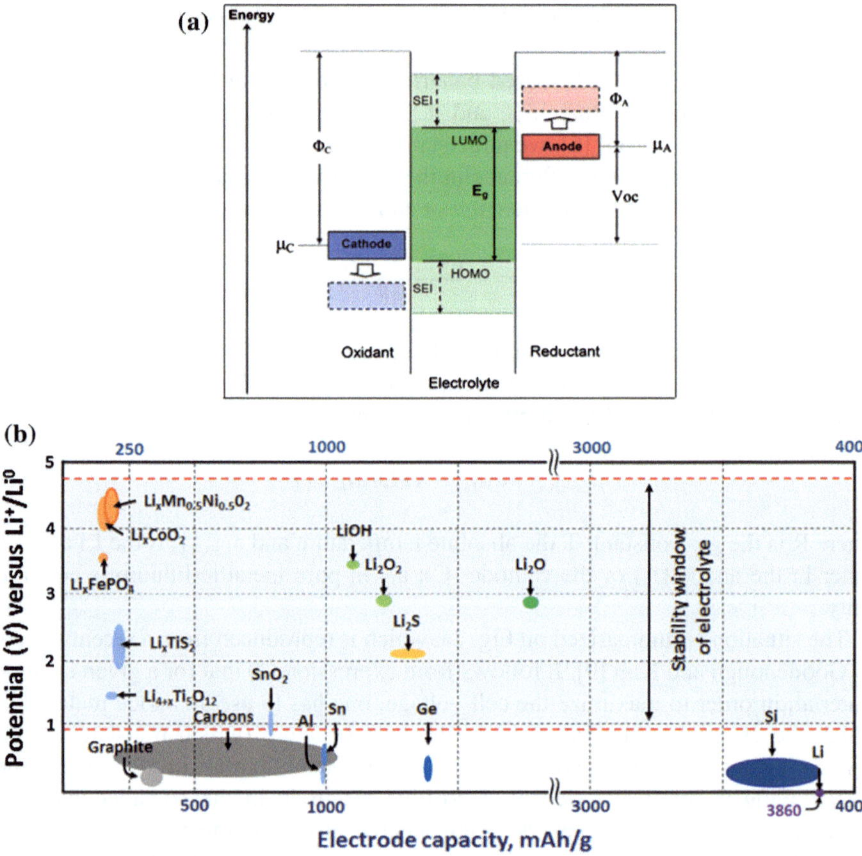

Fig. 2 **a** Energy diagram of a lithium cell in open circuit condition. Φ_A and Φ_C are the respective work functions of the anode and cathode. E_g is the energy window in which the electrolyte is thermodynamically stable. The stability domain of the electrolyte can be kinetically extended by the formation of a passivating film (the so-called solid electrolyte interphase—SEI) on each electrode. **b** Voltage (relative to Li^+/Li^0) and gravimetric capacity of several electrode materials. The light blue domain between the *dotted red lines* represents the thermodynamic stability window of the most common electrolyte used for Li-ion batteries: 1 M $LiPF_6$ in a 1:1 mixture of ethylene carbonate/diethyl carbonate. Capacity values for the Li-O_2 and Li-S systems are indicative [6, 7] as they are still debated [8]. Adapted and reprinted with permission from Ref. [9]

can be circumvented by the formation of a passivating film on the surface of the anode, which kinetically extends the stability domain of the electrolyte. This passivating film (usually called solid electrolyte interphase—SEI [10]) blocks electron transfer between the anode and electrolyte (hence preventing further electrolyte reduction), but is permeable to Li^+ ions, thus allowing their reduction at the anode when the battery is charged and their release in the electrolyte when the battery is discharged. The SEI usually forms naturally upon the first battery charge, where electrolyte decomposition is unavoidable. If the SEI is mechanically and chemically stable, there is no more electrolyte decomposition after the first battery charge and the anode is stabilized, even though its chemical potential lies outside the electrolyte stability window. The SEI composition depends mainly on the particular electrolyte and lithium salt used for battery fabrication and to a lesser extent on the anode material itself. For a graphite-based anode and a $LiFP_6$/EC/DEC electrolyte, the SEI is typically composed of lithium compounds (Li_2CO_3, LiF, Li_2O), polyolefins, semicarbonates, etc. [10–12]. Note that the stability of the SEI on graphite-based anodes is another reason for their large commercial success.

With the above remarks in mind, the third important parameter for a good anode is its ability to support/generate a stable SEI (which can necessitate the use of a particular electrolyte and the addition of some SEI promoters in the electrolyte [13]). For instance, the dimensional stability (small volume change) of the anode upon Li alloying/insertion, will guaranty the mechanical stability of the SEI on its surface. Actually, if the volume variation of the anode during Li insertion/alloying is too important (poor mechanical stability), the SEI will tend to locally break, thus exposing parts of the bare anode surface to the electrolyte, which will promote more local electrolyte decomposition, trapping more Li in some carbonates, fluorides, oxides or other Li-bearing decomposition products. Because the amount of cyclable Li is limited by the cathode content, continued Li trapping induces a rapid capacity fading as the battery charge/discharge cycling is carried on. Note that in any case, and even if the SEI is stable after the first battery charge, its formation irreversibly consumes some of the electrolyte and Li originally present in the battery casing, which induces an irreversible capacity loss.

The fourth parameter determining the usefulness of an anode is its cycle lifetime. This is the number of battery charge-discharge cycles it can withstand without significant degradation, i.e., without significant capacity loss. Intuitively, one feels that this parameter will depend on the magnitude of the capacity (the larger the capacity, the more Li can be packed, but the larger the swelling/straining/deformation of the host structure) and also on the depth of charge/discharge of the electrode. The Coulombic efficiency (CE), which is a closely related parameter, corresponds to the fraction of the prior charge that can be delivered during the following discharge. Of course, the CE has to be as close to 100 % as possible. For a 99.9 % value, corresponding to a 0.1 % only loss per charge-discharge cycle (and which could look like a good performance), the capacity will be reduced to ~82 % of its original value after 200 cycles and down to ~37 % after 1000 cycles. For a ~90 % capacity retention after 1000 cycles, one needs a 99.99 % CE.

A fifth important parameter is the charge/discharge rate of the anode, the so-called C-rate which has been defined earlier. The C-rate is controlled by several material as well as structural properties such as the Li diffusivity in the host material, the electrode structure (for a given coefficient of diffusion, the Li diffusion time is obviously shorter in nanosized grains), its electronic conductivity (which can vary with the state of charge), the quality of the contact with the current collector, etc. Typically, when a particular material is tested against the C-rate, various charge/discharge currents are applied to the anode. For instance, if a pure graphite anode is characterized, applying a 372 mA/g charge/discharge current will correspond to cycling the electrode at 1 C (or C/1) rate. In order to cycle the electrode at a C/10 rate, the experimenter will apply a 37.2 mA/g current, etc.

4 The Different Types of Li Reaction with Various Elements and Compounds

Lithium can alloy or react with a large number of elements or compounds, but in different ways. The major types of reactions found in electrochemical systems are (i) formation reactions, (ii) conversion reactions and (iii) insertion reactions [5]. When considering the possible reaction of lithium with other elements, it is useful to first have a look at the corresponding phase diagrams, even though most reactions do not occur at thermodynamic equilibrium.

A *formation (or alloying) reaction* can be represented by the following chemical equilibrium:

$$xLi + yM \Leftrightarrow Li_xM_y, \tag{4}$$

where usually the phase structure of the Li_xM_y reaction product is different from that of the parent element M. If the phase structure of the parent element is preserved, then the reaction product is just a solid solution ($y = 1$ and $x \leqq 1$); however, from a practical standpoint, solid solutions are not so interesting for Li storage since they are very often limited to at best a few atomic Li percent ($x \ll 1$), particularly at room temperature, which corresponds to small capacities. Examples of materials undergoing formation reactions are crystalline Si, Sn, Al and Sb.

For a *conversion (or displacement) reaction*, the equilibrium can be expressed as:

$$xLi + MN_y \Leftrightarrow Li_xN_y + M, \tag{5}$$

where the lithium displaces element M from phase MN_y, forming a new Li_xN_y phase. This type of reaction has first been exploited [14] and explained with transition metal oxides [15] (M = Co, Fe, Ni, Cu): $2Li + MO \Leftrightarrow Li_2O + M$, and it has been extended to transition metal sulfides, nitrides [16], fluorides [17] and phosphides. Note that the displaced element, M, can also react with Li according to an alloying reaction similar to (4). In the latter case however, reaction (5) is not reversible and there is a charge loss between the first lithiation and the following ones. Chapter 5 of this book presents displacement reactions for cathode materials.

Finally, *insertion reactions* involve the occupation of empty interstitial sites of the host crystal by Li ions. This type of reaction and the corresponding host anode materials are the object of Chap. 6 of this book (see also Ref. [18]). Graphite, with its layered structure, is the best example of a material very well adapted to such a reaction [18].

Before concluding this paragraph, and as briefly discussed below, we would like to emphasize that the type of reaction incurred by the anode material imposes the shape of the charge-discharge curve (e.g., in galvanostatic plots). Consider the Gibbs phase rule, which indicates the number of degrees of freedom, f, of a closed system under thermodynamic equilibrium: $f = C + 2 - P$, where C is the number of independent constituents and P the number of phases of the system. Actually, f is the number of intensive thermodynamic parameters that must be defined to completely describe the system. For instance, consider a simple formation reaction, where M is a pure element and Li_xM_y a compound exhibiting a crystal structure different from that of M. We have 2 constituents in our system, namely Li and M and 2 phases, which are M and Li_xM_y. According to the Gibbs phase rule, $f = 2$; so if the pressure (P) and the temperature (T) are kept constant, there are no degrees of freedom left for the system. Therefore, the chemical potential of Li in Li_xM_y has a fixed value and hence the electrode potential is constant as long as the 2 phases coexist. In other words, only the relative amounts of M and Li_xM_y change during lithiation. If the lithiation of M is represented in a voltage versus composition (or state of charge) graph, the curve will exhibit a long plateau as long as the 2 phases coexist. Note that if Li alloying does not change the structure of M (solid solution case), then we only have one phase in our system and $f = 3$; so at constant P and T, there is finally one degree of freedom in the system. Hence, the chemical potential of Li in M varies as lithiation progresses and so does the electrode potential; a sloping profile will be observed in a voltage versus composition graph. For a displacement reaction such as (5), there are 3 elements and 3 phases, so that the number of degrees of freedom is again zero (at constant P and T). Therefore, the voltage versus composition graph will exhibit a plateau as long as the 3 phases coexist.

Since the trend in the fabrication of battery electrodes is clearly towards the use of nanostructures, a word of caution is necessary, as the shape of voltage profiles can be altered by the particle size of their constituting materials [19]. For instance, the voltage-capacity curves for the lithiation–delithiation of Si at room temperature usually exhibit plateau or plateau-like regions (see below, Sect. 5.2). However, those voltage-capacity curves become sloped for nanosized Si [20]. The same type of behavior is observed for Sb-based alloys [21]. This can be explained by difficulties in nucleating a new phase at low dimensions (and high surface area), since the cost of overcoming the surface energy becomes too high. Sometimes, the grain size of the mother phase can even be smaller than the size of the critical nucleus for the nucleation of the new lithiated phase. Hence, the phase diagram can be modified by the extension of the solubility range of Li in solid solutions. Anatase Li_xTiO_2 is a clear example of this behavior [22].

5 Lithium Alloying with Selected Elements

Although we focus in this paragraph on Li alloying with simple elements of the periodic table, very often amorphous carbon is added to the particular element under scrutiny, either to prevent re-agglomeration of nanostructures (when such nanostructures are of concern) or to improve the electrical conductivity of the anode (especially when the particular element is a semiconductor). However amorphous carbon is usually inert towards Li and it does not participate in lithiation–delithiation reactions.

5.1 Column IV of the Periodic Table

Li interactions/alloys with elements from column IV of the periodic table are probably the most studied and the most interesting for battery anodes. This is particularly true for C, Si and Sn.

5.1.1 Carbon

Carbon, in its graphite allotropic form, is the mainly used material for Li-ion batteries anodes. However, carbon exists in many different forms, which result in Li reaction mechanisms other than the well-known intercalation. Capacities up to ~1100 mAh/g can be reached with carbonaceous materials. Although this topic and particularly Li intercalation in graphite is treated in Chap. 6, for the sake of completeness, we just recall here the major features of carbon-based anodes, particularly those using nanostructured carbons. Apart from graphite, sp^2-hybridized carbonaceous materials have been classified as soft carbons (graphitizable) or hard carbons (nongraphitizable), depending on the way their organic precursor decomposes upon pyrolysis in an inert gas [23]. For instance, Fig. 1 of Ref. [23] (not shown here) displays the Li capacity versus heat treatment temperature for a variety of soft and hard carbons. Clearly there are several possible interaction mechanisms which are briefly summarized in panels a–d of Fig. 3 [18, 23–27].

The general chemical reaction describing Li interaction with carbon-based materials can be expressed as:

$$6C + xLi^+ + xe^- \Leftrightarrow Li_xC_6 \qquad (6)$$

The most common situation is Li intercalation in graphite (Fig. 3a), which proceeds by the so-called staging phenomenon [18]. This staging phenomenon is described by an index, which indicates the number of unoccupied graphene layers between two Li-occupied layers [18, 26]. As already quoted, the maximum intercalation capacity is 372 mAh/g, which corresponds to x = 1 in reaction (6).

At this point, it is important to note that larger values of x can be obtained, up to 3 for some hard carbon materials (or mixtures of hard and soft carbons), the latter corresponding to the Li_3C_6 compound with a capacity of 1116 mAh/g. For instance, Zhou and coworkers [28] have used ordered mesoporous carbon (CMK-3) and they have obtained reversible Li capacities between 850 and 1100 mAh/g (corresponding to $2.3 < x < 3$). Unfortunately, their CMK-3 anodes exhibit a large hysteresis between charge and discharge, which results in a low energy efficiency

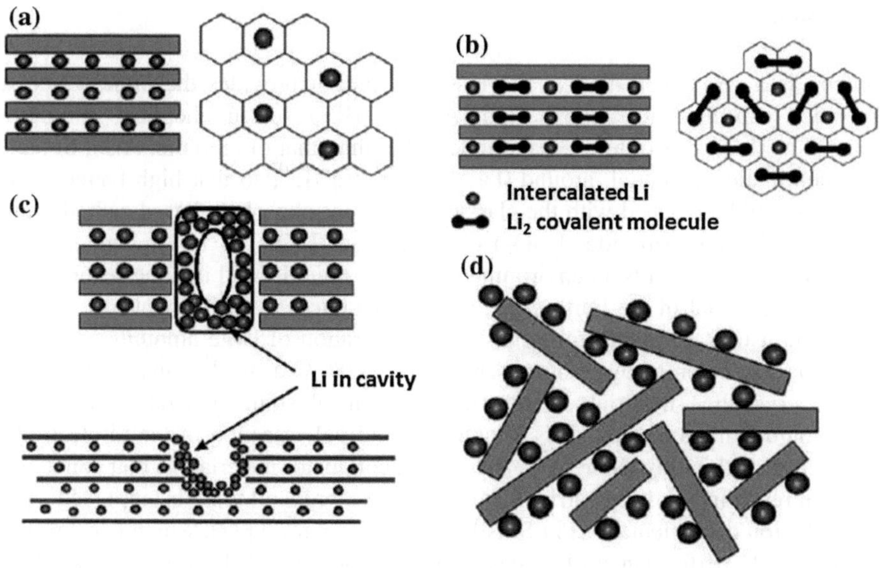

Fig. 3 Various carbonaceous materials and the way they store Li. **a** Li storage in graphite corresponding to LiC_6 at maximum capacity [18]. **b** Formation of Li_2 covalent molecules as proposed by Sato [24–26]. **c** Li storage in cavities and nanopores [25, 26]. **d** Li storage on both sides of isolated graphene sheets [25–27]. Reprinted with permission from Ref. [25]

for a full cell with an oxide cathode. In another example, Lahiri and coworkers have also obtained high capacity values (up to 1100 mAh/g) using multiwalled carbon nanotubes (MWNTs) coated with alumina [29]. Note however that MWNTs tend to suffer from Li-induced embrittlement [30] because when Li is inserted, the concentric and closed structure of MWNTs does not allow expansion of graphene sheets in the radial direction of MWNTs (which would correspond to the c-axis in graphite), thus inducing large stress levels. Of course, this embrittlement may have a deleterious influence upon anode ageing.

From a general point of view, disordered and partially graphitized carbons can yield larger capacities than graphite because they provide more sites for Li adsorption, including defect sites. For instance, the formation of Li clusters at nanocavities is shown on Fig. 3c. For graphene sheets, Li can adsorb on both sides, yielding higher capacities than graphite (Fig. 3d). Even if carbonaceous materials (other than graphite) induce problems such as a high irreversible capacity, a poor cycle lifetime and a low reaction potential, near 0 V (the latter being a potential safety issue because of possible Li deposition and dendrite formation), they are still actively studied because of their relatively simple synthesis methods. More details can be found in Chap. 6 of this book, as well as in Refs. [18] and [23–27]. Graphene applications in the field of energy storage are briefly highlighted in a recent topical paper [31].

5.1.2 Silicon

Silicon is a very interesting anode material since it provides the highest known specific capacity for room temperature (RT) electrochemical lithiation (>3600 mAh/g), one order of magnitude larger than that of graphite. Also, Si has a low delithiation potential, around 0.4 V against Li^+/Li^0, so that high battery voltages can still be reached with the classical oxide- or phosphate-based cathodes and also with lithium peroxide (Li_2O_2) or hydroxide (LiOH) if Li-Air cathodes are concerned. Finally, Si is an environment-friendly material and it is the second most abundant material in the Earth crust, which guarantees availability at low cost for future industrial use. Unfortunately, the incorporation of large amounts of lithium into Si is accompanied by a volume change of ~280 % as well as an amorphization of the crystalline material, so that the mechanical strain generated during the lithiation/delithiation steps leads to cracking and pulverization of the Si electrodes after a few cycles only. Moreover, Si is a semiconductor material, so that contacting it might result in Schottky barrier formation unless proper doping is employed. Also the diffusion coefficient of Li in Si is low ($\sim 10^{-13}$ cm^2/s) [32], which can strongly limit the rate performance of Si-based anodes. Despite all those drawbacks, Si is still one of the most studied materials, and several review papers have already been published over the past few years, concerning powders, thin films and more recently nanostructures [33–38], which indicates a rapidly moving research field of particular interest.

A number of structural studies, whether in situ or ex situ have revealed the phase transformations occurring in Si during alloying with Li. More specifically, in situ X-ray diffraction (XRD) analysis [39, 40], in situ transmission electron microscopy observations [41, 42], ex situ pair distribution function analysis [43] etc. have shown that crystalline Si (c-Si) converts to an amorphous phase (a-Li_xSi) during the first lithiation. The a-Li_xSi amorphous phase (where x ~3.5 is the most accepted value) usually crystallizes to the metastable c-$Li_{15}Si_4$ compound towards the end of lithiation, when the voltage goes below ~60 mV, thus reaching a capacity of 3579 mAh/g. This c-$Li_{15}Si_4$ polycrystalline phase exists over a range of compositions, so that Dahn and coworkers have labelled it $Li_{15\pm\delta}Si_4$ [40]. Hence the maximum capacity of a Si anode is ~3800 mAh/g corresponding to δ ~0.9 in overlithiated $Li_{15+\delta}Si_4$.

The first delithiation starts with an equilibrium between c-$Li_{15}Si_4$ and an amorphous Li_2Si phase; once all the Li extracted, the host Si is left amorphous (a-Si). For the second and following lithiation cycles, there is a first equilibrium between a-Si and another amorphous phase of composition $Li_{2.5}Si$ [44]. Once all the a-Si consumed, a third amorphous phase appears (with Li/Si > 2.5), in equilibrium with $Li_{2.5}Si$. When the lithiation voltage approaches zero, the c-$Li_{15}Si_4$ phase crystallizes again. The two-phase lithiation mechanisms of c-Si and a-Si are very similar, with a sharp phase boundary between c-Si and a-$Li_{3.5}Si$ or between a-Si and a-$Li_{2.5}Si$ [42, 44, 45]. Figure 4 summarizes the atomistic mechanism of Si lithiation, whether the starting material is crystalline or amorphous.

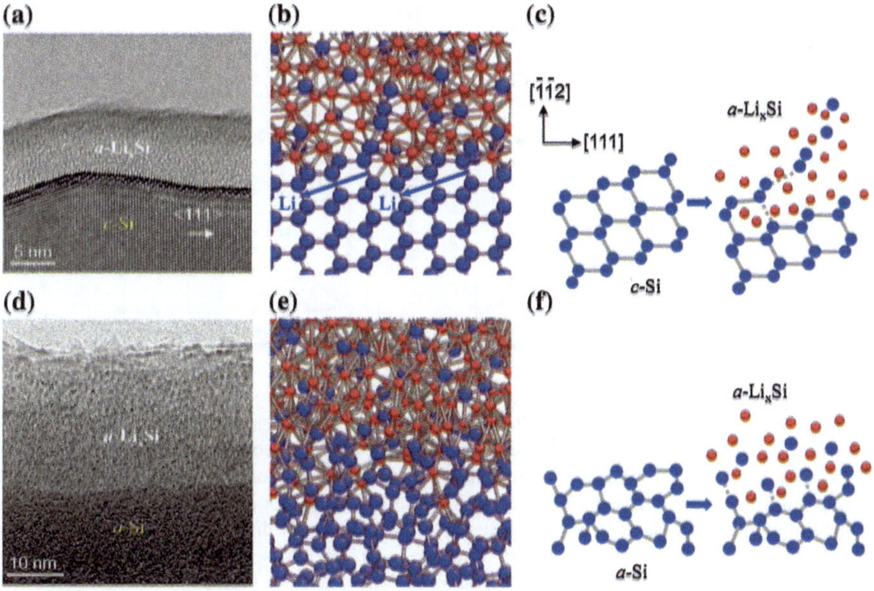

Fig. 4 Similarities of the two-phase lithiation mechanism of c-Si and a-Si. **a** High resolution transmission electron microscope (HRTEM) image showing the sharp phase boundary between c-Si and a-$Li_{3.5}$Si. **b** A snapshot from molecular dynamics simulation showing the atomic structures near the amorphous-crystal boundary. **c** The high Li concentration (*red atoms*) at the amorphous-crystal boundary softens the Si–Si covalent bonds, facilitating the peeling of Si atoms (*blue*) from the c-Si surface. **d** HRTEM image showing the sharp boundary between a-Si and a-$Li_{2.5}$Si (corresponding to the first sloping plateau of the galvanostatic lithiation curve of a-Si, see Fig. 5b, *red* or *blue curve*). **e** A snapshot from molecular dynamics simulation showing the atomic structures near the a-Si—a-$Li_{2.5}$Si interface. The a-Si consists of a continuous random network of Si atoms. **f** Schematics of the model for the formation of a-$Li_{2.5}$Si at the interface with a-Si. Here again, the high concentration of Li at the interface facilitates Si–Si bond breaking. Reprinted with permission from Ref. [44]. Copyright (2013) American Chemical Society

The second and following delithiation steps are similar to the first one. Maybe the charge-discharge behavior of a c-Si electrode can be summarized by referring to Fig. 5. Note however that the crystallization of the $Li_{15}Si_4$ phase does not take place in certain nanostructures, e.g., undoped nanowires [37, 41].

Although it is generally preferable to avoid phase changes to improve the lifetime and reliability of an electrode, whether crystallization of the $Li_{15}Si_4$ phase at the end of lithiation occurs or not is probably of little importance, since as pointed out by Liu and coworkers [46], the structure of the amorphous phase is very close to that of the crystal, which also explains the sudden crystallization of $Li_{15}Si_4$ when it takes place (atoms in the amorphous state are close to the positions they have in the crystal).

Finally, if instead of c-Si, amorphous Si is used as the starting material, the lithiation behavior is identical to that of the second cycle of c-Si (Fig. 5b, 2nd cycle in red).

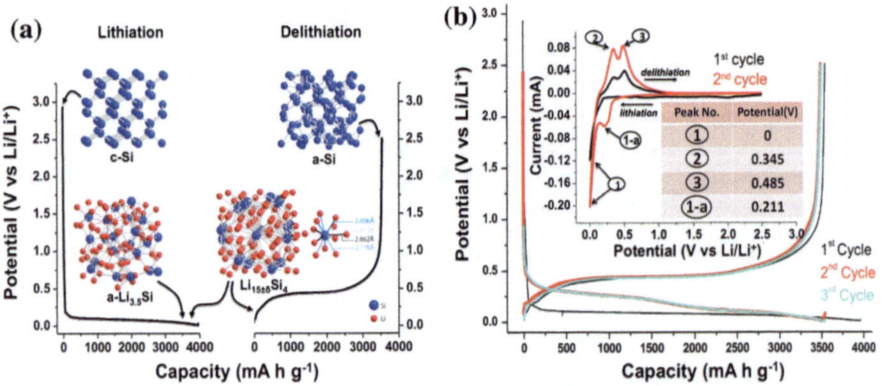

Fig. 5 Galvanostatic and cyclic voltammetry curves recorded on arrays of crystalline Si nanowires (NWs) mounted in a half-cell geometry with a Li counter electrode. **a** The first galvanostatic cycle, showing a slightly sloping plateau, starting at around 100 mV and corresponding to the equilibrium between c-Si and a-$Li_{3.5}Si$ (*black curve, left-hand side*). At the end of lithiation, when the voltage approaches 0 V, c-Si has totally disappeared and transformed to a-$Li_{3.75}Si$ that may suddenly crystallize into the c-$Li_{15}Si_4$ phase. Delithiation (*black curve, right-hand side*) produces a steep rise in potential, followed by a plateau at ~430 mV, which corresponds to an equilibrium between c-$Li_{15}Si_4$ (or a-$Li_{3.75}Si$) and a-Li_2Si. At the end of the delithiation step, when Li is totally extracted from the alloy, Si is left amorphous. The atomic structures of the various elements or alloys corresponding to particular alloying states have been constructed with the Crystal Maker software. **b** The three first galvanostatic curves of the Si NW array emphasizing the difference in lithiation behavior between the first cycle (c-Si, *black curve*) and the following ones (a-Si, *red* and *blue curves*). The 1st lithiation curve is of course identical to the one in (**a**), whereas the 2nd and 3rd lithiation curves show the 2 sloping plateaus which are typical of a-Si lithiation. Upon delithiation, the 3 curves are similar. The insert shows the 2 first cyclic voltammetry curves recorded on similar Si NW arrays. There is a single sharp peak for the first lithiation (*black curve*, peak marked as "*1*") corresponding to the one step, 2-phase lithiation of c-Si. Note that since the NWs are undoped, there is no peak corresponding to the crystallization of $Li_{15}Si_4$, which should occur below 80 mV. (See e.g. Ref. [37] for a more detailed discussion). The second lithiation produces 2 peaks (*red curve*, peaks marked as "*1*" and "*1-a*"), corresponding to the 2-step lithiation of a-Si. Upon delithiation, the curves for the first and second cycle are qualitatively similar, exhibiting 2 peaks (labeled 2 and 3) at approximately the same voltage. Reprinted with permission from Ref. [37]

To a large extent, the use of nanostructured Si (nanoparticles, nanowires, nanopillars etc.) can prevent cracking/pulverization and rapid degradation of Si-based anodes. For instance, c-Si nanoparticles do not crack upon full lithiation when their diameter is kept below ~150 nm [47]. Similarly, c-Si nanopillars keep their integrity unless their diameter is above ~300–360 nm [48, 49]. Note at this point that the lithiation of monocrystalline Si is highly anisotropic, the largest volume expansion occurring along ⟨110⟩ directions [48, 50, 51]. However, since c-Si becomes amorphous after the first charge-discharge cycle, this anisotropy should be of little importance for an anode, as long as cracking is prevented during the first lithiation. Particles with larger diameters can avoid cracking if a-Si (instead of c-Si) is the starting material [45].

Once the cracking problem solved, there still remain to stabilize the interface between Si and the electrolyte by forming a stable SEI (see Sect. 3 above). Actually, because of the large Si volume increase upon lithiation, the SEI tends to break, thus locally failing to protect the Si surface. As a consequence, the newly exposed Si surface keeps reacting with the electrolyte, trapping more Li in SEI products and inducing capacity fading. Furthermore, the SEI tends to thicken, which slows down Li exchanges, affecting the charge-discharge rate of the electrode. The problem is amplified by the use of nanostructures which exhibit large surface to volume ratios.

In order to solve the above-mentioned problem of SEI stability, various approaches have been explored during the past few years, aiming at keeping the mechanical advantages of nanostructures (i.e., avoid cracking), while preventing SEI breaking, still allowing Si to expand/contract. Schematically, those approaches have been applied to nanoparticles (0D materials), nanowire/nanotube-like structures (1D materials) and graphene-encapsulated nanostructures (2D-like materials). We do not pretend here to be exhaustive, but rather to illustrate some of the most promising approaches.

Brought along by the study of core-shell nanoparticles [52] (0D materials), hollowed structures have been designed, that are capable of accommodating Si expansion. Figure 6 shows an example of a hollow core-shell carbon-silicon (C-Si) structure, where oxidized Si nanoparticles are first coated with a carbon layer, followed by SiO_2 dissolution in hydrofluoric acid (HF) [53]. The resulting hollowed structure allows the Si to expand upon lithiation without breaking the C shell (Fig. 6a). Hence a stable SEI can be formed on this C shell, which does not experience any deformation. Such core-shell structures can be deposited on a current collector using the familiar slurry-type process, and because the overall core-shell nanoparticle is dimensionally stable during lithiation–delithiation, the electrical contact between neighboring particles is not affected by charge-discharge cycling. This hollowed core-shell structure has been recently improved by Cui and coworkers [54] who grouped together the C-Si core-shell nanoparticles into larger clusters, which are themselves surrounded by a thicker outer carbon layer which protects them from the electrolyte. The improved structure is inspired by a pomegranate fruit. An anode capacity of ~2500 mAh/g has been obtained (at C/20 rate), stable over 1000 charge-discharge cycles, using charge-discharge currents up to 3.7 mAh/cm^2, comparable to those of commercial Li-ion batteries [54].

Concerning 1D nanostructures, Si nanowires (NWs) have been largely investigated, including core-shell and coated structures (see Ref. [37] for a recent review, including the main synthesis methods for Si NWs). Porous Si NWs have shown interesting properties since the empty space inside the pores allows accommodation of some of the Si volume change. Using porous Si NWs etched from boron-doped Si wafers, Zhou and coworkers have reached 2000 charge-discharge cycles at a 0.5/1 C rate, keeping the reversible capacity over 1000 mAh/g [55].

Nanotube-like Si structures (Si NT) represent another interesting 1D approach, since the inside hollowed part of the structure can absorb some of the volume increase during lithiation, thus limiting the overall swelling of the host Si NT.

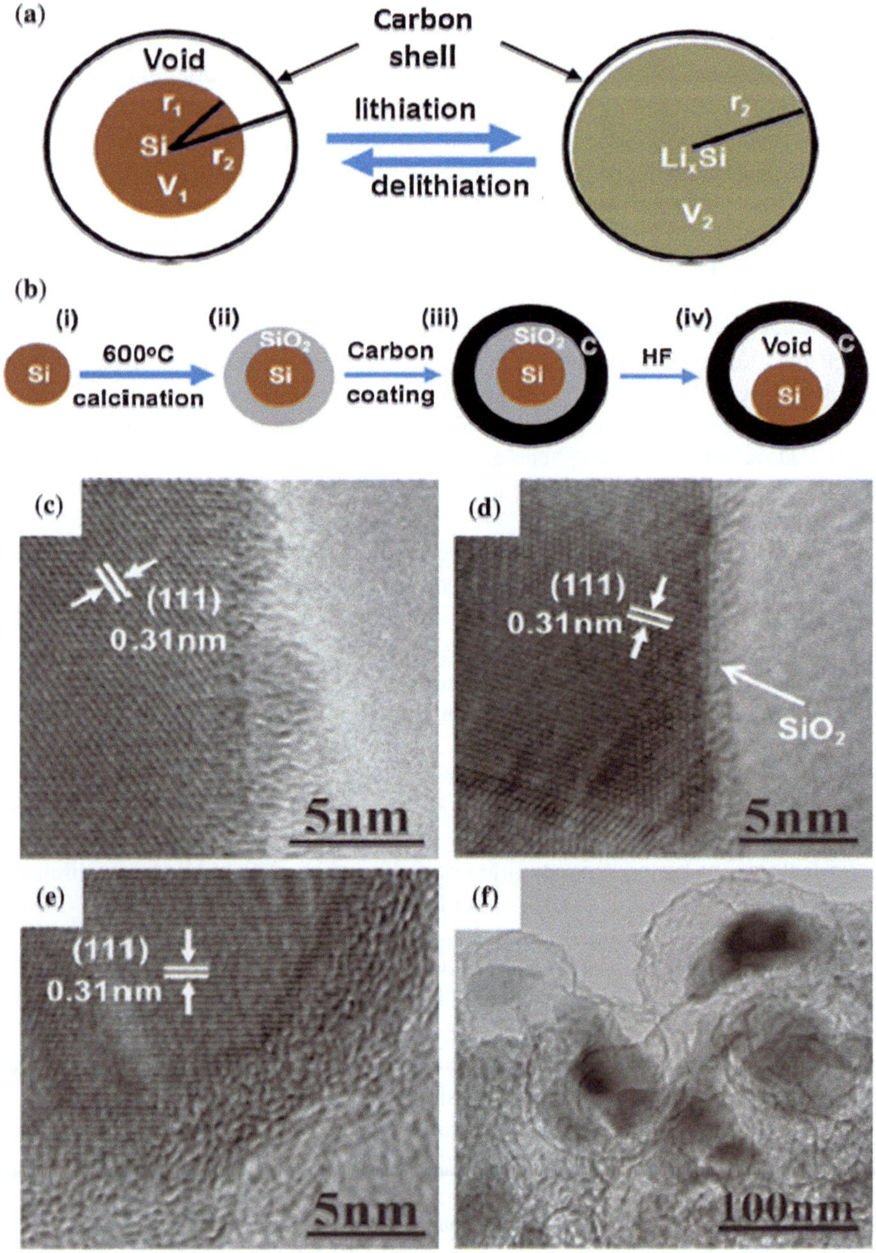

Fig. 6 Hollow Si-C core-shell nanostructure. **a** Schematic of the principle; Si can expand upon alloying with Li without breaking the carbon shell, that keeps its original dimension. **b** Schematic fabrication process, starting with Si nanoparticles. **c–f** Transmission electron microscope images of a pristine Si nanoparticle (**c**); a Si nanoparticle covered by a thin layer of SiO_2 (**d**); an oxidized nanoparticle (Si/SiO_2) coated with a thin carbon layer (**e**); some hollowed core-shell structures obtained after the removal of the SiO_2 layer (**f**). Reprinted with permission from Ref. [53]

Hence the SEI on the outer surface can be stabilized. Several groups have explored this route and interesting results have been obtained. Park and coworkers [56] have synthesized their nanotube-like Si structures inside porous anodic alumina templates and they have obtained a good stability upon charge-discharge cycling in a half cell, together with high charge-discharge rates of up to 5 C. They also built a complete battery, with a $LiCoO_2$ cathode, which exhibited a 89 % capacity retention after 200 charge-discharge cycles at 1 C rate. Song and coworkers [57] deposited a conformal Si shell on sacrificial ZnO NW cores and subsequently removed the ZnO using a moderate temperature (600 °C) reduction process; their mechanical model shows that free expansion at the inner surface of the Si NTs eases the volume increase during Si lithiation. Following those ideas, Cui and coworkers [58] synthesized their Si NTs by using sacrificial carbon fibers on which a CVD Si shell was deposited. The core carbon fibers were subsequently removed by a 500 °C thermal treatment in air which also provided a protective SiO_2 coating around the Si NTs. This outside SiO_2 coating (permeable to Li ions) served as a mechanical clamping layer, preventing expansion of the outer surface of the Si NT during lithiation, while the volume increase was accommodated by partially filling up the empty core of the Si NT. Because the outside surface of the Si nanotubes does not, or only slightly expands, the SEI which deposits on it is mechanically stable and serves as an efficient passivation layer. Hence, the stability of anodes based on those Si nanotubes is excellent, as 85 % of the original capacity is still retained after 6000 charge-discharge cycles at 12 C. The authors also report charging rates of up to 20 C which is one of the highest values reached for Si (if we except the results of Ref. [41] obtained inside the TEM in particular conditions). A rather severe drawback is that the areal Si mass is small, typically below 0.1 mg/cm^2. Also, because most of the Si NTs is empty, the volumetric capacity of the corresponding anodes is necessarily small.

A third type of approach consists in "wrapping" the Si nanostructures inside flexible graphene foils, resulting in 2D-like configurations. Actually, one can intuitively feel that graphene will provide a highly deformable electrical contact that can follow the expansion and contraction of the Si nanostructures as they are lithiated and delithiated. Graphene is the material of choice, since it is not only flexible, but also chemically inert and it exhibits a good electrical conductivity. Figure 7 schematically shows Si nanoparticles (NPs) embedded inside defect-decorated, porous graphene sheets [59].

Actually, although high temperature CVD processes on transition metal foils have been (and still are) developed for the synthesis of large area, electronic grade graphene [60], the simpler way to obtain defect-decorated graphene flakes/films for use in batteries or supercapacitors relies on solution-based processes involving the chemical exfoliation of graphite or graphite derivatives such as graphite oxide [61]. A typical preparation sequence for the fabrication of graphene films from liquid suspensions will first include the chemical oxidation of graphite particles with mixtures of strong acids and oxidants as in the Hummers method [62], which after ultrasonication in water will yield stable colloidal suspensions of hydrophilic graphite oxide (GO) sheets. GO which is a brittle and insulating material can be

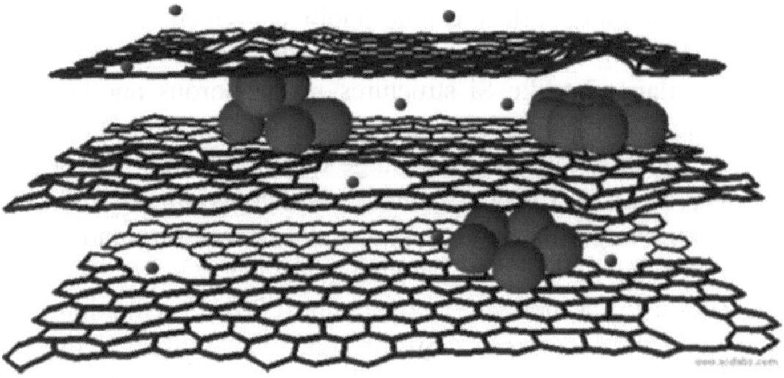

Fig. 7 A schematic drawing (not to scale) of a section of a composite electrode material constructed with a graphenic scaffold with in-plane carbon vacancy defects. The graphene sheets with these holey defects are displaced from each other for clarity. Electrochemically active components, for example, Si nanoparticles (*large spheres*), are sandwiched between graphene sheets, and these composites are structurally integrated with a 3-D graphenic network of interconnecting graphitic domains formed by reconstituting these graphene sheets. Li ions (*small spheres*) can diffuse easily across graphene sheets throughout the structure by passing through the in-plane vacancy defects. Reprinted with permission from Ref. [59]

spin-coated or spray-deposited or filtered and subsequently reduced back to defect-decorated graphene (actually called reduced graphite oxide, RGO), either chemically (e.g., using hydrazine hydrate [63]) or thermally [64]. Note that RGO is an ideal material since there is no need for a high crystalline quality graphene material for electrode applications and large lattice defect concentrations are even necessary to let the Li ions and electrolyte molecules permeate through the gas-tight carbon network [65]. However, making uniform composites of defect-decorated porous graphene sheets (e.g., RGO) and Si nanostructures is by no means an easy task. The simplest approach seems to make a water-based suspension with the Si nanostructures and GO sheets, spread/spray the suspension on the current collector (Cu foil) and then reduce the GO back to graphene. However, Si NPs/NWs are always oxidized, so that their OH-terminated surface tends to be negatively charged in solution. On the other hand, GO is also negatively charged because of the ionization of carboxylic acid and phenolic hydroxyl groups [66]. Hence the GO flakes tend to repel the Si nanostructures, which yields highly non-uniform composites. Zhou et al. [67] and Ye et al. [68] have used electrostatic self-assembly by modifying the surface charge of the Si particles and graphene sheets [68] with various polymers. Once the electrostatic assembly performed in water-based solutions, the composites are dried and fired at high temperature (≧500 °C) in a controlled reducing atmosphere. Sophisticated (and expensive) processes have also been developed for Si NWs, where they are first "wrapped" within a CVD graphene-like deposit and then further embedded within RGO sheets [69, 70]. Having said this, inspection of Fig. 7 tells us that if the electrolyte penetrates between the graphene sheets, then the SEI will form on Si NPs, electronically isolating them from the graphene. This will lead to rapid capacity fading. Moreover, most published papers show relatively low capacities for those graphene (or RGO)-Si composites, typically below 1500 mAh/g.

To briefly summarize this paragraph on Si, we emphasize that nanostructuring is indeed beneficial and that, thanks to clever engineering (including band gap engineering at the nanoscale to control the lithiation front[1] [71]), a number of reports show lifetimes over 1000 charge-discharge cycles with capacity values over 1250 mAh/g. However, for most approaches, increasing the "mass loading" (mg/cm^2) of anodes based on Si nanostructures is another challenge for the future. Typical mass loadings of several mg/cm^2 of active material, i.e., $\sim$10 times more than presently obtained should be reached. This low areal mass loading is a general drawback of nanomaterials, particularly when they are used in slurries where more inert binder and conductive additives are necessary. Finally, as with all industrial products, cost issues will probably determine the future of Si-based anodes and the simpler overall process will win in the end.

Several companies presently commercialize Si-based materials for anodes or Si-based anodes. For instance, Nexeon Ltd, a SME in the UK, has patented a nanostructured Si pillars technology (US patents no. 7402829, 7683359, and 7842535) which is supposed to overcome the problem of cycling, as well as provide larger capacities. SiNode, founded in 2012 through Northwestern University's NUvention program, is commercializing a patented composite of silicon nano-particles and RGO developed by Pr. Kung [59]. California Lithium Battery, (CalBattery) uses a process developed by Junbing Yang at Argone National Laboratory; again it is a composite of graphene and Si, where the Si is directly deposited on graphene by CVD from a chlorosilane compound [72]. Amprius which is a spin-off from Yi Cui's lab in Stanford also develops Si-based anodes.

5.1.3 Germanium

As silicon, germanium (Ge) crystallizes in the diamond structure and the Li-Ge system is analogous to the Li–Si system. However, probably because of its very high cost [73], Ge has received much less attention than Si. The relative abundance of Ge in the Earth's crust is about 1.5 ppm, when it is $\sim$27 % for Si. Ge gravimetric capacity is also lower than that of Si (1384 mAh/g for the $Li_{15}Ge_4$ compound), although its volumetric capacity is comparable to that of Si (7367 mAh/cm^3 for $Li_{15}Ge_4$, and 8334 mAh/cm^3 for $Li_{15}Si_4$).

Even though in some work performed on Ge films, several crystalline Li-Ge compounds were observed during lithiation [74], it seems that crystalline Ge is progressively amorphized during Li insertion (just as c-Si), at least for nanostructures [75, 76]. As with Si, in situ XRD and TEM analysis show that lithiated germanium crystallizes into cubic $Li_{15}Ge_4$ at the end of the full lithiation step, whether the starting material is crystalline [75, 76] or amorphous [77]. The delithiation of the c-$Li_{15}Ge_4$ compound yields a porous Ge amorphous phase [75, 76]. This porous structure tends to minimize the mechanical degradation during

[1] Liu et al. [71].

charge-discharge cycling [76]. After the first charge-discharge cycle, the material oscillates between c-$Li_{15}Ge_4$ at full lithiation and porous amorphous Ge when completely delithiated (again as with Si).

However, there are several major differences between Si and Ge. First, because of a smaller band gap, Ge is a better electronic conductor than Si. Second, the Li diffusivity in Ge is 2 orders of magnitude higher than that in Si [32]. Those two previous differences enable high-rate performance of Ge anodes [74, 78], particularly those made of nanotube-like structures [79]. The third difference is the toughness of Ge nanoparticles, that do not crack (for sizes between 100 and ~700 nm) during multiple lithiation–delithiation cycles, despite a ~260 % volume increase upon full lithiation [75]. The same behavior was observed for Ge NWs [76]. This has been explained by the isotropic lithiation of c-Ge, which contrast with that of c-Si [48–50]. Because of all the above properties and advantages over Si, there is a renewed interest in Ge [80–82], or Ge alloys [77, 83] in the literature. Recent work shows that anodes based on Ge NWs retain a capacity of 900 mAh/g after 1100 charge-discharge cycles, while showing excellent rate performance, even at 100 C [84]. To the best of our knowledge, there are no commercial developments for anodes based on Ge.

5.1.4 Tin

β-Sn (or white tin), which is the room temperature (300 K) stable allotrope crystallizes in the tetragonal structure. Below ~286 K (13 °C), β-Sn tends to transform into the α phase (grey tin) which adopts the diamond structure, like Ge and Si above in column IV of the periodic table. The tetragonal metallic β phase can be stabilized at low temperature by small amounts of Bi, Pb, Ag … The melting temperature of β-Sn is ~232 °C, much lower than those of Si (1414 °C) and Ge (938 °C).

The electrochemical lithiation of β-Sn has been reviewed by Winter and Besenhard in 1999 [85]. There are 9 different compounds likely to appear as the Li content in Sn increases, namely, Li_2Sn_5, LiSn, Li_3Sn_2 (doubtful existence), Li_7Sn_3, Li_5Sn_2, $Li_{13}Sn_5$, Li_7Sn_2, Li_4Sn (doubtful existence as well) and $Li_{22}Sn_5$. More recent in situ XRD characterizations of thin Sn films [86] show that β-Sn transforms to Li_2Sn_5, then to β-LiSn and finally to $Li_{22}Sn_5$. There is no amorphization of the starting element (in contrast with Si and Ge) since all the above compounds are crystalline. Figure 8 summarizes the various Sn-based phases that appear during the first lithiation–delithiation cycle observed on a thin Sn film. The fully lithiated $Li_{22}Sn_5$ compound corresponds to a Li/Sn ratio of 4.4 and a theoretical capacity of 993 mAh/g. Because of the large volume increase of ~360 % when going from Sn to $Li_{22}Sn_5$, and because of the successive phase transformations, anodes made of pure Sn (e.g., thin films [87]) rapidly disintegrate and fail.

Using nanostructures is usually a successful approach in order to cope with a large volume expansion; however, even 10 nm-diameter Sn nanoparticles (monodisperse) tend to crack upon alloying with Li, meaning that (again in contrast with Si) size reduction only is probably not a good option for solving the

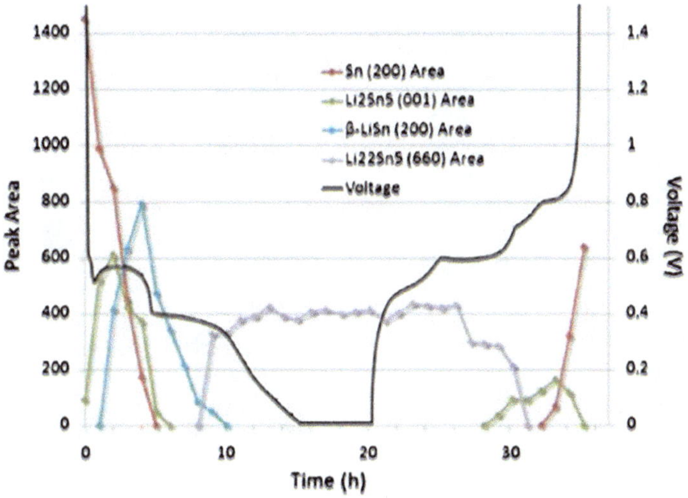

Fig. 8 In situ XRD characterization of a β-Sn thin film during the first lithiation–delithiation cycle. *Peak* areas for selected diffraction planes from each Sn-based phase are shown, together with voltage (against Li^0/Li^+) and degree of lithiation. Reprinted with permission from Ref. [86]. Copyright 2012, The Electrochemical Society

decrepitation/disintegration problem of Sn-based anodes [88]. The agglomeration of nanoparticles, preventing them from keeping their original nanometric size, has been invoked as one of the reasons for cracking [89].

In order to prevent agglomeration, different approaches have been investigated, which schematically consist in embedding the Sn nanoparticles into some kind of "binder". For instance, using 5 nm-diameter particles embedded in a nitrogen-doped porous carbon network, Zhu et al. have obtained a stable capacity of 722 mAh/g during 200 cycles at a current density of 200 mA/g [89]. In another approach, SnO_2 or other oxidized forms of Sn are used as the starting material [14, 90–92]. In this situation, the first lithiation of SnO_2 results in the irreversible formation of Sn nanoparticles embedded in a Li_2O matrix. This reaction is similar to a displacement reaction ($SnO_2 + 4Li \Leftrightarrow Sn + 2Li_2O$—see Sects. 4 and 7), but it is not reversible as Li cannot be extracted from Li_2O. However, the Sn particles thus created are stable and can be further cycled according to an alloying reaction ($xLi + Sn \Leftrightarrow Li_xSn$, with $0 < x < 4.4$). The mechanism of the formation of the Sn nanoparticles from SnO_2 decomposition has been studied using in situ TEM observations [93, 94]. Figure 9 is a schematic representation of the lithiation mechanism of a SnO_2 NW, as reconstituted from such in situ TEM analysis [94]. Li first diffuses along (020) crystallographic planes, which resembles an intercalation reaction. There is then short range lithiation of the surrounding SnO_2 matrix that becomes progressively amorphous, thus breaking Sn-O bonds and favoring the formation of a Li_2O matrix in which Sn nanoparticles precipitate. At a later stage of lithiation, the Sn nanoparticles transform into Li_xSn.

Finally, in yet another approach, Sony have used a quasi amorphous nano-structured Sn–Co–C alloy in their Nexelion battery, which was first commercialized

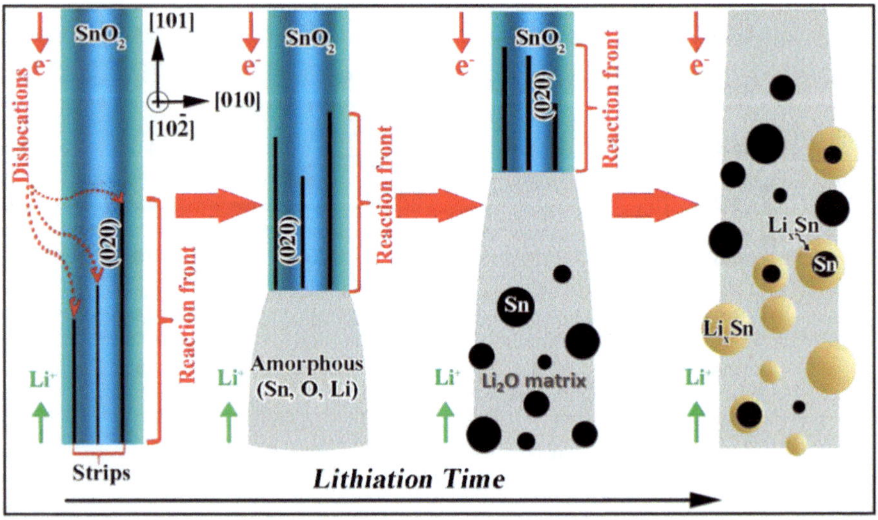

Fig. 9 Schematic of the lithiation mechanism of a crystalline SnO_2 nanowire, featuring the initial lithium-ion long-range diffusion mediated by dislocation activities inside the pristine lattice, followed by solid-state amorphization of the crystalline nanowire, nucleation of Sn particles, and finally the alloying of the Sn particles into Li_xSn. Reprinted with permission from Ref. [94]. Copyright 2013, American Chemical Society

in February 2005. The Nexelion battery exhibits a ~10 % increase in weight energy density and a 30 % increase in overall capacity compared to similar batteries built with graphite anodes. The capacity of the alloy, with composition around $Sn_{30}Co_{30}C_{40}$, can reach 700 mAh/g, i.e., about twice that of graphite [95]. This alloy, synthesized by a high energy mechanical milling process has been studied in more detail by Scrosati and coworkers [96], following previous combinatorial investigations by Dahn's group indicating the domain where a nanostructured quasi amorphous phase was obtained [95]. It seems that the CoSn-C alloy originally formed remains stable after 100 charge-discharge cycles [96].

To conclude this paragraph, there seem to be no real advantage of Sn-based electrodes over their Si counterparts, since Sn expands more upon full lithiation (inducing more mechanical damage), its capacity is lower than that of Si and it is more expensive than Si. Moreover, while Li insertion into Si tends to induce a brittle-to-ductile transition [97] which somehow delays cracking, the fully lithiated $Li_{22}Sn_5$ phase is brittle, which explains why it cracks even at low particle size [98].

5.1.5 Lead

Lead (Pb) crystallizes in the face centered cubic system and just like Sn, its melting temperature is low (327 °C) compared to those of Si and Ge. As far as Li-based battery anodes are concerned, Pb has received little attention, probably because of its toxicity and relative low capacity (see below). According to the phase diagram, Pb alloying with Li at room temperature produces 5 intermetallic compounds, namely αLiPb, Li_5Pb_2, Li_3Pb, $Li_{10}Pb_3$ and Li_4Pb [99]. However, in room temperature lithiation experiments, Huggins [100] identified 4 voltage plateaux for the Li_xPb alloys, at 0.601, 0.449, 0.374 and 0.292 V (over the range $0 < x < 4.5$) and reached the $Li_{4.4}Pb$ composition which would correspond to the $Li_{22}Pb_5$ compound, absent from the phase diagram. The capacity of this alloy is 568 mAh/g, which is a rather modest value compared to Sn or Si. Huggins' results have later been confirmed by Martos et al. [101], although their electrochemical characterizations yield somewhat different plateau voltages.

Because Pb-based anodes tend to degrade rapidly upon cycling (see e.g., Ref [101]), as with Sn, most studies have concentrated on decreasing particle size and dispersing them into a buffering inactive matrix. For instance, lead oxides have been investigated [101] and compared with tin oxide [102]. More recently, PbO-C composites have been studied [103], as well as Pb-sandwiched nanoparticles (i.e., SiC-Pb-C core-shell1-shell2 nanoparticles) [104], but the gravimetric capacity is rather low (~200 mAh/g after 200 charge-discharge cycles) although the volumetric capacity is about twice that of graphite, at 1580 mAh/cm^3 (i.e., ~60 % of the theoretical Pb volumetric capacity) after 600 cycles [104]. A $PbGeO_3$-graphene composite has also been tested and found to retain a ~600 mAh/g capacity after 50 charge-discharge cycles [105].

5.2 Column V of the Periodic Table

Elements from column V studied for their potential use in Li-ion batteries are essentially P, Sb and Bi. Data relative to As are relatively scarce in the literature, probably because of its high toxicity and also its relatively high atomic mass, which limits the gravimetric capacity. The most common allotrope of arsenic is grey or metallic As, which crystallizes in the rhombohedric system. In a small paragraph of their paper on Li intercalation in ZnSb, Park and Sohn have reported a capacity of 330 mAh/g (1300 mAh/cm^3) and ~100 % capacity retention after 300 cycles for an As/C composite anode [106].

5.2.1 Phosphorus

Phosphorus reacts with lithium to form Li_3P at complete lithiation, corresponding to a volume increase of ~300 % and a maximum capacity of 2595 mAh/g. However, among the three main phosphorus allotropes [107], the white variety tends to be unstable above 30 °C and only red and black phosphorous have been tested as anode materials. Red phosphorous (rP), which is commercially available, is usually amorphous and exhibits a low electrical conductivity, which induces a poor reversibility for the lithiation reaction [108]. Mixing rP with carbon (30.56 wt% of rP in the C-rP composite) has allowed Wang and coworkers [109] to obtain a good cycling stability, with a capacity retention of ~750 mAh/g after 55 cycles for the C-rP composite. Black phosphorous (bP) is a more interesting anode material (although difficult to synthesize), since it exhibits a much higher conductivity, even though it is a semiconductor material with a band gap of 0.3 eV [110]. Actually, bP is a layered material, just like graphite (except that the "phosphorene" layers are not planar but puckered) and bP with few layers is currently being investigated for thin film transistor applications [111]. At this point, it is interesting to note that Li intercalation in a half cell configuration is used to exfoliate bulk transition metal di-chalcogenides (MoS_2, WS_2 ...) and fabricate 2D layered materials which are studied for electronic applications [112, 113].

Park and Sohn [108] have mixed rP with carbon and transformed it into mixtures of bP + C by high energy mechanical milling (HEMM); they have obtained a capacity of ~600 mAh/g which remained stable after 100 charge-discharge cycles. By processing mixtures of bP and graphene, also using HEMM, Sun and coworkers [114] have synthesized a bP-graphene composite where phosphorus-carbon bonds are generated at the periphery of graphite's basal planes, providing stable connecting paths between the bP particles, even after complete lithiation/delithiation. An initial discharge capacity of ~1100 mAh/g was obtained for the composite, which decayed to ~800 mAh/g after 100 cycles at 0.2 C [114].

5.2.2 Antimony

Just like P, Sb can be lithiated up to Li_3Sb, which corresponds to a capacity of ~660 mAh/g. The volume expansion when going from Sb to Li_3Sb is about 135 %. Maybe Sb has been more investigated for Na storage in recent years [115, 116]. Using thin Sb films, Bagetto and coworkers have shown that during the first lithiation, 2 successive plateaus appear at 0.82 and 0.78 V, corresponding to the formation of Li_2Sb and Li_3Sb respectively. The latter is mainly cubic, although a small amount (~4 %) of the hexagonal allotrope is also found [116]. The first delithiation only shows one plateau (at 1.02 V), corresponding to the equilibrium between Li_3Sb and Sb [116]. This means that Li_2Sb is not re-formed during delithiation. The second lithiation shows two plateaus again, but the first one is now observed at ~0.89 V and probably still corresponds to the equilibrium between Sb and Li_2Sb [116], which has been seen before using in situ XRD [117].

One of the most recent work on Sb proposes to use monodisperse Sb nanocrystals with their size in the 10–20 nm range [118]. Interesting results have been obtained with the 20 nm nanocrystals, particularly concerning high rate cycling, up to 20 C, where the capacity retention is still ~80 % of the original one [118]. A drawback for Sb is the high voltage against Li (an average of ~0.85 V), which will reduce the corresponding voltage of any cell with an oxide cathode.

5.2.3 Bismuth

Bismuth (Bi) is not very common in the Earth's crust (~9 ppb) and consequently, tends to be expensive (~39 US $/100 g for pure element). The Bi–Li phase diagram shows that LiBi and Li_3Bi alloys can be formed [119], so that the maximum capacity is ~385 mAh/g, roughly equivalent to that of graphite. However, the volumetric capacity of Bi is ~3768 mAh/cm^3, about 4.5 times that of graphite.

Using thin films deposited by electroplating, Xianming and coworkers [120] have shown by in situ XRD that upon lithiation, LiBi first appears (~0.8 V), followed by Li_3Bi formation at ~0.72 V. During delithiation, LiBi first forms (although the separation between LiBi and Li_3Bi was not clearly observed), and at the end, only Bi is present in the electrode. The decomposition of Li_3Bi to LiBi and Bi occurs around 0.84 V. In a more recent work, Sohn's group at Seoul National University [121] have used nanosized Bi, Bi–C and Bi–Al_2O_3–C composites synthesized by high energy ball milling. They confirmed the results of Xianming in terms of the various phases appearing-disappearing during lithiation–delithiation. They also clearly show that anodes made of pure Bi degrade rapidly, whereas those using composite are more stable (61 % retention of the original capacity after 100 charge-discharge cycles); the most stable composite is the Bi-Al_2O_3-C one, with 74 % retention of the original capacity after 100 cycles. However, in this latter situation, the original capacity was only 450 mAh/g (~2400 mAh/cm^3). Also, as with Sb, the lithiation–delithiation voltages against Li are high (around 0.8 V on average), which will represent a penalty for a full cell with an oxide cathode.

5.3 Column III of the Periodic Table

Elements from column III which have been studied for Li-ion anodes are essentially aluminum (Al) and gallium (Ga). Although the boron-lithium phase diagram shows several compounds [122], Li poorly reacts with B at room temperature [123]. Actually, B or rather Li-B alloys have been studied for application in the so-called thermal batteries, which are primary devices working at high temperature (350–450 °C) with molten salt electrolytes [124, 125].

5.3.1 Aluminum

Aluminum is an interesting anode material since it can make a series of solid solutions and intermetallic compounds with Li (one of which recently discovered [126]), culminating at a capacity of 2235 mAh/g for the Li_9Al_4 compound of the equilibrium phase diagram. Moreover, Al is the 3rd most abundant element in the Earth's crust and it is non-toxic and relatively cheap. Actually, room temperature (RT) electrochemical alloying seems to be limited to the β-LiAl intermetallic compound, corresponding to a capacity of 993 mAh/g. β-LiAl crystallizes in the so-called Zintl structure (space group *Fd*-3 m), which can be described as two interpenetrating diamond lattices with each element occupying one of the lattices [127]. The lattice parameter of β-LiAl at stoichiometry is 6.37 Å [127] and its elemental cell accommodates 8 Al atoms, so that transforming Al (fcc structure, with a lattice parameter of 4.05 Å) into β-LiAl yields a ~95 % volume expansion per Al atom. Clearly the Al to β-LiAl transition upon electrochemical alloying with Li involves strong restructuring of the Al lattice.

The first studies concerning the electrochemical RT alloying of Li with an Al electrode were performed 4 decades ago [128]. Incidentally, Dey was the first one to point out the similarities between RT electrochemical alloying of Li and classical metallurgical alloying [128]. It was rapidly recognized that Li–Al alloys could replace pure lithium in secondary Li-ion battery anodes, thus circumventing the problem of Li dendritic growth upon battery charging [129]. Unfortunately, pre-lithiated Al electrodes were found to disintegrate after a few cycles, because of large volume variation and phase change upon lithiation–delithiation [130]. The use of Al thin films [131, 132] or micro- and nano-sized Al particles [133], or composites [134] did not improve significantly the cycling behavior. Perhaps one of the best results (50 % capacity retention after 100 charge-discharge cycles) was obtained by using $Al_{0.8}Cu_{0.2}$ alloys synthesized by sputtering [135].

Recently, one-dimensional Al nanostructures, such as nanowires [132] or nanorods [136] have been tested as battery anodes, but here again, the original capacity quickly fades away after a few cycles (even quicker than with thin films), which contrasts with other elements or alloys, where the use of such one-dimensional nanostructures was found to improve the cycling behavior (see the previous paragraphs). The reason why Al anodes based on nanowires or nanorods degrade rapidly has been elucidated in recent work, using in situ TEM observations of the behavior of Al nanowires during lithiation–delithiation cycles [137]. It was found that voids appear in the Al NWs during delithiation steps, which transform the original NWs into a series of isolated nanograins. Overall, the NWs retain their shape (although their length is observed to increase) because the Al grains are confined into a Li–Al–O glass tube, formed during the first Li alloying step, by lithiation of the native aluminum oxide film around each Al NW. Because the so-formed Al grains are separated by voids (even after the first delithiation step [137]) and thus electrically disconnected from the current collector, they cannot participate in the reduction of Li^+ ions and the subsequent Li alloying upon battery recharging, which induces a rapid capacity fade of the electrode. Figure 10 summarizes the situation.

Even thin film-type, all-solid-state Li-ion batteries with an Al anode lose 90 % of their original capacity after 100 cycles [138], a behavior in contrast to that observed with similar all-solid-state batteries using Si anodes [139]. The degradation mechanism of such Al-based all-solid-state batteries would be different from that of NWs as it would result from Li trapping in AlLi domains formed on the top surface of the Al anode. This trapping would be due to the lack of surface pathways allowing Li and Al to out-diffuse from the AlLi domains upon delithiation [138].

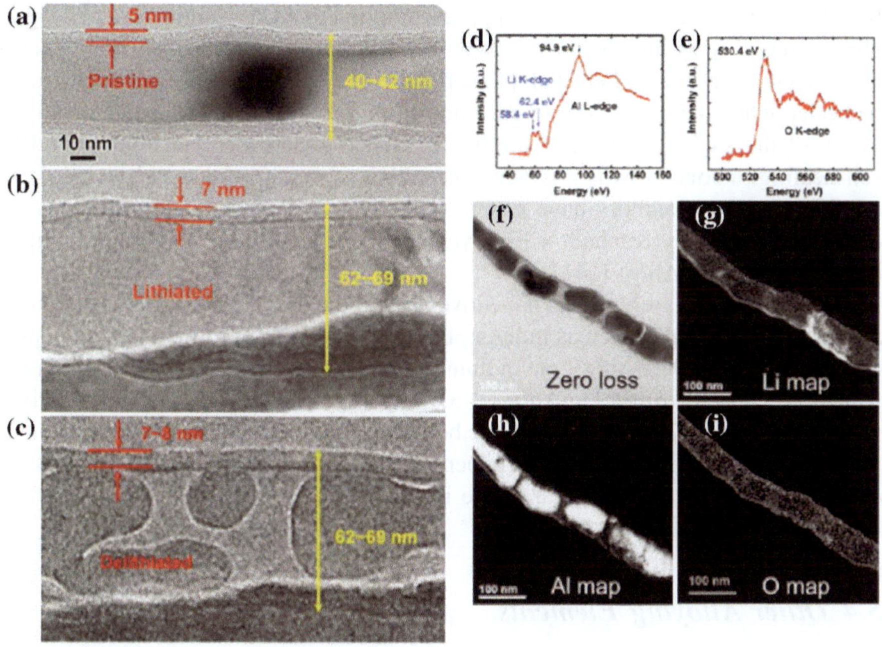

Fig. 10 Morphology evolution of the surface native oxide layer and Al NW during the first cycle, and EELS spectra as well as maps of Li, Al, and O elements of an Al NW after three electrochemical cycles. **a** A pristine Al NW with a surface Al_2O_3 layer of about 5 nm. **b** The fully lithiated NW, showing the thickness of surface layer was increased to about 7 nm. **c** The NW after the first delithiation, showing voids inside the Al NW and the thickness of the surface layer (about 7–8 nm), which did not change obviously from that after first lithiation. **d**, **e** Electron energy loss spectroscopy (EELS) spectra of Li–K, Al–L, and O–K edges with the pre-edge background subtracted, showing the presence of Li, Al, and O in the cycled NW. **f** Zero loss image showing that the pulverized nanoparticles were confined by a tube-like surface layer, like peas in a pea pod. **g**, **i** EELS maps of Li, Al, and O, respectively, indicating the nanoparticles in the tube were Al nanoparticles and the surface layer consisted of Li, Al, and O. The energy-filtered maps were obtained using a three-window technique. Reprinted with permission from Ref. [137]. Copyright 2011, American Chemical Society

In summary, although Al is an attractive electrode material, it seems that it does not withstand cycling, even when used in nanosized forms, which contrasts with other elements such as Si.

5.3.2 Gallium and Indium

The Ga–Li phase diagram shows several compounds, up to $GaLi_2$ [140], the latter corresponding to a specific capacity of $\sim$767 mAh/g, which is about twice that of graphite. However, the interest of Ga probably relies on the fact that it is in the liquid state above 29.8 °C. Based on the assumption that Ga would return to a liquid phase after alloying and de-alloying with Li (even though intermediate solid phases are formed in-between), Lee et al. [141] pointed out that any crack formation would be healed when the liquid is restored, since droplets would merge together under the action of the surface tension of the liquid. Hence a Ga-based electrode would be self-healing, a concept which is very interesting. Using a porous carbon matrix to confine Ga, those authors have observed a good cycling stability of their Ga-C electrode, since they reached ~100 cycles without significant loss of the original capacity (~400 mAh/g) [141].

On the other end, few papers have investigated the use of pure indium (In), and it seems that compounds such as indium phosphide [142] (InP, providing a reversible capacity of ~620 mAh/g) or indium oxide [143, 144] (In_2O_3, providing a reversible capacity of 883 mAh/g [143]) are more interesting starting materials, reacting with Li in 2 steps (the first one being irreversible), in a way similar to SnO_2 [145] (see above). Because In is an expensive metal (~750 $/kg in 2014), its use in commercial products in the near future is very unlikely.

5.4 Other Alloying Elements

5.4.1 Magnesium

The Mg–Li phase diagram is shown on Fig. 11. This diagram is the only one involving Li which exhibits both large solid solution domains and no intermetallic compounds (see Ref. [119], vol. 2, pp. 1487–1512). The interest of solid solutions is to avoid phase transformations and their deleterious effect on the cycling life time of the electrode.

However, Fig. 11 shows that the Mg-based α solid solution (hcp crystal structure) is limited to ~5 wt% Li (~15 at.%), which corresponds to a modest capacity of ~194 mAh/g. On the other hand, the β phase (bcc structure) extends from ~11 wt% Li (~30 at.% Li) to pure Li; Shi et al. [146] have calculated that cycling a Li–Mg alloy anode from $Li_{0.4}Mg_{0.6}$ to $Li_{0.6}Mg_{0.4}$ yielded a capacity of 642 mAh/g. The problem with using a β phase Li_xMg_{1-x} alloy anode is to avoid the delithiation below x ~0.3, which would induce the precipitation of the α phase. This implies some voltage limitations during battery discharge.

Kim and coworkers [147] have studied the lithiation of pure Mg. They have obtained a capacity of 3070 mAh/g for the first lithiation (probably going deep into the β phase region) and 2150 mAh/g for the first delithiation. However, they have noticed that the pure Mg electrode could not be lithiated at a current larger than 10 mA/g, which corresponds to a very small rate (~C/300). This poor rate capability has been attributed to the presence of a native oxide film on the surface of Mg or to the formation of an excessive SEI thickness. Alloys composed of inactive Mg_2Ni and Mg (global composition $Mg_{0.75}Ni_{0.25}$) reacted quite well with Li, but for pure Mg as well as Mg–Ni alloys, the initial capacity was almost lost after 10 cycles.

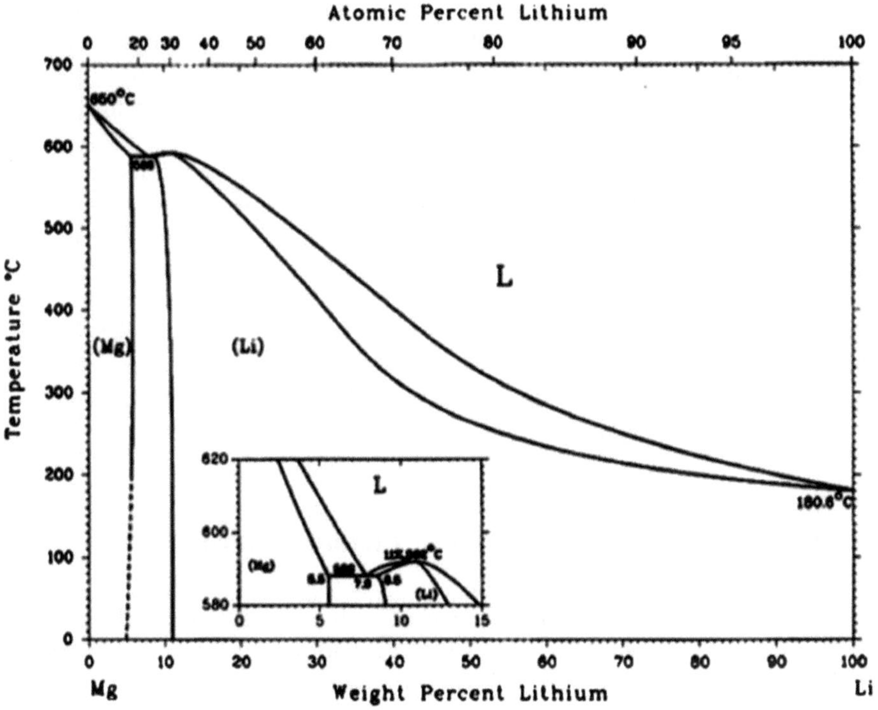

Fig. 11 The magnesium–lithium phase diagram. Note the large extension of the β solid solution

Better results were obtained by using a Mg-C composite anode [148], in particular concerning the C-rate. Altogether, it seems that the use of Mg has not been very successful so far.

5.4.2 Silver and Gold

Despite their high cost, noble metals (particularly silver [149]) have been considered for thin film-type batteries. The Ag-Li phase diagram shows several intermediate phases [119, vol. 1, p. 38], which culminate at the $AgLi_9$ composition [150]. However, upon RT lithiation–delithiation, only two reversible phases appear, namely AgLi, as well as an unknown and not yet identified phase (called phase II in Ref. [149]). Using 100 nm-thick sputtered Ag films in conjunction with a $Li_{1.2}Mn_{1.5}Ni_{0.5}O_4$ cathode, Taillades and Sarradin [149] have shown a capacity of ~25 μAh/cm^2 over 1000 cycles, at an average working voltage of 4.65 V. More recently, virus-templated Ag, Au and Ag_xAu_{1-x} alloy NWs have been synthesized and their electrochemical properties characterized [150]. For instance, 15 nm-diameter Ag NWs showed a reversible capacity of ~534 mAh/g (2nd cycle), which was reduced to 280 mAh/g after 10 cycles [150]. In this study, all nanowires, whether pure Ag, Au or AuAg alloys failed to maintain a stable capacity when tested up to 20 cycles [150]. The cycling properties can be improved when Ag is alloyed with Sn; for instance, an anode made of $Ag_{0.52}Sn_{0.48}$ (original reversible capacity of ~550 mAh/g) still exhibits a capacity of 200 mAh/g after 300 cycles [151].

Although there are few literature papers on the electrochemical alloying of Au with Li at RT, Au thin films have been shown to accommodate Li in the 0.02–0.5 V potential range with a poor reversibility [152, 153]. The galvanostatic charge-discharge curves obtained for such thin films of Au exhibit two voltage plateaus corresponding to the Li alloying process at 0.2 and 0.1 V, whereas Li removal from the alloy occurs in two steps at 0.18 and 0.4 V [152, 153]. Mesoporous Au films show a better cycling stability than their compact thin film counterparts, with ~16 % (at 80 mAh/g) of the original capacity retained after 30 charge-discharge cycles [154].

6 Intermetallic Compounds/Multi Element Alloys

Intermetallic and multi element alloys have also been investigated, where at least one element does not alloy with Li (e.g., Cu) and is used to buffer the strain/deformation induced by Li alloying with the other element(s). This is often called the mixed conductor matrix concept [155], designed to improve the cyclability of the electrode. The subject has recently been reviewed in detail by Zhang [156] and will not be developed further here. Figure 12 (from Ref. [156]) summarizes the main results from the literature for Si-based, Sn-based and Sb-based anodes. An example already quoted above is the use of an $Al_{0.8}Cu_{0.2}$ multilayer which improves the cyclability of Al-based anodes [135]. One of the drawbacks of this approach is the reduction of electrode capacity due to the addition of the inactive buffering element.

7 Conversion Anodes

A last type of Li anodes is based on the so-called conversion reactions, whose reversibility was fully understood by Poizot et al. in 2000 [15]. For instance, if we take a look at transition metal oxides, the conversion reaction can be written as:

$$2Li^+ + 2e^- + MO \underset{2}{\overset{1}{\leftrightarrows}} Li_2O + M, \tag{8}$$

where M = Fe, Ni, Co, Cu … The transition metal oxide reduction by Li (direction 1) is usually thermodynamically favorable (see e.g., an Ellingham diagram). However,

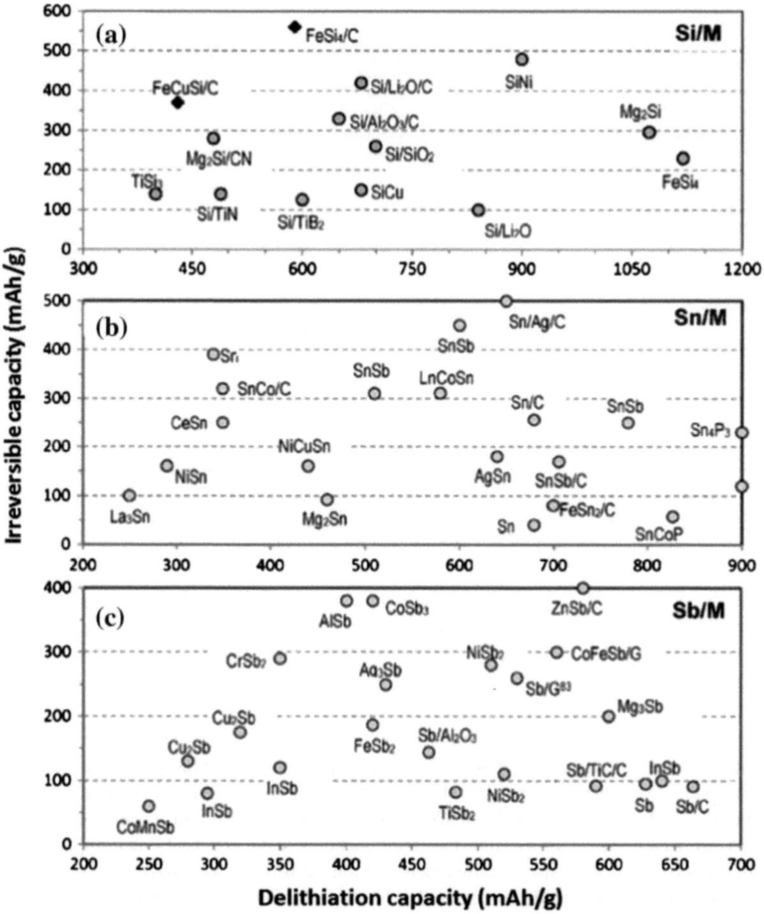

Fig. 12 The first-cycle irreversible capacity of various multi element alloy anodes. **a** Si-based. **b** Sn-based. **c** Sb-based. Reprinted with permission from Ref. [156]. Copyright 2011, American Chemical Society

the reverse reaction (direction 2, corresponding to Li extraction from Li_2O) cannot take place in normal conditions, since Li_2O is generally more stable than MO. Actually, the above reaction can be made reversible because of the formation of metal nanoparticles embedded in a Li_2O matrix during the first lithiation. Such nanoparticles then catalyze Li_2O decomposition when a reverse polarization is applied. Because the nanosize of the metal particles is maintained upon further charge-discharge cycling, the above reaction (8) is rendered reversible. The situation is summarized on Fig. 13.

Reaction (8) is distinct from the one seen in Sect. 5.1.4 concerning tin oxide, since once SnO_2 reduced to form Li_2O, the latter is stable and cannot be decomposed. Hence some of the Li is lost and there is a large irreversible capacity due to

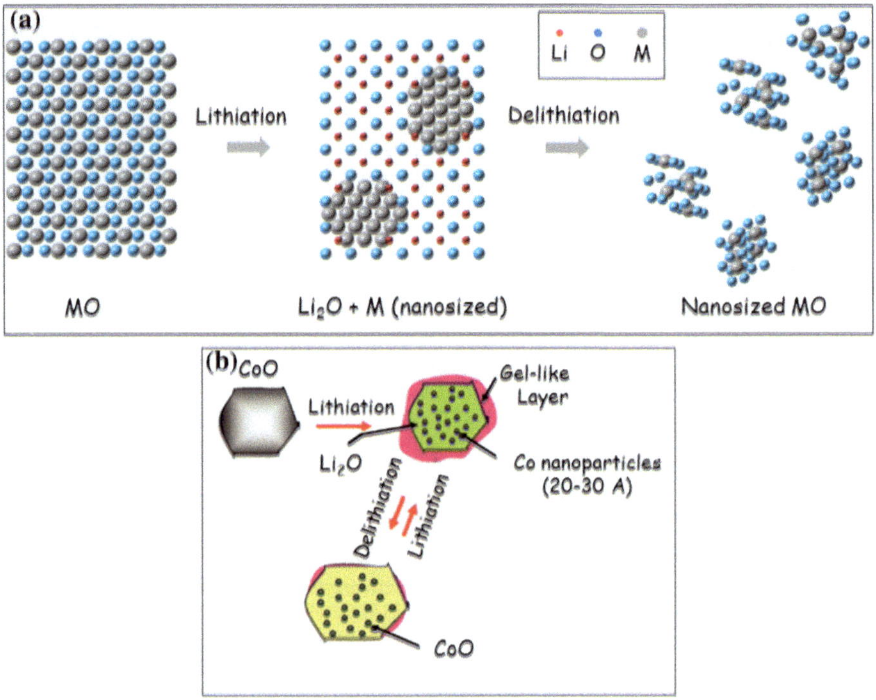

Fig. 13 Schematic mechanism of a conversion reaction based on a metal oxide MO. **a** The first cycle, showing the lithiation step yielding metal nanoparticles embedded in a Li_2O matrix and the delithiation step leaving nanosized metal oxide particles. **b** The overall reaction illustrated with cobalt oxide (*CoO*). Starting with large CoO particles, the first lithiation yields Co nanoparticles embedded in Li_2O (*horizontal*); the Co nanoparticles catalyze Li_2O decomposition and Li extraction when the polarization is reversed, which results in nanosized CoO. Thanks to the nanosized particles, the electrode can be cycled between CoO and Li_2O + Co as lithium is added or removed (*vertical*). Note however that the electrolyte decomposition (gel-like layer) is favored by the nanosized metal particles **b** is reprinted with permission from [157]

Li_2O formation during the first lithiation. Only the released Sn further participates to the lithiation–delithiation reactions.

Note that in reaction (8), and as already stated in Sect. 4, oxygen can be replaced by nitrogen [16], fluorine [17], sulfur [158], phosphorus [158] or even hydrogen [159]. The subject of conversion reactions has been thoroughly reviewed by Cabana et al. and Fig. 14 summarizes some of the electrode capacities obtained so far in the literature.

Even though they offer larger capacity values than graphite-based anodes, conversion electrodes suffer from an important hysteresis in voltage between charge and discharge, leading to poor energy efficiency and low battery voltages. This effect is illustrated on Fig. 15 taken from reference [157]. According to Oumellal and coworkers, if Co is taken as an example, the largest voltage hysteresis is observed for fluoride compounds ($\sim$1.1 V), followed by oxides ($\sim$0.9 V), then by

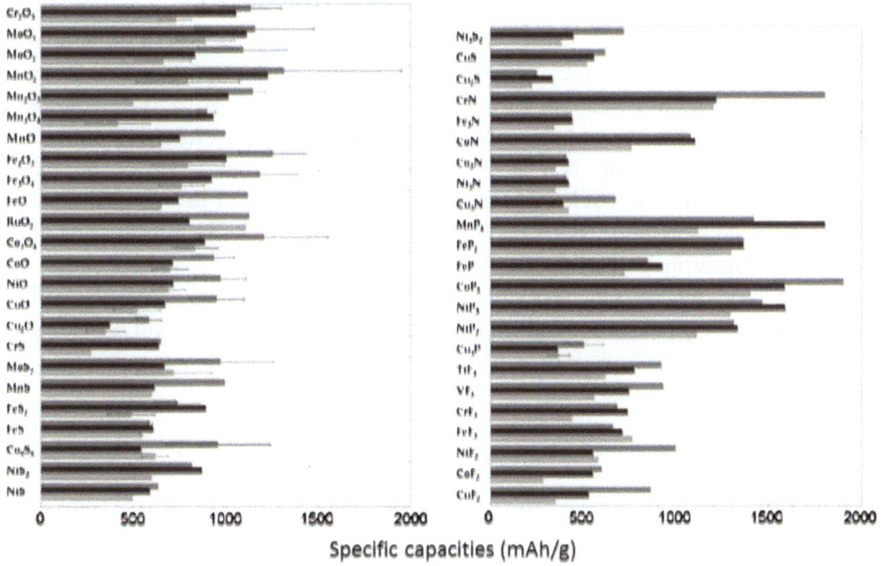

Fig. 14 Theoretical (*black bars*), first discharge (*dark grey*), and charge (*light grey*) specific gravimetric capacities of different compounds that react with lithium through a conversion reaction. The experimental capacities are taken from a series of reports for each compound. The "error" bars are provided as an indication of the dispersion of values observed in the bibliography and, thus, have no statistical meaning. Data for compounds with no bar have been taken from a single literature source. Reprinted with permission from [159]

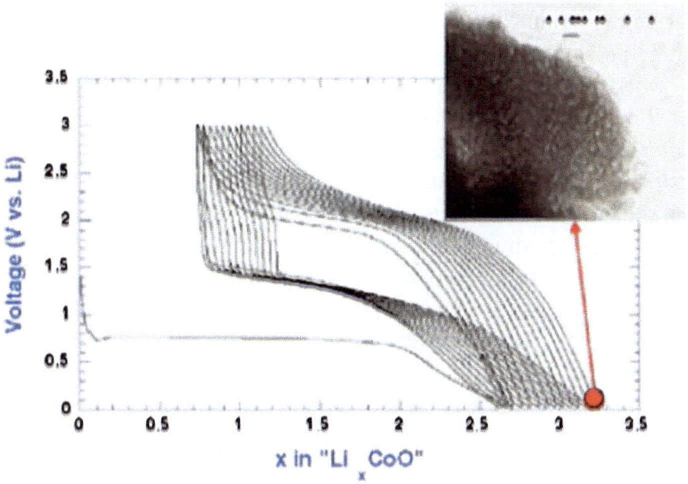

Fig. 15 Voltage composition curve for a CoO/Li cell cycled between 0.01 and 3 V. The *inset* shows a TEM micrograph of the nanostructured composite electrode at the end of the lithiation process. Note the large voltage hysteresis ($\sim$1 V) between lithiation and delithiation. Reprinted with permission from Ref. [157]

sulfides (0.7 V), then by nitrides ($\sim$0.6 V) and finally by phosphides ($\sim$0.4 V) and hydrides ($\sim$0.25 V) [159]. Moreover, the highly divided metal particles tend to favor excessive electrolyte decomposition and even if it is partially reversible [160], the formation of a thick SEI can partially block the exchange of Li ions. Moreover, the Li trapped in the SEI induces a large irreversible capacity loss. Recent in situ TEM observations on CuO NW conversion confirm the implication of the SEI in the ageing mechanism [161]. We also note a recent detailed in situ TEM observation of the conversion of FeF_2 nanoparticles resulting in 1–3 nm iron crystallites mixed in an amorphous LiF matrix [162].

In summary, although the work of Poizot [15] sparked intensive research efforts in the field of conversion anodes, it seems that no industrial and commercial developments have followed, in particular because of the large hysteresis between lithiation and delithiation.

8 Conclusions

Because they can offer much larger capacities than existing graphite-based insertion anodes, alloy and conversion anodes bear a real potential for applications in Li-based battery technology. Although the development of such anodes has been plagued by mechanical instabilities during charge-discharge cycling, the recent introduction of nanomaterials/nanostructures has enabled huge progress in this area. For instance, while the first attempts to use Si were largely unsuccessful, due to the rapid pulverization of the corresponding anodes, thanks to the use of nanostructures, there are now a number of published works showing charge-discharge stability over more than 1000 cycles, with typical electrode capacities over 1200 mAh/g, i.e. more than three times that of graphite-based anodes. Beyond laboratory work and scientific publications, there are also a number of start-up companies commercializing Si-based anodes.

Very often, nanostructuration is accompanied by mixing with some kind of amorphous/porous carbon, the latter preventing re-agglomeration and providing a conducting skeleton which improves the cycling rate of the electrode. Complex nanostructures such as core-shell nanoparticles or nanowires, hollow nanoparticles, nanotube-like structures and composites with graphene are being studied.

Although many elements and compounds have been investigated, few of them will experience further developments and industrialization in the future, since as with all mass-manufactured products, processing and material costs will prevail in the end. Abundant and non-toxic materials (e.g., Si) will be favored, together with simple, low temperature and repeatable manufacturing processes, with particular attention paid to their sustainable and environment-friendly aspects.

References

1. Brissot C et al (1998) In situ study of dendritic growth in lithium/PEO-salt/lithium cells. Electrochim Acta 43:1569–1574
2. Tarascon J-M, Armand M (2001) For a quick historical background on Li-based batteries. Issues and challenges facing rechargeable lithium batteries. Nature 414:359–367
3. Qi Y, Guo H, Hector LG Jr, Timmons A (2010) Threefold increase in the Young's Modulus of graphite negative electrode during lithium intercalation. J Electrochem Soc 157:A558–A566
4. In a traditional battery electrode process the Li-storage material (graphite for an anode) is incorporated into a slurry containing a binder (polymer) as well as a conductive additive (carbon-based powder). This slurry is subsequently deposited on the current collector of the battery electrode (usually a copper foil), typically by tape casting and dried in an oven before being used, in order to remove the solvents from the slurry. The anode is usually not lithiated when it is assembled in the battery and Li is provided by the cathode content
5. Huggins RA (2010) Energy storage. Springer, Heidelberg, pp 119–160
6. Christensen J, Albertus P, Sanchez-Carrera RS et al (2012) A critical review of Li/Air batteries. J Electrochem Soc 159:R1–R30
7. Ji X, Lee KT, Nazar LF (2009) A highly ordered nanostructured carbon-sulfur cathode for lithium-sulphur batteries. Nat Mater 8:500–506
8. Bruce PG, Freunberger SA, Hardwick LJ, Tarascon J-M (2012) Li–O_2 and Li–S batteries with high energy storage. Nat Mater 11:19–29
9. Goodenough JB, Kim Y (2010) Challenges for rechargeable Li batteries. Chem Mater 22:587–603
10. Peled E (1979) The electrochemical behavior of alkali and alkaline earth metals in nonaqueous battery systems-the solid electrolyte interphase model. J Electrochem Soc 126:2047–2051
11. Andersson AM et al (2003) Electrochemically lithiated graphite characterized by photoelectron spectroscopy. J Power Sour 119–121:522–527
12. Xu K, von Cresce A (2011) Interfacing electrolytes with electrodes in Li ion batteries. J Mater Chem 21:9849–9864
13. Möller K-C, Santner HJ, Kern W et al (2003) In situ characterization of the SEI formation on graphite in the presence of a vinylene group containing film-forming electrolyte additives. J Power Sour 119–121:561–566
14. Idota Y (1995) U.S. Pat. 5,478,671
15. Poizot P, Laruelle S, Grugeon S et al (2000) Nano-sized transition-metal oxides as negative-electrode materials for lithium-ion batteries. Nature 407:496–499
16. Poizot P, Laruelle S, Grugeon S, Tarascon J-M (2000) Rationalization of the low-potential reactivity of 3d-metal-based inorganic compounds towards Li. J Electrochem Soc 149: A1212–A1217
17. Li H, Richter G, Maier J (2003) Reversible formation and decomposition of LiF clusters using transition metal fluorides as precursors and their application in rechargeable Li batteries. Adv Mater 15:736–739
18. Winter M, Besenhard OJ, Spahr EM, Novák P (1998) Insertion electrode materials for rechargeable lithium batteries. Adv Mater 10:725–763
19. Zhang W-J (2011) Lithium insertion/extraction mechanism in alloy anodes for lithium-ion batteries. J Power Sour 196:877–885
20. Saint J et al (2007) Towards a fundamental understanding of the improved electrochemical performance of silicon-carbon composites. Adv Funct Mater 17:1765–1774
21. Park CM, Sohn HJ (2010) Electrochemical characteristics of $TiSb_2$ and Sb/TiC/C nanocomposites as anodes for rechargeable Li-ion batteries. J Electrochem Soc 157:A46–A51

22. Wagemaker M, Borghols WJH, Mulder FM (2007) Large impact of particle size on insertion reactions. A case for anatase Li_xTiO_2. J Am Chem Soc 129:4323–4327
23. Dahn JR, Zheng T, Liu Y, Xue JS (1995) Mechanisms for lithium insertion in carbonaceous materials. Science 270:590–593
24. Sato K, Noguchi M, Demachi A, Oki N, Endo M (1994) A mechanism of lithium storage in disordered carbons. Science 264:556–560
25. Su DS, Schlögl R (2010) Nanostructured carbon and carbon nanocomposites for electrochemical energy storage and applications. ChemSusChem 3:136–168
26. Kaskhedikar NA, Maier J (2009) Lithium storage in carbon nanostructures. Adv Mater 21:2664–2680
27. Zheng T, Xing W, Dahn JR (1996) Carbons prepared from coals for anodes of lithium-ion cells. Carbon 34:1501–1506
28. Zhou H, Zhu S, Hibino M, Honma I, Ichihara M (2003) Lithium storage in ordered mesoporous carbon (CMK-3) with highly reversible specific energy capacity and good cycling performance. Adv Mater 15:2107–2111
29. Lahiri I et al (2011) Ultrathin alumina-coated carbon nanotubes as an anode for high capacity Li-ion batteries. J Mater Chem 21:13621–13626
30. Liu Y et al (2011) Lithiation-induced embrittlement of multiwalled carbon nanotubes. ACS Nano 5:7245–7253
31. Liu J (2014) Charging graphene for energy. Nat Nanotechnol 9:739–740
32. Chou C-Y, Hwang GS (2014) On the origin of the significant difference in lithiation behavior between silicon and germanium. J Power Sources 263:252–258
33. Kasavajjula U, Wang C, Appleby AJ (2007) Nano- and bulk-silicon-based insertion anodes for lithium-ion secondary cells. J Power Sour 163:1003–1039
34. Teki R et al (2009) Nanostructured silicon anodes for lithium ion rechargeable batteries. Small 5:2236–2242
35. Szczech R, Jin S (2011) Nanostructured silicon for high capacity lithium battery anodes. Energy Environ Sci 4:56–72
36. Wu H, Cui Y (2012) Designing nanostructured Si anodes for high energy lithium ion batteries. Nano Today 7:414–429
37. Zamfir MR et al (2013) Silicon nanowires for Li-based battery anodes: a review. J Mater Chem A 1:9566–9586
38. Su X et al (2014) Silicon-based nanomaterials for lithium-ion batteries: a review. Adv Energy Mater 4:1300882 (23 pages)
39. Hatchard TD, Dahn JR (2004) In situ XRD and electrochemical study of the reaction of lithium with amorphous silicon. J Electrochem Soc 151:A838–A842
40. Li J, Dahn JR (2007) An in situ X-ray diffraction study of the reaction of Li with crystalline Si. J Electrochem Soc 154:A156–A161
41. Liu XH et al (2011) Ultrafast electrochemical lithiation of individual Si nanowire anodes. Nano Lett 11:2251–2258
42. Liu XH et al (2012) In situ atomic-scale imaging of electrochemical lithiation in silicon. Nat Nanotechnol 7:749–756
43. Key B, Morcrette M, Tarascon JM, Grey CP (2011) Pair distribution function analysis and solid state NMR studies of silicon electrodes for lithium ion batteries: understanding the (de) lithiation mechanisms. J Am Chem Soc 133:503–512
44. Wang JW et al (2013) Two-phase electrochemical lithiation in amorphous silicon. Nano Lett 13:709–715
45. McDowell MT et al (2013) In situ TEM of two-phase lithiation of amorphous silicon nanospheres. Nano Lett 13:758–764
46. Liu XH et al (2012) In situ TEM experiments of electrochemical lithiation and delithiation of individual nanostructures. Adv Energy Mater 2:722–741
47. Liu XH et al (2012) Size-dependent fracture of silicon nanoparticles during lithiation. ACS Nano 6:1522–1531

48. Lee SW et al (2012) Fracture of crystalline silicon nanopillars during electrochemical lithium insertion. Proc Natl Acad Sci USA 109:4080–4085
49. Ryu I, Choi JW, Cui Y, Nix WD (2011) Size-dependent fracture of Si nanowire battery anodes. J Mech Phys Solids 59:1717–1730
50. Yang H et al (2012) Orientation-dependent interfacial mobility governs the anisotropic swelling in lithiated silicon nanowires. Nano Lett 12:1953–1958
51. Goldman JL, Long BR, Gewirth AA, Nuzzo RG (2011) Strain anisotropies and self-limiting capacities in single-crystalline 3D silicon microstructures: models for high energy density lithium-ion battery anodes. Adv Funct Mater 21:2412–2422
52. Su L, Jing Y, Zhou Z (2011) Li ion battery materials with core–shell nanostructures. Nanoscale 3:3967–3983 and references therein
53. Li X et al (2012) Hollow core-shell structured porous Si-C nanocomposites for Li-ion battery anodes. J Mater Chem 22:11014–11017
54. Liu N et al (2014) A pomegranate-inspired nanoscale design for large-volume-change lithium battery anodes. Nat Nanotechnol 9:187–192
55. Ge M, Rong J, Fang X, Zhou C (2012) Porous doped silicon nanowires for lithium ion battery anode with long cycle life. Nano Lett 12:2318–2323
56. Park M-H et al (2009) Silicon nanotube battery anodes. Nano Lett 9:3844–3847
57. Song T et al (2010) Arrays of sealed silicon nanotubes as anodes for lithium ion batteries. Nano Lett 10:1710–1716
58. Wu H et al (2012) Stable cycling of double-walled silicon nanotube battery anodes through solid–electrolyte interphase control. Nat Nanotechnol 7:310–315
59. Zhao X, Hayner CM, Kung MC, Kung HH (2011) In-plane vacancy-enabled high-power Si–Graphene composite electrode for lithium-ion batteries. Adv Energy Mater 1:1079–1084
60. Zhang Y, Zhang L, Zhou C (2013) Review of chemical vapor deposition of graphene and related applications. Acc Chem Res 46:2329–2339
61. Park S, Ruoff RS (2009) Chemical methods for the production of graphenes. Nat Nanotechnol 4:217–224
62. Hummers WS, Offeman RE (1958) Preparation of graphitic oxide. J Am Chem Soc 80:1339
63. Stankovich S et al (2007) Synthesis of graphene-based nanosheets via chemical reduction of exfoliated graphite oxide. Carbon 45:1558–1565
64. Becerril HA et al (2008) Evaluation of solution-processed reduced graphene oxide films as transparent conductors. ACS Nano 2:463–470
65. Bunch JS et al (2008) Impermeable atomic membranes from graphene sheets. Nano Lett 8:2458–2462
66. Li D, Müller MB, Gilje S, Kaner RB, Wallace GG (2008) Processable aqueous dispersions of graphene nanosheets. Nat Nanotechnol 3:101–104
67. Zhou X, Yin Y-X, Wan L-J, Guo Y-G (2012) Self-assembled nanocomposite of silicon nanoparticles encapsulated in graphene through electrostatic attraction for lithium-ion batteries. Adv Energy Mater 2:1086–1090
68. Ye Yun-Sheng et al (2014) Improved anode materials for lithium-ion batteries comprise non-covalently bonded graphene and silicon nanoparticles. J Power Sources 247:991–998
69. Cho YJ et al (2011) Nitrogen-doped graphitic layers deposited on silicon nanowires for efficient lithium-ion battery anodes. J Phys Chem C 115:9451–9457
70. Wang X et al (2013) Adaptable silicon-carbon nanocables sandwiched between reduced graphene oxide sheets as lithium ion battery anodes. ACS Nano 7:1437–1445
71. Liu Y et al (2013) Tailoring lithiation behavior by interface and bandgap engineering at the nanoscale. Nano Lett 13:4876–4883
72. Ren J-G et al (2013) Silicon-Graphene composite anodes for high-energy lithium batteries. Energy Technol 1:77–84
73. About 1900 US $/kg in 2014, i.e., ~1000 times more expensive than metallurgical grade Si. http://www.financialpost.com/markets/data/commodity-cash_prices.html

74. Graetz J, Ahn CC, Yazami R, Fultz B (2004) Nanocrystalline and thin film germanium electrodes with high lithium capacity and high rate capabilities. J Electrochem Soc 151: A698–A702
75. Liang W et al (2013) Tough germanium nanoparticles under electrochemical cycling. ACS Nano 7:3427–3433
76. Liu XH et al (2011) Reversible nanopore formation in Ge nanowires during lithiation–delithiation cycling: an in situ transmission electron microscopy study. Nano Lett 11:3991–3997
77. Baggetto L, Notten PHL (2009) Lithium-ion (de)insertion reaction of germanium thin-film electrodes: an electrochemical and in situ XRD study. J Electrochem Soc 156:A169–A175
78. Abel PR et al (2013) Nanostructured Si(1-x)Gex for tunable thin film lithium-ion battery anodes. ACS Nano 7:2249–2257
79. Park M-H et al (2011) Germanium nanotubes prepared by using the Kirkendall effect as anodes for high-rate lithium batteries. Angew Chem Int Ed 50:9647–9650
80. Yuan F-W, Yang H-J, Tuan H-Y (2012) Alkanethiol-passivated Ge nanowires as high-performance anode materials for lithium-ion batteries: the role of chemical surface functionalization. ACS Nano 6:9932–9942
81. Gu M et al (2014) Bending-induced symmetry breaking of lithiation in germanium nanowires. Nano Lett 14:4622–4627
82. Liu J et al (2014) Ge/C nanowires as high-capacity and long-life anode materials for Li-ion batteries. ACS Nano 8:7051–7059
83. Bodnarchuk MI et al (2014) Colloidal tin-germanium nanorods and their Li-ion storage properties. ACS Nano 8:2360–2368
84. Kennedy T et al (2014) High-performance germanium nanowire-based lithium-ion battery anodes extending over 1000 cycles through in situ formation of a continuous porous network. Nano Lett 14:716–723
85. Winter M, Besenhard JO (1999) Electrochemical lithiation of tin and tin-based intermetallics and composites. Electrochim Acta 45:31–50
86. Rhodes KJ et al (2012) In situ XRD of thin film tin electrodes for lithium ion batteries. J Electrochem Soc 159:A294–A299
87. Beaulieu LY, Hatchard TD, Bonakdarpour A, Fleischauer MD, Dahn JR (2003) Reaction of Li with alloy thin films studied by in situ AFM. J Electrochem Soc 150:A1457–A1464
88. Xu L et al (2013) Monodisperse Sn nanocrystals as a platform for the study of mechanical damage during electrochemical reactions with Li. Nano Lett 13:1800–1805
89. Zhu Z et al (2014) Ultrasmall Sn nanoparticles embedded in nitrogen-doped porous carbon as high-performance anode for lithium-ion batteries. Nano Lett 14:153–157
90. Itoda Y, Kubota T, Matsufuji A, Maekawa Y, Miyasaka T (1997) Tin-based amorphous oxide: a high-capacity lithium-ion-storage material. Science 276:1395–1397
91. Courtney IA, Dahn JR (1997) Electrochemical and in situ X-ray diffraction studies of the reaction of lithium with tin oxide composites. J Electrochem Soc 144:2045–2051
92. Brousse T, Retoux R, Herterich U, Schleich DM (1998) Thin-film crystalline SnO_2-lithium electrodes. J Electrochem Soc 145:1–4
93. Huang JY et al (2010) In situ observation of the electrochemical lithiation of a single SnO_2 nanowire electrode. Science 330:1515–1520
94. Nie A et al (2014) Atomic-scale observation of lithiation reaction front in nanoscale SnO_2 materials. ACS Nano 7:6203–6211 and references therein
95. Dahn JR, Mar RE, Abouzeid A (2006) Combinatorial study of $Sn_{1-x}Co_x$ ($0 < x < 0.6$) and $[Sn_{0.55}Co_{0.45}]_{1-y}C_y$ ($0 < y < 0.5$) alloy negative electrode materials for Li-ion batteries. J Electrochem Soc 153:A361–A365
96. Hassoun J, Mulas G, Panero S, Scrosati B (2007) Ternary Sn–Co–C Li-ion battery electrode material prepared by high energy ball milling. Electrochem Comm 9:2075–2081
97. Zhao K et al (2011) Lithium-assisted plastic deformation of silicon electrodes in lithium-ion batteries: a first-principles theoretical study. Nano Lett 11:2962–2967

98. Wolfenstine J et al (2000) Experimental confirmation of the model for microcracking during lithium charging in single-phase alloys. J Power Sour 87:1–3
99. Massalski TB (1986) Binary alloy phase diagram, vol 2. American Society for Metals, Metal Park. Ohio 44073, p 1495
100. Wang J, King P, Huggins RA (1986) Investigations of binary lithium-zinc, lithium-cadmium and lithium-lead alloys as negative electrodes in organic solvent-based electrolyte. Solid State Ionics 20:185–189
101. Martos M, Morales J, Sanchez L (2003) Lead-based systems as suitable anode materials for Li-ion batteries. Electrochim Acta 48:615–621
102. Sandu I et al (2002) Comparison of the electrochemical behavior of SnO_2 and PbO_2 negative electrodes for lithium ion batteries. Ionics 8:27–35
103. Ng SH et al (2006) Spray pyrolyzed PbO-carbon nanocomposites as anode for lithium-ion batteries. J Electrochem Soc 153:A787–A793
104. Chen Z et al (2012) Pb-sandwiched nanoparticles as anode material for lithium-ion batteries. J Solid State Electrochem 16:291–295
105. Wang J et al (2014) In-situ one-step hydrothermal synthesis of a lead germanate-graphene composite as a novel anode material for lithium-ion batteries. Sci Rep 4(7030):1–7
106. Park CM, Sohn HJ (2010) Quasi-intercalation and facile amorphization in layered ZnSb for Li-ion batteries. Adv Mater 22:47–52
107. Schlesinger ME (2002) The thermodynamic properties of phosphorus and solid binary phosphides. Chem Rev 102:4267–4301
108. Park CM, Sohn HJ (2007) Black phosphorus and its composite for lithium rechargeable batteries. Adv Mater 19:2465–2468
109. Wang L et al (2012) Nano-structured phosphorus composite as high-capacity anode materials for lithium batteries. Angew Chem Int Ed 51:9034–9037
110. Akahama Y, Endo S, Narita S (1983) Electrical properties of black phosphorus single crystals. J Phys Soc Jpn 52:2148–2155
111. Li L et al (2014) Black phosphorous field-effect transistors. Nat Nanotechnol 9:372–377
112. Zeng Z et al (2012) An effective method for the fabrication of few-layer-thick inorganic nanosheets. Angew Chem Int Ed 51:9052–9056
113. Chhowalla M et al (2013) The chemistry of two-dimensional layered transition metal dichalcogenide nanosheets. Nat Chem 5:263–275
114. Sun J et al (2014) Formation of stable phosphorous-carbon bond for enhanced performance in black phosphorus nanoparticle-graphite composite battery anodes. Nano Lett 14:4573–4580
115. Darwiche A et al (2012) Better cycling performances of bulk Sb in Na-ion batteries compared to Li-ion systems: an unexpected electrochemical mechanism. J Am Chem Soc 134:20805–20811
116. Baggetto L et al (2013) Intrinsic thermodynamic and kinetic properties of Sb electrodes for Li-ion and Na-ion batteries: experiment and theory. J Mater Chem A 1:7985–7994
117. Hewitt KC, Beaulieu LY, Dahn JR (2001) Electrochemistry of InSb as a Li insertion host; problems and prospects. J Electrochem Soc 148:A402–A410
118. He M et al (2014) Monodisperse antimony nanocrystals for high-rate Li-ion and Na-ion battery anodes: nano versus bulk. Nano Lett 14:1255–1262
119. Massalski TB (ed) (1986) Binary alloy phase diagrams, vol 1. American society for metals, Metals park, Ohio 44073, p 514
120. Xianming W, Nishina T, Uchida I (2002) Lithium alloy formation at bismuth thin layer electrode and its kinetics in propylene carbonate electrolyte. J Power Sour 104:90–96
121. Park C-M et al (2009) Enhanced electrochemical properties of nanostructured bismuth-based composites for rechargeable lithium batteries. J Power Sour 186:206–210
122. Borgstedt HB, Guminski C (2003) The B-Li (Boron-Lithium) System. J Phase Equilib 24:572–574
123. Ding X et al (2013) Temperature-dependent lithium storage behavior in tetragonal boron (B50) thin film anode for Li-ion batteries. Electrochim Acta 87:230–235

124. Szwarc R et al (1982) Discharge characteristics of lithium-boron alloy anode in molten salt thermal cells. J Electrochem Soc 129:1168–1173
125. Sanchez P et al (1992) Preparation and characterization of lithium-boron alloys: electrochemical studies as anodes in molten salt media, and comparison with pure lithium-involving systems. J Mater Science 27:240–246
126. Puhakainen K et al (2010) A new phase in the system lithium–aluminum: characterization of orthorhombic Li_2Al. J Solid State Chem 183:2528–2533
127. Kishio K, Brittain JO (1979) Defect structure of β-LiAl. J Phys Chem Solids 40:933–940
128. Dey AN (1971) Electrochemical alloying of lithium in organic electrolytes. J Electrochem Soc 118:1547–1549
129. Rao BML, Francis RW, Christopher HA (1977) Lithium–aluminum electrode. J Electrochem Soc 124:1490–1492
130. Garreau M, Thevenin J, Fekir M (1983) On the processes responsible for the degradation of the aluminum-lithium electrode used as anode material in lithium aprotic electrolyte batteries. J Power Sour 9:235–238
131. Hamon Y et al (2001) Aluminum negative electrode in lithium ion batteries. J Power Sour 97–98:185–187
132. Hudak NS, Huber DL (2012) Size effects in the electrochemical alloying and cycling of electrodeposited aluminum with lithium. J Electrochem Soc 159:A688–A695
133. Lei X et al (2007) Effect of particle size on the electrochemical properties of aluminum powders as anode materials for lithium ion batteries. J Alloys Compounds 429:311–315
134. Lei X, Ma J (2009) Co_3O_4 coated Al composites as anode materials for lithium ion batteries. Mater Chem Phys 116:383–387
135. Wang CY et al (2008) Electrochemical properties of nanostructured $Al_{1-x}Cu_x$ alloys as anode materials for rechargeable lithium-ion batteries. J Electrochem Soc 155:A615–A622
136. Au M et al (2010) Free standing aluminum nanostructures as anodes for Li-ion rechargeable batteries. J Power Sour 195:3333–3337
137. Liu Y et al (2011) In situ transmission electron microscopy observation of pulverization of aluminum nanowires and evolution of the thin surface Al_2O_3 layers during lithiation–delithiation cycles. Nano Lett 11:4188–4194
138. Leite MS et al (2014) Insights into capacity loss mechanisms of all-solid-state Li-ion batteries with Al anodes. J Mater Chem A 2:20552–20559
139. Phan VP, Pecquenard B, Le Cras F (2012) High-performance all-solid-state cells fabricated with silicon electrodes. Adv Funct Mater 22:2580–2584
140. Okamoto H (2006) Gallium-Lithium phase diagram. J Phase Equilib Diff 27:200
141. Lee KT et al (2008) Liquid gallium electrode confined in porous carbon matrix as anode for lithium secondary batteries. Electrochem Solid-State Lett 11:A21–A124
142. Cui Y-H et al (2009) InP as new anode material for lithium ion batteries. Electrochem Comm 11:1045–1047
143. Zhou Y et al (2006) The electrochemistry of nanostructured In_2O_3 with lithium. J Power Sour 162:1373–1378
144. Kim D-W et al (2007) Highly conductive coaxial SnO_2-In_2O_3 heterostructured nanowires for Li-ion battery electrodes. Nano Lett 7:3041–3047
145. Li H, Huang XJ, Chen LQ (1999) Anodes based on oxide materials for lithium rechargeable batteries. Solid State Ionics 123:189–192
146. Shi Z et al (2001) Electrochemical properties of Li-Mg alloy electrodes for lithium batteries. J Power Sources 92:70–80
147. Kim H et al (2000) Electrochemical characteristics of Mg–Ni alloys as anode materials for secondary Li batteries. J Power Sour 90:59–63
148. Park C-M et al (2006) Enhancement of the rate capability and cyclability of an Mg-C composite electrode for Li secondary batteries. J Power Sour 158:1451–1455
149. Taillades G, Sarradin J (2004) Silver: high performance anode for thin film lithium ion batteries. J Power Sour 125:199–205

150. Lee YJ et al (2010) Biologically activated noble metal alloys at the nanoscale: for lithium ion battery anodes. Nano Lett 10:2433–2440
151. Yin J et al (2003) New Ag-Sn alloy anode materials for lithium-ion batteries. J Electrochem Soc 150:A1129–A1135
152. Taillades G et al (2002) Metal-based very thin film anodes for lithium ion microbatteries. Solid State Ionics 152–153:119–124
153. Laïk B et al (2008) Silicon nanowires as negative electrode for lithium-ion microbatteries. Electrochim Acta 53:5528–5532
154. Yuan L et al (2007) Mesoporous gold as anode material for lithium-ion cells. J New Mater Electrochem Syst 10:95–99
155. Boukamp BA, Lesh GC, Huggins RA (1981) All-solid lithium electrodes with mixed-conductor matrix. J Electrochem Soc 128:725–729
156. Zhang W-J (2011) A review of the electrochemical performance of alloy anodes for lithium-ion batteries. J Power Sour 196:13–24
157. Tarascon J-M et al (2005) New concepts for the search of better electrode materials for rechargeable lithium batteries. CR Chimie 8:9–15
158. Oumellal Y et al (2008) Metal hydrides for lithium-ion batteries. Nat Mater 7:916–921
159. Cabana J et al (2010) Beyond intercalation-based Li-ion batteries: the state of the art and challenges of electrode materials reacting through conversion reactions. Adv Mater 22:E170–E192
160. Grugeon S et al (2003) An update on the reactivity of nanoparticles Co-based compounds towards Li. Solid State Sci 5:895–904
161. Wang W et al (2012) Revealing the conversion mechanism of CuO nanowires during lithiation–delithiation by in situ transmission electron microscopy. Chem Comm 48:4812–4814
162. Wang F et al (2012) Tracking lithium transport and electrochemical reactions in nanoparticles. Nature Comm 3:1201. doi:10.1038/ncomms2185

Electrolytes for Lithium and Lithium-Ion Batteries

Libo Hu, Sheng Shui Zhang and Zhengcheng Zhang

In this chapter, new trends in the formulation of non-aqueous liquid electrolytes will be discussed. Novel solvents and salts used in Li-ion battery electrolytes are categorized and illustrated, and the progress in understanding the formation mechanism behind the solid-electrolyte interphase (SEI) is discussed.

1 Introduction

There has been an explosion of development in portable electronics over the last several decades. As a consequence, Li-ion battery technology has also evolved tremendously to meet the demand for higher energy density, longer cycle life, and lower cost. Now, the technology is moving into applications requiring even higher energy and power density, such as electric vehicles [10, 29, 102]. For this purpose, great advances have been made in both high-energy anode materials [49] and cathode materials [115].

On the anode side, silicon-based materials emerged as a super high-capacity anode for the Li-ion battery [111]. Other alloy anode materials [80, 137] and metal oxides [83, 112] also showed great promise. However, due to their different characteristics from the conventional carbonaceous anode, these anode materials raise new challenges for battery electrolytes. The ability to deliver higher capacity usually also means larger volume expansion during lithiation, and as a result, the SEI with conventional electrolyte is damaged/cracked due to repeated dramatic changes of the anode material morphology during cycling [37, 50].

L. Hu · Z. Zhang (✉)
Electrochemical Energy Storage Theme, Chemical Sciences and Engineering Division, Argonne National Laboratory, Lemont, IL 60439, USA
e-mail: zzhang@anl.gov

S.S. Zhang
U.S. Army Research Laboratory, Adelphi, MD 20783, USA
e-mail: Shengshui.zhang.civ@mail.mil

Z. Zhang and S.S. Zhang (eds.), *Rechargeable Batteries*,
Green Energy and Technology, DOI 10.1007/978-3-319-15458-9_8

On the cathode side, the introduction of high-energy materials has also presented new challenges. In conventional Li-ion batteries, the cathode/electrolyte interface is rarely a problem because the cut-off potential of conventional cathode materials (lower than 4 V vs. Li^+/Li) is usually far below the oxidative decomposition potential of the state-of-the-art (SOA) electrolyte. However, many of the new-generation cathode materials set cut-off potentials around 4.3 V, with the potential reaching above 4.5 V versus Li^+/Li [44]. The high-voltage cathode materials have even higher charging potentials, some close to 5 V versus Li^+/Li [19, 56, 87], which is well above the decomposition threshold of SOA electrolytes [129].

With both the anode and cathode chemistry becoming much more challenging, the electrolyte requirements for a high-energy Li-ion cell have never been more stringent. A single electrolyte formulation could never meet the demands of so many high-energy materials, so to accommodate different cell chemistries, researchers have developed numerous electrolyte formulations consisting of both conventional and novel organic solvents and lithium salts in combination with performance-enhancing additives. Besides the requirement of high performance, batteries used in transportation applications must also be safe. Conventional organic electrolytes are highly flammable, raising concerns over the safety of batteries with such electrolytes in electric vehicles and aircraft. Consequently, flame retardant additives and electrolyte formulations with overcharge protection have been investigated [14]. Some batteries used in niche applications demand specialty electrolytes. For example, the electrolytes for batteries used in oil well drilling need to withstand temperature over 200 °C as well as strong vibrations and shocks [40], and electrolytes for batteries used in satellites and space missions need to withstand extremely low temperatures [82]. In each of these situations, the electrolyte has to be tailored to the specific needs of the application.

Obviously, the properties of the electrolyte in Li-ion batteries are of crucial importance, and a few comprehensive general review articles have already been published on the subject [116, 117, 120, 133]. In this chapter, the recent advances and new trends in electrolyte research will be overviewed from a wide angle, with a focus on novel electrolytes and electrolyte additives for high-energy electrode materials.

2 Electrolyte Solvents for Non-aqueous Electrolyte

Due to the highly reductive nature of the lithiated anode material, the lithium and lithium ion battery electrolytes usually consist of a lithium salt and either a single aprotic organic solvent or a mixture of them [116] instead of the aqueous electrolytes used in many conventional primary and secondary batteries. Even with organic, aprotic electrolytes, there are still challenges associated with the electrolyte/electrode interface. Since the redox potential of carbonaceous anode (graphite), the most common anode materials for Li-ion batteries, is slightly above the redox potential of Li^+/Li, the charging potential of the graphite anode is usually much lower than the reduction potential of most organic solvents. As a result, the thermodynamically unstable organic electrolyte can only be stabilized kinetically on the anode surface through the generation of a protective passivation layer on the anode surface. This passivating layer can be formed by the reductive decomposition of the organic electrolyte or another component in the electrolyte. The passivation layer is often referred to as the solid-electrolyte interphase (SEI). Due to the extremely important role of the SEI layer, the ability to form a robust SEI (ionically conductive, resistant to degradation during cycling) becomes a governing factor to determine the suitability of a new electrolyte for the Li-ion chemistry.

Being an organic electrolyte, the organic solvent used in the electrolyte formulations is the most influential component to consider. The most prominent electrolyte solvent for the first-generation Li-ion battery is, without a doubt, ethylene carbonate (usually abbreviated as EC in the Li-ion battery literature) (Fig. 1). This cyclic carbonate possesses several desirable properties as an electrolyte solvent, including large dipole moment, good salt solubility, low vapor pressure and above all, the ability to form a robust SEI on the surface of the graphite anode [32]. The drawback of EC is that it has high melting point (34–37 °C) and high viscosity (1.90 cP at 40 °C), so it is conventionally used in combination with linear carbonates (Fig. 1) to increase the liquid range and ion mobility of the resulting electrolyte. The EC/linear carbonate formulations laid the foundation of modern Li-ion battery electrolyte technology and usually serve as the basis of various advanced formulations used in industry. Another well-known cyclic carbonate is propylene carbonate, abbreviated as PC, a solvent that was used in Li metal batteries until carbonaceous anodes became popular. Unfortunately, PC was found to co-intercalate with Li ions into the graphene layers causing exfoliation of the graphite structure, which deemed it inappropriate for Li-ion batteries [23]. Because of its destructive effect on carbonaceous anodes, to date it has been widely used as an electrolyte for the evaluation of anode SEI formation additives, which will be discussed in detail in Chap. 9.

The other major component in a liquid organic electrolyte is the lithium salt, which is the ionic source of the electrolyte. The salt is generally the lithium conjugate base of a super acid, which makes the anion anodically stable and easily dissociates from Li^+ in organic solvents. Compared with electrolyte solvents, the

Fig. 1 Common electrolyte solvents in SOA Li-ion batteries

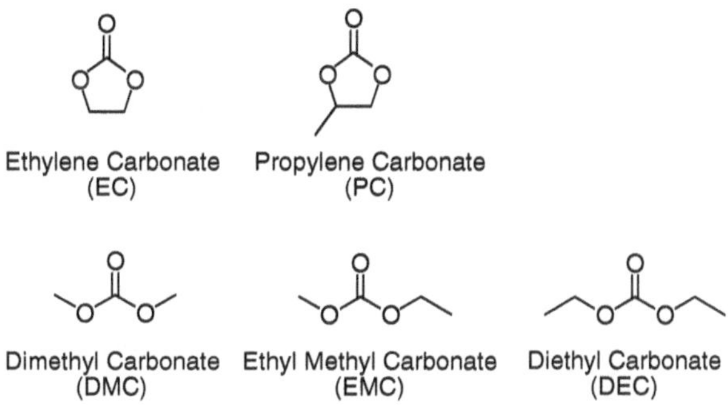

lithium salts generally receive less attention because the choice of salt is much more limited. The SOA lithium salt used in Li-ion batteries is lithium hexafluorophosphate, with the chemical formula $LiPF_6$. This salt has good solubility in the carbonate solvents mentioned above and good ion mobility, is resistant to reduction and oxidation, and can passivate the aluminum current collectors used in batteries [130]. It is the predominant lithium salt used in the EC-based electrolyte formulations and widely used in commercial Li-ion batteries. Other commonly used salts in the research labs include lithium perchlorate ($LiClO_4$), lithium tetrafluoroborate ($LiBF_4$), lithium hexafluoroarsenate ($LiAsF_6$), lithium trifluoromethanesulfonate ($LiSO_3CF_3$) and lithium salts based on fluorinated sulfonyl imide anions such as lithium bis(trifluoromethanesulfonyl)imide (LiTFSI). The negative charge in such anions (i.e., $TFSI^-$ or $[N(SO_2CF_3)_2]^-$) is extensively delocalized resulting in weak cation-anion interactions. Unfortunately, corrosion of aluminum current collectors occurs in liquid electrolytes containing LiTFSI.

With the development of high-energy electrode materials, the design of organic electrolytes has evolved accordingly. On the anode side, electrolytes that are compatible with novel high-capacity anode materials are in great demand. On the cathode side, high-voltage electrolytes using novel solvents are being investigated because conventional carbonate solvents are anodically unstable at the elevated charging potentials required to deliver higher energy. Furthermore, development of electrolytes with enhanced safety properties is also a focus of transportation technology research. In this part of the chapter, the organic solvents used in the cutting-edge Li-ion batteries will be discussed in detail, with a focus on novel solvents.

2.1 Fluorinated Carbonates

As stated above, carbonates are the main solvents in the current SOA Li-ion batteries due to their large dipole moment, good solubility of lithium salts, and ability to form a good SEI on carbonaceous anodes. However, conventional alkyl carbonate solvents experience extensive oxidative decomposition on the cathode side at higher potential (>4.3 V vs. Li^+/Li) [61, 129], which makes them unsuitable for high-energy cathode materials. With the exception of this disadvantage, carbonates generally have better properties than any other solvents used in Li-ion batteries. As a result, modified carbonates with electron withdrawing groups that raise the oxidation potential are promising targets of electrolyte research for high voltage high energy cathodes. Fluorination is one of the most popular methods to achieve a higher oxidation potential.

Theoretical calculations using density function theory (DFT) have shown a moderate to dramatic increase in the oxidation potential of organic carbonates compared with their non-fluorinated counterparts (Table 1) [138]. The calculated oxidation potentials are much higher than those measured on actual electrodes, but the results could serve as a source of comparison between different structures in the same model. The electrolytes formulated from these fluorinated carbonates showed

Table 1 Structure, oxidation potential, and energies for highest occupied molecular orbit (HOMO) and lowest unoccupied molecular orbit (LUMO) of carbonates (EC and EMC), ethers (EPE), fluorinated carbonates (F-AEC and F-EMC), and fluorinated ethers (F-EPE)

Molecule	Structure	Pox (V Theory)	HOMO (au)	LUMO (au)
EC		6.91 (6.83 open)	−0.31005	−0.01067
EMC	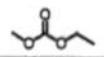	6.63	−0.29905	0.00251
EPE	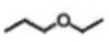	5.511	−0.26153	0.00596
F-AEC	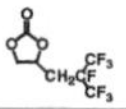	6.98	−0.31780	−0.01795
F-EMC	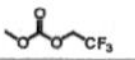	7.01	−0.31946	−0.00363
F-EPE	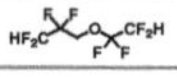	7.24	−0.35426	−0.00356

Reproduced with permission [138]. Copyright 2013 Royal Society of Chemistry

high stability on the 5-V Ni/Mn spinel cathode $LiNi_{0.5}Mn_{1.5}O_4$ (LNMO). However, the fluorinated cyclic carbonate F-AEC shown in Table 1 was only compatible with $Li_4Ti_5O_{12}$ (LTO) used as anode due to its inability to form a functional SEI on the graphite anode surface [138].

SEI formation on the graphite anode is a major concern with fluorinated carbonate electrolytes. It is generally agreed that fluorinated compounds reductively decompose and form an insulating layer on the anode surface, and the repeated decomposition on the anode results in higher interfacial impedance and irreversible capacity loss. One of the simplest and most used fluorinated carbonate as reported in the literature is fluoroethylene carbonate (4-fluoro-1,3-dioxolan-2-one), commonly abbreviated as FEC, or F-EC (Fig. 2). FEC is a well-known co-solvent and SEI formation additive for the graphite anode [67] and silicon anode [71]. Especially for the silicon anode, FEC is the singularly necessary electrolyte additive to mitigate capacity fading during repeated charge and discharge cycling. There are different views on the SEI formation mechanism of FEC as an anode additive, and

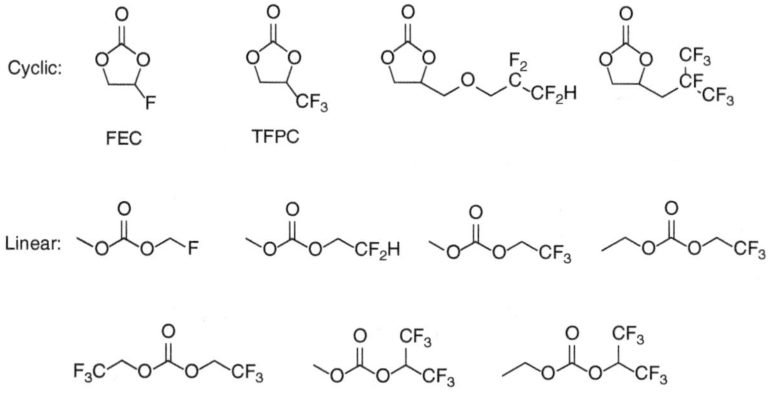

Fig. 2 A selection of fluorinated carbonates for high voltage Li-ion battery electrolytes

generally it falls into the vinylene carbonate (VC) mechanism [28] or vinyl fluoride mechanism [71]; their details will be discussed in Chap. 9. On the cathode side, FEC is reported to have better stability than EC on the LNMO cathode and enables greatly improved performance for LNMO/graphite full cells even at elevated temperature (55 °C) [45].

Other than FEC, the trifluoromethyl substituted ethylene carbonate (or fluorinated propylene carbonate) (TFPC, Fig. 2) also has received attention as a possible co-solvent and SEI formation additive. While most research suggests TFPC is not as effective as FEC in terms of capability of SEI formation on graphite anode, at least one study showed that TFPC is superior to FEC in suppressing the irreversible capacity of the graphite anode [106] (Fig. 3). While electron microscopy results indicated that TFPC is capable of forming a SEI on the graphite anode [48], TFPC has much higher production cost than FEC, so the interest in this molecule as a candidate for the commercial Li-ion battery is low.

Besides FEC and TFPC, more advanced structures of fluorinated cyclic carbonates have also been synthesized and evaluated in Li-ion batteries. Two examples of such molecules are given in Fig. 2. These molecules have traditionally been used in specialized applications such as coolants and supercapacitors, and their use in Li-ion batteries is relatively new [5, 6, 74, 78]. Regardless, these molecules have shown increased stability toward oxidation, better thermal stability, and low flammability.

Compared with fluorinated cyclic carbonates, less research has been conducted on fluorinated linear carbonates due to their limited availability. The structure variation of the fluorinated linear carbonates are much more diverse than that of

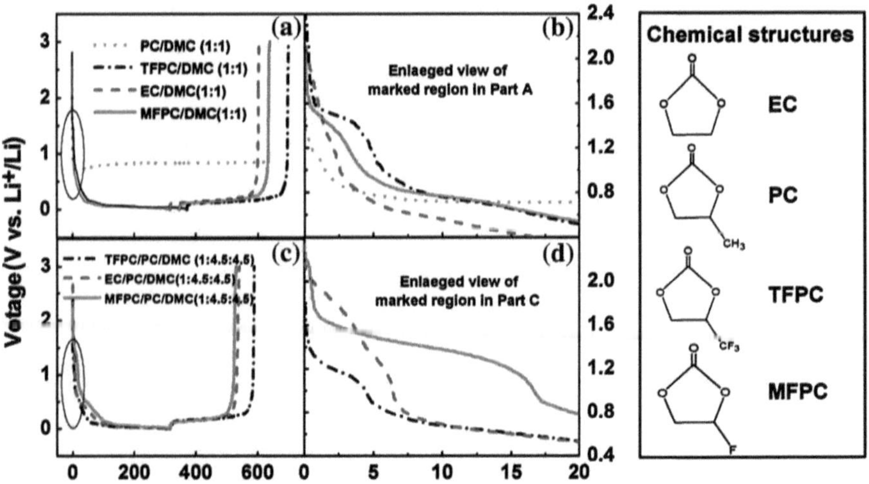

Fig. 3 Voltage profiles in the first cycle of Li/mesocarbon microbead (MCMB) half cells with 1 M $LiPF_6$ in bi-solvent and tri-solvent electrolytes containing EC, PC, FEC (labeled as MFPC in the graph), and TFPC. Reproduced with permission [106]. Copyright 2010 Elsevier

cyclic carbonates because the two groups on both side of the carbonate center can be introduced with various fluorinated alcohol precursors, and the synthesis chemistry is less complicated than that of the fluorinated cyclic carbonates. Consequently, fluorinated linear carbonates hold greater promise as electrolyte candidates for Li-ion batteries, with some examples of fluorinated linear carbonates shown in Fig. 2. Fluorination of linear carbonates is being investigated in order to improve both the oxidative stability and the melting point. Higher oxidation stability would enable use of these solvents with high-voltage cathodes. The methyl trifluoroethyl carbonate (FEMC) (asymmetric fluorinated EMC) has been shown to have much higher oxidation stability of the formulated electrolyte on the high voltage $LiNi_{0.5}Mn_{1.5}O_4$ [45, 138] than its non-fluorinated counterpart. The second purpose of fluorination is to lower both melting point and viscosity, which would enable the electrolyte for low temperature applications [93]. Fluorinated molecules behave differently than their non-fluorinated counterparts, and these different behaviors include decreased boiling and melting points and increased surface tension. Thus, fluorinated carbonates, especially linear carbonates, are excellent candidates for the low-temperature Li-ion batteries for space and deep-sea missions. Last but not least, the low-flammability to non-flammability of fluorinated compounds could potentially improve the safety of Li-ion batteries [5, 6].

In summary, the SOA carbonate-based electrolytes remain the most promising category of electrolyte for 4-V Li-ion batteries. However, to meet the requirements for the high-voltage high energy cells, fluorination plays a significant role in elevating the oxidation stability via F- and fluoroalkyl- substitution. Electrolyte researchers from academia, national labs and industry are actively working on this front. Widespread adoption of these fluorinated electrolytes may afford the performance, cost and safety of the high energy Li-ion cells for a wide range of applications.

2.2 *Sulfone-Based Electrolytes*

The research in sulfones electrolytes for Li-ion batteries dates back to the 1980s, even before the intercalation chemistry was studied [70]. However, since the emergence of the Li-ion chemistry, the interest in sulfones-based electrolytes declined due to the excellent performance of carbonate-based electrolytes for graphitic anodes. Sulfones, especially symmetric ones, have high boiling points and can be used to formulate electrolyte for batteries in high temperature applications. The pioneering studies in the sulfone-based electrolyte for Li-ion batteries were predominantly carried out by a research group led by professor Charles Austen Angell at Arizona State University [97, 98, 119], and nowadays more researchers are further exploring the potential of this group of solvents.

In recent years, the sulfone-based electrolytes have renewed interest because the new generation of cathode materials requires working potentials much higher than the oxidation stability limit of the conventional carbonate-based electrolytes.

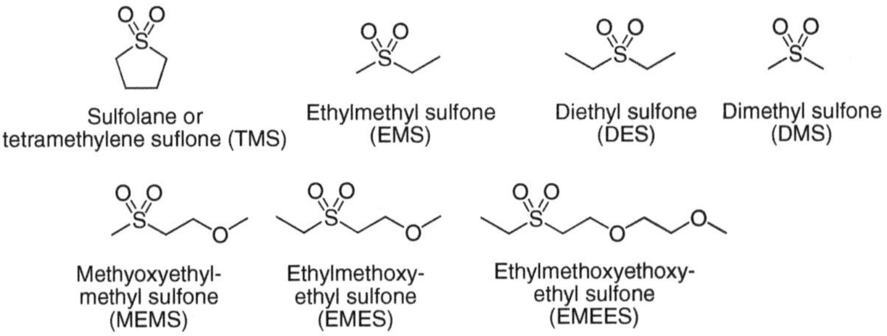

Fig. 4 Examples of commercially available and lab-synthesized sulfones

The sulfone functionality (−SO_2−) has the highest achievable oxidation state (+6) found in all sulfur-containing organic compounds, which makes it highly stable against oxidation. Based on quantum chemistry calculations and experimental data, sulfones generally have oxidation potentials above 5 V versus Li^+/Li (Fig. 4), even for the varieties with organic ether side chains [91]. Given proper formulation, they hold great promise as electrolyte solvents in high-voltage Li-ion batteries. However, from the earliest study of sulfone electrolytes in Li-ion batteries, it was found that many are not capable of forming an SEI on the graphite anodes [119]. The use of SEI formation additives or lithium salts that can form an SEI is common practice in many following studies [64, 99, 110].

A few commercially available sulfones and lab-synthesized sulfones that are suitable as electrolyte solvents are shown in Fig. 5. Most sulfones with simple structures suffer from high melting point and high viscosity, and consequently the formulated electrolytes generally suffer from lower ionic conductivity and subsequently low C-rate capability compared with carbonate-based electrolytes; To address this issue, in some designs, ether-type chains are incorporated into the structure [91].

One of the most widely examined sulfones is sulfolane (tetramethylene sulfone, abbreviated as TMS). Sulfolane has a melting point of 27.5 °C, and like EC it has to be mixed with a diluent to widen the liquid range and decrease the viscosity of the resulting electrolyte. Due to its high polarity, TMS can dissolve most of the common lithium salts and also dissolves less common lithium salts including lithium bis(oxalato)borate (LiBOB) [64] and lithium difluoro(oxalato)borate (LiDFOB) [110] which usually have low solubility in carbonate electrolytes. LiBOB and LiDFOB act as not only a lithium ion source in the electrolyte, but also passivate the graphite anode through the formation of a robust SEI. The utilization of such SEI-forming lithium salts eliminates the need of additional additives. TMS was reported to be stable on the high-voltage spinel cathode $LiNi_{0.5}Mn_{1.5}O_4$ when coupled with the $Li_4Ti_5O_{12}$ anode [1]. Mixed with ionic liquid, TMS can form a non-flammable electrolyte with a oxidation stability up to 4.8 V versus Li^+/Li, as examined on a $Li_{1.2}Ni_{0.2}Mn_{0.6}O_2$ cathode [114]. To improve the properties of the sulfone based electrolytes, investigations of fluorinated sulfones have been initiated

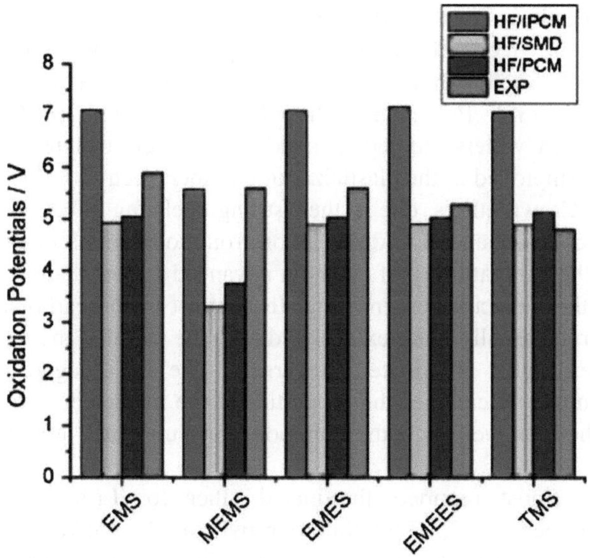

Fig. 5 Oxidation potentials calculated at the HF/6-31 + G (d, p) level for three solvation models: isodensity polarizable continuum model (IPCM), polarizable continuum model (PCM), and solvation model based on density (SMD). Experimental oxidation potentials (EXP) are from the literature. Reproduced with permission [91]. Copyright 2011 American Chemical Society

in order to further increase its voltage stability and the SEI formation capacity for high energy density lithium-ion batteries (Zhang, unpublished data).

In conclusion, sulfones show both great advantages and challenges as a new class of electrolyte solvents. The intrinsic oxidation stability makes them a good candidate for high-voltage Li-ion batteries, but the high melting point and viscosity of conventional sulfones make them inappropriate for high power applications. Future research should be focused on the functionalization of sulfones which makes them more suitable as electrolytes and/or electrolyte additives for new generation Li-ion battery.

2.3 Ether-Based Electrolytes

Due to the affinity of the oxygen atom to lithium cations, organic ethers have historically been investigated as electrolyte solvent or co-solvent to increase the ion mobility due to their low viscosity and acceptable polarity. Oligomeric linear ethers with repeating ethylene glycol units ($-CH_2CH_2O-$) such as dimethoxyethane, diglyme, triglyme, tetraglyme [26, 65, 103], and cyclic ethers such as tetrahydrofuran (THF) and 2-methyltetrahydrofuran (Me-THF) [2, 54] were investigated as the electrolyte components for Li metal batteries. However, aliphatic ethers suffer from low oxidation potentials [79], making them inappropriate for the majority of the SOA Li-ion chemistries.

Although the above-mentioned ethers may not be suitable for high-energy Li-ion batteries, they are frequently used in high-capacity batteries with relatively low operating voltage, such as lithium-sulfur (Li-S) batteries [141] and lithium-oxygen ($Li-O_2$) batteries [35]. Polyether with poly(ethylene oxide)(PEO) as a typical example has been widely studied as solid polymer electrolytes. Low molecular weight PEO is employed as the plasticizer of polymer electrolyte for Li-ion polymer batteries [96]. Crown ethers, due to their strong chelating effect, are used as electrolyte additives to enhance Li^+ dissociation from counterions in electrolytes [33].

To enable the compatibility of ethers in advanced Li-ion chemistry, fluorination is utilized in the molecular engineering. In contrast to the carbonates, the fluorination of ethers is usually more extensive due to the fact that the synthesis of such ethers involves highly fluorinated precursors. The high degree of fluorination sometimes completely changes the properties of the molecule. A few examples of fluorinated ethers as well as some common non-fluorinated ethers are shown in Fig. 6.

So far, the most reported fluorinated ether for Li-ion battery study is 1,1,2,2-tetrafluoroethyl-2,2,3,3-tetrafluoropropyl ether [6, 45, 78, 138]. This solvent is also listed in Table 1 for the molecule F-EPE. Compared with the non-fluorinated ethyl propyl ether, the calculated oxidation potential of F-EPE increases from 5.51 to 7.24 V, which is the most dramatic change among the fluorinated compounds listed in Table 1. The source of this dramatic increase in oxidation potential lies in the oxidation mechanism of ethers shown in Scheme 1. The source of the oxidative vulnerability of non-fluorinated ethers is that the C-H bond next to the ether oxygen is subject to hydrogen extraction to form radicals, which is then stabilized by the lone pair electron on the ether oxygen atom. The resulting radical subsequently traps an oxygen molecule to form a peroxide radical. In highly fluorinated ethers, especially those with fluorinated α carbon, the strong electron-withdrawing effect of the fluoroalkyl group pulls the electron density of the lone pair electrons on the oxygen atom away from the neighboring $-CH_2$ group, and the stabilizing effect is

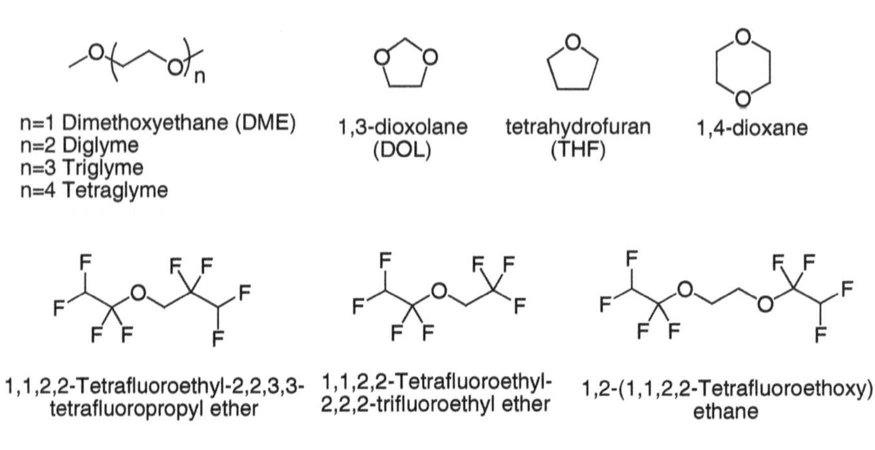

Fig. 6 Examples of non-fluorinated ethers and fluorinated ethers

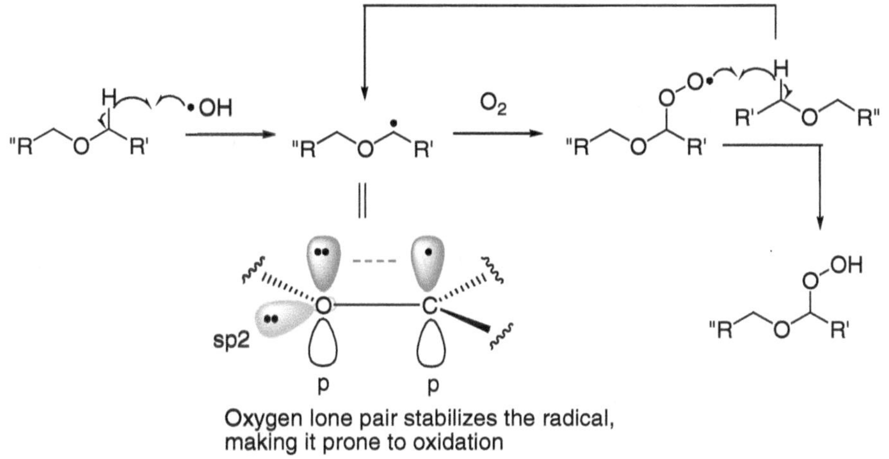

Scheme 1 Oxidation and peroxide formation mechanism of ether molecules

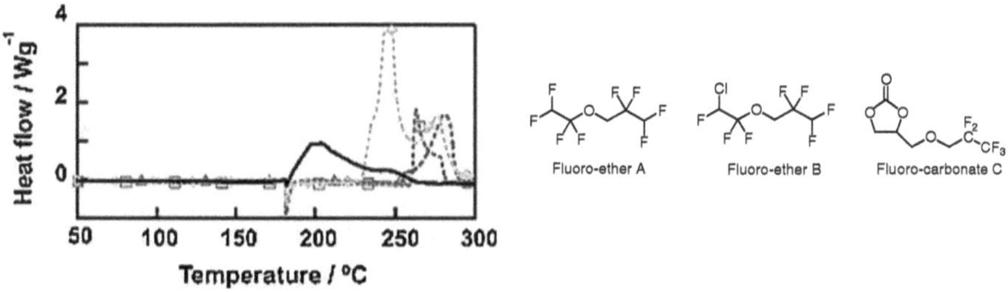

Fig. 7 Differential scanning calorimetry (DSC) curves for mixtures of metallic Li and EC/DMC (1:1 vol. ratio) or fluorine compound (**A**, **B** or **C**). *Solid black curve* EC/DMC; ---▲--- fluoro-ether **A**; ---□--- fluoro-ether **B**; ---◇--- fluoro-carbonate **C**. Reproduced with permission [78]. Copyright 2013 Elsevier

greatly reduced. This effect results in the much higher oxidation potential of fluorinated ethers, making them good candidates for high-voltage Li-ion batteries. In addition to the anodic stability, fluorinated ethers also possess wide liquid range, non-flammability, and enhanced thermal stability (Fig. 7) [78], all of which are necessary properties of Li-ion batteries designed for vehicles.

Besides F-EPE, many other fluorinated ethers are synthesized by the addition reaction of tetrafluoroethylene with fluorinated and non-fluorinated alcohols. Some of the new structures may have superior properties to F-EPE as electrolyte solvents for Li-ion batteries and beyond Li-ion batteries. Recently, F-EPE was investigated for the first time as the electrolyte solvent in the lithium-sulfur battery [12]. The new fluorinated electrolyte DOL/F-EPE-1.0 M LiTFSI suppressed the deleterious shuttling effect and improved the capacity retention and coulombic efficiency in Li-S cell tests. In addition, it was found to eliminate the self-discharge of lithium-sulfur batteries. Another fluorinated ether, bis(2,2,2-trifluoroethyl) ether (BTFE), was reported to mitigate self-discharge of Li-S cells having both low- and high-sulfur loading sulfur cathodes [13, 38]. This electrolyte significantly decreased self-discharge at elevated temperature, though differences in behavior of cells with high- and low-sulfur loading were also noted. Further investigation showed that this effect likely stems from the formation of a more robust protective film on the anode surface. New fluoroethers with unique structures are certainly worthy of exploration.

2.4 Ester and Lactone-Based Electrolytes

Esters and lactones (cyclic esters) can be considered as cousins of linear and cyclic carbonates, as their physical properties are alike. Figure 8 shows the chemical structures of some common esters and lactones. Due to the lower oxidation state of esters compared to that of carbonates, esters have generally inferior oxidation stability. For this reason, esters have not been widely studied as electrolyte solvents as the trend of Li-ion battery development is leading towards higher cell voltages. However, linear esters continue to be attractive as a diluent in electrolyte formulations due to their low viscosity and melting point which improve low temperature performance of the conventional Li-ion battery [95].

Among lactones, γ-butyrolactone (GBL) has been widely investigated as an alternative to cyclic carbonates and frequently used in many electrolyte compositions. Like EC, GBL has high dielectric constant, high solubility of lithium salts, high boiling point (b.p. 204 °C) and low melting point (−43.5 °C), making it more suitable for transportation application. Although the most compatible salt with carbonate solvents is $LiPF_6$, the optimal lithium salt for GBL is $LiBF_4$. It was first demonstrated by Japanese researchers from Toshiba that an electrolyte formulated with GBL/EC/$LiBF_4$ showed good performance in laminated thin-film Li-ion cells [100, 101]. A later paper [20] reported that among five salts ($LiBF_4$, $LiPF_6$, $LiAsF_6$, LiTFSI, and $LiClO_4$) tested with EC/GBL mixed solvents, only $LiBF_4$ offered reversible intercalation with capacity up to the theoretical value of graphite. After that, most studies involved only $LiBF_4$ as the default salt for GBL-based electrolytes. To further improve the stability and reduce the resistance of the electrolyte/anode interface, additives such as maleic anhydride [104], VC [53], FEC [52], and many others have been used to successfully enhance the performance of GBL-based electrolytes.

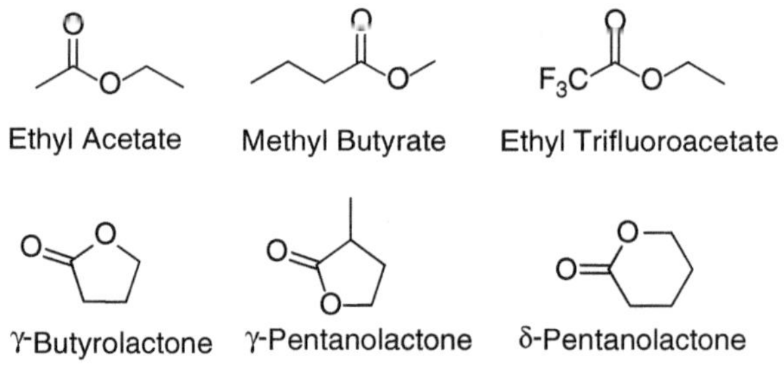

Fig. 8 Examples of esters and lactones

Another leading trend of GBL-based electrolyte research is associated with LiBOB as a lithium salt [118]. Being a fluorine-free salt, LiBOB has long been considered as a safer and greener alternative to the $LiPF_6$. However, LiBOB has low solubility in linear carbonates, so the resulting electrolyte has low conductivity compared with $LiPF_6$-based electrolytes. In contrast, LiBOB has high solubility in GBL and even linear esters such as ethyl acetate [118], which makes the LiBOB/ester system promising [11, 46]. An electrolyte consisting of solely GBL with LiBOB shows large irreversible capacity during formation [118], and additives [81] or mixing with carbonates [30] have been investigated to solve this issue. Not only a lithium ion source, LiBOB acts as a stable SEI formation additive as it is frequently used in carbonate and silicon-based electrolytes [9, 140].

While the majority study of GBL electrolyte have been focused on the 4 V electrode materials, at least one study reports that a binary mixture of GBL/Sulfolane results in an electrolyte that is more stable than the SOA carbonate electrolyte on the high-voltage spinel LNMO cathode [24]. The reported electrolyte also enables a higher discharge potential and comparable rate capability than the conventional electrolyte (Fig. 9). Beside the previously mentioned merits,

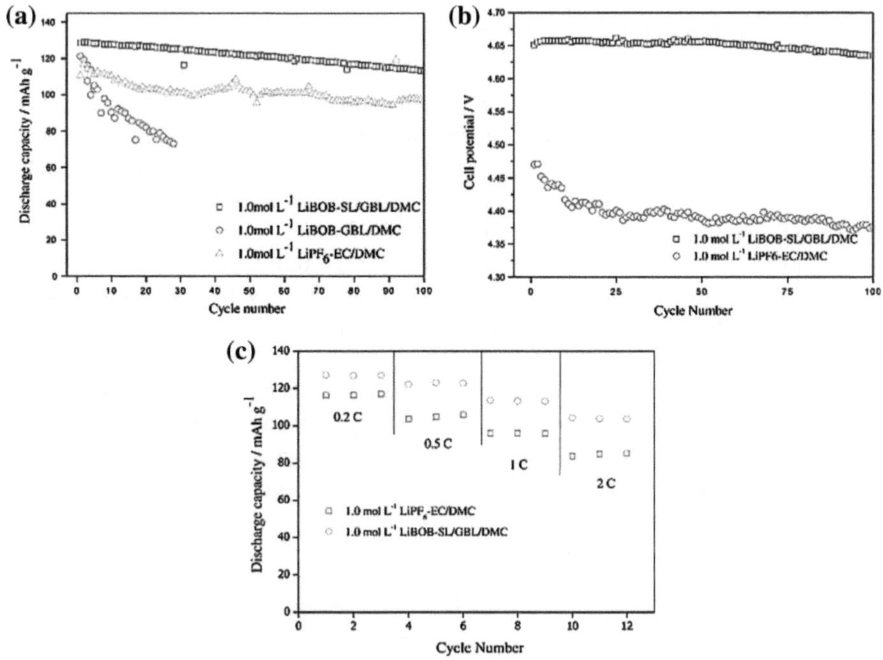

Fig. 9 **a** Cycle performance of $LiNi_{0.5}Mn_{1.5}O_4$/Li cells with different electrolytes cycled at 0.5 C discharge rate and room temperature. **b** Mean voltages of $LiNi_{0.5}Mn_{1.5}O_4$/Li cells with two electrolytes cycled at 0.5 C discharge rate and room temperature. **c** Discharge capacities of $LiNi_{0.5}Mn_{1.5}O_4$/Li cells cycled at different discharge rates and room temperature. Reproduced with permission [24]. Copyright 2013 Elsevier

GBL-based electrolyte has also shown superior thermal stability [18], especially when combined with the $LiBF_4$.

Linear esters, as mentioned above, are not widely used in Li-ion batteries due to their inferior anodic stability compared to linear carbonates. Although numerous patents for electrolyte formulations claim various ester structures as the co-solvents, literature reports on the esters are considerably fewer. Although esters oxidize much more easily than carbonates, at least one reference reported an electrolyte formulated from a mixture of sulfone and ethyl acetate showing better electrochemical properties than carbonate-based electrolyte in LNMO/Li half cells [108]. Due to their low viscosity and melting point, linear esters are considered to be candidates for low-temperature Li-ion batteries. Both non-fluorinated [94] and fluorinated esters [95] have been examined by scientists at the Jet Propulsion Laboratory as Li-ion battery electrolyte co-solvents for space missions.

In conclusion, the utilization of esters and lactones in Li-ion battery electrolytes is currently limited and is underexplored. The ease of synthesis and purification of esters should not be overlooked for potential cost reduction and more work on this group of solvents is needed.

2.5 Nitriles-Based Electrolytes

Due to the strong electron withdrawing ability of the cyanide functionality (−CN), organic nitriles are strongly polar solvents with high salt solubility, high oxidation potential, and low viscosity. The resulting electrolytes often have a wide electrochemical window and high conductivity. While nitriles have high stability toward oxidation, they are easily reduced on the anode. In fact, acetonitrile reacts with Li metal violently and is unable to form a stable passivation layer [86], which has led to a lack of interest in nitrile as an electrolyte solvent for Li-ion batteries. Other nitriles (Fig. 10) have shown potential as the electrolyte solvent of Li primary [39] and secondary batteries [105].

As the demand for high-voltage Li-ion batteries increased, the interest in the nitrile-based electrolyte was reignited due to its promising stability against oxidation, a similar case to the sulfone-based electrolyte described earlier. Conventional electrolyte additives [36] and novel additives tailored for the nitrile-based

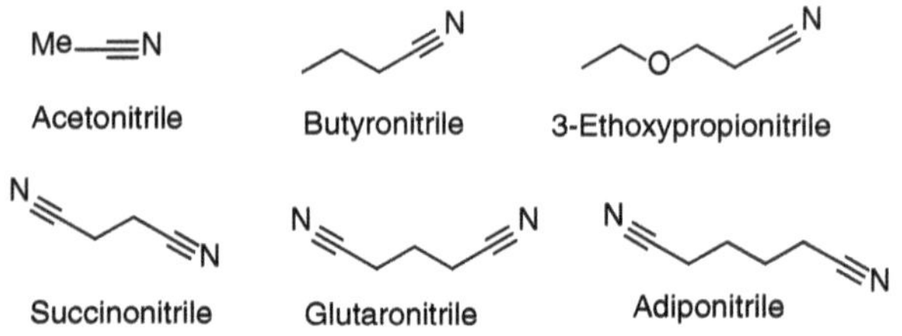

Fig. 10 Examples of organic nitriles used in Li-ion batteries

electrolytes [76, 77] were investigated to increase compatibility with the graphite anode. An electrolyte formulated with acetonitrile and LiBOB was studied for the purpose of physical properties, although no electrochemical evaluation was performed [42]. Recently, a ground-breaking discovery was reported by a Japanese group in which increasing the concentration of lithium bis(trifluoromethanesulfonyl)imide (LiTFSI) to 4.2 M suppressed the spontaneous reaction between acetonitrile and Li metal anode, and allowed for the reversible intercalation of Li ions into graphite (Fig. 11) [128]. The resulting electrolyte also demonstrated high ionic conductivity and C-rate capability. This discovery is very encouraging in terms of enabling the development of nitrile-based electrolytes, and it has also brought insight into the interaction between electrodes and electrolyte in the concentrated electrolyte solutions. More discussion on this topic will be centered on the salt LiTFSI in Sect. 3. On the cathode side, mono-functionalized nitriles [36] and di-functionalized nitriles (Fig. 10) were investigated on over-charged cathodes such as $LiCoO_2$ [3, 4] and Ni/Mn spinel LNMO [27] as high voltage electrolytes.

While nitriles may be similar to sulfones as an alternative solvent to carbonate for the high-voltage applications, nitriles are superior to sulfones due to lower viscosity and wider liquid range. More research is definitely needed to better understand and utilize the nitriles as the component in advanced Li-ion battery electrolytes.

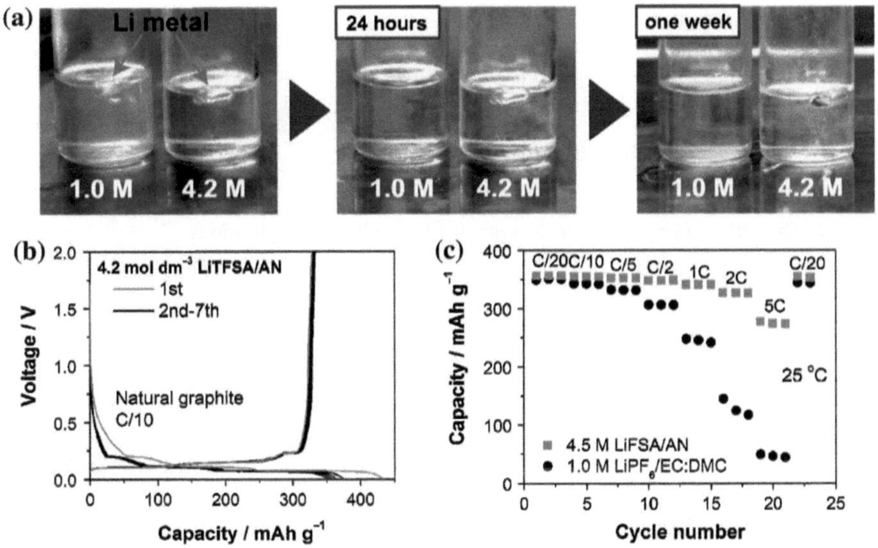

Fig. 11 **a** Reactivity of lithium metal foil and LiTFSI/acetonitrile solutions at 1.0 and 4.2 M concentrations at room temperature. **b** Charge-discharge curves of natural graphite/lithium metal cell with 4.2 M LiTFSI/acetonitrile electrolyte at C/10 rate. **c** Reversible capacity of natural graphite in the two electrolytes at various C-rates and 25 °C. Charge and discharge were conducted at the same C-rate without using a constant-voltage mode at both ends of charge and discharge, and the charge (lithium deintercalation) capacity was plotted. Reproduced with permission [128]. Copyright 2014 American Chemical Society

2.6 Organosilicon Compounds-Based Electrolytes

The application of organosilicon compounds in lithium batteries dates back to 1980s when polysiloxanes were studied as the matrix for the polymer electrolyte in Li metal batteries [31, 107]. However, the application of liquid organosilicon compounds as solvents for the Li-ion battery electrolyte is fairly recent, initiated by collaboration between Argonne National Laboratory and University of Wisconsin-Madison [7]. In this work, silanes and silyl ethers with ethylene glycol repeating units (Fig. 12) were used as the only solvent for LiBOB-based electrolyte. Such electrolytes showed tremendous improvement in the cycling performance over the $LiPF_6$ in EC/DEC electrolyte for the $LiNi_{0.8}Co_{0.15}Al_{0.05}O_2$ (NCA)/mesocarbon microbeads (MCMB) electrochemical couple. The cell showed no impedance rise over storage for one year at 80 % state of charge (SOC) at 55 °C (Fig. 13). Since then, this category of electrolytes have been widely applied to various Li-ion battery systems [57, 131, 132], and even to the emerging Li-O_2 battery [139].

Other groups also performed studies on this kind of solvent, but reports have been sporadic. Vinyl tris(2-methoxy)ethoxy silane [89] and ethyl tris(2-methoxy)

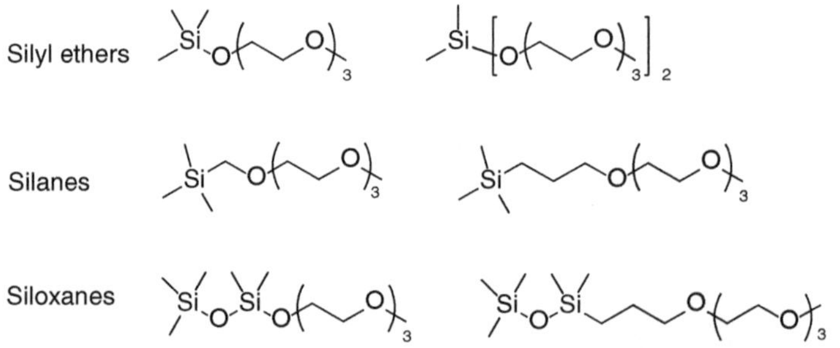

Fig. 12 Examples of organosilicon solvents for Li-ion battery applications

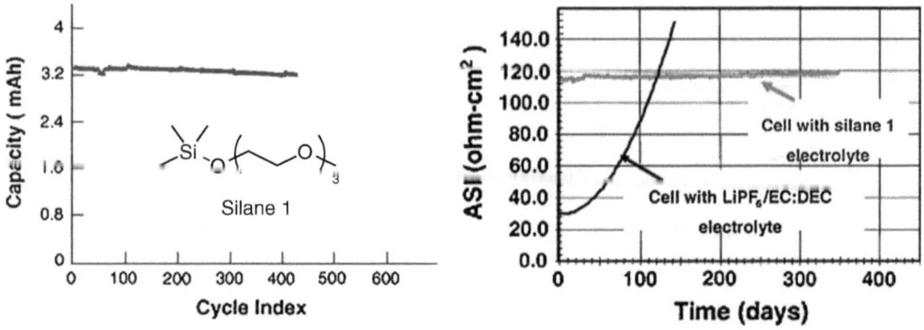

Fig. 13 Cycling performance and area specific impedance (ASI) of $MCMB/LiNi_{0.8}Co_{0.15}Al_{0.05}O_2$ cells with a 0.8 M LiBOB/silane or 1.0 M $LiPF_6$/EC:DEC (1:1) electrolyte. Reproduced with permission [7]. Copyright 2006 Elsevier

ethoxy silane [90] were investigated as an SEI formation additive and co-solvent for PC-based electrolyte and were shown to enable the intercalation of Li^+ ions into graphite anodes. Silanes were also shown to passivate the Li metal anode in Li-O_2 and Li-S batteries [72, 73].

Compared with other electrolyte solvents, organosilicon solvents are relatively new for battery applications. These solvents, especially those with short polyethylene glycol chains, have low toxicity and excellent biocompatibility compared to other solvents, which make them more suitable for medical applications [85]. Even with the electron-withdrawing effect of the terminal silyl group by back donating of 2p-orbital electrons of oxygen on the ether chain into the empty 3D-orbital of silicon, however, these solvents still have limited oxidation stability (usually less than 4.5 V) which restricts their applications in high-voltage Li-ion batteries.

2.7 Other Solvents

In addition to the organic solvents described above, there are quite a few more organic compounds considered as solvent candidates for Li-ion battery electrolytes. In many cases, these molecules are more appropriate as additives instead of solvents due to the inferior physical and electrochemical properties of the resulting electrolyte if such molecules are used in high concentration. These molecules include the non-fluorinated and fluorinated varieties of alkyl sulfates [113], alkyl sulfites [21, 59, 63], alkyl phosphates [66], and phosphazenes [84]. Some of them have been employed as SEI formation additives or flame retardant additives and will be discussed in the respective sections of Chap. 9.

3 Electrolyte Salts for Non-aqueous Electrolyte

While solvents certainly attract major attention, the role of salts has been often overlooked even though many cases showed that changing lithium salt vastly changes electrolyte performance. As mentioned in the introduction, the majority of lithium salts used in the SOA Li-ion battery is the lithium conjugate base of super acids such as HPF_6, HBF_4, and HSO_3CF_3. These counter-ions are non-coordinative and have high oxidation stability, which makes the respective lithium salts suitable as battery electrolyte solutes with high conductivities and wide electrochemical windows. However, most of these salts are corrosive to the cathode current collector materials, and each has its own drawbacks. For example, the most widely used salt, $LiPF_6$, has poor thermal stability and decomposes at temperatures as low as 70 °C in the presence of electrolyte solvents [92, 116]. Furthermore, it is very moisture sensitive. Other salts also have their drawbacks: $LiBF_4$ suffers from mediocre conductivity, $LiAsF_6$ is highly toxic, $LiClO_4$ is explosive, and LiTFSI is highly corrosive to aluminum, the standard material of the cathode current collector.

It seems that finding a "perfect" salt is much more difficult than finding a "perfect" solvent because the choices are very limited. Regardless, effort has been made to find less hazardous, more stable salts with better electrochemical properties and possibly the capability to form an SEI on the anode without needing the participation of solvent or additive molecules. The following discussion will address these characteristics individually in various lithium salts developed in the past decade.

3.1 Salts with SEI Formation Capability

The stabilization of the anode/electrolyte interface is one of the most pressing challenges in Li-ion batteries. Now, with the use of high voltage cathode materials, the oxidation stability of the electrolyte and the cathode/electrolyte interface are crucial to the long term cycling performance. While additives offer a solution by forming the passivation layer on electrode surface, lithium salts that can form SEI on both the anode and cathode are even more attractive. So far, some salts are already known to have such capability, independent of the solvents used. A few examples are shown in Fig. 14.

The most widely known film-forming salt is lithium bis(oxalato)borate, commonly abbreviated as LiBOB (Fig. 14). The fluorine free, non-corrosive, non-toxic (and thus environmentally friendly) features make this salt very appealing. The synthesis and solution properties of LiBOB were first reported by Prof. Angell's lab from Arizona State University in 2001, followed by the discovery of its unique capability in stabilizing the cycling performance of Li-ion battery [127]. LiBOB contains a very weakly coordinating anion because the negative charge on the anion is delocalized by eight oxygen atoms in the two oxalate groups. The solution of LiBOB in acetonitrile has a conductivity of 25.2 $mScm^{-1}$ at 25 °C, approaching that of aqueous solutions, and a LiBOB/PC electrolyte is anodically stable up to 4.5 V versus Li^{+}/Li [127]. The following year, the same group discovered that in a LiBOB/PC solution, Li^{+} can reversibly intercalate into and deintercalate out of graphite, which had not been observed with other salts (Fig. 15) [121]. This groundbreaking discovery led to the immense popularity of the salt and the interest in finding other SEI forming salts. Since then, LiBOB has been frequently used as an SEI formation additive for $LiPF_6$-based electrolyte, as well as the single salt in many novel electrolyte formulations [7, 9, 64]. Other than the ability to protect the

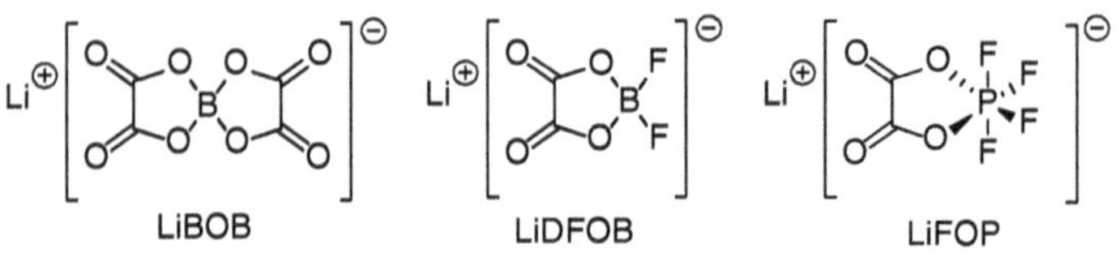

Fig. 14 Structures of LiBOB, LiDFOB, and LiFOP

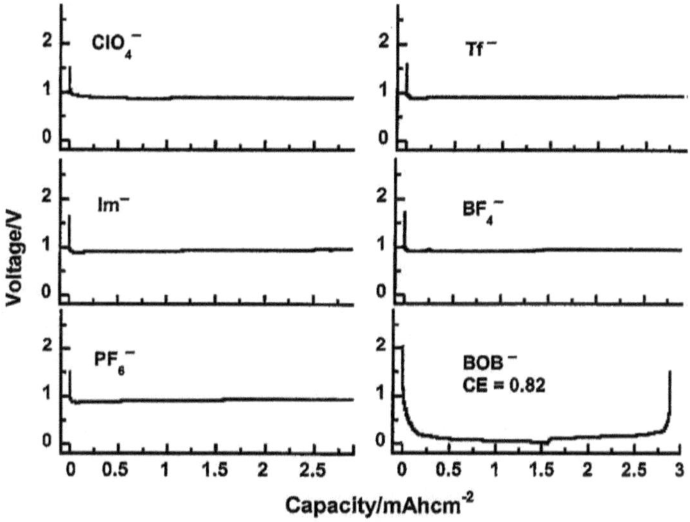

Fig. 15 The voltage profiles of the first discharge of Li/graphite half-cells employing an electrolyte with 1.0 M LiBOB dissolved in PC. For LiBOB/PC electrolyte, the first cycle is shown, with CE indicated in the graph. Reproduced with permission [121]. Copyright 2002 The Electrochemical Society

anode, LiBOB was also reported to be capable of forming a passivation film on the surface of the cathode, which mitigates the oxidative decomposition of electrolyte solvents on the high voltage cathodes [143, 122] and reduces the dissolution of transition metal ions from the cathode into the electrolyte [126]. Besides these desirable properties, both LiBOB and the SEI formed from it show excellent thermal stability [144].

One of the disadvantages of LiBOB is poor solubility in organic carbonates, especially in linear carbonates [118]. In order to increase the solubility, a structural analog of LiBOB, lithium difluoro(oxalato)borate, commonly abbreviated as LiDFOB, with a chemical structure consisting of half LiBOB and half $LiBF_4$, was developed in the U.S. Army Research Lab [22, 60, 134–136]. This salt has better solubility in linear carbonates than LiBOB while retaining similar SEI formation capability as LiBOB. Due to the similarities in chemical structure and properties, in many publications and patents LiDFOB and LiBOB are simultaneously studied/claimed as the independent lithium salt or electrolyte additive. Both LiBOB and LiDFOB seem to have become one of the "standard" additives for electrolyte formulation that have SEI formation issues. The current trend, as mentioned in Sect. 2, is to employ LiDFOB or LiBOB as the only salt for novel solvents that are not capable of forming SEI on their own [21, 59, 109, 131].

An analogue of LiDFOB is lithium tetrafluoro(oxalato)phosphate ($LiPF_4C_2O_4$, or LiFOP), in which phosphorous is the coordination center. This salt was initially synthesized by Prof. Brett Lucht's group at the University of Rhode Island. It has been demonstrated that the electrolyte prepared from this salt has comparable conductivity (9 $mScm^{-1}$ at 25 °C) with conventional $LiPF_6$-based electrolyte while its cell performance is also similar to the $LiPF_6$ [123, 125]. Moreover, LiFOP has much improved thermal stability than $LiPF_6$ [124]. The initial concern with this salt is whether it can form SEI on the graphite anode. This concern was ruled out by the same group, showing that LiFOP enables reversible intercalation of Li^+ ions with graphite electrode even with the problematic solvent PC, firm proof that LiFOP can form SEI by itself [142].

At present, it is quite evident that the oxalate-containing salts are able to form SEI on their own, because the oxalate moiety participates in the reduction reaction (1.5–1.7 V vs. Li^+/Li) during the charging of graphite anode. Other lithium salts with oxalate functional groups might have the same SEI-forming property.

While many novel lithium salts have been developed, many of them employed EC as a co-solvent so their stand-alone ability to form SEI is unknown. As mentioned earlier, the choice of lithium salts is very limited due to the bulky molecular size and/or less structural varieties, therefore the development of a new salt with SEI formation capability is much more challenging than the development of a new solvent with such capability.

3.2 Salts with Enhanced Thermal Stability

Among the available lithium salts, $LiPF_6$ is most commonly preferred for the Li-ion battery because it holds the best comprehensive properties compared with other salts, especially with respect to solubility, resultant ionic conductivity, and passivation of the Al current collector. However, $LiPF_6$ suffers from poor thermal stability especially in the solution state. When exposed to elevated temperatures, $LiPF_6$ decomposes and reacts with the organic solvents in the electrolyte generating HF gas, which causes rapid degradation of the cell chemistry [34, 62]. Although other salts mentioned earlier have better thermal stability than $LiPF_6$, alternative salts with higher thermal stability are in demand for safety reasons.

In addition to the exceptional SEI formation ability, LiBOB and LiDFOB have excellent thermal stability compared to $LiPF_6$. Other chelating boron-based salts with good thermal stability have also been reported. Similar in structure to LiDFOB, lithium difluoro(sulfato)borate ($LiBF_2SO_4$) (Fig. 16) has been tested at 80 °C in MCMB/Li and $LiFePO_4$/Li half-cells with improved performance [58]. Boron-based chelating salts such as lithium bis[1,2-benzenediolato(2)-O,O′]borate (LBBB) (Fig. 16), lithium bis[2,3-naphthalenediolato(2)-O,O′]borate (LBNB), and lithium bis[2,2′-biphenyldiolato(2)-O,O′]borate (LBBPB) also have good thermal stability, but the electrolyte solutions containing these salts showed much inferior conductivity due to their large anion size and narrow electrochemical window [15–17]. Besides the salts mentioned above, a multivalent salt with fluorinated

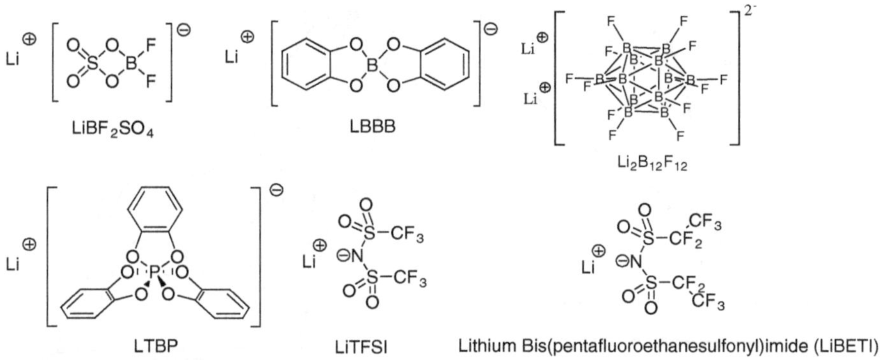

Fig. 16 A selection of thermally stable lithium salts with various structures

boron clusters, dilithium dodecafluorododecaborate ($Li_2B_{12}F_{12}$, where boron atoms occupy 12 vertices of a regular icosahedron, and each boron atom is connected to one fluorine atom) (Fig. 16) has shown enhanced storage and cycle life compared with $LiPF_6$ at 60 °C [8].

Although having P-F bonds in the structure, lithium tetrafluoro(oxalato)phosphate (LiFOP, Fig. 14) has much better thermal stability than $LiPF_6$. Another thermally stable phosphorous-based salt is a tri-chelated salt, lithium tris [1,2-benzenediolato(2)-O,O′]phosphate (LTBP, Fig. 16) [43], but it cannot be used in the Li-ion battery because of its insufficient electrochemical window related to the less stable phenol ligands.

The nitrogen-centered imide salts are good candidates for thermally stable lithium salts. The best known salt is LiTFSI (Fig. 16), which was found to be indefinitely stable at temperatures as high as 100 °C [25]. Unfortunately, this salt leads to corrosion of the aluminum current collector on the cathode [55], making it unsuitable for SOA Li-ion batteries. The larger imide analogue, lithium bis (perfluoroethanesulfonyl)imide (LiBETI, Fig. 16), exhibits much better passivation of aluminum, leading to improved stability [69, 75]. Recent progress in the understanding of the solvation properties of electrolyte solutions has brought new insight to LiTFSI salt. EC-based electrolytes with high LiTFSI concentrations, for example a 3:1 or 2:1 (molar ratio), did not corrode aluminum [68] (Fig. 17a). Moreover, the highly solvated solvent molecules makes the electrolyte solvent anodically and thermally stable (Fig. 17b). This finding suggests that it is feasible for researchers to reconsider utilizing thermally stable salts such as LiTFSI in a new way and this endeavor should be pursued for the safety benefits.

In addition to highly concentrated electrolytes, anions with longer perfluorinated chains also passivate aluminum while still providing good thermal stability to the resulting electrolyte [41, 47, 51, 55].

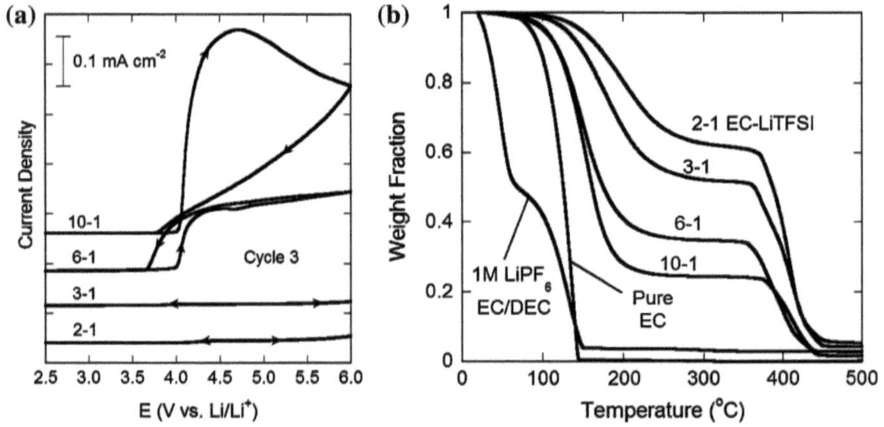

Fig. 17 **a** The 3rd cycle cyclic voltammetry scan of coin cells with Al foil as working electrodes, Li metal as counter/reference electrodes, and EC-LiTFSI (from 10:1 to 2:1) as electrolyte. **b** Thermal gravimetric analysis (5 °C min^{-1}) of 1.0 M $LiPF_6$ in EC/DEC (3:7 volume ratio), pure EC, and various EC-LiTFSI (from 10:1 to 2:1) electrolyte. Reproduced with permission [68]. Copyright 2014 Royal Society of Chemistry

3.3 Emerging Salts

There is new focus on developing lithium salts with fewer corrosion issues and less environmental impact. Most of the conventional lithium salts contain fluorine because it is among the smallest yet strongest electron withdrawing groups, and anions with fluorine substitution generally have high anodic stability. However, many fluorinated salts are corrosive and hazardous to the environment. Thermal runaway in batteries containing fluorinated salts may result in the release of toxic fumes containing HF, which is extremely dangerous as well as harmful to the atmosphere. As a result, developing fluorine-free salts marks the first step toward environmentally friendly batteries.

An alternative to fluorine as the electron-withdrawing group is the cyanide (CN) group. Recently synthesized cyanide salts include lithium dicyanamide (LDCA), lithium 4,5-dicyano-1,2,3-triazolate (LDCTA), and lithium tetracyanoborate ($LiB(CN)_4$) (Fig. 18) [88]. Although CN fits the goal of fluorine free, the oxidation potential of these anions is lower than that of fluorine-based salts, so the application of lithium cyanide salts is limited to low voltage cathodes such as $LiFePO_4$.

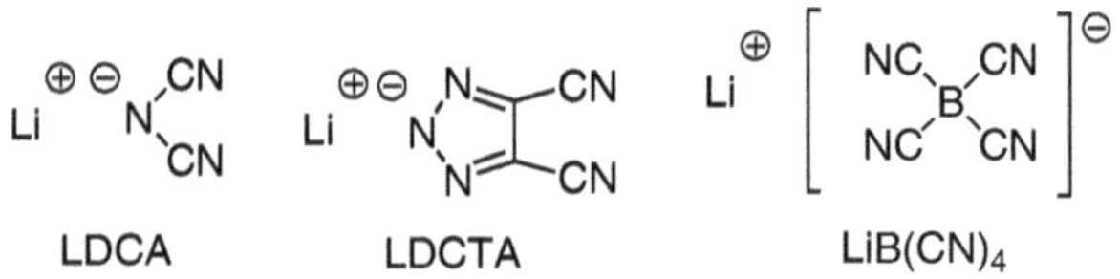

Fig. 18 Lithium salts based on cyanide functionality

4 Conclusion and Future Outlook

Li-ion battery chemistry is a mature technology that has been commercialized over the decades, although it is still far from reaching its limit. Lithium metal, which is considered as the "holy grail" of lithium battery anodes, is still unable to deliver the desired cell performance despite new progress made for effective suppression of the dendrite formation. Lithium transition metal oxide-based inorganic cathodes have been constantly achieving new levels of performance in terms of the capacity and voltage, while organic cathode materials have also started to show great promise. The development of new anode and cathode materials goes hand in hand with electrolyte development, as each new material with new challenges necessitates a compatible electrolyte. The number of patent application and journal publication on Li-ion battery electrolytes has increased exponentially in the past decade (Fig. 19), which is evidence of the burst of interest in this field. Novel electrolytes and electrolyte additives that are tailored to the various lithium-ion chemistries will be the main focus of future research and development. It is of vital importance that mechanistic studies using in situ characterization techniques and computational efforts should be carried out to support electrolyte development.

For emerging battery technologies such as Li-S, Li-O_2, Na-ion and magnesium-ion batteries, the knowledge achieved in the course of Li-ion battery electrolyte research will be able to provide insight and support for the development of new electrolytes for various electrochemical energy storage chemistries.

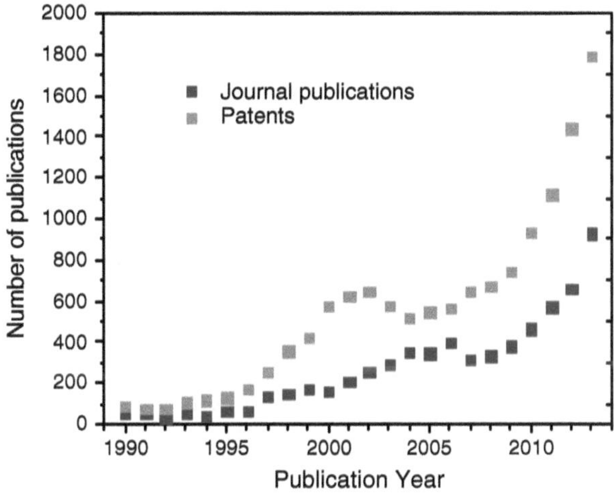

Fig. 19 Total numbers of journal publications and patents on the topic of Li-ion battery electrolyte by year searched by SciFinder. Copyright 2014 American Chemical Society

References

1. Abouimrane A, Belharouak I, Amine K (2009) Sulfone-based electrolytes for high-voltage Li-ion batteries. Electrochem Commun 11:1073–1076
2. Abraham KM, Goldman JL, Natwig DL (1982) Characterization of ether electrolytes for rechargeable lithium cells. J Electrochem Soc 129:2404–2409. doi:10.1149/1.2123556
3. Abu-Lebdeh Y, Davidson I (2009) High-voltage electrolytes based on adiponitrile for Li-ion batteries. J Electrochem Soc 156:A60–A65. doi:10.1149/1.3023084
4. Abu-Lebdeh Y, Davidson I (2009) New electrolytes based on glutaronitrile for high energy/power Li-ion batteries. J Power Sources 189:576–579. doi:10.1016/j.jpowsour.2008.07.113
5. Achiha T, Nakajima T, Ohzawa Y, Koh M, Yamauchi A, Kagawa M, Aoyama H (2009) Electrochemical behavior of nonflammable organo-fluorine compounds for lithium ion batteries. J Electrochem Soc 156:A483–A488. doi:10.1149/1.3111904
6. Achiha T, Nakajima T, Ohzawa Y, Koh M, Yamauchi A, Kagawa M, Aoyama H (2010) Thermal stability and electrochemical properties of fluorine compounds as nonflammable solvents for lithium-ion batteries. J Electrochem Soc 157:A707–A712. doi:10.1149/1.3377084
7. Amine K, Wang Q, Vissers DR, Zhang Z, Rossi NAA, West R (2006) Novel silane compounds as electrolyte solvents for Li-ion batteries. Electrochem Commun 8:429–433. doi:10.1016/j.elecom.2005.12.017
8. Arai J, Matsuo A, Fujisaki T, Ozawa K (2009) A novel high temperature stable lithium salt ($Li_2B_{12}F_{12}$) for lithium ion batteries. J Power Sources 193:851–854. doi:10.1016/j.jpowsour.2007.04.001
9. Aravindan V, Gnanaraj J, Madhavi S, Liu H-K (2011) Lithium-ion conducting electrolyte salts for lithium batteries. Chem Eur J 17:14326–14346. doi:10.1002/chem.201101486
10. Armand M, Tarascon JM (2008) Building better batteries. Nature 451:652–657. doi:10.1038/451652a
11. Azeez F, Fedkiw PS (2010) Conductivity of LIBOB-based electrolyte for lithium-ion batteries. J Power Sources 195:7627–7633. doi:10.1016/j.jpowsour.2010.06.021
12. Azimi N, Weng W, Takoudis C, Zhang Z (2013) Improved performance of lithium–sulfur battery with fluorinated electrolyte. Electrochem Commun 37:96–99. doi:10.1016/j.elecom.2013.10.020
13. Azimi N, Xue Z, Rago ND, Takoudis C, Gordin ML, Song J, Wang D, Zhang Z (2015) Fluorinated electrolytes for Li-S battery: suppressing the self-discharge with an electrolyte containing fluoroether solvent. J Electrochem Soc 162:A64–A68. doi:10.1149/2.0431501jes
14. Balakrishnan PG, Ramesh R, Prem Kumar T (2006) Safety mechanisms in lithium-ion batteries. J Power Sources 155:401–414. doi:10.1016/j.jpowsour.2005.12.002
15. Barthel J, Buestrich R, Gores HJ, Schmidt M, Wuhr M (1997) A new class of electrochemically and thermally stable lithium salts for lithium battery electrolytes: 4. Investigations of the electrochemical oxidation of lithium organoborates. J Electrochem Soc 144:3866–3870. doi:10.1149/1.1838103
16. Barthel J, Schmidt M, Gores HJ (1998) Lithium bis[5-fluoro-2-olato-1-benzenesulfonato (2-)-O, O']borate(1), a new anodically and cathodically stable salt for electrolytes of lithium-ion cells. J Electrochem Soc 145:L17–L20. doi:10.1149/1.1838265
17. Barthel J, Wuhr M, Buestrich R, Gores HJ (1995) New class of electrochemically and thermally stable lithium-salts for lithium battery electrolytes: 1. Synthesis and properties of lithium bis[1,2-Benzenediolato(2-)-O, O']borate. J Electrochem Soc 142:2527–2531. doi:10.1149/1.2050048
18. Belov D, Shieh D-T (2012) GBL-based electrolyte for Li-ion battery: thermal and electrochemical performance. J Solid State Electrochem 16:603–615. doi:10.1007/s10008-011-1391-y

19. Brutti S, Panero S (2013) Recent advances in the development of $LiCoPO_4$ as high voltage cathode material for Li-ion batteries. ACS Sym Ser 1140:67–99. doi:10.1021/bk-2013-1140.ch004
20. Chagnes A, Carre B, Willmann P, Dedryvere R, Gonbeau D, Lemordant D (2003) Cycling ability of gamma-butyrolactone-ethylene carbonate based electrolytes. J Electrochem Soc 150:A1255–A1261. doi:10.1149/1.1597882
21. Chen R, Zhu L, Wu F, Li L, Zhang R, Chen S (2014) Investigation of a novel ternary electrolyte based on dimethyl sulfite and lithium difluoromono(oxalato)borate for lithium ion batteries. J Power Sources 245:730–738. doi:10.1016/j.jpowsour.2013.06.132
22. Chen Z, Liu J, Amine K (2007) Lithium difluoro(oxalato)borate as salt for lithium-ion batteries. Electrochem Solid-State Lett 10:A45–A47. doi:10.1149/1.2409743
23. Chung G-C, Kim H-J, Yu S-I, Jun S-H, Choi J-W, Kim M-H (2000) Origin of graphite exfoliation: an investigation of the important role of solvent cointercalation. J Electrochem Soc 147:4391–4398. doi:10.1149/1.1394076
24. Cui X, Zhang H, Li S, Zhao Y, Mao L, Zhao W, Li Y, Ye X (2013) Electrochemical performances of a novel high-voltage electrolyte based upon sulfolane and γ-butyrolactone. J Power Sources 240:476–485. doi:10.1016/j.jpowsour.2013.04.063
25. Dominey LA, Koch VR, Blakley TJ (1992) Thermally stable lithium salts for polymer electrolytes. Electrochim Acta 37:1551–1554. doi:10.1016/0013-4686(92)80109-Y
26. Dudley JT, Wilkinson DP, Thomas G, LeVae R, Woo S, Blom H, Horvath C, Juzkow MW, Denis B et al (1991) Conductivity of electrolytes for rechargeable lithium batteries. J Power Sources 35:59–82. doi:10.1016/0378-7753(91)80004-h
27. Duncan H, Salem N, Abu-Lebdeh Y (2013) Electrolyte formulations based on dinitrile solvents for high voltage Li-ion batteries. J Electrochem Soc 160:A838–A848. doi:10.1149/2.088306jes
28. Etacheri V, Haik O, Goffer Y, Roberts GA, Stefan IC, Fasching R, Aurbach D (2012) Effect of fluoroethylene carbonate (FEC) on the performance and surface chemistry of Si-nanowire Li-ion battery anodes. Langmuir 28:965–976. doi:10.1021/la203712s
29. Etacheri V, Marom R, Elazari R, Salitra G, Aurbach D (2011) Challenges in the development of advanced Li-ion batteries: a review. Energy Environ Sci 4:3243–3262. doi:10.1039/c1ee01598b
30. Fan LZ, Xing TF, Awan R, Qiu WH (2011) Studies on lithium bis(oxalato)-borate/propylene carbonate-based electrolytes for Li-ion batteries. Ionics 17:491–494. doi:10.1007/s11581-011-0551-5
31. Fish D, Khan IM, Smid J (1986) Poly[(methoxyheptaethylene oxide)methylsiloxane]/lithium perchlorate complexes as solvent-free polymer electrolytes for high energy density storage devices. Polym Prepr (Am Chem Soc, Div Polym Chem) 27:325–326
32. Fong R, Von Sacken U, Dahn JR (1990) Studies of lithium intercalation into carbons using nonaqueous electrochemical cells. J Electrochem Soc 137:2009–2013. doi:10.1149/1.2086855
33. Foster DL, Wolfenstine J, Behl WK (1999) Tertiary polyamines as additives to lithium-ion battery electrolytes. Proc of Electrochem Soc 98–16:391–397
34. Gachot G, Grugeon S, Armand M, Pilard S, Guenot P, Tarascon J-M, Laruelle S (2008) Deciphering the multi-step degradation mechanisms of carbonate-based electrolyte in Li batteries. J Power Sources 178:409–421. doi:10.1016/j.jpowsour.2007.11.110
35. Girishkumar G, McCloskey B, Luntz AC, Swanson S, Wilcke W (2010) Lithium-air battery: promise and challenges. J Phys Chem Lett 1:2193–2203. doi:10.1021/jz1005384
36. Gmitter AJ, Plitz I, Amatucci GG (2012) High concentration dinitrile, 3-alkoxypropionitrile, and linear carbonate electrolytes enabled by vinylene and monofluoroethylene carbonate additives. J Electrochem Soc 159:A370–A379. doi:10.1149/2.016204jes
37. Goldman JL, Long BR, Gewirth AA, Nuzzo RG (2011) Strain anisotropies and self-limiting capacities in single-crystalline 3D silicon microstructures: models for high energy density lithium-ion battery anodes. Adv Funct Mater 21:2412–2422. doi:10.1002/adfm.201002487

38. Gordin M, Dai F, Chen S, Xu T, Song J, Tang D, Azimi N, Zhang Z, Wang D (2014) Bis (2,2,2-trifluoroethyl) ether as an electrolyte co-solvent for mitigating self-discharge in lithium-sulfur batteries. ACS Appl Mater Interfaces 6(11):8006–8010. doi:10.1021/am501665s
39. Guibert S, Cariou M, Simonet J (1988) Research of new solvents for lithium batteries: II. Behavior of aliphatic nitriles substituted by electron donating groups. Bull Soc Chim Fr 924–927
40. Guidotti RA, Reinhardt FW (2001) The performance of manganese oxides as a function of temperature in several molten-salt systems. ITE Lett Batteries New Technol Med 2:26–32
41. Han H, Guo J, Zhang D, Feng S, Feng W, Nie J, Zhou Z (2011) Lithium (fluorosulfonyl) (nonafluorobutanesulfonyl)imide (LiFNFSI) as conducting salt to improve the high-temperature resilience of lithium-ion cells. Electrochem Commun 13:265–268. doi:10.1016/j.elecom.2010.12.030
42. Han S-D, Allen JL, Boyle PD, Henderson WA (2012) Delving into the properties and solution structure of nitrile-lithium difluoro(oxalato)borate (LiDFOB) electrolytes for Li-ion batteries. ECS Trans 41:47–51. doi:10.1149/1.4717962
43. Handa M, Suzuki M, Suzuki J, Kanematsu H, Sasaki Y (1999) A new lithium salt with a chelate complex of phosphorus for lithium battery electrolytes. Electrochem Solid St 2:60–62. doi:10.1149/1.1390734
44. He P, Yu H, Li D, Zhou H (2012) Layered lithium transition metal oxide cathodes towards high energy lithium-ion batteries. J Mater Chem 22:3680–3695. doi:10.1039/c2jm14305d
45. Hu L, Zhang Z, Amine K (2013) Fluorinated electrolytes for Li-ion battery: an FEC-based electrolyte for high voltage $LiNi_{0.5}Mn_{1.5}O_4$/graphite couple. Electrochem Commun 35:76–79. doi:10.1016/j.elecom.2013.08.007
46. Huang J-Y, Liu X-J, Kang X-l YuZ-X, Xu T-T, Qiu W-H (2009) Study on γ-butyrolactone for LiBOB-based electrolytes. J Power Sources 189:458–461. doi:10.1016/j.jpowsour.2008.12.088
47. Huber B, Linder T, Hormann K, Froemling T, Sundermeyer J, Roling B (2012) Synthesis of novel lithium salts containing pentafluorophenylamido-based anions and investigation of their thermal and electrochemical properties. Z Phys Chem (Muenchen, Ger) 226:377–390. doi:10.1524/zpch.2012.0220
48. Inaba M, Kawatate Y, Funabiki A, Jeong SK, Abe T, Ogumi Z (1999) STM study on graphite/electrolyte interface in lithium-ion batteries: solid electrolyte interface formation in trifluoropropylene carbonate solution. Electrochim Acta 45:99–105. doi:10.1016/S0013-4686(99)00196-6
49. Ji L, Lin Z, Alcoutlabi M, Zhang X (2011) Recent developments in nanostructured anode materials for rechargeable lithium-ion batteries. Energy Environ Sci 4:2682–2699. doi:10.1039/c0ee00699h
50. Jung SC, Choi JW, Han Y-K (2012) Anisotropic volume expansion of crystalline silicon during electrochemical lithium insertion: an atomic level rationale. Nano Lett 12:5342–5347. doi:10.1021/nl3027197
51. Kamamura K, Umegaki T, Shiraishi S, Ohashi M, Takehara Z (2002) Electrochemical behavior of Al current collector of rechargeable lithium batteries in propylene carbonate with $LiCF_3SO_3$, $Li(CF_3SO_2)_2N$, or $Li(C_4F_9SO_2)(CF_3SO_2)N$. J Electrochem Soc 149:A185–A194. doi:10.1149/1.1433471
52. Kasnatscheew J, Schmitz RW, Wagner R, Winter M, Schmitz R (2013) Fluoroethylene carbonate as an additive for γ-butyrolactone based electrolytes. J Electrochem Soc 160: A1369–A1374. doi:10.1149/2.009309jes
53. Kinoshita S-C, Kotato M, Sakata Y, Ue M, Watanabe Y, Morimoto H, Tobishima S-I (2008) Effects of cyclic carbonates as additives to γ-butyrolactone electrolytes for rechargeable lithium cells. J Power Sources 183:755–760. doi:10.1016/j.jpowsour.2008.05.035
54. Koch VR, Young JH (1978) The stability of the secondary lithium electrode in tetrahydrofuran-based electrolytes. J Electrochem Soc 125:1371–1377. doi:10.1149/1.2131680

55. Krause LJ, Lamanna W, Summerfield J, Engle M, Korba G, Loch R, Atanasoski R (1997) Corrosion of aluminum at high voltages in non-aqueous electrolytes containing perfluoroalkylsulfonyl imides: new lithium salts for lithium-ion cells. J Power Sources 68:320–325. doi:10.1016/S0378-7753(97)02517-2
56. Kraytsberg A, Ein-Eli Y (2012) Higher, stronger, better… a review of 5 volt cathode materials for advanced lithium-ion batteries. Adv Energy Mater 2:922–939. doi:10.1002/aenm.201200068
57. Kusachi Y, Dong J, Zhang Z, Amine K (2011) Tri(ethylene glycol)-substituted trimethylsilane/lithium bis(oxalate)borate electrolyte for $LiMn_2O_4$/graphite system. J Power Sources 196:8301–8306. doi:10.1016/j.jpowsour.2011.06.033
58. Li S, Zhao W, Cui X, Zhang H, Wang X, Zhong W, Feng H, Liu H (2014) Lithium difluoro (sulfato)borate as a novel electrolyte salt for high-temperature lithium-ion batteries. Electrochim Acta 129:327–333. doi:10.1016/j.electacta.2014.02.090
59. Li S, Zhao W, Zhou Z, Cui X, Shang Z, Liu H, Zhang D (2014) Studies on electrochemical performances of novel electrolytes for wide-temperature-range lithium-ion batteries. ACS Appl Mater Interfaces 6:4920–4926. doi:10.1021/am405973x
60. Liu J, Chen Z, Busking S, Amine K (2007) Lithium difluoro(oxalato)borate as a functional additive for lithium-ion batteries. Electrochem Commun 9:475–479. doi:10.1016/j.elecom.2006.10.022
61. Lu DS, Xu MQ, Zhou L, Garsuch A, Lucht BL (2013) Failure mechanism of graphite/$LiNi_{0.5}Mn_{1.5}O_4$ cells at high voltage and elevated temperature. J Electrochem Soc 160:A3138–A3143. doi:10.1149/2.022305jes
62. Lux SF, Lucas IT, Pollak E, Passerini S, Winter M, Kostecki R (2012) The mechanism of HF formation in LiPF6 based organic carbonate electrolytes. Electrochem Commun 14:47–50. doi:10.1016/j.elecom.2011.10.026
63. Mao L, Li B, Cui X, Zhao Y, Xu X, Shi X, Li S, Li F (2012) Electrochemical performance of electrolytes based upon lithium bis(oxalate)borate and sulfolane/alkyl sulfite mixtures for high temperature lithium-ion batteries. Electrochim Acta 79:197–201. doi:10.1016/j.electacta.2012.06.102
64. Mao LP, Li BC, Cui XL, Zhao YY, Xu XL, Shi XM, Li SY, Li FQ (2012) Electrochemical performance of electrolytes based upon lithium bis(oxalate)borate and sulfolane/alkyl sulfite mixtures for high temperature lithium-ion batteries. Electrochim Acta 79:197–201. doi:10.1016/j.electacta.2012.06.102
65. Matsuda Y, Morita M, Yamashita T (1984) Conductivity of the lithium tetrafluoroborate ($LiBF_4$) mixed ether electrolytes for secondary lithium cells. J Electrochem Soc 131:2821–2827. doi:10.1149/1.2115416
66. Matsumoto K, Nakahara K, Inoue K, Iwasa S, Nakano K, Kaneko S, Ishikawa H, Utsugi K, Yuge R (2014) Performance improvement of Li ion battery with non-flammable TMP mixed electrolyte by optimization of lithium salt concentration and SEI preformation technique on graphite anode. J Electrochem Soc 161:A831–A834. doi:10.1149/2.091405jes
67. McMillan R, Slegr H, Shu ZX, Wang WD (1999) Fluoroethylene carbonate electrolyte and its use in lithium ion batteries with graphite anodes. J Power Sources 81:20–26. doi:10.1016/S0378-7753(98)00201-8
68. McOwen DW, Seo DM, Borodin O, Vatamanu J, Boyle PD, Henderson WA (2014) Concentrated electrolytes: decrypting electrolyte properties and reassessing Al corrosion mechanisms. Energ Environ Sci 7:416–426. doi:10.1039/c3ee42351d
69. Morita M, Shibata T, Yoshimoto N, Ishikawa M (2002) Anodic behavior of aluminum in organic solutions with different electrolytic salts for lithium ion batteries. Electrochim Acta 47:2787–2793. doi:10.1016/S0013-4686(02)00164-0
70. Morita M, Okada Y, Matsuda Y (1987) Lithium cycling efficiency on the aluminum substrate in blended sulfolane-ether systems. J Electrochem Soc 134(11):2665–2669. doi:10.1149/1.2100267

71. Nakai H, Kubota T, Kita A, Kawashima A (2011) Investigation of the solid electrolyte interphase formed by fluoroethylene carbonate on Si electrodes. J Electrochem Soc 158: A798–A801. doi:10.1149/1.3589300
72. Neuhold S, Schroeder DJ, Vaughey JT (2012) Effect of surface preparation and R-group size on the stabilization of lithium metal anodes with silanes. J Power Sources 206:295–300. doi:10.1016/j.jpowsour.2012.01.127
73. Neuhold S, Vaughey JT, Grogger C, Lopez CM (2014) Enhancement in cycle life of metallic lithium electrodes protected with Fp-silanes. J Power Sources 254:241–248. doi:10.1016/j.jpowsour.2013.12.057
74. Nishikawa D, Nakajima T, Ohzawa Y, Koh M, Yamauchi A, Kagawa M, Aoyama H (2013) Thermal and oxidation stability of organo-fluorine compound-mixed electrolyte solutions for lithium ion batteries. J Power Sources 243:573–580. doi:10.1016/j.jpowsour.2013.06.034
75. Naoi K, Mori M, Naruoka Y, Lamanna WM, Atanasoski R (1999) The surface film formed on a lithium metal electrode in a new imide electrolyte, lithium bis (perfluoroethylsulfonylimide) [LiN(C2F5SO2)2]. J Electrochem Soc 146:462–469. doi:10.1149/1.1391627
76. Oh B, Ofer D, Rempel J, Pullen A, Sriramulu S, Barnett B (2010) New and improved nonaqueous electrolyte components fo Li-ion batteries. In: Proceedings of power sources conference 44th, pp 170–173
77. Oh B, Ofer D, Singh SK, Sriramulu S, Barnett B (2008) Nitrile-based electrolytes for lithium-ion cells. Proceedings of power sources conference 43rd, pp 109–112
78. Ohmi N, Nakajima T, Ohzawa Y, Koh M, Yamauchi A, Kagawa M, Aoyama H (2013) Effect of organo-fluorine compounds on the thermal stability and electrochemical properties of electrolyte solutions for lithium-ion batteries. J Power Sources 221:6–13. doi:10.1016/j.jpowsour.2012.07.121
79. Ossola F, Pistoia G, Seeber R, Ugo P (1988) Oxidation potentials of electrolyte solutions for lithium cells. Electrochim Acta 33:47–50. doi:10.1016/0013-4686(88)80030-6
80. Park C-M, Kim J-H, Kim H, Sohn H-J (2010) Li-alloy based anode materials for Li secondary batteries. Chem Soc Rev 39:3115–3141. doi:10.1039/b919877f
81. Ping P, Wang Q, Sun J, Feng X, Chen C (2011) Effect of sulfites on the performance of LiBOB/γ-butyrolactone electrolytes. J Power Sources 196:776–783. doi:10.1016/j.jpowsour.2010.07.064
82. Ratnakumar BV, Smart MC, Huang CK, Perrone D, Surampudi S, Greenbaum SG (2000) Lithium ion batteries for Mars exploration missions. Electrochim Acta 45:1513–1517. doi:10.1016/s0013-4686(99)00367-7
83. Reddy MV, Subba Rao GV, Chowdari BVR (2013) Metal oxides and oxysalts as anode materials for Li ion batteries. Chem Rev (Washington, DC, US) 113:5364–5457. doi:10.1021/cr3001884
84. Rollins HW, Harrup MK, Dufek EJ, Jamison DK, Sazhin SV, Gering KL, Daubaras DL (2014) Fluorinated phosphazene co-solvents for improved thermal and safety performance in lithium-ion battery electrolytes. J Power Sources 263:66–74. doi:10.1016/j.jpowsour.2014.04.015
85. Rossi NAA, West R (2009) Silicon-containing liquid polymer electrolytes for application in lithium ion batteries. Polym Int 58:267–272. doi:10.1002/pi.2523
86. Rupich MW, Pitts L, Abraham KM (1982) Characterization of reactions and products of the discharge and forced over-discharge of Li/SO_2 cells. J Electrochem Soc 129:1857–1861. doi:10.1149/1.2124314
87. Santhanam R, Rambabu B (2010) Research progress in high voltage spinel $LiNi_{0.5}Mn_{1.5}O_4$ material. J Power Sources 195:5442–5451
88. Scheers J, Lim D-H, Kim J-K, Paillard E, Henderson WA, Johansson P, Ahn J-H, Jacobsson P (2014) All fluorine-free lithium battery electrolytes. J Power Sources 251:451–458. doi:10.1016/j.jpowsour.2013.11.042
89. Schroeder G, Gierczyk B, Waszak D, Kopczyk M, Walkowiak M (2006) Vinyl tris-2-methoxyethoxy silane: a new class of film-forming electrolyte components for Li-ion

cells with graphite anodes. Electrochem Commun 8:523–527. doi:10.1016/j.elecom.2006.01.021

90. Schroeder G, Gierczyk B, Waszak D, Walkowiak M (2006) Impact of ethyl tris-2-methoxyethoxy silane on the passivation of graphite electrode in Li-ion cells with PC-based electrolyte. Electrochem Commun 8:1583–1587. doi:10.1016/j.elecom.2006.07.030
91. Shao N, Sun XG, Dai S, Jiang DE (2011) Electrochemical windows of sulfone-based electrolytes for high-voltage Li-ion batteries. J Phys Chem B 115:12120–12125. doi:10.1021/jp204401t
92. Sloop SE, Pugh JK, Wang S, Kerr JB, Kinoshita K (2001) Chemical reactivity of PF_5 and $LiPF_6$ in ethylene carbonate/dimethyl carbonate solutions. Electrochem Solid-State Lett 4: A42–A44. doi:10.1149/1.1353158
93. Smart MC, Ratnakumar BV, Ryan-Mowrey VS, Surampudi S, Prakash GKS, Hub J, Cheung I (2003) Improved performance of lithium-ion cells with the use of fluorinated carbonate-based electrolytes. J Power Sources 119:359–367. doi:10.1016/S0378-7753(03)00266-0
94. Smart MC, Ratnakumar BV, Surampudi S (2002) Use of organic esters as cosolvents in electrolytes for lithium-ion batteries with improved low temperature performance. J Electrochem Soc 149:A361–A370. doi:10.1149/1.1453407
95. Smith KA, Smart MC, Prakash GKS, Ratnakumar BV (2008) Electrolytes containing fluorinated ester co-solvents for low-temperature Li-ion cells. ECS Trans 11:91–98. doi:10.1149/1.2938911
96. Song JY, Wang YY, Wan CC (1999) Review of gel-type polymer electrolytes for lithium-ion batteries. J Power Sources 77:183–197. doi:10.1016/s0378-7753(98)00193-1
97. Sun XG, Angell CA (2004) New sulfone electrolytes: part II. Cyclo alkyl group containing sulfones. Solid State Ionics 175:257–260. doi:10.1016/j.ssi.2003.11.035
98. Sun XG, Angell CA (2005) New sulfone electrolytes for rechargeable lithium batteries: part I. Oligoether-containing sulfones. Electrochem Commun 7:261–266. doi:10.1016/j.elecom.2005.01.010
99. Sun XG, Angell CA (2009) Doped sulfone electrolytes for high voltage Li-ion cell applications. Electrochem Commun 11:1418–1421. doi:10.1016/j.elecom.2007.05.020
100. Takami N, Ohsaki T, Hasebe H, Yamamoto M (2002) Laminated thin Li-ion batteries using a liquid electrolyte. J Electrochem Soc 149:A9–A12. doi:10.1149/1.1420704
101. Takami N, Sekino M, Ohsaki T, Kanda M, Yamamoto M (2001) New thin lithium-ion batteries using a liquid electrolyte with thermal stability. J Power Sources 97–98:677–680. doi:10.1016/S0378-7753(01)00699-1
102. Tarascon JM, Armand M (2001) Issues and challenges facing rechargeable lithium batteries. Nature 414:359–367. doi:10.1038/35104644
103. Tobishima S, Yamaki J, Yamaji A, Okada T (1984) Dialkoxyethane-propylene carbonate mixed electrolytes for lithium secondary batteries. J Power Sources 13:261–271. doi:10.1016/0378-7753(84)80034-8
104. Ufheil J, Baertsch MC, Wuersig A, Novak P (2005) Maleic anhydride as an additive to γ-butyrolactone solutions for Li-ion batteries. Electrochim Acta 50:1733–1738. doi:10.1016/j.electacta.2004.10.061
105. Wang Q, Pechy P, Zakeeruddin SM, Exnar I, Graetzel M (2005) Novel electrolytes for $Li_4Ti_5O_{12}$-based high power lithium ion batteries with nitrile solvents. J Power Sources 146:813–816. doi:10.1016/j.jpowsour.2005.03.157
106. Wang XJ, Lee HS, Li H, Yang XQ, Huang XJ (2010) The effects of substituting groups in cyclic carbonates for stable SEI formation on graphite anode of lithium batteries. Electrochem Commun 12:386–389. doi:10.1016/j.elecom.2007.12.041
107. Watanabe M, Nagano S, Sanui K, Ogata N (1987) Structure-conductivity relationship in polymer electrolytes formed by network polymers from poly[dimethylsiloxane-g-poly(ethylene oxide)] and lithium perchlorate. J Power Sources 20:327–332. doi:10.1016/0378-7753(87)80131-3

108. Watanabe Y, Kinoshita S-I, Wada S, Hoshino K, Morimoto H, Tobishima S-I (2008) Electrochemical properties and lithium ion solvation behavior of sulfone–ester mixed electrolytes for high-voltage rechargeable lithium cells. J Power Sources 179:770–779. doi:10.1016/j.jpowsour.2008.01.006
109. Wu F, Zhu Q, Li L, Chen R, Chen S (2013) A diisocyanate/sulfone binary electrolyte based on lithium difluoro(oxalato)borate for lithium batteries. J Mater Chem A 1:3659–3666. doi:10.1039/c3ta01182h
110. Wu F, Zhu QZ, Li L, Chen RJ, Chen S (2013) A diisocyanate/sulfone binary electrolyte based on lithium difluoro(oxalate)borate for lithium batteries. J Mater Chem A 1:3659–3666. doi:10.1039/C3TA01182H
111. Wu H, Cui Y (2012) Designing nanostructured Si anodes for high energy lithium ion batteries. Nano Today 7:414–429. doi:10.1016/j.nantod.2012.08.004
112. Wu HB, Chen JS, Hng HH, Wen Lou X (2012) Nanostructured metal oxide-based materials as advanced anodes for lithium-ion batteries. Nanoscale 4:2526–2542. doi:10.1039/c2nr11966h
113. Xia J, Sinha NN, Chen LP, Dahn JR (2014) A comparative study of a family of sulfate electrolyte additives. J Electrochem Soc 161:A264–A274. doi:10.1149/2.015403jes
114. Xiang J, Wu F, Chen RJ, Li L, Yu HG (2013) High voltage and safe electrolytes based on ionic liquid and sulfone for lithium-ion batteries. J Power Sources 233:115–120. doi:10.1016/j.jpowsour.2013.01.123
115. Xu B, Qian D, Wang Z, Meng YS (2012) Recent progress in cathode materials research for advanced lithium ion batteries. Mater Sci Eng R 73:51–65. doi:10.1016/j.mser.2012.05.003
116. Xu K (2004) Nonaqueous liquid electrolytes for lithium-based rechargeable batteries. Chem Rev 104:4303–4417. doi:10.1021/cr030203g
117. Xu K (2014) Electrolytes and interphases in Li-ion batteries and beyond. Chem Rev 114:11503–11618. doi:10.1021/cr500003w
118. Xu K (2008) Tailoring electrolyte composition for LiBOB. J Electrochem Soc 155:A733–A738. doi:10.1149/1.2961055
119. Xu K, Angell CA (2002) Sulfone-based electrolytes for lithium-ion batteries. J Electrochem Soc 149:L7–L7. doi:10.1149/1.1496104
120. Xu K, von Cresce A (2011) Interfacing electrolytes with electrodes in Li ion batteries. J Mater Chem 21:9849–9864. doi:10.1039/C0JM04309E
121. Xu K, Zhang S, Poese BA, Jow TR (2002) Lithium bis(oxalato)borate stabilizes graphite anode in propylene carbonate. Electrochem Solid-State Lett 5:A259–A262. doi:10.1149/1.1510322
122. Xu M, Tsiouvaras N, Garsuch A, Gasteiger HA, Lucht BL (2014) Generation of cathode passivation films via oxidation of lithium bis(oxalato) borate on high voltage spinel ($LiNi_{0.5}Mn_{1.5}O_4$). J Phys Chem C 118:7363–7368. doi:10.1021/jp501970j
123. Xu M, Xiao A, Li W, Lucht BL (2009) Investigation of lithium tetrafluorooxalatophosphate as a lithium-ion battery electrolyte. Electrochem Solid-State Lett 12:A155–A158. doi:10.1149/1.3134462
124. Xu M, Xiao A, Li W, Lucht BL (2010) Investigation of lithium tetrafluorooxalatophosphate [$LiPF_4(C_2O_4)$] as a lithium-ion battery electrolyte for elevated temperature performance. J Electrochem Soc 157:A115–A120. doi:10.1149/1.3258290
125. Xu M, Xiao A, Yang L, Lucht BL (2009) Novel electrolyte for lithium ion batteries: lithium tetrafluorooxalatophosphate ($LiPF_4C_2O_4$). ECS Trans 16:3–11. doi:10.1149/1.3123122
126. Xu MQ, Zhou L, Dong YN, Chen YJ, Garsuch A, Lucht BL (2013) Improving the performance of graphite/$LiNi_{0.5}Mn_{1.5}O_4$ cells at high voltage and elevated temperature with added lithium bis(oxalato) borate (LiBOB). J Electrochem Soc 160:A2005–A2013. doi:10.1149/2.053311jes
127. Xu W, Angell CA (2001) LiBOB and its derivatives: weakly coordinating anions, and the exceptional conductivity of their nonaqueous solutions. Electrochem Solid-State Lett 4:E1–E4. doi:10.1149/1.1344281

128. Yamada Y, Furukawa K, Sodeyama K, Kikuchi K, Yaegashi M, Tateyama Y, Yamada A (2014) Unusual stability of acetonitrile-based superconcentratede electrolytes for fast-charging lithium-ion batteries. J Am Chem Soc 136:5039–5046. doi:10.1021/ja412807w
129. Yang L, Ravdel B, Lucht BL (2010) Electrolyte reactions with the surface of high voltage $LiNi_{0.5}Mn_{1.5}O_4$ cathodes for lithium-ion batteries. Electrochem Solid-State Lett 13:A95–A97. doi:10.1149/1.3428515
130. Yazami R, Martinent A (2005) Fluorinated anions and electrode/electrolyte stability in lithium batteries. Elsevier, pp 173–194. doi:10.1016/b978-008044472-7/50036-7
131. Yong T, Wang J, Mai Y, Zhao X, Luo H, Zhang L (2014) Organosilicon compounds containing nitrile and oligo(ethylene oxide) substituents as safe electrolytes for high-voltage lithium-ion batteries. J Power Sources 254:29–32. doi:10.1016/j.jpowsour.2013.12.087
132. Zhang L, Lyons L, Newhouse J, Zhang Z, Straughan M, Chen Z, Amine K, Hamers RJ, West R (2010) Synthesis and characterization of alkylsilane ethers with oligo(ethylene oxide) substituents for safe electrolytes in lithium-ion batteries. J Mater Chem 20:8224–8226. doi:10.1039/c0jm01596b
133. Zhang SS (2006) A review on electrolyte additives for lithium-ion batteries. J Power Sources 162:1379–1394. doi:10.1016/j.jpowsour.2006.07.074
134. Zhang SS (2006) An unique lithium salt for the improved electrolyte of Li-ion battery. Electrochem Commun 8:1423–1428. doi:10.1016/j.elecom.2006.06.016
135. Zhang SS (2007) Electrochemical study of the formation of a solid electrolyte interface on graphite in a $LiBC_2O_4F_2$-based electrolyte. J Power Sources 163:713–718. doi:10.1016/j.jpowsour.2006.07.040
136. Zhang SS (2007) Lithium oxalyldifluoroborate as a salt for the improved electrolytes of Li-ion batteries. ECS Trans 3:59–68. doi:10.1149/1.2793577
137. Zhang W-J (2011) A review of the electrochemical performance of alloy anodes for lithium-ion batteries. J Power Sources 196:13–24. doi:10.1016/j.jpowsour.2010.07.020
138. Zhang Z, Hu L, Wu H, Weng W, Koh M, Redfern PC, Curtiss LA, Amine K (2013) Fluorinated electrolytes for 5 V lithium-ion battery chemistry. Energ Environ Sci 6:1806–1810. doi:10.1039/C3EE24414H
139. Zhang Z, Lu J, Assary RS, Du P, Wang H-H, Sun Y-K, Qin Y, Lau KC, Greeley J, Redfern PC, Iddir H, Curtiss LA, Amine K (2011) Increased stability toward oxygen reduction products for lithium-air batteries with oligoether-functionalized silane electrolytes. J Phys Chem C 115:25535–25542. doi:10.1021/jp2087412
140. Zhang Z, Dong J, West R, Amine K (2009) Oligo(ethylene glycol)-functionalized disiloxanes as electrolytes for lithium-ion batteries. J Power Sources 195(18):6062–6068. doi:10.1016/j.jpowsour.2007.12.067
141. Zheng M-S, Chen J-J, Dong Q-F (2012) The research of electrolyte on lithium/sulfur battery. Adv Mater Res (Durnten-Zurich, Switz) 476–478:1763–1766. doi:10.4028/www.scientific.net/AMR.476-478.1763
142. Zhou L, Lucht BL (2012) Performance of lithium tetrafluorooxalatophosphate (LiFOP) electrolyte with propylene carbonate (PC). J Power Sources 205:439–448. doi:10.1016/j.jpowsour.2012.01.067
143. Zhu Y, Li Y, Bettge M, Abraham DP (2013) Electrolyte additive combinations that enhance performance of high-capacity $Li_{1.2}Ni_{0.15}Mn_{0.55}Co_{0.1}O_2$-graphite cells. Electrochim Acta 110:191–199. doi:10.1016/j.electacta.2013.03.102
144. Zinigrad E, Larush-Asraf L, Salitra G, Sprecher M, Aurbach D (2007) On the thermal behavior of Li bis(oxalato)borate LiBOB. Thermochim Acta 457:64–69. doi:10.1016/j.tca.2007.03.00

Additives for Functional Electrolytes of Li-Ion Batteries

Libo Hu, Adam Tornheim, Sheng Shui Zhang and Zhengcheng Zhang

1 Introduction

The electrolyte is an indispensable element of Li-ion batteries. In normal operation, the electrolyte does not participate in electrochemical reactions but rather conducts ions to enable the electrode reactions on the cathode and anode. The electrolyte is typically composed of a lithium salt as the solute for lithium ions and a solvent or mixed solvent as the medium for ionic conduction. The performance of a Li-ion battery, measured by rate capability, cycling reversibility, cycling stability (life), and safety, is greatly affected by the properties of the electrolyte, including ionic conductivity, chemical compatibility, electrochemical stability, and flammability. In particular, both the chemical compatibility and electrochemical stability are more dependent on the kinetic stability than the thermodynamic stability. In such cases, the suitability of an electrolyte for the Li-ion battery is determined by the properties of the electrolyte-electrode interface, rather than those of the bulk electrolyte. Thus, the addition of small amount of foreign compounds may tremendously improve the battery's performance by altering the chemical kinetics at that interface. These compounds are referred to as electrolyte additives and their weight percentage in the electrolyte usually does not exceed 5 %. Due to their significant effect on the battery's performance, in recent years electrolyte additives have been intensively investigated for potential performance improvement of existing Li-ion batteries, or for specialized applications, such as operation at high voltage or extreme temperatures. In addition to the electrode/electrolyte interface, some bulk properties

L. Hu · A. Tornheim · Z. Zhang (✉)
Electrochemical Energy Storage Theme, Chemical Sciences and Engineering Division, Argonne National Laboratory, Lemont, IL 60439, USA
e-mail: zzhang@anl.gov

S.S. Zhang
U.S. Army Research Laboratory, Adelphi, MD 20783, USA
e-mail: Shengshui.zhang.civ@mail.mil

Z. Zhang and S.S. Zhang (eds.), *Rechargeable Batteries*,
Green Energy and Technology, DOI 10.1007/978-3-319-15458-9_9

of the electrolyte, such as electrochemical stability window and flammability, are also affected by the additive. In this chapter, electrolyte additives will be discussed in categories of the intended function in Li-ion batteries.

2 Anode Passivation Additives

The anode/electrolyte interface has been the focus of electrolyte research since the advent of the Li-ion battery. This interface is frequently referred to as solid-electrolyte interphase (SEI), a term that was originally used to describe the surface layer of metallic Li in primary Li batteries [100]. The SEI serves two purposes in Li-ion batteries. On one hand, it protects the electrolyte from electrochemical reduction on the charged (lithiated) anode surface, and on the other hand it protects the anode material from chemically reacting with the electrolyte. With graphitic anodes, the SEI protects the graphite layered structure from exfoliation caused by solvent co-intercalation [33]. In silicon anodes, the SEI acts as an elastic layer to maintain the structural integrity of Si particles during lithiation [60].

A functional SEI should provide good protection to the anode and electrolyte in addition to being highly ion-conductive. Moreover, the SEI should remain stable over repeated cycling and long storage, and should also be thermally stable when exposed to elevated temperatures. Although a conventional electrolyte based on $LiPF_6$ salt and ethylene carbonate (EC) solvent is capable of forming a SEI on graphitic anodes, the SEI formed from EC alone has relatively poor thermal stability [7]. The properties of this SEI can be improved with the addition of a small amount of additive. Since introduction of an additive usually does not change the bulk properties of the electrolyte, such as conductivity, viscosity, and liquid temperature range, the additive approach offers a cost-effective method for improving the SEI formation of a well-established electrolyte.

In this section, recent advancements in additives will be discussed for SEI formation at the anode, in which some additives are previously known compounds with new applications. Due to the proprietary nature of much of the additive research, public reports on the additives are rather limited. Moreover, a mechanistic understanding of the effect of a specific additive in the battery is frequently lacking in many cases. Regardless, the patterns between the structures and functions of known additives help identify new compounds to investigate.

2.1 SEI Additives for Carbonaceous Anodes

In Li-ion batteries, carbonaceous anodes are the most common commercial anode due to its superior reliability. In addition to graphite, a variety of non-graphitic carbon materials, such as carbon nanotubes [151], mesoporous carbon [177], non-graphitic hard carbons [19], and other composite carbon materials, [145] are suitable for the anode material. Of all these materials, graphite offers good energy density, good rate capability and has low manufacturing cost, which makes it the most widely used anode material. However, the layered structure of the graphite presents weak interplanar bonds, and such connections can be easily broken if solvent co-intercalation and subsequent electrochemical decomposition occur [33]. Luckily, a functional SEI prevents solvent co-intercalation, and graphite can reliably withstand reversible lithium intercalation.

In the early days, researchers were unsure of the function of the SEI formed on the graphite surface, as the only information available was the success of ethylene carbonate in the electrolyte. As a result, the design of the SEI-forming additives generally followed the guideline that the chemical structure must share some similarity with EC, which means either a cyclic carbonate derivative or a five-membered ring structure with other functionalities. Early examples of such additives include, but are not limited to, substituted catechol carbonate [129], ethylene sulfite [139], propylene sulfite [140], fluoroethylene carbonate (FEC) and vinylene carbonate (VC) [58] (Fig. 1). The mechanism involved with this kind of additive is believed to be similar to that for EC, which goes through reductive decomposition on the anode forming an insoluble SEI composed of inorganic and organic salts.

The outlier in this group is VC, which contains a double bond capable of polymerization by either a radical or anionic mechanism on the anode. This structure provides another mechanism of surface film formation besides the formation of insoluble lithium carbonates and lithium alkyl carbonates. Although the study on VC as an additive has more than 20 years of history, its working mechanism in batteries is still under debate. While some chemists believe VC is involved with radical/anionic polymerization, as there has been experimental proof of such a product [36], other chemists have proposed mechanisms that involve the interaction of VC with reduction products of EC, which agrees with the observation of gas products such as CO and CO_2 (Scheme 1) [126, 135]. Electrochemical processes can be very complicated, and in batteries they can be even more convoluted due to the heterogeneous surface chemistry of electrode materials. There is experimental evidence to support both of these mechanisms. Even though the success of VC is

Fig. 1 The first generation of SEI-forming additives that resemble EC in structure

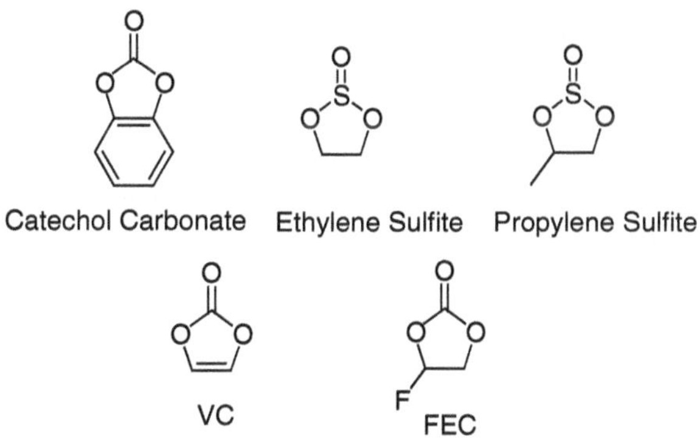

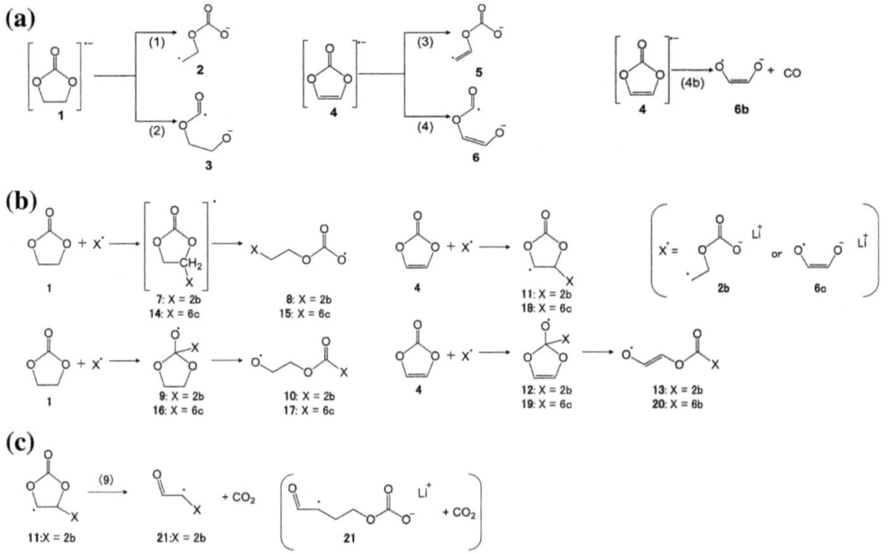

Scheme 1 **a** One-electron (*1e*) reductive decomposition reactions of EC and VC; **b** possible interactions between intact EC (*1*) and VC (*4*) molecules with the radical reduction products of EC (*2*) and VC (*6b*); **c** possible subsequent reaction after the radical attack of o_E−EC^- on VC for CO_2 evolution. Reproduced with permission [126]. Copyright 2013 American Chemical Society

not well understood, its discovery has led to the development of a group of additives featured with unsaturated bonds that are capable of polymerization on the anode surface. This strategy of additive design based on the success of VC led to the boom of additive research in the 2000s.

Presently, many organic additives fall into the category of polymerizable monomers. A non-exhaustive list includes esters (including carboxylic esters/carbonates and other inorganic esters such as phosphates, sulfates, and silicates) that are derived from vinyl and allyl alcohols [2, 54, 71], vinyl pyridine [63], acrylic acid nitrile [110], maleic acid derivatives [125, 131], vinyl sulfones [128], vinyl silanes [113], and isocyanates [65, 171] (Fig. 2). The synergistic effect of different unsaturated compounds used in various combinations has also been reported [1].

The reducible additives that do not contain polymerizable functionalities are also expanding, not limited to the five-membered ring structural analog. Examples of this group include substituted cyclic carbonates [52, 178], sulfates [41, 146], sulfites [123], sultones [59, 76], phosphates [98, 132], and possibly other compounds that have heteroatoms with high oxidation states (Fig. 2).

Boron-containing compounds, especially those with only three substitutions, are mostly known for their functions as anion receptors. However, some of these compounds are also capable of forming an SEI, making them bi-functional additives [78]. The lithium salts with tetra-substituted boron anions such as LiBOB and LiDFOB can serve as SEI-forming additives in electrolytes based on other salts

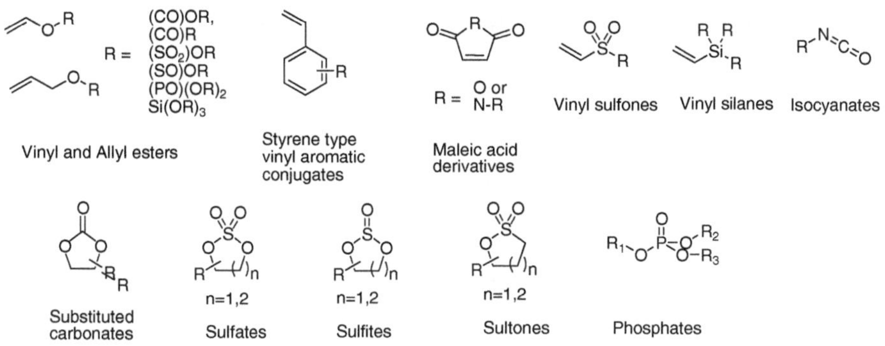

Fig. 2 Structural illustrations of polymerizable and reductive decomposable additives

such as $LiPF_6$ [82]. These additives often show synergistic effect when combined with other additives [6, 180]. A FTIR analysis showed that all the SEIs formed in the presence of these salts contain B-O molecular moieties, concluding that these boron-containing anions are participating in the SEI formation [181].

2.2 SEI Additives for Novel Anodes

Carbonaceous anodes cannot meet the demand for increasing energy density, so the development of high-capacity anode materials is of great interest to both academia and industry. However, high-capacity anode materials generally suffer from severe volume expansion during lithiation. This poses a new challenge to the stabilization of the anode/electrolyte interface, as graphite has a much lower degree of Li^+ intercalation (one Li per six carbon) and the resulting volume expansion is insignificant. With high-capacity anode materials such as silicon, the large volume expansion may become a critical factor accounting for the degradation of the formed SEI. As a result, new SEI-forming additives that can accommodate such large volume expansion need to be developed for these new materials.

So far, FEC is the most effective SEI-forming additive for all silicon anodes [23, 32, 35, 37, 39, 45, 103]. Although FEC is among the first group of additives studied for the graphite anode [58], the mechanism of how it improves the graphite anode remains unclear. The most acknowledged mechanism involves the elimination of HF from the FEC molecule triggered by a strong base, most likely ROLi, which is readily available on the charged anode, and the resulting HF reacts with the base to generate LiF. The elimination of HF from FEC generates VC as a product, and VC undergoes reductive decomposition or polymerization to form SEI on the anode (Scheme 2a) [23, 39]. Another mechanism adopts a reduction-first pathway (Scheme 2b) [94], in which the FEC molecule is reduced first to generate vinyl fluoride, which is then polymerized on the anode surface. The poly(vinyl fluoride) then undergoes elimination by base attack to generate polyacetylene and LiF.

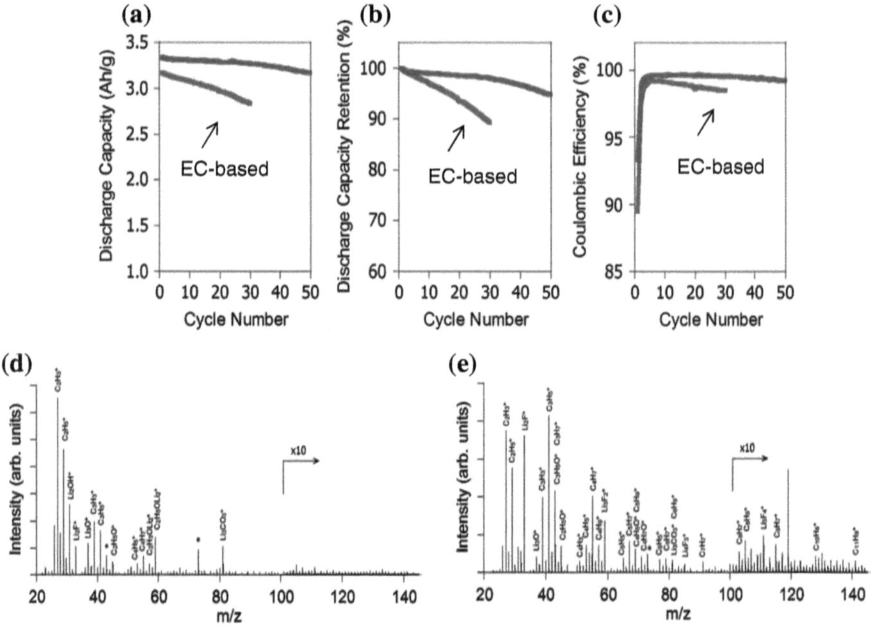

Fig. 3 Cycling performance of Si/Li cells using EC-based (*red line,* labeled) and FEC-based (*blue line*) electrolytes: **a** discharge capacity, **b** discharge capacity retention, and **c** coulombic efficiency. Also shown are TOF-SIMS surface spectra (positive secondary ion mode) for Si electrodes of the **d** EC-based and **e** FEC-based cells after the 30th cycle. Reproduced with permission [94]. Copyright 2011 The Electrochemical Society (color online)

This mechanism is proposed based on the characterization by time-of-flight secondary ion mass spectrometry (TOF-SIMS) (Fig. 3), which revealed polyene and LiF species on the anode. Nakai et al. also noted that FEC not only protected the electrolyte from further reductive decomposition on the Si anode, but also prevented the Si anode from being oxidized, in contrast to EC-based electrolytes. Besides these two mechanisms, other possibilities have been proposed on the basis of theoretical calculations [86].

Other than FEC, VC has also been shown to improve Si anode performance [21, 22, 35, 103], although in some cases the improvement is not as prominent as that of FEC. If FEC indeed protects the Si anode through the VC mechanism, then the LiF generated through the first elimination step may have some positive effect on the Si anode as well. Other additives that work on the Si anode include LiDFOB [35], succinic anhydride [80], alkoxysilanes [108, 118] and tris(pentafluorophenyl) borane [47].

The additive research for alloy type anodes is even scarcer because they have an even shorter history than that for the silicon anodes. So far, the additives for these anodes found in the literature are still conventional additives such as FEC, VC, and vinylethylene carbonate (VEC) [8].

Lithium titanate ($Li_4Ti_5O_{12}$, abbreviated as LTO) is a well-known intercalative anode material which offers high stability and reversibility (so-called "zero-strain

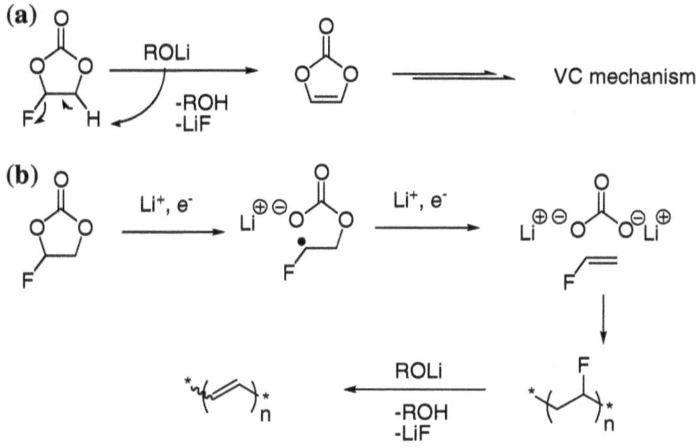

Scheme 2 Possible mechanisms of FEC SEI formation on Si anode

material") [106]. It was believed that no SEI would form on this anode because its working potential is around 1.5 V versus Li^+/Li, which is far above the decomposition potential of the alkyl carbonate solvents. However, a recent report has proven that an SEI is indeed formed on the LTO anode, and the SEI is also critical to the anode performance [49]. The same authors also noted that the addition of VC to the electrolyte enhanced the rate capability and cycle performance of LTO electrodes synthesized from lithium acetate. One issue associated with the LTO anode is the severe gassing in the initial formation process after cell assembly due to the reaction between the outer layer of the material and the liquid electrolyte [14]. While some researchers address this problem from the material side by using an inert coating on the LTO particles, an SEI that forms before the gassing reaction happens can serve a similar purpose [48].

To summarize this section, SEI-forming additives for the stabilization of anode/electrolyte interface are still primarily developed for graphitic materials. However, we fully expect additive research to include other anode materials as other novel additives become more mature in the future.

3 Cathode Passivation Additives

The cathode/electrolyte interface has received much less attention than the anode/electrolyte interface, mostly because carbonate electrolytes are electrochemically stable with conventional lithium transition oxide materials, such as $LiCoO_2$, $LiMn_2O_4$, and $LiFePO_4$. However, the cathode/electrolyte interface is actually still problematic. The catalytic effect of the transition metal oxide will lead to parasitic reactions of electrolyte on the surface of the cathode. Moreover, as some reactive species in the electrolyte (such as HF, PF_5 and POF_3 in $LiPF_6$-based electrolytes) can attack the cathode material, causing the dissolution of transition metal ions into the electrolyte. The resulting transition metal ions from this dissolution may migrate to the anode where they are reduced and deposited on the anode. The transition metal species on the anode will catalyze more decomposition of electrolyte forming thick and resistive SEI [11, 64, 166]. For high-voltage cathodes such as $LiNi_{0.5}Mn_{1.5}O_4$ (LNMO, 4.7 V) and $LiCoPO_4$ (4.8 V) the carbonate electrolytes are not thermodynamically stable on charged cathodes, and oxidative decomposition of electrolyte will occur if no protection is provided [85, 163]. As a result, cathode SEI additives are attracting more electrolyte researchers and will be of vital importance in the future as the high-voltage cathode materials are introduced into the high-energy high voltage batteries for electric vehicle applications.

3.1 Additives for 4-V Cathode Materials

The transition metal dissolution phenomenon has been identified in $LiMn_2O_4$ spinel as early as the mid-1990s [57, 147], and similar problems still plague high-energy cathode materials such as the high-voltage spinel $LiNi_{0.5}Mn_{1.5}O_4$ [101]. The problem was caused by the presence of protic solvent (water, alcohols, and acids) as impurities in the electrolyte or generated by oxidative decomposition of the organic carbonates [101, 130]. The proton donors can react with $LiPF_6$ in the electrolyte to generate HF, which reacts with the cathode material to cause metal dissolution. To solve this problem, researchers have primarily employed two tactics. The first is to scavenge the species that potentially produce protons such as water and HF with consumable additives, and the second is to form a protective layer on the cathode particle surface with sacrificial additives that blocks the direct contact of the electrolyte with the cathode while allowing Li ion diffusion.

The scavenger-type additives are not developed only for cathode specifically, as water and HF can negatively impact the anode as well. Regardless, some additives have been developed for this purpose but are only disclosed in patents and rarely appear in academic journals due to the commercial nature of additive research as mentioned in the introduction. A good summary of such compounds can be found in a review article [172] which highlights compounds that can scavenge water, such as *N,N′*-dicyclohexylcarbodiimide (DCC), or both water and HF, such as *N,N′*-diethylaminotrimethylsilane (Fig. 4). Similar compounds have appeared in later patents as well, and likely the Si-N bond containing compounds can all meet the

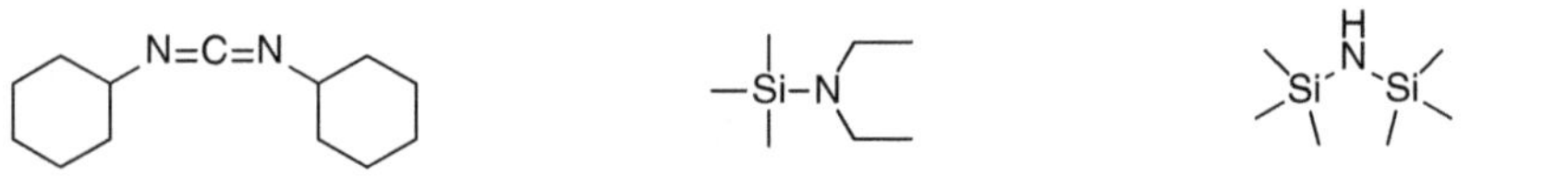

N,N'-diethylaminotrimethylsilane N,N'-diethylamino trimethylsilane Heptamethyldisilazane

Fig. 4 Examples of water/HF scavenger additives

purposes outlined above. An example is the use of heptamethyldisilazane (Fig. 4) alone [81, 144] or in combination with ethanolamine [143] to reduce Mn dissolution in the $LiMn_2O_4$ cathode during cycling and storage. In addition, nano-sized alumina has been reported to serve as an HF scavenger when dispensed in electrolyte by soaking [83]. It is likely that other metal oxides such as silica nanoparticles can offer similar functions and improve battery performance.

The film-forming additives have been more widely studied because of the various benefits a film on the cathode could provide. By preventing the reaction between the electrolyte and the cathode, the film not only suppresses the dissolution of metal ions in the cathode, but also protects the electrolyte from being oxidized on the cathode, a serious concern with high voltage cathodes. Several film-forming additives have been reported to enhance the thermal stability of the $LiMn_2O_4$ cathode and suppress Mn dissolution. As shown in Fig. 5, these additives include methylene methane disulfonate (MMDS) [182], 2-(triphenylphosphoranylidene) succinic anhydride [107], and prop-1-ene-1,3-sultone [75]. Figure 6 shows scanning electron microscopy images and performance data for both pristine and cycled $LiMn_2O_4$ cathode with and without the MMDS additive in ethylene carbonate/dimethyl carbonate/ethyl methyl carbonate (EC/DMC/EMC) electrolyte. Some of the film-forming additives (Fig. 5) have been shown to enhance the cycling performance of full Li-ion cells composed of a $LiMn_2O_4$ cathode and a graphite anode, such as tris(trimethylsilyl) borate [84] and tris(pentafluorophenyl) phosphine [155]. However, whether these additives reduce Mn dissolution in the electrolyte is not clear.

Besides the more problematic $LiMn_2O_4$ spinel cathode, additives have also been developed to enhance the performance of other cathode materials, such as $LiCoO_2$, $LiCo_xMn_yNi_{1-x-y}O_2$, and $LiFePO_4$. These additives mostly form a thin SEI layer on the cathode to suppress the buildup of decomposition products over cycling and enhance the capacity retention, especially at elevated temperatures. Many of the additives originally developed for the graphite anode have been found to be able to form films on the cathode as well, such as VC [10, 36, 142], VEC [77], FEC [99], LiBOB [124], and LiDFOB [179]. Other than these additives, the additives

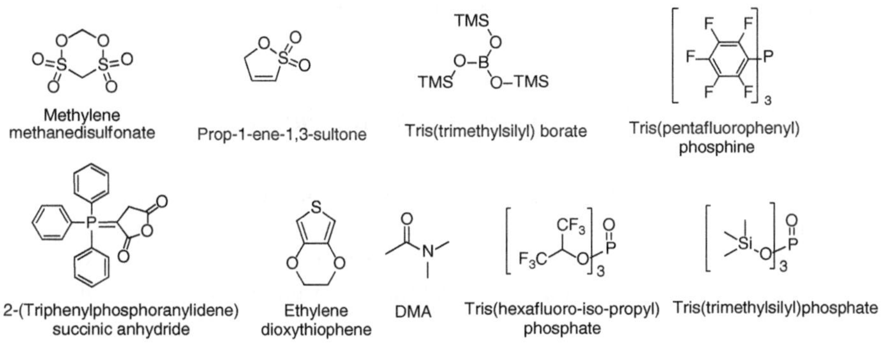

Fig. 5 Examples of film-forming additives for cathode SEI

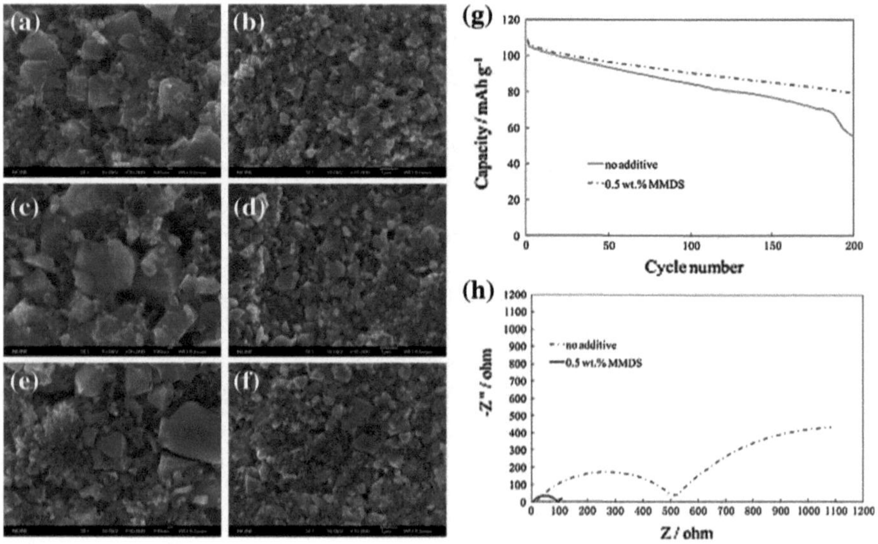

Fig. 6 SEM micrographs of $LiMn_2O_4$ electrodes before cycling **a** 30,000×, **b** 10,000×, and after 200 cycles at 55 °C in a 1 M $LiPF_6$ in EC/DMC/EMC (1:1:1, wt%) electrolyte without **c** 30,000×, **d** 10,000× and with 0.5 wt% MMDS additive **e** 30,000×, **f** 10,000×; **g** cycling performance of a Li/$LiMn_2O_4$ cell with and without 0.5 wt% MMDS additive at 55 °C. The cell was cycled at 1 C between 3.0 and 4.3 V; **h** electrochemical impedance spectra of the Li/$LiMn_2O_4$ cell with and without 0.5 wt% MMDS after 200 cycles at 55 °C. Reproduced with permission from [182]. Copyright 2014 Elsevier

designed specifically for cathodes have also been reported. An incomplete list (Fig. 5) includes those mentioned above for the $LiMn_2O_4$ spinel cathode, 2,5-dihydrofuran and γ-butyrolactone [161], ethylene dioxythiophene [72], dimethylacetamide [154], tris(hexafluoro-*iso*-propyl)phosphate [121], and tris(tri-methylsilyl)phosphate [167].

It is notable that the literature on cathode additives has become more abundant in recent years, showing a trend that the SEI additive research is gradually shifting from the anode side to both electrodes of the battery.

3.2 Additives for High-Voltage Cathode Materials

As mentioned above, cathode SEI additives have gained more attention recently, due primarily to the increasing interest of high-voltage cathode materials, defined as a cathode with a majority of capacity above 4.5 V in its voltage profile. So far the cathodes that have attracted the most attention from this group are LNMO spinel and $LiCoPO_4$.

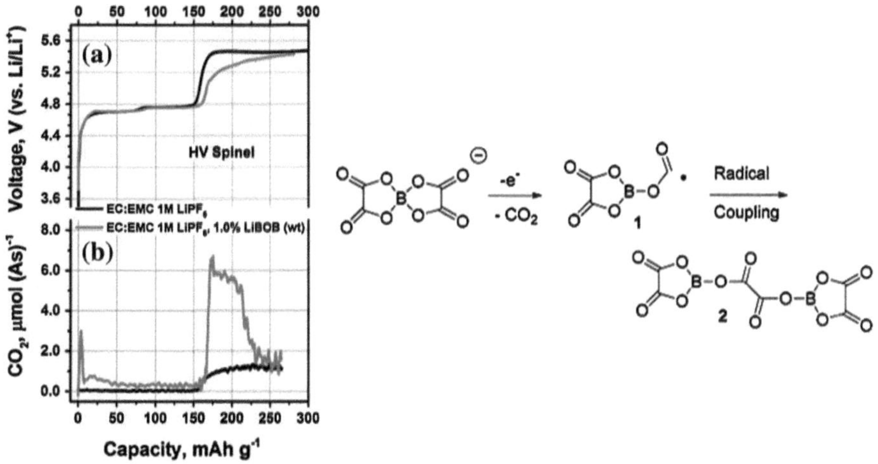

Fig. 7 Galvanostatic charge voltage profile (**a**) and CO_2 evolution rates (**b**) of $LiNi_{0.5}Mn_{1.5}O_4$ electrodes with electrolyte of EC/EMC (3/7, v/v) plus 1.0 M $LiPF_6$ without LiBOB additive (*black lines*) and with 1.0 wt% LiBOB additive (*red lines*). Proposed decomposition mechanism is shown on *right*. Reproduced with permission [157]. Copyright 2014 American Chemical Society (color online)

LNMO spinel is an attractive cathode candidate for next generation lithium-ion batteries as it offers high power capability with an operating voltage of 4.7 V and a capacity of 135 mA hg^{-1}. However, its commercialization is plagued by severe capacity fade, particularly at elevated temperatures, in full cells employing a graphite anode. [109]. The considerable efforts devoted to developing high voltage electrolyte, especially for this material, have been mentioned in Sect. 2.1 and 2.2 in Chap. 7, primarily with regard to the electrolyte solvents. Considerable efforts have been equally devoted to the additive. So far, the additive most studied for the LNMO cathode is LiBOB [46, 102, 157, 159, 162]. LiBOB has been reported to enhance the capacity retention of the cell cycled at room and high temperature, reduce self-discharge during storage, suppress interfacial impedance growth, and prevent Mn and Ni from dissolution into the electrolyte. The formation of a cathode passivation film from LiBOB additive was reported to be accompanied by CO_2 gas evolution, as proposed by a mechanism shown in Fig. 7 [157].

Besides LiBOB, other additives have been reported to protect the high voltage cathode and mitigate electrolyte decomposition, including LiDFOB [162], succinic anhydride [66, 122], 1,3-propane sultone [66], tris(hexafluoro-*iso*-propyl) phosphate [127], tris(pentafluorophenyl) phosphine [156], 3-hexylthiophene [3], 1,3-propanediolcyclic sulfate [41], dimethylacetamide [13], triethyl(2-methoxyethyl) phosphonium bis(trifluoromethylsulfonyl)imide [13], glutaric anhydride [16], 4-(trifluoromethyl)-benzonitrile [55], and 1-propylphosphonic acid cyclic anhydride [160]. The structures of the additives that are not listed in the previous text are shown in Fig. 8.

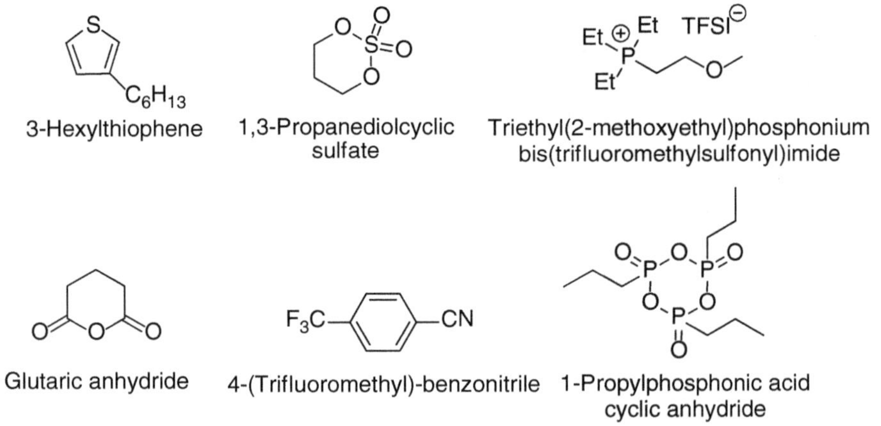

Fig. 8 Examples of electrolyte additives for high voltage cathodes

The additive research for $LiCoPO_4$ is sparser. Few examples include LiBOB [9], shown to mitigate capacity fade at 3 %, LiDFOB [53], which increased capacity retention at 40 cycles from 34 to 69 % at a 5.1 V cut-off potential, and thiophene [152], which increased capacity retention from 15 to 68 % after 30 cycles. Sharabi et al. [111] reported that FEC-containing electrolytes with a small amount (0.5–1 %) of trimethlyboroxine additive allowed for 90 % capacity retention after 100 cycles with a 5.2 V cut-off potential.

The field of high voltage electrolyte additive research is so nascent that most of the literature on this subject was published after 2010. There is no doubt that the demand for high energy density and high power density is pushing this research forward.

4 Additives for Improved Safety

While additives meant to improve the SEI certainly constitute a major portion of the additive research for Li-ion batteries, many other additives aim to improve different aspects of the Li-ion batteries, such as the safety characteristics, ionic conductivity of the electrolyte, and high/low temperature performance of the electrolyte. For example, researchers have developed redox shuttle and overcharge shutdown additives to protect the battery from overcharge and the resulting thermal runaway, flame retardant additives to reduce the flammability of the electrolyte, and anion receptors to enhance the ionic conductivity and increase the lithium transference number. These additives may not be critical to the cell performance, but may be very important and possibly necessary in commercial batteries.

4.1 Overcharge Protection Additives

Overcharge protection is of vital importance in a battery pack, as capacities of individual cells of a pack are not identical, meaning that uniform charging could lead to overcharge of cells with low capacity and even cause catastrophic thermal runaway. The strategies for overcharge protection generally employ one or both of the following methods: external circuit control, and electrolyte with a shutdown or redox shuttle additive [28].

Shutdown additives usually undergo an irreversible oxidative polymerization to form a non-conductive polymer layer that electrically blocks the current flow to protect the cell from being overcharged. The disadvantage of this strategy is that once the shutdown operation is activated, the cell is permanently disabled and the module that connects the damaged cell in series is consequently non-operational. For this reason, this type of overcharge protection additive received little interest. Classic examples of such additives are primarily aromatic compounds such as biphenyl [149], cyclohexylbenzene [56, 158], and xylene [44, 170].

The redox shuttle additives act as a reversible redox mediator between the cathode and anode when the voltage of the cell increases above a certain level. By selecting an appropriate redox shuttle additive, the overcharge protection could be set at slightly above the fully charged potential of the cathode [27]. The advantage of the redox shuttle additives is that they do not interfere with the cell normal operation the cell, so the performance of the battery pack is not compromised. However, no redox shuttle additive is completely reversible, especially in a real cell. The additives are gradually consumed over time and eventually cannot adequately protect against overcharge. Therefore, the chemical stability of a redox shuttle is an important criterion for the overcharge protection performance. Ideally, the redox shuttle additive should remain electrochemically inert during the normal operation of the battery, and should be strictly reversible when activated by overcharging. The concept of "redox shuttle" additives dates back to the era of lithium metal batteries and the very beginning of Li-ion batteries. The earliest studied compounds for redox shuttle additives were ferrocene family organometallic compounds [4, 20, 96], however, they were quickly phased out due to their low oxidation potentials.

In 2005, a series of hydroquinone ethers was first studied as redox shuttle additives by a group of scientists in Dalhousie University [17], which led to the growing survey of redox shuttles in this category. It was shown that 2,5-ditertbutyl-1,4-dimethoxybenzene (DDB), which can be reversibly oxidized at 3.9 V versus Li^+/Li with high rate of charge transfer, is well suited for the $LiFePO_4$ cathode (Fig. 9) [34]. Based on this result, modification to this molecule has led to the discovery of a series of high performance redox shuttles. A change in the structure, for example, adjusting the symmetry [42], using a fused ring system [136], or incorporating oligo (ethylene glycol) units [169] has been shown to increase the solubility of the shuttle additive and hence protect the cell from overcharging at high charging rate and extend the duration of protection. Introducing electron-withdrawing groups on the 1,4-dimethoxy groups [91] or directly onto the benzene ring [168] has been shown to

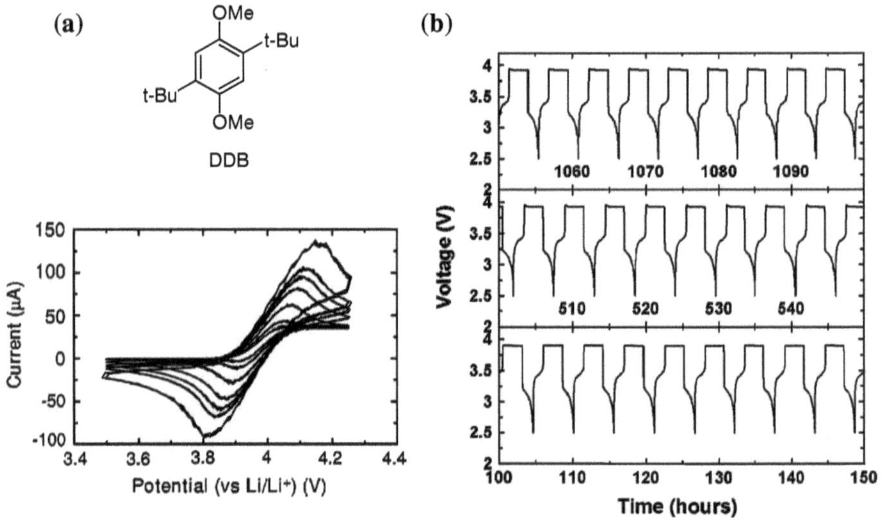

Fig. 9 **a** Cyclic voltammograms of 0.0992 M 2,5-ditertbutyl-1,4-dimethoxybenzene (DDB) additive in 0.5 M LiBOB plus PC:DMC:EC:DEC electrolyte at scanning rates ranging from 5 to 200 mV/s. **b** Potential versus time for selected 50-h portions of the charge-discharge of a $LiFePO_4$/graphite coin cell containing 0.2 M DDB. The cell was charged and discharged at a C/2 rate. The charge was lasted for 4 h, so the cell exhibited about 2.5 h of overcharge per cycle. Reproduced with permission [34]. Copyright 2005 The Electrochemical Society

increase the oxidation potential up to 4.8 V versus Li^+/Li, which is suitable for overcharge protection of 5-V cathode materials.

Other organic redox shuttles based on aromatic compounds with heteroatom substitutions include phenothiazine [18], triphenylamine [89], diarylamines with different substitutions [38], and 2-(pentafluorophenyl)-tetrafluoro-1,3,2-benzodioxaborole [29, 137]. Nitroxide radicals such as (2,2,6,6-Tetramethylpiperidin-1-yl) oxy (TEMPO) have been studied as a redox shuttle as well but showed inferior rate of charge transfer compared with DDB [90]. In addition, lithium borate cluster salts ($Li_2B_{12}F_{12}$) have also been reported to be suitable redox shuttle additives for 4-V lithium ion chemistry (Chen et al. [26]).

4.2 Flame Retardant Additives

Flame retardant additives are added to the electrolyte to inhibit or delay the onset of radical propagation reaction and prevent the spread of fire when a thermal runaway occurs in a battery. Flame retardant additives have been widely used in other areas such as the plastics and coating industry, and battery flame retardant additives are designed with the same principle. However, more stringent requirements are applied to the battery additives, as the additives must be electrochemically inert, and preferably would have a positive effect on the battery performance.

Flame retardant additives generally fall into two categories: halides and phosphorous compounds. In Li-ion batteries, the flame retardant additive design sometimes combines both functionalities into one molecule. The halides generally act by trapping hydrogen radicals to form more stable hydrogen halide, which terminates the propagation of the radical chain reaction. The mechanism behind phosphorous additives is generally more complicated; in addition to terminating the propagation of radical reaction, they also produce heavy phosphorus-oxygen compounds to form a non-flammable layer/cover that isolates the flame from air [112].

There are no reports on halogen compounds as flame retardant additives for Li-ion batteries, possibly due to the side effect of halogens (not including fluorine) acting as unwanted redox shuttles in batteries, with the exception of tetrabromobisphenol A [15]. Phosphorous additives may not be as effective as the halogenated ones, but they are less toxic and more environmentally friendly. Examples of such additives include alkyl phosphates [87, 115, 164], aryl phosphates [43, 117, 133, 176], mixed alkyl aryl phosphates [95, 114], alkyl phosphites [97, 164], alkyl phosphonates [148], phosphonamidate [51], phosphazenes [5, 40], and some other phosphorous compounds [141, 175]. Halogenated phosphorous additives have primarily been fluorinated ones, as many of the fluorinated alkyl solvents themselves are non-flammable. Examples include tris(2,2,2-trifluoroethyl) phosphite [95, 174], tris(2,2,2-trifluoroethyl) phosphate [134, 153], tris(pentafluorophenyl) phosphine [155], and bis(2,2,2-trifluoroethyl) methylphosphonate [165]. Chlorinated tris (2-chloroethyl) phosphate has also been reported to be effective in flame suppression and even enhanced the electrochemical performance of the cell [116]. Besides these organic additives, lithium salts with phosphoryl-rich flame-retardant counter-ions such as Li[B(DPC)(oxalato)], $Li[B(DPC)_2]$, $Li[B(DPC)F_2]$, and $Li[P(DPC)_3]$ (H2-DPC: tetraethyl (2,3-dihydroxy-1,4-phenylene) bis(phosphonate)) have also been reported as effective [105].

Other than the aforementioned compounds, flame retardant additives with other structures could be further explored. Many ionic liquids are known to be non-flammable, and ionic liquids could also be used as additives to lower the flammability of the electrolyte [12, 30]. Nitrogen-containing heterocyclic compounds with fluorinated substitutions have also been reported as new flame retardant additives [61, 62]. Finally, a silane compound, methyl phenyl bismethoxydiethoxysilane, has been reported to serve as both the SEI formation additive and flame retardant additive [79].

4.3 Ion Receptors

Ion receptors are developed to bind to the cations or anions of the lithium salt in the electrolyte to promote ion dissociation and, therefore, increase ionic conductivity of the electrolyte. Although both cation and anion receptors have been shown to improve the ionic conductivity of electrolytes, cationic receptors slow down the mobility of Li^+ and, hence, reduce the Li ion transference number. In contrast, anion receptors not only increase the ionic conductivity, but also increase the Li ionic transference number, which makes them much more preferable in Li-ion batteries.

Due to the negative effect on Li ionic transference number, the study of cation receptors only lasted a short period of time. Crown ethers, due to their ability to bind Li^+, were first studied as cation receptors [88, 92, 93]. The addition of crown ethers improved the solubility of lithium salts and ionic conductivity of the resulting electrolyte, especially with less polar solvents. However, these additives have negative effect on cycle life and the high toxicity of crown ethers also discouraged further investigation.

The concept of anion receptors first appeared in the literature before 2000 with two types of chemical structures, borane/borate with electron-withdrawing substitutions [67, 70, 119, 120, 173] and aza-ethers [68, 69]. From then on, the research has been focused on not only developing new anion receptors, but also expanding the use of known molecules.

As mentioned above, the primary purpose of anion receptors is to increase the ionic conductivity and Li ion transference number, although in some battery systems, the conductivity cannot be improved [73]. However, even in the early days of anion receptor research, researchers found other uses of these interesting molecules. To begin with, anion receptors can greatly increase the solubility of salts that are otherwise almost insoluble in organic electrolytes, such as LiF [70, 119, 138], Li_2O, and Li_2O_2 (Fig. 10) [150]. Electrolytes based on these salts that are traditionally

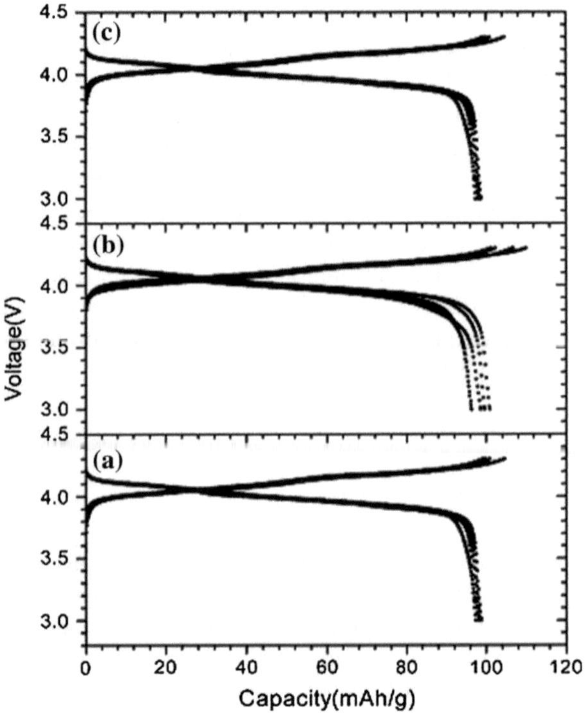

Fig. 10 Charge-discharge curves for Li/$LiMn_2O_4$ cells at room temperature, with various electrolytes: **a** 0.4 M tris(pentafluorophenyl) borane (TPFPB) and 0.2 M Li_2O in PC/DMC (1:1, v/v); **b** 0.4 M TPFPB and 0.2 M Li_2O_2 in PC/DMC (1:1, v/v); and **c** conventional $LiPF_6$-based electrolyte. The cells were cycled between 3 and 4.3 V at 0.2 C. Reproduced with permission [150]. Copyright 2008 Elsevier

insoluble in the organic solvents used in batteries have shown interesting properties. For example, the fluoride anion can be reversibly intercalated into graphite through a soluble fluoride-anion-receptor complex, making graphite as a cathode material. Based on this finding, a dual graphite intercalating cell with discharge voltage over 4.8 V has been realized [138]. Furthermore, anion receptors have been known to stabilize the metallic lithium anode in lithium metal batteries [31, 173]. They have also been reported to increase the thermal stability of the SEI layer [50], increase the power capability of batteries [24], and improve the overall performance of the cell [104]. Some anion receptors can also act as redox shuttles because of their reversible redox reaction at suitable potential ranges [25, 137].

5 Future Directions

Besides the above-mentioned additive categories, other additives aim for enhanced wetting of the electrolyte with separators and electrodes, inhibition of corrosion of electrode current collectors, and many other purposes [172]. In order to optimize the overall performance of the battery, two or more additives may be combined into a battery system. In this case, the compatibility of these additives in one battery system would require massive testing and optimization, and the introduction of new additives may change the whole picture again. These challenges call for a high-throughput combinatorial approach for additive and additive combination screening to be more efficient. While the requirements of labor and equipment might be impractical in academic institutions, pursuing such an approach is highly feasible in an industrial setting. Not surprisingly, companies are already utilizing this approach in developing new additives and have achieved impressive results in a short period of time [74]. It is expected that such an approach will be more widely adopted in material development and cell design in the future, and hopefully accelerate energy storage research in all respects.

References

1. Abe K, Miyoshi K, Hattori T, Ushigoe Y, Yoshitake H (2008) Functional electrolytes: synergetic effect of electrolyte additives for lithium-ion battery. J Power Sources 184:449–455. doi:10.1016/j.jpowsour.2008.03.037
2. Abe K, Yoshitake H, Kitakura T, Hattori T, Wang HY, Yoshio M (2004) Additives-containing functional electrolytes for suppressing electrolyte decomposition in lithium-ion batteries. Electrochim Acta 49:4613–4622. doi:10.1016/j.electacta.2004.05.016
3. Abouimrane A, Odom SA, Tavassol H, Schulmerich MV, Wu H, Bhargava R, Gewirth AA, Moore JS, Amine K (2013) 3-Hexylthiophene as a stabilizing additive for high voltage cathodes in lithium-ion batteries. J Electrochem Soc 160:A268–A271. doi:10.1149/2.039302jes
4. Abraham KM, Pasquariello DM, Willstaedt EB (1990) n-Butylferrocene for overcharge protection of secondary lithium batteries. J Electrochem Soc 137:1856–1857. doi:10.1149/1.2086817

5. Ahn S, Kim H-S, Yang S, Do JY, Kim BH, Kim K (2009) Thermal stability and performance studies of $LiCo_{1/3}Ni_{1/3}Mn_{1/3}O_2$ with phosphazene additives for Li-ion batteries. J Electroceram 23:289–294. doi:10.1007/s10832-008-9437-y
6. An Y, Zuo P, Du C, Ma Y, Cheng X, Lin J, Yin G (2012) Effects of VC-LiBOB binary additives on SEI formation in ionic liquid-organic composite electrolyte. RSC Adv 2:4097–4102. doi:10.1039/c2ra01040b
7. Andersson AM, Edstrom K (2001) Chemical composition and morphology of the elevated temperature SEI on graphite. J Electrochem Soc 148:A1100–A1109. doi:10.1149/1.1397771
8. Applestone D, Manthiram A (2012) Symmetric cell evaluation of the effects of electrolyte additives on Cu_2Sb-Al_2O_3-C nanocomposite anodes. J Power Sources 217:1–5. doi:10.1016/j.jpowsour.2012.05.119
9. Aravindan V, Cheah YL, Ling WC, Madhavi S (2012) Effect of LiBOB additive on the electrochemical performance of $LiCoPO_4$. J Electrochem Soc 159:A1435–A1439. doi:10.1149/2.024209jes
10. Aurbach D, Gamolsky K, Markovsky B, Gofer Y, Schmidt M, Heider U (2002) On the use of vinylene carbonate (VC) as an additive to electrolyte solutions for Li-ion batteries. Electrochim Acta 47:1423–1439. doi:10.1016/s0013-4686(01)00858-1
11. Aurbach D, Markovsky B, Salitra G, Markevich E, Talyossef Y, Koltypin M, Nazar L, Ellis B, Kovacheva D (2007) Review on electrode-electrolyte solution interactions, related to cathode materials for Li-ion batteries. J Power Sources 165:491–499. doi:10.1016/j.jpowsour.2006.10.025
12. Bae S-Y, Shim E-G, Kim D-W (2013) Effect of ionic liquid as a flame-retarding additive on the cycling performance and thermal stability of lithium-ion batteries. J Power Sources 244:266–271. doi:10.1016/j.jpowsour.2013.01.100
13. Bae S-Y, Shin W-K, Kim D-W (2014) Protective organic additives for high voltage $LiNi_{0.5}Mn_{1.5}O_4$ cathode materials. Electrochim Acta 125:497–502. doi:10.1016/j.electacta.2014.01.124
14. Belharouak I, Koenig GM Jr, Tan T, Yumoto H, Ota N, Amine K (2012) Performance degradation and gassing of $Li_4Ti_5O_{12}/LiMn_2O_4$ lithium-ion cells. J Electrochem Soc 159:A1165–A1170. doi:10.1149/2.013208jes
15. Belov DG, Shieh DT (2014) A study of tetrabromobisphenol A (TBBA) as a flame retardant additive for Li-ion battery electrolytes. J Power Sources 247:865–875. doi:10.1016/j.jpowsour.2013.08.143
16. Bouayad H, Wang Z, Dupre N, Dedryvere R, Foix D, Franger S, Martin JF, Boutafa L, Patoux S, Gonbeau D, Guyomard D (2014) Improvement of electrode/electrolyte interfaces in high-voltage spinel lithium-ion batteries by using glutaric anhydride as electrolyte additive. J Phys Chem C 118:4634–4648. doi:10.1021/jp5001573
17. Buhrmester C, Chen J, Moshurchak L, Jiang JW, Wang RL, Dahn JR (2005) Studies of aromatic redox shuttle additives for $LiFePO_4$-based Li-ion cells. J Electrochem Soc 152:A2390–A2399. doi:10.1149/1.2098265
18. Buhrmester C, Moshurchak L, Wang RL, Dahn JR (2006) Phenothiazine molecules-possible redox shuttle additives for chemical overcharge and overdischarge protection for lithium-ion batteries. J Electrochem Soc 153:A288–A294. doi:10.1149/1.2140615
19. Buiel E, Dahn JR (1999) Li-insertion in hard carbon anode materials for Li-ion batteries. Electrochim Acta 45:121–130. doi:10.1016/s0013-4686(99)00198-x
20. Cha CS, Ai XP, Yang HX (1995) Polypyridine complexes of iron used as redox shuttles for overcharge protection of secondary lithium batteries. J Power Sources 54:255–258. doi:10.1016/0378-7753(94)02079-i
21. Chen L, Wang K, Xie X, Xie J (2006) Enhancing electrochemical performance of silicon film anode by vinylene carbonate electrolyte additive. Electrochem Solid-State Lett 9:A512–A515. doi:10.1149/1.2338771
22. Chen L, Wang K, Xie X, Xie J (2007) Effect of vinylene carbonate (VC) as electrolyte additive on electrochemical performance of Si film anode for lithium ion batteries. J Power Sources 174:538–543. doi:10.1016/j.jpowsour.2007.06.149

23. Chen X, Li X, Mei D, Feng J, Hu MY, Hu J, Engelhard M, Zheng J, Xu W, Xiao J, Liu J, Zhang J-G (2014) Reduction mechanism of fluoroethylene carbonate for stable solid-electrolyte interphase film on silicon anode. Chemsuschem 7:549–554. doi:10.1002/cssc.201300770
24. Chen Z, Amine K (2006) Tris(pentafluorophenyl) borane as an additive to improve the power capabilities of lithium-ion batteries. J Electrochem Soc 153:A1221–A1225. doi:10.1149/1.2194633
25. Chen Z, Amine K (2007) Bifunctional electrolyte additive for lithium-ion batteries. Electrochem Commun 9:703–707. doi:10.1016/j.elecom.2006.11.002
26. Chen Z, Liu J, Jansen AN, GirishKumar G, Casteel B, Amine K (2010) Lithium borate cluster salts as redox shuttles for overcharge protection of lithium-ion cells. Electrochem Solid-State Lett 13:A39–A42. doi:10.1149/1.3299251
27. Chen Z, Qin Y, Amine K (2009) Redox shuttles for safer lithium-ion batteries. Electrochim Acta 54:5605–5613. doi:10.1016/j.electacta.2009.05.017
28. Chen Z, Qin Y, Amine K (2010) Chemical overcharge protection of lithium-ion cells. Nova Science Publishers, Inc., New York, pp 119–146
29. Chen ZH, Amine K (2007) Bifunctional electrolyte additive for lithium-ion batteries. Electrochem Commun 9:703–707. doi:10.1016/j.elecom.2006.11.002
30. Choi J-A, Sun Y-K, Shim E-G, Scrosati B, Kim D-W (2011) Effect of 1-butyl-1-methylpyrrolidinium hexafluorophosphate as a flame-retarding additive on the cycling performance and thermal properties of lithium-ion batteries. Electrochim Acta 56:10179–10184. doi:10.1016/j.electacta.2011.09.009
31. Choi N-S, Lee YM, Cho KY, Ko D-H, Park J-K (2004) Protective layer with oligo(ethylene glycol) borate anion receptor for lithium metal electrode stabilization. Electrochem Commun 6:1238–1242. doi:10.1016/j.elecom.2004.09.023
32. Choi N-S, Yew KH, Lee KY, Sung M, Kim H, Kim S-S (2006) Effect of fluoroethylene carbonate additive on interfacial properties of silicon thin-film electrode. J Power Sources 161:1254–1259. doi:10.1016/j.jpowsour.2006.05.049
33. Chung G-C, Kim H-J, Yu S-I, Jun S-H, Choi J-W, Kim M-H (2000) Origin of graphite exfoliation; an investigation of the important role of solvent cointercalation. J Electrochem Soc 147:4391–4398. doi:10.1149/1.1394076
34. Dahn JR, Jiang J, Moshurchak LM, Fleischauer MD, Buhrmester C, Krause LJ (2005) High-rate overcharge protection of $LiFePO_4$-based Li-ion cells using the redox shuttle additive 2,5-ditertbutyl-1,4-dimethoxybenzene. J Electrochem Soc 152:A1283–A1289. doi:10.1149/1.1906025
35. Dalavi S, Guduru P, Lucht BL (2012) Performance enhancing electrolyte additives for lithium ion batteries with silicon anodes. J Electrochem Soc 159:A642–A646. doi:10.1149/2.076205jes
36. El Ouatani L, Dedryvere R, Siret C, Biensan P, Reynaud S, Iratcabal P, Gonbeau D (2009) The effect of vinylene carbonate additive on surface film formation on both electrodes in Li-ion batteries. J Electrochem Soc 156:A103–A113. doi:10.1149/1.3029674
37. Elazari R, Salitra G, Gershinsky G, Garsuch A, Panchenko A, Aurbach D (2012) Li ion cells comprising lithiated columnar silicon film anodes, TiS2 cathodes and fluoroethylene carbonate (FEC) as a critically important component. J Electrochem Soc 159:A1440–A1445. doi:10.1149/2.029209jes
38. Ergun S, Elliott CF, Kaur AP, Parkin SR, Odom SA (2014) Controlling oxidation potentials in redox shuttle candidates for lithium-ion batteries. J Phys Chem C 118:14824–14832. doi:10.1021/jp503767h
39. Etacheri V, Haik O, Goffer Y, Roberts GA, Stefan IC, Fasching R, Aurbach D (2012) Effect of fluoroethylene carbonate (FEC) on the performance and surface chemistry of Si-nanowire Li-ion battery anodes. Langmuir 28:965–976. doi:10.1021/la203712s
40. Fei S-T, Allcock HR (2010) Methoxyethoxyethoxyphosphazenes as ionic conductive fire retardant additives for lithium battery systems. J Power Sources 195:2082–2088. doi:10.1016/j.jpowsour.2009.09.043

41. Felix Cheng J-H, Hy S, Rick J, Wang F-M, Hwang B-J (2013) Mechanistic basis of enhanced capacity retention found with novel sulfate-based additive in high-voltage Li-ion batteries. J Phys Chem C 117:22619–22626. doi:10.1021/jp409779x
42. Feng JK, Ai XP, Cao YL, Yang HX (2006) A highly soluble dimethoxybenzene derivative as a redox shuttle for overcharge protection of secondary lithium batteries. Electrochem Commun 9:25–30. doi:10.1016/j.elecom.2006.08.033
43. Feng JK, Cao YL, Ai XP, Yang HX (2008) Tri-(4-methoxyphenyl) phosphate: a new electrolyte additive with both fire-retardancy and overcharge protection for Li-ion batteries. Electrochim Acta 53:8265–8268. doi:10.1016/j.electacta.2008.05.024
44. Feng XM, Ai XP, Yang HX (2004) Possible use of methylbenzenes as electrolyte additives for improving the overcharge tolerances of Li-ion batteries. J Appl Electrochem 34:1199–1203. doi:10.1007/s10800-004-0771-8
45. Fridman K, Sharabi R, Elazari R, Gershinsky G, Markevich E, Salitra G, Aurbach D, Garsuch A, Lampert J (2013) A new advanced lithium ion battery: combination of high performance amorphous columnar silicon thin film anode, 5 V $LiNi_{0.5}Mn_{1.5}O_4$ spinel cathode and fluoroethylene carbonate-based electrolyte solution. Electrochem Commun 33:31–34. doi:10.1016/j.elecom.2013.04.010
46. Ha S-Y, Han J-G, Song Y-M, Chun M-J, Han S-I, Shin W-C, Choi N-S (2013) Using a lithium bis(oxalato) borate additive to improve electrochemical performance of high-voltage spinel $LiNi_{0.5}Mn_{1.5}O_4$ cathodes at 60 °C. Electrochim Acta 104:170–177. doi:10.1016/j.electacta.2013.04.082
47. Han G-B, Lee J-N, Choi JW, Park J-K (2011) Tris(pentafluorophenyl) borane as an electrolyte additive for high performance silicon thin film electrodes in lithium ion batteries. Electrochim Acta 56:8997–9003. doi:10.1016/j.electacta.2011.07.136
48. He Y-B, Li B, Liu M, Zhang C, Lv Lv, Yang C, Li J, Du H, Zhang B, Yang Q-H, Kim J-K, Kang F (2012) Gassing in $Li_4Ti_5O_{12}$-based batteries and its remedy. Sci Rep 2:00919. doi:10.1038/srep00913
49. He Y-B, Liu M, Huang Z-D, Zhang B, Yu Y, Li B, Kang F, Kim J-K (2013) Effect of solid electrolyte interface (SEI) film on cyclic performance of $Li_4Ti_5O_{12}$ anodes for Li ion batteries. J Power Sources 239:269–276. doi:10.1016/j.jpowsour.2013.03.141
50. Herstedt M, Stjerndahl M, Gustafsson T, Edstrom K (2003) Anion receptor for enhanced thermal stability of the graphite anode interface in a Li-ion battery. Electrochem Commun 5:467–472. doi:10.1016/s1388-2481(03)00106-1
51. Hu J, Jin Z, Zhong H, Zhan H, Zhou Y, Li Z (2012) A new phosphonamidate as flame retardant additive in electrolytes for lithium ion batteries. J Power Sources 197:297–300. doi:10.1016/j.jpowsour.2011.09.012
52. Hu L, Xue Z, Amine K, Zhang Z (2014) Fluorinated electrolytes for 5-V Li-ion chemistry: synthesis and evaluation of an additive for high-voltage $LiNi_{0.5}Mn_{1.5}O_4$/graphite cell. J Electrochem Soc 161:A1777–A1781. doi:10.1149/2.0141412jes
53. Hu M, Wei J, Xing L, Zhou Z (2012) Effect of lithium difluoro(oxalate)borate (LiDFOB) additive on the performance of high-voltage lithium-ion batteries. J Appl Electrochem 42:291–296. doi:10.1007/s10800-012-0398-0
54. Hu Y, Kong W, Li H, Huang X, Chen L (2004) Experimental and theoretical studies on reduction mechanism of vinyl ethylene carbonate on graphite anode for lithium ion batteries. Electrochem Commun 6:126–131. doi:10.1016/j.elecom.2003.10.024
55. Huang W, Xing L, Wang Y, Xu M, Li W, Xie F, Xia S (2014) 4-(Trifluoromethyl)-benzonitrile: a novel electrolyte additive for lithium nickel manganese oxide cathode of high voltage lithium ion battery. J Power Sources 267:560–565. doi:10.1016/j.jpowsour.2014.05.124
56. Iwayasu N, Honbou H, Horiba T (2011) Overcharge protection effect and reaction mechanism of cyclohexylbenzene for lithium ion batteries. J Power Sources 196:3881–3886. doi:10.1016/j.jpowsour.2010.12.082
57. Jang DH, Shin YJ, Oh SM (1996) Dissolution of spinel oxides and capacity losses in 4 V Li/LixMn2O4 coils. J Electrochem Soc 143:2204–2211. doi:10.1149/1.1836981

58. Jeong S-K, Inaba M, Mogi R, Iriyama Y, Abe T, Ogumi Z (2001) Surface film formation on a graphite negative electrode in lithium-ion batteries: atomic force microscopy study on the effects of film-forming additives in propylene carbonate solutions. Langmuir 17:8281–8286. doi:10.1021/la015553h
59. Jung HM, Park S-H, Jeon J, Choi Y, Yoon S, Cho J-J, Oh S, Kang S, Han Y-K, Lee H (2013) Fluoropropane sultone as an SEI-forming additive that outperforms vinylene carbonate. J Mater Chem A 1:11975–11981. doi:10.1039/c3ta12580g
60. Jung SC, Choi JW, Han Y-K (2012) Anisotropic volume expansion of crystalline silicon during electrochemical lithium insertion: an atomic level rationale. Nano Lett 12:5342–5347. doi:10.1021/nl3027197
61. Kam D, Kim K, Kim H-S, Liu HK (2009) Studies on film formation on cathodes using pyrazole derivatives as electrolyte additives in the Li-ion battery. Electrochem Commun 11:1657–1660. doi:10.1016/j.elecom.2009.06.020
62. Kim K, Ahn S, Kim H-S, Liu HK (2009) Electrochemical and thermal properties of 2,4,6-tris (trifluoromethyl)-1,3,5-triazine as a flame retardant additive in Li-ion batteries. Electrochim Acta 54:2259–2265. doi:10.1016/j.electacta.2008.10.043
63. Komaba S, Itabashi T, Ohtsuka T, Groult H, Kumagai N, Kaplan B, Yashiro H (2005) Impact of 2-vinylpyridine as electrolyte additive on surface and electrochemistry of graphite for C/LiMn2O4 Li-ion cells. J Electrochem Soc 152:A937–A946. doi:10.1149/1.1885385
64. Komaba S, Kumagai N, Kataoka Y (2002) Influence of manganese(II), cobalt(II), and nickel (II) additives in electrolyte on performance of graphite anode for lithium-ion batteries. Electrochim Acta 47:1229–1239. doi:10.1016/s0013-4686(01)00847-7
65. Korepp C, Kern W, Lanzer EA, Raimann PR, Besenhard JO, Yang M, Moeller KC, Shieh DT, Winter M (2007) Ethyl isocyanate—an electrolyte additive from the large family of isocyanates for PC-based electrolytes in lithium-ion batteries. J Power Sources 174:628–631. doi:10.1016/j.jpowsour.2007.06.140
66. Lee H, Choi S, Choi S, Kim H-J, Choi Y, Yoon S, Cho J-J (2007) SEI layer-forming additives for $LiNi_{0.5}Mn_{1.5}O_4$/graphite 5 V Li-ion batteries. Electrochem Commun 9:801–806. doi:10.1016/j.elecom.2006.11.008
67. Lee HS, Sun X, Yang XQ, McBreen J (2002) Synthesis and study of new cyclic boronate additives for lithium battery electrolytes. J Electrochem Soc 149:A1460–A1465. doi:10.1149/1.1513559
68. Lee HS, Sun X, Yang XQ, McBreen J, Callahan JH, Choi LS (2000) Synthesis of cyclic aza-ether compounds and studies of their use as anion receptors in nonaqueous lithium halide salts solution. J Electrochem Soc 147:9–14. doi:10.1149/1.1493856
69. Lee HS, Yang XQ, McBreen J, Okamoto Y, Choi LS (1995) A new family of anion receptors and their effect on ion pair dissociation and conductivity of lithium salts in non-aqueous solutions. Electrochim Acta 40:2353–2356. doi:10.1016/0013-4686(95)00192-h
70. Lee HS, Yang XQ, Xiang CL, McBreen J (1998) The synthesis of a new family of boron-based anion receptors and the study of their effect on ion pair dissociation and conductivity of lithium salts in nonaqueous solutions. J Electrochem Soc 145:2813–2818. doi:10.1149/1.1838719
71. Lee JT, Lin YW, Jan YS (2004) Allyl ethyl carbonate as an additive for lithium-ion battery electrolytes. J Power Sources 132:244–248. doi:10.1016/j.jpowsour.2004.01.045
72. Lee Y-S, Lee K-S, Sun Y-K, Lee YM, Kim D-W (2011) Effect of an organic additive on the cycling performance and thermal stability of lithium-ion cells assembled with carbon anode and $LiNi_{1/3}Co_{1/3}Mn_{1/3}O_2$ cathode. J Power Sources 196:6997–7001. doi:10.1016/j.jpowsour.2010.10.047
73. Lee YM, Seo JE, Choi N-S, Park J-K (2005) Influence of tris(pentafluorophenyl) borane as an anion receptor on ionic conductivity of $LiClO_4$-based electrolyte for lithium batteries. Electrochim Acta 50:2843–2848. doi:10.1016/j.electacta.2004.11.058
74. Li B, Bhat V, Shan J, Cheng G, Yang J-H, O'Neill C, Caldwell MA, Tong W, Kaye SS (2012) High throughput synthesis and screening for discovery of improved electrode materials for lithium-ion batteries. American Chemical Society, Washington DC, pp PETR-4

75. Li B, Wang Y, Rong H, Wang Y, Liu J, Xing L, Xu M, Li W (2013) A novel electrolyte with the ability to form a solid electrolyte interface on the anode and cathode of a $LiMn_2O_4$/graphite battery. J Mater Chem A 1:12954–12961. doi:10.1039/c3ta13067c
76. Li B, Xu M, Li B, Liu Y, Yang L, Li W, Hu S (2013) Properties of solid electrolyte interphase formed by prop-1-ene-1,3-sultone on graphite anode of Li-ion batteries. Electrochim Acta 105:1–6. doi:10.1016/j.electacta.2013.04.142
77. Li J, Yao W, Meng YS, Yang Y (2008) Effects of vinyl ethylene carbonate additive on elevated-temperature performance of cathode material in lithium ion batteries. J Phys Chem C 112:12550–12556. doi:10.1021/jp800336n
78. Li LF, Lee HS, Li H, Yang XQ, Huang XJ (2009) A pentafluorophenylboron oxalate additive in non-aqueous electrolytes for lithium batteries. Electrochem Commun 11:2296–2299. doi:10.1016/j.elecom.2009.10.015
79. Li LL, Li L, Wang B, Liu LL, Wu YP, van Ree T, Thavhiwa KA (2011) Methyl phenyl bis-methoxydiethoxysilane as bi-functional additive to propylene carbonate-based electrolyte for lithium ion batteries. Electrochim Acta 56:4858–4864. doi:10.1016/j.electacta.2011.02.117
80. Li Y, Xu G, Yao Y, Xue L, Zhang S, Lu Y, Toprakci O, Zhang X (2013) Improvement of cyclability of silicon-containing carbon nanofiber anodes for lithium-ion batteries by employing succinic anhydride as an electrolyte additive. J Solid State Electrochem 17:1393–1399. doi:10.1007/s10008-013-2005-7
81. Li Y, Zhang R, Liu J, Yang C (2009) Effect of heptamethyldisilazane as an additive on the stability performance of $LiMn_2O_4$ cathode for lithium-ion battery. J Power Sources 189:685–688. doi:10.1016/j.jpowsour.2008.08.075
82. Liu J, Chen Z, Busking S, Belharouak I, Amine K (2007) Effect of electrolyte additives in improving the cycle and calendar life of graphite/$Li_{1.1}[Ni_{1/3}Co_{1/3}Mn_{1/3}]_{0.9}O_2$ Li-ion cells. J Power Sources 174:852–855. doi:10.1016/j.jpowsour.2007.06.225
83. Liu JY, Liu N, Liu DT, Bai Y, Shi LH, Wang ZX, Chen LQ, Hennige V, Schuch A (2007) Improving the performances of $LiCoO_2$ cathode materials by soaking nano-alumina in commercial electrolyte. J Electrochem Soc 154:A55–A63. doi:10.1149/1.2388731
84. Liu Y, Tan L, Li L (2013) Tris(trimethylsilyl) borate as an electrolyte additive to improve the cyclability of $LiMn_2O_4$ cathode for lithium-ion battery. J Power Sources 221:90–96. doi:10.1016/j.jpowsour.2012.08.028
85. Lu DS, Xu MQ, Zhou L, Garsuch A, Lucht BL (2013) Failure mechanism of graphite/$LiNi_{0.5}Mn_{1.5}O_4$ cells at high voltage and elevated temperature. J Electrochem Soc 160:A3138–A3143. doi:10.1149/2.022305jes
86. Martinez de la Hoz JM, Balbuena PB (2014) Reduction mechanisms of additives on Si anodes of Li-ion batteries. Phys Chem Chem Phys: Ahead of Print. doi:10.1039/c4cp01948b
87. Matsumoto K, Nakahara K, Inoue K, Iwasa S, Nakano K, Kaneko S, Ishikawa H, Utsugi K, Yuge R (2014) Performance improvement of Li ion battery with non-flammable TMP mixed electrolyte by optimization of lithium salt concentration and SEI preformation technique on graphite anode. J Electrochem Soc 161:A831–A834. doi:10.1149/2.091405jes
88. Morita M, Hayashida H, Matsuda Y (1987) Effects of crown-ether addition to organic electrolytes on the cycling behavior of the TiS_2 electrode. J Electrochem Soc 134:2107–2111. doi:10.1149/1.2100833
89. Moshurchak LM, Buhrmester C, Dahn JR (2008) Triphenylamines as a class of redox shuttle molecules for the overcharge protection of lithium-ion cells. J Electrochem Soc 155:A129–A131. doi:10.1149/1.2816229
90. Moshurchak LM, Buhrmester C, Wang RL, Dahn JR (2007) Comparative studies of three redox shuttle molecule classes for overcharge protection of $LiFePO_4$-based Li-ion cells. Electrochim Acta 52:3779–3784. doi:10.1016/j.electacta.2006.10.068
91. Moshurchak LM, Lamanna WM, Bulinski M, Wang RL, Garsuch RR, Jiang J, Magnuson D, Triemert M, Dahn JR (2009) High-potential redox shuttle for use in lithium-ion batteries. J Electrochem Soc 156:A309–A312. doi:10.1149/1.3077578

92. Nagasubramanian G, Attia AI, Halpert G (1992) Effects of 12-crown-4 ether on the electrochemical performance of CoO_2 and TiS_2 cathodes in Li polymer electrolyte cells. J Electrochem Soc 139:3043–3046. doi:10.1149/1.2069030
93. Nagasubramanian G, Distefano S (1990) 12-Crown-4 ether-assisted enhancement of ionic-conductivity and interfacial kinetics in polyethylene oxide electrolytes. J Electrochem Soc 137:3830–3835. doi:10.1149/1.2086309
94. Nakai H, Kubota T, Kita A, Kawashima A (2011) Investigation of the solid electrolyte interphase formed by fluoroethylene carbonate on Si electrodes. J Electrochem Soc 158: A798–A801. doi:10.1149/1.3589300
95. Nam T-H, Shim E-G, Kim J-G, Kim H-S, Moon S-I (2008) Diphenyloctyl phosphate and tris (2,2,2-trifluoroethyl) phosphite as flame-retardant additives for Li-ion cell electrolytes at elevated temperature. J Power Sources 180:561–567. doi:10.1016/j.jpowsour.2008.01.061
96. Narayanan SR, Surampudi S, Attia AI, Bankston CP (1991) Analysis of redox additive-based overcharge protection for rechargeable lithium batteries. J Electrochem Soc 138:2224–2229. doi:10.1149/1.2085954
97. Nguyen DN, Park IJ, Kim JG (2012) Triethyl and tributyl phosphite as flame-retarding additives in Li-ion batteries. Met Mater Int 18:189–196. doi:10.1007/s12540-012-0025-y
98. Park I-J, Nam T-H, Kim J-G (2013) Diphenyloctyl phosphate as a solid electrolyte interphase forming additive for Li-ion batteries. J Power Sources 244:122–128. doi:10.1016/j.jpowsour. 2013.03.031
99. Park Y, Shin SH, Hwang H, Lee SM, Kim SP, Choi HC, Jung YM (2014) Investigation of solid electrolyte interface (SEI) film on $LiCoO_2$ cathode in fluoroethylene carbonate (FEC)-containing electrolyte by 2D correlation X-ray photoelectron spectroscopy (XPS). J Mol Struct 1069:157–163. doi:10.1016/j.molstruc.2014.01.041
100. Peled E (1979) The electrochemical-behavior of alkali and alkaline-earth metals in non-aqueous battery systems—the solid electrolyte interphase model. J Electrochem Soc 126:2047–2051. doi:10.1149/1.2128859
101. Pieczonka NPW, Liu ZY, Lu P, Olson KL, Moote J, Powell BR, Kim JH (2013) Understanding transition-metal dissolution behavior in LiNi0.5Mn1.5O4 high-voltage spinel for lithium ion batteries. J Phys Chem C 117:15947–15957. doi:10.1021/Jp405158m
102. Pieczonka NPW, Yang L, Balogh MP, Powell BR, Chemelewski K, Manthiram A, Krachkovskiy SA, Goward GR, Liu M, Kim J-H (2013) Impact of lithium bis(oxalate)borate electrolyte additive on the performance of high-voltage spinel/graphite Li-ion batteries. J Phys Chem C 117:22603–22612. doi:10.1021/jp408717x
103. Profatilova IA, Stock C, Schmitz A, Passerini S, Winter M (2013) Enhanced thermal stability of a lithiated nano-silicon electrode by fluoroethylene carbonate and vinylene carbonate. J Power Sources 222:140–149. doi:10.1016/j.jpowsour.2012.08.066
104. Qin Y, Chen Z, Lee HS, Yang XQ, Amine K (2010) Effect of anion receptor additives on electrochemical performance of lithium-ion batteries. J Phys Chem C 114:15202–15206. doi:10.1021/jp104341t
105. Rectenwald MF, Gaffen JR, Rheingold AL, Morgan AB, Protasiewicz JD (2014) Phosphoryl-rich flame-retardant ions (FRIONs): towards safer lithium-ion batteries. Angew Chem Int Ed 53:4173–4176. doi:10.1002/anie.201310867
106. Ronci F, Reale P, Scrosati B, Panero S, Rossi Albertini V, Perfetti P, di Michiel M, Merino JM (2002) High-resolution in-situ structural measurements of the $Li_{4/3}Ti_{5/3}O_4$ "zero-strain" insertion material. J Phys Chem B 106:3082–3086. doi:10.1021/jp013240p
107. Ryou M-H, Lee J-N, Lee DJ, Kim W-K, Choi JW, Park J-K, Lee YM (2013) 2-(triphenylphosphoranylidene)succinic anhydride as a new electrolyte additive to improve high temperature cycle performance of $LiMn_2O_4$/graphite Li-ion batteries. Electrochim Acta 102:97–103. doi:10.1016/j.electacta.2013.03.129
108. Ryu Y-G, Lee S, Mah S, Lee DJ, Kwon K, Hwang S, Doo S (2008) Electrochemical behavior of silicon electrodes in lithium salt solutions containing alkoxy silane additives. J Electrochem Soc 155:A583–A589. doi:10.1149/1.2940310

109. Santhanam R, Rambabu B (2010) Research progress in high voltage spinel $LiNi_{0.5}Mn_{1.5}O_4$ material. J Power Sources 195:5442–5451. doi:10.1016/j.jpowsour.2010.03.067
110. Santner HJ, Moller KC, Ivanco J, Ramsey MG, Netzer FP, Yamaguchi S, Besenhard JO, Winter M (2003) Acrylic acid nitrile, a film-forming electrolyte component for lithium-ion batteries, which belongs to the family of additives containing vinyl groups. J Power Sources 119–121:368–372. doi:10.1016/s0378-7753(03)00268-4
111. Sharabi R, Markevich E, Fridman K, Gershinsky G, Salitra G, Aurbach D, Semrau G, Schmidt MA, Schall N, Bruenig C (2013) Electrolyte solution for the improved cycling performance of $LiCoPO_4/C$ composite cathodes. Electrochem Comm 28:20–23. doi:10.1016/j.elecom.2012.12.001
112. Schartel B (2010) Phosphorus-based flame retardancy mechanisms—old hat or a starting point for future development? Materials 3:4710–4745. doi:10.3390/ma3104710
113. Schroeder G, Gierczyk B, Waszak D, Kopczyk M, Walkowiak M (2006) Vinyl tris-2-methoxyethoxy silane—a new class of film-forming electrolyte components for Li-ion cells with graphite anodes. Electrochem Commun 8:523–527. doi:10.1016/j.elecom.2006.01.021
114. Shim E-G, Nam T-H, Kim J-G, Kim H-S, Moon S-I (2008) Diphenyloctyl phosphate as a flame-retardant additive in electrolyte for Li-ion batteries. J Power Sources 175:533–539. doi:10.1016/j.jpowsour.2007.08.098
115. Shim E-G, Nam T-H, Kim J-G, Kim H-S, Moon S-I (2009) Effects of trioctyl phosphate and cresyl diphenyl phosphate as flame-retarding additives for Li-ion battery electrolytes. Met Mater Int 15:615–621. doi:10.1007/s12540-009-0615-5
116. Shim E-G, Park I-J, Nam T-H, Kim J-G, Kim H-S, Moon S-I (2010) Electrochemical performance of tris(2-chloroethyl) phosphate as a flame-retarding additive for lithium-ion batteries. Met Mater Int 16:587–594. doi:10.1007/s12540-010-0811-3
117. Smart MC, Krause FC, Hwang C, West WC, Soler J, Prakash GKS, Ratnakumar BV (2011) The evaluation of triphenyl phosphate as a flame retardant additive to improve the safety of lithium-ion battery electrolytes. ECS Trans 35:1–11. doi:10.1149/1.3646164
118. Song S-W, Baek S-W (2009) Silane-derived SEI stabilization on thin-film electrodes of nanocrystalline Si for lithium batteries. Electrochem Solid-State Lett 12:A23–A27. doi:10.1149/1.3028216
119. Sun X, Lee HS, Lee S, Yang XQ, McBreen J (1998) A novel lithium battery electrolyte based on lithium fluoride and a tris(pentafluorophenyl) borane anion receptor in DME. Electrochem Solid-State Lett 1:239–240. doi:10.1149/1.1390698
120. Sun X, Lee HS, Yang XQ, McBreen J (1999) Comparative studies of the electrochemical and thermal stability of two types of composite lithium battery electrolytes using boron-based anion receptors. J Electrochem Soc 146:3655–3659. doi:10.1149/1.1392529
121. Tan S, Zhang Z, Li Y, Li Y, Zheng J, Zhou Z, Yang Y (2013) Tris(hexafluoro-iso-propyl) phosphate as an SEI-forming additive on improving the electrochemical performance of the $Li[Li_{0.2}Mn_{0.56}Ni_{0.16}Co_{0.08}]O_2$ cathode material. J Electrochem Soc 160:A285–A292. doi:10.1149/2.066302jes
122. Tarnopolskiy V, Kalhoff J, Nadherna M, Bresser D, Picard L, Fabre F, Rey M, Passerini S (2013) Beneficial influence of succinic anhydride as electrolyte additive on the self-discharge of 5 V $LiNi_{0.4}Mn_{1.6}O_4$ cathodes. J Power Sources 236:39–46. doi:10.1016/j.jpowsour.2013.02.030
123. Tasaki K, Kanda K, Kobayashi T, Nakamura S, Ue M (2006) Theoretical studies on the reductive decompositions of solvents and additives for lithium-ion batteries near lithium anodes. J Electrochem Soc 153:A2192–A2197. doi:10.1149/1.2354460
124. Taubert C, Fleischhammer M, Wohlfahrt-Mehrens M, Wietelmann U, Buhrmester T (2010) LiBOB as electrolyte salt or additive for lithium-ion batteries based on LiNi0.8Co0.15Al0.05O2/graphite. J Electrochem Soc 157:A721–A728. doi:10.1149/1.3374666
125. Ufheil J, Baertsch MC, Wuersig A, Novak P (2005) Maleic anhydride as an additive to γ-butyrolactone solutions for Li-ion batteries. Electrochim Acta 50:1733–1738. doi:10.1016/j.electacta.2004.10.061

126. Ushirogata K, Sodeyama K, Okuno Y, Tateyama Y (2013) Additive effect on reductive decomposition and binding of carbonate-based solvent toward solid electrolyte interphase formation in lithium-ion battery. J Am Chem Soc 135:11967–11974. doi:10.1021/ja405079s
127. von Cresce A, Xu K (2011) Electrolyte additive in support of 5 V Li ion chemistry. J Electrochem Soc 158:A337–A342. doi:10.1149/1.3532047
128. Wagner R, Brox S, Kasnatscheew J, Gallus DR, Amereller M, Cekic-Laskovic I, Winter M (2014) Vinyl sulfones as SEI-forming additives in propylene carbonate based electrolytes for lithium-ion batteries. Electrochem Commun 40:80–83. doi:10.1016/j.elecom.2014.01.004
129. Wang C, Nakamura H, Komatsu H, Noguchi H, Yoshio M, Yoshitake H (1998) Suppression of electrochemical decomposition of propylene carbonate (PC) on a graphite anode in PC base electrolyte with catechol carbonate. Denki Kagaku oyobi Kogyo Butsuri Kagaku 66:286–292
130. Wang E, Ofer D, Bowden W, Iltchev N, Moses R, Brandt K (2000) Stability of lithium ion spinel cells III. Improved life of charged cells. J Electrochem Soc 147:4023–4028. doi:10.1149/1.1394013
131. Wang F-M, Yu M-H, Cheng C-S, Pradanawati SA, Lo S-C, Rick J (2013) Phenylenedimaleimide positional isomers used as lithium ion battery electrolyte additives: relating physical and electrochemical characterization to battery performance. J Power Sources 231:18–22. doi:10.1016/j.jpowsour.2012.12.093
132. Wang Q, Sun J, Chen C (2009) Improved thermal stability of graphite electrodes in lithium-ion batteries using 4-isopropyl phenyl diphenyl phosphate as an additive. J Appl Electrochem 39:1105–1110. doi:10.1007/s10800-008-9765-2
133. Wang Q, Sun J, Yao X, Chen C (2005) 4-Isopropyl phenyl diphenyl phosphate as flame-retardant additive for lithium-ion battery electrolyte. Electrochem Solid-State Lett 8: A467–A470. doi:10.1149/1.1993389
134. Wang W, Wang S, He Y, Yang X, Guo H (2013b) Tris(2,2,2-trifluoroethyl)phosphate (TFP) as flame-retarded additives for Li-ion batteries. Adv Mater Res (Durnten-Zurich, Switz) 787:40–45, 47. doi:10.4028/www.scientific.net/AMR.787.40
135. Wang Y, Nakamura S, Tasaki K, Balbuena PB (2002) Theoretical studies to understand surface chemistry on carbon anodes for lithium-ion batteries: how does vinylene carbonate play its role as an electrolyte additive. J Am Chem Soc 124:4408–4421. doi:10.1021/ja017073i
136. Weng W, Zhang Z, Redfern PC, Curtiss LA, Amine K (2011) Fused ring and linking groups effect on overcharge protection for lithium-ion batteries. J Power Sources 196:1530–1536. doi:10.1016/j.jpowsour.2010.08.049
137. Weng W, Zhang Z, Schlueter JA, Redfern PC, Curtiss LA, Amine K (2011) Improved synthesis of a highly fluorinated boronic ester as dual functional electrolyte additive for lithium-ion batteries. J Power Sources 196:2171–2178. doi:10.1016/j.jpowsour.2010.09.110
138. West WC, Whitacre JF, Leifer N, Greenbaum S, Smart M, Bugga R, Blanco M, Narayanan SR (2007) Reversible intercalation of fluoride-anion receptor complexes in graphite. J Electrochem Soc 154:A929–A936. doi:10.1149/1.2759841
139. Wrodnigg GH, Besenhard JO, Winter M (1999) Ethylene sulfite as electrolyte additive for lithium-ion cells with graphitic anodes. J Electrochem Soc 146:470–472. doi:10.1149/1.1391630
140. Wrodnigg GH, Wrodnigg TM, Besenhard JO, Winter M (1999) Propylene sulfite as film-forming electrolyte additive in lithium ion batteries. Electrochem Commun 1:148–150. doi:10.1016/s1388-2481(99)00023-5
141. Wu B, Pei F, Wu Y, Mao R, Ai X, Yang H, Cao Y (2013) An electrochemically compatible and flame-retardant electrolyte additive for safe lithium ion batteries. J Power Sources 227:106–110. doi:10.1016/j.jpowsour.2012.11.018
142. Wu H-C, Su C-Y, Shieh D-T, Yang M-H, Wu N-L (2006) Enhanced high-temperature cycle life of $LiFePO_4$-based Li-ion batteries by vinylene carbonate as electrolyte additive. Electrochem Solid-State Lett 9:A537–A541. doi:10.1149/1.2351954

143. Wu X, Li X, Wang Z, Guo H, Yue P, Zhang Y (2013) Improvement on the storage performance of $LiMn_2O_4$ with the mixed additives of ethanolamine and heptamethyldisilazane. Appl Surf Sci 268:349–354. doi:10.1016/j.apsusc.2012.12.095
144. Wu X, Wang Z, Li X, Guo H, Zhang Y, Xiao W (2012) Effect of lithium difluoro(oxalato) borate and heptamethyldisilazane with different concentrations on cycling performance of $LiMn_2O_4$. J Power Sources 204:133–138. doi:10.1016/j.jpowsour.2011.12.012
145. Wu YP, Rahm E, Holze R (2003) Carbon anode materials for lithium ion batteries. J Power Sources 114:228–236. doi:10.1016/s0378-7753(02)00596-7
146. Xia J, Sinha NN, Chen LP, Dahn JR (2014) A comparative study of a family of sulfate electrolyte additives. J Electrochem Soc 161:A264–A274. doi:10.1149/2.015403jes
147. Xia YY, Zhou YH, Yoshio M (1997) Capacity fading on cycling of 4 V Li/$LiMn_2O_4$ cells. J Electrochem Soc 144:2593–2600. doi:10.1149/1.1837870
148. Xiang HF, Xu HY, Wang ZZ, Chen CH (2007) Dimethyl methylphosphonate (DMMP) as an efficient flame retardant additive for the lithium-ion battery electrolytes. J Power Sources 173:562–564. doi:10.1016/j.jpowsour.2007.05.001
149. Xiao L, Ai X, Cao Y, Yang H (2004) Electrochemical behavior of biphenyl as polymerizable additive for overcharge protection of lithium ion batteries. Electrochim Acta 49:4189–4196. doi:10.1016/j.electacta.2004.04.013
150. Xie B, Lee HS, Li H, Yang XQ, McBreen J, Chen LQ (2008) New electrolytes using Li_2O or Li_2O_2 oxides and tris(pentafluorophenyl)borane as boron based anion receptor for lithium batteries. Electrochem Commun 10:1195–1197. doi:10.1016/j.elecom.2008.05.043
151. Xin S, Guo Y-G, Wan L-J (2012) Nanocarbon networks for advanced rechargeable lithium batteries. Acc Chem Res 45:1759–1769. doi:10.1021/ar300094m
152. Xing LY, Hu M, Tang Q, Wei JP, Qin X, Zhou Z (2012) Improved cyclic performances of $LiCoPO_4$/C cathode materials for high-cell-potential lithium-ion batteries with thiophene as an electrolyte additive. Electrochim Acta 59:172–178. doi:10.1016/j.electacta.2011.10.054
153. Xu K, Zhang SS, Allen JL, Jow TR (2002) Nonflammable electrolytes for Li-ion batteries based on a fluorinated phosphate. J Electrochem Soc 149:A1079–A1082. doi:10.1149/1.1490356
154. Xu M-Q, Hao L-S, Liu Y-L, Li W-S, Xing L-D, Li B (2011) Experimental and theoretical investigations of dimethylacetamide (DMAc) as electrolyte stabilizing additive for lithium ion batteries. J Phys Chem C 115:6085–6094. doi:10.1021/jp109562u
155. Xu M, Liang Y, Li B, Xing L, Wang Y, Li W (2014) Tris(pentafluorophenyl) phosphine: a dual functionality additive for flame-retarding and sacrificial oxidation on $LiMn_2O_4$ for lithium ion battery. Mater Chem Phys 143:1048–1054. doi:10.1016/j.matchemphys.2013.11.003
156. Xu M, Liu Y, Li B, Li W, Li X, Hu S (2012) Tris (pentafluorophenyl) phosphine: an electrolyte additive for high voltage Li-ion batteries. Electrochem Commun 18:123–126. doi:10.1016/j.elecom.2012.02.037
157. Xu M, Tsiouvaras N, Garsuch A, Gasteiger HA, Lucht BL (2014) Generation of cathode passivation films via oxidation of lithium bis(oxalato) borate on high voltage spinel ($LiNi_{0.5}Mn_{1.5}O_4$). J Phys Chem C 118:7363–7368. doi:10.1021/jp501970j
158. Xu MQ, Xing LD, Li WS, Zuo XX, Shu D, Li GL (2008) Application of cyclohexylbenzene as electrolyte additive for overcharge protection of lithium ion battery. J Power Sources 184:427–431. doi:10.1016/j.jpowsour.2008.03.036
159. Xu MQ, Zhou L, Dong YN, Chen YJ, Garsuch A, Lucht BL (2013) Improving the performance of graphite/LiNi/$_{0.5}Mn_{1.5}O_4$ cells at high voltage and elevated temperature with added lithium bis(oxalato) borate (LiBOB). J Electrochem Soc 160:A2005–A2013. doi:10.1149/2.053311jes
160. Yan G, Li X, Wang Z, Guo H, Xiong X (2014) Beneficial effects of 1-propylphosphonic acid cyclic anhydride as an electrolyte additive on the electrochemical properties of $LiNi_{0.5}Mn_{1.5}O_4$ cathode material. J Power Sources 263:231–238. doi:10.1016/j.jpowsour.2014.04.060

161. Yang L, Lucht BL (2009) Inhibition of electrolyte oxidation in lithium ion batteries with electrolyte additives. Electrochem Solid-State Lett 12:A229–A231. doi:10.1149/1.3238486
162. Yang L, Markmaitree T, Lucht BL (2011) Inorganic additives for passivation of high voltage cathode materials. J Power Sources 196:2251–2254. doi:10.1016/j.jpowsour.2010.09.093
163. Yang L, Ravdel B, Lucht BL (2010) Electrolyte reactions with the surface of high voltage $LiNi_{0.5}Mn_{1.5}O_4$ cathodes for lithium-ion batteries. Electrochem Solid St 13:A95–A97
164. Yao XL, Xie S, Chen CH, Wang QS, Sun JH, Li YL, Lu SX (2005) Comparative study of trimethyl phosphite and trimethyl phosphate as electrolyte additives in lithium ion batteries. J Power Sources 144:170–175. doi:10.1016/j.jpowsour.2004.11.042
165. Zeng Z, Jiang X, Wu B, Xiao L, Ai X, Yang H, Cao Y (2014) Bis(2,2,2-trifluoroethyl) methylphosphonate: an novel flame-retardant additive for safe lithium-ion battery. Electrochim Acta 129:300–304. doi:10.1016/j.electacta.2014.02.062
166. Zhan C, Lu J, Kropf AJ, Wu TP, Jansen AN, Sun YK, Qiu XP, Amine K (2013) Mn(II) deposition on anodes and its effects on capacity fade in spinel lithium manganate-carbon systems. Nat Commun 4:2437. doi:10.1038/Ncomms3437
167. Zhang J, Wang J, Yang J, NuLi Y (2014) Artificial interface deriving from sacrificial tris (trimethylsilyl)phosphate additive for lithium rich cathode materials. Electrochim Acta 117:99–104. doi:10.1016/j.electacta.2013.11.024
168. Zhang L, Zhang Z, Wu H, Amine K (2011) Novel redox shuttle additive for high-voltage cathode materials. Energy Environ Sci 4:2858–2862. doi:10.1039/c0ee00733a
169. Zhang L, Zhang ZC, Redfern PC, Curtiss LA, Amine K (2012) Molecular engineering towards safer lithium-ion batteries: a highly stable and compatible redox shuttle for overcharge protection. Energ Environ Sci 5:8204–8207. doi:10.1039/C2ee21977h
170. Zhang Q, Qiu C, Fu Y, Ma X (2009) Xylene as a new polymerizable additive for overcharge protection of lithium ion batteries. Chin J Chem 27:1459–1463. doi:10.1002/cjoc.200990245
171. Zhang SS (2006) Aromatic isocyanate as a new type of electrolyte additive for the improved performance of Li-ion batteries. J Power Sources 163:567–572. doi:10.1016/j.jpowsour.2006.09.046
172. Zhang SS (2006) A review on electrolyte additives for lithium-ion batteries. J Power Sources 162:1379–1394. doi:10.1016/j.jpowsour.2006.07.074
173. Zhang SS, Angell CA (1996) A novel electrolyte solvent for rechargeable lithium and lithium-ion batteries. J Electrochem Soc 143:4047–4053. doi:10.1149/1.1837334
174. Zhang SS, Xu K, Jow TR (2003) Tris(2,2,2-trifluoroethyl) phosphite as a co-solvent for nonflammable electrolytes in Li-ion batteries. J Power Sources 113:166–172. doi:10.1016/S0378-7753(02)00537-2
175. Zheng J, Li X, Yu Y, Feng X, Zhao Y (2014) Novel high phosphorus content phosphaphenanthrene-based efficient flame retardant additives for lithium-ion battery. J Therm Anal Calorim 117:319–324. doi:10.1007/s10973-014-3679-5
176. Zhou D, Li W, Tan C, Zuo X, Huang Y (2008) Cresyl diphenyl phosphate as flame retardant additive for lithium-ion batteries. J Power Sources 184:589–592. doi:10.1016/j.jpowsour.2008.03.008
177. Zhou H, Zhu S, Hibino M, Honma I, Ichihara M (2003) Lithium storage in ordered mesoporous carbon (CMK-3) with high reversible specific energy capacity and good cycling performance. Adv Mater (Weinheim, Ger) 15:2107–2111. doi:10.1002/adma.200306125
178. Zhu Y, Casselman MD, Li Y, Wei A, Abraham DP (2014) Perfluoroalkyl-substituted ethylene carbonates: novel electrolyte additives for high-voltage lithium-ion batteries. J Power Sources 246:184–191. doi:10.1016/j.jpowsour.2013.07.070
179. Zhu Y, Li Y, Bettge M, Abraham DP (2012) Positive electrode passivation by LiDFOB electrolyte additive in high-capacity lithium-ion cells. J Electrochem Soc 159:A2109–A2117. doi:10.1149/2.083212jes
180. Zhu Y, Li Y, Bettge M, Abraham DP (2013) Electrolyte additive combinations that enhance performance of high-capacity $Li_{1.2}Ni_{0.15}Mn_{0.55}Co_{0.1}O_2$ -graphite cells. Electrochim Acta 110:191–199. doi:10.1016/j.electacta.2013.03.102

181. Zhuang GV, Xu K, Jow TR, Ross PN (2004) Study of SEI layer formed on graphite anodes in PC/LiBOB electrolyte using IR spectroscopy. Electrochem Solid St 7:A224–A227. doi:10.1149/1.1756855
182. Zuo X, Wu J, Fan C, Lai K, Liu J, Nan J (2014) Improvement of the thermal stability of LiMn2O4/graphite cells with methylene methanedisulfonate as electrolyte additive. Electrochim Acta 130:778–784. doi:10.1016/j.electacta.2014.03.106

Phosphonium-Based Ionic Liquids

K. Tsunashima

1 Introduction

Lithium secondary batteries are energy storage devices that have been extensively studied and developed for a wide range of practical applications [1]. In recent years, not only charge-discharge performance of the batteries but also their safe performance has been high priorities for all uses of lithium secondary batteries due to the fact that the volatility and inflammability of organic electrolytes often cause serious safety problems. In this situation, room temperature ionic liquids (RTILs) have been regarded as the suitable safe electrolytes for lithium secondary batteries because of their unique physicochemical properties such as no measurable vapor pressure, high thermal stability, low flammability, etc. [2–4]. Most studies on RTIL-based electrolytes for lithium secondary batteries have been associated with nitrogen-based RTILs such as imidazolium, quaternary ammonium, pyridinium, pyrrolidinium derivatives (Fig. 1). It is well known that 1-ethyl-3-methylimidazolium (EMI) cation affords low-viscosity and low-melting RTILs together with a wide variety of anions, often being used as cationic components for the battery electrolytes [5, 6]. However, electrochemical stability of the EMI cation seems insufficient for the uses in the lithium battery system because the cathodic limiting potential is more positive than the redox potential of lithium (ca. +1.0 V vs. Li/Li^+). On the other hand, RTILs based on quaternary ammonium and the related cations have been intensively studied for lithium battery electrolytes since their cations generally have high cathodic stabilities [7–10], although the quaternary ammonium based RTILs tend to show somewhat higher viscosity and lower conductivity than those of EMI-based RTILs.

Recently, RTILs based on quaternary phosphonium cations (Fig. 2) have been receiving a great deal of attention as potential substitutes of the corresponding

K. Tsunashima (✉)
Department of Materials Science, National Institute of Technology,
Wakayama College, Wakayama, Japan
e-mail: tsunashima@wakayama-nct.ac.jp

Z. Zhang and S.S. Zhang (eds.), *Rechargeable Batteries*,
Green Energy and Technology, DOI 10.1007/978-3-319-15458-9_10

Fig. 1 Typical cations of RTILs used for lithium battery applications

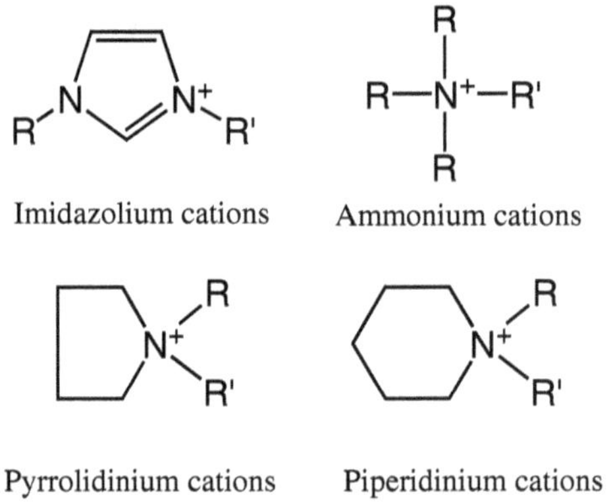

ammonium counterparts [11, 12]. In the pioneering study published by Bradaric et al. [11], it is suggested that phosphonium RTILs provide chemical and thermal stabilities in various applications. From this point of view, many attractive researches on the phosphonium RTILs have been made mainly toward the recyclable media for organic syntheses and solvent extractions [11, 12]. In most cases, the phosphonium RTILs reported therein are based on relatively large phosphonium cations such as tri-*n*-hexyl(tetradecyl)phosphonium cation ($P(C_6H_{13})_3(C_{14}H_{29})^+$, $P^+_{666(14)}$) derived from tri-*n*-hexylphosphine ($P(C_6H_{13})_3$). Such phosphonium RTILs tend to have high viscosities and the correspondingly low conductivities due to their bulky cation sizes, so that the electrochemical application using the phosphonium RTILs seems to be restricted. For this reason, the studies on lithium secondary batteries containing phosphonium IL electrolytes have also been rarely reported. Viscosity essentially influences the rate of mass transport in IL media, so that low viscosity is a major requirement and an advantage for the battery performance containing the IL electrolytes. In order to improve such transport properties of the phosphonium RTILs, we have developed a family of RTILs based on relatively small asymmetrical phosphonium cations together with bis(trifluoromethylsulfonyl) amide ($N(SO_2CF_3)_2^-$, TFSA) anion, characterizing their physicochemical and electrochemical properties for lithium battery electrolytes [13, 14]. The general aim of this chapter is to review our recent works on preparation, physicochemical characterizations and battery tests of phosphonium RTILs from the viewpoint of their possible application as lithium battery electrolytes.

Fig. 2 Quaternary phosphonium cations

2 Phosphonium Ionic Liquids and the Lithium Battery Application

2.1 Preparation of Phosphonium Ionic Liquids

Figure 3 shows the typical synthetic route to prepare the TFSA anion based phosphonium RTILs. We have chosen triethylphosphine ($P(C_2H_5)_3$) and tri(*n*-butyl)phosphine ($P(n\text{-}C_4H_9)_3$) as commercially available starting materials since these two phosphines can afford relatively small quaternary phosphonium cations. The preparation process of the phosphonium RTILs includes two steps: nucleophilic addition reactions of trialkylphosphines to alkyl halides to form precursor phosphonium halides and then aqueous ion exchange reactions of the precursor phosphonium halides with lithium bis(trifluoromethylsulfonyl)amide (Li-TFSA) to obtain TFSA anion based phosphonium RTILs.

In the first step, the nucleophilic addition reactions need to be controlled under inert gas atmosphere because both triethylphosphine and tri(*n*-butyl)phosphine are quite pyrophoric and easy to react with oxygen in air to form trialkylphosphine oxides. However, the nucleophilic addition reactions are relatively fast when compared to those of the corresponding trialkylamines [15]. Another important point is a selection of halides. The rates of the reaction with alkyl bromide are much

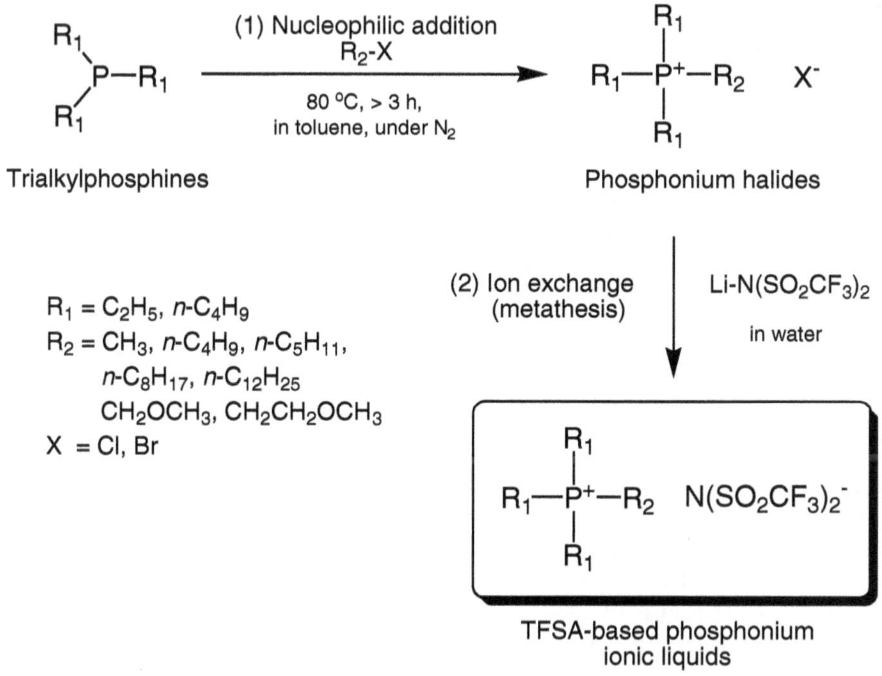

Fig. 3 Praparation of TFSA anion based phosphonium RTILs

faster than those with alkyl chloride; however, the choice of chlorides is more favorable for the efficiencies of the ion exchange reactions and the removal of residual halide anions from RTILs. In most cases, the phosphonium halides obtained are crystalline solids, so that the phosphonium halides can be easily purified by recrystallization.

The TFSA anion based phosphonium RTILs are easily synthesized by aqueous ion exchange reactions and can be separated from the aqueous phase because of the hydrophobicity of the phosphonium RTILs. The crude phosphonium RTILs obtained are purified by washing with pure water several times to remove residual halide anions and metal cation. The content of residual halides and metal cation can be analyzed by using an ion chromatography and an inductively coupled plasma spectroscopy (ICP), respectively. The content of residual halides can also be checked with the use of silver nitrate. The phosphonium RTILs are dried by high vacuum heating to remove residual water, and then are stored in an argon filled glove box to avoid the moisture absorption. When the RTILs are used as lithium battery electrolytes, very low contents of residual halides and water are required (preferably less than 10 ppm).

2.2 Properties of Phosphonium Ionic Liquids

The physicochemical properties of the TFSA anion based phosphonium RTILs we have obtained are summarized in Table 1 [13, 14]. Each phosphonium cations form low-melting salts together with TFSA anion because the packing of ions might be reduced by the asymmetrical cation structures. It is clearly found that the RTILs based on triethylalkylphosphonium cations tend to show low viscosities and high conductivities, which is due to the relatively small cation sizes. Even tri(*n*-butyl) alkylphosphonium cation based RTILs give relatively low viscosities when compared to that of the well-known traditional phosphonium RTIL ($P_{666(14)}$-TFSA, 450 mPa s at 20 °C [11]). Considerably low viscosities and high conductivities are observed in $P_{222(1O1)}$-TFSA (35 mPa s and 4.4 mS cm^{-1} at 25 °C) and $P_{222(2O1)}$-TFSA (44 mPa s and 3.6 mS cm^{-1} at 25 °C), which indicates that introduction of a methoxy group into the phosphonium cations drastically improves their transport property. These low-viscosity phosphonium RTILs seem to be advantageous for lithium battery electrolytes.

The comparison of physicochemical properties between phosphonium and the corresponding RTILs is very interesting. Table 2 lists the physicochemical properties of the corresponding ammonium RTILs for comparison [13]. If the intermolecular force is a major factor in the transport property, phosphonium RTILs having relatively large phosphorus atoms should give high viscosities and low conductivities because of increasing van der Waals interactions; however, the phosphonium RTILs listed in Table 1 show much lower viscosities and higher conductivities than those of the corresponding ammonium RTILs. Although the causes of this phenomenon still remain unclear at present, a presence of vacant

Table 1 Physicochemical properties of phosphonium RTILs based on TFSA anion

Phosohonium RTIL $[PR^1R^1R^1R^2]^+ N(SO_2CF_3)_2^-$			$T_m{}^a$/°C	d^b/g cm	η^c/mPa s	σ^d/mS cm	$T_{dec}{}^e$/°C	References
R^1	R^2	Abbreviation						
C_2H_5	n-C_5H_{11}	P_{2225}-TFSA	17	1.32	88	1.7	380	[13]
C_2H_5	n-C_8H_{17}	P_{2228}-TFSA	less than −50	1.26	129	0.98	400	[13]
C_2H_5	n-$C_{12}H_{25}$	$P_{222(12)}$-TFSA	13	1.21	180	0.47	408	[13]
C_2H_5	CH_2OCH_3	$P_{222(1O1)}$-TFSA	14	1.42	35	4.4	388	[13]
C_2H_5	$(CH_2)_2OCH_3$	$P_{222(2O1)}$-TFSA	10	1.39	44	3.6	404	[13]
n-C_4H_9	CH_3	P_{4441}-TFSA	16	1.28	207	0.42	379	[14]
n-C_4H_9	n-C_8H_{17}	P4448-TFSA	less than −50	1.18	250	0.27	373	[14]
n-C_4H_9	n-$C_{12}H_{25}$	$P_{444(12)}$-TFSA	16	1.13	303	0.18	383	[14]

[a]Melting point;
[b]Density at 25 °C;
[c]Viscosity at 25 °C;
[d]Condustivity at 25 °C;
[e]Thermal decomposition temperature (10 % weight loss)

Table 2 Physicochemical properties of the corresponding ammonium RTILs based on TFSA anion

Ammonium RTIL $[NR^1R^1R^1R^2]^+ N(SO_2CF_3)_2^-$			T_m [a]/°C	d [b]/g cm	η [c]/mPa s	σ [d]/mS cm	T_{dec} [e]/°C	References
R^1	R^2	Abbreviation						
C_2H_5	n-C_5h_{11}	N_{2225}-TFSA	less than −50	1.33	172	0.98	385	[13]
C_2H_5	n-C_8H_{17}	N_{2228}-TFSA	less than −50	1.28	217	0.56	397	[13]
C_2H_5	n-$C_{12}H_{25}$	$N_{222(12)}$-TFSA	7	1.22	316	0.19	390	[13]
C_2H_5	CH_2OCH_3	$N_{222(101)}$-TFSA	−3	1.44	69	3.0	287	[13]
C_2H_5	$(CH_2)_2OCH_3$	$N_{222(201)}$-TFSA	less than −50	1.40	85	2.2	384	[13]

[a]Melting point;
[b]Density at 25 °C;
[c]Viscosity at 25 °C;
[d]Condustivity at 25 °C;
[e]Thermal decomposition temperature (10 % weight loss)

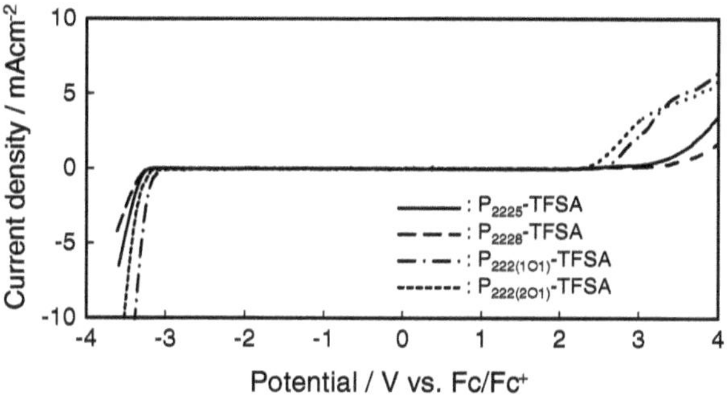

Fig. 4 Linear sweep voltammograms for pure TFSA anion based phosphonium RTILs at a glassy carbon electrode (0.008 cm^2) with a 5 mV s^{-1} potential sweep rate

d orbitals in the phosphorus atom seems likely to play an essential role in the ionic conduction.

The electrochemical stability is of particular importance when RTILs are used as electrolytic media for high voltage electrochemical devices like lithium secondary batteries. Figure 4 illustrates the linear sweep voltammograms measured in the pure phosphonium RTILs at a glassy carbon working electrode [13]. It is found that the potential windows, in which any decomposition currents are not observed, lie between ca. −3.1 and +2.3 V versus Fc/Fc$^+$. These potential windows seem comparable to those of the corresponding ammonium RTILs, although $P_{222(1O1)}$-TFSA and $P_{222(2O1)}$-TFSA show somewhat narrow potential windows due to the presence of a methoxymethyl group in the cation. This result allows us to infer that the phosphonium RTILs show the wide potential windows available for lithium battery electrolytes because the redox potential of lithium is around ca. −3.1 V versus Fc/Fc$^+$.

2.3 Phosphonium Ionic Liquids as Lithium Battery Electrolytes

The charge-discharge properties of the lithium battery cells containing the lithium-containing phosphonium and the corresponding ammonium RTILs are shown in Fig. 5 [16]. The cell containing the $P_{222(2O1)}$-TFSA electrolyte shows sufficient capacities for the charge and discharge at the first cycle (152 and 141 mAh g^{-1}, respectively) which are close to a theoretical capacity of Li_xCoO_2 ($0.5 < x < 1$, voltage range: 4.2–3.4 V). Furthermore, the charge and discharge capacities of the cell containing $P_{222(2O1)}$-TFSA are highly maintained even after 50 cycles when compared to those containing P_{2225}-TFSA and $N_{221(2O1)}$-TFSA. This result reveals the relatively high rechargeability and cyclability of the cell containing the $P_{222(2O1)}$-TFSA electrolyte.

The transport property (i.e., viscosity and conductivity) of the $P_{222(2O1)}$-TFSA electrolyte seems to be one of the major factors in the charge-discharge processes.

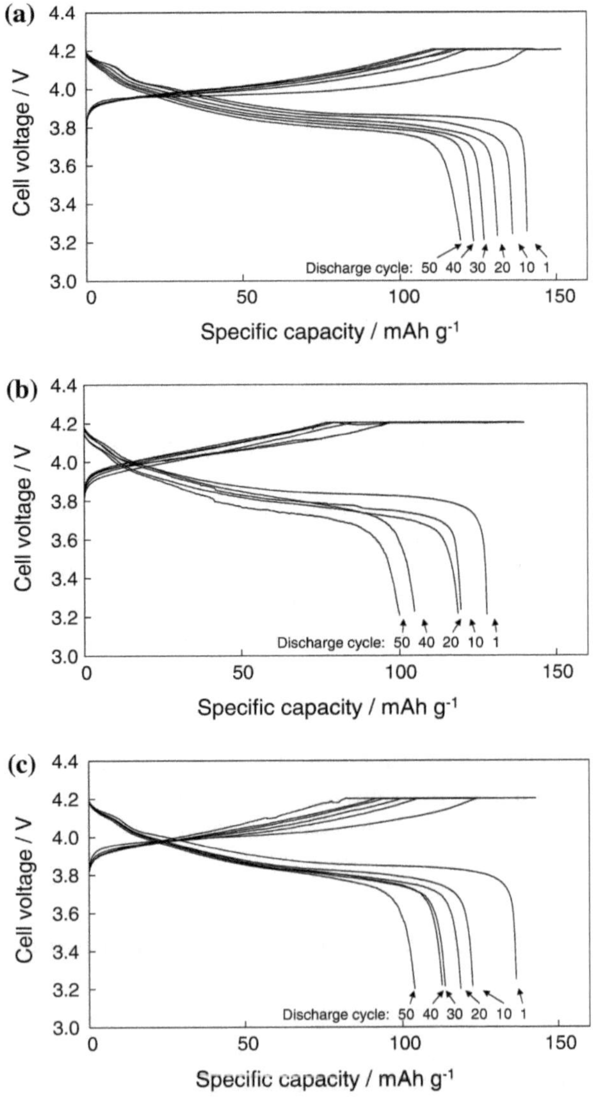

Fig. 5 Charge-discharge curves of the $LiCoO_2$ cathode based cells containing **a** $P_{222(2O1)}$-TFSA, **b** P_{2225}-TFSA and **c** $N_{221(2O1)}$-TFSA (DEME-TFSA) electrolytes (Li^+ concentration: 1.0 M). Cell type: CR2032, charge mode: CCCV (0.05 C, 25 h), discharge mode: CV (0.05 C), cathode: $LiCoO_2$ (8.5 mg cm^{-2}), anode: Li (1 cm^2)

Low viscosity of the electrolyte should contribute to increasing not only the mobility of lithium cation in the electrolyte but also the penetration of the electrolyte into the cathode and the separator materials. Therefore, the transport property of the $P_{222(2O1)}$-TFSA electrolyte should play a significant role in the charge-discharge properties, particularly the rate performance. In addition, as shown in Fig. 4, $P_{222(2O1)}$-TFSA has an appropriate electrochemical stability available for lithium battery electrolytes. Therefore, the relatively high charge-discharge performance observed seems to be attributed to the relatively high transport property and high electrochemical stability of $P_{222(2O1)}$-TFSA. Similar result has been observed in the lithium battery test of the

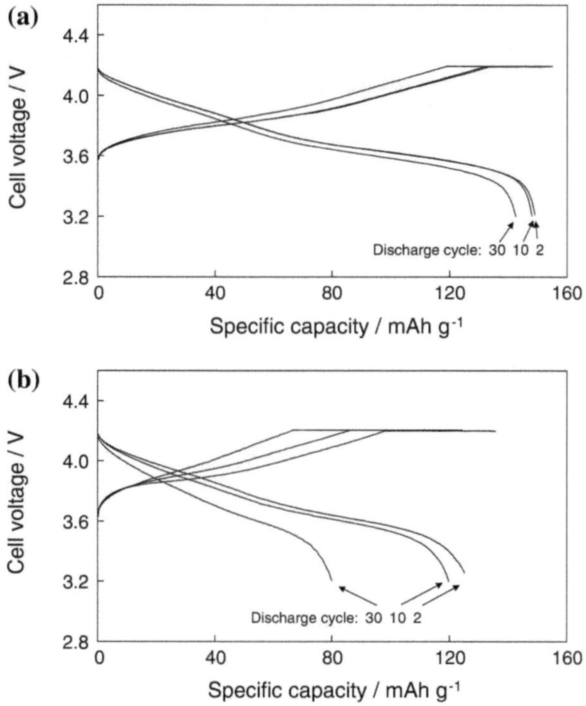

Fig. 6 Charge-discharge curves of the $LiNi_{0.8}Co_{0.1}Mn_{0.1}O_2$ cathode based cells containing **a** $P_{222(2O1)}$-TFSA and **b** P_{2225}-TFSA electrolytes (Li^+ concentration: 1.0 M). Cell type: CR2032, charge mode: CCCV (0.05 C, 25 h), discharge mode: CV (0.05 C), cathode: $LiNi_{0.8}Co_{0.1}Mn_{0.1}O_2$ (8.9 mg cm^{-2}), anode: Li (1 cm^2)

cell consisting of a $LiNiO_2$-based cathode ($LiNi_{0.8}Co_{0.1}Mn_{0.1}O_2$) and the phosphonium RTIL electrolytes as shown in Fig. 6 [17].

2.4 Phosphonium Ionic Liquids Mixed with Diluents

In the case of phosphonium RTILs showing relatively high viscosity and low conductivity (e.g. P_{4441}-TFSA), we have examined the electrolytic properties of the phosphonium RTILs mixed with a conventional organic electrolyte as a diluent to improve their transport property [18, 19]. Figure 7 represents the charge-discharge curves of lithium battery cells containing the phosphonium RTIL electrolytes mixed with the $LiPF_6$-ethylene carbonate (EC)-ethyl methyl carbonate (EMC) based organic electrolyte in comparison with that containing the corresponding ammonium RTIL electrolyte. It is shown that the cycling performance of the cells containing the mixed P_{2225}-TFSA and P_{4441}-TFSA electrolytes is superior to those of the other electrolytes, which suggests an improving effect of phosphonium RTILs on the charge-discharge performance. This result also suggests that both P_{2225}-TFSA and P_{4441}-TFSA can be regarded as effective additives to improve the charge-discharge cycling efficiency whereas P_{4444}-TFSA reduces the cycling performance. Especially, P_{4441}-TFSA shows the most remarkable improving effect.

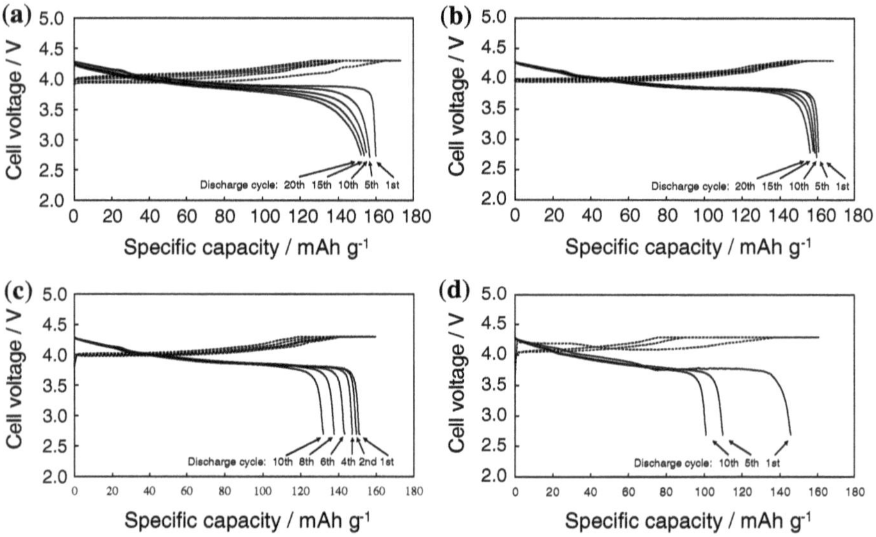

Fig. 7 Charge-discharge curves of the cells containing **a** P_{2225}-TFSA, **b** P_{4441}-TFSA, **c** P_{4444}-TFSA and **d** N_{4441}-TFSA electrolytes mixed with the 1.0 M $LiPF_6$/EC + EMC (1:2) electrolyte (1:1, v/v). Cell type: CR2032, charge mode: CCCV (0.5 C, 5 h), discharge mode: CV (0.2 C), cathode: $LiCoO_2$ (7.2 mg cm^{-2}), anode: Li (1 cm^2)

This improving effect clearly depends on the structures of the phosphonium cations because both P_{4441} and P_{2225} cations are relatively asymmetrical when compared to the symmetrical P_{4444} cation. Therefore, the RTILs having asymmetrical phosphonium cations seem to play an important role to improve the charge-discharge cycling performance. As a possibility, this effect might be attributed to the uniform formation of lithium on the anode surface, suppressing the unfavorable formation of needle-like lithium dendrites to accumulate dead lithium area on the surface.

As mentioned above, addition of organic solvent based electrolytes as diluents has been regarded as one of the potential methods to improve the transport property of RTIL electrolytes [20]. The molecular components added into RTILs can reduce the electrostatic interaction of RTILs, resulting in lower viscosity and the correspondingly higher conductivity. However, the addition of volatile organic solvents can spoil the significant advantage of the low flammable property of RTILs. On the other hand, it is known that phosphazene compounds act as nonflammable agents for organic solvent based electrolytes of lithium ion batteries [21–24]. For example, a diphenoxy-substituted phosphazene compound (DPPz, Fig. 8) is a liquid at ambient temperature, so that we have evaluated the eletrolytic properties of the mixed electrolytes containing the phosphazene compound as a suitable nonflammable diluent for RTIL-based lithium battery electrolytes [25]. Figure 9 shows the temperature dependences of viscosities of lithium-containing P_{2225}-TFSA and $P_{222(1O1)}$-TFSA electrolytes with and without DPPz. It is clearly found that the viscosities are considerably reduced by addition of DPPz, which means that DPPz

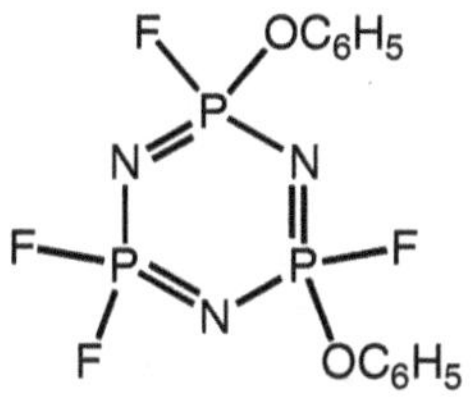

Fig. 8 Chemical structure of DPPz

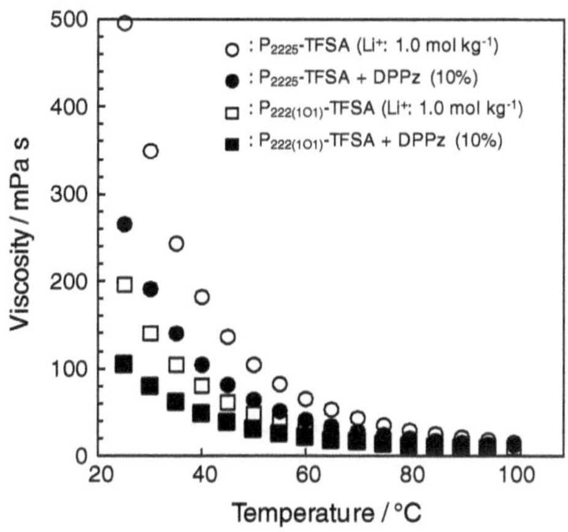

Fig. 9 Temperature dependence of viscosity of lithium-containing TFSA anion based phosphonium RTILs with and without DPPz

behaves as a diluent to improve the transport property. Figure 10 illustrates the cyclic voltamograms measured in the lithium-containing $P_{222(101)}$-TFSA electrolyte with DPPz in comparison with lithium-contaning $P_{222(101)}$-TFSA electrolyte. The characteristic large peak appearing around ca. −3 V versus Fc/Fc^{+} is observed in each voltammogram, showing that the cathodic deposition and anodic dissolution of metallic lithium considerably take place [26]. It is noted that the voltammetric response of the DPPz-containing $P_{222(101)}$-TFSA electrolyte is closely similar to that of the comparative $P_{222(101)}$-TFSA electrolyte. This result indicates that the redox reactions of lithium are not significantly interfared by the presence of DPPz. These results allow us to suggest that DPPz can be regarded as a nonflammable diluent for RTIL-based lithium battery electrolytes.

2.5 Thermal Stability Effect of Phosphonium Ionic Liquids

It is well known that organic phosphorus compounds often show high thermal stability, so that the phosphonium RTILs seem to be able to play a significant role in the thermal stability of lithium battery electrolytes. Figure 11 exhibits the thermogravimetric traces of several phosphonium RTILs in comparison with the

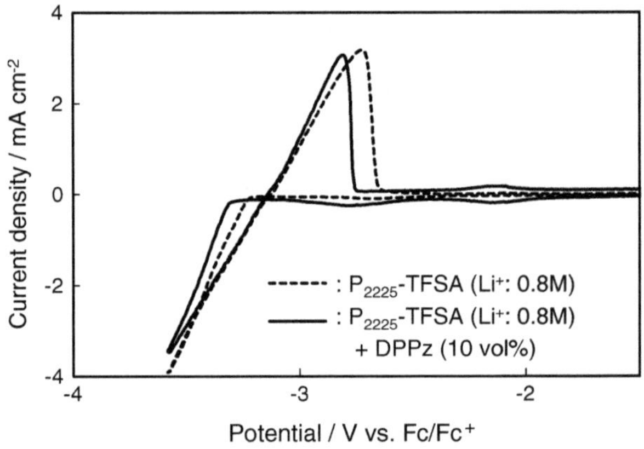

Fig. 10 First cycles of cyclic voltammograms measured in the DPPz-containing $P_{222(1O1)}$-TFSA (*solid line*) and lithium-containing (0.8 mol dm^{-3}) $P_{222(1O1)}$-TFSA (*dotted line*) electrolytes at a Ni working electrode (0.018 cm^2) with a 10 mV s^{-1} potential sweep rate

corresponding ammonium RTILs [13]. It is clearly seen that the phosphonium RTILs are thermally stable up to nearly 400 °C. Similar high thermal stabilities are observed in the phosphonium RTILs containing substituents such as vinyl and phenyl groups [27, 28]. As shown in Fig. 11b, $N_{222(1O1)}$-TFSA significantly reduces its thermal stability due to the presence of a methoxymethyl group in the cation; however, $P_{222(1O1)}$-TFSA maintains high thermal stability. It is worthwhile to note that each phosphonium RTIL also tend to give relatively slow gravimetric decrease when compared to the corresponding ammonium RTIL. These results indicate both high thermal stability and the mild thermal decomposition behavior of the phosphonium RTILs. The detailed thermal decomposition mechanism of phosphonium RTILs is unclear at present; however, unlike the corresponding ammonium systems, the decomposition processes seem to involve the formations of less volatile phosphorus-contain in species, such as organophosphate and polyphosphate derivatives, thereby making the decomposition reactions considerably mild. In addition, phosphorus-containing species thus formed might exhibit non-flammable property to contribute to the safe performance of lithium battery cells.

Figure 12 demonstrates the thermal decomposition behavior of the phosphonium RTIL consisting of dicyanamide ($N(CN)_2^-$, DCA) anion (P_{2225}-DCA) in comparison with the corresponding ammonium RTIL (N_{2225}-DCA) [29]. Generally, it is known that the thermal stability of DCA anion based RTILs is relatively low (<300 °C); however, as shown in Fig. 12, we have confirmed that the combination of phosphonium cations with DCA anion formed thermally stable phsophonium RTILs, which suggests that the thermal stability of DCA anion based RTILs can be enhanced by phosphonium cations. Although DCA anion based RTILs are not often used for lithium battery electrolytes, this result indicates that the choice of phosphonium cations should be advantageous to design thermally stable RTILs especially when thermally unstable anions are employed.

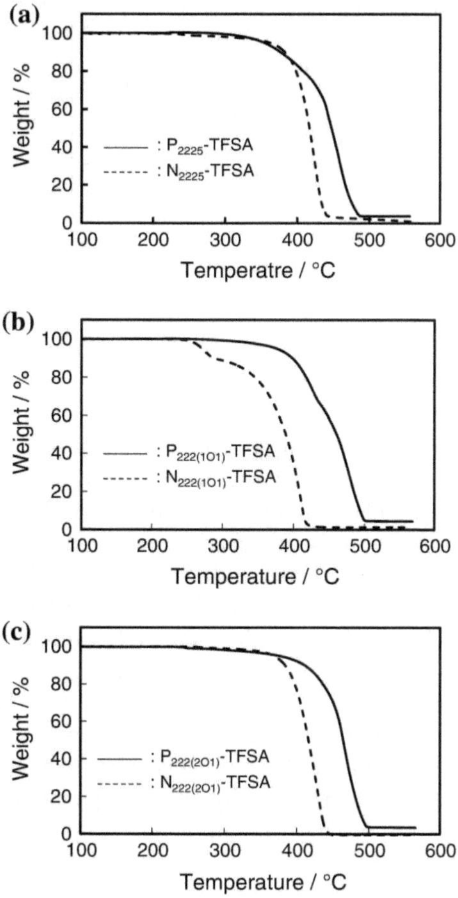

Fig. 11 Thermogravimetric traces for TFSA anion based phosphonium and the corresponding ammonium RTILs

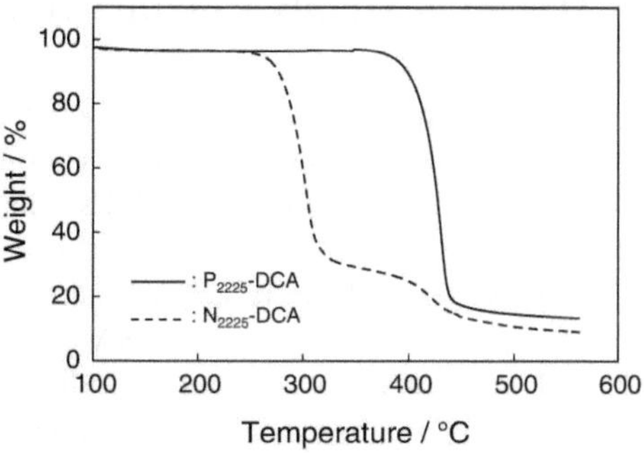

Fig. 12 Thermogravimetric traces for dicyanamide anion based phosphonium and the corresponding ammonium RTILs

We also have investigated the thermal stability effect of phosphonium RTILs on the electrode materials. Figure 13 depicts the DSC thermograms of $LiCoO_2$ cathodes charged in $LiPF_6$-EC-EMC based organic electrolytes with and without phosphonium RTILs as an example to demonstrate an observed thermal stability

Fig. 13 DSC thermograms of $LiCoO_2$ charged in 1.0 M $LiPF_6$/EC + EMC (1:2) electrolyte with and without TFSA anion based RTILs

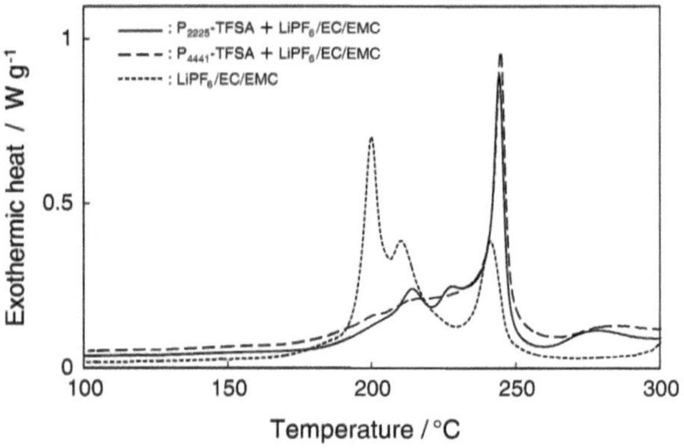

effect of the phosphonium system [19]. As shown by dotted line in Fig. 13, the $LiCoO_2$ cathode charged in the LiPF6-EC-EMC electrolyte with no additives is thermally decomposed around 198, 204 and 240 °C. In the thermograms for the $LiCoO_2$ cathodes charged in the mixed electrolyte containing the phosphonium salts (P_{2225}-TFSA and P_{4441}-TFSA, solid and dashed lines, respectively), the peak around 198 °C significantly vanishes whereas the peak around 240 °C grows larger, which means that the thermal decomposition points of $LiCoO_2$ cathodes are shifted to relatively high temperature regions. This means that the thermal decomposition temperatures of $LiCoO_2$ cathodes are shifted to relatively high temperature regions, indicating the thermal stabilization of $LiCoO_2$ by the phosphonium RTILs. Although the detailed mechanism remains still unclear at present, such a thermal stabilization effect by the phosphonium RTILs should contribute to the safety performance of lithium battery cells.

3 Design of Highly Conductive Phosphonium Ionic Liquids

We have demonstrated the physicochemical and battery electrolyte characterizations of the phosphonium RTILs based on the TFSA anion. As a result, it has been shown that the phosphonium RTILs exhibited relatively low viscosities and high conductivities when compared to those of the traditional phosphonium RTILs. Especially, the RTILs based on methoxy-substituted phosphonium cations, e.g. $P_{222(1O1)}$-TFSA and $P_{222(2O1)}$-TFSA give quite low viscosities (35 and 44 mPa s at 25 °C, respectively) and high conductivities (4.4 and 3.6 mS cm^{-1} at 25 °C, respectively) in the TFSA anion based phosphonium RTILs we have obtained. However, the transport property is not sufficient for lithium battery application because the conductivity of conventional organic electrolytes is generally ca. 10 mS cm^{-1} at 25 °C. In addition, RTILs increase their viscosities when lithium salts are added into RTILs due to the electrostatic interaction enhanced by the addition of lithium salts. Therefore, design of further highly conductive phosphonium RTILs should be inevitably required to improve the transport property of phosphonium RTIL electrolytes.

It is well known that the selection of anions frequently offers a drastic change in the various properties of RTILs. Recently, interests in RTILs based on a bis(fluorosulfonyl)amide anion ($N(SO_2F)_2^-$, FSA) have been increasing. One of the most promising applications of the FSA anion based RTILs is thought to be electrolytic media for lithium secondary batteries and electrochemical double layer capacitors because of their favorably high transport property [30–32]. Typical cationic components selected for this purpose include mainly the nitrogen-based cations such as imidazolium and *N*-heterocyclic ammonium based cations. On the other hand, the FSA anion based phosphonium RTILs have been rarely reported, so that the transport property of FSA anion based RTILs consisting of the quaternary phosphonium cations is of particular interest.

We have designed and successfully prepared a series of RTILs based on phosphonium cations together with the FSA anion from the viewpoint of lithium battery application [33, 34]. The preparation of FSA anion based phosphonium RTILs can be carried out according to the same procedure as the TFSA anion based phosphonium RTILs depicted in Fig. 3 because the FSA anion based phosphonium RTILs are also hydrophobic.

The physicochemical properties of FSA anion based phosphonium RTILs we have reported are summarized in Table 3 [33, 34]. In addition to the RTILs based on triethylalkylphosphonium cations, we have synthesized the RTILs based on trimethylalkylphosphonium cations derived from trimethylphosphine ($P(CH_3)_3$) since the phosphonium cations having smaller cation sizes are expected to give the phosphonium RTILs showing higher transport property. It should be noted that the FSA anion based phosphonium RTILs are obviously low-melting when compared to the corresponding TFSA anion based RTILs. For instance, the melting points (T_m) of P_{2225}-FSA, $P_{222(1O1)}$-FSA and $P_{222(2O1)}$-FSA (−35, −14 and −18 °C, respectively) are much lower than those of P_{2225}-TFSA, $P_{222(1O1)}$-TFSA and $P_{222(2O1)}$-TFSA listed in Table 1 (17, 14 and 10 °C, respectively). Even melting point of P_{2224}-FSA is −16 °C, despite the fact that the corresponding TFSA anion based RTIL (P_{2224}-TFSA) is a crystalline solid at room temperature (T_m: 55 °C) [13]. This low-melting behavior of the FSA anion based phosphonium RTILs is very important for the improvement of low temperature characteristics of RTIL based electrolytes.

The viscosities and conductivities of FSA anion based phosphonium RTILs at 25 °C are also listed in Table 3. It is worthwhile to note that the triethylalkylphosphonium RTILs based on FSA anion exhibits much lower viscosities and higher conductivities than those of the corresponding TFSA anion based phosphonium RTILs shown in Table 1. Further low viscosities and high conductivities are observed in the trimethylalkylphosphonium RTILs. For example, P_{1113}-FSA shows lower viscosity and higher conductivity (31 mPa s and 7.1 mS cm^{-1} at 25 °C) than those of P_{2224}-FSA (62 mPa s and 4.5 mS cm^{-1} at 25 °C), which means that the decrease in the cation size of the RTILs leads to a decrease in viscosity and the corresponding increase in conductivity. Particularly, $P_{111(1O1)}$-FSA having a methoxy group shows the lowest viscosity and highest conductivity in a series of the FSA

Table 3 Physicochemical properties of phosphonium RTILs based on FSA anion

Phosohonium RTIL $[PR^1R^1R^1R^2]^+ N(SO_2F)_2^-$			T_m [a]/°C	d [b]/g cm^{-3}	η [c]/mPa s	σ [d]/mS cm^{-1}	T_{dec} [e]/°C	References
R^1	R^2	Abbreviation						
CH_3	n-C_3H_7	P_{1113}-FSA	0	1.33	31	7.1	298	[34]
CH_3	CH_2OCH_3	$P_{111(1O1)}$-FSA	−6	1.39	21	8.9	262	[34]
C_2H_5	n-C_4H_9	P_{2224}-FSA	−16	1.26	62	4.5	313	[33]
C_2H_5	n-C_5H_{11}	P_{2225}-FSA	−35	1.24	70	3.0	309	[33]
C_2H_5	CH_2OCH_3	$P_{222(1O1)}$-FSA	−14	1.32	29	8.9	310	[33]
C_2H_5	$(CH_2)_2OCH_3$	$P_{222(2O1)}$-FSA	−18	1.31	37	5.7	285	[33]

[a]Melting point;
[b]Density at 25 °C;
[c]Viscosity at 25 °C;
[d]Condustivity at 25 °C;
[e]Thermal decomposition temperature (10 % weight loss)

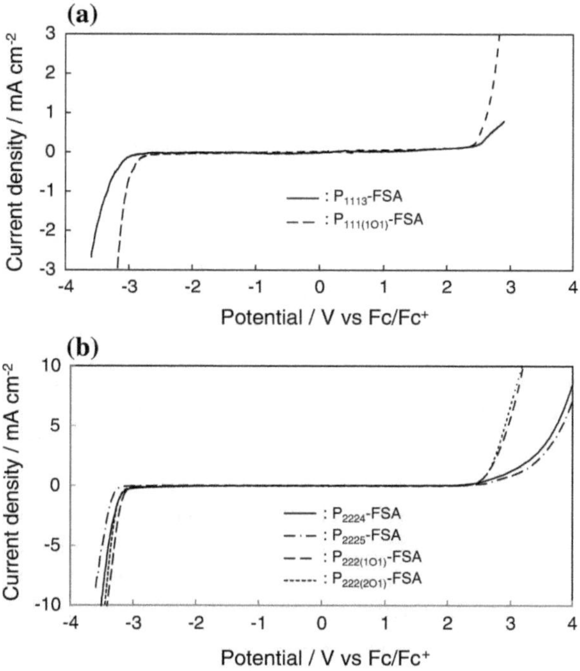

Fig. 14 Linear sweep voltammograms for pure FSA anion based phosphonium RTILs at a glassy carbon electrode (0.008 cm^2) with a 5 mV s^{-1} potential sweep rate

anion based phosphonium RTILs (21 mPa s and 8.9 mS cm^{-1} at 25 °C). Like $P_{222(1O1)}$-TFSA and $P_{222(2O1)}$-TFSA, the introduction of a electron-donating methoxy group into the phosphonium cation seems to reduce the positive charge of the phosphonium cation, thereby giving a drastic decrease in viscosity and the corresponding increase in conductivity. These findings allow us to indicate that the FSA anion based phosphonium RTILs can be classified into a new class of low viscosity and highly conductive RTILs.

The electrochemical stability of the FSA anion based phosphonium RTILs has been evaluated by the voltammetric measurement. Figure 14 illustrates the linear sweep voltammograms measured in pure FSA anion based phosphonium RTILs [33, 34]. The voltammetric patterns for the FSA anion based phosphonium RTILs are basically similar to those for the TFSA based phosphonium RTILs shown in Fig. 4, which means that the FSA anion based phosphonium RTILs have typically wide windows as electrochemically stable electrolytes. The anodic limit potentials for the FSA anion based phosphonium RTILs tend to be somewhat shifted to relatively negative potential region when compared to the case of TFSA anion based RTILs, which seems likely to be due to the anodic oxidation of the FSA anion.

In order to check the ability of the FSA anion based phosphonium RTILs as lithium battery electrolytes, the electrochemical behavior of lithium in the FSA anion based phosphonium RTILs has been investigated by using a cyclic voltammetry technique. Figure 15 displays the cyclic voltamograms of lithium measured in lithium-containing P_{1113}-FSA and $P_{111(1O1)}$-FSA electrolytes in comparison with

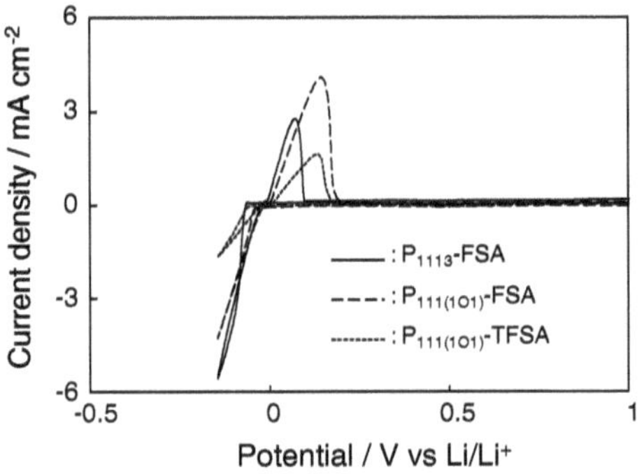

Fig. 15 First cycles of cyclic voltammograms measured in lithium-containing (0.5 mol kg^{-1}) P_{1113}-FSA (solid line), $P_{111(1O1)}$-FSA (*dashed line*) and $P_{111(1O1)}$-TFSA (*dotted line*) electrolytes at a Ni working electrode (0.018 cm^2) with a 10 mV s^{-1} potential sweep rate

the $P_{111(1O1)}$-TFSA electrolyte [34]. It is noteworthy that the voltammetric responses for the P_{1113}-FSA and $P_{111(1O1)}$-FSA electrolytes are much larger than those for the $P_{111(1O1)}$-TFSA electrolytes. Particularly, $P_{111(1O1)}$-FSA give the largest voltammetric response with a considerably high coulombic efficiency (97 %) estimated by the percentage of anodic and cathodic peak charges, which seems to be attributed to the very low viscosity and high electrochemical stability. This result suggests that the FSA anion based phosphonium RTILs can be regarded as promising RTIL electrolytes favorably used as electrolytic media for lithium secondary batteries.

4 Concluding Remarks

In this chapter, we have summarized the physicochemical and electrochemical properties of a variety of phosphonium RTILs recently reported. The battery performance of the cells containing the phosphonium RTIL electrolytes has been presented, as a demonstration of considerable effects of phosphonium RTILs on the charge-discharge cycling property. One of the characteristic features of the phosphonium RTILs is the thermal stability, which should contribute to the durability and the safe performance of lithium battery cells. Furthermore, it has been shown that combination of small phosphonium cations and the FSA anion gives low viscosity and highly conductive phosphonium RTILs, allowing the large and efficient redox behavior of lithium in the RTILs. In this context, the phosphonium RTILs can be regarded as promising candidates for lithium battery electrolytes.

Acknowledgments This work was partially supported by the Grant-in-Aid for Scientific Research (No. 22550131) from the Ministry of Education, Culture, Sports, Science and Technology, Japan, and by the Environment Research and Technology Development Fund

(3K123018) of the Ministry of the Environment, Japan. The author is grateful to Nippon Chemical Industrial Co., Ltd. and Professor Dr. M. Matsumiya, Yokohama National University, for collaboration in the development of the phosphonium based RTILs.

References

1. Tarascon J-M, Armand M (2001) Issues and challenges facing rechargeable lithium batteries. Nature 414:359–367
2. Ohno H (ed) (2005) Electrochemical aspects of ionic liquids. Wiley, New Jersey
3. MacFarlane DR, Forsyth M, Howlett PC, Pringle JM, Sun J, Annat G, Neil W, Izgorodina EI (2007) Ionic liquids in electrochemical devices and processes: managing interfacial electrochemistry. Acc Chem Res 40:1165–1173
4. Armand M, Endres F, MacFarlane DR, Ohno H, Scrosati B (2009) Ionic-liquid materials for the electrochemical challenges of the future. Nature Mater 8:621–629
5. Nakagawa N, Izuchui S, Kuwana K, Nukuda T, Aihara Y (2003) Liquid and polymer gel electrolytes for lithium batteries composed of room-temperature molten salt doped by lithium salt. J Electrochem Soc 150:A695–A700
6. Garcia B, Lavallee S, Perron G, Michot C, Armand M (2004) Room temperature molten salts as lithium battery electrolyte. Electrochim Acta 49:4583–4588
7. Sakaebe H, Matsumoto H (2003) *N*-Methyl-*N*-propylpiperidinium bis (trifluoromethanesulfonyl)imide (PP13–TFSI)—novel electrolyte base for Li battery. Electrochem Commun 5:594–598
8. Sato T, Maruo T, Markane S, Takagi K (2004) Ionic liquids containing carbonate solvent as electrolytes for lithium ion cells. J Power Sources 138:253–261
9. Sakaebe H, Matsumoto H, Tatsumi K (2005) Preparation of room temperature ionic liquids based on aliphatic onium cations and asymmetric amide anions and their electrochemical properties as a lithium battery electrolyte. J Power Sources 146:45–50
10. Seki S, Kobayashi Y, Miyashiro H, Ohno Y, Mita Y, Usami A, Terada N, Watanabe M (2005) Reversibility of lithium secondary batteries using a room-temperature ionic liquid mixture and lithium metal. Electrochem Solid-State Lett 8:A577–A578
11. Bradaric CJ, Downard A, Kennedy C, Robertson AJ, Zhou Y (2003) Industrial preparation of phosphonium ionic liquids. Green Chem 5:143–152
12. Fraser KJ, MacFarlane DR (2009) Phosphonium-based ionic liquids: an overview. Aust J Chem 62:309–321
13. Tsunashima K, Sugiya M (2007) Physical and electrochemical properties of low-viscosity phosphonium ionic liquids as potential electrolytes. Electrochem Commun 9:2353–2358
14. Tsunashima K, Sugiya M (2007) Physical and electrochemical properties of room temperature ionic liquids based on quaternary phosphonium cations. Electrochemistry 75:734–736
15. Henderson WA Jr, Buckler SA (1960) The nucleophilicity of phosphines. J Am Chem Soc 82:5794–5800
16. Tsunashima K, Yonekawa F, Sugiya M (2008) A lithium battery electrolyte based on a room-temperature phosphonium ionic liquid. Chem Lett 37:314–315
17. Tsunashima K, Yonekawa F, Sugiya M (2009) Lithium secondary batteries using a lithium nickelate-based cathode and phosphonium ionic liquid electrolytes. Electrochem Solid-State Lett 12:A54–A57
18. Tsunashima K, Yonekawa F, Kikuchi M, Sugiya M (2010) Tributylmethylphosphonium Bis (trifluoromethylsulfonyl)amide as an effective electrolyte additive for lithium secondary batteries. J Electrochem Soc 157:A1274–A1278
19. Tsunashima K, Yonekawa F, Kikuchi M, Sugiya M (2011) Effect of quaternary phosphonium salts in organic electrolyte for lithium secondary batteries. Electrochemistry 79:453–457

20. Bayley PM, Lane GH, Rocher NM, Clare BR, Best AS, MacFarlane DR, Forsyth M (2009) Transport properties of ionic liquid electrolytes with organic diluents. Phys Chem Chem Phys 11:7202–7208
21. Xu K, Ding MS, Zhang S, Allen JL, Jow TR (2002) An attempt to formulate nonflammable lithium ion electrolytes with alkyl phosphates and phosphazenes. J Electrochem Soc 149: A622–A626
22. Zhang Q, Noguchi H, Wang H, Yoshio M, Otsuki M, Ogino T (2005) Improved thermal stability of $LiCoO_2$ by cyclotriphosphazene additives in lithium-ion batteries. Chem Lett 34:1012–1013
23. Tsujikawa T, Yabuta K, Matsushita T, Matsushima T, Hayashi K, Arakawa M (2009) Characteristics of lithium-ion battery with non-flammable electrolyte. J Power Sources 189:429–434
24. Sazhin SV, Harrup MK, Gering KL (2011) Characterization of low-flammability electrolytes for lithium-ion batteries. J Power Sources 196:3433–3438
25. Tsunashima K, Taguchi H, Yonekawa F (2012) Influence of nonflammable diluents on properties of phosphonium ionic liquids as lithium battery electrolytes. ECS Trans 50:419–424
26. Tsunashima K, Sugiya M (2008) Electrochemical behavior of lithium in room-temperature phosphonium ionic liquids as lithium battery electrolytes. Electrochem Solid-State Lett 11: A17–A19
27. Tsunashima K, Niwa E, Kodama S, Sugiya M, Ono Y (2009) Thermal and transport properties of ionic liquids based on benzyl-substituted phosphonium cations. J Phys Chem B 113:15870–15874
28. Tsunashima K, Ono Y, Sugiya M (2011) Physical and electrochemical characterization of ionic liquids based on quaternary phosphonium cations containing a carbon–carbon double bond. Electrochim Acta 56:4351–4355
29. Tsunashima K, Kodama S, Sugiya M, Kunugi Y (2010) Physical and electrochemical properties of room-temperature dicyanamide ionic liquids based on quaternary phosphonium cations. Electrochim Acta 56:762–766
30. Ishikawa M, Sugimoto T, Kikuta M, Ishiko E, Kono M (2006) Pure ionic liquid electrolytes compatible with a graphitized carbon negative electrode in rechargeable lithium-ion batteries. J Power Sources 162:658–662
31. Matsumoto H, Sakaebe H, Tastumi K, Kikuta M, Ishiko E, Kono M (2006) Fast cycling of Li/$LiCoO_2$ cell with low-viscosity ionic liquids based on bis(fluorosulfonyl)imide $[FSI]^-$. J Power Sources 160:1308–1313
32. Seki S, Kobayashi Y, Miyashiro H, Ohno Y, Mita Y, Terada N (2008) Compatibility of *N*-methyl-*N*-propylpyrrolidinium cation room-temperature ionic liquid electrolytes and graphite electrodes. J Phys Chem C 112:16708–16713
33. Tsunashima K, Kawabata A, Matsumiya M, Kodama S, Enomoto R, Sugiya M, Kunugi Y (2011) Low viscous and highly conductive phosphonium ionic liquids based on bis (fluorosulfonyl)amide anion as potential electrolytes. Electrochem Commun 13:178–181
34. Tsunashima K, Sakai Y, Matsumiya M (2014) Physical and electrochemical properties of phosphonium ionic liquids derived from trimethylphosphine. Electrochem Commun 39:30–33

Solid-State Lithium Ion Electrolytes

C. Tealdi, E. Quartarone and P. Mustarelli

1 Introduction

Lithium-ion batteries are mature products in portable electronics, and are considered the most promising storage systems for automotive, and even for geographical electric grids [1]. It is also accepted that the current research on lithium-ion batteries will likely generate only an incremental improvement of the performances of these storage systems, whereas any quantum jump will require a substantial change of the cell chemistry, e.g. the development of Li-air or Li-sulphur technology [2]. On the other hand, state-of-the-art Li-ion batteries are based on liquid (or gel) organic electrolytes, which pose severe problems in terms of safety and cycle life. Therefore, the replacement of the currently used organic electrolytes with inorganic solid-state electrolytes (SSEs) is very appealing. First of all, SSEs can solve several concerns on capacity losses, cycle life, operation temperatures, safety and reliability better than the liquid ones [3]. In addition, they present other advantages such as absence of leakage and pollution, and better resistance to shocks and vibrations [4]. Finally, they are generally single ion-conductors (lithium transference number equal to one), which assure maximum electrochemical efficiency and the lowering of cell over-potentials [5]. Lithium SSEs can find application in other technological sectors where all-solid-state batteries and microbatteries are often mandatory, ranging from microelectronics to sensors in medical and military fields, biomedical devices, smart cards and other micro-devices (MEMS, NEMS), and also Radio Frequency Identification (RFID) tags. This clearly justifies the increasing interest towards SSEs, as shown in Fig. 1, which reports the number of papers in this topic (taken from ISI Web of Science).

C. Tealdi · E. Quartarone · P. Mustarelli (✉)
Department of Chemistry, Section of Physical Chemistry, University of Pavia, and INSTM,
Via Taramelli 12, 27100 Pavia, Italy
e-mail: piercarlo.mustarelli@unipv.it

Z. Zhang and S.S. Zhang (eds.), *Rechargeable Batteries*,
Green Energy and Technology, DOI 10.1007/978-3-319-15458-9_11

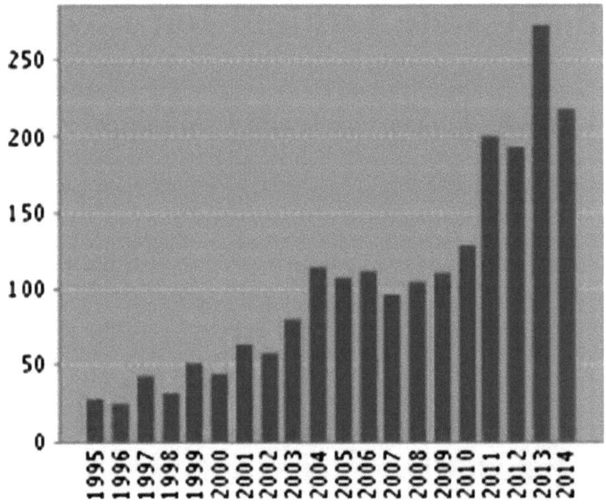

Fig. 1 Number of papers on solid-state lithium electrolytes (from ISI Web of Science, September 10, 2014)

Lithium solid-state electrolytes can be roughly divided into three main categories: (i) ceramic (CE), (ii) glasses (GL), (iii) solvent-free polymer electrolytes (SPEs). Indeed, the most appealing class is CE, which has been the object of recent good reviews [6–8]. These electrolytes can easily offer a relatively high conductivity (up to $10^{-3}\ \Omega^{-1}\ cm^{-1}$), and have the further advantage of a thermal expansion coefficient that can be made similar to that of ceramic electrode materials, so avoiding cracks and losses of contact during thermal cycling. An exceptionally high conductivity of 12 m $\Omega^{-1}\ cm^{-1}$ at room temperature was recently claimed for ceramic $Li_{10}GeP_2S_{12}$ [9], but this result has not yet been confirmed by independent research. GL electrolytes gained a great deal of attention during '80 and '90, chiefly because of their isotropic nature, which could allow direction-independent conductivity [10], absence of grain boundaries, ease preparation of thin films and wide attainable composition ranges [7]. At present, they are still investigated chiefly as model systems as far as concerns the relationships between local/medium range structure and transport properties [11], whereas possible technological applications are limited to anode-protective coatings for Li–O_2 batteries and electrolytes for some specific applications, e.g. in rechargeable batteries for intra-corporeal biomedical devices. In this chapter, GL electrolytes will be chiefly considered for their historical relevance. SPEs electrolytes are not generally reviewed as solid-state lithium conductors because of the polymeric—and partially amorphous—nature of the matrix [5]. On the other hand, they are practically solvent-free, and often made in form of (nano)composites with ceramic (nano)phases (SiO_2, TiO_2, layered silicates, etc.). Therefore, they will be considered in this chapter.

2 Ceramic Electrolytes

Crystalline inorganic electrolytes for Li-ion batteries can be divided into four main families of compounds, depending on their crystal structure: (1) A-site deficient perovskite-type Li-ion conductors; (2) Garnet-type Li-ion conductors; (3) NASICON-type Li-ion conductors; (4) LISICON-type Li-ion conductors.

These materials are generally prepared in bulk by solid-state reactions or sol-gel recipes. Recently, high-energy ball milling has gained interest both for inducing chemical reactivity and for particle size reduction. Thin films can be conveniently prepared by means of advanced techniques such as atomic layer deposition, whereas more conventional methods like r.f. magnetron sputtering are not well suited for these multiple-cations oxides. In the following, for each of these families we will present the main structural features, discuss the transport properties and illustrate the possible Li-ion conduction mechanism in the system.

2.1 A-Site Deficient Perovskite-Type Li-Ion Conductors

Compounds belonging to the solid solution of general formula $Li_{3x}La_{(2/3)-x}\Upsilon_{(1/3)-2x}TiO_3$ (with Υ indicating cationic vacancies on the A site) have been shown to possess a perovskite-type structure in a wide range of composition, extending approximately in the range $0.04 < x < 0.17$ [12]. Lithium ion conductivity in bulk materials of the lithium lanthanum titanate family (LLTO) was found to be extremely promising (of the order of $10^{-3}\ \Omega^{-1}\ cm^{-1}$ at room temperature), [13] largely dependent upon composition, and reaching a maximum of $1.1 \times 10^{-3}\ \Omega^{-1}\ cm^{-1}$ at room temperature for $x = 0.11$ [14]. Depending on composition and synthetic conditions, in addition to the perfectly cubic perovskite structure, tetragonal, orthorhombic and hexagonal distorted perovskite-type structures were also proposed to account for the structural features of this family. Indeed, the structural description of LLTO is still controversial, as also within a certain symmetry frequently more than one space group was proposed. Such an ambiguity mainly derives from the difficulties in identifying the Li positions, as well as the distribution of intrinsic cation vacancies. Figure 2 shows, as an example, the crystal structures of four polymorphs likely attributed to the LLTO family. As a rough guide, Li-poor compositions ($x < 0.1$) are generally reported to crystallize with orthorhombic symmetry (Fig. 2c), while for Li-rich compositions the tetragonal symmetry is preferentially reported (Fig. 2b) [15]. Both the systems are characterized by layers with large La-site occupancies, alternating with partially Li/La occupied layers characterized by a larger concentration of cationic vacancies [16]. The cubic symmetry (Fig. 2a) was obtained for specific compositions through quenching of the high temperature polymorph [17, 15 and references therein], whereas the hexagonal polymorph (Fig. 2d) was proposed for the $La_{0.5}Li_{0.5}TiO_3$ composition [18]. For a comprehensive structural survey, the reader is referred to specific review papers on the LLTO family (e.g. Ref. [4] and references therein).

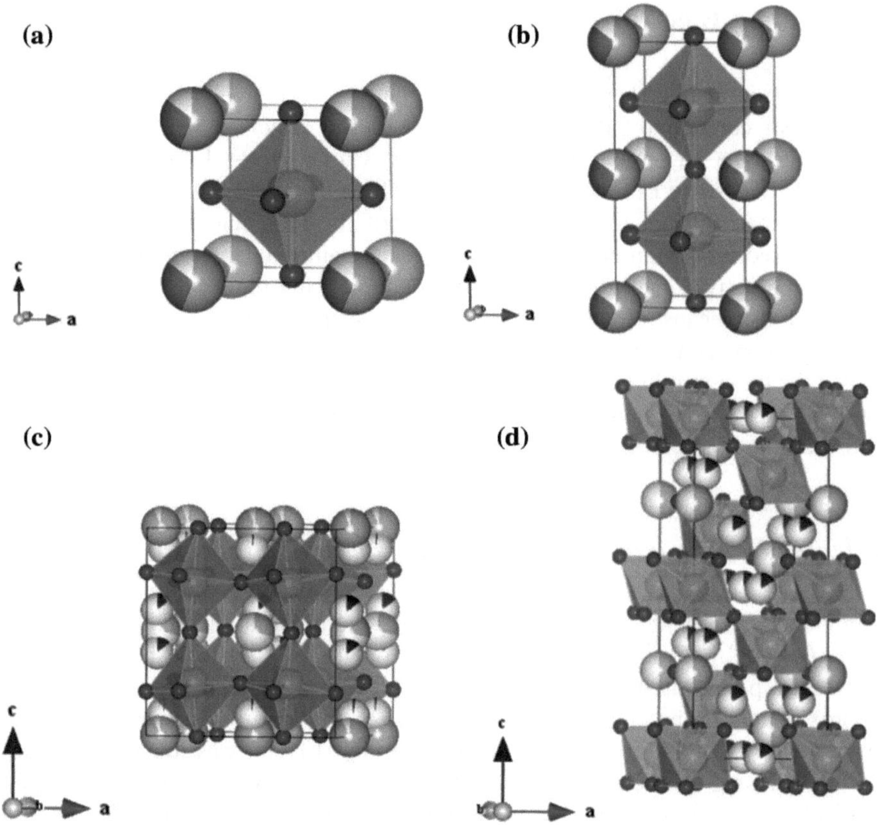

Fig. 2 Structural representation of the four main polymorphs of LLTO. **a** Cubic, space group P*m*3*m* [13]; **b** tetragonal, space group P4/*mmm* [78]; **c** orthorhombic, space group C*mmm* [79]; **d** hexagonal, space group R-3*c* [18]. Legend: *green*—lanthanum, *violet*—lithium, *pale blue*—titanium, *red*—oxygen, *grey*—cationic vacancies (Color figure online)

One important feature to note regarding the structural description of LLTO is that, for each polymorph, the crystallographic sites pertaining to La^{3+} and Li^{+} are characterized by a partial site occupancy, in addition to a certain number of vacant positions, depending on composition. The presence of such a structural disorder on the La/Li crystallographic sites and, above all, of cationic vacancies, is considered by most authors at the origin of the high conductivity values of LLTO at room temperature [17], as the most likely mechanism for Li diffusion in the system is suggested to be vacancy-mediated. A jump from a Li site into an adjacent vacant site requires that the Li ions pass through a square-type bottleneck as sketched in Fig. 3. Activation energy for Li ion migration in the system are correlated to the bottleneck size, i.e. the migration space available at the saddle point configuration.

The Arrhenius plot of conductivity for various LLTO compositions invariably shows a change in activation energy along with temperature [13, 17, 19 and references therein].

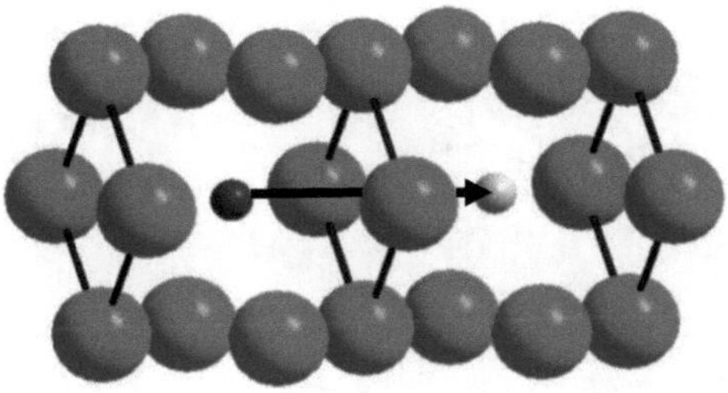

Fig. 3 Schematic representation of the square-type bottleneck for tetragonal LLTO. Ti and La ions have been omitted for clarity. Legend: *violet*—lithium, *red*—oxygen, *grey*—Li vacant site (Color figure online)

Two main regions can be identified: a low temperature region (below approximately 130 °C) following an Arrhenius-type behavior and a high temperature region where conductivity data are more correctly described by a Vogel-Tammann-Fulcher (VTF) type behavior. Such a deviation from linearity is attributed to a progressive tilting and rotation of the TiO_6 octahedra that may result in a progressive change with temperature of the available migration space for Li ion diffusion at the saddle point configuration. Activation energies for Li diffusion usually reported are around 0.3 eV, with differences due to composition and preparation methods [17]. The degree of ordering of the Li/La cations and vacancies strongly influences the transport properties of the materials, with higher conductivity generally found for the most disordered systems [20]. Interestingly, such a degree of order can be reversibly modified by careful temperature annealing in the 600–1150 °C range, thus partially allowing a modulation of the transport properties at a specific composition.

LLTO electrolytes present the advantage of being stable in dry and hydrated atmosphere and over a wide temperature range, while being characterized by a high electrochemical stability (>8 V) and almost pure ionic conductivity [16]. However, two main drawbacks affect the use of LLTO electrolytes: grain boundary resistance and electronic contribution to conductivity [7]. Grain boundary resistance can be improved through optimization of the sintering conditions, and a certain improvement of grain boundary contribution was reported in combination with the introduction of silica. Due to the presence of Ti^{4+}, LLTO is not particularly useful in combination with highly reducing negative electrodes (typically Li metal anode), as this would introduce a considerable electronic contribution to conductivity [21]. However, recent reports are showing its promising application as a separator for Li-air batteries [22].

2.2 Garnet-Type Li-Ion Conductors

Compounds exhibiting the general formula $Li_xLn_3M_2O_{12}$ (Ln = rare earth) belong to the large family of garnet-type oxides, crystallizing in a cubic structure, within

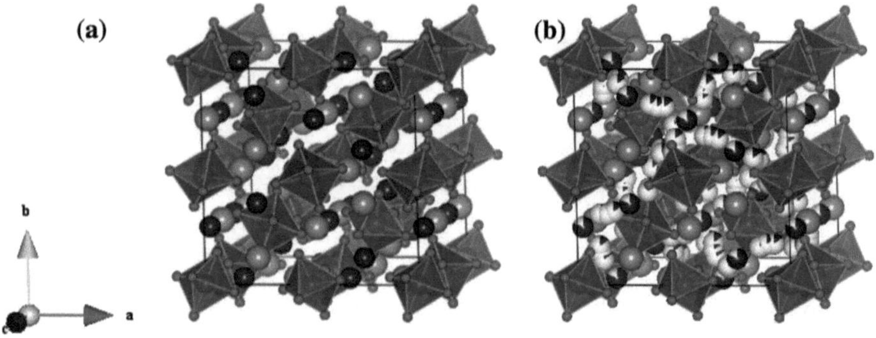

Fig. 4 Structural representation of the garnet-type structure (cubic, space group I*a*-*3d*) **a** stoichiometric garnet $Li_3La_3Ta_2O_{12}$; **b** Li-stuffed garnet $Li_5La_3Ta_2O_{12}$, showing the possible additional positions for Li ions and the partial site occupancy for Li ions on their respective sites [24]. Legend: *green*—lanthanum, *violet*—lithium, *blue*—tantalum, *red*—oxygen, *grey*—cationic vacancies (Color figure online)

the space group I*a*-*3d*, where the Ln, M and Li cations sit on distinct crystallographic sites. In particular, the Li environment is tetrahedral. When x is equal to 3, each crystallographic site is fully occupied and the compounds can be called stoichiometric garnets; when x is greater than 3, the compounds are known as Li-stuffed garnets and may exhibit high ionic conductivity [6].

The first studied Li-stuffed garnet-type compounds were $Li_5La_3M_2O_{12}$ (M = Nb, Ta) [23]. Such compounds deliver a bulk conductivity of about $10^{-6}\,\Omega^{-1}\,cm^{-1}$ at room temperature. Their crystal structure is schematically represented in Fig. 4, together with an example of the structure of stoichiometric garnet. In Li-stuffed garnets the Li ions additional with respect to the stoichiometric formula are distributed, with partial occupancy, over tetrahedral and distorted octahedral sites [24]. It is in particular the presence of vacant crystallographic sites at these additional Li positions that allows the migration of Li ions according to a vacancy-mediated hopping.

Various types of cationic isovalent and aliovalent substitutions are possible for this structure, and this allows also modulating the Li content within the system. Interestingly, an almost linear relationship between Li content and conductivity was found for $3 < x < 7$ [23, 25], although differences in transport properties may depend on both synthesis and sintering conditions. Indeed, the thermal history of the compound can affect the Li ions distribution among the additional Li crystallographic sites available for Li-stuffed garnets. The highest bulk conductivity in the family of garnet-type Li-ion conductors, of $10^{-3}\,\Omega^{-1}\,cm^{-1}$ at room temperature, was found for the $Li_{6.4}La_3Zr_{1.4}Ta_{0.6}O_{12}$ composition [23].

The garnet-type Li ion conductors generally show high ionic conductivity, wide electrochemical window and excellent stability, even in combination with metallic Li as the anode material, thanks to the low tendency of element such as Ta, Nb and Zr to change their oxidation state. This also ensures negligible electronic contributions to the conductivity.

2.3 NASICON-Type Li-Ion Conductors

Compounds of general formula $LiA_2(PO_4)_3$ (A = Ti, Zr, Ge, Hf) belong to the family of NASICON-type Li ion conductors. Most of the NASICON (Na SuperIonic CONductors) materials crystallize in a rhombohedral three-dimensional (3D) network structure (space group R-3c) characterized by large tunnels perpendicular to the *c* axis (Fig. 5a). Deviations form this symmetry were reported for lithium-excess compounds, e.g. $Li_3Ti_2(PO_4)_3$, in which two types of crystallographic sites are available for Li [26]. These two sites are partially occupied and such vacant sites represent a favourable condition to promote Li diffusion in the system (Fig. 5b). The degree of order/disorder of Li ions between the A1 and A2 sites, as well as their occupation, is dependent upon synthetic conditions and composition. Grain boundary resistance is high in polycrystalline samples but, in general, a good densification process will reduce porosity and improve Li ion conductivity [27].

Among the $LiA_2(PO_4)_3$ system, the compounds with Ti present high Li ion conductivity [7 and reference therein]. Partial substitution of Ti^{4+} by Al^{3+} results in an improved conductivity for the series $Li_{1+x}Al_xTi_{2-x}(PO_4)_3$, with the composition $Li_{1.3}Al_{0.3}Ti_{1.7}(PO_4)_3$ showing the best performances (about $3 \times 10^{-3}\ \Omega^{-1}\ cm^{-1}$).

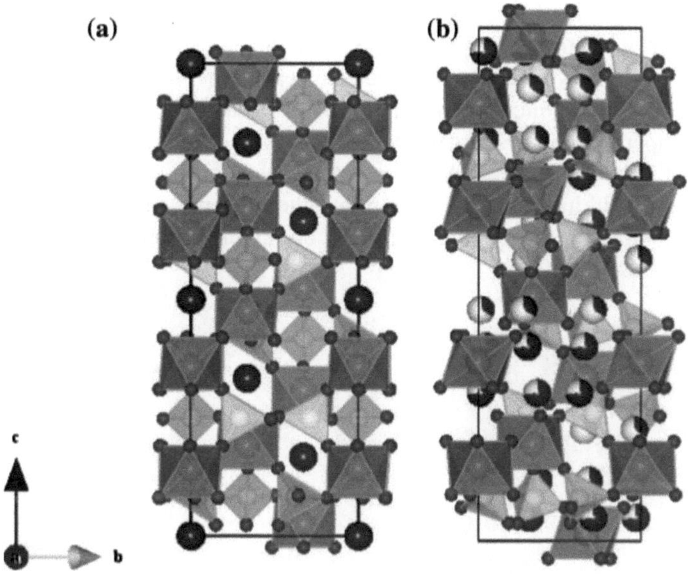

Fig. 5 Structural representation of the NASICON-type structure of **a** $LiTi_2(PO_4)_3$ (space group R-*3c*) showing the Li ions channels along the *a* axis; **b** $Li3Ti_2(PO_4)_3$ (space group R-*3*) showing the Li sites partially occupied [26]. Legend: *yellow*—phosporus, *violet*—lithium, *blue*—titanium, *red*—oxygen (Color figure online)

Just like the LLTO family of Li ion conductors, $LiTi_2(PO_4)_3$-based electrolytes are unstable towards Li metal anodes due to the facile reducibility of Ti^{4+} [21], but their excellent stability in water and air has prompted their application in all-solid-state lithium batteries as well as in lithium/air secondary batteries [28].

2.4 LISICON-Type Li-Ion Conductors

$Li_{14}ZnGe_4O_{16}$ is the main representative of the family of oxides known as LISICON (Lithium SuperIonic CONductors). The crystal structure of this family of compounds is a 3D skeleton where lithium ions can occupy four different crystallographic sites, two of them being interstitial sites where the Li ions are highly mobile. The conductivity of LISICON-type materials is generally poor, of the order of 10^{-7} Ω^{-1} cm^{-1} at room temperature, although it rapidly increases to 0.125 Ω^{-1} cm^{-1} at 300 °C [29]. Use of $Li_{14}ZnGe_4O_{16}$ is therefore impracticable because of its low conductivity at room temperature as well as its high reactivity towards Li metal.

The search for conductivity improvements in LISICON-type materials brought to the investigation of the thio-LISICON family of Li-ion conductors, in which the oxide ions are replaced by sulphur. The thio-LISICON family is represented by the general formula $Li_{4-x}A_{1-y}B_yS_4$ with A = Si, Ge and B = P, Al, Zn, Ga. The parent compound of this family is Li_4GeS_4 and its structure is based on isolated GeS_4 tetrahedra. In this structure (Fig. 6a) Li^+ ions are located in both octahedral and tetrahedral sites; LiS_6 octahedra are connected each other to form chains along the *b*-axis [30–32]. The conductivity of thio-LISICON materials exceeds 10^{-4} Ω^{-1} cm^{-1} at room temperature and is largely dependent upon composition. The highest conductivity of 2.2×10^{-3} Ω^{-1} cm^{-1} was reported for the system $Li_{3.25}Ge_{0.25}P_{0.75}S_4$, suggesting that lithium ion distribution and the presence of vacant sites on the Li sublattice compared to the parent Li_4GeS_4 compound positively affect the Li ion

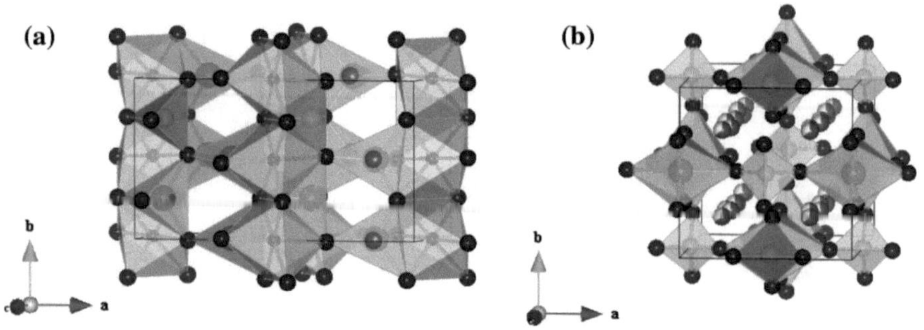

Fig. 6 **a** Crystal structure of Li_4GeS_4 [30–32]; **b** crystal structure of $Li_{10}GeP_2S_{12}$ [9] showing the channels running parallel to the *c* axis and occupied by the mobile Li ions on partially filled crystallographic sites. Legend: *yellow*—phosphorus, *cyano*—germanium, *violet*—lithium, *blue*—sulphur (Color figure online)

conduction in the structure. A similar effect was reported for the $Li_{4-x}Si_{1-x}B_xS_4$ system (B = Al, P), for which both the introduction of Li vacancies (Al doping) and Li interstitials (P doping) resulted in improved conductivities compared to Li_4SiS_4, with better performances for the Si-P solid solution.

As already stated, even higher values were claimed for the compound $Li_{10}GeP_2S_{12}$, presenting a conductivity of $1.2 \times 10^{-2}\ \Omega^{-1}\ cm^{-1}$ at room temperature [9]. $Li_{10}GeP_2S_{12}$, although closely related to the thio-LISICON family of conductor, present a completely different crystal structure (Fig. 6b) where Li ions with partial occupancy are present in channels running along the *c* crystallographic axis.

3 Glassy Electrolytes

3.1 Historical Framework

Warburg was the first one to report on ionic conduction in glasses at the beginning of the 20th century [33]. Glasses were also the first solid electrolytes to be used in thermochemical probes, and many cation-sensitive glass electrodes have been reported over the years. However, a characteristic of these glasses was their relatively high electrical resistivity at room temperature. The progresses in identifying glasses with higher ion conductivity were very slow until the '70s, probably because of the lack of studies about the local glass structure and the inability to separate carrier and mobility contributions to the conductivity.

The first examples of Fast Ion Conducting (FIC) glasses—at that time called "superionic"—were reported by Kunze in 1973 [34], and by Chiodelli et al. in 1974 [35]. Both these works were concerned with the transport of Ag^+ ions in AgI-Ag_2MO_4 (M = Se, Cr, W, Mo) glassy systems.

3.2 FIC Glasses Structure, Composition and Fabrication

Glasses are amorphous solids that present the glass transition, T_g, a quasi-thermodynamic phenomenon resembling a 2nd order phase transition. FIC glasses may be binary, ternary or even quaternary systems, including:

- one or two glass formers A_xB_y (A = Si, P, Ge, B; A = O, S);
- one glass modifier M_xA_y (M = Li, Na, K, Mg, Ag; A = O, S);
- one or more dopants M_xD_y (M = Li, Na, K, Mg, Ag; D = I, Cl, Br).

Here, Si, P, Ge and B are the classical glass formers individuated by Zachariasen chiefly on the basis of their ability to form coordination tetrahedra [36]. However, other metals usually forming different polyhedra (e.g. octahedra) can act as glass formers at least in certain composition ranges, e.g. MoO_2 in AgI–Ag_2MoO_4 [37].

Glasses were traditionally prepared by melt-quenching oxides, carbonates and/or nitrates at high temperature (even well above 1000 °C), and by pouring the melt on cold surfaces, or in liquid nitrogen. Glass-forming regions are normally reported in 2D and 3D ceramic phase diagrams. Glass-forming regions could be expanded by fast-quenching (up to 10^6 K/s) obtained by means of roller-quenchers [38]. During '90s sol-gel recipes were increasingly used to prepare silicate and later borate glasses [39]. Sol-gel techniques, in fact, allowed a further increase of the glass-forming regions and reduced the loss of light elements (e.g. Li, Na), thanks to the milder thermal conditions with respect to standard melt quenching.

During the last years, high-energy mechanical milling became a commonly used method to form amorphous materials and glasses [40]. This technique has two major advantages with respect to the other methods: the process is very simple and the synthesis can be performed at or near room temperature. Another advantage is that high-energy milling can induce chemical reactions and, therefore, make new compositions available.

3.3 Models of Ion Transport

Chandra et al. [11] recently published a thorough review on ion transport in FIC glasses. In most cases, ionic conduction is due to the motion of a single ionic species, either anion or cation. We may express the conductivity, σ, as the product of carrier charge concentration, n, and mobility, μ

$$\sigma = \mathrm{Ze}\, n\, \mu \tag{1}$$

where Ze is the charge on the conducting ion. The problem of modeling the ionic conductivity in glasses is that of understanding, and possibly separating, what determines n and μ. In the discussion of ionic conduction processes in glass, the models recurring up to the '90s were:

(i) *the strong electrolyte or Anderson-Stuart (A-S)*;
(ii) *the weak electrolyte (WE)*;
(iii) *the defect hypotheses.*

(i) The A-S model suggests that a substantial fraction of the ions is conducting. The activation energy, ΔE_A, is made of two terms

$$\Delta E_A = \Delta E_B + \Delta E_S \tag{2}$$

where ΔE_B is the electrostatic binding energy and ΔE_S is the so-called "strain energy", which accounts for the mechanical forces acting upon the ion as it expands the structure in order to move between sites [41]. Within some approximations, the A-S model gives

$$\Delta E_B = bZZ_0e^2/g(r + r_0) \quad (3)$$

$$\Delta E_S = 4pGr_0(r - r_D)^2 \quad (4)$$

where b and g are adjustable parameters, with the latter being set equal to the dielectric permittivity. Z and r and Z_0 and r_O are the charges and radii of cation and oxygen anion, respectively, r_D is the radius of the network constriction between the sites occupied by the cations, and G is the shear modulus of the glass.

The A-S model has been mainly applied to the borate glasses [42]. Here, ΔE_B decreases substantially with increasing alkali oxide, and this is probably due the cation jump distance, which decreases by decreasing the cation site separation, and thereby causing a greater overlapping of the coulomb orbitals.

(ii) WE theory was proposed by Ravaine and Souquet [43, 44]. They noted some similarities between the electrochemistry of glasses and aqueous systems, and hypothesized that when a modifier oxide or another dopant salt is added to a glassy matrix, alkali dissociation is the dominant energetic barrier experienced by the cation during the conduction process. Once a cation has been dissociated from its charge-compensating ion, it is free to migrate until it recombines. The dissociation takes place following simple chemical equilibria, e.g.

$$Li_2O \leftrightarrow Li^+ + OLi^- \quad (6)$$

where OLi^- is called the "dissociated anion site" and Li^+ is the "free" cation. In other words, whereas the A-S model supposes that the majority of cations do contribute to the conductivity, which is essentially determined by the mobility term, μ, the WE theory claims that only a small fraction of the ions are contributing, at a given time, to the conduction process. It should be pointed out that A-S is a "microscopic" approach, whereas WE is a "thermodynamic" one. In order to reconcile these two views, Martin and Angell [45] suggested these two models could be thought as the extremes of a more general approach, where both mobility and concentration cause conductivity variations. The two extreme models are applicable when one of the two barrier terms, migration (ΔE_S) or binding (ΔE_B), is the dominant one for a given glass: if $\Delta E_B > \Delta E_S$ the glass behaves like a weak electrolyte whereas, in contrast, the glass is a strong one. Since a strain energy barrier is always present because of the volume requirements of the migrating cations, the authors postulate the existence of metastable sites of "intermediate" energy to account for both dissociated and non-dissociated states.

(iii) Ionic transport in crystals usually involves migration of defects (vacancies, interstitial ions, interstitial pairs). Haven and Verkerk [46] argued that when the ionic motion involves defect mechanisms the experimental diffusion coefficient, D, is different (because of correlation effects) from the calculated one, D^*, which obeys the Nernst-Einstein equation

$$D^* = \sigma kT / ne^2 \tag{7}$$

Here, n is again the carrier concentration, and the other symbols have the usual meaning. The so-called "Haven ratio", D/D^*, gives information about the nature of the mechanisms involved in the conduction. In crystals the formation of defects can be treated in term of a quasi-chemical equilibrium, where concentrations and mobility of the defects may be controlled with doping. Whereas for a long time we believed that such a procedure could not be applied to glasses, since the ions cannot be considered as "foreign" to the matrix [47], Moynihan and Lesikar [48] applied the concept of doping to the Mixed Alkali Effect (the combination of two cations in a common glassy matrix, MAE), to explain the rapid falls in conductivity.

More recently, other microscopic and thermodynamic models were proposed to account for conductivity in FIC glasses. The most important of them are:

(a) *Random-site model*
Assuming that all the available ions as potentially mobile, Nassau et al. [49] proposed there should exist a wide distribution of alkali ion sites in glasses having differing local free energies. The activation energies of conductivity in these glasses will vary resembling the distribution of the local free energies. Glass and Nassau observed that the activation energy behaved as a linear function of Li content for many Li-based FICs, including $Li_2O{:}Al_2O_3$, $LiO_2{:}B_2O_3$, $Li_2O{:}Ga_2O_3$ etc.

(b) *Decoupling index model*
Angell [50] proposed the so called "decoupling index model" by introducing an index R_τ defined as the ratio of two relaxation times:

$$R_\tau = \tau_s / \tau_\sigma \tag{8}$$

where τ_s and τ_σ are the mechanical (=structure) and the electrical (=transport) relaxation times, respectively. Glasses with high R_τ values (e.g. the inorganic ones based on Zachariasen's glass formers) are termed "strong", whereas if R_τ is low they are called "fragile" (e.g. polymers and organic liquids like glycerol). For Ag^+ ion conducting iodomolybdate glasses, $R_\tau \sim 10^{14}$ [51], whereas for poorly conducting glasses, $R_\tau \sim 10^4$. Because there is such a large variation in R_τ, viscosity is indicative of analogous variations in ionic mobility and conductivity. In the lower temperature region (below the T_g), R_τ is generally higher than 10^{12}, which is indicative of the fact that the ion transport is decoupled from structural dynamics. R_τ changes very rapidly across the glass transition.

(c) *Cluster by-pass model*
Ingram et al. [52] suggested that in FIC glassy electrolytes ordered clusters (macro domains) with diameter in the range 2.5–5.0 nm will exist, within which preferred conducting pathways (a sort of connecting tissue) are imbedded. Fast ion conduction may be described as the migration of mobile ions through these preferred conducting pathways, which surround insulating (or less conducting) clusters. When the glass is quenched below T_g, the residual liquid, initially surrounding the clusters, gets solidified as a highly disordered phase forming conducting pathways. The presence of a few foreign cations on the surface of the clusters during cooling may cause blocks to preferred pathways. These blocks divert the current through the clusters. Cluster bypass model successfully provided simple explanations of transport in many FIC glasses, including those exhibiting mixed-alkali effect, high value of decoupling index, R_τ, as well as the curvature observed in Arrhenius plots of some AgI-rich glasses, which was explained as a result of continuous exchange of mobile ions between cluster and tissue region [53].

(d) *Dynamic structure model (DSM)*
The approach proposed by Bunde et al. [54, 55] is based on number of key features, the most important of them being: (i) at temperatures far below the glass transition temperature, the glassy structure is completely frozen; (ii) the mobile cations themselves are active enough to create/determine their own glass structures; (iii) ion transport in glasses takes place stepwise following a jump/hop mechanism. Within this model the Authors were able to explain some important findings about FICs, and namely:

- the anomalous dependence of ionic conductivity on the content of glass modifier. Up to 6 orders-of-magnitude increase were observed in the ionic conductivity of xNa$_2$O:(1 − x)B$_2$O$_3$ glass when x increased from 0.15 to 0.5;
- the strong decrease of the ionic conductivity as a consequence of MAE. For example, decreases of conductivity up to 10^4 times were observed in [xK$_2$O:(1 − x)Li$_2$O]:2SiO$_2$ for $x \cong 0.5$ with respect to the pure binary systems;
- other minor consequences of MAE, such as variation in diffusion coefficient of mobile cations, maxima of Haven's ratio, occurrence of internal friction peaks in the mechanical loss spectrum.

(e) *MIsmatch Generated Relaxation for the Accommodation and Transport of IONs (MIGRATION) model*
Recently, Funke and co-workers [56] explored the concept of ion hopping dynamics in disordered materials. Following their approach, structural and dynamic disorders are the key factors to understand ion conduction in FIC glasses. Here, ionic transport can no longer be described in terms of individual defects performing random walks in a static energy landscape, but rather in terms of a more challenging many-particle problem, with the mobile ions

interacting with each other and with their surrounding matrix. Each mobile ion is surrounded by its mobile neighbors, which create a cage-like potential for it. By ion initial hop, a mismatch is created and the actual position of the ion would be different from the position where it is expected for by the neighbors. According to the MIGRATION concept, the system operates in order to reduce the mismatch through two competing relaxations ways: single-particle and many-particles. In the single-particle route the ions hop backwards, whereas in the other route the neighbours rearrange themselves. The mismatch-generated relaxation process occurring along the many-particle route finally leads to an accommodation of the ion at the new site.

(f) *Graded percolation* (*GP*) *model*

This model was recently proposed by Mustarelli et al. [57] in order to describe the transport properties of AgI-based glasses. It is based on previous Reverse Monte Carlo (RMC) structural models of borate and molybdate glasses [58], coupled with bond valence calculations, which showed that the conductivity is related to the formation of "infinite pathways clusters" for transport, where the silver ions do experience a mixed iodine/oxygen (I/O) coordination. In GP model, the main factor affecting ionic conductivity is the mobility of the Ag^+ carriers, which is controlled by the Ag local environment. Ionic conductivity is explained in terms of a percolation between a low-conducting phase (purely oxygen-coordinated sites), and a high-conducting one (I/O coordinated sites). The percolation takes place along pathways (nano-channels) with fractal structure. The nature of the glass network (connectivity and dimensionality) plays a significant role only for low I/O values. This allowed explaining the transport and thermal anomalies observed in AgI-based borate and phosphate glasses.

3.4 Li-Based FIC Glasses

Lithium-ion conducting glasses can be divided into two big categories: oxides and sulphides. Lithium oxide glasses are more common and easy to prepare, however their conductivity generally does not exceed 10^{-8}–10^{-6} Ω^{-1} cm^{-1}, which is not enough for applications in high power density devices such as lithium batteries. Sulphide glasses can offer much higher conductivity levels (10^{-5}–10^{-3} Ω^{-1} cm^{-1}) due to the higher polarizability of sulphur ions, however they are very sensitive to moisture and do require inert ambient for preparation and handling. In particular, reaction with moisture can generate H_2S [21]. This problem can be partially solved, at the expenses of the conductivity level, by substituting a fraction of sulphur atoms with oxygens. From the point of view of the composition, for a better addressing of the glass forming regions, it is convenient to separately discuss binary and ternary glasses.

3.4.1 Li Binary Glasses

These are glasses of the systems $x\text{Li}_2\text{O}:(1 - x)\text{M}_y\text{O}_z$ and $x\text{Li}_2\text{S}:(1 - x)\text{M}_y\text{S}_z$ (M = Si, B, P, Ge, etc.). The conductivity normally increases by increasing the molar fraction, *x*, of the glass modifier Li_2O (Li_2S). Figures 7 and 8 report the behavior of the conductivity near the room temperature for some model lithium oxide and sulphide glassy systems, respectively.

From the data here reported it is clear that sulphide glasses display ion conductivity higher than oxide ones. It is also clear that the conductivity values strongly depend on the structure of the matrix, which is almost fully interconnected in the case of Si-based glasses, gets to be more open on B-ones and is chain-like in $0.5\text{Li}_2\text{M}:0.5\text{P}_2\text{M}_5$. A good discussion of the structure-properties relationships is reported in Ref. [10].

3.4.2 Li Ternary Glasses

Ternary systems may be obtained by adding a dopant (e.g.: LiI) to the binary glass. High conductivity levels in the range 10^{-3}–10^{-2} Ω^{-1} cm^{-1} at 300 °C were early reported for $\text{LiCl:Li}_2\text{O:B}_2\text{O}_2$ [59]. Sulphide glasses are among the best Li^+ conductors at room temperature. Figure 9 reports the room temperature conductivity for several compositions as a function of LiI content. It seems that conductivity tends to a limiting value of approximately 10^{-3} Ω^{-1} cm^{-1} as also reported for silver glasses doped with AgI [57].

Another way to improve the conductivity of glasses is to exploit the so-called "mixed anion" or "mixed former" effect, where two glass formers are mixed to give

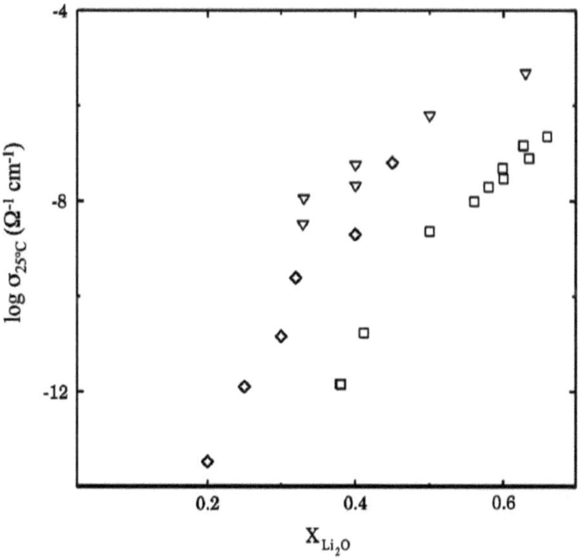

Fig. 7 Conductivity behaviour of glasses of the systems $x\text{Li}_2\text{O}:(1 - x)\text{M}_y\text{O}_z$. *Triangles* $\text{Li}_2\text{O:SiO}_2$; *rhombs* $\text{Li}_2\text{O:B}_2\text{O}_3$; *squares* $\text{Li}_2\text{O:P}_2\text{O}_5$. Graph redrawn from Ref. [10] (Color figure online)

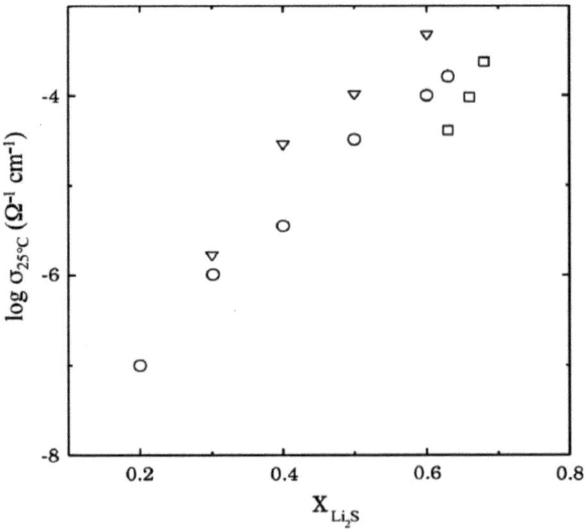

Fig. 8 Conductivity behaviour of glasses of the systems xLi$_2$S:(1 − x)M$_y$S$_z$. *Triangles* Li$_2$S:SiS$_2$; *circles* Li$_2$S:GeS$_2$; *squares* Li$_2$S:P$_2$S$_5$. Graph redrawn from Ref. [10] (Color figure online)

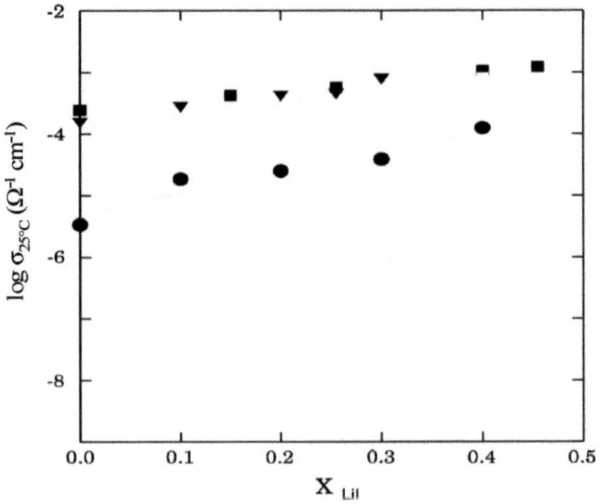

Fig. 9 Conductivity behaviour of glasses belonging to ternary systems xLiI:(1 − x)[Li$_2$S-M$_y$S$_z$] (M = Si, P, Ge). *Triangles* SiS$_2$; *circles* GeS$_2$; *squares* P$_2$S$_5$. Graph redrawn from Ref. [10] (Color figure online)

a ternary system, or even a quaternary one if a dopant is also used. Examining the alkali borophosphate families over a wide range of B/P ratios it was observed [60] that the ionic conductivity of these glasses is a complex function of composition and that, for the same alkali content, when P_2O_5 was substituted for B_2O_3 the conductivity passes through a maximum. The same results were later reported for sulphide glasses [61].

Among the most recent results, a conductivity value of $1.6 \times 10^{-3}\ \Omega^{-1}\ \mathrm{cm}^{-1}$ at 25 °C was reported for the glass 50Li$_2$S:17P$_2$S$_5$:33LiBH$_4$ [62]. Seino et al. [63] recently reported a value of $1.7 \times 10^{-2}\ \Omega^{-1}\ \mathrm{cm}^{-1}$ for the glass-ceramic composition

$70Li_2S{:}30P_2S_5$. However, the high values obtained on glass-ceramic systems are often due to grain boundary contributions and/or impurities (chiefly in the case of sulphides), and need a very careful experimental check.

4 Solid Polymer Electrolytes

With respect to ceramic and glassy materials, the polymer electrolytes show important benefits including flexibility, free-standing performances and an easier processability. In addition, the use of "dry" polymeric systems may limit some important drawbacks, related to the solvent leakage, the high Li reactivity and the consequent dendrites formation [4, 64]. They behave both as the separator and the electrolyte, also leading to more stable solid-state interfaces. In terms of battery safety, the presence of a polymer in the cell may guarantee higher thermal stability and thermal excursion up to 200 °C. Due to these promising aspects, SPEs are considered as a possible alternative to the liquid ones in Li- or Na-based batteries.

The concept of SPE dates back to 70s, when Armand firstly proposed a new ion conductor based on a lithium salt properly complexed by a polar and aprotic polymer matrix without the use of any liquid component (additives or liquid electrolytes) [65]. At the beginnings, the research on SPEs was exclusively focused on poly(ethyleneoxide) (PEO) as the complexing polymer [66]. Ever since, a lot of polymer/salt systems were deeply investigated, such as those based on PMMA, PAN, PVDF [66–69]. In principle, SPEs must satisfy some basic requirements: (i) ionic conductivity higher than 10^{-4} S/cm at room temperature, (ii) good thermal, chemical and mechanical stability, (iii) lithium transport number close to the unity, and (iv) compatibility with the electrodes and consequently wide electrochemical windows [67].

In spite of the wide spectrum of SPEs available in the literature, the preferred combinations are still those based on polyethylene oxide (PEO). PEO is a semi-crystalline polymer whose glass transition temperature T_g and melting point T_m are near −60 and 70 °C, respectively. It allows a better Li solvation due to its structural similarity to the crown-ethers and to the presence of ether oxygen in the structure, which confers some polarity ($\varepsilon_r = 8$) [4, 66–69]. Li transport takes place in the amorphous polymer domains above the glass transition temperature, T_g, via an oxygen-assisted ion hopping promoted by the long-range segmental motion along the polymer backbone (see Fig. 10).

The ionic conductivity of the PEO electrolytes, therefore, strictly depends on both crystallinity degree and T_g of the amorphous phase [70]. In particular, it is very low at room temperature ($<10^{-6}\ \Omega^{-1}\ \mathrm{cm}^{-1}$) where the crystalline fraction of the polymer is predominant, but it abruptly increases around the melting temperature reaching 1 $\Omega^{-1}\ \mathrm{cm}^{-1}$ above 80–90 °C, where the entire polymer is in a viscous liquid state. The research on SPEs ever pointed towards the enlargement of the amorphous fraction by means of several approaches. One of the most followed strategies was the choice of suitable Li salts and the identification of proper salt

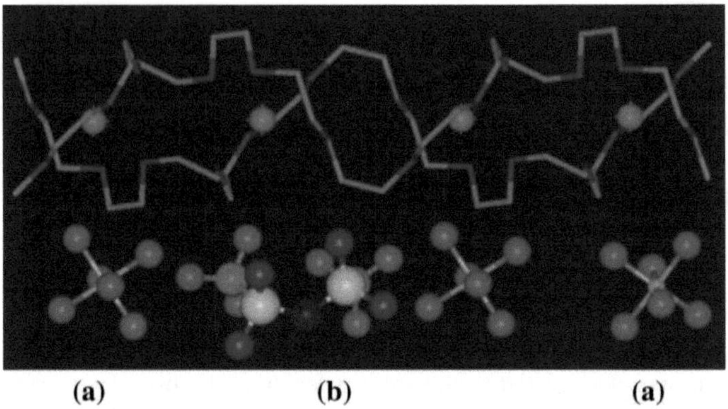

Fig. 10 Crystal structure of PEO_6:$(LiAsF_6)_{1-x}(LiTFSI)_x$. *Blue spheres* Li ions; *Pink structures* **a** AsF_6^- anion; *mixed colors structure* **b** $TFSI^-$ anion. Taken from Ref. [68] with the permission of the publisher (Color figure online)

Fig. 11 Conductivity versus temperature for some PEO: LiX complexes at different molar ratio, *n*. gl: Li_2O:$3B_2O_3$ glass. Unpublished data from the Authors (Color figure online)

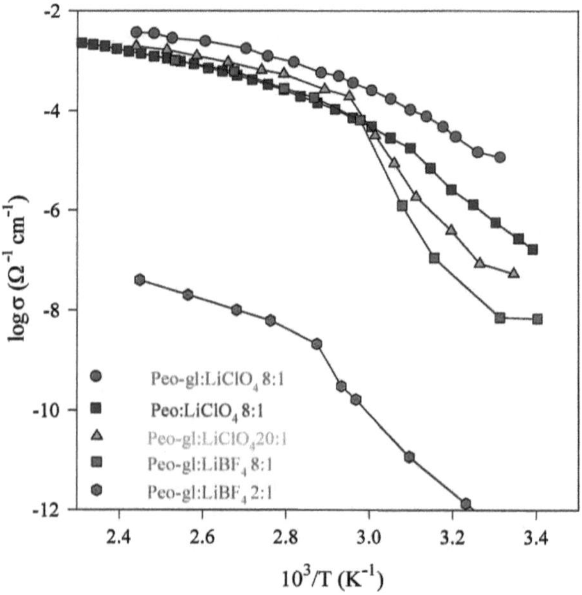

concentrations. Several PEO_n-LiX were explored in the past, by changing both the molar ratio *n*, expressed as mol_{EO}/mol_{Li}, and the anion X (where X = halide, ClO_4, CF_3SO_3, PF_6, TFSI, BF_4 etc.). Generally speaking, the presence of the salt increases the polymer amorphous fraction, that can reach 100 % for well-defined molar ratios, *n*, generally in the range 8–20, depending on the employed salt [4, 5]. In case of fully amorphous PEO-LiX, ionic conductivity exceeding $10^{-5}\ \Omega^{-1}\ cm^{-1}$ is generally observed, even at room temperature (see Fig. 11).

However, the absence of crystalline domains causes a drastic worsening of the mechanical performances (dimensional stability, filmability, free-standing properties). In addition, the amorphous phase is metastable and undergoes to crystallization within days or weeks, so causing drastic drops of conductivity. At higher salt content ($n < 6$), thermal and spectroscopic techniques evidence the presence of crystalline aggregates, which are responsible for the reduction of the conductivity values, because of the formation of ion pairs. In 2005, Bruce showed that some crystalline complexes (e.g. PEO-$LiAsF_6$) with $n = 6$ are arranged in unique cylindrical structures, where the lithium ions are not coordinated by the anions. Such complexes are not good ion conductors, but their conductivity remarkably improves when the AsF_6^- anions are partially replaced by the isovalent $TSFI^-$ ones [71].

SPEs are bi-ionic systems. This means that both anions and cations are mobile; unfortunately the bulk conductivity is primarily due to the anion mobility. In fact, lithium transference numbers, t^+_{Li}, as low as 0.05–0.3 are generally obtained in case of PEO-based electrolytes [61]. The higher mobility causes an accumulation of the anions around the electrodes, because they exchanges with Li^+ only. The resulting gradients in the salt concentration are responsible for the electrode polarization and do lead to worsening of performances. Some attempts to overcome this drawback were carried out during these last decades, as for instance the use of large and heavy anions (e.g. $TFSI^-$) or of anion receptors, like boron, linear and cyclic azo-ether compounds, or calix-arene derivatives [72], which behave similarly to the crown-ethers in case of cations. However, these strategies failed for different reasons: in some cases no substantial enhancement of t^+_{Li} was observed, in other ones the complexation of the anion caused a conductivity decrease. The more realistic strategy to forbid counter-ion diffusion is the anion anchoring to the polymer chains via chemical bonding to form single-ion conductors. The systems including charged groups along the backbone are known as polyelectrolytes [64, 73]. In the particular of case of batteries, this approach results in free cations with higher long-range mobility and consequently t^+_{Li} values very close to unit. Several classes of polymers were investigated as matrices for single-ion conduction, ranging from organic and/or inorganic polyanions, to polymeric ionic liquids and charged di- or tri-block copolymers [73].

The technological use of PEO-salt complexes may be possible if a proper compromise among the following factors is reached: (i) suitable ionic conductivities, (ii) good mechanical properties, and (iii) higher transference number. Many efforts were made to this aim. One explored way was the blending and/or cross-linking of PEO with other compatible polymers, as poly(acrylic acid) (PAA) and poly(methylmethacrilate) (PMMA), which improve the mechanical performances and increase the conductivity as well as the lithium transport number, by blocking the anion. In particular, polystyrene has been widely used to give a better dimensional stability to the polyether systems. In this case, the physical and chemical properties of the mixed system may be modulated by changing parameters as the ratio between the two polymers [4, 73]. Gomez et al. [74] for instance, found that the ionic conductivity of poly(styrene-block-ethylene oxide) copolymers increases with the molecular weight, due to local stresses in the block copolymer microdomains, which interfer with the capability of PEO to coordinate Li^+ cations.

The addition of ceramic particles to PEO-based electrolytes to form nano- or micro-composite systems is maybe the most interesting way to prevent the electrode degradation phenomena, to improve the chemical, thermal and mechanical stability and finally to reduce the tendency to crystallization. Many micro- and nano inorganic oxides were added to the polymer during the film casting, in particular insulating SiO_2, Al_2O_3 and TiO_2, but also superacid conducting zeolites, or lithium-based glasses, or more recently also $LiNbO_3$ or $BaTiO_3$ [75]. Besides the benefits discussed above, the fillers also improve the ionic transport and the SEI stability. The entity of these effects depends on the morphological features of the filler, namely particle dimension, which should be lower than 1 μm, and microstructure. An important advance in this field is the use of mesoporous fillers, like SBA-15 or MCM-41. It was found that the dispersion of such nanoporous fillers in PEO-$LiClO_4$ electrolytes leads to a conductivity enhancement three times higher than that obtained in case of microsized silica. This improvement has been interpreted in terms of a suppression of the PEO crystalline fraction, as proved by means of DSC measurements on the corresponding SiO_2-based composite electrolytes. The typical high surface area of the mesoporous materials favours the formation of a larger polymer/filler interphase, which has positive effect even on both electrochemical stability windows and transport numbers [75].

Some years ago, LIBOB was proposed as a new salt for PEO-based SPEs and nanocomposites with alumina and nanosized silica. This salt has a bulky anion with plasticizing properties, able to hinder the polymer crystallization. Ionic conductivity of about $10^{-5}\ \Omega^{-1}\ cm^{-1}$ were observed at 30 °C. The most interesting result was a transference number very close to 0.9 in case of a SPE based on PEO-PMMA block copolymers with high LIBOB loading ($n = 3$). These membranes also showed a wide electrochemical window exceeding 4.0 V and good interfacial stability with the lithium anode. A further advance in this field is the dissolution in PEO matrices of lithium borate salts, namely $Li–[CH_3(OCH_2CH_2)_nO]_3BC_3H_9$ containing a number of oxyethylene substituents, n, in the range 1–7. The salt with $n = 3$ shows an interesting conductivity of 2×10^{-5} S/cm at room temperature [5, 75].

Other polymers were also studied as alternative matrices for SPEs, all of them including ethylene oxide as basic unit, like for instance polyethylenoxide-methylethermethacrylate (PEOMA) and polyethylenglycol-alkylacrylate. However, no important improvements have been reported, and conductivity similar to that one observed in case of PEO-LiX complexes was typically observed [4, 73].

More recently, as a consequence of the use of ionic liquids (ILs) in lithium batteries, a new class of solid polymer electrolytes was proposed by Ohno and co. [76]. Basically, these SPEs, better known as polymeric ionic liquids (PILs), are prepared through the radical polymerization of ILs by properly combining different cations, anions and polymer backbones [77]. PILs show the potential advantage to combine the benefits of ionic liquids (high ionic conductivity, high thermal stability, non-flammability and low volatility) with those ones of polymers, namely mechanical stability, free-standing properties, safety, easier processing, packaging, etc. However, as in the case of other polyelectrolytes, the ionic conductivity is often too low for practical applications. In fact, whereas typical conductivity values obtained with the ionic liquids are around $10^{-3}\ \Omega^{-1}\ cm^{-1}$ at room temperature, after the polymerization process the conductivity dramatically falls, sometimes even of 2–4 orders-of-magnitude. Ohno found values 10^4 times lower for the system based on ethylvinylimidazolium-TFSI after the polymerization [76]. This phenomenon was rationalized in terms of a remarkable increase of T_g and of a reduction of the free mobile ions, because of the covalent bonding of the monomer components.

Finally, the most recent promising classes in the field of SPEs are the zwitterions, identified as "Li ion dissociation enhancers", namely ionic liquids where both the anion and cation are immobilized on the same structure. If dispersed into proper PEO-like polymers, the zwitterionic compounds may increase the ionic conductivity of the polymer electrolytes [80]. In case of addition of 1-butylimidazolium-3-(n-butanesulphonate) dissociator to Lithium methylacrylate copolymers, for instance, the ionic conductivity achieves a maximum close to $0.6\ m\ \Omega^{-1}\ cm^{-1}$ at 30 °C, in presence of a Li^+ mole fraction in the copolymer of about 0.050.

5 Conclusions

At present, the research on solid-state Li^+ conductors is chiefly focused on ceramic materials, which seem to offer the most room for improvements. Here, in addition to the investigation of structure/transport relationships, many issues must be considered: meso- and microstructure (particle dimensions, grain boundaries, thin films, etc.), chemical, electrochemical and mechanical stability. Glasses seem to have reached their maximum potential, having been explored as far as concerns composition, transport mechanisms, local and even medium range structure. Polymers (and composite) electrolytes may reserve some good news, chiefly in terms of chemical stability towards the electrodes, and electrochemical compliance with the recently proposed 5 V cathodes.

Magistris [10] concluded his review with these words: "Composite materials, dispersed phases or dynamically heterogeneous (solid–liquid) systems are little known and have enormous potential of optimization. They usually contain at least an amorphous, or highly disordered component, and their conductivity behaviour is strongly influenced by interfacial phenomena. If our skills in controlling the fine scale morphology of the "electrolyte system" will grow enough, we may dream of imitating a natural membrane, with its switchable and highly efficient ionic channels." About twenty years later this fine control has yet to be fully realized, and a great deal of attention must be devoted to bottom-up preparation techniques, such as atomic layer deposition, pulsed laser deposition, molecular self-assembling, etc.

References

1. DOE/EPRI 2013 Electricity Storage Handbook
2. Bruce PG, Freunberger SA, Hardwick LJ, Tarascon J-M (2012) Li–O_2 and Li–S batteries with high energy storage. Nat Mater 11:19–29
3. Hayashi A, Noi K, Sakuda A, Tatsumisago M (2012) Superionic glass-ceramic electrolytes for room-temperature rechargeable sodium batteries. Nat Commun 3:856–860
4. Fergus JW (2010) Ceramic and polymeric solid electrolytes for lithium-ion batteries. J Power Sour 195:4554–4569
5. Quartarone E, Mustarelli P (2011) Electrolytes for solid-state lithium rechargeable batteries: recent advances and perspectives. Chem Soc Rev 40:2525–2540
6. Cussen EJ (2010) Structure and ionic conductivity in lithium garnets. J Mater Chem 20:5167–5173
7. Cao C, Li Z-B, Wang X-L et al (2014) Recent advances in inorganic solid electrolytes for lithium batteries. Front Energy Res 25:1–10
8. Thangadurai V, Narayanan S, Pinzaru D (2014) Garnet-type solid state fast Li ion conductors for Li batteries: critical review. Chem Soc Rev 43:4714–4727
9. Kamaya N, Homma K, Yamakawa Y et al (2011) A lithium superionic conductor. Nat Mater 10:682–686
10. Magistris A (1993) Ionic conduction in glasses. In: Scrosati B, Magistris A, Mari CM, Mariotto G (eds) Fast ion transport in solids: proceedings of the NATO advanced research workshop, Belgirate, Italy. NATO Science Series E, vol 250. Kluwer, Dordrecht, pp 120–132, 20–26 September 1992
11. Chandra A, Bhatt A, Chandra A (2013) Ion conduction in superionic glassy electrolytes: an overview. J Mater Sci Technol 29:193–208
12. Belous AG, Novitskaya GN, Polyanetskaya SV et al (1987) Study of complex oxides with the composition $La_{2/3-x}Li_{3x}TiO_3$. Inorg Mater 23:412–415
13. Inaguma Y, Chen L, Itoh M et al (1993) High ionic conductivity in lithium lanthanum titanate. Solid State Commun 86:689–693
14. Kawai H, Kuwano J (1994) Lithium ion conductivity of A-site deficient perovskite solid-solution $La_{0.67-x}Li_{3x}TiO_3$. J Electrochem Soc 141:L78–L79
15. Gao X, Fisher CAJ, Kimura T et al (2013) Lithium atom and A-site vacancy distributions in lanthanum lithium titanate. Chem Mater 25:1607–1614
16. Bohnke O (2008) The fast lithium-ion conducting oxides $Li_{3x}La_{2/3-x}TiO_3$ from fundamentals to application. Solid State Ionics 179:9–15
17. Stramare S, Thangadurai V, Weppner W (2003) Lithium lanthanum titanates: a review. Chem Mater 15:3974–3990
18. Alonso JA, Sanz J, Santamaria J et al (2000) On the location of Li^+ cations in the fast Li-cation conductor $La_{0.5}Li_{0.5}TiO_3$ perovskite. Angew Chem Int Ed 39:619–621
19. Emery J, Buzare JY, Bohnke O et al (1997) Lithium-7 NMR and ionic conductivity studies of lanthanum lithium titanate electrolytes. Solid State Ionics 99:41–51
20. Harada Y, Hirakoso Y, Kawai H et al (1999) Order-disorder of the A-site ions and lithium ion conductivity in the perovskite solid solution $La_{0.67-x}Li_{3x}TiO_3$ ($x = 0.11$). Solid State Ionics 121:245–251
21. Knauth P (2009) Inorganic solid Li ion conductors: an overview. Solid State Ionics 180:911–916
22. Inaguma Y, Nakashima M (2013) A rechargeable lithium-air battery using a lithium ion-conducting lanthanum lithium titanate ceramics as an electrolyte separator. J Power Sour 228:250–255
23. Thangadurai V, Kaack H, Weppner WJF (2003) Novel fast lithium ion conduction in garnet-type $Li_5La_3M_2O_{12}$ (M = Nb, Ta). J Am Ceram Soc 86:437–440
24. Cussen EJ (2006) The structure of lithium garnets: cation disorder and clustering in a new family of fast Li^+ conductors. Chem Commun 4:412–413

25. Ramzy A, Thangadurai V (2010) Tailor-made development of fast Li Ion conducting garnet-like solid electrolytes. ACS Appl Mater Interfaces 2:385–390
26. Aatiq A, Menetrier M, Croguennec L et al (2002) On the structure of $Li_3Ti_2(PO_4)_3$. J Mater Chem 12:2971–2978
27. Aono H, Sugimoto E, Sadaoka Y et al (1991) Electrical property and sinterability of $LiTi_2(PO_4)_3$ mixed with lithium salt (Li_3PO_4 or Li_3BO_3). Solid State Ionics 47:257–264
28. Shimonishi Y, Zhang T, Imanishi N et al (2011) A study on lithium/air secondary batteries–stability of the NASICON-type lithium ion conducting solid electrolyte in alkaline aqueous solutions. J Power Sour 196:5128–5132
29. Robertson AD, West AR, Ritchie AG (1997) Review of crystalline lithium-ion conductors suitable for high temperature battery applications. Solid State Ionics 104:1–11
30. Kanno R, Hata T, Kawamoto Y, Irie M (2000) Synthesis of a new lithium ionic conductor, thio-LISICON lithium germanium sulfide system. Solid State Ionics 130:97–104
31. Murayama M, Kanno R, Kawamoto Y, Kamiyama T (2002) Structure of the thio-LISICON, Li_4GeS_4. Solid State Ionics 154–155:789–794
32. Murayama M, Kanno R, Irie M, Ito S, Hata T, No Sonoyama, Kawamoto Y (2002) Synthesis of new lithium ionic conductor Thio-LISICON—lithium silicon sulfides system. J Solid State Chem 168:140–148
33. Warburg E (1913) Über die Diffusion von Metallen in Glas. Ann Phys 40:327–334
34. Kunze D (1973) Silver ion conducting electrolyte with glass-like structure. In: Van Gool W (ed) Fast ion transport in solids. North Holland, Amsterdam, pp 405–408
35. Chiodelli G, Magistris A, Schiraldi A (1974) Some solid electrolyte cells. Electrochim Acta 19:655–656
36. Zachariasen WH (1932) The atomic arrangement in glass. J Am Chem Soc 54:3841–3851
37. Tomasi C, Mustarelli P, Magistris A (1998) Devitrification and metastability: revisiting the phase diagram of the system $AgI:Ag_2MoO_4$. J Solid State Chem 140:91–96
38. Barney ER, Hannon AC, Holland D, Winslow D, Rjial B, Affatigato M, Feller SA (2007) Structural studies of lead aluminate glasses. J Non-Cryst Solids 353:1741–1747
39. Mustarelli P, Quartarone E, Benevelli F (1997) A ^{11}B and ^{7}Li MAS-NMR study of sol-gel lithium triborate glass subjected to thermal densification. Mat Res Bull 32:679–687
40. Hayashi A, Hama S, Morimoto H et al (2001) Preparation of Li2S-P2S5 amorphous solid electrolytes by mechanical milling. J Am Ceram Soc 84:477–479
41. Anderson OL, Stuart DA (1954) Calculation of activation energy of ionic conductivity in silica glasses by classical methods. J Am Ceram Soc 37:573–780
42. Martin SW (1988) Conductivity activation energy relations in high sodium-content borate and aluminoborate glasses. J Am Ceram Soc 71:438–445
43. Ravaine D, Souquet JL (1977) A thermodynamic approach to ionic conductivity in oxide glasses. Part. 1. Correlation of the ionic conductivity with the chemical potential of constituents in binary alkali oxide glasses. Phys Chem Glasses 18:27–31
44. Ravaine D, Souquet JL (1978) A thermodynamic approach to ionic conductivity in oxide glasses. Part. 2. A statistical model for the variations of the chemical potential of constituents in binary alkali oxide glasses. Phys Chem Glasses 19:115–120
45. Martin SW, Angell CA (1986) Dc and ac conductivity in wide composition range Li2O-P_2O_5 glasses. J Non-Cryst Solids 83:185–207
46. Haven Y, Verkerk B (1965) Diffusion and electrical conductivity of sodium ions in sodium silicate glasses. Phys Chem Glasses 6:38–45
47. Charles RJ (1961) Polarization and diffusion in silicate glasses. J Appl Phys 32:1115–1126
48. Moynihan CT, Lesikar AV (1981) Weak-electrolyte models for the mixed alkali effectin glass. J Am Ceram Soc 64:40–46
49. Nassau K, Glass AM, Grasso M et al (1981) Quenched lithium-containing multiple sulphate glasses. J Non-Cryst Solids 46:45–58
50. Angell CA (1986) Recent developments in fast ion transport in glassy and amorphous materials. Solid State Ionics 18&19:72–88

51. Kawamura J, Shimoji M (1986) Ionic conductivity and glass transition in superionic conducting glasses $(Agi)_{1-x}(Ag_2MoO_4)_x$ ($x = 0.25, 0.30, 0.35$): I. Experimental results in the liquid and glassy states. J Non-Cryst Solids 88:281–294
52. Ingram MD, Mackenzie MA, Muller W et al (1988) Cluster and pathways: a new approach to ion migration in glass. Solid State Ionics 28–30:677–680
53. Senapati H, Parthasarathy G, Lakshmikumar SK et al (1983) Effect of pressure on the fast-ion conduction in silver iodide-silver oxide-molybdenum oxide glasses. Phil Mag B 47:291–297
54. Bunde A, Ingram MD, Maass P et al (1991) Mixed alkali effects in ionic conductors: a new model and computer simulations. J Non-Cryst Solids 131:1109–1112
55. Maass P, Bunde A, Ingram MD (1992) Ion transport anomalies in glasses. Phys Rev Lett 68:3064–3067
56. Funke K, Banhatti RD, Radha D (2006) Ionic motion in materials with disordered structures. Solid State Ionics 177:1551–1557
57. Mustarelli P, Tomasi C, Magistris A (2005) Fractal nanochannels as the basis of the ionic transport in AgI-based glasses. J Phys Chem B 109:17417–17421
58. St Adams, Swenson J (2000) Determining ionic conductivity from structural models of fast ionic conductors. Phys Rev Lett 84:4144–4147
59. Button DP, Tandon RP, Tuller HL et al (1981) Fast Li^+ conductance in chloroborate glasses II-diborates and metaborates. Solid State Ionics 5:655–658
60. Magistris A, Chiodelli G, Villa M (1985) Lithium borophosphate vitreous electrolytes. J Power Sour 14:87–91
61. Desphande V, Pradel A, Ribes M (1988) The mixed glass former effect in the $Li_2S{:}SiS_2{:}GeS$ system. Mat Res Bull 23:379–384
62. Yamauchi A, Sakuda A, Hayashi A, Tatsumisago M (2013) Preparation and ionic conductivities of $(100 - x)(0.75Li_2S{:}0.25P_2S_5){:}xLiBH_4$ glass electrolytes. J Power Sour 244:707–710
63. Seino Y, Ota T, Takada K, Hayashi A, Tatsumisago M (2014) A sulphide lithium superion conductor is superior to liquid ion conductors for use in rechargeable batteries. Energy Environ Sci 7:627–631
64. Maranas JK (2012) Polyelectrolytes for batteries: current state of understanding. In Page K (ed) Polymers for energy storage and delivery: polyelectrolytes and fuel cells. ACS Symposium Series. American Chemical Society, Washington
65. Armand M, Chabagno JM, Duclot MJ (1979). In: Vashishta P (ed) Fast ion transport in solids. North Holland, New York
66. Gray FM (1997) Polymer electrolytes. Roy Soc Chem, London
67. Arora P, Zhang Z (2004) Battery separators. Chem Rev 104:4419–4462
68. Armand MB, Bruce PG, Forsyth M et al (2011) Polymer electrolytes in energy materials. In: Bruce DW, O'Hare D, Walton RI (eds) Wiley, Chichester
69. Hollinan DT, Balsara NP (2013) Polymer electrolytes. Annu Rev Mater Res 43:503–525
70. Ratner MA, Shriver DF (1988) Ion transport in solvent-free polymers. Chem Rev 88:109–124
71. Christie AM, Lilley SJ, Staunton E et al (2005) Increasing the conductivity of crystalline polymer electrolytes. Nature 433:50–53
72. Mazor H, Golodnitsky D, Peled E et al (2008) A search for single-ion conducting polymer electrolyte: combined effect of anion trap and inorganic filler. J Power Sour 178:736–743
73. Quartarone E, Mustarelli P (2014) Polyelectrolytes for batteries, encyclopedia of polymeric nanomaterials. Springer, Berlin, pp 1–10
74. Gomez ED, Panday A, Feng EH et al (2009) Effect of ion distribution on conductivity of block copolymer electrolytes. Nano Lett 9:1212–1216
75. Quartarone E, Mustarelli P, Magistris A (1998) PEO-based composite electrolytes. Solid State Ionics 110:1–14
76. Ohno H (2007) Design of ion conductive polymers based on ionic liquids. Macromol Symp 249–250:551–556
77. Mecerreyes D (2011) Polymeric ionic liquids: broadening the properties and applications of polyelectrolytes. Progr Pol Sci 36:1629–1648

78. Fourquet JL, Duroy H, Crosnier-Lopez MP (1996) Structural and microstructural studies of the series $La_{2/3-x}Li_{3x}\square_{1/3-2x}TiO_3$. J Solid State Chem 127:283–294
79. Inaguma Y, Katsumata T, Itoh M et al (2002) Crystal structure of a lithium ion-conducting perovskite $La_{2/3-x}Li_{3x}TiO_3$ ($x = 0.05$). J Solid State Chem 166:67–72
80. Tiyapiboonchaiya C, Pringle JM, Sun J et al (2004) The zwitterion effect in high-conductivity polyelectrolytes materials. Nat Mater 3:29–32

Manufacture and Surface Modification of Polyolefin Separator

Zheng Xue, Zhengcheng Zhang and Sheng Shui Zhang

1 Introduction

Separator is a crucial component in the liquid electrolyte battery, which is placed between the positive and negative electrodes to prevent physical contact with each other while allowing free ionic transport and insulating electronic flow. In most cases the separator is a microporous polyolefin membrane that does not participate in any reactions inside the battery [1, 2]. Essentially, the separator is required to be chemically and electrochemically stable against the electrolyte and electrode materials, and mechanically strong to withstand the stress tension in the assembly and operation of battery. Structurally, the separator has sufficient porosity to absorb liquid electrolyte for necessary ionic conductivity. While playing the essential roles, the separator adds electric resistance and inactive volume, which adversely affect the battery's performances, in particular the energy density, power density, cycle life, and safety. For high energy and power densities, the separator is required to be very thin and highly porous while remaining mechanically strong. For safety, the separator should be able to shut down the current when the battery is accidently overheated, such as in an event of over-charging or short circuit, so that thermal runaway can be avoided. The shutdown function is generally obtained by structural design, for example, a multiplayer structure, in which at least one layer melts to close up the pores before the battery reaches thermal runaway temperature and the

Z. Xue · Z. Zhang
Chemical Sciences and Engineering Division,
Argonne National Laboratory, Lemont, IL 60439, USA
e-mail: zzhang@anl.gov

S.S. Zhang (✉)
Electrochemistry Branch, RDRL-SED-C, Sensors and Electron Devices Directorate,
U.S. Army Research Laboratory, Adelphi, MD 20783, USA
e-mail: shengshui.zhang.civ@mail.mil; shengshui@gmail.com

Z. Zhang and S.S. Zhang (eds.), *Rechargeable Batteries*,
Green Energy and Technology, DOI 10.1007/978-3-319-15458-9_12

others remain mechanically strong to isolate the electrodes from physical contact. In this chapter, the manufacture, properties and surface modifications of the polyolefin-based battery separators will be outlined.

2 Manufacture of Polyolefin Separator

Polyolefin materials used for the battery separator are based on a homopolymer or a blend of polyethylene (PE) and polypropylene (PP) in a number of combinations between high density polyethylene (HDPE) and ultrahigh molecular weight polyethylene (UHMWPE). The methods for manufacturing the microporous polyolefin membranes can be divided into the dry process and wet process. Both processes contain an extrusion step to produce a thin film, and employ one or more orientation steps to impart porosity and increase tensile strength. The membranes made by dry process show a distinct slit-pore microstructure, while those by wet process feature interconnected spherical or elliptical pores.

2.1 Dry Process

The dry process is based on a mechanical stretching principle to create microporous structure, which is generally composed of three steps: (1) extruding, (2) annealing, and (3) stretching. In the first step, the polyolefin resins are fed into an extruder where the resins are heated to above their melting temperature, followed by extruding the polyolefin melt through a shape-die and cooling down to form a precursor film. The morphology and orientation of the precursor film depend on the extruding conditions and the characteristics of polymer resins. The extruding conditions include melt temperature, die gap/shape, temperature, blowup ratio, and cooling speed. The parameters of polyolefin resins include M_w, M_w/M_n value, and polymeric ratio in the case of a blend. In the second step, the precursor film is annealed to form a well-oriented crystalline structure having lamellae arranged in rows with their long axis perpendicular to the machine direction (MD). Such structures are required for the formation of micropores in the next stretching step since only the stacked lamellae can be stretched out to create pores. In general, the annealing process with a control of the temperature and time is carried out at a temperature slightly lower than the melting temperature of the polyolefin resins.

In the third step, the annealed film is stretched to form porous structure at a temperature between the crystal dispersion and melting temperatures of the polyolefin resins, for example 100–130 °C for UHMWPE. Note that the crystal dispersion temperature is a parameter to describe the temperature characteristics of the dynamic viscoelasticity of a thermoplastic polymer. The temperature in this step is critical to determine the porosity. High temperature results in ineffective stretch due to the resulting loss in molecular orientation, whereas low temperature leads to break of the film due to the insufficient softness of the highly oriented crystalline film. The stretching process generally consists of a cold stretch, a hot stretch, and a relax. The cold stretch creates pore structure by stretching the film at a lower temperature with a faster strain rate, the hot stretch increases the size of the pores by further stretching the film at a higher temperature with a slower strain rate, and the relax reduces internal stress within the film by heating the film for a short time. The stretching can be divided into uniaxial stretch [3–8] and biaxial stretch [9–15]. In the former, the film is stretched only along MD, whereas in the latter the film is stretched in MD at lower temperature (70–110 °C), followed by stretching the film in transverse direction (TD) at higher temperature (110–140 °C, depending on the melting point of polymer). Variables of the stretching process include ratio, temperature, and speed of the stretch. The porosity of the final film depends on the morphology of the precursor film, annealing and stretching conditions, whereas tensile strength and its orientation of the final film depend on the ratio of the MD and TD stretches. Both pore shape and tensile strength of the uniaxially stretched film are strongly MD-oriented since the film is stretched only in MD. Significant difference between the MD and TD strengths may result in splitting problem, i.e. tearing as a result of puncture, which creates difficulties in handling the separators during battery assembly.

Figure 1 shows microstructure of a melt extruded HDPE film before and after stretch. Before stretch, the film shows a stacked lamellar morphology with the lamellae oriented along the TD (Fig. 1a). With uniaxial stretch, some of the stacked lamellae in the film separate to form microporous structure while others remain in their initial morphology. Therefore, the pores thus formed are distantly MD-oriented, being in almost perpendicular to the TD (Fig. 1b). In biaxial stretch, the MD-oriented film is further deformed along the TD to make the pores expend toward the TD, which as a result reduces the orientation of pores with an increase in the pore size. Therefore, the membranes made by the biaxial stretch are less orientation in the pore shape and tensile strength than those by the uniaxial stretch. Since the formation of pores by stretching is based on a lamellae-arranged crystalline structure, the dry

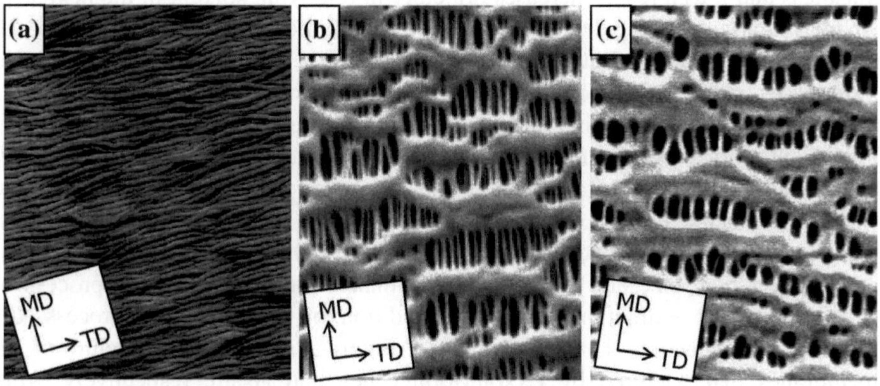

Fig. 1 Microstructure of a melt-extruded HDPE film before and after stretch **a** before stretch, **b** after uniaxial stretch, and **c** after biaxial stretch

process is only applicable to these polymers with high crystalline row structure and to the blends containing at least one of such polymers, such as PE-PP [16], polystyrene (PS)-PP [15, 17], and poly(ethylene terephthalate) (PET)-PP blends [15, 17]. When the polymers in the blend are immiscible with each other, the resulting microporous film exhibits two polymer phases interconnected throughout the film. For PE-PP blend membrane, thermal shutdown behavior was observed due to the different melting points between PE and PP phases [16]. That is, the PE phase melts down and closes up its pores in a temperature range of 135–140 °C, and the PP phase melts about 170 °C. For the cases of PS-PP and PET-PP blends (both in a 1:9 wt. ratio), the resulting membranes showed microcrack structures, which suggested that porous structures were formed by a cracking mechanism. That is, the minor PS or PET phase acts as a stress concentrator to induce the formation of cracks and the subsequent growth propagates the cracks into the major phase of blends [15].

2.2 Wet Process

The wet process is based on a phase separation principle to create porous structures, which is generally composed of the following steps: (1) mixing polymer resins, mineral oil, antioxidant, and other additives and heating to make a hot homogenous solution, (2) extruding the hot solution through a sheet die into a gel-like film, (3) extracting mineral oil and other additives to form microporous structures. In the first step, the dissolution is carried out above the melting temperature of the polyolefin resins and in most cases in a temperature range of 140–250 °C, depending on the types of polymers and solvents used. Upon the dissolution, the heated solution is extruded through a sheet die and immediately quenched to obtain a gel-like film. In this step, die gap determines the thickness of gel-like film, and the temperature, polymeric concentration, as well as quenching speed determine the size and shape of the oil droplets in the film. Slow quenching results in high thickness and it as a result increases difficulty in removing solvent in the next step, whereas rapid quenching decreases pore size and permeability. In the third step, the mineral oil and other additives are extracted using a volatile organic solvent, in most cases methylene chloride, to create porous structure. The wet process is applicable to both crystalline and amorphous polymers, and its resulting film is non-oriented for pore structure and mechanical strength. For semicrystalline polyolefin membranes, a stretching step can be added either before or after the extraction to achieve high porosity and large pore size [18–21]. It has been proven that the membranes produced by a process of stretching after extraction exhibit much larger pore size and air permeability than those produced by a process of stretching before extraction [21]. With addition of the stretching process, the resulting pores become somewhat oriented. Figure 2 shows microstructure of PE membranes manufactured by ExxonMobil/Tonen and Asahi, respectively, from which noticeable difference in the morphology is visible.

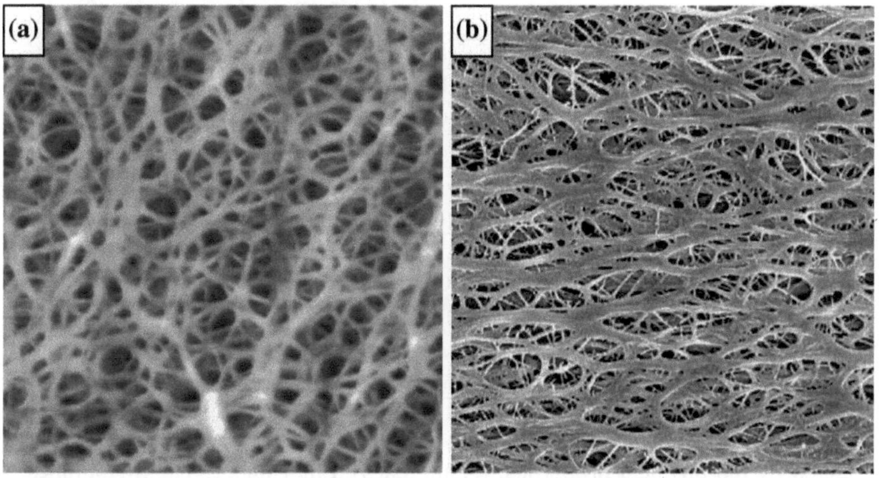

Fig. 2 Microstructure of PE membranes by wet process **a** Tonen Setela (porosity: 45 %) and **b** Asahi HiPore (porosity: 50 %)

In addition, removable inorganic fillers such as $CaCO_3$ and silica can be mixed with the polyolefin resins in the first step to increase the porosity of the membranes [16, 22, 23]. In such cases, additional step is needed to create the pores by removing inorganic fillers, for example an acid washing step for $CaCO_3$. On the other hand, the added inorganic fillers may be used to increase electrolyte wettability and retention if the fillers are stable against the battery components and their presence does not significantly reduce mechanical strength. A positive effect of the $CaCO_3$ fillers on the battery performance is that alkali $CaCO_3$ can neutralize acidic HF, which is inevitably present in the $LiPF_6$-based electrolytes. In addition, it was claimed [22] that leaving silica in the membrane could increase oxidation resistance and durability in the short circuit conditions. Regarding distribution of the pore size, Funaoka et al. [24] showed that addition of a hot solvent treatment before or after the extraction could result in the formation of the pore size decreasing gradually form the surface towards the center. For example, they dipped a regularly stretched PE membrane in a 120 °C-paraffin bath for 5 s to obtain a membrane having an average pore size of 8.2 μm in the vicinity of the one surface, 0.5 μm in the vicinity of the other surface, and 0.12 μm near the center. They considered that such structures favored increasing the electrolyte retention and battery safety.

2.3 Dry Process Versus Wet Process

As described above, the membranes made by the dry and wet processes show significant difference in the orientation of the pore structures. This difference is well

Table 1 Typical properties of the polyolefin separators by different processes

Manufacturer	Celgard	ExxonMobil/Tonen
Membrane name	Celgard 2400	Tonen-1
Process	Dry	Wet
Composition	PP	PE
Thickness, μm	25	25
Porosity, %	37	36
Gurley number, s	24	650
Tensile strength (MD), Kg cm^{-2}	1300	1500
Tensile strength (TD), Kg cm^{-2}	130	1300
Melting temperature, °C	163	135
Reference	[25]	[26]

reflected by the physical properties, as summarized in Table 2. The membranes by the dry process distinguish themselves by high MD-orientation in tensile strength and thermal shrinkage, and low Gurley number. Such orientations will not cause any adverse impacts on the battery performance as long as the separator is winded along MD in the battery assembly. The Gurley number, which is defined as the time required for a specific amount of air to pass through a specific area of separator under specific pressure, is a measure of the air permeability. When the porosity of the membranes is fixed, the Gurley number reflects tortuosity of the pores. As shown in Table 1, the membrane by wet process has larger Gurley number than that by dry process although they have the same thickness and porosity. This fact indicates that pores formed by wet process are more tortuous, which is an important feature to prevent the growth of lithium dendrites in the rechargeable Li and Li-ion batteries.

3 Correlation of Polyolefin Nature and Membrane Property

Battery separators are characterized by numerous properties, including material nature, membrane structural and functional properties. Material nature includes chemical stability, crystalline structure, hydrophilicity, thermal shrinkage, melting point, M_w and M_w/M_n of polyolefin materials. Structural properties include thickness, porosity, pore size, pore shape, pore tortuosity, and pore distribution. Functional properties include mechanical strength, electrical resistivity, air permeability, thermal shutdown, electrolyte wettability and retention. Many of the above properties are affected with each other and may be in a trade-off relationship. For example, the mechanical strength is affected in opposite manner by the thickness, porosity and permeability, as required by the battery performance.

3.1 Mechanical Property

From the viewpoint of material nature, both the M_w and M_w/M_n affect mechanical strength and thermal shrinkage of the membranes. Table 2 indicates their impacts on the mechanical strength for three microporous PE membranes having similar thickness and porosity. By comparing rows 1 and 2 and rows 2 and 3, respectively, one finds that high M_w and low M_w/M_n value are favorable for the mechanical strength. In the case of blends, the mechanical strength of the membranes can be controlled by adjusting the ratio of two polymers such as HDPE and UHMWPE for an HDPE-UHMWPE blend [21, 23, 24]. For example, Ihm et al. [21] studied the effect of the M_w and content of UHMWPE on the mechanical strength of HDPE-UHMWPE blend membranes made by the wet process under the same extruding, stretching and extracting conditions. They found that the mechanical strength was increased with the M_w and content of UHMWPE. Regarding the orientations of the pore structure and mechanical strength, they can be adjusted by changing the ratio of MD and TD stretches [27], or even eliminated by using wet process without stretching process [22, 23]. In the battery, however, no negative impact has been reported to be associated with the orientation of mechanical strength.

3.2 Thermal Property and Shutdown Function

The material nature determines the operation temperature range of the polyolefin separators. The highest temperature for the mechanical integrity of the membranes is limited by the melting point of polyolefin materials. Table 3 summarizes the typical physical properties of the PE and PP used for the battery separators. It is shown that depending on the density and M_w, melting point of PE is in a range of 98–135 °C, and that of PP in a range of 168–175 °C.

Based on the difference in the melting point of PE and PP, thermal shutdown separators can be made by laminating different PE and PP layers into a multiplayer membrane. In shutdown membrane, one or more PE layers serve as shutdown and at least one PP layer as the mechanically dimensional support. Typically, the PE-PP bilayer [14, 29] and PP-PE-PP trilayer [27, 30–36] designs have been widely adopted

Table 2 Impact of M_w and M_w/M_n on the mechanical strength[a]

Membrane code	1	2	3
M_w	10×10^6	5.0×10^5	5×10^5
M_w/M_n	12.0	11.0	6.9
Thickness, μm	25	25	25
Porosity, %	37	38	40
Tensile strength, kg/cm^2	770	650	940
Puncture strength, g	426	331	465

[a]Data was cited selectively from Table 1 of Ref [28]

Table 3 Typical physical properties of PE and PP[a]

	T_m, °C	T_g, °C	Tensile strength, psi	Density, g/cm^3
LDPE[b]	98–115	−25	1200–4550	0.917–0.932
LLDPE	122–124		1900–4000	0.918–0.940
HDPE	130–137		3200–4500	0.952–0.965
UHMWPE	125–135		5600	≥0.940
PP	168–175	−20	4500–6000	0.900–0.910

[a]Definitions: T_m crystalline melting temperature, T_g glass transition temperature. [b]Abbreviations: *LDPE* low density polyethylene

by separator manufacturers. With such structures, the PE layer is capable of melting and closing up the pores before the battery reaches such a temperature to initiate the thermal runaway. As a result of the closing up of the pores, the electrical resistance of the separator is dramatically increased, shutting down the current, whereas the PP layer still remains mechanical strength to isolate the electrodes from physical contact (internal circuit shorting). To evaluate the shutdown behavior, Venugopal et al. [37] studied the temperature dependence of three separators soaked with a 1 M $LiPF_6$ 3:2 EC/DEC electrolyte by measuring the resistance of membranes during heating and cooling. The results (Fig. 3) showed that the shutdown behavior of the PP-PE-PP separator was in excellent agreement with the melting point of PE, and that the membrane was able to withstand up to 165 °C, as determined by the melting point of PP. There is only ~30 °C temperature gap between the cell's shutdown and the melting down of PP layer, which may be sufficient for the protection of normal operations of small batteries. However, such a temperature gap is not sufficient for the protection of large batteries, especially in such conditions as nail penetration and overcharge at high current density. In these cases, the heat transfer in the large battery is not fast sufficiently so that the local temperature can easily exceed the melting point of PP layer before the battery is effectively shut down. As a result, the strongly oxidative cathode and the strongly reductive anode come into contact, and resulting chemical reactions initiate the thermal runaway.

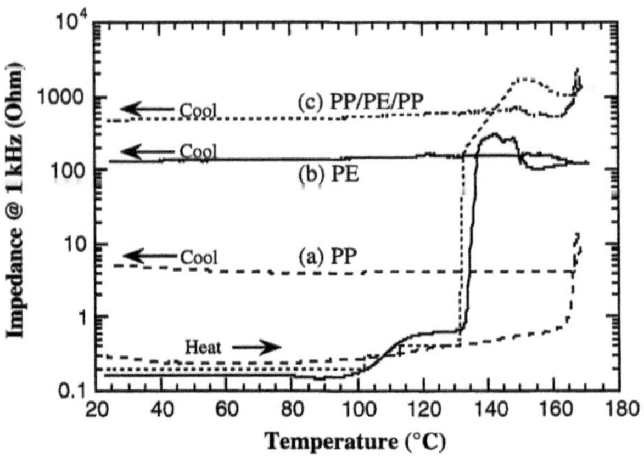

Fig. 3 Temperature dependence of the polyolefin membranes soaked with a 1 M $LiPF_6$ 3:2 EC/DEC electrolyte. Reprinted from Fig. 7 of Ref. [37], Copyright (1999), with permission from Elsevier

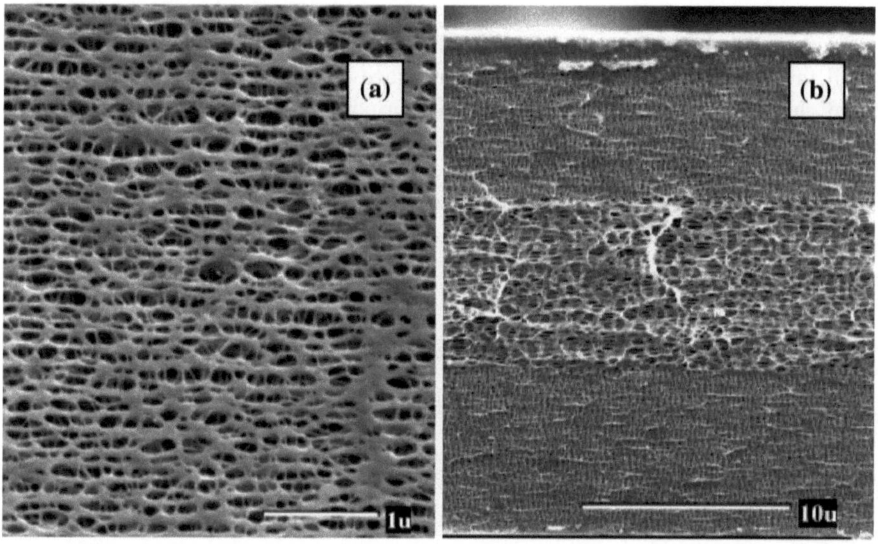

Fig. 4 Micrographs of Celgard PP-PE-PP trilayer membrane **a** Surface, and **b** cross section. Reprinted from Fig. 5 of Ref. [2], Copyright (2004), with permission from American Chemical Society

The multilayer shutdown separators usually are made by lamination of PE and PP layers together through calendaring, bonding, or welding [27, 34, 36]. Preferably, the cross-plied lamination where the axis of one ply is angularly biased relative to the axis of another ply can increase mechanical strength and puncture-resistance. Alternatively, the multilayer separators can be made through co-extrusion processes, in which all precursor layers are extruded simultaneously, and then are annealed and stretched to form a multilayer microporous structure [27, 34]. Figure 4 shows micrographs of a Celgard PP-PE-PP trilayer membrane. It is shown that the pores are more tortuous and that three layers have nearly equal thickness. In consideration of the impurities brought in by the resin materials, antioxidant, and other possible additives during extrusion, Kinouchi et al. [35] claimed that dispersing small amount of metal oxide particles into two PP layers of the PP-PE-PP trilayer membrane could effectively adsorb these impurities and consequently reduce their negative impacts on the battery performance.

4 Surface Modification

As a favorable separator, the microporous polyolefin membranes are required to absorb liquid electrolyte quickly in the electrolyte filling operation and retain the liquid electrolyte permanently for long-term operation of battery. Unfortunately, considerable difference in the polarity (dielectric constant) between the polyolefin material and electrolyte solvents makes it impossible. Beside this, the contact of the separator and electrode considerably affects the performance of batteries. Even small failure at the separator-electrode interface can result in uneven current distribution, which consequently causes the formation of lithium dendrites on the negative electrode in the Li and Li-ion batteries. On the other hand, the highest melting point of PP is only ~165 °C, beyond which the polyolefin membrane fails to separate the electrodes. In order to overcome these drawbacks, surface modification has been frequently practiced. According to the nature of modifications, these practices can be categorized into the physical modification and chemical modification, as discussed below.

4.1 Physical Modification

Physical modification generally fills the pores of membrane with a gel polymer electrolyte (GPE) or coats a functional layer onto the surface of membrane. The functional layer could be any of a surfactant layer, a dense polymer layer, a porous polymer layer, or a polymer composite layer. The physical modification not only increases the wettability and electrolyte retention of the membrane, but also considerably affects the structural and physical properties of the membrane, such as porosity, thickness, shrinkage, mechanical strength etc. The filling practice was earliest demonstrated in 1995 by Abraham et al., who encapsulated a GPE into the pores of Celgard membranes by first soaking the membrane with a liquid electrolyte containing tetraethylene glycol diacrylate (TEGDA) monomer and then photopolymerizing the monomer [38]. Alternatively, the GPE can be impregnated into the pores of membrane by immersing the membrane into a GPE solution and then evaporating the solvent. Without exception, the encapsulation or impregnation of a GPE greatly promotes the wettability and electrolyte retention of the microporous polyolefin membrane.

Treatment of polyolefin separators with a wetting agent (surfactant) is the simplest practice for the coating modification. To do this, the surfactant is dissolved into a volatile solvent to form a dilute surfactant solution, followed by either spraying the resulting solution onto the surfaces of separator or passing the separator through the surfactant solution bath, and lastly evaporating the solvent [39]. The amount of surfactant is generally small so that its presence hardly affects the thickness and porosity of the microporous membrane while significantly enhancing the hydrophility of the polyolefin membrane. Since the surfactant is subject to dissolving into the electrolyte, its improvement on the hydrophility is temporary and its stability in the battery must be considered for the suitability.

Most practices for the physical modification are the coating of a polymer or a polymer composite onto one or two surfaces of the microporous polyolefin membrane. In order to enhance the interfacial contact between the separator and electrode, a thin polymer layer that can be gelled by the liquid electrolyte can be coated onto the microporous membrane [40–46]. The suitable polymers for this purpose are poly (ethylene oxide) (PEO) [40, 41] and PEO-like polymer [42], polyvinylidene fluoride (PVDF) and its related copolymers such as poly(vinylidene fluoride-co-hexafluoropropylene) (PVDF-HFP) [43–45], and other electrolyte gellable polymers such as a copolymer of acrylonitrile and methyl methacrylate [46]. Upon gelled by the liquid electrolyte, the polymer becomes a GPE and it in turn serves as an adhesive to bond the separator and electrode together in addition to increasing the uptake and retention of the liquid electrolyte. In practice, the polymer layer can be applied to the microporous membrane by a means of dipping, spraying or coating with a GPE solution, followed by evaporation of the solvent. In assembly of battery, a heating step is preferably applied to promote the in situ formation of the GPE when the polymer-coated membrane is used as the separator. Due to the need and consumption for forming GPE, the polymer-modified separators are shown to have significantly high uptake of liquid electrolyte as compared with the pristine separator.

A negative effect of the above modification is that the resulting polymer layer is dense, which prevents the penetration of liquid electrolyte into the pores of microporous membrane. Therefore, the polymer layer with porous structure is highly desired so as to facilitate the electrolyte filling process. In such efforts, Jeong et al. [46] employed the phase inversion method to coat a microporous polymer layer. In their method, they first applied a dilute dimethylformamide (DMF) solution of an acrylonitrile-methyl methacrylate copolymer (ANMMA, AN: MMA = 84:16) onto the micropouous PE membrane with a thickness of 25 μm and a porosity of 40 %, and then immersed the coated PE membrane into a water bath to induce phase inversion. Due to the fast solvent exchange between the good-solvent (DMF) in the polymer solution film and the non-solvent (water) in the bath, the polymer layer with highly microporous structure is formed. They found that the Li-ion cell using the separator with a 14 μm polymer coating layer exhibited very stable capacity retention and excellent rate capability, which is attributed to the strong bonding between the separator and electrodes. In particular, a facile one-step phase inversion (coating process) is available when the polymer suits for a good-solvent with low boiling point and a non-solvent with high boiling point. By this manner, the polymer slowly precipitates out of the solution film to form porous structure with evaporation of the good-solvent. For example, the PVDF-HFP polymer can be dissolved into a hot mixed solvent of acetone (good-solvent) and water (non-solvent). Coating the hot solution onto a polyolefin separator and leaving the solvents to evaporate naturally in air leads to a porous PVDF-HFP layer coated separator. Moreover, the porosity of the polymeric coating layer can be controlled by adjusting the ratio of solvents. On the other hand, Kim et al. [45] introduced small amount of inorganic filler, SiO_2, into the polymer coating layer, and they found that the presence of the hydrophilic SiO_2 particles significantly reduced the contact angle of the liquid electrolyte and accelerated the penetration of the liquid electrolyte into the separator. Attributing to the improved interfacial contact, the Li-ion cell using such a polymer-modified separator exhibited excellent

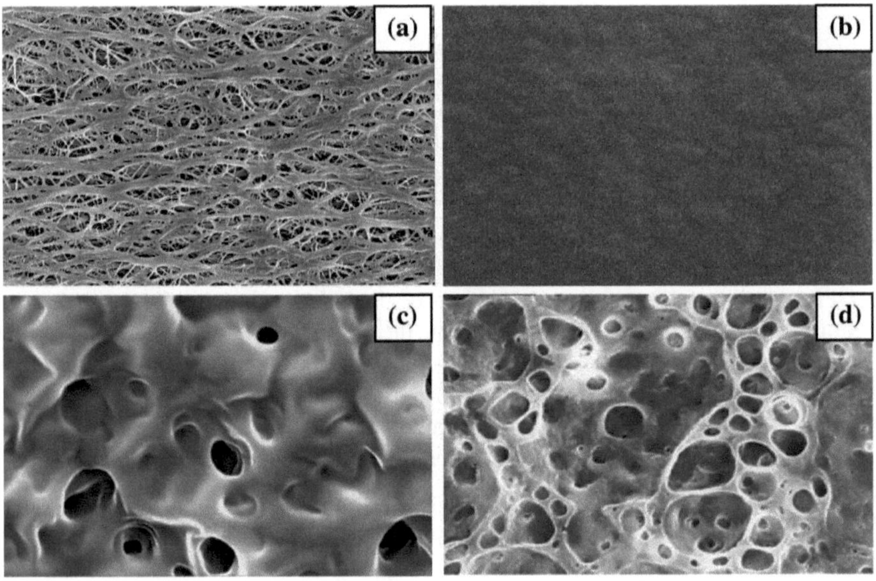

Fig. 5 Micrographs of the surface structure of a microporous PE membrane before and after polymer coating treatment **a** bare membrane, **b** with polymer coating by solution method, **c** with polymer coating by phase inversion, and **d** with polymer/silica coating by phase inversion. Reprinted from Fig. 4 and Fig. 2 of Refs. [42] and [45], respectively, Copyright (2006 and 2004, respectively), with permission from Elsevier

capacity retention and rate capability. As comparison, Fig. 5 displays the surface images of a microporous PE membrane before and after polymer coating by different methods. The differences in the surface morphology of the polymer layers coated by the solution and phase inversion methods are very clear.

At elevated temperatures, the GPE becomes fluid, losing the ability to separate the electrodes. Therefore, the modification of a polymer or a polymer composite cannot improve the safety of battery in addition to the performance improvement by increasing the uptake and retention of liquid electrolyte and promoting the interfacial contact between the separator and electrodes. In order to integrate the improvement of safety and the advantages of GPE into a single separator, Zhang et al. [47] proposed to significantly increase the content of SiO_2 in the polymer composite until the SiO_2 filler becomes the main component of the polymer composites. By this way, the SiO_2 filler not only increases the wettability and uptake of the liquid electrolyte, but also retains dimensional stability of the separator even when the separator melts down at elevated temperatures. This concept was demonstrated in a Li-S cell by coating a 13 μm 50:50 (wt.) PEO-SiO_2 composite layer onto one side of the Celgard membrane, showing that the Li/S cell with the modified separator had very stable capacity retention.

4.2 Chemical Modification

Chemical modification covalently grafts polar groups onto the surfaces or pore walls of the microporous polyolefin membrane, which turns the membrane from hydrophobic to hydrophilic [48–50]. Such modifications generally enhance the wettability and retention of the liquid electrolytes, especially those consisting of high content of ethylene carbonate, propylene carbonate or γ-butyrolactone, on the membrane without changing the structural parameters and physical properties of the membrane. The chemical grafting can be carried out by high-energy radiation induced polymerization [48–50]. During high-energy irradiation, the polyolefin is radiation-induced to cleave C-C or C-H bonds to generate polymeric radicals, and the resulting polymeric radicals subsequently initiate hydrophilic monomer to polymerize. In consideration of the high reactivity of the Li-ion battery electrode materials, only these monomers having no active protons, such as glycidyl methacrylate [49], methylmethacrylate [50] and oligo(ethylene glycol) [51], are suitable for the modification of the Li-ion battery separators. Due to the increased hydrophilicity, both uptake and retention of the liquid electrolytes on the modified separators showed great improvement, and the improving effectiveness varies with the degree of grafting and the type of monomers [49, 50].

Beside the functional groups, the strongly electron-withdrawing fluorine is also shown to increase the hydrophility of the polyolefin membrane. To verify this, Li et al. [52] directly fluorinated the PP separator under a F_2/N_2 gas atmosphere. Results showed that the fluorination did not change the size and shape of micropores, instead significantly introduced CF_x groups into the surfaces of PP separator. This treatment increased the surface energy of the PP separator from 37.8 to 47.7 m N/m, and accordingly led to about nine times increase in the alkali absorption ratio (AAR) without loss over 90 days storage at ambient temperature. Plasma grafting has also been used as the modifying carrier to coat the polyolefin separator with nanoparticles by first plasma treating the membrane in oxygen to create surface anchoring groups and then immersing the membrane into a dispersion of positively charged SiO_2 nanoparticles [53]. This treatment led to the nanoparticles electrostatically adsorbed onto the separator, which resulted in immediate wetting of the polyolefin separator by the highly polar ethylene carbonate and propylene carbonate that otherwise are unable to wet the pristine polyolefin separator.

5 Concluding Remarks

Separator is a crucial component of the liquid electrolyte battery, its essential role is to isolate the positive electrode and negative electrode from physical contact while permitting free flow of ions. The separator itself does not participate in any cell reactions, however, its properties significantly affect the performance and safety of the batteries. For high energy and power densities, the separator is required to be very thin and highly porous, while it adversely affects the safety and cycle life of the battery as a result of the reduced mechanical strength. Safety is the top priority for the rechargeable batteries, especially for applications in hybrid electric vehicle (HEV) and plug-in hybrid electric vehicle (PHEV) where both the high energy and high power are required. The PE-based multilayer shutdown separators are able to protect small batteries in a temperature range of 90–130 °C, depending on the M_w and M_w/M_n of the PE materials and the composition of the blend if a PE-PE blend is used. However, protection for large batteries still remains challenge. None of the polyolefin separators can be mechanically strong beyond the melting point (near 165 °C) of PP. Above the melting point, the polyolefin separator melts down, resulting in physical contact of the electrodes, which consequently initiates the highly energetic chemical reactions between the strongly oxidative cathode material and the strongly reductive anode material. Even near the softening temperature of polyolefin materials, the separators are subject to shrinking, which potentially results in the internal circuit shorting of battery. Future development of the battery separators should be made by balancing the performance, safety, and cost. The cost of the polyolefin separators is mainly in the production process, therefore, developing a more cost-effective process is essential for the cost reduction of polyolefin-based battery separators.

References

1. Linden D, Reddy TB (2002) Handbook of batteries, 3rd edn. McGraw-Hill, New York
2. Arora P, Zhang Z (2004) Battery separators. Chem Rev 104:4419–4462
3. Nagou S, Nakamura S (1988) Microporous film and process for production thereof. US Patent 4,791,144
4. Lee SY, Ahn BI, Im SG, Park SY, Song HS, Kyung YJ (2004) High crystalline polypropylene microporous membrane, multi-component microporous membrane and methods for preparing the same. US Patent 6,830,849
5. Higuchi H, Matsushita K, Ezoe H, Shinomura T (1995) Porous film, process for producing the same, and use of the same. US Patent 5,385,777
6. Johnson MB, Wilkes GL (2001) Microporous membranes of polyoxymethylene from a melt-extrusion process: (I) effects of resin variables and extrusion conditions. J Appl Polym Sci 81:2944–2963
7. Johnson MB, Wilkes GL (2002) Microporous membranes of polyoxymethylene from a melt-extrusion process: (II) effects of thermal annealing and stretching on porosity. J Appl Polym Sci 84:1762–1780
8. Johnson MB, Wilkes GL (2002) Microporous membranes of isotactic poly (4-methyl-1-pentene) from a melt-extrusion process. I. Effects of resin variables and extrusion conditions. J Appl Polym Sci 83:2095–2113
9. Hashimoto A, Yagi K, Mantoku H (2000) Porous film of high molecular weight polyolefin and process for producing same. US Patent 6,048,607
10. Xu M, Hu S, Guan J, Sun X, Wu W, Zhu W, Zhang X, Ma Z, Han Q, Liu S (1992) Polypropylene microporous film. US Patent 5,134,174
11. Chu F, Yamaoka T (1996) Polychloroprene 'popcorn' growth kinetics measured by the McBain balance. Polymer 37:537–539

12. Fisher HM, Wensley CG (2002) Polypropylene microporous membrane for battery separator. US Patent 6,368,742
13. Jacoby P, Bauer CW, Clingman, SR, Tapp, WT (1994) Oriented polymeric microporous films US Patent 5,317,035
14. Yu WC, Geiger MW (1996) Shutdown, bilayer battery separator. US Patent 5,565,281
15. Chandavasu C, Xanthos M, Sirkar KK, Gogos CG (2000) Preparation of microporous films from immiscible blends via melt processing. J Plast Film Sheeting 16:288–300
16. Higuchi H, Matsushita K, Ezoe M, Shinomura T (1995) Porous film, process for producing the same, and use of the same. US Patent 5,385,777
17. Chandavasu C, Xanthos M, Sirkar KK, Gogos C (2004) Preparation of microporous films from immiscible blends via melt processing and stretching. US Patent 6,824,680
18. Kaimai N, Takita K, Kono K, Funaoka H (2000) Method of producing highly permeable microporous polyolefin membrane. US Patent 6,153,133
19. Takita K, Kono K, Takashima T, Okamoto K (1991) Microporous polyolefin membrane and method of producing same. US Patent 5,051,183
20. Jacoby P, Bauer CW, Clingman SR, Tapp WT (1993) Oriented polymeric microporous films. US Patent 5,176,953
21. Ihm DW, Noh JG, Kim JY (2002) Effect of polymer blending and drawing conditions on properties of polyethylene separator prepared for Li-ion secondary battery. J Power Sources 109:388–393
22. Doi Y, Fugii O, Kaneko S, Hanamura T (1980) Microporous film battery separator. US Patent 4,210,709
23. Nishimura Y, Segawa T (2006) Separator for metal halogen cell. US Patent 7,081,321
24. Funaoka H, Takita K, Kaimai N, Kobayashi S, Kono K (2003) Microporous polyolefin film and process for producing the same. US Patent 6,666,969
25. Data sheet of Celgard microporous membranes, http://www.celgard.com/
26. Kimishima K, Yaniyama Y, Lesniewski ME, Brant P, Kono K (2006) Advanced automotive battery and ultracapacitor conference (AABC—06), Baltimore
27. Callahan RW, Call RW, Harleson KJ, Yu TH (2003) Battery separators with reduced splitting propensity. US Patent 6,602,593
28. Kaimai N, Takita K, Kono K (1998) Method of producing of microporous polyolefin membrane. US Patent 5,830,554
29. Lundquist JT, Lundsager CB, Palmer NI, Troffkin HJ, Howard J (1987) Battery separator. US Patent 4,650,730
30. Lundquist JT, Lundsager CB, Palmer N, Troffkin HJ (1988) Battery separator. US Patent 4,731,304
31. Yu WC, Dwiggins CF (1997) Methods of making cross-ply microporous membrane battery separator, and the battery separators made thereby. US Patent 5,667,911
32. Yu WC (1997) Shutdown, trilayer battery separator. US Patent 5,691,077
33. Yu TH (2000) Trilayer battery separator. US Patent 6,080,507
34. Yu WC (2005) Continuous methods of making microporous battery separators. US Patent 6,878,226
35. Kinouchi M, Akazawa T, Oe T, Kogure R, Kawabata K, Nakakita Y (2003) Battery separator and lithium secondary battery. US Patent 6,627,346
36. Spotnitz RM (2001) Trilayer battery separator. US Patent 6,180,280
37. Venugopal G, Moore J, Howard J, Pendalwar S (1999) Characterization of microporous separators for lithium-ion batteries. J Power Sources 77:34–41
38. Abraham KM, Alamgir M, Hoffman DK (1995) Polymer electrolytes reinforced by Celgard membranes. J Electrochem Soc 142:683–687
39. Taskier HT (1982) Hydrophilic polymer coated microporous membranes capable of use as a battery separator. US Patent 4,359,510
40. Kim DW, Ko JM, Chun JH, Kim SH, Park JK (2001) Electrochemical performances of lithium-ion cells prepared with polyethylene oxide-coated separators. Electrochem Commun 3:535–538

41. Kim DW, Noh KA, Chun JH, Kim SH, Ko JM (2001) Highly conductive polymer electrolytes supported by microporous membrane. Solid State Ionics 144:329–337
42. Oh JS, Kang YK, Kim DW (2006) Lithium polymer batteries using the highly porous membrane filled with solvent-free polymer electrolyte. Electrochim Acta 52:1567–1570
43. Eschbach FO, Oliver M (1997) Gel electrolyte bonded rechargeable electrochemical cell and method of making same. US Patent 5,681,357
44. Hamano K, Shiota H, Shiraga S, Aihara S, Yoshida Y, Murai M, Inuzuka T (1999) Lithium ion secondary battery and method of fabricating thereof. US Patent 5,981,107
45. Kim JY, Kim SK, Lee SJ, Lee SY, Lee HM, Ahn S (2004) Preparation of micro-porous gel polymer for lithium ion polymer battery. Electrochim Acta 50:363–366
46. Jeong YB, Kim DW (2004) Effect of thickness of coating layer on polymer-coated separator on cycling performance of lithium-ion polymer cells. J Power Sources 128:256–262
47. Zhang SS, Tran DT (2013) How a gel polymer electrolyte affects performance of lithium/sulfur batteries. Electrochim Acta 114:296–302
48. Gineste JL, Pourcelly G (1995) Polypropylene separator grafted with hydrophilic monomers for lithium batteries. J Membrane Sci 107:155–164
49. Ko JM, Min BG, Kim DW, Ryu KS, Kim KM, Le YG, Chang SH (2004) Thin-film type Li-ion battery, using a polyethylene separator grafted with glycidyl methacrylate. Electrochim Acta 50:367–370
50. Gao K, Hu GX, Yi TF, Dai CS (2006) PE-g-MMA polymer electrolyte membrane for lithium polymer battery. Electrochem Acta 52:443–449
51. Man C, Jiang P, K-w Wong, Zhao Y, Tang C, Fan M, Lau WM, Mei J, Li S, Liu H, Hui D (2014) Enhanced wetting properties of a polypropylene separator for a lithium-ion battery by hyperthermal hydrogen induced cross-linking of poly(ethylene oxide). J Mater Chem A 2:11980–11986
52. Li B, Gao J, Wang X, Fan C, Wang H, Liu X (2014) Surface modification of polypropylene battery separator by direct fluorination with different gas components. Appl Surf Sci 290:137–141
53. Fang J, Kelarakis A, Lin YW, Kang CY, Yang MH, Cheng CL, Wang Y, Giannelis EP, Tsai LD (2011) Nanoparticle-coated separators for lithium-ion batteries with advanced electrochemical performance. Phys Chem Chem Phys 13:14457–14461

Microstructurally Composed Nanoparticle Assemblies as Electroactive Materials for Lithium-Ion Battery Electrodes

Evan Uchaker and Guozhong Cao

1 Summary

Lithium-ion batteries are a well-established technology that has seen steady gains in performance based on materials chemistry as well as microstructure design and assembly over the past several decades. There are many material selections available when designing and assembling the device such as electro-active species, additives, and particle size/morphology to name a few. Many of the research proclamations focusing on the advantages of nanosized electrodes have yet to find commercial application, and considerable improvements in energy density and stability are still necessary in order to achieve energy storage parity. Therefore, the design and use of kinetically stabilized nanostructures should be considered. Over the past several years, significant studies have been conducted examining the synthesis and performance of heterogeneous structures. While heterogeneous structures typically refer to the combination of two or more materials, in this case it refers to architectures displaying more than one size scale (i.e., micro/nano). A great deal of recent efforts have focused on the formation and understanding of nanoparticle superstructures with a vast range of architectures. The design of microstructurally composed nanoparticle assemblies would, for instance, possess the

E. Uchaker · G. Cao (✉)
Department of Materials Science and Engineering, University of Washington, Seattle, WA, USA
e-mail: gzcao@u.washington.edu

G. Cao
Beijing Institute of Nanoenergy and Nanosystems, Chinese Academy of Sciences, Beijing 100083, China

G. Cao
School of Materials Science and Engineering, Dalian University of Technology, Dalian 116023, China

Z. Zhang and S.S. Zhang (eds.), *Rechargeable Batteries*,
Green Energy and Technology, DOI 10.1007/978-3-319-15458-9_13

structural and chemical stability of microsized electrodes while exploiting the beneficial properties associated with nanosized electrodes and their large specific surface area.

2 Introduction

Batteries, as devices for chemically storing energy, demonstrate high conversion efficiency, relatively high energy density, long cycle life, and zero exhaust release. They are ideal power sources for portable devices, automobiles, and backup power supplies. Unfortunately, considerable improvements and advancements in rechargeable battery materials and technologies are still necessary in order to achieve energy sustainability; namely, the energy density, power density, and stability must be vastly enhanced in order to make smart grid and fully electric vehicle technologies an attainable reality.

The development and implementation of nanostructured electrode materials has been well documented over the past decade; many excellent reviews concerning the benefits of nanostructured materials for electrochemical energy conversion and storage have already been published, to which the reader is referred [1–3]. Given all the unique qualities and benefits of nanostructures, the synthesis and characterization of nanostructured electrode materials of various chemistries have been extensively investigated [4–6]. However, nanomaterials as a whole suffer from several basic limitations that restrict their performance in energy storage applications [7]. Conversely, microstructurally composed nanoparticle assemblies can potentially circumvent the thermodynamic instability, undesired side reactions, high processing costs, and potential nano-toxicity effects associated with nanoparticle usage [8–10]. Herein, this chapter strives to offer a succinct outline of the undesired characteristics and limitations associated with nanosized electrode materials, provide a brief discussion of the benefits of microstructurally composed nanoparticle assemblies, and then summarize the findings of studies employing nano-based micro-assemblies for electrochemical energy storage systems, primarily Li-ion batteries.

3 Nanostructured Li-Ion Battery Electrodes

3.1 Advantages of Nanostructured Electrodes

Despite being an overall mature technology, battery design and performance have changed drastically over the past decade with the development and implementation of nanomaterials [8, 11–13]. The nanoscale size reduction leads to enhancements of the Li-ion battery intercalation capability by increasing the specific surface area for interfacial Faradic reactions, improving the mass and charge diffusion paths, and modifying the surface thermodynamics which can facilitate phase transitions and eliminate miscibility gaps [14]. However, the most dominant and substantial advantages imparted by nanostructuring is drawn from enhancements in the kinetics and corresponding diffusivities because the characteristic time for diffusion is proportional to the square of the diffusion path length [15]. The reduction of the electrode particle size into the nanometer-based regime tremendously improves the intercalation kinetics and ultimately enhances the overall degree of intercalation. Similar to the effects of transport on Li-ions, electron transport is also enhanced in nanometer sized electro-active particles [13].

The benefits of nanostructured metal oxide electrodes have been well documented throughout the literature. Wagemaker et al. [16] compared the lithium ion intercalation performance of anatase TiO_2 micro- and nano-particles. The size of the anatase TiO_2 nanoparticles was incrementally altered in order to investigate the nano-size effects. It was determined that the Li-ion solubility was greater with decreasing particle size, and increased from 0.55 to 0.7 Li-ion per formula unit when the micron-sized TiO_2 particles were replaced with nanoparticles smaller than 40 nm. The discrepancy in the quantity of inserted Li-ion was ascribed to a more homogeneous, and Li-rich phase in the nanoparticles compared with the microparticles that had coexisting Li-rich/poor phases because they were too large for homogeneous Li-ion penetration [17]. Ganapathy et al. [18] also revealed that the limiting Li-ion diffusion rate between two-phase boundary electrode particles has an activation energy of approximately 50 kJ mol^{-1}, but nanomaterials smaller than 40 nm did not have the phase boundary and had an activation energy of 10–25 kJ mol^{-1}. Hu et al. [19] also corroborated the enhanced electrochemical effects of nanostructuring TiO_2 as a Li-ion battery electrode material. They revealed that the discrepancy in the quantity of intercalated Li-ion could be attributed to dissimilar diffusion along the *ab*-plane, and that Li-ion surface storage on the nanoparticles is energetically more favorable than bulk insertion.

The development of silicon based anodes for Li-ion batteries has also received a considerable amount of interest over the past several years. The Li–Si binary system indicates that an individual silicon atom can accommodate the insertion of 4.4 lithium ions, yielding one of the highest known theoretical charge capacities; additionally, the discharge potential of silicon is quite low, and it is already a highly processed material. Although this theoretical capacity value is eleven times larger than that of the current commercially used graphite anodes, the use of silicon as an anode material is limited because the insertion of 4.4 Li-ions is accompanied by a 400 % volume expansion of the lattice. This tremendous expansion results in cracking and disintegration of the electrode, with active material loss via reduced electronic contact, giving way to severe capacity fade [20]. Such has been the case for past studies involving Si bulk films and micrometer-sized particles, where capacity fading and short battery lifetimes were attributed to the pulverization and loss of electrical contact between the active material and the current collector.

Fortunately, the use of Si nanostructures such as nanowires can alleviate these issues because there is a critical particle size below which fracture will not propagate [21, 22].

While Li-ion surface storage and increased fracture resistance are performance benefits inherent to many nanostructured electrode materials, an even more complex interaction has been observed in nano-sized $LiFePO_4$ because of its incomplete miscibility gap (region in a phase diagram in which two phases with nearly the same structure have no solubility in one another) [23, 24]. Meethong et al. [25] showed that the miscibility gap in pristine $LiFePO_4$ diminishes with decreasing nanoparticle size, suggesting that the miscibility gap completely disappears below some critical value. The enhanced solubility of Li-ions in nano-sized $LiFePO_4$ can be ascribed to the presence of multiple phases in the individual particles. The coexistence of two crystallographic phases within one particle leads to a domain-wall-related energy penalty, which is determined by the strain prompted from the difference in the lattice parameters and can destabilize the two-phase coexistence in smaller particles. Nanostructures have been utilized to increase the equilibrium compositions and effectively reduce the lattice mismatch between coexisting phases. Therefore, the energy gain associated with phase separation will drop with decreasing particle size and the miscibility gap will gradually diminish [26]. Experimentally, the miscibility gap began to decrease while the solubility increased starting with nanoparticles approximately 100 nm in diameter, and showed a strong effect in particles smaller than 35 nm [27]. Changes in the miscibility gap were ascribed to particle size associated modifications in the molar free energy of mixing for either one or both phases. The associated particle size effects were also speculated to emerge from the relative contributions of the particle-matrix surface energy and stress, and the compatibility between stresses in the coherently interfaced two-phase particles [25].

Studies have also revealed that the implementation of nanostructured electrode materials can result in the initiation of new lithium storage mechanisms. These effects typically manifest either via a pseudocapacitive storage mechanism that accommodates lithium ions on the surface/interface of the particles below a critical particle size or through a conversion mechanism that involves the formation and decomposition of at least two separate phases [28–31]. The pseudocapacitive mechanism is more pronounced because of the more prominent role of surfaces and grain interfaces in nanomaterials. Reversible conversion reactions based on the reduction and oxidation of metal nanoparticles can ensue between binary compounds comprised of some second or third period element, a transition metal oxide, and metallic lithium [32–37]. Nanoparticles are extremely effective toward this means because of their large specific surface area that is very active toward the decomposition of the lithium binary compound. Furthermore, reduction of some micrometer sized materials to the nanoscale has been shown to activate or enable reversible electrode reactions that would otherwise not take place, typically materials with low Li-ion diffusion coefficients.

3.2 Shortcomings of Nanostructured Electrodes

One of the most prominent features of nanomaterials is their large specific surface area that has a correspondingly higher surface energy than bulk contemporaries which can allow for tailoring of the chemical potential [28]. As previously mentioned, this large and energetically relevant surface area can contribute to a pseudocapacitive Li-ion storage mechanism [38–41]. However, there are several shortcomings and limitations that can be categorized as direct consequences of this additional free surface energy.

Diminished voltage plateaus and curved voltage profiles have both been attributed to the strain and interfacial energy from coexisting lattice terminations within electrode nanoparticles. These differing structural environments in the near-surface region result in a distribution of the redox potentials; varying structural environments depend on the distance from the surface and the surface orientation itself, substantiating that size and morphology play an important role in dictating particle stability [42]. Hence, the irreversible capacity can be larger for smaller particles with a surface area greater than some critical value where the optimal particle size is dictated by the voltage window [43]. Moreover, too high of Li-ion intercalation at the surface leads to passivation of the surface region by surface reconstruction or mechanical failure, and accompanying irreversible capacity loss is inevitable [44, 45]. Thus, there exists an optimal particle range for nanomaterials where cracks resulting from non-surface related volume expansion will not propagate, and surface reconstruction will not occur [16].

Expanding upon this, the formation of the solid electrolyte interphase (SEI) is even less predictable in nanomaterials because of their high surface area and corresponding high surface energy that increases the potential for secondary or side reactions along the surface area of the particle. The SEI layer is a film that forms on the surface of the electrode through the decomposition of the electrolyte into lithium containing organics and inorganics in an attempt to reduce the contact area and thereby prevent further electrolyte decomposition [46]. The formation of a stable SEI is often critical because it can act as an interfacial stabilizer and improve the capacity retention of the electrode over long cycle lifetimes at the expense of a large irreversible capacity drop because its ionic and electronic conductivities are relatively low and it effectively intercepts and stores Li-ions that would otherwise be intercalated by the electrode [47]. However, a non-uniform, thick, and unstable SEI can have unfavorable impact on electrode performance, and is more likely to be present in nanomaterials rather than their bulk counterparts. Past studies have clearly demonstrated the deleterious effects of SEI formation on nanostructured electrodes. It was confirmed that a thicker SEI would form on $LiCoO_2$ as the particle size decreased [48]. The thicker layers negated the positive effects of nanostructuring the electrode material by severely impeding Li-ion diffusion, and diminishing the electrode performance at higher current densities. Similar results were seen for $LiMn_2O_4$ where the SEI thickness on a cycled electrode increased in a near linear fashion with the number of cycles, and the cycle stability was the performance parameter most impacted [49].

Another issue concerning the size-stability of nanoparticles is the proclivity of the electro-active metal ion to dissolve in the electrolyte, resulting in cyclical capacity degradation [50]. The majority of cases documenting the dissolution of the electro-active metal ion have focused on $LiMn_2O_4$, where Mn-ions in the electrode are easily dissolved into the electrolyte by acids generated via oxidation of the solvent molecules, but the process can occur in any Li_xMO_y phase [51–53]. Migration of the dissolved species to the anode can have fatal effects on the functionality of the cell; their precipitation on the anode terminal destroys the passivation layer on the negative electrode and can lead to the formation of a plated layer [54]. However, dissolution can be prevented by applying an oxide coating on the nanoparticles to decrease the surface area and eliminate any side reactions [55]. The electrochemical operating voltage range for nanostructured electrode materials narrows during electro-active material dissolution.

Electronic and ionic conductivity have long been limiting factors to the development of advanced electrode materials for Li-ion battery [56]. These barriers stem from several fundamental issues: the metal oxide materials typically employed as cathodes have relatively low electronic conductivity, there exists a complex relationship between the electronic and ionic conductivities in electrodes, and phase transformations upon lithiation can change the conductive properties. Additionally, it has been well established that nanomaterials suffer from much greater resistance than their corresponding bulk material; nanowires, for example, typically display resistivity values that are $\sim$20 % greater than what is seen in the bulk regime, and this discrepancy can extend up to several orders of magnitude [57]. The reason for the drastic increase in the resistivity of nanomaterials is due to the surface scattering of electrons as a result of the sheer increase in the relative surface area and the number of grain boundaries that are strongly dependent on the particle size and morphology [58]. These physical parameters negatively impact the conduction mechanisms because of the scattering of electrons at particle surfaces and grain boundaries. These effects are also more pronounced in the electrode materials with smaller mean particle size. To compensate for these effects, conductive additives such as carbon are typically added to electrodes, but such incorporations decrease the already low packing density of nanostructured electrode materials even further.

Tortuosity is one of the most important parameters to characterize a porous medium, and reflects the reduction in transport within the electrode due to the complex porous structure comprised of active particles, binder, and conductive carbon [59–61]. Complex, tortuous nanostructures can lead to decreased effective electrolyte conductivity and diffusivity for porous electrodes by limiting transport in the electrolyte phase. The concept of electrode tortuosity (τ) is used along with electrode porosity (ε) as a measure for the decrease in effective electrolyte conductivity and diffusivity due to the structure of the electrode within the confines of the porous electrode description; the tortuosity of a material should decrease as the porosity increases, and Bruggeman suggested a quantitative relationship where tortuosity is inversely proportional to the square root of porosity [62, 63]; note that this simplification is only appropriate for describing the packing of equally sized spheres [64]. Tortuous transport paths result in a higher macroscopic ohmic resistance, which in turn hinders the delivered energy and power density [65]. Based on geometrical constraints, nanoparticle based electrodes would have less effective packing with a large amount of small pores that would ultimately hinder electrolyte penetration and thus negatively affect the tortuosity, conductivity, and diffusivity.

An additional limitation stemming from the use of nanomaterials relates to their packing density or mass loading which is the amount of active material per unit area of the electrode. There are several characteristics that must be strictly controlled in order to assure a homogeneous and ideal electrode mass loading; such features include the physical composition, packing density to maximize active material content, and open porosity to access the electrolyte. Ultimately, the desired mass loading is dictated by and should reflect the diffusion coefficient of lithium ions, electrical conductivity of particles, and electronic continuity to guarantee charge exchange to the current collectors. The standard thickness and mass loading for a typical commercial electrode is 50 μm and 20 mg cm^{-2}, respectively [66]. However, the overwhelming majority of commercially available cells employ microsized electrode materials because the mass loading of nanosized particles is far less, typically 1–2 mg cm^{-2} [67, 68]. Given that the density of a nanomaterial is generally less than the same material formed from micrometer-sized particles, the volumetric energy density is effectively limited in nanomaterials because there is a larger ratio of components not participating in the electrochemical reaction (such as binding agents or conductive additives) [8, 13]. In fact, carbon black additives can constitute 80–98 % of the composite electrode surface area as governed by the average particle size [69].

Processing conditions are also heavily impacted by the high reactivity of nanoparticles. Many nanomaterials exhibit poor stability when exposed to ambient atmosphere; such is the case for olivine $LiFePO_4$ due to the high reactivity of Fe^{2+} in the presence of atmospheric oxygen and water. This undesired reaction under ambient exposure is known to be more pronounced for smaller particles because of their increased surface/exposure area, and can result in spontaneous lithium and surface oxidation [70]. It has also been revealed that nanostructured electrode materials with poor adherence to the current collector will agglomerate during cycling; nano-SnSb undergoes successive agglomeration during Li-ion insertion and extraction, and experiences quick capacity fade as a result [71]. Inactive $LiMO_y$ phases with transition metal cations of lower oxidation are formed from redox reactions with solution species [54]; moreover, these compounds can be spontaneously delithiated under ambient conditions involving reactions with CO_2 [72]. Thus, the application of some electro-active materials may be limited due to the high processing costs associated with avoiding these secondary reactions and stability issues.

4 Nanoparticle Assemblies as Li-Ion Battery Electrodes

The controlled assembly and arrangement of nanoparticles across extended and varying length scales is a critical element in the design of integrated materials with advanced functions. There is strong motivation to use nanoparticles as building blocks because of the uniquely extensive effects that are only observed in nanosized systems. The interaction between nanoparticles in spatially assembled micro-arrangements can dictate some of the properties, thereby broadening the potential applications for which these materials can be utilized.

Structured micro-assemblies built from nanoparticle primary subunits can be classified into several categories: arranged, hierarchical, and oriented nanoparticle assemblies. In addition to this, one-, two-, and three-dimensional systems exist under each of these divisions. Methods of preparation can differ vastly depending on the material being synthesized or processed. There is hardly a universal method that can be applied for the majority of chemistries, but there are several general approaches involving wet chemical techniques that are commonly used. Examples of commonly used techniques include: solvothermal, sol–gel processing, surfactant-assisted synthesis, or solvent-controlled synthesis [73].

The following sections include examples of microsized nanoparticle assemblies. The covered examples are not exhaustive but meant to provide the reader with a broad summary of the available methods and considerations.

4.1 Arranged Nanoparticle Assemblies

Self-assembled nanostructures with particular morphology and novel properties have been the subject of intense research over the past several decades due to their novel properties and tunable functions that can be utilized in a wide range of applications [74–76]. The use of organic surfactants and polymers in material synthesis has been a popular method of achieving morphological control, where the polymer or foreign ions typically work to regulate and stabilize the surface of the nanoparticles or directly serve as a templating agent [77].

One of the more facile approaches, because of its ease of processing and lack of template, that falls under this scope is the polyol method. Utilizing the polyol technique, hollow vanadyl glycolate nanostructured microspheres were synthesized via a highly scalable and template free polyol induced solvothermal process [78]. Subsequent calcination transformed the precursor material into vanadium pentoxide, a well-studied transition metal oxide. The vanadyl glycolate nanoparticles were synthesized through a self-seeding process, and then aggregated around N_2 bubbles formed during the reaction that acted as "quasi-micelles" due to the large polarization discrepancy between nitrogen and water. The proposed formation mechanism provides a firm understanding of the processes leading to the observed hollow microsphere morphology.

Fig. 1 **a, b** SEM and **c** rate performance of the V_2O_5 microsphere electrode at 1, 3, 6, 12, and 20 C of hollow V_2O_5 nanostructured microspheres. Reprinted with permission from Uchaker et al. [78]. Copyright 2013 American Chemical Society

As shown by the SEM micrographs in Fig. 1, the hollow microspheres had an average outer diameter of approximately 3.0 μm and an inner diameter of approximately 750 nm, and were composed of aggregated nanoparticles. The size of the individual nanoparticles that formed the hierarchical aggregate nanostructure ranged from 60–110 nm, and was confirmed with a calculated average size of 77 nm as determined by applying the Scherrer equation to the full width at half-maximum of the (001) peak from the XRD spectrum. The thermally treated material was tested as cathode for Li-ion battery, and showed excellent cycle stability and high rate performance; specifically, the discharge capacities of the V_2O_5 microspheres at current densities of 300 (1 C), 900 (3 C), 1800 (6 C), 3600 (12 C), and 6000 (20 C) mA g^{-1} were 241, 165, 127, 96, and 78 mAh g^{-1}, respectively (Fig. 1). The exceptional electrochemical performance was attributed to the relatively thin walled structure that ensured fast phase penetration between the electrolyte and active material, and shortened lithium-ion migration distance. The prolonged cycling stability is ascribed to the inherent morphological void that can readily accommodate volume expansion and contraction upon cycling.

Similarly, Cao et al. [79] synthesized hollow V_2O_5 microspheres composed of nanorods formed via self-assembly. Hollow microsphere precursors of vanadyl glycolate composed of assembled nanorods were first synthesized utilizing the polyol process, and were then transformed to V_2O_5 with a simple annealing step. No change in the morphology was observed upon annealing. The individual nanorods comprising the structure were approximately 200 nm in diameter and varied in length, while the overall spherical structures were approximately 2 μm in diameter. The structures were successfully able to intercalate up to one Li-ion per V_2O_5 formula unit, and showed moderate capacity retention up to 15 cycles with an initial discharge capacity of 286 mAh g^{-1} [79]. It was determined that interconnecting nanoparticles between the self-assembled nanorods lead to a reduction in the Li-ion diffusion distance, thereby increasing the kinetics of the material and the overall electrochemical performance. It has since been revealed that hollow structures can result in favorable enhancements in capacity retention with cycling due to the presence of the inner cavity that can accommodate cycling induced strain due to the de/intercalation of Li-ion [80].

Fig. 2 **a**, **b** SEM images of LNMO under different magnifications; **c** capacity retention versus cycle number of the LNMO cycled between 4.8 and 2.0 V at various specific currents ranging from 25 to 500 mA g^{-1} [84]

Lithium rich layered oxides are promising cathode materials on account of their, typically, much higher capacity (>250 mAh g^{-1}) and potential for reduced cost [81]. Despite these advantages, they suffer from intrinsic poor rate capability and cycle stability, primarily because lithium extraction is accompanied with Li_2O oxidation in the initial charge followed by substantial transition metal ion migration into the lithium sites [82]. This oxidation and subsequent irreversible migration event is stabilized by surface structural rearrangement that effectively impedes Li-ion transport [83]. To combat these issues, micro/nano $Li_{1.2}Ni_{0.2}Mn_{0.6}O_2$ (LNMO) cathode materials were processed using a hydrothermal reaction followed by calcination [84]. Rod-like, micro-sized oxalate precursor matrices are formed during the facile hydrothermal synthesis, as depicted in Fig. 2. Subsequent decomposition of the oxalate precursor is accompanied with large amounts of gas release, leading to the formation of cubic or spherical shaped nanoparticles as the assembly units of the rod-like matrix. The release of gas during calcination also leads to the formation of a fairly porous structure, which can ease electrolyte penetration.

The rod-like structures (1–3 μm in width and 3–10 μm in length) are assembled from nanoparticles with average sizes of 20–150 nm. When electrochemically tested, this material retains a high discharge capacity of 212 mAh g^{-1} (97 % of the initial capacity) after 30 cycles at 1 C, and exhibits an initial discharge capacity of 198 mAh g^{-1} at 2 C. The superior performance is attributed to the merits of the assembled micro/nano structures; namely, the ability to reduce secondary reactions occurring between the electrode and the electrolyte because of the high surface area of the not completely isolated nanoparticles. These effects ultimately improve the capacity retention, as demonstrated by the high discharge capacities of 244, 225, 212, and 186 mAh g^{-1} at 0.2, 0.5, 1, and 2 C current densities for 30 cycles at each rate. The capacity retention values of 94, 91, 97 and 94 %, respectively, at each current density step attest to the outstanding current rate and cycling performance of the micro/nano material. These findings support the use of micro-sized nanoparticle assemblies given that the issues of poor rate capability and cycle stability are circumvented with this material design.

The manipulation of arranged microscaled nanoparticle assemblies has also proven to be a productive route toward improved battery performance.

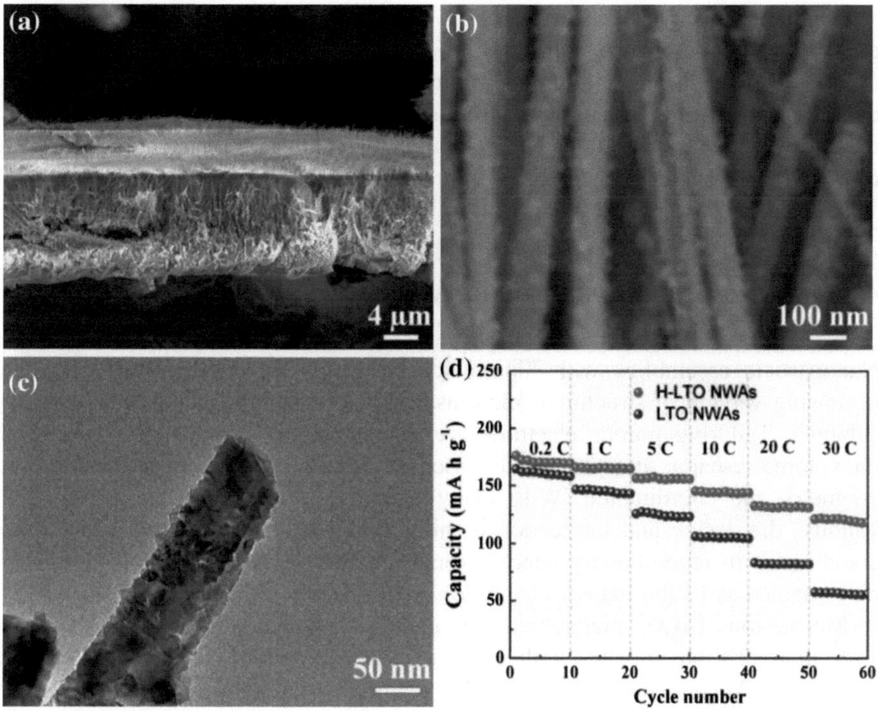

Fig. 3 **a**, **b** Cross-sectional SEM and **c** TEM images of H-LTO nanowire arrays; **d** Comparison of the specific discharge capacities for the LTO and H-LTO nanowire arrays at various current densities. Reprinted with permission from Shen et al. [85]. Copyright 2012 John Wiley & Sons, Inc

$H_2Ti_2O_5{\cdot}H_2O$ nanowire arrays with an average diameter and length of 80 nm and 8 μm, respectively, were directly grown on Ti foil through an alkali hydrothermal process combined with an ion-exchange process, as shown in Fig. 3. This precursor material was then chemically lithiated and annealed under Ar/H_2 gas at various temperatures to convert the precursor into hydrogenated spinel $Li_4Ti_5O_{12}$ (LTO). Heat treatment in hydrogen may cause the loss of oxygen in LTO, resulting in the formation of both oxygen vacancies and trivalent titanium ions that can enhance the electronic conductivity [85]. The presence of hydrogen was confirmed through XPS analysis of the O 1s peak which was altered by the presence of Ti-OH bonds.

Under this design, the hydrogenated LTO nanowire arrays possess large specific surface area that facilitates the fast transfer of Li-ions. The presence of Ti^{3+} ions can facilitate the electronic conductivity by allowing electron hopping from Ti^{3+} to Ti^{4+}. The oxygen vacancies do not contribute to electronic conductivity, at least not directly, as they support ionic conduction. However, the presence of oxygen vacancies leads to more void space inside the crystal lattice, and thus may allow easier migration/diffusion of lithium ions. Comparison of H-LTO nanowire arrays with pristine LTO nanowire arrays demonstrates the value of hydrogenation in terms of their high capacity (173 mAh g^{-1} at 0.2 C), excellent rate capability (121 mAh g^{-1} at 30 C), and good cyclic stability (~5 % capacity loss after 100 cycles at a rate of 5 C), as shown in Fig. 3. Ultimately, the manipulation of the nanowire assembly contributed to an effective ambipolar diffusion of Li-ion and electrons into and out of the electro-active species in the H-LTO nanowire architecture - enabling remarkable rate capability, cycling performance, and improved stability.

4.2 Hierarchical Nanoparticle Assemblies

Nanoparticle assemblies with hierarchical structure are another intriguing and promising version of structured micro-assemblies built from nanoparticle primary subunits. Tailoring porous electrode architecture at the mesoscale is largely an unexplored research direction. Such structures are often three-dimensional in their geometry and coordination. With many of such structures having proven their validity, the design and fabrication of hierarchically structured porous electrodes could result in revolutionary energy storage systems. Several examples and their performance as Li-ion battery electrodes are discussed.

Mesoporous Co_3O_4 hierarchical nanobundles were produced through a facile hydrothermal strategy paired with subsequent annealing [86]. Microscopy revealed that the Co_3O_4 nanobelts are composed of many irregular pores uniformly dispersed with an average diameter of approximately 15–25 nm. The overall nanobundle consists of assembled nanobelts with an average diameter of about 200 nm and length of approximately 5 μm. The formation of the porous structure in the Co_3O_4 nanobelts can be attributed to the thermally driven contraction process during the thermal decomposition of the precursor accompanied with the release of CO_2 and H_2O. Examination of the electrode after 60 cycles (ending in the discharged state) revealed that the original morphology of the Co_3O_4 is perfectly retained. This retentive quality was reflected in the electrochemical results. The reversible specific capacity was stable at 1667 mAh g^{-1} at a current density of 0.1 A g^{-1} after 60 cycles. Even at high current densities of 1 and 5 A g^{-1}, the Co_3O_4 electrode still could deliver a remarkable discharge capacity of 1264 and 603 mAh g^{-1}, respectively, for 100 cycles.

Spinel $CoMn_2O_4$ porous nanosheets were assembled into hierarchical microspheres using solvothermal synthesis followed by calcination [87]. The diameter of the overall microspheres varies from 4–6 μm, and the nanosheets are relatively thick at approximately 350 nm, as displayed in Fig. 4, but this in turn effectively ensures architectural stability.

The hierarchical architecture with spherical morphology possesses relatively low surface energy, which results in significantly less self-aggregation during the charge/discharge process. As a result, $CoMn_2O_4$ hierarchical microspheres can cycle relatively stable with high capacity and rate capability despite their overall tremendous size (Fig. 4). Similar morphologies and enhanced electrochemical

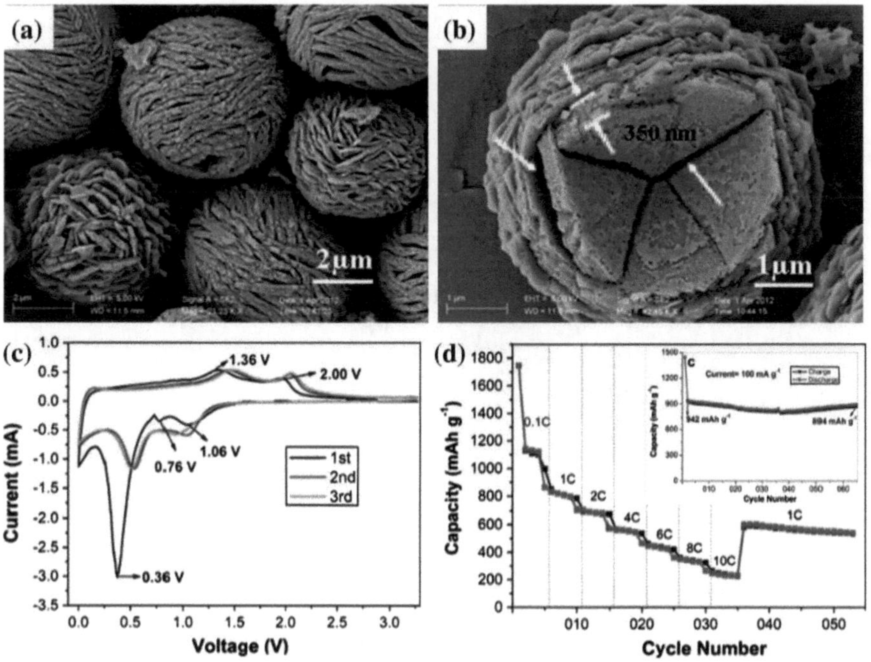

Fig. 4 **a**, **b** SEM images, **c** first three consecutive CVs, and **d** cycling performance at various current rates (1 C = 900 mA g^{-1}) of the $CoMn_2O_4$ hierarchical microspheres. Reprinted with permission from Macmillan Publishers Ltd: Scientific Reports 2: 986 © 2012 [87]

performances have been reported for an array of electrode material candidates, including $LiFePO_4$, Mn_2O_3, TiO_2, Fe_3O_4, V_2O_5, and Co_3O_4 [88–93].

MoS_2-polyaniline (PANI) nanowires were synthesized through the hydrothermal conversion of MoO_x/PANI nanowires using thiourea, and the end product is shown in Fig. 5 [94]. The two-dimensionally MoS_2 nanostructures grew anisotropically due to the crystallographically layered structure held by van der Waals interactions. The MoO_x/PANI nanowire precursor provides an excellent one-dimensional template for hierarchical MoS_2/PANI conversion and growth because of the soft and flexible PANI chains that buffer volume expansion and thus preserve the overall morphology during the sulfidation process. The end result was a 200–500 nm diameter nanowire uniformly coated with ultrathin, irregularly shaped MoS_2 nanosheets approximately 10 nm thick.

Comparing the cyclic voltammograms for the MoS_2/PANI and MoS_2 microparticles clearly revealed that the polymer incorporated design can more actively and reversibly store Li-ions. In addition to a higher specific capacity in the initial cycle, the MoS_2/PANI nanowires show better cycling performance than the MoS_2 microparticles, as well as rate capability (Fig. 5). Based on the Nyquist impedance results shown in Fig. 5c, the MoS_2/PANI experienced significantly decreased charge-transfer resistance at the electrode/electrolyte interface when compared with

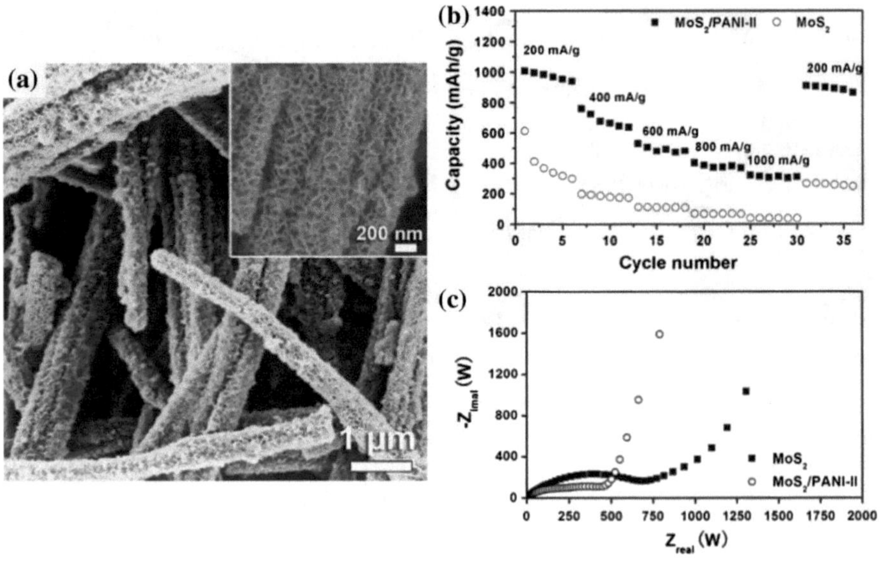

Fig. 5 **a** SEM images of the MoS_2/PANI nanowires; **b** rate performances (0.01–3.0 V at 100 mA g^{-1}), and **c** Nyquist plots (100 kHz–10 mHz) of the MoS_2/PANI nanowires and commercial MoS_2 microparticles. Reprinted with permission from Yang et al. [94]. Copyright 2012 John Wiley & Sons, Inc

the MoS_2 microparticles, indicating the greatly decreased charge-transfer resistance at the electrode/electrolyte interface. The improved electrochemical performance of the MoS_2/PANI assemblies was ultimately attributed to nano-confinement of the MoS_2 primary units which promote kinetics because of the shortened diffusion paths, while the primary architecture at the micrometer scale effectively avoids aggregation of the active nanomaterials and facilitates the transport of electrons and ions. Similar results have been reported for Fe_2O_3 except that the adopted morphology was spherical instead of wire based [95].

Similar to these structures, columnar-shaped hierarchical SiO_x nanoconifers were directly self-organized on metallic $NiSi_x$ nanowires through chemical vapor deposition [96]. The SiO_x nanoconifers act as an active element for alloying/de-alloying with Li-ions because of their substantially higher surface areas, and the metallic $NiSi_x$ template serves both as efficient electron transport pathway and as rigid mechanical support. The deposited template wires have an average length and diameter of 5 μm and 170 nm, respectively, with a uniform 30 nm thick $NiSi_x$ core. The SiO_x shells do not conformally deposit on the $NiSi_x$ template and instead adopt a conifer shape where periodic SiO_x columns are self-assembled with a high packing density, and is a result of the limited number of nucleation sites. After the first cycle, the SiO_x nanoconifer cell shows a very high coulombic efficiency of over 97 % for the subsequent cycles. After 20 cycles, the capacity is measured to be ~1375 mAh g^{-1}, which corresponds to ~80 % of the initial capacity. Then, its

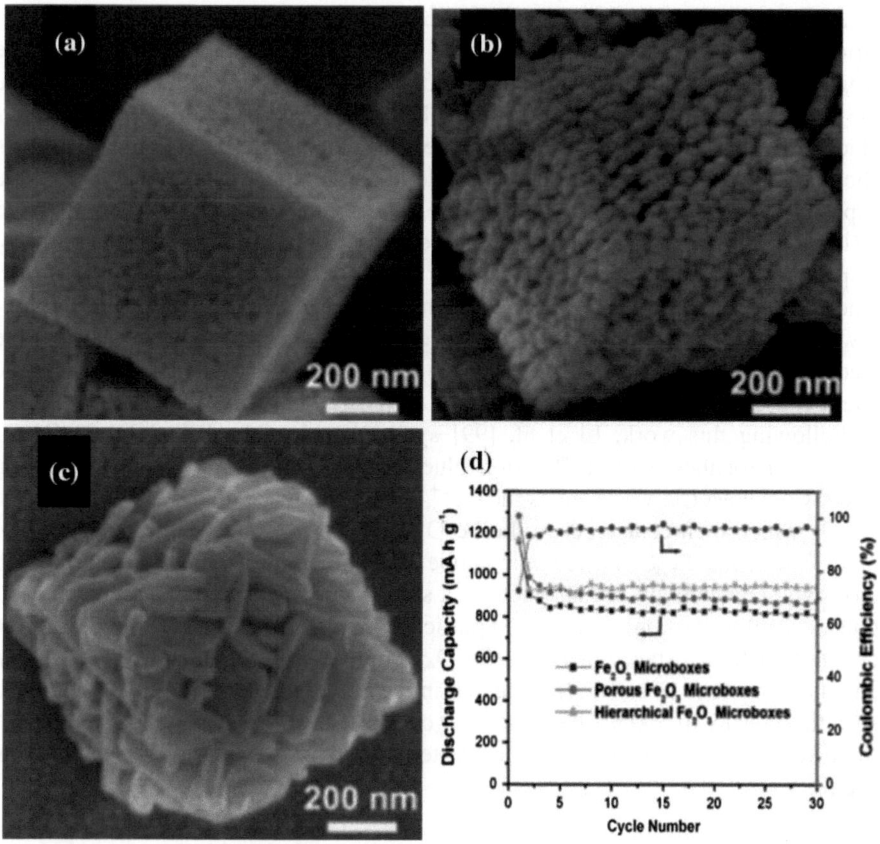

Fig. 6 SEM images of the **a** relatively dense, **b** highly porous, and **c** hierarchically structured Fe_2O_3 microboxes; cycling performance over the voltage range of 0.01–3.0 V at a current density of 200 mA g^{-1}. Reprinted with permission from Zhang et al. [98]. Copyright 2012 American Chemical Society

capacity is gradually degraded to about 800 mAh g^{-1} up to the 100th cycle. For comparison, SiO_x bulk exhibits very low capacity around 500–600 mAh g^{-1}, and its capacity drastically drops after a few cycles. It is speculated that the regularly spaced voids among individual SiO_x columns accommodate cycling induced volume expansion and reduce the Li-ion diffusion pathways through facile electrolyte penetration. In this regard, the columnar shape provides great advantages in terms of structural stability as well as enhanced kinetics. Analogous morphologies and results have been seen in Cu-Si-Al_2O_3 cable arrays as well [97].

Zhang et al. [98] synthesized Fe_2O_3 microboxes by controlling the thermal treatment conditions under which Prussian blue ($Fe_4[Fe(CN)_6]_3$) precursors were processed. Using this method yielded three different structures: relatively dense iron oxide shelled microboxes with a large interior cavity, highly porous microboxes constructed from enlarged Fe_2O_3 nanoparticles ($\sim$70 nm), and hierarchically structured shelled microboxes consisting of Fe_2O_3 nanoplatelets (200 nm edge length with a thickness of $\sim$40 nm); all of which are displayed in Fig. 6. Despite the primary nanoparticle size being smaller for the porous microbox, the hierarchical microbox actually displayed a higher and more stable discharge capacity. As shown in Fig. 6, the hierarchical Fe_2O_3 microboxes exhibited the highest reversible capacity of 945 mAh g^{-1} in the 30th cycle, while comparable capacities of 802 and 871 mAh g^{-1} were also obtained for the dense and porous Fe_2O_3 microboxes, respectively. This finding further verifies the impact and importance that hierarchical structuring can have on electrochemical performance. Specifically, the hollow structure and the porous shell architecture facilitate penetration of the electrolyte and transport of Li-ion ions in the electrode and efficiently buffers the stress caused by volume variation during the charge-discharge process.

Following this work, Li et al. [99] synthesized double-shelled Fe_2O_3/Co_3O_4 hollow microcubes using a Prussion blue-supported annealing chemical reaction route. Cobalt acetate was used as the Co^{2+} source and reacted with OH^- to generate Co_3O_4 which then coalesced on the Fe_2O_3 hollow microcube shell acting as template. The robust Fe_2O_3 hollow microcube at the inner layer not only displays good electronic conductivity but also acts as a stable support for the hierarchical Co_3O_4 outside shell consisting of nanosized particles. Impedance analysis revealed that the charge transfer resistance of Fe_2O_3/Co_3O_4 is much lower than that of pure Co_3O_4, and the composite electrode recorded an initial coulombic efficiency of 74.4 % and specific capacity of 500 mAh g^{-1} after 50 cycles at a current density of 100 mA g^{-1}; these results are a factor of 3 times better than that of pure Co_3O_4 nanoparticle sample.

4.3 Oriented Nanoparticle Assemblies

There has been considerable effort carried out regarding the synthesis and characterization of ordered nanoparticle superstructures with a vast range of architectures, also known as mesocrystals. As initially described by Cölfen and Antonietti, mesocrystals, short for mesoscopically structured crystals, are crystallographically oriented nanoparticle superstructures [75, 100, 101]. For some time, mesocrystals were only studied in biomineral materials, but current research efforts have shifted to the development of mesocrystalline organic molecules, metal oxides, and other functional materials [102].

To date, novel nanoparticle arrangement has been achieved with colloidal crystals and supercrystals or nanocrystal superlattices; however, individual nanoparticle orientation is not taken into account under these systems. Similar to these systems, mesocrystals can be classified by their high degree of crystallinity and high porosity, but differentiate themselves with subunit (individual nanoparticle) alignment along a crystallographic register. These highly desirable properties are due in part to mesocrystal formation mechanisms, which are still poorly understood, and make mesocrystals the ideal material candidates for catalysis, sensing, and energy storage and conversion applications. Porous materials with large specific surface areas have been shown to enhance the performance of lithium-ion battery electrode materials because of the more prevalent and uniform pores that ease intercalation by decreasing the Li-ion diffusion distance and pathways. Thus, mesocrystalline materials could be exceedingly beneficial when used as electrode for electrochemical energy storage devices, particularly Li-ion batteries, because of the inherent and uniform porosity associated with the well-defined nanoparticle orientation [103].

4.3.1 Anodic Materials

The majority of the work on anodic electrode materials has focused on TiO_2 and its derivatives because of its proven safety, good cyclic stability, and high rate performance [104–106]. Initial reports of TiO_2 mesocrystals focused on the topotactic conversion of NH_4TiOF_3 [107, 108], and were followed with the synthesis of rutile TiO_2 through the use of organic additives [109, 110]. Hong et al. [111] were among the first to develop rutile TiO_2 mesocrystals using an additive free, low temperature synthesis process. It is worth note that lithiation in microsized rutile TiO_2 is limited to a negligible amount because of anisotropic Li-ion diffusion that perturbs the rutile framework which successfully blocks Li-ions from occupying thermodynamically preferential octahedral sites [112, 113]. There are no such effects in nanoparticle rutile TiO_2 because of the short diffusion length, and significantly more Li-ions can be reversibly incorporated [114].

TiO_2 nanowires were first synthesized by hydrothermal growth, and were then dispersed in an aqueous acid solution to yield mesocrystal bundles through homoepitaxial aggregation. The individual nanowires measured several hundred nanometers long and were 3–5 nm in diameter, while the overall bundles had a diameter of approximately 70 nm. Corresponding selected area electron diffraction (SAED) patterns verified the single crystal characteristics of the overall bundle. Higher resolution TEM images further confirm the presence of homoepitaxially oriented nanowires as well as their growth along the [001] direction.

The TiO_2 mesocrystals demonstrated good rate performance and delivered a large stable capacity of 200 mAh g^{-1} at 0.5 C; furthermore, the mesocrystals had a stable reversible capacity of 171.3 mAh g^{-1} after 100 cycles at 1 C. The enhanced lithium storage properties were ultimately attributed to the mesoscopic structure of the rutile TiO_2 mesocrystals, which offered a large specific surface area of 38.5 m^2 g^{-1} and short transport distance as evinced by the low volume of mesopores. Both of these factors would facilitate lithium-ion intercalation at the interface, shorten both the mass and charge transport distances, and better accommodate any volume change accompanied with lithium-ion intercalation.

A mechanism for the formation of the rutile TiO_2 mesocrystals in the absence of polymer additives was proposed to occur through the homoepitaxial aggregation of hydrogen titanate single crystal nanowire subunits. It is proposed that Brownian motion would bring two titanate nanowires in contact which would then be held together by weak van der Waals forces. The attractive force between two hydrogen titanate nanowires would be rather weak since the Hamaker constant of water is more than an order of magnitude larger than that of hydrogen titanate, giving rise to the formation of mesocrystals. Additionally, the acidic conditions would ensure a relatively slow condensation reaction between the two nanowire surfaces, allocating for better crystal orientation prior to aggregation. It should also be noted that aggregation along the [001] direction provided the maximum attachment among neighboring nanowires, and the greatest reduction in surface energy. It is also proposed that the slow phase transformation at low temperature accounted for the stability of the mesocrystal against fusing into single crystal. Interestingly, Wulff-shaped octahedral and nanorod-like nanoporous mesocrystals composed of ultrathin rutile TiO_2 nanowires were produced when the surfactant sodium dodecyl benzene sulfonate (SDBS) was introduced into the reaction system [115]. The morphology of the mesocrystals depended largely on the content of SDBS, specifically the titanate/SDBS ratio, resulting in rutile mesocrystals with different shapes. Again, the mesocrystals formed through homoepitaxial self-assembly of nanocrystallites, but this time with the assistance of the SDBS additive.

As opposed to rutile TiO_2, Ye et al. [116] synthesized ellipsoidal anatase TiO_2 mesocrystals approximately 380 nm in length and 280 nm in diameter through the solvothermal reaction of acetic acid and tetrabutyl titanite (TBT), the end result of which is shown in Fig. 7. The mesocrystalline structures displayed a seemingly rough surface and were composed of oriented nanoparticles ranging between 10–20 nm in diameter. SAED confirmed the single crystalline characteristics of the mesocrystal and revealed that the elongated ellipsoid direction corresponded to the [001] direction, as observed in the inset of Fig. 7c. Time dependent growth investigations revealed that the mesocrystals formed through a series of several chemical

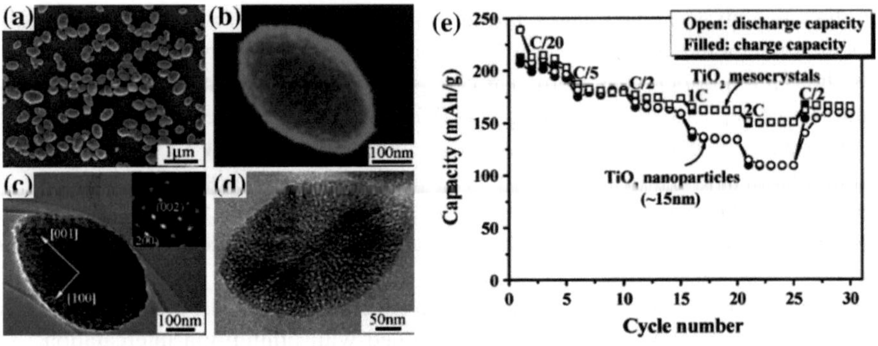

Fig. 7 **a**, **b** SEM and **c**, **d** TEM images, and **e** electrochemical performance of nanoporous anatase TiO_2 mesocrystals. Reprinted with permission from Ye et al. [116]. Copyright 2011 American Chemical Society

reactions involving ligand exchange, esterification, hydrolysis-condensation, and finally transient metamorphosis.

It is desirable to develop additive-free solution routes considering that the organic additives usually employed in mesocrystal synthesis are too costly for wide-spread application. Towards this means, the acetic acid solvent played multiple roles during the mesocrystal formation process. The acetic acid acted as a chemical modifier and lowered the reactivity of TBT while reacting with it to form metastable precursors for the slow release of soluble titanium-containing species so that nascent anatase nanocrystals were continuously being formed. Acetic acid reacted with TBT to yield butyl acetate that operated as a templating agent during oriented aggregation while the nanocrystals were stabilized by acetic acid, thus preventing single crystal formation and eventually leading to mesocrystal formation along the [001] direction through the oriented attachment mechanism.

The mesocrystalline nature of the particles was most likely achieved through the oriented attachment mechanism and infusion between the TiO_2 nanoparticles, leading to a decrease of the interfacial nucleation sites for the rutile phase. The formation of the TiO_2 mesocrystals through the oriented attachment mechanism also decreased the number of grain boundaries existing between individual TiO_2 nanoparticles, thus resulting in facile electronic conduction and fast Li-ion transport between the electrolyte and the mesocrystal electrode. The uniformly dispersed nanopores throughout the structure of the mesocrystals can facilitate contact with the electrolyte, hence increasing the electrode/electrolyte interface which favors fast Li-ion transport. These speculations were validated when the mesocrystal TiO_2 anode was compared to a TiO_2 anode composed of nanoparticles of approximately the same size as the mesocrystalline subunits, approximately 15 nm (Fig. 7e). The capacity of the mesocrystals and nanocrystals were comparable at lower current densities. However, the mesocrystalline anode considerably outperformed its nanoparticle counterpart at higher current rates, 152 mAh g^{-1} compared to 115 mAh g^{-1} at 2 C (1 C = 170 mA g^{-1}), respectively, while demonstrating an overall good rate capability. The mesocrystals also demonstrated good cycle stability, with approximately 74.2 % of the initial 205 mAh g^{-1} discharge capacity retained after 60 cycles at 1 C.

Expanding upon their previous work, Hong et al. [117] went on to control the morphology of both rutile and anatase TiO_2 mesocrystals through the use of different counterions during synthesis. Dumbbell shaped rutile TiO_2 mesocrystals were obtained when an HCl aqueous solution was used (Fig. 8a/b), while truncated-octahedral anatase TiO_2 mesocrystals were obtained when the aqueous solution was H_2SO_4 based (Fig. 8c/d). The dumbbell shaped rutile TiO_2 superstructures were approximately 800 nm in length, and assembled from thin nanowires with average diameters of 3–5 nm grown along the [001] direction. Nitrogen adsorption analysis measured the specific surface area at 12.6 m^2 g^{-1} and revealed that the sample had a low volume of mesopores.

The samples treated with an aqueous H_2SO_4 solution produced truncated-octahedral anatase TiO_2 mesocrystals measuring 25–50 nm that were made up of 3–5 nm sized nanoparticles. SAED was used to determine that the nanoparticle

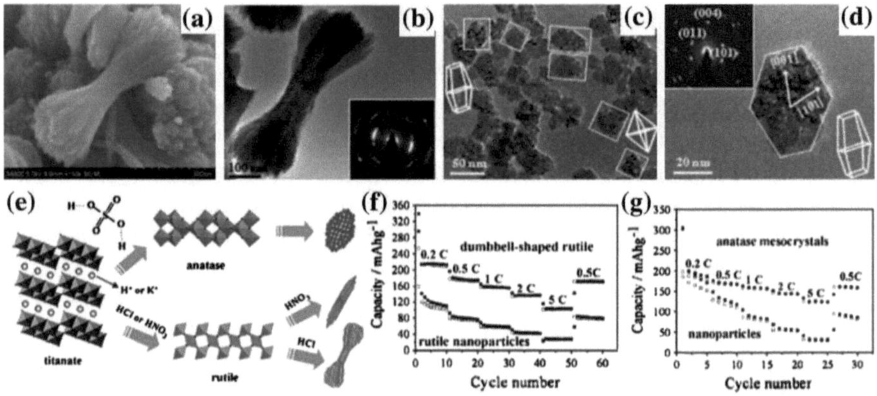

Fig. 8 **a** SEM and **b** TEM images of the rutile TiO_2 dumbbell mesocrystals, **c** and **d** TEM images of the truncated-octahedral anatase TiO_2 mesocrystals, **e** schematic of the proposed formation mechanism of TiO_2 superstructures with different morphology and phase. The insets in **b** and **d** are the corresponding SAED patterns. Rate capability of the **f** dumbbell shaped rutile and **g** truncated-octahedral anatase TiO_2 mesocrystals and comparison with similarly sized commercially available particles of the same phase (open—discharge, filled—charge). Reprinted with permission from Hong et al. [117]. Copyright 2012 John Wiley & Sons, Inc

subunits were highly oriented along the [101] direction and elongated along the [001] direction, seemingly adopting a Wulff-shaped formation corresponding to the principles of single crystal growth [115, 118]. These results indicate that the truncated-octahedral anatase TiO_2 mesocrystals were a single crystal intermediate. The Brunauer–Emmett–Teller (BET) nitrogen adsorption derived surface area and total pore volume was 225 $m^2\ g^{-1}$ and 0.43 $cm^3\ g^{-1}$, respectively. The anatase mesocrystals had a narrow micropore size distribution of approximately 0.5 nm, while the average mesopore was approximately 18 nm.

A proposed formation mechanism for the counterion controlled TiO_2 superstructures is depicted in Fig. 8e. It has been well documented that the titanate precursor has a layered structure made up of edge sharing TiO_6 octahedra with H^+ or K^+ occupying the interlayer spaces. The TiO_6 octahedra will rearrange under acidic conditions into either anatase, which is assembled from four edge and corner sharing octahedra, or rutile, which is assembled from two edge and corner sharing octahedra. Based on geometrical considerations, linear chains can only form rutile-type nuclei whereas skewed chains can only form anatase-type nuclei [119]. Thus, the titanate nanowire precursor kept its morphology when subjected to acidic conditions except when exposed to $[SO_4^{-}]$ because of its steric hindrance effects. Therefore, the titanate nanowires readily decomposed to nanoparticles in the presence of H_2SO_4; in either case, the mesocrystals formed through oriented aggregation.

The galvanostatic charge-discharge performance as a function of current density for both the rutile and anatase mesocrystals was compared to nanoparticles of their corresponding material and size; both mesocrystal specimens demonstrated their advantages. At a discharge rate of 5 C, the rutile mesocrystals had a discharge capacity of 103 mAh g^{-1}, while the rutile nanoparticles only exhibited 28 mAh g^{-1}. Equivalent differences in specific capacity were observed at lower current densities as well, attesting to the overall superiority of the mesocrystal material as revealed in Fig. 8f. Similarly, the anatase mesocrystals exhibited a discharge capacity of 124 mAh g^{-1}, while the anatase nanoparticles only exhibited 32 mAh g^{-1} at a discharge rate of 5 C (Fig. 8g). For the anatase mesocrystal, discrepancies in the capacity were more pronounced at higher discharge rates and less noticeable at lower current density. The improved Li-ion intercalation rate performance was attributed to the intrinsic characteristics of the mesoscopic TiO_2 superstructures, which had a single crystal-like and porous nature that facilitated fast electron transport and relieved the strain from volumetric change.

Xu et al. [120] developed an electrochemical synthesis route that produced leaf-like CuO mesocrystals without the aid of any surfactants. Copper foils were simply immersed in an aqueous solution of $NaNO_3$ to which a constant voltage was applied. The resulting mesocrystals adopted a two-dimensional needle-like morphology, and were designated 'nanoleaves', as shown in Fig. 9a. The CuO nanoleaves were approximately 50 nm wide and several hundreds of nanometers long. Moreover, the nanoleaves were composed of many small particles (Fig. 9b). Figure 9c displays an HRTEM image of a single nanoleaf, while Fig. 9d reveals the corresponding SAED pattern. The single crystal like quality of the SAED pattern indicates that the nanoparticles making up the nanoleaf share the same crystallographic orientation. Time resolved deposition/growth analysis uncovered that the CuO mesocrystals form through the oriented attachment of small nanocrystals. It is speculated that the CuO mesocrystals formed through the dehydration of $Cu(OH)_2$, and subsequently self-assembled along a mutual orientation. The leaf-like morphology was adopted because of defects and surface energy constraints, which lead to different growth rates along the width and length of the nanoleaves. However, a more lucid and detailed formation mechanism has yet to be divulged.

Upon investigating the electrochemical performance of the CuO mesocrystal, it was established that the CuO mesocrystals exhibit a high initial discharge capacity of 1063 mAh g^{-1} and a reversible capacity of 674 mAh g^{-1}, which is close to the theoretical capacity of 670 mAh g^{-1} as shown in Fig. 9e. The cycle stability is also displayed in Fig. 9e, and slight capacity fading was observed over the course of 30 cycles. Comparatively, the CuO microflake electrode deteriorated quite quickly and had a reversible capacity less than 200 mAh g^{-1} after 30 cycles. The improved electrochemical characteristics of the CuO mesocrystalline nanoleaves were ultimately attributed to their high surface area, confined thickness, and oriented single crystal domains.

There are several other material systems that have been studied and not included for the sake of brevity; some such systems are: SnO, SnO_2, Fe_2O_3, and Co_3O_4 [121–126].

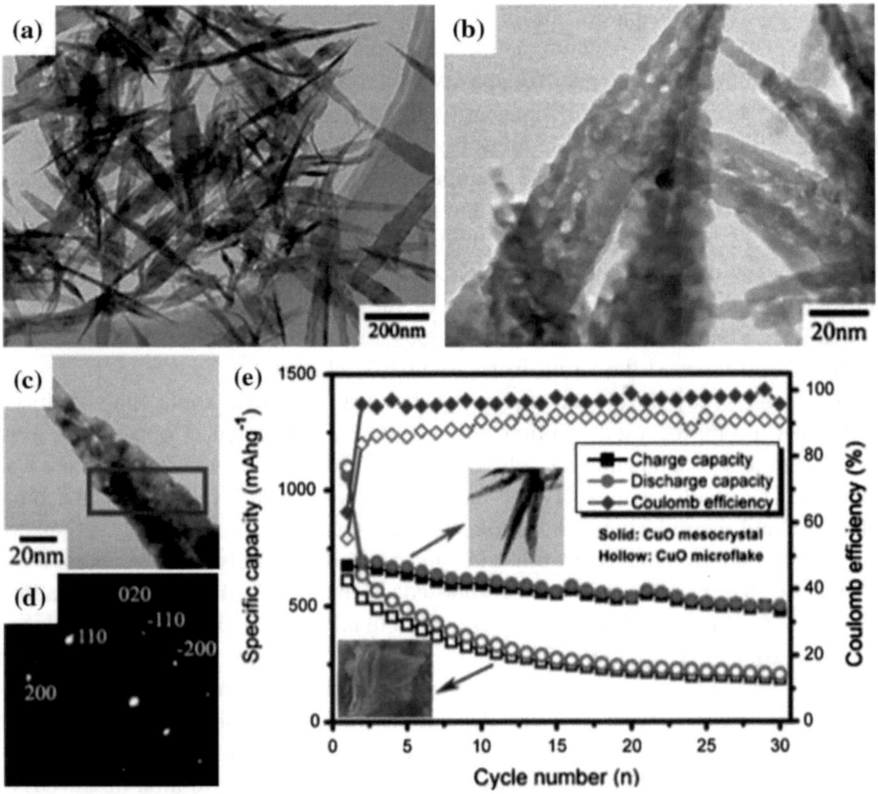

Fig. 9 TEM images of the leaf-like CuO mesocrystals at **a** low and **b** high magnification; **c** TEM image and **d** SAED pattern (from the highlighted area) of an individual CuO nanoleaf. **e** The cycle performance of the leaf-like CuO mesocrystal compared with CuO microflake. Reproduced from Ref. [120] with permission from The Royal Chemistry Society

4.3.2 Cathodic Materials

Compared to their anodic partners, there are considerably fewer studies concerning cathodic electrode materials; it is probably attributable to the fact that cathodic materials are typically complex oxides. The synthesis of nanostructured primary units and crystallographic alignment of complex oxides are understandably more challenging and difficult to control. Specifically, it is common for cathodic materials to contain lithium in the as-prepared state. Provided the qualities of lithium, its stoichiometry under most synthetic processing methods is problematic to govern.

Monoclinic nanostructured metastable vanadium dioxide, VO_2(B), was synthesized from the reduction of commercially available V_2O_5 utilizing additive-free solution-based processing methods where oxalic acid served as both the reducing and chelating agent due to its disposition to serve as a bidentate ligand [127]. The size and morphology of the as-prepared VO_2(B) material was examined using SEM,

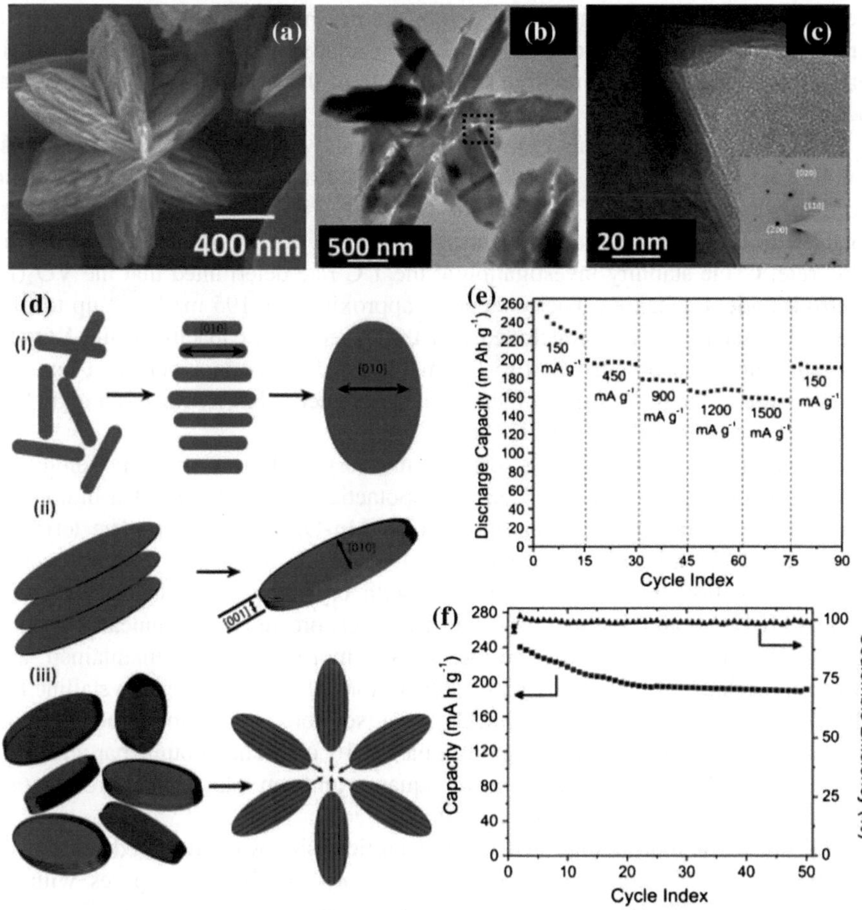

Fig. 10 **a** Representative SEM image of the as-synthesized vanadium dioxide mesocrystals showing the stacked nanosheets composing each arm; **b** and **c** TEM images of a cross-sectioned VO_2(B) mesocrystal and corresponding SAED pattern (inset). **d** Proposed formation mechanism of the VO_2(B) mesocrystals; **e** discharge-rate capability at various current densities and **f** cycle stability and Coulombic efficiency over 50 cycles at 150 mA g^{-1}. Reprinted with permission from Uchaker et al. [127]. Copyright 2013 John Wiley & Sons, Inc

the results of which are displayed in Fig. 10a. A typical VO_2(B) sample was mainly composed of uniform micron sized ellipsoidal stars with an average height and width of approximately 2.8 and 1.5 μm, respectively. Each mesocrystal was composed of six arms that are self-assembled from stacked nanosheets 20–60 nm thick and radially aligned with respect to the center of the structure. Cross-sectional TEM images of an ultramicrotomed sample (Fig. 10b/c) approximate the overall thickness of each hierarchically nanostructured arm at 200–300 nm. The inset of Fig. 10c displays the SAED pattern for an arm region, the single crystalline nature of which confirms mesocrystal formation. Indexing the SAED pattern revels that the individual $VO_2(B)$ nanosheets exhibit preferential growth along the [010] direction, indicating that the (010) plane has a relatively high stacking rate that is realized by considering the relative stacking rate of the octahedra at various crystal faces [128].

The rate capability of the $VO_2(B)$ mesocrystals was tested at incremental discharge rates, for 15 cycles at each increment, starting at 150 mA g^{-1} (1 C) and terminating at 1500 mA g^{-1}, the results of which are displayed in Fig. 10e. Most notably, the $VO_2(B)$ electrodes demonstrated a high capacity of 158 mAh g^{-1} at the 10 C rate. Cycle stability investigation at the 1 C rate determined that the $VO_2(B)$ electrodes adopt a stable capacity value of approximately 195 mAh g^{-1} up to fifty cycles. In essence, it was concluded that the exceptional capacity of the $VO_2(B)$ superstructures can be attributed to the mesocrystalline arrangement that leads to the exposure of the (001) facet, which has a lower energy barrier for faster and more efficient Li-ion intercalation [129, 130].

Dang et al. [131] established a selective mesocrystal formation methodology for various manganese oxides based on the topotactic transformation of a biomimeticaly prepared $MnCO_3$ precursor through an Mn_5O_8 intermediate. The term topotaxy describes all solid state reactions that lead to a material with respective crystal orientation that can be correlated with crystal orientations in the initial product [101]. In brief, the mesocrystalline $MnCO_3$ precursor was annealed to form the Mn_5O_8 intermediate, while the mesocrystal morphology was maintained, and was then reacted hydrothermally with LiOH to form the various mesocrystalline Li–Mn–O compounds. The initial $MnCO_3$ precursor took on a rhombohedral morphology with an edge length of approximately 10 μm. The subunit particles that made up the $MnCO_3$ precursor were square, uniform platelets approximately 500 nm in size, as revealed in Fig. 11a/b.

The precursor morphology and subunit particle size was preserved during the intermediate Mn_5O_8 topotactic transformation process; however, pores with an average size of 100 nm were formed due to the decomposition of $MnCO_3$, specifically the volume reduction attributed to the removal of CO and CO_2 (Fig. 11c/d). More notably, different Li–Mn–O compounds could be acquired by adjusting the quantity of LiOH used during the hydrothermal process—specifically, $LiMnO_2$, $LiMn_2O_3$, and $LiMn_2O_4$. Too high of annealing temperature resulted in a randomly oriented Mn_2O_3 product which also failed to adopt nanoparticle orientation following hydrothermal treatment. A schematic diagram of the various formation pathways is illustrated in Fig. 11e.

The shape and size of the overall $MnCO_3$ precursor was maintained through the hydrothermal process, and imparted unto the $LiMnO_2$, $LiMn_2O_3$, and $LiMn_2O_4$ products. However, the size and morphology of the subunit nanoparticles comprising the superstructures differed significantly. For instance, $LiMn_2O_4$ prepared with only 12 wt% more LiOH exhibited both nanosheet and nanoparticle morphology with average sizes of 100 and 250 nm, respectively. The packing density and surface roughness of the mesocrystal composing nanoparticles could also be tailored by adjusting the amount of LiOH used, based on SEM images.

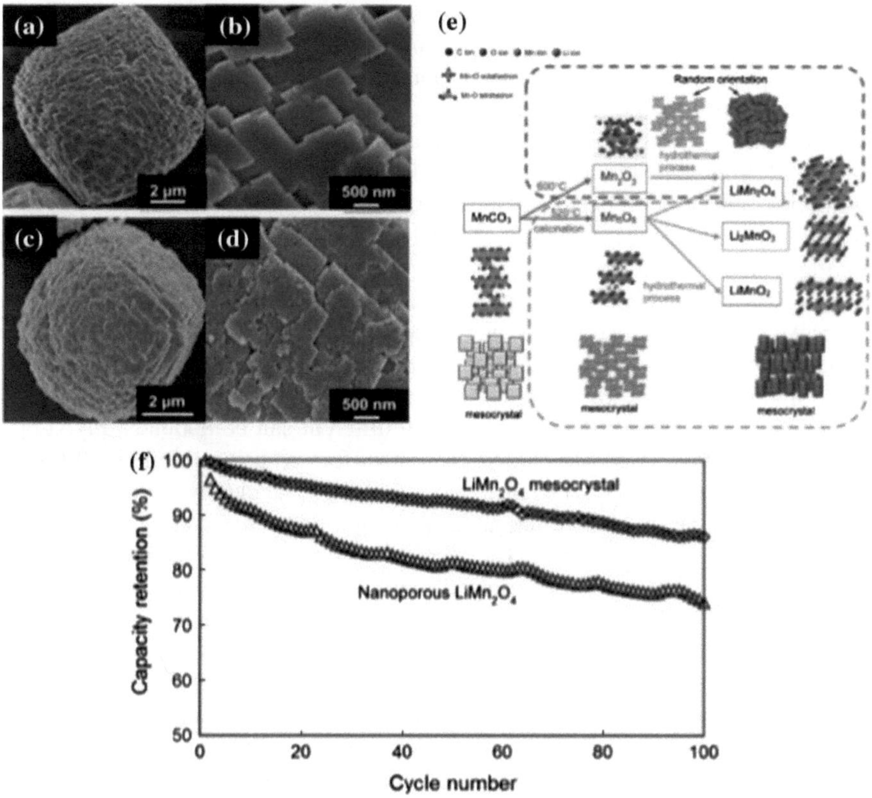

Fig. 11 SEM images of the **a**, **b** $MnCO_3$ and **c**, **d** Mn_5O_8 precursor mesocrystals. **e** Transformation from $MnCO_3$ precursors to lithium manganates via manganese oxides where the final crystal phase depends on the [LiOH]/[Mn_5O_8] molar ratio during the hydrothermal process. (**f**) Cycle stability in terms of capacity retention for the $LiMn_2O_4$ mesocrystals compared with nanoporous $LiMn_2O_4$ particles between cutoff voltages of 3.1–4.4 V and at a current density of 100 mA g^{-1}. Reproduced from Ref. [131] with permission from The Royal Chemistry Society

The lithium intercalation capability of the $LiMn_2O_4$ mesocrystals were then investigated, and compared to that of porous, single-crystalline $LiMn_2O_4$ nanoparticles. The typical two-stepped plateau of the $LiMn_2O_4$ cubic spinel, corresponding to two different Li-ion intercalation processes, was obtained and the specific capacity was 98 mAh g^{-1} at 100 mA g^{-1}. It is worth mentioning that the half-cells were cycled between 3.1 and 4.4 V, which is a relatively high electrochemical window. A comparison of the capacity retention of the $LiMn_2O_4$ mesocrystals and randomly oriented nanoparticles, when cycled at 100 mA g^{-1}, is shown in Fig. 11f. After 100 cycles, the mesocrystalline $LiMn_2O_4$ retained 86 % of its initial capacity, while the nanoporous specimen was only able to maintain 74 % of its preliminary value. Additionally, a composite electrode prepared from equal parts

$LiMnO_2$ and $LiMn_2O_3$ demonstrated exemplary capacity at various current densities, most notably 192 mAh g^{-1} at 10 mA g^{-1}. The enhanced electrochemical performance of the mesocrystals was attributed to their high degree of crystallinity and exposure of specific surface facets.

Cao et al. [132] successfully developed an approach to regulate the morphology of Mn and Ni based carbonate precursors. Highly uniform and oriented particulate assemblies yielded well-designed superstructures with high tap density and high voltage capability. $LiMn_{1.5}Ni_{0.5}O_4$ (LMNO) adopts the spinel structure while offering a high operating voltage and rate capability, but its morphology is difficult to control because of the high synthesis temperatures often required. Manganese and nickel containing carbonates were effectively synthesized by exploiting the gradual production of NH_3 and CO_2 stemming from the decomposition of urea during hydrothermal treatment.

The morphology was controlled by using different salt compounds; Fig. 12a, c show the carbonate based Mn and Ni precursors formed using metal chlorides and metal sulfates, respectively. The inset picture of Fig. 12a clearly demonstrates that the mesocrystalline microcube had an edge length of approximately 10 μm and a tap density as high as 1.7 g cm^{-3}. Closer inspection reveals that the microcubes are in fact built up from layers of oriented and stacked square nanosheets approximately 200 nm in dimension. Time resolved hydrothermal growth investigations suggested that $MnCO_3$ precipitated out preferentially, and that the Ni^{2+} ions took some time to fully precipitate into $NiCO_3$. However, the full precipitation of $NiCO_3$ coincided with radical change in the sample morphology from random and irregular to oriented and hierarchical. The carbonate based microspheres, Fig. 12c, had an average diameter of 15 μm but were also composed of nanoparticle subunits, albeit less uniform than the microcube counterpart. As expected, based on geometrical considerations, the microspheres had a higher tap density of 1.9 g cm^{-3}.

Elemental mapping of each morphology proved that $MnCO_3$ and $NiCO_3$ were successfully co-precipitated throughout each particle; no detectable degree of segregation was noticed. XRD confirms the presence of $MnCO_3$ and $NiCO_3$ in both the mesocrystalline microcubes and microspheres, although the peaks are more intense for the microcubes indicating a higher degree of crystallinity. The narrow size distribution of both the superstructure morphologies is ascribed to Ostwald ripening. Heat treatment of these materials with LiOH generated LMNO.

The electrochemical performance of the LMNO mesocrystals was compared to that of irregularly shaped, commercially available material when cycled between 3.4 and 5 V. Figure 12e reveals that both the mesocrystalline microspheres and microcubes exhibited superior cycle stability with essentially no capacity fade over fifty cycles; the commercial LMNO sample demonstrated moderately lower initial capacity and experienced an 11 % capacity fade under the same conditions. It is possible that the improvement in cycle stability might be attributed to the uniformity in particle size and geometrical robustness of the mesocrystal particles. Overall, the study validated the superiority of the mesocrystal materials in terms of higher volumetric capacity and stability.

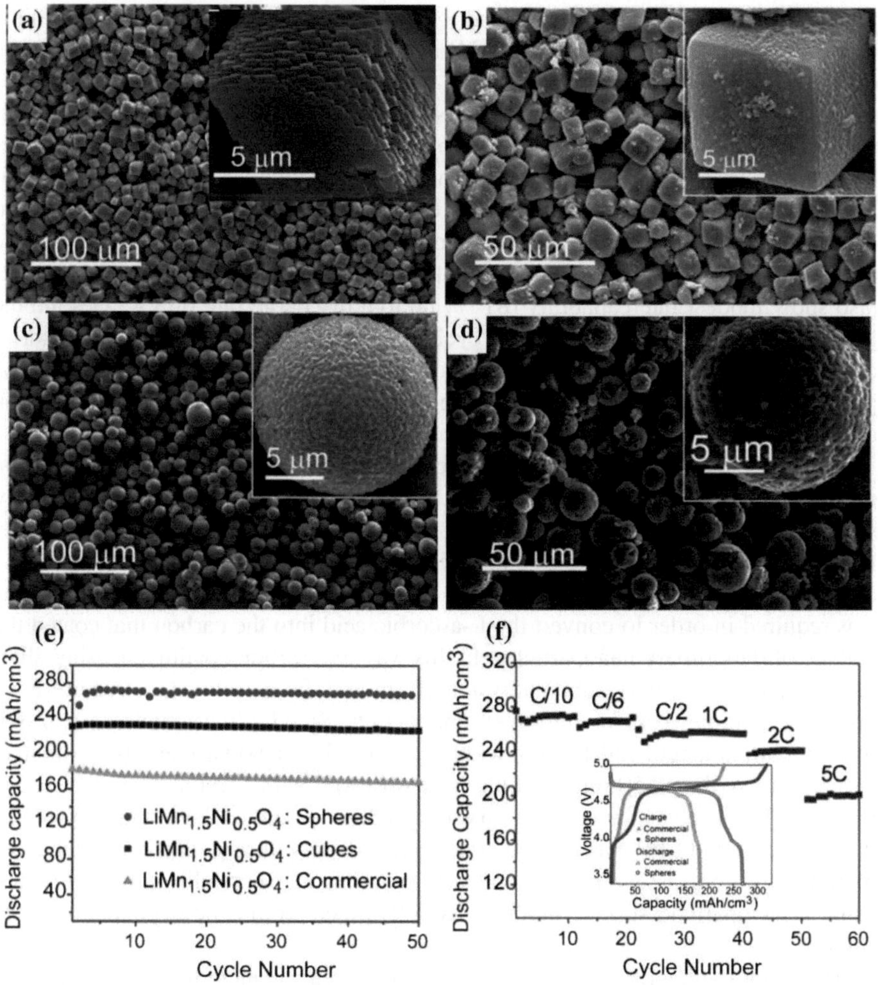

Fig. 12 SEM micrographs of mesocrystalline microcube **a** carbonate precursor and **b** LMNO final product; mesocrystalline microsphere **c** carbonate precursor and **d** LMNO final product. **e** Comparative cycle performance of LMNO mesocrystalline microcubes, microspheres, and commercial nanoparticles. **f** Rate capability of $LiMn_2O_4$ mesocrystalline microspheres; inset: first dis/charge profiles of the LMNO mesocrystalline microcubes, microspheres, and commercial nanoparticles at the 0.1 C current rate. Reproduced from Ref. [132] with permission from The Royal Chemistry Society

Lithium iron phosphate ($LiFePO_4$ or LFP) has attracted much attention in both research and industrial communities because of its high theoretical capacity of 170 mAh g^{-1}, stable discharge rate at a voltage of ~3.4 V, low cost due to abundance of raw materials, and excellent thermal and chemical stability [133]. Yet, LFP has despairingly low electrical conductivity that limits its application in high power devices but could potentially be mitigated by the use of mesocrystalline electrode materials. Bilekca et al. [134] synthesized $LiFePO_4$ mesocrystals via an extremely efficient microwave-assisted route. The versatility of this synthesis method was illustrated by its ability to synthesize $LiMnPO_4$ under similar conditions. The as obtained product was composed of slightly elongated nanoparticles ($\sim$100 nm) that nearly fused together into discs approximately 1 μm in diameter. The LFP mesocrystals were tested as cathode material for Li-ion battery, and exhibited an initially moderate capacity of 150 mAh g^{-1} that was extremely stable upon cycling from 2.0 to 4.5 V at a rate of 340 mA g^{-1} (2 C). The LFP mesocrystals also show modest rate capability (81 %) up to a discharge rate of 8 C. Despite the lack of a carbon coating and modest cycling performance, the study did not clarify or expand upon the relationship between the unique structural features and the performance of the LFP mesocrystals and only suggested that the implementation of mesocrystals as electrode material may be the linking factor [134].

In an attempt to compensate for the low electronic conductivity of LFP, carbon coated LFP mesocrystals were later studied by Xia et al. [135], where they were synthesized using a solvothermal and post-growth annealing route. A nearly identical technique, except for the inclusion of surfactant during synthesis, yielding an analogous product has been reported by Chen et al [136]. The annealing route was required in order to convert the L-ascorbic acid into the carbon that coated the surface of the primary nanoparticles. In this way, the carbon coating encasing all of the nanoparticle primary subunits was uniform and the mesocrystal structure itself provided a framework for improved conduction. Spindle-like structures approximately 2 μm in length and 1 μm in width, composed of nanoparticles approximately 100 nm in length and 50 nm in width, were observed with SEM and are depicted in Fig. 13. TEM analysis revealed the mesocrystalline nature of the obtained product; it was determined LFP was formed almost immediately during reaction and that the morphology progressed through standard non-classical crystallization techniques using a nanoparticle surface modifier as determined from time dependent growth studies.

The spindle-like LFP mesocrystals demonstrated an incredibly stable cycling stability of approximately 157 mAh g^{-1} over the course of 50 cycles at the 0.1 C rate, with an overall drop in capacity of only 3 %. Conversely, electrochemical impedance spectroscopy revealed that the charge transfer resistance (R_{ct}) decreased from 57.7 to 28.7 Ω after the 50 cycles; the LFP mesocrystals also exhibited a moderate rate capability that terminated at 117 mAh g^{-1} at a rate of 5 C [135]. The electrochemical performance was attributed to the porous structure that shortened the Li-ion diffusion length, enhanced electrolyte penetration, and increased the electrochemical reaction surface, thus alleviating electrode polarization. The incorporation of the uniform carbon coating throughout the structure also aided the electrochemical capability by enhancing the electronic conductivity.

In a separate study, $LiFePO_4$ dumbbell-like mesocrystals, as shown in Fig. 14, were prepared by a simple solvothermal process in the presence of polyvinyl propylene (PVP) [137]. SEM imaging shows that the dumbbell-like microstructures

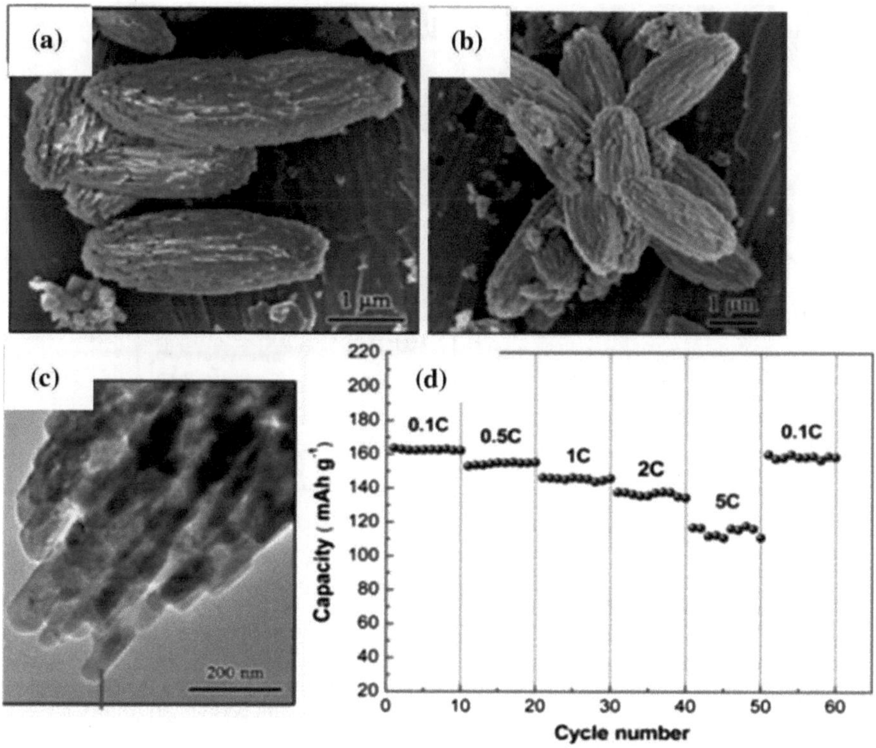

Fig. 13 SEM images of spindle-like LFP structure obtained for a growth period of **a** 20 h, **b** pH of 10; **c** HRTEM image. **d** The charge/discharge profiles of the spindle-like LFP architecture in the voltage range from 2.5 to 4.2 V at various current rates. Reproduced from Ref. [135] with permission from Elsevier

have length ranging from 1.5 to 2 μm and that they are hierarchically constructed with two-dimensional nanoplates that are approximately 300 nm in length and 50 nm thick. Without PVP, irregular and non-uniform plates with a large size were formed; reduced amounts of PVP resulted in three-dimensionally hierarchical microstructures formed from nanoplates, but with a wide size distribution. Synthesis with other surfactants failed to produce oriented LFP mesocrystals but rather irregular and non-uniform disc-like particles of large size.

On the basis of TEM time dependent investigations, a formation mechanism behind the dumbbell shaped LFP mesocrystals was proposed. The process was stated to carry out over several different stages that started with initial nucleation and growth of rectangular nanocrystals that would then aggregate and orientedly attach into large rectangular shaped structures due to PVP related effects. With prolonged reaction time the previously formed structures consisting of nanoplates demonstrated a preference to tilt at both ends because of lattice tension or surface interaction in the edge areas, resulting in the formation of the bowed-out assembly.

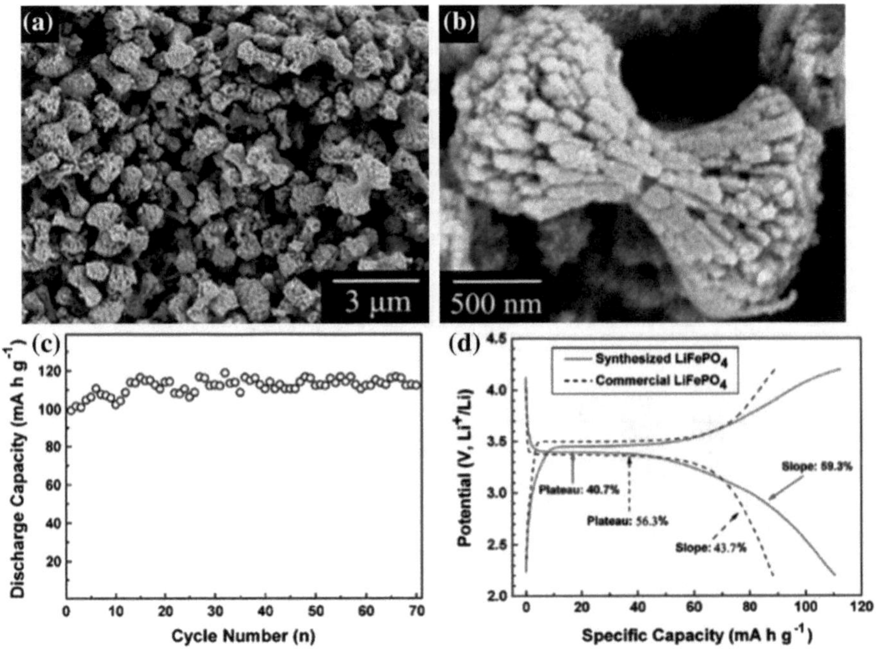

Fig. 14 **a**, **b** SEM images, **c** cycle stability, and **d** comparison with commercially available material of the $LiFePO_4$ dumbbell-like mesocrystals. Reprinted with permission from Yang et al. [137]. Copyright 2009 American Chemical Society

Li_3PO_4, whose presence was detected using XRD, would then dissolve and recrystallize to $LiFePO_4$ where it would collect on the edges of the pre-existing $LiFePO_4$ nanoplates. Thus, is can be surmised that dumbbell $LiFePO_4$ mesocrystal formation is based on a dissolution-recrystallization process with eventually complete phase transformation.

Prior to electrochemical testing, the PVP was converted to carbon through annealing; the inherent amount of carbon present was measured to be 1 wt%. At a current rate of 0.03 C, the dumbbell mesocrystals had an initial discharge capacity of 100 mAh g^{-1} that improved to 110 mAh g^{-1} after 70 cycles. The lack of capacity fade indicates that the material is extremely stable. The capacity of the dumbbell mesocrystals was also higher than that of commercially available LFP while showing less polarization and a longer discharge plateau (Fig. 14d). The authors reasoned that the disparity between the two materials may exist because of a pseudocapacitive effect in the mesocrystalline LFP imparted by its nanosized component. Similarly shaped $LiFePO_4$ mesocrystals formed through several other routes have been reported as well [138, 139].

5 Concluding Remarks

Ensembles and novel arrangements of nanoparticles can collectively exhibit properties vastly different than individual nanoparticles or bulk materials. Micro-sized nanoparticle assemblies are a promising class of hierarchically nanostructured solid materials that coincidentally have many inherent traits that are desired for Li-ion battery electrodes. Such characteristics include: large surface area, high porosity, small primary subunit size, large overall assembly size, and high degree of crystallinity. A great deal of the research efforts, to date, concerning micro/nano assemblies have focused on understanding their formation mechanisms, but newer works reviewing their application are starting to emerge. In terms of battery performance, the porous structure inherent to micro/nano assemblies can facilitate contact with the electrolyte thereby leading to fast Li-ion transport mechanics while at the same time accommodating lithiation induced volume expansion and strain giving way to improved cycling stability.

For mesocrystals, the oriented arrangement of nanoparticle subunits can eliminate the grain boundaries between adjacent particles thus offering much better charge and mass transport, and ultimately better rate capability. Provided this unique combination of nanoparticle properties and order combined with a microscopic or even macroscopic size, mesocrystals have strong potential as active materials for lithium-ion battery electrodes. These assemblies possess the structural and chemical stability of microsized electrodes while exploiting the beneficial properties associated with nanosized electrodes and their large reactive surface area.

Micro/nano assemblies have also demonstrated great potential for applications in catalysis, sensors, and electronics. Other applications are yet to be explored where huge surface area and good charge and mass transfer properties are required or the hierarchical structure can offer synergistic advantages; for example, nanostructure-based solar cells, such as dye-sensitized solar cells or quantum dots solar cells, would greatly benefit from the applications of such micro/nano assemblies, as they would offer charge transfer properties similar to single crystals, large surface area as seen for nanoparticles, and desirable light scattering.

Acknowledgments Part of this work was financially supported by the National Science Foundation (NSF, CMMI-1030048) and the University of Washington TGIF grant. This material is based in part upon work supported by the State of Washington through the University of Washington Clean Energy Institute.

References

1. Guo Y-G, Hu J-S, Wan L-J (2008) Nanostructured materials for electrochemical energy conversion and storage devices. Adv Mater 20(15):2878–2887. doi:10.1002/adma.200800627
2. Liu D, Cao G (2010) Engineering nanostructured electrodes and fabrication of film electrodes for efficient lithium ion intercalation. Energy Environ Sci 3(9):1218–1237. doi:10.1039/b922656g

3. Scrosati B, Garche J (2010) Lithium batteries: status, prospects and future. J Power Sources 195(9):2419–2430. doi:10.1016/j.jpowsour.2009.11.048
4. Marom R, Amalraj SF, Leifer N, Jacob D, Aurbach D (2011) A review of advanced and practical lithium battery materials. J Mater Chem 21(27):9938–9954. doi:10.1039/c0jm04225k
5. Ji L, Lin Z, Alcoutlabi M, Zhang X (2011) Recent developments in nanostructured anode materials for rechargeable lithium-ion batteries. Energy Environ Sci 4(8):2682–2699. doi:10.1039/c0ee00699h
6. Wang Y, Li H, He P, Hosono E, Zhou H (2010) Nano active materials for lithium-ion batteries. Nanoscale 2(8):1294–1305. doi:10.1039/c0nr00068j
7. Koziej D, Lauria A, Niederberger M (2014) 25th anniversary article: metal oxide particles in materials science: addressing all length scales. Adv Mater 26(2):235–257
8. Guo YG, Hu JS, Wan LJ (2008) Nanostructured materials for electrochemical energy conversion and storage devices. Adv Mater 20(15):2878–2887. doi:10.1002/adma.200800627
9. Nel A, Xia T, Madler L, Li N (2006) Toxic potential of materials at the nanolevel. Science 311(5761):622–627. doi:10.1126/science.1114397
10. Colvin VL (2003) The potential environmental impact of engineered nanomaterials. Nat Biotechnol 21(10):1166–1170. doi:10.1038/nbt875
11. Heiligtag FJ, Niederberger M (2013) The fascinating world of nanoparticle research. Mater Today 16(7–8):262–271. doi:10.1016/j.mattod.2013.07.004
12. Bruce PG, Scrosati B, Tarascon J-M (2008) Nanomaterials for rechargeable lithium batteries. Angew Chem Int Ed 47(16):2930–2946. doi:10.1002/anie.200702505
13. Arico AS, Bruce PG, Scrosati B, Tarascon J-M, van Schalkwijk W (2005) Nanostructured materials for advanced energy conversion and storage devices. Nat Mater 4(5):366–377
14. Meethong N, Huang HYS, Carter WC, Chiang YM (2007) Size-dependent lithium miscibility gap in nanoscale $Li_{1-x}FePO_4$. Electrochem Solid-State Lett 10(5):A134–A138. doi:10.1149/1.2710960
15. Liu D, Cao GZ (2010) Engineering nanostructured electrodes and fabrication of film electrodes for efficient lithium ion intercalation. Energy Environ Sci 3(9):1218–1237. doi:10.1039/b922656g
16. Wagemaker M, Borghols WJH, Mulder FM (2007) Large impact of particle size on insertion reactions. A case for anatase Li_xTiO_2. J Am Chem Soc 129(14):4323–4327. doi:10.1021/ja067733p
17. Wagemaker APMKFMM M (2002) Equilibrium lithium transport between nanocrystalline phases in intercalated TiO_2 anatase. Nature 418:397–399
18. Ganapathy S, van Eck ER, Kentgens AP, Mulder FM, Wagemaker M (2011) Equilibrium lithium-ion transport between nanocrystalline lithium-inserted anatase TiO_2 and the electrolyte. Chemistry 17(52):14811–14816. doi:10.1002/chem.201101431
19. Hu YS, Kienle L, Guo YG, Maier J (2006) High lithium electroactivity of nanometer-sized rutile TiO_2. Adv Mater 18(11):1421–1426. doi:10.1002/adma.200502723
20. Kasavajjula U, Wang CS, Appleby AJ (2007) Nano- and bulk-silicon-based insertion anodes for lithium-ion secondary cells. J Power Sources 163(2):1003–1039. doi:10.1016/j.jpowsour.2006.09.084
21. Chan CK, Peng H, Liu G, McIlwrath K, Zhang XF, Huggins RA, Cui Y (2008) High-performance lithium battery anodes using silicon nanowires. Nat Nano 3(1):31–35. doi:http://www.nature.com/nnano/journal/v3/n1/suppinfo/nnano.2007.411_S1.html
22. Li X, Gu M, Hu S, Kennard R, Yan P, Chen X, Wang C, Sailor MJ, Zhang J-G, Liu J (2014) Mesoporous silicon sponge as an anti-pulverization structure for high-performance lithium-ion battery anodes. Nat Commun 5. doi:10.1038/ncomms5105
23. Yamada A, Koizumi H, Nishimura S-i, Sonoyama N, Kanno R, Yonemura M, Nakamura T, Kobayashi Y (2006) Room-temperature miscibility gap in Li_xFePO_4. Nat Mater 5(5):357–360. doi:http://www.nature.com/nmat/journal/v5/n5/suppinfo/nmat1634_S1.html

24. Wagemaker M, Ellis BL, Lützenkirchen-Hecht D, Mulder FM, Nazar LF (2008) Proof of supervalent doping in olivine $LiFePO_4$. Chem Mater 20(20):6313–6315. doi:10.1021/cm801781k
25. Meethong N, Huang H-YS, Carter WC, Chiang Y-M (2007) Size-dependent lithium miscibility gap in nanoscale Li[sub 1 − x]FePO[sub 4]. Electrochem Solid-State Lett 10(5): A134. doi:10.1149/1.2710960
26. Kobayashi G, S-i Nishimura, Park M-S, Kanno R, Yashima M, Ida T, Yamada A (2009) Isolation of solid solution phases in size-controlled Li_xFePO_4 at room temperature. Adv Funct Mater 19(3):395–403. doi:10.1002/adfm.200801522
27. Wagemaker M, Singh DP, Borghols WJ, Lafont U, Haverkate L, Peterson VK, Mulder FM (2011) Dynamic solubility limits in nanosized olivine $LiFePO_4$. J Am Chem Soc 133 (26):10222–10228. doi:10.1021/ja2026213
28. Balaya P, Bhattacharyya AJ, Jamnik J, Zhukovskii YF, Kotomin EA, Maier J (2006) Nano-ionics in the context of lithium batteries. J Power Sources 159(1):171–178. doi:10.1016/j.jpowsour.2006.04.115
29. Liu Y, Liu D, Zhang Q, Yu D, Liu J, Cao G (2011) Lithium iron phosphate/carbon nanocomposite film cathodes for high energy lithium ion batteries. Electrochim Acta 56 (5):2559–2565. doi:10.1016/j.electacta.2010.11.050
30. Cabana J, Monconduit L, Larcher D, Palacín MR (2010) Beyond intercalation-based li-ion batteries: the state of the art and challenges of electrode materials reacting through conversion reactions. Adv Mater 22(35):E170–E192. doi:10.1002/adma.201000717
31. Lin F, Nordlund D, Weng T-C, Zhu Y, Ban C, Richards RM, Xin HL (2014) Phase evolution for conversion reaction electrodes in lithium-ion batteries. Nat Commun 5. doi:10.1038/ncomms4358
32. Godshall NA, Raistrick ID, Huggins RA (1980) Thermodynamic investigations of ternary lithium-transition metal-oxygen cathode materials. Mater Res Bull 15(5):561–570. doi:http://dx.doi.org/10.1016/0025-5408(80)90135-X
33. Kaun TD, Nelson PA, Redey L, Vissers DR, Henriksen GL (1993) High temperature lithium/sulfide batteries. Electrochim Acta 38(9):1269–1287. doi:http://dx.doi.org/10.1016/0013-4686(93)80057-7
34. Badway F, Mansour AN, Pereira N, Al-Sharab JF, Cosandey F, Plitz I, Amatucci GG (2007) Structure and electrochemistry of copper fluoride nanocomposites utilizing mixed conducting matrices. Chem Mater 19(17):4129–4141. doi:10.1021/cm070421g
35. Poizot P, Laruelle S, Grugeon S, Dupont L, Tarascon JM (2000) Nano-sized transition-metal oxides as negative-electrode materials for lithium-ion batteries. Nature 407(6803):496–499
36. Débart A, Dupont L, Poizot P, Leriche J-B, Tarascon JM (2001) A transmission electron microscopy study of the reactivity mechanism of tailor-made CuO particles toward lithium. J Electrochem Soc 148(11):A1266–A1274. doi:10.1149/1.1409971
37. Grugeon S, Laruelle S, Dupont L, Tarascon JM (2003) An update on the reactivity of nanoparticles Co-based compounds towards Li. Solid State Sci 5(6):895–904. doi:http://dx.doi.org/10.1016/S1293-2558(03)00114-6
38. Dylla AG, Henkelman G, Stevenson KJ (2013) Lithium insertion in nanostructured TiO_2(B) architectures. Acc Chem Res 46(5):1104–1112. doi:10.1021/ar300176y
39. Zukalova M, Kalbac M, Kavan L, Exnar I, Graetzel M (2005) Pseudocapacitive lithium storage in TiO2(B). Chem Mater 17(5):1248–1255. doi:10.1021/cm048249t
40. Jiang Y-M, Wang K-X, Zhang H-J, Wang J-F, Chen J-S (2013) Hierarchical $Li_4Ti_5O_{12}/TiO_2$ composite tubes with regular structural imperfection for lithium ion storage. Scientific reports 3. doi:10.1038/srep03490
41. Zhu K, Wang Q, Kim J-H, Pesaran AA, Frank AJ (2012) Pseudocapacitive lithium-ion storage in oriented anatase TiO_2 nanotube arrays. J Phys Chem C 116(22):11895–11899. doi:10.1021/jp301884x
42. Wagemaker M, Mulder FM, Van der Ven A (2009) The role of surface and interface energy on phase stability of nanosized insertion compounds. Adv Mater 21(25–26):2703–2709. doi:10.1002/adma.200803038

43. Borghols WJH, Wagemaker M, Lafont U, Kelder EM, Mulder FM (2009) Size effects in the $Li_{4+x}Ti_5O_{12}$ spinel. J Am Chem Soc 131(39):17786–17792
44. Jung S-K, Gwon H, Hong J, Park K-Y, Seo D-H, Kim H, Hyun J, Yang W, Kang K (2014) Understanding the degradation mechanisms of $LiNi_{0.5}Co_{0.2}Mn_{0.3}O_2$ cathode material in lithium ion batteries. Adv Energy Mater 4(1):n/a-n/a. doi:10.1002/aenm.201300787
45. Lin F, Markus IM, Nordlund D, Weng T-C, Asta MD, Xin HL, Doeff MM (2014) Surface reconstruction and chemical evolution of stoichiometric layered cathode materials for lithium-ion batteries. Nat Commun 5. doi:10.1038/ncomms4529
46. Ratnakumar BV, Smart MC, Surampudi S (2001) Effects of SEI on the kinetics of lithium intercalation. J Power Sources 97–98:137–139
47. Béguin F, Chevallier F, Vix-Guterl C, Saadallah S, Bertagna V, Rouzaud JN, Frackowiak E (2005) Correlation of the irreversible lithium capacity with the active surface area of modified carbons. Carbon 43(10):2160–2167. doi:10.1016/j.carbon.2005.03.041
48. Jo M, Hong Y-S, Choo J, Cho J (2009) Effect of $LiCoO_2$ cathode nanoparticle size on high rate performance for Li-ion batteries. J Electrochem Soc 156(6):A430–A434. doi:10.1149/1.3111031
49. Lei JL, Li LJ, Kostecki R, Muller R, McLarnon F (2005) Characterization of SEI layers on $LiMn_2O_4$ cathodes with in situ spectroscopic ellipsometry. J Electrochem Soc 152(4):A774–A777. doi:10.1149/1.1867652
50. Pieczonka NPW, Liu Z, Lu P, Olson KL, Moote J, Powell BR, Kim J-H (2013) Understanding transition-metal dissolution behavior in $LiNi_{0.5}Mn_{1.5}O_4$ high-voltage spinel for lithium ion batteries. J Phys Chem C 117(31):15947–15957. doi:10.1021/jp405158m
51. Jang Dong H, Shin Young J, Oh SM (1996) Dissolution of spinel oxides and capacily losses in 4 V $Li/Li_xMn_2O_4$ cells. J Electrochem Soc 143:2204–2211
52. Jang DH, Oh SM (1997) Electrolyte effects on spinel dissolution and cathodic capacity losses in 4 V $Li/Li_xMn_2O_4$ rechargeable cells. J Electrochem Soc 144:3342–3348
53. Markevich E, Salitra G, Aurbach D (2005) Influence of the PVdF binder on the stability of $LiCoO_2$ electrodes. Electrochem Commun 7(12):1298–1304. doi:10.1016/j.elecom.2005.09.010
54. Kumagai N, Komaba S, Kataoka Y, Koyanagi M (2000) Electrochemical behavior of graphite electrode for lithium ion batteries in Mn and Co additive electrolytes. Chem Lett 29 (10):1154–1155. doi:10.1246/cl.2000.1154
55. Yi T-F, Zhu Y-R, Zhu X-D, Shu J, Yue C-B, Zhou A-N (2009) A review of recent developments in the surface modification of $LiMn_2O_4$ as cathode material of power lithium-ion battery. Ionics 15(6):779–784. doi:10.1007/s11581-009-0373-x
56. Balke N, Jesse S, Morozovska AN, Eliseev E, Chung DW, Kim Y, Adamczyk L, Garcia RE, Dudney N, Kalinin SV (2010) Nanoscale mapping of ion diffusion in a lithium-ion battery cathode. Nat Nano 5(10):749–754. doi:http://www.nature.com/nnano/journal/v5/n10/abs/nnano.2010.174.html#supplementary-information
57. Ou MN, Harutyunyan SR, Lai SJ, Chen CD, Yang TJ, Chen YY (2007) Thermal and electrical transport properties of a single nickel nanowire. Physica Status Solidi (B) 244 (12):4512–4517. doi:10.1002/pssb.200777114
58. Cao G, Wang Y (2011) Characterization and properties of nanomaterials. Nanostructures and nanomaterials, vol world scientific series in nanoscience and nanotechnology, 2nd edn. World Scientific Publishing Co. Pte. Ltd., Singapore. doi:10.1142/9789814340571_0008
59. Ebner M, Chung D-W, Garcia RE, Wood V (2013) Tortuosity Anisotropy in Lithium-Ion Battery Electrodes. Adv Energy Mater 4(5). doi:10.1002/aenm.201301278
60. Vijayaraghavan B, Ely DR, Chiang Y-M, García-García R, García RE (2012) An analytical method to determine tortuosity in rechargeable battery electrodes. J Electrochem Soc 159(5): A548–A552. doi:10.1149/2.jes113224
61. Newman JS, Tobias CW (1962) Theoretical analysis of current distribution in porous electrodes. J Electrochem Soc 109(12):1183–1191. doi:10.1149/1.2425269

62. Ebner M, Geldmacher F, Marone F, Stampanoni M, Wood V (2013) X-Ray tomography of porous, transition metal oxide based lithium ion battery electrodes. Adv Energy Mater 3 (7):845–850. doi:10.1002/aenm.201200932
63. Ferguson TR, Bazant MZ (2012) Nonequilibrium thermodynamics of porous electrodes. J Electrochem Soc 159(12):A1967–A1985. doi:10.1149/2.048212jes
64. Newman J, Tiedemann W (1975) Porous-electrode theory with battery applications. AlChE J 21(1):25–41. doi:10.1002/aic.690210103
65. Tiedemann W, Newman J (1975) Maximum effective capacity in an ohmically limited porous electrode. J Electrochem Soc 122(11):1482–1485. doi:10.1149/1.2134046
66. Aurbach D, Markovsky B, Rodkin A, Cojocaru M, Levi E, Kim H-J (2002) An analysis of rechargeable lithium-ion batteries after prolonged cycling. Electrochim Acta 47(12):1899–1911. doi:10.1016/S0013-4686(02)00013-0
67. Sakamoto JS, Dunn B (2002) Vanadium oxide-carbon nanotube composite electrodes for use in secondary lithium batteries. J Electrochem Soc 149(1):A26–A30. doi:10.1149/1.1425791
68. Bresser D, Paillard E, Copley M, Bishop P, Winter M, Passerini S (2012) The importance of "going nano" for high power battery materials. J Power Sources 219:217–222. doi:10.1016/j.jpowsour.2012.07.035
69. Syzdek J, Marcinek M, Kostecki R (2014) Electrochemical activity of carbon blacks in $LiPF_6$-based organic electrolytes. J Power Sources 245:739–744. doi:10.1016/j.jpowsour.2013.07.033
70. Martin JF, Yamada A, Kobayashi G, Nishimura S, Kanno R, Guyomard D, Dupré N (2008) Air exposure effect on $LiFePO_4$. Electrochem Solid-State Lett 11(1):A12. doi:10.1149/1.2801016
71. Li H, Shi LH, Wang Q, Chen LQ, Huang XJ (2002) Nano-alloy anode for lithium ion batteries. Solid State Ionics 148(3–4):247–258. doi:10.1016/s0167-2738(02)00061-9
72. Martha SK, Markevich E, Burgel V, Salitra G, Zinigrad E, Markovsky B, Sclar H, Pramovich Z, Heik O, Aurbach D, Exnar I, Buqa H, Drezen T, Semrau G, Schmidt M, Kovacheva D, Saliyski N (2009) A short review on surface chemical aspects of Li batteries: a key for a good performance. J Power Sources 189(1):288–296. doi:10.1016/j.jpowsour.2008.09.084
73. Niederberger M, Pinna N (2009) Metal oxide nanoparticles in organic solvents. Engineering materials and processes. Springer, Berlin
74. Antonietti M, Ozin GA (2004) Promises and problems of mesoscale materials chemistry or why meso? Chem-A Eur J 10(1):28–41. doi:10.1002/chem.200305009
75. Colfen H, Antonietti M (2005) Mesocrystals: inorganic superstructures made by highly parallel crystallization and controlled alignment. Angew Chem Int Ed 44(35):5576–5591. doi:10.1002/anie.200500496
76. Liu J, Cao G, Yang Z, Wang D, Dubois D, Zhou X, Graff GL, Pederson LR, Zhang J-G (2008) Oriented nanostructures for energy conversion and storage. Chemsuschem 1(8–9):676–697. doi:10.1002/cssc.200800087
77. Kim F, Connor S, Song H, Kuykendall T, Yang PD (2004) Platonic gold nanocrystals. Angewandte Chemie-Int Ed 43(28):3673–3677. doi:10.1002/anie.200454216
78. Uchaker E, Zhou N, Li Y, Cao G (2013) Polyol-Mediated solvothermal synthesis and electrochemical performance of nanostructured V_2O_5 hollow microspheres. J Phys Chem C 117(4):1621–1626. doi:10.1021/jp310641k
79. Cao AM, Hu JS, Liang HP, Wan LJ (2005) Self-assembled vanadium pentoxide (V_2O_5) hollow microspheres from nanorods and their application in lithium-ion batteries. Angew Chem Int Ed 44(28):4391–4395. doi:10.1002/anie.200500946
80. Liu J, Xue D (2010) Hollow nanostructured anode materials for Li-ion batteries. Nanoscale Res Lett 5(10):1525–1534. doi:10.1007/s11671-010-9728-5
81. Ito A, Li D, Sato Y, Arao M, Watanabe M, Hatano M, Horie H, Ohsawa Y (2010) Cyclic deterioration and its improvement for Li-rich layered cathode material Li [Ni0.17Li0.2Co0.07Mn0.56]O2. J Power Sources 195(2):567–573. doi:10.1016/j.jpowsour.2009.07.052

82. Luo D, Li G, Fu C, Zheng J, Fan J, Li Q, Li L (2014) A new spinel-layered Li-rich microsphere as a high-rate cathode material for Li-Ion batteries. Adv Energy Mater:n/a-n/a. doi:10.1002/aenm.201400062
83. Johnson CS, Li N, Lefief C, Vaughey JT, Thackeray MM (2008) Synthesis, characterization and electrochemistry of lithium battery electrodes: $_xLi_2MnO_3\ _{(1-x)}LiMn_{0.333}Ni_{0.333}Co_{0.333}O_2$ ($0 \le x \le 0.7$). Chem Mater 20(19):6095–6106. doi:10.1021/cm801245r
84. Zhang L, Wu B, Li N, Mu D, Zhang C, Wu F (2013) Rod-like hierarchical nano/micro $Li_{1.2}Ni_{0.2}Mn_{0.6}O_2$ as high performance cathode materials for lithium-ion batteries. J Power Sources 240:644–652. doi:10.1016/j.jpowsour.2013.05.019
85. Shen L, Uchaker E, Zhang X, Cao G (2012) Hydrogenated $Li_4Ti_5O_{12}$ nanowire arrays for high rate lithium ion batteries. Adv Mater 24(48):6502–6506. doi:10.1002/adma.201203151
86. Xiao Y, Hu C, Cao M (2014) High lithium storage capacity and rate capability achieved by mesoporous Co_3O_4 hierarchical nanobundles. J Power Sources 247:49–56. doi:10.1016/j.jpowsour.2013.08.069
87. Hu L, Zhong H, Zheng X, Huang Y, Zhang P, Chen Q (2012) $CoMn_2O_4$ Spinel hierarchical microspheres assembled with porous nanosheets as stable anodes for lithium-ion batteries. Sci Rep 2. doi:http://www.nature.com/srep/2012/121217/srep00986/abs/srep00986.html#supplementary-information
88. Wei W, Chen D, Wang R, Guo L (2012) Hierarchical $LiFePO_4$/C microspheres with high tap density assembled by nanosheets as cathode materials for high-performance Li-ion batteries. Nanotechnology 23(47):475401
89. Hu L, Sun Y, Zhang F, Chen Q (2013) Facile synthesis of porous Mn_2O_3 hierarchical microspheres for lithium battery anode with improved lithium storage properties. J Alloys Compd 576:86–92. doi:10.1016/j.jallcom.2013.04.146
90. Lan T, Liu Y, Dou J, Hong Z, Wei M (2014) Hierarchically porous TiO_2 microspheres as a high performance anode for lithium-ion batteries. J Mater Chem A 2(4):1102–1106. doi:10.1039/c3ta14178k
91. Wang B, Wu HB, Zhang L, Lou XW (2013) Self-supported construction of uniform Fe_3O_4 hollow microspheres from nanoplate building blocks. Angew Chem Int Ed 52(15):4165–4168. doi:10.1002/anie.201300190
92. Pan A, Wu HB, Yu L, Lou XW (2013) Template-free synthesis of VO_2 hollow microspheres with various interiors and their conversion into V_2O_5 for lithium-ion batteries. Angew Chem Int Ed 52(8):2226–2230. doi:10.1002/anie.201209535
93. Tao F, Gao C, Wen Z, Wang Q, Li J, Xu Z (2009) Cobalt oxide hollow microspheres with micro- and nano-scale composite structure: fabrication and electrochemical performance. J Solid State Chem 182(5):1055–1060. doi:10.1016/j.jssc.2009.01.030
94. Yang L, Wang S, Mao J, Deng J, Gao Q, Tang Y, Schmidt OG (2013) Hierarchical MoS_2/polyaniline nanowires with excellent electrochemical performance for lithium-ion batteries. Adv Mater 25(8):1180–1184. doi:10.1002/adma.201203999
95. Jeong J-M, Choi BG, Lee SC, Lee KG, Chang S-J, Han Y-K, Lee YB, Lee HU, Kwon S, Lee G, Lee C-S, Huh YS (2013) Hierarchical hollow spheres of Fe_2O_3@polyaniline for lithium ion battery anodes. Adv Mater 25(43):6250–6255. doi:10.1002/adma.201302710
96. Song K, Yoo S, Kang K, Heo H, Kang Y-M, Jo M-H (2013) Hierarchical SiO_x nanoconifers for Li-ion battery anodes with structural stability and kinetic enhancement. J Power Sources 229:229–233. doi:10.1016/j.jpowsour.2012.12.002
97. Cao F-F, Deng J-W, Xin S, Ji H-X, Schmidt OG, Wan L-J, Guo Y-G (2011) Cu-Si nanocable arrays as high-rate anode materials for lithium-ion batteries. Adv Mater 23(38):4415–4420. doi:10.1002/adma.201102062
98. Zhang L, Wu HB, Madhavi S, Hng HH, Lou XW (2012) Formation of Fe_2O_3 microboxes with hierarchical shell structures from metal-organic frameworks and their lithium storage properties. J Am Chem Soc 134(42):17388–17391. doi:10.1021/ja307475c
99. Li Z, Li B, Yin L, Qi Y (2014) Prussion blue-supported annealing chemical reaction route synthesized double-shelled Fe_2O_3/Co_3O_4 hollow microcubes as anode materials for lithium-ion battery. ACS Appl Mater Interfaces 6(11):8098–8107. doi:10.1021/am500417j

100. Niederberger M, Colfen H (2006) Oriented attachment and mesocrystals: Non-classical crystallization mechanisms based on nanoparticle assembly. PCCP 8(28):3271–3287. doi:10.1039/b604589h
101. Song R-Q, Cölfen H (2010) Mesocrystals—ordered nanoparticle superstructures. Adv Mater 22(12):1301–1330. doi:10.1002/adma.200901365
102. Meldrum FC, Coelfen H (2010) Crystallization and formation mechanisms of nanostructures. Nanoscale 2(11):2326–2327. doi:10.1039/c0nr90029j
103. Wang Y, Cao GZ (2006) Synthesis and enhanced intercalation properties of nanostructured vanadium oxides. Chem Mater 18(12):2787–2804. doi:10.1021/cm052765h
104. Chen JS, Tan YL, Li CM, Cheah YL, Luan Deyan, Srinivasan Madhavi FYC, Boey LA, Archer Lou XW (2010) Constructing hierarchical spheres from large ultrathin anatase TiO_2 nanosheets with nearly 100 % exposed (001) facets for fast reversible lithium storage. J Am Chem Soc 132:6124–6130
105. Yang Z, Choi D, Kerisit S, Rosso KM, Wang D, Zhang J, Graff G, Liu J (2009) Nanostructures and lithium electrochemical reactivity of lithium titanites and titanium oxides: a review. J Power Sources 192(2):588–598. doi:10.1016/j.jpowsour.2009.02.038
106. Ren Y, Hardwick LJ, Bruce PG (2010) Lithium intercalation into mesoporous anatase with an ordered 3D pore structure. Angew Chem Int Ed 49(14):2570–2574. doi:10.1002/anie.200907099
107. Zhou L, Boyle DS, O'Brien P (2007) Uniform NH_4TiOF_3 mesocrystals prepared by an ambient temperature self-assembly process and their topotaxial conversion to anatase. Chem Commun 2:144–146. doi:10.1039/b611476h
108. Zhou L, Smyth-Boyle D, O'Brien P (2008) A facile synthesis of uniform NH_4TiOF_3 mesocrystals and their conversion to TiO_2 mesocrystals. J Am Chem Soc 130(4):1309–1320. doi:10.1021/ja076187c
109. Zhang D, Li G, Wang F, Yu JC (2010) Green synthesis of a self-assembled rutile mesocrystalline photocatalyst. CrystEngComm 12(6):1759–1763. doi:10.1039/b922477g
110. Liu S-J, Gong J-Y, Hu B, Yu S-H (2009) Mesocrystals of rutile TiO_2: mesoscale transformation, crystallization, and growth by a biologic molecules-assisted hydrothermal process. Cryst Growth Des 9(1):203–209. doi:10.1021/cg800227x
111. Hong Z, Wei M, Lan T, Jiang L, Cao G (2012) Additive-free synthesis of unique TiO_2 mesocrystals with enhanced lithium-ion intercalation properties. Energy Environ Sci 5 (1):5408–5413. doi:10.1039/c1ee02551a
112. Ohzuku T, Takehara Z, Yoshizawa S (1979) Nonaqueous lithium/titanium dioxide cell. Electrochim Acta 24(2):219–222. doi:10.1016/0013-4686(79)80028-6
113. Kavan L, Fattakhova D, Krtil P (1999) Lithium insertion into mesoscopic and single-crystal TiO_2 (rutile) electrodes. J Electrochem Soc 146(4):1375–1379. doi:10.1149/1.1391773
114. Jiang C, Honma I, Kudo T, Zhou H (2007) Nanocrystalline rutile TiO_2 electrode for high-capacity and high-rate lithium storage. Electrochem Solid-State Lett 10(5):A127–A129. doi:10.1149/1.2712041
115. Hong Z, Wei M, Lan T, Cao G (2012) Self-assembled nanoporous rutile TiO_2 mesocrystals with tunable morphologies for high rate lithium-ion batteries. Nano Energy 1(3):466–471. doi:10.1016/j.nanoen.2012.02.009
116. Ye J, Liu W, Cai J, Chen S, Zhao X, Zhou H, Qi L (2011) Nanoporous anatase TiO(2) mesocrystals: additive-free synthesis, remarkable crystalline-phase stability, and improved lithium insertion behavior. J Am Chem Soc 133(4):933–940. doi:10.1021/ja108205q
117. Hong Z, Xu Y, Liu Y, Wei M (2012) Unique ordered TiO(2) superstructures with tunable morphology and crystalline phase for improved lithium storage properties. Chemistry 18 (34):10753–10760. doi:10.1002/chem.201200515
118. Yang HG, Sun CH, Qiao SZ, Zou J, Liu G, Smith SC, Cheng HM, Lu GQ (2008) Anatase TiO_2 single crystals with a large percentage of reactive facets. Nature 453(7195):638–641. doi:10.1038/nature06964
119. Zheng YQ, Shi ER, Chen ZZ, Li WJ, Hu XF (2001) Influence of solution concentration on the hydrothermal preparation of titania crystallites. J Mater Chem 11(5):1547–1551

120. Xu M, Wang F, Ding B, Song X, Fang J (2012) Electrochemical synthesis of leaf-like CuO mesocrystals and their lithium storage properties. RSC Advances 2(6):2240. doi:10.1039/c2ra01119k
121. An Z, Zhang J, Pan S, Yu F (2009) Facile template-free synthesis and characterization of elliptic alpha-Fe_2O_3 superstructures. J Phys Chem C 113(19):8092–8096. doi:10.1021/jp9004168
122. Ning J, Jiang T, Men K, Dai Q, Li D, Wei Y, Liu B, Chen G, Zou B, Zou G (2009) Syntheses, characterizations, and applications in lithium ion batteries of hierarchical SnO nanocrystals. J Phys Chem C 113:14140–14144
123. Chen S, Wang M, Ye J, Cai J, Ma Y, Zhou H, Qi L (2013) Kinetics-controlled growth of aligned mesocrystalline SnO_2 nanorod arrays for lithium-ion batteries with superior rate performance. Nano Research 6(4):243–252. doi:10.1007/s12274-013-0300-3
124. Duan X, Mei L, Ma J, Li Q, Wang T, Zheng W (2012) Facet-induced formation of hematite mesocrystals with improved lithium storage properties. Chem Commun 48(100):12204–12206. doi:10.1039/c2cc36620g
125. Ma J, Teo J, Mei L, Zhong Z, Li Q, Wang T, Duan X, Lian J, Zheng W (2012) Porous platelike hematite mesocrystals: synthesis, catalytic and gas-sensing applications. J Mater Chem 22(23):11694–11700. doi:10.1039/c2jm30216k
126. Su D, Dou S, Wang G (2014) Mesocrystal Co_3O_4 nanoplatelets as high capacity anode materials for Li-ion batteries. Nano Research 7(5):794–803. doi:10.1007/s12274-014-0440-0
127. Uchaker E, Gu M, Zhou N, Li Y, Wang C, Cao G (2013) Enhanced intercalation dynamics and stability of engineered micro/nano-structured electrode materials: vanadium oxide mesocrystals. Small 9(22):3880–3886
128. Liu JF, Li QH, Wang TH, Yu DP, Li YD (2004) Metastable vanadium dioxide nanobelts: Hydrothermal synthesis, electrical transport, and magnetic properties. Angew Chem Int Ed 43(38):5048–5052. doi:10.1002/anie.200460104
129. Liu Y, Uchaker E, Zhou N, Li J, Zhang Q, Cao G (2012) Facile synthesis of nanostructured vanadium oxide as cathode materials for efficient Li-ion batteries. J Mater Chem 22 (46):24439–24445. doi:10.1039/c2jm34078j
130. Sun CH, Yang XH, Chen JS, Li Z, Lou XW, Li C, Smith SC, Lu GQ, Yang HG (2010) Higher charge/discharge rates of lithium-ions across engineered TiO_2 surfaces leads to enhanced battery performance. Chem Commun 46(33):6129–6131. doi:10.1039/c0cc00832j
131. Dang F, Hoshino T, Oaki Y, Hosono E, Zhou H, Imai H (2013) Synthesis of Li-Mn-O mesocrystals with controlled crystal phases through topotactic transformation of MnCO(3). Nanoscale 5(6):2352–2357. doi:10.1039/c3nr33767g
132. Cao A, Manthiram A (2012) Shape-controlled synthesis of high tap density cathode oxides for lithium ion batteries. Phys Chem Chem Phys 14:6724–6728. doi:10.1039/c2cp40209b
133. Zhou N, Liu Y, Li J, Uchaker E, Liu S, Huang K, Cao G (2012) Synthesis and characterization of high power $LiFePO_4$/C nano-plate thin films. J Power Sources 213:100–105. doi:10.1016/j.jpowsour.2012.04.022
134. Bilecka I, Hintennach A, Djerdj I, Novak P, Niederberger M (2009) Efficient microwave-assisted synthesis of $LiFePO_4$ mesocrystals with high cycling stability. J Mater Chem 19(29):5125–5128. doi:10.1039/b909545d
135. Xia Y, Zhang WK, Huang H, Gan YP, Tian J, Tao XY (2011) Self-assembled mesoporous $LiFePO_4$ with hierarchical spindle-like architectures for high performance lithium ion batteries. J Power Sources 196(13):5651–5658. doi:10.1016/j.jpowsour.2011.02.044
136. Chen M, Teng F, Li G, Shi H, Wang J, Xu M, Ji X, Lu T, Lv Y, S-i Mho (2012) Self-assembly of highly uniform $LiFePO_4$ hierarchical nanostructures by surfactant molecules in a new mixture medium. Ionics 18(6):541–547. doi:10.1007/s11581-012-0703-2
137. Yang H, Wu X-L, Cao M-H, Guo Y-G (2009) Solvothermal synthesis of $LiFePO_4$ hierarchically dumbbell-like microstructures by nanoplate self-assembly and their application as a cathode material in lithium-ion batteries. J Phys Chem C 113(8):3345–3351. doi:10.1021/jp808080t

138. Popovic J, Demir-Cakan R, Tornow J, Morcrette M, Su DS, Schloegl R, Antonietti M, Titirici M-M (2011) LiFePO(4) mesocrystals for lithium-ion batteries. Small 7(8):1127–1135. doi:10.1002/smll.201002000
139. Zhou N, Wang H-Y, Uchaker E, Zhang M, Liu S-Q, Liu Y-N, Cao G (2013) Additive-free solvothermal synthesis and Li-ion intercalation properties of dumbbell-shaped $LiFePO_4$/C mesocrystals. J Power Sources 239:103–110. doi:10.1016/j.jpowsour.2013.03.136

2D and 3D Imaging of Li-Ion Battery Materials Using Synchrotron Radiation Sources

Ulrike Boesenberg and Ursula E.A. Fittschen

1 Introduction

Improving the performance of Li-ion batteries in terms of e.g. power density, life time and stability is mandatory to promote applications beyond consumer electronics such as electric vehicles or large scale energy storage. Careful characterization of the involved materials regarding composition, phases and their interactions with electrolyte and/or charge carriers is essential for understanding the functionality and performance of the battery. X-ray based studies on the solid state characteristics of materials/electrodes used in Li-ion batteries are very resourceful for understanding performance of Li-ion cell systems. Recently a full beamline was dedicated to the study of rechargeable batteries at Spring-8 in Japan [1]. X-ray diffraction (XRD) and X-ray absorption spectroscopy (XAS) provide knowledge on phase changes and redox activities and belong to the canon of standard techniques for bulk material characterization.

Common battery electrodes are typically hierarchically structured. They are usually composites of materials where particulates of active and conductive material are embedded into a binder to form a porous structure. One can immediately envision a whole number of challenges concerning issues like charge and mass transport or interface reactions. Figure 1 shows the complex physical microstructural characteristics of a composite electrode [2].

On the material level the performance and capacity are linked to the physical microstructure and local chemistry. Insertion of Li ions in intercalation materials generally requires a slight adaptation of the crystal lattice to accommodate for the

U. Boesenberg
Deutsches Elektronen-Synchrotron, Notkestrasse 85, Hamburg 22607 Germany

U.E.A. Fittschen (✉)
Chemistry Department, Washington State University, Pullman, WA 99164, USA
e-mail: ursula.fittschen@wsu.edu

Z. Zhang and S.S. Zhang (eds.), *Rechargeable Batteries*,
Green Energy and Technology, DOI 10.1007/978-3-319-15458-9_14

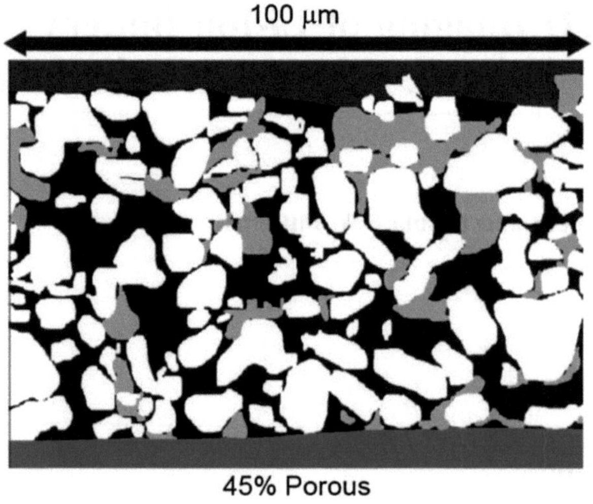

Fig. 1 Segmented FIB/SEM images of electrode films with varying porosities. *White* represents active material domains, *light gray* represents carbon domains, *black* represents pore domains, and the top *dark gray* and bottom *gray* represent the exterior volume and current collector respectively. Reprinted with permission from reference [2]. Reproduced with the permission of the Electrochemical Society

Li-ion and requires a change of oxidation state to compensate for the additional positive charge, i.e. the Fe in $FePO_4$ will undergo valence change from Fe^{3+} to Fe^{2+}. The high level performance of battery electrodes is thus quite a complex matrix of chemistry and morphology on a range of length scales from nm to cm.

And indeed, inhomogeneities in elemental distribution, defects in the atomic lattice and heterogeneity in physical structure, such as changes in porosity and tortuosity on the mesoscale, are considered key factors for degradation and capacity fade [3–5]. Heterogeneities can easily lead to inhomogeneous current distribution and inaccessible capacity. Capacity loss is the major contribution to aging of Li ion batteries. For an accurate description and modeling of the aging process, experimentally obtained data is essential [6]. Deeper understanding of degradation mechanisms may lead to improved electrode design and thus result in an increase of life-time.

Not only aging and degradation is affected by the microstructural characteristics of an electrode. Parameters such as particle and/or agglomerate size, porosity, layer thickness etc. can also make the difference in the application as high power or high energy cells. This for example described for graphitic anodes in Ender et al. [7]. To fully understand the correlation between morphology and chemistry, spatial characterization is therefore needed over a large range from the nanoscale all the way to the electrode level.

This chapter will highlight Li-ion battery research using 2D and 3D synchrotron based X-ray imaging techniques and show examples from characterizations of full electrodes as well as single particles. We will give a brief introduction into the available techniques and designated set-ups to make room for the results that provide spatial resolved insight into battery material chemistry and physical microstructure. For the details of the contrast methods and the X-ray optics we would like to refer the reader designated textbooks or review articles where these are developed in greater detail.

2 X-Ray Micro Probes

In situ micro analysis on Li ion batteries has been reviewed recently focusing on electron microscopy and touching also other microscopic probes as well [8]. Electron microprobes offer spatial resolution of a few nano meters and less. However in situ and *operando* capabilities are limited due to the low depth of information and the restriction of the probe environment.

Although much less common, X-rays are an efficient probe for spatially resolved measurements generally referred to as X-ray microscopy or X-ray micro- and nanoprobe where data can be obtained in 2 or 3 dimensions (tomography). Their high penetration depth and good spectral resolution allow access to relate physical structures to properties with reasonable spatial resolution. Commonly down to 20 nm at 3rd generation synchrotron facilities and recently even down to 5 nm using ptychography with soft X-rays ($E < 2$ keV) [9, 10]. In recent years, multiple groups around the world have focused on imaging Li-ion batteries and their materials on a range of length scales. Synchrotron light sources are often preferred for their higher photon flux in comparison to lab sources. This drastically reduces imaging times and facilitates higher spatial resolution.

2.1 X-Ray Imaging Contrasts Available for 2D/3D Imaging

The three most common X-ray contrast methods used for spatially resolved measurements are absorption, fluorescence and diffraction/scattering. X-ray absorption spectroscopy measurements are a special case of utilizing the absorption contrast, due to the strong contrast variation in the vicinity of an absorption edge. The selection of a suitable instrument and contrast method strongly depends on the relevant scientific question. In the following we will briefly introduce a few fundamental characteristics of X-rays and highlight the three most common methods to obtain contrast.

- Absorption contrast images in general show physical distribution of matter such as morphology e.g. particle assembly in an anode, similar to optical and electron microscope images. Special cases are
  - phase contrast for soft matter and
  - energy resolved imaging for chemical speciation.

- Diffraction contrast images show changes in crystal structure. Crystallinity of the material is a prerequisite. A special case is coherent diffraction imaging, where the far field diffraction pattern is collected which allows conclusions on the distribution of matter (morphology) with higher resolution than the optics provided. This method is independent of the crystallinity of the material.
- Fluorescence contrast allows to access the distribution of elements depending on the limits of detection for the respective elemental lines. In general the probe is more sensitive for high Z elements having low limits of detection in the ppb range.

2.1.1 Absorption Contrast

X-ray absorption contrast the most commonly applied contrast method is frequently used in daily live for e.g. taking radiographs for medical reasons. Absorption imaging visualizes matter according to its density and absorption cross sections.

Interaction of X-rays with matter is based on the interaction of high energy photons with predominantly the core electrons of the respective atoms.

The X-ray attenuation is a measure to describe the X-ray interaction with a respective element or compound. Overall it decreases with rising energy of the photons and increases with increasing atomic number. This results in rather large penetration depth and thus enables studies of volumes representative for the regarded material. In comparison to other spatially resolved characterization techniques such as electron microscopy, X-ray microscopy is therefore generally considered a bulk technique, unless special experimental setups or low energy X-rays are employed to e.g. remain surface sensitive.

Within the illuminated sample volume, the absorption follows the Beer-Lambert law $I(x) = I_0 e^{-\mu(E)\rho x}$. Where I is the transmitted intensity after a path of the length x (e.g. the thickness of the sample). I_0 represents the intensity at $x = 0$, while $\mu(E)$ corresponds to the energy dependent mass attenuation coefficient and ρ the density of the material. Changes in the density of the material by e.g. defects (cracks, voids) or phase boundaries will therefore lead to changes in the absorption and can image morphology and phase distribution (phases with heavier elements vs. lighter elements).

Phase contrast uses information concerning changes in the phase of an X-ray beam that passes through an object. Images are created similar to phase contrast used in optical microscopy. Accordingly, the beam's phase shift caused by the sample is not measured directly, but is transformed into variations in intensity, which then can be recorded by the detector [11–13]. It can be better understood thinking in terms of wave optics rather than ray optics. Typically this method enhances edges of sample features. Phase contrast is most commonly used for imaging materials that show poor absorption contrast i.e. biological matter, due to the low absorption cross sections of the light elements but has recently been applied to battery materials [14, 15].

The weak interaction of X-rays with matter causes quasi none-destructiveness in comparison to other probes such as ions and electrons and thus allows for *in situ* or *operando* analysis. Nonetheless, radiation induced damage can become an issue especially for experiments which have a long exposure, or high radiation dose per area (focused beam) and can be non-neglectable.

Microscopes in the hard X-ray regime ($E > 2$ keV) are often operated in air due to the high penetration power of the X-rays and can have a working distance of a few centimeters, thus facilitating *in situ* or *operando* setups in comparison to e.g. transmission electron microscopy or the sub-mm working distance in microscopes using soft X-rays ($E < 2$ keV).

2.1.2 X-Ray Absorption Spectroscopy

As described earlier the absorption contrast of a given material decreases in general with higher energies. However, each element has distinct and characteristic so called "absorption edges", where a core electron is excited and the transmission of the material drastically drops as the attenuation coefficient of the element rises. The energy of the absorption edge of an element varies with the oxidation state due to the change in electron binding. In general the edge is shifted slightly (a few eV) but distinctly to higher values for oxidized material in comparison to the metallic state.

An advantage of using synchrotron light sources in comparison to laboratory based sources is the much higher photon flux and brilliance, which allows to tune the energy of the incoming radiation with an excellent spectral resolution and offering still high flux on the sample.

Scanning the energy provides means for acquiring absorption edge spectra and allows for chemical speciation (e.g. determination of the oxidation state of an element or possibly the compound) based on element/compound specific absorption properties.

Measurements of transmission (or fluorescence) directly across the absorption edge as a function of energy are generally called XANES for X-ray absorption near edge structure. Extended absorption measurements up to about 1000 eV above the energy of the absorption edge are called EXAFS for extended X-ray absorption fine structure (Fig. 2a). Bulk EXAFS using large X-ray beams can reveal information on nearest neighbors and distances to surrounding atoms. However, requirements on the number of data points and signal-to-noise ratio are much more stringent than for XANES and are therefore not commonly applied in high resolution imaging. Please see e.g. [16, 17] for a more detailed description and explanation.

The variation of the absorption edge with the oxidation state is especially interesting for active materials in Li-ion batteries, since the oxidation state can be used as a measure for the state of charge (SOC) see e.g. [18, 19]. In Fig. 2b the changes in the Ni K-edge -XANES spectrum at different SOC of $LiNi_{0.5}Mn_{1.5}O_4$ composite electrodes are exemplarily shown. Ni is the redox active element in this compound. The non-cycled electrode represents the pristine state (discharged) and

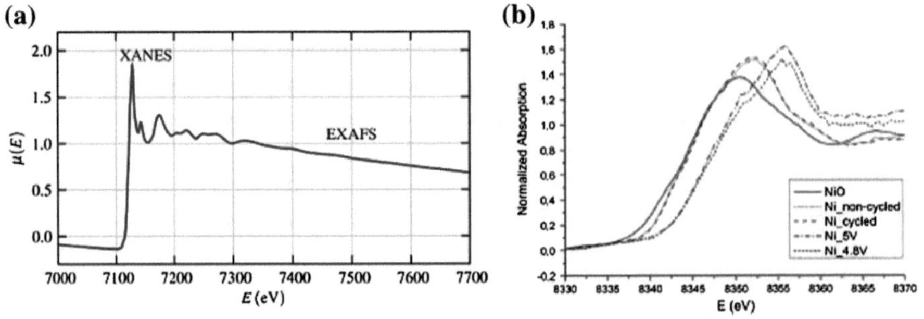

Fig. 2 **a** XANES and EXAFS at the Fe K-edge and **b** normalized XANES spectra of the Ni K edge of $LiNi_{0.5}Mn_{1.5}O_4$ at different SOCs and a Ni^{2+} reference. Reprinted with permission from Menzel et al. [68]. Copyright (2012) by Elsevier

the cycled electrode the discharged state, whereas the electrode extracted at 5 and 4.8 V represent charged states.

Spatially resolved XANES measurements can be collected by acquisition of a number of 2D images at multiple energies in the vicinity of the absorption edge. Another approach is to acquire a full energy spectrum per sample point before the sample is moved to the next spot of interest.

The acquisition of multiple 2D image to obtain spatial resolved species information is illustrated in Fig. 3 (1)–(3) [20]. It illustrates how the intensity distribution changes for a stack of 2D image with changing energy and how the species information is then reconstructed for each pixel. Usually reference compounds are used to fit the obtained spectra for the determination of the phase distribution. This can then be extended for 3D data collection, where the sample is rotated and such a phase map is determined for each angle. This dataset can then be reconstructed and rendered for display of 3D phase distribution.

Analysis software like e.g. the TXMwizard [21] enables stacking and alignment routines for a set of 2D images from which then single pixel XANES or element specific information can be extracted and further analyzed, e.g. by linear combination fitting of suitable standards or principal component analysis [21].

Absorption edges in general but most commonly those of 3D elements, which are typically the redox active elements in materials for Li-ion batteries, can also be used to extract elemental information. Imaging below and above an absorption edge is often done to obtain element specific information in transmission geometry. However, this can become rather time consuming for e.g. tomography, because the measurement has to be repeated at multiple energies. Zone plates and lenses have to be adjusted to maintain the focus at each energy while mirror systems can be achromatic. Alignment and resolution correction of the multiple datasets may have to be performed to obtain comprehensive information.

Preparation and mounting of suitable sample cells and electrode samples often poses a challenge for X-ray microscopy. Microscopes operating in the hard X-ray

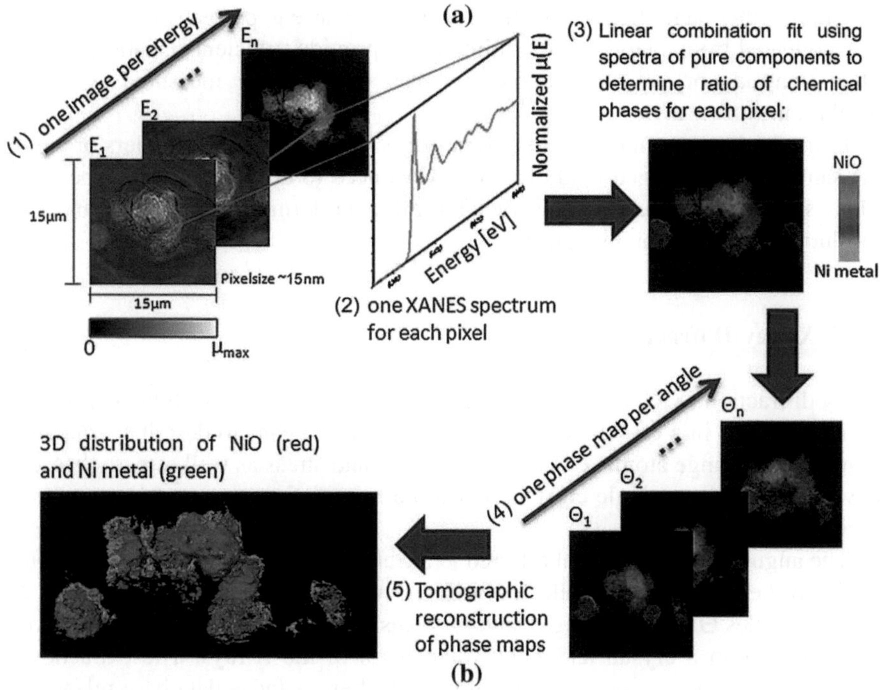

Fig. 3 Principles of data processing for three-dimensional XANES microscopy. (1) One image is acquired in absorption contrast at each energy in the XANES scan. (2) XANES are constructed from each pixel plotting normalized absorption versus energy. (3) XANES from each pixel is fit to create a chemical phase map. (4) The sample is rotated to collect a 3D dataset and a phasemap is determined for each angle and (5) tomographic reconstruction and rendering for 3D phase distribution. Reprinted with permission from [20]. Copyright 2011

regime bear the great advantage of operating with working distance usually in the few mm to cm range, while microscopes in the soft X-ray regime operating at high resolutions (<50 nm) have typically sub-mm working distances. Further, the latter instruments generally operate under slight vacuum conditions to avoid absorption of the X-rays by air.

Nonetheless, besides the geometrical limits, absorption effects by the sample environment and materials are non-neglectable in both energy regimes. Those need to be considered carefully already when the cell from which the material is to be extracted is designed.

For example; capturing reaction progress in single particles or secondary particles, the concentration of the active material needs to be low, well defined and yet provide sufficient contrast for the X-rays. For such cells, the electrochemical response is often weak, but hopefully metastable intermediate states can be caught before they can relax during harvesting of the electrodes.

On the other hand, larger scale information such as e.g. porosity and tortuosity is best collected from a large region of interest to provide sufficient statistics, which is often contradictory to the sample volumes that can be measured with high resolution.

From the instrumental side, long term drifts of stages, sample holders or simply the sample in the imaged field of view (FOV) need to be avoided or compensated. This issue becomes especially crucial for longer term measurements and high resolution imaging (tomography).

2.1.3 X-Ray Diffraction

X-ray diffraction (XRD) bases on the interaction of the X-rays with the periodically arranged atoms in a crystal lattice and thus provides means to describe structure in terms of long range atomic order, phases, strain and stress as well as particle size in powder materials or single crystals. For more information we to refer to textbooks such as [22].

The angular-resolved signal referred to as reflections obtained from diffraction of an X-ray beam is usually collected with a 1 or 2 dimensional detector. The diffraction angles Θ follow Bragg's law: $n\lambda = 2d\sin\Theta$ where d is the distance between lattice planes in a crystal and λ the wavelength of the X-rays. The extraction of information and processing of diffraction data often includes data base referencing or Rietveld refinement to determine e.g. exact lattice parameters, phase fractions or site occupation of atoms.

In Fig. 4 a scheme of a micro-XRD set up using a 2D CCD detector is depicted as used by Reinsberg et al. [23]. A single bounce capillary is used to produce a low-divergence beam focus. The reflections can be radially integrated to give 1D powder diffraction patterns.

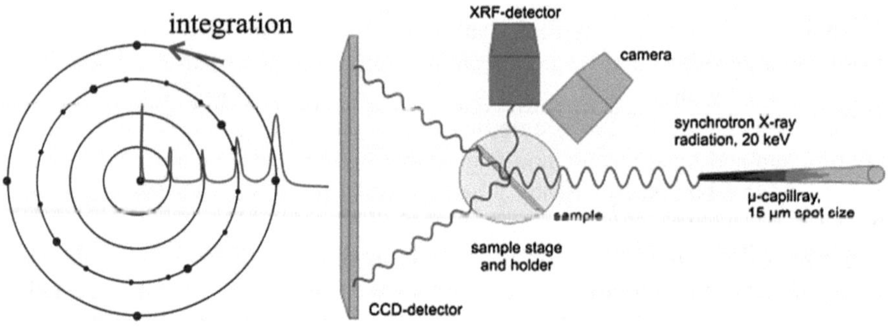

Fig. 4 Schematic of a scanning setup using diffraction contrast. The sample is scanned through a focused beam and 2D diffraction patterns and X-ray fluorescence spectra are recorded simultaneously. Reprinted with permission from Klaus Reinsberg

Spatially resolved XRD relies on the crystallinity of the material. However, if the beam is small in comparison to the imaged grains in the sample, they can appear as a single crystals. In general, the data analysis can be quite complex for multidimensional datasets.

2.1.4 X-Ray Fluorescence

X-ray fluorescence (XRF) allows for elemental imaging. It is directly correlated to X-ray absorption. In X-ray absorption usually a core electron (e.g. K-shell) interacts with a photon from the excitation radiation and parts from the atom. The vacancy is then filled by an electron from outer shells (e.g. L or M). The energy from the transition (difference of the final and the initial state of this electron) can be emitted as fluorescence photon.

The energy resolved spectrum in general allows to identify each element by its specific lines. The intensity is correlated to the concentration of the element.

XRF is therefore a suitable technique to provide qualitative and quantitative elemental information. Furthermore, synchrotron XRF and also other set ups like total reflection XRF (TXRF) are sensitive to low concentrations and can detect elements on the 10 ppb concentration level.

The spatial resolution can be achieved either by scanning the sample through a focused excitation beam or using an array detector together with an optic (full field).

For each pixel a full spectrum is acquired and then analyzed by fitting a model spectrum to the peaks. For reliable quantification often preliminary information on specific characteristics of the sample such as thickness or layered structure are included to account for e.g. self-absorption effects. In Fig. 5 a schematic workflow is shown. Figure 5a illustrates the acquisition of full spectra for each pixel using a focused beam. In (b) the fitting of one of these spectra is shown. Figure 5c shows how the image is obtained by plotting the integrated intensity (and optionally corrected) for each pixel. Figure 5d shows a schematic set-up for a full-field XRF detector in combination with a one dimensionally focused beam.

Inter-correlation of elements or morphology can be extracted using PCA or cluster analysis. Please see e.g. [24] for more details on the method.

CCD energy dispersive array detectors have been used for full field micro-XRF for different applications [25]. A prototype pnCCD form IFG/pnSensor is may be the most advanced instrumentation available to date. Because the complete spectrum is to be acquired in each pixel the optics used in front of the detector need to be fairly achromatic, which limits the choice of available optics and thus the achievable spatial resolution. At the moment capillary optics are mostly used either to guide the photons to the detector array (1:1) or provide slight magnification (e.g. 1:8). In case of 1:1 optics allow the acquisition of about 50 μm spatially resolved full energy spectra of an $12.7 \times 12.7\ \text{mm}^2$ area [26, 27]. From a set-up as sketched in Fig. 5d), elemental distribution in space (2 or 3D) can be extracted. An alternative to tomography and full field XRF to obtain 3D information using X-ray fluorescence is the implementation of confocal micro-XRF. Here, one optic is used

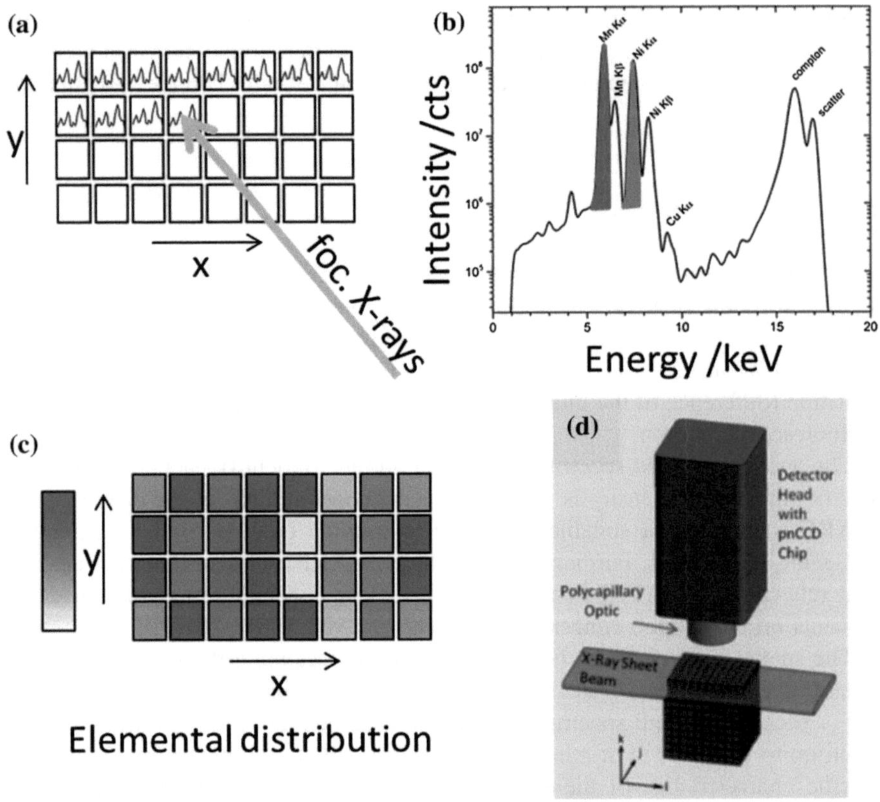

Fig. 5 Principle of Micro-XRF (**a**) the sample is scanned through the focused beam in 2 or 3 dimensions. A spectrum is acquired for each pixel, (**b**) Each individual spectrum is analyzed and fitted using appropriate routines, (**c**) 2 or 3D information on the elemental distribution or correlations between multiple elements can be displayed, (**d**) schematic setup of a full field micro-XRF detector in combination with a sheet beam, reprinted with permission from reference [25]. Copyright 2014 Royal Chemical Society

to focus the primary beam and another optic is applied to the detector to discriminate fluorescence photons in depth. Because multiple elements and therefore multiple fluorescence lines at a variety of energies need to be detected, the implemented optics need to be achromatic. Typically capillary optics are used and resolutions in the range of 10 μm can be obtained.

The advantage of X-ray fluorescence (XRF) imaging over scanning below and above the respective absorption edge of multiple elements is to correlated information on multiple elements within a single measurement. Additionally, limits of detection for the respective elements are generally lower than in transmission geometry.

Accordingly, acquisition of the full X-ray fluorescence spectrum at each pixel is especially interesting for cathodes material containing multiple transition metals or materials containing trace concentrations of transition metals and other high Z elements.

Recent developments of suitable detectors can make full use of the high flux available at 3rd generation synchrotrons and thus open the door to scanning large areas in the square millimeter to square centimeter range with sub-micrometer resolution in a few hours, XANES mapping using the fluorescence detector or even fluorescence tomography. Such a setup can be e.g. found at the PETRAIII storage ring at DESY in Hamburg, the ESRF in Grenoble, APS in Chicago, AS in Melbourne and others.

Nonetheless, sub 100 nm resolution poses restriction to sample size and environment. Accordingly, larger samples may not be fitted into the setup and *in situ* applications may be restricted. In addition to this limitation careful evaluation of absorption and shading effects is required to ensure for reliable results especially with thick samples.

2.2 Instrumentation: Full Field and Scanning X-Ray Microscopy

In microscopy regardless if X-rays, electrons or optical light is used, two general modes can be distinguished, which are

- the full field microscopy, where the object is fully illuminated and an array detector provides spatial resolution, usually coupled to an optic and
- scanning microscopy, where the object is scanned through a focused beam and the signal is analyzed for each pixel. An optic is required to focus the probing beam.

A detailed review on general X-ray microscopy can be found e.g. in [13, 28].

Multiple kinds of X-ray optics exist to modify the beam. Robust and highest resolution under working conditions of 20–30 nm can be obtained using Fresnel-Zone plates, although their efficiency is comparatively low. The development of highly efficient zone plates in the medium energy range (2–12 keV) has enabled measurements under ambient conditions. The larger penetration depth of hard X-rays promoted the development of sophisticated sample environments for e.g. in situ set ups [29].

There has also been a tremendous development of designing and manufacturing X-ray focusing lenses such as compound refractive lenses. Highly efficient in terms of flux are mirrors aligned in Kirkpatrick-Baez geometry. Ref. [30] includes a detailed review on the recent developments of X-ray optics for microscopy.

Most X-ray microscopes -regardless of fullfield or scanning type- work in transmission geometry using the contrast of the X-ray absorption in the material. However, emerging technologies which use other contrast methods such as fluorescence [31], phase contrast [14], diffraction [32, 33] and diffractive imaging techniques (the latter allows for spatial resolution no longer limited by the optical elements in the beam [34–36]) are also applied for studies of battery materials.

Examples for X-ray microscopes both in the soft and hard X-ray regime can be found at almost every synchrotron in the world with further developments with regard to measurement time, stability and resolution constantly taking place. A scheme of such a set-up is depicted in Fig. 6 as an example from a set up at the European Synchrotron Radiation Facility (ESRF) [37]. The monochromatic beam coming from an undulator source is focused using a zone plate objective lens and an aperture. The sample is raster scanned in 2 or 3 dimensions through the beam, often interferrometrically controlled. Stages ensure the accuracy of positioning. In this example both the fluorescence signal and transmission signal are detected simultaneously.

On the other hand, using the unfocused or only slightly collimated beam, optics such as Fresnel-Zone plates, can be placed behind the sample to reproduce an image in higher magnification on CCD detector, so called full field transmission X-ray microscopy (FFTXM).

This technique bears the advantage of imaging a larger field of view (FOV) at once, depending on the achieved resolution of the integrated optics. Because information from about 1 M pixel can be obtained simultaneously the 2D information may be acquired much faster than in the scanning mode, provided that the contrast is strong.

Usually, tomography or 2D XANES which require the scan of the rotation or the energy across an absorption edge can be performed in a much more time efficient way, when using full field set ups. The spatial resolution ranges from around 30 nm up to several μm, depending on the optics.

On the other hand, FFTXM is obviously limited in coupling with contrast methods, where 2dimensional data is acquired for each pixel such as diffraction or ptychography (diffractive imaging).

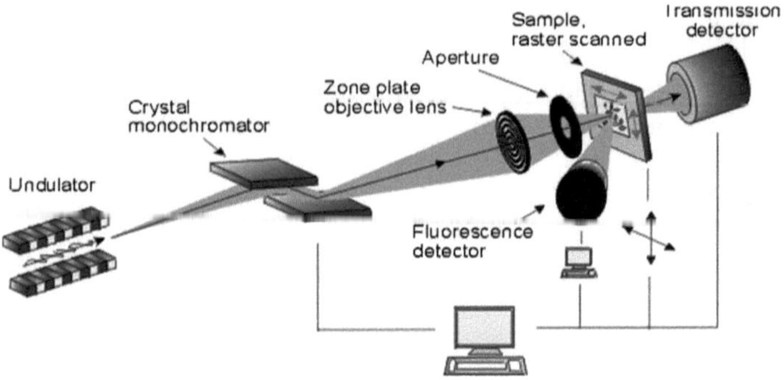

Fig. 6 Schematic setup of a scanning setup at an undulator beamline at ID21 at ESRF. Reprinted with permission from reference [37]. Copyright 2010 American Chemical Society

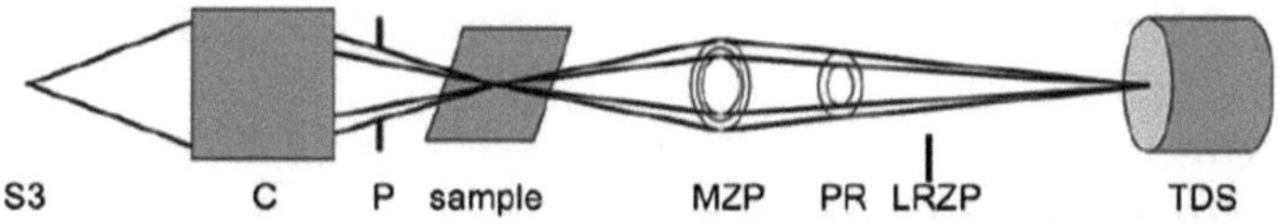

Fig. 7 Schematic of the transmission X-ray microscope at SSRL located at a wiggler beamline, the microscope setup consisting of a capillary condenser (C), the sample stage, pinhole (P), Zone plate (MZP) and area detector (TDS). A phase ring (PR) can be used to provide Zernike phase contrast. Reprinted with permission from reference [38]. Copyright Microscopy Society of America 2010

One example for a full-field Microscope operating in transmission geometry and using Fresnel-Zone plates to obtain *ca.* 30 nm resolution can be found at Beamline 6-2, SSRL Stanford, USA. [38] Fig. 7 sketches the general setup for such an instrument. The instrument is based on the design of an Xradia (now Zeiss) microscope for use with a laboratory X-ray source. An elliptical capillary condenser focuses the monochromatic X-rays onto the sample. The condenser provides a hollow cone illumination of the sample over the field of view (FOV). An objective lens, consisting of a micro- zone plate, images the transmitted X-rays onto a transmission detector system, yielding either a $\sim 30\ \mu m^2$ FOV with 10 × objective or $\sim 14\ \mu m^2$ FOV with 20 × objective.

The implementation of a rotation stage enables tomographic imaging and thus 3 dimensional reconstruction. An extensive review on three dimensional techniques for imaging the physical micro structure of matter over a large range of length scales can be found e.g. in [39].

To overcome the issue of a limited FOV of FF microscopes -while maintaining the high spatial resolution- multiple images are acquired with a slight overlap and stitched together using appropriate routines [9, 40, 41]. Stitching is very successful for 2D imaging, also under operando conditions. However, it becomes time consuming for tomography, especially when multiple tomograms with variation of the beam energy are needed [see Fig. 3 (4) and (5)].

The Figures of merit of the analysis in terms of sensitivity, spatial resolution and time resolution will greatly depend on the respective beam line and set up e.g. source, optics, and detectors.

In a round-robin project Kanitpanyacharoen et al. [42] have compared the characteristics of X-ray tomographic microscope instruments at multiple Synchrotrons such as the Swiss Light Source (SLS), the Advanced Photon Source (APS) and the Advanced Light Source (ALS) which produced comparable results.

3 X-Ray Micro Probes in Li-Ion Battery Research

3.1 Imaging and Spatially Resolved XANES (XAFS) in Transmission Geometry

Most X-ray microscopes and spatially resolved imaging techniques especially with the option of tomography and thus 3D imaging utilize the absorption contrast of a material. In the following section we illustrate the results obtained for Li-ion batteries by a few selected examples from this vastly growing research area.

The research focuses on two major size ranges addressing different functionality aspects. First, there is a high level of interest on the general morphology of full electrodes. Data obtained on porosity and tortuosity is used to identify the bottleneck of transportation mechanisms. These parameters are further implemented in models since they are regarded keys to accurately describe aging and degradation mechanism in electrodes.

On the other hand describing and visualizing the phase transformation in single particles is mandatory to fully understand reaction mechanisms, i.e. core-shell reaction mechanism or the crystallographic orientation of the reaction front within a single particle. The canonical example is $LiFePO_4$ for which the exact mechanism of phase transformation is still extensively discussed [43]. The latter aspect often combines absorption contrast with energy selective measurements (XANES) to i.e. monitor the oxidation state and thus the reaction progress.

3.1.1 Electrode Level

Maybe, the first 3D imaging of a large volume (hundreds of micrometer) capturing the morphology of a commercial graphitic anode [44] and cathode [45] was achieved using an X-ray microscope (laboratory instrument) [18]. Those experiments give insight to the large scale porosity and physical macrostructure of the electrode and can be e.g. performed with X-rays penetrating through the intact steel housing of the cell, see Fig. 8 [46]. Geometrical parameters such as porosity, tortuosity, surface area and pore and particle size distribution were extracted. For example Channagiri et al. elaborated on the increase in porosity measured in commercial $LiFePO_4$ electrodes with increasing number of cycles and where thus able to put these changes and effects into numbers [47]. On the other hand Ender et al. have used X-ray nanotomography to visualize the differences in electrode morphology between high power and high energy graphitic anodes [7]. They observed increased porosity by almost a factor of two for the high power anodes which relates to a lower tortuosity and a narrower particle size distribution.

Rather than focusing on the physical macrostructure, Chen-Wiegart et al. [48] have pointed their efforts at elucidating the effect of cycling on the single particle level and the chemistry of different materials. To adapt the sample to the limited FOV of high resolution microscopes they have used a focused ion beam (FIB) to

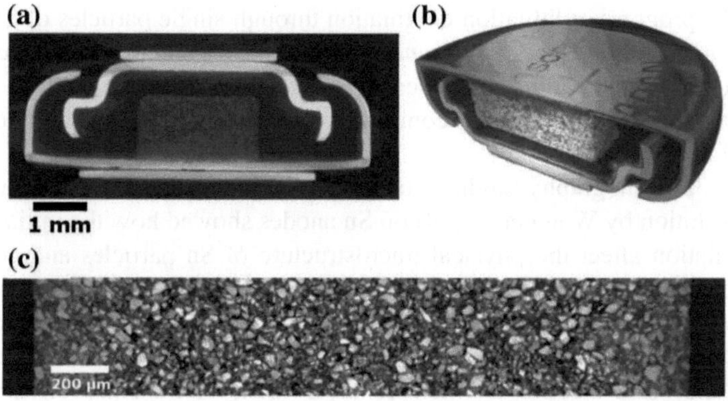

Fig. 8 **a** A virtual diametrical slice through the battery showing the cell architecture. **b** A cut-away image of the reconstructed cell, showing the manganese oxide cathode attached to the positive (uppermost) battery terminal. **c** At higher resolution, the virtual slice shows the manganese oxide electrode in the charged state. Reprinted with permission from reference [46]. Copyright 2014 Wiley VHC

cut a cone of a mixed electrode containing particles of $LiCoO_2$ and lithium nickel manganese cobalt oxide (NMC), which could be well identified from their difference in absorption contrast [48]. The sample was cycled multiple times. They found cracking dominantly in NMC particles, which showed a larger number of "higher curvature" features than the $LiCoO_2$. [48] Understanding these changes is relevant. Cracking may lead to alteration of the integration of the particle into the conductive network. Cracks can also create new surfaces for side reactions with the electrolyte. Both effects directly affect the cycle life and available capacity of the electrode.

The origin of cracks and other morphological changes is often found on the molecular level e.g. when insertion of Li ions causes a change of volume.

Because volume changes of the active material during battery operation is considered a significant barrier for the long term functionality of alloying and conversion materials, these systems have received considerable attention for X-ray microscopy. On these systems using SnO [49, 50], Ge [51], CuO [52] or even lithium-sulfur [41] also one of the first *in situ* imaging in absorption contrast was performed. For example the Ge particles showed volume extensions up to 400 % and could only be measured with ~30 nm resolution by stitching multiple FOV next to each other [51].

Ebner et al. have used model electrodes with SnO for studies under operando conditions in 3dimensions. From the 3D tomographic reconstructions they deduce particle size distribution, porosity and tortuosity for several materials [5, 53]. They propose a new and efficient way to gather such information from 2D images. In their studies they witnessed variation in onset and rate of core-shell lithiation for different particles. They also monitored crack initiation and growth along preexisting defects, and irreversible distortion of the electrode. Careful evaluation of the changes in absorption contrast further enabled Ebner et al. [4] to quantitatively analyze the progress of lithiation/delithiation through single particles of SnO within the working electrode under *operando* conditions. Detailed knowledge of these structural features and aging processes are essential for adequate models describing the cycling and aging process of composite electrodes and opens the door to new, improved electrode designs.

In situ 3D tomography studies covering a large field of view 40 μm and ca. 30 nm resolution by Wang et al. [54] on Sn anodes showed how the initial lithiation and delithiation affect the physical microstructure of Sn particles and correlate to mechanical degradation.

Since low atomic number elements have a low cross section of interaction with high energy X-rays, imaging materials such as graphite (anode), binder, carbon black or the solid electrolyte interface layer (SEI) in transmission geometry is problematic due to the low contrast. Unfortunately, using lower energy photons for imaging is generally hampered, due to the increasing air absorption and limited penetration.

Promising strategies regarding this issue were followed recently. Eastwood et al. and Lin et al. have made use of the phase contrast to successfully image a graphite electrode with sub-μm resolution [14] or to understand the dissolution and re-deposition of polysulfides in lithium sulfur cells [15]. Another approach followed by Zielke et al. was to model structures corresponding to low atomic number elements in combination with FFTXM tomography reconstructions [55].

3.1.2 XANES Imaging

Elucidation of the phase transformation in single particles shows the impressive capabilities of high resolution X-ray microscopy coupled with XANES and/or elemental contrast and allows for unprecedented geometric and chemical insight into the active materials of an electrode. For example, 3D XANES imaging of NiO particles in a conversion reaction with Li captured the reaction propagation along defects and cracks [20].

Despite extensive studies the phase transformation in single particles of $LiFePO_4$ upon de-lithiation is still one of the most discussed mechanisms [43]. This is best studied by *in situ* or *operando* studies, to overcome the open question of material relaxation during the harvesting, but. Those pose a tremendous challenge on the experimenter. Very well defined particle shapes were investigated at specific states of delithiation by Boesenberg et al. highlighting the importance of correlating microstructural properties and chemical state/phase distribution, see Fig. 9 [56]. The extremely high defect density in the central part of the studied individual particles limited the phase transformation to the mostly intact outer edge of the particles. This is in agreement with a more recent study in higher resolution by Shapiro et al. using soft X-ray ptychography [10].

Obtaining representative results for analysis on the individual particle level is a general problem. To probe multiple particles sections of a $LiFePO_4$ electrode stacked in the vertical direction has been prepared using FIB (at an intermediate state of charge).

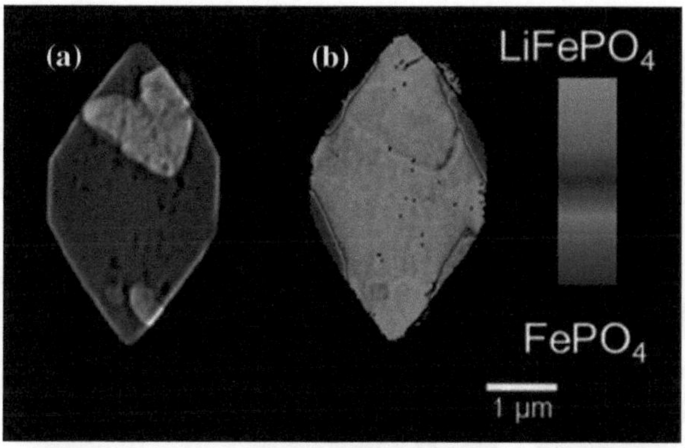

Fig. 9 (a) FF TXM image of a selected crystal in a sample with nominal composition $Li_{0.74}FePO_4$ collected at 7080 eV. (b) Chemical phase map obtained by LC fitting of XANES data at each pixel. Reprinted with permission from reference [56]. Copyright 2013 American Chemical Society

These specimens have then been spectroscopically imaged using a STXM. In combination with TEM images the phase distribution over many particles was illustrated. It was found that the stage of charge was fairly homogeneous distributed over the complete depth of the electrode (from current collector to the separator). Regarding the individual particles, it was found that particles in general appeared either charged or discharged. The percentage of charged to discharged particles in relation to their sized changed from ca 67 % discharged in smaller particles (around 200 nm) and around 60 % for larger particles (>250 nm) [57]. 2D imaging with chemical contrast spanning from single particle to multiple particles of $LiFePO_4$ under *operando* conditions were recently realized at BNL [58, 59]. Here evolution of charging in individual particles in the μm size range could be visualized.

Crucial information on the transport and degradation mechanism of NCM materials can also be obtained by X-ray microscopy. Full-field imaging below and above the absorption edge of Mn, Co and Ni on the secondary particles of NMC materials is able to capture elemental distribution in the cycled state [60]. Accordingly, migration and segregation can be monitored. This is illustrated in Fig. 10 (a–c). On the left hand side the individual elemental distributions for Mn, Co and Ni of a secondary particle are shown after cycling. In d) all three elements are correlated. A cross section (e) illustrates the increase in Mn concentration in comparison to Ni and Co at the edge of the secondary particle after cycling. This indicates ongoing elemental migration during the cycling of these layered materials. Similarly, this method was used to illustrate the Ni distribution in single particle of $LiNi_{0.75}Co_{0.10}Mn_{0.15}O_2$ with a core shell concentration gradient by Sun et al. [61].

But not only intercalation materials were characterized by XANES mapping, a recent study by Wang et al. shows an in situ study of the conversion reaction in CuO [52]. They capture both the evolution of morphology as well as the phase distribution at selected voltages revealing a core shell lithiation-delithiation mechanism.

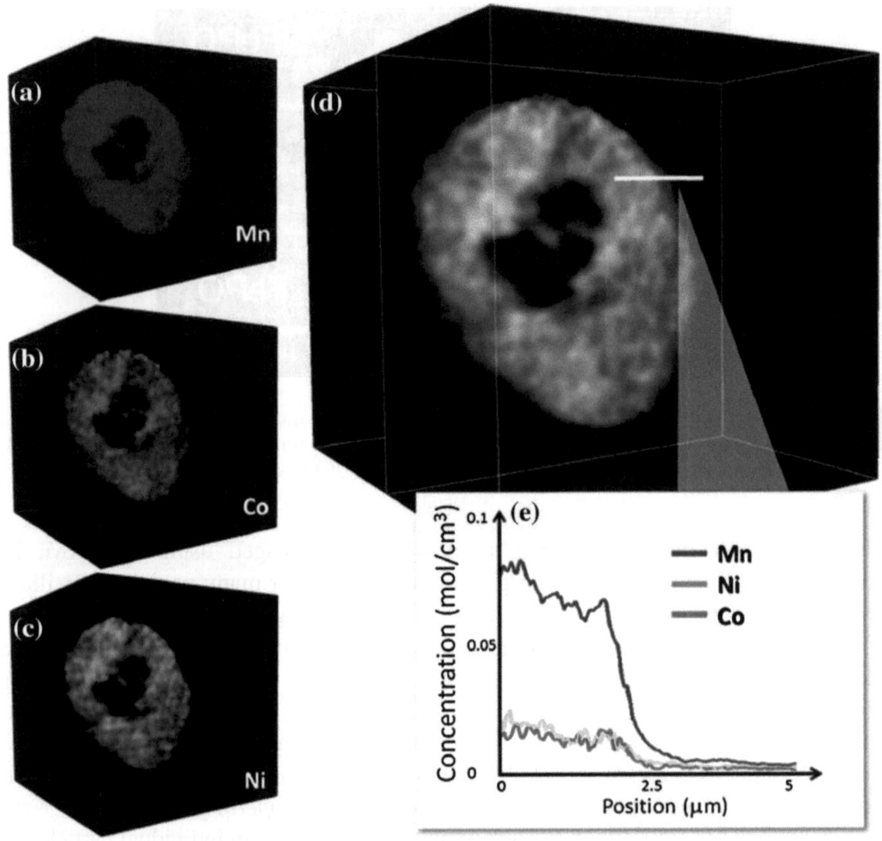

Fig. 10 Panel a-c Distribution of TMs (Mn, Co, and Ni, respectively) over a single selected slice (one voxel thickness) through the 3D volume of the 1x-cyled electrode. Panel d: Overlay of all TMs over the same slice. Panel e: Elemental concentration profile over the line (5 μm in length) indicated in panel d. Reprinted with permission from reference [60]. Copyright 2014 American Chemical Society

3.2 Micro-Diffraction

Bulk X-ray diffraction, thus averaging the X-ray diffraction signal over a large amount of sample ($\sim mm^3$), *ex situ* or under *operando* conditions can be considered a standard technique using laboratory based or synchrotron based sources. Countless findings rely on the precise description of the atomic order obtained from Laue diffraction patterns to describe structure, phases, reaction pathways, strain, disorder etc.

Although spatially resolved Laue diffraction has been established even in 3 dimensions [33, 62, 63], its application to describe the distribution of atomic order (such as phases) throughout the electrode is much less common. A study by

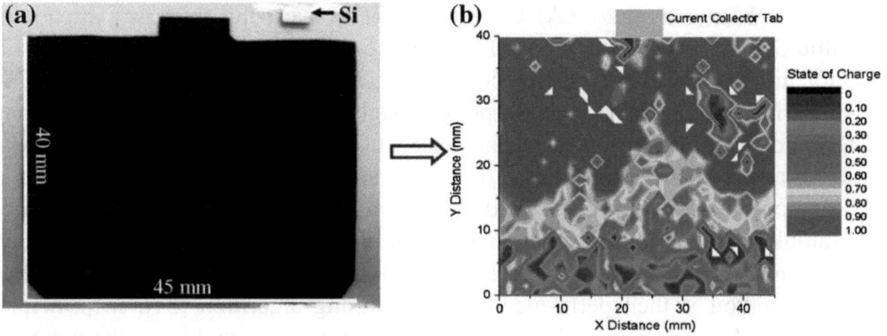

Fig. 11 **a** Optical image of the 40 mm × 45 mm rectangular LiFePO4 electrode. **b** FePO4 phase concentration profile of the prismatic electrode at 50 % SOC (charged at 2 A g-1). Reprinted with permission from reference [64]. Copyright 2010 American Chemical Society

Liu et al. [64] using micro-diffraction has illustrated nonuniform distribution of state of charge (SOC) by capturing the inhomogeneous distribution of state of charge in electrodes of $LiFePO_4$. They found those to occur horizontally over the length and width of an electrode (Fig. 11) as well as vertically over the depth of an electrode probing from the current collector the electrode-separator interface (Fig. 12). In Fig. 11 the state of charge distribution at 50 % charging in an electrode, with the current collecting contact at one side of the square electrode is shown. It demonstrates how inhomogeneities in charge are found on a relatively large scales of millimeters. For each pixel a powder diffraction pattern was integrated from the 2D image obtained by the CCD detector.

In depth-profiling shows further a significant difference between the state of charge at the current collector and the electrode-separator interface by a change (Fig. 12e) from 50 to 70 % in the state of charge over a depth of 50 μm [64].

To describe the electrochemical lithiation of the layered oxysulfide $Sr_2MnO_2Cu_{3.5}S_3$ Roberts et al. [65] have used micro-diffraction in combination

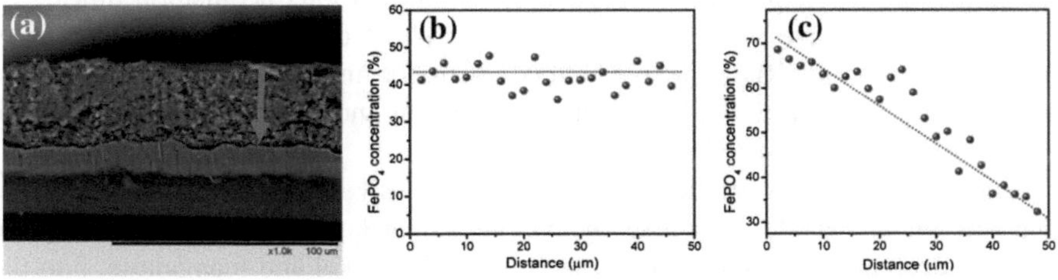

Fig. 12 **a** SEM image. The approximate locations of the vertical and horizontal scans are superimposed. FePO4 phase concentration versus scan distance along, **c** the vertical direction at 50 % SOC (charged at 20 mA g-1). FePO4 phase concentration versus scan distance along, **e** the vertical direction of the LiFePO4 electrode at 50 % SOC (charged at 3 A g-1). Reprinted with permission from reference [64]. Copyright 2010 American Chemical Society

with micro-XRF and micro-XANES. Micro XRD was capable to show extrusion of metallic Cu at an early state of preparation of the active material, where the XANES spectra still resembled the pristine condition.

Smearing the borders to absorption imaging, coherent diffraction imaging (CDI) and ptychography have recently been used to obtain extremely high resolution images of battery materials [10, 66, 67] in the sub 10 nm range. The method is based on the collection of far field diffraction patterns where the illumination of the sample in the confined coherent beam is highly redundant, thus overlapping. Iterative reconstruction of the object's transmission function and the wave field of the illumination is then performed using a phasing algorithm [34]. In principle, ptychography can overcome the limitations of X-ray optics and bears the potential of extreme resolutions coupled with excellent chemical speciation using the local complex refractive index. The potential for Li-ion battery materials was illustrated by Shapiro et al. [10], who have imaged a single particle of partially delithiated $LiFePO_4$ with chemical speciation. They were able to correlate the formation of cracks to the c-direction of the crystal and with the phase transformation. The observed phase distribution around the crack led to the conclusion, that the reaction is solid-state limited and kinetically controlled by microstructural defects.

3.3 *Micro X-Ray Fluorescence*

Using a micro-XRF approach, elemental distribution for a number of elements can be studied simultaneously. Acquiring the full XRF spectrum at each pixel is especially interesting for cathodes material containing multiple transition elements like e.g. lithium-nickel-manganese-oxide. Because of the low detection limits of the method, XRF is also a suitable method to detect and localize trace elements e.g. impurities or contamination deposited on the electrodes.

Benefits of micro-XRF regarding investigations on elemental migration, as part of the degradation process in a Li-Ion cell have just been recognized by the community. Accordingly, there are just a limited number of studies using this technique due to the previously limited resolution and long acquisition times.

Robert et al. have illustrated the beneficial effect of combination of spatially resolved XRF with XRD for active battery electrodes under in operando conditions, by detecting and correlating the distribution of Cu and Mn in layered oxysulfide $Sr_2MnO_2Cu_{3.5}S_3$ particles [65].

The authors recent studies using confocal X-ray fluorescence on elemental deposits on the anode in course of cycling vs. nickel doped manganese spinel indicate deep penetration of both transition metals, nickel and manganese, into the pores of the anode structure [68]. Figure 13 shows elemental distribution of Ni (green) and Mn (red) on a cycled graphite anode vs. nickel doped manganese spinel. The current collector is indicated by Ga (blue), which is an impurity present in the Al current collector.

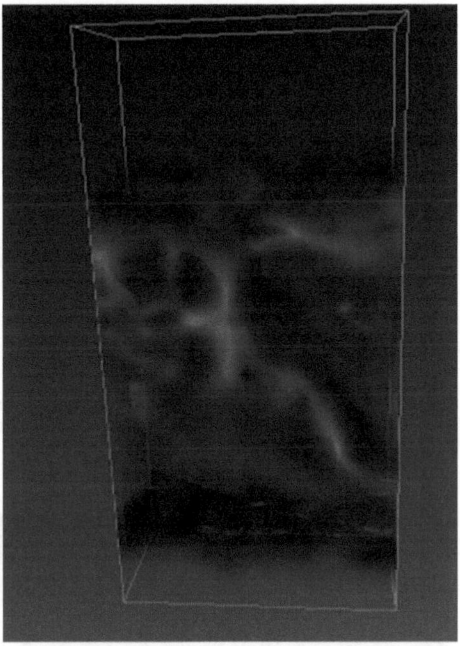

Fig. 13 Ni (gr), Ga (bl) Mn (r): 3D fluorescence images over an area of 100 × 100 × 200 μm³ with the current collector at the bottom, Ni and Mn in the bulk material obtained by confocal XRF [68]

A confocal micro-XRF XANES setup was also employed to investigate the oxidation states at the cathode/electrolyte interface [69]. This set up allows to study the oxidation state of both Ni and Mn in situ at different depth levels. It was shown, that aged (over 50 charge/discharge cycles) nickel doped manganese spinels cathodes change from 30 % Mn^{3+} and 70 %Mn^{4+} to ca. 100 % Mn^{4+} at the surface region (ca. 20 μm) although Ni is the redox active element of the material. The pristine oxygen deficient spinel fd3 m (Mn oxidation state 3.7) and the "Mn^{4+} only" spinel $P4_332$ oxidation state was used as reference. Mn^{2+} was not considered, because if present it readily dissolves in the electrolyte. Therefore, if at all present in the solid fraction should be a minor contribution. Nonetheless, XANES gives an average valence of all the chemical species present in the probed volume. Hence, it needs careful consideration of the references. Changes in the oxidation state of non-redox active elements implies further changes (chemical or structural) in the material to remain charge balanced, and can thus reveal information about aging and degradation processes. In Fig. 14 the measurement points in different depth are indicated by red dots.

Nonetheless, these are only snapshots at selected points. To capture elemental and valence distribution in a representative area/volume a rapid measurement technique is essential. The authors have recently made use of the ultrafast scanning setup available at P06, DESY, Hamburg to investigate elemental distribution on the electrode level of nickel doped manganese spinel full cells after cycling [70].

Though sub μm spatial resolution are easily achievable at this set up, lower resolution was favored to image the complete sample. The data was collected with

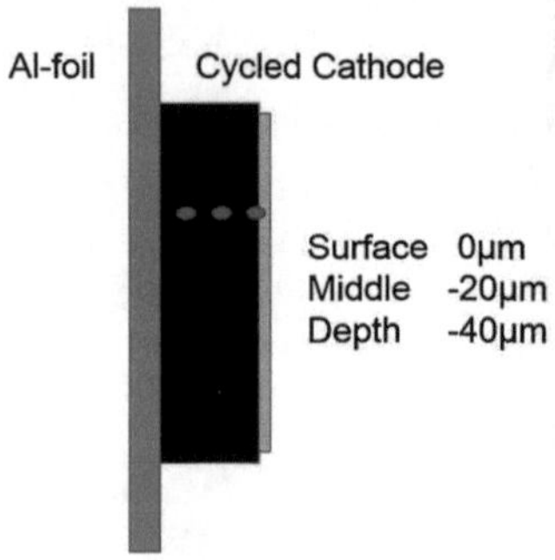

Material and position	Edge shift (Δ E (eV))	Mn valence	rel. Cont. MnIV
Aged electrode electrode/separator Interface	(13.1)	+4.0	100%
Aged electrode ca. -20µm from interface	(12.3)	+3.9	90%
Aged electrode ca. -40µm from interface	(11.9)	+3.8	80%
Pristine material	(12.0)	+3.8	80%
Reference MnIV	(12.9)	+4.0	100%
Reference MnIII	(8.5)	+3.0	0%

Fig. 14 Sketch of the measurement points for confocal XANES measurements at the electrode/separator interface and in depth and the results showing valence change of Mn at the surface after aging (according to Menzel et al. [69])

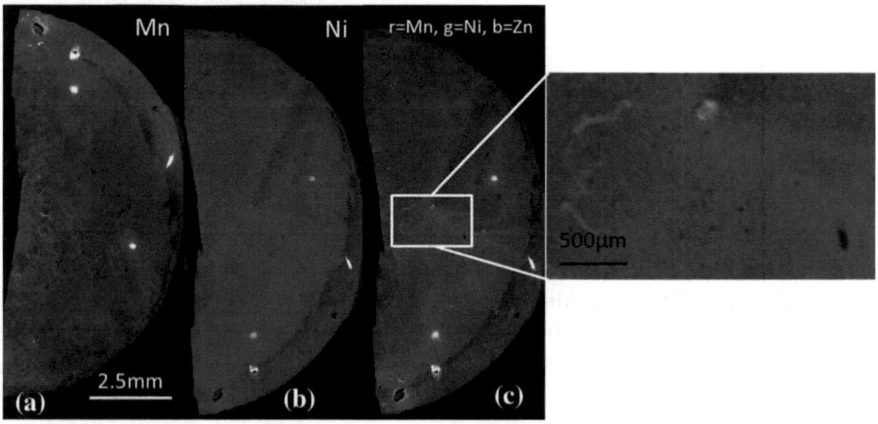

Fig. 15 µ-XRF images of nickel doped manganese spinel electrodes cycled in full cells versus graphitic anodes. (**a**) and (**b**) elemental distribution of Mn and Ni, respectively and (**c**) a correlated rgb image of Mn(r), Ni (g) and Zn(b). The data was collected at PETRAIII, DESY, Hamburg Germany using a Maia detector with 2 µm pixel size and a dwell time of 1 ms per pixel

a 2 µm step size and 1 ms exposure time per pixel, thus the acquisition was finished within a 4 h. Figure 15 shows the individual elemental distribution of Mn and Ni (a and b, respectively) and the correlated distribution in an rgb image with Mn(r), Ni (g) and Zn (b). Zn is a low concentration contamination possibly from the protective foil around the sample to avoid air exposure. The experiment revealed few but significant inhomogeneities both in the Ni and Mn distribution over the full electrode and regions with distorted stoichiometry. The example illustrates also the importance of knowledge on multiple length scales from the nano to the microscale. With a small FOV of a few 10 µm it would be a challenge to find the regions of interest (inhomogeneities) and estimate their quantity.

4 Concluding Remarks

Spatial characterization on all relevant size levels will truly help to understand the ongoing processes and functionality in active battery materials and the electrodes made thereof. Although the here presented examples are focused on conventional intercalation materials because of the good access to the transition metal absorption edges, the door is wide open to study novel battery concepts utilizing the conventional as well as more sophisticated contrast methods. Especially spatially resolved spectra-microscopy in the hard X-ray regime is a powerful tool to characterize ongoing phase transformations of hierarchical materials but the experiments are always a delicate balance between spatial resolution, measurement time and the imaged region. While highest spatial resolution is best achieved using transmission electron microscopy (TEM), X-rays are especially valuable when it comes to chemical speciation because of the high spectral resolution and larger penetration depth. X-ray microscopes around the world span an amazing range of length scales from the cm (with sub μm resolution) to the nm with many available contrasts to be chosen for the respective problem to be studied. While especially the more sophisticated contrast methods such as diffraction or diffractive imaging, phase contrast and XRF require some expert knowledge for data analysis, they can also provide unique information. Regarding the development in this field, this chapter can give only a non-comprehensive insight into the wide range of available tools using X-ray microscopy and their applications in research for lithium batteries.

References

1. Tanida H, Fukuda K, Murayama H et al (2014) RISING beamline (BL28XU) for rechargeable battery analysis. J Synchrotron Radiat 21:268–272. doi:10.1107/S1600577513025733
2. Stephenson DE, Walker BC, Skelton CB et al (2011) Modeling 3D microstructure and ion transport in porous li-ion battery electrodes. J Electrochem Soc 158:A781. doi:10.1149/1.3579996
3. Harris SJ, Lu P (2013) Effects of inhomogeneities—nanoscale to mesoscale—on the durability of li-ion batteries. J Phys Chem C 117:6481–6492
4. Ebner M, Marone F, Stampanoni M, Wood V (2013) Visualization and quantification of electrochemical and mechanical degradation in Li ion batteries. Science 342:716–720. doi:10.1126/science.1241882
5. Ebner M, Chung D-W, Garcia ER, Wood V (2014) Toruosity anisotropy in lithium-ion battery electrodes. Adv Energy Mater 4:1301278
6. Thiedmann R, Stenzel O, Spettl A et al (2011) Stochastic simulation model for the 3D morphology of composite materials in Li–ion batteries. Comput Mater Sci 50:3365–3376. doi:10.1016/j.commatsci.2011.06.031
7. Ender M, Joos J, Weber A, Ivers-Tiffée E (2014) Anode microstructures from high-energy and high-power lithium-ion cylindrical cells obtained by X-ray nano-tomography. J Power Sources 269:912–919

8. Shao M (2014) In situ microscopic studies on the structural and chemical behaviors of lithium-ion battery materials. J Power Sources 270:475–486. doi:10.1016/j.jpowsour.2014.07.123
9. Andrews JC, Weckhuysen BM (2013) Hard X-ray spectroscopic nano-imaging of hierarchical functional materials at work. ChemPhysChem 14:3655–3666
10. Shapiro DA, Yu Y-S, Tyliszczak T et al (2014) Chemical composition mapping with nanometre resolution by soft X-ray microscopy. Nat Photonics 1–5. doi:10.1038/nphoton.2014.207
11. Fitzgerald R (2000) Phase -sensitive X-ray imaging. Phys Today 23:23–26
12. Burvall A, Lundström U, Takman Pac et al (2011) Phase retrieval in X-ray phase-contrast imaging suitable for tomography. Opt Express 19:10359–10376
13. Schroer CG, Cloetens P, Rivers M et al (2004) High-resolution 3D imaging microscopy using hard X-rays. MRS Bull 29:157–165
14. Eastwood DS, Bradley RS, Tariq F et al (2014) The application of phase contrast X-ray techniques for imaging Li-ion battery electrodes. Nucl Instrum Methods Phys Res Sect B Beam Interact with Mater Atoms 324:118–123. doi:10.1016/j.nimb.2013.08.066
15. Lin C-N, Chen W-C, Song Y-F et al (2014) Understanding dynamics of polysulfide dissolution and re-deposition in working lithium-sulfur battery by in-operando transmission X-ray microscopy. J Power Sources 263:98–103
16. Bunker G (2010) Introduction to XAFS: a practical guide to X-ray absorption fine structure spectroscopy, 1st edn. Cambridge University Press, Cambridge
17. Koningsberger DC, Prins R (1988) X-Ray absorption: principles, applications, techniques of EXAFS, SEXAFS and XANES. Wiley-Interscience, New York
18. Shearing P, Wu Y, Harris SJ, Brandon N (2011) In situ X-Ray spectroscopy and imaging of battery materials. Electrochem Soc Interface 20:43–47
19. McBreen J, O'Grady WE, Pandya KI (1988) EXAFS: a new tool for the study of battery and fuel cell materials. J Power Sources 22:323–340. doi:10.1016/0378-7753(88)80027-2
20. Meirer F, Cabana J, Liu Y et al (2011) Three-dimensional imaging of chemical phase transformations at the nanoscale with full-field transmission X-ray microscopy. J Synchrotron Radiat 18:773–781. doi:10.1107/S0909049511019364
21. Liu Y, Meirer F, Williams Pa et al (2012) TXM-wizard: a program for advanced data collection and evaluation in full-field transmission X-ray microscopy. J Synchrotron Radiat 19:281–287. doi:10.1107/S0909049511049144
22. Dinnebier RE, Billinge SJL (2008) Powder diffraction—theory and practice. RSC Publishing
23. Reinsberg K-G, Schumacher C, Zastrow S et al (2013) Investigation on the homogeneity of pulsed electrochemically deposited thermoelectric films with synchrotron μ-XRF, μ-XRD and μ-XANES. J Mater Chem A 1:4215–4220. doi:10.1039/c3ta01480k
24. Janssens KHA, Adams FCV, Rindby A (1999) Microscopic X-ray fluorescence analysis. Wiley, New York
25. Radtke M, Buzanich G, Curado J et al (2014) Slicing—a new method for non destructive 3D elemental sensitive characterization of materials. J Anal At Spectrom 29:1339–1344. doi:10.1039/C4JA00085D
26. Scharf O, Ihle S, Ordavo I et al (2011) Compact pnCCD-based X-ray camera with high spatial and energy resolution: a color X-ray camera. Anal Chem 83:2532–2538
27. Boone MN, Garrevoet J, Tack P et al (2014) High spectral and spatial resolution X-ray transmission radiography and tomography using a color X-ray camera. Nucl Instrum Methods Phys Res A 735:644–648. doi:10.1016/j.nima.2013.10.044
28. Falcone R, Jacobsen C, Kirz J et al (2011) New directions in X-ray microscopy. Contemp Phys 52:293–318. doi:10.1080/00107514.2011.589662
29. Gonzalez-Jimenez ID, Cats K, Davidian T et al (2012) Hard X-ray nanotomography of catalytic solids at work. Angew Chem Int Ed Engl 51:11986–11990. doi:10.1002/anie.201204930
30. Sakdinawat A, Attwood D (2010) Nanoscale X-ray imaging. Nat Photonics 4:840–848. doi:10.1038/nphoton.2010.267

31. De Jonge MD, Vogt S (2010) Hard X-ray fluorescence tomography–an emerging tool for structural visualization. Curr Opin Struct Biol 20:606–614. doi:10.1016/j.sbi.2010.09.002
32. De Nolf W, Janssens K (2009) Micro X-ray diffraction and fluorescence tomography for the study of multilayered automotive paints. Surf Interface Anal 42:411–418
33. Larson BC, Yang W, Ice GE et al (2002) Three-dimensional X-ray structural microscopy with submicrometre resolution. Nature 415:887–890. doi:10.1038/415887a
34. Schropp A, Hoppe R, Patommel J et al (2012) Hard x-ray scanning microscopy with coherent radiation: beyond the resolution of conventional X-ray microscopes. Appl Phys Lett 100:253112. doi:10.1063/1.4729942
35. Dam HF, Andersen TR, Pedersen EBL et al (2014) Enabling flexible polymer tandem solar cells by 3D ptychographic imaging. Adv Energy Mater n/a–n/a. doi:10.1002/aenm.201400736
36. Holler M, Diaz A, Guizar-Sicairos M et al (2014) X-ray ptychographic computed tomography at 16 nm isotropic 3D resolution. Sci Rep 4:3857. doi:10.1038/srep03857
37. Cotte M, Susini J, Dik J, Janssens K (2010) Synchrotron-based X-ray absorption spectroscopy for art conservation: looking back and looking forward. Acc Chem Res 43:705–714
38. Andrews JC, Almeida E, Van Der Meulen MCH et al (2010) Nanoscale X-Ray microscopic imaging of mammalian mineralized tissue. Microsc Microanal 16:327–336
39. Cocco AP, Nelson GJ, Harris WM et al (2013) Three-dimensional microstructural imaging methods for energy materials. Phys Chem Chem Phys 15:16377–16407. doi:10.1039/c3cp52356j
40. Liu Y, Meirer F, Wang J et al (2012) 3D elemental sensitive imaging using transmission X-ray microscopy. Anal Bioanal Chem 404:1297–1301. doi:10.1007/s00216-012-5818-9
41. Nelson J, Misra S, Yang Y et al (2012) In operando X-ray diffraction and transmission X-ray microscopy of lithium sulfur batteries. J Am Chem Soc 134:6337–6343. doi:10.1021/ja2121926
42. Kanitpanyacharoen W, Parkinson DY, De Carlo F et al (2013) A comparative study of X-ray tomographic microscopy on shales at different synchrotron facilities: ALS, APS and SLS. J Synchrotron Radiat 20:172–180. doi:10.1107/S0909049512044354
43. Yuan L-X, Wang Z-H, Zhang W-X et al (2011) Development and challenges of LiFePO4 cathode material for lithium-ion batteries. Energy Environ Sci 4:269. doi:10.1039/c0ee00029a
44. Shearing PR, Howard LE, Jørgensen PS et al (2010) Characterization of the 3-dimensional microstructure of a graphite negative electrode from a Li-ion battery. Electrochem Commun 12:374–377. doi:10.1016/j.elecom.2009.12.038
45. Shearing PR, Brandon NP, Gelb J et al (2012) Multi length scale microstructural investigations of a commercially available Li-ion battery electrode. J Electrochem Soc 159:A1023–A1027
46. Eastwood DS, Yufit V, Gelb J et al (2014) Lithiation-induced dilation mapping in a lithium-ion battery electrode by 3D X-Ray microscopy and digital volume correlation. Adv Energy Mater 4:1300506
47. Channagiri Sa, Nagpure SC, Babu SS et al (2013) Porosity and phase fraction evolution with aging in lithium iron phosphate battery cathodes. J Power Sources 243:750–757. doi:10.1016/j.jpowsour.2013.06.023
48. Chen-Wiegart YK, Liu Z, Faber KT et al (2013) 3D analysis of a LiCoO2–Li (Ni1/3Mn1/3Co1/3)O2 Li-ion battery positive electrode using x-ray nano-tomography. Electrochem Commun 28:127–130. doi:10.1016/j.elecom.2012.12.021
49. Chao S-C, Yen Y-C, Song Y-F et al (2011) In situ transmission X-ray microscopy study on working SnO anode particle of Li-ion batteries. J Electrochem Soc 158:A1335–A1339
50. Chao S-C, Yen Y-C, Song Y-F et al (2010) A study on the interior microstructures of working Sn particle electrode of Li-ion batteries by in situ X-ray transmission microscopy. Electrochem Commun 12:234–237. doi:10.1016/j.elecom.2009.12.002
51. Weker JN, Liu N, Misra S et al (2014) In situ nanotomography and operando transmission X-ray microscopy of micron-sized Ge particles. Energy Environ Sci 7:2771. doi:10.1039/C4EE01384K

52. Wang J, Chen-Wiegart YK, Wang J (2013) In situ chemical mapping of a lithium-ion battery using full-field hard X-ray spectroscopic imaging. Chem Commun (Camb) 49:6480–6482. doi:10.1039/c3cc42667j
53. Ebner M, Geldmacher F, Marone F et al (2013) X-Ray tomography of porous, transition metal oxide based lithium ion battery electrodes. Adv Energy Mater 3:845–850
54. Wang J, Chen-Wiegart YK, Wang J (2014) In situ three-dimensional synchrotron X-ray nanotomography of (de)lithiation processes in tin anodes. Angew Chem Int Ed 53:4460–4464
55. Zielke L, Hutzenlaub T, Wheeler DR et al (2014) A combination of X-ray tomography and carbon binder modeling: reconstructing the three phases of LiCoO2 Li-ion battery cathodes. Adv Energy Mater 4:1301617
56. Boesenberg U, Meirer F, Liu Y et al (2013) Mesoscale phase distribution in single particles of LiFePO4 following lithium deintercalation. Chem Mater 25:1664–1672. doi:10.1021/cm400106k
57. Chueh WC, El Gabaly F, Sugar JD et al (2013) Intercalation pathway in many-particle LiFePO (4) electrode revealed by nanoscale state-of-charge mapping. Nano Lett 13:866–872. doi:10.1021/nl3031899
58. Wang J, Chen-Wiegart YK, Wang J (2014) In operando tracking phase transformation evolution of lithium iron phosphate with hard X-ray microscopy. Nat Commun 5:1–10. doi:10.1038/ncomms5570
59. Chen-Wiegart YK, Wang J, Wang J (2013) Development of in situ full field spectroscopic imaging analysis and application on Li-ion battery using transmission X-ray microscopy. Proceediongs of the SPIE 8851, X-Ray Nanoimaging Instruments Methods. p 88510C
60. Yang F, Liu Y, Martha SK et al (2014) Nanoscale morphological and chemical changes of high voltage lithium-manganese rich NMC composite cathodes with cycling. Nano Lett 14:4334–4341. doi:10.1021/nl502090z
61. Sun Y-K, Chen Z, Noh H-J et al (2012) Nanostructured high-energy cathode materials for advanced lithium batteries. Nat Mater 11:942–947. doi:10.1038/nmat3435
62. Poulsen HF, Jensen DJ, Vaughan GBM (2004) Three-dimensional X-Ray diffraction microscopy using high-energy X-Rays. MRS Bull 29:166–169
63. Bleuet P, Welcomme E, Dooryhée E et al (2008) Probing the structure of heterogeneous diluted materials by diffraction tomography. Nat Mater 7:468–472. doi:10.1038/nmat2168
64. Liu J, Kunz M, Chen K et al (2010) Visualization of charge distribution in a lithium battery electrode. J Phys Chem Lett 1:2120–2123. doi:10.1021/jz100634n
65. Robert R, Zeng D, Lanzirotti A et al (2012) Scanning X-ray fluorescence imaging study of lithium insertion into copper based oxysulfides for Li-Ion batteries. Chem Mater 24:2684–2691
66. Singer A, Ulvestad A, Cho H et al (2014) Noequilibrium structural dynamics of nanoparticles in LiNi1/2Mn3/2O4 cathode under operando conditions. Nano Lett, ASAP
67. Ulvestad A, Singer A, Cho H-M et al (2014) Single Particle Nanomechanics in operando batteries via lensless strain mapping. Nano Lett. doi:10.1021/nl501858u
68. Fittschen U, Boesenberg U, Falk M et al (2014) Confocal XRF imaging of elemental deposition of Mn, Ni and Cu on the graphite anode in cycled LiNi0.5Mn1.5O4 /graphite full cells. Anka Annual Report
69. Menzel M, Schlifke A, Falk M et al (2013) Surface and in-depth characterization of lithium-ion battery cathodes at different cycle states using confocal micro-X-ray fluorescence-X-ray absorption near edge structure analysis. Spectrochim Acta Part B At Spectrosc 85:62–70. doi:10.1016/j.sab.2013.04.001
70. Boesenberg U, Falk M, Fittschen UEA et al (2015) Correlation between chemical and morphological heterogeneities in $LiNi_{0.5}Mn_{1.5}O_4$ spinel composite electrodes for lithium-Ion batteries determined by Micro-X-ray Fluorescence Analysis, Chemistry of Materials, doi:10.1021/acs.chemmater.5b00119

Hazard Characterizations of Li-Ion Batteries: Thermal Runaway Evaluation by Calorimetry Methodology

Yih-Wen Wang and Chi-Min Shu

1 Introduction

As society becomes more dependent on electronic equipment, the development of capable energy storing systems continues to grow at a rapid rate supported by the crucial demand all over the world. The most familiar products are consumer electronic devices such as computers, camcorders, cellular phones, power portable tools, industrial movement of materials handling equipment, and automotive applications. Rechargeable batteries have a lower total cost of use and fewer environmental impacts than the other disposable ones. As an energy-storage device, rechargeable batteries are characterized by high power density, flat discharge profiles, sound low temperature performance, addition to their rechargeable-ability. More recently, rechargeable batteries have received renewed interest as a power source for electric vehicles (EVs) and hybrid electric vehicles (HEVs). The advantage of the efforts and benefits is the automotive industry achieving high voltage energy storage systems. In addition, large-scale applications of rechargeable battery packs in EVs to reduce CO_2 or other pollutants' emissions are considered a green system from an environmentally friendly power source [1, 2]. With the rapid development of new energy vehicles, power battery industries swiftly have become a popular investment.

As significant performance improvements have been made with rechargeable battery systems, such as the lead-acid, the nickel-cadmium, nickel-hydrogen, lithium-ion batteries (LIBs), polymer Li-ion batteries and so on, have been introduced

Y.-W. Wang (✉)
Department of Occupational Safety and Health, College of Public Health, China Medical University, Taichung, Taiwan, ROC
e-mail: evenwang@mail.cmu.edu.tw

C.-M. Shu
Department of Safety, Health, and Environmental Engineering, National Yunlin University of Science and Technology, Yunlin, Taiwan, ROC

Z. Zhang and S.S. Zhang (eds.), *Rechargeable Batteries*,
Green Energy and Technology, DOI 10.1007/978-3-319-15458-9_15

into commercial use under advanced development [1, 3]. Much of the development work on new systems has been supported by the need for high-performance batteries for portable consumer electronic applications and automotive vehicles. The typical performance, characteristics and applications of progressive secondary batteries are summarized in Table 1.

The chemistry, performance, cost and safety characteristics vary among the different commercial rechargeable batteries. The LIB is one of the most popular battery types so that more than 90 % of battery packs use Li-ion cells for portable electronics, such as cellular phones, laptop computers and camcorders, and they have one of the best energy-to-weight ratios (maximum power density), large temperature working range, better life performance such as no memory effect and a slow loss of charge when not in use, and wide availability. The development trends of the battery industry guide its investment strategy, and provide a reference for venture capital institutions. Lead-acid batteries account for half the demand of mainstream secondary batteries. LIBs have emerged in the last decade to capture over half of the sales value of the secondary consumer market. The market share of LIBs, which are mainly used for smart phones, tablet computer, transportation vehicles, mobile power pack, wearable electronics, stand-by applications, and un-interruptable power system (UPS), was 15.6 billion USD in 2013. The penetration rate of global LIB capacity shipped for use in electric motors is moving forward [4]. Accompanying the needs of increasing battery capacity, the development of large-scale LIBs has continued. Systems adopting the load of a great quantity of cells could improve not only the heat dissipation but also simplify the management of the battery. However, in the face of large-scale battery development, stability and uniformity of the cell materials affect significantly the chemical, electrochemical and safety performance. Therefore, understanding the properties and chemical/electrochemical hazards of the cell components is the primary purpose. In addition, specifications for cell thermal abuse have raised some safety issues for a large number of consumers in electric products. LIBs are known for high power capacity, high charging and discharging efficiency, and long life span. They are now commonly used in consumer, industrial and military electronics applications, especially digital cameras, mobile phones and notebooks. LIBs can be categorized by different anode materials, such as Li-Co, Li-Mn, and $LiMPO_4$ (M = Fe, Mn, etc.), which have become popular in recent years and graphite is used as cathode material. The electrolytes in LIBs typically consist of lithium salts, such as $LiPF_6$, $LiBF_4$ and $LiClO_4$, in an organic solvent that conducts lithium-ions, acting as a carrier between the cathode and the anode when the battery carries an electric current through an external circuit. Li-Co batteries, which account for the majority on market, should have circuit protection to ensure safety. On top of that, the rechargeable battery packs have suffered fire or explosive accidents that caused shortages in global supply, and the manufacturers are now thus actively seeking new development and inherently safer design of cell materials.

Table 1 Typical performance, characteristics and applications of select rechargeable batteries

	Lead-acid battery	Nickel-cadmium battery	Nickel-hydrogen battery	Lithium-ion battery	Polymer Li-ion battery
Commodification	1890	1956	1990	1991	1999
Cell voltage (V)	2	1.25	1.25	3.6	3.6
Power density (Wh/kg)	30–50	50–80	60–120	110-160	100–130
Cycle life (cycles)	200–300	500	500	500–1000	500
Charge time (h)	8–16	1	2–4	2–4	2–4
Operating temperature (°C)	−20–60	−40–60	−20–60	−20–60	0–60
Pollution	Yes	Yes	No	No	No
Applications	Automobile starting, lighting and ignition, and large backup power supplies	Industrial, and emergency power applications, communication equipment	Applications including HEVs and EVs	Portable and consumer electronic equipment, electric vehicles, and space applications	Portable and consumer electronic equipment, electric vehicles, and space applications
Characteristics	Popularity and dominant position are its low cost with good performance	Long cycle life, little maintenance, high discharge rates and low temperatures	Long cycle life, long life, high energy density, can tolerate overcharge, and environmentally acceptable and recyclable	Excellent cycle life, high specific energy, and high energy density	Safer than LIB, high specific energy, and high energy density
Limitations	Relatively low cycle life, limited energy density, poor charge retention and storability, and lead pollution	Lower energy density and more expensive	High cost, long cycle life at low depth of discharge, self-discharge proportional to H_2 pressure and temperature	Lower charge rate and high initial cost	Relatively low cycle life to LIBs

Both electrical disconnection of the electroactive species and electronic short circuits must also be avoided. In addition, thermal abuse should be taken into account in safe battery design. The organic solvent or polymer electrolytes that are typically used in LIBs are not stable in the presence of lithium activities while using elemental anodes in contact with electrolytes containing cationic groups. Safety issues related to thermal runaway involve a complicated process involving chemistry, material properties and engineering design, and should be considered from electrode materials and electrolyte to cell design. However, many different electrode materials, electrolytes and battery types show different thermal runaway behaviors, and these exact reactions at different thermal runaway stages are under investigation, depending on the component materials, cathode, anode and electrolyte. The reactions also are dominated by the state of charge, discharging rate, etc. To disclose all of these reactions requires our sustained research.

In recent years, numerous fires or explosive incidents have been caused by LIBs, because of their short-circuiting, crashes, or thermal instabilities worldwide. It can be seen that the fires or explosions are caused by thermal abuse of the battery pack, which is where the runaway reactions were triggered. It is commonly thought that the LIBs failures are related to the flammability of the electrolyte, the rate of charge and/or discharge, and the engineering of the battery pack. It can rupture, ignite, or explode when exposed to high temperature or short-circuiting. The adjacent cells may also then heat up and fail, in some cases, causing the entire battery to ignite or rupture. Nevertheless, numerous battery failures or explosions with electronic products have happened worldwide, and faulty LIB packs may be the reason. Table 2 lists some accidents researched from the Internet and newspapers. Many laptop brands, such as Sony, HP and Dell, have recalled their products to rule out potential explosive hazards, and even Apple has investigated explosive hazards in its iPhone smartphone. LIBs' potential unstable properties also threaten the EVs. For these reasons incidents of overheating and of batteries catching fire during device usage have caused safety concerns on LIBs. The thermal abuses in the electrode-electrolyte reactions occurr at elevated temperatures under conditions of heating, crushing, or short-circuiting [5].

Li-ion batteries, unlike other rechargeable batteries, have a potentially hazardous flammable electrolyte with a powerful oxidizer, and require strict safer design during charging or operation. This chapter will verify the safety characteristics and hazards of commercial LIBs. Overcharge, external heating, short-circuit/crush are the main reasons to result in explosion or rupture of LIBs due to electrochemical reactions and thermal behaviors that could compromise unsafe operations. Accompanied with broad and new applications, however, large-scale LIBs should consider safety more seriously. Safety test standards and testing methodologies are introduced to cell design and advanced materials.

Table 2 Case of fire or explosion incidents caused by LIBs

Date	Cell manufacturer	Product	Brand	Failure causes/results
August 2006	Sony	Laptop computer	Dell, Apple, and Lenovo	Recall of 960 million LIBs
September 2006	Sony	Laptop computer	Lenovo, China	Recall of 526,000 LIBs
December 2006	Sanyo	Mobile phone	Mitsubishi electric, Japan	Recall of ca. 1.3 million of the battery packs
March 2007	Sanyo	Laptop computer	Lenovo, China	Recall of 205,000 9-cell lithium-ion batteries
April 2007	Sanyo	Mobile phone	NTT and DoCoMo, Japan	Recall of 1.3 million cellphone batteries
June 2007	NEC	Mobile phone	KDDI, Japan	Recall of 69,000 mobile phones batteries
July 2007	Sony	Laptop computer	Toshiba, Japan	Recall of 1400 battery packs
August 2007	Panasonic	Mobile phone	Nokia, Finland	Recall of 46 million mobile phone batteries
March 2008	NEC	Mobile phone	Au, Japan	Cause three burn injuries
August 2009	n.a.	Mobile phone	Apple, USA	Apple recalls on first generation iPhones
July 2010	Sony	Laptop computer	Sony, Japan	Recall of 535,000 Vaio laptops
April 2011	BYD	EV taxi	Zotye, China	LIB short-circuiting and ignition
May 2011	LG	Laptop computer	HP, USA	Recall of 54,000 mobile phone batteries
November 2011	A123	PHEV	GM, USA	Chevrolet Volt plug-in hybrid caught fire after a routine crash test
April 2012	n.a.	GM motors	GM, USA	Explosion at General Motors tech center battery laboratory and injures two workers

(continued)

Table 2 (continued)

Date	Cell manufacturer	Product	Brand	Failure causes/results
July 2012	Samsung	Mobile phone	Samsung, Korea	Rid the phone of water damage and that act caused the explosion
November 2012	n.a.	Tablet	Google, USA	Nexus 7 began smoking while plugged in and charging on its original factory charger
January 2013	Yuasa	Airplane	Boeing, USA	At least four aircraft suffered from electrical system problems stemming from LIBs
July 2013	Samsung	Mobile phone	Samsung, Korea	Crush and fie
October 2013	Panasonic	EV	Tesla, USA	Hitting a minor part of the road barrier which ignited the car's fuel cell
October 2013	Panasonic	EV	Tesla, USA	LIB packs crash and fire
November 2013	Panasonic	EV	Tesla, USA	LIB packs crash and fire
February 2014	Panasonic	EV	Tesla, USA	Fire
March 2014	Elentec	Mobile phone	Samsung, Korea	LIB packs expanded
April 2014	Bosch	EV	Nissan, Japan	Charging unit damaged a Leaf EV and led to a fire
April 2014	n.a.	Laptop computer	Leveno, china	Voluntarily recalled certain lithium-ion batteries
July 2014	Panasonic	EV	Tesla, USA	Stolen Model S split in half, catches fire after massive crash

n.a. not available

2 Chemical/Electrochemical Reaction with Rechargeable Batteries

The applications of LIB have attracted wide attention, from small cells in electronic products to large-scale devices in power tools and EVs. However, the chemistry, electrochemistry and thermal instability of LIB components must also be considered. A cell consists of cathode (positive electrode) and anode (negative electrode) immersed in a dissociated salt electrolyte solution in which Li^+, H^+ ions, etc. transfer between the two electrodes will cause an electrochemical reaction. These materials characteristics require a better understanding of material issues so as to master their physical, electrochemical/chemical properties, thermal stability and more specifically reactivity. The cathode materials in LIBs, such as $LiCoO_2$, $LiMnO_2$, $Li[Ni_xCo_yMn_z]O_2$ and $LiFePO_4$, are thermally unstable and release oxygen at elevated temperatures, which induces an autocatalytic reaction with the electrolytes [5].

2.1 Theory of Electrochemical Power and Energy

Power density is the ratio of the power available from a battery to its volume (W/liter). Specific power generally refers to the ratio of power to mass (W/kg). Comparison of electrical power (P) to cell mass is more common. Defined by voltage (V) and current (I) [1],

$$P = VI \tag{1}$$

Since

$$V = IR \tag{2}$$

$$P = I^2R \tag{3}$$

Power also can be described by energy (E) emitted per unit of time (t) [6]:

$$E = \int_0^{\Delta t} IV(t)dt = \int_0^{Q} V(q)dq \tag{4}$$

where Q is the total charge per unit weight (Ah kg^{-1}) or per volume (Ah L^{-1}) transferred by the current I = dq/dt on charge/discharge. The features of merit of a rechargeable cell are its density (specific and volumetric) of stored energy, its output power for a given charge/discharge current, and its calendar life.

$$P(q) = V(q)I \tag{5}$$

Thus

$$P = Et = VIt = QV \tag{6}$$

When a current flows through a device it induces Joule heating. In a LIB, the electrical resistance consists of the resistance of the positive and negative electrodes, electrolyte and separator. In each region, the current passes through different phases, hence Joule heating should be considered in all the phases. Joule heating is always positive and contributes to a rise in temperature. The reaction directions for charge and discharge cycles are opposite to each other, thus entropy change is endothermic during charge cycle and exothermic during discharge cycle. To enhance the energy density of a rechargeable battery with solid electrodes to where it can compete with the internal combustion engine, it will be necessary to find a way to raise V while retaining a large cathode Q at high currents I, such as the layered oxides $LiCoO_2$ and $LiNiO_2$ having an intrinsic voltage limit. Here we address the safety issues related to strategies for individual rechargeable battery cells. The management of the battery modules becomes more complex, as does the cost, the larger the number of cells required for a given battery application [7].

In an electrochemical cell, reactions essentially take place at two sites in the device. The overall reaction in the cell is given by addition of these two half-cell reactions [1].

$$aA + bB \leftrightarrow cC + dD \tag{6}$$

The change in the standard free energy ΔG^0 of this reaction is expressed as,

$$\Delta G^0 = -nFE^0 \tag{7}$$

where F is a constant known as the Faraday (96,487 coulombs), and E^0 the standard electromotive force.

When conditions are other than in the standard state, the voltage E of a cell is given by the Nernst equation,

$$E = E^0 - \frac{RT}{nF} \ln \frac{a_C^c a_D^d}{a_A^a a_B^b} \tag{8}$$

where a is the activity of relevant species, R is the gas constant, and T is the absolute temperature.

The change in the standard free energy G^0 of a cell reaction is the driving force which enables a battery to deliver electrical energy to an external circuit. The measurement of the electromotive force, incidentally, also makes available data on changes in free energy, entropies and enthalpies together with activity coefficients, equilibrium constants, and solubility products.

The reaction heat generation when the LIB is undergoing thermal runaway, mainly including solid electrolyte interface (SEI) decomposition, electrode reaction with electrolyte, and electrode decomposition. For the individual reaction, the heat generation can be expressed as Eq. (9) [7],

$$Q = \frac{dH}{dt} = \Delta H M^{n} A \exp\frac{-E_a}{RT} \tag{9}$$

where H is the heat of reaction, M is the mass of reactant, n is the reaction order, A is a pre-exponential factor, and E_a is the activation energy. The total heat generation is the summarized values of all the reactions.

2.2 Electrochemical Reaction Mechanisms of LIB

The three participants in the electrochemical reactions in a LIB are the anode, cathode and electrolyte. A LIB is also known as a swing battery or rocking chair battery as two-way movement of lithium-ions between anode and cathode through the electrolyte evolves during charge and discharge process. When a lithium-based battery is discharging, the lithium-ions are extracted from the anode and inserted into the cathode, while electrons flow in through an external electrical circuit. The reverse process occurs when the cell is discharged. Electrical energy can only be extracted if electrons flow through a closed external circuit. The overall reaction has its limits. Overdischarge will supersaturate lithium metal oxide, leading to the production of lithium oxide and often to unrein highly energetic reactions. The more lithium the electrodes can take in, the more total energy the battery can store, and the longer it can last. Liquid electrolytes in LIBs generally consist of Li salts in an organic solvent, which conduct Li^+ ions, acting as carrier between the cathode and the anode when a battery passes an electric current through an external circuit. Table 3 shows the batteries are based on the graphite (C)/$LiPF_6$ in ethylene carbonate–dimethyl carbonate ((EC–DMC)/$LiMO_2$ sequence and operate on a process. Unfortunately, organic solvents are easily decomposed on anodes during charging. However, when appropriate organic solvents are used as the electrolyte, the solvent is decomposed in initial charging and forms a solid layer called the solid electrolyte

Table 3 General chemical reaction mechanisms of the LIBs

Reaction	Mechanism	
Cathode half-reaction	$LiMO_2 \leftrightarrow Li_{1-x}MO_2 + xLi + xe^-$	(10)
Anode half-reaction	$nC + xLi^+ + xe^- \leftrightarrow Li_xC_n$	(11)
Full cell reaction	$LiMO_2 + nC \leftrightarrow Li_{1-x}MO_2 + Li_xC_n$	(12)

Note Where M is Co, Ni, Fe, or W, etc., the cathode materials may be $LiCoO_2$, $LiNiO_2$, $LiMn_2O_4$, $LiFeO_2$, or $LiWO_2$, etc., and the anode materials may be $LixC_6$, TiS_2, WO_3, NbS_2, or V_2O_5, etc.

interphase (SEI), which is electrically insulating yet sufficiently conductive to Li ions. Furthermore, there is a constant risk of electrolyte decomposition caused by high temperatures or overcharging [7, 8].

The electrodes of LIBs undergo exothermic reactions with the cell electrolyte when exposed to elevated temperatures. For example, layered Li_xCoO_2 is used as the cathode presently in most of the lithium-ion cells due to its easy synthesis and excellent reversible lithium extraction/insertion properties. $LiCoO_2$ is one of the most common commercially used cathode materials offering good capacity and high voltage. The proposed autocatalytic reaction mechanisms of a Li_xCoO_2 cathode are as follows [5, 9–11]:

$$Li_xCoO_2 \rightarrow xLiCoO_2 + \frac{(1-x)}{3}Co_3O_4 + \frac{(1-x)}{3}O_2 \tag{13}$$

$$Co_3O_4 \xrightarrow{400K,\,Solvent} 3CoO + 0.5O_2 \tag{14}$$

$$Co_3O_4 + 0.5C \xrightarrow{400K} 3CoO + 0.5CO_2 \tag{15}$$

$$CoO \xrightarrow{473K,\,Solvent} Co + 0.5O_2 \tag{16}$$

$$2LiCoO_2 + CO_2 \xrightarrow{520K,\,Solvent} Li_2CO_3 + 2Co + 0.5O_2 \tag{17}$$

$$2.5O_2 + C_3H_4O_3 \xrightarrow{\Delta} 3CO_2 + 2H_2O + Heats \tag{18}$$

The reaction in a solvent predominantly composed of Li_2CO_3 from decomposition of lithium alkyl carbonates and the combustion of solvent reacting with the released oxygen may generate the heat of the cell. Because of the cathode-electrolyte reactions, the generation of heat can be calculated from the battery exposed to a high temperature environment. The autocatalytic reaction of $Li_{0.5}CoO_2$ in the presence of EC solvent to proceed to CoO is as follows [7, 9],

$$Li_{0.5}CoO_2 + 0.1C_3H_4O_3 \rightarrow 0.5LiCoO_2 + 0.5CoO + 0.3CO_2 + 0.2H_2O \tag{19}$$

Graphite remains the pre-eminent anode material for lithium-ion batteries because of its good performance. Thermal reactions at salt-based SEIs proceed via surface salt decomposition and yield mainly LiF; the reaction in predominantly solvent based SEIs proceeds via decomposition of lithium-alkyl carbonates to Li_2CO_3. The reaction mechanism of anode for lithium-ion cell is expressed as follows [12, 13]:

- Thermal decomposition of lithium alkyl carbonate

$$2Li + (CH_2OCO_2Li)_2 \rightarrow 2Li_2CO_3 + C_2H_4 \tag{20}$$

$$(CH_2OCO_2Li)_2 \rightarrow Li_2CO_3 + C_2H_4 + CO_2 + 0.5O_2 \tag{21}$$

$$ROCO_2Li \rightarrow R^{\cdot} + H^{\cdot} + LiCO_3 \tag{22}$$

$$R^{\cdot} + O_2 \rightarrow ROO^{\cdot} + H^{\cdot} \rightarrow ROOH \tag{23}$$

- Decomposition of $LiPF_6$ and thermal reaction of PF_5 with intercalated Li ions

$$LiPF_6 \rightarrow LiF + PF_5 \tag{24}$$

$$PF_5 + Li \rightarrow Li_{1-x}PF_{5-x} + xLiF \tag{25}$$

- Reactions of PF_5 with dimethyl carbonate (DEC), which HF is significantly lower the stability of solutions

$$\begin{aligned} PF_5 + C_5H_{10}O_3 &\rightarrow C_2H_5OCOOPF_4 + C_2H_4 + HF \\ &\rightarrow C_2H_5OCOOPF_4 + C_2H_5F \end{aligned} \tag{26}$$

$$C_2H_5OCOOPF_4 \rightarrow PF_3O + C_2H_5F + CO_2 \tag{27}$$

- Thermal reaction of HF with Li_2CO_3 in SEI layer

$$2HF + Li_2CO_3 \rightarrow 2LiF + H_2CO_3 \tag{28}$$

- Thermal reaction of HF with intercalated Li ions

$$-CH_2{-}CF_2- \rightarrow -CH{=}CF- + HF \tag{29}$$

$$HF + Li \rightarrow LiF + 0.5H_2 \tag{30}$$

In practice, organic solvents are easily decomposed on the anodes during charging and are not stable in the presence of high lithium activities. There is also a potential risk of electrolyte decomposition as a result of high temperatures or overcharging. This decomposition is a common problem when using elemental lithium anodes in contact with electrolytes containing organic groups.

2.3 Adiabatic Calorimeter Testing

Heat cannot be directly measured instead of by using temperature measurement. Some calorimeters are able to directly measure the heat release rate or thermal power. There are different ways to classify the calorimetric and thermo-analytic methods. The energy release from chemical reactions is a key parameter for chemical research and related industries. When energy is generated by an exothermically induced reaction and the heat transfer to the outer surface is lesser than the generated amount, thermal runaway could occur and might cause fire or explosion accidents. Thermal stability screening is the cornerstone of reactivity screening. The most common tests employed for initial thermal stability screening are differential scanning calorimetry (DSC) and differential thermal analysis (DTA). Thermal analysis is a technique in which a physical property of a substance is measured as a function of temperature while the substance is subjected to a controlled temperature program [14, 15]. Figure 1 shows a DSC signal during the exothermic reaction of energetic chemicals.

If the initial screening tests indicate exothermic activity, then more sophisticated adiabatic thermal testing, such as by the use of accelerating rate calorimeter (ARC) and vent sizing package 2 (VSP2), can provide better quantitative data on apparent onset temperature, reaction enthalpy, instantaneous heat production as a function of temperature, maximum temperature and pressure excursions as a consequence of runaway, and additional data useful for facility design and operation [16, 17]. There are many reasons for explosions of hazardous substances. Runaway exothermic reactions are one of the most common and these can be avoided if reactions are understood. It should be remembered that specific materials can decompose exothermically so that there is a temperature at which a thermal runaway and potential subsequent explosion can be initiated. Heat from reaction will increase in rate with temperature. For a safety issue, the temperature-pressure-time (T-P-t) data of substance obtained under adiabatic conditions is needed. If the thermal stability screening tests by DSC/DTA show that a substance of interest is thermally unstable close to the temperature range of large-scale storage or handling (using criteria provided in the description of the respective screening test), further

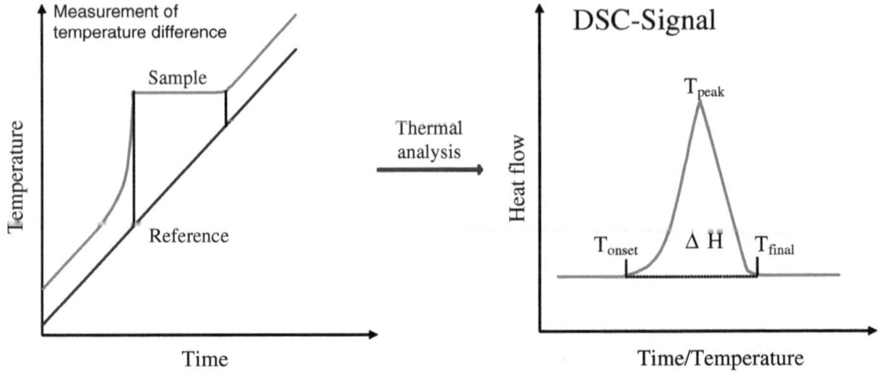

Fig. 1 Typical DSC thermogram curve by thermal analysis

Table 4 The experimental data measured from adiabatic calorimeter

Basic data	Model data
Onset temperature	Temperature of no return
Temperature versus time	Activation energy
Pressure versus time	Order of reaction
Self-heating rate	Arrhenius frequency factor
Pressure rising rate	Heat of reaction
Time to maximum adiabatic rate	Specific heat

investigations by specific, more accurate, more sensitive and more dedicated tests are recommended. In particular, ARC and VSP2 are often used within the context of thermal stability screening. The adiabatic calorimeter is used for screening purposes due to the higher-quality and more useful results. Adiabatic calorimeters could be simulating exactly a real-life scenario, pressure measurement, discriminating between runaway reaction, giving high quality thermal data, and taking whole devices such as batteries. Other thermal stability tests such as isothermal storage tests and gas evolution tests are addressed in the context of a consequence analysis. Table 4 lists the experimental data from adiabatic calorimeter. Quantities that can be determined by thermal stability tests include:

1. Precise thermal stability (including autocatalysis and inhibitor depletion, chemical in compatibility, etc.)
2. Maximum operating temperature to avoid exothermic decomposition
3. Maximum rate of reaction
4. Runaway behavior
5. Adiabatic temperature rise
6. Gas evolution (mass flux) during decomposition and/or runaway reaction
7. Behaviors of the material under external heat load such as an external fire
8. Induction time to the start of exothermic reaction; e.g., when inhibitor depletion is involved
9. Incompatibility with metals, additives, and contaminants.

To investigate the heat generation and gas evolution during decomposition and runaway, the adiabatic calorimeter simultaneously records the temperature and the pressure trajectories, which are usually proportional to the gas evolution during reaction or decomposition. The accelerating rate calorimeter provides the temperature-time and pressure-time data for tested materials under adiabatic conditions. The adiabatic calorimeter is particularly well-suited for determining a useful onset temperature T_0 for exothermic activity. Due to its greater sensitivity, the T_0 determined by the calorimeter will nearly always be significantly lower than T_0 values measured by DSC/DTA.

Generally, thermal runaway occurs when an exothermic reaction goes out of control, that is the reaction rate increases due to an increase in temperature, causing a further increase in temperature and hence a further increase in the reaction rate, which possibly results in an explosion. The temperature of an energetic chemical is determined by the temperature balance between the amount of heat generated and that dissipated by the system. The heat generation follows the exponential function and the heat dissipation keeps the linear function, as depicted in Fig. 2.

An actual test to depict thermal runway reactions is in the plots often referred to as Semenov plots, as drawn in Fig. 3. The curved line represents the heat generation due to an exothermic reaction with exponential function assuming Arrhenius law, while the straight lines represent the heat removal which is a linear function with Newton's law of cooling at different coolant temperatures. The temperature of the coolant can be sufficiently low (case of line B) or insufficiently, like in case C where thermal control is not possible under any circumstances. Line B has one tangent

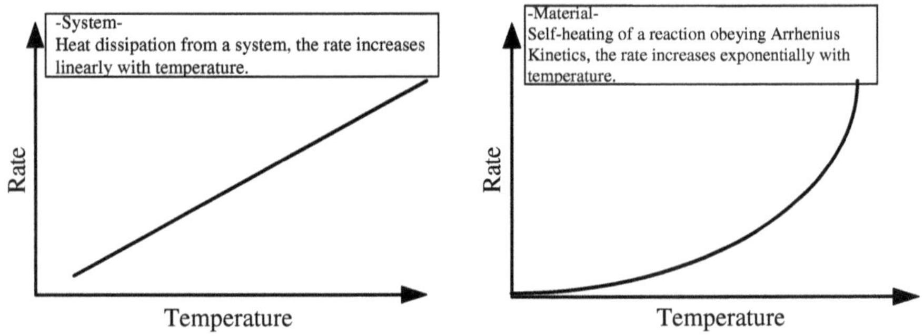

Fig. 2 Comparison of heat generation of an energetic chemical and heat dissipation of the surrounding system

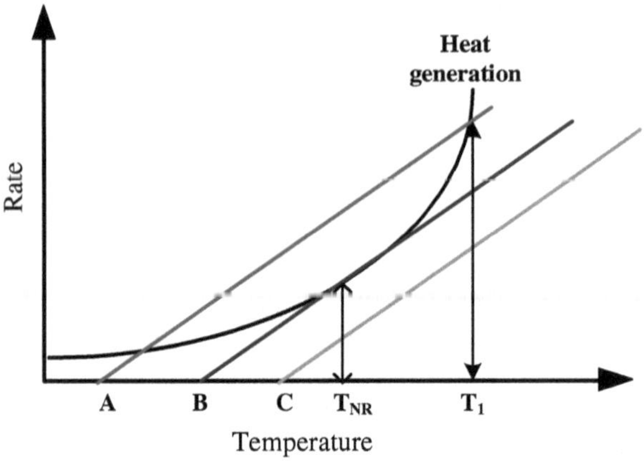

Fig. 3 Self-heating of a reaction and heat dissipation from a system, at three ambient temperatures, *A*, *B* and *C*. *A* can control the sample to temperature T_1, *B* is at the critical temperature T_{NR}, and *C* cannot control the thermal runaway

point with curved line, this point is a critical point, while heat removal is equal to heat generation, and thus, this critical equilibrium temperature is called the temperature of no return (T_{NR}). The LIB can be regarded as a reaction system, in which heat is generated by the electrochemical reactions between its compounds. And then, under different working and boundary conditions, when the battery temperature reaches to the T_{NR}, the thermal runaway will inevitably occur [7, 18, 19].

Calorimetry methodology is important for the design and testing of inherently safer batteries. Table 5 displays the testing needs of battery via various calorimeters. Once the properties of the cell components are known then the components need to be evaluated in combination with each other. An adiabatic calorimeter is often used when conducting multiphase compatibility studies. Testing full cells is an important aspect of battery safety evaluation, that include,

1. Thermal stability test.
2. Charging and discharging.
3. Designing and testing of protective devices, such as PTC (positive temperature coefficient) current limiting devices, CID (current interrupt device) which shuts off power based on internal pressure, and vent system.
4. Test uses and misuse scenarios.

Table 5 Battery safety testing needs by various calorimeters

Testing subject	Calorimeter	Data
Components	DSC, DTA, TGA[a], ARC, and VSP	Compare the relative hazards of various active materials for the purpose of thermal screening:
		1. Onset temperature, enthalpy, maximum heating rate 2. Time-temperature history effects onset temperature 3. Gas pressure generation rate, maximum gas pressure, total gas 4. Materials' properties: heat capacity, vaporization and thermal transport characteristics
Mixtures	DSC, DTA, ARC, VSP, and C80[b]	Compare the relative stability of pressure rise of electrolyte reactivity and cathode decomposition at higher temperature
Battery packs	ARC	Uncharged, charged, overcharged, and cycling: Adiabatic onset temperature, enthalpy, maximum heating rate
		Pressure rising rate, maximum pressure, vent opening pressure and temperature heat output during normal operations

[a]Thermogravimetric analyzer
[b]Reaction, isothermal and scanning calorimeter

3 Thermal Runaway Reaction of LIBs by Calorimetry Methodology

The chemical and electrochemical reactions of battery abuse are complex and dependent on the temperature and power density, which can be increased by electrical heating during use, by chemical reactions, and by external thermal effects. Heat accumulation is therefore more likely to be a problem when exothermic reactions reinforce the heat generated by the current flow. Nonetheless, thermal runaway is still possible during the charging of these batteries if they are subjected to overcharging or elevated temperatures. Heat generation increases exponentially with temperature while heat removing only dissipates linearly. Insufficient heat removing or incorrect venting devices could cause thermal runaway, fire, or explosion [20]. For example, the overcharge can be severe as additional energy is added to the cell. The cell current is forced through the cell up to some limiting voltage. Heat is generated by electrochemical reactions and by current flowing through the cell. If overheated or overcharged, LIBs may suffer thermal runaway and in case of some unpredictable accidents. Faulty chargers can affect the safety of the battery because they can destroy the battery's protection circuit [21].

In extreme cases this can lead to combustion. Short-circuiting will cause the cell to overheat and possibly to catch fire. Adjacent cells may then overheat and fail, possibly causing the battery packs to ignite or rupture. In the event of a fire, nail and crush are somewhat similar hazards for a battery. External damage allows the battery packs to discharge uniformly and generate heat locally due to short-circuit of the battery [21, 22]. Heat is generated by current flowing through the cell, and by current flowing through the nail. Initially the nail is positioned outside of the battery wall and, when the test begins, is forced through the battery wall and into the battery at a constant speed. As the nail moves forward, forming direct shorts between adjacent electrode pairs, the current flowing through the nail itself decreases.

In practice, thermal abuse conditions, such as over charge/discharge, short circuits, and exposure to higher temperature, accelerate the rate of heat generation in a LIB. The thermal abuses in the electrode-electrolyte reactions of the LIB occur at elevated temperatures under conditions of heating, crushing, or short-circuiting. The cathode materials in LIBs, such as $LiCoO_2$, $LiMnO_2$, $Li[Ni_xCo_yMn_z]O_2$ and $LiFePO_4$, are thermally unstable and release oxygen at elevated temperatures, which induces an autocatalytic reaction with the electrolytes [23]. A thermal runaway reaction that leads to an explosion is the result of an exothermic reaction of the active materials that causes a temperature increase and gas generation. The runaway reaction initially proceeds slowly, but it accelerates exponentially as the temperature increases until the process becomes dramatically out of control. The exothermic onset temperatures, the rate of heat production and the rate of pressure development at runaway are needed to be characterized for inherently hazardous chemicals. The calorimetric methodology is an important and effective measure to analyze the thermal hazards of an unstable substance. The use of adiabatic systems has the advantage that no heat loss is allowed from the sample, and as a result, its thermal behavior on a real-scale size can be predicted. A high heat of reaction is not on its own sufficient to make an exothermic reaction hazards. A large quantity of gas released during a fast self-heating process causes the true danger associated with an explosive reaction. In a closed vessel, such as the steel casing of a battery, the high pressure generated by the gas evolution can ruin the container. Adiabatic calorimeters have been used for the measurement of temperature (T)–pressure (P)–time (t) profiles. The low thermal inertia of the adiabatic calorimeter is useful for extrapolating beyond the peak temperature of the experimental trials. Thermal runaway is one of the failure causes in rechargeable batteries. Many researches have been conducted to evaluate the hazards of battery components and to find safer alternative materials. We listed the researches related to the thermal hazards of LIBs by various calorimetry methodology to briefly describe the potential properties of LIB in Table 6.

Many factors influence the operational characteristics, capacity, energy output and performance of a battery. Bcause of the many possible interactions, these effects can be presented only as generalizations, and the influence of each factor is usually greater under more stringent operating conditions. There are many causes to the LIB safety problems, such as:

1. Flammable and ignitable electrolyte.
2. Thermally unstable cathode.
3. Overcharge sensitive cathode/anode.
4. Thin fragile separator.
5. High sensitivity to metallic elements.

It is commonly recognized that the thermal instability, fire or explosion of LIBs are related to the flammability of the electrolyte, the rate of charge/discharge, and the electrochemical potential of the battery pack. In general, thermal runaway is initiated at an exothermic reaction out of control, that is, the reaction rate increases exponentially with the increasing temperature. Furthermore, a thermal runaway reaction that leads to an explosion is the result of an exothermic reaction of the active materials that causes a temperature increase and gas generation [7, 45]. The thermal abuses in the electrode-electrolyte reactions of the LIB occurred at elevated temperatures under conditions of heating, crushing, or short-circuiting. The cathode materials in LIBs, such as $LiCoO_2$, $LiMnO_2$, $Li[Ni_xCo_yMn_z]O_2$ and $LiFePO_4$, are thermally unstable and release oxygen at elevated temperatures, which induce an autocatalytic reaction with the electrolytes. The organic solvents are easily decomposed on the anodes during charging and are not stable in the presence of high lithium activities. The electrochemical reactions of battery abuse are intriguing and dependent on the temperature, which can be increased by electrical heating during use, by chemical reactions, and by external thermal effects. The electrolytes used in LIBs are less stable in the presence of high lithium compositions, which is a common problem when using elemental lithium negative electrodes in contact with electrolytes containing an organic or polymeric cationic group [23].

Table 6 Related research of LIBs by calorimetry methodologies (selected)

Authors	Calorimeters	Cell components	Related thermal hazard research of LIBs	Reference
Biensan et al.	DSC	Full cell, cathode, and electrolyte	Safety of lithium-ion cells is mainly related to thermal reactivity of the components. Adequately substituted nickel based materials appear as the most stable positive materials. Electrolyte is mainly concerned through its reactivity with lithium of Li_xC_6 and with O_2 generated by the positive materials decomposition. The sorting in terms of energy and kinetics shows that both electrodes contribute to the thermal instability of lithium-ion cells	[24]
MacNeil et al.	DSC and ARC	Cathode and electrolyte	1. The thermal decomposition of $Li_{0.5}CoO_2$ was studied by ARC. The reaction of $Li_{0.5}CoO_2$ with ethylene EC:DEC solvent initiates at temperatures as low as 130 °C, which is much lower than the decomposition temperature of $Li_{0.5}CoO_2$ itself, and reduction to CoO occurs. That caused the reducing power of the solvent and the heat generated by this reaction is consistent with that expected from the combustion of the solvent by the oxygen liberated during the decomposition of the solid 2. The combustion of the solvents reduces $Li_{0.5}CoO_2$ to $LiCoO_2$ and Co_3O_4. In the presence of electrolyte salt, there is a dramatic inhibiting effect that is not only fully understood, but could be due to a coating of the electrode particles by salt decomposition products.	[25]
Baba et al.	DSC and TGA	Cathode, and electrolyte	It is well known that charged Li_xCoO_2 is metastable, and that oxygen evolution has been observed at temperatures above 200 °C. The DSC profile of $Li_{0.49}CoO_2$ showed two exothermic peaks, one beginning at 190 °C and the other beginning at 290 °C and the maximum exothermic heat from 190 to 230 °C based on cathode weight was ca. 420 J/g, and that from 230 to 300 °C was ca. 1000 J/g	[26]
Spotnitz et al.	DSC, ARC	Cathode, electrolyte, SEI, Lithium salt	These kinetic parameters can be used to simulate DSC and ARC experiments. The simulations indicate that the negative electrode binder plays a relatively unimportant role in thermal runaway.	[27]
Yamaki et al.	DSC	Cathode and electrolyte	1. The thermal stability of electrolytes with Li_xCoO_2 cathode or lithiated carbon anode showed two exothermic peaks, one beginning at 190 °C and the other beginning at 290 °C	[28]

(continued)

Table 6 (continued)

Authors	Calorimeters	Cell components	Related thermal hazard research of LIBs	Reference
			2. DSC revealed that the thermal stability of electrochemically lithiated graphite with 1 M LiPF6/EC + DMC and PVdF-binder has shown a mild heat generation starting from 130 °C with a small peak at 140 °C and continued until a sharp exothermic peak appeared at 280 °C	
Jiang et al.	ARC	Full cells	1. Accelerating rate calorimeter (ARC) has been used to compare the thermal stability of three different cathode materials, $LiCoO_2$, $Li[Ni_{0.1}Co_{0.8}Mn_{0.1}]O_2$, and $LiFePO_4$, in EC/DEC solvent and in 1.0 M $LiPF_6$ EC/DEC or 0.8 M LiBoB EC/DEC electrolytes 2. The cathode materials were charged to 4.2 V versus Li metal before analysis. In EC/DEC solvent, the onset temperatures for self-sustained exothermic reactions are 150, 220 and 310 °C for $LiCoO_2$, $Li[Ni_{0.1}Co_{0.8}Mn_{0.1}]O_2$ and $LiFePO_4$ (all charged to 4.2 V), respectively 3. That graphite/LiBoB-based electrolyte/LiFePO4 Li-ion cells will be very abuse-tolerant	[29]
Roth et al.	DSC	Full cells	DSC was used to measure the thermal interactions between several binder materials and representative anode carbons both in the presence of cell electrolyte. The temperature and magnitude of the exothermic reactions were measured up to 375 °C and average enthalpy values were obtained over several duplicate samples to allow good quantitative comparison of the material reactions. The exothermic anode reactions were sensitive to the state of charge and presence of electrolyte	[30]
Maleki et al.	ARC	Full cells	1. Thermal stability of Li-ion cells and their cathodes and anodes was evaluated as a function of SOC using ARC 2. Results indicated the cell consisting of Sn-$LiCoO_2$ cathode and meso-carbon micro-fiber (MCMF) anode offers a higher thermal stability than the cell consisting of $LiCoO_2$ cathode and graphite anode	[31]
Roth et al.	DSC, ARC	Full cell, electrolytes, anode, cathode	1. The thermal behavior of commercial and prototype cells was measured under varying conditions of cell composition, age and state-of-charge (SOC)	[32]

(continued)

Table 6 (continued)

Authors	Calorimeters	Cell components	Related thermal hazard research of LIBs	Reference
			2. The thermal runaway behavior of full cells was measured along with the thermal properties of the cell components 3. Gas generation and gas composition over the temperature range were measured corresponding to the thermal runaway regime 4. Characterization of cell thermal abuse tolerance and an understanding of the mechanisms that result in cell thermal runaway	
Fey et al.	DSC	Cathode	Commercial $LiCoO_2$ cathode samples were coated with cobalt oxides by a wet chemical coating process. The thermal behavior of the fully charged $LiCoO_2$ reported that the quantity of heat generated by the reaction of $LixCoO_2$ with the electrolyte was independent of the amount of the electrolyte	[33]
Wang et al.	C80	Electrolyte	A C80 calorimeter was used to study the thermal behaviors of electrolyte. C80 results show that alone shows one exothermic peak, which is attributed to the solid electrolyte interphase (SEI) decomposition, Li-electrolyte reaction as well as new SEI film formation, new SEI film decomposition, and Li with PVDF/other products reactions	[34]
Wang et al.	ARC	Full cells	The high temperature reactions of 1 M $LiPF_6$ EC:DEC and $LiCoO_2$, Li $(Ni_{1/3}Co_{1/3}Mn_{1/3})O_2$ (NCM) or $Li(Ni_{0.8}Co_{0.15}Al_{0.05})O_2$ (NCA) charged to 4.2 V and 4.4 V, respectively, were studied by ARC. The reactivity of the samples depends on the electrolyte:active material ratio used during ARC testing. Electrode materials charged to 4.4 V are more reactive than the electrode materials charged to 4.2 V	[35]
Belov et al.	DSC	Full cell, anode, cathode and separator	1. Overcharge of a 720 mAh LIBi-ion battery was evaluated using soft overcharge technique which follow the overcharge process gradually to find and analyze the stage when cell conditions become most critical and irreversible. DSC tests were used to characterize battery material at different overcharge stage 2. It was shown that the thermal runaway will happen when the heat generation is not equal to the heat dissipation during the battery operated at an overcharge condition	[36]

(continued)

Table 6 (continued)

Authors	Calorimeters	Cell components	Related thermal hazard research of LIBs	Reference
Veluchamy et al.	DSC and TGA	Cathode	1. Thermal behavior of Li_xCoO_2 cathode material from cells charged to different voltages had been analyzed, using TGA and DSC 2. The mass loss appearing between 60 and 125 °C in TGA and the exothermic peaks with 4.9 and 7.0 J g^{-1} in DSC around 75 and 85 °C for the Li_xCoO_2 cathodes of 4.20 and 4.35 V cells was explained based on SEI film-break down	[37]
Chang et al.	DSC and TGA	Cathode	1. Both DSC and TGA data revealed that the ZnO coating is effective in retarding the decomposition reaction of $LiCoO_2$ with the electrolyte, thereby leading to a decrease in the amount of heat generated 2. The effect of ZnO coating on the thermal stability of charged cathodes was examined by DSC. The extent of the exothermic reaction is much less and the onset temperature for thermal decomposition is much higher for all ZnO-coated $LiCoO_2$ materials than for bare $LiCoO_2$ 3. The weight-loss profile significantly varies with the coating amount. As is widely known, the large weight losses that occur at approximately 250 °C correspond to oxygen liberated from the degraded $LiCoO_2$ layered structure at high temperatures	[38]
Gabrisch et al.	TGA	Cathode	Layered $LixCoO_2$ transforms to a mixture of stoichiometric $LiCoO_2$ and Co_3O_4 at elevated temperature. The thermal decomposition of $LixCoO_2$ by characterizing its decomposition products after short and long-term isothermal annealing at 350 °C	[39]
Lee	DSC and TGA	Cathode and electrolyte	The electrolyte is composed of ethylene carbonate, propylene carbonate, and dimethyl carbonate. Direct observation of gas species resulting from the reactions is beneficial in understanding the reaction mechanisms. The reduction of $LixCoO_2$ to $LiCoO_2$ and Co_3O_4 is triggered at 190 °C and completed at 400 °C	[40]
Jhu et al.	VSP	Full cells	VSP2 was used to measure the thermal hazard and runaway characteristics of commercial 18650 lithium-ion batteries. The tests evaluated the thermal hazard characteristics, such as initial exothermic temperature (T0), self-heating rate (dT dt^{-1}), pressure-rise rate (dP dt^{-1}), pressure temperature profiles, maximum	[41]

(continued)

Table 6 (continued)

Authors	Calorimeters	Cell components	Related thermal hazard research of LIBs	Reference
			temperature, and pressure which were observed by adiabatic calorimetric methodology via VSP2 using customized test cells	
Jhu et al.	VSP2	Full cells	Thermal abuse behaviors relating to adiabatic runaway reactions in commercial 18650 $LiCoO_2$ batteries were studied by VSP2. The self-heating reactions were applied the thermal analytical model combining with Arrhenius equation	[5]
Jhu et al.	VSP	Full cells	The thermal stabilities of charged 18650 Li-ion batteries with $LiCoO_2$ and LiMnCoNi, were characterized via VSP2. The $LiCoO_2$ batteries proved to be more hazardous than the LiMnCoNi ones	[42]
Eddahech et al.	ARC	Full cells	A series of tests are conducted using an accelerating rate calorimeter to promote an adiabatic environment. Cell heat capacity is identified and the overall heat generated is quantified. Cell entropy is measured, using both potentiometric and calorimetric methods. The part of reversible reaction in the overall thermal behavior is determined during charge-discharge tests and compared to joule losses. The influence of the SOC variation and the impact of charge-discharge current rate on battery heat generation are highlighted	[43]
Kim et al.	ARC	Cathode and electrolyte	The effect of lithium bis(oxalato)borate (LiBOB), vinylene carbonate (VC) and succinonitrile (SN) electrolyte additives and $LiPF_6$ salt on the reactivity between electrolyte and charged positive electrode material was investigated by ARC. The results shown here suggest that additives and $LiPF_6$ salt can play a different role in thermal stability depending on the positive electrode material	[44]

The calorimetric application of thermal hazards analysis for energetic LIBs has proven to be a useful alternative technique because of the requirement of temperature measurements. The ARC experimental results of the thermal runaway test showed that the temperature of the exothermic reaction was near 100 °C. As the exothermic reaction progressed beyond 150 °C, the reaction rate accelerated, and the battery disintegrated. At this point, the battery components were ejected from the casing, which could lead to burns or even explosions [9, 23].

According to the studies reviewed, the porous separator breaks down at temperatures above 120 °C and then induces an internal short circuit [29]. A thermal runaway reaction of LIBs occurs after the self-generated heating reaches a critical temperature and is usually accompanied by an explosion. Therefore, the safety issues of LIBs should not be ignored, and their safe use should be ensured. After cycling, the LIBs had charging voltages of 3.7 and 4.2 V on commercially available 18650 LIB with nominal capacity of 2600 mAh, and were then used to test the thermal runaway reaction in an adiabatic calorimeter [5]. The VSP2 trials of the LIBs provided T-P-t profiles for the runaway reactions taking place under thermal adiabatic conditions. A charged 18650 cylinder cell exothermally initiated at approximately 129.1 °C, and between 120 and 160 °C, the heat accumulated and triggered a thermal runaway reaction. Beyond 160 °C, the violently rapid, self-heating reaction resulted in a thermal explosion accompanied by high temperature and pressure, as shown in Figs. 4 and 5: The ARC tests for the CR2325 coin cell of Li_xCoO_2 (4.1–4.3 V), in which the maximum temperature and self-heating rate

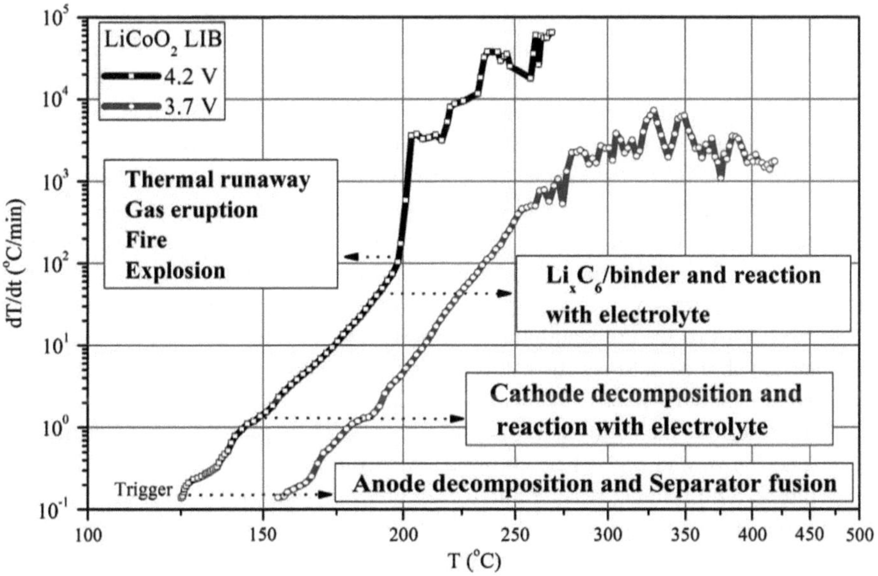

Fig. 4 Self-heating rate profiles of the commercial 18650 LIBs by a VSP2 adiabatic test is outlining four stages of thermal runaway

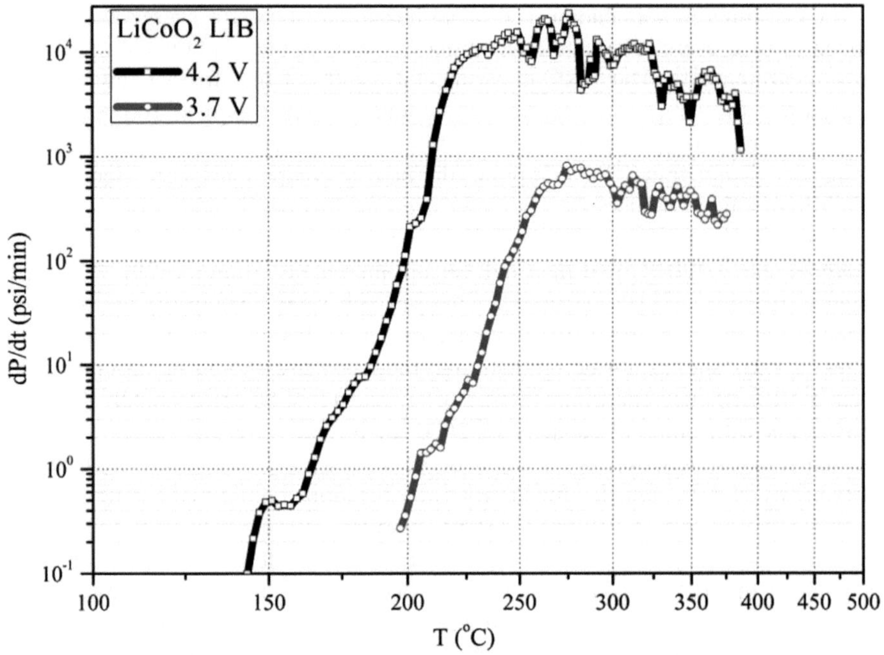

Fig. 5 Pressure rise rate for the thermal decomposition of the commercial 18650 charged (4.2 V) and uncharged (3.7 V) LIBs

were below 300 °C and 100 °C min^{-1}, respectively [23]. Furthermore, the liquid electrolytes in the LIBs conducting lithium-ions, acted as a carrier between the anodes and cathodes when the battery passed an electrical current through an external circuit. The exothermic temperature range of the commercial electrolyte solutions was determined to be 170–330 °C by the ARC tests [46, 47]. Richard and Dahn used an ARC to determine the thermal abuse behaviors of the battery electrolyte and of $Li_{1+x}Mn_{2-x}O_4$ in the electrolyte; the self-heating rate was initially detected at 130 and 190 °C, respectively. The self-heating at 50 °C increased swiftly after 100 °C for the 18650 LIB as measured by the ARC tests. Without sufficient heat removal in a battery, a runaway reaction can occur that may eventually be followed by autocatalytic decomposition or a thermal explosion [48]. If LIBs are capable of auto-reactive decomposition at elevated temperatures and if by charging the battery packs are heated so that the electrodes accelerate toward a runaway reaction, then the energetic substance is an explosive. The thermal runaway reaction of a charged LIB showed a maximum self-heating rate, a high final temperature, a high final pressure and a high pressure rise rate. Hence, to understand the thermal runaway hazards of the LIB, we summarized the experimental data to assess the degree of hazard of the energetic LIBs.

The thermal reactive hazards of the charged batteries exponentially increased with temperature, while the uncharged batteries did not have such a sharp trend.

Thus, we determined that with an increasing charge level, there were more battery hazards. Comparison of the thermal decomposition at the charge cells shows which experienced catastrophic gas generation. Hence, the impact of the charge level on the thermal runaway reaction needs to be taken into account when considering the safety of LIBs. The time required to reach maximum temperature trajectories is a practical measure for comparing the various LIBs. The temperature induction and the heat accumulation of the LIBs mainly resulted in thermal runaway reactions. When the temperature gradually increased to approximately 120 °C, the LIBs experienced solid-electrolyte interphase (SEI) decomposition. Moreover, the separator diaphragm fused at approximately 130 °C, causing the electrodes to shut down. Then cathode oxidation by the electrolyte of the LIBs initiated a self-heating reaction and heat accumulation from 130 to 160 °C in the experimental trials. Finally, the swiftly increasing temperature and pressure resulted in a thermal explosion. The heating rate increased dramatically until the temperature reached above 180 °C, and the sharp exothermic heat generation and gas liberation could exacerbate potential damage to the LIBs [5]. In our experiments, high temperatures caused the LIBs to suffer thermal runaway and cell rupture if inadvertently mistreated. Roth et al. who reviewed the electrochemical reactions of LIB, state that SEI typically decomposes at 90–130 °C, intercalated lithium reacts with the electrolyte at 90–290 °C (electrolyte decomposes at 200–300 °C), and positively active materials decompose and react with the solvent at 150–500°C [32, 37].

We investigated the initial reaction of LIBs under adiabatic conditions by calorimeters. The use of the adiabatic system has the advantage that practically no heat loss is allowed from the battery, and therefore, the behavior of a real large-scale battery pack can be estimated. We obtained the essential parameters of the thermal hazard via adiabatic calorimeter, such as the onset temperature (T_0), self-heating rate (dT/dt), pressure rise rate (dP/dt), maximum temperature (T_{max}) and pressure (P_{max}). Finally, the self-reactive heats of various LIBs were measured via the adiabatic calorimetry methodology to rank the thermal hazards. Then, the dynamic scanning calorimeter and adiabatic calorimeter were used to analyze and dynamically scan various kinds of chemical components inside the LIB under charge, discharge and overcharge conditions to determine their thermal behavior. The worst case of LIB under exothermic behavior and heat accumulation inside the cell without sufficient heat dissipation were determined by combining the data from calorimetry methodology. Our purpose is to enhance LIB being designed and produced under the principle of inherently safer concept. To achieve such purpose, measurements of exothermic onset temperature, peak power, heat of decomposition, adiabatic self-heat rate, pressure-rising rate, maximum temperature of reaction system, maximum pressure, adiabatic time to maximum rate, etc., were performed to establish the thermodynamics of exothermic behavior in LIB. Figure 6 shows the main thermal hazards by the internal and external damages during 100–200 °C that result in uncontrolled runaway. Under an unexpected abuse situation, the heat generated inside the cell at 100 °C will initiate the organic electrolyte decomposition. The fusion of the separator when the temperature reaches 150 °C will cause

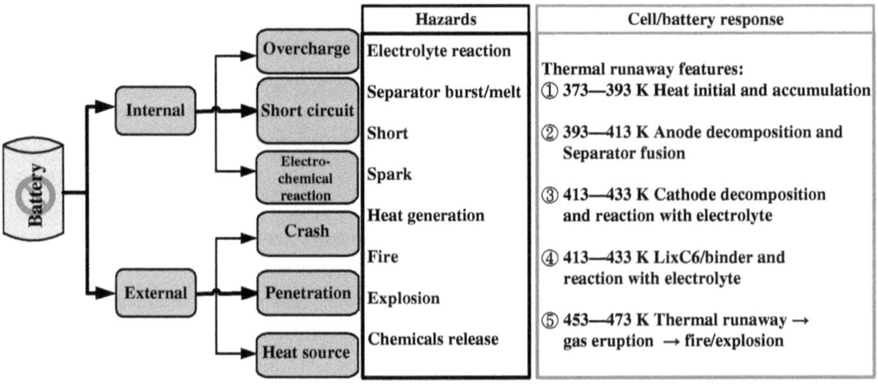

Fig. 6 Thermal runaway due to various abuse conditions of LIBs

the short circuit of the cell. Furthermore, the electrodes decompose to release oxygen at 200 °C. The reaction of a cell changes with the increasing temperature. Thermal runaway includes the thermal decomposition of the electrolyte, oxidation reaction of cathode and electrolyte, reduction reaction of anode and electrolyte, thermal cracking of cell electrodes, and separator fusion. Figure 7 and Table 7 demonstrate the internal short circuit and exothermic reactions of a LIB. The giant enthalpy inside the cell is the reaction of cathode and electrolyte. The thermal stabilities of the cell components, such as electrode, electrolyte and separator, are an important safe issue to develop high power density LIBs [5, 49, 50].

Table 7 The activation process of the battery failure, thermal runaway causes the rapid release of energy

Use/misuse/abuse			Thermal runaway and cell failure
Heat accumulation in cell	→	Exothermic activity on electrochemical reaction	• Rapid rise pressure and temperature of a cell/battery • Fire or explosion
↖		↙	
Acceleration/active action of further reaction			

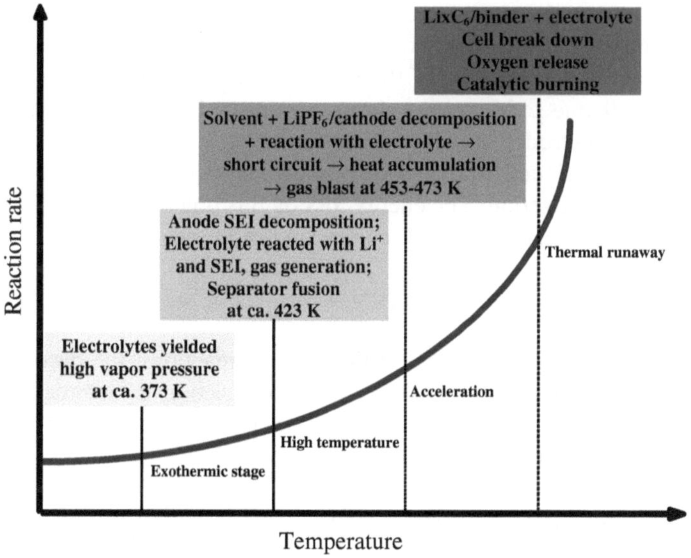

Fig. 7 Thermal runaway progress and internal reaction of a lithium-ion cell

4 Thermal Hazard Evaluation and Analytical Equations of LIBs

Li-ion batteries are potentially vulnerable to abuse conditions due to their complex chemical and electrochemical reactions. Electrical abuse (overcharge and external short circuit) may be mitigated by controlling circuitry. Consequences of short circuit caused by mechanical abuse such as nail and crush, etc. or manufacturing issues will depend on the cell design, manufacturing quality as well as nature of the event. Most internal potential hazards result in poor battery thermal performance which may trigger runaway reaction. The internal short circuit of a cell can induce an energetic exothermic process and cause significant self-heating reaction inside the cell [51]. The calorimetry methodology gives a quantitative test concerning the thermal potential of a LIB, and also provides the information related to the reaction rate profiles of the enthalpy of a cell and the parameters that affect the volatility of the cell components/materials. We summarized the thermodynamic/kinetic relations to describe the reactivity of the energetic LIBs. The heat generating mechanisms have been explained as to the reasons that enhance thermal instability at elevated temperature. With the knowledge of the mechanisms of cell internal electrochemical reaction, from quantitative measurement of the LIBs subjected to the thermal abuse it is possible to propose new strategies to improve thermal stability on LIBs. Performance and thermal analysis in combination are practical methods in optimizing the design of scaled-up cells and batteries for power applications.

The temperature of a short circuit cell will only increase if the generated heats are insufficiently removed. In a short circuit cell, heat is initially generated by the electrical current in the cell (Eq. 1). Richard et al. proposed that the electrical power of a lithium cell due to the short circuit at elevated temperatures the chemical reactions at the electrodes could generate heat (P_{elec}) [48]. When electric current flows through the cell, cell voltage V deviates from open-circuit potential V_0 due to electrochemical polarization. The energy loss by this polarization dissipates as heat. Equation (31) is exothermic at both charge and discharge cycles [7].

$$P_{elec} = I\,(V - V_0) = I^2R \tag{31}$$

The combination of the heat generation by two reactions, the decomposition of the metastable components of the SEI and the reaction of intercalated Li with electrolyte to form stable SEI, caused the cell temperature to increase. They can be expressed as a heat generating function as follows [52]:

$$P_{chem} = [(\frac{\Delta H_1}{C_p} A_1 \exp\frac{-E_{a1}}{k_B T})x_f^n + (\frac{\Delta H_2}{C_p} A_2 \exp\frac{-E_{a2}}{k_B T})x_i \exp\frac{-z}{z_0}]C_p m \tag{32}$$

where the heats of reaction for the formation of stable SEI from metastable SEI and for the formation of stable SEI from lithium and solvent reaction are represented by H_1 and H_2, respectively. C_p is the specific heat of the cell. E_i and A_i are the activation energy and the frequency factor for the different reactions. The amount of lithium intercalated into the carbon is represented by x_i, and the lithium in the metastable SEI is represented by x_f. The thickness of the SEI layer is related to the amount of lithium present in both stable and metastable SEI components, z. n is the reaction order, k_B is Boltzmann's constant and T is the temperature of the cell.

In an adiabatic experiment, no heat is lost to the surroundings, and all of the liberated reaction energy is used for the self-heating reaction of LIBs. The adiabatic thermal explosion is important for practical discussions of safety because it represents the most dangerous case. The self-heating of a LIB by an exothermic reaction can be described by the heat of reaction (ΔH), total heat capacity, cell mass (m_{cell}) and adiabatic temperature rise (ΔT_{ad}) of the LIB These quantities are related as follows [23, 53, 54]:

$$\Delta H = m_{cell} \times C_p \times \Delta T_{ad} \tag{33}$$

The heat of reaction of a cell is determined by the adiabatic calorimetric experiments, which indicate the dramatic exothermic hazards of charged Li-ion cells at elevated temperatures.

In a runaway reaction, the overpressure in a vessel is normally due to the heat of reaction, which is released by the runaway of the energetic substances. The adiabatic exothermic behavior was more quantitatively shown in the VSP2 trials, which provided temperature-pressure-time profiles for the runaway reactions. A LIB typically forms materials on the anode, depending upon the ease of decomposition by the electrochemical reactions. Without sufficient heat removal, a runaway reaction can occur, that may eventually be followed by auto-ignition or a thermal explosion. The runaway hazards of a LIB can be recognized by analyzing the adiabatic trajectories obtained from VSP2 calorimetric trials. When the potential exists for an adiabatic runaway reaction, the temperature and the pressure trajectories of the reaction can be used as a proper measure of the magnitude of the thermal hazard of the LIB using the VSP2 adiabatic calorimetric methodology. The potential thermal hazards of LIBs may cause problems during various kinds of applications. For example, the heat of reaction for the charged LIBs as determined by VSP2 experiments is calculated in Table 8 and Fig. 8.

Finally, We applied the simple analytical equation of self-heating rate by an exothermic reaction, which for an adiabatic temperature rise can be written as [5, 7, 53]:

$$\frac{dT}{dt} = \Delta T_{ad} A \exp\left(\frac{-E_a}{k_B T}\right)(1 - \mu)^n \tag{34}$$

Table 8 Calculation for energy of reaction on 18650 cell from VSP tests

Worked example: From adiabatic calorimetric tests a fully charged 18650 cell is known to have caused a thermal runaway reaction. A reaction become dominant above 125 °C and reached 578 °C after 70 min. The cell blasted finally by means of the thermal abuse failure
Question: Calculate the energy of reaction of a cell after thermal runaway under adiabatic conditions
Data: 1. The mass of a cell is 46.0 g 2. Apparent onset temperature (T_0) is 125 °C, and the maximum temperature (T_m) is reached at 578 °C 3. The specific heat of cell (C_p), which includes the cell and its can, is approximated 0.73 J g °C^{-1} 4. Temperature versus time profile by adiabatic calorimeter is delineated in Fig. 7
Solution: 1. The energy of reaction is directly linked with the cell failure, that is, the potential of destruction of a runaway. Where a reactive cell cannot exchange energy with its surroundings, adiabatic condition prevail. The temperature rise is proportional to the energy released 2. The experimental data, the adiabatic temperature rise (ΔT_{ad}), is a convenient way to assess the severity of a runaway reaction. It can be calculated by the following, $\Delta T_{ad} = T_m - T_0 = 578 - 125 = 453\,°C$ 3. The specific heat of cell (C_p) is calculated by the following equation: $C_p = \frac{\sum_i m_i C_{pi}}{\sum_i m_i} = \frac{(C_{cell} m_{cell} + C_{can} \times m_{can})}{m_{cell} + m_{can}}$ 4. These values can be calculated by substituting the mass of the cell, specific heat of cell and the adiabatic rise into Eq. (33). In this example, the energy of reaction is given by the following expression: $\Delta H = m_{cell} \times C_p \times \Delta T_{ad}$ $= 46\,(g) \times 0.73 (J/g\,°C) \times 453 (°C) = 15{,}211.74\,J$

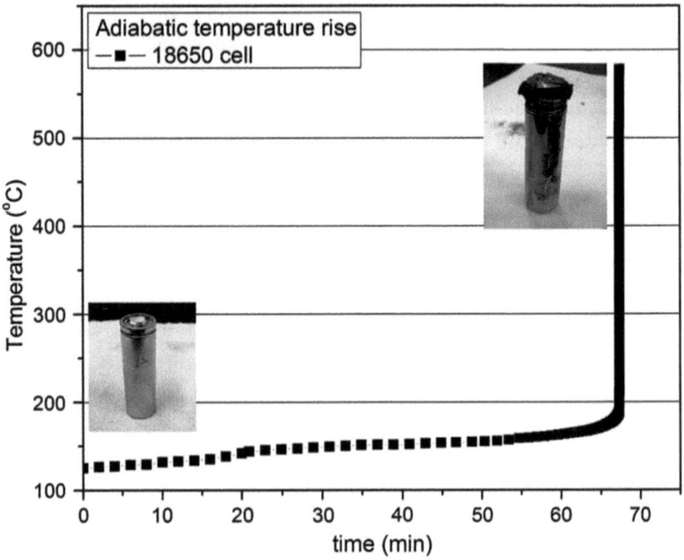

Fig. 8 Temperature-time profiles of 18650 lithium-ion cell after thermal runaway

$$\ln\frac{dT}{dt} = \ln \Delta T_{ad} + \ln A - \frac{E_a}{k_B T} + \ln(1-\mu)^n \tag{35}$$

where μ is the degree of conversion. If A is much greater than ΔT_{ad} and $(1-\mu)^n$, then Eq. (34) can be simplified as:

$$\ln\frac{dT}{dt} \approx \ln \Delta n_{ad} A - \frac{E_a}{k_B T} \tag{36}$$

By plotting the natural logarithm of the self-heating rate versus the inverse of temperature, the activation energy (E_a) of LIBs can be calculated from the slope, and the frequency factor (A) of the simplified Arrhenius equation is determined from the intercept of the fitted line.

It is better to measure the self-heating rate of 18650 LIBs using a VSP2 adiabatic calorimeter. The experimental data can then be used to determine E_a and A with an Arrhenius plot, which is shown in Fig. 9. The fitted lines were used to calculate the kinetic parameters of the thermal explosion for various LIBs. These results agree with the values of E_a and A of Li-ion coin cells that were found using ARC tests by MacNeil et al. [23] and Argue et al. [55]. Wang et al. summarized that the reaction heat generation is the total result of all possible reactions when the cell is undergoing thermal runaway, mainly including SEI decomposition, electrodes reaction with electrolyte, electrodes decomposition. For the thermal runaway reaction, the heat generation can be expressed as Q_{cell} [5, 7, 56, 57]:

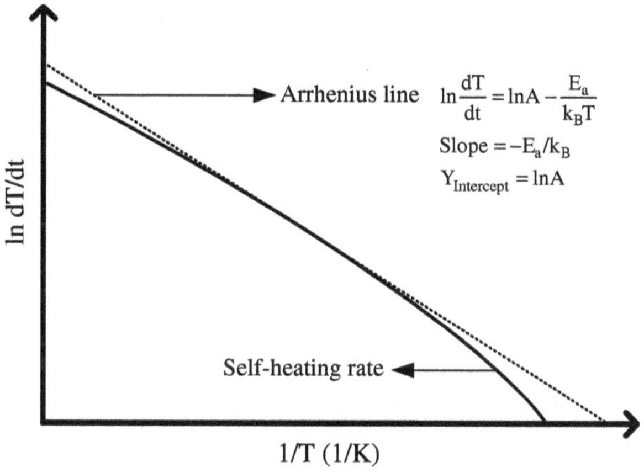

Fig. 9 Characteristic self-heating rate for an energetic substance by Arrhenius equation

$$Q_{cell} = \Delta H m^{n} A \exp\frac{-E_a}{k_B T} \quad (37)$$

From calorimetry methodology, Richard et al. calculated the power generation of the cell (P_{cell}) by ARC measurement as the following:

$$P_{cell} = C_p m \frac{dT}{dt} \quad (38)$$

We then substituted Eq. (34) into Eq. (38) to obtain the simplified expression for the heat of reaction of lithium-ion cell.

$$P_{cell} = C_p m \Delta T_{ad} A \exp(\frac{-E_a}{k_B T})(1 - \mu)^{n} \quad (39)$$

Based on the observations of these studies, the self-heating rate of LIBs increased exponentially with the temperature. The results of the adiabatic runaway reaction experiments agreed with those of the calorimetric methodology. Violent thermal decomposition and explosions due to runaway reactions pose serious safety hazards that must be considered when using LIBs.

The total heat generation as a function with temperature increase, P(T), is the electrical power due to the short-circuit, P_{elec}, and the chemically generated power due to the reaction of the cell, $P_{chem,}$ are added, but then the power loss to the environment, $P_{out,}$ is subtracted.

$$P(T) = P_{elec} + P_{chem} + P_{out} \quad (40)$$

where P_{out} is the main thermal exchanges between the cell surface and environments and includes convection heat transfer (Q_{conv}) and radiation (Q_{radi}). When the cell/battery temperature (T_{cell}) exceeds the ambient temperature (T_{amb}), the convection starts to dissipate the heat. When in high temperature cells/batteries, radiation plays an important role and should be considered. Convective and radiative heat flux out to ambient is evaluated in the following Eq. (35) [7]:

$$P_{out} = Q_{conv} + Q_{radi} = hA(T_{cell} - T_{amb}) + \varepsilon\sigma A(T^4_{cell} - T^4_{amb}) \tag{41}$$

where h is a convection heat transfer coefficient, A is the area of the cell, ε is the emissivity of the cell surface and σ is the Stefan–Boltzmann constant.

Li-ion batteries, unlike other rechargeable batteries, have potentially hazardous and pressurized materials and energy density, and require strict quality control during manufacturing. Calorimetry methodology is useful for the design and determining of inherent safer LIBs, although there is a limitation of battery pack size which can be tested. A faulty battery can affect the safety of the battery because it can destroy the battery's protection circuit to cause a serious fire or explosion. Calorimetry methodology has previously been shown to give useful qualitative tests concerning the cell components and thermal abuse on the relativity of electrochemical reaction. The analytical equations precisely consider the complicated process involving material properties, chemistry, and engineering to achieve a comprehensive thermal analysis. Based on this model, some important phenomena such as the thermal response profiles and the temperature distribution on the surface can be simulated precisely. The model is coupled with electrochemical reaction and thermal response to study in detail the temperature field distribution and evolution inside cell. It also considers the geometrical features to simulate an oven test, which is significant in larger cells for electric vehicle application. Safety improvements can more from various technology improvements, such as thermally stable cathode, non-flammable electrolyte, improved separator, coated active materials and increase formation and quality assurance of the battery packs.

5 Summary

Li-ion batteries are being developed for high-power applications in HEVs and EVs currently being designed for traffic transportation. These cells offer superior performance in terms of power and energy density over current cell chemistries. Cells using this chemistry are the basis of battery systems for both gasoline and fuel cell based hybrids. However, the safety of these cells needs to be understood and improved for eventual widespread commercial application in power system. Short-circuiting a battery will cause the cell to accumulate heat and possibly to catch thermal runaway. Adjacent cells within the battery pack may overheat and fail, possibly causing the entire battery to ignite or rupture and damage the products. The energy content of an electrical and chemical reaction in case of fire is often serious. In about 2010, large LIBs were introduced in place of other batteries to power systems on some aircraft; some serious lithium-ion battery fires, or smoke, occurred on the Boeing 787 passenger aircraft, which did not cause crashes but had the potential to do so [58, 59]. In the US, the Federal Aviation Administration (US FAA) Office of Security and Hazardous Materials Safety (ASH) has striven to enhance safety in air transportation by preventing hazardous materials accidents and incidents aboard aircraft. 144 air incidents involving smoke, fire, extreme heat or explosion of the batteries carried as cargo or baggage have been recorded since 1991–2014 [60].

Batteries are sources of energy and when used properly will deliver their energy in a safe manner. There are instances, however, when a battery may vent, rupture, or even explode if it is abused. The design of the battery should include protective devices and other features which can prevent or minimize the problem. Table 9 lists the strategies of battery safety improvement. These conditions may cause an internal pressure increase within the cells, resulting in an activation of the vent device or a rupture or explosion of the battery. There are a number of means to minimize the possibilities of these occurrences and Fig. 10 shows the safety mechanism of the LIB pack [61]. The thermal behavior of commercial and prototype cells has been measured under varying conditions of cell composition, age

Table 9 Strategies of battery safety improvement

Battery/cell monitoring	• Operates at battery level
	• Monitoring and software control of cell
	• Balancing
	• Calorimetry test on battery packs
Inherent safer materials	• Contribution at cell-level
	• Development of improved safety in anode, cathode, separators and electrolytes
	• Calorimetry test on materials
Safety devices	• Work at cell and battery-level
	• Include PTC, CID, vent, fuses, or interlock, etc. are designed to intervene before runaway occurs
	• Pressure relief and calorimetry studies

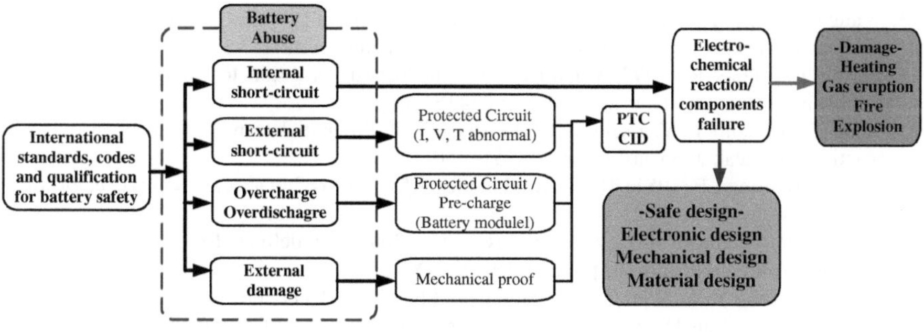

Fig. 10 Safety mechanism of lithium-ion battery pack

and SOC. The thermal runaway behavior of batteries has been measured along with the thermal properties of the cell components. We have also described gas generation and gas composition over the temperature range corresponding to the thermal runaway regime. These studies have allowed characterization of cell thermal abuse tolerance and an understanding of the mechanisms that result in cell thermal runaway. Therefore, thermal analysis and management is very important to improve the electrochemical and safety performance in batteries.

Acknowledgments This work was supported financially by China Medicine University (CMU), Taiwan, ROC (grant no. CMU102-N-09).

References

1. Reddy TB, Linden D (2002) Handbook of batteries, 3rd edn. McGraw-Hill, New York
2. Julien C (2000) Design considerations for lithium batteries. In: Julien C, Stoynov Z (eds) Materials for lithium-ion batteries. Kluwer Academic Publishers, The Netherlands, pp 1–20
3. Industrial Economics & Knowledge Center (IEK). Sept 2011. http://ieknet.itri.org.tw. Assessed June 2014
4. Industrial Economics & Knowledge Center (IEK). Spring 2014. http://www.materialsnet.com.tw. Assessed June 2014
5. Jhu C-Y, Wang Y-W, Shu C-M, Chang J-C, Wu H-C (2011) J Hazard Mater 192:99–107
6. Goodenough JB, Park K-S (2013) The Li-ion rechargeable battery: a perspective. J Am Chem Soc 135:1167–1176
7. Wang Q, Ping P, Zhao X, Chu G, Sun J, Chen C (2012) Thermal runaway caused fire and explosion of lithium ion battery. J Power Sources 208:210–224
8. Ozawa K (2009) Lithium ion rechargeable batteries. Wiley-VCH, Germany
9. MacNeil DD, Dahn JR (2002) The reactions of $Li_{0.5}CoO_2$ with non-aqueous solvents at elevated temperatures. J Electrochem Soc 149:912–919
10. Jiang J, Dahn JR (2004) ARC studies of the thermal stability of three different cathode materials: $LiCoO_2$; $Li[Ni_{0.1}Co_{0.8}Mn_{0.1}]O_2$; and $LiFePO_4$, in $LiPF_6$ and LiBoB EC/DEC electrolytes. Electrochem Commun 6:39–43
11. Zhang SS (2007) A review on the separators of liquid electrolyte Li-ion batteries. J Power Sources 164:351–364
12. Choi NS, Profatilova IA, Kim SS, Song EH (2008) Thermal reactions of lithiated graphite anode in $LiPF_6$-based electrolyte. Thermochimica Acta 480 (1–2):10–14
13. Wang Q, Sum J, Yao X, Chen C (2005) Thermal stability of $LiPF_6$/EC + DEC electrolyte with charged electrodes for lithium ion batteries. Thermochim Acta 437:12–16
14. Barton J, Rogers R (1997) Chemical reaction hazards: a guide to safety, 2nd edn. Gulf Publishing Company, Houston. ISBN 0-88415-274-X
15. Ando T, Fujimoto Y, Morisaki S (1991) Analysis of differential scanning calorimetric data for reactive chemical. J Hazard Mater 28:251–280
16. Tou JC, Whiting LF (1981) The thermokinetic performance of an accelerating rate calorimeter. Thermochimica Acta 48:21–42
17. Center for Chemical Process Safety (CCPS) (2005) Guidelines for chemical reactivity evaluation and application to process design, NY, USA
18. Jhu C-Y, Wang Y-W, Wen C-Y, Chiang C-C, Shu C-M (2011) Self-reactive rating of thermal runaway hazards on 18650 lithium-ion batteries. J Therm Anal Calorim 106:159–163

19. Jhu C-Y, Wang Y-W, Wen C-Y, Shu C-M (2012) Thermal runaway potential of $LiCoO_2$ and $Li(Ni_{1/3}Co_{1/3}Mn_{1/3})O_2$ batteries determined with adiabatic calorimetery methodology. Appl Energy 100:127–131
20. Maleki H, Deng G, Anani A, Howard J (1999) Thermal stability studies of Li-ion cells and components. J Electrochem Soc 146(9):3224–3229
21. Spotnitz R, Franklin J (2003) Abuse behavior of high-power, lithium-ion cells. J Power Sources 113:81–100
22. Power, Safety and handling guidelines for Electrochem Lithium Batteries. Electrochem Commercial Power, Division of Greatbatch, Ltd. MA, USA, 2006
23. MacNeil DD, Christensen L, Landuect J, Paulsen JM, Dahn JR (2000) An autocatalytic mechanism for the reaction of $LixCoO_2$ in electrolyte at elevated temperature. J Electrochem Soc 147:970–979
24. Biensan Ph, Simon B, Peres JP, de Guibert A, Broussely M, Bodet JM, Perton F (1999) On safety of lithium-ion cells. J Power Sources 81–82:906–912
25. MacNeil DD, Dahn JR (2001) The reaction of charged cathodes with nonaqueous solvents and electrolytes: I. $Li_{0.5}CoO_2$. J Electrochem Soc 148(11):A1205–A1210
26. Okada S, Yamaki J-I (2002) Thermal stability of $LixCoO_2$ cathode for lithium ion battery. Solid State Ionics 148:311–316
27. Spotnitz R, Franklin J (2003) Abuse behavior of high-power, lithium-ion cells. J Power Sources 113(1):81–100
28. Yamaki J-I, Baba Y, Katayama N, Takatsuji H, Egashira M, Okada S (2003) Thermal stability of electrolytes with $LixCoO_2$ cathode or lithiated carbon anode. J Power Sources 121:789–793
29. Jiang J, Dahn JR (2004) ARC studies of the thermal stability of three different cathode materials: $LiCoO_2$; $Li[Ni_{0.1}Co_{0.8}Mn_{0.1}]O_2$; and $LiFePO_4$, in $LiPF_6$ and LiBoB EC/DEC electrolytes. Electrochem Commun 6:39–43
30. Roth EP, Doughty DH, Franklin J (2004) DSC investigation of exothermic reactions occurring at elevated temperatures in lithium-ion anodes containing PVDF-based binders. J Power Sources 134:222–234
31. Maleki H, Howard JN (2004) Role of the cathode and anode in heat generation of Li-ion cells as a function of state of charge. J Power Sources 137:117–127
32. Roth EP, Crafts CC, Doughty DH, McBreen J (2004) Advanced technology development program for lithium-ion batteries: thermal abuse performance of 18650Li-ion cells. Sandia National Laboratories, SAND2004-0584, USA, March 2004
33. Fey GT-K, Lin YY, Kumar TP (2005) Enhanced cyclability and thermal stability of $LiCoO_2$ coated with cobalt oxides. Surf Coat Technol 191:68–75
34. Wang Q, Sun JH, Yao XL, Chen CH (2006) Micro calorimeter study on the thermal stability of lithium-ion battery electrolytes. J Loss Prev Process Ind 19:561–569
35. Wang Y, Jiang J, Dahn JR (2007) The reactivity of delithiated Li $(Ni_{1/3}Co_{1/3}Mn_{1/3})O_2$, Li $(Ni_{0.8}Co_{0.15}Al_{0.05})O_2$ or $LiCoO_2$ with non-aqueous electrolyte. Electrochem Commun 9:2534–2540
36. Belov D, Yang M-H (2008) Investigation of the kinetic mechanism in overcharge process for Li-ion battery. Solid State Ionics 179(27–32):1816–1821
37. Veluchamy A, Doh C-H, Kim D-H, Lee J-H, Shin H-M, Jin B-S, Kim H-S, Moon S-I (2009) Thermal analysis of $LixCoO_2$ cathode material of lithium ion battery. J Power Sources 189:855–858
38. Chang WY, Choi J-W, Im J-C, Lee J-K (2010) Effects of ZnO coating on electrochemical performance and thermal stability of $LiCoO_2$ as cathode material for lithium-ion batteries. J Power Sources 195:320–326
39. Gabrisch H, Kombolias M, Mohanty D (2010) Thermal decomposition of $Li\mathit{x}CoO_2$ monitored by electron energy loss spectroscopy and magnetic susceptibility measurements. Solid State Ionics 181:71–78
40. Lee SH, Jung JM, Ok JH, Park C-H (2010) Thermal studies of charged cathode material ($Li\mathit{x}CoO_2$) with temperature-programmed decomposition–mass spectrometry. J Power Sources 195:5049–5051

41. Jhu C-Y, Wang Y-W, Wen C-Y, Chiang C-C, Shu C-M (2011) Self-reactive rating of thermal runaway hazards on 18650 lithium-ion batteries. J Therm Anal Calorim 106:159–163
42. Jhu C-Y, Wang Y-W, Wen C-Y, Shu C-M (2012) Thermal runaway potential of LiCoO2 and Li(Ni1/3Co1/3Mn1/3)O2 batteries determined with adiabatic calorimetry methodology. Appl Energy 100:127–131
43. Eddahech A, Briat O, Vinassa J-M (2016) Thermal characterization of a high-power lithium-ion battery: potentiometric and calorimetric measurement of entropy changes. 61:432–439
44. Kim G-Y, Dahn JR (2014) ARC studies of the effects of electrolyte additives on the reactivity of delithiated $Li_{1-x}[Ni_{1/3}Mn_{1/3}Co_{1/3}]O_2$ and $Li_{1-x}[Ni_{0.8}Co_{0.15}Al_{0.05}]O_2$ positive electrode materials with electrolyte. J Electrochem Soc 161(9):A1394–A1398
45. Torabi F, Esfahanian V (2011) Study of thermal–runaway in batteries: I. Theoretical study and formulation. J Electrochem Soc 158:A850–A858
46. MacNeil DD, Christensen L, Landuect J, Paulsen JM, Dahn JR (2000) An autocatalytic mechanism for the reaction of Li_xCoO_2 in electrolyte at elevated temperature. J Electrochem Soc 147:970–979
47. Gnanaraj JS, Zinigrad E, Asraf L, Gottlieb HE, Sprecher M, Aurbach D, Schmidt M (2003) The use of accelerating rate calorimetry (ARC) for the study of the thermal reactions of Li-ion battery electrolyte solutions. J Power Sources 119–121:794–798
48. Richard MN, Dahn JR (1999) Predicting electrical and thermal abuse behaviours of practical lithium-ion cells from accelerating rate calorimeter studies on small samples in electrolyte. J Power Sources 79:135–142
49. Belov D, Hung CJ, Shieh DT (2009) Influence of electrolyte and separator on the safety of lithium-ion batteries. Ind Mater 275:66–76 (in Chinese)
50. Cheng CS (2008) The thermal analyses of the lithium-ion battery materials. Ind Mater 264:118–122 (in Chinese)
51. Maleki H, Howard JN (2009) Internal short circuit in Li-ion cells. J Power Sources 191:568–574
52. Richard MN, Dahn JR (1999) Accelerating rate calorimetry study on the thermal stability of lithium intercalated graphite in electrolyte. Modeling the results and predicting differential scanning calorimeter curves. J Electrochem Soc 146(6):2078–2084
53. Wang Y-W, Duh Y-S, Shu C-M (2006) Evaluation of adiabatic runaway reaction and vent sizing for emergency relief from DSC calorimetry. J Therm Anal Calorim 85:225–234
54. Stoessel F (2008) Thermal safety of chemical processes: risk assessment and process design. ISBN: 978-3-527-31712-7
55. Argue S, Davidson IJ, Ammundsen B, Paulsen J (2003) A comparative study of the thermal stability of Li1–xCoO2 and Li3–xCrMnO5 in the presence of 1 M LiPF6 in 3:7 EC/DEC electrolyte using accelerating rate calorimetry. J Power Sources 119:664–668
56. Ottaway M (2009) Lithium batteries, highly energetic materials: The varied use of adiabatic calorimetry to aid safety and battery development. In: 37th North American thermal Analysis Society (NATAS) annual conference, USA
57. Wang QS, Ping P, Sun JH (2010) Catastrophe analysis of cylindrical lithium ion battery. Nonlinear Dynam 61:763–772
58. Kanellos M (2013) Can anything tame the battery flames? 15 Aug 2006. Retrieved 14 June 2013
59. http://en.wikipedia.org/wiki/Lithium_ion_battery#cite_note-100, Guardian newspaper: Heathrow fire on Boeing Dreamliner started in battery component, 2013
60. http://www.faa.gov/about/office_org/headquarters_offices/ash/ash_programs/hazmat/
61. Cheng CS, Yang CR, Hsu JM (2009) High safety materials STOBA for lithium-ion batteries. Ind Mater 275:77–82 (in Chinese)

Li-Ion Battery Pack and Applications

Michael S. Mazzola and Masood Shahverdi

1 Introduction

An accurate battery pack model is of significant importance for electric-vehicle drivetrain design and simulation. It is not uncommon to see simple resistance battery models used in vehicle simulations or energy storage system simulations [1, 2] even involving fast dynamics in vehicle power delivery. In contrast to the view that vehicle system level simulation does not require highly accurate battery models [3], a high fidelity battery pack model is critical for the vehicle simulation because the drivetrain power management, the motor/generator control, AC/DC and DC/DC converter design and control, the battery pack state of power (SOP) management, etc. are highly dependent on the accurate prediction of the battery power and battery state of charge (SOC). This is true largely because of the dynamics of the battery current when a real-world battery electric vehicle (BEV), hybrid electric vehicle (HEV), or plug-in hybrid electric vehicle (PHEV) is modeled with realistic drive cycles. As a result, simple models are not capable of predicting the dynamic responses of the battery pack, which can limit the validity of the entire simulation, especially if controller performance is included.

Battery packs usually consist of hundreds of battery cells connected in series and parallel, including battery packs made up of several battery modules, with each battery module containing several battery cells in series, parallel, or series–parallel configuration. Much battery modeling work has been reported at the battery cell level [4–9], with less work discussing battery models at the battery pack level. The work of integrating cell models into pack models may end up being done by the system level designer or power electronics designer who do not have expertise in batteries. Going from battery cell model to battery pack model is not simply aggregating cell models to

M.S. Mazzola (✉) · M. Shahverdi
Department of Electrical and Computer Engineering, Mississippi State University, Box 9571, Mississippi, MS 39762, USA
e-mail: mazzola@ece.msstate.edu

Z. Zhang and S.S. Zhang (eds.), *Rechargeable Batteries*,
Green Energy and Technology, DOI 10.1007/978-3-319-15458-9_16

make a pack model, because not only could it introduce unnecessary computational requirements for the system simulation, but also because some phenomena that can easily be observed in the battery pack are ignored [10]. Significant fidelity loss will occur if inadequate attention is paid to the battery pack behavior, as opposed to cell-level modeling alone. Thus it is worth investigating the construction of a battery pack *behavior* model separately from the cell model.

A high level of model accuracy (less than 1.11 % error) for a 360-V, 21.3-kWh lithium-ion battery pack has been achieved by correlating the bandwidth of the battery model with the bandwidth of the battery application [11]. Since a battery is a continuous nonlinear system which involves complex reactions between anode and cathode no natural exponential moments should be expected because of the nonlinearity of the underlying processes. The electrical analogue battery model, which has been the subject of many modeling papers addressing different formats and different chemistries, is actually a truncated multi-term exponential series in which exponential terms are used to approximate the battery behavior. Without linearization there are no natural time constants in these models arising from physical or chemical analysis, and thus it can be argued that the limited number of time constants available in the battery model should be based on the users' simulation objectives [12]. This relationship between achievable model bandwidth and application need is ignored in most scenarios. While much work has been done seeking natural moments for the exponential terms [13–15], in reality the model parameters (including the time constants that define the bandwidth of the battery model) are estimated to give the "best fit" to an arbitrary load stimulus during the battery test. Large modeling errors may result. An alternative is to base the bandwidth of the model on the bandwidth of the battery application; and when this is done a priori of the parameter extraction, then significantly higher fidelity can be achieved for an electrical analogue battery model of the same order of dynamic approximation. In short, the bandwidth of the battery pack model is chosen to be the bandwidth of the actual battery pack application [12]. A description of the general model is found in Sect. 2, with an example of parameter extraction in Sect. 3. The application in hybrid electric vehicles of two Li-ion behavioral models of varying assigned bandwidth is covered in two examples in Sect. 4. Section 5 concludes the chapter with a summary.

2 Electrical Analogue Battery Model

The electrical analogue battery model [16–18] (Fig. 1) includes two parts. The left part is the SOC estimator, where battery SOC is calculated based on direct coulomb-counting. The right part is the circuit behavior representation of a physical battery, which can be conveniently solved using circuit simulation software such as MATLAB/Simulink and CADENCE/Pspice. The bridge between the two parts is the state of charge—open circuit voltage (OCV) mapping. The OCV estimation depends on the SOC estimation, as the OCV on the circuit part is found by correlating the OCV with the estimated SOC on the mapping.

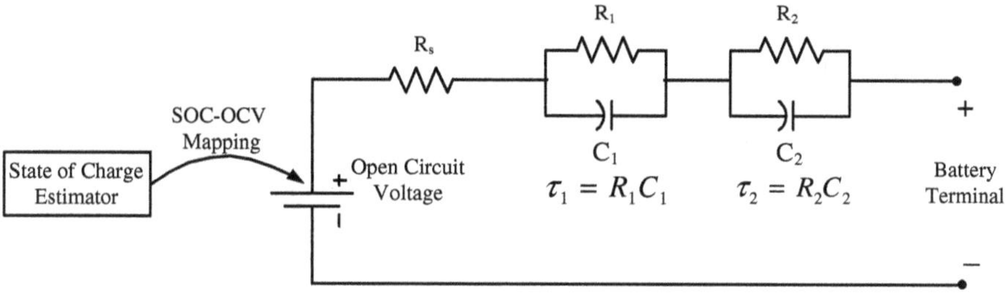

Fig. 1 The electrical analogue battery model

Based on Kirchhoff's current law and voltage law, a discrete time mathematical description of the electrical analogue battery model is derived as state update Eq. (1) and output Eq. (2).

$$\begin{bmatrix} V_{c1,k} \\ V_{c2,k} \\ SOC_k \end{bmatrix} = \begin{bmatrix} -\frac{\Delta t}{R_1C_1} & 0 & 0 \\ 0 & -\frac{\Delta t}{R_2C_2} & 0 \\ 0 & 0 & 0 \end{bmatrix} \begin{bmatrix} V_{c1,k-1} \\ V_{c2,k-1} \\ SOC_{k-1} \end{bmatrix} + \begin{bmatrix} \frac{1}{C_1} \\ \frac{1}{C_2} \\ \frac{1}{C} \end{bmatrix} i_{k-1}\Delta t + \mathbf{v}_{k-1} \quad (1)$$

where the state variables $V_{c1,k-1}$, $V_{c2,k-1}$, and SOC_k are capacitor C_1 voltage, capacitor C_2 voltage, and battery SOC, respectively. R_s, R_1, C_1, R_2, and C_2 are model parameters which need to be identified prior to using this model and will be discussed in Sect. 4. C is a constant representing battery capacity. i_{k-1} is the input current at time instant $k-1$. Parameter $\mathbf{v}_{k-1}$ is white Gaussian noise with zero mean and covariance $\mathbf{Q}$. The sampling interval is Δt. The measurement equation is

$$V_k = V_{ocv,k}(SOC_k) + R_s i_{k-1} + V_{c1,k} + V_{c2,k} + \mathbf{n}_k \quad (2)$$

where $V_{ocv,k}$ is the OCV and V_k is the battery terminal voltage. Variable $\mathbf{n}_k$ is the measurement noise which is assumed to be white Gaussian noise with zero mean and covariance $\mathbf{R}$.

The mapping of OCV on SOC is reflected by (3), where $V_{ocv,k}$ is an eighth order polynomial equation in SOC_k.

$$\begin{aligned} V_{ocv,k} = {} & a_8SOC_k^8 + a_7SOC_k^7 + a_6SOC_k^6 + a_5SOC_k^5 \\ & + a_4SOC_k^4 + a_3SOC_k^3 + a_2SOC_k^2 + a_1SOC_k + a_0 \end{aligned} \quad (3)$$

3 Battery Model Parameter Extraction

The circuit parameters in (1), (2), and the SOC-OCV mapping in (3) need to be estimated before applying the model.

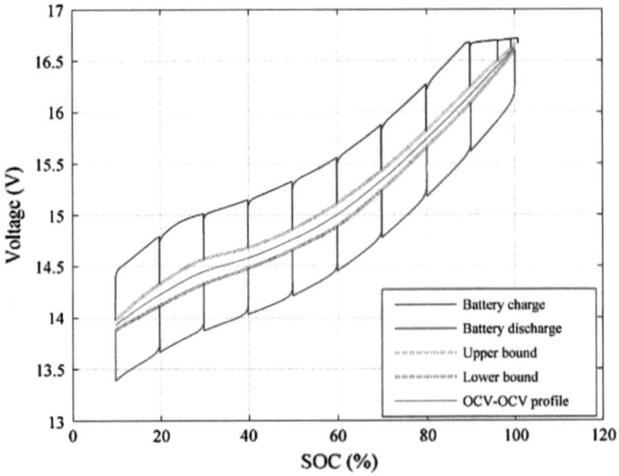

Fig. 2 SOC-OCV profile extraction on a 6.8 Ah Ultralife UBBL10 lithium-ion battery module

Table 1 Coefficients for the polynomial equation

a_8	−139.1
a_7	481.664
a_6	−555.94
a_5	123.455
a_4	230.917
a_3	−185.97
a_2	50.1971
a_1	−2.5346
a_0	13.872

Battery SOC-OCV profile was extracted from a battery pulse charging/discharging test [19]. The pulse length was taken as charging/discharging the battery by 10 % SOC and there was 1 min rest period after each pulse. After combing the charging test voltage and discharging test voltage together, the SOC-OCV profile was extracted as the red line in Fig. 2. The detailed procedure can be found in [19]. After the SOC-OCV profile was extracted, a further battery test with long rest period (24 h) after each pulse with fewer pulses was conducted to correct the initial SOC-OCV extraction based on the fact that battery terminal voltage should reach the true open-circuit voltage indicating the true SOC in about 24 h [19]. The coefficients for the eighth order polynomial equations representing the SOC-OCV profile are shown in Table 1.

A battery test with the test profile in Fig. 3 was conducted on the battery. Mixed charging and discharging current with different current rates were performed to sufficiently excite the battery for model parameter extraction. The errors between the measured terminal voltage and the model output terminal voltage were

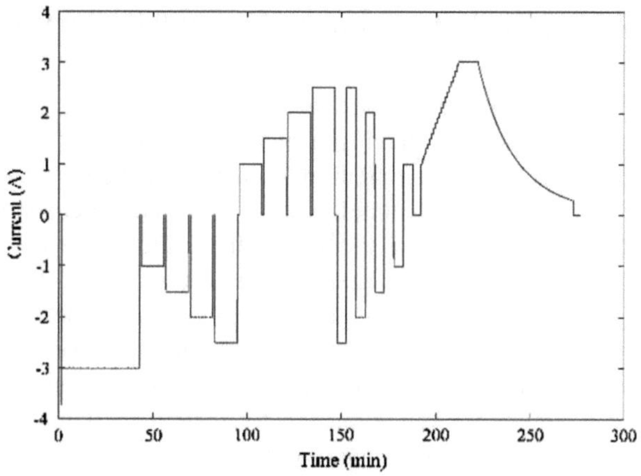

Fig. 3 Battery test current profile (positive direction is taken as when the battery is being charged) (after [11])

Table 2 Estimated parameters

R_S (Ω)	R_1 (Ω)	C_1 (kF)	R_2 (Ω)	C_2 (kF)
0.145	0.032	1.852	0.044	47.730

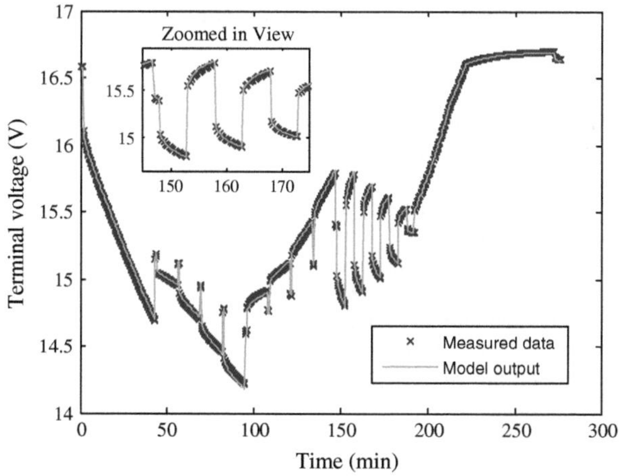

Fig. 4 Results of the battery terminal voltage estimation against experimental data showing the fine agreement when the model bandwidth is assigned so as to cover the application's natural bandwidth (after [11])

calculated and minimized by varying the circuit parameters using the Sequential Quadratic Programming (SQP) algorithm. The extracted parameters are shown in Table 2. Figure 4 shows the battery terminal voltage estimation results with the extracted parameters. Accurate off-line battery terminal voltage estimation results can be observed.

4 Application of Large-Format Battery Behavioral Model

To evaluate the performance of a large-format Li-Ion battery in an automotive electrical system, having a behavioral model of a battery at the module level is required. In the simulation domain, the behavioral battery model should be a flexible electrical model, and in a mathematical structure that does not rely on battery chemistry in the sense that partial differential equations are solved. For different types of batteries like Lithium-Ion (Li-Ion), Nickel Metal Hydride (NiMH), or lead acid, the model parameters should be updated based on specification of a battery such as State of Charge verses Open Circuit Voltage (SOC vs. OCV) curve, and internal impedances. The behavioral model also needs to indicate SOC which is a key parameter for system level analysis. Depending on the required fidelity, internal impedances can be represented by either a single resistance or a resistance pulse one or more series RC networks. The RC network specification determines the fidelity of the model for higher bandwidth studies. In low bandwidth analysis the simplified model in Fig. 5 gives reasonable accuracy. In this model the internal impedance is estimated by a single resistance only, where the numerical effect is to represent the difference between terminal voltage and open-circuit voltage as having no delayed response. The method given in Sects. 2 and 3 will work for both this case as well as those cases where the delay must be considered, such as when a battery management or powertrain controller's performance is part

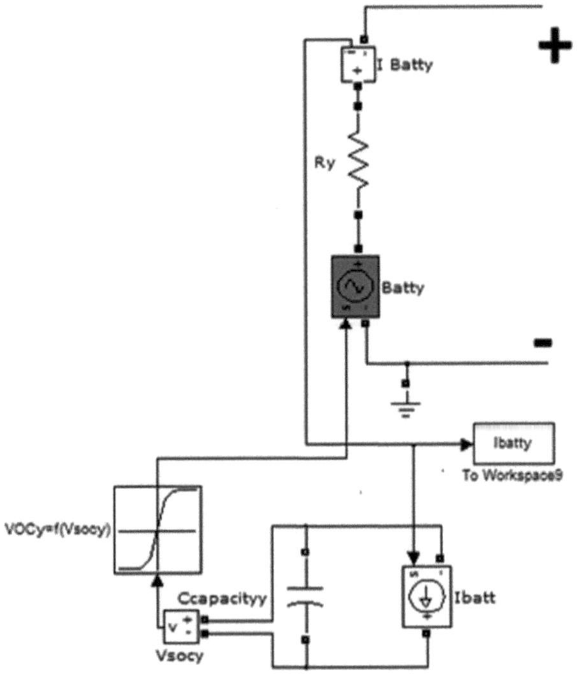

Fig. 5 Simplified behavioral battery pack model. The model includes VOC-SOC nonlinear relationship and a single series resistance to model the difference between open-circuit voltage (related to SOC) and terminal voltage when the battery is loaded by current during charging or discharging

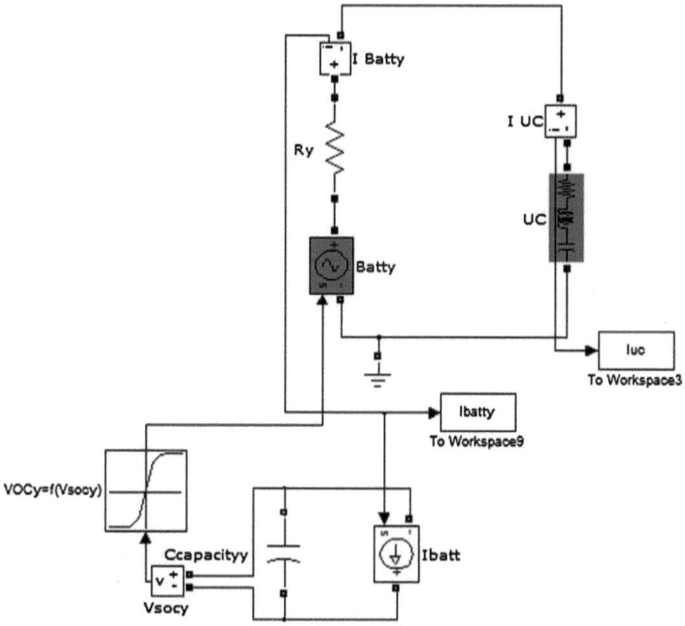

Fig. 6 Simplified hybrid energy storage pack behavioral model. A first order UC model in parallel with the battery forms a hybrid energy storage model

of the problem being simulated. Here a capacitor is responsible for indicating the state of charge in the battery. The voltage across the capacitor is equivalent to SOC of the battery. A lookup table creates the non-linear relationship between SOC and OCV. If the VOC-SOC curve and the single resistor of the pack is extracted by the method in Sect. 2 the pack model can be directly derived. This has the advantage of avoiding errors associated with an ideal cell model that does not include the distribution in cell performance that can make building the pack model built by scaling up the cell level model unreliable. Ultra-capacitors can be modeled in pack level by using a first order series RC network. Figure 6 displays the pack level UC model in parallel with pack level battery model in Simulink.

4.1 Battery Behavioral Model in the Powertrain of a Series Hybrid Electric Vehicle

The application of the battery pack model is illustrated by studying a design example in the electrified vehicle area. In this section, the battery pack model is the key part of the powertrain of a series hybrid electric vehicle. A unique (for passenger duty) vehicle is studied in the Autonomie simulation environment [20]. Then, several Energy Storage System (ESS) options, which are combinations of battery strings and/or UC strings, are nominated and considered in an analysis of alternatives. Since 1-Hz standard drive schedules for city and highway driving condition published by US Environmental Protection Agency (EPA) are used as a simplification, so then it is assumed that higher frequency loading does not affect this analysis. Therefore, the bandwidth of this analysis is limited to 1 Hz. This is a practical example of why the battery model can be limited to the bandwidth of the simulation application assumptions.

4.1.1 Reference Vehicle Specifications

Figure 7 illustrates the architecture of the reference vehicle in which the Subaru EE20 gasoline engine [21] is the main source of energy coupled with one axial flux permanent magnet (AFPM) generator, the YASA-400 [22]. The HESS is charged by the generator and operates as an intermediate energy buffer for a traction motor, another YASA-400. The power rating and other specifications of the main components of the power train are shown. In Autonomie, each component is defined by a behavioral model in the Simulink environment and Autonomie couples these models together to run an overall model of the vehicle over selected drive cycles (Fig. 8). An existing series-hybrid mid-size vehicle model is modified for this study. For engine modeling, a validated Prius model available in Autonomie is scaled up to 110 kW to reflect the ratings of the Subaru engine. For the generator and motor, the YASA-400 efficiency map data and other specifications of the machine are extracted and used for modifying an available machine model in Autonomie. Next, a modified control strategy is applied in state flow environment (Fig. 9). Three fixed

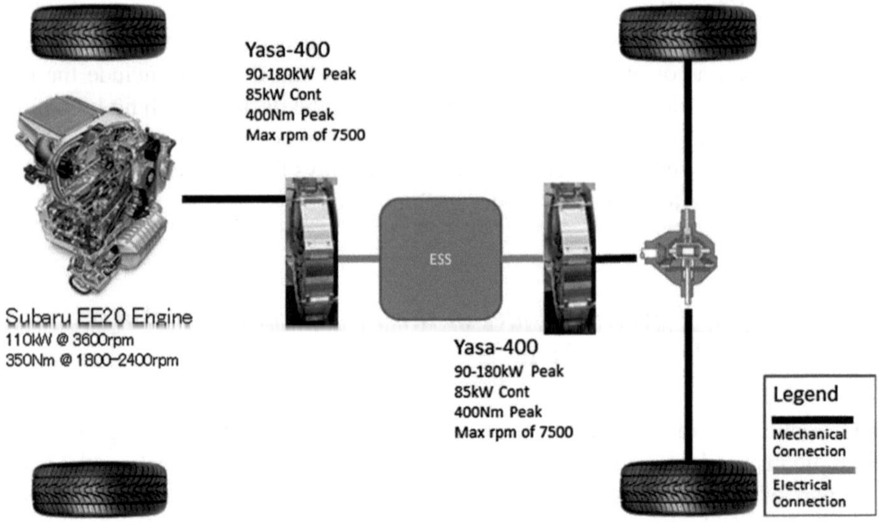

Fig. 7 The series-hybrid architecture of the reference vehicle

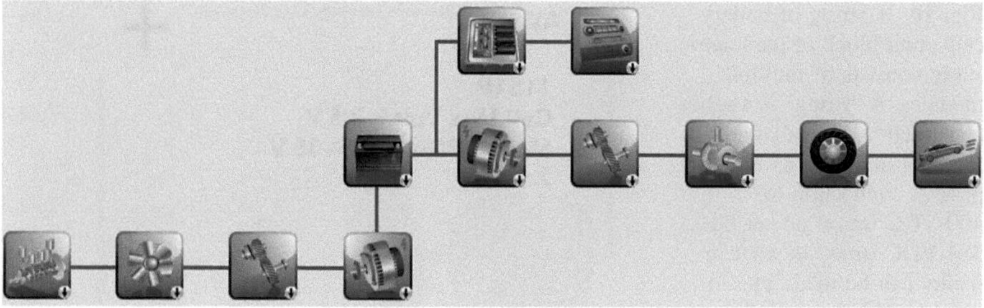

Fig. 8 Series architecture of the reference vehicle in autonomie

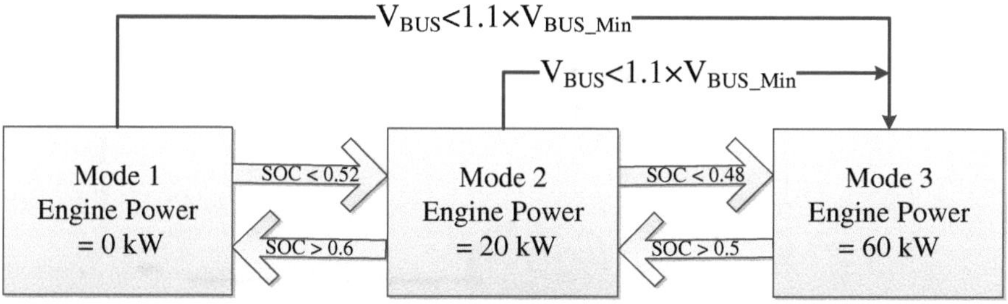

Fig. 9 Control strategy state flow diagram

engine operating points are used for the engine control strategy. They are (1) an idle mode (no load); (2) a high efficiency operating point, and (3) a wide open throttle (WOT) operating point. One of the three operating points is selected by the controller depending on battery state of charge (SOC). The idle operating point is selected when the battery SOC is high enough to provide energy to the propulsion motor for a while. If the SOC falls below the first low threshold, mode two is selected to recharge the battery to the first high threshold; in this way the engine operates at an efficient or high Brake Specific Fuel Consumption (BSFC) operating point. If the SOC goes below a second low threshold, mode three is activated to supply the load with maximum power from the engine which simultaneously charges the battery back to a second high threshold.

4.1.2 Parametric Study for Selecting Light and Efficient ESS Using Battery Pack Model

The cost of the battery in a Hybrid Electric Vehicle (HEV) is still high, so any solution for selecting a smaller capacity, lighter battery can reduce the total cost of the vehicle. A Hybrid ESS (HESS) which ends up with a minimum battery size is called a *battery light* solution. For minimizing the size, one method is to use optimization algorithms. However, because of power-bus voltage constraints in the

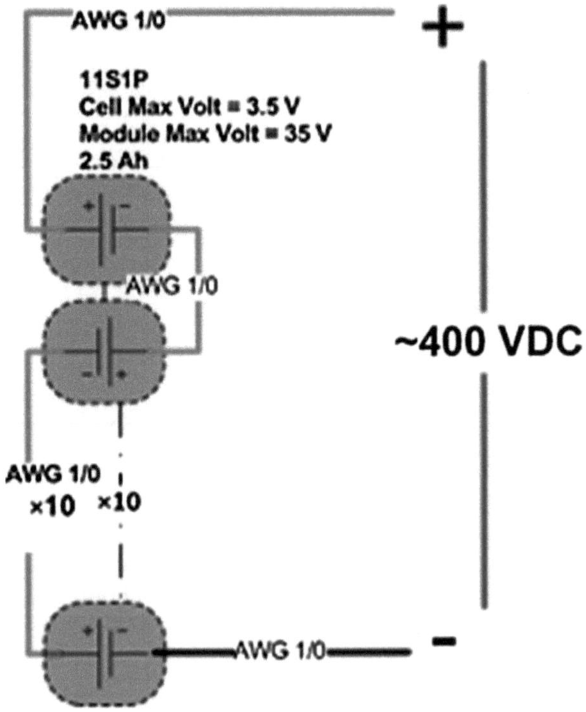

Fig. 10 A string of battery cells: unit block of parametric study consists of multiple modules. A "string" is a series construction of available modules and/or cells to allow passive connection to 400-VDC (max) power bus. 384 VDC (max) is what in reality can be made by considered battery modules. All HESS options are constrained to integer numbers of strings

passive configuration, the size of the battery or the UC string is not a continuously variable parameter which can be set as an optimization variable, but a discrete variable instead. One efficient method for testing for alternatives is to carry out a parametric study over available options which meet the voltage constraints. The upper range of dc voltage of the motor controller recommended for the YASA-400 is 400 V. This is the limiting factor which determines the maximum high line voltage of the ESS. 400 VDC is the voltage at which maximum motor peak power of 93 kW is achieved. Therefore, the nominal battery voltage of a string should be close to this value to meet the peak power of the vehicle acceleration requirement. To create a 400 VDC unit block (string) of battery or UC options, multiple modules are required to be connected in series (Fig. 10). In this study, a parametric study has been carried out for finding the best available ESS solution. The basic idea is to vary the number of parallel strings and compare fuel economies (MPG) of the vehicle. Using every possible practical combination of strings from Table 3, three possible configurations of battery only, UC only, and combined battery and UC are compared. Two Gen4 motor/generator controllers from Sevcon [23] are used to control both the YASA motor and generator. Battery or UC behavioral Simulink models in the Autonomie library are updated with information from Table 3. In the next two paragraphs, a commercially available string of UC and the commercially available strings of batteries are examined respectively.

Table 3 One string of alternative energy storage devices

Specifications	UC#1	ANR#1	AMP#1	NiMH#1
Cell maximum voltage (V)	2.83	3.5	3.4	1.4
Cell nominal voltage (V)	2.67	3.3	3.3	1.2
Number of cells in modules	6.0	10.0	6.0	6.0
Number of modules in series	24.0	11.0	19.0	45.0
Total number of cells	150.0	110.0	114.0	270.0
Nominal voltage (V)	384.0	363.0	376.2	324.0
Maximum voltage (V)	407.7	381.2	384.0	380.7
Estimated cells mass (kg)	26.7	10.5	51	66
Estimated cells volume (L)	21.3	6.0	33.8	26.6
Energy capacity (kWh)	–	0.9	5.3	2.1
Continuous power (kW)	33.3	18.0	48	32
Power in 1 s (kW)	<230	43.6	–	–

One String of UC: Referring to the bus voltage constraint, one string of UC can be sized. From data available in UC manufacturers' data sheets, 1 Hz is the assumed bandwidth limitation of UC's. Therefore, the largest demand at 1 Hz is considered as a reference for sizing one string of 400 VDC UC. Among standard drive cycles, US06 is the most aggressive with harsh accelerations and decelerations. Therefore, the power draw from the motor associated with the US06 drive cycle is used to size the UC bank, which contains a peak power of almost 100 kW.

Maxwell UC cells are used in this study, so the manufacturer's instruction for finding the number of cells and the capacitance of cells is used. In (4), W_{UC} is the energy requirement which in this case is 30 Wh. U_r is the high line voltage set to 384 VDC because the 384 VDC is the maximum voltage that can be built with both available battery modules (100 % charged) and available UC modules (100 % charge). This maximum voltage is the largest that the passive connection of available components can achieve while remaining below the maximum voltage of 400 VDC imposed by the motor drive. K, 0.5, is the factor which determines the ratio of low line to high line voltage. This value is recommended by the manufacturer [24].

$$\mathbf{W}_{\mathbf{UC}} = \frac{\mathbf{C}_0\mathbf{U}_{\mathbf{r}}^2(1 - \mathbf{K}^2)}{2(3600)} \tag{4}$$

1.95 F is computed for C_0; however, considering the manufacturer's recommended margin and also based on available cells (650 F, 2.67 V), overall capacitance of 4.33 F is the closest minimized option. 150 650-F cells are used in series to create one string of 384-V UC. In Table 3, additional specifications of the UC are listed.

One String of Batteries: Lithium-ion chemistry is recognized as one of the best solutions for today's and future EV's and HEV's because of high energy and power density. Also, nickel metal hydride (NiMH) batteries are widely used by manufacturers [25]. There are several manufacturers which make Li-ion and NiMH cells; however, the number of options available in the form of modules or packages which have the cell balancing function built-in are limited. Selected for consideration in this parametric study are the NiMH prismatic cells from Panasonic used in the Prius, the ANR cylindrical cells from A123, and the AMP14 prismatic cells from A123 because they come in packaging with balancing specifications. For each of these three options the specifications of the lightest 384 VDC string are shown in Table 3. The lightest strings have the lowest costs, and obviously the lowest capacity, so if more power or energy is required one or more strings with the same specifications must be added in parallel. Therefore, battery strings listed in Table 3 are the smallest units which fulfill the voltage requirement. Having identical voltage ratings for the battery and the UC strings allow them to work in parallel using the full SOC range of both.

4.1.3 HESS Parametric Study Results

Figure 11 shows the result of the parametric study performed by comparing the combined fuel economies of the vehicle with different ESSs. Combined fuel economy is defined as a weighted average of 55 % city driving (modeled with the UDDS drive cycle) and 45 % highway driving (modeled with the HWFET drive cycle). On the horizontal axis the type and number of strings are shown; as an example "ANR#2UC#1" represents two parallel strings of ANR cylindrical cells from A123 in parallel with one string of UC from Maxwell. In the ANR case, it is shown that when the number of parallel strings goes up, the fuel economy goes up as well until a peak happens at four strings. After this point, the negative impact of added weight on fuel economy outweighs the benefits of extra power and as a result a reduction in combined fuel economy is observed. This trend is clearly seen with the first three UC options.

However, the option with the highest fuel economy is not necessarily the best solution because the additional cost of adding a string must be considered. Therefore, there is always a tradeoff between increasing fuel economy and minimizing the cost and weight of the HESS. From Fig. 11, it can be argued that it is not worth investing money for additional strings for achieving only one MPG of additional fuel economy. Also, given the minimal AER strategy, a lighter, less costly battery option is desired if fuel economy is not affected too much. Attending to this argument, four attractive options are identified numbered one through four in Fig. 11.

4.1.4 HEV Energy Storage Design Example—Concluding Remarks

The final option selected by a vehicle designer would depend on the results of additional detailed design that may include augmenting the battery model with additional bandwidth capability, as discussed earlier in the chapter. This may be

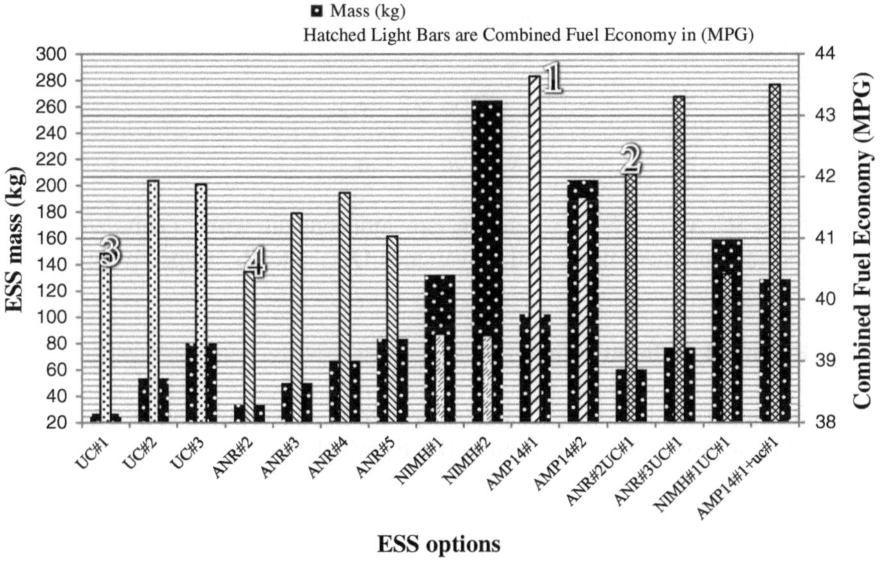

Fig. 11 Combined fuel economy for possible ESS options for the reference vehicle 100-kW powertrain

needed, for instance, when considering controller dynamics. One such example follows in the next section where an adaptive filter for real-time state-of-charge-estimation requires the additional dynamic simulation fidelity provided by the behavioral model augmented with two additional exponential modes not used in the HEV design example in this section (Fig. 11).

4.2 Battery Behavioral Model for Real-Time Battery State-of-Charge Estimation

Battery state of charge (SOC) estimation is a key issue for a battery management system (BMS), especially with the rapid development and commercialization of hybrid electric vehicles (HEVs) and plug-in hybrid electric vehicles (PHEVs) [26]. Accurate SOC estimation is crucial for determining the optimal operation mode of the PHEV or HEV, so that the best overall performance can be achieved. Compared with the batteries in low power applications, e.g., laptops or cell phones, batteries in high power applications encounter faster dynamics and more cycles of charging/discharging until the battery SOC can be recalibrated [26–28]. Therefore, it is more difficult to get an accurate SOC estimation in high power applications. In this section, two kinds of Gaussian approximation filters, the extended Kalman filter and the quadrature based Gaussian approximation filters, are briefly reviewed. They will be used to estimate the SOC.

Consider a class of nonlinear discrete-time dynamical systems and the measurement equation described by:

$$\mathbf{x}_k = \boldsymbol{f}(\mathbf{x}_{k-1}) + \mathbf{v}_{k-1} \tag{5}$$

$$\mathbf{y}_k = \boldsymbol{h}(x_k) + \mathbf{n}_k \tag{6}$$

where $\mathbf{x}_k \in \mathbf{R}^n$; $\mathbf{y}_k \in \mathbf{R}^m$, $\mathbf{v}_{k-1}$ and $\mathbf{n}_k$ are independent white Gaussian process noise and measurement noise with covariance $\mathbf{Q}_{k-1}$ and $\mathbf{R}_k$, respectively.

4.2.1 Extended Kalman Filter

Given the initial estimate state $\hat{\mathbf{x}}_0$ and covariance $\mathbf{P}_0$, the estimation of states can be obtained by the extended Kalman filter (EKF), which consists of the prediction and the update steps:

Prediction:

$$\hat{\mathbf{x}}_{k|k-1} = \boldsymbol{f}\left(\hat{\mathbf{x}}_{k-1|k-1}\right) \tag{7}$$

$$\mathbf{P}_{k|k-1} = \mathbf{F}_{k-1}\mathbf{P}_{k-1|k-1}\mathbf{F}_{k-1}^{\mathrm{T}} + \mathbf{Q}_{k-1} \tag{8}$$

where $\hat{\mathbf{x}}_{k-1|k-1}$ and $\mathbf{P}_{k|k-1}$ are the state estimate and the covariance at time $k-1$, respectively. $\mathbf{F}_{k-1}$ is the Jacobian matrix of $\boldsymbol{f}$ evaluated at $\hat{\mathbf{x}}_{k-1|k-1}$.

Update:

$$\mathbf{K}_k = \mathbf{P}_{k|k-1}\mathbf{H}_k^{\mathrm{T}}\left(\mathbf{H}_k\mathbf{P}_{k|k-1}\mathbf{H}_k^{\mathrm{T}} + \mathbf{R}_k\right) \tag{9}$$

$$\hat{\mathbf{x}}_{k|k} = \hat{\mathbf{x}}_{k|k-1} + \mathbf{K}_k\left(\mathbf{y}_k - \boldsymbol{h}\left(\hat{\mathbf{x}}_{k|k-1}\right)\right) \tag{10}$$

$$\mathbf{P}_{k|k} = (\mathbf{I} - \mathbf{K}_k\mathbf{H}_k)\mathbf{P}_{k|k-1} \tag{11}$$

where $\mathbf{H}_k$ and $\mathbf{K}_k$ are the Jacobian matrix of $\boldsymbol{h}$ evaluated at $\hat{\mathbf{x}}_{k|k-1}$ and the Kalman gain at time k, respectively.

Since the EKF is based on the first-order Taylor series expansion, the accuracy and stability of the EKF may not be sufficient for many applications with large uncertainties. Many quadrature-based Gaussian approximation filters can be used in the same filtering framework to improve the performance of the EKF.

4.2.2 Quadrature Based Gaussian Approximation Filters

Assuming that the probability density function (PDF) of the states is Gaussian, the Gaussian approximation filters can be obtained as follows [29–31]. Note that only the mean and covariance need to be calculated and the filtering algorithm also consists of two steps:

Prediction:

$$\hat{\mathbf{x}}_{k|k-1} = \sum_{i=1}^{N_p} W_i \boldsymbol{f}(\xi_i) \tag{12}$$

$$\mathbf{P}_{k|k-1} = \sum_{i=1}^{N_p} W_i \left(\boldsymbol{f}(\xi_i) - \hat{\mathbf{x}}_{k|k-1}\right)\left(\boldsymbol{f}(\xi_i) - \hat{\mathbf{x}}_{k|k-1}\right)^{\mathrm{T}} + \mathbf{Q}_{k-1} \tag{13}$$

where N_p is the total number of quadrature points, W_i is the point weight, and ξ_i is the transformed point obtained from the covariance decomposition, i.e.

$$\mathbf{P}_{k-1|k-1} = \mathbf{S}\mathbf{S}^{\mathrm{T}} \tag{14}$$

$$\xi_i = \mathbf{S}\gamma_i + \hat{\mathbf{x}}_{k-1|k-1} \tag{15}$$

where γ_i is the quadrature point to approximate the filtering integral with respect to the normal distribution $N(\mathbf{x}; \mathbf{0}; \mathbf{I}_n)$ with mean 0 and covariance $\mathbf{I}_n$. n is the state dimension.

Update:

$$\hat{\mathbf{x}}_{k|k} = \hat{\mathbf{x}}_{k|k-1} + \mathbf{L}_k(\mathbf{y}_k - \mathbf{z}_k) \tag{16}$$

$$\mathbf{P}_{k|k} = \mathbf{P}_{k|k-1} - \mathbf{L}_k \mathbf{P}_{xz}^{\mathrm{T}} \tag{17}$$

where

$$\mathbf{L}_k = \mathbf{P}_{xz}(\mathbf{R}_k + \mathbf{P}_{zz})^{-1} \tag{18}$$

$$\mathbf{z}_k = \sum_{i=1}^{N_p} W_i \boldsymbol{h}(\tilde{\xi}_i) \tag{19}$$

$$\mathbf{P}_{xz} = \sum_{i=1}^{N_p} W_i \left(\tilde{\xi}_i - \hat{\mathbf{x}}_{k|k-1}\right)\left(\boldsymbol{h}(\tilde{\xi}_i) - \mathbf{z}_k\right)^{\mathrm{T}} \tag{20}$$

$$\mathbf{P}_{zz} = \sum_{i=1}^{N_p} W_i \big(\boldsymbol{h}(\tilde{\xi}_i) - \mathbf{z}_k\big)\big(\boldsymbol{h}(\tilde{\xi}_i) - \mathbf{z}_k\big)^{\mathrm{T}} \tag{21}$$

where $\tilde{\xi}_i$ is the transformed point obtained from the decomposition of the predicted covariance, i.e.

$$\mathbf{P}_{k|k-1} = \widetilde{\mathbf{S}}\widetilde{\mathbf{S}}^{\mathrm{T}} \tag{22}$$

$$\tilde{\xi}_i = \widetilde{\mathbf{S}}\boldsymbol{\gamma}_i + \hat{\mathbf{x}}_{k|k-1} \tag{23}$$

γ_i and W_i can be chosen according to the Gauss-Hermite quadrature (GHQ) rule [29, 30] or the Unscented Transformation (UT) [31]. The corresponding filters using GHQ and UT for estimation are Gauss-Hermite quadrature filter (GHQF) and Unscented Kalman filter (UKF), respectively.

The points and weights for UT are given as follows.

For the UT with $2n + 1$ points [31], γ_i and W_i are given by

$$\begin{cases} \gamma_1 = [0, 0, \cdots 0, 0]^{\mathrm{T}}; & W_1 = \frac{2\kappa}{2(n+\kappa)} \\ \gamma_i = \sqrt{n+\kappa}\,\mathbf{e}_{i-1}; & W_i = \frac{1}{2(n+\kappa)}, \quad 2 \le i \le n+1 \\ \gamma_i = -\sqrt{n+\kappa}\,\mathbf{e}_{i-n-1}; & W_i = \frac{1}{2(n+\kappa)}, \quad n+2 \le i \le 2n+1 \end{cases} \tag{24}$$

where $\mathbf{e}_{i-1}$ is the unit vector in $\mathbf{R}^n$ with the (i − 1)th element being 1 and κ is a tuning parameter with the suggested optimal value $\kappa = 3 - n$ for Gaussian distributions [31]. The UT is exact for all polynomials of the form $x_1^{i_1} x_2^{i_2} \cdots x_n^{i_n}$ with $1 \le i_1 + \cdots + i_n \le 3$ [30].

For the univariate GHQ rule with m quadrature points, γ_i and W_i can be calculated as follows [29]. If $m = 1$, then $\gamma_1 = 0$ and $W_1 = 1$. If $m > 1$, first construct a symmetric tri-diagonal matrix J with zero diagonal elements and $J_{i,i+1} = J_{i+1,i} = \sqrt{i/2}$, $1 \le i \le m - 1$. Then the quadrature point γ_i is calculated by $\gamma_i = \sqrt{2}\varepsilon_i$, where ε_i is the ith eigenvalue of J. The corresponding W_i is calculated by $W_i = (\mathbf{v}_i)_1^2$ where $(\mathbf{v}_i)_1$ is the first element of the ith normalized eigenvector of J. The univariate GHQ rule with m points is exact up to the $(2m - 1)$th order of polynomials [30].

The multivariate GHQ rule extends the univariate m-point set to the n-dimensional point set by the tensor product rule [29, 30]. It is exact for all polynomials of the form $x_1^{i_1} x_2^{i_2} \cdots x_n^{i_n}$ with $1 \le i_j \le 2m - 1$ [30]. However, the total number of points $N_p = m^n$ increases exponentially with the dimension n. Hence, it is hard to use for high dimensional problems. To alleviate this problem, the sparse Gauss-Hermite quadrature can be used [32]. In this paper, the conventional Gauss-Hermite quadrature is used since the dimension of this problem is three.

Since the EKF uses the first order Taylor expansion to approximate the non-linear function, the approximation can be accurate only to the first order. But the UKF approximation accuracy can be up to the 3rd order polynomials because it uses the Unscented Transformation. Using a more advanced quadrature rule for approximation, the GHQF can achieve arbitrarily higher order estimation accuracy than the UKF.

4.2.3 Performance Testing and Results

The real-time battery SOC estimation approach has been tested in simulation using the experimentally determined battery model described in Sect. 3 for a 6.8 Ah Ultralife UBBL10 Lithium-ion battery module. The test current is shown in Fig. 3. For the model represented by (1) and (2), the process noise and measurement noise are assumed to be white Gaussian and are added to the battery model. Thus, this battery model is assumed to accurately represent the true system within the bandwidth of the original current excitation (i.e., Fig. 3) used for parameter extraction. To test the robustness of the filters to uncertain initial conditions, an initial value of 40 % SOC was given to the model instead of the true value of 100 % SOC. The initial values for $\hat{V}_{c1,0}$ and $\hat{V}_{c2,0}$ are set to be 0.05 and 0.05 V instead of the true values 0 and 0 V, respectively. The initial covariance is set to be $\mathbf{P}_0 = \mathrm{diag}([0.01, 0.01, 0.1])$. The process noise covariance is $Q = 1 \times 10^{-8}$ and the measurement noise is $R = 1 \times 10^{-4}$.

The SOC estimation from running one time is shown in Fig. 12 with a zoomed view in Fig. 13. The SOC estimation with the EKF diverged from the true SOC and thus was not plotted here. Figure 14 shows the SOC estimation errors.

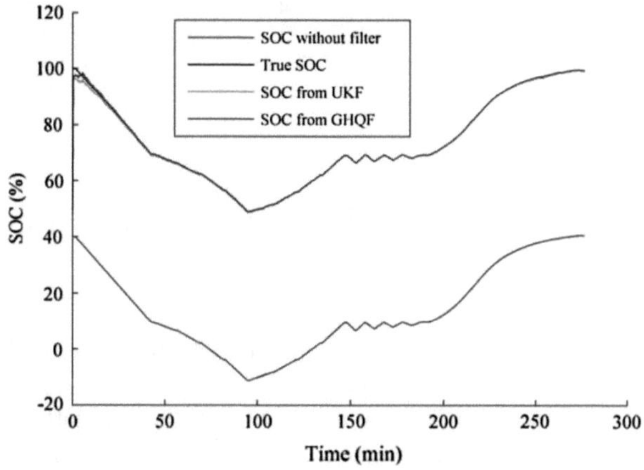

Fig. 12 SOC estimation results from running the simulation one time

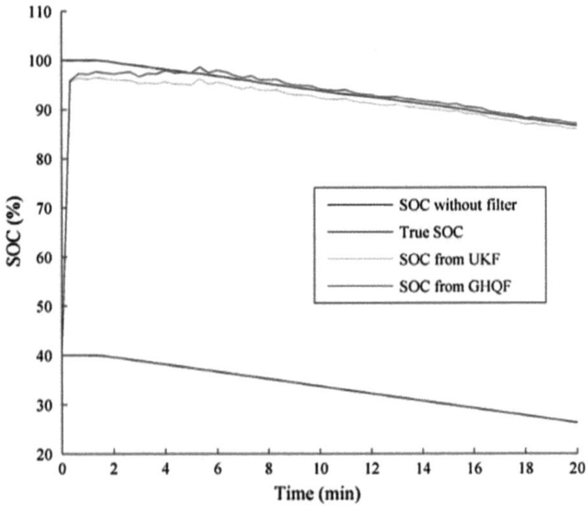

Fig. 13 Zoomed view of the SOC estimation in the first 20 min of the simulation

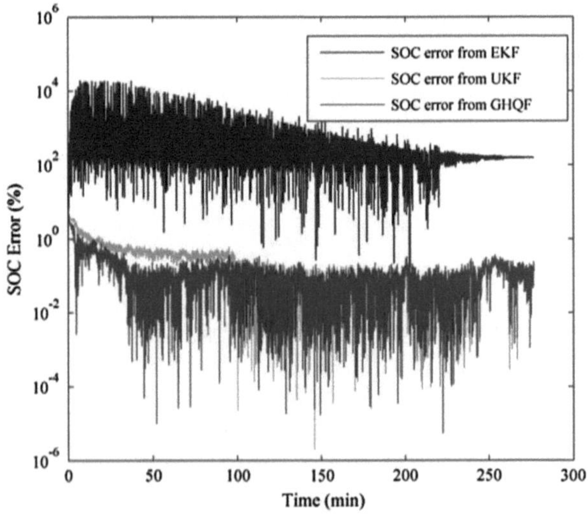

Fig. 14 SOC estimation errors versus time for the three different filters

The results shown in Fig. 15 are generated from 50 runs of the simulation, where the root-mean-square error (RMSE) of the 50 runs is used as the criteria for comparison. Since these are 50 runs of the simulation rather than 50 battery tests, no battery aging related issues are involved or considered at this point.

From these results it can be observed that the estimated SOC quickly converges to the true SOC. The errors from both GHQF and UKF quickly drop to 1 % and stay below 0.2 % in most parts. The GHQF converges faster than the UKF and proves to be the best estimator compared with the UKF and EKF in the battery SOC estimation problem.

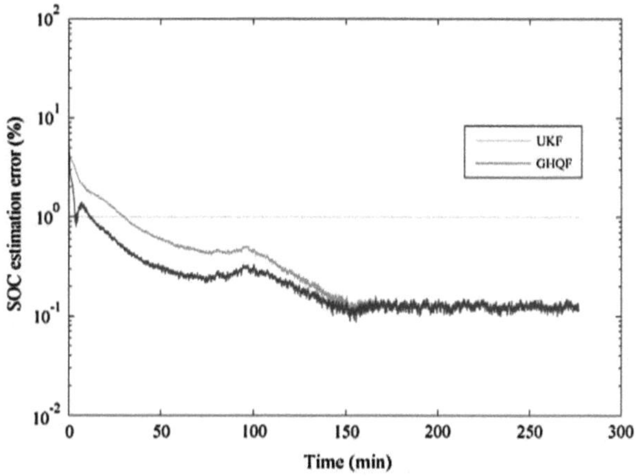

Fig. 15 Root-mean-square error of SOC for the UKF and GHQF after 50 simulation runs

4.2.4 Real-Time SOC Estimation Filters—Concluding Remarks

The Gauss-Hermite quadrature filter was introduced with an otherwise accurate open-loop behavioral battery model to estimate battery SOC in real-time. The ability of the filter to self-start was tested by giving the model poor initial values. The EKF and the UKF were built on the same battery model for comparison. The results show that the GHQF is capable of starting with large deviated initial conditions to quickly converge to the true SOC. SOC estimations from the EKF diverged from the true SOC indicating that EKF is not capable of being used on the selected type of electrical analogue battery model with poor initial conditions. The UKF is able to provide satisfactory estimation accuracy but it converges to the true SOC comparatively more slowly than the GHQF. SOC estimation errors for GHQF and UKF remains between 0.13 and 1 % of SOC in most parts, with the error from UKF slightly higher than that from GHQF. Therefore, when both the self-starting capability and estimation accuracy are considered, the GHQF excels in comparison to commonly reported EKF and UKF for real-time SOC estimation with the common electrical-analogue battery model, because the GHQF is known to be more capable of handling high order polynomial based models [30, 32].

5 Summary

This chapter has examined a well-known behavioral model (i.e., Fig. 1) used with Li-ion batteries as well as other chemistries. The need for accuracy in dynamic simulation as well as real-time application in battery management systems varies widely. Physics based models are essential to the design of batteries, but are a burden in systems engineering and in detailed design of applications that use batteries. The large format Li-ion battery has many promising emerging applications, and thus there is a need for useful models of large-format Li-ion batteries. But the use of single-cell behavioral models as a means for directly scaling to models of large format batteries has some disadvantages. This chapter presented an emerging technique for directly modeling a large-format battery with a flexible electrical-analogue battery behavioral model. A general procedure for extracting the parameters of the model is discussed, and additional references to the literature are given for the details should the reader choose to practice the method. A key concept in this approach is to consider the bandwidth of the model and that of the simulation application to be tied. Thus, the available degrees of freedom in the form of the time constants of the dynamic portion of the behavioral model are selected a priori to cover accurately the bandwidth of the application. The result is accurate dynamic modeling of the terminal voltage of the battery while maintaining accurate track of the key internal variable, which is the state-of-charge of the battery.

Two examples are presented in this chapter. The first is a system level parametric study of the available battery technologies to satisfy a design problem with a series hybrid electric vehicle. At the level of the study, which involved comparing battery weight, cost, and the resulting predicted vehicle fuel economy, a dynamic model with independent state variables was not required. In this case, a single resistance was used to model the difference between the terminal voltage and the open-circuit voltage of the battery during dynamic current fluctuations. In such a model a single resistor is often thought to represent the internal "dc" resistance of the battery; but this is misleading. In fact, in a real battery, the terminal voltage continues to change for many hours even after the battery current has ceased flowing. In the bandwidth-centric dynamic behavioral modeling reported in [11, 12], this resistance actually models all dynamic changes in the battery terminal voltage that are completed faster than the characteristic time scale of the current, in other words, the resistor is less a physical resistance and more a "catch all" for battery dynamics that occur faster than that included in the model. The value of this limiting case in the first example is that it allowed the controller algorithm to be stimulated by the capacity-limited bandwidth of two different battery types and in synergy with the capacity-limited bandwidth of the ultracapacitor so that relative vehicle (i.e., system) performance could be judged in terms of the fuel efficiency of the engine.

However, internal dynamics is one of the more common causes for battery state-of-charge misestimation. So, in the second example, it was shown that adaptive filters could be very effective at estimating and tracking the internal state-of-charge of a large format battery with the benefit of the presented behavioral model. The behavioral model was augmented over that in the first example with two RC tank circuits to add internal state variables. The time constants of the RC circuits were selected not based on an unlikely extraction of natural exponential moments inside the battery reflecting independent physical processes, but instead the time constants were assigned a priori (but not the R's and C's themselves, see Sect. 3) to cover the expected application bandwidth to the necessary accuracy. The justification for this approach is that the presented behavioral model represents in fact a truncated exponential series that approximate the actual dynamics of the battery (linear or non-linear) to be expected over the application bandwidth.

The result was impressive performance in converging to the true SOC after starting from an arbitrary initial condition, with the Gauss-Hermite quadrature filter (GHQF) performing the best in this respect. After relaxation, both the Unscented Kalman Filter (UKF) and the GHQF filter maintained accurate track of the SOC (less than 1 % error) during dynamic variations in battery charging and discharging in the presence of Gaussian distributed noise. The conclusion taken from both examples is that matching the expected dynamics of the simulation/application environment with a flexible-bandwidth behavioral model of a large-format Li-ion battery is an effective and economical engineering method.

Acknowledgments The authors acknowledge the contributions made by Dr. Jian Wei Li to this chapter while a graduate student at Mississippi State University prior to his employment in industry.

References

1. Zhang W, Dong D, Cvetkovic I, Lee FC, Boroyevich D (2011) Lithium-based energystorage management for DC distributed renewable energy system. In: Energy conversion congress andexposition (ECCE), pp 3270–3277
2. Chen Y, Li G, Zhang F (2011) Simulation study on control strategy for a hybrid electric vehicle with batteryand ultracapacitor. In: International conference on electric information and control engineering (ICEICE), pp 2601–2604
3. Tremblay O, Dessaint LA, Dekkiche AI (2007) A generic battery model for the dynamic simulation of hybridelectric vehicles. In: IEEE vehicle power and propulsion conference, pp 284–289
4. Chen M, Rincon-Mora GA (2006) Accurate electrical battery model capable of predicting runtime and IVperformance. IEEE Trans Energy Convers 21:504–511
5. Gao L, Liu S, Dougal RA (2002) Dynamic lithium-ion battery model for system simulation. IEEE TransCompon Packag Technol 25:495–505
6. Schweighofer B, Raab KM, Brasseur G (2003) Modeling of high power automotive batteries by the use of anautomated test system. IEEE Trans Instrum Meas 52:1087–1091
7. Hussein AAh, Kutkut N, Batarseh I (2011) A hysteresis model for a lithium battery cell with improvedtransient response. In: Applied power electronics conference and exposition (APEC), pp 1790–1794
8. Zhang J, Ci S, Sharif H, Alahmad M (2010) An enhanced circuit-based model for single-cell battery. In:Applied power electronics conference and exposition (APEC), pp 672–675
9. Abu-Sharkh S, Doerffel D (2004) Rapid test and non-linear model characterisation of solid-state lithium-ionbatteries. J Power Sources 130:266–274
10. Shafiei A, Momeni A, Williamson SS (2011) Battery modeling approaches and management techniquesfor plug-in hybrid electric vehicles. In: Vehicle power and propulsion conference (VPPC), pp 1–5
11. Li J, Mazzola M, Gafford J, Jia B, Xin M (2012) Bandwidth based electrical-analogue battery modeling forbattery modules. J Power Sources 218:331–340
12. Li J, Mazzola M (2013) Accurate battery pack modeling for automotive applications. J Power Sources237:215–228
13. Watrin N, Bouquain D, Blunier B, Miraoui A (2011) Energy-source-sizing methodology for hybrid fuel cellvehicles based on statistical description of driving cycles. In: Vehicle power and propulsion conference (VPPC), pp 1–5

14. Zheng M, Qi B, Du X (2009) In: 4th IEEE conference on industrial electronics and applications (ICIEA), pp2867–2871
15. Tsang KM, Sun L, Chan WL (2010) Dynamic model for characteristics of Li-ion battery on electric vehicle.Energy Convers Manag 51:2857–2862
16. Chen M, Rincon-Mora GA (2006) Accurate electrical battery model capable of predicting runtime and IVperformance. IEEE Trans Energy Convers 21:504–511
17. Kroeze RC, Krein PT (2008) Electrical battery model for use in dynamic electric vehicle simulations. In:Power electronics specialists conference, pp 1336–1342
18. Hu Y et al (2009) A technique for dynamic battery model identification in automotive applications usinglinear parameter varying structures. Control Eng Pract 17:1190–1201
19. Abu-Sharkh S, Doerffel D (2004) Rapid test and non-linear model characterisation of solid-state lithiumionbatteries. J Power Sources 130:266–274
20. Autonomie, Argonne National Laboratory (2014) http://www.autonomie.net. Accessed 26 Sept 2014
21. Boxeer—Subaru Diesel Engine (2014) http://www.boxeer.com. Accessed 26 Sept 2014
22. YASA Motors (2014) http://www.yasamotors.com/products. Accessed 26 Sept 2014
23. Sevcon (2014) http://www.sevcon.com/ac-controllers/gen-4-size-8.aspx. Accessed 26 Sept 2014
24. Miller JM et al (2009) Why hybridization of energy storage is essential for future hybrid, plug-in and batteryelectric vehicles. In: Energy Conversion Congress and Exposition. IEEE, pp 2614–2620
25. Burke AF (2007) Batteries and ultracapacitors for electric, hybrid, and fuel cell vehicles. Proc IEEE 95(4):806–820
26. Pang S et al (2001) Battery state-of-charge estimation. In: Proceedings of the American controlconference, pp 1644–1649
27. Piller S et al (2001) Methods for state-of-charge determination and their applications. J Power Sources 96(2001):113–120
28. Vasebi A et al (2007) A novel combined battery model for state-of-charge estimation in lead-acid batteriesbased on extended Kalman filter for hybrid electric vehicle applications. J Power Sources 174:30–40
29. Arasaratnam I et al (2007) Discrete-time nonlinear filtering algorithms using Gauss–Hermite quadrature. Proc IEEE 95:953–977
30. Ito K, Xiong K (2000) Gaussian filters for nonlinear filtering problems. IEEE Trans Autom Control 45:910–927
31. Julier S et al (2000) A new method for the nonlinear transformation of means and covariances in filters andestimators. IEEE Trans Autom Control 45:77–482
32. Jia B et al (2011) Sparse Gauss-Hermite quadrature filter with application to spacecraft attitude estimation. J Guidance Control Dyn 34:367–379

High Voltage Cathode Materials

Christian M. Julien, Alain Mauger, Karim Zaghib and Dong Liu

1 Introduction

Energy production, storage and utilization are basic problems in industrialized countries from the economical, environmental, political and technological points of view. On the crucial technological aspects is energy saving and storage for powering diverse devices such as microelectronics (cellar phones, tablers, laptops), green transportation (hybrid electric vehicles (HEVs), electric vehicles (EVs), bikes) and smart grid. The development of inexpensive high-power-density secondary (rechargeable) electrochemical cells such as Li-ion batteries (LiBs) would also have a highly positive environmental impact through the replacement of gasoline motors. Intensive studies have been devoted to electrode materials for LiBs. Indeed, their larger gravimetric and volumetric energy densities make them good candidates to alleviate the environmental and energy pressures [1, 2]. These batteries are commonly identified by the name of the positive electrode (currently named cathode) element, i.e. a transition-metal insertion compound consisting of a host structure for Li^+ ions and the operating voltage of the redox potential. Three classes of materials can be considered: the 3-V cathodes, e.g. MnO_2, $LiFePO_4$, the 4-V cathodes, e.g. $LiCoO_2$, $LiCo_{1-y-z}Ni_yAl_zO_2$, $LiNi_{1/3}Mn_{1/3}Co_{1/3}O_2$, $LiMn_2O_4$, and the 5-V cathodes that are considered in this chapter. With high voltage, the battery delivers an energy density higher than 500 Wh kg^{-1} at an industrial scale

C.M. Julien (✉)
PHENIX, Université Pierre et Marie Curie, Paris, France
e-mail: Christian.Julien@upmc.fr

A. Mauger
IMPMC, Université Pierre et Marie Curie, Paris, France

K. Zaghib · D. Liu
IREQ, Varennes, QC, Canada

Z. Zhang and S.S. Zhang (eds.), *Rechargeable Batteries*,
Green Energy and Technology, DOI 10.1007/978-3-319-15458-9_17

requesting fast discharge rates over many charge/discharge cycles with good thermal stability [3].

Several reviews have been published on electrode materials for LiBs [2, 4–8]. Ellis et al. [9] provided an overview of the major developments in the area of cathode materials but did not mention the 5-volt systems. Fergus [10] has critically exposed developments in cathode materials for LiBs including a comparison of the performance characteristics of the promising cathode materials and approaches for improving their performances. Zaghib et al. [11] gave an overview of olivines in lithium batteries for green transportation and energy storage. Concerning the high-voltage materials, several reports have been dedicated to the spinel $LiNi_{0.5}Mn_{1.5}O_4$ (denoted LNM) and its derivatives [12–15]. Recently, Kraytsberg et al. [14] outlined and analyzed the achievements, and challenges in the field of high-voltage cathode materials for Li-ion cells.

In this chapter we give the state of the art in the understanding of the properties of the 5-V cathode materials. Owing to the progress in this field, these compounds are promising active cathode elements for the next generation of Li-ion batteries to improve the technology of the energy storage and electric transportation. This paper is organized as follows. In Sect. 2, the preliminary considerations are dedicated to the principles governing LiBs and the electron energies in the electrodes. Section 3 addresses physico-chemical and electrochemical properties of the $LiNi_{0.5}Mn_{1.5}O_4$ spinel oxides and their related doped parents. Section 4 is devoted to other 5-V cathodes materials with olivine, inverse spinel, fluorovanadate and fluorophosphate structures, and Sect. 5 gives some concluding remarks.

2 Electron Energy in LiBs

The operating potential of a cell is limited by the open-circuit voltage V_{oc}, which is the potential difference across terminals of the battery when no current is being drawn

$$V_{oc} = -\frac{1}{nF}(\mu_A - \mu_C), \tag{1}$$

where $(\mu_A - \mu_C)$ is the difference in the electrochemical potential of the anode (A) and the cathode (C), n is the number of electrons involved in the chemical reaction of the cell, and F is the Faraday's constant. For high-voltage cathode that Is a semiconductor in nature, it implies to consider the Fermi level as discussed by Goodenough et al. [2]. If the active transition-metal cation contains a localized d-electron manifold, the manifold acts as a redox couple, e.g. $Ni^{2+/4+}$ in LNM. Successive redox couples are separated by an on-site effective Coulomb correlation energy U that can be large when augmented by either a crystal-field splitting or an intra-atomic exchange splitting [1]. However, when the Fermi energy E_{FC} of the cathode material approaches the top of the anion p bands of the host, the p-d covalent mixing may transform the correlated d electrons at E_{FC} into band

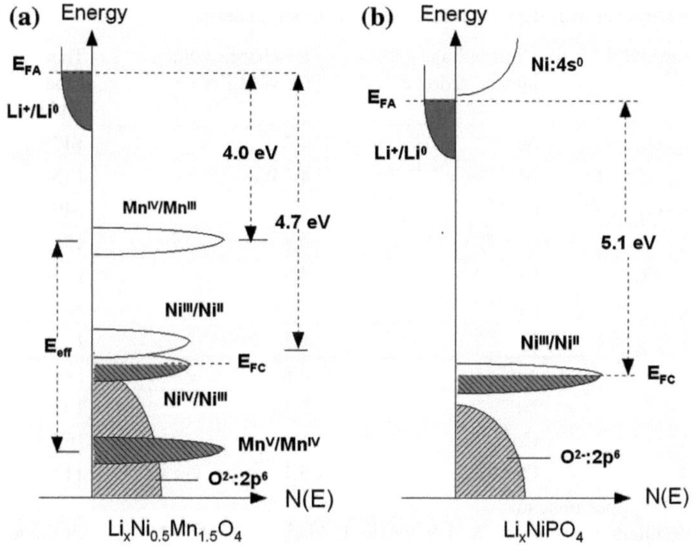

Fig. 1 Schematic density of states and Fermi energies for **a** $Li_xNi_{0.5}Mn_{1.5}O_4$ spinel and **b** Li_xNiPO_4 olivine cathodes. The origin of energies is chosen at the Fermi energy of lithium metal

electrons occupying one-electron states. In the absence of a crystal-field splitting of the *d* orbitals at E_{FC}, which is the case for Ni(IV) to Ni(II), the one-electron states are not separated by any on-site energy *U* and there is no step in the voltage of the battery. E_{FC} is moved from one formal valence state to another upon the reduction or oxidation of the host.

Figure 1 reports a comparison of the schematic density of states and Fermi energies for (a) $Li_xNi_{0.5}Mn_{1.5}O_4$ spinel and (b) Li_xNiPO_4 olivine cathodes. Access to Ni(III) and (Ni(IV) valence states is possible in the spinel case, while the Mn^{5+}/Mn^{4+} couple lies well-below the top of the O 2*p* bands; the solid electrolyte interface (SEI) layer formed at voltage V > 4.3 V is self-limiting and Li-permeable. On the other hand, the properties of Li_xNiPO_4 (LNP) are greatly influenced by the counteraction of the $(PO_4)^{3-}$ polyanion on the Ni^{2+}/Ni^{3+} couple that is pinned at the top of the O 2*p* bands, which provides an intrinsic voltage limit for this cathode. Table 1 lists the electrochemical data of the high-voltage cathode materials.

3 Spinel Frameworks

3.1 *Structure of* $LiNi_{0.5}Mn_{1.5}O_4$

$LiNi_{0.5}Mn_{1.5}O_4$ crystallizes in two possible crystallographic structures: the face-centered spinel (S.G.$Fd\overline{3}mm$) and the simple cubic phase (S.G. $P4_332$) named as "ordered spinel". The difference in the diffraction patterns of the cubic $P4_332$

Table 1 Electrochemical data for high-voltage cathode materials

Cathode material	High-voltage plateau redox couple	Discharge voltage[a] (V vs. Li^+/Li^0)	Theoretical capacity ($mAh\ g^{-1}$)
$LiNi_{0.5}Mn_{1.5}O_4$	$Ni^{2+/4+}$	4.7	147
$LiNi_{0.45}Mn_{1.45}Cr_{0.1}O_4$	$Ni^{2+/4+}/Cr^{3+/4+}$	4.7/4.8	145
$LiCr_{0.5}Mn_{1.5}O_4$	$Cr^{3+/4+}$	4.8	149
$LiCrMnO_4$[b]	$Cr^{3+/4+}$	4.8	151
$LiCu_{0.5}Mn_{1.5}O_4$	$Cu^{2+/3+}$	4.9	147
$LiCoMnO_4$[b]	$Co^{3+/4+}$	5	147
$LiFeMnO_4$[b]	$Fe^{3+/4+}$	5.1	148
$LiNiVO_4$	$Ni^{2+/3+}$	4.8	148
$LiNiPO_4$	$Ni^{2+/3+}$	5.1	167
$LiCoPO_4$	$Co^{2+/3+}$	4.8	167
Li_2CoPO_4F	$Co^{2+/4+}$	5.1	115

[a]Voltage of the upper plateau
[b]A partial delithiation occurs at ca. 4 V versus Li^0/Li^+ due to the $Mn^{3+/4+}$ redox couple

symmetry is characterized by additional weak Bragg lines located at 2θ = 15.3, 39.7, 45.7 and 57.5° [16–18] due to the ordering of the Ni and Mn cations. The primitive unit cell in the cubic system ($P4_332$) resulting from the 1:3 cation ordering is illustrated in Fig. 2. The cubic cell parameter Å falls from a = 8.243 Å for $LiMn_2O_4$ to a = 8.1685 Å for $LiNi_{0.5}Mn_{1.5}O_4$. In the ordered phase, the larger Ni^{2+} ions (ionic radius 0.69 Å) occupy only the 4b sites that give more room than the 16d sites of the normal spinel structure [19]. The cation distribution in the $P4_332$

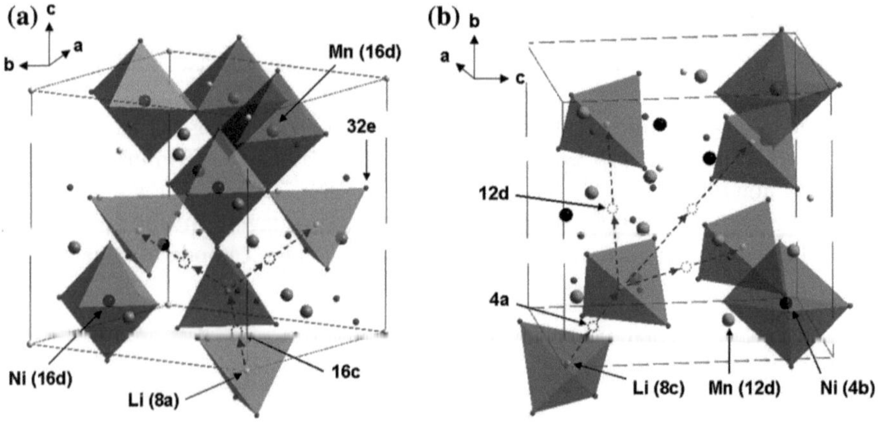

Fig. 2 Schematic representation of the structure of $A[B_2]O_4$ spinel lattices. **a** the smallest (primitive) cubic unit cell of normal spinel ($Fd\bar{3}m$) and **b** the unit cell of the 1:3 ordered spinel ($P4_332$). The structure is composed of alternating octants of AO_4 tetrahedra and B_4O_4 cubes to build the fcc unit cell [28]

symmetry is then Li on 8*c*, Ni on 4*b*, Mn on 12*d*, and O(1) and O(2) oxygen ions occupy the 24*e* and 8*c* Wyckoff positions, respectively. The net result is thus a significant optimisation of space occupation leading to a reduced unit cell volume. In addition, Strobel et al. [20] suggested that the occurrence of cation ordering corresponds to the larger valence difference $\Delta Z(Mn - M) \geq 2$. Moreover, it has been reported that the lattice parameters depends on the synthetic route [21–26]. Similar results have been reported for $LiMg_{0.5}Mn_{1.5}O_4$ [27]. Magnetic measurements have evidenced the ferrimagnetic ordering at $Tc = 129$ K in LMN. When no impurity phase (NiO, $Li_yNi_{1-y}O$) is detected, the ferrimagnetic behavior is an intrinsic property trivially due to the collinear ferrimagnetic ordering in which both the Ni sublattice and the Mn sublattice are ferromagnetic [28, 29] and spin-polarized in the opposite direction.

It has been pointed out that phase-pure $LiNi_{0.5}Mn_{1.5}O_4$ is difficult to synthesize due to impurities such as NiO and/or $Li_yNi_{1-y}O$ usually exist [29]. Mukai and Sugiyama identified the $Li[Ni_{1/2}Mn_{3/2}]O_4$ ($P4_332$) from dc-susceptibility measurements [30]. As an alternative approach, the off-stoichiometric material $LiNi_{0.5}Mn_{1.5}O_{4-\delta}$ has been synthesized in the disordered structure [9, 31–35]. The small voltage plateau at ca. 4 V for $LiNi_{0.5}Mn_{1.5}O_{4-\delta}$ is attributed to the Mn^{3+}/Mn^{4+} couple due to the existence of a small amount of Mn^{3+} ions resulting from the charge compensation of oxygen loss. The crystal structure of $LiMn_{1.5}Ni_{0.5}O_4$ has been determined by various techniques including XRD [35, 36], neutron diffraction [36, 37], XAFS [38], XANES [39], ex situ electron diffraction [36], in situ synchrotron X-ray absorption (XAS) [40], Raman scattering spectroscopy [19, 25, 41], ex situ Fourier transform infrared (FTIR) spectroscopy [25, 41–43] and in situ FTIR [44]. Raman and FTIR spectroscopy have been proven to be effective tools in differentiating the ordered versus disordered LNM structures; the sample with cationic ordering is known to exhibit characteristic infrared bands at 650, 465 and 430 cm^{-1} [43].

3.2 Origin of the High Voltage

The electrochemical reaction in $Li_xNi_yMn_{2-y}O_4$ has been described by several groups [2, 17, 45, 46]. Using ultraviolet photoelectron spectroscopy, Gao et al. [45] studied the top of the valence band of $Li_xNi_yMn_{2-y}O_4$ for a series of samples with $0.0 < y < 0.5$. A partial density of states attributed to Ni 3*d* electrons is located about 0.5 eV above that of the Mn 3*d* e_g electrons. When $y = 0$, the voltage plateau of Li// $LiMn_2O_4$ is located at 4.1 V. As *y* increases, the capacity associated to the 4.1 V plateau decreases as $1 - 2y$ Li per formula unit and a new plateau at 4.7 V appears. The capacity associated to the 4.7 V plateau increases as 2*y* Li per formula unit, so that the total capacity of the samples (the sum of the contributions from the 4.1 and 4.7 V plateau) is constant. This is taken as evidence that the oxidation state of Ni in these samples is +2, and therefore they can be written as $Li^+Ni_y^{2+}Mn_{1-2y}^{3+}Mn_{1+y}^{4+}O_4^2$.

The 4.1 V plateau is related to the oxidation of Mn^{3+} to Mn^{4+} and the 4.7 V plateau to the oxidation of Ni^{2+} to Ni^{4+}. The reason for this behavior is shown in Fig. 3.

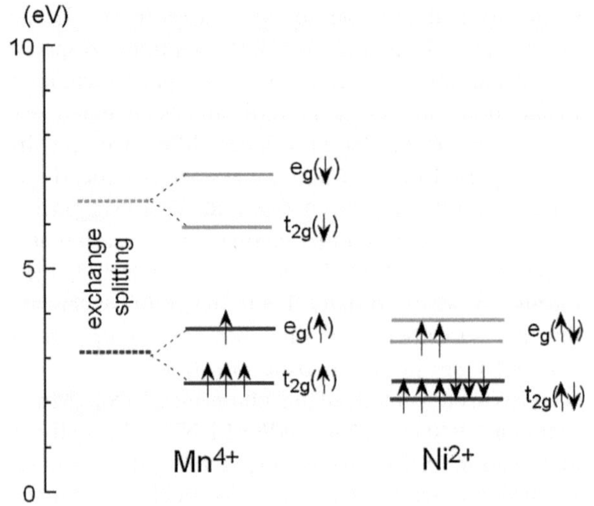

Fig. 3 Schematic diagram showing the 3*d* electronic levels of Mn^{3+} and Ni^{2+} in $LiNi_yMn_{2-y}O_4$ spinel. It is noted that Ni^{2+} favors the low-spin configuration in $LiNi_yMn_{2-y}O_4$

The crystal field splits the 3*d* levels of Mn and Ni octahedrally coordinated with oxygen into e_g and t_{2g} levels [45]. For Mn^{3+}, among the four $3d^4$ electrons with majority spin (↑) three electrons are on $t_{2g}(\uparrow)$ and one electron is on $e_g(\uparrow)$. In the low-spin configuration, the $3d^8$ electrons of Ni^{2+} have six electrons on the $t_{2g}(\uparrow\downarrow)$ level and two electrons on the $e_g(\uparrow\downarrow)$ level. As an electron is removed from Mn^{3+}, it is removed from Mn $e_g(\uparrow)$, which has an electron binding energy at around 1.5–1.6 eV, and this is on the 4.1-V plateau. When there are no more electrons left on Mn $e_g(\uparrow)$ (all Mn are oxidized to Mn^{4+}), electrons are removed from Ni $e_g(\uparrow\downarrow)$ which has an electron binding energy of about 2.1 eV, and the voltage plateau moves up to 4.7 V because of the increased energy needed to remove electrons.

3.3 Synthesis of LNM Spinels

$LiMn_{1.5}Ni_{0.5}O_4$ particles, because of their easy preparation, and their economical and environmental advantages, have been prepared by various synthetic methods with different morphologies and sizes. Generally, LNM powders prepared by solid-state reaction (SSR) have a large deficiency in oxygen, which results in the appearance of a wide voltage plateau at 4.1 V versus Li^0/Li^+ and the presence of impurity phase such as NiO or/and $Li_yNi_{1-y}O$. It is thus detrimental to the electrochemical properties. The amount of impurity phase can be reduced by annealing the sample in high O_2 pressure (at ca. 2 MPa) [30]. Spinel oxides were synthesized using carbonates [47], mixture of Li_2Co_3, NiO and electrolytic MnO_2 as reactants [48, 49]. Modified solid-state reactions include one-step process [50], mechanical activated SSR [49, 51], and ball milling [52]. LNM powders were also grown by low-temperature SSR using excess of oxalic acid [53], two-step and three-step synthesis [54]. Ohzuku et al. [55, 56] reported that an optimum condition to prepare a LNM material is a two-step solid state reaction, i.e. crystallization at 1000 °C followed by oxidation at 700 °C.

Molten salt synthesis based on the use of salts with low melting point has been reported to be one of the simplest techniques for preparing ceramic materials [57]. Well-faced LNM crystals were synthesized by molten salt technology using stoichiometric amount of LiOH, $Ni(OH)_2$, and γ-MnOOH (2:1:3) calcined at 700–1000 °C [57]. Micro-sized LNM crystals in plate and octahedral shapes were synthesized by molten-salt method in LiCl and LiCl-KCl fluxes; the main surface facets on the plates were determined to be (112) crystal planes [58]. Recently, microscale LNM was synthesized using nanothorn structured MnO_2 mixed with Li and Ni precursors [59].

Many approaches have been used to synthesize LNM by wet-chemical methods using aqueous or alcoholic solutions providing high-purity spinel phase. These techniques include combustion method [26, 60], sucrose-aided combustion method [61], co-precipitation method (CPM) [62], ultrasonic CPM [63], two-dryness CPM [50], chloride-ammonia CPM [64], emulsion drying [5, 65], oxalate co-precipitation process [66]. The LNM materials prepared by co-precipitation from acetate solution by oxalic acid and annealing at 900 °C in air had the preferred disordered spinel structure [67]. Idemoto et al. [31] showed that the oxygen content in samples obtained by sol-gel method and annealed under a high pressure of oxygen increases in comparison with that of samples prepared by the solid-state method. For the $LiNi_{0.5}Mn_{1.5}O_4$ sample prepared by combustion method using urea as fuel with a urea/Li molar ratio of 0.57 and a heat treatment temperature of 900 °C, the particle-size distribution fell in a narrow range of 1–2 μm [68]. Cao and Manthiram proposed a novel synthetic approach using urea as a reservoir for CO_2 to control the formation of LNM and optimize its morphology for high tap density [69]. A modified Pechini method utilized metal nitrates dissolved in distilled water and added drop-wise to citric acid–ethylene glycol (1:4 molar ratio); this solution heated to 140 °C enables the chelation (reaction of functional carboxyl group of acid with metal ions) process for the esterification of acid with ethylene glycol. The final product is composed by primary crystallites of 50–70 nm size with a surface area of 15.6 $m^2\ g^{-1}$ [70].

Spray pyrolysis methods are also commonly used including the flame-type method that can be used for mass production of cathode materials [71]. Other derivative methods include spray drying [72], ultrasonic spray [24], ultrasonic spray pyrolysis [73] spray pyrolysis associated with internal combustion-type [74]. Internal combustion type spray pyrolysis apparatus was used to prepare LNM cathode materials under the form of spherical particles with an average size of about 2 μm after calcination at 800 °C. The morphology of the particles changed from spherical to an irregular shape at temperatures higher than 900 °C [75].

The use of polymers containing oxygen-based functional groups [polyethylene glycol (PEG), poly(methyl methacrylate) (PMMA)] has been studied to synthesize highly crystalline nanometric LNM. Mechanical activation of hydrated salts in the presence of oxalic acid and the polymer followed by heating at 800 °C for a few minutes is sufficient to obtain pseudopolyhedral particles ranging from 60–80 nm in size [76]. LNM powders are also synthesized by polymer-pyrolysis growth [13], radiated-polymer sol-gel synthesis [54], poly(methyl methacrylate)-assisted method [77], polymer-assisted synthesis [14].

3.4 Structural Modification of LNM Spinels

The influence of the structure modification on electrochemical properties of high-voltage spinel cathodes has been reported in numerous works (see [78] for a review). Manthiram et al. [79] have pointed out the beneficial effect of cation (M'') substitution or doping in spinel oxides, allowing them to deliver the capacity around a nearly flat voltage of 5 V. The higher operating potential is attributed to a binding of the M'':3d e_g electrons by at least 0.5 eV higher than that of the Mn: 3d e_g electrons [17]. This can be achieved by appropriate double doping with elements such as Cr, Fe, Co, which have larger bonding energy than that of Mn–O. Among the numerous dopant elements, chromium is the most popular [41, 79, 80]. The lithium-rich the spinel $Li_{1+\delta}Ni_{0.5}Mn_{1.5}O_4$ was also prepared. The extra lithium is used in this material to convert Mn^{III} to Mn^{IV} state, thus minimizing the impact of Jahn–Teller (JT) distortion [55].

As a result, the superior performances of these cathodes are attributed to the smaller lattice parameter differences among the three cubic phases formed during the charge-discharge process. For $LiCr_{2y}Ni_{0.5-y}Mn_{1.5-y}O_4$ ($0 < y \leq 0.2$) powders synthesized by a sucrose-aided combustion method and heated at 900 °C, the most remarkable result was obtained in the range $y \leq 0.1$, where a very high capacity retention at 55 °C was observed [56]. Park et al. [81] found that Cr doping ($y = 0.05$) provided a wider plateau during charge–discharge tests by suppressing the Mn^{3+} oxidation that increases the structural stability by reduction of the JT distortion.

Other dopants for $LiNi_{0.5}Mn_{1.5}O_4$ are the fluorine and the sulfur. When part of oxygen atoms is replaced by fluorine, discharge capacitance generally decreases, but stability under cycling grows considerably [82–86]. Besides the effects of the structure, particle size and cation substitution, there are some works relative to the substitution of a small amount of fluorine for oxygen atoms in $LiNi_{0.5}Mn_{1.5}O_{4-z}F_z$ ($0.05 \leq z \leq 0.2$) [83]. The electrochemical measurements show that stable cycling performance can be obtained when the fluorine concentration z is higher than 0.1, but the specific capacity is decreased and 4 V plateau capacity resulting from a conversion of Mn^{4+}/Mn^{3+} remains. Sun et al. [87, 88] studied the effect of sulfur substitution on the structural integrity of spinel frameworks synthesized by wet chemistry. The capacity loss for the $LiAl_{0.24}Mn_{1.76}O_{3.98}S_{0.02}$ electrode was attributed to the presence of the rock-salt phase Li_2MnO_3 detected at the surface of the oxysulfide particles in the electrode cycled in the 4-V region at 80 °C [87].

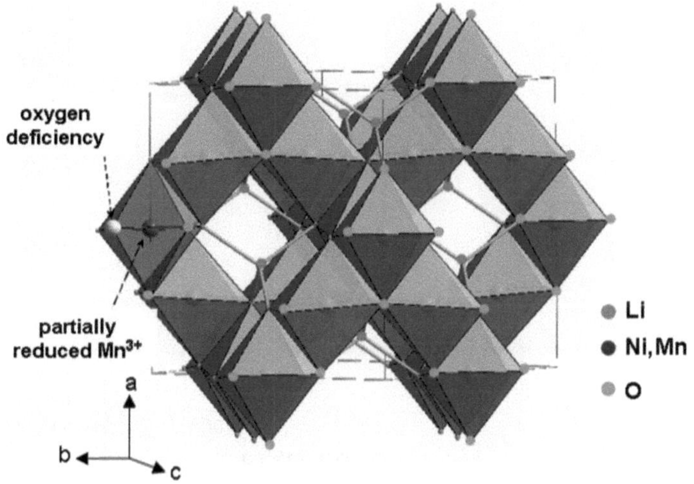

Fig. 4 Schematic representation of the spinel structure with oxygen vacancy

3.5 Oxygen Deficiency

More oxygen loss in the LNM framework leads to more Mn^{3+} generated to keep the electric neutrality, and the larger ionic radii of Mn^{3+} (0.645 Å) compared to Mn^{4+} (0.530 Å) results in the larger cell volume (Fig. 4). Moreover, less Mn^{3+} can reduce the dissolution of manganese, decrease the polarization, and increase the cycle performance, therefore, lead to a significant contribution to the comprehensive electrochemical performances of materials [89]. The partial replacement of Ni and Mn by Cr in $LiNi_{0.5}Mn_{1.5}O_4$ is an effective way to alleviate this problem [18]. In addition, Cr^{3+}, which has only three electrons in its 3*d* energy level; therefore no JT distortion is retained. Zheng et al. [90] demonstrated that the presence of an appropriate amount of oxygen deficiency and/or Mn^{3+} is critical to accelerate the Li^+ ion transport within the crystalline structure, which is beneficial to enhance the electrochemical performance of LNM offering high rate capability of 96 mAh g^{-1} at 10 °C.

3.6 LNM Versus Cr-Doped LNM

The two-electron redox reaction Ni^{2+}/Ni^{3+} and Ni^{3+}/Ni^{4+} for the $LiNi_xMn_{2-x}O_4$ cathodes have been investigated by several spectroscopies [45, 91]. With the increase of x in $LiNi_xMn_{2-x}O_4$, the 5-V plateau capacity rises and finally the 4-V plateau disappears basically at $x = 0.5$. Figures 5 and 6 display, respectively, the typical charge/discharge profiles measured at C/12 rate of LNM//Li cells for pristine and Cr-doped cathode samples and the related differential capacity curves,

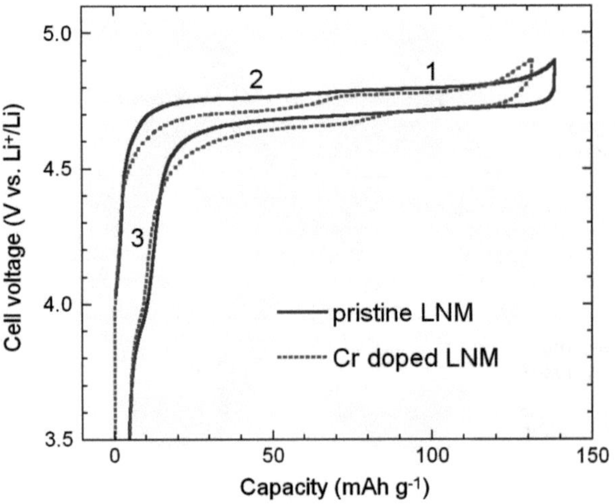

Fig. 5 Typical charge-discharge profiles (3rd cycle) measured at C/12 rate of LNM//Li cells for pristine $LiNi_{0.5}Mn_{1.5}O_4$ and Cr-doped $LiCr_{0.1}Ni_{0.45}Mn_{1.45}O_4$ samples (adapted from Ref. [39])

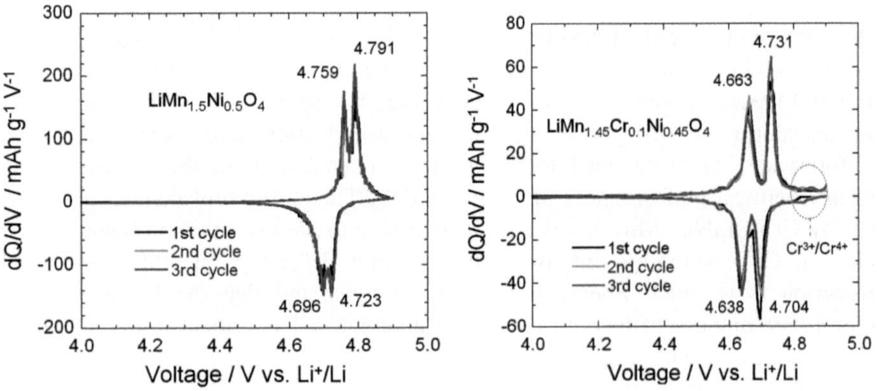

Fig. 6 Differential capacity curves, dQ/dV versus V, of $LiMn_{1.5}Ni_{0.5}O_4$ and $LiMn_{1.45}Cr_{0.1}Ni_{0.45}O_4$ spinel materials. The values at the peaks are given in volt (adapted from Ref. [39])

dQ/dV versus V that provide more details on the electrochemical behavior [92]. These figures show that the electrochemical properties of LNM spinels are intimately connected to cation ordering on the lattice spinel lattice. Thus the doping onto Ni site has the effect of inhibiting the ordering. However, Shin et al. [93] have pointed out the existence of cation-ordered domains. The two dQ/dV peaks in the 4.7-V region for either the oxidation process (charge) and the reduction process (discharge) indicate the two-step redox reaction. The narrower voltage separation

($\Delta V = \sim 30$ mV) between the two oxidation peaks of the pristine LNM compared to those ($\Delta V = \sim 70$ mV) of the Cr-doped LNM is reminiscent of the higher degree of cation ordering in the pristine sample.

Lithium insertion in $Li_{1+x}Mn_{1.5}Ni_{0.5}O_4$ occurs in the 3-V region as a two-phase system with the coexistence of the cubic $LiMn_{1.5}Ni_{0.5}O_4$ and the tetragonal $Li_2Mn_{1.5}Ni_{0.5}O_4$ phase [91]. Micrometer-sized LNM single crystals synthesized by SSR at the relatively low temperature of 700 °C showed an initial discharge capacity of 130 mAh g^{-1} between 4.95 and 3.00 V at the 3 C rate [94]. Kanamura et al. [95] obtained a discharge capacity higher than 130 mA h g^{-1} and 100 % rechargeability with the coulombic efficiency about 97 % and average discharge potential 4.75 V versus Li^+/Li. The electrochemical performance has been evaluated as a function of the deviation of the ratio Ni/Mn concentrations from the ideal composition 1/3/ [96–99] and oxygen deficiency [100]. The electrochemical properties are not very sensitive to a Ni excess. On another hand, the loss of rechargeable capacity is remarkable when the Ni/Mn ratio is smaller then 1/3, a result that is in agreement with the fact that the optimization of the electrochemical properties of $LiNi_{0.5}Mn_{1.5}O_4$ spinel requires minimization of the nickel deficiency [101].

3.7 Phase Diagram of $Li_xNi_{0.5}Mn_{1.5}O_4$

The phase evolution during lithium de-intercalation/intercalation within the LNM spinel phase has been studied using both ex situ [69, 102–104] and in situ [35, 39, 42, 105] X-ray diffraction techniques. Conflicting results have been reported. Alcantara et al. [104] investigated the mechanism of lithium extraction-insertion around 5 V by recording ex situ XRD of the LNM electrodes. They found that the XRD patterns of $Li_xNi_{0.5}Mn_{1.5}O_4$ are still indexed in a single cubic phase during the first charge of the battery up to $x = 0.5$. Further lithium extraction caused the reflections to split and two cubic phases with different unit cell parameter are detectable from $Li_{0.45}Ni_{0.5}Mn_{1.5}O_4$ to the complete extraction $x = 0$. The second phase shows a contraction of the unit cell volume, similarly to that found in the deinserted phases emerging from $LiMn_2O_4$. In the ordered LMN, three distinct cubic phases were detected related to each of the Ni^{2+}, Ni^{3+}, and Ni^{4+} oxidation states in the ordered LMN material. These phases transformed from one to another by means of two distinct two-phase regions that correspond to the wide voltage plateaus observed in the charge-discharge curves [36]. In contrast, the disordered LMN material showed a small shift of the XRD spectrum to higher angles without any clearly identified two-phase region until the end of charge is approached; consistently, the plateaus in the voltage profiles are less pronounced [36]. On the other hand, a structural change has been reported upon delithiation of disordered $Li_xMn_{1.5}Ni_{0.5}O_4$ at $x = 0.5$ with loss of the glide symmetry [106]. However, a different phase diagram has been reported for this same disordered LMN, with a solid solution for large values of x, followed with a two-phase region at $x \sim 0.6$,

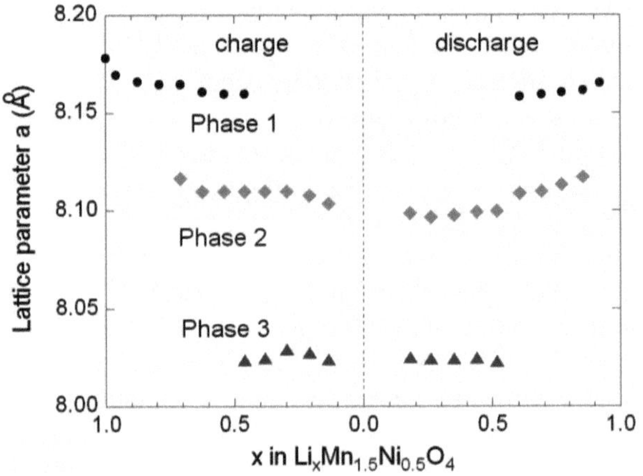

Fig. 7 Variation of the lattice parameter of the different cubic phases in the LNM sample as a function of the Li concentration x during the charge (**a**) and discharge (**b**) during cycling at C/24 rate (adapted from Ref. [105])

due to the onset of a second cubic phase, phase II, then another two-phase region is observed at $x < 0.4$, in which phase II coexists with another cubic phase, phase III [42]. Rhodes et al. [35] have found that the three phases coexist in a finite range of concentrations. However, the authors did not specify if their measurements were made on ordered or disordered LMN. We can simply note that their result is in agreement with the phase diagram found by Wang et al. [42] for LMN in the ordered phase, since the three phases are found to coexist only in this case. Moreover, the phase diagram is reported to depend on the morphology of the particles [107]. The phase transitions were reversible, meaning that migration of cations is highly reversible during lithiation/delithiation.

Recent report has shown that the phase diagram for both samples $Li_xMn_{1.5}Ni_{0.5}O_4$ and $Li_xMn_{1.45}Ni_{0.45}Cr_{0.10}O_4$ showed the existence of three phases that form alternatively solid solutions and two-phase regions [105]. The results shown in Fig. 7 have been understood on a basis of a model that takes strain effects into account, also explaining the fact that the phase diagram is sample dependent, and hence the different results reported in the literature. The analysis of the phase diagram confirms the faster dynamics of the Li-insertion/de-insertion in the Cr-doped sample, evidenced by the improved capacity retention at high C-rates. The other benefit of the Cr-substitution is the increase of the stability of the lattice. The drawback is a decrease in the energy density that is not due to a loss of capacity, but a smaller redox potential of the nickel versus Li^+/Li [105].

3.8 Surface Modified by Coating

LNM materials demonstrate a significant capacity loss at elevated temperature $T > 50$ °C, which is a critical environment for HEV and EV applications. Researches have shown that an extended cycling of the Li//LNM cells at elevated temperature resulted in the local Mn and Ni dissolution. The factors that limit the capacity retention of LNM at high temperature (~60 °C) have been identified [108]. The most important failure involves the surface of the cathode material where decomposition of the nonaqueous electrolyte occurs because of the catalytic effect of the transition-metal ions. The product of electrolyte oxidation is formed on the surface of the cathode material and acts as an electrically insulating layer that could decrease the electronic conductivity and prevent the easy transportation of ions. Surface analysis of aged $LiNi_{0.5}Mn_{1.5}O_4$ electrode operating at 60 °C involves the formation of LiF, C-F and $P\text{-}F_x$ species and results in local Mn and Ni dissolution and transformation of the active material to $\lambda\text{-}MnO_2$ [109]. In an ionic liquid solvent at elevated temperature (55 °C) the surface film is enriched by inorganic fluorinated species and becomes thicker with cycling, resulting in a continued electrode polarization and ultimately cell failure [110]. The effect of different membranes and aluminum current collectors on the initial coulombic efficiency of LNM//Li have been investigated, and the cycling performance at different rates and temperatures and the storage performance at 60 °C for a week discussed [111].

Efforts to inhibit the electrode-electrolyte reactions have included coating the surface with substance that would block them suppressing the surface disproportionation reaction $2Mn^{3+} \rightarrow Mn^{2+}Mn^{4+}$ [1]. Fu et al. [112] have reviewed the importance of surface structural features of electrode materials for their electrochemical performance in Li batteries; both anode and cathode materials were considered. Many studies indicated that a surface coating approach is an effective method to improve the electrochemical performance of $LiNi_{0.5}Mn_{1.5}O_4$ [108, 113–117]. So far, surface modification of 5 V-LNM includes formation of oxide coated layer (OCL), noble metal layer or phosphate layer. Among the varieties of coating components, Al_2O_3 is the most popular [108, 113]. To illustrate the effect of coating, the discharge capacity at 50th cycle versus C-rate of $LiMn_{1.42}Ni_{0.42}Co_{0.16}O_4$ before and after coating with 2 wt% Al_2O_3 is displayed in Fig. 8.

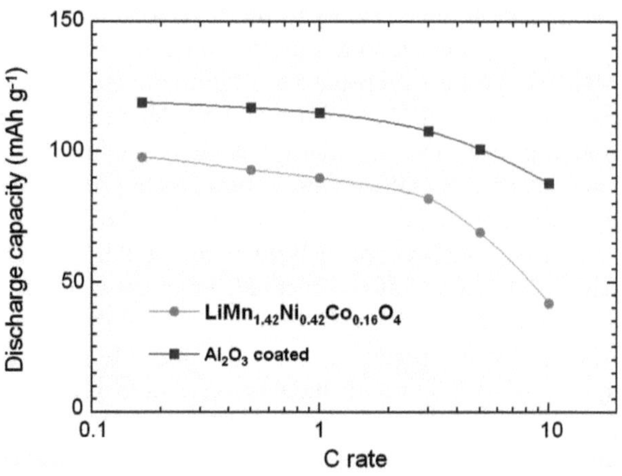

Fig. 8 Discharge capacity of $LiMn_{1.42}Ni_{0.42}Co_{0.16}O_4$ at 50th cycle as a function of C-rate before and after coating with 2 wt% Al_2O_3 (adapted from Ref. [113])

Cho et al. [118] attributed the positive effect of the OCL on the cycling performance to the increased fracture toughness of the coated material, which would suppress the lattice stress by constraining the LNM particles against lattice changes. However, Chen and Dahn et al. [119] suggested that OCL does not suppress the lattice parameter evolution and that the enhancement of electrochemical performance is independent of the fracture toughness of the coating oxide. Appapillai et al. [120] pointed out that the better cycling performance is largely due to the differences in the surface microstructure rather than structural instability. In the same spirit, Liu and Manthiram [108] argued that improvement in the electrochemical properties of surface modified Co-doped LNM is due preferentially to a much more stable surface chemistry which avoids the development of SEI layer acting as a protection shell between the active material and the electrolyte.

4 Inverse Spinel Cathodes

In 1993, Dahn discovered that both $LiNiVO_4$ and $LiCoVO_4$ belong to the 5-V class cathode with an inverse spinel structure [121]. These compounds were synthesized through solid-state reaction between their respective nitrates and oxides at a very high temperature, ca. 800 °C. In this lattice (S.G. *Fd*3*m*), the pentavalent vanadium is located on the tetrahedral 8*a* sites (1/4, 1/4, 1/4), Li and Ni(or Co) are distributed on the octahedral 16*d* sites (1/2,1/2,1/2), the distribution being disordered and the positions of oxygen are in 32*e* sites (z,z,z). Using in situ XRD, it has been proved that the reaction occurring at potential as high as 4.8 V versus Li^+/Li for the compound $LiNiVO_4$ is a deintercalation/intercalation process involving the removal/insertion of Li^+. The redox couple to operate at the plateau over 4.5 V is $Ni^{2+/3+/4+}$ providing a theoretical capacity 148 mAh g^{-1}. If all the Li^+ ions were located in octahedral 16*d* sites, no Li^+ conduction would be expected through the tetrahedral site-empty octahedral site pathway. The striking difference in cell voltage between $LiNiVO_4$ (4.8 V) and its analogue $LiCoVO_4$ (4.2 V) implies that the transition-metal atoms in the inverse spinel structure play a critical role. Fey et al. [122] have reported the presence of a NiO impurity peak at 43.5° in their $LiNiVO_4$ product prepared via a high-temperature solid-state reaction (cubic lattice parameter a = 8.222 Å), while Prabaharan et al. [123] have demonstrated that free-impurity $LiNiVO_4$ was synthesized at temperatures as low as 500 °C using the aqueous glycine–nitrate combustion process (a = 8.2198 Å). The particle size of the product annealed at 500 °C was found to be 38 nm.

Various methods have been used for the synthesis of $LiNiVO_4$ (named after LNV). The conventional method, i.e. solid-state reaction, provides big crystallite size and poor stoichiometry [124–126]. Fey et al. [127] prepared samples by reacting $LiNiO_2$ precursor and V_2O_3 or V_2O_5 in air at 700 °C for 2 h. Wet-chemical synthesis techniques include sol-gel route [126–128], gel-combustion synthesis [127, 129, 130], combustion synthesis using gelatine as fuel [131], sol-gel via citrate [132], sol-gel via oxalate [133], glycerol-assisted gel combustion synthesis

[129], starch-assisted combustion method [78], urea-assisted combustion synthesis [134], precipitation method [135], aqueous glycine-nitrate combustion process [122]. LNV particles, sized 50–70 nm, prepared by a citric acid polymeric process in addition with urea were coated by Al_2O_3 derived from (methoxyethoxy) acetate-alumoxane. The galvanostatic cycling studies suggest that 0.5 wt% coating obtained from a (methoxyethoxy) acetate-alumoxane precursor enhanced the cycle stability by a factor two compared with the pristine LNV [398]. Hydrothermal technique can be used to obtain much smaller particles (few nanometers only) prepared at low temperature (200 °C) [126, 127, 136–138]. Synthesis of Li- and La-doped LNV has been also reported [139] but this doping did not lead to significant improvement. Stoichiometric and non-stoichiometric LNV thin films were formed by using rf-sputtering by varying the partial pressure of oxygen [140]. Stoichiometric sputtered LNV with a partial pressure of oxygen (1 %) gives the best electrochemical properties, and has obtained a capacity of 1300 mAh g^{-1} during the first discharge cycle in the range 3–0.2 V at low C-rate.

Electrochemical tests of coin cells cycled at 0.2 mA cm^{-2} in the voltage range between 4.8 and 3.0 V versus Li^+/Li show at the first discharge capacity of 98 and 83 mAh g^{-1}, for the two samples, respectively. Pristine and Al-doped LNV compounds were synthesized by starch-assisted combustion followed by a thermal treatment at 700 °C for 3 h. The discharge capacity at the 1st cycle of pristine LNV is 90 mAh g^{-1}, while the capacity of $LiNiAl_{0.1}V_{0.9}O_4$ has increased to 119 mAh g^{-1} [141]. Chitra et al. [130] showed that LNV inverse spinel prepared by sol-gel synthesis via oxalate route and fired at 800 °C has domain size of 52 nm. Two well-defined potential regions were observed in the charge–discharge curves of $Li//Li_xNiVO_4$ cell cycled in the compositional domain $0.2 \le x \le 1.0$. The first region extends up to $x = 0.65$ and the second one region up to $x = 0.2$. This latter x-value is considered as the maximum faradaic yield for Li^+ extraction in Li_xNiVO_4. The deintercalation process of Li^+ ions from $LiNiVO_4$ cathode materials leads to the quantitative oxidation of Ni^{II} into Ni^{III} and the non-oxidation of V atoms beyond the 5 + state. The upper high voltage plateau at 4.7 V versus Li^+/Li is attributed to the two-phase domain $NiVO_4$ + $Li_{0.65}NiVO_4$ (Fig. 9a). Phase transitions are clearly observed in the derivative capacity dQ/dV versus V plot (Fig. 9b). The oxidation-reduction peaks observed at 4.82 and 4.68 V, respectively, are associated with a single electron transfer process during lithium intercalation–deintercalation process.

The structural properties of the as-prepared and electrochemically of Li_xNiVO_4 ($1 > x > 0.56$) were investigated by XRD and Raman spectroscopy. Only little effect of the Li-extraction on the crystal structure could be detected with XRD, whereas Raman revealed a softening of the V–O bonds and severer distortion of the $[VO_4]$ tetrahedron with Li-extraction and also a decrease of the electronic conductivity at $x = 0.56$ [132]. Rietveld refinement of the $Li_{0.56}NiVO_4$ electrode (charge state) showed that the lattice parameter of the delithiated cathode material almost did not change with respect to that of fresh $LiNiVO_4$, indicating a rigid structure of the LNV lattice (a = 8.2178(4) Å at x(Li) = 1). This could be attributed to the co-existence of Li and Ni on the 16d site because the stronger Ni–O bonds may

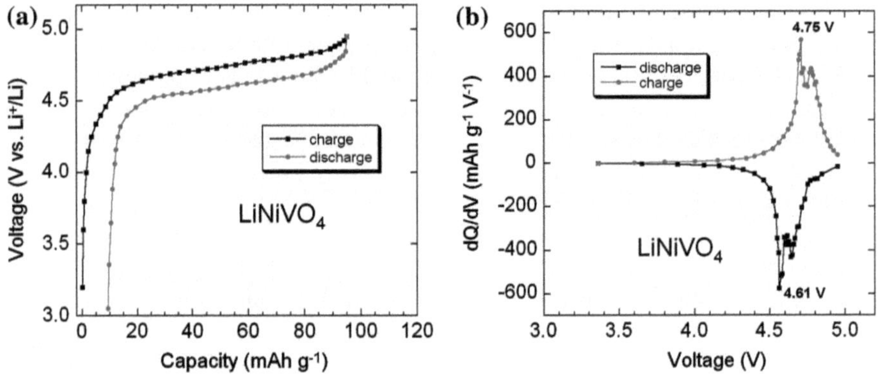

Fig. 9 **a** Charge-discharge profile of Li//$LiNiVO_4$ cell at C/5 in the potential range 3.0–4.9 V. **b** Derivative capacity dQ/dV as a function of the cell voltage

restrain the shrinkage of the material lattice during Li insertion/extraction. However, the peak intensities decreased with lithium extraction, and the diffraction peaks became broader and broader. This indicates progressive formation of defects in the material lattice as a result of Li^+ extraction [132].

5 Polyanionic Frameworks

The third family of 5-V cathode materials is based on poly-anionic frameworks with the olivine and olivine-related structures. Recently, a short review of this class of materials, as cathodes for advanced lithium-ion batteries, has been published [14]. Since the discovery of the electrochemical activity of Li$M$$PO_4$ (M = Fe, Mn) olivines with $Fe^{3+/2+}$ and $Mn^{3+/2+}$ redox potentials 3.5 and 4.1 V versus Li^+/Li, respectively and theoretical capacity 170 mAh g^{-1} [7], there were many attempts to develop cathode materials with higher intercalation voltages over 4.5 V that deliver high specific energy density as high as 800 Wh kg^{-1}. Candidates for high-voltage electrodes are polyanionic materials containing either nickel or cobalt (for structure description, see Ref. [142]).

5.1 *Olivine Frameworks*

Li$M$$PO_4$ compounds crystallize in the olivine structure belonging to the orthorhombic symmetry (S.G. *Pnma*). According to this structure, the lithium ions are distributed along channels. The simplified picture for the ionic conductivity is then a 1D diffusion of the lithium ions along these channels. A common feature to all the olivine materials is their poor electronic conductivity, so that bare particles have very

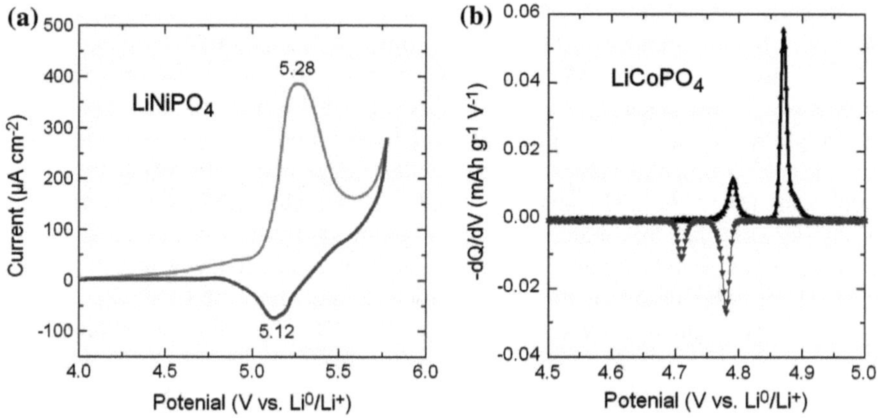

Fig. 10 Electrochemical features of olivine frameworks. **a** Cyclic voltammogram of Li//$LiNiPO_4$. **b** Incremental capacity −dQ/dV of Li//$LiCoPO_4$

poor electrochemical properties (Fig. 10). Therefore, the active element of the cathode is always a nano-composite C-Li$M$$PO_4$ that designates the nanoparticle with its carbon coat [143]. On the other hand, the ability to suppress thermal runaway of Li$M$$PO_4$ olivine frameworks is attributed to the high covalent feature of the P-O bonds in the tetrahedral (PO_4) units, which stabilizes the olivine structure and prevents oxygen release from the charged (delithiated) olivine materials up to 600 °C. However, it is still controversial for the delithiated state of $LiCoPO_4$ [144–146].

Various techniques were used for the growth of $LiNiPO_4$ (LNP) and $LiCoPO_4$ (LCP) isostructural olivines. Synthetic methods include solid state reaction with a final heat treatment at 775 °C for 48 h in argon [145, 147–154], ball milling mixing carbon with the precursors [155], freeze-drying process assisted by formic acid [156], polyvinyl-pyrrolidone assisted sol-gel route [157], precipitation method [158], Pechini method [159, 160], polyol method using 1,2 propanediol and ethylene glycol [161], thin film deposition [158]. LCP powders were prepared by an original solid-state synthesis procedure based on the use of an alternative cobalt-containing precursor $CoNH_4PO_4$ and a lithium excess synthesis with carbon black as temporal dispersing agent, later eliminated as CO_2 [162]. The sol-gel technique in ethylene glycol [163, 164] appeared to be a simple method to prepare submicron-sized carbon-coated LNP and LCP with only little dispersion in the size distribution. Bramnik et al. [148] reported the effect of different synthesis routes on Li extraction–insertion from $LiCoPO_4$; samples prepared by SSR method at high temperature demonstrated unsatisfactory electrochemical performance [149], although some improvement was observed by a synthetic approach based on the precursor $NH_4CoPO_4{\cdot}H_2O$ [162]. Surface modification of LCP particles was obtained via a thin layer of Al_2O_3 deposited (~10 nm) by a sputtering method [158] or via a thin layer of $LiFePO_4$ (~4 nm) by SSR method [153], and will be discussed below.

Recently, several researchers reported the electrochemical features of LiMPO$_4$ (M = Ni, Co) compounds (Fig. 10). Wolfenstine and Allen [146] determined the Ni^{3+}/Ni^{2+} redox potential in $LiNiPO_4$ between 5.1 and 5.3 V. To overcome the problems with low electrolyte stability, a 1 mol L^{-1} $LiPF_6$ in tetramethylene sulfone electrolyte was used because of its high oxidative stability, around 5.8 V versus Li^+/Li. These experimental values are in excellent agreement with recent theoretical predictions [165, 166]. The absence of redox peaks when LNP was heated under argon suggests that this material has a very low intrinsic electrical conductivity. This result is in good agreement with the electrical conductivity results of Rissouli et al. [167], who observed that the electrical conductivity of $LiNiPO_4$ was 2–3 decades lower than that for $LiCoPO_4$ and $LiMnPO_4$. Therefore, additional treatments such as carbon coating are mandatory for LNP to exhibit Li insertion/deinsertion. Magnetic anisotropy in Li-phosphates and the origin of their magneto-electric properties have been investigated [168–172]. Magnetic properties of LCP and LNP show that antiferromagnetic M-O-M superexchange interactions couple the spins in planes parallel to (100) [169]. Local environment and bonding strength of the cations were studied by Raman and FTIR [172–174]. The voltammetry displayed an oxidation peak at ~5.28 V and a reduction peak at ~5.12 V (Fig. 10a). Recently, Dimesso et al. [175] have reported the preparation of LNP and LCP by a Pechini assisted sol-gel process that provides material exhibiting redox peaks at ~5.2 and ~4.9 V versus Li^+/Li. Mg-substituted LNP/graphitic carbon foams composite was also synthesized by the same method, showing a discharge capacity of 126 mAh g^{-1} at C/10 rate by substituting 0.2 Mg for Ni [176].

In the early work by Amine et al. [177], it is demonstrated that Li can be reversibly removed from LCP at an average voltage of 4.8 V versus Li^+/Li with only a small contraction in the unit cell volume of the olivine lattice and the formation of a second olivine-like phase upon Li extraction from Li_xCoPO_4 with limited $\Delta x = 0.42$ lithium per formula unit. The electrochemical properties of $LiCoPO_4$ have been studied as a function of several parameters to improve the discharge capacity and capability. In particular, LNP-LCP mixing to obtain solid solutions as cathodes [178], carbon coating [179, 180] was prepared; oxygen partial pressure effects on the discharge capacity [181], structural stability upon delithiation [182] have been investigated. Okada et al. [152] have shown that $LiCoPO_4$ exhibited the highest 4.8 V discharge plateau of 100 mAh g^{-1} after initial charging to 5.1 V giving an energy density of 480 Wh kg^{-1} comparable to that of $LiCoO_2$. The electrical conductivity of LCP and mixed (Co, Ni), (Co, Mn) compounds [167, 168] has been measured, and also that of doped LCP as well [183] aiming to improve it. LCP, like LNP, has low electron conductivity, so that its use as the cathodic material is possible only in the case of synthesis of the $LiCoPO_4$/C composite [155, 184]. Such composites can be discharged at the potentials of 4.7–4.8 V. and their initial discharge capacity is close to the theoretical one, 167 mAh g^{-1}. However, their cycleability is very low, because the decomposition of liquid electrolyte occurs under charging in the potential range of 4.8–5.1 V simultaneously with oxidation of Co^{2+} to Co^{3+}. Several groups have studied the phase transitions occurring upon lithium insertion-extraction of $LiCoPO_4$ [151, 185]. A two phase mechanism was confirmed by in situ synchrotron diffraction [148]. An amorphization of the phosphate was observed after electrochemical or chemical oxidation [149, 182]. Based on X-ray absorption spectroscopy measurements, Nagayama et al. [185] suggested f a hybridization effect between the Co 3*d* and O 2*p* orbitals and the polarization effect introduced by Li ions [150]. Bramnik et al. [151] revealed the appearance of two orthorhombic phases upon electrochemical Li extraction. The$LiCoPO_4$ and the Li deficient phases, $Li_{0.7}CoPO_4$ and $CoPO_4$ phases are responsible for the two voltage plateaus at 4.8 and 4.9 V versus Li^+/Li. The hedgehog-like $LiCoPO_4$ microstructures in the size of about 5–8 μm are composed of large numbers of nanorods in diameter of ca. 40 nm and length of ca. 1 μm, which are coated with a carbon layer of ca. 8 nm in thickness by in situ carbonization of glucose during the solvothermal reaction. As a 5-V positive electrode material for rechargeable lithium battery, the hedgehog-like $LiCoPO_4$ delivers an initial discharge capacity of 136 mAh g^{-1} at 0.1 C rate and retains its 91 % after 50 cycles [184]. Surface modification of particles provides a satisfactory cycleability for $LiCoPO_4$ to be used as a 5-V cathode material [153, 158]. The capacity retention of Al_2O_3-coated LCP was 105 mAh g^{-1} after 50 cycles at $T = 55$ °C at low C-rate in $LiBF_4$ electrolyte [158]. Jang et al. [153] claimed that $LiFePO_4$-coated LCP particles prepared by SSR method (~100–150 nm) show improved battery performance with an initial discharge capacity of 132 mAh g^{-1} but did neither mention the C-rate nor the current density. The main effect of the $LiFePO_4$ coating is then the increase of the capacity retention. After 15 cycles at low C-rate, the capacity has almost vanished without the coating, while it is still about 90 mAh g^{-1} after coating. This result, however, suggests that the coating with Al_2O_3 is more efficient.

The ability to suppress thermal runaway of Li*M*PO_4 olivine frameworks is attributed to the high covalent feature of the P-O bonds in the tetrahedral (PO_4) units, which stabilizes the olivine structure and prevents oxygen release from the charged (delithiated) olivine materials up to 600 °C. This is still controversial in LCP and LNP olivine lattices. In particular, a thermal instability has been reported in the charged (i.e. delithiated) state of $LiCoPO_4$ [144]. Both olivine-like phases Li_zCoPO_4 (z = 0.6) and $CoPO_4$ appearing during the delithiation of LCP are unstable upon heating, and decompose readily in the range 100–200 °C. The decomposition of lithium-poor phases leads to gas evolution and the crystallization of $Co_2P_2O_7$. Incorporation of lithium bis(oxalato)borate (LiBOB) as additive in conventional electrolyte solutions enhances the electrochemical performance of LCP electrode [154]. Recent investigations of $LiCoPO_4$—3D carbon nanofiber composites and plate-like $LiCoPO_4$ nanoparticles have demonstrated their capability as high-voltage cathode materials [186, 187].

5.2 *Fluorovanadates*

A new class of high–voltage cathode materials consists in fluorinated lithium phosphates [188–193]. Fluorides are expected to display high-voltage because the electronegativity of fluor is greater than that of oxygen [192]. Lithiated vanadium fluorovanadate ($LiVPO_4F$) is characterized by the discharge potential of 4.5 V. However, discharge capacity at a relative low discharge rate, ca. 1 C, does not exceed 130 mAh g^{-1} [188]. Of much greater interest are fluorinated cobalt phosphates. The theoretical capacity of such materials is twice higher than that of lithiated iron and cobalt phosphates and the process of lithium insertion/extraction occurs at the potentials of 4.8–5.1 V. The fluorinated cobalt phosphate realizes the combination of high discharge capacity and high voltage allows assuming that an experimental performance close to theoretical could be achieved, as the specific energy consumption of such materials would be at least 2.5–3 times higher than specific energy consumption of lithiated cobalt oxide.

5.3 Fluorophosphates

Fluorophosphates of general formula A_2MPO_4F (A = Li, Na and M = Fe, Mn, Co, Ni) crystallize in three structure types, which differ in the connectivity of (MO_4F_2) octahedra: face-shared (Na_2FePO_4F), edge-shared (Li_2MPO_4F, M = Co, Ni) and corner-shared (Na_2MnPO_4F) [190, 194–199]. Both $LiCoPO_4$ and Li_2CoPO_4F (isostructural with Li_2NiPO_4F) crystallize in the orthorhombic system (S.G. *Pnma*, $Z = 8$). Nevertheless, there are remarkable differences between the structures from a crystallographic point of view. $LiCoPO_4$ has CoO_6 octahedra, LiO_6 octahedra and PO_4 tetrahedra. In contrast, Li_2CoPO_4F has CoO_4F_2 octahedra instead of CoO_6 octahedra. In addition, Li_2CoPO_4F has two kinds of Li sites, 4*c* and 8*d* [194]. It was confirmed that Li_2CoPO_4F is a new class of 5-V cathode materials similar to $LiCoPO_4$. A considerable theoretical upper limit of approximately 310 mAh g^{-1} is expected for Li_2CoPO_4F and Li_2NiPO_4F (Fig. 11)

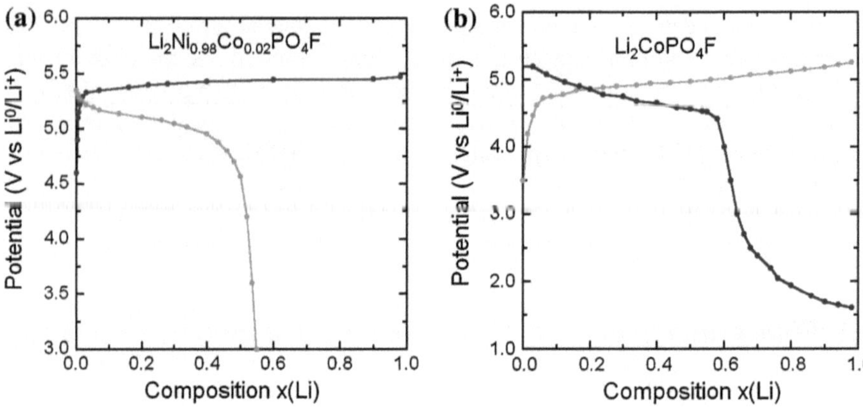

Fig. 11 Electrochemical features of fluoro-phospates. **a** Li_2NiPO_4F and **b** Li_2CoPO_4F

The theoretical estimation of the intercalation voltage of ~4.9 V for Li_2CoPO_4F cathode is in good agreement with the voltage plateau observed at ca. 5 V [194]. A fault of both Li_2CoPO_4F and lithiated cobalt phosphate is a high irreversible capacity (especially in the first cycles), which is related to decomposition of electrolyte at high anodic potentials. Experimentally, $LiNiPO_4F$ discharge voltage is demonstrated to be close to 5.3 V [480]. Khasanova et al. [190] have investigated the electrochemical performance and structural properties of the high-voltage cathode material Li_2CoPO_4F. The cyclic voltammetry and coulometry under potential step mode in the voltage range 3.0–5.1 V versus Li revealed a structural transformation at potentials above 4.8 V. This transformation occurring upon Li-extraction appears to be irreversible: the subsequent Li-insertion does not result in restoration of the initial structure, but takes place within a new "modified" framework. According to the structure refinement this modification involves the mutual rotations of (CoO_4F_2) octahedra and (PO_4) tetrahedra accompanied by the considerable unit cell expansion which is expected to enhance the Li transport upon subsequent cycling. The new framework demonstrates a reversible Li-insertion/extraction in a solid-solution regime with stabilized discharge capacity at around 60 mAh g^{-1}. The Li_2CoPO_4F is prepared by a two step solid state method, followed by the application of wet coating containing various amounts of ZrO_2. Among the samples, the 5 wt% ZrO_2 coated Li_2CoPO_4F material shows the best performance with an initial discharge capacity of up to 144 mAh g^{-1} within the voltage range of 2–5.2 V versus Li at 10 mA g^{-1} [195].

The redox potential of the fluorosufates phases $LiMSO_4F$ (M = Co, Ni, Mn) phases are expected to exhibit redox potentials of 4.25, 4.95 and 5.25 V for the corresponding sulfates. Barpanda et al. [196] succeeded in preparing the $Li(Fe_{1-x}M_x)SO_4F$ solid solutions solely when using the corresponding monohydrate solid solutions as precursors. Neither of these phases was shown to present any electrochemical activity up to 5 V. Evidently, Co^{2+}/Co^{3+}, Ni^{2+}/Ni^{3+} and Mn^{2+}/Mn^{3+} redox reaction does not occur within the explored galvanostatic cycling potential window, with all the capacity solely coming from the Fe^{2+}/Fe^{3+} redox reaction. Subsequently, the electrochemical activity of pure $LiMSO_4F$ phases were tested at different rates (C/2–C/10) by cycling up to 5 V using an aluminum plunger for cathode. Therefore, even at such high voltages, which are prone to electrolyte oxidation/decomposition, the activity of the corresponding M^{2+}/M^{3+} redox couples did not be triggered. Among the Co-based polyanionic insertion compounds $Li_2CoP_2O_7$ [197] is considered as 4.9 V cathode. This pyrophosphate crystallizes in the monoclinic structure (S.G. $P2_1/c$), in which Li occupied five sites, two are tetrahedrally coordinated, one forms bipyramidal sites, and two Li share them occupancy with Co bipyramids. The material synthesized using a two-step solid-state method delivered a discharge capacity of ca. 80 mAh g^{-1}at C/20 rate.

6 Concluding Remarks

Many attempts have been made since a decade to increase the voltage of the Li-ion batteries up to 5 V. This is a challenge, because many fulfillments must be achieved before a commercialization at a wide scale can be envisioned. (1) The cathode element must be stable, and support many cycles to have a reasonable cycling life. (2) The anode must have the same property. (3) The raise of temperature up to 60 ° C should not degrade the performance. (4) The electrolyte should not oxidize even at such a voltage, and an electrolyte with a higher HOMO (say 6 V vs. Li) would be helpful. (5) The thermal stability is mandatory. The progress has been constant since a decade, but all the problems have not been solved, yet [200–204]. This review focused on the electrode, and mainly on the cathode elements. Among them, two materials have emerged as the most promising candidates: the $LiMn_{0.5}Mn_{1.5}O_4$ spinel and the $LiCoPO_4$ olivine. The poor cycleability of $LiCoPO_4$, however, is a severe disadvantage. Even if coating, with Al_2O_3 in particular, is an improvement, 10 % loss after only 50 cycles is still far from the performance required for penetration on the market. The cycleability of $LiMn_{0.5}Mn_{1.5}O_4$ is better, at least at room temperature. It is now possible to maintain the capacity of this material over 100 cycles without significant loss at room temperature, in particular when the particles are coated with a noble metal or a phosphate layer, or some other layers that have been reviewed here. In addition, the C-rate and current density available have significantly improved by decreasing the size of the particles to the nano-range. The state-of-the-art is thus a multi-composite nanoparticle made of a core region in the 5-V element, surrounded by a coat, few nanometers in thickness, of a protective element. The drawback to obtain such a high-tech particle is that the synthesis is a two-step process in which the core region is synthesized first, and then the coating is made, which makes the whole synthesis process expensive. It is difficult, however, to avoid it, because of the problem of the electrolyte: the problem (2) mentioned above. The state-of-the-art (SOA) electrolytes that are used in Li-ion batteries today do not support the 5 V potential. As a first step, unsymmetrical sulfones were proposed as an electrolyte system that can resist oxidation beyond 5.0 V. However, intrinsic shortcomings of sulfone as a major electrolyte component, including its failure to form a protective layer on graphitic anode, slow Li ion kinetics, and poor electrode active matcrial utilization caused by high viscosity. More recently [198], new additives to SOA electrolytes have been proposed, based on phosphate ester with highly fluorinated alkyl side arms in SOA non-aqueous electrolytes. In particular, the addition of 1 % Tris(1,1,1,3,3,3-hexafluoro-iso-pro pyl)phosphate (HFiP) to the SOA electrolyte improves very efficiently the cycling life of the $LiNi_{0.5}Mn_{1.5}O_4$ cathode. The capacity fade of the half-cell with Li anode is only about 10 % at the 130th cycle at a slow C-rate that has not been specified. Nevertheless, the results for the full-cell with graphite anode are not as good: the initial discharge capacity about 0.85 mAh is lower than with the SOA electrolyte; after 500 cycles, the capacity is 0.5 mAh, comparable with 0.4 mAh with the SOA electrolyte. This result shows that the cycling life at this stage is not limited by the cathode, but instead by the formation of the SEI with the carbon anode. One solution is to replace the graphite by LTO. Indeed, we have reported in this review prior works showing that the choice of LTO as the active anode element improves considerably the cycling life of the battery. The drawback, however, is that we do not have a 5-V battery anymore, because the redox potential 1.77 V of the titanate ($Li_4Ti_5O_{12}$) with respect to Li^+/Li is lost in the process. We are thus left with a 3-V battery, which, however, is sufficient for many applications, provided that safety concerns do not disqualify it: we meet the problem (5) above mentioned. The basic difficulty on the intrinsic safety of the battery comes from the fact that the partial pressure of the oxygen increases very fast with the voltage in oxides of transition metals [199]. This inevitable thermodynamic law implies an increase of the risk of thermal runaway due to the cathode element with the voltage, especially if the anode is graphitic. In this case, oxygen can be released from the cathode will react with the carbon at the anode side to generate carbonic gas. This reaction is exothermic, at the origin of the thermal runaway. We have mentioned earlier in this review that this problem is a debate for the cobalt phosphate in the delithiated phase. In the case of $LiMn_{0.5}Mn_{1.5}O_4$, this problem can be overcome, at least partly, by coating the 5-V particles by a protective layer. Therefore, the first role of the protective layer of the active particles is to embed the core region made of the 5-V element to keep the oxygen inside the particles. The inconvenience is that this protective layer is usually inert thus implying a loss of energy density, which however, remains small when the thickness layer is only few nanometers. The exception is the $LiFePO_4$ coating: this material is electrochemically active, and has a growing success as the cathode element of Li-ion batteries working at 3.4 V. Moreover, it takes its outstanding thermal stability from the fact that the oxygen is tightly bound to the phosphorous to build PO_4 units. As a result, $LiFePO_4$ is very stable up to 5.3 V, and can thus efficiently be used as a protective layer for $LiMn_{0.5}Mn_{1.5}O_4$ against loss of oxygen. Another difficulty met with $LiMn_{0.5}Mn_{1.5}O_4$ is the general problem of the Mn-spinels, namely the dissolution of Mn in the electrolyte, which reduces the life of the battery. This loss of Mn can be under control at room temperature, but increases very much with temperature, so that the capacity of $LiMn_{0.5}Mn_{1.5}O_4$ decreases dramatically with temperature, while everybody wants a battery that works up to 60 °C, a temperature that can be reached during the summertime in vehicles, for instance: the problem (3) above mentioned. Again, the coat of the $LiMn_{0.5}Mn_{1.5}O_4$ particles should also be protective against the loss of Mn, at it is for the loss of oxygen, but very few experiments have been made so far to test this effect upon cycling at 60 °C. Nevertheless, this discussion shows that, while many efforts have been made to try different forms of doping, the most promising route to the operational and safe 5-V batteries is the synthesis of multi-composite nanoparticles as the active cathode element [200–204]. However, a concomitant work is mandatory to pursue the progress on electrolytes to find electrolytes that not only do not oxidize at such a voltage, but also solve the problem of the solid-electrolyte interface on the carbon surface of the anode: all the five problems we have mentioned are still pending challenges.

References

1. Goodenough JB (2002) Oxides cathodes. In: Advances in lithium-ion batteries. Kluwer Academic/Plenum, New York pp 135–154
2. Goodenough JB, Kim Y (2010) Challenges for rechargeable Li batteries. Chem Mater 22:587–603
3. Zaghib K, Dubé J, Dallaire A, Galoustov K, Guerfi A, Ramanathan M, Benmayza A, Prakash J, Mauger A, Julien CM (2012) Enhanced thermal safety and high power performance of carbon-coated $LiFePO_4$ olivine cathode for Li-ion batteries. J Power Sour 219:36–44
4. Mizushima K, Jones PC, Wiseman PJ, Goodenough JB (1980) Li_xCoO_2 ($0 < x<1$): a new cathode material for batteries of high energy density. Mater Res Bull 15:783–789
5. Ohzuku T, Makimura Y (2001) Layered lithium insertion material of $LiCo_{1/3}Ni_{1/3}Mn_{1/3}O_2$ for lithium-ion batteries Chem Lett 30:642–643
6. Thackeray MM, David WIF, Bruce PG, Goodenough JB (1982) Lithium insertion into manganese spinels. Mater Res Bull 18:461–472
7. Padhi AK, Nanjundaswamy KS, Goodenough JB (1997) Phospho-olivines as positive-electrode materials for rechargeable lithium batteries. J Electrochem Soc 144:1188–1194
8. Whittingham MS (2004) Lithium batteries and cathode materials. Chem Rev 104:4271–4301
9. Ellis BL, Lee KT, Nazar LF (2010) Positive electrode materials for Li-ion and Li-batteries. Chem Mater 22:691–714
10. Fergus JW (2010) Recent developments in cathode materials for lithium ion batteries. J Power Sour 195:939–954
11. Zaghib K, Mauger A, Julien CM (2012) Overview of olivines in lithium batteries for green transportation and energy storage. J Solid State Electrochem 16:835–845
12. Santhanam R, Rambabu B (2010) Research progress in high voltage spinel $LiNi_{0.5}Mn_{1.5}O_4$ material. J Power Sour 195:5442–5451
13. Liu GQ, Wen L, Liu YM (2010) Spinel $LiNi_{0.5}Mn_{1.5}O_4$ and its derivatives as cathodes for high-voltage Li-ion batteries. J Solid State Electrochem 14:2191–2202
14. Kraytsberg A, Ein-Eli Y (2012) Higher, stronger, better. A review of 5 volt cathode materials for advanced lithium-ion batteries. Adv Energy Mater 2:922–939
15. Liu D, Han J, Dontigny M, Charest P, Guerfi A, Zaghib K, Goodenough JB (2010) Redox behaviors of Ni and Cr with different counter cations in spinel cathodes for Li-ion batteries. J Electrochem Soc 157:A770–A775
16. Shin Y, Manthiram A (2003) Origin of the high voltage (>4.5 V) capacity of spinel lithium manganese oxides. Electrochim Acta 48:3583–3592
17. Obrovac MN, Gao Y, Dahn JR (1998) Explanation for the 4.8-V plateau in $LiCr_xMn_{2-x}O_4$. Phys Rev B: Condens Matter 57:5728–5733
18. Gryffroy D, Vaudenberghe RE (1992) Cation distribution, cluster structure and ionic ordering of the spinel series $LiNi_{0.5}Mn_{1.5-x}Ti_xO_4$ and $LiNi_{0.5-y}Mg_yMn_{1.5}O_4$. J Phys Chem Solids 53:777–784
19. Amdouni N, Zaghib K, Gendron F, Mauger A, Julien CM (2006) Structure and insertion properties of disordered and ordered $LiNi_{0.5}Mn_{1.5}O_4$ spinels prepared by wet chemistry. Ionics 12:117–126
20. Strobel P, Ibarra-Palos A, Anne M, Poinsignon C, Crisci A (2003) Cation ordering in $Li_2Mn_3MO_8$ spinels: structural and vibration spectroscopy studies. Solid State Sci 5:1009–1018
21. Ariyoshi K, Iwakoshi Y, Nakayama N, Ohzuku T (2004) Topotactic two-phase reactions of $Li[Ni_{1/2}Mn_{3/2}]O_4$ ($P4_332$) in nonaqueous lithium cells. J Electrochem Soc 151:A296–A303
22. Kanamura K, Hoshikawa W, Umegaki T (2002) Electrochemical characteristics of $LiNi_{0.5}Mn_{1.5}O_4$ cathodes with Ti or Al current collectors. J Electrochem Soc 149:A339–A345

23. Ohzuku T, Takeda S, Iwanaga M (1999) Solid-state redox potentials for $Li[Me_{1/2}Mn_{3/2}]O_4$ (Me: 3d-transition metal) having spinel-framework structures: a series of 5 volt materials for advanced lithium-ion batteries. J Power Sour 81–82:90–94
24. Okada M, Lee YS, Yoshio M (2000) Cycle characterizations of $LiM_xMn_{2-x}O_4$ (M = Co, Ni) materials for lithium secondary battery at wide voltage region. J Power Sour 90:196–200
25. Dokko K, Mohamedi M, Anzue N, Itoh T, Uchida I (2002) In situ Raman spectroscopic studies of $LiNi_xMn_{2-x}O_4$ thin film cathode materials for lithium ion secondary batteries. J Mater Chem 12:3688–3693
26. Takahashi K, Saitoh M, Sano M, Fujita M, Kifune K (2004) Electrochemical and structural properties of a 4.7 V-class $LiNi_{0.5}Mn_{1.5}O_4$ positive electrode material prepared with a self-reaction method. J Electrochem Soc 151:A173–A177
27. Ooms FGB, Kelder EM, Schoonman J, Wagemaker M, Mulder FM (2002) High-voltage $LiMg_{\delta}Ni_{0.5-\delta}Mn_{1.5}O_4$ spinels for Li-ion batteries. Solid State Ionics 152–153:143–153
28. Blasse G (1966) Ferromagnetism and ferrimagnetism of oxygen spinels containing tetravalent manganese. J Phys Chem Solids 27:383–389
29. Amdouni N, Zaghib K, Gendron F, Mauger A, Julien CM (2007) Magnetic properties of $LiNi_{0.5}Mn_{1.5}O_4$ spinels prepared by wet chemical methods. J Magn Magn Mater 309:100–105
30. Mukai K, Sugiyama J (2010) An indicator to identify the $Li[Ni_{1/2}Mn_{3/2}]O_4$ ($P4_332$): dc-susceptibility measurements. J Electrochem Soc 157:A672–A676
31. Idemoto Y, Narai H, Koura N (2003) Crystal structure and cathode performance dependence on oxygen content of $LiMn_{1.5}Ni_{0.5}O_4$ as a cathode material for secondary lithium batteries. J Power Sour 119–121:125–129
32. Park SH, Oh SW, Myung ST, Sun YK (2004) Mo^{6+}-doped $Li[Ni_{(0.5+x)}Mn_{(1.5-2x)}Mo_x]O_4$ spinel materials for 5 V lithium secondary batteries prepared by ultrasonic spray pyrolysis. Electrochem Solid-State Lett 7:A451–A454
33. Moorhead-Rosenberg Z, Shin DW, Chemelewski KR, Goodenough JB, Manthiram A (2012) Quantitative determination of Mn^{3+} content in $LiMn_{1.5}Ni_{0.5}O_4$ spinel cathodes by magnetic measurements. Appl Phys Lett 100:213909-1–213909-5
34. Idemoto Y, Narai H, Koura N (2002) Oxygen content and electrode characteristics of $LiMn_{1.5}Ni_{0.5}O_4$ as a 5 V class cathode material for lithium secondary battery. Electrochemistry 70:587–589
35. Rhodes K, Meisner R, Kim Y, Dudney N, Daniel C (2011) Evolution of phase transformation behavior in $Li(Mn_{1.5}Ni_{0.5})O_4$ cathodes studied by in situ XRD. J Electrochem Soc 158:A890–A897
36. Kim JH, Myung ST, Yoon CS, Kang SG, Sun YK (2004) Comparative study of $LiNi_{0.5}Mn_{1.5}O_{4-\delta}$ and $LiNi_{0.5}Mn_{1.5}O_4$ cathodes having two crystallographic structures: *Fd*-3 *m* and $P4_332$. Chem Mater 16:906–914
37. Bhaskar A, Bramnik NN, Senyshyn A, Fuess H, Ehrenberg H (2010) Synthesis, characterization, and comparison of electrochemical properties of LiM0.5Mn1.5O4 (M = Fe Co, Ni) at different temperatures. J Electrochem Soc 157:A689–A695
38. Terada Y, Yasaka K, Nishikawa F, Konishi T, Yoshio M, Nakai I (2001) In situ XAFS analysis of Li(Mn, M)2O4 (M = Cr Co, Ni) 5 V cathode materials for lithium-ion secondary batteries. J Solid State Chem 156:286–291
39. Wen W, Kumarasamy B, Mukerjee S, Auinat M, Ein-Eli Y (2005) Origin of 5 V electrochemical activity observed in non-redox reactive divalent cation doped $LiM_{0.5-x}Mn_{1.5+x}O_4$ ($0 \leq x \leq 0.5$) cathode materials in situ XRD and XANES spectroscopy studies. J Electrochem Soc 152:A1902–A1911
40. Mukerjee S, Yang XQ, Sunb X, Lee SJ, McBreen J, Ein-Eli Y (2004) In situ synchrotron X-ray studies on copper–nickel 5 V Mn oxide spinel cathodes for Li-ion batteries. Electrochim Acta 49:3373–3382
41. Liu D, Lu Y, Goodenough JB (2010) Rate properties and elevated-temperature performances of $LiNi_{0.5-x}Cr_{2x}Mn_{1.5-x}O_4$ ($0 \leq 2x \leq 0.8$) as 5 V cathode materials for lithium-ion batteries. J Electrochem Soc 157:A1269–A1273

42. Wang L, Li H, Huang X, Baudrin E (2011) A comparative study of *Fd*-3 *m* and $P4_332$ $LiNi_{0.5}Mn_{1.5}O_4$. Solid State Ionics 193:32–38
43. Julien CM, Gendron F, Amdouni N, Massot M (2006) Lattice vibrations of materials for lithium rechargeable batteries. VI: Ordered spinels. Mater Sci Eng B 130:41–48
44. Matsui M, Dokko K, Kanamura K (2010) Surface layer formation and stripping process on $LiMn_2O_4$ and $LiNi_{1/2}Mn_{3/2}O_4$ thin film electrodes. J Electrochem Soc 157:A121–A129
45. Gao Y, Myrtle K, Zhang MJ, Reimers JN, Dahn JR (1996) Valence band of $LiNi_xMn_{2-x}O_4$ and its effects on the voltage profiles of $LiNi_xMn_{2-x}O_4$/Li electrochemical cells. Phys Rev B: Condens Matter 54:16670–16675
46. Shin Y, Manthiram A (2004) Factors influencing the capacity fade of spinel lithium manganese oxides. J Electrochem Soc 151:A204–A208
47. Patoux Q, Daniel L, Bourbon C, Lignier H, Pagano C, Le Cras F, Jouanneau S, Martinet S (2009) High voltage spinel oxides for Li-ion batteries: From the material research to the application. J Power Sour 189:344–352
48. Fang HS, Wang ZX, Li XH, Guo HJ, Peng WJ (2006) Exploration of high capacity $LiNi_{0.5}Mn_{1.5}O_4$ synthesized by solid-state reaction. J Power Sources 153:174–176
49. Chen ZY, Ji S, Linkov V, Zhang JL, Zhu W (2009) Performance of $LiNi_{0.5}Mn_{1.5}O_4$ prepared by solid-state reaction. J Power Sour 189:507–510
50. Miao C, Shi L, Chen G, Dai D (2012) Preparation of precursor of $LiNi_{0.5}Mn_{1.5}O_4$ with high density. Adv Mater Res 463–464:881–884
51. Liu G, Qi L, Wen L (2006) Synthesis and electrochemical performance of $LiNi_xMn_{2-x}O_4$ spinel as cathode material for lithium ion batteries. Rare Met Mater Eng 35:299–302
52. Fang HS, Wang ZX, Yin ZL, Li XH, Guo HJ, Peng WJ (2005) Effect of ball milling and electrolyte on properties of high-voltage $LiNi_{0.5}Mn_{1.5}O_4$ spinel. Trans Nonferrous Met Soc Chin (English) 15:1429–1432
53. Fang HS, Li LP, Li GS (2007) A low-temperature reaction route to high rate and high capacity $LiNi_{0.5}Mn_{1.5}O_4$. J Power Sour 167:223–227
54. Xu HY, Xie S, Ding N, Liu BL, Shang Y, Chen CH (2006) Improvement of electrochemical properties of $LiNi_{0.5}Mn_{1.5}O_4$ spinel prepared by radiated polymer gel method. Electrochem Acta 51:4352–4357
55. Arunkumar TA, Manthiram A (2005) Influence of lattice parameter differences on the electrochemical performance of the 5 V spinel LiMn1.5 − yNi0.5 − zMy + zO4 (M = Li, Mg, Fe Co, and Zn). Electrochem Solid-State Lett 8:A403–A405
56. Aklalouch M, Amarilla JM, Saadoune I, Rojo JM (2011) LiCr0.2Ni0.4Mn1.4O4 spinels exhibiting huge rate capability at 25 and 55 C: analysis of the effect of the particle size. J Power Sour 196:10222–10227
57. Kim JH, Myung ST, Sun YK (2004) Molten salt synthesis of $LiNi_{0(5}Mn_{1(5}O_4$ spinel for 5 V class cathode material of Li-ion secondary battery. Electrochim Acta 49:219–227
58. Chen G, Hai B, Shukla AK, Duncan H (2012) Impact of $LiMn_{1.5}Ni_{0.5}O_4$ crystal surface facets. ECS Symp Abstr700
59. Lim SJ, Ryu WH, Kim WK, Kwon HS (2012) Electrochemical performance of $LiNi_{0.5}Mn_{1.5}O_4$ cathode material fabricated from nanothorn sphere structured MnO_2. ECS Symp Abstr953
60. Zhao ZQ, Ma JF, Tian H, Xie LJ, Zhou J, Wu PW, Wang YG, Tao JT, Zhu XY (2005) Preparation and characterization of nano-crystalline $LiNi_{0.5}Mn_{1.5}O_4$ cathode material by the soft combustion reaction method. J Am Ceram Soc 88:3549–3552
61. Chen J, Cheng F (2009) Combination of lightweight elements and nanostructured materials for batteries. Acc Chem Res 42:713–723
62. Fan YK, Wang JM, Ye XB, Zhang JQ (2007) Physical properties and electrochemical performance of $LiNi_{0.5}Mn_{1.5}O_4$ cathode material prepared by a co-precipitation method. Mater Chem Phys 103:19–23
63. Yi TF, Hu XG (2007) Preparation and characterization of sub-micro $LiNi_{0.5-x}Mn_{1.5+x}O_4$ for 5 V cathode materials synthesized by an ultrasonic-assisted co-precipitation method. J Power Sour 167:185–191

64. Ohzuku T, Ariyoshi K, Yamamoto S (2002) Synthesis and characterization of $Li[Ni_{1/2}Mn_{3/2}]O_4$ by two-step solid state reaction. J Ceram Soc Jpn 110:501–505
65. Myung ST, Komaba S, Kumagai N, Yashiro H, Chung HT, Cho TH (2002) Nano-crystalline $LiNi_{0.5}Mn_{1.5}O_4$ synthesized by emulsion drying method. Electrochim Acta 47:2543–2549
66. Zhao Q, Ye N, Li L, Yan F (2010) Oxalate coprecipitation process synthesis of 5 V cathode material $LiNi_{0.5}Mn_{1.5}O_4$ and its performance. Rare Met Mater Eng 39:1715–1718
67. Liu D, Han J, Goodenough JB (2010) Structure, morphology, and cathode performance of $Li_{1-x}[Ni_{0.5}Mn_{1.5}]O_4$ prepared by coprecipitation with oxalic acid. J Power Sour 195:2918–2923
68. Yang K, Su J, Zhang L, Long Y, Lv X, Wen Y (2012) Urea combustion synthesis of $LiNi_{0.5}Mn_{1.5}O_4$ as a cathode material for lithium ion batteries. Particuology 10:765–770
69. Cao A, Manthiram A (2012) Controlled synthesis of high tap density $LiMn_{1.5}Ni_{0.5}O_4$ with tunable shapes. ECS Symp Abstr699
70. Kunduraci M, Amatucci GG (2006) Synthesis and characterization of nanostructured 4.7 V $Li_xMn_{1.5}Ni_{0.5}O_4$ spinels for high-power lithium-ion batteries. J Electrochem Soc 153: A1345–A1352
71. Yamada M, Dongying B, Kodera T, Myoujin K, Ogihara T (2009) Mass production of cathode materials for lithium ion battery by flame type spray pyrolysis. J Ceram Soc Jpn 117:1017–1020
72. Wu HM, Tu JP, Chen XT, Shi DQ, Zhao XB, Cao GS (2006) Synthesis and characterization of abundant Ni-doped $LiNi_xMn_{2-x}O_4$ (x = 0.1–0.5) powders by spray-drying method. Electrochim Acta 51:4148–4152
73. Park SH, Oh SW, Yoon CS, Myung ST, Sun YK (2005) $LiNi_{0.5}Mn_{1.5}O_4$ showing reversible phase transition on 3 V region. Electrochem Solid-State Lett 8:A163–A167
74. Ogihara T, Kodera T, Myoujin K, Motohira S (2009) Preparation and electrochemical properties of cathode materials for lithium ion battery by aerosol process. Mater Sci Eng, B 161:109–114
75. Kojima M, Mukoyama I, Myoujin K, Kodera T, Ogihara T (2009) Mass production and battery properties of $LiNi_{0.5}Mn_{1.5}O_4$ powders prepared by internal combustion type spray pyrolysis. Key Eng Mater 388:85–88
76. Sigala C, Guyomard D, Verbaere A, Piffard Y, Tournoux M (1995) Positive electrode materials with high operating voltage for lithium batteries: $LiCr_yMn_{2-y}O_4$ (0 < y<1). Solid State Ionics 81:167–170
77. Arrebola JC, Caballero A, Hernan L, Morales J (2008) PMMA-assisted synthesis of $Li_{1-x}Ni_{0.5}Mn_{1.5}O_{4-\delta}$ for high-voltage lithium batteries with expanded rate capability at high cycling temperatures. J Power Sources 180:852–858
78. Kalyani P, Kalaiselvi N, Muniyandi N (2003) An innovative soft-chemistry approach to synthesize $LiNiVO_4$. Mater Chem Phys 77:662–668
79. Liu J, Manthiram A (2009) Understanding the improved electrochemical performances of Fe-substituted 5 V spinel cathode $LiMn_{1.5}Ni_{0.5}O_4$. J Phys Chem C 113:15073–15079
80. Zhong GB, Wang YY, Yu YQ, Chen CH (2012) Electrochemical investigations of the LiNi0.45M0.10Mn1.45O4 (M = Fe Co, Cr) 5 V cathode materials for lithium ion batteries. J Power Sour 205:385–393
81. Park SB, Eom WS, Cho WI, Jang H (2006) Electrochemical properties of $LiNi_{0.5}Mn_{1.5}O_4$ cathode after Cr doping. J Power Sour 159:679–684
82. Amatucci GG, Pereira N, Zheng T, Tarascon JM (2001) Failure mechanism and improvement of the elevated temperature cycling of $LiMn_2O_4$ compounds through the use of the $LiAl_xMn_{2-x}O_{4-z}F_z$ solid solution. J Electrochem Soc 148:A171–A182
83. Oh SW, Park SH, Kim JH, Bae YC, Sun YK (2006) Improvement of electrochemical properties of $LiNi_{0.5}Mn_{1.5}O_4$ spinel material by fluorine substitution. J Power Sour 157:464–470
84. Xu XX, Yang J, Wang YQ, Wang JL (2007) $LiNi_{0.5}Mn_{1.5}O_{3.975}F_{0.05}$ as novel 5-V cathode material. J Power Sour 174:1113–1116
85. Du GD, NuLi Y, Yang J, Wang J (2008) Fluorine-doped $LiNi_{0.5}Mn_{1.5}O_4$ for 5 V cathode materials of lithium-ion battery. Mater Res Bull 43:3607–3613

86. Wu X, Zong X, Yang Q, Jin Z, Wu H (2001) Electrochemical studies of substituted spinel $LiAl_yMn_{2-y}O_{4-z}F_z$ for lithium secondary batteries. J Fluorine Chem 107:39–44
87. Sun YK, Park GS, Lee YS, Yoshio M, Nahm KS (2001) Structural changes (degradation) of oxysulfide $LiAl_{0.24}Mn_{1.76}O_{3.98}S_{0.02}$ spinel on high-temperature cycling. J Electrochem Soc 148:A994–A998
88. Sun YK, Oh SW, Yoon CS, Bang HJ, Prakash J (2006) Effect of sulfur and nickel doping on morphology and electrochemical performance of $LiNi_{0.5}Mn_{1.5}O_{4-x}S_x$ spinel material in 3-V region. J Power Sour 161:19–26
89. Xi N, Zhong B, Chen M, Yin K, Li L, Liu H, Guo X (2013) Synthesis of $LiCr_{0.2}Ni_{0.4}Mn_{1.4}O_4$ with superior electrochemical performance via a two-step thermo polymerization technique. Electrochim Acta 97:184–191
90. Zheng J, Xiao J, Yu X, Kovarik L, Gu M, Omenya F, Chen X, Zhang JG (2012) Enhanced Li^+ ion transport in $LiNi_{0.5}Mn_{1.5}O_4$ through control of site disorder. Phys Chem Chem Phys 14:13515–13521
91. Amine K, Tukamoto H, Yasuda H, Fujita Y (1996) A New three-volt spinel $Li_{1+x}Mn_{1.5}Ni_{0.5}O_4$ for secondary lithium batteries. J Electrochem Soc 143:1607–1613
92. Liu D, Hamel-Paquet J, Trottier J, Barray F, Gariépy V, Hovington P, Guerfi A, Mauger A, Julien CM, Goodenough JB, Zaghib K (2012) Synthesis of pure phase disordered $LiMn_{1.45}Cr_{0.1}Ni_{0.45}O_4$ by a post-annealing method. J Power Sour 217:400–406
93. Shin DW, Bridges CA, Huq A, M. Paranthaman MP, Manthiram A (2012) Role of cation ordering and surface segregation in high-voltage spinel $LiMn_{1.5}Ni_{0.5-x}M_xO_4$ (M = Cr, Fe, and Ga) cathodes for lithium-ion batteries. Chem Mater 24:3720–3731
94. Takahashi Y, Sasaoka H, Kuzuo R, Kijima N, Akimoto J (2006) A low-temperature synthetic route and electrochemical properties of micrometer-sized $LiNi_{0.5}Mn_{1.5}O_4$ single crystals. Electrochem Solid-State Lett 9:A203–A206
95. Kanamura K, Hoshikawa W, Umegaki T (2001) Preparation and evaluation of new cathode materials for rechargeable lithium battery with 5 V. J Japn Soc Powder Met 48:283–287
96. Maeda Y, Ariyoshi K, Kawai T, Sekiya T, Ohzuku T (2009) Effect of deviation from Ni/Mn stoichiometry in $Li[Ni_{1/2}Mn_{3/2}]O_4$ upon rechargeable capacity at 4.7 V in nonaqueous lithium cells. J Ceram Soc Jpn 117:1216–1220
97. Yoshio M, Konishi T, Todorov YM, Noguchi H (2000) Electrochemical behavior of nonstoichiometric $LiMn_{2-x}Ni_xO_4$ as a 5-V cathode material. Electrochemistry 68:412–414
98. Xia H, Meng YS, Lu L, Ceder G (2007) Electrochemical properties of nonstoichiometric $LiNi_{0.5}Mn_{1.5}O_{4-\delta}$ thin-film electrodes prepared by pulsed laser deposition. J Electrochem Soc 154:A737–A743
99. Pasero D, Reeves N, Pralong V, West AR (2008) Oxygen nonstoichiometry and phase transitions in $LiMn_{1.5}Ni_{0.5}O_{4-\delta}$. J Electrochem Soc 155:A282–A291
100. Jin YC, Lin CY, Duh JG (2012) Improving rate capability of high potential $LiNi_{0.5}Mn_{1.5}O_{4-x}$ cathode materials via increasing oxygen non-stoichiometries. Electrochim Acta 69:45–50
101. Wu X, Kim SB (2002) Improvement of electrochemical properties of $LiNi_{0.5}Mn_{1.5}O_4$ spinel. J Power Sour 109:53–57
102. Wu HM, Tu JP, Yuan YF, Li Y, Zhao XB, Cao GS (2005) Electrochemical and ex situ XRD studies of a $LiMn_{1.5}Ni_{0.5}O_4$ high-voltage cathode material. Electrochim Acta 50:4104–4108
103. Kim JH, Yoon CS, Myung ST, Prakash J, Sun YK (2004) Phase transitions in $Li_{1-\delta}Ni_{0.5}Mn_{1.5}O_4$ during cycling at 5 V. Electrochem Solid-State Lett 7:A216–A220
104. Alcántara R, Jaraba M, Lavela P, Tirado JL (2002) Optimizing preparation conditions for 5 V electrode performance, and structural changes in $Li_{1-x}Ni_{0.5}Mn_{1.5}O_4$ spinel. Electrochim Acta 47:1829–1835
105. Zhu W, Liu D, Trottier J, Gagnon C, Mauger A, Julien CM, Zaghib K (2013) In-situ XRD study of the phase evolution in un-doped and Cr-doped $Li_xMn_{1.5}Ni_{0.5}O_4$ ($0.1 \le x \le 0.1$) 5-volt cathode materials. J Power Sour 242:236–243
106. Kim JH, Pieczonka NPW, Li Z, Wu Y, Harris S, Powell BR (2013) Understanding the capacity fading mechanism in $LiNi_{0.5}Mn_{1.5}O_4$/graphite Li-ion batteries. Electrochim Acta 90:556–562

107. Hai B, Shukla AK, Duncan H, Chen G (2013) The effect of particle surface facets on the kinetic properties of $LiMn_{1.5}Ni_{0.5}O_4$ cathode materials. J Mater Chem A 1:759–769
108. Sun YK, Yoon CS, Oh IH (2003) Surface structural change of ZnO-coated $LiNi_{0.5}Mn_{1.5}O_4$ spinel as 5 V cathode materials at elevated temperatures. Electrochim Acta 48:503–506
109. Aurbach D, Markovsky B, Talyosef Y, Salitra G, Kim HJ, Choi S (2006) Studies of cycling behavior, ageing, and interfacial reactions of $LiNi_{0.5}Mn_{1.5}O_4$ and carbon electrodes for lithium-ion 5-V cells. J Power Sour 162:780–789
110. Mun J, Yim T, Park K, Ryu JH, Kim YG, Oh SM (2011) Surface film formation on $LiNi_{0.5}Mn_{1.5}O_4$ electrode in an ionic liquid solvent at elevated temperature. J Electrochem Soc 158:A453–A457
111. Wu W, Li X, Wang Z, Guo H, Wang J, Xue P (2013) Comprehensive reinvestigation on the initial coulombic efficiency and capacity fading mechanism of $LiNi_{0.5}Mn_{1.5}O_4$ at low rate and elevated temperature. J Solid State Electrochem. doi: 10.1007/s10008-012-1963-5
112. Fu LJ, Liu H, Li C, Wu YP, Rahm E, Holze R, Wu HQ (2006) Surface modifications of electrode materials for lithium ion batteries. Solid State Sci 8:113–128
113. Liu J, Manthiram A (2009) Understanding the improvement in the electrochemical properties of surface modified 5 V $LiMn_{1.42}Ni_{0.42}Co_{0.16}O_4$ spinel cathodes in lithium-ion cells. Chem Mater 21:1695–1707
114. Liu J, Manthiram A (2009) Kinetics study of the 5 V spinel cathode $LiMn_{1.5}Ni_{0.5}O_4$ before and after surface modifications. J Electrochem Soc 156:A833–A838
115. Kobayashi Y, Miyashiro H, Takei K, Shigemura H, Tabuchi M, Kageyama H, Iwahori T (2003) 5 V class all-solid-state composite lithium battery with Li_3PO_4 coated $LiNi_{0.5}Mn_{1.5}O_4$. J Electrochem Soc 150:A1577–A1582
116. Arrebola J, Caballero A, Hernan L, Morales J, Castellon ER, Ramos-Barrado JR (2007) Effects of coating with gold on the performance of nanosized $LiNi_{0.5}Mn_{1.5}O_4$ for lithium batteries. J Electrochem Soc 154:A178–A184
117. Fan Y, Wang J, Tang Z, He W, Zhang J (2007) Effects of the nanostructured SiO_2 coating on the performance of $LiNi_{0.5}Mn_{1.5}O_4$ cathode materials for high-voltage Li-ion batteries. Electrochim Acta 52:3870–3875
118. Cho J, Kim YJ, Kim TJ, Park B (2001) Zero-strain intercalation cathode for rechargeable Li-ion cell. Ang Chem Int Ed 40:3367–3369
119. Chen Z, Dahn JR (2002) Effect of a ZrO_2 Coating on the structure and electrochemistry of Li_xCoO_2 when cycled to 4.5 V. Electrochem Solid-State Lett 5:A213–A216
120. Appapillai AT, Mansour AN, Cho J, Shao-Horn Y (2007) Microstructure of $LiCoO_2$ with and without "$AlPO_4$" nanoparticle coating: combined STEM and XPS studies. Chem Mater 19:5748–5757
121. Fey GTK, Li W, Dahn JR (1994) $LiNiVO_4$: A 4.8 volt electrode material for lithium cells. J Electrochem Soc 141:2279–2282
122. Fey GTK, Dahn JR, Zhang M, Li W (1997) The effects of the stoichiometry and synthesis temperature on the preparation of the inverse spinel $LiNiVO_4$ and its performance as a new high voltage cathode material. J Power Sour 68:549–552
123. Prabaharan SRS, Michael MS, Radhakrishna S, Julien C (1997) Novel low-temperature synthesis and characterization of $LiNiVO_4$ for high-voltage Li-ion batteries. J Mater Chem 7:1791–1796
124. Fey GTK, Perng WB (1997) A new preparation method for a novel high voltage cathode material: $LiNiVO_4$. Mater Chem Phys 47(1997):279–282
125. Rissouli K, Benkhouja K, Touaiher M, Ait-Salah A, Jaafari K, Fahad M, Julien C (2005) Structure and conductivity of lithiated vanadates LiMVO4 (M = Mn Co, Ni). J Phys IV France 123:265–269
126. Lu CH, Liou SJ (1998) Preparation of submicrometer $LiNiVO_4$ powder by solution route for lithium ion secondary batteries. J Mater Sci Lett 17:733–735
127. Fey GTK, Huang DL (1999) Synthesis, characterization and cell performance of inverse spinel electrode materials for lithium secondary batteries. Electrochim Acta 45:295–314

128. Cao X, Xie L, Zhan H, Zhou Y (2008) Rheological phase synthesis and characterization of $LiNiVO_4$ as a high voltage cathode material for lithium ion batteries. J New Mater Electrochem Syst 11:193–198
129. Vivekanandhan S, Venkateswarlu M, Satyanarayana N (2004) Glycerol-assisted gel combustion synthesis of nano-crystalline $LiNiVO_4$ powders for secondary lithium batteries. Mater Lett 58:1218–1222
130. Chitra S, Kalyani P, Yebka B, Mohan T, Haro-Poniatowski E, Gangadharan R, Julien C (2000) Synthesis, characterization and electrochemical studies of $LiNiVO_4$ cathode material in rechargeable lithium batteries. Mater Chem Phys 65:32–37
131. Subramania A, Angayarkanni N, Karthick SN, Vasudevan T (2006) Combustion synthesis of inverse spinel $LiNiVO_4$ nano-particles using gelatine as the new fuel. Mater Lett 60:3023–3026
132. Li X, Wei YJ, Ehrenberg H, Liu DL, Zhan SY, Wang CZ, Chen G (2009) X-ray diffraction and Raman scattering studies of Li^+/e^--extracted inverse spinel $LiNiVO_4$. J Alloys Compd 471:L26–L28
133. Lai QY, Lu JZ, Liang XL, Yan FY, Ji XY (2001) Synthesis and electrochemical characteristics of Li-Ni vanadates as positive materials. Intern J Inorg Mater 3:381–385
134. Palanichamy K (2011) On the modified inverse spinel-$LiCo(PO_4)_x(VO_4)_{1-x}$ as cathode for rechargeable lithium batteries. Ionics 17:391–397
135. Fey GTK, Chen KS (1999) Synthesis, characterization, and cell performance of $LiNiVO_4$ cathode materials prepared by a new solution precipitation method. J Power Sour 81–82:467–471
136. Lu CH, Liou SJ (2000) Hydrothermal preparation of nanometer lithium nickel vanadium oxide powder at low temperature. Mater Sci Eng, B 75:38–42
137. Phuruangrat A, Thongtem T, Thongtem S (2007) Preparation and characterization of nano-crystalline $LiCoVO_4$ and $LiNiVO_4$ used as cathodes for lithium ion batteries. J Ceram Proc Res 8:450–452
138. Phuruangrat A, Thongtem T, Thongtem S (2007) Characterization of nano-crystalline $LiNiVO_4$ synthesized by hydrothermal process. Mater Lett 61:3805–3808
139. Wang GX, Zhong S, Bradhurst DH, Dou SX, Liu HK (1999) Rare earth element (La) doped $LiNiVO_4$ as cathode material for secondary lithium ion cells. Mater Sci Forum 315–317:105–112
140. Reddy MV, Pecquenard B, Vinatier P, Levasseur A (2007) Cyclic voltammetry and galvanostatic cycling characteristics of $LiNiVO_4$ thin films during lithium insertion and re/de-insertion. Electrochem Commun 9:409–415
141. Kalyani P, Kalaiselvi N, Renganathan NG (2005) $LiNiM_xV_{1-x}O_4$ (M = Co, Mg and Al) solid solutions—prospective cathode materials for rechargeable lithium batteries. Mater Chem Phys 90:196–202
142. Zaghib K, Mauger A, Goodenough JB, Gendron F, Julien CM (2009) Positive electrode: lithium iron phosphate. In: Garche J (ed) Encyclopedia of electrochemical power sources. Elsevier Science Amsterdam 5:264–296
143. Julien CM, Mauger A, Ait-Salah A, Massot M, Gendron F, Zaghib K (2007) Nanoscopic scale studies of $LiFePO_4$ as cathode material in lithium-ion batteries for HEV application. Ionics 13:395–411
144. Bramnik NN, Nikolowski K, Trots DM, Ehrenberg H (2008) Thermal stability of $LiCoPO_4$ cathodes. Electrochem Solid-State Lett 11:A89–A93
145. Herle PS, Ellis B, Coombs N, Nazar LF (2004) Nano-network electronic conduction in iron and nickel olivine phosphates. Nat Mater 3:147–152
146. Wolfenstine J, Allen J (2005) Ni^{3+}/Ni^{2+} redox potential in $LiNiPO_4$. J Power Sour 142:389–390
147. Minakshi M, Sharma N, Ralph D, Appadoo D, Nallathamby K (2011) Synthesis and characterization of $Li(Co_{0.5}Ni_{0.5})PO_4$ cathode for Li-ion aqueous battery applications. Electrochem Solid-State Lett 14:A86–A89

148. Bramnik NN, Bramnik KG, Baehtz C, Ehrenberg H (2005) Study of the effect of different synthesis routes on Li extraction–insertion from $LiCoPO_4$. J Power Sour 145:74–81
149. Bramnik NN, Bramnik KG, Buhrmester T, Baehtz C, Ehrenberg H, Fuess H (2004) Electrochemical and structural study of $LiCoPO_4$-based electrodes. J Solid State Electrochem 8:558–564
150. Nakayama M, Goto S, Uchimoto Y, Wakihara M, Kitayama Y (2004) Changes in electronic structure between cobalt and oxide ions of lithium cobalt phosphate as 4.8-V positive electrode material. Chem Mater 16:3399–3401
151. Bramnik NN, Nikolowski K, Baehtz C, Bramnik KG, Ehrenberg H (2007) Phase transition occurring upon lithium insertion-extraction of $LiCoPO_4$. Chem Mater 19:908–915
152. Okada S, Sawa S, Egashira M, Yamaki JI, Tabuchi M, Kageyama H, Konishi T, Yoshino A (2001) Cathode properties of phospho-olivine $LiMPO_4$ for lithium secondary batteries. J Power Sour 97–98:430–432
153. Jang IC, Lim HH, Lee SB, Karthikeyan K, Aravindan V, Kang KS, Yoon WS, Cho WI, Lee YS (2010) Preparation of $LiCoPO_4$ and $LiFePO_4$ coated $LiCoPO_4$ materials with improved battery performance. J Alloys Compd 497:321–324
154. Aravindan V, Cheah YL, Chui Ling WC, Madhavi S (2012) Effect of LiBOB additive on the electrochemical performance of $LiCoPO_4$. J Electrochem Soc 159:A1435–A1439
155. Rabanal ME, Gutierrez MC, Garcia-Alvarado F, Gonzalo EC, Arroyoy de Dompablo ME (2006) Improved electrode characteristics of olivine–$LiCoPO_4$ processed by high energy milling. J Power Sour 160:523–528
156. Koleva V, Zhecheva E, Stoyanova R (2010) Ordered olivine-type lithium-cobalt and lithium-nickel phosphates prepared by a new precursor method. Eur J Inorg Chem 26:4091–4099
157. Kandhasamy S, Pandey A, Minakshi M (2012) Polyvinyl-pyrrolidone assisted sol-gel route $LiCo_{1/3}Mn_{1/3}Ni_{1/3}PO_4$ composite cathode for aqueous rechargeable battery. Electrochim Acta 60:170–176
158. Eftekhari A (2004) Surface modification of thin-film based $LiCoPO_4$ 5 V cathode with metal oxide. J Electrochem Soc 151:A1456–A1460
159. Deniard P, Dulac AM, Rocquefelte X, Grigorova V, Lebacq O, Pasturel A, Jobic S (2004) High potential positive materials for lithium-ion batteries: transition metal phosphates. J Phys Chem Solids 65:229–233
160. Prabu M, Selvasekarapandian S, Kulkarni AR, Karthikeyan S, Hirankumar G, Sanjeeviraja C (2011) Structural, dielectric, and conductivity studies of yttrium-doped $LiNiPO_4$ cathode materials. Ionics 17:201–207
161. Karthickprabhu S, Hirankumar G, Maheswaran A, Sanjeeviraja C, Daries-Bella RS (2013) Structural and conductivity studies on $LiNiPO_4$ synthesized by the polyol method. J Alloys Compd 548:65–69
162. Lloris JM, Pérez-Vicente C, Tirado JL (2002) Improvement of the electrochemical performance of $LiCoPO_4$ 5 V material using a novel synthesis procedure. Electrochem Solid-State Lett 5:A234–A237
163. Yang J, Xu JJ (2006) Synthesis and characterization of carbon-coated lithium transition metal phosphates LiMPO4 (M = Fe, Mn Co, Ni) prepared via a nonaqueous sol-gel route batteries, fuel cells, and energy conversion. J Electrochem Soc 153:A716–A723
164. Gangulibabu N, Bhuvaneswari D, Kalaiselvi N, Jayaprakash N, Periasamy P (2009) CAM sol-gel synthesized $LiMPO_4$ (M = Co, Ni) cathodes for rechargeable lithium batteries. J Sol-Gel Sci Technol 49:137–144
165. Zhou F, Cococcioni M, Kang K, Ceder G (2004) The Li intercalation potential of LiMPO4 and LiMSiO4 olivines with M = Fe, Mn Co, Ni. Electrochem Commun 6:1144–1148
166. Howard WF, Spotnitz RM (2007) Theoretical evaluation of high-energy lithium metal phosphate cathode materials in Li-ion batteries. J Power Sour 165:887–891
167. Rissouli K, Benkhouja K, Ramos-Barrado JR, Julien C (2003) Electrical conductivity in lithium orthophosphates. Mater Sci Eng B 98:185–189
168. Goñi A, Lezama L, Barberis GE, Pizarro JL, Arriortua MI, Rojo T (1996) Magnetic properties of the $LiMPO_4$ (M = Co, Ni) compounds. J Magn Magn Mater 164:251–255

169. Santoro RP, Segal DJ, Newnham RE (1966) Magnetic properties of $LiCoPO_4$ and $LiNiPO_4$. J Phys Chem Solids 27:1192–1193
170. Kornev I, Bichurin M, Rivera JP, Gentil S, Schmid H, Jansen AGM, Wyder P (2000) Magnetoelectric properties of $LiCoPO_4$ and $LiNiPO_4$. Phys Rev B Condens Matter 62:12247–12253
171. Yamauchi K, Picozzi S (2010) Magnetic anisotropy in Li-phosphates and origin of magnetoelectricity in $LiNiPO_4$. Phys Rev B: Condens Matter 81:024110
172. Julien CM, Mauger A, Zaghib K, Veillette R, Groult H (2012) Structural and electronic properties of the $LiNiPO_4$ orthophosphate. Ionics 18:625–633
173. Fomin VI, Gnezdilov VP, Kurnosov VS, Peschanskii AV, Yeremenko AV, Schmid H, Rivera JP, Gentil S (2002) Raman scattering in a $LiNiPO_4$ single crystal. Low Temp Phys 28:203–209
174. Shang SL, Wang Y, Mei ZG, Hui XD, Liu ZK (2012) Lattice dynamics, thermodynamics, and bonding strength of lithium-ion battery materials LiMPO4 (M = Mn, Fe Co, and Ni): a comparative first-principles study. J Mater Chem 22:1142–1149
175. Dimesso L, Jacke S, Spanheimer C, Jaegermann W (2012) Investigation on $LiCoPO_4$ powders as cathode materials annealed under different atmospheres. J Solid State Electrochem 16:3911–3919
176. Dimesso L, Spanheimer C, Jaegermann W (2013) Effect of the Mg-substitution on the graphitic carbon foams—$LiNi_{1-y}Mg_yPO_4$ composites as possible cathodes materials for 5 V applications. Mater Res Bull 48:559–565
177. Amine K, Yasuda H, Yamachi M (2000) Olivine $LiCoPO_4$ as 4.8-V electrode material for lithium batteries. Electrochem Solid-State Lett 3:178–179
178. Wolfenstine J, Allen J (2004) $LiNiPO_4$–$LiCoPO_4$ solid solutions as cathodes. J Power Sour 136:150–153
179. Ni J, Gao L, Lu L (2013) Carbon coated lithium cobalt phosphate for Li-ion batteries: Comparison of three coating techniques. J Power Sour 221:35–41
180. Wolfenstine J, Read J, Allen J (2007) Effect of carbon on the electronic conductivity and discharge capacity $LiCoPO_4$. J Power Sour 163:1070–1073
181. Wolfenstine J, Lee U, Poese B, Allen J (2005) Effect of oxygen partial pressure on the discharge capacity of $LiCoPO_4$. J Power Sour 144:226–230
182. Wolfenstine J, Poese B, Allen J (2004) Chemical oxidation of $LiCoPO_4$. J Power Sour 138:281–282
183. Wolfenstine J (2006) Electrical conductivity of doped $LiCoPO_4$. J Power Sour 158:1431–1435
184. Wang F, Yang J, Li YN, Wang J (2011) Novel hedgehog-like 5 V $LiCoPO_4$ positive electrode material for rechargeable lithium battery. J Power Sour 196:4806–4810
185. Nakayama M, Goto S, Uchimoto Y, Wakihara M, Kitayama Y, Miyanaga T, Watanabe I (2005) X-ray absorption spectroscopic study on the electronic structure of $Li_{1-x}CoPO_4$ electrodes as 4.8 V positive electrodes for rechargeable lithium ion batteries. J Phys Chem B 109:11197–11203
186. Dimesso L, Spanheimer C, Jaegermann W, Zhang Y, Yarin AL (2013) $LiCoPO_4$—3D carbon nanofiber composites as possible cathode materials for high voltage applications. Electrochim Acta 97:38–42
187. Devaraju MK, Rangappa D, Honma I (2012) Controlled synthesis of plate-like $LiCoPO_4$ nanoparticles via supercritical method and their electrode property. Electrochim Acta 85:548–553
188. Reddy MV, Subba-Rao GV, Chowdari BVR (2010) Long-term cycling studies on 4 V-cathode lithium vanadium fluorophosphates. J Power Sour 195:5768–5774
189. Okada S, Ueno M, Uebou Y, Yamaki JI (2005) Fluoride phosphate Li_2CoPO_4F as a high-voltage cathode in Li-ion batteries. J Power Sour 146:565–569
190. Khasanova NR, Gavrilov AN, Antipov EV, Bramnik KG, Hibst H (2011) Structural transformation of Li_2CoPO_4F upon Li-deintercalation. J Power Sour 196:355–360

191. Stroukoff KR, Manthiram A (2011) Thermal stability of spinel $Li_{1.1}Mn_{1.9-y}M_yO_{4-z}F_z$ (M = Ni, Al, and Li, $0 \leq y \leq 0.3$, and $0 \leq z \leq 0.2$) cathodes for lithium ion batteries. J Mater Chem 21:10165–10170
192. Koyama Y, Tanaka I, Adachi H (2000) New fluoride cathodes for rechargeable lithium batteries. J Electrochem Soc 147:3633–3636
193. Dutreilh M, Chevalier C, El-Ghozzi M, Avignant D, Montel JM (1999) Synthesis and crystal structure of a new lithium nickel fluorophosphates Li_2NiFPO_4 with an ordered mixed anionic framework. J Solid State Chem 142:1–5
194. Nagahama M, Hasegawa N (2010) Okada S (2010) High voltage performances of Li_2NiPO_4F cathode with dinitrile-based electrolytes. J Electrochem Soc 157:A748–A752
195. Amaresh S, Karthikeyan K, Kim KJ, Kim MC, Chung KY, Cho BW, Lee YS (2013) Facile synthesis of ZrO_2 coated Li_2CoPO_4F cathode materials for lithium secondary batteries with improved electrochemical properties. J Power Sour. doi:10.1016/j.jpowsour.2012.12.010
196. Barpanda P, Recham N, Chotard JN, Djellab K, Walker W, Armand M, Tarascon JM (2010) Structure and electrochemical properties of novel mixed $Li(Fe_{1-x}M_x)SO_4F$ (M = Co, Ni, Mn) phases fabricated by low temperature ionothermal synthesis. J Mater Chem 20:1659–1668
197. Kim H, Lee S, Park YU, Kim H, Kim J, Jeon S, Kang K (2011) Neutron and X-ray diffraction study of pyrophosphate-based $Li_{2-x}MP_2O_7$ (M = Fe, Co) for lithium rechargeable battery electrodes. Chem Mater 23:3930–3937
198. Xu KC, Cresce AVW (2012) Electrolytes in support of 5 V Li-ion chemistry. Patent appl number: 20120225359
199. La Mantia F, Huggins RA, Cui Y (2013) Oxidation processes on conducting carbon additives for lithium-ion batteries. J Appl Electrochem 43:1–17
200. Fang HS, Wang ZX, Li XH, Guo HJ, Peng WJ (2006) Low temperature synthesis of $LiNi_{0.5}Mn_{1.5}O_4$ spinel. Mater Lett 60:1273–1275
201. Liu YJ, Liu ZY, Chen XH, Chen L (2012) Synthesis and performance of $LiNi_{0.5}Mn_{1.5}O_4$ cathodes. J Central South Univ (Sci and Technol) 43:4248–4252
202. Julien C, Massot M, Pérez-Vicente C (2000) Structural and vibrational studies of $LiNi_{1-y}Co_yVO_4$ ($0 \leq y \leq 1$) cathodes materials for Li-ion batteries. Mater Sci Eng B 75:6–12
203. Minakshi M, Singh P, Appadoo D, Martin DE (2011) Synthesis and characterization of olivine $LiNiPO_4$ for aqueous rechargeable battery. Electrochim Acta 56:4356–4360
204. Chevrier VL, Ong SP, Armiento R, Chan MKY, Ceder G (2010) Hybrid density functional calculations of redox potentials and formation energies of transition metal compounds. Phys Rev B 82:075122

Non-aqueous Metal–Oxygen Batteries: Past, Present, and Future

Maxwell D. Radin and Donald J. Siegel

1 What Is the Motivation for High Energy-Density Batteries?

A metal-oxygen battery (sometimes referred to as a 'metal-air' battery) is a cell chemistry in which one of the reactants is gaseous oxygen, O_2. Oxygen enters the cell typically in the positive electrode—perhaps after being separated from an inflow of air—and dissolves in the electrolyte. The negative electrode is typically a metal monolith or foil. Upon discharge, metal cations present in the electrolyte react with dissolved oxygen and electrons from the electrode to form a metal-oxide or metal-hydroxide discharge product. In some chemistries the discharge product remains dissolved in the electrolyte; in other systems it precipitates out of solution, forming a solid phase that grows in size as discharge proceeds. In secondary metal-oxygen batteries the recharge process proceeds via the decomposition of the discharge phase back to O_2 and dissolved metal cations. In light of the processes associated with discharge and charging, reversible metal-oxygen batteries with solid discharge products are often referred to as precipitation-dissolution systems, a category that also includes lithium–sulfur batteries.

The interest in metal-oxygen chemistries follows from their very high theoretical energy densities. Figure 1 summarizes the gravimetric and volumetric energy densities for several metal-oxygen couples, and compares these to the theoretical energy density of a conventional lithium-ion battery. On the basis of these energy densities, it is clear that many metal-oxygen systems hold promise for surpassing the state-of-the-art Li-ion system.

Achieving this goal, however, remains a significant challenge when factors beyond energy density are accounted for: cycle life, round-trip efficiency, and cost

M.D. Radin · D.J. Siegel (✉)
University of Michigan, Ann Arbor, USA
e-mail: djsiege@umich.edu

Z. Zhang and S.S. Zhang (eds.), *Rechargeable Batteries*,
Green Energy and Technology, DOI 10.1007/978-3-319-15458-9_18

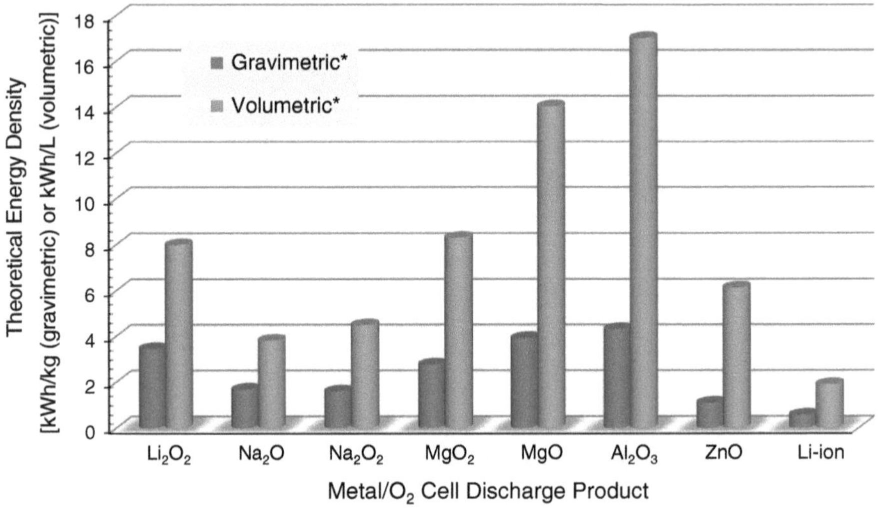

Fig. 1 Theoretical energy densities for several metal-oxygen couples compared to a conventional Li-ion battery. The abscissa indexes the discharge product(s) associated with each metal composition

must also be considered. The present chapter serves as a primer for new researchers interested in tackling these challenges. We begin with a brief history of metal-oxygen batteries, followed by a deep-dive into arguably the most ambitious secondary metal-oxygen chemistry, the non-aqueous Li–O_2 system. The current status of Li–O_2 cell performance is summarized which an emphasis on capacity, rate capability, cycle life, and efficiency. Subsequent sections review (i.) operating mechanisms, (ii.) challenges and failure modes, and (iii.) novel concepts for improving performance. We conclude with a brief discussion of non-lithium-based systems.

2 The History of Metal-Oxygen Batteries

2.1 Overview of Metal-Oxygen Batteries

The long history of metal-oxygen batteries is often unappreciated. To our knowledge, the earliest written description of a metal-oxygen battery is Vergnes' aqueous Zn-air battery from 1860 [1]. This design is in some respects remarkably similar to today's advanced metal-oxygen cells, which frequently employ porous carbon positive electrodes and noble-metal catalysts [2]. Figure 2 shows Vergnes' design, containing a zinc metal anode and a porous platinized coke positive electrode. The overall reaction in these cells is $Zn + \frac{1}{2}O_2 \rightarrow ZnO$. Zn-air batteries matured into a practical energy storage technology in the early 20th century [3], and as of the early

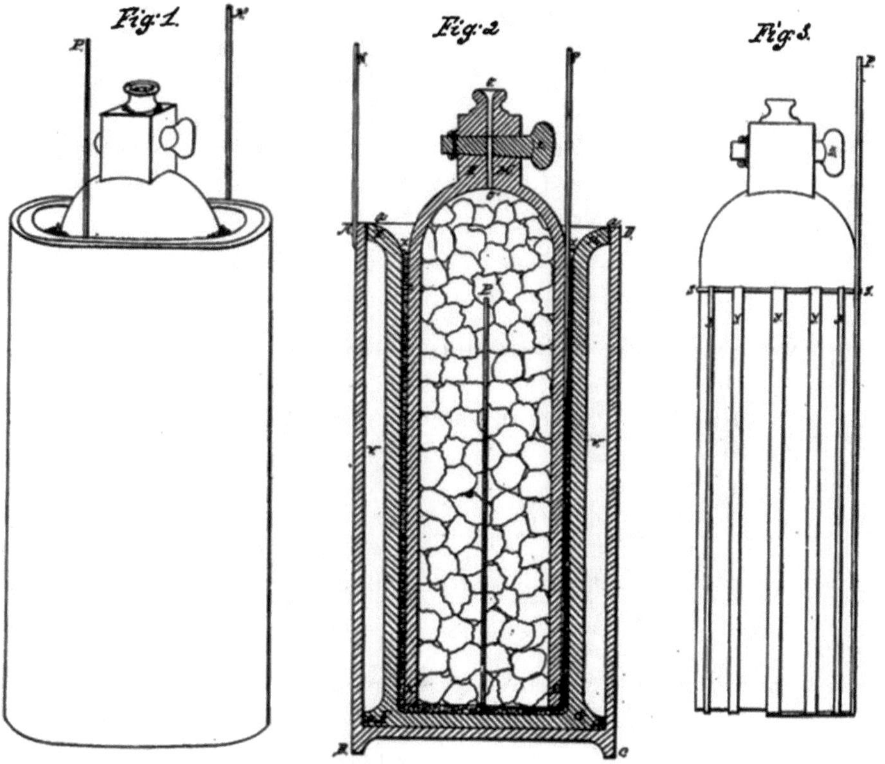

Fig. 2 Vergnes' 1860 Zn-air battery design, taken from Ref. [1]

21st century still remain the most prominent metal-oxygen chemistry. Industrially produced primary Zn-air cells exhibit high energy densities [4] and are employed in a number of applications, such as hearing aids.

Over the years, many other metal-oxygen couples have also been considered. In Tables 1 and 2, we enumerate reports of operating metal-oxygen cells, as well as oxygen cells based on the oxidation of several non-metals (C, H, and Si). The references cited in these tables are not intended to capture all of the work done on each metal-oxygen couple, but rather to highlight reviews and representative experiments. Table 1 shows cells that employ aqueous electrolytes, including composite electrolytes (i.e., the combination of a solid electrolyte in addition to an aqueous electrolyte). Table 2 shows non-aqueous chemistries, which are subdivided into high- and low-temperature. While all metal-oxygen chemistries can in principle be mechanically recharged (by replacing the metal anode and/or electrolyte), in Tables 1 and 2 we denote only those that are *electrochemically* rechargeable as 'secondary batteries.' Although Tables 1 and 2 are limited to couples in which a single element is oxidized, it should be noted that so-called 'direct' fuel cells involve the oxidation of compounds with multiple elements, including sugars [5], methanol [6], formic acid [7], and borohydrides [8].

Table 1 Summary of oxygen couples with aqueous electrolytes

	Reaction	Type of electrolyte	
		Non-composite	Composite
H	$H_2 + \frac{1}{2}O_2 \rightarrow H_2O$	Secondary [4, 6]	
Li	$Li + \frac{1}{4}O_2 + \frac{1}{2}H_2O \rightarrow LiOH$	Primary [11]	Secondary [15]
Na	$Na + \frac{1}{4}O_2 + \frac{1}{2}H_2O \rightarrow NaOH$		Primary [157]
Mg	$Mg + \frac{1}{2}O_2 + H_2O \rightarrow Mg(OH)_2$	Primary [4, 158]	
Ca	Unknown	Primary [159]	
V	$4H^+ + O_2 + 4V^{2+} \rightarrow 2H_2O + 4V^{3+}$		Secondary [160]
Mo	$Mo + \frac{3}{2}O_2 + H_2O \rightarrow H_2MoO_4$	Primary [161]	
	$Mo + \frac{3}{2}O_2 + 2KOH \rightarrow K_2MoO_4 + H_2O$	Primary [161]	
Fe	$Fe + \frac{1}{2}O_2 + H_2O \rightarrow Fe(OH)_2$	Secondary [4]	
	$3Fe(OH)_2 + \frac{1}{2}O_2 + H_2O \rightarrow Fe_3O_4 + 4H_2O$	Secondary [4]	
Zn	$Zn + \frac{1}{2}O_2 \rightarrow ZnO$	Secondary [4]	
Cd	Unknown	Secondary [162]	
Al	$Al + \frac{3}{4}O_2 + \frac{3}{2}H_2O \rightarrow Al(OH)_3$	Primary [4, 163]	Secondary [164]
Si	$Si + O_2 + 2H_2O \rightarrow Si(OH)_4$	Primary [165]	
Sn	$Sn + O_2 + 2KOH + 2H_2O \rightarrow K_2Sn(OH)_6$	Primary [166]	

Table 2 Summary of couples with non-aqueous electrolytes

	Reaction	Type of cell demonstrated	
		Low-temperature	High-temperature
H	$H_2 + \frac{1}{2}O_2 \rightarrow H_2O$	Secondary [6]	Secondary [6]
Li	$2Li + O_2 \rightarrow Li_2O_2$	Secondary [2, 15, 16]	
	$FeSi_2Li_x + \frac{x}{4}O_2 \rightarrow \frac{x}{2}Li_2O + FeSi_2$		Secondary [12]
Na	$Na + O_2 \rightarrow NaO_2$	Secondary [144, 145]	
	$2Na + O_2 \rightarrow Na_2O_2$	Secondary [145, 146]	
K	$K + O_2 \rightarrow KO_2$	Secondary [147]	
Mg	$Mg + \frac{1}{2}O_2 \rightarrow MgO$	Secondary [148]	Primary [167]
Ca	$2CaSi + \frac{1}{2}O_2 \rightarrow CaO + CaSi_2$		Secondary [168]
Mo	$Mo + O_2 \rightarrow MoO_2$		Secondary [169]
W	$W + O_2 \rightarrow WO_2$		Secondary [170]
Fe	$Fe + \frac{1}{2}O_2 \rightarrow FeO$ $3Fe + 2O_2 \rightarrow Fe_3O_4$		Secondary [171]
Al	Unknown	Secondary [150]	
C	$C + O_2 \rightarrow CO_2$		Primary [172]
Si	$Si + O_2 \rightarrow SiO_2$	Primary [173]	

Here 'low-temperature' refers to cells that operate below 100 °C and 'high-temperature' to those that operate above 100 °C

2.2 History of Li–O_2 Technology

The birth of the modern non-aqueous Li–O_2 battery is generally considered to be the 1996 demonstration of a room-temperature secondary cell by Abraham and Jiang [9]. While this development was a breakthrough, the history of earlier Li–O_2 batteries is often overlooked. To the best of our knowledge the first investigation of the Li–O_2 couple dates back to 1966 [10]. Although this early study also employed non-aqueous electrolytes—including propylene carbonate, today's preeminent Li-ion solvent—the design pursued was a 'moist' Li–O_2 system: the oxygen supply was saturated with water vapor. Interestingly, even this preliminary study identified some of the issues that remain critical for modern Li–O_2 cells, such as the formation of lithium carbonate and the role of impurities.

Other Li–O_2 designs emerged later. Primary Li–O_2 cells with aqueous electrolytes received considerable attention in the 1970s [11], and moisture-free high-temperature secondary cells were developed in the 1980s [12]. However, Abraham and Jiang's 1996 study represents the first demonstration of a moisture-free room temperature secondary Li–O_2 cell [9], and is therefore a key development in the history of Li–O_2 batteries. An amusing historical note is that the development of this cell was not intentional, but was instead a serendipitous discovery caused by the leakage of oxygen from a syringe into a sealed lithium-graphite cell [13].

Since 1996, research on non-aqueous Li–O_2 cells grown immensely. This has also led to the development of related chemistries, including true Li-air cells [14] (i.e., using ambient air rather than pure oxygen) and also reversible aqueous Li–O_2 cells [15]. It is not possible to summarize all of the studies performed to date. Instead, we strive to summarize and unify the key lessons, observations, and hypotheses that have been presented in the literature. For additional details beyond those presented here, the reader is encouraged to explore other reviews of the field [2, 15–20].

3 State of the Art

3.1 Current Status—Current, Capacity, Cycle Life, Efficiency

Much of the research on non-aqueous Li–O_2 batteries has focused on improving four critical aspects of performance: rate capability, capacity, voltaic efficiency, and cycle life. Some state-of-the-art Li–O_2 cells have been demonstrated to perform adequately with regard to these measures individually, but none have performed satisfactorily in all four simultaneously. This is because rate capability, capacity, voltaic efficiency, and cycle life are highly interdependent, often in surprising ways. Some interdependencies manifest as tradeoffs in performance; examples include:

1. Higher discharge rates reduce maximum capacity due to electrical passivation issues and/or oxygen transport limitations, as discussed in Sects. 3.3.1 and 3.3.2.

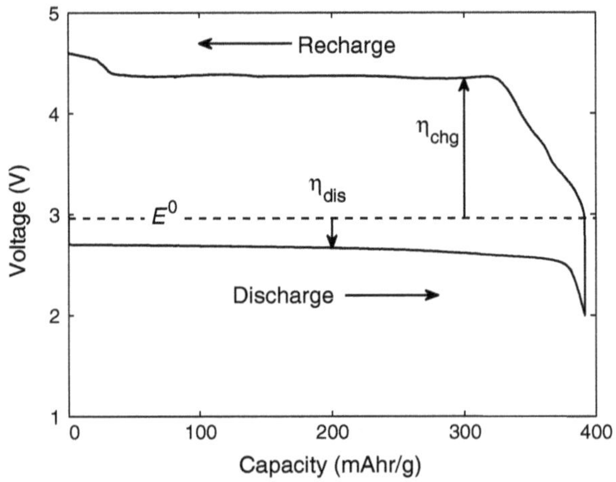

Fig. 3 Potential profile from a galvanostatic discharge/charge cycle of a parallel-electrode, aprotic Li–O_2 battery with a porous carbon positive electrode, Li metal anode, and LiTFSI/DME electrolyte at a current of 0.2 mA/cm^2. Data courtesy of L. Griffith, Monroe Research Group

2. Curtailing the discharge capacity increases cycle life and voltaic efficiency [21, 22].
3. Higher discharge rates (at fixed capacity) may *improve* voltaic efficiency, as the discharge product morphologies produced at high currents can exhibit lower charging overpotentials than the morphologies produced at low currents [23].

Figure 3 shows the potential profile from a galvanostatic discharge/charge cycle of a typical non-aqueous Li–O_2 cell. Several features shown here are typical for Li–O_2 cells. First, discharge proceeds at a constant voltage close to the theoretical cell potential E^0 for the formation of Li_2O_2. Discharge then terminates with a rapid drop in potential ('sudden death'), possibly due to oxygen transport limitations or electrical passivation. The recharge potential profile is more complex and exhibits several distinct stages with high recharge overpotential η_{chg}, resulting in low voltaic efficiencies.

We next summarize the performance of state-of-the-art Li–O_2 cells and compare to performance targets. We note that comparing capacities and currents across different experiments is non-trivial because different authors adopt different normalization schemes [24, 25]. For example, many studies employing carbon-black electrodes report capacities normalized to the mass of the carbon black. Underlying this convention is the notion that the capacity ought to be proportional to the mass of the carbon black. This can lead to misleading conclusions in cases where the gas diffusion layer (GDL) or current collector contributes significantly to capacity; one study found that carbon-black mass normalization can inflate the capacity of a typical Li–O_2 electrode by as much as an order of magnitude [24]. The carbon-mass-normalization convention can also lead to misleading conclusions in cases where only a small fraction of the carbon black is utilized due to oxygen transport limitations, as discussed in Sect. 3.3.2. Furthermore, this convention does not allow for meaningful comparison between electrodes where the mass of the catalyst and binder is significant, or to carbon-free electrodes.

To facilitate meaningful comparisons between experiments and with performance targets, we recommend the following protocol for reporting currents and capacities:

1. Capacities and currents should be reported normalized both to the mass and to the positive electrode's geometric area, because battery pack performance depends both on the current/capacity per mass and per geometric area. (At a minimum, one should supply enough data to allow readers to convert between normalization schemes.)
2. When normalizing to mass, the masses of all positive electrode components that scale with the loading (e.g., binders and catalysts) should be included because these contributions to the total mass are important from the perspective of battery system design.
3. If one excludes the mass of the GDL/current collector from the mass normalization, then one must verify that its contribution to capacity (per area or per cell) is negligible compared to the contribution from the active materials [23, 24]. Note that it is not sufficient to show that the capacity per mass of the GDL is small compared to the capacity per mass of the active material, because the mass of the GDL often greatly exceeds that of the active material.

The tradeoff between current and capacity is illustrated in Fig. 4, which shows the capacities and rates obtained in various Li–O_2 cells reported in the literature, normalized to the geometric area of the electrode. Additionally, the current densities

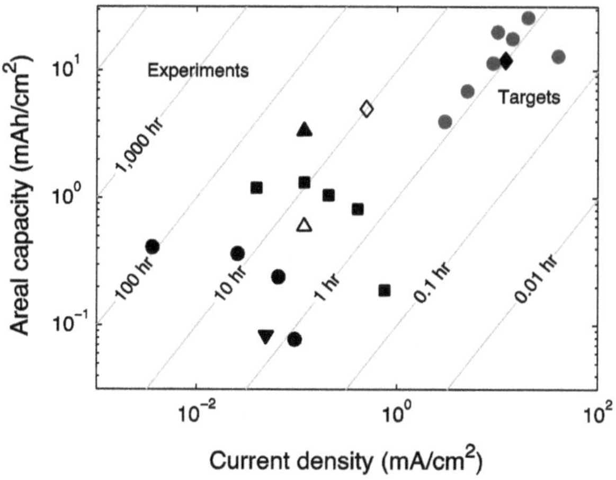

Fig. 4 Reported capacities for galvanostatic operation of Li–O_2 cells from various experiments [28, 29, 34, 155, 156] during the first discharge (*black solid symbols*) and in cells which can be cycled many times (*black open symbols*). The *red symbols* indicate the currents and capacities assumed in hypothetical battery designs [15, 26, 27]. *Diagonal lines* identify the time required for discharge (Color figure online)

and capacities assumed in several hypothetical designs for practical Li–O_2 batteries [15, 26, 27] are shown.

One of the fundamental reasons why current cell designs fall short of the areal performance targets [15, 26, 27] is electrode thickness: while experiments often consider electrodes of thickness ~10 μm, proposed battery designs have assumed much larger thicknesses of 150–300 μm. A practical Li–O_2 battery requires that the electrode be fairly thick so as to minimize the mass and volume penalties associated with the inactive components (e.g., separators, electrolyte, current collectors, packaging). However, the full utilization of thick electrodes is likely limited by oxygen transport, as discussed in Sect. 3.3.2. Thus the development of a practical Li–O_2 battery will require either a solution to the oxygen transport problem within thick electrodes, or a battery pack design that achieves high system-level performance with thin electrodes.

The gap between experiments and targets is smaller on a mass basis. The mass-specific capacity targets of ~1000 mAh/g (including the mass of catalysts and binders) [26, 27] are routinely achieved at fairly high rates (~1 h discharge), sometimes even over many cycles [28]. Although mass-capacities significantly higher than this have been reported [29, 30], from the perspective of system design there is limited benefit to increasing the gravimetric capacity beyond ~1000 mAh/g. As the capacity increases beyond this value, the gravimetric capacity ultimately becomes limited by the mass of the discharge product [27]: the theoretical capacity of the Li–O_2 couple is $1168\,\mathrm{mAh/g_{Li_2O_2}}$ [31].

In the next section, we summarize the key observations and theories regarding the operating mechanisms of Li–O_2 cells. Possible origins of these performance limitations are also described. It is important to keep in mind that different mechanisms may dominate under different operating conditions. For example, it has been shown that the current density [23], positive electrode material/architecture [32, 33], and system cleanliness [34–36] can play a significant role in the reaction mechanisms.

3.2 Proposed Mechanisms

3.2.1 The Discharge Product

The first step in understanding the performance of Li–O_2 batteries is understanding the discharge product. It is often presumed that the discharge product is bulk crystalline Li_2O_2; however, this is probably too simplistic an assumption, as there is now good evidence that the discharge product can have a complex morphology, microstructure, and composition.

Morphology. A number of different discharge product morphologies have been reported, including disks [23, 37], films [37, 38], needles [39], and hollow spheres [40]. Biconcave disks (similar to red blood cells) are among the most commonly observed morphologies, as shown in Fig. 5. (This morphology is often referred to as a 'toroid'; however, these particles are not strictly speaking toroids because they

Fig. 5 SEM image of biconcave Li_2O_2 disks in a Li–O_2 cell. From Adams et al. [23]

lack a hole that runs through the center of the disk.) Figure 6 shows the basic structure of a typical Li_2O_2 disk, which consists of a stack of relatively flat crystallites. The disks are highly textured (i.e., the misorientation between crystallites is small), with the {0001} axis being aligned with the central axis of the disk. In some cases the regions between the plates appear to be filled with components of the electrolyte [36], but in others it has been suggested that the inter-plate regions contain a distinct phase or grain boundary region [41]. This second phase could be, for example, amorphous Li_2O_2 or a lithium-deficient compound such as $Li_{2-x}O_2$. Note that it can be the case that multiple distinct morphologies appear concurrently in the positive electrode of a single cell; for example, large biconcave disks and small particles have been observed simultaneously [37].

The morphology of the discharge product has been suggested to influence discharge capacity and recharge overpotentials [23, 42–44]; therefore an understanding of the factors which control morphology may enable the design of cells with improved performance. It has been reported that low current densities and high water concentrations (hundreds to thousands of ppm) both promote the growth of biconcave disks [23, 37, 45–47]. Similar biconcave disks have also been observed in the precipitation of silicates [48] and corn starch [49], suggesting that there may be a common growth mechanism. It has also been reported that the characteristic size of these particles decreases with increasing current densities, and that at sufficiently high rates the deposit forms a conformal film rather than discrete particles [23, 37, 45, 47]. However, it has been suggested that the putative conformal films produced at high currents are in fact carpets of nano-scale needles [39]. Additionally, several experiments have concluded that the support and/or catalyst can strongly influence discharge product morphology [42–44].

A concrete picture of the discharge product growth mechanism remains elusive. A continuum-scale growth model has been proposed to explain the transition from particle to film with increasing current [50], and a separate model has been proposed to explain the increase in disk size with increasing water concentration [36].

Crystallinity. A growing number of experiments have suggested that amorphous Li_2O_2 can be present in the discharge product [23, 32, 33, 40]. The formation of an

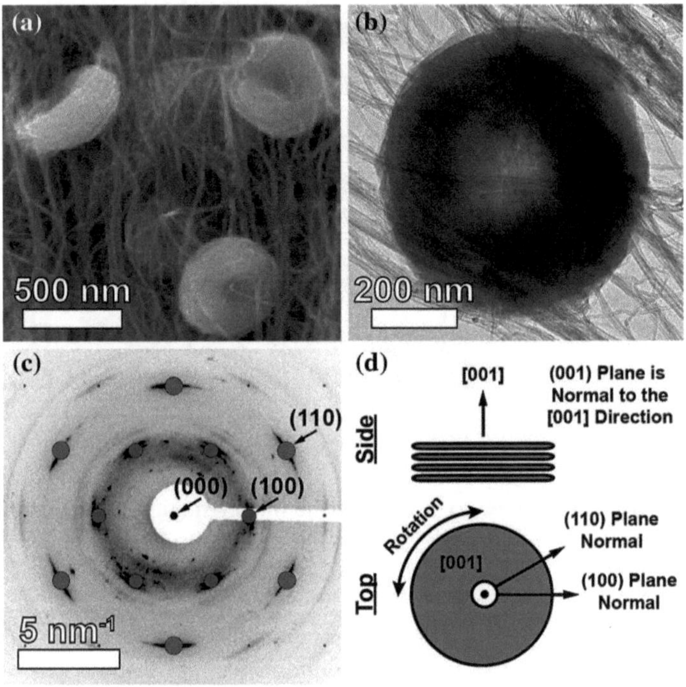

Fig. 6 Morphology of a biconcave Li_2–O_2 disk on a carbon nanotube support: **a** SEM, **b** bright-field TEM, **c** electron diffraction pattern, **d** schematic of microstructure. From Mitchell et al. [37]

amorphous deposit is consistent with Ostwald's rule, which states that unstable phases tend to precipitate before stable phases [51–53]. It has been reported that higher discharge rates [23], as well as certain catalysts, can promote the formation of amorphous Li_2O_2 [32, 33].

Several experimental [23, 32, 33] and computational [32, 54] studies have suggested that amorphous Li_2O_2 is easier to recharge than amorphous Li_2O_2, perhaps due to improved electron or Li-ion transport properties. If correct, this would suggest that Li–O_2 electrode designs (or operating conditions) that promote the formation of amorphous Li_2O_2 may yield superior performance.

Composition. Although the discharge product is primarily thought of as Li_2O_2, deviations from this composition have been proposed. One recurring theme is the occurrence of superoxide ions, O_2^-, in the discharge product [55]. The presence of a superoxide component perhaps should not be a surprise, given that it is known that other alkali metals form mixed peroxide-superoxide phases [56]. It remains unclear where exactly the superoxide component resides in the discharge product. It has been suggested to represent a surface species [57, 58], an oxygen-rich phase located in the inter-plate regions [41], or to be associated with the presence of point defects such as hole polarons [59, 60].

Relatively few studies have found evidence for Li_2O in the discharge product [61, 62]. Although Li_2O has a higher theoretical specific energy density than Li_2O_2 (5200 vs. 3505 Wh/kg [61]), it may not be a desirable discharge product for secondary Li–O_2 batteries because the electrochemical oxidation of Li_2O is more difficult than that of Li_2O_2 [63, 64].

It has been recognized that Li–O compounds are not the only phases present in the discharge product. Side reactions (i.e., reactions involving decomposition of the salt, solvent, or positive electrode) have been observed to produce other compounds, such as lithium carbonate, lithium acetate, lithium formate, and lithium fluoride [65, 66]. The products of these side reactions can comprise a substantial fraction of the discharge product; one experiment found that in a typical Li–O_2 cell with an ethereal solvent, the yield of Li_2O_2 was at best 91 % of the theoretical amount expected from coulometry [65]. It is important to note that in addition to the precipitated side reaction products, there may be additional soluble side reaction products. Side reactions are discussed in more detail in Sect. 3.3.4.

3.2.2 Discharge/Recharge Mechanisms

A number of different discharge and recharge mechanisms have been proposed, as illustrated in Fig. 7. It is important to keep in mind that different mechanisms may predominate depending on the experimental conditions (e.g., rate, electrolyte, electrode/catalyst, temperature, depth of discharge, and cleanliness).

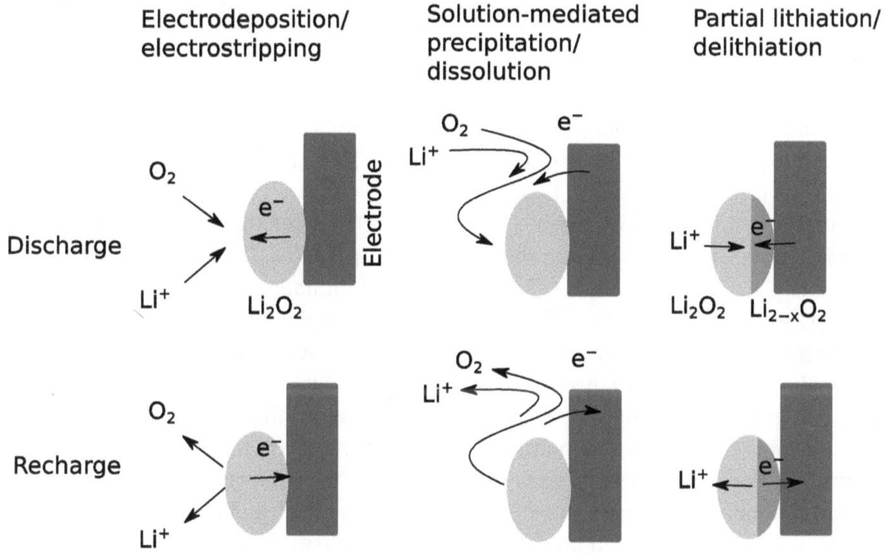

Fig. 7 Possible mechanisms for discharge/recharge in a Li–O_2 cell. As discussed in the text, in the case of solution-mediated mechanisms, there are many possible intermediate species; the central column illustrates a scenario where the intermediate species is molecular Li_2O_2

Electrodeposition/electrostripping. In some experiments it has been suggested that the growth/dissolution of a film occurs via the electrochemical deposition/stripping of Li_2O_2. In such a mechanism, electron transport presumably would occur through the growing deposit. It has been suggested that this could occur via electron tunneling [38, 67] or hole polaron hopping [59, 60, 68]. Limitations associated with charge transport are discussed in Sect. 3.3.1.

Solution-mediated precipitation/dissolution. The growth of large particles has been proposed to occur via a solution-mediated precipitation process, which would allow charge-transport limitations through the particles to be bypassed [23, 69–71]. For example, one proposed discharge mechanism is that O_2 is reduced on the positive electrode surface to form LiO_2: $Li^+ + O_2 + e^- \rightarrow LiO_2$. The LiO_2 could then diffuse in the electrolyte (or perhaps along the positive electrode surface), and then precipitate out via a disproportionation reaction: $2LiO_2 \rightarrow Li_2O_2 + O_2$. Such a mechanism requires that there be an intermediate species (be it LiO_2 or something else) that is either at least sparingly soluble or capable of rapid surface diffusion. Additives which solubilize such intermediates have been suggested to play a role in the dynamics of discharge product precipitation [46, 47].

A solution-mediated process (such as the reverse of the above reactions) could also occur during recharge. For example, it has been proposed that impurities present as contaminants or by-products of electrolyte decomposition may serve as the soluble intermediate species [35]. These impurities in effect function as redox mediators, or perhaps transform Li_2O_2 into a more soluble species. For example, a small amount of protons has been suggested to enable a recharge mechanism that begins with the transformation of Li_2O_2 into H_2O_2 via a single-displacement reaction, $Li_2O_2 + 2H^+ \rightarrow H_2O_2 + 2Li^+$ [35]. H_2O_2, being more soluble than Li_2O_2, could then diffuse to the electrode and be electrochemically oxidized via the reaction $H_2O_2 \rightarrow 2H^+ + O_2 + 2e^-$, yielding a net reaction of $Li_2O_2 \rightarrow 2Li^+ + O_2 + 2e^-$.

Partial lithiation/delithiation. The partial delithiation of the discharge product has been suggested to be the first step of recharge [60, 72–74]. This could occur as a two-phase reaction [72, 74]: $Li_2O_2 \rightarrow Li_{2-x}O_2 + xLi^+ + xe^-$. The equilibrium potential for this reaction when $x = 1$ has been calculated from first-principles methods to be 0.3–0.4 V above the equilibrium potential for the full oxidation of Li_2O_2, $Li_2O_2 \rightarrow O_2 + 2Li^+ + 2e^-$ [72]. Partial lithiation/delithiation could also occur as a solid solution [60, 73]. Although the two-phase pathway is predicted to be thermodynamically more stable than the solid-solution pathway [72], the fact that high currents and small particle sizes can suppress phase separation in Li-ion battery materials [75] suggests that one cannot rule out the solid-solution pathway for Li–O_2 based on thermodynamics alone. Note that even if a delithiation process occurs (either via a two-phase or solid-solution pathway), the intermediate lithium-deficient phase may not be readily observable if recharge occurs one particle at a time (i.e., via a 'domino cascade' mechanism) [75].

3.3 Challenges/Failure Modes

3.3.1 Charge Transport Within the Discharge Product

Charge transport through the discharge product has been thought to limit the performance of Li–O_2 cells in many circumstances [38, 76–79]. The presence of a passivating layer on the positive electrode would shut down electrochemical activity, potentially leading to limitations in capacity, voltaic efficiency, and rate capability. Although the charge-transport mechanism(s) at play are not well understood, several possibilities have been proposed:

1. *Electron tunneling.* In thin films (<5 nm), electron tunneling has been suggested to be the dominant charge-transport mechanism [38, 67]. This mechanism has been suggested to account for sudden death behavior, which would occur when the film thickness exceeds the distance over which electron tunneling can readily occur.
2. *Hole polaron hopping.* Experiments and first-principles modeling have found that hole polarons are the dominant electronic charge carrier in Li_2O_2 [59, 60, 80]. Polaron hopping has also been suggested to account for sudden death behavior. In this scenario, sudden death would occur when the deposit thickness exceeds the thickness of space-charge layers associated with the Li_2O_2/electrolyte and Li_2O_2/electrode interfaces [81].
3. *Li-ion vacancy diffusion.* Experiments and first-principles modeling have found that Li-ion vacancies are the dominant Li defect in Li_2O_2 [60, 80]. The role of Li-ion vacancies is different from that of electronic charge carriers because ionic defects cannot readily cross the interface between the discharge product and electrode support [74]. That is, at the Li–O_2 equilibrium potential, the amount of Li which can be inserted into (or deinserted from) the positive electrode support typically represents only a small fraction of the amount of Li in the discharge product. Thus the support can be thought of as an ion-blocking electrode.

The relative importance of these mechanisms may vary depending on the conditions (discharge product morphology, temperature, current density, etc.). For example, some studies have speculated that charge transport in Li_2O_2 could be enhanced at extended defects, such as surfaces [57, 58, 82], grain boundaries [83], amorphous regions [32, 54], or interfaces [81, 84].

3.3.2 Oxygen Transport in the Electrolyte

In many cell designs the low solubility and diffusivity of oxygen in the electrolyte can limit discharge capacity [25, 76, 85–89]. In this case, only the region of the positive electrode near the gas inlet is utilized. Sluggish oxygen transport can be further compounded by pore-clogging, i.e., the obstruction of oxygen-diffusion pathways by the discharge product [87]. Oxygen transport limitations can lead to a sudden drop in voltage during a galvanostatic discharge (sudden death) [87–89].

Improvements in oxygen transport may be required in order to fully utilize the thick electrodes required to meet performance targets, as discussed in Sect. 3.1. While tailoring the pore network of the electrode (as discussed in Sect. 3.4.1) can improve oxygen transport, it cannot overcome the fundamental limits determined by the solubility and diffusivity (i.e., permeability) of oxygen in the electrolyte [88]. Strategies for extending these fundamental limits are discussed in Sect. 3.4.5.

3.3.3 Kinetics

A number of studies have examined the kinetics of Li–O_2 cells. Systematic experiments have found that both the discharge and recharge kinetics are facile [79]. Several computational studies have explored mechanisms for the layer-by-layer deposition/stripping of Li_2O_2. The 'thermodynamic overpotentials' associated with layer-by-layer deposition/stripping were found to be small (<0.2 V), and it was suggested on this basis that kinetics would be fast [90]. (Note, however, that thermodynamic overpotentials can only be compared qualitatively to the overpotentials observed in experiments; for example, the thermodynamic overpotentials do not account for the density of reactive sites (e.g., step edges or kinks) or the exchange currents associated with different reaction steps.) A few other first-principles studies concluded that the kinetics of layer-by-layering deposition/stripping was slow, and would limit cell performance [91, 92]. The differences among conclusions in the literature result primarily not from differences among atomistic calculations, but rather from differing interpretations of the computational results—that is, how the energies for various reaction steps relate to the current-voltage relationship.

3.3.4 Degradation

Most experiments on Li–O_2 systems prior to 2010 used electrolytes developed for Li-ion batteries, employing carbonate solvents such as propylene carbonate (PC), ethylene carbonate (EC), and dimethyl carbonate (DMC). These solvents were natural choices, as they had been widely successful for Li-ion batteries; some even refer to PC as 'the new water' due to its widespread use for Li-ion electrochemistry [93]. In 2010 the Li–O_2 community began to recognize that carbonate solvents are in fact highly unstable in Li–O_2 cells [94–96]. Therefore, studies prior to 2010 must be regarded with caution, since electrolyte degradation, rather than Li–O electrochemistry, is thought to dominate carbonate-containing Li–O_2 cells.

It is now recognized that solvent stability is a critical issue for Li–O_2 batteries [2, 15, 97], and furthermore it has been observed that the salt [98–100], carbon support [101, 102], and binder [103] can also react irreversibly. Side reactions can lead to poor cyclability due to the loss of electrolyte and accumulation of side-reaction products [97, 102, 104]. Furthermore, the oxidation of side-reaction products during recharge can result in high charging overpotentials [101, 102].

Quantitative measurements [e.g., via differential electrochemical mass spectroscopy (DEMS)] are critical for distinguishing reversible cycling from parasitic processes [97]. This is demonstrated in Fig. 8, which shows the cycling of a typical Li–O_2 cell. Figure 8a shows that the cell is cyclable; in fact the capacity increases during the first few cycles. However, the cycling does not represent the reversible formation of Li_2O_2. Figure 8b and c show that the amount of oxygen released during discharge is less than the amount consumed during discharge. Furthermore, Fig. 8d shows that the ratio of electrons transferred to oxygen released during recharge deviates dramatically from the value which would be expected for the oxidation of Li_2O_2, $2e^-/O_2$. Thus despite the apparent cycability of this cell, the chemistry is dominated by side reactions.

Much work presently is being done to design Li–O_2 cells with sufficient stability for a practical battery. The stability of the solvent, salt, and support/catalyst are interdependent [102, 105]; thus the challenge is to find a combination of these that are sufficiently stable. Carbonate solvents have been abandoned in favor of ethers, ionic liquids, and other solvent classes [19, 106]. Although an improvement over carbonates, even these solvents exhibit some degree of degradation [65, 66]. For example, a typical ethereal electrolyte with a carbon positive electrode was found to exhibit an Li_2O_2 yield of at most 91 % [65]. Improved stability has been reported for certain combinations, such as $LiClO_4$/DMSO with a nanoporous gold positive electrode [105].

Since the number of possible salt/solvent/electrode combinations is large, a mechanistic understanding of degradation processes will be important for

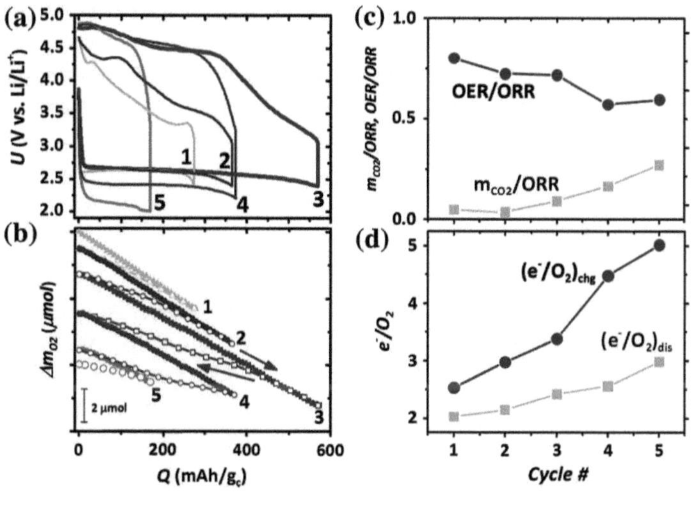

Fig. 8 **a** Potential profiles for five discharge/charge cycles of a Li–O_2 cell with Li-TFSI/DME electrolyte. **b** Oxygen consumption/evolution during discharge/charge. **c** Ratio of the amount of oxygen and CO_2 released during charge to the amount of oxygen consumed during discharge. **d** Ratio of the number of electrons transferred at the positive electrode to number of oxygen molecules consumed/evolved during discharge/charge. From McCloskey et al. [97]

identifying combinations with high stability. A summary of some of the solvent degradation processes that have been proposed is presented below. (Less effort has been invested in a mechanistic understanding of salt [99, 107], support, and binder stability, although these are clearly critical issues.)

1. *Chemical attack by electrochemical intermediates.* Chemical attack by intermediates of the oxygen reduction reaction during discharge can cause substantial degradation in some solvents. In particular, attack by superoxide (O_2^-) radicals is thought to be the main source of decomposition in carbonate solvents [108]. Others have suggested that oxidation intermediates could also lead to solvent degradation during recharge. In particular, 'nascent' oxygen evolved during recharge has been speculated to attack the solvent [97, 109]. Here 'nascent' refers to oxygen released in a highly reactive form, such as atomic oxygen or O_2 molecules in the singlet state.
2. *Auto-oxidation.* Organic solvents can undergo auto-oxidation (chemical reaction with molecular O_2). This has been hypothesized to contribute to solvent degradation in Li–O_2 cells [110, 111]. The importance of auto-oxidation may not be visible in typical experiments, whose time scales (days) are much shorter than those required for a practical automotive battery (years).
3. *Chemical attack by the discharge product.* Another solvent degradation mechanism is the chemical reaction between the solvent and the discharge product. A few experiments have sought to probe this [97, 100], and atomistic studies have examined solvent degradation on Li_2O_2 clusters [112] and surfaces [113].
4. *Electrochemical oxidation.* In addition to the chemical degradation processes listed above, electrochemical processes can also lead to solvent degradation. Many common solvents exhibit minimal oxidation up to ~4 V versus Li/Li^+ on carbon electrodes. However, it has been suggested that solvent oxidation is enhanced by Li_2O_2 [97]. Additionally, some oxygen-reduction catalysts used in Li–O_2 cells also catalyze solvent oxidation [114, 115].

3.3.5 Impurities

The presence of even small amounts of impurities have been suggested to have a substantial effect on cell performance [34, 35, 46, 47, 116–118]. We focus on H_2O, as this appears to be the most problematic and well-documented contaminant. It has been observed that even small amounts of water can influence Li–O_2 cells in complex ways:

1. The presence of water at concentrations as low as a few hundred ppm can significantly increase discharge capacity [21, 34, 47, 116].
2. Water can influence the discharge product morphology [21, 46, 47], and it has been found that concentrations of water in the hundreds to thousands of ppm can promote the formation of biconcave disks; see Sect. 3.2.1.

3. Water can significantly reduce the cyclability of Li–O_2 cell [21, 116].
4. The presence of water can result in the formation of LiOH, instead of, or in addition to, Li_2O_2 [14, 21].

The mechanisms by which water interacts with cell operation are not well understood. Water can react with Li metal in the negative electrode, as discussed in Sect. 3.3.6 [116]. Furthermore, water has been suggested to act as a mediator or solubilizing agent during discharge and/or recharge [35, 46, 47], as discussed in Sect. 3.2.2.

The minimization of water contamination in a Li–O_2 cell is non-trivial task [116], and it may be impractical to reduce the water concentration below ~10 ppm. However, it is not clear whether the complete elimination of water is necessary, or even desirable, for a practical Li–O_2 battery [46].

The effects of contamination by CO_2 have also been explored. It has been observed that CO_2 will react with the discharge product to form Li_2CO_3 [118]. Because of the high potentials required to oxidize Li_2CO_3 (and accompanying electrolyte decomposition), exposure to CO_2 should be minimized in secondary Li–O_2 cells. Unsaturated oxygenated hydrocarbons present as impurities in industrially produced ethers have also been found to be reactive in Li–O_2 cells [117].

3.3.6 Negative Electrode

Although graphite is the negative electrode of choice for commercial Li-ion batteries, the full benefit of the high specific capacity of the Li–O_2 positive electrode can only be realized when it is paired with a high specific capacity negative electrode. For this reason, nearly all Li–O_2 experiments to date have employed Li-metal negative electrodes [119]. Consequently, we focus in the remainder of this section of Li-metal negative electrodes. Of course, practical Li–O_2 cells could also take advantage of alternative negative electrodes, such as Si alloys [119].

We can divide challenges for Li-metal negative electrodes in Li–O_2 batteries into two categories: (i) challenges that are intrinsic to Li-metal electrodes (and which arise regardless of what positive electrode is used), and (ii) challenges that are specific to Li–O_2 systems. Among the intrinsic challenges for Li-metal electrodes, dendrite formation and coulombic efficiency are perhaps the most prominent. Intrinsic challenges for Li-metal electrodes will not be discussed here, as these have been reviewed elsewhere [4, 120].

Regarding challenges that are specific to Li–O_2 batteries, it has been recognized that the crossover of molecular oxygen, contaminants, and/or soluble side reaction products from the positive electrode may have a deleterious effect [116, 121, 122]; however, one study found that the presence of O_2 can promote the formation of a stable solid-electrolyte interphase (SEI) in DMSO [123]. The high reactivity of Li metal may require that the negative electrode of a practical Li–O_2 battery be protected with a solid electrolyte or SEI.

Note that most Li–O_2 experiments do not distinguish between limitations of the positive and negative electrodes, and in some cases it has been found that the Li-metal negative electrode contributes significantly to cyclability limitations and cell impedance [116, 124, 125]. A few studies have used Li_xFePO_4 instead of Li metal [126, 127]. This configuration would not be used in a practical battery because the equilibrium potential for the lithiation of Li_xFePO_4 is above the equilibrium Li–O_2 potential; however, the use of a highly stable electrode such as Li_xFePO_4 can allow one to isolate the phenomena specific to the O_2 electrode [126, 127].

3.4 Novel Concepts

Numerous new designs and materials have been developed in the years since Abraham and Jiang's development of the modern non-aqueous Li–O_2 battery. It is not possible to discuss them all here; rather we highlight a few novel concepts.

3.4.1 Advanced Positive Electrodes

While a 'baseline' Li–O_2 positive electrode consists of carbon black and binder, gains in performance have been reported using more advanced designs. Many studies have sought to modify the structure of the carbon at the nano-, micro-, or macro-scale [17, 18, 128]. Some experiments have explored carbon-free electrodes. Nanoporous gold [105] and titanium carbide [126] electrodes with DMSO-based electrolytes have been shown to have improved performance over carbon electrodes. The addition of new materials such as oxides and transition metals [2, 17, 18] has also been examined. These additions are frequently referred to as 'catalysts', but this terminology is misleading given these materials most likely do not function as conventional electrocatalysts. The term 'promoter' has been suggested as a more general term to describe materials which improve performance [42, 129].

3.4.2 Redox Mediators

The use of redox mediators has been shown to reduce charging overpotentials, presumably by bypassing charge transport limitations in the discharge product [127, 130]. The idea is that a soluble species that undergoes a reversible redox reaction at a potential near the Li–O_2 redox potential would be able to ferry electrons from the electrode to the discharge product. This mechanism assumes facile charge transfer from the mediator to the electrode. Such a mechanism may be incompatible with achieving high capacities: electron transfer will be blocked if the (insulating) discharge product covers the cathode support.

3.4.3 All Solid-State Designs

A somewhat different approach to the Li–O_2 chemistry is the all solid-state design. In this case the positive electrode is a porous material with the ability to support both lithium-ion and electron transport. Lithium and oxygen would react to form solid Li_2O_2 or Li_2O in the pores. Solid-state Li–O_2 cells have been demonstrated using composite cathodes comprised of carbon and LAGP, a fast ion conductor [15, 131]. Potential advantages of a solid-state design include improved safety and cyclability by avoiding the need for a liquid electrolyte that could degrade.

3.4.4 Hybrid Li-Ion/Li–O_2 Insertion Electrodes

A novel type of Li–O_2 battery has been suggested in which both Li and O are accommodated into a transition metal oxide host [132, 133]. This has been referred to as a 'hybrid' Li-ion/Li–O_2 battery because it contains elements of both chemistries: Li^+ ion insertion into a host as well as the reduction of oxygen. One example of this chemistry is the reaction $LiFeO_2 + 4Li^+ + 4e^- + O_2 \rightarrow Li_5FeO_4$, which can be thought of as the addition of Li_2O to Fe_2O_3. Although the presence of the transition metal oxide lowers the maximum theoretical energy density compared to a 'traditional' Li–O_2 battery that forms Li_2O_2, some hybrid Li-ion/Li–O_2 chemistries have been predicted to a have an energy density competitive with Li_2O_2. A recent high-throughput computational study identified several candidate hybrid Li-ion/Li–O_2 chemistries based on capacity, voltage, and band gap [134].

3.4.5 Other Concepts

Metal-oxygen batteries bear a resemblance to some biological systems, where the reduction of oxygen is used as an energy source. Some of the challenges of non-aqueous Li–O_2 batteries have also been encountered by nature, including 'oxidative stress' (deleterious reactions involving reactive species such as superoxide radicals, peroxides, and singlet oxygen [135]) and the sluggish transport of O_2 in fluids.

Nature's solutions to these challenges may inspire improvements in Li–O_2 cell design. For example, biological systems mitigate oxidative stress by antioxidants: molecules that scavenge reactive species [136]. One study applied this concept to Li–O_2 cells, and found that synthetic melanin additives led to improved cyclability [137]. Oxygen transport limitations are addressed in nature via the use of oxygen-binding proteins (e.g., hemoglobin) that improve O_2 solubility and also forced convection through the cardiovascular system [135]. These concepts can be applied to Li–O_2 cells: one study found that oxygen-binding perfluorinated additives improved discharge capacity [138]. No reports, to the best of our knowledge, have employed forced convection in Li–O_2 cells; however, continuum-scale models have predicted that the use of forced convection could significantly improve Li–O_2 discharge capacity [139].

Other approaches for improving oxygen transport in Li–O_2 cells have also been explored. The conceptually simplest approach is simply to increase the partial pressure of oxygen gas, which has been demonstrated to increase capacity significantly [140, 141]. Another concept for improving oxygen transport is the use of two immiscible liquids: one that facilitates Li-ion transport, and another that facilitates oxygen transport. A recent study demonstrated that the use of perfluorinated carbon liquids in this manner can significantly improve the capacity of Li–O_2 cells [142]. A similar concept is the use of additives to improve the solubility of Li_2O_2 and reaction intermediates [143].

4 Other Metal-Oxygen Chemistries

At this point our discussion has focused primarily on the Li–O_2 system, as among non-aqueous metal-oxygen chemistries, this system has received by far the most attention in the literature. A few recent studies, however, have begun to examine secondary room-temperature non-aqueous systems based on other alkali and alkaline-earth metals such as sodium [144–146], potassium [147], magnesium [148, 149], and aluminum [150], as shown in Table 2. The high abundance of these elements is often provided as a motivation for these systems, although projections indicate that the worldwide supply of Li is adequate for the next century [151]. Below we discuss some other potential advantages and disadvantages of these systems compared to Li–O_2.

Although necessarily sacrificing some gravimetric performance, the heavier Na- and K-based systems are noteworthy for two reasons: First, under some operating conditions, they appear to form a superoxide (NaO_2 or KO_2) discharge product, rather than peroxide. Second, the overpotentials observed during charging of these superoxides are very small in comparison to those for Li_2O_2 or Na_2O_2. One may therefore argue that what these cells lack in specific capacity is partially compensated for by an increase in voltaic efficiency. More importantly, if the formation of a superoxide discharge product is indeed responsible for higher efficiency, then a potential pathway for improving the Li–O_2 system may be at hand: by stabilizing a lithium superoxide (LiO_2) discharge phase one may realize high capacity and efficiency (i.e., low recharge overpotentials) simultaneously. However, such an approach may be challenging: although a superoxide component has been observed in the Li–O_2 discharge product, bulk LiO_2 is apparently unstable under ambient temperatures and pressures [152, 153]. The fact that NaO_2 and KO_2 are more stable than LiO_2 has been attributed to the smaller size of the Li^+ cation [56].

In addition to alkali-metal-based systems, recent studies have recently reported secondary non-aqueous metal-oxygen cells using magnesium [148, 149] and aluminum [150] negative electrodes. These chemistries are noteworthy because the theoretical gravimetric and volumetric energy densities of cells that discharges to MgO or Al_2O_3 surpass the energy densities of a cell that discharges to Li_2O_2, Fig. 1.

However, further quantitative measurements will be required to definitively determine to what extent MgO/Al_2O_3 formation occurs in these cells. Furthermore, the challenges facing these systems appear to be even greater than for Li–O_2 because of the difficulties in finding electrolytes compatible with a Mg- or Al-metal negative electrode [154].

5 Concluding Remarks

Metal-oxygen batteries have been known for more than 150 years. Despite this long history, new twists on this well-known chemistry have continued to emerge up to the present day. Arguably the most exciting and rapid developments have occurred in only the past five years, coinciding with the demonstration of non-aqueous, reversible systems that in some cases exhibit extremely high energy densities in a laboratory setting. In particular, research into the Li–O_2 chemistry has progressed rapidly, and has been successful in pinpointing the primary challenges that must be overcome for a reversible Li–O_2 battery to become commercially viable. Key amongst these are: electrolyte stability, efficient transport (of oxygen within the electrolyte and electronic charge carriers within the discharge product), and implementing a high-capacity metal negative electrode. Although breakthroughs are needed in all three areas, a prudent strategy would be to focus first on realizing a reversible metal negative electrode. Such a technology could also be translated (perhaps with minimal additional development) to other, more mature systems such as those based on conventional Li-ion or lithium–sulfur technology, potentially 'killing *several* birds with one stone.' Success in this area will likely hinge upon development of a solid electrolyte capable of suppressing dendrite formation, while allowing for high ionic conductivity.

Should these cell-level challenges be overcome, another set of challenges for non-aqueous metal-oxygen batteries loom at the system level. These include the engineering of an efficient balance of plant that would either store oxygen on board (in a closed system), or separate it from an incoming flow of air (open system). Table 3 summarizes projected energy densities at the system level for an automotive scale Li–O_2 battery from three recent studies. While there is a wide range in the projected values, it is clear that the mass and volume associated with the system

Table 3 Projected system-level energy densities for non-aqueous Li–O_2 batteries

Institution	Gravimetric energy density (Wh/kg)	Volumetric energy density (Wh/L)
JCESR [26]	200–500	300–450
Bosch [15]	650–850	550–950
Ford [27]	640	600

incur a large penalty based on the much higher theoretical densities reported in Fig. 1. Minimizing these penalties will require novel engineering solutions.

Acknowledgments The authors gratefully acknowledge financial support from the U.S. National Science Foundation, grant no. CBET-1351482.

References

1. Vergnes M (1860) Improvement in the construction of voltaic gas-batteries. US Patent 28317
2. Shao Y, Ding F, Xiao J et al (2013) Making Li-air batteries rechargeable: material challenges. Adv Funct Mater 23:987–1004. doi:10.1002/adfm.201200688
3. Heise GW, Schumacher EA (1932) An air-depolarized primary cell with caustic alkali electrolyte. J Electrochem Soc 62:383–391. doi:10.1149/1.3493794
4. Reddy TB, Linden D (2011) Handbook of batteries. McGraw-Hill Companies, New York
5. Moehlenbrock MJ, Minteer SD (2008) Extended lifetime biofuel cells. Chem Soc Rev 37:1188–1196. doi:10.1039/b708013c
6. Srinivasan S (2006) Fuel cells: from fundamentals to applications. Springer, New York
7. Yu X, Pickup PG (2008) Recent advances in direct formic acid fuel cells (DFAFC). J Power Sources 182:124–132. doi:10.1016/j.jpowsour.2008.03.075
8. Ma J, Choudhury NA, Sahai Y (2010) A comprehensive review of direct borohydride fuel cells. Renew Sustain Energy Rev 14:183–199. doi:10.1016/j.rser.2009.08.002
9. Abraham KM, Jiang Z (1996) A polymer electrolyte-based rechargeable lithium/oxygen battery. J Electrochem Soc 143:1–5. doi:10.1149/1.1836378
10. Toni JEA, McDonald GD, Elliott WE (1966) Lithium-moist air battery. Fort Belvoir, Virginia
11. Blurton KF, Sammells AF (1979) Metal/air review batteries: their status and potential—a review. J Power Sources 4:263–279. doi:10.1016/0378-7753(79)80001-4
12. Semkow KW, Sammells AF (1987) A lithium oxygen secondary battery. J Electrochem Soc 134:2084–2085. doi:10.1149/1.2100826
13. Abraham KM (2008) A brief history of non-aqueous metal-air batteries. ECS Trans 3:67–71. doi:10.1149/1.2838193
14. Zhang T, Zhou H (2013) A reversible long-life lithium-air battery in ambient air. Nat Commun 4:1817. doi:10.1038/ncomms2855
15. Imanishi N, Luntz AC, Bruce P (2014) The lithium air battery: fundamentals. Springer, Berlin
16. Lu J, Li L, Park J-B et al (2014) Aprotic and aqueous Li–O_2 batteries. Chem Rev 114:5611–5640. doi:10.1021/cr400573b
17. Wang J, Li Y, Sun X (2013) Challenges and opportunities of nanostructured materials for aprotic rechargeable lithium–oxygen batteries. Nano Energy 2:443–467. doi:10.1016/j.nanoen.2012.11.014
18. Li Q, Cao R, Cho J, Wu G (2014) Nanostructured carbon-based cathode catalysts for nonaqueous lithium–oxygen batteries. Phys Chem Chem Phys. doi:10.1039/C4CP00225C
19. Balaish M, Kraytsberg A, Ein-Eli Y (2014) A critical review on lithium-air battery electrolytes. Phys Chem Chem Phys 16:2801–2822. doi:10.1039/c3cp54165g
20. Yuan J, Yu J-S, Sundén B (2015) Review on mechanisms and continuum models of multi-phase transport phenomena in porous structures of non-aqueous Li-Air batteries. J Power Sources 278:352–369. doi:10.1016/j.jpowsour.2014.12.078
21. Guo Z, Dong X, Yuan S et al (2014) Humidity effect on electrochemical performance of Li–O_2 batteries. J Power Sources 264:1–7. doi:10.1016/j.jpowsour.2014.04.079

22. Trahan MJ, Mukerjee S, Plichta EJ et al (2012) Studies of Li-air cells utilizing dimethyl sulfoxide-based electrolyte. J Electrochem Soc 160:A259–A267. doi:10.1149/2.048302jes
23. Adams BD, Radtke C, Black R et al (2013) Current density dependence of peroxide formation in the Li–O_2 battery and its effect on charge. Energy Environ Sci 6:1772. doi:10.1039/c3ee40697k
24. Geaney H, O'Connell J, Holmes JD, O'Dwyer C (2014) On the use of gas diffusion layers as current collectors in Li–O_2 battery cathodes. J Electrochem Soc 161:A1964–A1968. doi:10.1149/2.0021414jes
25. Hougton R, Gouty D, Allinson J et al (2012) Monitoring the location of cathode-reactions in Li–O_2 batteries. J Electrochem Soc 162:A3126–A3132. doi:10.1149/2.0191502jes
26. Gallagher KG, Goebel S, Greszler T et al (2014) Quantifying the promise of lithium–air batteries for electric vehicles. Energy Environ Sci. doi:10.1039/c3ee43870h
27. Adams J, Karulkar M (2012) Bipolar plate cell design for a lithium air battery. J Power Sources 199:247–255. doi:10.1016/j.jpowsour.2011.10.041
28. Jung H-G, Hassoun J, Park J-B et al (2012) An improved high-performance lithium-air battery. Nat Chem 4:579–585. doi:10.1038/nchem.1376
29. Mitchell RR, Gallant BM, Thompson CV, Shao-Horn Y (2011) All-carbon-nanofiber electrodes for high-energy rechargeable Li–O_2 batteries. Energy Environ Sci 4:2952–2958. doi:10.1039/c1ee01496j
30. Sun B, Huang X, Chen S et al (2014) Porous graphene nanoarchitectures: an efficient catalyst for low charge-overpotential, long life, and high capacity lithium–oxygen batteries. Nano Lett 14:3145–3152. doi:10.1021/nl500397y
31. Kwabi DG, Ortiz-Vitoriano N, Freunberger Sa et al (2014) Materials challenges in rechargeable lithium–air batteries. MRS Bull 39:443–452. doi:10.1557/mrs.2014.87
32. Lu J, Lei Y, Lau KC et al (2013) A nanostructured cathode architecture for low charge overpotential in lithium–oxygen batteries. Nat Commun 4:2383. doi:10.1038/ncomms3383
33. Yilmaz E, Yogi C, Yamanaka K et al (2013) Promoting formation of noncrystalline Li_2O_2 in Li–O_2 battery with RuO_2 nanoparticles. Nano Lett 13:4679–4684. doi:10.1021/nl4020952
34. Meini S, Piana M, Tsiouvaras N et al (2012) The effect of water on the discharge capacity of a non-catalyzed carbon cathode for Li–O_2 batteries. Electrochem Solid-State Lett 15:A45–A48. doi:10.1149/2.005204esl
35. Meini S, Solchenbach S, Piana M, Gasteiger Ha (2014) The role of electrolyte solvent stability and electrolyte impurities in the electrooxidation of Li_2O_2 in Li–O_2 batteries. J Electrochem Soc 161:A1306–A1314. doi:10.1149/2.0621409jes
36. Aetukuri NB, McCloskey BD, García JM et al (2014) Solvating additives drive solution-mediated electrochemistry and enhance toroid growth in non-aqueous Li–O_2 batteries. Nat Chem 7:50–56. doi: 10.1038/nchem.2132
37. Mitchell RR, Gallant BM, Shao-Horn Y, Thompson CV (2013) Mechanisms of morphological evolution of Li_2O_2 particles during electrochemical growth. J Phys Chem Lett 4:1060–1064. doi:10.1021/jz4003586
38. Viswanathan V, Thygesen KS, Hummelshøj JS et al (2011) Electrical conductivity in Li_2O_2 and its role in determining capacity limitations in non-aqueous Li–O_2 batteries. J Chem Phys 135:214704. doi:10.1063/1.3663385
39. Griffith LD, Sleightholme AES, Mansfield JF et al (2015) Correlating Li/O_2 Cell Capacity and Product Morphology with Discharge Current. ACS Appl Mater Interfaces 7:7670–7678. doi:10.1021/acsami.5b00574
40. Jung H-G, Kim H-S, Park J-B et al (2012) A transmission electron microscopy study of the electrochemical process of lithium–oxygen cells. Nano Lett 1:2–4. doi:10.1021/nl302066d
41. Zhai D, Wang H-H, Yang J et al (2013) Disproportionation in Li–O_2 batteries based on a large surface area carbon cathode. J Am Chem Soc 135:15364–15372. doi:10.1021/ja403199d
42. Xia C, Waletzko M, Peppler K, Janek J (2013) Silica nanoparticles as structural promoters for oxygen cathodes of lithium–oxygen batteries. J Phys Chem C 117:19897–19904. doi:10.1021/jp407011d

43. Xu J-J, Wang Z-L, Xu D et al (2013) Tailoring deposition and morphology of discharge products towards high-rate and long-life lithium–oxygen batteries. Nat Commun 4:2438. doi:10.1038/ncomms3438
44. Lu J, Cheng L, Lau KC et al (2014) Effect of the size-selective silver clusters on lithium peroxide morphology in lithium–oxygen batteries. Nat Commun 5:4895. doi:10.1038/ncomms5895
45. Xia C, Waletzko M, Chen L et al (2014) Evolution of Li_2O_2 growth and its effect on kinetics of Li–O_2 batteries. ACS Appl Mater Interfaces 6:12083–12092. doi:10.1021/am5010943
46. Schwenke KU, Metzger M, Restle T et al (2015) The influence of water and protons on Li_2O_2 crystal growth in aprotic Li–O_2 cells. J Electrochem Soc 162:A573–A584. doi:10.1149/2.0201504jes
47. Aetukuri NB, McCloskey BD, García JM et al (2014) Solvating additives drive solution-mediated electrochemistry and enhance toroid growth in non-aqueous Li–O_2 batteries. Nat Chem. doi:10.1038/nchem.2132
48. Kosma Va, Beltsios KG (2013) Simple solution routes for targeted carbonate phases and intricate carbonate and silicate morphologies. Mater Sci Eng, C 33:289–297. doi:10.1016/j.msec.2012.08.042
49. Felker FC, Kenar JA, Fanta GF, Biswas A (2013) Comparison of microwave processing and excess steam jet cooking for spherulite production from amylose–fatty acid inclusion complexes. Starch 65:864–874. doi:10.1002/star.201200218
50. Horstmann B, Gallant B, Mitchell R et al (2013) Rate-dependent morphology of Li_2O_2 growth in Li–O_2 batteries. J Phys Chem Lett 4:4217–4222
51. Morse JW, Casey WH (1988) Ostwald processes and mineral paragenesis in sediments. Am J Sci 288:537–560
52. Feenstra TP, De Bruyn PL (1981) The Ostwald rule of stages in precipitation from highly supersaturated solutions: a model and its application to the formation of the nonstoichiometric amorphous calcium phosphate precursor phase. J Colloid Interface Sci 84:66–72
53. Ostwald W (1897) Studien über die Umwandlung fester Körper. Z Phys Chem 22:289–330
54. Tian F, Radin MD, Siegel DJ (2014) Enhanced charge transport in amorphous Li_2O_2. Chem Mater 26:2952–2959. doi:10.1021/cm5007372
55. Lau KC, Lu J, Luo X et al (2014) Implications of the unpaired spins in Li–O_2 battery chemistry and electrochemistry: a minireview. Chempluschem 80:336–343. doi:10.1002/cplu.201402053
56. Vannerberg N-G (1962) Peroxides, superoxides, and ozonides of the metals of groups Ia, IIa, and IIb. Prog Inorg Chem. Wiley, Hoboken, pp 125–197
57. Lu J, Jung H-J, Lau KC et al (2013) Magnetism in lithium–oxygen discharge product. ChemSusChem 6:1196–1202. doi:10.1002/cssc.201300223
58. Radin MD, Rodriguez JF, Tian F, Siegel DJ (2012) Lithium peroxide surfaces are metallic, while lithium oxide surfaces are not. J Am Chem Soc 134:1093–1103. doi:10.1021/ja208944x
59. Ong SP, Mo Y, Ceder G (2012) Low hole polaron migration barrier in lithium peroxide. Phys Rev B 85:081105. doi:10.1103/PhysRevB.85.081105
60. Radin MD, Siegel DJ (2013) Charge transport in lithium peroxide: relevance for rechargeable metal-air batteries. Energy Environ Sci 6:2370–2379. doi:10.1039/c3ee41632a
61. Trahan MJ, Jia Q, Mukerjee S et al (2013) Cobalt phthalocyanine catalyzed lithium–air batteries. J Electrochem Soc 160:A1577–A1586. doi:10.1149/2.118309jes
62. Thapa AK, Hidaka Y, Hagiwara H et al (2011) Mesoporous β-MnO_2 air electrode modified with Pd for rechargeability in lithium–air battery. J Electrochem Soc 158:A1483. doi:10.1149/2.090112jes
63. Xu W, Xu K, Viswanathan VV et al (2011) Reaction mechanisms for the limited reversibility of Li–O_2 chemistry in organic carbonate electrolytes. J Power Sources 196:9631–9639. doi:10.1016/j.jpowsour.2011.06.099

64. Meini S, Tsiouvaras N, Schwenke KU et al (2013) Rechargeability of Li–air cathodes pre-filled with discharge products using an ether-based electrolyte solution: implications for cycle-life of Li–air cells. Phys Chem Chem Phys 15:11478–11493. doi:10.1039/c3cp51112j
65. Mccloskey BD, Valery A, Luntz AC et al (2013) Combining accurate O_2 and Li_2O_2 assays to separate discharge and charge stability limitations in nonaqueous Li–O_2 batteries. J Phys Chem Lett 4:2989–2993. doi:10.1021/jz401659f
66. Freunberger S, Chen Y, Drewett NE et al (2011) The lithium-oxygen battery with ether-based electrolytes. Angew Chem Int Ed Engl 50:8609–8613. doi:10.1002/anie.201102357
67. Luntz AC, Viswanathan V, Voss J et al (2013) Tunneling and polaron charge transport through Li_2O_2 in Li–O_2 batteries. J Phys Chem Lett 4:3494–34997. doi:10.1021/jz401926f
68. Garcia-Lastra JM, Myrdal JSG, Christensen R et al (2013) DFT+U study of polaronic conduction in Li_2O_2 and Li_2CO_3: implications for Li–Air batteries. J Phys Chem C 117:5568–5577. doi:10.1021/jp3107809
69. Lu Y-C, Gallant BM, Kwabi DG et al (2013) Lithium–oxygen batteries: bridging mechanistic understanding and battery performance. Energy Environ Sci 6:750. doi:10.1039/c3ee23966g
70. Safari M, Adams BD, Nazar LF (2014) Kinetics of oxygen reduction in aprotic Li–O_2 cells: a model-based study. J Phys Chem Lett 5:3486–3491. doi:10.1021/jz5018202
71. Xue K, Mcturk E, Johnson L et al (2015) A comprehensive model for non-aqueous lithium air batteries involving different reaction mechanisms 162:614–621. doi:10.1149/2.0121504jes
72. Kang S, Mo Y, Ong SP, Ceder G (2013) A facile mechanism for recharging Li_2O_2 in Li–O_2 batteries. Chem Mater 25:3328–3336. doi:10.1021/cm401720n
73. Gallant BM, Kwabi DG, Mitchell RR et al (2013) Influence of Li_2O_2 morphology on oxygen reduction and evolution kinetics in Li–O_2 batteries. Energy Environ Sci 6:2518. doi:10.1039/c3ee40998h
74. Radin MD, Monroe CW, Siegel DJ (2015) How dopants can enhance charge transport in Li_2O_2. Chem Mater 27:839–847. doi:10.1021/cm503874c
75. Malik R, Abdellahi A, Ceder G (2013) A critical review of the Li insertion mechanisms in $LiFePO_4$ electrodes. J Electrochem Soc 160:A3179–A3197. doi:10.1149/2.029305jes
76. Albertus P, Girishkumar G, McCloskey B et al (2011) Identifying capacity limitations in the Li/Oxygen battery using experiments and modeling. J Electrochem Soc 158:A343. doi:10.1149/1.3527055
77. Das SK, Xu S, Emwas A-H et al (2012) High energy lithium-oxygen batteries—transport barriers and thermodynamics. Energy Environ Sci 5:8927. doi:10.1039/c2ee22470d
78. Lu Y-C, Shao-Horn Y (2013) Probing the reaction kinetics of the charge reactions of nonaqueous Li–O_2 batteries. J Phys Chem Lett 4:93–99. doi:10.1021/jz3018368
79. Viswanathan V, Nørskov JK, Speidel A et al (2013) Li–O_2 kinetic overpotentials: Tafel plots from experiment and first-principles theory. J Phys Chem Lett 4:556–560. doi:10.1021/jz400019y
80. Gerbig O, Merkle R, Maier J (2013) Electron and ion transport in Li_2O_2. Adv Mater 25:3129–3133. doi:10.1002/adma.201300264
81. Radin MD (2014) First-principles and continuum modeling of charge transport in Li–O_2 batteries. University of Michigan, Ann Arbor
82. Radin MD, Tian F, Siegel DJ (2012) Electronic structure of Li_2O_2 0001 surfaces. J Mater Sci 47:7564–7570. doi:10.1007/s10853-012-6552-6
83. Geng WT, He BL, Ohno T (2013) Grain boundary induced conductivity in Li_2O_2. J Phys Chem C 117:25222–25228. doi:10.1021/jp405315k
84. Zhao Y, Ban C, Kang J et al (2012) P-type doping of lithium peroxide with carbon sheets. Appl Phys Lett 101:023903. doi:10.1063/1.4733480
85. Zhu D, Zhang L, Song M et al (2013) Intermittent operation of the aprotic Li–O_2 battery: the mass recovery process upon discharge interval. J Solid State Electrochem 17:2539–2544. doi:10.1007/s10008-013-2116-1

86. Sahapatsombut U, Cheng H, Scott K (2013) Modelling the micro–macro homogeneous cycling behaviour of a lithium–air battery. J Power Sources 227:243–253. doi:10.1016/j.jpowsour.2012.11.053
87. Nimon VY, Visco SJ, De Jonghe LC et al (2013) Modeling and experimental study of porous carbon cathodes in Li–O_2 cells with non-aqueous electrolyte. ECS Electrochem Lett 2:A33–A35. doi:10.1149/2.004304eel
88. Liu J, Monroe CW (In preparation)
89. Chen XJ, Bevara VV, Andrei P et al (2014) Combined effects of oxygen diffusion and electronic resistance in Li–Air batteries with carbon nanofiber cathodes. J Electrochem Soc 161:A1877–A1883. doi:10.1149/2.0721412jes
90. Hummelshøj JS, Luntz AC, Nørskov JK (2013) Theoretical evidence for low kinetic overpotentials in Li–O_2 electrochemistry. J Chem Phys 138:034703. doi:10.1063/1.4773242
91. Mo Y, Ong S, Ceder G (2011) First-principles study of the oxygen evolution reaction of lithium peroxide in the lithium–air battery. Phys Rev B 84:205446. doi:10.1103/PhysRevB.84.205446
92. Lee B, Seo D-H, Lim H-D et al (2014) First-principles study of the reaction mechanism in sodium–oxygen batteries. Chem Mater 26:1048–1055. doi:10.1021/cm403163c
93. Leung K (2013) Electronic structure modeling of electrochemical reactions at electrode/electrolyte interfaces in lithium ion batteries. J Phys Chem C 117:1539–1547. doi:10.1021/jp308929a
94. Mizuno F, Nakanishi S, Kotani Y et al (2010) Rechargeable Li–Air batteries with carbonate-based liquid electrolytes. Electrochemistry 78:403–405
95. McCloskey B, Bethune D, Shelby R et al (2011) Solvents' critical role in nonaqueous lithium–oxygen battery. J Phys Chem Lett 2:1161–1166
96. Laino T, Curioni A (2012) A new piece in the puzzle of lithium/air batteries: computational study on the chemical stability of propylene carbonate in the presence of lithium peroxide. Chem—Eur J 18:3510–3520. doi:10.1002/chem.201103057
97. McCloskey BD, Bethune DS, Shelby RM et al (2012) Limitations in rechargeability of Li–O_2 batteries and possible origins. J Phys Chem Lett 3:3043–3047
98. Veith GM, Nanda J, Delmau LH, Dudney NJ (2012) Influence of lithium salts on the discharge chemistry of Li–Air cells. J Phys Chem Lett 3:1242–1247. doi:10.1021/jz300430s
99. Du P, Lu J, Lau KC et al (2013) Compatibility of lithium salts with solvent of the non-aqueous electrolyte in Li–O_2 batteries. Phys Chem Chem Phys 15:5572–5581. doi:10.1039/c3cp50500f
100. Younesi R, Hahlin M, Bjo F et al (2013) Li–O_2 battery degradation by lithium peroxide (Li_2O_2): a model study. Chem Mater 25:77–84. doi:10.1021/cm303226g
101. Mccloskey BD, Speidel A, Scheffler R et al (2012) Twin problems of interfacial carbonate formation in nonaqueous Li–O_2 batteries. J Phys Chem Lett 3:997–1001. doi:10.1021/jz300243r
102. Ottakam Thotiyl MM, Freunberger SA, Peng Z, Bruce PG (2013) The carbon electrode in nonaqueous Li–O_2 cells. J Am Chem Soc 135:494–500. doi:10.1021/ja310258x
103. Nasybulin E, Xu W, Engelhard MH et al (2013) Stability of polymer binders in Li–O_2 batteries. J Power Sources 243:899–907. doi:10.1016/j.jpowsour.2013.06.097
104. Shui J-L, Wang H-H, Liu D-J (2013) Degradation and revival of Li–O_2 battery cathode. Electrochem Commun 34:45–47. doi:10.1016/j.elecom.2013.05.020
105. Peng Z, Freunberger SA, Chen Y, Bruce PG (2012) A reversible and higher-rate Li–O_2 battery. Science 80(337):563–566. doi:10.1126/science.1223985
106. Kar M, Simons TJ, Forsyth M, MacFarlane DR (2014) Ionic liquid electrolytes as a platform for rechargeable metal–air batteries: a perspective. Phys Chem Chem Phys 16:18658–18674. doi:10.1039/C4CP02533D
107. Lau KC, Lu J, Low J et al (2014) Investigation of the decomposition mechanism of lithium bis(oxalate)borate (LiBOB) salt in the electrolyte of an aprotic Li–O_2 battery. Energy Technol 2:348–354. doi:10.1002/ente.201300164

108. Bryantsev V (2011) Computational study of the mechanisms of superoxide-induced decomposition of organic carbonate-based electrolytes. J Phys Chem Lett 2:379–383
109. Beyer H, Meini S, Tsiouvaras N et al (2013) Thermal and electrochemical decomposition of lithium peroxide in non-catalyzed carbon cathodes for Li–air batteries. Phys Chem Chem Phys 15:11025–11037. doi:10.1039/c3cp51056e
110. Bryantsev VS, Faglioni F (2012) Predicting autoxidation stability of ether- and amide-based electrolyte solvents for Li-air batteries. J Phys Chem A 116:7128–7138. doi:10.1021/jp301537w
111. Zhu D, Zhang L, Song M et al (2013) Solvent autoxidation, electrolyte decomposition, and performance deterioration of the aprotic Li–O_2 battery. J Solid State Electrochem 17:2865–2870. doi:10.1007/s10008-013-2202-4
112. Assary RS, Lau KC, Amine K et al (2013) Interactions of dimethoxy ethane with Li_2O_2 clusters and likely decomposition mechanisms for Li–O_2 batteries. J Phys Chem C 117:8041–8049
113. Laino T, Curioni A (2013) Chemical reactivity of aprotic electrolytes on a solid Li_2O_2 surface: screening solvents for Li–air batteries. New J Phys 15:095009. doi:10.1088/1367-2630/15/9/095009
114. McCloskey B, Scheffler R, Speidel A et al (2011) On the efficacy of electrocatalysis in nonaqueous Li–O_2 batteries. J Am Chem Soc 133:18038–18041
115. Harding JR, Lu Y, Shao-horn Y (2012) Evidence of catalyzed oxidation of Li_2O_2 for rechargeable Li–Air battery applications. Phys Chem Chem Phys 14:10540–10546. doi:10.1039/c2cp41761h
116. Cho MH, Trottier J, Gagnon C et al (2014) The effects of moisture contamination in the Li–O_2 battery. J Power Sources 268:565–574. doi:10.1016/j.jpowsour.2014.05.148
117. Schwenke KU, Meini S, Wu X et al (2013) Stability of superoxide radicals in glyme solvents for non-aqueous Li–O_2 battery electrolytes. Phys Chem Chem Phys 15:11830–11839. doi:10.1039/c3cp51531a
118. Gowda SR, Brunet A, Wallraff GM, Mccloskey BD (2013) Implications of CO_2 contamination in rechargeable nonaqueous Li–O_2 batteries. J Phys Chem Lett 4:276–279
119. Li F, Zhang T, Zhou H (2013) Challenges of non-aqueous Li–O_2 batteries: electrolytes, catalysts, and anodes. Energy Environ Sci 6:1125–1141. doi:10.1039/c3ee00053b
120. Xu W, Wang J, Ding F et al (2014) Lithium metal anodes for rechargeable batteries. Energy Environ Sci 7:513. doi:10.1039/c3ee40795k
121. Assary RS, Lu J, Du P et al (2012) The effect of oxygen crossover on the anode of a Li–O_2 battery using an ether-based solvent: insights from experimental and computational studies. ChemSusChem 6:51–55. doi:10.1002/cssc.201200810
122. Shui J-L, Okasinski JS, Kenesei P et al (2013) Reversibility of anodic lithium in rechargeable lithium-oxygen batteries. Nat Commun 4:2255. doi:10.1038/ncomms3255
123. Roberts M, Younesi R, Richardson W et al (2014) Increased cycling efficiency of lithium anodes in dimethyl sulfoxide electrolytes for use in Li–O_2 batteries. ECS Electrochem Lett 3: A62–A65. doi:10.1149/2.007406eel
124. Adams J, Karulkar M, Anandan V (2013) Evaluation and electrochemical analyses of cathodes for lithium-air batteries. J Power Sources 239:132–143. doi:10.1016/j.jpowsour.2013.03.140
125. Yoon DH, Park YJ (2014) Characterization of real cyclic performance of air electrode for Li–Air batteries. J Electroceram. doi:10.1007/s10832-014-9937-x
126. Ottakam Thotiyl MM, Freunberger SA, Peng Z et al (2013) A stable cathode for the aprotic Li–O_2 battery. Nat Mater 12:1–7. doi:10.1038/nmat3737
127. Chen Y, Freunberger SA, Peng Z et al (2013) Charging a Li–O_2 battery using a redox mediator. Nat Chem 5:489–494. doi:10.1038/NCHEM.1646
128. Tan P, Shyy W, An L et al (2014) A gradient porous cathode for non-aqueous lithium–air batteries leading to a high capacity. Electrochem Commun 46:111–114. doi:10.1016/j.elecom.2014.06.026

129. Black R, Lee J-H, Adams B et al (2013) The role of catalysts and peroxide oxidation in lithium-oxygen batteries. Angew Chem Int Ed Engl 52:392–396. doi:10.1002/anie.201205354
130. Kim DS, Park YJ (2014) Effect of multi-catalysts on rechargeable Li–Air batteries. J Alloys Compd 591:164–169. doi:10.1016/j.jallcom.2013.12.208
131. Li F, Kitaura H, Zhou H (2013) The pursuit of rechargeable solid-state Li–Air batteries. Energy Environ Sci 6:2302. doi:10.1039/c3ee40702k
132. Thackeray MM, Chan MKY, Trahey L et al (2013) Vision for designing high-energy, hybrid Li Ion/Li–O_2 cells. J Phys Chem Lett 4:3607–3611
133. Trahey L, Karan NK, Chan MKY et al (2012) Synthesis, characterization, and structural modeling of high-capacity, dual functioning MnO_2 electrode/electrocatalysts for Li–O_2 cells. Adv Energy Mater 3:75–84. doi:10.1002/aenm.201200037
134. Kirklin S, Chan M, Trahey L et al (2014) High-throughput screening of high-capacity electrodes for hybrid Li-ion/Li–O_2 cells. Phys Chem Chem Phys 16:22073–22082. doi:10.1039/C4CP03597F
135. Popel AS (1989) Theory of oxygen transport to tissue. Crit Rev Bioeng 17:257–321
136. Kohen R, Nyska A (2002) Oxidation of biological systems: oxidative stress phenomena, antioxidants, redox reactions, and methods for their quantification. Toxicol Pathol 30:620–650. doi:10.1080/0192623029016672
137. Kim BG, Kim S, Lee H, Choi JW (2014) Wisdom from the human eye: a synthetic melanin radical scavenger for improved cycle life of Li–O_2 battery. Chem Mater 26:4757–4764
138. Wang Y, Zheng D, Yang X-Q, Qu D (2011) High rate oxygen reduction in non-aqueous electrolytes with the addition of perfluorinated additives. Energy Environ Sci 4:3697. doi:10.1039/c1ee01556g
139. Li XL, Huang J, Faghri A (2014) Modeling study of a Li–Air battery with an active cathode. Energy 1–12. doi:10.1016/j.energy.2014.12.062
140. Nemanick EJ, Hickey RP (2014) The effects of O_2 pressure on Li–O_2 secondary battery discharge capacity and rate capability. J Power Sources 252:248–251. doi:10.1016/j.jpowsour.2013.12.016
141. Zhang Y, Zhang H, Li J et al (2013) The use of mixed carbon materials with improved oxygen transport in a lithium–air battery. J Power Sources 240:390–396. doi:10.1016/j.jpowsour.2013.04.018
142. Balaish M, Kraytsberg A, Ein-Eli Y (2013) Realization of an artificial three-phase reaction zone in a Li–Air battery. ChemElectroChem n/a–n/a. doi:10.1002/celc.201300055
143. Li C, Fontaine O, Freunberger SA et al (2014) Aprotic Li–O_2 battery: influence of complexing agents on oxygen reduction in an aprotic solvent. J Phys Chem C 118:3393–3401. doi:10.1021/jp4093805
144. Hartmann P, Bender CL, Vračar M et al (2013) A rechargeable room-temperature sodium superoxide (NaO_2) battery. Nat Mater 12:228–232. doi:10.1038/nmat3486
145. Ha S, Kim J-K, Choi A et al (2014) Sodium-metal halide and sodium-air batteries. ChemPhysChem 15:1971–1982. doi:10.1002/cphc.201402215
146. Liu W, Sun Q, Yang Y et al (2013) An enhanced electrochemical performance of a sodium-air battery with graphene nanosheets as air electrode catalysts. Chem Comm 49:1951–1953. doi:10.1039/c3cc00085k
147. Ren X, Wu Y (2013) A low-overpotential potassium-oxygen battery based on potassium superoxide. J Am Chem Soc 135:2923–2926. doi:10.1021/ja312059q
148. Shiga T, Hase Y, Yagi Y et al (2014) Catalytic cycle employing a TEMPO—anion complex to obtain a secondary Mg–O_2 battery. J Phys Chem Lett 5:1648–1652. doi:10.1021/jz500602r
149. Shiga T, Hase Y, Kato Y et al (2013) A rechargeable non-aqueous Mg–O_2 battery. Chem Commun (Camb) 49:9152–9154. doi:10.1039/c3cc43477j
150. Revel R, Audichon T, Gonzalez S (2014) Non-aqueous aluminium-air battery based on ionic liquid electrolyte. J Power Sources 272:415–421. doi:10.1016/j.jpowsour.2014.08.056

151. Gruber PW, Medina PA, Keoleian GA et al (2011) Global lithium availability: a constraint for electric vehicles? J Ind Ecol 15:760–775. doi:10.1111/j.1530-9290.2011.00359.x
152. Sangster J, Pelton A (1992) The Li–O (lithium-oxygen) system. J Phase Equilibria 13:296–299
153. Lau KC, Curtiss LA, Greeley J (2011) Density functional investigation of the thermodynamic stability of lithium oxide bulk crystalline structures as a function of oxygen pressure. J Phys Chem C 115:23625–23633. doi:10.1021/jp206796h
154. Muldoon J, Bucur CB, Gregory T (2014) Quest for nonaqueous multivalent secondary batteries: magnesium and beyond. Chem Rev 114:11683–11720. doi:10.1021/cr500049y
155. Lu YC, Kwabi DG, Yao KPC et al (2011) The discharge rate capability of rechargeable Li–O_2 batteries. Energy Environ Sci 4:2999–3007. doi:10.1039/c1ee01500a
156. Kim BG, Kim H-J, Back S et al (2014) Improved reversibility in lithium-oxygen battery: understanding elementary reactions and surface charge engineering of metal alloy catalyst. Sci Rep 4:4225. doi:10.1038/srep04225
157. Hayashi K, Shima K, Sugiyama F (2013) A mixed aqueous/aproticssodium/air cell using a NASICON ceramic separator. J Electrochem Soc 160:A1467–A1472. doi:10.1149/2.067309jes
158. Downing BW (2012) Metal—air technology. Electrochem Technol Energy Storage Convers. doi:10.1002/9783527639496.ch6
159. Cooper JF (1977) High performance metal/air fuel cells. OSTI ID: 7084912. doi:10.2172/7084912
160. Hosseiny SS, Saakes M, Wessling M (2011) A polyelectrolyte membrane-based vanadium/air redox flow battery. Electrochem Commun 13:751–754. doi:10.1016/j.elecom.2010.11.025
161. Walker CW, Walker J (2012) Molybdenum/air battery and cell design. 2: US Patent 8148020 B2
162. Wagner OC (1969) Secondary cadmium-air cells. J Electrochem Soc 116:693. doi:10.1149/1.2412023
163. Egan DR, Ponce de León C, Wood RJK et al (2013) Developments in electrode materials and electrolytes for aluminium–air batteries. J Power Sources 236:293–310. doi:10.1016/j.jpowsour.2013.01.141
164. Mori R (2014) A novel aluminium–air secondary battery with long-term stability. RSC Adv 4:1982. doi:10.1039/c3ra44659j
165. Zhong X, Zhang H, Liu Y et al (2012) High-capacity silicon-air battery in alkaline solution. ChemSusChem 5:177–180. doi:10.1002/cssc.201100426
166. Jiang R (2007) Combinatorial electrochemical cell array for high throughput screening of micro-fuel-cells and metal/air batteries. Rev Sci Instrum 78:072209. doi:10.1063/1.2755439
167. Inoishi A, Ju Y-W, Ida S, Ishihara T (2013) Mg-air oxygen shuttle batteries using a ZrO_2-based oxide ion-conducting electrolyte. Chem Commun (Camb) 49:4691–4693. doi:10.1039/c3cc40880a
168. Pujare NU, Semkow KW, Sammells AF (1988) A calcium oxygen secondary battery. J Electrochem Soc 135:260–261. doi:10.1149/1.2095574
169. Zhao X, Gong Y, Li X et al (2013) A new solid oxide molybdenum–air redox battery. J Mater Chem A 1:14858. doi:10.1039/c3ta12726e
170. Zhao X, Li X, Gong Y et al (2013) A high energy density all solid-state tungsten-air battery. Chem Commun (Camb) 49:5357–5359. doi:10.1039/c3cc42075b
171. Zhao X, Li X, Gong Y et al (2014) A novel intermediate-temperature all ceramic iron–air redox battery: the effect of current density and cycle duration. RSC Adv 4:22621. doi:10.1039/c4ra02768j
172. Desclaux P, Nürnberger S, Stimming U (2010) Direct carbon fuel cells. In: Steinberge-Wilckens R, Lehnert W (eds) Innovations in fuel cell technologies. The Royal Society of Chemistry, Cambridge, p 190–211 doi:10.1039/9781849732109-00190
173. Cohn G, Starosvetsky D, Hagiwara R et al (2009) Silicon–air batteries. Electrochem Commun 11:1916–1918. doi:10.1016/j.elecom.2009.08.015

Oxygen Redox Catalyst for Rechargeable Lithium-Air Battery

Sheng Shui Zhang and Zhengcheng Zhang

1 Introduction

Li-air battery consists of a Li anode and a porous air cathode, in which the air cathode may be a pure carbon, a catalyst-loaded carbon, or a mixture of carbon and catalyst. In discharge, Li metal oxidizes into Li^+ ion and oxygen from air reduces into peroxide or oxide anion, which is instantly combined with Li^+ ions from the electrolyte to form insoluble Li_2O_2 and Li_2O. The pores in the air cathode serve as the gas diffusion channels to enable free flow of oxygen and accommodate insoluble oxygen reduction products (Li_2O_2 and Li_2O). The rechargeablility of Li-air battery is based on the reversible oxygen reduction reaction (ORR) and oxygen evolution reaction (OER) on the air cathode, which are the keys to guarantee the power capability, energy efficiency and reversibility of Li-air battery. The ORR and OER are affected by many factors, including oxygen mass transport in the cathode, reaction kinetics as well as reaction product removal from the cathode. Of them, the ORR and OER often are the rate-determining step due to their slow kinetics. In order to make the Li-air battery practical, an electrocatalyst is necessary for facilitating the ORR and OER. In this chapter, the current efforts on the oxygen redox catalyst in Li-air battery will be summarized and discussed.

S.S. Zhang (✉)
Electrochemistry Branch, RDRL-SED-C, Sensors and Electron Devices Directorate, U.S. Army Research Laboratory, Adelphi, MD 20783, USA
e-mail: shengshui.zhang.civ@mail.mil

Z. Zhang
Chemical Sciences and Engineering Division, Argonne National Laboratory, Lemont, IL 60439, USA
e-mail: zzhang@anl.gov

Z. Zhang and S.S. Zhang (eds.), *Rechargeable Batteries*,
Green Energy and Technology, DOI 10.1007/978-3-319-15458-9_19

2 Oxygen Redox Process in Li/Air Battery

The ORR and OER are two opposite processes, discussing one makes understanding of the other. Therefore, in this section the ORR in Li-air battery will be discussed. The ORR is a multiple process consisting of a series of electrochemical reductions and/or chemical disproportionations, as described Fig. 1, in which the stability of MO_2 and M_2O_2 intermediates is greatly affected by the type of cation (M^+) and solvents [1, 2]. Since superoxide (MO_2) is instable, which quickly reduces or disproportionates into peroxide (M_2O_2), the first reduction step of oxygen is often referred as to a two-electron reduction. The final ORR products vary with the type of the catalyst and electrolyte so that the overall ORR may be a four-electron reduction, a two-electron reduction, or a mixed reaction of two- and four-electron reductions. In an aqueous electrolyte, oxygen first undergoes a two-electron reduction to form a peroxide anion, which immediately combines protons from water to form a hydrogen peroxide (H_2O_2), as described by Eq. 1.

$$O_2 + 2H_2O + 2e \rightarrow H_2O_2 + 2OH^- \quad E^0 = -0.065\ V \tag{1}$$

Resulting H_2O_2 may further reduce or disproportionate into H_2O, as expressed by Eqs. 2 and 3, respectively.

$$H_2O_2 + 2e \rightarrow 2OH^- \quad E^0 = 0.867\ V \tag{2}$$

$$H_2O_2 \rightarrow H_2O + {}^1\!/\!{}_2O_2 \tag{3}$$

$$O_2 + 2H_2O + 4e \rightarrow 4OH^- \quad E^0 = 0.401V \tag{4}$$

Since H_2O_2 is miscible (soluble) with the aqueous electrolyte, Eqs. 2 and 3 occur on the liquid-solid (i.e., the solution-catalyst) interface and the reaction products can quickly diffuse into the electrolyte solution. Therefore, Eq. 1 may be readily followed by Eq. 2 and/or Eq. 3. Combination of Eqs. 1 and 2 or Eqs. 1 and 3 leads to Eq. 4, which shows that the net ORR is a four-electron reduction. In an aqueous electrolyte Li-air battery, the ORR may end at a two-electron reduction, a four-electron reduction or a mixture of two- and four-reductions, depending on the type and the activity of catalyst towards Eqs. 2 and 3.

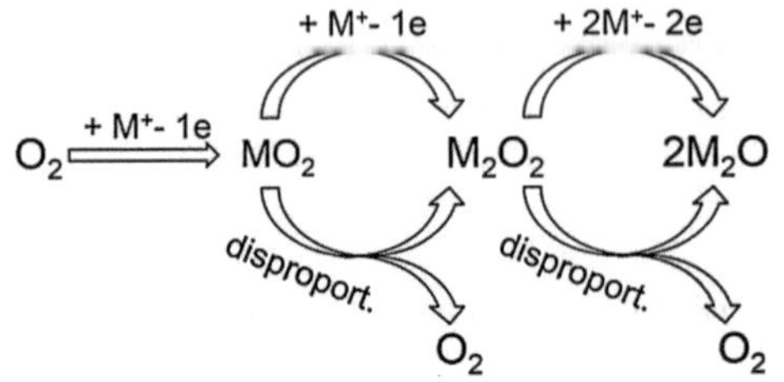

Fig. 1 General reduction processes of oxygen, where M^+ is the cation contained in the electrolyte

In the same manner, oxygen in a non-aqueous electrolyte first reduces into a peroxide anion, which instantly combines Li^+ ions from the electrolyte to form a Li_2O_2 (Eq. 5). Resulting Li_2O_2 may subsequently undergo further reduction or disproportionation to form more stable Li_2O, as described by Eqs. 6 and 7, respectively.

$$O_2 + 2Li^+ + 2e \rightarrow Li_2O_2 \quad 2.959 \text{ V vs. Li/Li}^+ \tag{5}$$

$$Li_2O_2 + 2Li^+ + 2e \rightarrow 2Li_2O \quad 2.867 \text{ V vs. Li/Li}^+ \tag{6}$$

$$Li_2O_2 \rightarrow Li_2O + {}^1\!/\!{}_2O_2 \tag{7}$$

Since Li_2O_2 is insoluble in the non-aqueous electrolyte, it accumulates on the surface of catalyst sites. As a result, Eqs. 6 and 7 take place on the solid-solid (i.e., the Li_2O_2-catalyst) interface. Due to the electrically insulting nature of Li_2O_2 and the insolubility of both the reactant and products, Eqs. 6 and 7 are kinetically slow and inefficient. Therefore, the major product of ORR in a non-aqueous electrolyte is Li_2O_2 in accompany with small amount of Li_2O, and the ORR is dominated by the two-electron reduction.

3 Role of Electrocatalyst in Li-Air Battery

The function of a catalyst is to reduce the reaction activation energy (ΔE_a) while not changing the free energy (ΔG^0), as shown in Fig. 2. In the other words, a catalyst increases the ORR and OER kinetics but unable to increase the open-circuit potential of the battery, which is reflected by a reduction of the over-potential in the operation of a Li-air battery, as indicated in Fig. 3. By reducing the over-potential, the catalyst moves the potential of ORR higher (Fig. 3a), which accordingly increases the discharge voltage of a Li-air battery (Fig. 3b). On contrary, by reducing the over-potential, the catalyst lowers the potential of OER, which accordingly reduces the charge voltage of a Li-air battery. As a result, the use of a catalyst narrows the voltage gap between the discharge and charge processes (Fig. 3b), which corresponds to an increase in the charge-discharge energy efficiency.

4 Requirements for Oxygen Redox Catalyst

A rechargeable Li-air battery requires that the ORR and OER be highly reversible. In the aqueous electrolyte, the ORR products (H_2O_2 and H_2O as shown in Eqs. 1–3) are miscible (soluble) with the electrolyte solution, which makes the OER reversible. The aqueous electrolyte Li-air battery can share the same catalyst as those used in the alkaline fuel cells and metal-air batteries, which have been intensively

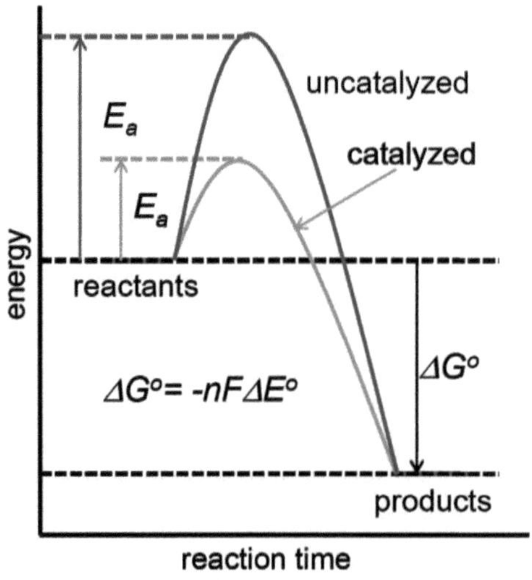

Fig. 2 Reduction of reaction activation energy by a catalyst

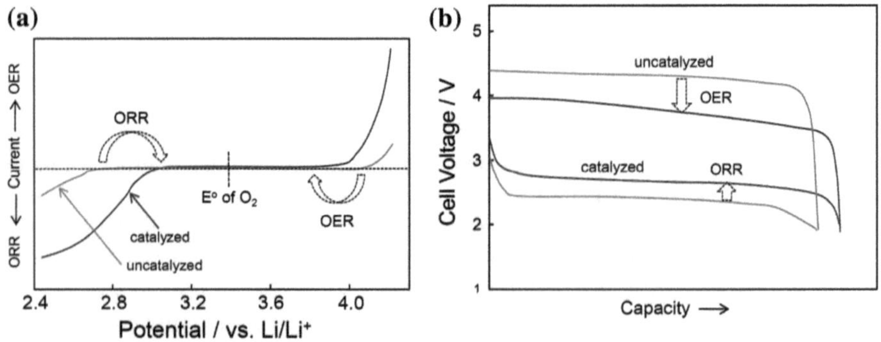

Fig. 3 Role of a catalyst in a Li-air battery. **a** ORR and OER polarization curves, and **b** charge-discharge voltage curves

investigated and widely reviewed [3–8]. In the non-aqueous electrolyte system, the insolubility of the ORR products (Li_2O_2 and Li_2O) makes the ORR much more complicated although the common knowledge about the ORR catalyst can be shared with the aqueous electrolyte system. In particular, the Li_2O is considered to be irreversible because its oxidation over-potential is so high that the oxidation potential of Li_2O is far higher than the electrochemical window of most organic solvents. Therefore, Eqs. 6 and 7 should be avoided or at least minimized for the good reversibility of Li-air battery. In the other words, a catalyst that enables two-electron ORR while being inactive towards Eqs. 6 and 7 is necessary for the non-aqueous electrolyte rechargeable Li-air battery. Among many requirements, the following properties are in particular important for the non-aqueous electrolyte rechargeable Li-air batteries.

1. High catalytic activity towards two-electron ORR: It is required for effectively reducing the over-potential of ORR and OER to enable the high power capability and high energy cycling efficiency.
2. Inactivity to the reduction and disproportionation (i.e., Eqs. 6 and 7) of Li_2O_2: For high reversibility, the formation of Li_2O should be avoided or at least minimized because of the inability of Li_2O to be oxidized to Li_2O_2 or O_2.
3. High porosity and high surface area: In order to obtain high specific capacity, high porosity is required to accommodate the insoluble ORR product (Li_2O_2), and high surface area to increase the ORR area.
4. Durability: For long cycle life, the catalyst is required to be chemically stable under the Li-air battery conditions, such as attack by alkaline and dissolution by solvent.
5. Cost: To make the Li-air commercially acceptable, the material and process should be inexpensive.

5 Category of Oxygen Redox Catalyst

As shown by Eqs. 1 and 5, the ORRs in the aqueous alkaline electrolyte and in the non-aqueous electrolyte share the same two-electron reduction. This means that all catalysts showing the catalytic activity towards the ORR in alkaline fuel cells and metal-air batteries are theoretically suitable for the non-aqueous electrolyte Li-air batteries. Base on the chemical composition of the materials, these catalysts can be briefly classified into the following categories: (1) porous carbon and doped carbon materials, (2) transition metal oxides, nitrides, and sulfides, (3) marcocyclic transition metal complexes, (4) non-precious metals and alloys, (5) precious metals and alloys, (6) organic redox mediators.

5.1 Porous Carbon and Doped Carbon Materials

All carbon materials show more or less catalytic activity towards the ORR in aqueous alkaline media [3, 4]. It is widely accepted that the ORR on carbon surface undergoes a two-electron reduction as suggested by Eqs. 1 and 5. Garten and Weiss [9] proposed that the O_2 reduction mainly occurs on the edges of graphite planes and the surface quinine groups are responsible for the O_2 reduction to peroxide. Depending on the surface chemistry of carbon materials, some carbon materials are shown to catalyze the disproportionation of H_2O_2 [10]. Therefore, modification of surface function groups can control the final ORR products. On the other hand, the architecture structure or porosity of carbon materials is important for accommodating large amount of Li_2O_2, which is required by the high specific capacity. Most of recent efforts have been focused on increasing the specific capacity of Li-air battery by making the carbon materials more porous and smaller [11–14].

Electronic defects in the carbon materials are known to favor the chemical adsorption of oxygen molecules and hence increase the catalytic activity of ORR. Therefore, doping of hereoatoms has been frequently adopted to improve the catalytic activity of the carbon materials. Suitable hereoatoms include sulfur (S), nitrogen (N), phosphorus (P), and boron (B), of which the S, N, and P are electron-rich atoms and B is electron-deficient atom as compared with carbon atom [15, 16]. The hereoatoms are doped either through pyrolyzing a hereoatom-containing organic precursor or through heat-treating carbon materials with a hereoatom source. The heteroatoms are usually located at the edge or defect sites, which create the negatively or positively charge sites that are favorable for the adsorption of oxygen. Wang et al. [17] have shown that doping of sulfur promotes the nucleation of Li_2O_2 on carbon, which results in better Li_2O_2 crystals and improved performance. Co-doping of two or more hereoatioms often leads to synergistic effects on the ORR [18–21]. As indicated in Fig. 4, the synergistic effect not only reduces the ORR over-potential, but also increases the ORR current. In particular, Liu et al. [21] showed that the synergistic effect by the co-doping of S and N makes a carbon fiber higher catalytic activity and better stability than the commercial Pt/C catalyst. Therefore, the doped carbon materials offer an excellent approach for the development of metal-free and low cost ORR catalysts.

5.2 Transition Metal Oxides, Nitrides, and Sulfides

It seems that all transition metal oxides have some catalytic effects towards the ORR. Metal oxides based on a wide range of transition metals have been investigated as the ORR catalyst for rechargeable Li-air battery. Typical transition metals include

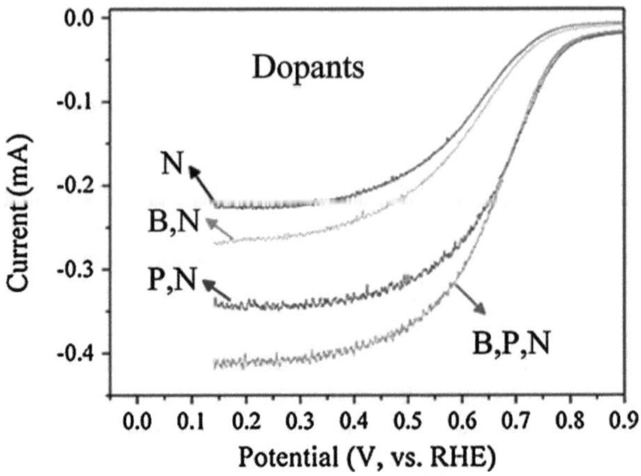

Fig. 4 Synergistic effect of co-doping on the ORR catalytic activity of graphene. ORR was performed in a 1 M $HClO_4$ electrolyte. Reprinted with permission from Ref. [18]

manganese (MnO_2, Mn_2O_3, or Mn_3O_4) [22–25], iron (Fe_2O_3, or Fe_3O_4) [26–28], cobalt (Co_2O_3) [28, 29], nickel (NiO) [30, 31], vanadium (V_2O_5) [32], and ruthenium (RuO_2) [33], which have been reviewed by several recent articles [11–14]. The transition metal oxides catalyze the ORR generally through a two-electron reduction, followed by catalyzing the disproportionation (decomposition) of peroxide [34]. There is a close correlation between the ORR catalytic activity in the Li-air battery and the ability to catalyze decomposition of H_2O_2 in an aqueous solution [35]. By comparing a few metal oxides, Giordani et al. [35] found that α-MnO_2 decomposes H_2O_2 fastest and it has the lowest over-potential in the charge process of a Li-air battery, as indicated in Fig. 5 [35]. Therefore, evaluating the rate of H_2O_2 decomposition offers a simple tool for the screening of ORR catalyst for the rechargeable Li-air battery.

RuO_2 is among the excellent catalysts for the Li-air battery. Sun et al. [33] showed that the RuO_2 nanocrystals not only significantly reduce the over-potentials of the ORR and OER but also offer extremely high capacity due to their excellent ORR catalytic activity and large specific surface area and porosity. It is shown in Fig. 6 that the Li-air cell catalyzed by the RuO_2 nanocrystals has an as low as 0.37 V voltage hysteresis (gap) between the discharging and charging processes, and a reversible capacity of about 9800 mAh/g.

Perovskite oxides, which feature the high electronic/ionic conductivity, fast oxygen ion mobility and excellent ORR catalytic activity, have been long studied as the cathode catalyst in solid oxide fuel cells. The perovskite oxides exhibit exceptional ORR catalytic activity because their electronic states favor the chemical adsorption of oxygen to the metal [36]. Recently, many publications have shown that such materials are also suitable for the ORR catalyst of Li-air batteries and outperform the conventional transition metal oxides [37–42]. It has been reported that the perovskite $Ba_{0.5}Sr_{0.5}Co_{0.8}Fe_{0.2}O_{3-x}$ catalyzes the OER at least an order of magnitude higher than the state-of-the-art iridium oxide catalyst in alkaline media [37]. The perovskite oxides typically make the Li-air batteries lower ORR and OER over-potentials and higher rate capability. For example, Xu et al. [40] reported that

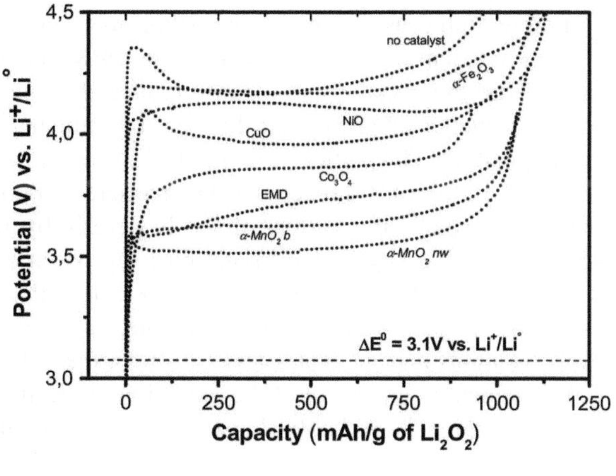

Fig. 5 Voltage curve of the first charge of Li/Li_2O_2 cells with different transition metal oxide catalysts in a 1 M $LiPF_6$ propylene carbonate electrolyte when charged at 70 mA/g, at 25 °C and P = 1 atm. Reprinted with permission from Ref. [35]

Fig. 6 Discharge-charge voltage curves of the first two cycles for Ru-carbon black (*solid line*) and carbon black electrode (*dash line*) at 200 mA/g. Reprinted with permission from Ref. [33]

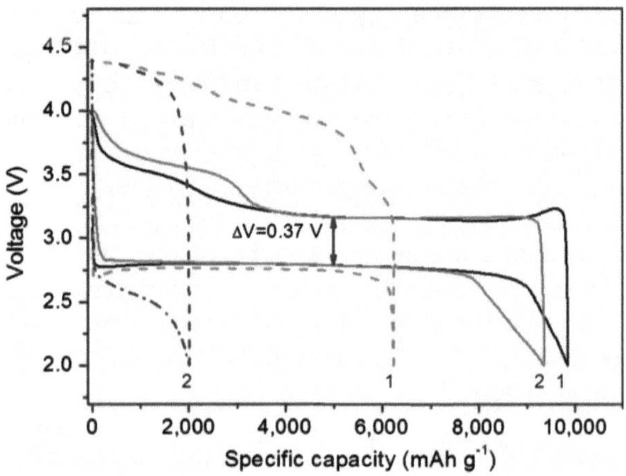

the perovskite $La_{0.75}Sr_{0.25}MnO_3$ raises the discharge voltage about 30–50 mV higher and drives the charge voltage about 200 mV lower for a Li-air cell as compared with the pristine KB carbon. Probably due to their low activity or even inactivity to catalyze the reduction and disproportionation of Li_2O_2, the perovskite oxides also enhance the ORR-OER reversibility, which make the Li-air battery cycling longer.

Most of the recent efforts on the perovskite oxide catalysts have been focused on the syntheses of novel pore architectures and highly conducting networks for the purpose to increase the specific capacity and power capability of Li-air battery. Using nanotechnologies, variety types of nano-structured perovskite oxide catalysts have been synthesized and shown much improved performance in terms of the specific capacity, energy cycling efficiency, and cycle number. Typical examples are $La_{0.75}Sr_{0.25}MnO_3$ nanotubes [40], 3D ordered macroporous $LaFeO_3$ [41], and hierarchical mesoporous $La_{0.5}Sr_{0.5}CoO_{2.91}$ nanowires [42], among which the hierarchical mesoporous $La_{0.5}Sr_{0.5}CoO_{2.91}$ nanowires exhibit an ultrahigh capacity of over 11,000 mAh/g [42].

Being similar as the oxides, transition metal nitrides [43–45] and sulfides [46] also show catalytic activity towards the ORR and OER in Li-air batteries. It is shown that these catalysts not only reduce the over-potentials, but also enhance battery's energy efficiency and lifespan by reducing or eliminating the electrolyte decomposition. The particle size of nitrides was shown to greatly affect the catalytic activity. Li et al. [43] showed that TiN particles supported on Vulcan XC-72 carbon reduce the charge-discharge voltage hysteresis by 60 mV in micro-size and by 450 mV in nano-size as the Li-air cell is cycled at a 50 mA/g (vs. carbon). Some results showed that the nitrides are more active than the oxides in catalyzing the ORR in Li-air battery. For example, a Fe/N/C composite gives lower over-potentials and generates less CO_2 than the α-MnO_2/carbon catalyst [44]. The CO_2 is generally a product of the oxidation of electrolyte solvents, the reduction of CO_2 generation is probably due to the shortened life of the lithium surperoxide

(LiO_2) intermediate by the Fe/N/C catalysist, which reduces the opportunity to oxidize the electrolyte solvents.

5.3 Marcocyclic Transition Metal Complexes

Transition metal N_4-macrocyclic complexes have been long known to catalyze ORR in aqueous electrolytes. Typically, the transition metals are Fe, Co, Ni, and Cu, and the N_4-macrocyclic ligands are the derivatives of phthalocyanine- or porphyrin-like compounds. The catalytic activity of these materials is affected by the center metal ion and ligand in the complexes. Jahnke et al. [47] proposed that the ORR is catalyzed through a "redox catalysis" mechanism. In the ORR process, oxygen first oxidizes the center metal ion (for example, M^{2+}) to higher valent metal ion (M^{3+}) followed by an electrochemical reduction to the original valance (M^{2+}) on the electrode, whereas the resulting peroxide anion combines protons from the aqueous electrolyte or Li^+ ions from the non-aqueous electrolyte to form H_2O_2 or Li_2O_2, which is either the final product or is further reduced to H_2O or Li_2O, depending on the transition metal complex used. The transition metal macrocyclic complex was first used in a non-aqueous based gel polymer electrolyte Li-air cell in 1996 by Abraham et al. [48] who used a cobalt phthalocyanine as the ORR catalyst and found the final ORR product was Li_2O_2, followed by further investigation [49].

The transition metal macrocyclic complexes are chemically instable with the alkaline ORR products (Li_2O_2 and Li_2O) and potentially dissolve into the non-aqueous electrolyte. Therefore, thermal treatment has shown to enhance the stability and catalytic activity of these complexes. In heat-treatment, the complex is first loaded onto high-surface-area carbon particles and then heated in a temperature range from 450 to 900 °C under an inert atmosphere [50]. The chemical composition of the final product depends on the heating temperature, generally from $M\text{-}N_4$ complex below 450 °C through $M\text{-}N_2$/MO mixture at 450–800°C and metal oxide (MO) at 800–900 °C to metal (M) above 1000 °C [3]. In an effort to develop the heated complexes as the ORR catalyst of Li-air battery, Ren et al. loaded a Fe and Cu mixed phthalocyanine complex onto Ketjenblack carbon and heated the mixture at 800–900 °C in argon atmosphere to get a FeCu/C catalyst [50]. When discharged at 0.2 mA/cm^2, the Li-air cell with the resultant FeCu/C catalyst shows a 0.2 V higher discharge voltage and an about 100 mAh/g lower specific capacity as compared with that using the baseline carbon, as shown in Fig. 7a, in which the higher discharge voltage is attributed to the reduced over-potential and the lower capacity to the reduced porosity of supporting carbon as a result of the filling by the heated complex into the carbon pores. Meanwhile, the FeCu/C catalyst induces a second lower discharge voltage plateau at ~1.7 V (see Fig. 7a), which was identified to be a catalyst-induced reduction of electrolyte solvents [50, 51]. After discharging, Zhang et al. [51] observed that the open-circuit voltage of the discharged Li-air cell recovered faster when the FeCu/C catalyst was used, as shown by Fig. 7b. They contributed this phenomenon to the subsequent disproportionation

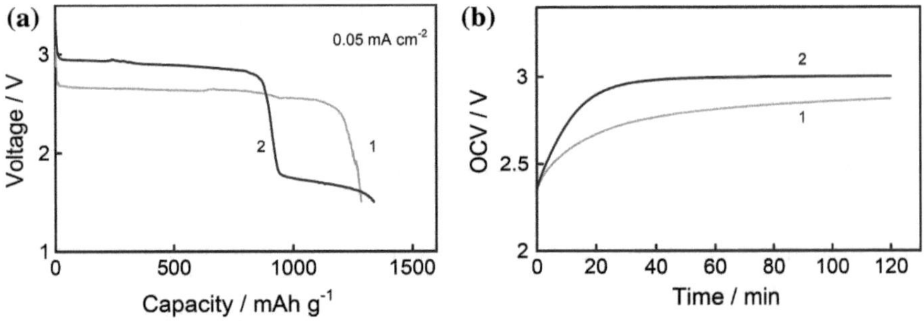

Fig. 7 Catalytic effect of FeCu/C catalyst in comparison with pristine carbon. **a** discharge voltage curve, and **b** open-circuit voltage recovery after discharging to a capacity of 600 mAh/g. Reprinted with permission from Refs. [50, 51], respectively

of Li_2O_2 catalyzed by the FeCu/C catalysis, and concluded that this type of catalysts may not be suitable for the rechargeable Li-air battery because the disproportionation produces irreversible Li_2O.

5.4 Transition Metals and Alloys

Transition metals, especially silver, show the ORR catalytic activity in aqueous alkaline media [52–54]. This is understandable because in oxygen atmosphere, the transition metals are oxidized into metal oxides, and in this case the transition metals function as an ORR catalyst as their corresponding oxides do. Therefore, the mechanism and catalytic activity of the transition metals can be referred as to the transition metal oxides.

5.5 Precious Metals and Alloys

The ORR on a Pt electrode in an acidic media has been intensively studied, and it is considered to be a multiple process generally consisting of: (1) O_2 adsorbs on the Pt catalytic sites, (2) the adsorbed O_2 reduces either through a two-electron reduction to H_2O_2 or through a four electron reduction to H_2O, (3) H_2O_2, if formed, is further reduced or disproportioned to H_2O. Step 3 is so fast that the H_2O_2 is rarely detected and the overall ORR on the Pt electrode can be taken as a four-electron reduction. Other precious metals such as Au and Pd also show the similar ORR catalytic activity, and the alloys of two or more precious metals are often explored for the synergistic effect of the ORR catalytic activity. In the Li-air battery, the ORR process is a little different from the aqueous acidic solution due to the insolubility of the Li_2O_2 immediate. Upon the reduction of oxygen, the formed Li_2O_2 accumulates on the surface of the catalytic sites, on which the Li_2O_2 is further reduced or disproportionated into Li_2O. Therefore, the ORR in the non-aqueous electrolyte is dominated by the four-electron reduction and the major ORR product is Li_2O.

In the non-aqueous electrolyte, it was found that the ORR catalytic activity is strongly correlated to the adsorption energy of oxygen on the surface of precious metal particles, and that the catalytic activity varies in an order of Pd > Pt > Ru ~ Au among a few common precious metals [55]. Asymmetric catalytic activities on the ORR and OER in the Li-air battery have been frequently observed. In comparison with carbon, Lu et al. concluded that the Au/C composite only catalyzes the ORR [56] whereas the Pt/C composite only catalyzes the OER [57]. However, the Pt-Au alloy catalyzes both the ORR and OER by increasing the energy cycling efficiency from 57 to 73 % [58], showing an excellent the synergistic effect. The asymmetric catalysis was also reported on the Pd/C composite by Xu et al. [59] who observed that the Pd/C composite preferentially catalyzes the OER, as shown by Fig. 8. It is shown that hollow spherical carbon (HSC) catalyzes both the ORR and OER, as compared with the pristine carbon paper (CP). With the incorporation of nano-structured Pd particles, the potential of the OER is further reduced whereas the potential of the ORR remains unchanged, as shown by comparing the P-HSC and HSC in Fig. 8. The asymmetric catalysis on the ORR and OER may be associated with the removal of reduction products. In the ORR, the ORR products (Li_2O_2 and Li_2O) are insoluble in the non-aqueous electrolyte, they accumulate on the surface of catalyst and block the access of external oxygen to the catalytic sites. Thus, the diffusion of external oxygen into the catalytic sites may become the rate-determining step, which determined the over-potential of ORR. On contrary, in the OER, the formed oxygen can be readily removed from the catalytic sites, which is not possible to be the rate-determining step. Therefore, the asymmetric catalysis is a common characteristic of the non-aqueous rechargeable Li-air batteries, which can also be observed from other catalyst systems.

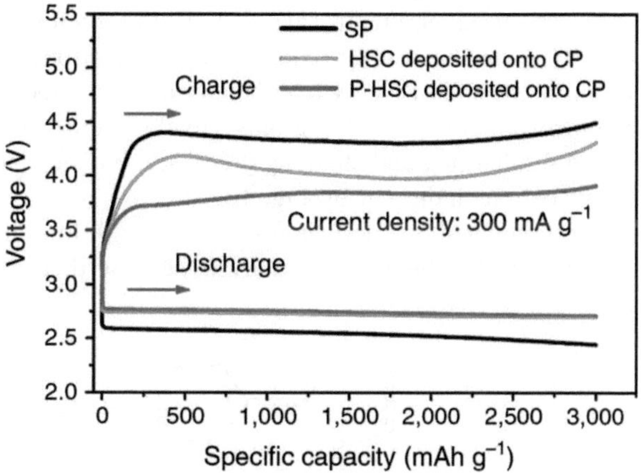

Fig. 8 First discharge and charge curves of Li-O_2 cells at a current density of 300 mA/g and a specific capacity limit of 3000 mAh/g. Reprinted with permission from Ref. [59]

5.6 Organic Redox Mediators

All ORRs on the catalysts described above are heterogeneous reactions, in which the gaseous or adsorbed oxygen is reduced on the surface of solid catalyst. The accumulation of insoluble ORR products on the catalytic sites gradually invalidates the catalytic activity. Therefore, moving the reaction from the solid surface to the solution phase must be a very effective approach for promoting the ORR and OER in Li-air battery. Based on this concept, Zheng et al. [60] proposed an anion acceptor, tris(penftafluorophenyl) borane, as the Li^+ ionic complexing additive, and found that the additive dramatically increases the solubility of Li_2O_2 through the Lewis acid-base interaction between boron and peroxide anion. As a result of the dissolution of Li_2O_2, the kinetic of the OER of Li_2O_2 at a carbon electrode is significantly enhanced. The similar concept was pursued by Li et al. [61], who examined a series of bis- and tris-imidazoliums as the cationic complexing agent, and concluded that the cationic complexing agents increase the solubility of Li_2O_2 and hence promote the OER kinetics of Li_2O_2 in the non-aqueous electrolyte.

Another approach for moving the oxygen redox reaction from the solid surface to the solution phase is the use of an organic redox mediator (RM), in which the redox potentials of RM match those of the ORR and ORR, and the RM in both the reduced and oxidized states is soluble in the non-aqueous electrolyte. The RM catalyzes the ORR through a redox mechanism, which can be described by Eqs. 8 and 9, whereas the OER proceeds in the opposite direction.

$$RM + 2Li^+ + O_2 \rightarrow RM^{2+} + Li_2O_2 \tag{8}$$

$$RM^{2+} + 2e \rightarrow RM \tag{9}$$

With the RM, O_2 first oxidizes the RM in the solution to produce Li_2O_2, and then the resulting RM^{2+} is electrochemically reduced to its original state at the electrode. An excellent example for this approach was demonstrated by Chen et al. [62], who used tetrathiafulvalene (TTF, as shown in Fig. 9) as the RM and showed that TTF can be readily oxidized at the electrode and the oxidized TTF in turn oxidizes the solid Li_2O_2 while the oxidized TTF reducing back to TTF. As a result, the incorporation of TTF not only greatly reduces the round-trip voltage hysteresis of the discharging and charging processes, but also increases the reversibility of the Li-air cell, as indicated by Fig. 10.

Quinone and its derivatives (see Fig. 9 for chemical structure) offer great opportunity for this type of catalysts because their redox potentials can be readily

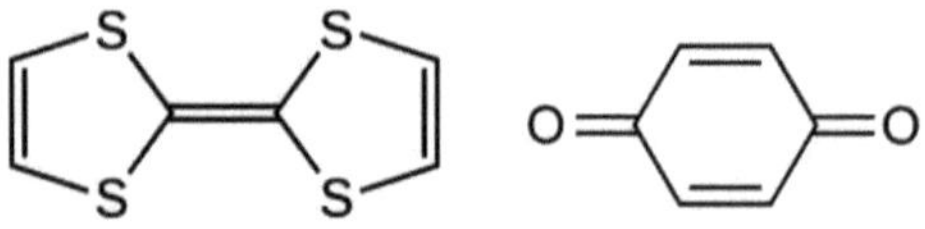

Fig. 9 Chemical structure of tetrathiafulvalene (*left*) and quinine (*right*)

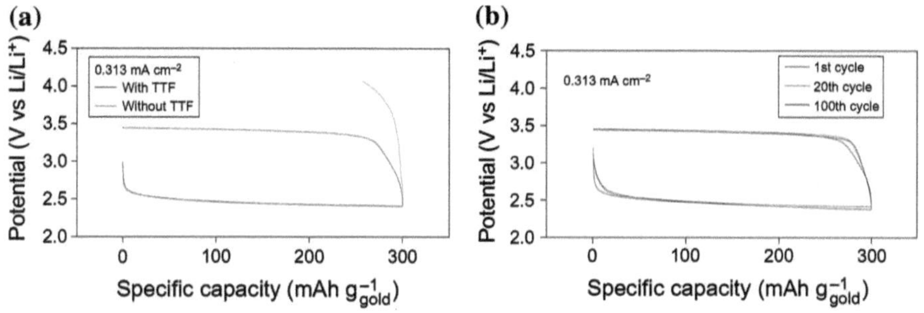

Fig. 10 Improvement of cycling reversibility (**a**) and stability (**b**) of a Li-air cell by a TTF redox mediator. Reprinted with permission from Ref. [62]

adjusted to match the ORR and OER by modifying the substituting groups. This type of compounds have been widely investigated as the metal-free ORR catalyst in various pH aqueous media [63–65], however, up to date there are no publications available about their applications in the non-aqueous electrolyte Li-air batteries. The RM-assisted ORR generally proceeds through the two-electron reduction, and the RM is catalytically inactive to the reduction and disproportionation of Li_2O_2. These features are of great importance for enabling the high reversibility and long cycle life of rechargeable Li-air battery.

6 Concluding Remarks

Li-air battery is an integrated system of many mechanical and electrochemical components, its performance is affected by many factors, including O_2 mass transport in the gas diffusion layer and air cathode, electrolyte's ionic conductivity and chemical stability against the ORR products and intermediates, and O_2 electrochemical kinetics at the cathode. Porosity of the air cathode is the key to the high specific capacity for accommodating large amount of insoluble ORR products and the high power capability for fast O_2 mass transport. Chemical stability of the electrolyte against the ORR products and intermediates is the key to the high reversibility and long life span. Oxygen redox catalyst is only one of parts, which facilitates the kinetics of O_2 electrochemical kinetics at the cathode. The insoluble and electrically insulating Li_2O is considered to be irreversible in the non-aqueous electrolyte because of its extremely high over-potential that drives the oxidation potential beyond the electrochemical window of organic electrolytes. Solid-state catalysts, especially the transition or precious metal containing catalysts, are more or less catalytically active towards the reduction or disproportionation of Li_2O_2, which produces the irreversible Li_2O. In order to develop rechargeable Li-air batteries with the high reversibility and long life span, a catalyst that catalyzes the two-electron ORR and is free of the transition or precious metal is highly demanded. The organic RMs, which catalyze two-electron ORR and do not contain any transition and precious metals, would be a kind of very promising catalysts for the non-aqueous electrolyte rechargeable Li-air battery.

References

1. Laoire CO, Mukerjee S, Abraham KM, Plichta EJ, Hendrickson MA (2009) Elucidating the mechanism of oxygen reduction for lithium-air battery applications. J Phys Chem C 113:20127–20134
2. Laoire CO, Mukerjee S, Abraham KM, Plichta EJ, Hendrickson MA (2010) Influence of nonaqueous solvents on the electrochemistry of oxygen in the rechargeable lithium-air battery. J Phys Chem C 114(19):9178–9186
3. Yeager E (1984) Electrocatalysts for O_2 reduction. Electrichem Acta 29:1527–1537
4. Yeager E (1986) Dioxygen electrocatalysis: mechanisms in relation to catalyst structure. J Mol Catal 38:5–25
5. Wang B (2005) Recent development of non − platinum catalysts for oxygen reduction reaction. J Power Sources 152:1–15
6. Bezerra CWB, Zhang L, Lee K, Liu H, Marques ALB, Marques EP, Wang H, Zhang J (2008) A review of Fe–N/C and Co–N/C catalysts for the oxygen reduction reaction. Electrochim Acta 53:4937–4951
7. Bidault F, Brett DJL, Middleton PH, Brandon NP (2009) Review of gas diffusion cathodes for alkaline fuel cells. J Power Sources 187:39–48
8. Neburchilov V, Wang H, Martin JJ, Qu W (2010) A review on air cathodes for zinc–air fuel cells. J Power Sources 195:1271–1528
9. Garten VA, Weiss PE (1955) Quinone − hydroquinone character of activated carbon and carbon black. Austral J Chem 8:68–95
10. Taylor RJ, Humffray AA (1975) Electrochemical studies on glassy carbon electrodes. J Electroanal Chem 64:63–84
11. Christensen J, Albertus P, Sanchez-Carrera RS, Lohmann T, Liedtke B, Kozinsky R, Ahmed J, Kojic A, Ahmed J (2012) A critical review of Li/air batteries. J Electrochem Soc 159(2):R1–R30
12. Shao Y, Park S, Xiao J, Zhang JG, Wang Y, Liu J (2012) Electrocatalysts for nonaqueous lithium-air batteries: status, challenges, and perspective. ACS Catal 2:844–857
13. Imanishi N, Yamamoto O (2014) Rechargeable lithium-air batteries: characteristics and prospects. Mater Today 17:24–30
14. Wang ZL, Xu D, Xu JJ, Zhang XB (2014) Oxygen electrocatalysts in metal–air batteries: from aqueous to nonaqueous electrolytes. Chem Soc Rev 43:7746–7786
15. Kicinski W, Szala M, Bystrzejewski M (2014) Sulfur-doped porous carbons: synthesis and applications. Carbon 68:1–32
16. Daems N, Sheng X, Vankelecom IFJ, Pescarmona PP (2014) Metal-free doped carbon materials as electrocatalysts for the oxygen reduction reaction. J Mater Chem A 2:4085–4110
17. Li Y, Wang J, Li X, Geng D, Banis MN, Tang Y, Wang D, Li R, Sham TK, Sun X (2012) Discharge product morphology and increased charge performance of lithium-oxygen batteries with graphene nanosheet electrodes: the effect of sulphur doping. J Mater Chem 22:20170–20174
18. Choi CH, Park SH, Woo SI (2012) Binary and ternary doping of nitrogen, boron, and phosphorus into carbon for enhancing electrochemical oxygen reduction activity. ACS Nano 6 (8):7084–7091
19. Liang J, Jiao Y, Jaroniec M, Qiao SZ (2012) Sulfur and nitrogen dualdoped mesoporous graphene electrocatalyst for oxygen reduction with synergistically enhanced performance. Angew Chem Int Ed 51:11496–11500

20. Xu J, Dong G, Jin C, Huang M, Guan L (2013) Sulfur and nitrogen codoped, few-layered graphene oxide as a highly efficient electrocatalyst for the oxygen. ChemSusChem 6:493–499
21. Liu Z, Nie H, Yang Z, Zhang J, Jin Z, Lu Y, Xiao Z, Huang S (2013) Sulfur-nitrogen co-doped three-dimensional carbon foams with hierarchical pore structures as efficient metal-free electrocatalysts for oxygen reduction reactions. Nanoscale 5(8):3283–3288
22. Ogasawara T, Debart A, Holzapfel M, Novak P, Bruce PG (2006) Rechargeable Li_2O_2 electrode for lithium batteries. J Am Chem Soc 128(4):1390–1393
23. Crisostomo VMB, Ngala JK, Alia S, Dobley A, Morein C, Chen CH, Shen X, Suib SL (2007) New synthetic route, characterization, and electrocatalytic activity of nanosized manganite. Chem Mater 19(7):1832–1839
24. Cheng H, Scott K (2010) Carbon-supported manganese oxide nanocatalysts for rechargeable lithium-air batteries. J Power Sources 195(5):1370–1374
25. Benbow EM, Kelly SP, Zhao L, Reutenauer JW, Suib SL (2011) Oxygen reduction properties of bifunctional α-manganese oxide electrocatalysts in aqueous and organic electrolytes. J Phys Chem C 115(44):22009–22017
26. Minowa H, Hayashi M, Hayashi K, Kobayashi R, Takahashi K (2013) Mn-Fe-based oxide electrocatalysts for air electrodes of lithium–air batteries. J Power Sources 244:17–22
27. Zhang Z, Zhou G, Chen W, Lai Y, Li J (2014) Facile synthesis of Fe_2O_3 nanoflakes and their electrochemical properties for Li-air batteries. ECS Electrochem Lett 3(1):A8–A10
28. Debart A, Bao J, Armstrong G, Bruce PG (2007) An O_2 cathode for rechargeable lithium batteries: the effect of a catalyst. J Power Sources 174(2):1177–1182
29. Yang Y, Sun Q, Li YS, Hong H, Fu ZW (2013) A CoO_x/carbon double-layer thin film air electrode for nonaqueous Li-air batteries. J Power Sources 223:312–318
30. Zhao G, Zhang L, Pan T, Kening K (2013) Preparation of NiO/multiwalled carbon nanotube nanocomposite for use as the oxygen cathode catalyst in rechargeable Li-O_2 batteries. J Solid State Electrochem 17(6):1759–1764
31. Liu WM, Gao TT, Yang Y, Sun Q, Fu ZW (2013) A hierarchical three-dimensional $NiCo_2O_4$ nanowire array/carbon cloth as an air electrode for nonaqueous Li-air batteries. Phys Chem Chem Phys 15(38):15806–15810
32. Lim SH, Kim DH, Byun JY, Kim BK, Yoon WY (2013) Electrochemical and catalytic properties of V_2O_5/Al_2O_3inrechargeable Li-O_2 batteries. Electrochim Acta 107:681–685
33. Sun B, Munroe P, Wang G (2013) Ruthenium nanocrystals as cathode catalysts for lithium-oxygen batteries with a superior performance. Sci Rept 3:2247. doi:10.1038/srep02247
34. Mao L, Zhang D, Sotomura T, Nakatsuc K, Koshiba N, Ohsaka T (2003) Mechanistic study of the reduction of oxygen in air electrode with manganese oxides as electrocatalysts. Electrochim Acta 48:1015–1021
35. Giordani V, Freunberger SA, Bruce PG, Tarascona JM, Larcher D (2010) H_2O_2 decomposition reaction as selecting tool for catalysts in Li-O_2 Cells. Electrochem Solid-State Lett 13(12):A180–A183
36. Suntivich J, Shao-Horn Y (2013) Trend in oxygen reduction reaction on transition metal oxide surfaces. ECS Trans 58:715–726
37. Suntivich J, May KJ, Gasteiger HA, Goodenough JB, Shao-Horn Y (2011) A perovskite oxide optimized for oxygen evolution catalysis from molecular orbital principles. Science 34 (6061):1383–1385
38. Fu ZH, Lin XJ, Huang T, Yu AS (2012) Nano-sized $La_{0.8}Sr_{0.2}MnO_3$ as oxygen reduction catalyst in nonaqueous Li/O_2 batteries. J Solid State Electrochem 16:1447–1452
39. Yang W, Salim J, Li S, Sun C, Chen L, Goodenough JB, Kim Y (2012) Perovskite $Sr_{0.95}Ce_{0.05}CoO_{3-\delta}$ loaded with copper nanoparticles as a bifunctional catalyst for lithium-air batteries. J Mater Chem 22:18902–18907
40. Xu JJ, Xu D, Wang ZL, Wang HG, Zhang LL, Zhang XB (2013) Synthesis of perovskite-based porous $La_{0.75}Sr_{0.25}MnO_3$ nanotubes as a highly efficient electrocatalyst for rechargeable lithium-oxygen batteries. Angew Chem Int Ed 52:3887–3890

41. Xu JJ, Wang ZL, Xu D, Meng FZ, Zhang XB (2014) 3D Ordered macroporous $LaFeO_3$ as efficient electrocatalyst for Li-O_2 batteries with enhanced rate capability and cyclic performance. Energy Environ Sci 7:2213–2219
42. Zhao Y, Xu L, Mai L, Han C, An Q, Xu X, Liu X, Zhang Q (2012) Hierarchical mesoporous perovskite $La_{0.5}Sr_{0.5}CoO_{2.91}$ nanowires with ultrahigh capacity for Li-air batteries. PNAS 109:19569–19574
43. Li F, Ohnishi R, Yamada Y, Kubota J, Domen K, Yamada A, Zhou HS (2013) Carbon supported TiN nanoparticles: an efficient bifunctional catalyst for non-aqueous Li-O_2 batteries. Chem Commun 49:1175–1177
44. Shui J, Karan NK, Balasubramanian M, Li S, Liu DJ (2012) Fe/N/C composite in Li-O_2 battery: studies of catalytic structure and activity toward oxygen evolution reaction. J Am Chem Soc 134:16654–16661
45. Dong SM, Chen X, Zhang KJ, Gu L, Zhang LX, Zhou XH, Li LF, Liu ZH, Han PX, Xu HX, Yao JH, Zhang CJ, Zhang XY, Shang CQ, Cui GL, Chen LQ (2011) Molybdenum nitride based hybrid cathode for rechargeable lithium-O_2 batteries. Chem Commun 47:11291–11293
46. Reeve RW, Christensen PA, Hamnett A, Haydock SA, Roy SC (1998) Methanol tolerant oxygen reduction catalysts based on transition metal sulfides. J Electrochem Soc 145:3463–3471
47. Jahnke H, Schonborn M, Zimmermann G (1976) Organic dye stuffs as catalysts for fuel cells. Top Cur Chem 61:133–181
48. Abraham KM, Jiang Z (1996) A polymer electrolyte-based rechargeable lithium/oxygen battery. J Electrochem Soc 143:1–5
49. Trahan MJ, Jia Q, Mukerjee S, Plichta EJ, Hendrickson MA, Abraham KM (2013) Cobalt phthalocyanine catalyzed lithium-air batteries. J Electrochem Soc 160(9):A1577–A1586
50. Ren XM, Zhang SS, Tran D, Read J (2011) Oxygen reduction reaction catalyst on lithium/air battery discharge performance. J Mater Chem 21(27):10118–10125
51. Zhang SS, Ren XM, Read J (2011) Heat-treated metal phthalocyanine complex as an oxygen reduction catalyst for non-aqueous electrolyte Li/air batteries. Electrochim Acta 56(12):4544–4548
52. Lee HK, Shim JP, Shim MJ, Kim SW, Lee JS (1996) Oxygen reduction behavior with silver alloy catalyst in alkaline media. Mater Chem Phys 45:238–242
53. Wu CY, Wu PW, Lin P, Li YY, Lin YM (2007) Silvercarbon nanocapsule electrocatalyst for oxygen reduction reaction. J Electrochem Soc 154:B1059–B1062
54. Marschilok AC, Zhu SL, Milleville CC, Lee SH, Takeuchi ES, Takeuchi KJ (2011) Electrodes for nonaqueous oxygen reduction based upon conductive polymer-silver composites. J Electrochem Soc 158(3):A223–A226
55. Lu YC, Gasteiger HA, Shao-Horn Y (2011) Catalytic activity trends of oxygen reduction reaction for nonaqueous Li-air batteries. J Am Chem Soc 133:19048–19051
56. Lu YC, Gasteiger HA, Parent MC, Chiloyan V, Shao-Horn Y (2010) The influence of catalysts on discharge and charge voltages of rechargeable Li-oxygen batteries. Electrochem Solid State Lett 13(6):A69–A72
57. Lu YC, Gasteiger HA, Crumlin E, McGuire R, Shao-Horn Y (2010) Electrocatalytic activity studies of select metal surfaces and implications in Li-air batteries. J Electrochem Soc 157(9): A1016–A1025
58. Lu YC, Xu ZC, Gasteiger HA, Chen S, Hamad-Schifferli K, Shao-Horn Y (2010) Platinum gold nanoparticles: a highly active bifunctional electrocatalyst for rechargeable lithium-air batteries. J Am Chem Soc 132(35):12170–12171
59. Xu JJ, Wang ZL, Xu D, Zhang LL, Zhang XB (2013) Tailoring deposition and morphology of discharge products towards high-rate and long-life lithium-oxygen batteries. Nat Commun 4:2438. doi:10.1038/ncomms3438
60. Zheng D, Lee HS, Yang ZQ, Qu D (2013) Electrochemical oxidation of solid Li_2O_2 in non-aqueous electrolyte using peroxide complexing additives for lithium-air batteries. Electrochem Commun 28:17–19

61. Li C, Fontaine O, Freunberger SA, Johnson L, Grugeon S, Laruelle S, Bruce PG, Armand M (2014) Aprotic Li-O_2 battery: influence of complexing agents on oxygen reduction in an aprotic solvent. J Phys Chem C 118:3393–3401
62. Chen Y, Freunberger SA, Peng Z, Fontaine O, Bruce PG (2013) Charging a Li-O_2 battery using a redox mediator. Nat Chem 5:489–494
63. Sarapuu A, Helstein K, Vaik K, Schiffrin DJ, Tammeveski K (2010) Electrocatalysis of oxygen reduction by quinones adsorbed on highly oriented pyrolytic graphite electrodes. Electrochim Acta 55(22):6376–6382
64. Valarselvan S, Manisankar P (2011) Electrocatalytic reduction of oxygen at glassy carbon electrode modified by polypyrrole/anthraquinones composite film in various pH media. Electrochim Acta 56(20):6945–6953
65. Wang A, Bonakdarpour A, Wilkinson DP, Gyenge E (2012) Novel organic redox catalyst for the electroreduction of oxygen to hydrogen peroxide. Electrochim Acta 66:222–229

Aqueous Lithium-Air Batteries

O. Yamamoto and N. Imanishi

1 Introduction

High specific energy density batteries are attracting increased attention as possible power sources for electric vehicles (EVs) [1–7]. However, the driving range of commercialized EVs with the current lithium-ion batteries is considerably lower than that of the vehicles with internal combustion (IC) engines, because the energy density for lithium-ion battery packs of around 100 Wh kg^{-1} is too low compared with the IC engine. The energy density of a battery that is comparable to that of an IC engine would be approximately 700 Wh kg^{-1}. Thus, batteries with seven times higher energy density than conventional lithium-ion batteries should be developed for EV applications. The calculated energy density for a lithium-ion battery with a carbon anode and a $LiCoO_2$ cathode is 387 Wh kg^{-1}. To extend the driving range of EVs to be comparable with IC engine vehicles, a new battery system with a calculated energy density several times higher than that of the lithium-ion system is required. There are few options; however, rechargeable lithium-air batteries are the most promising system because they have fair higher energy density and a lower materials cost than other battery systems. Recently, many research groups have been developing rechargeable lithium-air (or oxygen) batteries.

At present, two types of the rechargeable lithium-air (or oxygen) batteries have been developed; non-aqueous and aqueous systems. The non-aqueous lithium-air system consists of a lithium anode, non-aqueous electrolyte and a carbon/catalyst air electrode. The specific energy densities of the non-aqueous system calculated from reaction (1) and an open-circuit voltage (OCV) of 2.96 V [1] are respectively

O. Yamamoto (✉) · N. Imanishi
Graduate School of Engineering, Mie University, Tsu, Mie, Japan
e-mail: Yamamoto@chem.mie-u.ac.jp

Z. Zhang and S.S. Zhang (eds.), *Rechargeable Batteries*,
Green Energy and Technology, DOI 10.1007/978-3-319-15458-9_20

3458 Wh kg^{-1} and 4172 Wh L^{-1} for the discharge state (including oxygen) and 11,430 Wh kg^{-1} and 6104 Wh L^{-1} for the charged state (excluding oxygen).

$$2Li + O_2 = Li_2O_2. \quad (1)$$

The energy density of 11,430 Wh kg^{-1} for the charged state is close to the energy density of gasoline (ca. 13,000 Wh kg^{-1}). The calculated energy densities per unit mass and per unit volume of the non-aqueous lithium-air system are approximately 10 and 6 times higher than those of lithium-ion batteries, respectively. The rechargeable aqueous lithium-air batteries consist of a water-stable lithium electrode, aqueous electrolyte, and a carbon based air electrode. The concept of the water-stable lithium metal electrode was proposed by Visco et al. in 2004 [8]. The water-stable lithium electrode adopts a water-stable NASICON-type lithium-ion conducting solid electrolyte of $Li_{1+x+y}Al_x(Ti,Ge)P_{4-y}Si_yO_{12}$ (O-LATP), which was supplied from Ohara Ltd., Japan, as a protective layer that covers and isolates the lithium metal from direct contact with the aqueous electrolyte. The air electrode has been extensively studied for zinc-air batteries and fuel cells, and well stablished electrodes have been proposed. A polyethylene based electrolyte [9] and Li_3N [8] are used as an interlayer between the lithium metal and O-LATP to prevent direct contact because O-LATP is unstable in contact with lithium metal [10]. The key component in the aqueous lithium-air battery is the water-stable lithium electrode, and especially the water-stable lithium conducting solid electrolyte. For the aqueous system, water molecules are involved in the redox reaction at the air electrode:

$$4Li + 6H_2O + O_2 = 4(LiOH{\cdot}H_2O). \quad (2)$$

The OCV is dependent on the concentration of OH^- in the solution and is 3.90 V in a neutral solution [9]. However, the concentration of OH^- increases with the discharge depth. A saturated aqueous solution of LiOH is reached at 5.3 M L^{-1}, which corresponds to a discharge depth of ca. 5 %. The reaction product is deposited on the air electrode and/or in the electrolyte as $LiOH{\cdot}H_2O$. O-LATP is unstable in saturated LiOH aqueous solution [11]; however, the electrolyte is stable in an aqueous solution of saturated LiOH with saturated LiCl [12]. The OCV with an aqueous solution of saturated LiOH with saturated LiCl at room temperature was reported to be around 3.0 V [13]. The specific energy densities of the aqueous lithium-air battery calculated from reaction (2) and an OCV of 3.0 V are 1917 Wh kg^{-1} and 2895 Wh L^{-1} for the discharged state, and 2369 Wh kg^{-1} and 2010 Wh L^{-1} for the charged state. The mass and volume energy densities of the aqueous lithium-air battery are lower than those of the non-aqueous battery.

However, non-aqueous lithium-air battery systems have some severe problems that still need to be addressed, such as lithium corrosion by water when operated in the air (short shelf-life) [14], decomposition of the electrolyte during the discharge and charge processes (poor cycle performance) [15–17], high charge and discharge cell voltage difference (low energy conversion efficiency) [18], and decomposition of the air electrode carbon material during the charging process (short shelf-life)

[19]. Successful results for non-aqueous lithium-air cells have been obtained under pure oxygen [3, 5]. Thus, an oxygen selecting membrane should be developed for non-aqueous lithium-air batteries for EV applications. Oxygen selecting membranes for lithium-air batteries have been reported, but the selectivity is not perfect [20]. Protection of the lithium metal anode from water is the most critical point for long-life stability of non-aqueous lithium-air batteries, as pointed out by Armand and Tarascon [2]. These severe problems, except for carbon decomposition, would be removed with the use of an aqueous electrolyte and a water-stable lithium electrode. In this chapter, the characteristics of water-stable lithium conducting solid electrolytes for the water-stable lithium electrode of aqueous lithium-air batteries are introduced and the cell performance with a water-stable lithium electrode is examined.

2 Water-Stable Lithium Conducting Solid Electrolytes for Aqueous Lithium-Air Batteries

The water-stable lithium electrode is essential for the aqueous lithium-air battery and could also be necessary in non-aqueous lithium-air batteries to protect the negative electrode from reaction with moisture in ambient air. The key material for the water-stable lithium metal electrode is a water-stable lithium conducting solid electrolyte. Various types of solid lithium conducting electrolytes have been reported, such as the layered-type Li_3N structure [21], Li_2S-based glass [22], LISICON-type $Li_{14}Zn(GeO_4)_4$ [23], $Li_{10}GeP_2S_{12}$ [24], NASICON-type $Li_{1+x}A_xTi_{2-x}(PO_4)_3$ (A = Al, Fe, Sc) (LATP) [25], perovskite-type $La_{2/3-x}Li_{3x}TiO_3$ (LLTO) [26], and garnet-type $Li_7La_3Zr_2O_{12}$ (LLZ) [27]. The highest lithium conductivity of 1.2×10^{-2} S cm^{-1} at room temperature was reported for $Li_{10}GeP_2S_{12}$, which is higher than that of conventional liquid electrolytes for lithium-ion batteries. However, at present, only two types of lithium-ion conducting solid electrolytes, LATP and LLZ, have been reported that are unstable in water, but stable in aqueous saturated LiOH with saturated LiCl. LLTO was used as the protective layer for an aqueous lithium-air cell that had stable cell discharge performance at a low current density of 0.05 mA cm^{-2} in a 0.5 M LiOH aqueous solution for approximately 10 days [28]. However, the long term stability of LLTO in an aqueous electrolyte solution for lithium-air batteries has not been reported yet.

2.1 NASICON-Type Lithium-Ion Conducting Solid Electrolyte

NASICON (Na Super Ionic Conductor)-type solid ion conductors were discovered in 1976 by Goodenough and co-workers [29]. The general formula of this structure is $Na_{1+x}Zr_2(P_{1-x}Si_xO_4)_3$, which has three-dimensional conduction path ways. The NASICON-type high lithium-ion conductivity solid electrolytes of $Li_{1+x}A_xTi_{2-x}(PO_4)_3$ (A = Al, Cr, Ga, Fe, Sc, In, Lu, Y, and La) were reported by Aono et al. [25, 30]. The highest conductivity of 7×10^{-4} S cm^{-1} at 25 °C was observed for $Li_{1.4}Al_{0.4}Ti_{1.6}(PO_4)_3$. Fu [31] reported higher lithium-ion conductivity of 1.3×10^{-3} S cm^{-1} at room temperature for the $Li_{1+x}Al_xTi_{1-x}(PO_4)_3$ glass ceramics. The glass ceramic was prepared by quenching the Li_2O–Al_2O_3–TiO_2–P_2O_5 glass from 1450 °C and sintering at 550 °C. Ohara Inc. Japan has commercialized a water-impermeable lithium conducting glass ceramic of $Li_{1+x+y}Al_x(Ti,Ge)_{2-x}Si_yP_{3-y}O_{12}$ (O-LATP). The electrical conductivity of the O-LATP plate (0.15 mm thick) is 1×10^{-4} S cm^{-1} at room temperature. The stability of O-LATP in aqueous solutions has been extensively studied by Imanishi and colleagues [11]. Figure 1 shows impedance profiles of O-LATP samples with sputtered Au electrodes that were immersed in various aqueous solutions. The impedance profiles of O-LATP immersed in 1 M $LiNO_3$ and 1 M LiCl for 3 weeks was no different from that of the pristine O-LATP. The electrical conductivity of O-LATP immersed in distilled water was slightly decreased. However, the electrical conductivity estimated from the impedance profiles of O-LATP immersed in aqueous solutions of 1 M LiOH and 0.1 M HCl was significantly decreased. The protective layer for water-stable lithium electrodes in aqueous lithium-air batteries should be stable in saturated LiOH aqueous solution. Shimonishi et al. [12] reported that O-LATP is stable in an aqueous solution of saturated LiOH with saturated LiCl. Figure 2 shows the change in impedance over time for O-LATP immersed in a solution saturated with LiOH and LiCl at 50 °C. The total conductivity of O-LATP immersed in this solution was increased after 3 weeks and no change in conductivity was observed for 3 months. It was concluded that O-LATP is stable in a solution of saturated LiOH and saturated LiCl. O-LATP is thus acceptable for use the protective layer in water-stable lithium electrodes for aqueous lithium-air batteries.

The O-LATP glass ceramics is water-impermeable, but the conductivity is slightly low for a high power density EV battery. The resistance of a 0.15 mm thick O-LATP plate is 1500 Ω cm^2. The IR drop at 0.5 mA cm^{-2} is as high as 0.75 V, which corresponds to a 25 % cell voltage drop. Therefore, to reduce the IR drop, the thickness of the electrolyte should be reduced and/or the electrical conductivity should be improved. High power density is an important requirement for EV batteries, in addition to high energy density. Therefore, higher lithium-ion conductivity solid electrolytes with excellent mechanical properties should be developed to realize a high power density aqueous lithium-air battery. The conductivity target at the operation temperature is 10^{-3} S cm^{-1} [32]; the IR drops for 0.15 and 0.075 mm thick electrolytes is 0.75 and 0.375 V at 5 mA cm^{-2}, respectively. Many conductivity data for the $Li_{1+x}A_xGe_yTi_{2-y}(PO_4)_3$ system have been reported in the last 25 years. The conductivity is highly dependent on the preparation method because this type of electrolyte has a high grain boundary resistance. An extremely high conductivity of 4.62×10^{-3} S cm^{-1} at 25 °C was reported for the $Li_{1.5}Al_{0.5}Ge_{1.5}(PO_4)_3$ glass ceramic [33]; however, the electrical conductivity at 25 °C of the same composition glass ceramic was reported to be 4×10^{-4} S cm^{-1} by

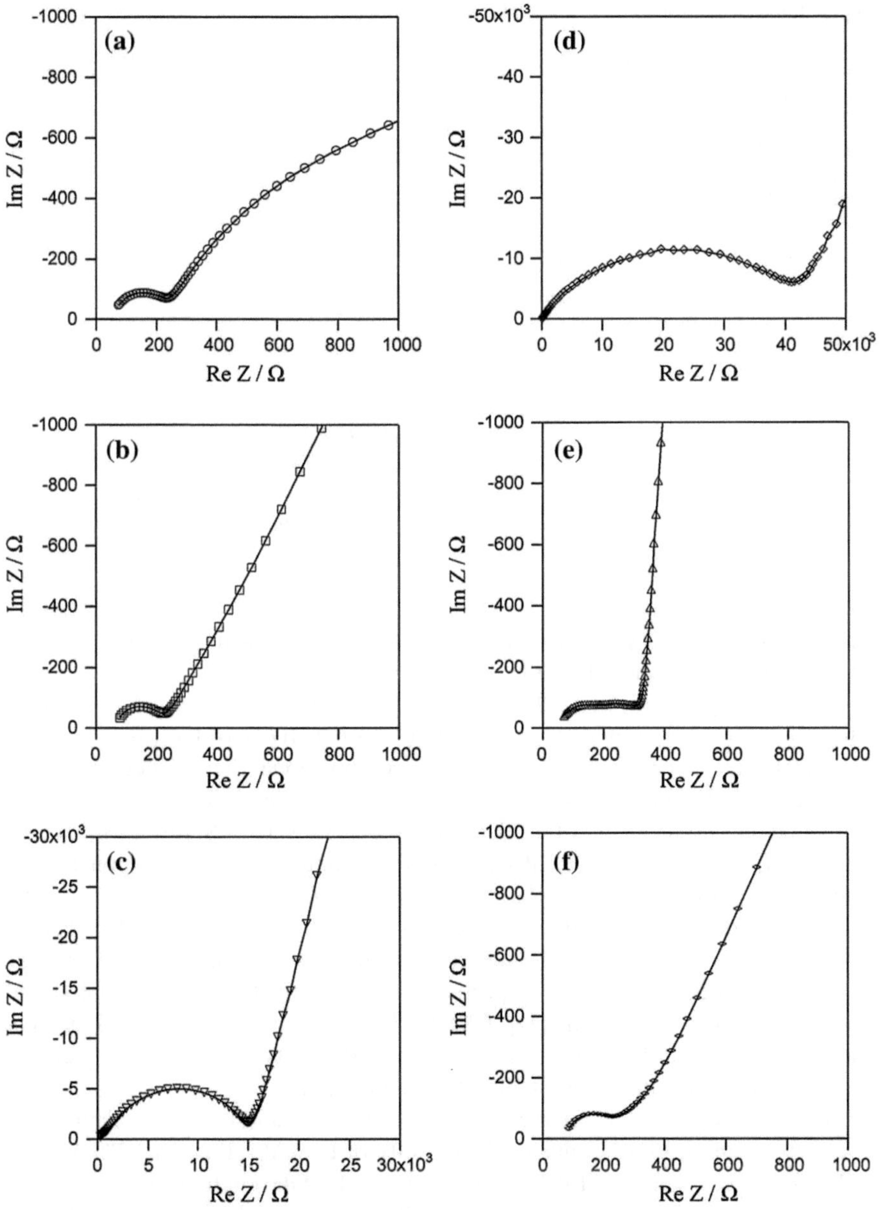

Fig. 1 Impedance profiles measured at 25 °C for **a** pristine O-LATP and O-LATP immersed in **b** 1 M $LiNO_3$ for 3 weeks, **c** 0.1 M HCl for 3 weeks, **d** 1 M LiOH for 8 months, **e** distilled water for one month, and **f** 1 M LiCl for 3 weeks (from Ref. [11])

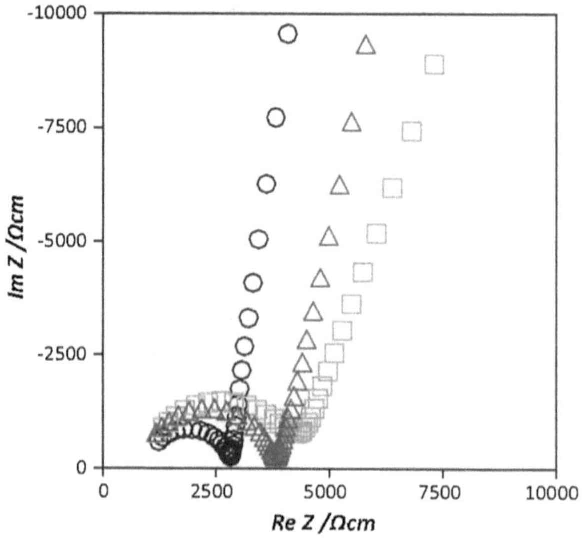

Fig. 2 Impedance profiles measured at 25 °C for O-LATP plate immersed in 11.6 M LiCl with 5.1 M LiOH at 50 °C for 3 weeks (*square*), 3 months (*triangle*), and pristine O-LATP (*circle*) (from Ref. [12])

Fu [34] and 7.25×10^{-4} S cm^{-1} by Wen and coworkers [35]. The conductivity of glass ceramics may be dependent on the preparation method and the effect of ageing on the conductivity should also be considered.

The $Li_{1+x}Al_xTi_{2-x-y}Ge_y(PO_4)_3$ solid solution has been examined by many researchers [36–38]. Zhang et al. reported the conductivity of the $Li_{1.4}Al_{0.4}Ti_{1.6-x}Ge_x(PO_4)_3$ (x = 0–1.6) system [38], and the compositional dependence of the bulk, total and grain boundary conductivities are shown in Fig. 3. The highest total and bulk conductivity at 25 °C of 1.29×10^{-3} and 2.35×10^{-3} S cm^{-1}, respectively, were observed for $Li_{1.4}Al_{0.4}Ti_{1.4}Ge_{0.2}(PO_4)_3$. The bulk conductivity is much higher than that of 9.43×10^{-4} S cm^{-1} for $Li_{1.4}Al_{0.4}Ti_{1.27}Ge_{0.33}(PO_4)_3$ reported by Wen and coworkers [37]. The activation energies for bulk and grain boundary conduction in $Li_{1.4}Al_{0.4}Ti_{1.4}Ge_{0.2}(PO_4)_3$ were estimated from the temperature dependence to be 20.2 and 55.0 kJ mol^{-1}, respectively. These activation energies are slightly lower than those for $Li_{1.4}Al_{0.4}Ti_{1.6}(PO_4)_3$. The conductivity of this lithium conducting solid electrolyte thus satisfies the conductivity target of the protective electrolyte layer for the water-stable lithium electrode.

The other requirement for the protective layer is stability in an aqueous electrolyte with a high LiOH content. The stability of $Li_{1.4}Al_{0.4}Ti_{1.4}Ge_{0.2}(PO_4)_3$ stored in distilled water, saturated LiCl solution, saturated LiOH solution, and a solution of saturated LiOH with saturated LiCl at 50 °C for one week was examined. X-ray diffraction (XRD) analysis of $Li_{1.4}Al_{0.4}Ti_{1.4}Ge_{0.2}(PO_4)_3$ immersed in these solutions revealed no impurity phases and no change in the diffraction peaks from before immersion in these solutions. However, the impedance profiles did reveal a change after immersion in these solutions, as shown in Fig. 4. The grain boundary resistance of $Li_{1.4}Al_{0.4}Ti_{1.4}Ge_{0.2}(PO_4)_3$ immersed in distilled water was increased from 12.2 to 552 Ω cm^2, while the bulk resistance increased slightly. The increase

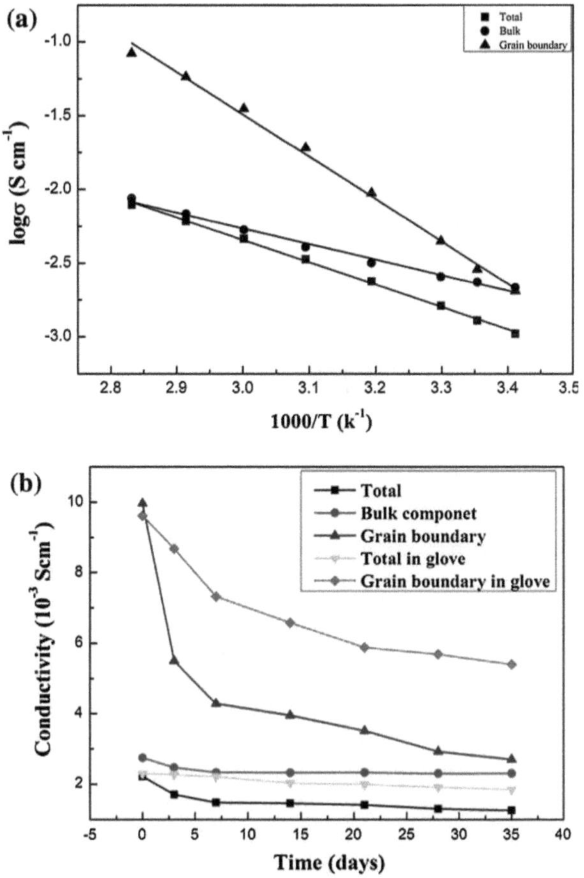

Fig. 3 **a** Temperature dependence of bulk, grain boundary, and total conductivity of $Li_{1.4}Al_{0.4}Ti_{1.44}Ge_{0.16}(PO_4)_3$, and **b** bulk, grain boundary, and total conductivity of $Li_{1.4}Al_{0.4}Ti_{1.6-x}Ge_x(PO_4)_3$ at 25 °C as a function of aging period, where glove means the samples were stored in glove box (from Ref. [38])

in the grain boundary resistance could be explained by an ion exchange reaction between Li^+ and H^+, as observed in $Li_7La_3Sn_2O_{12}$ [39]. An increase of grain boundary resistance was also observed for $Li_{1.4}Al_{0.4}Ti_{1.4}Ge_{0.2}(PO_4)_3$ immersed in saturated LiOH aqueous solution. However, there was no significant change in the bulk and grain boundary resistances of $Li_{1.4}Al_{0.4}Ti_{1.4}Ge_{0.2}(PO_4)_3$ after immersion in a solution of saturated LiCl or in a solution of saturated LiOH with saturated LiCl. $Li_{1.4}Al_{0.4}Ti_{1.4}Ge_{0.2}(PO_4)_3$, which has high lithium-ion conductivity and is stable in saturated LiOH with saturated LiCl, is thus a promising candidate for the protective layer of a water-stable lithium electrode.

2.2 Garnet-Type Lithium-Ion Conducting Solid Electrolyte

A garnet-type lithium-ion conducting solid electrolyte is another candidate for the protective layer of a water-stable lithium electrode. In 2007, Weppner and

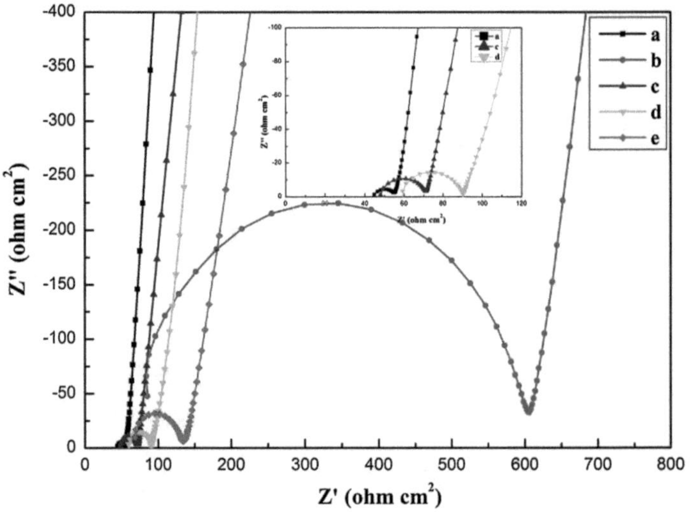

Fig. 4 Impedance profiles measured at 25 °C for (*a*) Au/pristine $Li_{1.4}Al_{0.4}Ti_{1.44}Ge_{0.16}(PO_4)_3$/Au, and Au/$Li_{1.4}Al_{0.4}Ti_{1.44}Ge_{0.16}(PO_4)_3$/Au immersed in (*b*) water, (*c*) saturated LiCl aqueous solution, (*d*) saturated LiCl and LiOH aqueous solution, and (*e*) saturated LiOH aqueous solution at 50 °C for one week (from Ref. [38])

co-workers reported that the garnet-type $Li_7La_3Zr_2O_{12}$ (LLZ) with lithium excess exhibited a high lithium-ion conductivity of 2.44×10^{-4} S cm^{-1} at 25 °C was stable in molten lithium metal and was not moisture sensitive [27]. This compound is quite attractive for the protective layer in a water-stable lithium electrode for lithium-air batteries. The NASICON-type lithium-ion conductor is unstable in contact with lithium metal [10]; therefore, an interlayer that conducts lithium ions and is stable in contact with lithium metal should be used to avoid direct contact of the lithium metal with the NASICON-type solid electrolyte. Visco et al. [8] and Zhang et al. [9] have used Li_3N and polyethylene oxide (PEO) based polymer electrolytes, respectively. Recently, the crystal structure [40], phase stability [41], Zr-site substituted LLZ [42–44], and the stability of LLZ in aqueous solution [45] have been reported. The tetragonal structure of the nominal $Li_7La_3ZrO_3$ composition is stable [46] and transforms to the cubic structure at 645 °C [41]. The electrical conductivity of tetragonal $Li_7La_3Zr_2O_{12}$ is as low as 1.63×10^{-6} S cm^{-1} at 25 °C. The high temperature cubic phase was stabilized at room temperature by the substitution of Al^{3+} for Li^+ [43], and of Ta^{5+} [44] and Nb^{5+} [42] for Zr^{4+}. The electrical conductivity of LLZ is dependent on the preparation method and dopants. The highest lithium-ion conductivity of 1×10^{-3} S cm^{-1} at 25 °C was reported for $Li_{6.4}La_3Zr_{1.4}Ta_{0.6}O_{12}$ by Goodenough and coworkers [44], which is comparable with that for $Li_{1.4}Al_{0.4}Ge_{0.2}Ti_{1.4}(PO_4)_3$ [38].

The stability of the high lithium-ion conducting garnet-type Al doped $Li_7La_3Zr_2O_{12}$ and $Li_{7-x}La_3Zr_{2-x}M_xO_{12}$ (M = Nb, Ta) in contact with lithium metal was examined by Imanishi and co-workers. The impedance of a

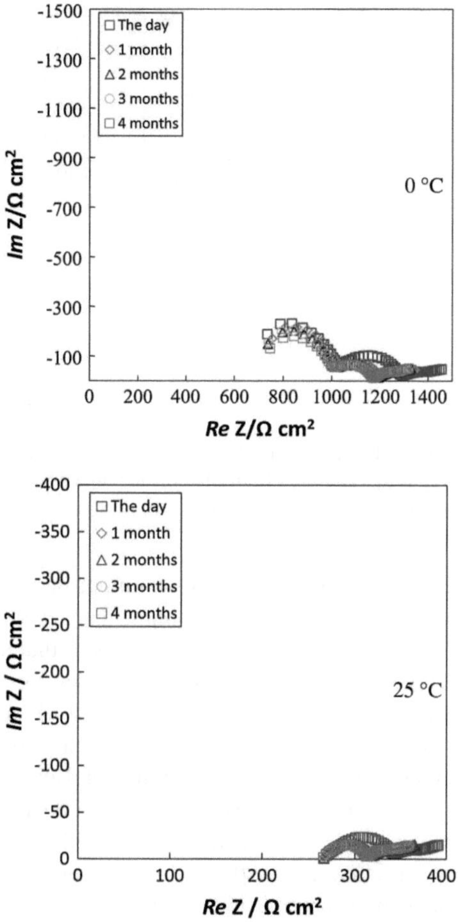

Fig. 5 Impedance profiles for $Li/Li_{6.75}La_3Zr_{1.75}Ta_{0.25}O_{12}/Li$ measured at 0 and 25 °C for various room temperature storage times (from Ref. [48])

$Li/Li_{6.75}La_3Zr_{1.75}Nb_{0.25}O_{12}/Li$ cell was gradually increased with the storage period at room temperature. The increase could be explained by the reduction of Nb by lithium [47]. In addition, $Li_{6.75}La_3Zr_{1.75}Ta_{0.25}O_{12}$ was stable in contact with lithium metal. Figure 5 shows impedance profiles for the $Li/Li_{6.75}La_3Zr_{1.75}Ta_{0.25}O_{12}/Li$ cell with respect to the storage time at room temperature. The impedance profiles show two semicircles; those in the high and low frequency ranges correspond to the grain boundary resistance and interface resistance, respectively. The grain boundary resistance showed no change with ageing and the interface resistance was decreased slightly for two weeks and then remained constant for 4 months. Therefore, $Li_{6.75}La_3Zr_{1.75}Ta_{0.25}O_{12}$ is stable in contact with lithium metal.

The stability of $Li_{6.75}La_3Zr_{1.75}Ta_{0.25}O_{12}$ in an aqueous solution of saturated LiOH with 10 M LiCl was also examined. Figure 6 shows the change in the impedance of $Li_{6.75}La_3Zr_{1.75}Ta_{0.25}O_{12}$ with the storage period in saturated LiOH with 10 M LiCl using a H-type cell. The sintered pellet of $Li_{6.75}La_3Zr_{1.75}Ta_{0.25}O_{12}$ with a relative

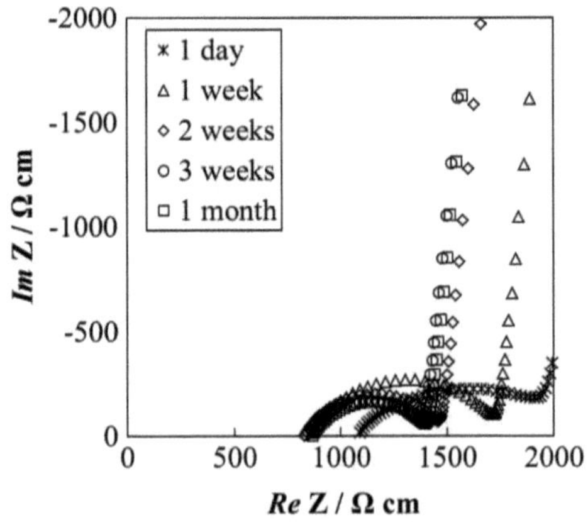

Fig. 6 Impedance profiles for the Pt/saturated LiOH/10 M LiCl aqueous solution/$Li_{6.75}La_3Zr_{1.75}Ta_{0.25}O_{12}$/saturated LiOH/10 M LiCl aqueous solution/Pt cell at room temperature as a function of storage time (from Ref. [48])

density of 96.7 % was water-impermeable. No significant change in the impedance was observed for one month. The estimated electrical conductivity of $Li_{6.75}La_3Zr_{1.75}Ta_{0.25}O_{12}$ was comparable with that measured using the sputtered Au electrode. Therefore, the interface resistance between $Li_{6.75}La_3Zr_{1.75}Ta_{0.25}O_{12}$ and the aqueous solution of saturated LiOH with 10 M LiCl is low and $Li_{6.75}La_3Zr_{1.75}Ta_{0.25}O_{12}$ is stable in saturated LiOH with 10 M LiCl. In view of the electrical conductivity, and the stability in contact with lithium metal and in saturated LiOH with 10 M LiOH aqueous solution, $Li_{6.75}La_3Zr_{1.75}Ta_{0.25}O_{12}$ is an attractive candidate for the protective layer of water-stable lithium electrodes for lithium-air batteries. The lithium metal can be used in direct contact with LLZ. However, lithium dendrite formation during the lithium deposition process is a serious problem between lithium and LLZ. Short-circuits of Li/LLZ/Li cells were reported for a short polarization period at low current density [47–49]. Figure 7 shows the change in cell voltage with the polarization time at 0.5 mA cm^{-1} for a Li/$Li_{7-x}La_3Zr_{2-x}Ta_xO_{12}$ (x = 0.25, 0.5, 0.7)/Li cell at 25 °C. A steady-state cell voltage was obtained for a short period, and the cell resistance calculated from the cell voltage just after polarization was comparable with that estimated from impedance profiles of the cell. Abrupt drops in cell voltages after polarization for a short period were observed for all cells. This may be due to short-circuit by the formation of lithium dendrites. Similar abrupt cell voltage drops were observed for Li/Al_2O_3 doped $Li_7La_3Zr_2O_{12}$/Li [49] and Li/$Li_{6.75}La_3Zr_{0.25}Nb_{0.25}O_{12}$/Li [47] cells. $Li_{6.75}La_3Zr_{1.75}Ta_{0.25}O_{12}$ with the highest relative density of 96.7 % showed the shortest short-circuit period. There is no clear dependence of the short-circuit period on the relative density of LLZ, but the lithium-ion diffusion kinetics at grain boundaries may be a key factor, because $Li_{6.75}La_3Zr_{1.75}Ta_{0.25}O_{12}$ with a high grain boundary conductivity of 2.5 × 10^{-3} S cm^{-1} at 25 °C had a shorter short-circuit period than $Li_{6.3}La_3Zr_{1.3}Ta_{0.7}O_{12}$ with a lower grain boundary conductivity of 1.37 × 10^{-3} S cm^{-1} at 25 °C. Therefore, the grain boundaries of

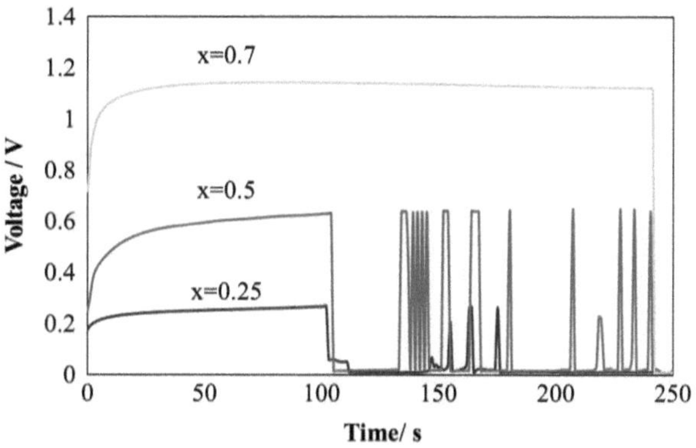

Fig. 7 Cell voltage versus polarization period curves for the Li/$Li_{7-x}La_3Zr_{2-x}Ta_xO_{12}$ (x = 0.25, 0.5, and 0.7)/Li cell at 0.5 mA cm^{-2} and 25 °C (from Ref. [48])

water-impermeable LLZ with high lithium-ion conductivity should be improved to suppress lithium dendrite formation during lithium deposition on lithium at high current density. In Table 1, the electrical conductivity and stability of the lithium-ion conducting solid electrolyte candidates for the protective layer of water-stable lithium electrodes are summarized.

2.3 *Water-Stable Lithium Metal Electrode*

At present, the only acceptable water-stable lithium-ion conducting solid electrolyte is the NASICON-type electrolyte of doped $LiTi_2(PO_4)$ (LATP). However, LATP is unstable in contact with lithium; therefore, an interlayer between lithium and LATP is required that is stable with lithium. Visco's group at Polyplus has used a thin Li_3N layer on LATP as an interlayer material that is stable in contact with lithium but unstable in water [8]. Stevens et al. of EDF have used Lipon ($Li_{3-x}(P,N)O_4$) as a protective layer, which was sputtered on the LATP plate [50]. The preparation methods for the Li_3N and Lipon interlayers are somewhat complicated and not convenient for the preparation of large size electrodes. Some groups have used a gel-type lithium-ion conducting electrolyte [51] and the organic electrolyte used for conventional lithium-ion batteries [52]; however, these electrolytes show significant lithium dendrite formation at high current density [53]. Lithium conducting PEO-based polymer electrolytes are another promising candidate for the interlayer. However, a disadvantage of the PEO based polymer electrolyte is the low lithium-ion conductivity at room temperature. Imanishi and co-workers have extensively studied the lithium conducting PEO based polymer electrolyte as the interlayer between lithium metal and LATP. The water-stable lithium electrode proposed by Imanishi et al. consists of lithium, $PEO_{18}Li(CF_3SO_2)_2N$(LiTFSI), and O-LATP [9]. Generally, the interface resistance between lithium and a polymer electrolyte is as high as several 100 Ω/cm^2 at 60 °C [54]. The addition of oxide fillers has been effective to reduce the interface resistance. For example, the stability

Table 1 Electrical conductivity and physical properties of water stable lithium-ion solid conductors

Materials	Conductivity (S cm^{-1})		Stability with Li	Stability			Reference
	25 °C	60 °C		Atmosphere	In H_2O	In sat. LiOH/LiCl	
$Li_{1+x+y}Al_x(Ti,Ge)_{2-x}Si_yP_{3-y}O_{12}$ (ohara plate)	1×10^{-4}	7.5×10^{-4}	Poor	Good	Poor	Excellent	[11]
$Li_{1.3}Al_{0.3}Ti_{1.7}(PO_4)_3$	7×10^{-4}	1.7×10^{-3}					[30]
$Li_{1.4}Al_{0.4}Ti_{1.4}Ge_{0.2}(PO_4)_3$	1.29×10^{-3}	4.3×10^{-3}	Poor	Poor	Poor	Excellent	[38]
$Li_7La_3Zr_2O_{12}^a$	2.74×10^{-4}	8.1×10^{-4}					[27]
$Li_{6,75}La_3Zr_{1.75}Ta_{0.25}O_{12}^a$	5.2×10^{-4}		Excellent	Poor	Poor	Excellent	[48]
$Li_{6.75}La_3Zr_{1.75}Nb_{0.25}O_{12}^a$	6.13×10^{-4}		Poor	Poor	Poor	Excellent	[47]
$Li_{0.29}La_{0.57}TiO_3$	5×10^{-4}		Poor			Stable in 0.5 M LiOH	[28]

[a]small amount of Al (0.01–0.5 atomic ratio in $Li_7La_3Zr_2O_3$) included

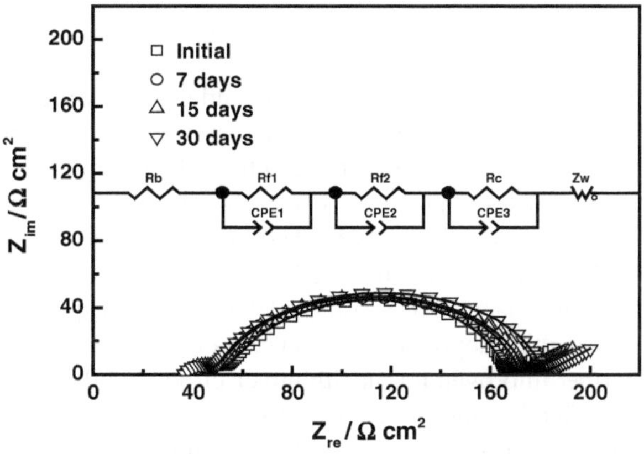

Fig. 8 Time dependence of impedance profiles for Li/PEO$_{18}$Li(CF$_3$SO$_2$)$_2$-10 wt % BaTiO$_3$/O-LATP/1 M LiCl aqueous solution/Pt at 60 °C (from Ref. [9])

of a water-stable lithium metal electrode of Li/PEO$_{18}$LiTFSI-10 wt% BaTiO$_3$/LATP in a 1 M aqueous solution of LiCl at 60 °C was examined and the change in the impedance profiles as a function of the storage period are shown in Fig. 8 [9]. The impedance profiles show a small semicircle in the high frequency range and a large semicircle in the low frequency range; the former corresponds to the grain boundary resistance of O-LATP and the polymer electrolyte, and the latter to the resistance of a passivation film, the interface resistance between the polymer electrolyte and O-LATP, and the charge transfer resistance. The cell resistance was less than 200 Ω cm^2 at 60 °C and slightly increased with the storage period.

Lithium dendrite formation at the water-stable lithium electrode is an important issue for the development of aqueous lithium-air batteries. In the early stage of lithium rechargeable batteries with aprotic electrolytes, lithium metal was mostly used as an anode. However, these batteries have not been commercialized due to lithium dendrite formation. Therefore, such aprotic electrolytes may be not suitable for the interlayer between lithium metal and LATP. However, lithium dendrite formation can be suppressed with the use of a polymer electrolyte. The growth of lithium dendrites in a Li/polymer electrolyte/Li cell has been extensively studied by Brissot, Rosso and colleagues [55, 56] using direct in situ observation technology and simultaneous cell potential evaluation. The mechanism for dendrite formation has been investigated by Chazalviel [57] and Leger et al. [58]. According to Chazalviel, a positive space charge appears in the vicinity of the negative electrode when the ionic concentration falls to zero, which indicates a local space charge near the lithium electrode and instability of the interface, such as expected with dendrite growth. This happens after a time τ_s (Sand time), which varies as the square of the current density J, according to the following equation:

$$\tau_s = \pi e^2 D[(\mu_a + \mu_c)/\mu_a]^2 C_o^2/(4J^2), \quad (3)$$

where e is the electronic charge, C_o is the initial concentration, μ_a and μ_c are the anionic and cationic mobility, respectively, and D is the ambipolar diffusion coefficient:

$$D = (\mu_a D_c + \mu_c D_a)/(\mu_a + \mu_c), \quad (4)$$

where D_c and D_a are the anionic and cationic diffusion constants, respectively. The Sand time τ_s, corresponds to the dendrite formation onset time, t_o. Brissot et al. [55] reported that the dendrite growth in Li/PEO_{20}LiTFSI corresponded to the model proposed by Chazalviel in the current density range of 0.02–0.3 mA cm^{-2}. The dendrite onset time for typical polymer electrolytes such as PEO_{18}LiTFSI is as short as 10 h at 1.0 mA cm^{-1} and 125 h at 0.1 mA cm^{-2} at 60 °C, which corresponds to specific weight capacities of 870 and 1025 mAh g^{-1}, respectively, including the weight of the 10 μm thick copper foil current collector. Thus, the specific weight capacity should be further improved for the lithium electrode in lithium-air batteries. Liu et al. [59] have examined lithium dendrite formation for a composite polymer electrolyte of PEO_{18}LiTFSI and nano-SiO_2 and/or ionic liquid of N-butyl-N-methyl pyrrolidium-TFSI (PP13TFSI). The dendrite formation onset time was extended by the addition of these materials; the onset times at 1.0 mA cm^{-2} were 10–15 h for the PEO_{18}LiTFSI-10 wt% SiO_2, 17 h for PEO_{18}LiTFSI-1.44PP13TFSI and 21 h for PEO_{18}LiTFSI-1.44PP13TFSI-10 wt% SiO_2. The specific capacity at 1.0 mA cm^{-2} for PEO_{18}LiTFSI-1.44PP13TFSI-10 wt% SiO_2 was 1458 mAh g^{-1} (including the weight of the 10 μm thick copper foil current collector), which is an attractive value for the lithium anode. However, the specific capacity should be improved at high current density. Liu et al. [59] measured τ_s as a function of the current density in the range of 0.1–1.0 mA cm^{-2} for Li/PEO_{18}LiTFSI-PP13TFSI/Li. The log τ_s versus log current density curve had good linearity in the range of 0.1 to 1.0 mA cm^{-2}, as shown in Fig. 9. The slope of 1.25 estimated from the curve is lower than that from Eq. (4). In this current range, the dendrite growth could not be explained by the Chazalviel model, which indicates that local fluctuations of current density may be an important factor, as indicated by Rosso et al. [60].

Wang et al. [61] proposed a new composite polymer electrolyte of PEO_{18}LiTFSI and tetraethylene glycol dimethyl ether (TEGDME). The lithium-ion conductivity

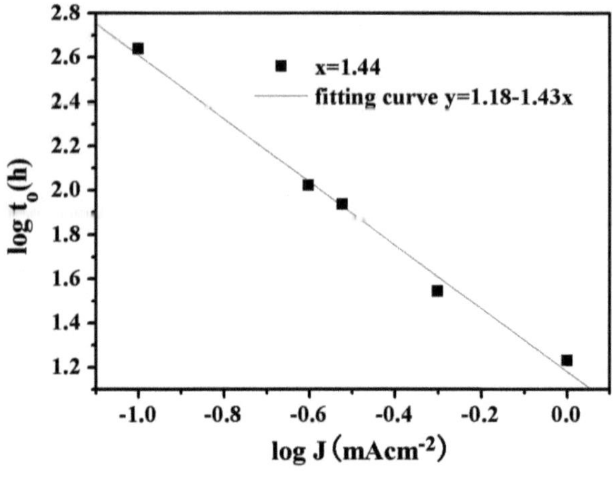

Fig. 9 Log current density (J) versus log dendrite formation onset time (t_o) curve for Li/PEO_{18}LiTFSI-1.44PP13TFSI/Li at 60 °C (from Ref. [59])

and salt diffusion coefficient of $PEO_{18}LiTFSI$ were significantly enhanced by the addition of TEGDME. The lithium-ion conductivity for $PEO_{18}LiTFSI$ of 1.24×10^{-4} S cm^{-1} and the salt diffusion coefficient of 0.36×10^{-7} cm^2 s^{-1} at 60 °C were respectively increased to 8.91×10^{-4} S cm^{-1} and 3.37×10^{-7} cm^2 s^{-1} by the addition of two moles of TEGDME. The Li/$PEO_{18}LiTFSI$-2.0 TEGDME/O-LATP/saturated LiCl aqueous solution/Pt, air cell showed a steady cell OCV of 3.48 V, which is comparable to calculated OCV. The cell resistance of 84 Ω cm^2 at 60 °C was significantly lower than that of 539 Ω cm^2 for Li/$PEO_{18}LiTFSI$/O-LATP/1 M LiCl aqueous solution/Pt, air cell [62], and also lower than that of 118 Ω cm^2 for Li/$PEO_{18}LiTFSI$-10 wt% $BaTiO_3$/O-LATP/1 M LiCl aqueous solution/Pt, air cell [32]. The lithium dendrite formation was examined using the Li/$PEO_{18}LiTFSI$-2.0 TEGDME/O-LATP/saturated LiCl aqueous solution/Pt, air cell at 60 °C and a constant current density of 1.0 mA cm^{-2}, where the thickness of the composite polymer electrolyte was 100 μm, and platinum with platinum black was used as the air and reference electrodes. The cell impedance was measured every 4 h during polarization and Fig. 10a shows the potential versus polarization time curve. The lithium electrode potential increased suddenly after 25 h polarization. This potential increase may be due to lithium deposition on O-LATP by lithium dendrite formation, which would result in the formation of a high resistance layer by the reaction of lithium and O-LATP [10]. Figure 10b shows the impedance profiles as a function of the polarization period. The resistance of the water-stable lithium

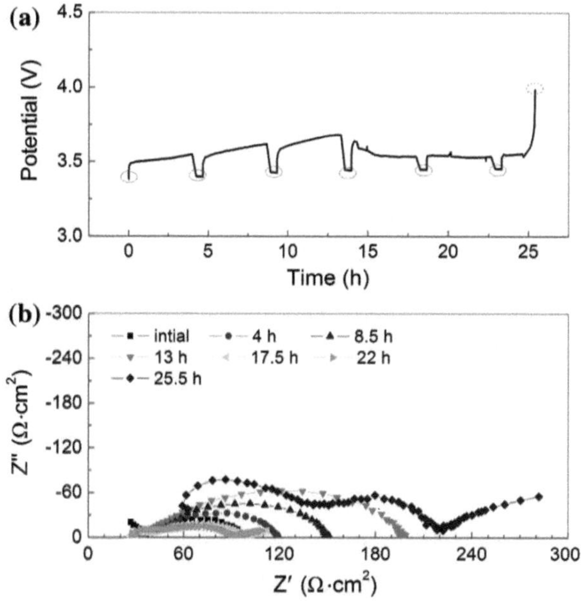

Fig. 10 **a** Charge profiles for the Li/$PEO_{18}LiTFSI$-2G4/LATP/saturated LiCl aqueous solution/Pt, air cell at 1 mA cm^{-2} and 60 °C, and **b** impedance profiles after each polarization period. The cell voltage and impedance were measured using a platinized platinum air reference electrode (from Ref. [61])

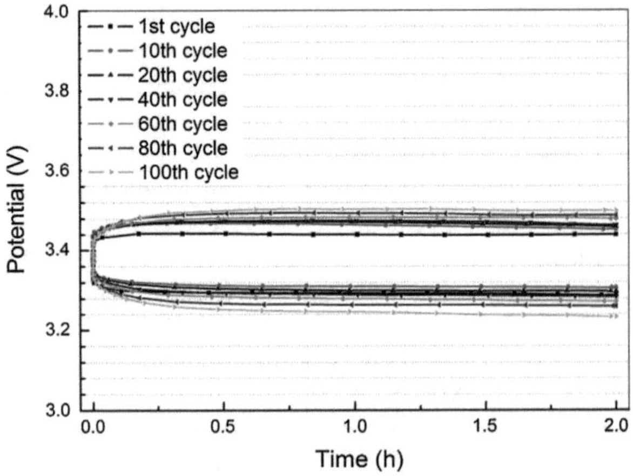

Fig. 11 Charge-discharge profiles for the Li/PEO_{18}LiTFSI-2G4(100 μm)/LATP/saturated LiCl aqueous solution/Pt, air cell at 1.0 mA cm^{-2} and 60 °C (from Ref. [61])

electrode increased gradually with the polarization period up to 13 h and then decreased with further increase of the polarization period up to 22 h. The decrease of the electrode resistance may be due to lithium dendrite formation. However, the short circuit period of 24 h is approximately 2.5 times longer than that measured for the Li/PEO_{18}LiTFSI/O-LATP/10 M LiCl-4 mM LiOH aqueous solution/Pt, air cell [63]. The cyclability of the Li/PEO_{18}LiTFSI-2.0 TEGDME/O-LATP/saturated LiCl aqueous solution/Pt, air cell for lithium deposition and stripping was measured at 60 °C and at a constant current density of 1.0 mA cm^{-2}, where the current was passed for 2 h. Figure 11 shows the voltage versus time profiles. After 100 cycles, the respective overpotentials at 1.0 mA cm^{-2} for lithium deposition and stripping increased slightly from 0.1 to 0.15 V and 0.08 to 0.13 V. The advantage of this composite electrolyte is the low and stable resistance of the solid electrolyte interface (SEI) formed between lithium and the composite electrolyte, and the overpotentials at a high current density of 4.0 mA cm^{-2} were as low as 0.4 V for lithium deposition and 0.3 V for lithium stripping [61]. However, lithium dendrite formation at high current densities should be suppressed further.

3 Cell Performance of Aqueous Lithium-Air Cells

The cell reaction for Li/H_2O/air is:

$$4Li + 2H_2O + O_2 = 4LiOH. \tag{5}$$

Self-discharge is high at OCV and low current density discharge due to the corrosion reaction of:

$$2Li + 2H_2O = 2\,LiOH + H_2. \tag{6}$$

The corrosion reaction in 3–5 M LiOH can be vertically eliminated if the lithium electrode is anodically polarized by approximately 300 mV [64]. The current efficiency is governed by the ratio of the two competing reactions. If lithium is in contact with a lithium conducting electrolyte that is stable at a potential lower than −3.040 V versus NHE, then water is not decomposed. The water-stable NSAICON-type lithium-ion conducting solid electrolyte of LATP is used for aqueous lithium-air batteries [8, 9]. However, this solid electrolyte is unstable in contact with lithium metal; therefore, a lithium conducting interlayer should be used to protect from the direct contact of LATP with lithium metal. Solid electrolytes such as Li_3N [8, 65] and Lipon [50], the liquid electrolyte of $LiClO_4$ in ethylene carbonate (EC)/dimethyl carbonate (DEC) [52] and PEO-based polymer electrolyte [9] can be used as the interlayer. The lithium-ion conductivity of Li_3N is as high as 10^{-3} S cm^{-1} at room temperature [20]; however, the preparation of Li_3N thin film on LATP is somewhat complicated. The conductivity of Lipon is too low at room temperature; therefore, an extremely thin film prepared by sputtering [50] could be used. The polymer electrolyte also exhibits poor conductivity at room temperature, so that a cell with the polymer electrolyte interlayer would have to be operated at above 50 °C. Lastly, for use of the liquid electrolyte, the suppression of lithium dendrite formation during lithium deposition should be improved.

Figure 12 shows a schematic diagram of the laboratory test cell presented by Visco et al. of Polyplus, USA, and the measured discharge performance [65], where LATP (50 μm thick) was used for the water-stable lithium-ion solid conductor and Li_3N for the interlayer between lithium and LATP. The lithium-ion conductivity of LATP was 2×10^{-4} S cm^{-1}. The discharge performance at high current densities was considerably better than that for the non-aqueous lithium air cells [4, 6]. However, the charging performance for this cell was not reported. PolyPlus reported the short range cyclic performance of a beaker-type lithium-air cell [8], where excellent cycling performance with low overpotentials was observed in the current density range of 1–10 mA cm^{-2}. An advantage of the aqueous system is that the reaction product of LiOH is soluble in the aqueous electrolyte, while the reaction product of Li_2O_2 is insoluble in the non-aqueous electrolyte.

The solubility of LiOH is 5.3 M L^{-1} and saturation of LiOH in the electrolyte is reached at ca. 5 % discharge depth. Stevens et al. of EDF, France, [50] proposed the use of an anion exchange membrane. The anion exchange membrane prevents the reaction of LiOH and CO_2 in air. They also proposed a positive bi-electrode, which consists of an air electrode for oxygen reduction with a secondary oxygen evolution electrode. The bi-electrode prevents degradation of the air electrode from carbon oxidation. Carbon oxidation has been observed for both the non-aqueous lithium-oxygen cell [19] and the aqueous lithium-air cell [66]. A schematic diagram of the cell proposed by Stevens et al. [50] is shown in Fig. 13. A protective layer of Lipon (0.5–2 μm thick) was deposited on one side of O-LATP by sputtering. The composite air electrode was prepared using an air electrode supplied by

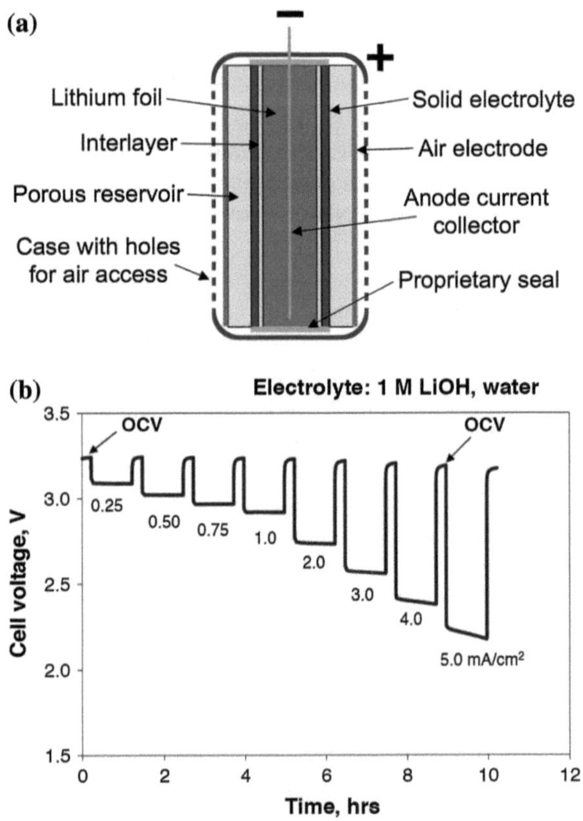

Fig. 12 **a** Schematic diagram of Polyplus aqueous lithium-air laboratory cell and **b** discharge performance as a function of discharge rate (from Ref. [65])

Electric Fuel Ltd. which was modified by the inclusion of an anionic polymeric membrane. The polymeric membrane was composed of interpenetrated network of two polymers. A polycationic cross-linked polyepichlorohydrin was used as the ionic network and poly(hydroxyl ethyl methacrylate) was used as the structural polymer to provide mechanical stability and reduced swelling. The cyclic performance of the cell using a saturated aqueous solution of LiOH and untreated ambient air is shown in Fig. 13b. Relatively high capacities were obtained; however, on cycling the lithium metal formed a porous or columnar deposit that increased in volume and caused a loss of contact between the lithium metal and O-LATP [50]. The lifetime of this composite air electrode when used with untreated air in 5 M or saturated LiOH aqueous solution was increased from 10 h without the anion exchange membrane to 1000 h.

Inaguma et al. of Gakushuin University, Japan, have reported an aqueous lithium-oxygen cell with a perovskite-type lithium conducting solid electrolyte of $La_{0.75}Li_{0.29}TiO_3$ (LLTO), where a carbon tube closed at one end (thickness of 2 mm) was used as an electrolyte separator [67]. A schematic diagram of the test cell is shown in Fig. 14a. A conventional aprotic electrolyte of 1.0 M $LiClO_4$ in EC/DEC

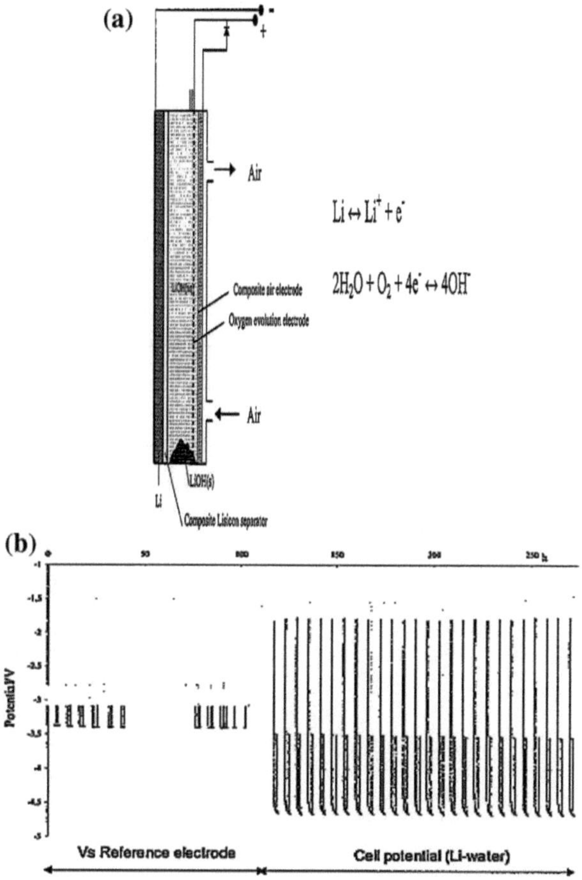

Fig. 13 **a** Schematic diagram of the EDF aqueous lithium-air cell and **b** cycling performance at 2 mA cm^{-2} and 20 °C (from Ref. [50])

was used for the interlayer between lithium metal and LLTO, and a 0.5 M LiOH aqueous solution was used as the electrolyte. Figure 14b shows the cyclic performance of the cell at 1 mA and room temperature. The current density calculated from the surface area of the porous carbon tube air electrode was 0.05 mA cm^{-2}. The high overpotential for the oxygen evolution reaction (OER) (charging process) may be attributed to the use of the porous carbon tube without a catalyst.

Imanishi and co-workers of Mie University, Japan, [68] have examined the charge discharge performance of an aqueous lithium-air cell with O-LATP and saturated LiOH with 10 M LiCl aqueous solution with a third electrode for the OER, as proposed by Stevens [50]. Rechargeable lithium-air batteries require a bi-functional air electrode for the oxygen reduction reaction (ORR) and the OER. When an air electrode is used for the OER, it is exposed to a highly corrosive potential. Arai et al. [69] examined a high surface area carbon black (Ketjen black; KB) electrode for the ORR and OER in aqueous 8 M KOH and observed electrode deterioration during the OER, which they claimed was due to a loss of the electrochemically active surface

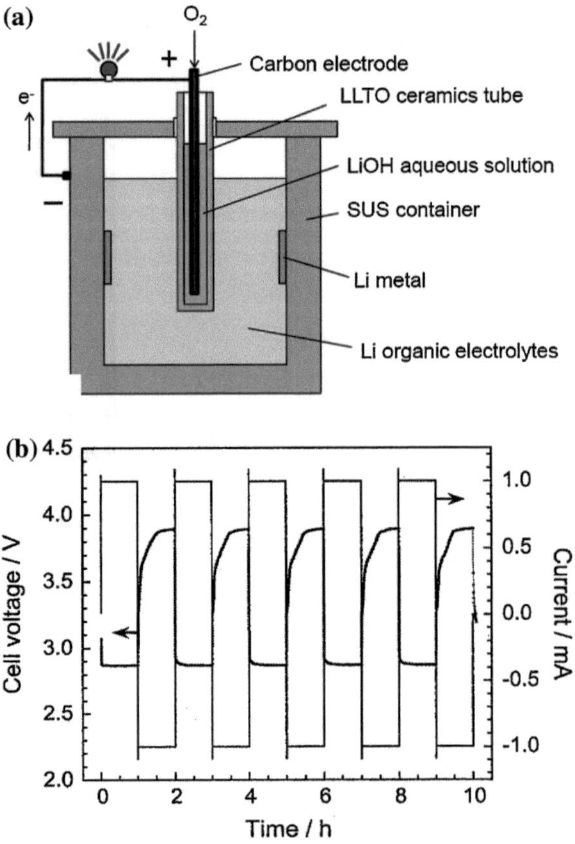

Fig. 14 **a** Schematic diagram of the Gakushuin aqueous lithium-air cell with LLTO and **b** cycling performance at 1 mA (0.05 mA cm^{-1}) and room temperature (from Ref. [67])

area of the electrode, mainly due to carbon corrosion. Ohkuma et al. [66] reported the electrode performance for the ORR and OER on a KB air electrode for an aqueous lithium-air cell, where an aqueous solution with saturated LiOH and 10 M LiCl was used as the electrolyte. Steady electrode performance for the ORR was observed on the high surface area carbon black without a catalyst at 2.0 mA cm^{-2} and at room temperature. The overpotential for the OER was slightly higher than that for the ORR and gradually increased with the polarization period. Analysis of the gas in the cell after polarization above 0.4 V versus NHE revealed the evolution of a small amount of CO during the OER by the decomposition of carbon in the electrode. The decomposition of carbon during the OER was also observed in the non aqueous lithium-air cell [19]. Thus, suppression of carbon decomposition during the charging process is one of the most important challenges to develop an acceptable rechargeable lithium-air battery for EVs. Figure 15 shows a schematic diagram of a test cell with an auxiliary electrode for the OER, which was prepared by pressing oxide powder onto a Ti mesh. A conventional liquid electrolyte of $LiClO_4$ in EC-DEC was used as the interlayer between lithium and O-LATP, and the electrolyte was an aqueous solution of saturated LiOH with 10 M LiCl. As shown in the reaction (2),

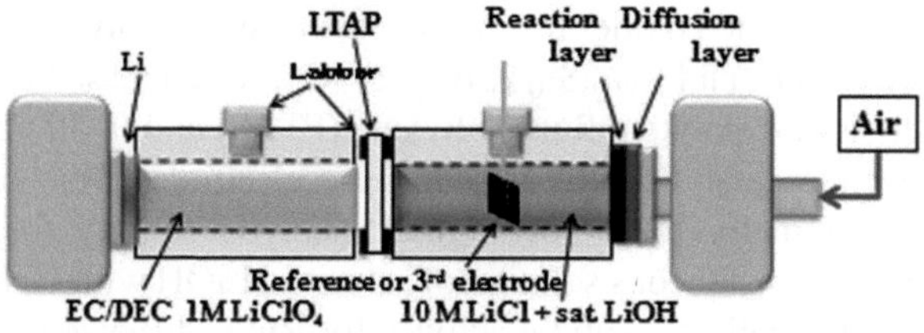

Fig. 15 Schematic diagram of the Mie University test cell (from Ref. [68])

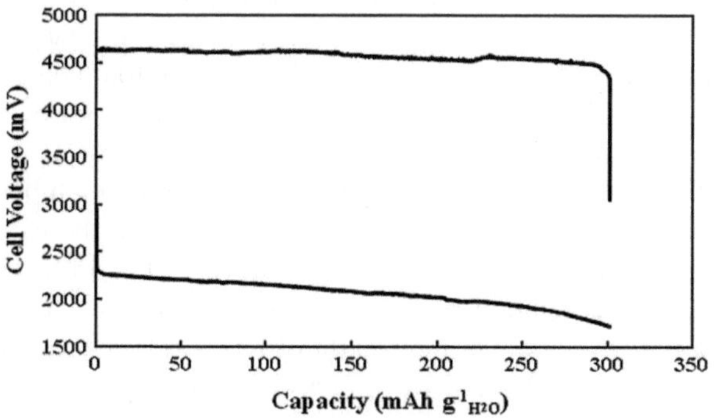

Fig. 16 Charge-discharge curves for the Li/1 M $LiClO_4$ in EC-DEC/O-LATP/saturated LiOH with 10 M LiCl/KB, air cell using a Pt third electrode for the OER at 0.64 mA cm^{-2} and 25 °C (from Ref. [68])

water is consumed with the cell reaction and the reaction product of $LiOH{\cdot}H_2O$ is deposited on the air electrode and/or in the electrolyte. High utility of the active material (water) is required to obtain a high specific energy density. Figure 16 shows charge and discharge curves for the Li/1 M $LiClO_4$ in EC-DEC/O-LATP/saturated LiOH with10 M LiCl aqueous solution/KB, air cell at 25 °C where platinum and platinum black as the 3rd electrode was used and the current densities for discharge and charge were 0.88 and 1.0 mA cm^{-2}, respectively. The loading of water and KB were 0.38 and 0.0255 g, respectively. The cell was successfully charged and discharged up to 300 mAh g^{-1} of water, which corresponds to a 30 % discharge depth for the capacity of the loaded water, and the air electrode capacity per gram of KB corresponds to 3600 mAh. The utility of water for the cell reaction is slightly lower than that for a Cd/NiOOH rechargeable cell, in which concentrated KOH aqueous solution is used as the electrolyte and water is involved in the cell reaction and the utility of water is 30–50 %. The energy density of aqueous lithium-air batteries is dependent on the utility of water; therefore, the water utility should be improved for this type cell. The charge voltage was as high as 4.5 V at 1.0 mA cm^{-2}. This high charging voltage may be due to the poor catalytic activity of the platinum/platinum black electrode for the OER at room temperature in the aqueous solution of saturated LiOH with 10 M LiCl. Metal electrodes such as Ag and Ni, which exhibit excellent catalytic activity for the OER in alkaline solution [70, 71], were unstable in the solution with LiCl. Selected oxides that are stable in the electrolyte with LiCl, such as MnO_2, RuO_2, $NiCo_2O_2$, and $La_{0.6}Sr_{0.4}Co_{0.2}Fe_{0.8}O_3$, were pressed onto the Ti mesh (100 mesh) with 4 wt% polytetrafluoroethylene (PTFE) and 0–50 vol.% Ti powder (15–30 μm), and the catalytic activities of these electrodes were compared using linear sweep voltammetry [68]. The highest catalytic activity for the OER was observed for RuO_2 in an aqueous solution of saturated LiOH with 10 M LiCl. Yilmaz et al. [72] reported excellent catalytic activity for RuO_2 dispersed on carbon nanotubes for the OER in non-aqueous lithium-oxygen cells.

The long-term cycling performance of the Li/1 M $LiClO_3$ in EC-DEC/O-LATP/saturated LiOH with 10 M LiCl/KB, air cell using a RuO_2 3rd electrode for the OER was tested at 0.64 mA cm^{-2} and 25 °C. A steady discharge cell voltage of 2.5 V and charge cell voltage of 3.9 V were observed for four charge and discharge cycles. At the 5th cycle, the discharge voltage decreased significantly to 1.57 V after discharge for 7 mAh, where the total operation period was 75 h. The cyclic performance of the cell was improved by passing air through a tube with soda lime to remove CO_2. The degradation in cell performance could be explained by contamination with CO_2 in the air, as observed by Stevens et al. [50]. The test cell was operated under pure oxygen flow and the cycling performance at 0.64 mA cm^{-2} and 25 °C is shown in Fig. 17, where the weight of KB was 0.0040 g and excess water loading. The cell was discharged to the specific capacity of the air electrode at 2000 mAh g^{-1} of the air electrode. Good cyclic performance was observed until the 8th cycle and then the discharge cell voltage was gradually decreased with cycling. The degradation of the discharge performance may be due to poor cyclability of the lithium electrode. The formation of many fine lithium particles was observed in the 1 M $LiClO_4$/EC-DEC electrolyte, which suggests lithium dendrite formation in the liquid electrolyte [73]. The polymer electrolyte is a good candidate for the interlayer between lithium metal and LATP, but the formation of lithium dendrites at high current density should be further suppressed, as discussed in the previous section of this chapter. To improve the cycling performance of the aqueous lithium-air cell,

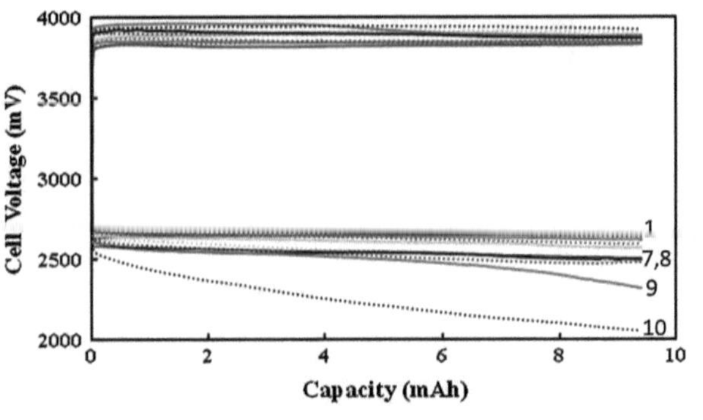

Fig. 17 Cycle performance for the Li/1 M $LiClO_4$ in EC-DEC/O-LATP/saturated LiOH with 10 M LiCl/KB, O_2 cell using a RuO_2 third electrode for the OER at 0.64 mA cm^{-2} and 25 °C (from Ref. [68])

CO_2 should also be removed from the air and a lithium dendrite formation-free interlayer or an alternative lithium electrode, such as $Li_{4.4}Si$ [74], should be developed.

4 Concluding Remarks

The aqueous lithium-air system is expected to be developed for a rechargeable battery with a high specific energy density because of its high theoretical energy density of 1910 Wh kg^{-1}, which is four times higher than that of conventional lithium-ion batteries. Furthermore, the aqueous lithium-air system has the potential to be charged and discharged at a high current density with a low overpotential because the reaction product is soluble in the electrolyte. However, many materials and technical problems must be resolved to develop practical batteries with the expected high performance. The lithium dendrite formation-free water-stable lithium electrode is a key component for aqueous lithium-air batteries, in which the interlayer between lithium metal and a solid lithium-ion conducting electrolyte is the most challenging research target. We are waiting to discover an appropriate water- and lithium-stable lithium-ion conducting solid electrolyte, such as the garnet-type solid electrolyte. A thin solid electrolyte with high lithium-ion conductivity and excellent mechanical properties should be developed to obtain a high energy density battery; the target thickness is less than 100 μm. A carbon-free air electrode or stable carbon for the OER are also important considerations for a long-life cell. At present, the use of the 3rd electrode for the OER is a possible solution to prevent degradation of the air electrode with cycling.

The specific energy density from the mass of lithium, oxygen, carbon and water is calculated to be approximately 1000 Wh kg^{-1} for 50 % water utility and 700 Wh kg^{-1} for 30 % water utility based on the discharge voltage of 2.8 V and air electrode capacity of 2000 mA g^{-1} of carbon. The calculated energy density is significantly dependent on water utilization. We should design a cell with a high water utility.

References

1. Girishkumar G, McCloskey B, Luntz AC, Swanson S, Wiche W (2010) Lithium-air battery: promise and challenges. J Phys Chem Lett 1:2193
2. Armand M, Tarascon M (2008) Building better batteries. Nature 451:652
3. Padbury R, Zjhang X (2011) Lithium-oxygen batteries—limiting factors that affect performance. J Power Sources 196:4436
4. Bruce PG, Freunberger SA, Hardwick LJ, Trascon M (2012) Li-O_2 and Li-S batteries with high energy storage. Nat Mater 11:19
5. Wang J, Yongliang L, Xueliang S (2013) Challenges and opportunities of nanostructured materials for aprotic rechargeable lithium-air batteries. Nano Energy 2:433

6. Imanishi N, Yamamoto O (2014) Rechargeable lithium-air batteries: characteristics and prospects. Mater Today 17:24
7. Kuwabi DG, Ortiz-Vitoriano N, Freunberger SA, Chen Y, Imanishi N, Bruce PG, Shao-Hon Y (2014) Materials challenges in rechargeable lithium-air batteries. MRS Bull 39:443
8. Visco SJ, Nimon E, Katz B, De Jonghe LC, Chu M-Y (2004) Lithium metal aqueous batteries. 12th International meeting on lithium batteries. Abstract # 53, Nara, Japan
9. Zhang T, Imanishi N, Hasegawa S, Hirano A, Xie J, Takeda Y, Yamamoto O, Sammes N (2009) Water-stable lithium anode with the three-layer construction for aqueous lithium-air secondary batteries. Electrochem Solid State Lett 12:A132
10. Imanishi N, Hasegawa S, Zhang T, Hirano A, Takeda Y, Yamamoto O (2008) Lithium anode for lithium-air batteries. J Power Sources 185:1392
11. Hasegawa S, Imanishi N, Zhang T, Xie J, Hirano A, Takeda Y, Yamamoto O (2009) Study on lithium/air secondary batteries-Stability of NASICON-type lithium ion conducting glass-ceramics with water. J Power Sources 189:371
12. Shimonishi Y, Zhang T, Imanishi N, Im D, Lee D-J, Hirano A, Takdeda Y, Yamamoto O, Sammes N (2011) A study on lithium/air batteries-Stability of the NASICON-type lithium ion conducting solid electrolyte in alkaline aqueous solutions. J Power Sources 196:5128
13. Imanishi N, Yamamoto O (2014) Polymer electrolyte for lithium-air batteries. In: Scrosati B, Abraham KM, Schalkwijk W, Hassoun J (eds) Lithium batteries: advanced technology and applications. Wiley and ECS, p 217
14. Kuboki T, Okuyama T, Ohsaki T, Takami N (2005) Lithium-air batteries using hydrophobic room temperature ionic liquid electrolyte. J Power Sources 146:766
15. Mizuno F, Nakanishi S, Kotani Y, Yokoishi S, Iba H (2010) Rechargeable Li-air batteries with carbonate-based liquid electrolyte. Electrochemistry 78:403
16. Freungberg SA, Chen Y, Pemg Z, Griffin JM, Hardwick LJ, Bard F, Novik P, Bruce PG (2011) Reaction in the rechargeable lithium-O_2 battery with alkyl carbonate electrolyte. J Am Chem Soc 133:8040
17. McCloskey BD, Bethune DS, Shelby RM, Girishkumar G, Luntz AC (2011) Solvents' critical role in nonaqueous lithium-oxygen battery electrochemistry. J Phys Chem Lett 2:1161
18. Ogasswara T, Debart A, Holzapfel M, Novak P, Bruce PG (2006) Rechargeable Li_2O_2 electrode for lithium batteries. J Am Chem Soc 128:1390
19. Thotiyl MMO, Freunberger SA, Peng Z, Bruce PG (2013) The carbon electrode in nonaqueous Li-O_2 cells. J Am Chem Soc 135:494
20. Zhang J, Xu W, Liu W (2010) Oxygen selective immobilized liquid membranes for operation of lithium-air batteries in ambient air. J Power Sources 195:7438
21. Alen UA, Rabenau A, Talat GH (1977) Ionic conductivity in Li_3N single crystal. Appl Phys Lett 30:631
22. Mizuno F, Hayashi A, Tadanaga K, Tatsumisago M (2006) New high ion conductive crystal precipitated Li_2S-P_2S_5 glass. Adv Mater 17:918
23. Hong HY-P (1978) Crystal structure and ionic conductivity of $Li_{14}Zn\ (GeO_4)_4$ and other Li^+ super ion conductor. Mat Res Bull 13:117
24. Kamaya N, Honma K, Yamakawa Y, Nakamura T, Hirayama M, Kanno R, Yomemura M, Kamiyama T, Kudo Y, Hama S, Kawamoto K, Mitsui A (2011) A lithium superior conductor. Nat Mater 10:686
25. Aono H, Sugimoto E, Sadaoka Y, Imanaka N, Adachi G (1989) Ionic conductivity of the lithium titanium phosphate ($Li_{1+x}M_xTi_{2-x}(PO_4)_3$, M = Al, Sc, Y and La) system. J Electrochem Soc 136:590
26. Inaguma Y, Chen I, Ito M, Nakamura T, Uchida T, Wakihar M (1993) High ionic conductivity in lithium lanthanum titanate. Solid State Commun 86:689
27. Murugan R, Tangadurai V, Weppner W (2007) Fast lithium ion conduction in garnet-type $Li_7La_3Zr_2O_{12}$. Angew Chem Int Ed 48:7778
28. Inaguma Y, Nakashima M (2013) A rechargeable lithium-air battery using a lithium ion-conducting lanthanum lithium tiatanate ceramics as an electrolyte separator. J Power Sources 229:250

29. Goodenough JB, Hong HYR, Kafalas JA (1976) Fast Na^+ ion transport in skeleton structure. Mater Res Bull 11:203
30. Aono H, Sugimoto E, Sadaoka Y, Imanaka N, Adachi G (1990) Ionic conductivity of solid electrolyte based on lithium titanium phosphate. J Electrochem Soc 137:1023
31. Fu J (1997) Superionic conductivity of glass-ceramics in the system $Li_2O–Al_2O_3–TiO_2–P_2O_5$. Solid State Ionics 96:195
32. Zhang T, Imanishi N, Takeda Y, Yamamoto O (2011) Aqueous lithium/air rechargeable batteries. Chem Lett 40:668
33. Thokchom JS, Kummer R (2008) The effect of crystallization parameter on the ionic conductivity of a lithium aluminum germanium phosphate glass-ceramic. J Power Sources 185:191
34. Fu J (1997) Fast Li^+ ion conducting glass-ceramics in the system $Li_2O–Al_2O_3–GeO_2–P_2O_5$. Solid State Ionics 104:191
35. Xu X, Wen Z, Wu X, Yang X, Gu Z (2007) Lithium-ion-conducting glass-ceramics of $Li_{1.5}Al_{0.5}Ge_{1.5}(PO_4)_3–xLi_2O$ (x = 0.0–0.2) with good electrical and electrochemical properties. J Am Ceram Soc 90:2802
36. Maldonado P, Losilla ER, Lara MM, Arand MAG, Brugue S, Mouahid FE, Zahir M (2003) High lithium ionic conductivity in the $Li_{1+x}Al_xGe_yTi_{2-x-y}(PO_4)_3$ NASICON series. Chem Mater 15:1870
37. Xu X, Wen Z, Gu Z, Xu X, Lin Z (2004) Lithium ion conductivity glass ceramics in the system $Li_{1.4}Al_{0.4}(Ge_{1-x}Ti_x)_{1.6}(PO_4)_3$ (x = 0–1.0). Solid State Ionics 171:207
38. Zhang P, Matsui M, Hirano A, Takeda Y, Yamamoto O, Imanishi N (2013) Water-stable lithium ion conducting solid electrolyte of the $Li_{1.4}Al_{0.4}Ti_{1.6-x}Ge_x(PO_4)_3$ system (x = 0–1.0) with NASICON-type structure. Solid State Ionics 253:175
39. Galven C, Fourquet J-L, Crosnier-Lopez M-P, Le Berre F (2011) Instability of the lithium garnet $Li_7La_3Sn_2O_{12}$; Li^+/H^+ exchange and structural study. Chem Mater 23:1892
40. Geiger CA, Alekseev E, Lazic B, Fisch M, Armbuster T, Langner R, Fechtalkor M, Kim N, Pettkea T, Weppner W (2011) Crystal chemistry and stability of "$Li_7La_3Zr_2O_{12}$" garnet: a first ion conductor. Inorg Chem 50:1089
41. Larraz G, Orera A, Sanjuan ML (2013) Cubic phases of garnet-type $Li_7La_3Zr_2O_{12}$. J Mater Chem A 1:11419
42. Ohta S, Kobayashi T, Asaoka T (2011) High lithium ionic conductivity in the garnet-type oxide $Li_{1-x}La_3(Zr_{2-x}Nb_x)O_{12}$ (x = 0–2). J Power Sources 196:3342
43. Ragasamy E, Wolfenstien J, Sakamoto J (2011) The role of Al and Li concentration on the formation of cubic garnet solid electrolyte of nominal composition $Li_7La_3Zr_2O_{12}$. Solid State Ionics 206:28
44. Li Y, Han T, Wang C-A, Xie H, Goodenough JB (2012) Optimizing Li^+ conductivity in a garnet framework. J Mater Chem 22:15357
45. Shimonishi Y, Toda A, Zhang T, Hirano A, Imanishi N, Yamamoto O, Takeda Y (2011) Synthesis of garnet-type $Li_{7-x}La_3Zr_2O_{12-0.5x}$ and its stability in aqueous solutions. J Power Sources 183:48
46. Awaka J, Kijima N, Hayakawa H, Akimoto J (2009) Synthesis and structural analysis of tetragonal $Li_7La_3Zr_2O_{12}$ with the garnet-related type structure. J Solid State Chem 182:2040
47. Ishiguro K, Nakata Y, Matsui M, Uechi I, Takeda Y, Yamamoto O, Imanishi N (2013) Stability of Nb-doped cubic $Li_7La_3Zr_2O_{12}$ with lithium metal. J Electrochem Soc 160:A1690
48. Ishiguro K, Nemori H, Sumahiro S, Nakata Y, Sudo R, Matsui M, Takeda Y, Yamamoto O, Imanishi N (2014) Ta-doped $Li_7La_3Zr_2O_{12}$ for water-stable lithium electrode of lithium-air batteries. J Electrochem Soc 161:A668
49. Sudo R, Nakata Y, Ishiguro K, Matsui M, Hirano A, Takeda Y, Yamamoto O, Imanishi N (2014) Interlayer behavior between garnet-type lithium-conducting solid electrolyte and lithium metal. Solid State Ionics 262:151
50. Stevens P, Toussaint G, Caillon G, Viaud P, Vinatier P, Cantau C, Ficher O, Sarrazin C, Mallouki M (2010) Development of a lithium air rechargeable battery. ECS Trans 28:1

51. Wang X, Hou Y, Zhu Y, Wu Y, Halze R (2013) An aqueous rechargeable lithium battery using coated Li meal as anode. Sci Rep 3:1401 (2013)
52. Wang Y, Zhou H (2010) A lithium-air battery with a potential to continuously reduce by O_2 from air for delivering energy. J Power Sources 195:358
53. Park HE, Hong CH, Yoon WY (2008) The effect of internal resistance on dendrite growth on lithium metal electrode in lithium secondary batteries. J Power Sources 178:765
54. Barghini MC, Mastragostino M, Passerini S, Scrosati B (1996) Electrochemical properties of polyethylene oxide-Li[$(CF_3SO_2)_2N$]-gamma-$LiAlO_2$ composite polymer electrolyte. J Electrochem Soc 142:2118
55. Brissot C, Rosso M, Chazalviel J-N, Lascude S (1998) In situ study of dendritic growth in lithium/PEO-salt/lithium cells. Electrochim Acta 43:1569
56. Rosso M, Brissot C, Ladcand S, Dolle M, Sammier L, Tarascon J-M, Bouchet R, Lascand S (2006) Dendrite short-circuit and fuse effect on Li/polymer/Li cells. Electrochem. Acta 51:5334
57. Chazalviel J-N (1990) Electrochemical aspects of the generation of ramified metallic electrodeposition. Phys Rev A 42:7355
58. Leger C, Elezgaray J, Argoul F (1998) Dynamical characterization of one-dimensional stationary growth regimes in diffusion-limited electrodeposition processes. Phys Rev E.58:7700
59. Liu S, Imanishi N, Zhang T, Hirano A, Takeda Y, Yamamoto O, Yang J (2010) Lithium dendrite formation in Li/polyethylene oxide-lithium bis(trifluoromethanesulfonyl)imide and N-methy-N-propylpiperidinum bis(trifluoromethanesulfonyl)imide/Li cells. J Electrochem Soc 157:A1092
60. Rosso M, Gobron T, Brissot C, Chzalviel J-N, Lascaud S (2001) Onset of dendritic growth in lithium polymer cells. J Power Sources 97–98:804
61. Wang H, Matsui M, Takeda Y, Yamamoto O, Im D, Lee DJ, Imanishi N (2013) Interface properties between lithium metal and a composite polymer electrolyte of $PEO_{18}Li(CF_3SO_2)_2N$-tetraethylene glycol dimethyl ether. Membrane 3:298
62. Zhang T, Imanishi N, Hasegawa S, Hirano A, Xie J, Takeda Y, Yamamoto O, Sammes N (2008) Li/polymer electrolyte/water-stable lithium-conducting glass ceramics for lithium-air secondary batteries with an aqueous electrolyte. J Electrochem Soc 155:A965
63. Wang H, Im D, Lee DJ, Matsui M, Takdeda Y, Yamamoto O, Imanishi N (2013) A composite polymer electrolyte protect layer between lithium and water stable ceramics for aqueous lithium-air batteries. J Electrochem Soc 160:A728
64. Littouer EL, Tsai KC (1977) Corrosion of lithium in alkaline solution. J Electrochem Soc 124:850
65. Visco SJ, Nimon N, Katz B, Jonghe LCD, Chu MY (2006) The development of high energy density lithium/air and lithium/water batteries with no self-discharge. The 210th Electrochemical Society meeting, Cancun, Mexico, Abstract # 389, 2006
66. Ohkuma H, Uechi I, Matsui M, Takeda Y, Yamamoto O, Imanishi N (2014) Stability of carbon electrodes for aqueous lithium-air secondary batteries. J Power Sources 245:942
67. Inaguma Y, Nakashima M (2013) A rechargeable lithium-air battery using a lithium ion-conducting lanthanum lithium titanate ceramics as a electrolyte separator. J Power Sources 228:250
68. Sunahiro S, Matsui M, Takeda Y, Yamamoto O, Imanishi N (2014) Rechargeable aqueous lithium-air batteries with an auxiliary electrode for the oxygen evolution. J Power Sources 262:338
69. Arai H, Muller S, Haas O (2000) AC impedance analysis of bifunctional air electrode for metal air batteries. J Electrochem Soc 146:3584
70. Bronoel G, Reby J (1980) Mechanism of oxygen evolution in basic medium at a nickel electrode. Electrochem Acta 23:973
71. Carlsson L, Ojefors L (1980) Bifunctional air electrode for metal-air batteries. J Electrochem Soc 127:525

72. Yilmaz E, Yogi C, Yamanaka K, Ohta T, Byon HR (2013) Promoting formation of noncrystal Li_2O_2 in the Li-O_2 battery with RuO_2 nano particles. Nano Lett 13:4679
73. Peled E (1979) The electrochemical behavior of alkaline and alkaline earth metal in nonaqueous battery systems-the solid electrolyte interphase model. J Electrochem Soc 133:2047
74. Teranishi R, Mizukoshi F, Si Q, Kawakubo M, Matusi M, Takeda Y, Yamamoto O, Imanishi N (2015) Silicon anode for rechargeable lithium-air batteries. J Power Sources 273:538

Lithium-Sulfur Battery

Shuli Li and Zhan Lin

1 Introduction

Energy storage systems have been considered as an effective way to conquer global warming by replacing the traditional fossil fuels with low-carbon energies [1, 2]. Rechargeable (or secondary) batteries such as nickel-metal hydride (Ni-MH), lead acid (Pd-acid), lithium-ion (Li-ion), lithium-air (Li-air) and lithium-sulfur (Li-S) batteries are reversible and highly durable energy storage devices, and they can electrochemically store electric energy and deliver stored chemical energy back into electricity repeatedly [3]. Among them, Li-ion batteries are the most popular rechargeable batteries, since they can supply high specific energy and power densities with long cycle life, fast charging ability, and no memory effect [4–6]. However, Li-ion batteries are far from satisfaction in terms of energy density when they are expanding their traditional applications from portable electronics to large-scale emerging applications such as electric vehicles and large-scale grids [7, 8]. Li-S batteries offer the theoretical specific energy five times higher than that of Li-ion batteries (2600 vs. 500 Wh kg^{-1}), and they have been considered as one of alternatives for large-scale emerging applications [9, 10]. Moreover, sulfur is abundant in nature with the advantages of low-cost, good safety, and environmental friendliness [11].

A typical Li-S cell consists of a lithium metal anode and a sulfur-carbon composite cathode with an organic liquid electrolyte in between, which is shown in Fig. 1. The Li-S chemistry is based on the following conversion reaction:

$$16\text{Li} + \text{S}_8 \leftrightarrow 8\text{Li}_2\text{S} \quad E^o = 2.20 \text{ V vs. Li/Li}^+ \tag{1}$$

S. Li · Z. Lin (✉)
Key Laboratory of Biomass Chemical Engineering of Ministry of Education, College of Chemical and Biological Engineering, Zhejiang University, Hangzhou 310027, People's Republic of China
e-mail: zhanlin@zju.edu.cn

Z. Zhang and S.S. Zhang (eds.), *Rechargeable Batteries*,
Green Energy and Technology, DOI 10.1007/978-3-319-15458-9_21

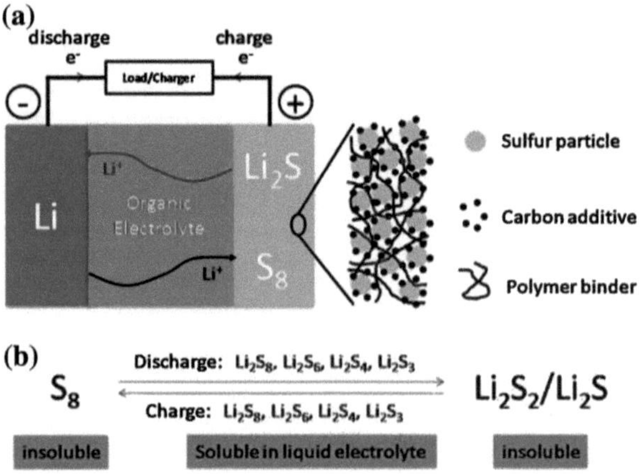

Fig. 1 **a** Charge (*red*)/discharge (*black*) process involved in a typical Li-S cell. **b** Charge/discharge process involving the formation of soluble lithium polysulfides (Li_2S_x, x = 2–8) and insoluble Li_2S_2/Li_2S [11]. Reprinted with permission; © 2013 American Chemical Society

Assuming complete conversion, the sulfur cathode has a theoretical specific capacity of 1675 mAh g^{-1}; and the whole Li-S cell demonstrates a theoretical energy of 2500 Wh kg^{-1} or 2800 Wh L^{-1}.

A typical discharge profile in a Li-S cell is shown in Fig. 2 [12, 13]. During discharge, sulfur is reduced by a two-electron reduction process to form polysulfide intermediates (Li_2S_x, x = 2–8) until lithium sulfide (Li_2S) is formed at the end. Two voltage plateaus at 2.4 and 2.0 V, which correspond to the reduction of long chain lithium polysulfides (Li_2S_x, 4 ≤ x) and short chain lithium polysulfides (Li_2S_x,

Fig. 2 A typical discharge profile in a traditional Li-S cell [15]. Reprinted with permission; © 2013 Nature Publishing Group

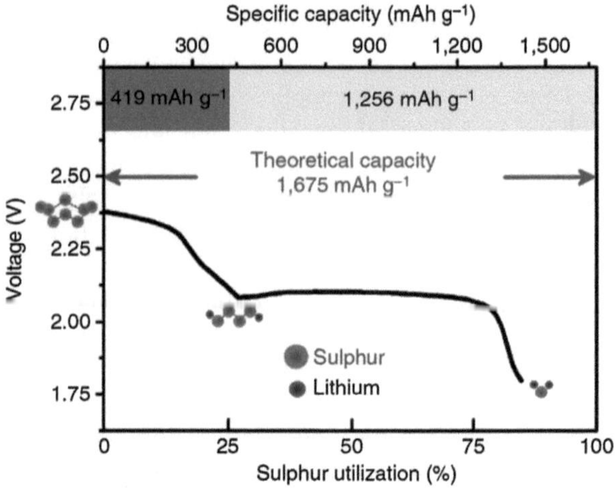

$x \leq 4$), are observed. The detailed discharge mechanism always goes through two stages [7, 10, 14]:

Stage I: Elemental sulfur reacts with Li to form long chain soluble Li_2S_x ($4 \leq x$).

$$S_8 + 2Li^+ + 2e^- \rightarrow Li_2S_8 \tag{2}$$

$$Li_2S_8 + 2Li^+ + 2e^- \rightarrow 2Li_2S_4 \tag{3}$$

Stage II: The Li_2S_4 was further reduced to form insoluble Li_2S_2 and Li_2S.

$$Li_2S_4 + 2Li^+ + 2e^- \rightarrow 2Li_2S_2 \tag{4}$$

$$Li_2S_2 + 2Li^+ + 2e^- \rightarrow 2Li_2S \tag{5}$$

Although the study on the Li-S batteries has been ongoing over three decades, they have been hindered from commercialization due to several practical challenges [16]. First, the large volume change of 76 % in the sulfur electrode upon cycling from S to Li_2S, making the cathode pulverize and lose electrical contact. As a result, the utilization fraction of active material decreases and fast capacity fade is observed [11, 17, 18]. The second problem is the low conductivity of sulfur (e.g., the electronic conductivity of S is 5×10^{-30} S cm^{-1} at 25 °C) and its discharge intermediates during cycling, which result in unstable electrochemical contact within sulfur cathodes. Carbon materials and/or conducting polymers are used to improve the electronic conductivity of the sulfur cathode, while the ionic conductivity is enhanced by using liquid electrolyte. However, liquid electrolyte with high solubility of lithium polysulfides leads to the third challenge in conventional Li-S batteries, i.e., the polysulfide shuttle. The polysulfide shuttle migrates soluble sulfur species from the cathode to the anode, where they react with lithium anode and form a passivation layer on its surface. The polysulfide shuttle leads to the loss of active materials, corrosion of lithium anode (e.g., the insoluble Li_2S accumulates on the Li anode), and low coulombic efficiency [8, 19].

Impressive breakthroughs in designing novel electrodes for advanced Li-S batteries have been made in terms of addressing the above challenges, i.e., (1) the improvement in the electronic and ionic nature of sulfur, (2) the mitigation in volume change of sulfur cathode, and (3) the strategies in conquering the polysulfide shuttle [18]. Until now, the advances in the Li-S batteries and its promising future have aroused great interest in the field; several good review articles on Li-S batteries have been published [20–23]. Based on great advances in the Li-S battery field especially the developments in recent two years [24, 25], herein in this book chapter we will review and discuss most recent developments in exploring different components in both traditional liquid cells and recent solid cells for longevity of high specific energy Li-S batteries.

2 The Liquid Cells

In traditional Li-S liquid cells, in order to enhance electronic conductivity of sulfur and its discharge products, carbon materials with/without conducting polymers are always used. In the meantime, liquid electrolytes supply the ionic conductivity for sulfur electrode during cycling [26, 27]. However, use of liquid electrolytes always leads to the polysulfide shuttle, which is a tough challenge associated with Li-S liquid cells [28]. In this section, we look back on the most recent advances in electrochemical performance of traditional Li-S cells from different cell components, i.e., the sulfur-based cathode, the lithium anode, and the liquid electrolyte.

2.1 The Sulfur-Based Cathode

The electronic conductivity of S is relatively low, and carbons or conducting polymers are widely used for good utilization and excellent cycling of sulfur electrodes [29, 30]. High surface area and/or porous structure also benefit sulfur adoption and prevent polysulfides from dissolving into electrolytes. Based on compositions and structures, here we discuss the sulfur-based cathode in the following three categories: (1) carbon-sulfur composites, (2) conducting polymer-sulfur composites, and (3) prelithiated sulfur composites.

2.1.1 Carbon-Sulfur Composites

Shortly after the discovery of mesoporous carbon (CMK-3)-sulfur cathode composites [31], many research activities on nanostructured carbon materials for high-performance Li-S batteries have been conducted [32–57].

Porous carbons Porous carbons always supply high surface and good electrical conductivity, which are essential criteria for accommodating active sulfur species in the cathode as well as enhancing cathode conductivity [58]. Qu et al. [59] reported a highly ordered nitrogen-rich mesoporous carbon (HNMC) by pyrolysis of biomass waste in a template process. The sulfur-HNMC (53.3 wt% sulfur) composite cathode exhibited an initial discharge capacity of 1209 mAh g^{-1} and retained at 600 mAh g^{-1} after 200 cycles at 1 C. The high rate capability and long term stability are enhanced by encapsulating sulfur into HNMC structure. Hierarchical porous carbon (HPC) with tailored pore structure was synthesized using a versatile approach based on ZnO nanoparticles by Kaskel et al. (Fig. 3a) [60]. The resulting materials show high pore volume (up to 3.9 cm^3 g^{-1}) and high specific surface area (up to 3060 m^2 g^{-1}). The HPC/sulfur composite cathode exhibited a capacity of >1200 mAh g^{-1}/sulfur (>750 mAh g^{-1}/electrode) at a high sulfur loading of $\geq$3 mg cm^{-2}.

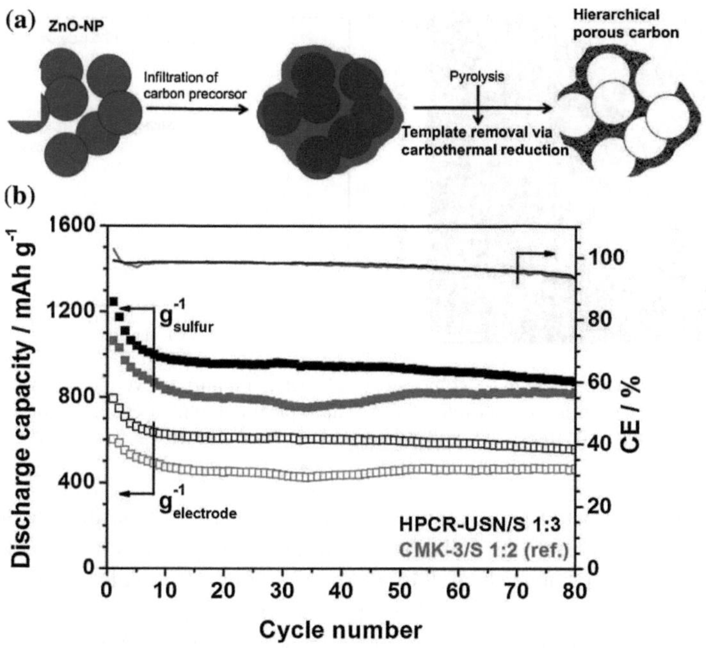

Fig. 3 **a** Synthesis schematic of hierarchical porous carbon. **b** Cycling performance of the HPCR-USN/S sample compared to CMK-3/S (reference sample) [60]. Reprinted with permission; © 2014 WILEY-VCH Verlag GmbH & Co. KGaA, Weinheim

Graphene or graphene oxides Graphene has aroused much interest since its first discovery in 2005 [61], due to its two-dimensional structure, high surface area, and good electronic conductivity. In Li-S cells, graphene has been used for the improvement in the conductivity of sulfur materials and as the reservoir for polysulfide species [62–67]. For example, Liu et al. reported a novel graphene-sulfur-carbon nanofibers (G-S-CNFs) multilayer and coaxial nanocomposite as cathode materials for Li-S batteries. These nanocomposite electrodes delivered a reversible capacity of 694 mAh g^{-1} at C/10 and 313 mAh g^{-1} at 2C. The cathode made with G-S-CNFs was able to maintain ~ 273 mAh g^{-1} even after 1500 cycles at 1C, representing an extremely low decay rate (0.043 % per cycle after 1500 cycles). Such improved rate capability and cycle stability are attributed to the unique coaxial architecture of the nanocomposite, in which graphene and CNFs enable electrodes with improved electrical conductivity, better ability to trap soluble polysulfide intermediates and accommodate volume expansion/shrinkage of sulfur during cycling [68]. Gao et al. reported a reduced graphene oxide-sulfur composite aerogel with a compact self-assembled graphene oxide skin as cathode materials for Li-S batteries [69]. This electrode was further modified by an atomic layer deposition of ZnO or MgO for a free-standing electrode to prevent the polysulfide shuttle (Fig. 4a). The ZnO modified G-S electrode with 55 wt% sulfur

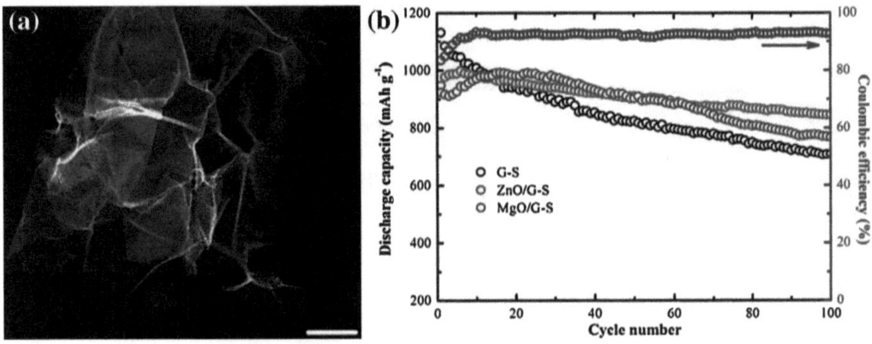

Fig. 4 **a** STEM image of ZnO/G-S composite (scale bars, 100 nm). **b** Cycling performances of the composite electrodes at a constant rate of C/5. The typical coulombic efficiency of the ZnO/G-S electrode is shown [69]. Reprinted with permission; © 2015 The Royal Society of Chemistry

loading showed a better performance than that of the MgO modified electrode. The ZnO modified G-S electrode delivered an initial discharge capacity of 998 mAh g^{-1} at C/5 and a high capacity of 846 mAh g^{-1} after 100 cycles with coulombic efficiencies over 92 % (Fig. 4b). The graphene oxide skin and metal oxide coating are natural and artificial barriers to constrain polysulfides within the cathode. The resulting Li-S batteries exhibited good cycling stability with high coulombic efficiency.

Hybrid carbons Porous carbons and graphene are always in amorphous states with sp^3-hybridized C–C bonding, and they exhibit relatively poor electrical conductance. The combination of highly conductive carbon nanomaterials with porous carbons or graphene forms hybrid carbon materials, which can help improve the conductivity of sulfur cathode and prevent the polysulfide shuttle. Multiwall nanotubes (MWNTs) were encapsulated into hollow porous carbon nanotubes to prepare a tube-in-tube carbon nanostructure cathode (S-TTCN) for Li-S cells by Zhao et al. [70]. The S-TTCN composite cathode delivered a discharge capacity of 918 mAh g^{-1} after 50 cycles with the columbic efficiency of ~98 % at 500 mA g^{-1}. Based on the overall composite mass, this electrode delivered a high capacity of 652 mAh g^{-1} after 50 cycles. The MWNTs improve the electronic conductivity and capacity, while porous carbon layers inhibit the polysulfide dissolution [71]. The aligned carbon nanotube/graphene (CNT/G) hybrid materials as a 3D conducting framework for sulfur accommodation were reported by Wang et al. (Fig. 5a). Based on its superior structure, the S-CNT/G (1:1 in weight ratio) nanocomposite achieved an initial discharge capacity of 1048 mAh g^{-1} at 1 C with a capacity fade as low as ~0.04 % per cycle over 200 cycles. The CNT network acts as a skeleton to form a binder-free, highly conductive, and flexible cathode, while 2D graphene sheet improves restriction for the polysulfide shuttle. Moreover, the CNT/G hybrid carbon enables sulfur species closely attach to conductive

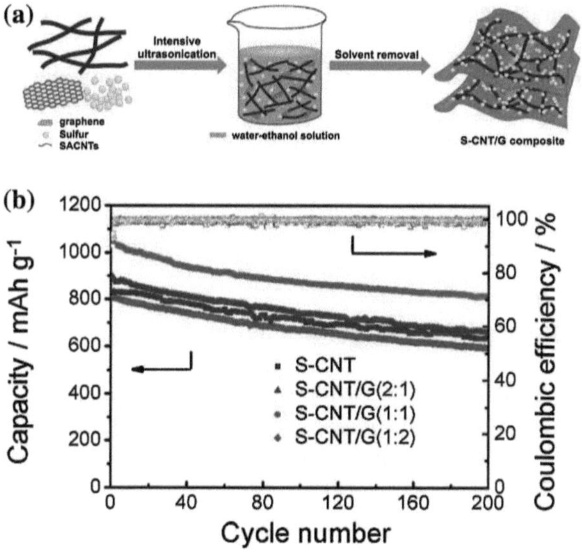

Fig. 5 **a** Schematic illustration of the synthesis procedure of the S-CNT/G composite. **b** Cycling performance of S-CNT and S-CNT/G at 500 mA g^{-1} [71]. Reprinted with permission; © 2015 The Royal Society of Chemistry

framework, which greatly enhance the electronic conductivity and thereby the utilization of active materials.

2.1.2 Conducting Polymer Sulfur Composites

Polymer materials are always synthesized at temperatures below 100 °C, which make it a feasible approach to protect sulfur cathode. Conducting polymers have been used to fabricate polymer-sulfur composites for the conductivity enhancement of sulfur species [72]. The functional groups and unique chain structure of conducting polymers further effectively confine sulfur species within the cathode [73, 74]. Many conducting polymers have been sulfur cathodes, including polyaniline (PANI) [29, 75–77] and polypyrrole (PPy) [78, 79]. For example, Zhang et al. synthesized PPy nanowire through a surfactant mediated approach, and further prepared sulfur-PPy composites by heating the mixture of element sulfur and PPy nanowire together [79]. The PPy nanowires serve as conductive additive, distribution agent and absorbing agents for active sulfur materials. The initial discharge capacity of sulfur-PPy composites was 1222 mAh g^{-1} and the remaining capacity is 570 mAh g^{-1} after 20th cycles. Yolk-shell PANI-coated sulfur composite cathodes were synthesized by Zhou et al. for Li-S batteries [80]. The method is using a heating vulcanization of a PANI-sulfur core-shell structure. Resulting sulfur cathode exhibited a stable capacity of 765 mAh g^{-1} at C/5 after 200 cycles, owing to internal void space inside the polymer shell for the accommodation of sulfur dissolution. Recently, conducting polymers (CPs)/graphene oxide (GO)@sulfur composites were synthesized via a facile one-pot route by Zhang et al. (Fig. 6a) [81]. The poly(3,4-ethylenedioxythiophene) (PEDOT) and polyaniline (PANI) polymers were used by interface polymerization on GO sheets. The PEDOT/GO@S composites exhibited a discharge capacity of 800 mAh g^{-1} after 200 cycles at C/5, which is much higher than that of PANI/GO@S composites (599 mAh g^{-1}) and PANI@S (407 mAh g^{-1}). The conducting polymers serve as a conductive matrix and an adsorbing agent, while GO physically and chemically confines polysulfides within the cathode.

2.1.3 Prelithiated Sulfur Composites

Prelithiated sulfur, also known as lithium sulfide (Li_2S) was also used as a sulfur cathode because it possesses a high theoretical capacity of 1166 mAh g^{-1} [82–84]. Li_2S has a higher melting point than that of sulfur (938 vs. 120 °C), which allows for various heat treatments for sulfur protection. Due to lithium source, the Li_2S cathode can avoid direct use of a metallic lithium anode and match with high capacity Si or Sn anodes for improved energy density [85, 86]. Bulk Li_2S (also micro-Li_2S) was always considered as an electrochemically inactive material, since it has electronic and ionic conductivities as low as 10^{-14} and 10^{-13} S cm^{-1}, respectively. When carbon materials were used to improve its conductivity, it was successfully used as cathode material for Li-S cells. For example, Fu et al. constructed a sandwiched

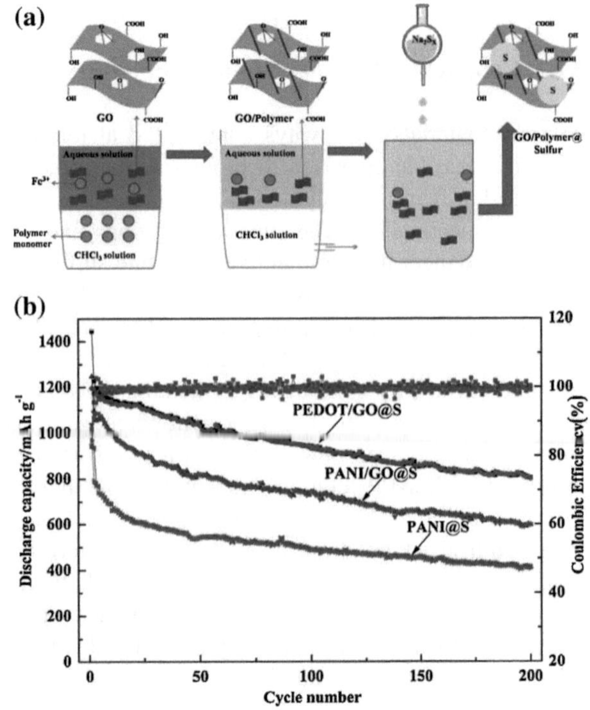

Fig. 6 **a** Synthesis schematic of the conducting polymers/GO@S composites. **b** Cyclic performance of PEDOT/GO@S, PANI/GO@S and PANI@S cathodes and the corresponding coulombic efficiency at C/5 [81]. Reprinted with permission; © 2015 Elsevier

sulfur cathode by using two layers of self-weaving, binder-free carbon nanotube films with micro-Li_2S powder in between [87]. Though carbon nanotube films improved the conductivity of micro-Li_2S powder, a high overpotential of 4 V was still needed at the first charge to overcome energy barrier. After activation, high electronic resistivity and low lithium ion diffusivity were conquered and no further overpotential was observed. A high capacity of 838 mAh g^{-1} at C/10 was achieved for Li_2S, which corresponds to 72 % of the theoretical maximum. Recently, Lin et al. reduced the micro-Li_2S to nano-Li_2S by using a solution-based reaction at the room temperature (Fig. 7a) [88, 89]. Carbon-coated NanoLi$_2$S (NanoLi$_2$S@carbon) composites were then designed, which consist of Li_2S nanoparticles as the core and a carbon coating as the shell (Fig. 7b). The carbon shell prevents the NanoLi$_2$S core from directly contacting the liquid electrolyte, which improves the performance of Li-S cells to provide longer cycle life and high sulfur utilization (Fig. 7c). The cyclability of Li-S cells is further enhanced by mixing the core-shell NanoLi$_2$S@carbon composites with graphene oxide, which chemically immobilizes polysulfides in the cathode through their functional groups. The resulting Li-S cell shows an initial specific discharge capacity of 1263 mAh g^{-1} (normalized to sulfur) at the C/10 rate and a capacity retention of 65.4 % after 200 cycles (Fig. 7c).

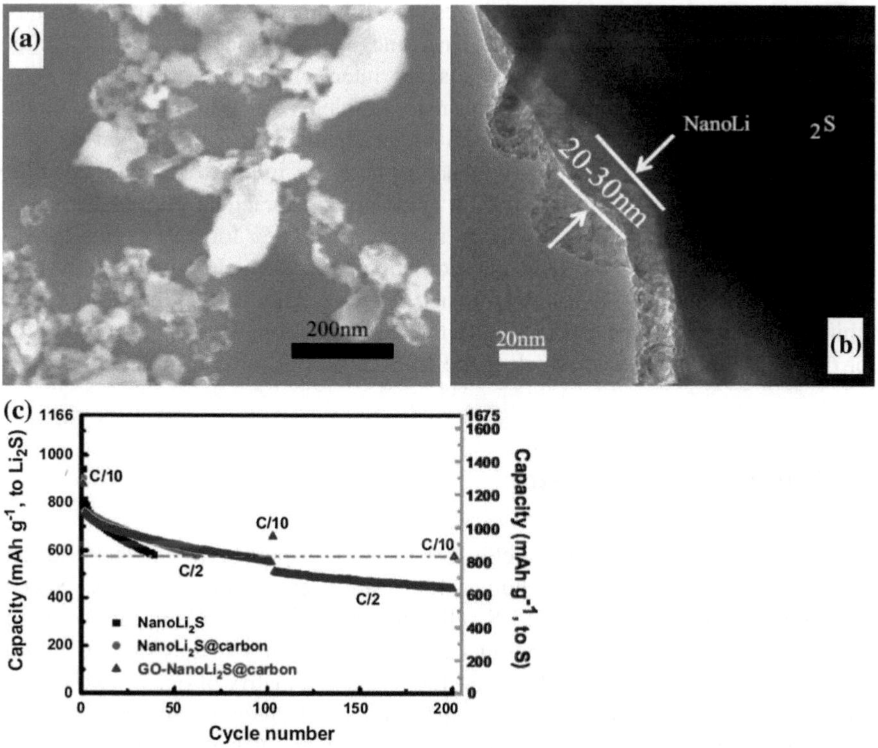

Fig. 7 **a** The SEM image of synthesized NanoLi$_2$S. **b** TEM image of core-shell NanoLi$_2$S@carbon composites. **c** Cycling comparisons of NanoLi$_2$S, NanoLi$_2$S@carbon, and GO-NanoLi$_2$S@carbon cathodes, respectively, at the C/2 rate [88]. Reprinted with permission; © 2014 Elsevier

2.2 The Lithium Anode

Although metallic lithium anode has a theoretical capacity of 3860 mAh g^{-1}, it suffers from poor cycling performance due to lithium dendrite formation. The lithium dendrite reacts with polysulfides, resulting in poor cycling performance. The lithium anode was protected by surface modifications, e.g., the use of lithium nitrate ($LiNO_3$), to minimize or eliminate dendrite formation [90]. During cycling, a passivation layer forms on Li electrodes, which consist of Li_xNO_y and Li_xSO_y species that come from reduction of $LiNO_3$ and oxidation of sulfur species by $LiNO_3$ [91]. This passivation layer diminishes reaction of polysulfide species with metallic lithium, subsequently improving the coulombic efficiency and cyclability of Li-S cells. As comparison, phosphorous pentasulfide (P_2S_5) gained attention for its ability to form a passivation layer as well as increase the solubility of Li_2S [92]. P_2S_5 led to the formation of a highly protective passivation layer on the metallic lithium surface (Fig. 8a), which functioned similarly to the passivation layer formed using $LiNO_3$. Moreover, P_2S_5 combined with Li_2S to form Li_2S/P_2S_5 complex that dissolves in organic electrolyte, which greatly promoted its electrochemical reversibility. The resulting Li-S cells demonstrated a discharge capacity of 900 mAh g^{-1} over 40 cycles with coulombic efficiency of ≥90 % at C/10 (Fig. 8b).

Though effective by adding additives, the lithium anode still suffers unstable electrodeposition, a consequence to form uneven electrodeposits on the lithium surface with inevitable defects. Nanoscale interfacial engineering could be a promising strategy to tackle the intrinsic problems of lithium metal anodes. Recently, Cui et al. coated the lithium metal anode with a monolayer of interconnected amorphous hollow carbon nanospheres to improve its cycling ability (Fig. 9) [93]. Lithium dendrites do not form up to a practical current density of 1 mA cm^{-2}. The coulombic efficiency improves to 99 % for 150 cycles. This is significantly better than the bare unmodified samples, which usually show rapid coulombic efficiency decay in fewer than 100 cycles. This strategy helps isolate the lithium metal depositions and facilitates the SEI formation. This opens a possibility for practical lithium anode, which make Li-S cells applicable in the near future.

2.3 The Liquid Electrolytes

Traditionally electrolytes provide effective Li-ion transport between electrodes and work as a charge-transfer medium within sulfur-containing cathodes. Special requirements for electrolytes in Li-S cells include low viscosity and low solubility of sulfur species. A common Li-S electrolyte consists of a Li salt such as lithium triflate ($LiCF_3SO_3$), LiTFSI, $LiPF_6$, and $LiClO_4$, and a matrix of one or two organic solvents. Based on liquid electrolytes used in Li-S cells, here we divide and discuss them into four categories: (1) ether-based electrolytes, (2) carbonate-based electrolytes, (3) ionic liquid-based electrolytes, and (4) other new liquid systems.

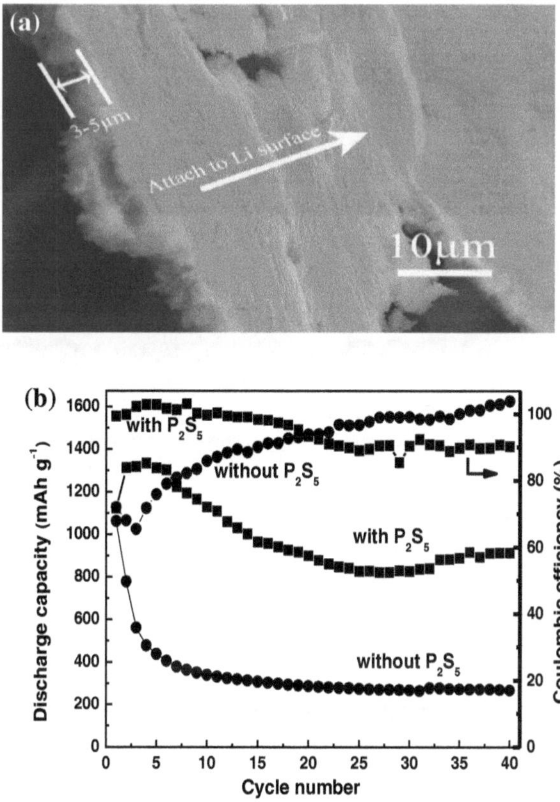

Fig. 8 **a** The cross-section of the passivation layer peeled off from the lithium anode. **b** Cycling performance of Li-S batteries with/without P_2S_5 [92]. Reprinted with permission; © 2013WILEY-VCH Verlag GmbH & Co. KGaA, Weinheim

2.3.1 Ether-Based Electrolytes

Ethers are stable with reduced sulfur species during Li-S battery cycling. The most commonly used ethers are 1,3-dioxolane (DOL), 1,2-dimethoxyethane (DME), and tetra(ethylene glycol) dimethyl ether (TEGDME) [94–97]. An early study into the optimization of the solvent ratio of TEGDME and DOL was studied by Chang et al. [98]. According to the solubility of polysulfides in liquid electrolytes, a 1:2 volume ratio of TEGDME:DOL gave the best discharge capacity when $LiCF_3SO_3$ was employed. The charge-discharge characteristics of Li-TEGDME-S cells at room temperatures were investigated by Ryu et al. [99]. The performance was improved by adding DOL and methyl acetate into the TEGDME solutions. The optimum Li-S cell had an initial discharge capacity of 1342 mAh g^{-1}. However, the capacity fading become serious after 20 cycles. Choi et al. further worked on finding better mixtures of ether solvents by comparing the performance of single and binary solvents. The systems with TEGDME, DME/DEGDME (diethylene glycol dimethyl ether) and DIOX/TEGDME revealed the beneficial effect of DIOX, which is attributed to the low viscosity and improved compatibility with lithium (Fig. 10) [100].

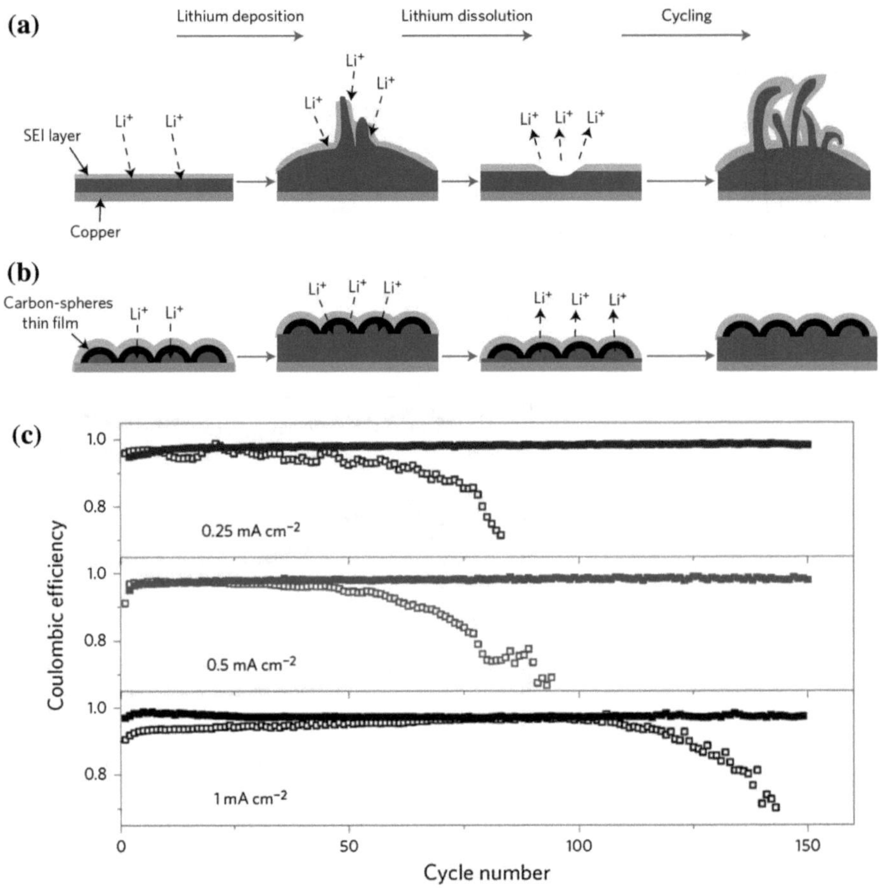

Fig. 9 Schematic diagrams of the different Li anode structures of **a** a thin film of SEI layer forms quickly on the surface of deposited Li (*blue*). Volumetric changes during the Li deposition process can easily break the SEI layer, especially at high current rates, and **b** modifying the Cu substrate with a hollow carbon nanosphere layer creates a scaffold for stabilizing the SEI layer. The volumetric change of the Li deposition process is accommodated by the flexible hollow-carbon-nanosphere coating. **c** Comparison of cycling performances of the hollow carbon nanosphere-modified electrode (*solid symbols*) and the control Cu electrode (*hollow symbols*) at different current rates. The amount of Li deposited in each cycle is 1 mAh cm^{-2} [93]. Reprinted with permission; © 2014 Nature Publishing Group

2.3.2 Carbonate-Based Electrolytes

In traditional Li-ion batteries, liquid electrolytes are based on organic carbonates, such as ethylene carbonate (EC), propylene carbonate (PC), dimethyl carbonate (DMC) and diethyl carbonate (DEC). Carbonate solvents underwent a variety of reactions with reduced sulfur species. These were recently confirmed by XAS analysis; as a result, this class of solvents is no longer considered for use in Li-S

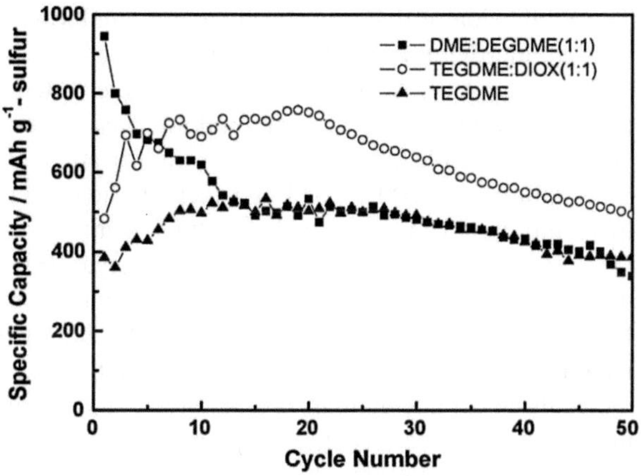

Fig. 10 Comparison of cycle performance of Li-S cells at room temperature with 1 M $LiCF_3SO_3$ in the electrolytes of TEGDME, TEGDME/DIOX (1:1, v/v) and DME/DEGDME (1:1, v/v) [100]. Reprinted with permission; © 2006 Elsevier

batteries [101]. However, when sulfur is confined in microporous carbons, the carbonate-based electrolytes were successfully applied in the Li-S cells. Wang et al. prepared conductive polymer/sulfur composites by heating the mixture of polyacrylonitrile (PAN) and sublimed sulfur. PAN was dehydrogenated by sulfur forming a conductive main chain, while –CN functional group form a thermally stable heterocyclic compound in which sulfur was confined. These composites was tested as cathode for Li-S cells, while the electrolyte system consisted of polyvinylidenefluoride-hexafluoropropylene (PVdF-HFP) colloid system that contains 1 M $LiPF_6$ in EC/DMC (1:1 in volume ratio). The composite cathode exhibited an initial discharge capacity of 850 mAh g^{-1}, while the capacity remained above 600 mAh g^{-1} after 50 cycles [102, 103].

2.3.3 Ionic Liquid-Based Electrolytes

Ionic liquids (ILs) possess the properties of non-volatile and non-flammable nature with relatively large electrochemical stability windows. Compared to the conventional organic electrolytes, ILs are much safer electrolytes in terms of avoiding the safety issues of leakage and/or gassing. However, ILs have higher viscosity than that of traditional organic liquid electrolytes, which resulting in lower conductivities/mobility. They are always used as additives to the traditional liquid electrolytes due to a substantially high cost. With the development of organic/IL hybrid electrolytes, the resulting cell properties may compensate cost differences. As mentioned above, the short cycle-life and low energy efficiency of Li-S batteries is rooted in the polysulfide shuttle phenomenon that results from the use of liquid electrolyte. In addition, stable long-term cycling of metallic liquid anode in an organic liquid electrolyte is a long-standing problem due to the formation of an unstable solid electrolyte interphase and the dendritic growth of lithium deposition.

The application of ILs in Li-S cells promise tailoring the solubility of polysufides as well as improving cycling of lithium anodes.

Recently, a series of IL electrolytes were tested for their applications in Li-S cells. Traditionally, the TFSI anion dominates the anion part of the ILs for the Li-S electrolytes, while typical cation examples are including the 1-butyl-3- methyl-imidazolium (BMIM), 1-ethyl-3-methylimidazolium (EMIM), 1-butyl-1-methy lpyrrolidinium (PYR14), and 1-butyl-1-methylpiperidinium (PiP14) in Fig. 11 [18]. As in the traditional liquid electrolyte systems, the physical properties determine the solubility power; charge distribution, polarity, viscosity and so forth. In the IL systems, however, the permittivity is largely independent of the combination of cations and anions, while variation in cations and anions affects the molecular level interactions, type/strength, and solvation. Due to unique properties, the ILs were studied as effective liquid electrolytes for the Li-S cells.

Yuan et al. first reported the operation of a Li-S cell with an IL electrolyte [104], where a mixture of 1 M LiTFSI and N-methyl-N-butyl-piperidinium TFSI was employed. The ionic liquid was used to suppress the polysulfide dissolution and thereby reduced capacity loss. The initial discharge capacity was higher than 1000 mAh g^{-1}; however, the cycling performance fell away dramatically over 10 cycles. The significant progress in Li-S cells when using the ILs was from the Cairns group [105, 106], who investigated the performance of mixtures of poly (ethylene glycol) dimethyl ether (PEGDME) with N-methyl-N-butyl-pyrrolidinium (PYR_{14}) TFSI. Song et al. reported an ionic liquid-based electrolyte consists of 1 mM g^{-1} LiTFSI in (n-methyl-(n-butyl) pyrrolidiniumbis(trifluoromethanesulfonyl) imide (PYR_{14}TFSI)/polyethylene glycol dimethyl ether (PEGDME) mixture

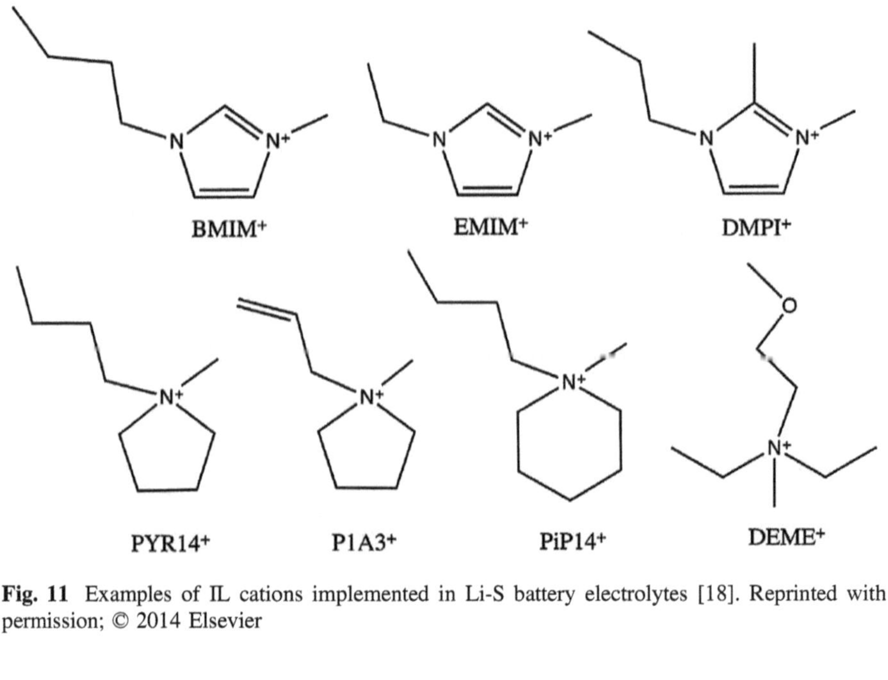

Fig. 11 Examples of IL cations implemented in Li-S battery electrolytes [18]. Reprinted with permission; © 2014 Elsevier

(1:1, by weight) for sulfur-graphene oxide nanocomposite cathode [107]. The Li-S cell showed an excellent cycle life of 1500 cycles with a high specific capacity of ~740 mAh g^{-1} and a low decay rate of 0.039 % per cycle, which confirm the important role of ionic liquid-based electrolyte for Li-S cells.

2.3.4 Other Liquid Electrolytes

Based on electrolyte systems mentioned-above, traditional liquid electrolytes in Li-S batteries always use the lithium salt of LiTFSI in single or combination of solvents,

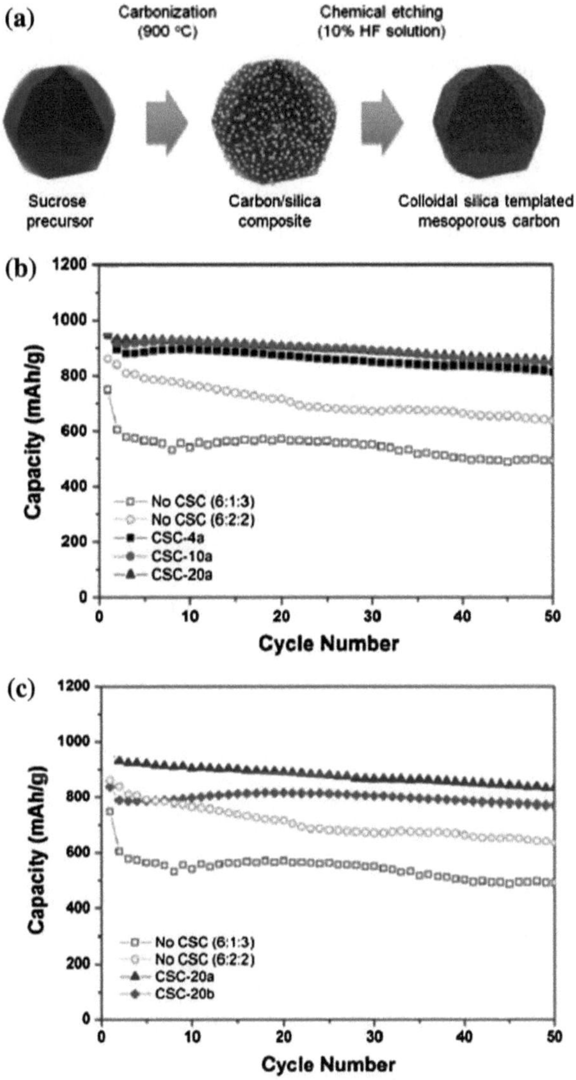

Fig. 12 **a** A schematic of the preparation of CSC as a functional additive for use in Li-S batteries: (i) mixing of sucrose and colloidal silica template in water, (ii) carbonization at 900 °C under Ar atmosphere, (iii) removal of silica templates by chemical etching with 10 % HF solution. **b** Cyclic performance of Li-S batteries employing CSC additives with different pore sizes and the same pore volume: CSC-4a (*filled square*), CSC-10a (*filled circle*), and CSC-20a (*filled triangle*). **c** Cyclic performance of Li-S batteries employing CSC additives with different pore volumes and the same pore size: CSC-20a (*filled triangle*) and CSC-20b (*filled diamond*) [113]. Reprinted with permission, © 2013 Elsevier

including 1,3-dioxolane(DOL) [98, 108], dimethoxyethane (DME) [100], tetraethylene glycol dimethyl ether (TEGDME) [99], polyethylene glycol dimethyl ether (PEGDME) [106], tetrahydrofuran (THF) [109], and ILs [104, 110, 111]. Except for the above solvents, other liquid electrolyte systems were also developed for the cycling performance of Li-S cells. For example, partially fluorinated ethers own superior properties including low melting point, high oxidation potential, and low flammability [112]. When used in Li-S cells, sulfur cathode delivered an initial discharge capacity of 1195 mAh g^{-1} and a reversible capacity of 836 mAh g^{-1} after 100 cycles. Partially fluorinated ether electrolytes introduced the formation of stable thin film on the lithium metal surface, which protect lithium anode from dissolved polysulfides. Carbon-based materials also show their potential as superior additives in Li-S batteries. For example, mesoporous carbon, which was templated by colloidal silica with particles of 4, 10 and 20 nm, was prepared from sucrose as additives CSC-4a, CSC-10a, CSC-20a, and CSC-20b for Li-S batteries (Fig. 12) [113]. The total pore volumes of the CSC-4a, CSC-10a, CSC-20a samples were the same, which was determined by the amount of each colloidal silica template. The results demonstrated that the additives improved the cycling performance of batteries significantly, which was affected by the total pore volume rather than pore size. Total pore volume determines the amount of soluble Li-polysulfides that were confined within CSC mesopores and sulfur utilization. As a result, it was proposed that the total pore volume should reach a certain value in order to absorb Li-polysulfide massively, regardless of pore size for future development of high-energy Li-S batteries.

3 The Solid Cells

There have been great improvements in cycling performance of Li-S liquid cells mentioned-above; however, the polysulfide shuttle and lithium dendrite formation are still tough challenges. When replaced by solid electrolytes, the polysulfide shuttle can be completely eliminated and the stable cycling of metallic lithium anodes can be achieved [114, 115]. In this section, we will discuss and highlight most recent breakthroughs in all-solid Li-S cells, including solid electrolytes and electrode materials.

3.1 Solid Electrolytes

A good solid electrolyte always has excellent ionic conductivity at room temperature, good compatibility with metallic lithium, and a wide electrochemical window (up to 5 V vs. Li/Li^+). Different kinds of solid electrolytes, including polymer electrolytes [1, 116–118], thio-LISICON electrolytes [119], and Li_2S/P_2S_5-based electrolytes [120, 121], have been widely studied for all-solid rechargeable lithium batteries. In this section, we will discuss and highlight most recent breakthroughs in solid electrolytes for all-solid Li-S cells, including polymer electrolytes and non-polymer electrolytes.

3.1.1 Polymer Electrolytes

Solid electrolytes always show very low conductivities at room temperature, which greatly inhibit their electrochemical performance. Polymer electrolytes with specific compositions were designed specifically for desired battery performance. Cairns et al. first characterized Li-S cells with three different polymer electrolytes; PEO/LiTFSI, poly(ethylene-methylene oxide) (PEMO) with LiTFSI, and a composite made by mixing fumed silica with PEGDME and LiTFSI [118]. Among three polymer electrolytes, the PEO cells yielded high initial discharge capacities of 1600 mAh g^{-1}. However, a rapid fall of discharge capacity to 200 mAh g^{-1} was observed after 20 cycles, and the cell needs to be operated at 104 °C to meet ionic conductivity requirement. Recently, Liang et al. reported the solid Li-S cell based on polymer electrolyte of $PEO_{18}Li(CF_3SO_2)_2N$-10 wt% SiO_2 [122]. This polymer electrolyte demonstrated the conductivity of 5×10^{-4} S cm^{-1} at 70 °C. When applied in all solid lithium batteries, sulfur-mesoporous carbon sphere (S-OMC) composite cathode showed a reversible discharge capacity of 800 mAh g^{-1} after 25 cycles at 70 °C (Fig. 13). This is attributed to good conductivity of OMC matrix and PEO polymer electrolyte at enhanced temperature.

3.1.2 Non-polymer Electrolytes

Although the inherent flexibility and low cost of polymer-based electrolytes are attractive features, their application in solid Li-S batteries are still limited by the

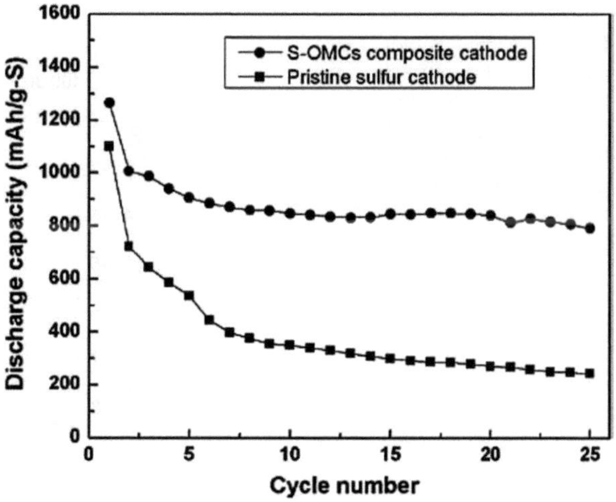

Fig. 13 Cycling performance of the Li-S cell with S-OMCs composite and pristine sulfur as cathode, 0.1 mA cm^{-2} at 70 °C [122]. Reprinted with permission, © 2010 Elsevier

requirement for operation at above room temperatures (e.g., 70 °C). With the discovery of solid electrolytes with ionic conductivities comparable to that of liquid electrolytes, all-solid rechargeable lithium batteries promise to be the next breakthrough for electric energy storage [123]. A serial of glass-ceramic solid electrolyte based on Li_2S-P_2S_5 have been reported by Hayashi and Tatsumisago et al. [124–129]. They first demonstrated that ball-milling of $80Li_2S$-$20P_2S_5$ (mol%) glasses gave rise to exceptionally high room temperature conductivity of 10^{-3} S cm^{-1}. This was attributed to the formation and stabilization of the high-temperature phase Li_7PS_6. Later on, they also described superionic analogues of thio-LISICON ($Li_{3.25}Ge_{0.25}P_{0.75}S_4$), which was deployed in energy storage devices. For example, the discharge capacity above 850 mAh g^{-1} was maintained in Li-S cells for 200 charge-discharge cycles at 1.3 mA g^{-1} at ambient temperature [121]. Moreover, lithium thiophosphates also were studied intensively as promising solid electrolytes for all-solid lithium batteries. Recently, the partial substitution of phosphorus (P) atoms by germanium (Ge) atoms in lithium thiophosphate showed an unprecedented high ionic conductivity of 1.2×10^{-2} S cm^{-1}, a value comparable to those of the liquid electrolytes used in Li-ion batteries [123]. Unfortunately, the Ge doping sacrifices the chemical compatibility of the lithium thiophosphate with metallic lithium. Liu et al. improved the chemical stability of lithium thiophosphates by the synthesis of nanostructured Li_3PS_4. This solid electrolyte showed the ionic conductivity of 1.6×10^{-4} S cm^{-1} at room temperature, which is 3 orders of magnitude higher than that of bulk lithium thiophosphate (Fig. 14) [130]. Both nano-sized framework and high surface area contributed to the enhancement of ionic conductivity, which make Li_3PS_4 a good solid electrolyte for all-solid Li-S cells [131, 132]. More recently, Sahu et al. reported arsenic substitution of Li_4SnS_4 solid electrolyte with high conduction and excellent air stability [133]. The solid electrolyte of composition $Li_{3.833}Sn_{0.833}As_{0.166}S_4$ has a high ionic conductivity of 1.39 mS cm^{-1} at 25 °C. When taking the high lithium-ion transference number into

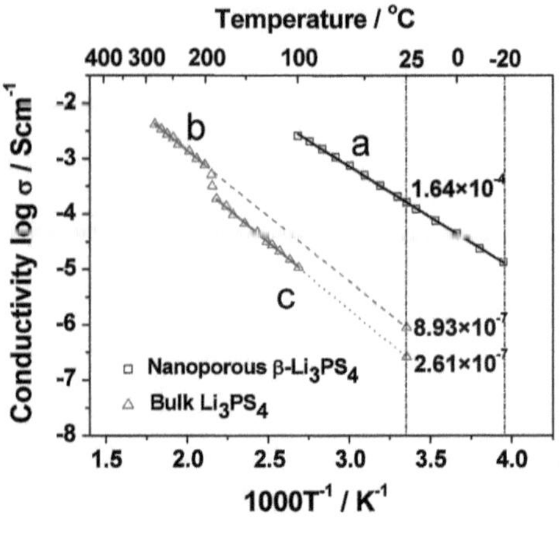

Fig. 14 Arrhenius plots for nanoporous β-Li_3PS_4 (*line a*), bulk β-Li_3PS_4 (*line b*), and bulk γ-Li_3PS_4 (*line c*) [130]. Reprinted with permission; © 2013 American Chemical Society

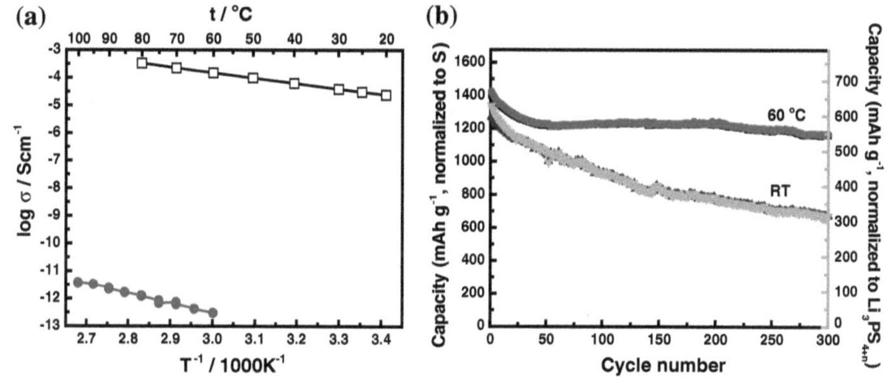

Fig. 15 **a** Temperature dependency of ionic conductivity of lithium polysulfidophosphates (Li_3PS_{4+5}) and (*open square*, Ea = 0.37 eV) and Li_2S (*filled circle*, Ea = 0.74 eV). **b** The cycling performance of Li_3PS_{4+5} at the rate of C/10 at room temperature and 60 °C (*pink/red* charge, *black/blue* discharge) [132]. Reprinted with permission; © 2013WILEY-VCH Verlag GmbH & Co. KGaA, Weinheim

consideration, this phase conducts lithium ions as well as carbonate-based liquid electrolytes.

3.2 Electrode Materials

As well documented, seamless transport of ions and transfer of electrons are basic requirements for all-solid Li-S batteries. The electronic conductivity of sulfur electrode is always improved by adding carbon; however, it is hard to enhance its ionic conductivity through simple mixing of solid electrolytes with sulfur [134, 135]. Recently, Lin et al. improved the ionic conductivity of lithium sulfide (Li_2S) by two orders of magnitude, through reacting elemental S with lithium triethylborohydride ($LiEt_3BH$) [131]. As mentioned above, Li_2S has a favorably high theoretical capacity of 1166 mAh g^{-1} and can be used as a pre-lithiated cathode to avoid the direct use of metallic lithium as the anode. The ionic conductivity of Li_2S was further improved by four orders of magnitude by coating it with solid electrolyte of Li_3PS_4, thus rendering the material a lithium superionic sulfide (LSS). The LSS cathode had an initial discharge capacity of 848 mAh g^{-1} and stabilized at 594 mAh g^{-1} after 100 cycles at 60 °C. Lin et al. further reported the first series of sulfur-rich compounds, i.e., lithium polysulfidophosphate (Li_3PS_{4+n}, $0 < n < 9$), with high ionic conductivities and excellent electrochemical reversibility for all-solid Li-S batteries [132]. These Li_3PS_{4+n} compounds are Li-ion conductors with room-temperature ionic conductivity in the range of 10^{-4} to 10^{-6} S cm^{-1}. When n = 5, for example, the ionic conductivity of Li_3PS_{4+5} is 3.0×10^{-5} S cm^{-1}, which is about 8 orders of magnitude higher than that of bulk Li_2S (Fig. 15a).

The Li_3PS_{4+5} cathode showed an initial discharge capacity of 1272 mAh g^{-1} (based on the incorporated sulfur content) (Fig. 15b) with 100 % coulombic efficiency after a few initial cycles at room temperature. Even better cycling performance was observed at 60 °C. The initial capacity was >1400 mAh g^{-1}, and a high capacity of 1200 mAh g^{-1} was maintained after 300 cycles. The increased ionic conductivity of the sulfur-rich cathode and the elimination of the polysulfide shuttle are responsible for improved cycling performance.

4 Conclusions and Perspectives

Li-S batteries have been considered as one of the promising candidates for the electrification of vehicles since they have the theoretical energy density 5 times greater than Li-ion batteries. Though promising, the tough challenges of the insulating properties of sulfur species, the polysulfide shuttle, and the lithium dendrite formation and safety concerns are still preventing them from practical applications. In the past few years, significant efforts have been dedicated to traditional Li-S liquid cells for improved cyclability. However, many issues still need further exploration before full commercialization of Li-S batteries in liquid electrolytes: (1) A deeper investigation and broader understanding of electrochemical reaction mechanisms need to be studied for different liquid electrolytes; (2) New liquid electrolyte compositions need be designed that improve cycle life by retaining polysulfides within positive electrodes; (3) Good additives need to be explored to improve stability and solve safety concerns when cycling metallic lithium anodes; (4) Safe and efficient cycling of metallic lithium anode in liquid electrolytes are highly needed; and (5) The anode with high capacity needs to be explored further.

By moving from the traditional liquid cells to all-solid cells, the polysulfide shuttle can be eliminated and high-energy Li-S chemistry with an intrinsically safe cell design can be enabled. With the emergence of solid electrolytes with ionic conductivities comparable to that of liquid electrolytes, all-solid Li-S batteries promise to be the next breakthrough for electric energy storage. Nevertheless, several issues need to be explored for the possible applications of all-solid Li-S batteries: (1) Novel solid electrolytes with required ionic conductivities at room temperature need to be designed; (2) The electrochemical windows and compatibility with metallic lithium of solid electrolytes need to be improved; (3) Interfacial resistance within the electrode and between the electrolyte and electrode need to be reduced, and (4) The in situ studies on the interfacial reaction need to conducted.

In summary, significant progresses have been made in improving cycling performance and understanding mechanisms of both traditional liquid and all solid Li-S batteries. Since Li-S batteries supply theoretical energy density 5 times higher than Li-ion batteries, they are one most promising energy storage system for electric vehicles and smart grids. With continued great interest and further extensive research in the battery community, we believe that Li-S batteries could be a next-generation breakthrough in advanced energy storage for large-scale grids and vehicular transportation.

References

1. Winter M, Brodd RJ (2004) Chem Rev 104:4245
2. Bruce PG, Hardwick LJ, Abraham KM (2011) MRS Bull 36:506
3. Cavaliere S, Subianto S, Savych I, Jones DJ, Roziere J (2011) Energy Environ Sci 4:4761
4. Liu C, Li F, Ma LP, Cheng HM (2010) Adv Mater 22:E28
5. Cairns EJ, Albertus P (2010) Annu Rev Chem Biomol 1:299
6. Zhang WJ (2011) J Power Sources 196:13
7. Yang Y, Zheng GY, Cui Y (2013) Chem Soc Rev 42:3018
8. He XM, Pu WH, Ren JG, Wang L, Wang JL, Jiang CY, Wan CR (2007) Electrochim Acta 52:7372
9. Li Q, Zhang ZA, Zhang K, Fang J, Lai YQ, Li J (2014) J Power Sources 256:137
10. Peled E, Gorenshtein A, Segal M, Sternberg Y (1989) J Power Sources 26:269
11. Manthiram A, Fu YZ, Su YS (2013) Acc Chem Res 46:1125
12. Cheon SE, Ko KS, Cho JH, Kim SW, Chin EY, Kim HT (2003) J Electrochem Soc 150: A796
13. Cheon SE, Ko KS, Cho JH, Kim SW, Chin EY, Kim HT (2003) J Electrochem Soc 150: A800
14. Mikhaylik YV, Akridge JR (2004) J Electrochem Soc 151:A1969
15. Su YS, Fu YZ, Cochell T, Manthiram A (2013) Nat Commun 4:1331
16. Zhang SS (2013) J Power Sources 231:153
17. Yin YX, Xin S, Guo YG, Wan LJ (2013) Angew Chem Int Ed 52:13186
18. Scheers J, Fantini S, Johansson P (2014) J Power Sources 255:204
19. Evers S, Nazar LF (2013) Acc Chem Res 46:1135
20. Bruce PG, Freunberger SA, Hardwick LJ, Tarascon JM (2012) Nat Mater 11
21. Ellis BL, Lee KT, Nazar LF (2010) Chem Mater 22:691
22. Marom R, Amalraj SF, Leifer N, Jacob D, Aurbach D (2011) J Mater Chem 21:9938
23. Dunn B, Kamath H, Tarascon JM (2011) Science 334:928
24. Wang J, Xin HLL, Wang DL (2014) Part Part Syst Char 31:515
25. Fedorkova A, Orinakova R, Cech O, Sedlarikova M (2013) Int J Electrochem Sci 8:10308
26. Li WY, Zheng GY, Yang Y, Seh ZW, Liu N, Cui Y (2013) Proc Natl Acad Sci 110:7148
27. Wei SC, Zhang H, Huang YQ, Wang WK, Xia YZ, Yu ZB (2011) Energy Environ Sci 4:736
28. Shim J, Striebel KA, Cairns EJ (2002) J Electrochem Soc 149:A1321
29. Zhang SC, Zhang L, Wang WK, Xue WJ (2010) Synthetic Met 160:2041
30. Wang C, Chen JJ, Shi YN, Zheng MS, Dong QF (2010) Electrochim Acta 55:7010
31. Ji XL, Lee KT, Nazar LF (2009) Nat Mater 8:500
32. Xu GL, Xu YF, Fang JC, Peng XX, Fu F, Huang L, Li JT, Sun SG (2013) Acs Appl Mater Inter 5:10782
33. Yu LH, Brun N, Sakaushi K, Eckert J, Titirici MM (2013) Carbon 61:245
34. Xi K, Cao S, Peng XY, Ducati C, Kumar RV, Cheetham AK (2013) Chem Commun 49:2192
35. Wang XF, Fang XP, Guo XW, Wang ZX, Chen LQ (2013) Electrochim Acta 97:238
36. Tao XY, Chen XR, Xia Y, Huang H, Gan YP, Wu R, Chen F, Zhang WK (2013) J Mater Chem A 1:3295
37. Wang DW, Zeng QC, Zhou GM, Yin LC, Li F, Cheng HM, Gentle IR, Lu GQM (2013) J Mater Chem A 1:9382
38. Ye H, Yin YX, Xin S, Guo YG (2013) J Mater Chem A 1:6602

39. Zhao SR, Li CM, Wang WK, Zhang H, Gao MY, Xiong X, Wang AB, Yuan KG, Huang YQ, Wang F (2013) J Mater Chem A 1:3334
40. Hwang TH, Jung DS, Kim JS, Kim BG, Choi JW (2013) Nano Lett 13:4532
41. Lu ST, Cheng YW, Wu XH, Liu J (2013) Nano Lett 13:2485
42. Park MS, Yu JS, Kim KJ, Jeong G, Kim JH, Yim T, Jo YN, Hwang U, Kang S, Woo T, Kim H, Kim YJ (2013) Rsc Adv 3:11774
43. Yang X, Zhang L, Zhang F, Huang Y, Chen YS (2014) ACS Nano 8:5208
44. Ma XZ, Jin B, Xin PM, Wang HH (2014) Appl Surf Sci 307:346
45. He G, Mandlmeier B, Schuster J, Nazar LF, Bein T (2014) Chem Mater 26:3879
46. Ding B, Yuan CZ, Shen LF, Xu GY, Nie P, Zhang XG (2013) Chem Eur J 1013:19
47. Wang B, Wen YF, Ye DL, Yu H, Sun B, Wang GX, Hulicova-Jurcakova D, Wang LZ (2014) Chem Eur J 20:5224
48. Wang JL, Yin LC, Jia H, Yu HT, He YS, Yang J, Monroe CW (2014) ChemSusChem 7:563
49. Chung SH, Manthiram A (2014) Electrochem Commun 38:91
50. Zhou L, Lin XJ, Huang T, Yu AS (2014) Electrochim Acta 116:210
51. Huang JQ, Peng HJ, Liu XY, Nie JQ, Cheng XB, Zhang Q, Wei F (2014) J Mater Chem A 2:10869
52. Qu YH, Zhang ZA, Wang XW, Lai YQ, Liu YX, Li J (2013) J Mater Chem A 1:14306
53. Jin KK, Zhou XF, Zhang LZ, Xin X, Wan GH, Liu ZP (2013) J Phys Chem C 117:21112
54. Cheng XB, Huang JQ, Zhang Q, Peng HJ, Zhao MQ, Wei F (2014) Nano Energy 4:65
55. Brun N, Sakaushi K, Yu LH, Giebeler L, Eckert J, Titirici MM (2013) Phys Chem Chem Phys 15:6080
56. Lu ST, Chen Y, Wu XH, Wang ZD, Lv LY, Qin W, Jiang LX (2014) Rsc Adv 4:18052
57. Tang JJ, Yang J, Zhou XY (2013) Rsc Adv 3:16936
58. Xin S, Gu L, Zhao NH, Yin YX, Zhou LJ, Guo YG, Wan LJ (2012) J Am Chem Soc 134:18510
59. Qu Y, Zhang Z, Zhang X, Ren G, Lai Y, Liu Y, Li J (2015) J Mater Chem A 84:399
60. Strubel P, Thieme S, Biemelt T, Helmer A, Oschatz M, Brückner J, Althues H, Kaskel S (2015) Adv Funct Mater 25:287
61. Novoselov KS, Geim AK, Morozov SV, Jiang D, Katsnelson MI, Grigorieva IV, Dubonos SV, Firsov AA (2005) Nature 438:197
62. Zhang YG, Zhao Y, Konarov A, Gosselink D, Soboleski HG, Chen P (2013) J Power Sources 241:517
63. Lu LQ, Lu LJ, Wang Y (2013) J Mater Chem A 1:9173
64. Lin TQ, Tang YF, Wang YM, Bi H, Liu ZQ, Huang FQ, Xie XM, Jiang MH (2013) Energ. Environ Sci 6:1283
65. Huang JQ, Liu XF, Zhang Q, Chen CM, Zhao MQ, Zhang SM, Zhu WC, Qian WZ, Wei F (2013) Nano Energy 2:314
66. Yan Y, Yin YX, Xin S, Guo YG, Wan LJ (2012) Chem Commun 48:10663
67. Li NW, Zheng MB, Lu HL, Hu ZB, Shen CF, Chang XF, Ji GB, Cao JM, Shi Y (2012) Chem Commun 48:4106
68. Ji LW, Rao MM, Zheng HM, Zhang L, Li YC, Duan WH, Guo JH, Cairns EJ, Zhang YG (2011) J Am Chem Soc 133:18522
69. Yu M, Wang A, Tian F, Song H, Wang Y, Li C, Hong J-D, Shi G (2015) Nanoscale 7:5292
70. Zhao Y, Wu W, Li J, Xu Z, Guan L (2014) Adv Mater 26:5113
71. Sun L, Kong W, Jiang Y, Wu H, Jiang K, Wang J, Fan S (2015) J Mater Chem A 3:5305
72. Yin LC, Wang JL, Yang J, Nuli YN (2011) J Mater Chem 21:6807
73. Xiao LF, Cao YL, Xiao J, Schwenzer B, Engelhard MH, Saraf LV, Nie ZM, Exarhos GJ, Liu J (2012) Adv Mater 24:1176
74. Wu F, Chen JZ, Chen RJ, Wu SX, Li L, Chen S, Zhao T (2011) J Phys Chem C 115:6057
75. Zhang Z, Li Q, Lai YQ, Li J (2014) J Phys Chem C 118:13369
76. Jin J, Wen ZY, Ma GQ, Lu Y, Rui K (2014) Solid State Ionics 262:170
77. Aleksandrov IA, Abramchuk SS, Solodovnikov SP, Zezin SB, Shevchenko VG, Aleksandrov AI (2012) Polym Sci Ser 54:407

78. Wang W, Li GC, Wang Q, Li GR, Ye SH, Gao XP (2013) J Electrochem Soc 160:A805
79. Zhang YG, Bakenov Z, Zhao Y, Konarov A, The NLD, Malik M, Paron T, Chen P (2012) J Power Sources 208:1
80. Zhou WD, Yu YC, Chen H, DiSalvo FJ, Abruna HD (2013) J Am Chem Soc 135:16736
81. Wang X, Zhang Z, Yan X, Qu Y, Lai Y, Li J (2015) Interface polymerization synthesis of conductive polymer/graphite oxide@sulfur composites for high-rate lithium-sulfur batteries. Electrochem Acta 155:54
82. Yang Y, Zheng GY, Misra S, Nelson J, Toney MF, Gui Y (2012) J Am Chem Soc 134:15387
83. Yang ZC, Guo JC, Das SK, Yu YC, Zhou ZH, Abruna HD, Archer LA (2013) J Mater Chem A 1:1433
84. Jeong S, Bresser D, Buchholz D, Winter M, Passerini S (2013) J Power Sources 235:220
85. Yang Y, McDowell MT, Jackson A, Cha JJ, Hong SS, Cui Y (2010) Nano Lett 10:1486
86. Hassoun J, Sun YK, Scrosati B (2011) J Power Sources 196:343
87. Fu YZ, Su YS, Manthiram A (2014) Adv Energy Mater 4:1300655
88. Lin Z, Nan C, Ye Y, Guo J, Zhu J, Cairns EJ (2014) Nano Energy 9:408
89. Nan C, Lin Z, Liao H, Song M-K, Li Y, Cairns EJ (2014) J Am Chem Soc 136:4659
90. Xiong SZ, Kai X, Hong XB, Diao Y (2012) Ionics 18:249
91. Aurbach D, Pollak E, Elazari R, Salitra G, Kelley CS, Affinito J (2009) J Electrochem Soc 156:A694
92. Lin Z, Liu ZC, Fu WJ, Dudney NJ, Liang CD (2064) Adv Funct Mater 1013:23
93. Zheng G, Lee SW, Liang Z, Lee H-W, Yan K, Yao H, Wang H, Li W, Chu S, Cui Y (2014) Nat Nanotech 9:618
94. Zhao Y, Wu WL, Li JX, Xu ZC, Guan LH (2014) Adv Mater 26:5113
95. Wu FX, Kim H, Magasinski A, Lee JT, Lin HT, Yushin G (2014) Adv Energy Mater 4. doi:10.1002/aenm.201400196
96. Sun L, Li MY, Jiang Y, Kong WB, Jiang KL, Wang JP, Fan SS (2014) Nano Lett 14:4044
97. Rong JP, Ge MY, Fang X, Zhou CW (2014) Nano Lett 14:473
98. Chang DR, Lee SH, Kim SW, Kim HT (2002) J Power Sources 112:452
99. Ryu HS, Ahn HJ, Kim KW, Ahn JH, Cho KK, Nam TH, Kim JU, Cho GB (2006) J Power Sources 163:201
100. Choi JW, Kim JK, Cheruvally G, Ahn JH, Ahn HJ, Kim KW (2007) Electrochim Acta 52:2075
101. Gao J, Lowe MA, Kiya Y, Abruna HD (2011) J Phys Chem C 115:25132
102. Wang JL, Yang J, Wan CR, Du K, Xie JY, Xu NX (2003) Adv Funct Mater 13:487
103. Wang JL, Yang J, Xie JY, Xu NX (2002) Adv Mater 14:963
104. Yuan LX, Feng JK, Ai XP, Cao YL, Chen SL, Yang HX (2006) Electrochem Commun 8:610
105. Rao MM, Song XY, Cairns EJ (2012) J Power Sources 205:474
106. Shin JH, Cairns EJ (2008) J Power Sources 177:537
107. Song M-K, Zhang Y, Cairns EJ (2013) Nano Lett 13:5891
108. Peled E, Sternberg Y, Gorenshtein A, Lavi Y (1989) J Electrochem Soc 136:1621
109. Yamin H, Gorenshtein A, Penciner J, Sternberg Y, Peled E (1988) J Electrochem Soc 1045:135
110. Park JW, Ueno K, Tachikawa N, Dokko K, Watanabe M (2013) J Phys Chem C 117:20531
111. Ueno K, Park JW, Yamazaki A, Mandai T, Tachikawa N, Dokko K, Watanabe M (2013) J Phys Chem C 117:20509
112. Weng W, Pol VG, Amine K (2013) Adv Mater 25:1608
113. Park MS, Jeong BO, Kim TJ, Kim S, Kim KJ, Yu JS, Jung Y, Kim YJ (2014) Carbon 68:265
114. Agrawal RC, Pandey GP (2008) J Phys D-Appl Phys 41
115. Quartarone E, Mustarelli P (2011) Chem Soc Rev 40:2525
116. Bouchet R, Maria S, Meziane R, Aboulaich A, Lienafa L, Bonnet JP, Phan TNT, Bertin D, Gigmes D, Devaux D, Denoyel R, Armand M (2013) Nat Mater 12:452
117. Jeddi K, Ghaznavi M, Chen P (2013) J Mater Chem A 1:2769
118. Marmorstein D, Yu TH, Striebel KA, McLarnon FR, Hou J, Cairns EJ (2000) J Power Sources 89:219

119. Nagao M, Imade Y, Narisawa H, Kobayashi T, Watanabe R, Yokoi T, Tatsumi T, Kanno R (2013) J Power Sources 222:237
120. Youngsik K, Arumugam N, Goodenough JB (2008) Chem Mater 20:470
121. Nagao M, Hayashi A, Tatsumisago M (2011) Electrochim Acta 56:6055
122. Liang XA, Wen ZY, Liu Y, Zhang H, Huang LZ, Jin J (2011) J Power Sources 196:3655
123. Kamaya N, Homma K, Yamakawa Y, Hirayama M, Kanno R, Yonemura M, Kamiyama T, Kato Y, Hama S, Kawamoto K, Mitsui A (2011) Nat Mater 10:682
124. Hakari T, Nagao M, Hayashi A, Tatsumisago M (2014) Solid State Ionics 262:147
125. Nagao M, Hayashi A, Tatsumisago M (2011) J Power Sources 196:6902
126. Hayashi A, Ohtsubo R, Nagao M, Tatsumisago M (2010) J Mater Sci 45:377
127. Ohtomo T, Hayashi A, Tatsumisago M, Kawamoto K (2013) J Non-Cryst Solids 364:57
128. Ito Y, Sakuda A, Ohtomo T, Hayashi A, Tatsumisago M (2013) Solid State Ionics 236:1
129. Muramatsu H, Hayashi A, Ohtomo T, Hama S, Tatsumisago M (2011) Solid State Ionics 182:116
130. Liu ZC, Fu WJ, Payzant EA, Yu X, Wu ZL, Dudney NJ, Kiggans J, Hong KL, Rondinone AJ, Liang CD (2013) J Am Chem Soc 135:975
131. Lin Z, Liu ZC, Dudney NJ, Liang CD (2013) ACS Nano 7:2829
132. Lin Z, Liu Z, Fu W, Dudney NJ, Liang C (2013) Angew Chem Int Ed 52:7460
133. Sahu G, Lin Z, Li JC, Liu ZC, Dudney N, Liang CD (2014) Energy Environ Sci 1053:7
134. Hayashi A, Ohtomo T, Mizuno F, Tadanaga K, Tatsumisago M (2003) Electrochem Commun 5:701
135. Takeuchi T, Kageyama H, Nakanishi K, Tabuchi M, Sakaebe H, Ohta T, Senoh H, Sakai T, Tatsumi K (2010) J Electrochem Soc 157:A1196

Why Grignard's Century Old Nobel Prize Should Spark Your Curiosity

Claudiu B. Bucur, Thomas Gregory and John Muldoon

1 Introduction

Without a doubt, the Holy Grail in battery research is the development of post lithium ion batteries (PLiB) with two or three times the energy density of lithium–ion [34]. One strategy towards PLiBs may require shifting from alloys to pure metal anodes and from insertion to conversion cathodes such as sulfur or air [2]. In order to grasp the full potential of a magnesium metal anode, here is a short background of the roadblocks encountered with lithium metal anodes. Li metal anode is a superior candidate because of high volumetric capacity (2062 mAh cm^{-3}) and very negative reduction potential (−3.04 V vs. SHE). Attempts at commercializing a rechargeable battery containing a Li metal anode have been dissuaded by the inherent instability of Li metal, especially during charging. Unfortunately, batteries containing Li metal anodes suffer from thermal runaway, which is characterized by temperatures quickly rising above the melting point of metallic Li (180.5 °C). This inherent safety issue can be triggered by dendrite formation on the Li metal anode which can form an internal short-circuit in the battery, leading to rapid uncontrolled discharge and internal heating [57]. In an effort to mitigate this intrinsic instability of batteries containing a Li metal anode, research shifted towards lithiated graphite and alloy-based anodes containing Si and Sn. Consequently, one of the major challenges in the commercialization of PLiBs containing a Li metal anode will be the suppression of dendritic growth; this is not a trivial task. Recent approaches to stabilize the lithium metal anode and avoid dendrite formation include the development of new electrolytes or additives which promote more uniform lithium

C.B. Bucur · J. Muldoon (✉)
Toyota Research Institute of North America, 1555 Woodridge Avenue,
Ann Arbor, MI 48105, USA
e-mail: john.muldoon@tema.toyota.com

T. Gregory
Borealis Technology Solutions LLC, Midland, MI 48642, USA

Z. Zhang and S.S. Zhang (eds.), *Rechargeable Batteries*,
Green Energy and Technology, DOI 10.1007/978-3-319-15458-9_22

electrodeposition. Noteworthy approaches include the work of Cui who reports a flexible, interconnected, hollow amorphous carbon nanosphere coating with the aim of realizing a chemically stable and mechanically strong interfacial layer in the highly reducing environment provided by the Li metal anode [61]. Also, Ding et al. [13] proposes a positively charged electrostatic shield composed of small concentrations of selected cations such as cesium and rubidium which inhibits the formation of lithium dendrites. Replacement of liquid electrolytes with Li–ion conducting solid electrolytes is another strategy for suppressing dendrite growth [46]. In contrast to Li metal, magnesium (Mg) metal anodes are not plagued by dendritic formation [35]. Additionally, Mg is safer than Li when exposed to air, more abundant in the earth crust (currently resulting in a cost <5 % of that of Li metal), provides double the volumetric capacity of Li and has a standard reduction potential of −2.356 V versus SHE, which is 700 mV less negative than Li (Fig. 1). Lithium anode operation is enabled by the formation of a passive layer via reduction of the electrolyte to form a solid electrolyte interface which is blocking to electrons but not to the alkali metal cations. This allows transport of lithium ions between the anode and electrolyte without substantial degradation of either component. On the other hand Mg has a unique electrochemistry, the formation of passivating layers on Mg metal by reduction of water, electrolyte or air is a formidable roadblock which prohibits its reversible deposition [34]. For example in carbonate solvents containing currently commercial ionic salts such as magnesium bis (trifluoromethane sulfonyl)-imide or magnesium perchlorate; reduction of these electrolytes results in the formation of an electronically blocking surface layer which does not conduct magnesium ions and inhibits deposition. However, efficient Mg deposition in ethereal solutions of Grignard reagents has been known since

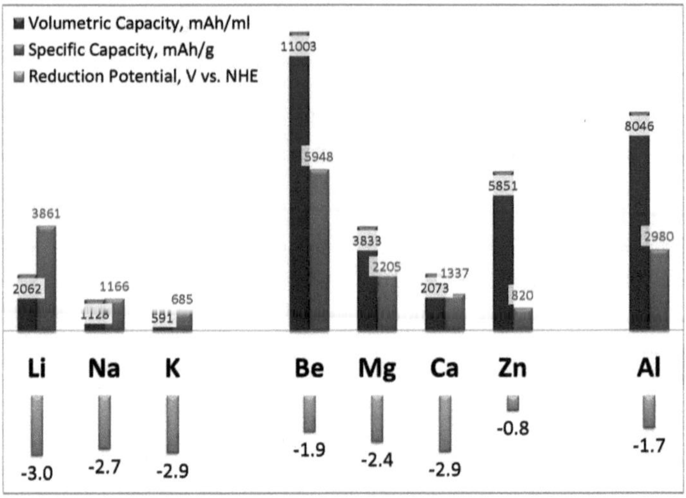

Fig. 1 Capacities and reductive potentials for various metal anodes. Reprinted with permission from Muldoon et al. [34]. Copyright 2014 American Society of Chemistry

1927 [15]. Unfortunately they are unsuitable Mg battery electrolytes due to therefore low conductivity and narrow electrochemical window (<2 V vs. Mg). A key to successful secondary Mg battery development is therefore the discovery of novel electrolytes.

Another strategy towards doubling the energy density of Li-ion batteries is to increase the capacity of cathodes by achieving two electron reduction at a single metal redox center [56]. This approach has to overcome the challenge of inserting two lithium cations for every transition metal center in the cathode material. This is not trivial due to the volume expansion of the intercalate which can lead to severe distortion of the crystal lattice and pulverization of the solid. Interestingly, two electron reduction can be achieved by insertion of a single divalent magnesium cation which has a similar radius to a monovalent lithium cation (ionic radius of Li^+ is 0.9 Å, and of Mg^{2+} is 0.86 Å), thus circumventing the excessive volume expansion problem. However, the higher charge density of magnesium cations poses solid state diffusion challenges due to strong interactions between the Mg cations and anionic constituents of the cathode material, potentially leading to rate and capacity limitations. This book chapter will highlight some of the major discoveries in both the area of electrolyte and cathode research and some of the future challenges which must be overcome. If the reader wishes more comprehensive reviews and perspectives, the following references are suggested: [5, 34, 44, 58]. Since the pioneering work by Gregory et al. [19] at the Dow Chemical Company in the early 1980s and the demonstration of a rechargeable, long cycle life but low energy density magnesium battery by Aurbach et al. [4], Aurbach et al. [4] the field of magnesium battery was largely dormant until fairly recently. Wide recognition for magnesium as a post lithium ion battery was catalyzed by research at Toyota's R&D labs [23]. The promise of a high energy rechargeable magnesium battery has stimulated intense research from industry, government labs and academics including Samsung, Pellion Technologies, JCESR and RISING to name a few.

1.1 Electrolytes

One of the obstacles in developing high voltage rechargeable magnesium batteries is moving beyond the oxidative stability of Grignards. Current strategies for developing modern magnesium electrolytes with high oxidative stabilities are deeply rooted in the early chemistries investigated by Connor et al. [12], Brenner [8], Gregory et al. [19], Mayer [32] and di Noto. It is important for the reader to grasp these discoveries and how they laid the foundation for the synthesis of modern magnesium electrolytes. In 1957 Connor et al. investigated the electrodeposition of magnesium from magnesium borohydride generated in situ from the reaction of $MgBr_2$ and $LiBH_4$. In 1971, Brenner evaluated electroplating from magnesium decaborane generated from the reaction of a lithium deacaborane complex with $MgCl_2$ in THF. In 1980s Gregory et al. synthesized magnesium organoborates such as $Mg(B(C_4H_9)_4)_2$. In addition, Gregory also spiked alkyl

Grignards with aluminum trichloride to enhance the electrochemical plating. In the same year, Mayer also reported the deposition of magnesium from trialkylaluminums and dialkylmagnesium. In the 1990s di Noto demonstrated magnesium electrolytes generated from the reaction of δ-$MgCl_2$ and aluminum trichloride. While these discoveries paved the way for modern magnesium electrolytes, this early work did not yield reversible magnesium electrolytes with high oxidative stability.

Based on the work of Gregory it was assumed that the character of the Lewis acid was an important factor in enhancing the voltage stability and that transmetallation was key in the formation of the electrochemical species. Derived from this premise, Aurbach et al. [5] investigated how the combination/ratio of dialkylmagnesium to Lewis acids affected the coulombic efficiency and oxidative stability of electrolytes. Aurbach discovered that the reaction of one equivalent of dibutylmagnesium and two equivalents of ethylalumiunm dichloride yielded an in situ generated magnesium organohaloaluminate electrolyte with the given formula of $Mg(AlCl_2BuEt)_2$ (DCC). This electrolyte exhibits an electrochemical stability window of 2.2 V and a coulombic efficiency of 100 % for the reversible deposition of magnesium. (Figure 2) To characterize the structure of DCC, single crystals were precipitated from a THF solution by the addition of hexane. However, the redissolved crystal $(Mg_2(\mu\text{-}Cl)_3{\cdot}6THF)(EtAlCl_3)$ in THF showed no reversible magnesium depositing. Aurbach concluded that the cleavage of the relatively weak Al-C bond in DCC via β-hydrogen elimination was the cause of the low voltage stability. The electrochemical window of DCC was enhanced by the substitution of the alkyl group with an aromatic group which excluded β-hydrogen elimination as a possible oxidation route. Aurbach's second generation magnesium organohaloaluminate electrolyte called APC was generated in situ via the reaction of one equivalent of aluminum trichloride with 2 equivalents of phenylmagnesium chloride [42]. This displayed a significantly broader electrochemical window of 3.3 V versus Mg on a Pt working electrode and a coulombic efficiency of 100 % (Figure 2). Aurbach demonstrated the first long cycle life rechargeable magnesium battery using the DCC electrolyte. This magnesium battery showed impressive cycle life of greater than 3500 cycles, low capacity fade, negligible self-discharge and a wide operating temperature. Unfortunately, this prototype battery was only considered as a replacement for Ni–Cd or lead acid battery because of the high formula weight and the low voltage of the Chevrel phase $Mg_xMo_3S_4$ insertion cathode.

In situ generated magnesium organohaloaluminates are complex "soup" mixtures operating under dynamic equilibria, containing corrosive and nucleophilic components. For example, in situ APC contains PhMgCl and Ph_2Mg. In addition, they do not allow varying solvents; they often need electrochemical conditioning before reaching optimal performance (such as high coulombic efficiency and low overpotential of deposition) and are difficult to store (their performance changes over time). By contrast, electrolytes for Li–ion batteries are prepared by simply dissolving salts of lithium ($LiPF_6$, LiTFSI, $LiClO_4$) in organic solvents (ethers, carbonates or ionic liquids). Because the salt and solvent are distinct components, this allows for electrolyte optimization by varying the salt:solvent combination and

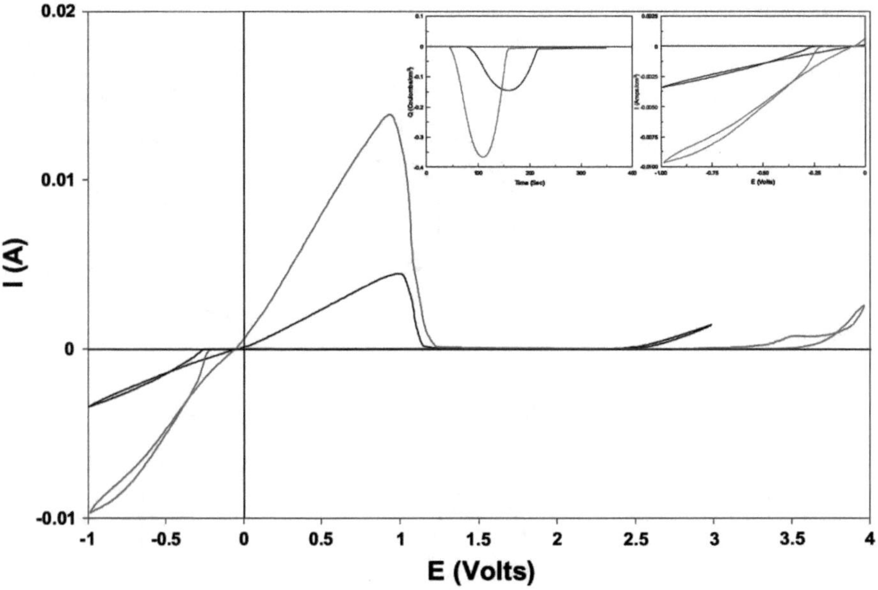

Fig. 2 Comparison between the voltammetric behavior of THF solutions containing 0.25 M of the reaction product between 1:2 $MgBu_2$ and $AlCl_2Et$ (DCC) and 0.4 M of the reaction product between 1:2 $AlCl_3$ and PhMgCl (designated APC) as indicated. 25 mV s^{-1}, Pt wire working electrode, 25 °C. *Right insert* Enlargement of the cyclic voltammograms near 0 V versus Mg R.E., comparing the over-potential for Mg deposition in the two solutions. *Left insert* The charge balance upon typical Mg deposition-dissolution cycles in these solutions (100 % cycling efficiency of Mg deposition). Reprinted with permission from Aurbach et al. [6]. Copyright 2007 WILEY-VCH Verlag GmbH & Co. KGaA, Weinheim

ratio. One of the "Holy Grails" of magnesium battery research is to develop an electrolyte solution based on low molecular weight, high conductivity magnesium salts which are highly soluble in a variety of solvents over a wide temperature range. Recently, our group has demonstrated that crystallization results in an improvement in both the oxidative stability and the coulombic efficiency of deposition/dissolution over the in situ generated magnesium organohaloaluminates. For example, the in situ generated electrolyte formed from the reaction of a 3:1 mixture of the Hauser base hexamethyldisilazide magnesium chloride (HMDSMgCl) and $AlCl_3$ shows an oxidative stability of 2.5 V versus Mg (which is identical to the oxidative stability of HMDSMgCl). The crystallized electrolyte $(Mg_2(\mu\text{-}Cl)_3{\cdot}6THF)(HMDS_nAlCl_{4-n})$ (n = 1,2) 1 (Fig. 3) shows a voltage stability of 3.2 V versus Mg and a coulombic efficiency of 100 % [23] (Fig. 4). Demonstration of the desired electrochemical activity of crystallized electrolyte GEN1 prompted the synthesis and experimental investigation of crystallized products obtained from in situ generated magnesium organohaloaluminates previously reported by Aurbach et al. It was found that, in contrast to previous reports on crystallized DCC, crystallized $(Mg_2(\mu\text{-}Cl)_3{\cdot}6THF)(EtAlCl_3)$ obtained from the

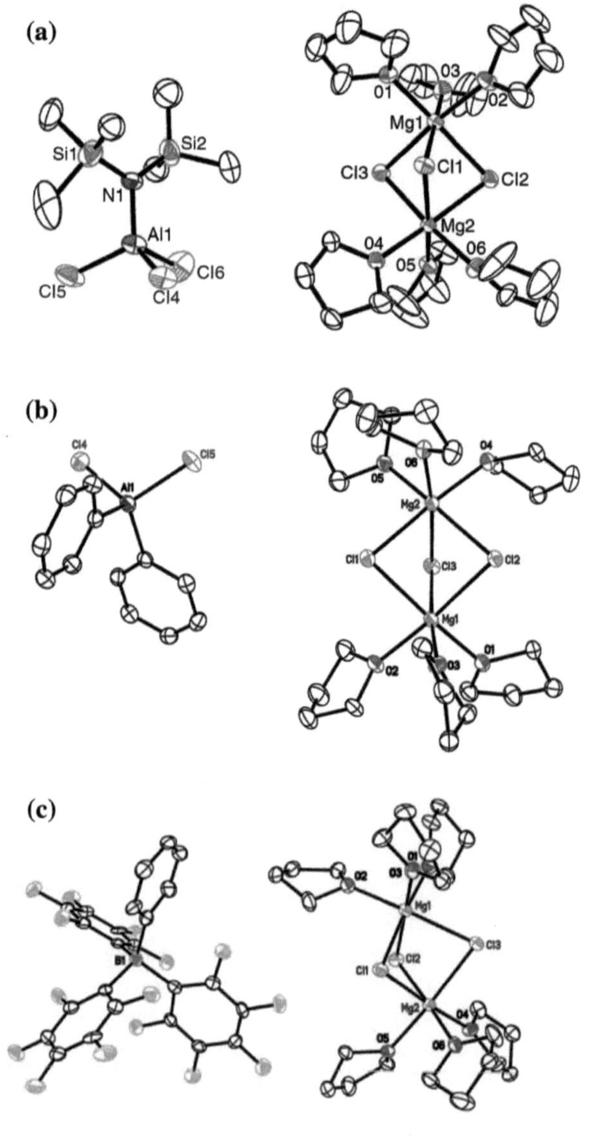

Fig. 3 **a** ORTEP plot (25 % thermal probability ellipsoids) of crystallized product, $(Mg_2(\mu\text{-}Cl)_3{\cdot}6THF)$ $(HMDSAlCl_3)$ (GEN1). Hydrogen atoms, THF of crystallization and second component of disorder are omitted for clarity. Reprinted with permission from Kim et al. [23]. Copyright 2011 Nature Publishing Group. **b** ORTEP plot (25 % thermal probability ellipsoids) of $(Mg_2(\mu\text{-}Cl)_3{\cdot}6THF)$ (Ph_2AlCl_2) (APC), 50 % thermal probability ellipsoids, hydrogen atoms omitted for clarity. Reprinted with permission from Muldoon et al. [35]. Copyright 2012 Royal Society of Chemistry. **c** ORTEP plot (25 % probability ellipsoids) of $(Mg_2(\mu\text{-}Cl)_3{\cdot}6THF)$ (B $(C_6F_5)_3Ph)$ (GEN3), hydrogen atoms omitted for clarity. Reprinted with permission from Muldoon et al. [34]. Copyright 2014 American Society of Chemistry

reaction of n-butylmagnesium and ethylaluminum chloride was, in fact, electrochemically active (Fig. 4). Similarly, the electrochemical activity of the crystallized reaction product between PhMgCl and $AlCl_3$ at a 2:1 ratio was also confirmed. Crystallization of the in situ generated electrolyte APC yields a compound with the general formula $(Mg_2(\mu\text{-}Cl)_3{\cdot}6THF)(Ph_nAlCl_{4-n})$ (n = 1–4). The predominant product of this reaction was identified as $(Mg_2(\mu\text{-}Cl)_3{\cdot}6THF)(Ph_2AlCl_2)$ by single X-ray diffraction (Fig. 3). All crystallized magnesium organohaloaluminates share

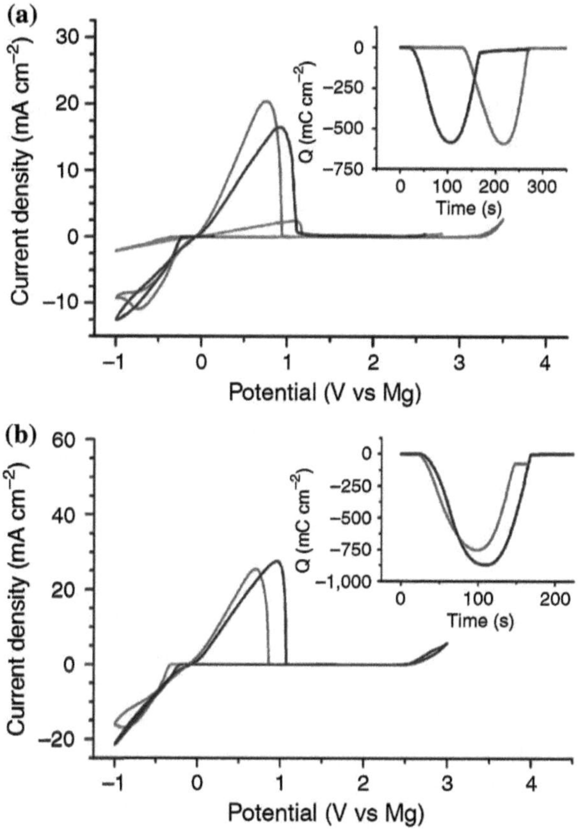

Fig. 4 **a** Cyclic voltammograms of HMDSMgCl (*green*), the reaction product generated in situ from a 3:1 mixture of HMDSMgCl to $AlCl_3$ (*blue*), and the crystal obtained from a 3:1 mixture of HMDSMgCl to $AlCl_3$ (designated GEN1) (*red*). Inset shows the charge balance during the deposition and the subsequent dissolution of Mg. Reprinted with permission from Kim et al. [23]. Copyright 2011 Nature Publishing Group. **b** Cyclic voltammograms of 0.4 M THF solution of the reaction product generated in situ from a 2:1 mixture of Bu_2Mg to $EtAlCl_2$ (DCC) (*blue*), and the crystal obtained from a 2:1 mixture of Bu_2Mg to $EtAlCl_2$ (crystallized DCC) (*red*). Scan rate is 0.025 V s^{-1}. Reprinted with permission from Kim et al. [23]. Copyright 2011 Nature Publishing Group

the general cation $(Mg_2(\mu\text{-}Cl)_3{\cdot}6THF)^+$ and their redox stability is determined by their unique anions.

Extending the oxidative stability beyond that of APC would require functionalizing the phenyl group with electron withdrawing groups such as fluorine or –CF_3. Based on our results and the literature, fluorinated aryl metal reagents and fluorinated aluminates are explosive and should be handled with caution [7]. In addition, the non-fluorinated trialkyl and triarylaluminum Lewis acids are highly reactive with air and water and can spontaneously result in uncontrollable fires and should be handled by experienced organometallic chemists. In contrast, triphenylboron Lewis acids are air and moisture stable and the corresponding fluorinated borates are non-explosive. The crystallized magnesium organoborate ($Mg_2(\mu\text{-}Cl)_3{\cdot}6THF)(B(C_6F_5)_3Ph)$ GEN3 (Fig. 3) synthesized from the reaction of trispentafluorophenyl borate and PhMgCl shows an oxidative stability in excess of 3.7 V versus Mg and is the highest voltage stability recorded to date for a magnesium battery electrolyte [35].

Grignards are highly air sensitive and corrosive in nature and must be handled under an inert atmosphere. The use of $MgCl_2$ as a Lewis base in the synthesis of magnesium electrolytes was first proposed by the early work of di Noto. While the conductivities of these electrolytes were examined, the detailed electrochemical performance (such as oxidative stability and overpotential of deposition) was not reported. Recent efforts to mitigate the hazardous nature of organometallic-based electrolytes include synthesis of a magnesium aluminum chloro complex (MACC) consisting of inorganic salts by Aurbach and Pellion Technologies [14]. In situ generated MACC is obtained from the reaction of $MgCl_2$ and $AlCl_3$ in THF or glyme and exhibit reversible deposition of magnesium as well as an oxidative stability of 3.1 V versus Mg. It was reported that electrochemical "conditioning" is needed to lower the overpotential for magnesium deposition and improve the coulombic efficiency [49] (Fig. 5). Using Pt as the working electrode and Mg foil as the counter, the electrolyte is cycled in a closed cell between −1.2 V and 2.8 V at a slow scan rate of 1 mV/sec (DCC, APC and GEN1 can routinely cycle at 25 mV/sec with 99–100 % coulombic efficiencies). Approximately 15 cycles for each 10 ml solution are required to obtain coulombic efficiencies of <99 % and $<$ −0.25 V overpotential for magnesium deposition. Although the "conditioning" process is not fully understood, it has been proposed that it contributes to the elimination of water impurities present in MACC which can account for the sluggish deposition/dissolution of magnesium. A significant limitation of these inorganic electrolytes is a dramatically lower current density for magnesium electrodeposited than can be obtained from magnesium organohaloaluminates. This may be due to the reduced solution conductivity and the absence of the organoaluminate anion. Recently, Liu et al. [29] have prepared electrolytes which can achieve oxidative stabilities of 3.4 V versus Mg and coulombic efficiencies of up to 100 % by the reaction of $MgCl_2$ with aluminum Lewis acids such as $AlEtCl_2$ and $AlPh_3$. An electrochemically active product is not isolated by crystallization but the $(Mg_2(\mu\text{-}Cl)_3{\cdot}6THF)^+$ dimer is found in the electrolyte solution by NMR and X-ray diffraction studies. High current densities are reported for the electrolyte solution $MgCl_2$–$AlPh_3Cl$ and a 100 % coulombic efficiency for the deposition of magnesium is reported for 250 cycles on a platinum working electrode for the deposition of pure and non-dendritic magnesium.

Building on Connor's discovery, the use of $Mg(BH_4)_2$ as a magnesium battery electrolyte has also been recently demonstrated by Mohtadi et al. [33]. The current densities of magnesium deposition from glyme are dramatically enhanced by the addition of $LiBH_4$ but the oxidative stability is low (<1.5 V vs. Mg), similar to that of Grignards. In contrast to the impure deposits reported by Connor, Mohtadi reports pure magnesium electrodeposition. Early reports of magnesium deposition

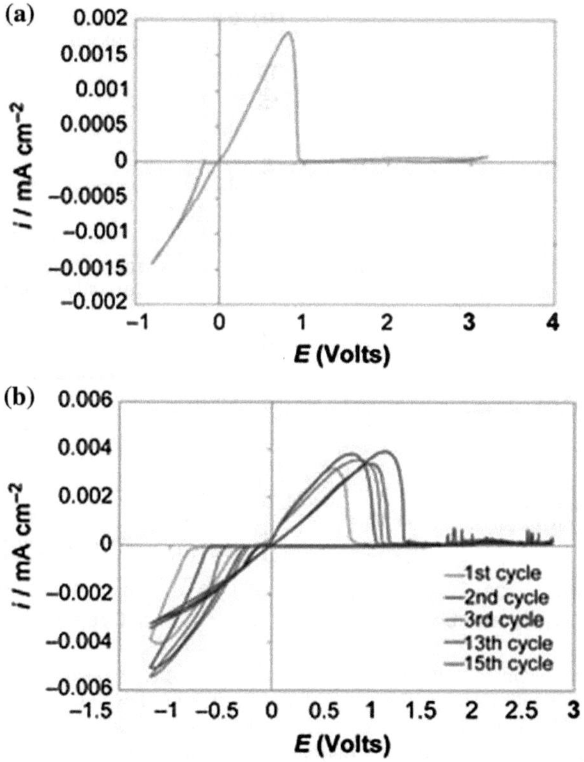

Fig. 5 Inorganic salt-based electrolyte solutions. **a** Cyclic voltammogram of conditioned $MgCl_2$–$AlCl_3$/THF (MACC; THF = tetrahydrofuran) solution. The MACC solution exhibits electrochemical performance on par with all phenyl complex solutions, with the added benefit of not containing the highly hazardous Grignard reagents. **b** Selected cycles during the conditioning procedure. Although not yet understood, the conditioning procedure highly enhances the electrochemical performance of MACC solutions, especially the reversibility and overpotential for deposition. Reprinted with permission from Shterenberg et al. [49]. Copyright 2014 Materials Research Society

from boron based compounds came from Brenner who evaluated electroplating from magnesium decaborane clusters. More recently, Mohtadi also demonstrated enhanced oxidative stabilities (3.2 V vs. Mg) over $Mg(BH_4)_2$ by a magnesium decaborane cluster (1-(1,7-carboranyl) magnesium chloride) dissolved in THF [10].

While the nondendritic deposition of magnesium holds the promise for the utilization of magnesium metal anodes, research efforts have also been directed towards coupling of commercially available magnesium salts ($Mg(TFSI)_2$ or $Mg(ClO_4)_2$) with alloy anodes such as Sn or Bi [3, 50]. High rate performance of a Bi electrode made from nanotubes [47] dramatically improved the discharge rates previously reported by Arthur et al. Figure 6 summarizes the cyclic voltammetry of magnesium insertion and de-insertion of the bismuth nanotubes as well as cycling stability and rate performance. This work emphasizes the importance of

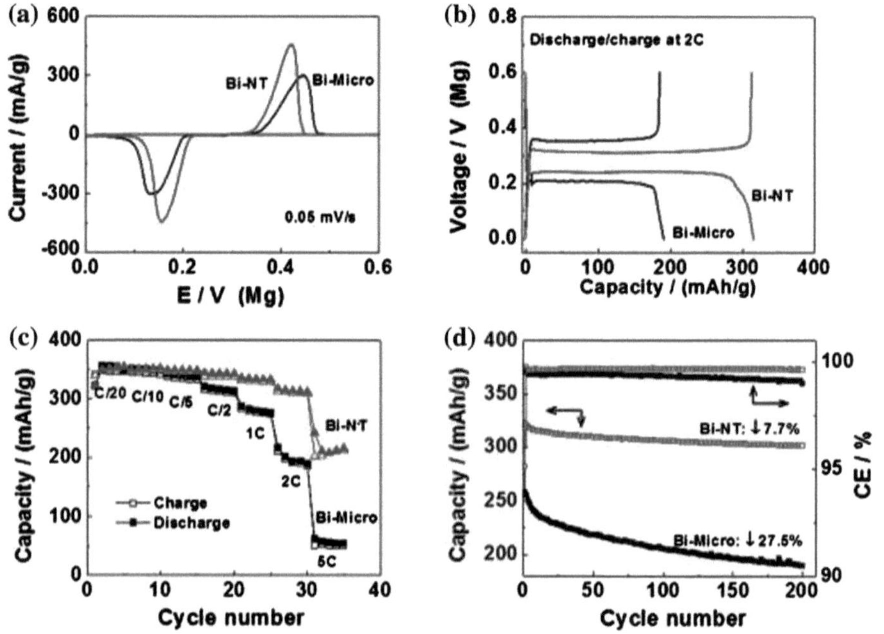

Fig. 6 **a** Cyclic voltammograms of Mg insertion/deinsertion in bismuth; **b** discharge/charge profile of an Mg–Bi cell; **c** rate performance of an Mg–Bi cell; **d** cycling stability and Coulombic efficiency (CE) of bismuth electrode for reversible Mg insertion/deinsertion. Cell configuration: Mg/0.1 M $Mg(BH_4)_2$ − 1.5 M $LiBH_4$ − diglyme/Bi. Reprinted with permission from Shao et al. [47]. Copyright 2014 American Chemical Society

nanotechnology to enhance the rate performance of alloy anodes. Surprisingly, the compatibility of $Mg(TFSI)_2$ with magnesium metal is still under debate. A few reports demonstrate the sluggish reversible deposition of magnesium from $Mg(TFSI)_2$ dissolved in glyme (less than 0.1 M $Mg(TFSI)_2$ can be dissolved in THF). The oxidative stability of $Mg(TFSI)_2$ has been reported below 3.2 V versus Mg and its deposition of magnesium is plagued by high overpotentials (varying from −0.5 to −1.0 V vs. Mg) and low coulombic efficiencies which vary drastically with the deposition rate.

1.2 Corrosion

Ideally, the current collector material used for cathode studies should be completely inert in the electrolyte solution otherwise currents observed during charging could be due to oxidation of the current collector rather than of the cathode active material. Unfortunately, magnesium organohaloaluminates electrolytes with high oxidative stabilities (above 2.2 V vs. Mg) are corrosive towards widely used materials such as stainless steel, aluminum and nickel [35]. Platinum and carbon are inert and preferred when testing cathodes. In order to test cathodes operating at voltages higher than 2.2 V versus Mg, corrosivity of magnesium electrolytes has to be mitigated. Corrosion of metals can be indicated by current hysteresis during cyclic oxidative stability tests. It is important to observe the current during the cathodic (reverse) scan because false "high" oxidative stabilities are often claimed when only the anodic (forward) scan is reported. For example GEN1 and crystallized APC show an oxidative stability of 3.2 V versus Mg on a stainless steel working electrode on the first cathodic scan. However, high hysteresis currents are observed during the reverse scan down to 2.2 V versus Mg. The second and following anodic scans always show an oxidative stability around 2.2 V versus Mg on a stainless steel working electrode (Figure 7). A plausible explanation for the corrosion observed using magnesium organohaloaluminates is the presence of chlorides in the both the cation and anion components. To examine the effect of the anion, the magnesium organoborate electrolyte ($Mg_2(\mu\text{-}Cl)_3{\cdot}6THF$) ($BPh_4$) GEN2 with a halogen free anion BPh_4^- was synthesized and tested electrochemically. GEN2 exhibits reduced corrosivity (2.6 V vs. Mg on either stainless steel or platinum electrodes) with no hysteresis. To examine the effect of the cation, the magnesium organoborate $Mg(BArF)_2$ GENX was synthesized [36]. GENX does not contain the chlorinated magnesium dimer $(Mg_2(\mu\text{-}Cl)_3{\cdot}6THF)^+$ cation and it's

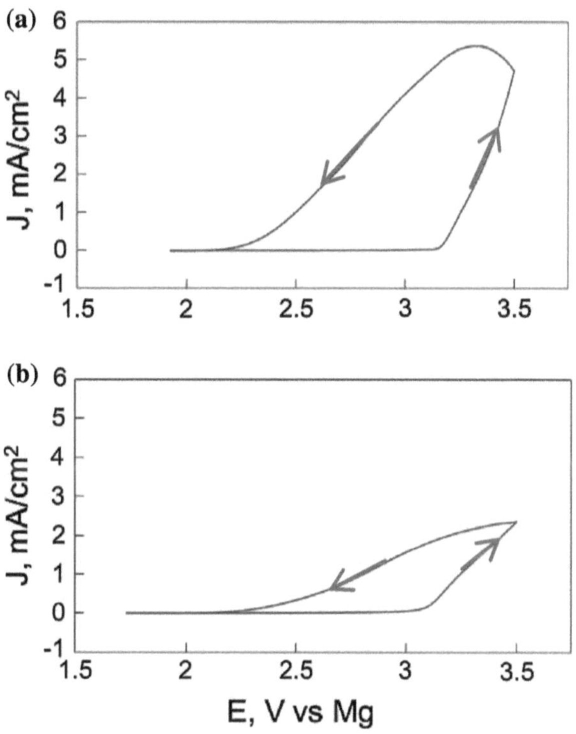

Fig. 7 Cyclic voltammograms of ($Mg_2(\mu\text{-}Cl)_3{\cdot}6THF$) ($Ph_2AlCl_2$) (APC) (**a**) and ($Mg_2(\mu\text{-}Cl)_3{\cdot}6THF$) ($HMDSAlCl_3$) (GEN1) (**b**) on a stainless steel working electrode with an area of 0.02 cm^2 at a temperature of 21 °C. Counterclockwise arrows designate hysteresis. Scan rate is 25 mV/s and the counter and reference electrode are both magnesium

$BArF^-$ anion is chlorine free. As a result, it exhibits very high oxidative stability (more than 4.5 V vs. Mg) on a stainless steel working electrode. However, reduction of the anion precludes reduction of magnesium and blocks its deposition. GENX cannot function acceptably as an electrolyte in a rechargeable magnesium battery but serves as a useful model compound to evaluate the corrosivity of magnesium organoborates at high potentials (above 3.0 V vs. Mg). These experiments appear to validate the hypothesis that magnesium organoborates which do not contain chlorides in either the cation or anion should exhibit reduced corrosivity.

Magnesium organoborates GEN2 based on the magnesium dimer $(Mg_2(\mu\text{-}Cl)_3{\cdot}6THF)^+$ show no indication of corrosivity to stainless steel due to the absence of hysteresis during cyclic voltammograms. In order to enhance the oxidative stability of GEN2 the use of fluorinated Lewis acids such as $B(C_6F_5)_3$ was considered. The reaction of $B(C_6F_5)_3$ with PhMgCl resulted in the formation of GEN3 which has an oxidative stability of 3.7 V versus Mg on a platinum working electrode. Unfortunately, during the reverse scan back to OCV, GEN3 exhibits hysteresis and exhibits a stability of 2.2 V versus Mg on a stainless steel working electrode which is indicative of corrosion. Corrosivity of GEN3 is surprising given the noncorrosive nature of GEN2 and their analogous structure which contains the same cation. Is the fluorinated anion of GEN3 responsible for corrosion or are there other factors at fault (such as purity of starting material or water contamination)? This is still an open ended question in our laboratory which we are presently investigating. If the magnesium dimer $(Mg_2(\mu\text{-}Cl)_3{\cdot}6THF)^+$ is the reason for corrosion, it would be desirable to have a universal synthetic method capable of converting the dimer to the naked magnesium such as $(Mg{\cdot}6THF)^{2+}$. One approach we deem universal is an ion exchange pathway where the magnesium dimer of GEN3 is converted to the potassium salt (Fig. 8) which can be readily converted to the silver salt and then exchanged to the naked magnesium salt of GEN3 (Fig. 8). The only requirement for this chemistry is that the anion is water stable. One of the challenges with this chemistry is the aqueous workup for the formation of the potassium salt which can introduce detrimental water with negative effects on the electrochemistry of the electrolyte. It is also possible to directly convert the potassium salt to the naked magnesium but one of the advantages of the silver intermediate is a reduction in the water adducts carried over to the naked magnesium salt. Other multivalent salts can be readily obtained using this universal ion exchange pathway. One of the promises of a naked magnesium electrolyte is a non-corrosive nature. However, magnesium electrolytes with a dimer cation have significantly higher solubilities and can accommodate double the moles of magnesium per mole of salt. In addition, it has not been proven beyond reasonable doubt that the magnesium dimer is responsible for the corrosive nature of magnesium organoborates.

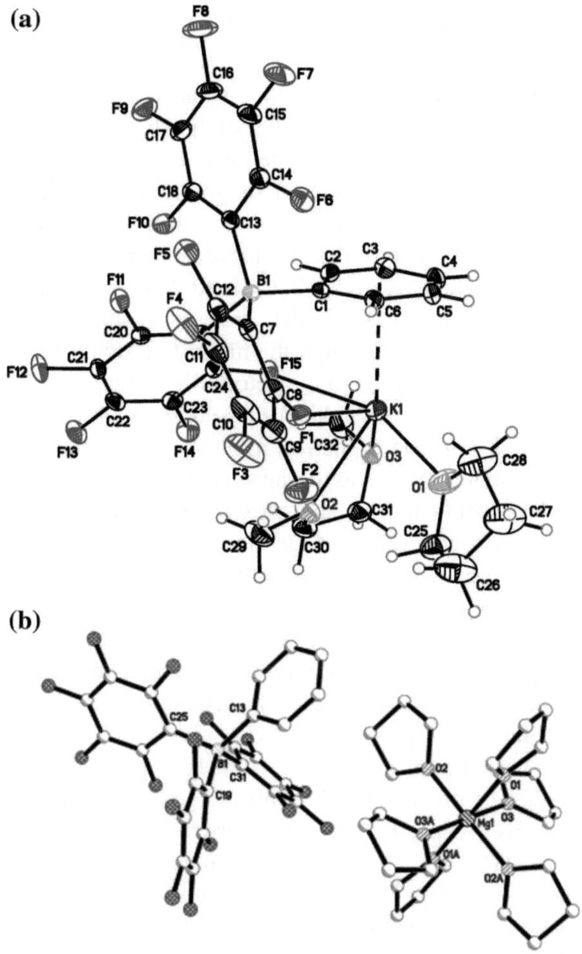

Fig. 8 **a** ORTEP plot for potassium salt analog of $(Mg_2(\mu\text{-}Cl)_3{\cdot}6THF)(B(C_6F_5)_3Ph)$ (GEN3); **b** ORTEP plot for naked Mg salt analog of GEN3. Hydrogen atoms are omitted for clarity. There is a 25 % thermal probability for the ellipsoids

2 Cathode

Electrochemical intercalation and conversion reactions have been investigated in the quest to discover practical cathodes for secondary magnesium batteries. Magnesium intercalation cathodes in particular are attractive due to the potential for a 2-electron reduction at the metal center with insertion of a single ion with size similar to univalent Li^+. As with Mg anodes, the electrolyte chemistry also plays a substantial role in the behavior of candidate cathode materials. A comprehensive survey of Mg battery cathode technology can be found in the review by Muldoon et al. [34] this chapter summarizes key discoveries in this field.

2.1 Intercalation Cathodes

Evaluation of Mg^{2+} intercalation into a variety of oxide, sulfide, and boride host materials was a focus of early studies by Gregory et al. [19], Bruce et al. [9] and Novák et al. [37]. Both chemical (via reaction between the host material and compounds such as dibutyl magnesium) and electrochemical intercalation techniques were demonstrated as being useful to screen cathode candidates. Materials which showed promise in these screening experiments included TiS_2, ZrS_2, γ-MnO_2, RuO_2, Co_3O_4, V_2O_5, and V_6O_{13}. In particular, Novák and Disilvestro demonstrated that V_2O_5 delivered promising first cycle capacity ($\sim$170 mAh/g) and reversibility, although the highest capacity was achieved when 1 M water was added to the $Mg(ClO_4)_2$/acetonitrile electrolyte; rapid capacity fade was also observed. It was speculated that preferential solvation of the Mg^{2+} ions by water molecules instead of the AN molecules aided the electrochemical insertion reaction and resulted in increased capacity, perhaps by co-intercalating with the Mg^{2+} ion and shielding it from the host material. Incorporation of water into the cathode material itself instead of the electrolyte also proved beneficial; V_2O_5 xerogel containing up to about 3 water molecules per V_2O_5 also demonstrated first cycle capacities of about 170 mAh/g in both dry AN and ionic liquid-based electrolytes [38]. Rapid capacity fade was typically observed when using wet electrolytes and the incompatibility of water with Mg metal and electrolytes which have been shown to be capable of reversible Mg deposition/dissolution makes use of this technique problematic in practical batteries.

These promising results for Mg^{2+} intercalation into V_2O_5 led to numerous studies of this compound with the goal of demonstrating reversible behavior with improved capacity retention. Particularly noteworthy is the work of Gershinsky et al. [16] who investigated the fundamental properties of electrochemical intercalation of Mg^{2+} into 200 nm films of V_2O_5 at very low current density in an attempt to negate the effects of slow solid-state diffusion of Mg^{2+}. Highly reversible electrochemical intercalation of Mg^{2+} into the V_2O_5 thin films was observed with a first cycle capacity of about 180 mAh/g, decreasing to a stable value of 150 mAh/g (corresponding to 0.5 mol Mg per mole V_2O_5) after 25 cycles. A similar capacity for Mg^{2+} intercalation into nanoparticulate V_2O_5 was obtained by Amatucci et al. [1]. Jiao et al. [22] synthesized Cu-doped vanadium oxide nanotubes which outperformed undoped nanotubes. Such results using thin film and nanostructured cathodes illustrate the performance improvement possible when the effect of slow solid state ionic diffusion is minimized.

Gershinsky et al. also addressed mechanistic aspects of Mg^{2+} insertion into V_2O_5. Cyclic voltammograms exhibited similar insertion and half-wave potentials for both Mg^{2+} and Li^+ insertion, thus supporting a mechanism involving reduction of the vanadium metal centers. A color change from yellow to faint green upon Mg^{2+} insertion was also indicative of the desired 2-electron V^{5+}/V^{3+} transition. Raman and XRD analysis of V_2O_5 at different stages of Mg^{2+} insertion/deinsertion by Gershinsky et al. and Yu and Zhang [59] indicated structural reversibility.

Variations of vanadium oxide evaluated as potential Mg battery cathodes include hydrated vanadium bronzes ($MV_3O_8.xH_2O$ where M = Li, Na, K, $Ca_{0.5}$, $Mg_{0.5}$, and $V_2O_5.yH_2O$ where y is typically up to 3) by Novák et al. [38] V_2O_5 aerogels by Le et al. [26] and V_2O_5/sulfur composites by Inamoto et al. [21] V_2O_5/S/metal oxide composites in particular delivered very high specific capacities up to about 300 mAh/g, although S solubility in the electrolyte was noted and the electrolyte used did not support Mg electrode reversibility; evaluation of these electrodes with electrolytes that are capable of reversible Mg electrode operation and reduce S solubility appears to be of merit.

For neat V_2O_5, the foregoing studies support a reversible capacity of approximately 150 mAh/g. Slow solid-state diffusivity of Mg^{2+} due to its high charge density and strong interaction with the host material appears to limit cell performance which can be ameliorated by addition of water to the electrolyte. However, this has a negative impact on Mg electrode performance and other techniques should be developed to achieve similar performance. Modification of the basic V_2O_5 structure may also improve its electrochemical performance. Fundamental differences in the intercalation behavior of Li^+ and Mg^{2+} into V_2O_5 are illustrated in Fig. 9 which shows that divalent Mg^{2+} intercalation yields higher specific capacity than univalent Li^+, but slower Mg^{2+} solid state diffusion results in significant voltage hysteresis and polarization.

Low cost MnO_2 is an attractive battery cathode candidate, and positive results obtained with γ-MnO_2 vs. other crystalline forms of this compound in early screening studies led to investigation of various MnO_2 crystal structures [Hollandite (2 × 2 tunnel), OMS-5 (4 × 2 tunnel) and Birnessite (layered)] by Rasul et al. [43] Hollandite delivered the highest capacity ($\sim$140 mAh/g) but slow Mg^{2+} ion diffusion was noted. Hydrothermally synthesized composite Hollandite/acetylene black exhibited higher capacity than the sol-gel material which was attributed to improved stability due to the presence of K^+ and H_2O in the structure. Improved cycling performance due to shielding of the charge on the inserted cations with water molecules in the hydrothermally synthesized material was consistent with behavior of other cathode materials, although some Mg^{2+} trapping was seen. The performance of commercial Hollandite MnO_2 was evaluated by Zhang et al. [60] using 0.2 M GEN1/THF electrolyte, and a first cycle discharge capacity of 280 mAh/g MnO_2 at 36 $\mu A/cm^2$ was measured, corresponding to a loading of about 0.47 Mg per Mn which indicates near full discharge to $MgMn_2O_4$. However, significant capacity fade was observed (<100 mAh/g after 6 cycles) which was attributed to Jahn-Teller distortion leading to crystal structure instability and Mn dissolution from the cathode material. XRD and XPS spectra of the positive electrode at different states of charge during both charge and discharge were indicative of reversible redox reactions of the MnO_2. MnO_2 particle size also affected first cycle capacity: 280 mAh/g was measured with 20 nm average particle size MnO_2 versus 170 mAh/g for 100 nm particles, which could also indicate slow Mg^{2+} diffusion. Mg^{2+} trapping in the MnO_2 structure was also supported by XPS and XAS studies of the MnO_2 electrode at varying state of charge, leading to the observed irreversible capacity loss. The authors speculated that improved cycling

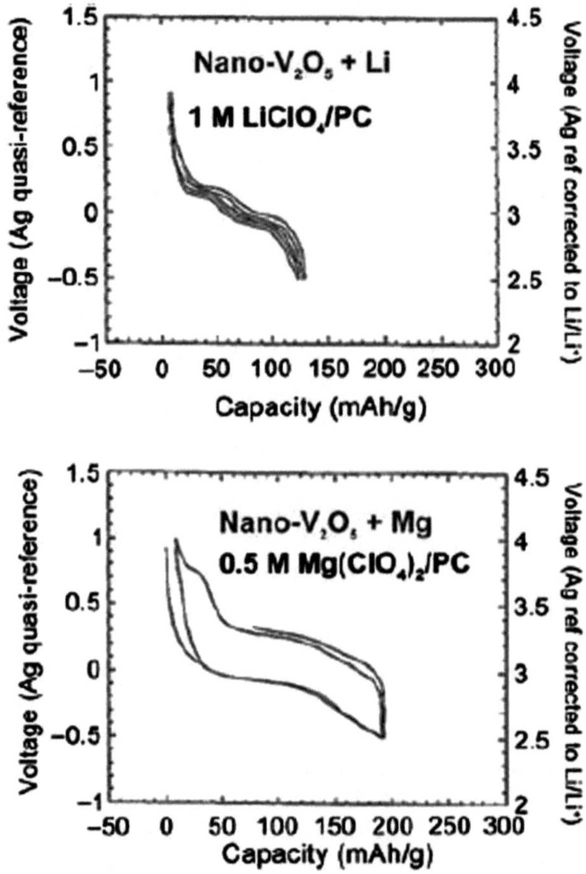

Fig. 9 Cyclic voltammograms of electrochemical behavior of nano-V_2O_5 during cation intercalation. Reprinted with permission from Shterenberg, et al. [49]. Copyright 2014 Materials Research Society

performance might be obtained by stabilizing the tunnel structure; techniques which promote improved performance of manganese oxide-based cathode materials in Li-ion batteries may also reduce capacity fade and Mn dissolution in Mg batteries.

Other metal oxide cathode materials which have been evaluated for Mg batteries include MoO_3, RuO_2 and Co_3O_4. Spahr et al. [51] achieved a first cycle discharge capacity in an ionic liquid electrolyte of about 150 mAh/g for MoO_3 (corresponding to an intercalant stoichiometry of $Mg_{0.38}MoO_3$) but rapid capacity fade was observed. Use of a wet organic electrolyte, however, yielded 210 mAh/g in the first cycle and stabilized around 160 mAh/g after about 8 cycles. Rate capability was not studied. Fundamental characteristics of Mg^{2+} intercalation into MoO_3 thin film cathodes were studied by Gershinski et al. [16] a reversible capacity of 210 mAh/g with 95 % coulombic reversibility over the first 10 cycles was reported. Mg^{2+} intercalation into MoO_3 appeared to suffer from greater kinetic limitations than V_2O_5 as evidenced by a greater charge/discharge voltage hysteresis. Koch et al. [24] and Sutto and Duncan [52] evaluated electrochemical Mg^{2+} intercalation into RuO_2.

A first cycle capacity of 101 mAh/g obtained by Sutto and Duncan implies a discharged stoichiometry of $Mg_{0.25}RuO_2$; capacity fade was attributed to poor reversibility. Sutto and Duncan also measured a capacity of 74 mAh/g for Mg^{2+} intercalation into Co_3O_4 in ionic liquid electrolytes, corresponding to a composition of $Mg_{0.33}Co_3O_4$ [53]. As with RuO_2, rapid capacity fade was attributed to difficulty in deintercalating Mg^{2+} although Co_3O_4 exhibited improved reversibility. Ichitsubo et al. [20] utilized co-precipitation to synthesize $MgCo_2O_4$ (spinel structure) and $Mg_{0.67}Ni_{1.33}O_2$ (rock salt structure) in an attempt to produce higher voltage cathode materials. The cobalt spinel material exhibited an initial open circuit voltage of 3.5 V versus Mg in an electrolyte of 1 M $Mg(ClO_4)_2$/AN; however voltage decay was seen and was attributed to structural relaxation and migration of Mg from the interior of the particles to the surface. The magnesium nickelate exhibited a potential of 3.5–3.8 V versus Mg although specific capacities were not reported.

Chevrel phase $M_xMo_6T_8$ (where M = metal and T = S, Se, or a combination of both) has received considerable attention as an intercalation cathode for rechargeable Mg batteries. These compounds consist of cubic anion frameworks enclosing octahedral clusters of Mo atoms and can be described as stacks of Mo_6T_8 blocks. The Mo_6 clusters can sustain charge injection of up to 4 electrons, resulting in a theoretical capacity of up to 2 Mg^{2+} ions per Chevrel phase unit (129 mAh/g Mo_6S_8). Aurbach et al. [4] first utilized Mo_3S_4 as a reversible cathode material in the cell Mg|DCC|Mo_3S_4. Fig. 10 illustrates the structure and electrochemical performance of this material.

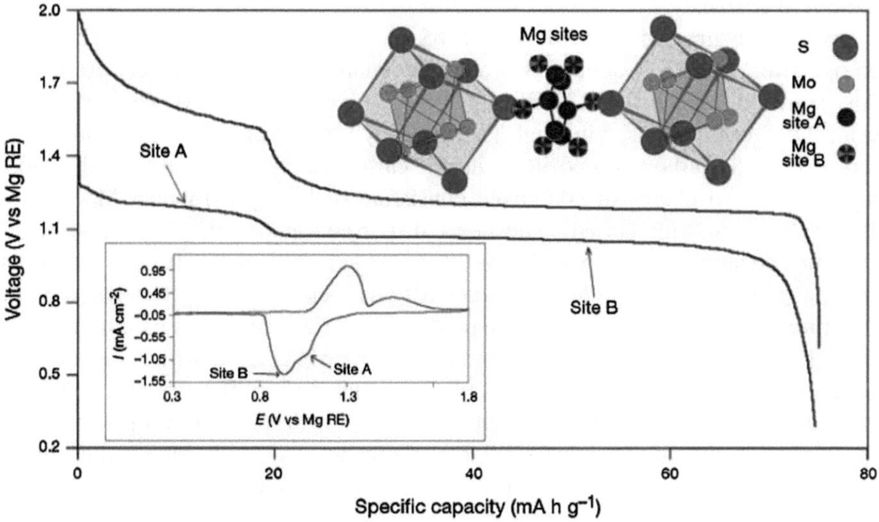

Fig. 10 Electrochemical characteristics of Mg^{2+} intercalation into Mo_3S_4. Cell discharge current density was 0.3 mA/cm^2 and CV scan rate was 0.05 mV/sec. Note the different response to intercalation into site A versus site B. Reprinted with permission from Aurbach et al. [4]. Copyright 2000 Nature Publishing Group

The Site A and Site B portions of the discharge curve represent Mg^{2+} intercalation into the inner and outer sites in the Chevrel phase material, corresponding to intercalation of the first and second Mg^{2+} ions per Mo_6S_8. Coin cells constructed with these materials delivered impressive performance: >2000 discharge/charge cycles at 100 % depth of discharge at practical rates (0.1–1 mA/cm^2) over a wide temperature range of −20 to + 80 °C with <15 % capacity fade. While the low cell voltage is not conducive to construction of high energy density batteries, it appears that this chemistry may be applicable to long cycle life Mg batteries for applications such as stationary energy storage. A key feature of these Chevrel phase compounds is relatively rapid Mg^{2+} diffusivity, attributed to a high concentration of vacant sites with relatively short distances between these sites. The octahedral Mo clusters with delocalized electrons also permit rapid redistribution of the two electrons which provide charge balance for each Mg^{2+} ion inserted. Some irreversible capacity loss attributed to Mg^{2+} ion trapping in the Chevrel phase materials was noted, however. Elevated temperature and substitution of up to 2 S atoms by Se in Mo_6S_8 resulted in improved performance. Lancry et al. [25] noted that the surface of Chevrel phase materials was stable in electrolytes capable of reversible Mg deposition/dissolution and did not form surface films. This could enable the use of nanoparticulate material to improve electrode kinetics via increased specific surface area and reduced diffusion distances without increasing electrolyte decomposition. Chevrel-type materials with stoichiometries such as Mo_6T_6 and $Mo_{15}T_{19}$ and cluster-containing Mo oxides such as $NaMo_4O_6$ and $Mn_{1.5}Mo_8O_{11}$ have been proposed to improve cathode voltage and capacity; other metal clusters which may be useful in this application include Fe_6, Co_6, and Cr_6, although electrochemical data to support this are not available.

Exfoliated "graphene-like" layered MoS_2 prepared by Liang et al. [28] achieved an operating voltage of about 1.8 V and a reversible capacity of about 170 mAh/g. An estimated capacity of 223 mAh/g for fully exfoliated MoS_2 was determined using DFT calculations. Liu et al. [30] used hydrothermal synthesis to produce a series of MoS_2/carbon composites with the carbon interspersed between the MoS_2 layers. A microspheres of composite consisting of highly exfoliated MoS_2 (0.5–1 μm diameter with curled nanosheets dispersed on the surface, 46.04 % C) exhibited superior capacity which was attributed to a synergistic effect of the carbon coating (which improves electronic conductivity) and the highly exfoliated MoS_2 structure (which improves electrolyte penetration). TiS_2 nanotubes prepared by Tao et al. [54] achieved a discharged composition of $Mg_{0.49}TiS_2$ with good capacity retention, while polycrystalline TiS_2 only intercalated about 0.2 mol Mg, consistent with previously results. Small-grain bulk TiS_2 with platelet morphology was speculated to possibly exhibit similar performance. As with MoS_2, this invites speculation that exfoliation of bulk TiS_2 could produce a material with high capacity.

Polyanion compounds such as phospho-olivines have been commercialized as positive electrode materials in lithium ion batteries and may represent a promising class of materials for magnesium batteries. Electrochemical investigation of $MgMnSiO_4$ was conducted by NuLi et al. [39, 40] using DCC electrolytes; at C/20 rate, a discharge capacity of 253.8 mAh/g was reported, corresponding to utilization of 0.81 Mg per unit formula. Magnesium cobalt silicate ($MgCoSiO_4$) synthesized by a solvothermal technique by Zheng et al. [62] was found to produce mesoporous material which exhibited better rate capability, higher specific capacity, and a higher discharge voltage than material obtained with high temperature solid state and molten salt syntheses, presumably due to its larger specific surface area. Li et al. [27] used molten salt techniques to synthesize $MgFeSiO_4$ but only achieved capacities of about 110 mAh/g for material synthesized at 900 °C, similar to molten salt synthesized $MgCoSiO_4$. Thus, solvothermal synthesis of $MgFeSiO_4$ could be an interesting avenue to explore in search of improved performance. $Mg_{0.5}Ti_2(PO_4)_3$ prepared by sol-gel techniques by Makino et al. [31] demonstrated intercalation of up to one Mg^{2+} ion per host unit (136 mAh/g), corresponding to reduction of Ti from IV to III valence. Substitution of Ti with various cations caused variations in unit cell volume and capacity. It would be instructive to prepare and evaluate these phosphates using the solvothermal method found to be optimum for silicate cathodes and an electrolyte capable of reversible Mg electrochemistry. The electronic conductivity of polyanion Mg compounds is likely to be similar to corresponding poorly conductive Li compounds, thus these cathode materials would likely need to use techniques such as nanostructured morphology and carbon coatings to reduce the impact of poor conductivity and ionic diffusion.

Inorganic Mg^{2+} intercalation cathodes tend to experience relatively slow Mg^{2+} diffusion and low intrinsic conductivity which negatively impacts both rate capability and effective capacity at acceptable charge and discharge rates. Weaker intermolecular forces in redox-active organic materials may permit higher Mg^{2+} mobility. Organosulfur compounds which can undergo redox reactions involving cleavage and recombination of S–S bonds such as 2,5-dimercapto-1,3,4-dithiazole (DMcT) and other organic materials such as poly-2,2'-dithiodianiline, [41] poly (2,2,6,6-tetramethylpiperidinyloxy-4-yl methacrylate, [11] and 2,5-dimethoxy-1,4-benzoquinone (DMBQ; Fig. 11a) have been evaluated as Mg cathode materials. DMBQ was determined to undergo 2-electron redox reactions with the insertion and deinsertion of one Mg^{2+} ion per DMBQ molecule and achieved a first cycle discharge capacity of about 250 mAh/g, remaining at about 200 mAh/g after 5 cycles. [45] The high relative capacity of DMBQ merits evaluation with an electrolyte capable of reversible Mg deposition and dissolution.

2.2 Conversion Cathodes

Sulfur has a high theoretical capacity of 1671 mAh/g and is thus an attractive cathode material, however successful demonstration of secondary Mg/S batteries was not achieved until non-nucleophilic electrolytes such as GEN1 were discovered; a first cycle discharge capacity of 1200 mAh/g S (vs. 1675 mAh/g theoretical capacity based on a 2-electron reduction to form MgS) was achieved by Kim et al. [23] using such electrolytes. Cell voltage was <1 V, however. Significant

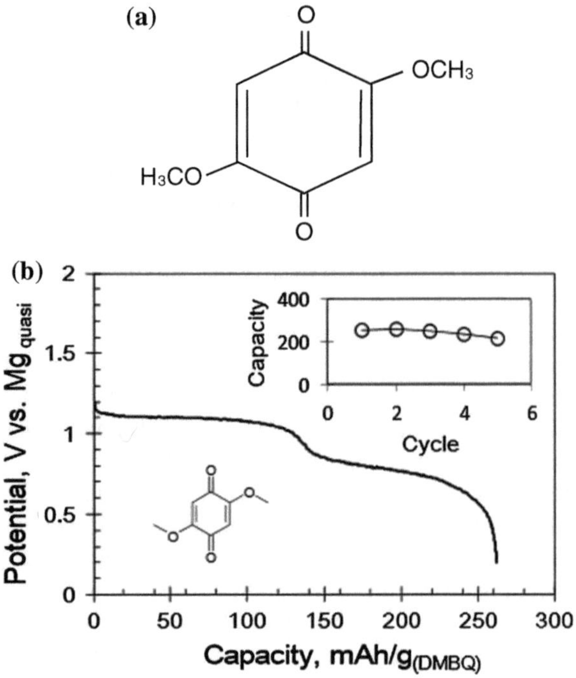

Fig. 11 **a** DMBQ **b** Discharge behavior of the cell Mg|0.5 M $Mg(ClO_4)_2$-γ-butyrolactone| DMBQ-acetylene black-PTFE (discharge rate not specified). Reprinted with permission from Sano et al. [45]. Copyright 2012 Chemical Society of Japan

charge/discharge capacity mismatch was attributed to a shuttling mechanism caused by soluble polysulfide species in the electrolyte similar to that observed in Li/S batteries. Post-mortem cell analysis revealed a yellow discoloration of the separator, indicative of the presence of polysulfide species; this was also supported by XPS analysis of the cathode. Thus, reduction in polysulfide solubility, perhaps via use of proper electrolyte additives, could potentially improve performance of Mg–S batteries although low cell voltage may still preclude practical implementation of this chemistry.

Primary and mechanically rechargeable Mg/air batteries have been demonstrated but electrically rechargeable Mg–O_2 batteries have received little attention. Conceptual Mg/O_2 batteries are similar to those of rechargeable Li/O_2 batteries, with either aqueous electrolytes with a metal ion conducting membrane to protect the negative electrode from corrosion by the electrolyte or non-aqueous electrolytes capable of supporting reversible oxygen electrochemistry. The lack of a Mg^{2+} conducting solid which is stable in contact with aqueous electrolytes hampers development of aqueous Mg/O_2 batteries. Conventional non-aqueous Mg/O_2 batteries may not be feasible since the probable discharge product, MgO, is considered to be electrochemically irreversible at ambient temperature and is non-conductive and insoluble in organic electrolytes. Shiga et al. [48] demonstrated a catalytic cycle for recharge of a Mg–O_2 battery using an I_2/DMSO complex as shown in Fig. 12a. Figure 12b shows electrochemical results.

Fig. 12 **a** Catalytic cycle for $Mg–O_2$ battery recharge **b** Charge/discharge curves for the I_2 mediated Mg/O_2 battery. Reprinted with permission Shiga et al. [48]. Copyright 2013 Royal Society of Chemistry

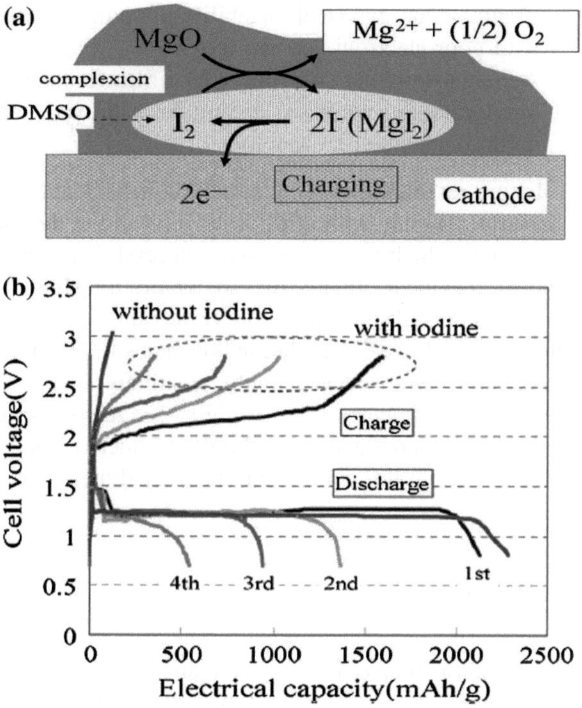

Little work on metal/metal halide conversion cathodes for use in rechargeable Mg batteries has been reported. Solubility of the cathode material in the electrolyte is a potential issue for these materials. Graphite fluorides have been evaluated in high specific energy primary but not secondary Mg batteries; MgF_2 was identified as a discharge product and thus reversibility is questionable [17]. WS_2 can function as a conversion cathode in primary Mg batteries with solid polymer electrolytes but again reversibility was not determined [18]. High temperature molten salt Mg/conversion cathode batteries have also been evaluated, but low cell voltage, poor rate capability, and high polarization were typically noted [55].

Direct comparison of the aforementioned results is difficult due to the use of widely varying experimental conditions. However, a few commonalities and conclusions may be inferred from the existing body of work:

1. Slow solid state diffusion of Mg^{2+} ions in many of the cathode materials leads to increased electrode polarization and ion trapping, resulting in low capacity and poor rate capability.
2. Possible co-intercalation of water and solvent molecules has been inferred, which may help shield the charge on the Mg^{2+} ion but can also cause severe structural distortion as well as decompose within the cathode material. Water incompatibility with the anode is another issue.

3. Chemical and physical instability of some intercalates has been observed along with poor electronic conductivity.
4. Apparent minimal cathode surface film formation allows the use of nanostructured cathode materials leading to improved performance, but at the cost of reduced volumetric capacity.

Future research involving structural modifications of intercalation cathodes may be fruitful. Doping with other metal ions and/or the use of multimetal compositions, techniques which have been very successful in lithium ion battery cathode development, may be useful avenues to explore for Mg batteries and blending of different cathode materials may also prove advantageous. Fundamental studies involving elucidation of the intercalation mechanism, including identification of the exact intercalating Mg species, could positively impact both cathode and electrolyte development. Irreversibility of high voltage conversion reactions due to formation of stable, non-conductive discharge products hampers use of such cathodes. Such thermodynamic limitations may be very difficult to overcome.

3 Summary

One would perceive the major advantages of magnesium as an anode are a higher volumetric capacity and the apparent lack of dendrite formation during charging which is highly dependent on the electrolyte and overcomes major safety and performance challenges encountered with the use of a lithium metal anode. However, the solid state chemistry and electrochemistry of magnesium is substantially more complex than that of the monovalent alkali metals due its higher charge density, which leads to greater difficulty in identifying and implementing practical battery chemistry. As highlighted here, the major developments for a high energy multivalent metal battery are rooted in material discovery of both high energy cathodes and noncorrosive electrolytes. The most significant issues for the development of magnesium batteries stems from the compatibility of the electrolyte with both the anode and the cathode and the discovery of high energy cathodes capable of undergoing multiple electron transfers to the same metal center with rapid diffusion of magnesium ion in the solid state. Nanotechnology may offer an opportunity to help reduce the impact of slow solid state diffusion as well as structural distortion during intercalation/deintercalation but also tends to reduce the volumetric capacity of the cathode structure. It is important to note that formation of passivation layers on the cathode and the anode by oxidation or reduction of the electrolyte might impede the search for cathodes capable of intercalating magnesium. Such trade-offs must be identified and optimized in the quest to develop practical rechargeable magnesium batteries.

References

1. Amatucci GG, Badway F, Singhal A et al (2001) Investigation of yttrium and polyvalent ion intercalation into nanocrystalline vanadium oxide. J Electrochem Soc 148:A940–A950. doi:10.1149/1.1383777
2. Armand M, Tarascon J-M (2008) Building better batteries. Nature 451:652–657. doi:10.1038/451652a
3. Arthur TS, Singh N, Matsui M (2012) Electrodeposited Bi, Sb and $Bi_{1-x}Sb_x$ alloys as anodes for Mg-ion batteries. Electrochem Commun 16:103–106. doi:10.1016/j.elecom.2011.12.010
4. Aurbach D, Lu Z, Schechter A et al (2000) Prototype systems for rechargeable Magnesium batteries. Nature 407:724–727. doi:10.1038/35037553
5. Aurbach D, Weissman I, Gofer Y, Levi E (2003) Nonaqueous Magnesium electrochemistry and its application in secondary batteries. Chem Rec 3:61–73. doi:10.1002/tcr.10051
6. Aurbach D, Suresh GS, Levi E, Mitelman A, Mizrahi O, Chusid O, Brunelli M (2007) Progress in rechargeable Magnesium battery technology. Adv Mater 19:4260–4266. doi:10.1002/adma.200701495
7. Bochmann M, Sarsfield MJ (1998) Reaction of AlR_3 with $[CPh_3][B(C_6F_5)_4]$: facile degradation of $[B(C_6F_5)_4]^-$ by transient "$[AlR_2]^+$."Organometallics 17:5908–5912. doi: 10.1021/om980400j
8. Brenner A (1971) Note on the electrodeposition of Magnesium from an organic solution of a Magnesium–Boron complex. J Electrochem Soc 118:99. doi:10.1149/1.2407964
9. Bruce PG, Krok F, Nowinski J et al (1991) Chemical intercalation of Magnesium into solid hosts. J Mater Chem 1:705–706. doi:10.1039/JM9910100705
10. Carter TJ, Mohtadi R, Arthur TS et al (2014) Boron clusters as highly stable Magnesium-battery electrolytes. Angew Chem Int Ed 53:3173–3177. doi:10.1002/anie.201310317
11. Chen Q, NuLi YN, Guo W, Yang J, Wang JL, Guo YG (2013) PTMA/Graphene as a novel cathode material for rechargeable Magnesium batteries. Acta Phys Chim Sin 29:2295–2299. doi:10.3866/PKU.WHXB201309241
12. Connor JH, Reid WE, Wood GB (1957) Electrodeposition of metals from organic solutions v. electrodeposition of Magnesium and Magnesium alloys. J Electrochem Soc 104:38–41. doi:10.1149/1.2428492
13. Ding F, Xu W, Graff GL et al (2013) Dendrite-free Lithium deposition via self-healing electrostatic shield mechanism. J Am Chem Soc 135:4450–4456. doi:10.1021/ja312241y
14. Doe RE, Han R, Hwang J et al (2013) Novel, electrolyte solutions comprising fully inorganic salts with high anodic stability for rechargeable Magnesium batteries. Chem Commun 50:243–245. doi:10.1039/C3CC47896C
15. Gaddum LW, French HE (1927) The electrolysis of grignard solutions1. J Am Chem Soc 49:1295–1299. doi:10.1021/ja01404a020
16. Gershinsky G, Yoo HD, Gofer Y, Aurbach D (2013) Electrochemical and spectroscopic analysis of Mg^{2+} intercalation into thin film electrodes of layered oxides: V_2O_5 and MoO_3. Langmuir 29:10964–10972. doi:10.1021/la402391f
17. Giraudet J, Claves D, Guérin K et al (2007) Magnesium batteries: towards a first use of graphite fluorides. J Power Sources 173:592–598. doi:10.1016/j.jpowsour.2007.04.067
18. Gregory TD, Hoffman RJ (1987) Electrochemical generator utilizing solid polymer-salt complex. United States Patent: 4702974
19. Gregory TD, Hoffman RJ, Winterton RC (1990) Nonaqueous electrochemistry of Magnesium: applications to energy storage. J Electrochem Soc 137:775–780. doi:10.1149/1.2086553
20. Ichitsubo T, Adachi T, Yagi S, Doi T (2011) Potential positive electrodes for high-voltage Magnesium–Ion batteries. J Mater Chem 21:11764–11772. doi:10.1039/C1JM11793A
21. Inamoto M, Kurihara H, Yajima T (2013) Vanadium Pentoxide-based composite synthesized using microwave water plasma for cathode material in rechargeable Magnesium batteries. Materials 6:4514–4522. doi:10.3390/ma6104514

22. Jiao L-F, Yuan H-T, Si Y-C et al (2006) Synthesis of Cu0.1-doped vanadium oxide nanotubes and their application as cathode materials for rechargeable Magnesium batteries. Electrochem Commun 8:1041–1044. doi:10.1016/j.elecom.2006.03.043
23. Kim HS, Arthur TS, Allred GD et al (2011) Structure and compatibility of a Magnesium electrolyte with a sulphur cathode. Nat Commun 2:427. doi:10.1038/ncomms1435
24. Koch VR, Nanjundiah C, Orsini M (1995) Rechargeable Magnesium power cells. NASA Technical Support Package MSC-22293
25. Lancry E, Levi E, Gofer Y, Levi MD, Aurbach D (2005) The effect of milling on the performance of a Mo_6S_8 Chevrel phase as a cathode material for rechargeable Mg batteries. J Solid State Electrochem 9:259–266. doi:10.1007/s10008-004-0633-7
26. Le DB, Passerini S, Coustier F et al (1998) Intercalation of polyvalent cations into V_2O_5 aerogels. Chem Mater 10:682–684. doi:10.1021/cm9705101
27. Li Y, Nuli Y, Yang J et al (2011) $MgFeSiO_4$ prepared via a molten salt method as a new cathode material for rechargeable Magnesium batteries. Chin Sci Bull 56:386–390. doi:10.1007/s11434-010-4247-4
28. Liang Y, Feng R, Yang S et al (2011) Rechargeable Mg batteries with graphene-like MoS_2 cathode and ultrasmall Mg nanoparticle anode. Adv Mater 23:640–643. doi:10.1002/adma.201003560
29. Liu T, Shao Y, Li G et al (2014) A facile approach using $MgCl_2$ to formulate high performance Mg^{2+} electrolytes for rechargeable Mg batteries. J Mater Chem A 2:3430–3438. doi:10.1039/C3TA14825D
30. Liu Y, Jiao L, Wu Q et al (2013) Sandwich-structured graphene-like MoS_2/C microspheres for rechargeable Mg batteries. J Mater Chem A 1:5822–5826. doi:10.1039/C3TA10786H
31. Makino K, Katayama Y, Miura T, Kishi T (2001) Electrochemical insertion of Magnesium to $Mg_{0.5}Ti_2(PO_4)_3$. J Power Sources 99:66–69. doi:10.1016/S0378-7753(01)00480-3
32. Mayer A (1990) Electrodeposition of Aluminum, Aluminum/Magnesium Alloys, and Magnesium from organometallic electrolytes. J Electrochem Soc 137:2806–2809. doi:10.1149/1.2087078
33. Mohtadi R, Matsui M, Arthur TS, Hwang S-J (2012) Magnesium Borohydride: from Hydrogen storage to Magnesium battery. Angew Chem Int Ed 51:9780–9783. doi:10.1002/anie.201204913
34. Muldoon J, Bucur CB, Gregory T (2014) The quest for non-aqueous multivalent secondary batteries: Magnesium and beyond. Chem Rev (accepted)
35. Muldoon J, Bucur CB, Oliver AG et al (2012) Electrolyte roadblocks to a Magnesium rechargeable battery. Energy Environ Sci 5:5941–5950. doi:10.1039/C2EE03029B
36. Muldoon J, Bucur CB, Oliver AG et al (2013) Corrosion of Magnesium electrolytes: chlorides—the culprit. Energy Environ Sci 6:482–487. doi:10.1039/C2EE23686A
37. Novák P, Desilvestro J (1993) Electrochemical insertion of Magnesium in metal oxides and sulfides from aprotic electrolytes. J Electrochem Soc 140:140–144. doi:10.1149/1.2056075
38. Novák P, Scheifele W, Joho F, Haas O (1995) Electrochemical insertion of Magnesium into hydrated Vanadium bronzes. J Electrochem Soc 142:2544–2550. doi:10.1149/1.2050051
39. NuLi Y, Yang J, Li Y, Wang J (2010) Mesoporous Magnesium Manganese silicate as cathode materials for rechargeable Magnesium batteries. Chem Commun 46:3794–3796. doi:10.1039/C002456B
40. NuLi Y, Yang J, Wang J, Li Y (2009) Electrochemical intercalation of Mg^{2+} in Magnesium Manganese Silicate and its application as high-energy rechargeable Magnesium battery cathode. J Phys Chem C 113:12594–12597. doi:10.1021/jp903188b
41. NuLi Y, Guo Z, Liu H, Yang J (2007) A new class of cathode materials for rechargeable Magnesium batteries: organosulfur compounds based on Sulfur–Sulfur bonds. Electrochem Commun 9:1913–1917. doi:10.1016/j.elecom.2007.05.009
42. Pour N, Gofer Y, Major DT, Aurbach D (2011) Structural analysis of electrolyte solutions for rechargeable Mg batteries by stereoscopic means and DFT calculations. J Am Chem Soc 133:6270–6278. doi:10.1021/ja1098512

43. Rasul S, Suzuki S, Yamaguchi S, Miyayama M (2011) Microstructural effects on the Mg–ion intercalation mechanism in MnO_2/Acetylene black composite cathodes for Magnesium–Ion rechargeable batteries. Abstract 615, 220th ECS Meeting, Boston
44. Saha P, Kanchan Datta M, Velikokhatnyi OI, Manivannan A, Alman D, Kumta PN (2014) Rechargeable Magnesium battery: current status and key challenges for the future. Prog Mat Sci 66:1–86. doi:10.1016/j.pmatsci.2014.04.001
45. Sano H, Senoh H, Yao M, Sakaebe H, Kiyobayashi T (2012) Mg^{2+} storage in organic positive-electrode active material based on 2,5-Dimethoxy-1,4-benzoquinone. Chem Lett 41:1594–1596. doi:10.1246/cl.2012.1594
46. Schaefer JL, Lu Y, Moganty SS, Agarwal P, Jayaprakash N, Archer LA (2012) Electrolytes for high-energy lithium batteries. Appl Nanosci 2:91–109. doi:10.1007/s13204-011-0044-x
47. Shao Y, Gu M, Li X et al (2014) Highly reversible Mg insertion in nanostructured Bi for Mg Ion batteries. Nano Lett 14:255–260. doi:10.1021/nl403874y
48. Shiga T, Hase Y, Kato Y et al (2013) A rechargeable non-aqueous Mg–O_2 battery. Chem Commun 49:9152–9154. doi:10.1039/C3CC43477J
49. Shterenberg I, Salama M, Gofer Y et al (2014) The challenge of developing rechargeable Magnesium batteries. MRS Bull 39:453–460. doi:10.1557/mrs.2014.61
50. Singh N, Arthur TS, Ling C et al (2012) A high energy-density tin anode for rechargeable Magnesium–Ion batteries. Chem Commun 49:149–151. doi:10.1039/C2CC34673G
51. Spahr ME, Novák P, Haas O, Nesper R (1995) Electrochemical insertion of Lithium, Sodium, and Magnesium in Molybdenum(VI) oxide. J Power Sources 54:346–351. doi:10.1016/0378-7753(94)02099-O
52. Sutto TE, Duncan TT (2012) Electrochemical and structural characterization of Mg ion intercalation into RuO_2 using an ionic liquid electrolyte. Electrochim Acta 79:170–174. doi:10.1016/j.electacta.2012.06.099
53. Sutto TE, Duncan TT (2012) Electrochemical and structural characterization of Mg ion intercalation into Co_3O_4 using ionic liquid electrolytes. Electrochim Acta 80:413–417. doi:10.1016/j.electacta.2012.07.050
54. Tao Z-L, Xu L-N, Gou X-L, et al (2004) TiS2 nanotubes as the cathode materials of Mg–ion batteries. Chem Commun 18:2080–2081. doi: 10.1039/B403855J
55. Uhler EF (1963) Investigation of new cathode–anode couples for secondary batteries using molten salt electrolytes. U.S. Air Force Systems Command Technical Documentary Report ASD-TDR-63–115
56. Whittingham MS (2004) Lithium batteries and cathode materials. Chem Rev 104:4271–4302. doi:10.1021/cr020731c
57. Xu K (2004) Nonaqueous liquid electrolytes for Lithium-based rechargeable batteries. Chem Rev 104:4303–4418. doi:10.1021/cr030203g
58. Yoo HD, Shterenberg I, Gofer Y, Gershinsky G, Pour N, Aurbach D (2013) Mg rechargeable batteries: an on-going challenge. Energy Environ Sci 6:2265–2279. doi:10.1039/C3EE40871J
59. Yu L, Zhang X (2004) Electrochemical insertion of magnesium ions into V_2O_5 from aprotic electrolytes with varied water content. J Colloids Interface Sci 278:160–165. doi: 10.1016/j.jcis.2004.05.028
60. Zhang R, Yu X, Nam K-W et al (2012) α-MnO_2 as a cathode material for rechargeable Mg batteries. Electrochem Commun 23:110–113. doi:10.1016/j.elecom.2012.07.021
61. Zheng G, Lee SW, Liang Z et al (2014) Interconnected hollow carbon nanospheres for stable Lithium metal anodes. Nat Nanotechnol 9:618–623. doi:10.1038/nnano.2014.152
62. Zheng Y, NuLi Y, Chen Q et al (2012) Magnesium Cobalt silicate materials for reversible Magnesium Ion storage. Electrochim Acta 66:75–81. doi:10.1016/j.electacta.2012.01.037

Organic Cathode Materials for Rechargeable Batteries

Ruiguo Cao, Jiangfeng Qian, Ji-Guang Zhang and Wu Xu

1 Introduction

Lithium-ion batteries based on intercalation chemistry have dominated the battery market in the applications for portal electronics and electric vehicles since Sony released the world's first commercial lithium-ion battery products in 1991 [1]. In conventional lithium-ion batteries, lithium intercalation compounds such as lithium transition metal oxides (e.g. $LiCoO_2$) or lithium transition metal phosphates (e.g. $LiFePO_4$) are used as the cathode materials where lithium ions are de-intercalated from the cathode during charge process and reversibly intercalated into the cathode during discharge process, while graphite carbon is widely used as the anode material in which the lithium ions are intercalated during charge process and de-intercalated during discharge process of the batteries. Such inorganic cathode materials (e.g. $LiCoO_2$ and $LiFePO_4$) can provide a typical specific capacity of 140–170 mAh g^{-1} [2–4]. Lithium- and manganese-rich (LMR) layered-structure cathode materials deliver a much higher specific capacity ($\sim$250 mAh g^{-1}) than traditional cathode materials such as $LiFePO_4$ and $LiCoO_2$ [5–8]. However, significant challenges, including voltage fade and limited cycle life in LMR cathodes still remain to be overcome prior to their large-scale market penetration [9–11].

After more than two decades of persistent research and development on lithium-ion batteries, the specific capacities of the electrode materials and the energy densities (both gravimetric and volumetric) of the batteries at cell level are approaching their intrinsic theoretical values [9, 12]. On the other hand, due to the limited solid-state diffusion coefficient, the intercalation/de-intercalation mechanism for most of the present inorganic cathode materials limits their high-rate performance for high-power applications [13]. Another concern on inorganic cathode materials comes from the

R. Cao · J. Qian · J.-G. Zhang · W. Xu (✉)
Energy and Environment Directorate, Pacific Northwest National Laboratory, Richland, WA 99354, USA
e-mail: wu.xu@pnnl.gov

Z. Zhang and S.S. Zhang (eds.), *Rechargeable Batteries*,
Green Energy and Technology, DOI 10.1007/978-3-319-15458-9_23

limited resource and high cost of transition metals especially cobalt. On the other hand, most of inorganic cathode materials are produced from mining process which is not renewable. To further develop greener rechargeable batteries, organic cathode materials based on conversion reactions are alternative candidates for high-capacity and low-cost batteries. The advantages of organic cathode materials stem from the nature of organic chemistry and these cathode materials could be potentially generated from renewable natural resources and easily tuned through changing the chemical environment surrounded the redox centers [14]. The flexible nature and high energy density of the organic cathode materials also have advantages over heavy inorganic materials to satisfy the need for the future wearable electronics [15–17].

The idea for organic cathode materials was initiated in the late 1960s, which is even earlier than the existence of the intercalated inorganic cathode materials [2, 18]. Although organic cathode materials have slightly longer history than that of the intercalation inorganic cathode materials, the research progress for organic electrode materials began to lag behind their inorganic counterparts when layered metal oxides were found to have good reversibility of insertion/de-insertion by lithium ions. Since then, inorganic cathode materials have taken the center stage in lithium-ion battery research, because they could provide higher cell voltage, lithium source and, more importantly, longer cycle life. With the extensive development of various inorganic cathode materials in the last several decades, the specific energy of these cathode materials is expected to reach their theoretical limits in the next decade. Therefore, the pursue for alternative cathode materials, including organic cathode materials, has attracted increased attention in recent years [19–22]. With increased research efforts, certain organic cathode materials have demonstrated electrochemical performance comparable or ever superior to the conventional inorganic cathode materials in terms of energy density, power density and cycling stability [21].

Generally, organic electrode materials can be categorized into different types based on their electrochemically active groups involved in the redox reactions, including conjugated carbonyl compounds, conducting polymers, organosulfides and free radicals. Conjugated carbonyl compounds were first discovered for energy storage materials in 1969 [18]. Carbonyl compounds stabilized with a variety of different substituent R-groups, i.e. R-(C=O)-R′, undergo a reversible redox reaction through an enolization mechanism forming R-($C^{\cdot}$-O^{-})-R′. Quinone is one of the most well-known examples that have carbonyl groups and exhibit fast reversible redox reactions. Besides quinones, many other conjugated carbonyl compounds can also be used to accommodate lithium ions through a reversible redox reaction. Benefit from their small unit molecular weight and fast reaction kinetics, conjugated carbonyl compounds can usually achieve high specific capacity and high power density. In 1977, some of the conducting polymers were found to exhibit remarkably high conductivities when subjected to a variety of chemical dopings [23]. After that, conducting polymers, such as polyacetylene, polyphenylene, polypyrrole, polythiophene, and polyaniline, have been extensively investigated for charge storage in lithium batteries [19]. The advantages of conducting polymers as electrode materials for charge storage include their high conductivity, flexibility and easy processibility. The drawback is the relatively low capacity restricted by the doping level. Organosulfides can be functionalized as organic electrode materials taking advantage of the S–S bond which can be broken and rebuilt reversibly. Two electrons are involved in the redox reaction of the S–S bond so that organosulfides are expected to deliver a higher capacity than conducting polymers. However, the high dissolution and slow kinetics of organosulfides greatly impede their electrochemical performances and limit their practical applications. On the other hand, some of the organic cathodes based on free radicals have found their applications in cases for extremely high rate but low energy-density requirements.

Besides the applications in lithium batteries, the implementation of organic electrode materials for sodium batteries has attracted significant interests [24–30]. Comparing to lithium batteries, sodium batteries are desirable to provide low-cost energy storage devices because sodium has much rich natural abundance, easy access to sources and, consequently, low price [30, 31]. Sodium has a suitable redox potential ($E^{\circ}_{(Na/Na^{+})} = -2.71$ V vs. standard hydrogen electrode, or 0.33 V above that of Li/Li^{+}) and similar intercalation chemistry to lithium, which make it possible to be a substituent of lithium for battery applications. However, although the studies on lithium-metal and lithium-ion batteries provide the valuable experiences and examples for researches on sodium-metal and sodium-ion batteries, the differences in size and bonding strength between lithium ions and sodium ions lead to different thermodynamic and/or kinetic properties for sodium batteries than for lithium batteries [13]. Most studied organic electrode materials for sodium batteries include conjugated carbonyl compounds with C=O functionalities which can accommodate sodium ions via an enolization reaction through C=O bond and sodium. Conducting polymers have also been proposed to act as electrochemically active materials used for electrode components in organic sodium batteries.

This chapter will review the recent advances on the development of organic electrode materials for their applications in three main areas, including rechargeable lithium batteries, sodium batteries and redox flow batteries. Four kinds of organic cathode materials, including conjugated carbonyl compounds, conducting polymers, organosulfides and free radical polymers, will be discussed in terms of their electrochemical performances in these three battery systems. Fundamental issues related to the synthesis-structure-activity correlations of these organic cathodes, as well as the working principles in these energy storage systems and their capacity fading mechanisms will also be discussed.

2 Organic Cathode Materials for Lithium-Based Energy Storage Systems

The organic materials in rechargeable lithium batteries have several advantages over inorganic electrode materials. First, organic electrode materials are usually lighter than their inorganic counterpart so that they could deliver higher gravimetric energy densities than the latter. Second, organic electrode materials are more flexible to be adapted in the electrode processing for the future wearable electronics. Third, taking advantage of versatility of organic reactions, the property of electrochemical redox reaction of organic electrode materials could be tuned by changing the chemical structure of functional groups. In addition, the adoption of organic electrode materials in lithium-based batteries can aid in the design of greener electrochemical storage devices and simplify the battery recycle processes. From the viewpoint of sustainability and renewability, organic electrode materials could be potentially extracted or made from renewable natural resources (biomass). For example, the oxocarbon salt $Li_2C_6O_6$ containing carbonyl groups as redox active center can be synthesized through potentially low cost processes and free of toxic solvents and by enlisting the use of natural organic sources (CO_2-harvesting entities) [32]. In this section, we will introduce four types of organic cathode materials for lithium-based batteries: conjugated carbonyl compounds, free radical polymers, conducting polymers, and organosulfide compounds.

2.1 Conjugated Carbonyl Compounds

The first attempt to use organic cathode materials for lithium batteries started 40 years ago when Williams et al. [18] proposed dichloroisocyanuric acid, a positive N-chloro compound with carbonyl functionalities, as a high energy density material for a primary lithium battery. Later, Alt et al. [33] investigated the possibility of quinone compounds with carbonyl functionalities as secondary (i.e. rechargeable) battery cathode materials. Since then, lots of carbonyl compounds have been investigated for cathode or anode materials in lithium batteries [21, 34–41]. The redox mechanism of carbonyl functionalities can be generalized as an enolization reaction and a reverse reaction of the carbonyl group (Fig. 1) [21, 22, 42]. When stabilized by conjugation, the redox system is most often reversible through a one-electron reduction to form a radical monoanion and has fast reaction kinetics [42]. Quinone-type structures are the most common organic molecules with carbonyl groups as the redox center. Some small organic molecules based on conjugated carbonyl groups with their theoretical specific capacity values are shown

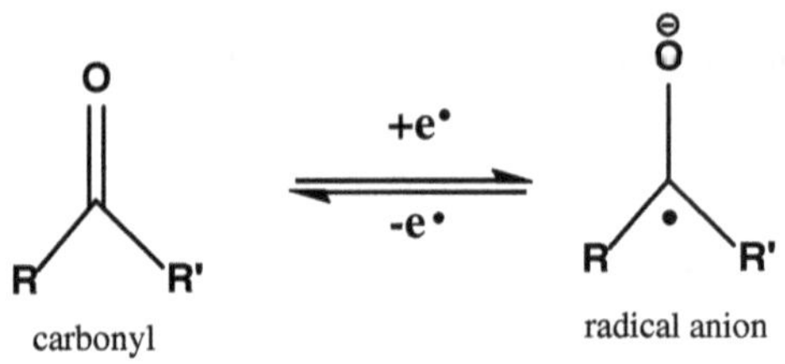

Fig. 1 Typical redox reaction mechanism for an organic cathode material with a carbonyl redox center

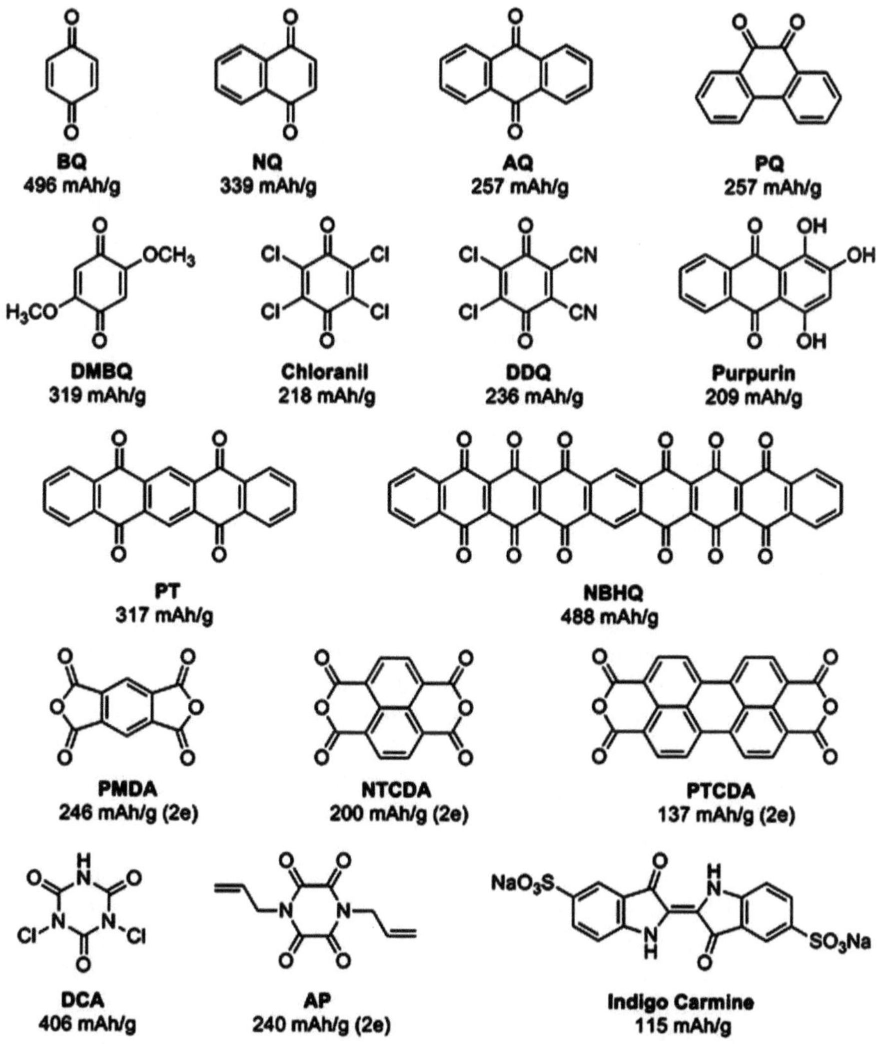

Fig. 2 Typical small organic molecules based on conjugated carbonyl groups proposed as positive electrode materials with their theoretical specific capacity values. Reprinted with permission from Ref. [21]. Copyright (2013) Royal Society of Chemistry

in Fig. 2 [11]. Owing to their multiple electron redox reaction nature and relatively light molecular weight per unit, conjugated carbonyl compounds usually deliver high specific capacity (typically more than 200 mAh g^{-1}) for lithium storage. Based on their high specific capacity and fast reaction kinetics, as well as the structure diversity, conjugated carbonyl compounds are believed to be the most promising organic electrode materials for cathode application in rechargeable lithium batteries.

The major problem of small quinone-type molecules as electrode materials in non-aqueous lithium batteries stems from the solubility of active materials in organic electrolytes, usually leading to poor cycling stability. In order to address this issue, one successful strategy is to link small molecules with quinone groups to form larger molecular sizes and accordingly to reduce the solubility in organic electrolytes. An insoluble oligomer with a larger molecular size and a higher molecular weight, nonylbenzo-hexaquinone (NBHQ), has been investigated in organic electrolytes, and a good cycling performance up to 500 cycles with a practical working potential was achieved [36, 43]. However, the low specific capacity and slow reaction kinetics of NBHQ makes it less promising in practical battery applications. Another strategy to tackle the dissolution problem of quinone-type cathode materials is to incorporate small quinone-type molecules into mesoporous pores to improve the retention of active materials on electrodes through confinement effect [44]. However, even though the small molecules were incorporated in mesoporous carbon materials, the cycling stability still suffered from the slow dissolution leading to irreversible capacity degradation.

Another approach to prevent the dissolution of small quinone species have been thoroughly studied through polymerization of quinone-type monomers [37, 45, 46]. For example, the monomer of 5-amino-1,4-naphthoquinone (ANQ) was polymerized through electrochemical or chemical process to form a redox polymer poly (5-amino-1,4-naphthoquinone) (PANQ), which delivered a capacity of 300 Ah kg^{-1} in the non-aqueous electrolyte and retained 200 Ah kg^{-1} after 17 cycles [46]. Han et al. [37] developed a relatively stable positive electrode through polymerization of 3,4,9,10-perylene-tetracarboxylicacid-dianhydride (PTCDA). A mixture of PTCDA and excess sulfur could form a PTCDA sulfide polymer through calcination at 400–500 °C under argon atmosphere. The resulted carbonyl polymer compounds with thioether bonds between PTCDA units showed a discharge capacity of 131 mAh g^{-1} at a current density of 100 mA g^{-1} and exhibited excellent cycling performance with almost no degradation after 250 cycles. The improved cycling stability can be attributed to the improvement of the electronic conductivity and the prevention of dissolution of active materials (Fig. 3). It is noteworthy that the drawback for the PTCDA sulfide polymer is its relatively low energy density.

Polyimides can be synthesized by simple polycondensation from dianhydrides and diamines, which contain rich carbonyl functionalities and can potentially deliver high specific capacities. Song et al. have synthesized a series of typical polyimides and studied their electrochemical energy storage performance [47]. It was found that the polyimide based on 1,4,5,8-naphthalenetetracarboxylic dianhydride (NTCDA) showed the best cycling performance while the PTCDA-based polyimide delivered the highest energy density. Another finding was that the NTCDA-based polyimide tended to improve the discharge voltage and to reduce the voltage gap between charge and discharge plateaus. In another study, density functional theory (DFT) calculations indicated that the six-membered cyclic 1,2-diketones served as an excellent core structure because of the high redox energy change resulted from the favorable coordination of the oxygen atoms to lithium and the aromaticity of the reduced form [48]. Pyrene-4,5,9,10-tetraone units, which contains two

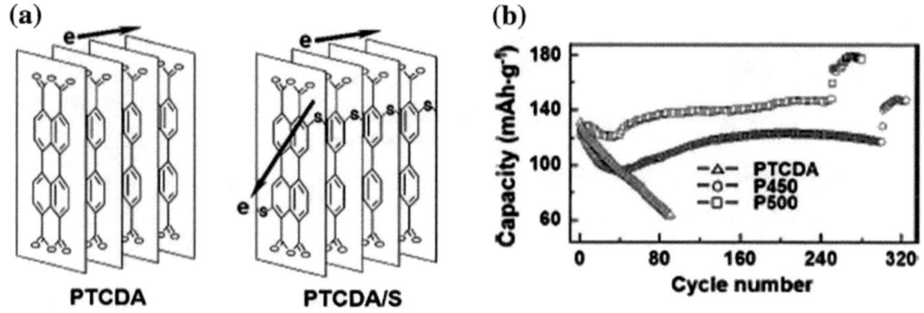

Fig. 3 **a** Schematic diagram showing the contribution of thioether bonds to the electronic conductivity. **b** Discharge capacity versus cycle number (current density: 100 mAh g^{-1}) for PTCDA and the sulfide polymers (P450 and P500 represent the sulfide polymers formed under argon atmosphere at 450 and 500 °C, respectively). Reprinted from Ref. [37] with permission from WILEY-VCH

six-membered-ring 1,2-diketones units, was anchored onto poly-(methacrylate) resulting in a polymer (PPYT) with porous structure. PPYT showed excellent charge-discharge ability based on a reversible lithium insertion/de-insertion mechanism and exhibited surprisingly outstanding cell performance with high capacity and excellent cyclability, as well as good rate capability.

Despite of polymerization, immobilization of small carbonyl compounds on certain solid substrates through chemical or physical bonding can also prevent the dissolution of active materials from electrodes. For example, a quinone derivative of calix[4]arene immobilized onto the surface of silica nanoparticles can significantly mitigate the dissolution of small organic molecules into organic electrolytes [49]. However, the partial loss of capacity is a big problem for this strategy. The improvement of energy and power density can be realized by substituting a non-conductive substrate with a conductive material. Recently, Lee et al. [50] demonstrated that active organic materials with carbonyl redox centers were immobilized onto conductive scaffolds through non-covalent bonding resulting in a self-standing and flexible nanohybrid organic electrode (Fig. 4). The nanohybrid organic electrodes exhibited surprisingly high capacity retention with nearly no capacity degradation after 100 cycles. More recently, a binder-free and flexible polyimide/single-walled carbon nanotube (PI/SWCNT) composite electrode (Fig. 5) has been developed as the cathode for lithium batteries, which showed very good cycling stability and rate capability [51]. The PI/SWCNT was synthesized by in-situ polymerization giving an extremely light weight electrode with significantly improved electronic conductivity. The morphology of polyimide on carbon nanotubes can be controlled to form nanoflake structures deposited on one side of the SWCNT film. The nanostructure of polyimide nanoflakes coupled with good conductive SWCNT leads to a good rate capability and a promising cycling stability in lithium batteries.

Anion-based organic salts with permanent negative charges generally show much poorer solubility than their neutral organic counterparts owing to their high

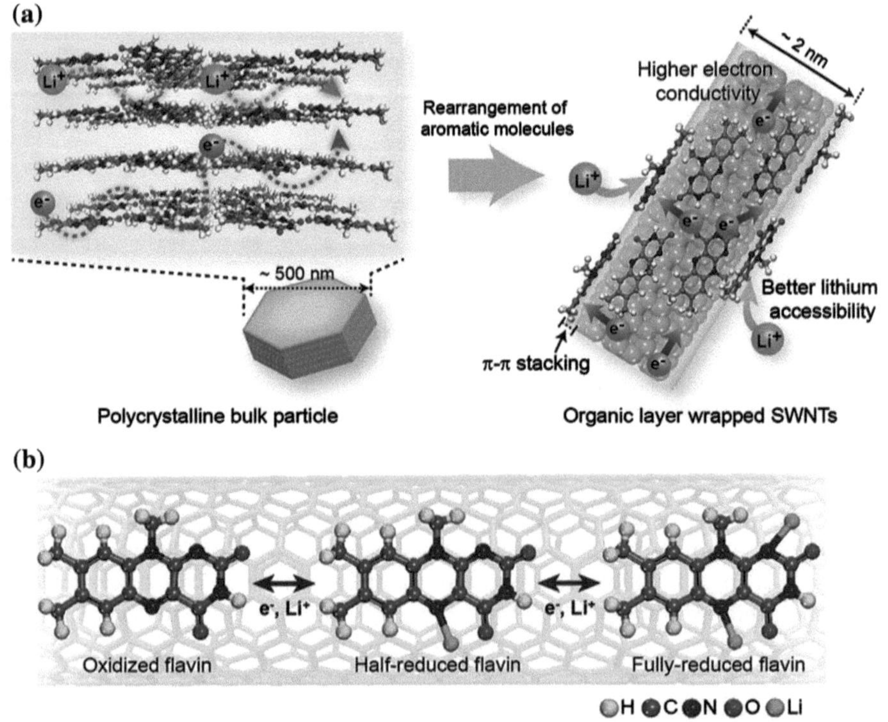

Fig. 4 a Nanohybridization strategy of aromatic redox molecules with SWCNTs for high-performance rechargeable lithium batteries. The organic nanolayer on SWCNTs formed via disassembly of crystalline structure of lumiflavin (LF) upon the immobilization with SWCNTs can have higher electron conductivity and better lithium accessibility than the polycrystalline bulk particle. **b** The energy storage mechanism of flavin in lithium rechargeable batteries via two lithium-ion coupled two electron transfer reaction. Reprinted from Ref. [50] with permission from WILEY-VCH

polarities via salt formation [52]. Therefore, lithiation of small organic cathode materials provide a possible way to decrease the solubility of organic cathode materials. Several lithiated organic salts with carbonyl groups, such as hydroxyl quinones [32, 53–55], conjugated dicarboxylates [56, 57] and dianhydride derivatives [39, 58, 59], have been investigated as electrode materials for lithium-ion batteries. A lithiated carbonyl compound, namely $Li_2(C_6H_2O_4)$, has been investigated as a positive active material in lithium-ion batteries [53]. This organic salt has an initial discharge capacity of 176 mAh g^{-1} and a Coulombic efficiency of 93.2 % in the first cycle. More importantly, $Li_2(C_6H_2O_4)$ exhibited significant improvement in terms of cycling stability comparing to small organic compounds. The use of lithiated redox organic molecules containing electrochemically active C=O functionalities, such as lithiated oxocarbon salts, has been proposed by Poizot and coworkers [54]. The key material is the tetralithium salt of tetrahydroxybenzoquinone ($Li_4C_6O_6$), which can be both oxidized to $Li_2C_6O_6$ and reduced to $Li_6C_6O_6$.

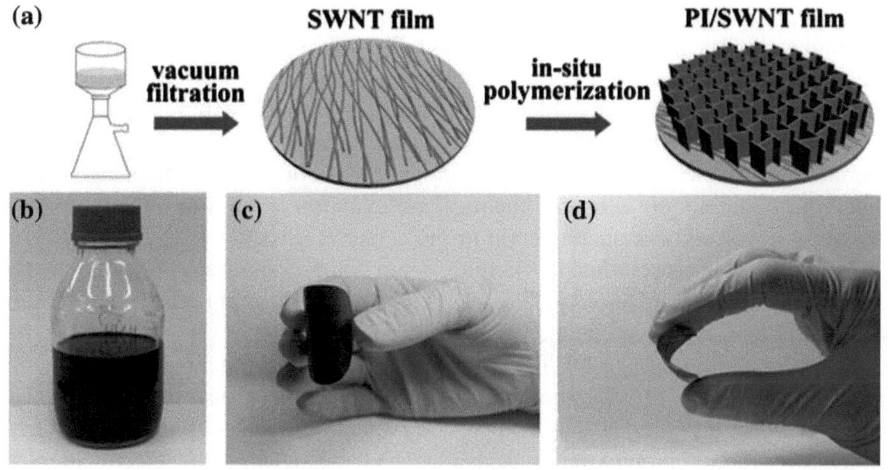

Fig. 5 **a** Schematic of the preparation process of PI/SWCNT film; photographs of **b** SWCNT aqueous dispersion; **c** SWCNT film; and **d** PI/SWCNT film. Reprinted from Ref. [51] with permission from WILEY-VCH

The $Li_4C_6O_6$ compound exhibited good performance with a specific capacity of ~ 200 mAh g^{-1} at an average potential of 1.8 V, showing a capacity decay of only 10 % after 50 cycles.

Taking advantage of the structural diversity of organic chemistry, the electrochemical properties of organic electrode materials can be easily tuned to give higher specific capacity or higher voltage gain, as well as better cycling stability. In order to achieve higher specific capacity, smaller molecular weight of the active center and less "electrochemically dead matter" are desirable to build stable structure of electrochemically active molecules [42]. The dilemma existing here is that large active groups with "dead matter" as supporting parts are good to achieve reasonable structural stability and reversible redox reaction. Therefore, the balance between capacity loss and stability gain needs to be considered when a practical organic electrode material is successfully designed. For example, comparing to naphthoquinone (NQ) and anthraquinone (AQ), benzoquinone (BQ) has a higher theoretical specific capacity of 496 mAh g^{-1} owing to its smaller molecular weight. However, NQ and AQ usually show better reversible redox reaction and more stable cycling performance because of their more conjugated structure and poorer solubility in organic electrolytes. Therefore, the holistic consideration on electrochemically active structure can lead to a successful molecular design. Poizot and coworkers designed a pyromellitic diimide dilithium salt by following an integration of some criteria: (i) having a restricted five-membered heterocycle to decrease the "electrochemically dead matter", (ii) connecting to a single phenyl group in a planar structure to stabilize the as-produced radical in reduction, and (iii) including permanent negative charges (i.e., organic anion rather than neutral molecular structure) [58]. The latter aims to decrease the redox potential compared to that of the tetraketopiperazine as well as the solubility issue. Taking benefit of the rational molecular design, the as-synthesized pyromellitic diimide dilithium salt exhibited quite stable electrochemical performance by delivering a two-electron redox reaction mechanism and a reasonable specific capacity of 200 mAh g^{-1}.

In addition to improving the specific capacity and the cycling stability through rational design, the redox potential of conjugated carbonyl compounds can also be tuned through the multiplicity of chemical combination at the molecular level [60]. By using a well-known phenomenon in molecular electrochemistry that a positive potential shift exists when switching from para- to ortho-position in the quinone/hydroquinone moiety, Poizot and coworkers reported that dilithium (2,3-dilithium-oxy)-terephthalate (Li_4-*o*-DHT), which is simply the ortho-regioisomer of Li_4-*p*-DHT, demonstrated a positive potential shift of nearly 300 mV (Fig. 6) [61]. In addition, an interesting electrochemical performance was observed with fast kinetics and good capacity retention upon cycling. The chemical structure variation around redox active centers can change the electrochemical performance, which may provide the capability of finely tuning organic electrode materials to give desirable performance. Recently, Wan et al. studied the effects of anthraquinone with and without SO_3Na-functional groups on electrochemical performance of lithium ion batteries [62]. It was found that the SO_3Na-functional groups were electrochemically inactive during the electrochemical process. Even though the specific capacity decreased due to the addition of electrochemically inactive species, as a trade-off, the structural stability and the cycling performance were significantly improved. Introducing functional groups to the quinone molecules can not only open a new approach to solve the dissolution problems of organic cathode materials but also provide a valuable guidance for tuning other properties such as specific capacities and redox potentials.

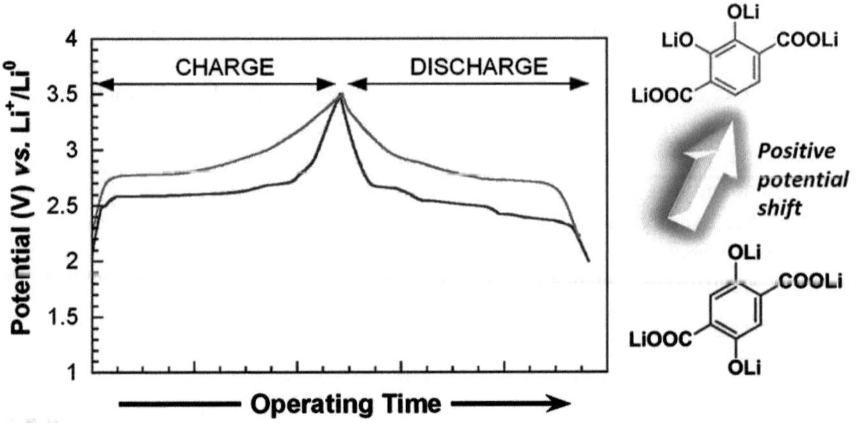

Fig. 6 Dilithium (2,3-dilithium-oxy)-terephthalate (Li_4-*o*-DHT), which is simply the ortho-regioisomer of Li_4-*p*-DHT, demonstrated a positive potential shift of nearly 300 mV. Reprinted with permission from Ref. [61]. Copyright (2014) American Chemical Society

Besides the dissolution problem of organic electrode materials in electrolytes which leads to poor cycling stability, another disadvantage of small organic molecules is their non-conductive nature. To address this issue, electrochemical active materials with carbonyl groups need to be incorporated with a large portion of conductive carbon in the electrode preparation process. Recently, CNTs and graphene have attracted extensive attention in a variety of research field because of their outstanding physical and chemical properties. Oxidization of CNTs or graphene can generate various oxygen-containing groups, such as carbonyl, carboxyl, hydroxyl, epoxide, etc. Therefore, oxidized CNTs and graphene have the potential to be used as electrochemical active materials in lithium-ion batteries. Shao-Horn and coworkers proposed the possibility of using the surface-functionalized carbon materials as electrochemical active cathode materials [63–65]. By using additive-free, functionalized layer-by-layer multi-walled carbon nanotubes (LBL-MWCNTs) as active materials, the electrode could store lithium up to a reversible gravimetric capacity of ~200 mAh g^{-1}-electrode, deliver a power of 100 kW kg^{-1}-electrode and provide a lifetime in excess of thousands of cycles [63]. This excellent performance is attributed to the fast redox reaction of surface oxygen-containing groups on electrodes (Fig. 7). In their follow-on research, the role of oxygen functional groups for lithium storage has been investigated in the electrodes with oxidized CNTs/graphene mixtures as active materials [65]. By tuning the surface oxygen chemistry, faradaic contributions from electrochemically active carbonyl, carboxylic and ester groups could provide gravimetric energies of ~230 Wh kg^{-1}. Since CNTs and graphene can provide a platform to undergo

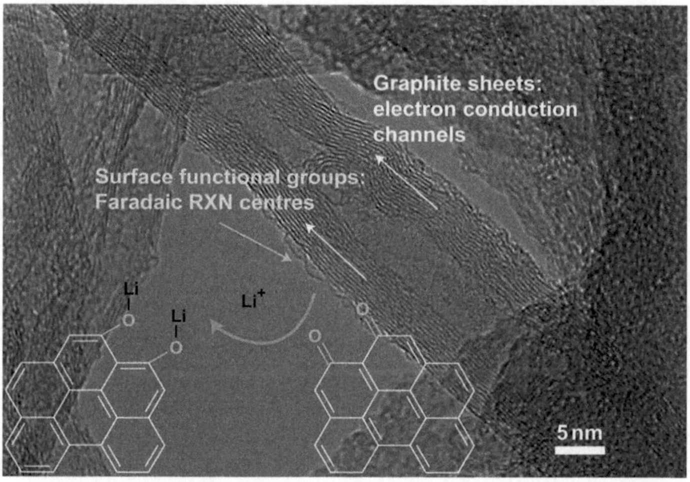

Fig. 7 Schematic of the energy storage mechanism of LBL-MWCNT electrodes. Faradaic reactions between surface oxygen functional species (*orange arrows*) and Li schematically illustrated on an HRTEM image of the LBL-MWCNT electrodes. Intact graphite layers inside the MWCNTs (*white arrows*) are indicated as electron conduction channels. Reprinted from Ref. [63] with permission from Macmillan Publishers (Color figure online)

versatile chemistry through organic functionalization [66–68], it is reasonable to expect that modification of the surface functional groups on carbon may allow the tuning of redox potentials and increase the efficiency by reducing the voltage difference during charge and discharge processes.

2.2 Free Radical Polymers

Many kinds of radicals are stable enough to be isolated at room temperature [69]. Some stable organic radicals can be oxidized and reduced electrochemically through reversible redox reactions with fast kinetics so that they could be good candidates for battery applications, which are called organic radical batteries (ORBs) [70–72]. Stable organic radicals often need to be anchored on the polymer chains or other substrates to commit their reversible redox functions. Free radical polymers are an important family of organic electrode materials, which utilize polymers with defined electrochemically active moieties as electrode materials. Most radical polymers consist of two main components including a robust radical pendant and a backbone polymer. A polymer bearing stable organic radicals as electrochemically active pendant groups can display striking performance, such as fast charging/discharging rate, stable cell voltage and good processibility. The first known application of stable radicals for energy storage systems was proposed by Nakahara et al. in 2002 [70]. A stable nitroxyl polyradical, poly (2,2,6,6-tetramethylpiperidinyloxy methacrylate) (PTMA), was used as the cathode material in rechargeable lithium batteries. The batteries based on PTMA exhibited a discharge capacity of 77 Ah kg^{-1} at an average discharge voltage of 3.5 V and demonstrated a surprisingly high cycling stability up to 500 cycles. After that, more and more efforts have been put onto this promising material and most of the research attention mainly focused on nitroxyl radicals, such as 2,2,6,6-tetramethylpiperidine-1-oxy (TEMPO) and 2,2,5,5-tetramethylpyrrolidine - 1-oxyl (PROXYL) [71, 73–82].

The research work on radical-based lithium batteries was mainly conducted by Nakahara and coworkers [70, 71, 83–85], and then followed by Nishide and coworkers [74, 80, 86–100]. The efforts have focused on polymer design, electrode optimization and search for robust radicals. A polymer backbone is indispensable to bear organic radicals and prevent their dissolution from electrodes. The requirements for polymer backbones include stability, insolubility, processibility, as well as light molecular weight. Various polymer backbones have been used to bear radical pendants, including polymethacrylate (PMA) [70, 71, 76, 83, 84, 101–107], polyacetylene (PAc) [73, 108] polynorbornene (PNB) [73, 86–88, 109, 110], poly (vinyl ether) (PVE) [85, 111, 112], polystyrene (PS) [88, 93, 113], polyether (PEG) [74, 82], cellulose [81], DNA [77] etc. Because of the easiness of polymerization method, PMA and PNB are the most popularly used polymer backbones. Interestingly, DNA can be used as scaffold to form DNA-lipid complexes containing TEMPO radicals, which can enable organic radical batteries [77].

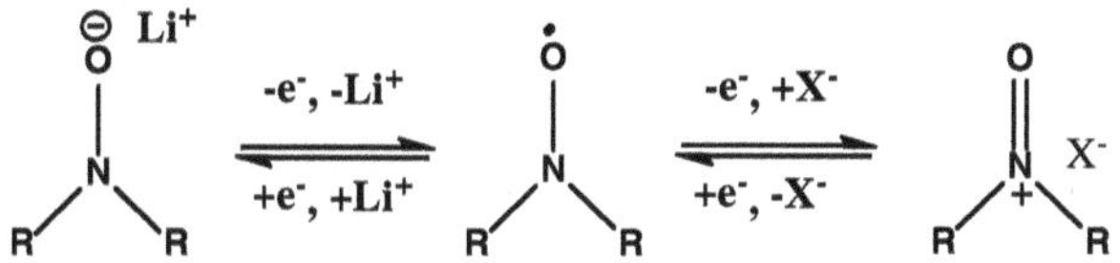

Fig. 8 Schematic of n-doping and p-doping mechanism of nitroxide compounds. Reprinted from Ref. [22] with permission from WILEY-VCH

The TEMPO-containing DNA-lipid complexes displayed reversible two-stage charge/discharge processes with a discharge capacity of 40–60 Ah kg^{-1}.

A nitroxide radical can be reversibly n-doped to aminooxy anion and p-doped to oxoammonium cation (Fig. 8) [22]. Based on the bipolar property of nitroxide radicals, it is possible to construct an all-organic battery in which the cathode and the anode contain the same or different nitroxide radicals with suitable redox potentials [96]. For example, a total organic polymer-based rechargeable device with a symmetric configuration has been developed by using poly(nitronilnitroxylstyrene) as the p- and n-type bipolar redox-active radical polymer [93]. The polymer offered the charge capacity of 82 and 91 mAh g^{-1} for the n-type and p-type redox reactions, respectively. Such an all-organic-based battery with the symmetric configuration, which consists of the same redox-active polymer for both anode and cathode, delivered a charge capacity of 44 mAh g^{-1} at a plateau voltage of 1.3 V, as well as a good cycle life (more than 250 cycles). In addition, this "rocking-chair type" battery could also be enabled by using different redox-active polymers as anode and cathode based on the potential gap between two different redox couples. To help the selection of suitable redox-active polymer couples, the potentials of various redox active radicals (Fig. 9) and synthetic approaches towards organic radical bearing polymers have been summarized in a review paper written by Schubert and coworkers [72].

Even though the organic radical itself has fast electrochemical redox kinetics, most of the polymer backbones are non-conductive thus leading to poor rate capability for the nitroxide radical batteries. To mitigate the drawback of low electronic conductivity, large amount of conductive carbon additives are usually required to make a satisfactory electrode. The electrochemical performance of redox-active radicals also greatly depends on the electrode structure where radical polymers may aggregate during cycling hence degrading the capacity. To tackle this problem, SWCNTs were employed to form a composite with radical polymer PTMA (Fig. 10) [91]. SWCNTs were wrapped by the radical polymer at molecular level to provide a high dispersibility of the SWCNTs and resulted in the charge propagation on the PTMA, thus providing a quantitative charge-storage capacity at a high rate. Recently, Choi et al. developed a method to terminally modify the radical polymer of redox-active nitroxide and then to graft the modified polymer uniformly onto MWCNTs [99]. The radical polymer-grafted MWCNTs allowed a high dispersibility in organic solvents and fadeless redox ability of nitroxide. The radical polymer wrapped MWCNTs exhibited quantitative and high rate redox

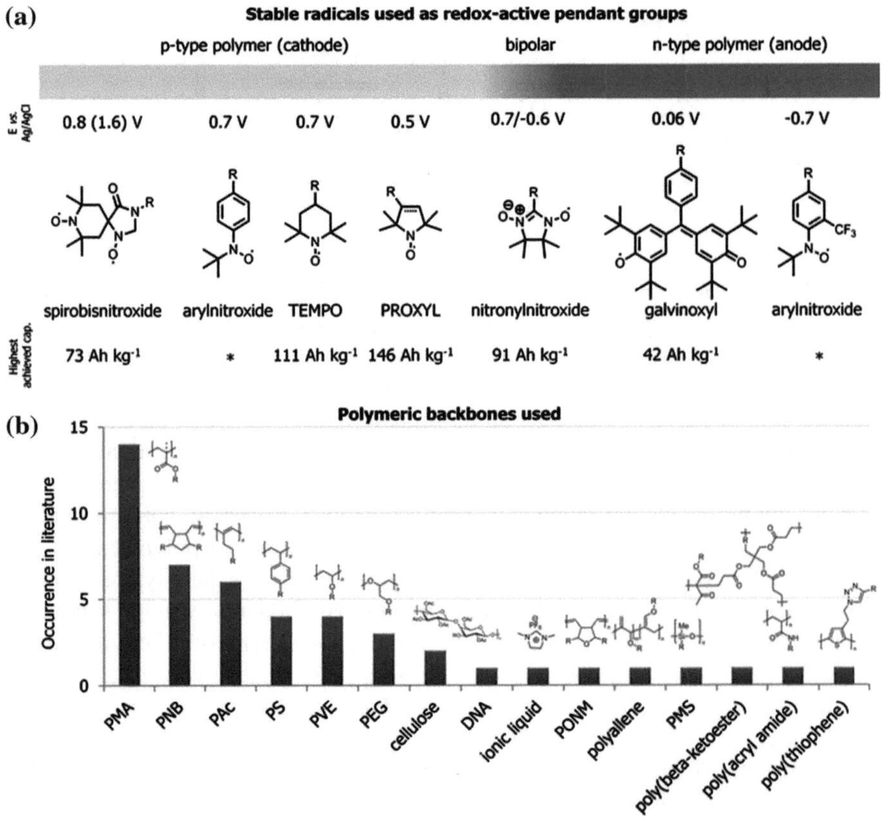

Fig. 9 **a** Stable organic radicals used as redox-active pendant groups in ORBs and their potentials as well as capacities. **b** Commonly employed polymeric backbones for hosting the organic radicals shown in (**a**). Reprinted from Ref. [72] with permission from WILEY-VCH

properties of nitroxide suitable as a cathode active material. The findings indicate that nano-scale homogeneous wrapping can prevent the aggregation of radical polymers and allow the rapid counterion accommodation leading to quantitative and rapid charge propagation.

2.3 Conducting Polymers

Conventional polymers such as plastics, rubbers, etc. which take advantage of strength, flexibility, elasticity, stability and mouldability typically exhibit significant resistance to electrical conduction and are either dielectrics or insulators. Since the discovery of conductive polyacetylene in 1977 [23], conducting polymers have received significant attentions from both science and engineering communities and

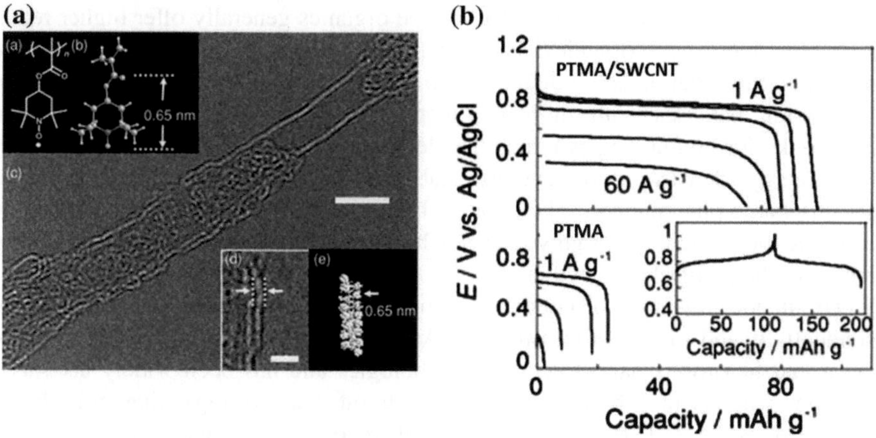

Fig. 10 **a** Structure of a radical polymer-wrapped SWCNT. **b** Discharging curves of the PTMA/SWCNT (4 wt%) composite and PTMA at the current density of 1, 3, 10, 30, and 60 A g^{-1}. *Inset* Charging–discharging curves of the PTMA/SWCNT composite at 1 A g^{-1} in acetonitrile containing 0.1 M $TBAClO_4$. Reprinted from Ref [91] with permission from WILEY-VCH

their applications in a wide range of fields including electronic devices and energy storage have been explored owing to their unique properties including good electronic conductivity and reversible redox activity [16, 19]. The attempts to apply conducting polymers for battery applications had first been proposed by Nigrey et al. and followed by a continuation of publications from many research groups [19, 114]. Conducting polymers are usually used as positive electrode materials in the p-doping oxidation state, which exchange anions during the charging and discharging processes. Therefore, the salt concentration in the electrolyte changes with the charge state of the conducting polymers, which is unfavorable for the specific energy and the operation of the battery [19]. The general redox chemistry of a p-type conducting polymer in a non-aqueous electrolyte can be described as below:

$$[\mathrm{Polymer}^{n+}][\mathrm{A}^-]_n + \mathrm{ne}^- \leftrightarrow \mathrm{Polymer} + \mathrm{nA}^-$$

As the cathode is discharged, the oxidized p-type polymer doped with anions takes up electrons and releases dopant anions into the electrolyte. Meanwhile, lithium metal anode releases lithium ions into the electrolyte and gives up electrons [115]. The conducting polymers that are most studied for battery applications include polyacetylene (PAc), polypyrrole (PPy), polythiophene (PTh), polyaniline (PAn) and polyparaphenylene (PPP). The theoretical specific capacities of the conducting polymers are comparable to those of metal oxide cathodes normally used in lithium-ion batteries. The capacities of conducting polymers are determined by the doping level which is defined by the ratio of doped units to all structural units, expressed as x in $[(P^{x+})(xA^-)]_n$ $(0 \le x \le 1)$. Typically, the reported capacities of p-type conducting polymers, such as polyanilines and polypyrroels, fall in the range of 80–150 mAh g^{-1} [115]. The p-type organics generally offer higher redox potentials than n-type organics. The battery with conducting polymers as cathode materials can offer a cell voltage up to ~4.0 V, comparable to conventional lithium-ion batteries with inorganic metal oxides as positive electrodes. The drawbacks for conducting polymers as electrode materials include their relatively low energy density, poor cycling stability, low Coulombic efficiency and self-discharge phenomenon. The poor cycling stability is caused by the irreversible overoxidation reaction at high voltages, which leads to capacity degradation during repeated charge/discharge cycling.

The major advantage of conducting polymers arises from their high electronic conductivity and ease of processibility. PAn that has several oxidation states can be fabricated in various nanostructural morphologies and has been widely used as a cathode material in lithium batteries, which offer an average charge/discharge voltage of ~3.5 V [116–119]. Cheng et al. reported a template method to synthesize PAn nanotubes and nanofibers and investigated the performance in Li|PAn rechargeable batteries [117]. In that work, anodic aluminum oxide (AAO) membranes that have cylindrical pore arrays were used as the template to synthesize 1-D structured PAn. Comparing to the Li|PAn cell with commercially available doped PAn powders, the cells with as-synthesized 1-D PAn nanotubes exhibited a high energy density of 227 Wh kg^{-1}. However, PAn gradually loses its electrochemical activity during cycling because of the irreversible oxidation even at moderately oxidizing potentials (3.5 V vs. Li/Li^{+}). To address this issue, the polyaniline:poly(2-acrylamido-2-methyl-1-propanesulfonic acid) (PAn: PAAMPSA) complex had been incorporated into an LBL electrode and the as-resulted electrode exhibited excellent electrochemical performance in terms of capacity and cycling stability [120]. After 1000 cycles between 1.5 and 4.5 V, no significant decrease was found for the LBL electrode fabricated with the PAn/PAAMPSA composite. In order to fabricate free-standing and flexible energy storage electrodes, hybrid aerogels consisting of interpenetrating SWCNT/PAn nanoribbons have been prepared and employed in lithium batteries (Fig. 11) [121]. Thin PAn nanoribbons (thickness of 10–100 nm, width of 50–1000 nm, and length of 10–20 μm) were formed within the network after polymerization of aniline. The lithium battery with optimized SWCNT/PAn nanoribbon aerogels showed high capacity of 185 mAh g^{-1} and good cycling stability. The improved electrochemical performance can be ascribed to the synergistic effects of efficient ion/electron transport within the 3D carbon nanotubes network, shortened ion diffusion distance, optimized strain relaxation from nanoribbons and nanotubes, and effective penetration of electrolyte within interconnected nanopores in the electrode.

Polypyrrole (PPy) is another intensively studied conducting polymer used for energy storage applications. PPy has been synthesized from the polymerization of pyrrole by either electrochemical or chemical oxidation method [16, 19]. Since it is difficult to n-dope PPy, p-type PPy as positive electrode material is the common choice reported in lithium batteries. Typically, the specific energy of PPy falls in the range of 80–390 Wh kg^{-1} with the open-circuit voltage of 3–4 V. One of the main drawbacks of PPy as cathode material is its relatively low theoretical capacity.

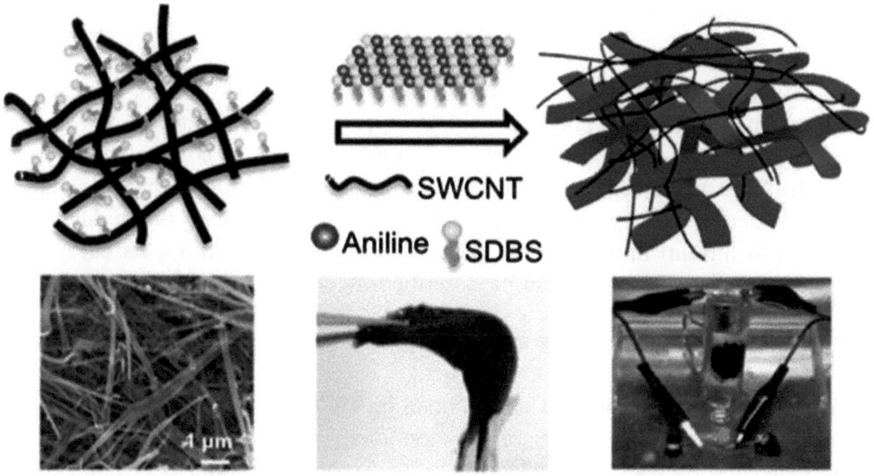

Fig. 11 Hybrid aerogels from SWCNT/PAn nanoribbons as free-standing, flexible energy storage electrodes. Reprinted with permission from Ref. [121]. Copyright (2014) American Chemical Society

Recently, several approaches have been developed to improve the specific energy of PPy. The most common strategy is the modification of PPy with redox active functionalities or incorporation of redox active molecules [122–126]. Park et al. proposed to improve the specific capacity of PPy by chemically and/or physically attaching a suitable redox couple to the polymer backbone, which should in general be a $3d^n/3d^{n+1}$ pair of the first-row transition elements [122]. As an example of the chemical bonding of a redox couple to the polymer, ferrocene groups were covalently anchored to the PPy backbone. The PPy/ferrocene polymer film with a uniform thickness was electrodeposited onto a stainless-steel mesh and the film delivered a specific capacity of up to 65 mAh g^{-1} with a short voltage plateau near 3.5 V versus Li/Li^+, comparing to the specific capacity of 20 mAh g^{-1} for a pure PPy cathode. Kong et al. incorporated cobalt and iron into PPy to form coordination complexes which exhibited high lithium storage capacity and cycling stability [124, 125]. The characterizations and DFT calculations indicated that the coordination complexes have multilayer structures in which the strong and stable intralayer metal-N coordination permits the material to possess high specific capacity, high cycling stability and outstanding rate capability. Recently, Zhou et al. reported the dramatic capacity enhancement of PPy by doping with $Fe(CN)_6^{4-}$ anions [123]. The $Fe(CN)_6^{4-}$ doped PPy material could deliver not only a greatly enhanced redox capacity of 140 mAh g^{-1}, three time higher than undoped PPy, but also had an excellent cycling stability. More recently, a high charge capacity redox molecule, 1,4-benzoquinone, has been used to increase the energy density of PPy electrode [126]. PPy electrodes doped with 1,4-benzoquinone exhibited specific charge capacities as high as 104 mAh g^{-1} compared to 50 mAh g^{-1} for the reference PPy electrodes.

2.4 Organosulfide Compounds

It is well known that the S–S bond contained in the organosulfides or thiolates can be reversibly broken and rebuilt, in which the two-electron reaction is involved to provide higher capacity than the doping/undoping reactions of conducting polymers [21, 22]. The redox reaction mechanism for organosulfides with S–S bond in lithium batteries undergoes the reversible reactions between the reduction of disulfides to lithium thio salts and the oxidation of lithium thio salts back to the corresponding disulfides and can be described as:

$$RSSR + 2e^- + 2Li^+ \leftrightarrow 2LiSR$$

where R is an organic moiety. The S–S bond in RSSR can be progressively cleaved during discharge, leading ultimately to monomeric anions, and then the monomer anion can be subsequently oxidized back to form the original S–S bond during charge. Even though organosulfide compounds potentially could offer higher capacity due to the multi-electron redox reactions, they suffer from major drawbacks including the slow reaction kinetics and the poor cycling stability. Generally, organosulfide compounds can be categorized into three types: dimeric organosulfides, main-chain-type organosulfur polymers and side-chain-type organosulfur polymers.

Tetraethylthiuram disulfide (TETD), one of dimeric organosulfides, was first proposed by Visco and DeJonghe [127] as an energy storage material in high temperature sodium batteries. Later, Liu et al. [128–130] pushed forward these efforts by following a series of researches to thoroughly examine the reaction mechanism and the kinetic behavior of a diverse group of organodisulfide/thiolate redox couples. Because of the high solubility of RSSR and RS^- in organic electrolytes, it is very difficult to achieve satisfactory cycling performance in non-aqueous rechargeable lithium batteries. Therefore, the early studies on this kind of organosulfides were mainly concentrated on high-temperature lithium batteries with solid-state electrolytes.

In order to enhance the specific capacity of organosulfide compounds, a series of organosulfur polymers with disulfides bonds in the main chain have been investigated [129, 130]. However, due to the problems of the dissolution and slow kinetics, these compounds were only studied in solid-state lithium batteries at elevated temperatures (typically 80–130 °C). Oyama et al. introduced a new composite cathode for high energy density rechargeable lithium batteries, which consisted of 2,5-demercapto-1,3,4-thiadiazole (DMcT), PAn and gel electrolyte [131]. The lithium cell had a cell voltage over 3.0 V and exhibited the energy density of 303 Wh kg^{-1} at a current density of 0.1 mA cm^{-2} at room temperature. DMcT-based organosulfides thereafter attracted much attention among all organosulfur compounds for cathode materials in lithium batteries owing to their high theoretical capacity (362 mAh g^{-1}) [132–141]. However, because of its slow kinetics of redox reactions, DMcT usually offers poor rate capability and not all of

DMcT incorporated in the electrode is reactive so it is unfavorable for practical applications in lithium/lithium-ion batteries. The redox behavior of the dimercaptan can be accelerated greatly when coupled with electrocatalysts, such as PAn, poly (3,4-ethylenedioxythiophene) (PEDOT) and copper(II) salt [132, 133, 137, 140, 142]. For example, the incorporation of PAn can enable DMcT to reach its theoretical capacity limit which exceeds that of the intercalated inorganic electrodes in commercial lithium-ion batteries [132]. Furthermore, PEDOT was also found to exhibit dramatic electrocatalytic activity toward the redox reaction of DMcT. Abruña et al. have contributed great efforts to the studies on the electrocatalytic effect of conducting polymers toward the redox reactions of thiadiazole-based compounds through both experiments and calculations [137, 139, 140, 143–145]. It was suggested that PEDOT could be an effective oxidative electrocatalyst for any aromatic organosulfur compound whose highest occupied molecular orbital (HOMO) energy level was in between the neutral PEDOT's HOMO level and the cationic PEDOT's singly-occupied molecular orbital (SOMO) level (Fig. 12) [139]. Similarly, a compound with a lowest unoccupied molecular orbital (LUMO) energy level close to the HOMO level of PEDOT could exhibit reductive electrocatalysis by PEDOT.

In addition to DMcT, other organosulfur compounds based on dimercaptothiophene and its derivatives with a variety of functional groups (electron-donating or electron-withdrawing groups) have been synthesized and investigated as cathode materials in lithium batteries [139, 144]. However, the composite cathodes, consisting of a thiolate compound such as DMcT and a conducting polymer, often exhibit poor charge/discharge cyclability due to dissolution of the reduction products (typically monomers) of the disulfide polymer when an organic liquid electrolyte is used [137, 139, 140]. Therefore, the battery performance with DMcT-based composite electrode is still far from practical use because of the intrinsic dissolution problem and slow kinetics.

Side-chain-type organosulfur polymers typically consist of conducting polymer as the main chain and S–S bond or polysulfide bond as the side chain. The feature of side-chain-type organosulfur polymers avoids backbone cleavage when the S-S bond is broken during charge/discharge processes and is expected to offer better cycling stability comparing to main-chain-type organosulfur polymers. Naoi et al. firstly developed a new class of conducting polymers with S–S bonds as side chain linkers to interconnect between two moieties of aniline [146]. 2,2′-Ditiodianiline (DTDA) was used to conduct electrochemical polymerization to form an electroactive poly(DTDA) thin film. The electrochemically polymerized poly(DTDA) has several advantages because of its high theoretical energy density, faster kinetics and higher electrical conductivity over an organosulfur cathode when used in lithium batteries. The as-prepared poly(DTDA) offered a specific capacity of 270 Ah kg^{-1} and an energy density of 675 Wh kg^{-1}. Following this concept, polymers with similar structures containing interconnected S–S bonds have been developed for energy storage in lithium batteries [147, 148]. Although poly(DTDA) contains a conductive polymer as main chain, the dislocation of the two main chains during discharge and the dissolution of the discharged polymer main chain resulted in low

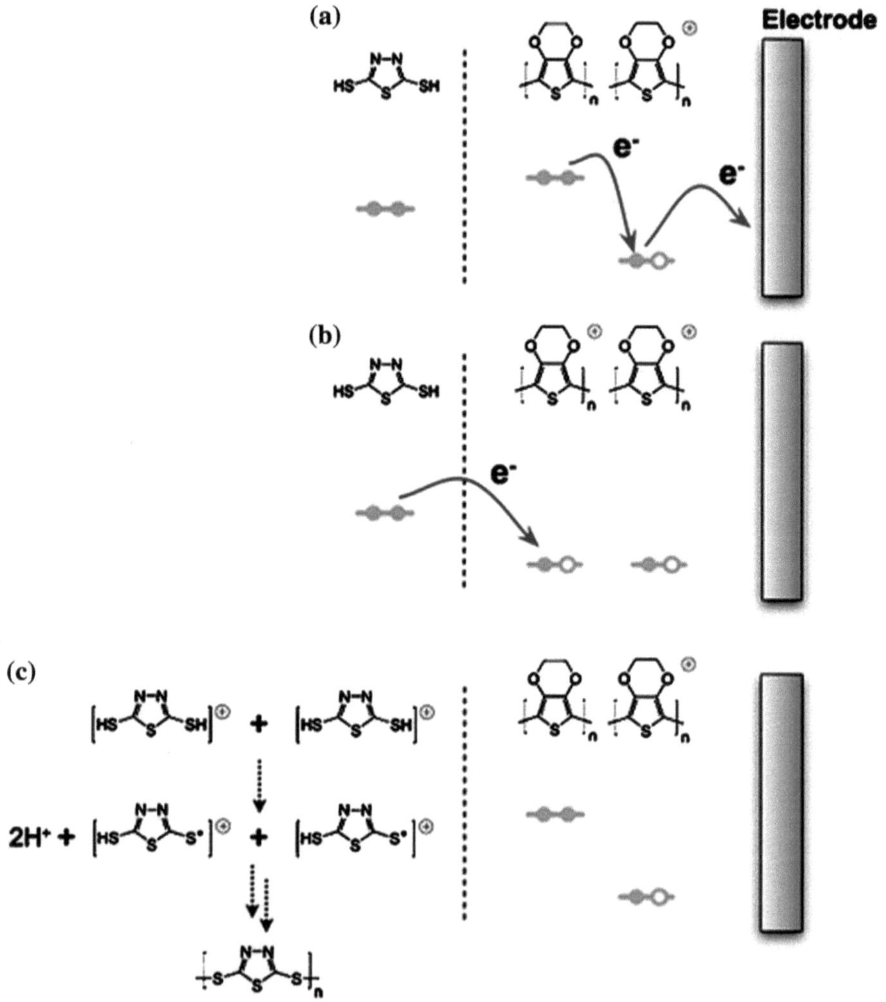

Fig. 12 Schematic of the electrocatalytic cycle between DMcT and a PEDOT film-coated electrode showing the oxidative (charge) and reducing (discharge) processes. **a** Neutral EDOT (monomer unit within the PEDOT film) species is electrochemically oxidized, forming $[EDOT]_n^+$ followed by (**b**) oxidation of the DMcT monomer (or oligomer), regenerating neutral EDOT species followed by (**c**) proton-coupled polymerization of DMcT monomers (or oligomers) into oligomers/polymers. Reprinted with permission from Ref. [139]. Copyright (2007) Royal Society of Chemistry

recombination efficiency, consequently leading to a poor cycling stability. To overcome this problem, other type polymer with S–S bond as the side groups of polymer chain has been developed [149–152]. Because of the S–S bond occurred on the side of same polymer chain, the recombination of S–S bond during charge has been improved which results in better cycling performance.

3 Organic Cathode Materials for Sodium Batteries

Even though lithium batteries are characterized by high specific energy, high efficiency and long cycle life which make them dominant in the battery market for portable electronics, scaling up the lithium battery technology for power systems in stationary energy storage applications such as smart grid is still facing significant challenges [153]. The potential large-scale production of lithium-ion batteries for vehicle applications will eventually drive the cost of lithium higher in the near future because of the limited lithium resource which in turn makes lithium-ion batteries unlikely to meet the fast-growing global demands. Taking advantage of the high content level in the earth crust and the low cost of sodium when compared to lithium, rechargeable sodium batteries have been recognized recently as an alternative to rechargeable lithium batteries as energy storage systems for smart grid and renewable energy power plants [13, 30, 154]. However, most of inorganic intercalation oxides for lithium storage cannot be directly used for sodium storage because of their different cation radius sizes, which hampers the development of sodium-ion batteries. In contrast, organic cathode materials are not sensitive to the insertion ion radius and therefore can be directly transplanted to sodium storage systems.

In early research, tetracyanoethylene (TCNE) had been used as a cathode material to circumvent the corrosion problems prevalent in many existing electrochemical couples using the Na/β″-alumina half cells [24]. TCNE exhibited reversible electrochemical characteristics with a potential over 3.0 V versus Na/Na^+ and an energy density of ~620 Wh kg^{-1}. Recently, most of attention has been focused on carbonyl compounds owing to their favorable electrochemical performance for sodium ions. Despite of carbonyl compounds, other organic electrode materials such as conducting polymers and free radicals have also been characterized for sodium storage.

Unlike lithium-ion batteries using graphite as anode, one of the key barriers for sodium-ion batteries is the lack of stable anode materials which can insert sodium ions reversibly at relatively low potentials. Therefore, many efforts have been made to search for suitable organic anode materials for sodium-ion batteries. Park et al. reported on the first construction of high performance sodium-ion batteries using a small organic molecule, disodium terephthalate ($Na_2C_8H_4O_4$), and its derivatives [27]. Disodium terephthalate can reversibly de-insert two sodium cations. Moreover, the substituents on a phenyl ring (amino-, bromo-, and nitro- functional groups) and region-isomerism of dicarboxylates (ortho-, meta- and para- positions) can affect the thermodynamic and kinetic properties of disodium terephthalate and its derivatives. Almost at the same time, Zhao et al. [28] reported similar work by using disodium terephthalate as high performance anode materials for sodium-ion batteries. Later, the sodium insertion in a series of carboxylate based anode materials were characterized by Abouimrane et al. [26] and the full sodium batteries were assembled and tested. It demonstrated that organic conjugated carboxylates can be used as promising anode materials for sodium-ion batteries.

Besides as anode materials, conjugated carbonyl compounds can also be used as cathode materials in sodium batteries. Zhang et al. synthesized a series of dianhydride-based polyimides (PI), which bear various commercially available dianhydrides such as pyromellitic dianhydride (PMDA), 1,4,5,8-naphthalenetetra carboxylic dianhydride (NTCDA), and perylene 3,4,9,10-tetracarboxylic dianhydride (PTCDA), and reported their performances as cathode materials in sodium batteries [31]. Among all the polymers, PTCDA-based PI exhibited a high power density of ~21 kW kg^{-1}, a specific energy of 285 Wh kg^{-1}, and an excellent cycling stability of 87.5 % capacity retention after 5000 cycles. The superior performance for PTCDA-based PI to PDMA- and NTCDA-based PI analogues can be contributed to its low solubility in electrolytes and the lowest LUMO energy. In addition, indigo carmine (IC), a water-soluble organic molecule widely used as a food dye, has been proved to reversibly store not only lithium but also sodium via a redox reaction, and can be used as the cathode material in sodium batteries [155]. The IC electrode exhibited a discharge capacity of 106 mAh g^{-1} with an average potential of 1.8 V versus Na/Na^+ for the first cycle. Also, the IC electrode showed good cycling stability with a capacity drop of 20 mAh g^{-1} after 40 cycles. Recently, PTCDA, a well-known organic pigment has been used as a cathode material in sodium batteries [156]. In the voltage range of 1–3 V, PTCDA delivered a high capacity of 145 mAh g^{-1} at 10 mA g^{-1}, an excellent rate capability (91 mAh g^{-1} at 1000 mA g^{-1}), and a good cycling stability for 200 cycles. Another example of carbonyl groups for sodium storage has been proposed by using functionalized graphite nanoplatelets [157]. The XPS spectra results suggested that the sodium ions were chemically bound to C=O functional groups by breaking the double bonds and forming single bonds between carbon and oxygen.

Recently, all-organic sodium-ion batteries have been reported by using organic compounds as both anode and cathode materials. Deng et al. used p-doped polytriphenylamine as cathode and n-type redox-active poly(anthraquinonyl sulphide) as anode in their study. Such an all-organic sodium-ion battery had a voltage output of 1.8 V, a considerable specific energy of 92 Wh kg^{-1}, a high rate capability with 60 % capacity release at 16C rate (or 3200 mA g^{-1}), and an excellent cycling stability with 85 % capacity retention after 500 cycles at 8C rate [158]. Later, Wang et al. reported an all-organic sodium-ion battery with tetrasodium salt of 2,5-dihydroxyterephthalic acid ($Na_4C_8H_2O_4$), which was prepared through a one-pot method, as both anode and cathode. $Na_4C_8H_2O_4$ delivered compatible and stable capacities of ~180 mAh g^{-1} with excellent cycling stability as either anode in 0.1–1.8 V or cathode in 1.6–2.8 V versus Na/Na^+ [159]. The use of $Na_4C_8H_2O_4$ as the single initial active material for both anode and cathode in the simple all-organic sodium-ion battery provided an average operation voltage of 1.8 V, a practical energy density of about 65 Wh kg^{-1}, a good rate capability up to 5C rate and a stable cycling stability for 100 cycles. It is worth to note that this type of all-organic "rocking-chair" sodium-ion batteries can be realized from renewable, sustainable and synthetically modifiable organic compounds and polymeric materials.

Conducting polymers have also been proposed to store sodium as positive electrodes in sodium batteries. Most of the research on conducting polymers for sodium batteries has been done by Yang et al. [160–162]. An aniline-nitroaniline copolymer was prepared through a chemical oxidative polymerization and the as-prepared product was investigated as a cathode material for sodium batteries [160].

The electrode with the aniline-nitroaniline copolymer provided a reversible capacity of 180 mAh g^{-1} at an average potential of ~3.2 V versus Na/Na^+ and remained 173 mAh g^{-1} after 50 cycles, indicating a good cycling stability. Polypyrrole doped with redox-active $Fe(CN)_6^{-4}$ was found to deliver a specific capacity of 135 mAh g^{-1} in sodium battery with good rate capability and cycling stability [161]. Another polypyrrole functionalized with ionizable sodium sulfonate has been developed by the same group and it was found that the self-doping polymer could reversibly undergo sodium insertion-extraction reaction with a specific capacity of 85 mAh g^{-1} [162].

4 Organic Cathode Materials for Redox Flow Batteries

With increasing demand for electricity, renewable energy resources have been growing their popularity to a significant portion in the electric grid during the past several years [163]. However, the intermittent nature of renewable energy, such as solar and wind, requires energy storage systems to seamlessly integrate the variable output renewable generation into smart grid to fulfill a set of ancillary services including frequency regulation, cold start, contingency reserves and peak load curtailment [163]. Although existing energy storage is dominated by pumped hydroelectric, there is the recognition that batteries which can reversibly convert electrical power to chemical energy can address the energy storage need more easily, flexibly and economically [164]. Among all the potential battery technologies, redox flow batteries (RFBs) show the most promising potential to meet the requirements for stationary electrical energy storage because of their unique features such as moderate cost, modularity, transportability and flexible operation [163, 165–168]. Unlike solid-electrode batteries which can only maintain discharge at peak power for a short period of time and thus have difficulties to fully regulate renewable energy output, RFBs have a significant advantage in this regard which can independently scale the power and energy components of the system by maintaining all of the electro-active species in fluid form [169, 170]. A RFB can reversibly convert electrical energy into chemical energy through two soluble redox couples contained in external electrolyte tanks sized in accordance with application requirements. Generally, inorganic redox couples in aqueous electrolytes are dominant in RFB chemistries, which is limited by the electrochemical stability of water that is only stable within a small potential window (typically 1.2–1.6 V) beyond which water electrolysis occurs [171]. To broaden the voltage output of conventional aqueous RFBs, non-aqueous RFBs in which redox couples in non-aqueous electrolytes are employed to store energy have attracted increasing interests which promise to enable higher energy density for commercially-viable energy technology. Redox-active organic molecules in non-aqueous RFBs offer a series of merits including a wider electrochemical window, a larger energy density and a higher energy efficiency. Nevertheless, the research of organic redox couples for RFBs is still in its early state, which has attracted a great deal of research interests recently for the selection of suitable organic redox couples [170,172–175].

Inspired by organic electrode materials in lithium-ion batteries, Li et al. reported a novel all-organic RFB employing 2,2,6,6-tetramethyl-1-piperidinyloxy (TEMPO) and N-methyl-phthalimide as cathode and anode active materials, respectively [172]. $NaClO_4$ dissolved in the anhydrous acetonitrile was used as the supporting electrolyte. Stable charge/discharge plateaus and relatively high Coulombic efficiency (90 %) were achieved for the first 20 cycles. In another work, Brushett et al. reported a similar all-organic RFB employing 2,5-di-tert-butyl-1,4-bis (2-methoxyethoxy)benzene as high potential redox compound and a variety of molecules derived from quinoxaline as low potential active materials and the common electrolytes used in conventional lithium-ion batteries as the supporting electrolyte [171]. It was found that the choice of electrolyte and of substituent groups could have a significant impact on the electrochemical performance of active materials. The charge/discharge characteristics were investigated in a proof-of-principle coin cell flow battery which delivered a moderate voltage of <2.0 V and showed a stable electrochemical performance over the course of 30 discharge-charge cycles.

Though it is easy to find organic molecules as high potential redox compound for non-aqueous RFBs, the low potential redox compounds usually have difficulties to reach low enough redox potentials to take advantage of the full utilization of positive electrode materials [176]. To achieve higher voltage output and greater energy density, metal-based anodes have been studied to pair a high potential redox compound in RFBs [173, 177–183]. Wang et al. [178] developed a non-aqueous, hybrid metal-organic RFB based on tailored anthraquinone as cathode active material and lithium metal as anode. The molecule of choice for this work was 1,5-bis(2-(2-(2-methoxyethoxy)ethoxy)ethoxy)anthracene-9,10-dione (15D3GAQ) which had higher solubility in polar solvents than anthraquinone owing to the functional groups. The cell with the tailored molecules demonstrated an energy density of ~82 % and a specific discharge energy density similar to those of aqueous RFBs. Recently, in another work done by the same group, an ionic-derivatized ferrocene (Fc1N112-TFSI) compound has been designed to approach an increased solubility which was 20 times higher than its precursor in the supporting electrolyte [184]. Because of the modification of the molecular structure of ferrocene which is evidenced by NMR and DFT calculation (Fig. 13a), the solvent molecules would have preferential positions within the solvation sphere resulting in a higher solubility. Besides the higher solubility, the modified ferrocene exhibited more positive redox potential in comparison with its pristine counterpart (Fig. 13b). The Li|Fc1N112-TFSI flow cell was evaluated using LiTFSI-based electrolyte containing 5 wt% FEC, which exhibited the capacity retention of 99.95 % per cycle through the 100 cycles (Fig. 13c). More recently, Zhao et al. demonstrated a promising rechargeable battery with a redox flow cathode in which an environmentally friendly, low-cost ferrocene/ferrocenium molecular redox couple was employed as an active material [183]. The cell showed about 95 % energy efficiency and about 90 % capacity retention after 250 full charge/discharge cycles.

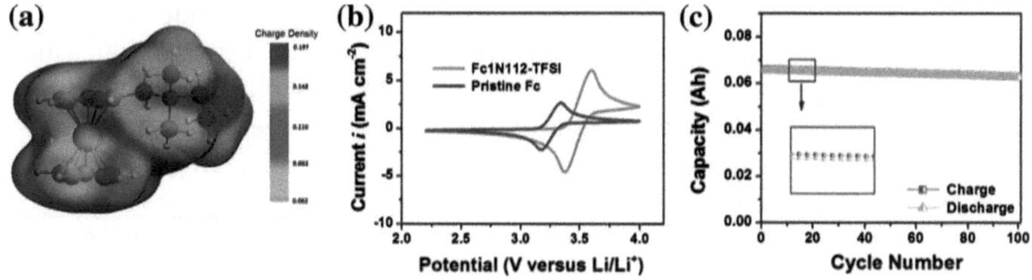

Fig. 13 **a** Charge density around Fc1N112 cation calculated using B3LYP level theory and the TZ2P basis set using the ADF 2013 package; **b** CV scans of 0.04 M pristine ferrocene and 0.1 M Fc1N112-TFSI in a 1.0 M LiTFSI supporting electrolyte in EC/PC/EMC; **c** cycling performance (charge and discharge capacities with respect to cycle number) of the Li|Fc1N112-TFSI flow cell using 0.1 M Fc1N112-TFSI in 1.0 M LiTFSI with 5 wt% FEC at 3.5 mA cm^{-2}. Reprinted from Ref. [184] with permission from WILEY-VCH

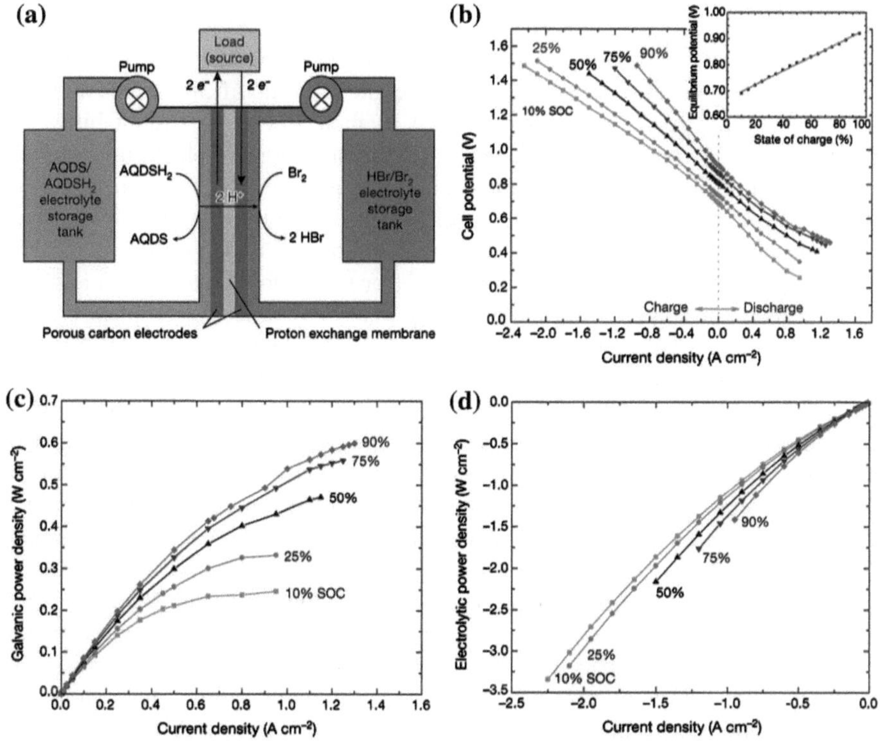

Fig. 14 **a** Schematic cell configuration of redox flow cell at discharge mode. $AQDSH_2$ refers to the reduced form of AQDS. **b** Cell potential versus current density at five different states of charge. **c** Galvanic power density versus current density for the same SOCs. **d** Electrolytic power density versus current density. Reprinted from Ref. [170] with permission from Macmillan Publishers

Like aforementioned organic cathode materials in solid-electrode lithium batteries, tunable molecular structure is their important feature through which their electrochemical properties such as redox potential and solubility can be changed. The active organic materials in flow batteries espouse the same dogma. For example, quinone based molecules are common active species employed in flow batteries and their structure can be readily tuned to achieve favorable electrochemical properties [170, 174, 178]. With tuned small organic molecules called 9,10-anthraquinone-2,7-disulphonic acid (AQDS), a team of Harvard scientists demonstrated that AQDS underwent extremely rapid and reversible two-electron two-proton reduction in sulphuric acid [170]. An aqueous flow battery with the quinone/hydroquinone couple as anode and the Br_2/Br^- redox couple as cathode yielded a peak power density exceeding 0.6 W cm^{-2} at 1.3 A com^{-2} (Fig. 14).

5 Summary and Outlook

Organic redox-active molecules are perceived as promising electrode materials for energy storage devices owing to their light weight, easy modification, and ability to undergo multiple redox processes, as well as their low cost, sustainability and greener nature in comparison with the conventional inorganic electrode materials. Especially, tunability of organic molecular structure can take advantage of rich organic chemistry and facilitate the development of more stable organic electrode materials with greater capacity, improved stability, and non-solubility (for solid-electrode batteries) or higher solubility (for redox flow batteries). In addition, with the increasing interests in the emerging field of wearable electronics, such as roll-up screens, conformable active radio-frequency identification tags, wearable sensors and implantable medical devices, organic electrode materials with flexible nature are perfect candidates to fulfill the applications in foldable electronics. More importantly, organic electrode materials have the potential to be manufactured by using biomass as precursors through a greener process comparing to inorganic electrode materials which are usually produced by less environmentally-friendly ways that generate a great deal of waste water and gases. However, while organic electrode materials possess the notable advantages and attract extraordinary research interests, their wide adoption by battery community still faces big challenges. Many difficulties remain in developing suitable organic electrode materials with satisfactory performance in comparison with their inorganic counterparts. We believe that the following research directions are important to the development of next generation organic electrode materials for different energy storage applications.

1. Organic electrode materials often suffer from their thermally unstable nature of organic molecules leading to short cycling life. Rational design of the organic molecular structure is critical to improve its chemical and physical stability.
2. Organic electrode materials have relatively low packing or tap density in comparison with inorganic electrode materials. As a result, it is difficult to achieve high volumetric energy density of the batteries when employing light

organic electrode materials. Therefore, it is necessary to investigate the self-assembly mechanism of organic electrode materials to increase their packing densities. Another strategy is to develop the organic/inorganic composite electrode materials to take the advantage of the merits from both organic and inorganic materials.

3. Organic electrode materials used in solid-electrode-based batteries exhibit another drawback of appreciable solubility in conventional organic electrolytes which leads to poor cycling stability. To address this problem, several approaches can be taken to prevent the solubility of organic electrode materials, such as structure modification, polymerization, substrate anchoring, as well as selection of suitable electrolytes.
4. Organic electrode materials used in redox flow batteries still have low solubility in non-aqueous electrolytes and exhibit low energy density. This problem can be addressed by structural modifications with ionization or high polarizable groups, and selection of appropriate electrolytes.
5. Low conductivity of most organic molecules usually requires quite a large amount of conductive carbons during the electrode processing to guarantee the electrochemical performance of organic active materials. Another strategy is to develop the composite structure with highly conductive support materials such as graphene and carbon nanotubes.
6. It is important to search for "greener" approaches to manufacture organic electrode materials.
7. Computational calculations are useful to the development of favorable organic electrode materials with high energy density and cycling stability.
8. Finally, the in-situ microscopic characterizations (such as in-situ transmission electron microscopy and in-situ nuclear magnetic resonance) can be powerful tools to reveal the details of the electrochemical processes occurred in the electrode materials and help guide the further development of organic electrode materials.

References

1. Yoshino A (2012) The birth of the lithium-ion battery. Angew Chem Int Ed 51(24):5798–5800
2. Whittingham MS (2004) Lithium batteries and cathode materials. Chem Rev 104(10):4271–4302
3. Goodenough JB, Kim Y (2009) Challenges for rechargeable Li batteries†. Chem Mater 22 (3):587–603
4. Manthiram A (2011) Materials challenges and opportunities of lithium ion batteries. J Phys Chem Lett 2(3):176–184
5. Lu Z, Dahn JR (2002) Understanding the anomalous capacity of $Li/Li[Ni_xLi_{(1/3-2x/3)}Mn_{(2/3-x/3)}]O_2$ cells using in situ X-ray diffraction and electrochemical studies. J Electrochem Soc 149(7):A815–A822
6. Thackeray MM, Johnson CS, Vaughey JT et al (2005) Advances in manganese-oxide 'composite' electrodes for lithium-ion batteries. J Mater Chem 15(23):2257–2267
7. Thackeray MM, Kang S-H, Johnson CS et al (2007) Li_2MnO_3-stabilized $LiMO_2$ (M=Mn, Ni, Co) electrodes for lithium-ion batteries. J Mater Chem 17(30):3112–3125

8. Sun Y-K, Lee M-J, Yoon CS et al (2012) The role of AlF_3 coatings in improving electrochemical cycling of Li-enriched nickel-manganese oxide electrodes for Li-ion batteries. Adv Mater 24(9):1192–1196
9. Choi N-S, Chen Z, Freunberger SA et al (2012) Challenges facing lithium batteries and electrical double-layer capacitors. Angew Chem Int Ed 51(40):9994–10024
10. Yu H, Zhou H (2013) High-energy cathode materials (Li_2MnO_3–$LiMO_2$) for lithium-ion batteries. J Phys Chem Lett 4(8):1268–1280
11. Zheng J, Gu M, Xiao J et al (2013) Corrosion/Fragmentation of layered composite cathode and related capacity/voltage fading during cycling process. Nano Lett 13(8):3824–3830
12. Tarascon JM, Armand M (2001) Issues and challenges facing rechargeable lithium batteries. Nature 414(6861):359–367
13. Hong SY, Kim Y, Park Y et al (2013) Charge carriers in rechargeable batteries: Na ions vs. Li ions. Energy Environ Sci 6(7):2067–2081
14. Renault S, Gottis S, Barres A-L et al (2013) A green Li-organic battery working as a fuel cell in case of emergency. Energy Environ Sci 6(7):2124–2133
15. Jeong G, Kim Y-U, Kim H et al (2011) Prospective materials and applications for Li secondary batteries. Energy Environ Sci 4(6):1986–2002
16. Nyholm L, Nyström G, Mihranyan A et al (2011) Toward flexible polymer and paper-based energy storage devices. Adv Mater 23(33):3751–3769
17. Gwon H, Hong J, Kim H et al (2014) Recent progress on flexible lithium rechargeable batteries. Energy Environ Sci 7(2):538–551
18. Williams DL, Byrne JJ, Driscoll JS (1969) A high energy density lithium/dichloroisocyanuric acid battery system. J Electrochem Soc 116(1):2–4
19. Novák P, Müller K, Santhanam KSV et al (1997) Electrochemically active polymers for rechargeable batteries. Chem Rev 97(1):207–282
20. Stolar M, Baumgartner T (2013) Organic n-type materials for charge transport and charge storage applications. PCCP 15(23):9007–9024
21. Song Z, Zhou H (2013) Towards sustainable and versatile energy storage devices: an overview of organic electrode materials. Energy Environ Sci 6(8):2280–2301
22. Liang Y, Tao Z, Chen J (2012) Organic electrode materials for rechargeable lithium batteries. Adv Energ Mater 2(7):742–769
23. Shirakawa H, Louis EJ, MacDiarmid AG et al (1977) Synthesis of electrically conducting organic polymers: halogen derivatives of polyacetylene, $(CH)_x$. J Chem Soc, Chem Commun 16:578–580
24. Ratnakumar BV, Stefano S, Williams RM et al (1990) Organic cathode materials in sodium batteries. J Appl Electrochem 20(3):357–364
25. Dai Y, Zhang Y, Gao L et al (2010) A sodium ion based organic radical battery. Electrochem Solid-State Lett 13(3):A22–A24
26. Abouimrane A, Weng W, Eltayeb H et al (2012) Sodium insertion in carboxylate based materials and their application in 3.6 V full sodium cells. Energy Environ Sci 5(11):9632–9638
27. Park Y, Shin D-S, Woo SH et al (2012) Sodium terephthalate as an organic anode material for sodium ion batteries. Adv Mater 24(26):3562–3567
28. Zhao L, Zhao J, Hu Y-S et al (2012) Disodium terephthalate ($Na_2C_8H_4O_4$) as high performance anode material for low-cost room-temperature sodium-ion battery. Adv Energ Mater 2(8):962–965
29. Chihara K, Chujo N, Kitajou A et al (2013) Cathode properties of $Na_2C_6O_6$ for sodium-ion batteries. Electrochim Acta 110:240–246
30. Palomares V, Casas-Cabanas M, Castillo-Martinez E et al (2013) Update on Na-based battery materials. A growing research path. Energy Environ Sci 6(8):2312–2337
31. Wang H-G, Yuan S, Ma D-l et al (2014) Tailored aromatic carbonyl derivative polyimides for high-power and long-cycle sodium-organic batteries. Adv Energ Mater 4(7):1301651
32. Chen H, Armand M, Demailly G et al (2008) From biomass to a renewable $Li_XC_6O_6$ organic electrode for sustainable Li-ion batteries. ChemSusChem 1(4):348–355

33. Alt H, Binder H, Köhling A et al (1972) Investigation into the use of quinone compounds-for battery cathodes. Electrochim Acta 17(5):873–887
34. Ohzuku T, Wakamatsu H, Takehara Z et al (1979) Nonaqueous lithium/pyromellitic dianhydride cell. Electrochim Acta 24(6):723–726
35. Si Tobishima, Ji Yamaki, Yamaji A (1984) Cathode characteristics of organic electron acceptors for lithium batteries. J Electrochem Soc 131(1):57–63
36. Pasquali M, Pistoia G, Boschi T et al (1987) Redox mechanism and cycling behaviour of nonylbenzo-hexaquinone electrodes in Li cells. Solid State Ionics 23(4):261–266
37. Han X, Chang C, Yuan L et al (2007) Aromatic carbonyl derivative polymers as high-performance Li-ion storage materials. Adv Mater 19(12):1616–1621
38. Geng J, Bonnet J-P, Renault S et al (2010) Evaluation of polyketones with N-cyclic structure as electrode material for electrochemical energy storage: case of tetraketopiperazine unit. Energy Environ Sci 3(12):1929–1933
39. Walker W, Grugeon S, Mentre O et al (2010) Ethoxycarbonyl-based organic electrode for Li-batteries. J Am Chem Soc 132(18):6517–6523
40. Liang Y, Zhang P, Chen J (2013) Function-oriented design of conjugated carbonyl compound electrodes for high energy lithium batteries. Chem Sci 4(3):1330–1337
41. Hernández-Burgos K, Burkhardt SE, Rodríguez-Calero GG et al (2014) Theoretical studies of carbonyl-based organic molecules for energy storage applications: the heteroatom and substituent effect. J Phys Chem C 118(12):6046–6051
42. Poizot P, Dolhem F (2011) Clean energy new deal for a sustainable world: from non-CO_2 generating energy sources to greener electrochemical storage devices. Energy Environ Sci 4 (6):2003–2019
43. Boschi T, Pappa R, Pistoia G et al (1984) On the use of nonylbenzo-hexaquinone as a substitute for monomeric quinones in non-aqueous cells. J Electroanal Chem Interfacial Electrochem 176(1–2):235–242
44. Lei Z, Wei-kun W, An-bang W et al (2011) A MC/AQ parasitic composite as cathode material for lithium battery. J Electrochem Soc 158(9):A991–A996
45. Xu W, Read A, Koech PK et al (2012) Factors affecting the battery performance of anthraquinone-based organic cathode materials. J Mater Chem 22(9):4032–4039
46. Häringer D, Novák P, Haas O et al (1999) Poly(5-amino-1,4-naphthoquinone), a novel lithium-inserting electroactive polymer with high specific charge. J Electrochem Soc 146 (7):2393–2396
47. Song Z, Zhan H, Zhou Y (2010) Polyimides: promising energy-storage materials. Angew Chem Int Ed 49(45):8444–8448
48. Nokami T, Matsuo T, Inatomi Y et al (2012) Polymer-bound pyrene-4,5,9,10-tetraone for fast-charge and -discharge lithium-ion batteries with high capacity. J Am Chem Soc 134 (48):19694–19700
49. Genorio B, Pirnat K, Cerc-Korosec R et al (2010) Electroactive organic molecules immobilized onto solid nanoparticles as a cathode material for lithium-ion batteries. Angew Chem Int Ed 49(40):7222–7224
50. Lee M, Hong J, Kim H et al (2014) Organic nanohybrids for fast and sustainable energy storage. Adv Mater 26(16):2558–2565
51. Wu H, Shevlin SA, Meng Q et al (2014) Flexible and binder-free organic cathode for high-performance lithium-ion batteries. Adv Mater 26(20):3338–3343
52. Burkhardt SE, Bois J, Tarascon J-M et al (2012) Li-carboxylate anode structure-property relationships from molecular modeling. Chem Mater 25(2):132–141
53. Xiang J, Chang C, Li M et al (2007) A novel coordination polymer as positive electrode material for lithium ion battery. Cryst Growth Des 8(1):280–282
54. Chen H, Armand M, Courty M et al (2009) Lithium salt of tetrahydroxybenzoquinone: toward the development of a sustainable Li-ion battery. J Am Chem Soc 131(25):8984–8988
55. R-h Zeng, X-p Li, Y-c Qiu et al (2010) Synthesis and properties of a lithium-organic coordination compound as lithium-inserted material for lithium ion batteries. Electrochem Commun 12(9):1253–1256

56. Armand M, Grugeon S, Vezin H et al (2009) Conjugated dicarboxylate anodes for Li-ion batteries. Nat Mater 8(2):120–125
57. Walker W, Grugeon S, Vezin H et al (2011) Electrochemical characterization of lithium 4,4 [prime or minute]-tolane-dicarboxylate for use as a negative electrode in Li-ion batteries. J Mater Chem 21(5):1615–1620
58. Renault S, Geng J, Dolhem F et al (2011) Evaluation of polyketones with N-cyclic structure as electrode material for electrochemical energy storage: case of pyromellitic diimide dilithium salt. Chem Commun 47(8):2414–2416
59. Kim DJ, Je SH, Sampath S et al (2012) Effect of N-substitution in naphthalenediimides on the electrochemical performance of organic rechargeable batteries. RSC Adv 2(21):7968–7970
60. Zhu X-Q, Wang C-H (2010) Accurate estimation of the one-electron reduction potentials of various substituted quinones in DMSO and CH_3CN. J Org Chem 75(15):5037–5047
61. Gottis S, Barrès A-L, Dolhem F et al (2014) Voltage gain in lithiated enolate-based organic cathode materials by isomeric effect. ACS Appl Mat Interfaces 6(14):10870–10876
62. Wan W, Lee H, Yu X et al (2014) Tuning the electrochemical performances of anthraquinone organic cathode materials for Li-ion batteries through the sulfonic sodium functional group. RSC Adv 4(38):19878–19882
63. Lee SW, Yabuuchi N, Gallant BM et al (2010) High-power lithium batteries from functionalized carbon-nanotube electrodes. Nat Nano 5(7):531–537
64. Lee SW, Gallant BM, Lee Y et al (2012) Self-standing positive electrodes of oxidized few-walled carbon nanotubes for light-weight and high-power lithium batteries. Energy Environ Sci 5(1):5437–5444
65. Byon HR, Gallant BM, Lee SW et al (2013) Role of oxygen functional groups in carbon nanotube/graphene freestanding electrodes for high performance lithium batteries. Adv Funct Mater 23(8):1037–1045
66. Georgakilas V, Otyepka M, Bourlinos AB et al (2012) Functionalization of graphene: covalent and non-covalent approaches, derivatives and applications. Chem Rev 112 (11):6156–6214
67. Ambrosi A, Chua CK, Bonanni A et al (2014) Electrochemistry of graphene and related materials. Chem Rev 114(14):7150–7188
68. Chua CK, Pumera M (2013) Covalent chemistry on graphene. Chem Soc Rev 42(8):3222–3233
69. Hicks RG (2007) What's new in stable radical chemistry? Org Biomol Chem 5(9):1321–1338
70. Nakahara K, Iwasa S, Satoh M et al (2002) Rechargeable batteries with organic radical cathodes. Chem Phys Lett 359(5–6):351–354
71. Nishide H, Iwasa S, Pu Y-J et al (2004) Organic radical battery: nitroxide polymers as a cathode-active material. Electrochim Acta 50(2–3):827–831
72. Janoschka T, Hager MD, Schubert US (2012) Powering up the future: radical polymers for battery applications. Adv Mater 24(48):6397–6409
73. Katsumata T, Satoh M, Wada J et al (2006) Polyacetylene and polynorbornene derivatives carrying TEMPO. Synthesis and properties as organic radical battery materials. Macromol Rapid Commun 27(15):1206–1211
74. Suga T, Yoshimura K, Nishide H (2006) Nitroxide-substituted polyether as a new material for batteries. Macromol Symp 245–246(1):416–422
75. Lee SH, Kim J-K, Cheruvally G et al (2008) Electrochemical properties of new organic radical materials for lithium secondary batteries. J Power Sources 184(2):503–507
76. Bugnon L, Morton CJH, Novak P et al (2007) Synthesis of poly(4-methacryloyloxy-TEMPO) via group-transfer polymerization and its evaluation in organic radical battery. Chem Mater 19(11):2910–2914
77. Qu J, Morita R, Satoh M et al (2008) Synthesis and properties of DNA complexes containing 2,2,6,6-tetramethyl-1-piperidinoxy (TEMPO) moieties as organic radical battery materials. Chem Eur J 14(11):3250–3259

78. Kim J-K, Matic A, Ahn J-H et al (2012) Preparation and application of TEMPO-based di-radical organic electrode with ionic liquid-based polymer electrolyte. RSC Adv 2 (27):10394–10399
79. Xu L, Yang F, Su C et al (2014) Synthesis and properties of novel TEMPO-contained polypyrrole derivatives as the cathode material of organic radical battery. Electrochim Acta 130:148–155
80. Oyaizu K, Kawamoto T, Suga T et al (2010) Synthesis and charge transport properties of redox-active nitroxide polyethers with large site density. Macromolecules 43(24):10382–10389
81. Qu J, Khan FZ, Satoh M et al (2008) Synthesis and charge/discharge properties of cellulose derivatives carrying free radicals. Polymer 49(6):1490–1496
82. Oyaizu K, Suga T, Yoshimura K et al (2008) Synthesis and characterization of radical-bearing polyethers as an electrode-active material for organic secondary batteries. Macromolecules 41(18):6646–6652
83. Nakahara K, Iriyama J, Iwasa S et al (2007) High-rate capable organic radical cathodes for lithium rechargeable batteries. J Power Sources 165(2):870–873
84. Nakahara K, Iriyama J, Iwasa S et al (2007) Cell properties for modified PTMA cathodes of organic radical batteries. J Power Sources 165(1):398–402
85. Suguro M, Iwasa S, Kusachi Y et al (2007) Cationic polymerization of poly(vinyl ether) bearing a TEMPO radical: a new cathode-active material for organic radical batteries. Macromol Rapid Commun 28(18–19):1929–1933
86. Suga T, Konishi H, Nishide H (2007) Photocrosslinked nitroxide polymer cathode-active materials for application in an organic-based paper battery. Chem Commun 17:1730–1732
87. Oyaizu K, Ando Y, Konishi H et al (2008) Nernstian adsorbate-like bulk layer of organic radical polymers for high-density charge storage purposes. J Am Chem Soc 130(44):14459–14461
88. Suga T, Ohshiro H, Sugita S et al (2009) Emerging N-type redox-active radical polymer for a totally organic polymer-based rechargeable battery. Adv Mater 21(16):1627–1630
89. Ibe T, Frings RB, Lachowicz A et al (2010) Nitroxide polymer networks formed by Michael addition: on site-cured electrode-active organic coating. Chem Commun 46(20):3475–3477
90. Yoshihara S, Isozumi H, Kasai M et al (2010) Improving charge/discharge properties of radical polymer electrodes influenced strongly by current collector/carbon fiber interface. J Phys Chem B 114(25):8335–8340
91. Choi W, Ohtani S, Oyaizu K et al (2011) Radical polymer-wrapped SWNTs at a molecular level: High-rate redox mediation through a percolation network for a transparent charge-storage material. Adv Mater 23(38):4440–4443
92. Oyaizu K, Sukegawa T, Nishide H (2011) Dual dopable poly(phenylacetylene) with nitronyl nitroxide pendants for reversible ambipolar charging and discharging. Chem Lett 40(2):184–185
93. Suga T, Sugita S, Ohshiro H et al (2011) p- and n-type bipolar redox-active radical polymer: toward totally organic polymer-based rechargeable devices with variable configuration. Adv Mater 23(6):751–754
94. Suga T, Takeuchi S, Nishide H (2011) Morphology-driven modulation of charge transport in radical/ion-containing, self-assembled block copolymer platform. Adv Mater 23(46):5545–5549
95. Yoshihara S, Katsuta H, Isozumi H et al (2011) Designing current collector/composite electrode interfacial structure of organic radical battery. J Power Sources 196(18):7806–7811
96. Takeo S, Hiroyuki N (2012) Redox-active radical polymers for a totally organic rechargeable battery. In: Polymers for energy storage and delivery: polyelectrolytes for batteries and fuel cells, vol 1096. ACS symposium series. American Chemical Society, pp 45–53
97. Chae IS, Koyano M, Oyaizu K et al (2013) Self-doping inspired zwitterionic pendant design of radical polymers toward a rocking-chair-type organic cathode-active material. J Mater Chem A 1(4):1326–1333

98. Chae IS, Koyano M, Sukegawa T et al (2013) Redox equilibrium of a zwitterionic radical polymer in a non-aqueous electrolyte as a novel Li^+ host material in a Li-ion battery. J Mater Chem A 1(34):9608–9611
99. Choi W, Endo S, Oyaizu K et al (2013) Robust and efficient charge storage by uniform grafting of TEMPO radical polymer around multi-walled carbon nanotubes. J Mater Chem A 1(9):2999–3003
100. Sukegawa T, Kai A, Oyaizu K et al (2013) Synthesis of pendant nitronyl nitroxide radical-containing poly(norbornene)s as ambipolar electrode-active materials. Macromolecules 46(4):1361–1367
101. Nakahara K, Iriyama J, Iwasa S et al (2007) Al-laminated film packaged organic radical battery for high-power applications. J Power Sources 163(2):1110–1113
102. Kim J-K, Cheruvally G, Choi J-W et al (2007) Effect of radical polymer cathode thickness on the electrochemical performance of organic radical battery. Solid State Ionics 178(27–28):1546–1551
103. Guo W, Yin Y-X, Xin S et al (2012) Superior radical polymer cathode material with a two-electron process redox reaction promoted by graphene. Energy Environ Sci 5(1):5221–5225
104. Kim Y, Jo C, Lee J et al (2012) An ordered nanocomposite of organic radical polymer and mesocellular carbon foam as cathode material in lithium ion batteries. J Mater Chem 22 (4):1453–1458
105. Lin H-C, Li C-C, Lee J-T (2011) Nitroxide polymer brushes grafted onto silica nanoparticles as cathodes for organic radical batteries. J Power Sources 196(19):8098–8103
106. Hung M-K, Wang Y-H, Lin C-H et al (2012) Synthesis and electrochemical behaviour of nitroxide polymer brush thin-film electrodes for organic radical batteries. J Mater Chem 22 (4):1570–1577
107. López-Peña HA, Hernández-Muñoz LS, Cardoso J et al (2009) Electrochemical and spectroelectrochemical properties of nitroxyl radical species in PTMA, an organic radical polymer. Influence of the microstructure. Electrochem Commun 11(7):1369–1372
108. Qu J, Katsumata T, Satoh M et al (2007) Synthesis and charge/discharge properties of polyacetylenes carrying 2,2,6,6-tetramethyl-1-piperidinoxy radicals. Chem Eur J 13 (28):7965–7973
109. Katsumata T, Qu J, Shiotsuki M et al (2008) Synthesis, characterization, and charge/discharge properties of polynorbornenes carrying 2,2,6,6-tetramethylpiperidine-1-oxy radicals at high density. Macromolecules 41(4):1175–1183
110. Dai Y, Zhang Y, Gao L et al (2011) Electrochemical performance of organic radical cathode with ionic liquid based electrolyte. J Electrochem Soc 158(3):A291–A295
111. Koshika K, Sano N, Oyaizu K et al (2009) An aqueous, electrolyte-type, rechargeable device utilizing a hydrophilic radical polymer-cathode. Macromol Chem Phys 210(22):1989–1995
112. Koshika K, Sano N, Oyaizu K et al (2009) An ultrafast chargeable polymer electrode based on the combination of nitroxide radical and aqueous electrolyte. Chem Commun 7:836–838
113. Suga T, Pu Y-J, Kasatori S et al (2007) Cathode- and anode-active poly(nitroxylstyrene)s for rechargeable batteries: p- and n-type redox switching via substituent effects. Macromolecules 40(9):3167–3173
114. Nigrey PJ, MacInnes D, Nairns DP et al (1981) Lightweight rechargeable storage batteries using polyacetylene, (CH)x as the cathode-active material. J Electrochem Soc 128(8):1651–1654
115. Mike JF, Lutkenhaus JL (2013) Electrochemically active polymers for electrochemical energy storage: opportunities and challenges. ACS Macro Letters 2(9):839–844
116. Jeon J-W, Ma Y, Mike JF et al (2013) Oxidatively stable polyaniline:polyacid electrodes for electrochemical energy storage. PCCP 15(24):9654–9662
117. Cheng F, Tang W, Li C et al (2006) Conducting poly(aniline) nanotubes and nanofibers: controlled synthesis and application in lithium/poly(aniline) rechargeable batteries. Chem Eur J 12(11):3082–3088

118. Zhao R-R, Zhu L-M, Qian J-F, Yang H-X (2012) Poly(aniline/o-nitroaniline): a high capacity cathode material for lithium ion batteries. J Electrochem 18(4):310–313
119. Zhang H, Cao G, Wang Z et al (2008) High-rate lithium-ion battery cathodes using nanostructured polyaniline/carbon nanotube array composites. Electrochem Solid-State Lett 11(12):A223–A225
120. Jeon J-W, O'Neal J, Shao L et al (2013) Charge storage in polymer acid-doped polyaniline-based layer-by-layer electrodes. ACS Appl Mat Interfaces 5(20):10127–10136
121. Ge D, Yang L, Honglawan A et al (2014) In situ synthesis of hybrid aerogels from single-walled carbon nanotubes and polyaniline nanoribbons as free-standing, flexible energy storage electrodes. Chem Mater 26(4):1678–1685
122. Park KS, Schougaard SB, Goodenough JB (2007) Conducting-polymer/iron-redox-couple composite cathodes for lithium secondary batteries. Adv Mater 19(6):848–851
123. Zhou M, Qian J, Ai X et al (2011) Redox-active $Fe(CN)_6^{4-}$-doped conducting polymers with greatly enhanced capacity as cathode materials for Li-ion batteries. Adv Mater 23(42):4913–4917
124. Guo B, Kong Q, Zhu Y et al (2011) Electrochemically fabricated polypyrrole–cobalt–oxygen coordination complex as high-performance sithium-storage materials. Chem Eur J 17 (52):14878–14884
125. Mao Y, Kong Q, Guo B et al (2011) Polypyrrole-iron-oxygen coordination complex as high performance lithium storage material. Energy Environ Sci 4(9):3442–3447
126. Arcila-Velez MR, Roberts ME (2014) Redox solute doped polypyrrole for high-charge capacity polymer electrodes. Chem Mater 26(4):1601–1607
127. Visco SJ, DeJonghe LC (1988) Ionic conductivity of organosulfur melts for advanced storage electrodes. J Electrochem Soc 135(12):2905–2909
128. Liu M, Visco SJ, De Jonghe LC (1990) Electrode kinetics of organodisulfide cathodes for storage batteries. J Electrochem Soc 137(3):750–759
129. Liu M, Visco SJ, De Jonghe LC (1991) Novel solid redox polymerization electrodes: all-solid-state, thin-film, rechargeable lithium batteries. J Electrochem Soc 138(7):1891–1895
130. Liu M, Visco SJ, De Jonghe LC (1991) Novel solid redox polymerization electrodes: electrochemical properties. J Electrochem Soc 138(7):1896–1901
131. Sotomura T, Uemachi H, Takeyama K et al (1992) New organodisulfide—polyaniline composite cathode for secondary lithium battery. Electrochim Acta 37(10):1851–1854
132. Oyama N, Tatsuma T, Sato T et al (1995) Dimercaptan-polyaniline composite electrodes for lithium batteries with high energy density. Nature 373(6515):598–600
133. Oyama N, Pope JM, Sotomura T (1997) Effects of adding copper(II) salt to organosulfur cathodes for rechargeable lithium batteries. J Electrochem Soc 144(4):L47–L51
134. Oyama N, Tatsuma T, Sotomura T (1996) Disulfide-polyaniline composite cathodes for rechargeable batteries with high energy density. Macromol Symp 105(1):85–90
135. Oyama N, Hatozaki O (2000) Lithium polymer battery with high energy density. Macromol Symp 156(1):171–178
136. Oyama N (2000) Development of polymer-based lithium secondary battery. Macromol Symp 159(1):221–228
137. Kiya Y, Hutchison GR, Henderson JC et al (2006) Elucidation of the redox behavior of 2,5-dimercapto-1,3,4-thiadiazole (DMcT) at poly(3,4-ethylenedioxythiophene) (PEDOT)-modified electrodes and application of the DMcT-PEDOT composite cathodes to lithium/lithium ion batteries. Langmuir 22(25):10554–10563
138. Canobre SC, Davoglio RA, Biaggio SR et al (2006) Performance of a polyaniline(DMcT)/ carbon fiber composite as cathode for rechargeable lithium batteries. J Power Sources 154 (1):281–286
139. Kiya Y, Henderson JC, Hutchison GR et al (2007) Synthesis, computational and electrochemical characterization of a family of functionalized dimercaptothiophenes for potential use as high-energy cathode materials for lithium/lithium-ion batteries. J Mater Chem 17(41):4366–4376

140. Kiya Y, Iwata A, Sarukawa T et al (2007) Poly[dithio-2,5-(1,3,4-thiadiazole)] (PDMcT)–poly(3,4-ethylenedioxythiophene) (PEDOT) composite cathode for high-energy lithium/lithium-ion rechargeable batteries. J Power Sources 173(1):522–530
141. NuLi Y, Guo Z, Liu H et al (2007) A new class of cathode materials for rechargeable magnesium batteries: organosulfur compounds based on sulfur–sulfur bonds. Electrochem Commun 9(8):1913–1917
142. T-y Chi, Li H, X-w Li et al (2013) Synthesis and electrochemical performance of hierarchically porous carbon-supported PDMcT–PANI composite for lithium-ion batteries. Electrochim Acta 96:206–213
143. Rodríguez-Calero GG, Lowe MA, Kiya Y et al (2010) Electrochemical and computational studies on the electrocatalytic effect of conducting polymers toward the redox reactions of thiadiazole-based thiolate compounds. J Phys Chem C 114(13):6169–6176
144. Gao J, Lowe MA, Conte S et al (2012) Poly(2,5-dimercapto-1,3,4-thiadiazole) as a cathode for rechargeable lithium batteries with dramatically improved performance. Chem Eur J 18 (27):8521–8526
145. Rodríguez-Calero GG, Lowe MA, Burkhardt SE et al (2011) Electrocatalysis of 2,5-dimercapto-1,3,5-thiadiazole by 3,4-ethylenedioxy-substituted conducting polymers. Langmuir 27(22):13904–13909
146. Naoi K, Ki Kawase, Mori M et al (1997) Electrochemistry of poly(2,2′-dithiodianiline): a new class of high energy conducting polymer interconnected with S-S bonds. J Electrochem Soc 144(6):L173–L175
147. Su Y-Z, Dong W, Zhang J-H et al (2007) Poly[bis(2-aminophenyloxy)disulfide]: a polyaniline derivative containing disulfide bonds as a cathode material for lithium battery. Polymer 48(1):165–173
148. Su Y-Z, Niu Y-P, Xiao Y-Z et al (2004) Novel conducting polymer poly[bis(phenylamino) disulfide]: synthesis, characterization, and properties. J Polym Sci, Part A: Polym Chem 42 (10):2329–2339
149. Deng S-R, Kong L-B, Hu G-Q et al (2006) Benzene-based polyorganodisulfide cathode materials for secondary lithium batteries. Electrochim Acta 51(13):2589–2593
150. Li J, Zhan H, Zhou L et al (2004) Aniline-based polyorganodisulfide redox system of high energy for secondary lithium batteries. Electrochem Commun 6(6):515–519
151. Cho J-S, Sato S, Takeoka S et al (2001) Synthesis of disulfide-containing aniline and copolymerization with aniline. Macromolecules 34(9):2751–2756
152. Uemachi H, Iwasa Y, Mitani T (2000) Preparation and charge-discharge properties of a novel organosulfur polymer, Poly(p-phenylene thiuret), for battery applications. Chem Lett 29 (8):946–947
153. Scrosati B, Garche J (2010) Lithium batteries: status, prospects and future. J Power Sources 195(9):2419–2430
154. Pan H, Hu Y-S, Chen L (2013) Room-temperature stationary sodium-ion batteries for large-scale electric energy storage. Energy Environ Sci 6(8):2338–2360
155. Yao M, Kuratani K, Kojima T et al (2014) Indigo carmine: an organic crystal as a positive-electrode material for rechargeable sodium batteries. Sci Rep 4
156. Luo W, Allen M, Raju V et al (2014) An organic pigment as a high-performance cathode for sodium-ion batteries. Adv Energ Mater 4(15):1400554
157. Kim H, Park Y-U, Park K-Y et al (2014) Novel transition-metal-free cathode for high energy and power sodium rechargeable batteries. Nano Energy 4:97–104
158. Deng W, Liang X, Wu X et al (2013) A low cost, all-organic Na-ion battery based on polymeric cathode and anode. Sci Rep 3:2671
159. Wang S, Wang L, Zhu Z et al (2014) All organic sodium-ion batteries with $Na_4C_8H_2O_6$. Angew Chem Int Ed 53(23):5892–5896
160. Zhao R, Zhu L, Cao Y et al (2012) An aniline-nitroaniline copolymer as a high capacity cathode for Na-ion batteries. Electrochem Commun 21:36–38
161. Zhou M, Zhu L, Cao Y et al (2012) Fe(CN)6-4-doped polypyrrole: a high-capacity and high-rate cathode material for sodium-ion batteries. RSC Adv 2(13):5495–5498

162. Zhu L, Shen Y, Sun M et al (2013) Self-doped polypyrrole with ionizable sodium sulfonate as a renewable cathode material for sodium ion batteries. Chem Commun 49(97):11370–11372
163. Yang Z, Zhang J, Kintner-Meyer MCW et al (2011) Electrochemical energy storage for green grid. Chem Rev 111(5):3577–3613
164. Dunn B, Kamath H, Tarascon J-M (2011) Electrical energy storage for the grid: a battery of choices. Science 334(6058):928–935
165. Ponce de León C, Frías-Ferrer A, González-García J et al (2006) Redox flow cells for energy conversion. J Power Sources 160(1):716–732
166. Skyllas-Kazacos M, Chakrabarti MH, Hajimolana SA et al (2011) Progress in flow battery research and development. J Electrochem Soc 158(8):R55–R79
167. Leung P, Li X, Ponce de Leon C et al (2012) Progress in redox flow batteries, remaining challenges and their applications in energy storage. RSC Adv 2(27):10125–10156
168. Weber A, Mench M, Meyers J et al (2011) Redox flow batteries: a review. J Appl Electrochem 41(10):1137–1164
169. Rugolo J, Aziz MJ (2012) Electricity storage for intermittent renewable sources. Energy Environ Sci 5(5):7151–7160
170. Huskinson B, Marshak MP, Suh C et al (2014) A metal-free organic-inorganic aqueous flow battery. Nature 505(7482):195–198
171. Brushett FR, Vaughey JT, Jansen AN (2012) An all-organic non-aqueous lithium-ion redox flow battery. Adv Energ Mater 2(11):1390–1396
172. Li Z, Li S, Liu S et al (2011) Electrochemical properties of an all-organic redox flow battery using 2,2,6,6-tetramethyl-1-piperidinyloxy and N-methylphthalimide. Electrochem Solid-State Lett 14(12):A171–A173
173. Zhao Y, Si S, Liao C (2013) A single flow zinc/polyaniline suspension rechargeable battery. J Power Sources 241:449–453
174. Bachman JE, Curtiss LA, Assary RS (2014) Investigation of the redox chemistry of anthraquinone derivatives using density functional theory. J Phys Chem A 118(38):8852–8860
175. Liu P, Y-l Cao, Li G-R et al (2013) A solar rechargeable flow battery based on photoregeneration of two soluble redox couples. ChemSusChem 6(5):802–806
176. Wang W, Luo Q, Li B et al (2013) Recent progress in redox flow battery research and development. Adv Funct Mater 23(8):970–986
177. Lu Y, Goodenough JB (2011) Rechargeable alkali-ion cathode-flow battery. J Mater Chem 21(27):10113–10117
178. Wang W, Xu W, Cosimbescu L et al (2012) Anthraquinone with tailored structure for a nonaqueous metal-organic redox flow battery. Chem Commun 48(53):6669–6671
179. Lu Y, Goodenough JB, Kim Y (2011) Aqueous cathode for next-generation alkali-on batteries. J Am Chem Soc 133(15):5756–5759
180. Wang Y, He P, Zhou H (2012) Li-redox flow batteries based on hybrid electrolytes: at the cross road between Li-ion and redox flow batteries. Adv Energ Mater 2(7):770–779
181. Zhao Y, Wang L, Byon HR (2013) High-performance rechargeable lithium-iodine batteries using triiodide/iodide redox couples in an aqueous cathode. Nat Commun 4:1896
182. Fan FY, Woodford WH, Li Z et al (2014) Polysulfide flow batteries enabled by percolating nanoscale conductor networks. Nano Lett 14(4):2210–2218
183. Zhao Y, Ding Y, Song J et al. (2014) Sustainable electrical energy storage through the ferrocene/ferrocenium redox reaction in aprotic electrolyte. Angew Chem Int Ed 53 (41):11036–11040
184. Wei X, Cosimbescu L, Xu W et al (2015) Towards high-performance nonaqueous redox flow electrolyte via ionic modification of active species. Adv Energ Mater 5(1):1400678

Recent Developments and Trends in Redox Flow Batteries

Liang Su, Jeffrey A. Kowalski, Kyler J. Carroll and Fikile R. Brushett

1 Introduction

A major challenge of the 21st century is the development of efficient and sustainable means of energy conversion, distribution, and storage on a global scale. Presently, fossil fuel technologies make up the backbone of our energy economy, notably transportation and bulk generation for electrification. The International Energy Agency (IEA) reported that, in 2012, of the total energy consumed worldwide, around 70.8 % was derived from fossil fuel sources (oil, natural gas, coal, etc.), with the majority of energy consumption and demand growth coming from developing nations [1]. In the United States, the 2011 Energy Information Administration (EIA) annual review reported that around 78 % of the total energy consumed was derived from fossil fuel sources (i.e., oil, coal, and natural gas) and the two largest consumption sectors are electricity (38 %) and transportation (27 %), both of which are dominated by fossil fuel sources [2, 3]. However, in the future, this energy mix will not be feasible [4]. Rising population and continuing economic growth in the developing world are projected to double global energy consumption by 2050 [3]. Non-renewable fossil fuel reserves, which took millennia to accumulate, are finite and rapidly disappearing. Moreover, the continued and increasing generation of anthropogenic carbon dioxide (CO_2) from fossil fuel combustion will likely have negative implications for the global climate [5]. Analysis by the Intergovernmental Panel on Climate Change (IPCC) indicated that to stabilize the atmospheric concentration of CO_2 at 350–400 ppm (near its current level), global CO_2 emissions would need to be tapered by 2050 to a level of 20–50 % of the 2000 emissions [6]. Thus a tremendous need exists for scientific and technological advances to address these grand challenges, sparking worldwide

L. Su · J.A. Kowalski · K.J. Carroll · F.R. Brushett (✉)
Department of Chemical Engineering, Massachusetts Institute of Technology, Cambridge, MA 02139, USA
e-mail: brushett@mit.edu

Z. Zhang and S.S. Zhang (eds.), *Rechargeable Batteries*,
Green Energy and Technology, DOI 10.1007/978-3-319-15458-9_24

investment in low-carbon/carbon-neutral power generation, carbon capture and storage, and system-wide energy efficiency [7].

Stationary energy storage systems (ESS) will play a pivotal role in the widespread integration of renewable, non-dispatchable energy sources (e.g., solar photovoltaic (PV), wind) and in the improvement of energy efficiency of the electric grid [8, 9]. Indeed, increased energy storage assets can enhance energy security, reduce carbon emissions, and introduce new revenue streams for a range of stakeholders [10, 11]. In cases where no transmission or distribution constraints exist, grid-connected ESS are not required to be co-located with the energy source, providing the flexibility to optimize storage performance characteristics and minimize costs. Where constraints exist (e.g., developing economies, island nations), coupling energy storage with local generation resources (e.g., solar PV, fuel cells, micro-turbines) can enable the development of robust micro-grids. Services provided by ESS and their remuneration create the incentive for adopting energy storage. These services can be broadly classified as bulk energy, ancillary, transmission and distribution (T&D), renewables integration, and customer energy management (Table 1) [12, 13]. However, despite this promise, only $\sim$ 2.5 % of total electric production in the US relies on grid energy storage with the principal barrier to widespread installation being the system cost [14]. Therefore, the development of cost competitive energy storage technologies with validated reliability and safety is of paramount importance to the continued evolution of the electric power sector in the US and worldwide.

While a range of ESS options exist to meet the aforementioned services, no single technology is suitable for all applications. Present storage technologies vary in their performance characteristics, level of technological maturity, and, most importantly, cost. The applicability of different technologies based on their associated attributes is shown in Fig. 1, an adaptation from a recent Sandia National Laboratory report [15]. The storage technology comparison shown divides the applications roughly into three broad segments based on the discharge time and system power requirements: uninterruptible power supply, T&D grid support-load shifting, and bulk management. Both cost and bulk-storage-relevant attributes have resulted in pumped hydroelectric storage (pumped hydro) being responsible for over 97 % of the worldwide energy storage capacity (ca. 127 GW) [16]. Pumped hydro however suffers from constraints arising from geographical settings, licensing, environmental regulations, and uncertainty in long-term electric markets [10, 17]. Though they presently constitute a significantly smaller installed capacity (ca. 100 MW), electrochemical energy storage technologies have a number of desirable characteristics including high power/energy density, high round-trip efficiency, rapid response time, and terrain-independence. However, the economics of grid storage are challenging. In a 2013 report, the United States Department of Energy (DOE) has outlined a near-term system capital cost goal of $250/kWh for grid storage systems with a long term reduction to $150/kWh [10]. Other DOE programs have proposed more aggressive cost targets. For example, the Advanced Research Projects Agency-Energy (ARPA-E) Grid-Scale Rampable Intermittent Dispatchable Storage (GRIDS) program has set a target of less than $100/kWh for

Table 1 Key characteristics of storage systems for selected energy services adapted from Ref. [13]

Services	Size (MW)	Discharge duration	Cycles (typical)	Response time	Output (electricity 'e', thermal 't')
Bulk energy services					
Seasonal storage	500–2000	d–mo	1–5/y	d	e, t
Arbitrage	100–2000	8–24 h	0.25–1/d	>1 h	e
Ancillary services					
Frequency regulation	1–2000	1–15 min	20–40/d	1 min	e
Load following	1–2000	15 min–1 d	1–29/d	<15 min	e, t
Voltage support	1–40	1–60 s	10–100/d	0.001–1 s	e
Black start	0.1–400	1–4 h	<1/y	<1 h	e
Spinning reserve	10–2000	15 min–2 h	0.5–2/d	<15 min	e
Non-spinning reserve	10–2000	15 min–2 h	0.5–2/d	>15 min	e
Transmission and distribution infrastructure services					
Transmission and distribution congestion relief	10–500	2–4 h	0.14–1.25/d	>1 h	e, t
Transmission and distribution investment deferral	1–500	2–5 h	0.75–1.25/d	>1 h	e, t
Renewable and other integration services					
Variable Supply Resource Integration	1–400	1 min–h	0.5–2/d	<15 min	e, t
Waste Heat Utilization	1–10	1–24 h	1–20/d	<10 min	t
Combined Heat and Power	1–5	min–h	1–10/d	<15 min	t
Customer energy management services					
Demand shifting and peak reduction	0.001–1	min–h	1–29/d	<15 min	e, t
Off-grid	0.001–	3–5 h	0.75–1.5/d	<1 h	e, t

new storage technologies [18]. Further, the Joint Center for Energy Storage Research (JCESR), a DOE-funded Energy Innovation Hub, aims to develop battery prototypes that, when scaled to manufacturing, are projected to reach battery price levels to enable widespread market adoption (e.g., \$100 per useable kWh) [19]. These stringent targets reflect the low cost of electricity in the United States but market penetration may be realized in global regions with higher electricity costs (e.g., Europe, Japan [1]) or through niche applications where higher system costs can be offset by the value of provided services.

Redox flow batteries (RFBs) have attracted considerable academic and industrial interest based on their favorable combination of performance, cost, and safety. Over the past few years, several comprehensive reviews have been published on RFBs

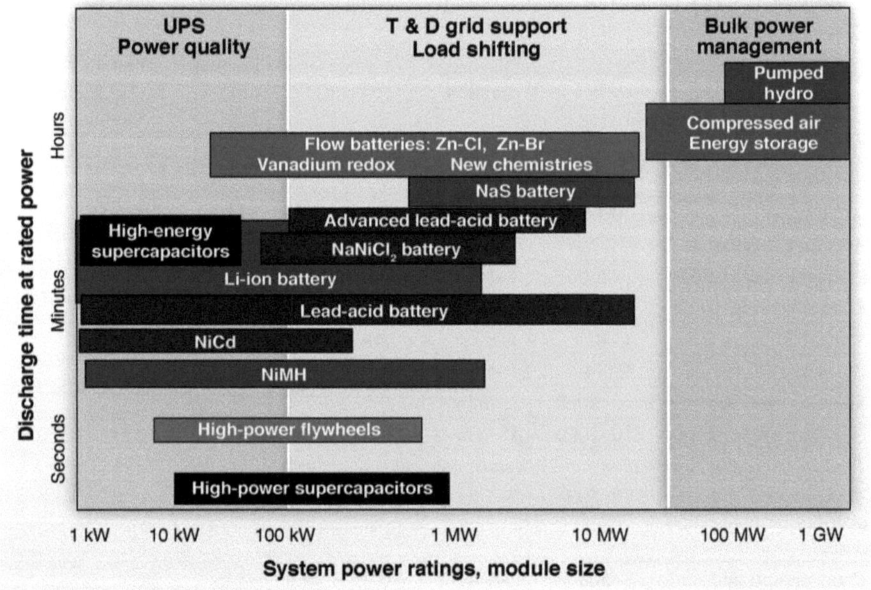

Fig. 1 Comparison of select storage technologies and associated addressable services broadly divided into three categories: uninterruptible power supply (UPS)-power quality, transmission and distribution (T&D) grid support-load shifting, and bulk power management. When roughly compared to Table 1, UPS-power quality is comparable to customer energy management services, T&D grid support-load shifting to ancillary services, T&D infrastructure services, renewable and other integration services, while bulk power management can be related to bulk energy services. Figure adapted from Ref. [15]. Reproduced with permission. Copyright 2010 Electric Power Research Institute

with detailed assessments of individual components (i.e., membranes [20–22] and electrodes/bipolar plates [23–25]) and of utility for stationary ESS applications [26–31]. For the sake of completeness, we will briefly review several topics covered in previous literature and, where appropriate, we will highlight in-depth reports for the interested reader. However, our goal is to discuss emerging, potentially transformative, strategies for enhancing RFB technologies through molecular design, electrolyte development, and cell-level engineering. In particular, we will survey the non-aqueous RFB literature which, to date, has been underrepresented in critical reviews. To this end, following a short overview of RFB basics and key performance metrics (Sect. 2), we highlight recent advances in redox active materials for aqueous and non-aqueous RFBs (Sect. 3) and new cell configurations to enhance common chemistries and enable the emerging redox couples and flowable formulations (Sect. 4). Finally, we conclude with future research directions and key challenges for RFB technologies (Sect. 5).

2 Overview of Redox Flow Batteries

A redox flow battery (RFB) is a rechargeable electrochemical device that utilizes the reversible redox reactions of two soluble electroactive species for energy storage. Figure 2 shows a simplified schematic of a flow battery. The system includes electrochemical reactors, storage vessels, circulation pumps, a heat exchanger, and power conditioning equipment. The positive and negative electrolytes (also referred to as the catholyte and anolyte) are fed to one or more electrochemical reactors, where the active species are oxidized or reduced to alternately charge or discharge the battery. Within an electrochemical cell, each electrolyte reacts on the corresponding positive or negative electrodes (also referred to as the cathode or anode), which are separated by either an ion-selective membrane or a nanoporous separator. These generic reactions are shown below (assuming a 1-electron transfer process):

$$\begin{array}{lll}
\text{Positive electrode:} & X^{n+} + e^- \rightleftharpoons X^{(n-1)+} & E_1^0 \\
\text{Negative electrode:} & Y^{m+} + e^- \rightleftharpoons Y^{(m-1)+} & E_2^0 \\
\text{Total reaction:} & X^{n+} + Y^{(m-1)-} \rightleftharpoons X^{(n-1)+} + Y^{m+} & E_{cell} = E_1^0 - E_2^0
\end{array}$$

As compared to enclosed batteries, RFB architecture has several compelling features including:

- Independent power (W) and energy (Wh) due to separate tank and electroreactor configurations, which enables modular flexibility.
- Long life cycle with deep charge capabilities as storage is based on facile solution-phase redox reactions where the electrodes serve as a source or sink of electrons. Indeed, Skyllas-Kazacos et al. reported that durability generally exceeds 5000 deep cycles for flow batteries [31].
- Superior safety as reactants are contained in physically separated tanks with a relatively small volume in close proximity within the reactor stack.

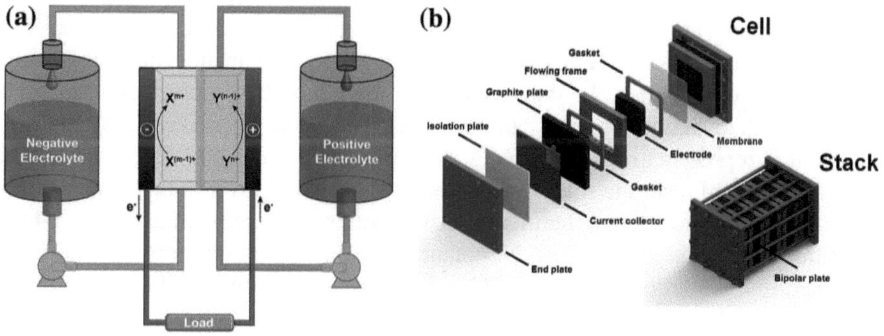

Fig. 2 **a** Schematic of a redox-flow cell (discharge mode). **b** Exploded view of components in a redox flow cell and an assembled redox flow battery with 4 stacked cells

- Simplified manufacturing and lower cell-to-cell variability due to simple electrode and stack construction. Further, replacement costs are likely to be lower.
- High active-to-inactive materials ratios are possible, especially for long duration storage where the battery cost asymptotically approaches the cost of redox solutions.

However, these advantages must be balanced against low energy density and increased system complexity. The energy densities of common aqueous flow systems, such as all-vanadium redox (VRB), are lower than those of enclosed batteries such as lead-acid and lithium-ion. Fortunately, this deficiency can be tolerated in many stationary applications. Flow battery systems require pumps to circulate the electrolytes, resulting in parasitic losses and complicating independent operation on the grid. Shunt currents between electrochemical cells within the stack also lead to efficiency losses. In this chapter, we will not contemplate these system-level considerations but direct the interested reader to reference [32] for the estimation of system efficiency in the presence of shunt current and pumping loss.

At present, RFBs are broadly considered too expensive for widespread deployment. While materials cost reductions can be expected through manufacturing scale and learning by doing, and additional contributions to price will decrease through competition and volume, new redox couples, electrolytes, and reactor configurations are likely needed to meet the DOE cost targets. As battery cost ($/kWh) are proportional to materials costs ($/kg) and inversely proportional to materials requirements (kWh/kg), two approaches can be taken to lower the system cost: cheaper storage materials and higher energy density (less material required). As we hope to highlight in this chapter, both of these avenues are being pursued by the research community.

3 Redox Active Compounds for Flow Cells

A wide range of electroactive materials have been investigated for use in RFBs. These materials can be broadly classified as inorganic compounds, which are far more prevalent at this point, and organic compounds, which are emergent. Moreover, their performance and durability have been evaluated in aqueous and non-aqueous electrolytes. In this section we highlight both select common and newly-identified redox couples and the resulting cell chemistries. While this treatment is representative rather than exhaustive, the ideal redox couple should be cost-effective and have the following properties: excellent reversibility and stability (both chemical and electrochemical), fast kinetics and mass transfer, and high solubility in relevant electrolytes. Note that only solution-phase redox reactions are discussed in this section. Dissolution/deposition reactions are introduced in Sect. 4.

3.1 Redox Active Compounds in Aqueous Flow Cells

3.1.1 Inorganic Redox Active Compounds

The application of inorganic electroactive compounds in aqueous RFBs have been the subject of the vast majority of the literature to date. Table 2 summarizes the standard electrode potentials of common redox couples while Table 3 highlights prominent cell chemistries based on combinations of these redox couples. Of these chemistries, iron–chromium (ICB) [33], polysulfide–bromide (PSB) [14], and all-vanadium (VRB) [31] systems have yielded industry-level demonstrations (order of 100 kW–10 MW). Below, these RFB chemistries are introduced in some detail with key advantages, disadvantages, and challenges highlighted.

Iron–Chromium Redox Flow Batteries

As the first modern RFB, the Fe–Cr system was proposed by NASA in 1973 and triggered considerable research activity on electrodes, membranes, and catalysts for optimizing performance and durability throughout the 1980s [43]. Furthermore, a 1 kW/13 kWh ICB system with 8 × 39-cell stacks was prototyped as the energy

Table 2 Electrochemical series for redox flow batteries [34]

Electrode reaction	E^0 (V)
$Li^+ + e^- \rightleftharpoons Li$	−3.040
$2H_2O + 2e^- \rightleftharpoons H_2 + 2OH^-$	−0.828
$Zn^{2+} + 2e^- \rightleftharpoons Zn$	−0.762
$Cr^{3+} + e^- \rightleftharpoons Cr^{2+}$	−0.407
$S_4^{2-} + 2e^- \rightleftharpoons 2S_2^{2-}$	−0.265
$V^{3+} + e^- \rightleftharpoons V^{2+}$	−0.255
$Pb^{2+} + 2e^- \rightleftharpoons Pb$	−0.126
$TiOH^{3+} + H^+ + 2e^- \rightleftharpoons Ti^{3+} + H_2O$	−0.055
$2H^+ + 2e^- \rightleftharpoons H_2$	0.000
$O_2 + 2H_2O + 4e^- \rightleftharpoons 4OH^-$	0.401
$I_3^- + 2e^- \rightleftharpoons 3I^-$	0.536
$Fe^{3+} + e^- \rightleftharpoons Fe^{2+}$	0.771
$VO_2^+ + 2H^+ + e^- \rightleftharpoons VO^{2+} + H_2O$	0.991
$Br_2 + 2e^- \rightleftharpoons 2Br^-$	1.807
$O_2 + 4H^+ + 4e^- \rightleftharpoons 2H_2O$	1.229
$PbO_2 + 4H^+ + 2e^- \rightleftharpoons Pb^{2+} + 2H_2O$	1.455
$Mn^{3+} + e^- \rightleftharpoons Mn^{2+}$	1.542
$Ce^{4+} + e^- \rightleftharpoons Ce^{3+}$	1.720

Here, E^0 represents the standard electrode potential

Table 3 Summary of major aqueous redox flow battery chemistries

Chemistry	Cell voltage (V)	Electrolyte (positive/negative)	Current density (mA/cm^2)	Efficiency	Reference
Fe/Ti	0.826	1 M $FeCl_3$ + 0.5 M HCl	≤25	CE ≈ 100 %	[35]
		1 M $TiCl_3$ + 6 M HCl		EE = 40 %	
Fe/V	1.026	1.25 M $FeCl_2$ + 1.25 M VCl_3 + 2.3 M HCl	50	CE = 97 %	[36]
		1.25 M $FeCl_2$ + 1.25 M VCl_3 + 2.3 M HCl		EE = 78 %	
Fe/Cr	1.178	1.25 M $FeCl_2$ + 2.3 M HCl + 1.25 M $CrCl_3$	40	CE = 97 %	[37]
		1.25 M $FeCl_2$ + 2.3 M HCl + 1.25 M $CrCl_3$		EE = 73 %	
VRB	1.246	1.5 M $VOSO_4$ + 2 M H_2SO_4	40	CE = 90 %	[38]
		1.5 M $VOSO_4$ + 2 M H_2SO_4		EE = 73 %	
Bromide/polysulfide	1.352	4.0 M NaBr	40	EE = 64 %	[39]
		1.3 M Na_2S_4			
Polyhalide/V	~1.43	1 M NaBr + 1 M HCl	20	CE = 83 %	[40]
		1 M VCl_3 + 1.5 M HCl		EE = 66 %	
Mn/V	1.797	0.3 M Mn(II) + 5 M H_2SO_4	20	CE = 69 %	[41]
		0.3 M V(III) + 5 M H_2SO_4		EE = 63 %	
Ce/V	1.975	0.5 M Ce(III) + 1 M H_2SO_4	22	CE = 87 %	[42]
		0.5 M V(III) + 1 M H_2SO_4			

Here, CE and EE represent coulombic efficiency and energy efficiency, respectively

storage device for a photovoltaic array in the early 1980s [44]. The electrode reactions are[1]:

$$\begin{array}{lll} \text{Positive electrode:} & Fe^{3+} + e^- \rightleftharpoons Fe^{2+} & E^0 = 0.771\,V \\ \text{Negative electrode:} & Cr^{3+} + e^- \rightleftharpoons Cr^{2+} & E^0 = -0.407\,V \\ \text{Total reaction:} & Fe^{3+} + Cr^{2+} \rightleftharpoons Fe^{2+} + Cr^{3+} & E = 1.178\,V \end{array}$$

Hydrochloric acid and an cation or anion-selective membrane are commonly employed in the system. While the ferrous/ferric (Fe^{2+}/Fe^{3+}) redox couple demonstrates facile kinetics on carbonaceous surfaces, the kinetics of the chromous/chromic (Cr^{2+}/Cr^{3+}) couple are fairly sluggish, requiring a relatively high overpotential to drive the reaction. Thus, the hydrogen evolution reaction (HER) inevitably becomes a concern at these low potentials, resulting in the parasitic losses on the negative electrode. A gold-lead bimetallic electrocatalyst has been shown to effectively mitigate the problem as lead suppresses HER while gold catalyzes the redox reaction of Cr^{2+}/Cr^{3+} [44]. However, the incorporation of gold is undesirable as it dramatically

[1]All electrode potential values are referenced to the standard hydrogen electrode unless otherwise stated.

increases the capital cost. Moreover, for these systems, rebalancing the system due to side reactions and active species crossover (through the membrane) is necessary for long-term use, which adds to the overall operating expenses.

Polysulfide-Bromide Redox Flow Batteries

The PSB is an attractive chemistry due to the low cost, high abundance, and high solubility of both polysulfide and bromine in aqueous electrolytes. Considered applicable in MW/MWh class installations, this chemistry has been extensively studied since the earlier 1990s with several large-scale demonstration undertaken by various companies (e.g., Regenesys Technology) [14]. The electrode reactions are:

Positive electrode:	$Br_3^- + 2e^- \rightleftharpoons 3Br^-$	$E^0 = +1.087\,V$
Negative electrode:	$S_4^{2-} + 2e^- \rightleftharpoons 2S_2^{2-}$	$E^0 = -0.265\,V$
Total reaction:	$Br_3^- + 2S_2^{2-} \rightleftharpoons 3Br^- + S_4^{2-}$	$E = 1.352\,V$

A cation exchange membrane is typically used with a sodium ion shuttling between the positive and negative electrolytes containing NaBr and Na_2S_4, respectively. Nickel [45] and cobalt [39] have been found to catalyze the negative electrode reaction. Further, the use of activated carbon/polyolefin composite electrodes has been shown to increase charging voltages, from 1.7 to 2.1 V, due to bromine adsorption [31]. However, as with ICBs, species crossover over the operating life dramatically decreases the efficiency and capacity of the cell. Moreover, side reactions may lead to the formation of toxic gases (i.e., Br_2, H_2S) and insoluble insulating precipitants. Therefore, the system requires stringent maintenance to separate the electrolyte and to recover the reactants, which substantially increases the operating cost. Despite the favorable kinetics, the mass transfer overpotential on the positive electrode limits the discharge performance of the battery [46]. Scamman and coworkers developed a numerical model for PSBs, and showed that there would still be a net loss for arbitrage applications even if operated under optimum conditions. However, if the total capital cost were reduced by 20–30 % or the buy and sell price difference for electrical energy were increased, the system could become economically feasible [47].

All Vanadium Redox Flow Batteries

As arguably the most well-known RFB chemistry, VRBs take advantage of the four oxidation states of vanadium within the stability window of water. This enables operation with the same element as an electroactive species as both negative and positive electrolytes and limits concerns about solution crossover and the associated permanent deleterious effects (e.g., capacity fade, irreversible side reactions). Since the initial electrochemical studies of the V(IV)/V(V) and the V(II)/V(III) redox couples in 1985 [48, 49] and the first demonstration of an all-vanadium redox flow cell in 1986 [50] by Skyllas-Kazacos and co-workers, VRBs have been the focus of intensive research, development, demonstration, and deployment activities for more than two decades. These efforts have culminated in several pilot plant scale or utility scale installations worldwide [31]. The electrode reactions of a VRB are:

Positive electrode:	$VO_2^+ + 2H^+ + e^- \rightleftharpoons VO^{2+} + H_2O$	$E^0 = +0.991\ V$
Negative electrode:	$V^{3+} + e^- \rightleftharpoons V^{2+}$	$E^0 = -0.255\ V$
Total reaction:	$VO_2^+ + V^{2+} + 2H^+ \rightleftharpoons VO^{2+} + V^{3+} + H_2O$	$E = 1.246\ V$

This system typically uses sulfuric acid as the electrolyte with a proton exchange membrane. While a porous separator could be used, for high efficiency operation, ion-selective membranes are generally preferred as vanadium crossover leads to losses in coulombic efficiency. At present, Nafion is the membrane of choice as V(V) is a powerful oxidizing agent, which can attack cheaper hydrocarbon-based ion selective membranes [21]. The redox reactions of different vanadium species have displayed reversibility and high activity on carbon based electrodes. Moreover, Li et al. discovered the catalytic effects of bismuth nanoparticles on V(II)/V(III) [51] and of niobium oxide nanorods on both V(II)/V(III) and V(IV)/V(V) [52], which have been shown to further enhance the energy efficiency of the VRB by more than 10 %.

One challenge of VRBs is the low energy density (<25 Wh/L), which is, at least partially, restricted by the low solubility of $VOSO_4$ (<1.7 M), which is sensitive to both acid concentration (3–4 M H_2SO_4) and solution temperature (10–40 °C) [53]. Recently, Li et al. developed sulfate-chloride mixed acid electrolyte that solubilizes V(V) in the form of $VO_2Cl(H_2O)_2$ up to 2.5 M, which corresponds to a 70 % improvement of the battery capacity compared to the conventional sulfate based electrolyte [54]. A 1 kW/1 kWh prototype VRB system using the mixed acid electrolyte has been demonstrated [55] and the underlying patents are being licensed by several companies. A second challenge for VRB technologies is the high cost of vanadium, reportedly 43 % of total cost for 1 MW/4 MWh mixed acid system [32]. This may set a floor for the potential system costs despite reductions enabled by mass production and engineering improvements. Alternative approaches are exploring the use of bromine polyhalide positive electrolytes, which leverages the higher solubility of vanadium in the presence of halides and the lower cost of bromine-based electrolytes [30, 31]. However, the aforementioned challenges with ICB and PSB systems; namely species crossover and bromine toxicity, are re-introduced.

3.1.2 Organic Redox Active Molecules

Recently, electroactive organics have been explored as active materials in aqueous RFBs and may open new avenues to meet the stringent grid storage cost targets. Indeed, organic molecules may offer several advantages over inorganic molecules. First, they are often comprised of earth-abundant elements (carbon, hydrogen, oxygen, nitrogen, sulfur, etc.), and thus, their cost and availability is less constrained by the production and reserves of key elements (e.g., vanadium). Moreover, while transition metals must be mined and purified from ore, organic materials may be synthesized in sustainable fashion using green chemistry routes [56]. Second, while inorganic redox couples are restricted by the known periodic table, a broader array of organic redox-active molecules are available, allowing for the realization of new redox couples. Third, key electrochemical and physical properties (e.g., redox potential, solubility) can be tailored via modification of the redox moiety or the surrounding molecular structure. Further, a number of organics are known to undergo two electron transfer (e.g., quinones), which may lead to higher intrinsic capacities and thus, higher cell energy densities. These design degrees of freedom allow for a wide array of potential storage materials to be contemplated and for high throughput computational tools to be leveraged to drive the discovery and development process. However, as with any emerging materials, a number of unanswered questions exist including organic stability in acidic electrolytes and the organic solubility in aqueous media. Below, we highlight recent developments in organic aqueous RFBs.

Huskinson et al. studied the electrochemistry of 9,10-anthraquinone-2,7-disulphonic acid (AQDS) and demonstrated its applicability as the low potential compound in a AQDS-bromide aqueous redox flow cell [57]. AQDS undergoes a reversible, 2-electron transfer process on the glassy carbon electrode. Excitingly, the reported reaction kinetics were one to three orders of magnitude faster than many common redox couples Fe^{2+}/Fe^{3+}, Cr^{2+}/Cr^{3+}, VO^{2+}/VO_2^{+}, V^{2+}/V^{3+}, Br_2/Br^-, and S_4^{2-}/S_2^{2-} on similar carbon surfaces. The electrode reactions are:[2]

Positive electrode:	$Br_2 + 2e^- \rightleftharpoons 2Br^-$	$E^0 = +1.087\,V$
Negative electrode:	$AQDS + 2H^+ + 2e^- \rightleftharpoons AQDSH_2$	$E^0 = +0.213\,V$
Total reaction:	$Br_2 + AQDSH_2 \rightleftharpoons 2Br^- + AQDS + 2H^+$	$E = 0.874\,V$

Promising cycling performance and capacity retention was obtained using a benchtop cell (Fig. 3). Specifically, the cell was operated at a large current density (0.5 A/cm^2) with a high current efficiency ($\sim$99 %) and low capacity degradation (0.78 %/cycle). Since energy density is directly proportional to the number of electrons transferred by the redox active compound, the 2-electron transfer redox event of AQDS is highly favorable in the application of redox flow batteries. Moreover, via DFT-informed molecular design, the researchers were able to identify a more promising second generation AQDS derivative with one hydroxyl group on each benzene ring (1,8-dihydroxy-9,10-anthraquinone-2,7-disulphonic acid). Compared to AQDS, the derivative displayed a lower redox potential (consequently, a higher cell voltage) and higher kinetic rate constant, which not only produces a more favorable low potential compound, but also validates the applicability of computationally directed research for RFBs. Based on this work, the authors further explored the long-term cycling performance of the $AQDS/Br_2$ cell

[2]Determined by CV of 1 mM AQDS + 1 M H_2SO_4 on a glassy carbon electrode versus SHE.

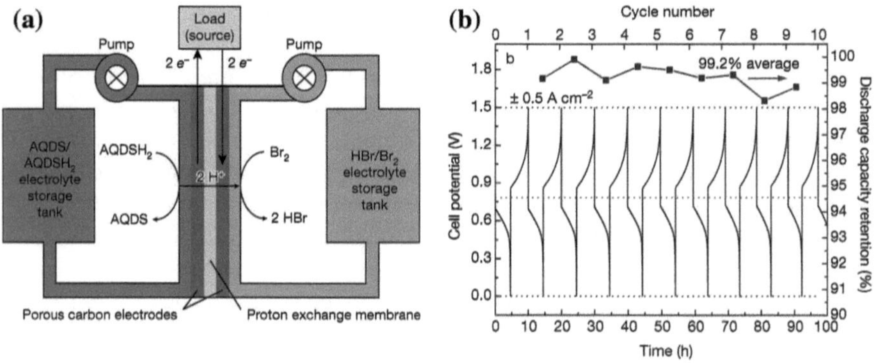

Fig. 3 **a** Cell schematic. Discharge mode is shown; the *arrows* are reversed for electrolytic/charge mode. $AQDSH_2$ refers to the reduced form of AQDS. **b** Constant-current cycling (0.5 A/cm^2) at 40 °C using a 3 M HBr + 0.5 M Br_2 solution on the positive side and a 1 M AQDS + 1 M H_2SO_4 solution on the negative side. Discharge capacity retention is indicated for each cycle [57]. Reproduced with permission. Copyright 2014 Nature

containing 0.1 L AQDS (1 M) and H_2SO_4 (1 M) on the negative side and 0.12 L Br_2 (0.5 M) and HBr (3 M) on the positive side with a Nafion 115 membrane separating the electrolytes [58]. The cell showed good stability at 100 % state-of-charge and 0.75 A/cm^2 over 750 cycles with an average discharge capacity retention and current efficiency as high as 99.84 and 98.35 %, respectively.

Beyond anthraquinone, the fundamental electrochemistry of other quinonoid compounds such as benzoquinone and naphthoquinone in aqueous systems have been thoroughly studied [59, 60], paving the way for their new applications in RFBs. Huskinson et al. reported the cyclic voltammetry results of 1,4-benzoquinone in neutral solution with high reversibility on both glassy carbon and platinum disc electrodes [61]. Moreover, they also studied the cycling performance of 0.1 M benzoquinone as the positive electrolyte couple with hydrogen reduction reaction with a theoretical cell voltage of ~0.70 V in acidic electrolyte using I-V polarization curve. The authors pointed out that the cell performance was primarily restricted by mass transport of benzoquinone due to its limited solubility whereas higher flow rate improved the peak power density and limiting current density. Xu et al. investigated the electrochemical behavior of 4,5-dihydroxybenzene-1,3-disulfonic acid disodium salt (Tiron) as a high potential compound in the application of aqueous redox flow batteries. Constant current electrolysis (vs. a lead counter electrode) was performed on Tiron (0.05 M) in a H_2SO_4 solution (3 M) [62]. The proposed reaction mechanism involves a reversible 2-electron transfer process followed by the addition of a hydroxyl group and a subsequent reversible 2-electron transfer process, which is also known as an ECE (electrochemical–chemical–electrochemical) process (Fig. 4). The reversible 2-electron transfer redox reaction between compound 3 and compound 4 at 0.96 V makes this redox couple an ideal high potential compound for aqueous redox flow batteries. Tiron, herein, serves as a starting material due to its commercial availability and low price.

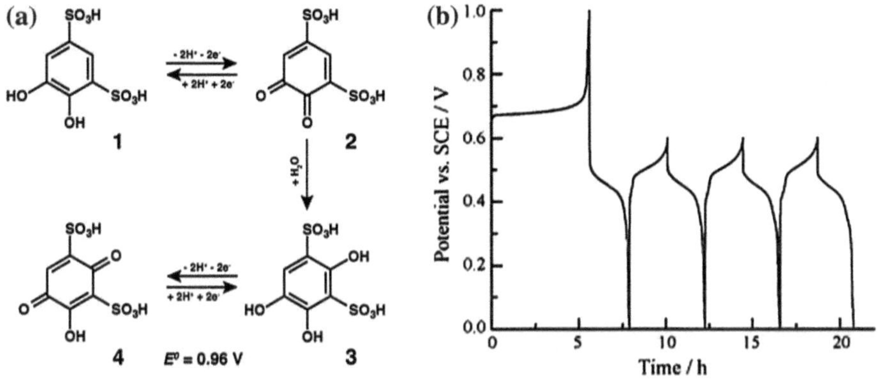

Fig. 4 **a** The mechanism of the redox reaction of 4,5-dihydroxybenzene-1,3-disulfonic acid. **b** Electrolytic characterization of tiron in 3 M H_2SO_4 at 298 K with a lead counter electrode [62]. Reproduced with permission. Copyright 2010 Electrochimica Acta

Combining Tiron and AQDS as the high potential and low potential redox active compounds, respectively, Yang et al. reported an all organic, aqueous redox flow cell with an open circuit voltage (OCV) of ~0.75 V [100 % state of charge (SOC)] [63]. A demonstration cell was operated using 0.2 M Tiron (positive side), 0.2 M AQDS (negative side), and 1 M H_2SO_4 (electrolyte) at a current density of 8 mA/cm^2 and a flow rate of 1000 mL/min, and displayed high current efficiency and acceptable capacity retention. In this paper, the authors also pointed out that the strong interaction between the ionized quinonoid species (R-SO_3^-) and water results in the low diffusion coefficient of the quinone based redox active molecules. Corroborating this statement, computational analysis of the I-V relationship revealed that lowering mass transfer resistance is critical to achieving a high operating current density in this type of redox flow cell.

3.2 Redox Active Compounds in Non-aqueous Flow Cells

As compared to their aqueous counterparts, non-aqueous RFBs are in their infancy. Transitioning from aqueous to non-aqueous electrolytes offers a wider window of electrochemical stability that enables cell operation at higher voltages. Indeed, depending on the salt, solvent, and electrode material, the stability window for non-aqueous electrolytes can span greater than 4 V [64]. Higher cell voltage leads to higher energy density, and typically to higher efficiency as well. These benefits promise to reduce cost of energy and to lessen system footprints that may enable specific applications such as energy storage in an urban environment. In addition, a broader selection of redox materials may be available due to either the wider potential window or the variety of available non-aqueous solvents. However, this promise must be balanced with the challenges associated with non-aqueous electrolyte, including increased cost, reduced ionic conductivity, and other unfavorable physical properties (e.g., flammability). Understanding and balancing these competing factors will be key to determining the true value of non-aqueous RFBs. Below, we highlight recent research activities in the investigation of redox active compounds for non-aqueous flow cells.

3.2.1 Metal-Centered Coordination Complexes

As a natural extension of the studies conducted on transition metals for aqueous RFBs, metal-centered coordination complexes have been investigated for use in non-aqueous systems. The structure of a metal-centered coordination complex can be expressed as $[ML_n]^{z+}$ where M is a transition metal center (i.e., ruthenium, iron, vanadium, chromium, manganese, nickel, etc.). L is a ligand [i.e., acetylacetonate ($acac^-$), 2,2′-bipyridine (bpy), phenanthroline (phen), etc.], n is the number of ligands (typically n = 3 for bidentate ligands), and z is the total charge that combines the valence of the metal center and the charge of the ligand. Because an entire compound must be neutral, a counter ion is necessary if z is not zero. Using $[Ru(bpy)_3]$, $[Ru(acac)_3]$, and $[Ru(phen)_3]$ as examples, Fig. 5a–c depict the bonding environment of the ruthenium metal center as well as 2-D structures of 2,2′-bipyridine, acetylacetonate, and phenanthroline, respectively. While the metal center provides the electrochemical activity of $[ML_n]$, the coordinating ligands plays a pivotal role in determining a number of key parameters of the complex, such as the solubility, diffusivity, reversibility, and redox potential of each electron transfer event. Moreover, non-innocent (redox active) ligands may store additional charge thus increasing the intrinsic capacity of the complex [65]. A multi-step redox process with a sufficient potential difference enables the use of the same redox active compound as both the high potential and the low potential compound, which alleviates the concern of cross contamination. To date, most systems employ ACN and a non-coordinating salt [e.g., tetraethylammonium tetrafluoroborate ($TEABF_4$)] as the solvent and the supporting salt, respectively, with an anion exchange

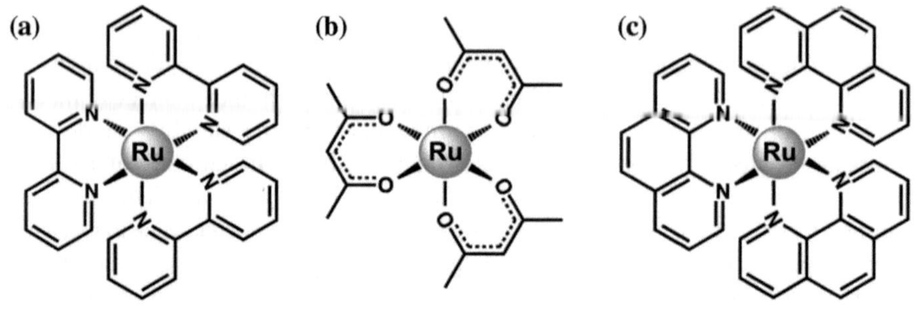

Fig. 5 The molecular structures of (A) $[Ru(bpy)_3]$, (B) $[Ru(acac)_3]$, and (C) $[Ru(phen)_3]$

Table 4 Non-aqueous redox flow batteries based on metal-centered coordination complex

Reactions	Electrolyte	Cell voltage (V)	Reference
$[Mn(IV)(acac)_3]^+ + e^- \rightleftharpoons Mn(III)(acac)_3$ $Mn(III)(acac)_3 + e^- \rightleftharpoons [Mn(II)(acac)_3]^-$	0.5 M $TEABF_4$ Acetonitrile	1.1	[66]
$[Ru(IV)(acac)_3]^+ + e^- \rightleftharpoons Ru(III)(acac)_3$ $Ru(III)(acac)_3 + e^- \rightleftharpoons [Ru(II)(acac)_3]^-$	0.05 M $TEABF_4$ Acetonitrile	1.77	[67]
$[Co(III)(acacen)]^+ + e^- \rightleftharpoons Co(II)(acacen)$ $Co(II)(acacen) + e^- \rightleftharpoons [Co(I)(acacen)]^-$	0.1 M $TEABF_4$ Acetonitrile	2.0	[68]
$[V(IV)(acac)_3]^+ + e^- \rightleftharpoons V(III)(acac)_3$ $V(III)(acac)_3 + e^- \rightleftharpoons [V(II)(acac)_3]^-$	0.1 M $TEABF_4$ Acetonitrile	2.2	[69]
$[Fe(III)(bpy)_3]^{3+} + e^- \rightleftharpoons [Fe(II)(bpy)_3]^{2+}$ $[Ni(II)(bpy)_3]^{2+} + 2e^- \rightleftharpoons Ni(0)(bpy)_3$	0.05 M $TEABF_4$ Propylene carbonate	2.3	[70]
$[Ru(III)(bpy)_3]^{3+} + e^- \rightleftharpoons [Ru(II)(bpy)_3]^{2+}$ $[Ru(II)(bpy)_3]^{2+} + e^- \rightleftharpoons [Ru(I)(bpy)_3]^+$	0.1 M $TEABF_4$ Acetonitrile	2.6	[71]

membrane (AEM) to separate the positive and negative electrolytes. Table 4 summarizes several reported metal-centered coordination complexes and the cell voltage of the proposed non-aqueous redox chemistry.

Ruthenium-Centered Coordination Complex

In 1988, Matsuda et al. demonstrated the first non-aqueous redox flow cell using $[Ru(bpy)_3]^{2+}/[Ru(bpy)_3]^{3+}$ and $[Ru(bpy)_3]^{+}/[Ru(bpy)_3]^{2+}$ as the high potential and low potential couples, respectively, with an OCV of 2.6 V [71]. A static H-cell polarization experiment identified the efficiency/performance limiting reaction as $[Ru(bpy)_3]^{+}/[Ru(bpy)_3]^{2+}$, which might be due to a side reaction and/or the low stability of $[Ru(bpy)_3]^{+}$. A flow cell cycling experiment demonstrated the charge-discharge capability of the cell with a 1.2 V discharge voltage, a 40 % coulombic efficiency and a 47 % utilization efficiency at 5 mA/cm^2. Although the performance metrics of this cell were far from practical, this seminal work initiated the application of metal-centered coordination complexes to non-aqueous redox flow cells and triggered ensuing research interest in this area. More recently, Chakrabarti et al. investigated ruthenium acetylacetonate ($Ru(acac)_3$) that displayed two 1-electron transfer processes with ~1.77 V separation according to the cyclic voltammetry [67]. The charge/discharge performance of $Ru(acac)_3$ was studied in the H-cell configuration containing 40 mL $Ru(acac)_3$ (0.1 M), $TEABF_4$ (1 M), and ACN in each compartment. At the charge and discharge rate of 1 and 0.5 mA, respectively, the first cycle gave an energy efficiency of 74 % and the second cycle gave an energy efficiency of 57 %. Based on this work, a redox flow cell using Ru $(acac)_3$ as the active materials with two 2.5-L storage tanks and 5 cm × 5 cm active area was demonstrated [72]. For 0.02 M $Ru(acac)_3$, the optimum power output was determined to be 35 mW at the discharge current density of 2.1 mA/cm^2 and the voltage efficiency is 52.1 % (at 100 % SOC). Increasing the concentration of Ru $(acac)_3$ to 0.1 M increased the current density and power output by a factor of 5 (at the same velocity) due to the improved mass transfer of the active species.

Vanadium-Centered Coordination Complexes

Thompson et al. investigated a series of metal acetylacetonates [$M(acac)_3$] (where M = V [73], Cr [74], and Mn [66]) for their potential application in non-aqueous redox flow batteries. Amongst these beta-diketonate candidates, [$V(acac)_3$] has been systematically studied to optimize the cell performance with respect to the electrode surface, the impurity concentration, and the electrolyte. Specifically, the influence of the electrode surface (glassy carbon, platinum, and gold) on the electrochemistry of $[V(III)(acac)_3]^0/[V(IV)(acac)_3]^+$ (positive electrolyte) and $[V(II)(acac)_3]^-/[V(III)(acac)_3]^0$ (negative electrolyte) was examined [75]. The positive electrode reaction exhibits fast and reversible kinetics on all electrode surfaces. In contrast, the negative electrode reaction is surface-selective and the kinetic constant of $[V(II)(acac)_3]^-/[V(III)(acac)_3]^0$ on a gold electrode is about 6.5 times higher than that on a glassy carbon electrode. As the main source of impurities, oxygen and water deteriorate the performance of [$V(acac)_3$] based, non-aqueous redox flow cells [69]. Oxygen may degrade the electrolyte and block the reduction reaction on the negative electrode while water can not only impede the kinetics on the negative electrode but also hinder the positive electrode reaction by forming vanadyl acetylacetonate [$VO(acac)_2$]. In addition, a range of solvents (ACN, dimethylformamide, hexane, tetrahydrofuran, dimethylcarbonate) and salts [tetraethylammonium tetrafluoroborate, tetrabutylammonium tetrafluoroborate, tetrabutylammonium hexafluorophosphate,(1-butyl, 3-methyl)imidazolium bis(trifluoromethanesulfonyl) imide] were screened aiming at improving the conductivity of the electrolyte and the solubility of [$V(acac)_3$] [76]. According to their results, ACN and $TEABF_4$ is still the optimal combination for [$V(acac)_3$]. However, even under carefully controlled conditions, the charge/discharge performance of [$V(acac)_3$] (in a stirred H-cell) was still not desirable with coulombic and energy efficiencies of 70 and 35 %, respectively at 50 % SOC and a discharge current of 0.014 mA/cm^2.

Recently, vanadium-centered coordination complexes using the non-innocent redox active ligand—maleonitriledithiolate [$(mnt)^{2-}$] were investigated for their potential application in non aqueous RFBs [65]. It has also been demonstrated that the reduction of $[V(mnt)_3]^{2-}$ adds one electron to the vanadium (IV) center and the oxidation of $[V(mnt)_3]^{2-}$ removes one electron from the $(mnt)^{2-}$ ligand. Therefore, non-innocent ligands may store additional electrochemical energy beyond what can be accessed from the metal center. In addition, improved complex stability can be expected when electron transfer takes place on the ligand rather than involving metal-ligand bonding. The redox potential of V(III)/V(IV) is tunable by the selection of cation in the supporting electrolyte, which might be attributed to the decreased relative electrostatic stabilization of $[V(mnt)_3]^{2-}$ and the increased steric shielding of the cationic charge. The cycling performance using 0.02 M $TEA_2[V(mnt)_3]$ in 0.1 M tetrabutylammonium hexafluorophosphate ($TBAPF_6$)/ACN as both the positive and the negative electrolyte with a battery separator (Tonen) demonstrated the rechargeability of the non-aqueous redox flow cell with ~90 % coulombic efficiency and good capacity retention, although the energy efficiency is <45 %.

Fe, Co, Ni Based Coordination Complexes

Recently, promising results were reported by Mun et al. using $[Fe(bpy)_3]$ and $[Ni(bpy)_3]$ as the high potential and low potential compounds, respectively [70]. The 1-electron transfer $[Fe(II)(bpy)_3]^{2+}/[Fe(III)(bpy)_3]^{3+}$ takes place at +0.65 V versus Ag/Ag^+ and the 2-electron transfer $[Ni(bpy)_3]^0/[Ni(II)(bpy)_3]^{2+}$ takes place at −1.66 V versus Ag/Ag^+, which results in the nominal cell voltage of 2.31 V. The redox stability of both reactions in $TEABF_4$ (0.05 M)/PC was verified by a 100 cycle CV on a glassy carbon electrode. The advantageous electrochemical properties of the redox active compounds were further validated by the constant current (0.8 mA/cm^2) cycling in a flow cell containing 0.4 M $[Fe(bpy)_3](BF_4)_2$ and 0.2 M $[Ni(bpy)_3](BF_4)_2$ in $TEABF_4$ (0.5 M)/PC, and an AEM. The OCV of the cell was 2.2 V and the coulombic and energy efficiencies of the cell were maintained at 90.4 % and 81.8 %, respectively, throughout the cycling. The utilization efficiency (based on the first charging phase) was 92.8 % of the theoretical capacity. However, the capacity of the cell decayed over time, which might be due to the crossover of redox active species. An improved capacity retention was observed by replacing $[Ni(bpy)_3]$ with $[V(acac)_3]$ in a non-flowing cell [77]. Besides the non-aqueous chemistries discussed above, several other combinations such as $[Ni(phen)_3]$-$[Fe(phen)_3]$, $[Co(bpy)_3]$-$[Fe(bpy)_3]$, $[Co(phen)_3]$-$[Fe(phendione)_3]$[3] have been evaluated as well, and their performances were similar to that of the $[Fe(bpy)_3]$-$[Ni(bpy)_3]$ cell [77].

Ferrocene Derivatives

Ferrocene is a well-known redox model compound for its superior electrochemical reversibility and stability in non-aqueous electrolytes, and has been widely adopted as an internal standard for electrochemical measurements [78]. However, ferrocene's low solubility (e.g., <0.1 M in carbonate-based solvent) restricts its use in RFBs for grid energy storage. Recently, Wei et al. described a structural modification strategy to greatly enhance the solubility of ferrocene [79]. Figure 6a shows the structure of the ferrocene derivative (Fc1N112-TFSI). As evidenced experimentally and computationally by nuclear magnetic resonance (NMR) and DFT, respectively, the

[3] phendione stands for 1,10-phenanthroline-5,6-dione.

quaternary ammonium cation accentuates the interaction between the ionic pendant and the polarizable oxygen atoms on the carbonate solvent molecules. As a result, the as-prepared ionic-derivatized ferrocene compound can be dissolved in a carbonate based electrolyte [1.2 M LiTFSI in the mixture of ethylene carbonate (EC)/ propylene carbonate (PC)/ethyl methyl carbonate (EMC) (4:1:5 by weight)] up to 0.85 M—more than 20 fold higher than the solubility of pristine ferrocene in the same electrolyte. Besides advancing the solvation chemistry, the quaternary ammonium cation also induces the electron withdrawing effect on the ferrocene moiety, which increases the redox potential of the ferrocene derivative by 0.23 V as compared to pristine ferrocene. Stable half-cell cycling performance of 0.1 M Fc1N112-TFSI with a lithium foil negative electrode is shown in Figs. 6b and c, further validating the electrochemistry of the compound. Furthermore, using a hybrid lithium-graphite negative electrode and appropriate additive (15 wt% fluoroethylene carbonate) in the electrolyte, the cycling performance of 0.8 M Fc1N112-TFSI was demonstrated at 1.5 mA/cm^2, resulting in 90 % coulombic efficiency, 76 % energy efficiency, 0.93 %/cycle degradation of discharge capacity.

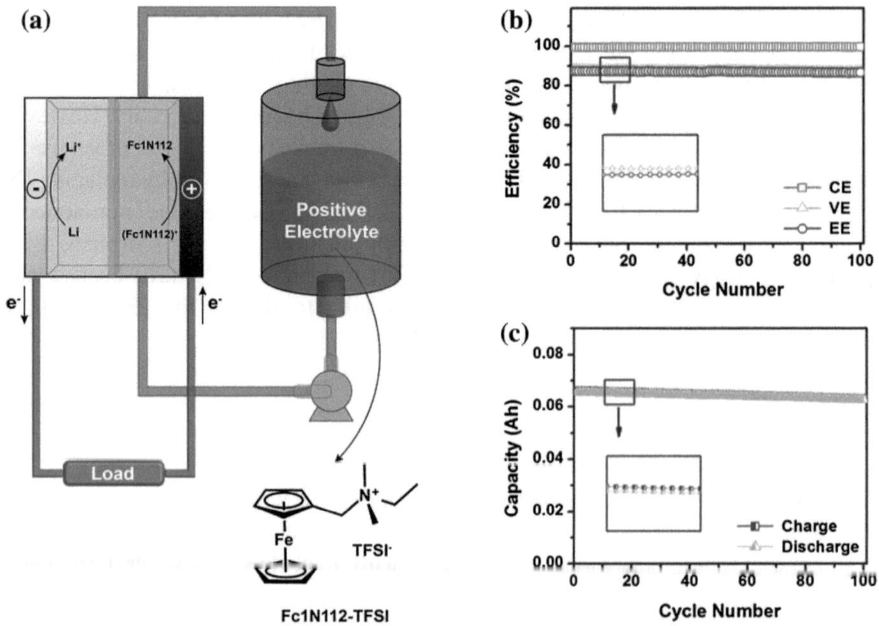

Fig. 6 **a** Cell schematic (discharge mode) and the molecular structure of Fc1N112-TFSI. **b** Coulombic (CE), voltage (VE), and energy (EE) efficiencies and **c** capacities of the half-cell test using 0.1 M Fc1N112-TFSI in 1.0 M LiTFSI in EC/PC/EMC with 5 wt% FEC at a current rate of 3.5 mA/cm^2 and a flow rate of 40 mL/min [79]. Reproduced with permission. Copyright 2014 Advanced Energy Materials

3.2.2 Organic Redox Active Molecules

Metal-centered coordination complexes are usually bulky, resulting in large molecular weights, and consequently, if not offset by multiple electron transfer events, small specific capacities (Mol. Cap. = nF/3.6*MW [Ah/kg]). In this regard, redox active organic molecules may offer the advantage of lower molecular weights and multi-electron transfer. Further, given the variety of non-aqueous solvents, redox active organic molecules may achieve higher solubilities, with fewer molecular modifications, than in aqueous electrolytes. For example, sulfonic groups (MW = 80 g/mol) are often attached to improve the organic solubility in aqueous electrolyte but these modifications can lower intrinsic capacity and may lead to less favorable properties (e.g., slower diffusion). In contrast, numerous functional groups are accessible for engineering the redox-active compounds in non-aqueous electrolytes, which enables rationally designed molecular modifications towards improved performances such as enabling multi-electron transfer, increasing solubility, improving the electrochemical reversibility, and shifting the redox potential.

Organic Molecules

Over the past decade, a wide variety of redox active organic materials have been investigated for use as solid electrodes in rechargeable Li-metal and Li-ion cells as well as for performance-enhancing solution-phase additives for advanced batteries [56, 80, 81]. For Li-metal and Li-ion cells, active material insolubility is desirable without tethering the discrete monomers to a polymeric backbone (lowering electrode capacity). In general, this can be pursued through enhanced polarity, π-π stacking, or reversible polymerizations. Despite these efforts, dissolution remains a key performance-limiting challenge for organic materials, especially at different charge states. A number of highly engineered redox active molecules have been developed as redox shuttles for overcharge protection in advanced batteries including phenothiazines, nitroxide-based radicals (vide infra), and substituted di-methoxybenzenes [82]. A number of these organic materials can be employed for non-aqueous RFB applications.

Brushett et al. reported the proof-of-concept study of an all-organic, lithium-ion based, non-aqueous redox flow cell using 2,5-di-tert-butyl-1,4-bis(2-methoxyethoxy)benzene (DBBB) and 2,3,6-trimethylquinoxaline (TMQ) in 0.2 M $LiBF_4$/PC with a theoretical OCV of 1.54 V (Fig. 7a) [83]. A wealth of knowledge from the development and optimization of electrolytes for Li-ion batteries [64, 84] can be leveraged to inform the design and construction of Li-ion based non-aqueous RFBs, which further enriches the selection of solvent and salts. As a product of molecular engineering, DBBB was originally designed for the overcharge protection of lithium-ion batteries [85], which requires a high redox potential, high electrochemical reversibility, moderate solubility, and high chemical compatibility. All of these properties are in perfect alignment with the criteria of redox active compounds in non-aqueous RFBs. Moreover, it has been determined

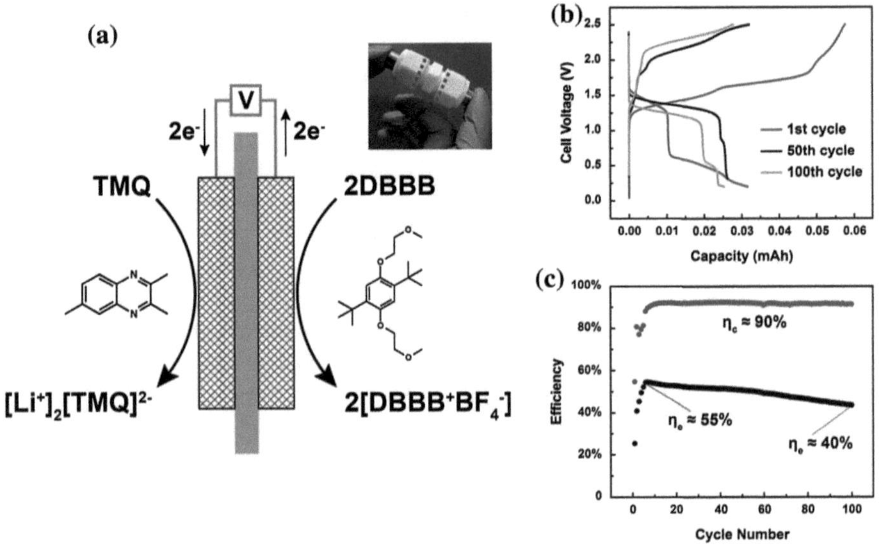

Fig. 7 **a** The image of Swagelok® cell and the corresponding reactions on the positive and negative electrodes during charging. **b** The 1st, 50th, and 100th constant current charge-discharge profiles. **c** The coulombic and energy efficiency over 100 cycles

that the rate constant and diffusivity of DBBB are comparable to those of ferrocene [86]. Quinoxaline, a commercially available compound with a molecular weight of only 130 g/mol, has displayed very high solubility (~7 M in most solvents) and promising electrochemical behavior in 0.2 M $LiBF_4$/PC. The addition of methyl groups on the conjugated rings not only enhances the reversibility of electron transfer, but also lowers the redox potential to a more negative value. The charge–discharge of the proposed system was demonstrated for 100 cycles in a non-flowing, Swagelok® cell configuration with lithiated Nafion 117 as the membrane, and displayed average coulombic and energy efficiencies of ~90 and ~50 %, respectively (Fig. 7b, c). Although the cell performance is far from being optimized, this archetypical work represents the baseline of organic redox active molecules for non-aqueous RFBs, and more importantly, provides a guideline for the future development of energy storage strategy in this category with higher energy density. To this end, more recent efforts have focused on the optimization of the positive electrolyte in terms of improving the conductivity and the solubility of DBBB in a range of carbonate based solvents and their mixtures [propylene carbonate (PC), ethylene carbonate (EC), dimethyl carbonate (DMC), diethyl carbonate (DEC)] with commonly used lithium salts (LiTFSI, $LiBF_4$, $LiPF_6$) in a high-throughput, automated screening platform [86]. The best combination of solubility and conductivity (at 25 °C) was obtained as 0.6 M (DBBB) and 5.25 mS/cm, respectively, in 0.6 M $LiPF_6$ in a ternary mixture of PC:EC:DMC (0.08:0.12:0.8). In addition, higher solubility of DBBB can be expected in ether based solvent due to the presence of methoxymethyl group in the solute and solvent.

The superior electrochemistry of anthraquinone derivatives has not only found use in aqueous system as discussed in Sect. 3.1.2, but also attracts attention in non-aqueous RFBs. Wang et al. reported the structural modification of pristine anthraquinone by adding two triethylene glycol monomethyl ether groups onto the 1 and 5 carbons [87]. The adapted anthraquinone derivative showed a significantly improved solubility in non-aqueous polar solvent such as PC. This modification strategy towards the enhancement of solubility has recently been rationalized by DFT calculations [88]. Cyclic voltammetry of this compound displayed two 1-electron transfer processes centered at 2.82 and 2.50 V versus Li/Li^+ in 1 M $LiPF_6$/PC. The step-wise redox activity was also captured galvanostatically in a non-flowing, half-cell measurement. Moreover, this anthraquinone derivative based non-aqueous redox flow cell displayed high-energy efficiency (~82 %) and relative stable discharge energy density (with an average degradation rate of 0.08 mAh/L/cycle) in the cycling experiment at 0.1 mA/cm^2 for 40 cycles.

Using DFT calculations, Hernández-Burgos et al. investigated the effect of heteroatoms and substituents on the redox potential and stability of quinone based and other carbonyl based organic redox active molecules for energy storage applications [89]. As a rule, electron-withdrawing groups result in a positive shift of the redox potential while electron-donating groups result in a negative shift of the redox potential. Higher potential compounds tend to be less stable than the lower potential compounds, and the anionic species are predicted to be more stable than the neutral species. Furthermore, molecular engineering can also be realized by replacing the C atom with a heteroatom such as N, O, or S, especially for a five-membered ring, which will unburden the capacity penalty imposed by the increased molecular weight from the substituent. The same group also formulated a unique method to tune the redox potential of a carbonyl based organic redox active molecule by exploring the effect of cations in the supporting electrolyte as observed by RDV and verified by DFT [90]. Taking 1,2-di(thiophen-2-yl)ethane-1,2-dione (DTED) as an example, in the presence of a non-coordinating salt (tetrabutylammonium perchlorate), there are two 1-electron transfer processes observed at −1.36 and −2.20 V versus Ag/Ag^+. In the presence of a lithium salt ($LiClO_4$), the first reduction reaction slightly increases to −1.35 V and the second one noticeably increases to −1.80 V versus Ag/Ag^+, which can be explained by the stabilization effect of Li^+ on the anionic DTED due to its smaller size than that of $(TBA)^+$. In the presence of a magnesium salt [$Mg(ClO_4)_2$], the two 1-electron transfer processes merge to one 2-electron transfer process at dramatically higher potential (−1.11 V vs. Ag/Ag^+), which can be attributed to an even smaller ion with multiple charges. Therefore, this provides a flexible strategy for adjusting the redox potential of electroactive species based on the choice of supporting electrolyte ions according to their role as a high potential or low potential compound. While organic active materials in non-aqueous environments have the potential to enable high energy density storage, several key questions about the stability and solubility of active materials remain, particularly in their charged states. Furthermore, while a broad array of materials are available, archetypical redox families and resulting cell chemistries with common ion exchange have yet to be developed and exploited. Establishing and testing these baseline systems is a critical next step.

Stable Radicals

Stable radicals are sub-valent compounds with an unpaired electron, which is stabilized via steric protection of the radical center and/or delocalization of the unpaired electron throughout the molecular structure. An important feature of stable organic radicals is the oxidation and reduction reactions producing closed-shell cations and anions via simple one-electron transfer outer-sphere reactions with no broken or formed chemical bonds. Often, but not always, these reactions are very fast and reversible. Certain classes of stable radicals, most notably nitroxides, are both durable and reactive under specific conditions, and thus are of interest as charge storage materials. Indeed, significant efforts by Nishide and co-workers have focused on developing organic radical batteries based on stable radical monomers tethered to polymeric backbones [91, 92]. Efforts by Buhermester et al. and Nakahara et al. have focused on understanding and exploiting nitroxide-based radicals, typically derivatives of 2,2,6,6-tetramethyl-1-piperidinyloxy (TEMPO), as active materials in non-aqueous electrolytes (e.g., for overcharge protection) [93, 94]. More recently, stable radicals have been explored for RFB applications. Li et al. proposed a non-aqueous redox flow cell utilizing TEMPO and, a redox-active organic molecule, N-methylphthalimide (NMP), as the high potential and low potential species, respectively, in $NaClO_4$ (1 M) in ACN [95]. The measured rate constants of TEMPO and NMP are on the order of 10^{-1} and 10^{-2} cm/s, respectively, both of which are higher than that of ferrocene ($\sim 10^{-3}$ cm/s); the diffusion coefficients of TEMPO and NMP are in the order of 10^{-5} cm^2/s (10-fold higher than ferrocene). A constant current cycling in a non-flowing cell kept a high coulombic efficiency of 90 % for 20 cycles with an average charging and discharging voltages of 1.65 and 1.36 V, respectively. More recently, Wei et al. demonstrated a hybrid non-aqueous RFB using TEMPO as the positive electrolyte and lithium metal as the negative electrode [96]. As TEMPO is quite soluble in carbonate-based solvents (e.g., >2.0 M actives in the presence of supporting salt), high energy flow cells were proposed and demonstrated (>100 Wh/L). Initial results are very promising and further chemistry development may lead to a standard positive electrolyte system for non-aqueous RFBs.

4 New Configuration of Redox Flow Batteries

From Fe^{2+}/Fe^{3+} in aqueous electrolytes to organic radicals in non-aqueous electrolytes, the development of novel redox active compounds over the past 30 years has greatly enriched the options for the electrochemistry in flow batteries. However, the cell design and operating principles of redox flow cells have remained, for the most part, similar to the original configuration proposed by Thaller in 1974 [97]. New paradigms that incorporate unique flow patterns, electrode geometries, and electrolyte compositions are critical for the next generation flow batteries. This section highlights recent advances in these areas and describes emerging flow cell technologies.

4.1 Cell Architecture and Flow Channel Configuration

In general, the conventional electro-reactor architecture of a redox flow cell utilizes flow-through graphite felt electrodes coupled with graphite plate current collectors. Alternatively, redox flow cells can adopt design paradigms from more mature electrochemical technologies, such as proton exchange membrane fuel cells, where advances in cell engineering have led to high-performance reactor stacks. As shown in Fig. 8a, a graphite plate is in close contact with a thinner electrode such as carbon paper. Compared to the flow-through configuration, different flow patterns such as parallel, serpentine, and interdigitated flow, can be engraved on the graphite plate (Fig. 8b). The compression of the carbon electrode can be controlled by incompressible gaskets with varying thicknesses. The improved performance of this zero-gap cell architecture with a serpentine flow channel has been verified by Aaron et al. for VRB in terms of high peak power density (557 mW/cm^2, 1 M vanadium and 4 M H_2SO_4, 50 mL reservoirs) compared to other cell architectures under similar experimental conditions [98].

Recently, Darling and Perry systematically investigated the influence of a flow channel configuration on the cell parameters of the high-potential redox couple [V (IV)/V(V)] in VRBs [99]. Specifically, pressure drop, ohmic resistance, and the

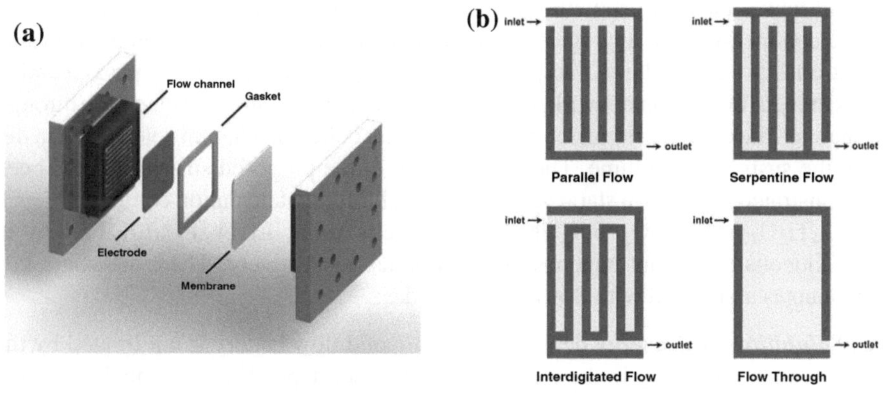

Fig. 8 **a** Exploded view of components in a redox flow cell with a flow channel. **b** Different flow channel configurations

limiting current were measured with respect to a combination of the electrode and channel configurations. The limiting current was measured at a constant voltage (0.4 V) and reflects the mass transfer limitation of the configuration. For the flow through configuration, the electrode was inserted in the flow channel whereas in the interdigitated and the parallel flow, electrodes were fixed on top of, but still in close contact, with the flow channel. The key results are that the interdigitated-flow carbon paper outperforms the conventional flow-through graphite felt in terms of pressure drop, which is directly related to pumping losses, and an increased cell resistance, which is directly related to ohmic loss, with equally efficient mass transfer. Therefore, the interdigitated flow configuration with carbon paper will be beneficial for high viscosity and low conductivity systems such as non-aqueous redox flow cells. While these experiments were conducted in a "half-cell" configuration, the results can lead to system level engineering and enable more combinatorial approaches towards enhancing flow battery performance.

4.2 Semi-solid Flow Batteries

In conventional redox flow cells, a flowing electrolyte carries redox active compounds to the surface of a static electrode that is in contact with a current collector to enable the electron transfer. Recently, Chiang and co-workers revolutionized this paradigmatic operating mode by introducing the concept of semi-solid flow batteries (Fig. 9a), which combines the features of high energy density lithium ion batteries and the advantage of scalability and flexibility from conventional redox flow cells [100]. The conventional static porous carbon electrode is replaced by a percolating network of nanoscale conductors that serve as the "dynamic electrode" in the semi-solid flow cell. In detail, electrochemically active compounds (redox, intercalation, etc.) and Ketjenblack carbon nanoparticles (<2 vol.%) are dispersed in the supporting electrolyte (salt + solvent) under sonication (Fig. 9b). This leads to the formation of a fractal nanoscale network within the electrolyte, which enables both electronic and ionic conduction. The aggregated carbon nanoparticles form a continuous flowable electrodes at a low volume fraction with the storage materials densely packed and in proximity to the carbon network. By pumping the suspension-based electrode through a channel, the carbon network contacts the current collector, enabling electron transfer into the suspension volume. The applicability of the proposed suspension has been demonstrated not only in static half-cells using lithium intercalation compounds such as $Li_4Ti_5O_{12}$ / $LiCoO_2$ but also in flow cells using $Li_3Ti_2(PO_4)_3$/$LiFePO_4$, in non-aqueous [100] and aqueous [101] systems, respectively. There are some key advantages associated with flowable electrodes:

- ***Solubility***—The energy density of a semi-solid flow battery is not limited by the solvation chemistry in the electrolyte, but rather depends on the rheology of the suspension (the suspension must still be flowable). For example, 20 vol.% $LiCoO_2$ (ρ = 5.06 g/cm^3, MW = 97.9 g/mol) can be easily dispersed in the

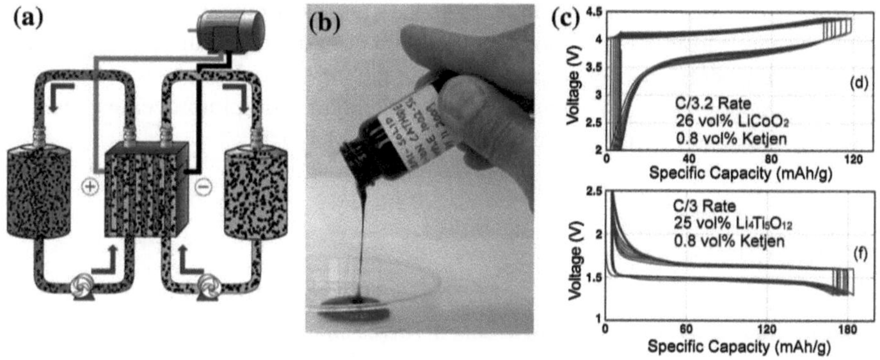

Fig. 9 **a** Scheme: semi-solid flow cell (SSFC) system using flowing lithium-ion cathode and anode suspensions could enable new models such as transportation 'fuels' tuned for power versus range, or cold versus warm climates, with flexible refueling and recycling options. **b** Fluid semi-solid suspension containing $LiCoO_2$ powder as the active material and Ketjenblack as the dispersed conductive phase, dispersed in alkyl carbonate electrolyte. **c** Galvanostatic charge/discharge curves for semi-solid suspensions having 26 vol.% $LiCoO_2$ (LCO) dispersed in 1.3 M $LiPF_6$ in an alkyl carbonate blend and 25 vol.% $Li_4Ti_5O_{12}$ (LTO) dispersed in 1 M LiPF6 in dimethyl carbonate [100]. Reproduced with permission. Copyright 2011 Advanced Energy Materials

suspension, corresponding to 10.2 M electrochemically active compound, which is much higher than any other maximum achievable solubility of redox active compounds in aqueous or non-aqueous systems [100].

- ***Surface area***—For the commonly used graphite felt (6 μm diameter fiber, 3 mm thickness), the specific surface area is ~0.33 m^2/g and the fraction of the fiber is ~6 vol.%. Therefore, the surface area of graphite felt in a cell is ~0.04 m^2/cm^3. For the carbon nanoparticles, the specific surface area is ~1400 m^2/g. Therefore, 1.5 vol.% Ketjenblack results in the surface area of the "nano-conductor" to be as large as ~42 m^2/cm^3 [102].
- ***Pumping efficiency***—Intuitively, the high viscosity suspension (~700 cP) would seem harder to pump than a solution (~50 cP for VRBs). However, if solution flow through a porous media is consider, as might be expected in an operating flow cell, the "wired-up" suspension becomes more attractive. For example, Fan et al. showed that for same flow channel length and diameter, the pressure drop of a suspension-based electrode through an open channel is actually three orders of magnitude lower than that of a solution-based electrolyte through a porous media (e.g., carbon electrode) with a 10 μm pore diameter [102].

Semi-solid flow cells with "wired" up electrolytes are an intriguing concept which can be applied to both soluble and insoluble chemistries [100, 101]. However, the unique properties that enable desirable electrolyte-level properties present new challenges for cell and system design. First, the electrolytes display non-Newtonian viscoelasticity properties (i.e., Herschel-Bulkley behavior), which confounds both traditional cell design and pumping. Second, the electronic conductivity of the solution may lead to parasitic losses due to shunt currents throughout the system. Third, the impact of suspended particulates in the fluid may lead to increased pump and component wear (e.g., abrasion). Continued engineering improvements will be required to realize the full potential of this novel approach.

4.3 Hybrid Redox Flow Batteries

By definition, a hybrid redox flow cell involves plating and stripping of a redox active metal such as zinc, lithium, and lead, which generally serves as the negative electrode of a hybrid redox flow cell with the corresponding metal cation existing in the electrolyte. Given the high solubility in its discharged state (metal cations) and the zero mobility in its charged state (metal), the energy density limitation and the crossover concern of hybrid redox flow cells are not imposed by the low potential side. However, the concomitant penalty associated with the metal negative electrode is the formation of dendrites, which can puncture the membrane or separator, and consequently, short-circuit the battery.

4.3.1 Zinc-Based Redox Flow Batteries

A classic example of a hybrid redox flow battery is the zinc-halogen system, such as zinc-bromine, which can be traced back to 1885 [103], and was systematically re-evaluated as a secondary battery in 1964 [104]. Similarly, zinc–chlorine was investigated in the late 1970s and 1980s but was found to be environmentally hazardous (chlorine gas formation) and more complicated in a system level than the zinc-bromine chemistry. More recently, Lim et al. introduced a circulating aqueous electrolyte to the zinc-bromine battery (ZBB), which dramatically improved the performance of the system [105]. The electrode reactions of a ZBB are:

Positive electrode:	$Br_2 + e^- \rightleftharpoons 2Br^-$	$E^0 = +1.087\,V$
Negative electrode:	$Zn^{2+} + 2e^- \rightleftharpoons Zn$	$E^0 = -0.762\,V$
Total reaction:	$Br_2 + Zn \rightleftharpoons 2Br^- + Zn^{2+}$	$E = 1.849\,V$

The most attractive feature of ZBBs is the low cost of both zinc and bromine. In addition, the electrode polarization is very small due to the fast kinetics on both the positive and negative electrodes [105]. The circulation of electrolyte not only greatly improves the coulombic efficiency of the cell, but also alleviates the formation of dendrites on the negative electrode. However, self-discharge, resulting from the crossover of Br_2, is the primary concern of ZBBs, which can be aggravated by the formation of tribromide ions (Br_3^-) and other polybromide ions in the positive electrolyte. In practice, complexing agents such as quaternary ammonia salts are usually added to the positive electrolyte, which bind to these polybromide ions and form a new, low-solubility liquid phase that is separated from the electrolyte [106]. In addition, a periodic stripping cycle, which aims to completely remove zinc from the negative electrode (100 % depth-of-discharge), is recommended to ensure a smooth zinc deposition during normal operation [14]. Commercially available ZBBs are designed for a 20-year service life with a modular 50 kWh unit and scalable to 500 kWh in a single enclosure and up to 4 enclosures (2 MWh) from a single point of system connection [107].

Besides the zinc-halogen systems, other zinc-based hybrid chemistries have been explored. One example is the alkaline zinc-ferricyanide redox flow cell that was demonstrated by Adams in 1979 [108]. The advantage of this zinc-ferricyanide redox flow cell includes high efficiency, high cell voltage, and low toxicity [109]. The electrode reactions are:

$$\begin{array}{lll} \text{Positive electrode:} & [Fe(CN)_6]^{3-} + e^- \rightleftharpoons [Fe(CN)_6]^{4-} & E^0 = +0.358\,V \\ \text{Negative electrode:} & [Zn(OH)_4]^{2-} + 2e^- \rightleftharpoons Zn + 4OH^- & E^0 = -1.199\,V \\ \text{Total reaction:} & 2[Fe(CN)_6]^{3-} + Zn + 4OH^- \rightleftharpoons 2[Fe(CN)_6]^{4-} + [Zn(OH)_4]^{2-} & E = 1.557\,V \end{array}$$

The system utilizes zinc and carbon as the negative and positive electrodes, respectively. On discharge, zinc is converted to a zincate ion ($[Zn(OH)_4]^{2-}$) which is then converted and stored in the negative electrolyte tank as ZnO. Similarly, sodium ferro- and ferricyanides are stored as precipitates in the positive electrolyte tank. This unique design of solid storage in the electrolyte tanks results in a smaller footprint, but in a more complex system management and temperature control.

Another example is the zinc-cerium redox flow cell developed by Clarke et al. in 2004 [110]. The electrode reactions are:

$$\begin{array}{lll} \text{Positive electrode:} & Ce^{4+} + e^- \rightleftharpoons Ce^{3+} & E^0 = +1.720\,V \\ \text{Negative electrode:} & Zn^{2+} + 2e^- \rightleftharpoons Zn & E^0 = -0.7618\,V \\ \text{Total reaction:} & 2Ce^{4+} + Zn \rightleftharpoons 2Ce^{3+} + Zn^{2+} & E = 2.482\,V \end{array}$$

Methanesulfonic acid is preferentially used as the electrolyte on both sides to increase the solubility of the metal cations [111]. Zinc-cerium chemistries enable a very high cell voltage. However, the concomitant parasitic reactions such as HER and OER must be avoided or minimized to achieve acceptable cell efficiencies.

4.3.2 Soluble Lead-Acid Flow Batteries

An extension of hybrid redox flow batteries is the "double hybrid" soluble lead-acid flow batteries (SLFBs) where deposition and dissolution of redox active compounds are involved in both high potential and low potential electrode reactions. Pletcher et al. explored the concept of SLFBs in 2004 [112], and have reported their systematic study in a series of papers [113–120]. The electrode reactions of a SLFB are:

Positive electrode:	$PbO_2 + 4H^+ + 2e^- \rightleftharpoons Pb^{2+} + 2H_2O$	$E^0 = +1.455\,V$
Negative electrode:	$Pb^{2+} + 2e^- \rightleftharpoons Pb$	$E^0 = -0.126\,V$
Total reaction:	$PbO_2 + Pb + 4H^+ \rightleftharpoons 2Pb^{2+} + 2H_2O$	$E = 1.581\,V$

As shown in Fig. 10, the electrolyte consists of lead (II) methansulfonate (up to 2 M) in a biodegradable methanesulfonic acid (1.5 M). Charging the cell results in metallic lead (Pb^0) being deposited at the negative electrode and lead (IV) dioxide at the positive electrode; discharging the cell results in Pb^{2+} being re-dissolved into the electrolyte from both electrodes. SLFBs differ from traditional lead-acid batteries in that, rather than residing on the electrodes in the form of $PbSO_4$, the lead methansulfonate is dissolved in methanesulfonic acid, which enables the flow functionality of SLFBs. Furthermore, the single redox active molecule (Pb^{2+}) existing in the electrolyte contributes two very attractive features of SLFBs compared to other redox flow batteries. (1) There is no need for an ion-selective membrane in the cell, and consequentially, (2) only one electrolyte storage tank is required. These key features allow for a simplified cell design and decreased footprint of the system, and in turn, greatly reduces the construction and maintenance cost of a SLFB.

However, similar to other hybrid redox flow systems where liquid-solid reactions occur, SLFBs suffer from Pb^0 dendrite formation on the negative electrode

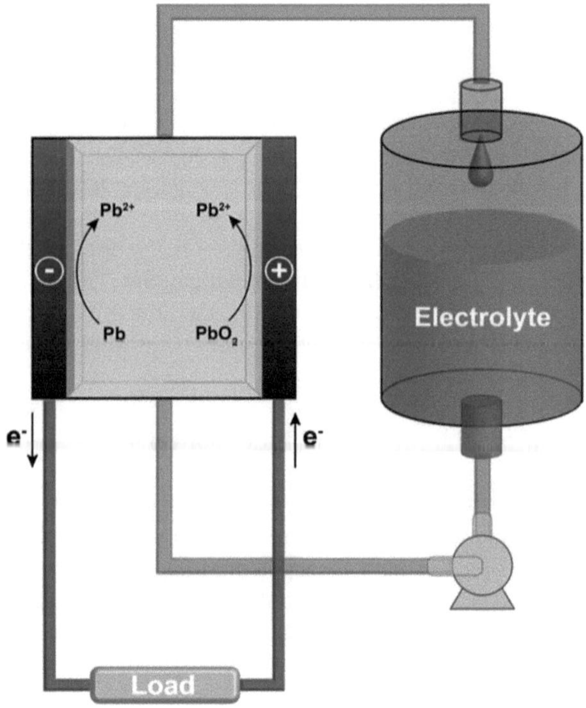

Fig. 10 Schematic of a soluble lead-acid redox flow battery

and more severely, irreversible PbO_2 polymorph (both α- and β-PbO_2) formation on the positive electrode, which dictates the low current density applied in SLFBs (typically <20 mA/cm^2). Recently, Verde et al. demonstrated that increased mass transfer results in the formation nanoscale PbO_2 and Pb^0 on the positive and negative electrodes, respectively, which effectively mitigates the polymorph and dendrite issues during charging [121]. In a beaker-cell configuration with sufficient stirring, their soluble lead-acid flow cell achieved an energy efficiency of ~79 % for 2000 cycles at 20 mA/cm^2. Continued efforts are needed in the development of novel flow reactor architectures that allow for faster mass transport of Pb ions across the electrode surfaces to minimize depletion layer thickness while keeping the pumping cost low.

4.3.3 Lithium-Aqueous Flow Batteries

Lithium metal is an extremely attractive negative electrode material due to its high theoretical specific capacity (3861.9 mAh/g), low electrochemical potential (−3.04 V vs. SHE), and possibility of coupling with a charged or discharged positive electrode material [122, 123]. However, despite significant efforts over the past 40 years, continued challenges with safety and efficiency during prolonged and/or high rate battery operation prevent the deployment of advanced Li batteries. However, emerging needs for high energy batteries for transportation applications (e.g., Li–metal oxide, Li–S, Li–O_2) continued to drive research and development. Indeed, over the past five years, progress has been made in enhancing cyclability and safety through the mechanically robust and chemically-stable solid-electrolyte membranes, based on ceramics, polymers or hybrid materials [124]. These advances may also be leveraged to develop hybrid flow systems where the lithium metal negative electrode is paired with a flowing positive electrolyte.

Lu and Goodenough demonstrated a hybrid redox flow cell using a lithium negative electrode in static non-aqueous electrolyte and a carbon positive electrode in circulating aqueous electrolyte with 0.1 M $K_3Fe(CN)_6$ as the redox active molecule [126]. The key design component is a sub-millimeter thick water stable lithium-ion conducting ceramic, $Li_{1+x+y}Al_xTi_{2-x}Si_yP_{3-y}O_{12}$, which enables Li-ion transfer between the two compartments. To avoid chemical reduction of the solid electrolyte on the lithium metal, two pieces of electrolyte soaked polypropylene separator (Celgard) were used. This concept can be extended to a wide range of aqueous redox couples (Fig. 11) many of which have been explored by different groups [127]. Recently, Zhao and Byon described a hybrid lithium-iodine redox flow cell and systematically investigated the influence of the concentration of redox active molecule, flow rate, discharge current, and temperature on the cell performance [128]. Figure 12a illustrates the configuration of the system. Lithium ion conductive Li_2O-Al_2O_3-TiO_2-P_2O_5 (LATP) solid electrolyte was employed to separate the positive and negative flowing electrolyte, and a layer of glass fiber was employed to separate the solid electrolyte and lithium metal. The electrode reactions are:

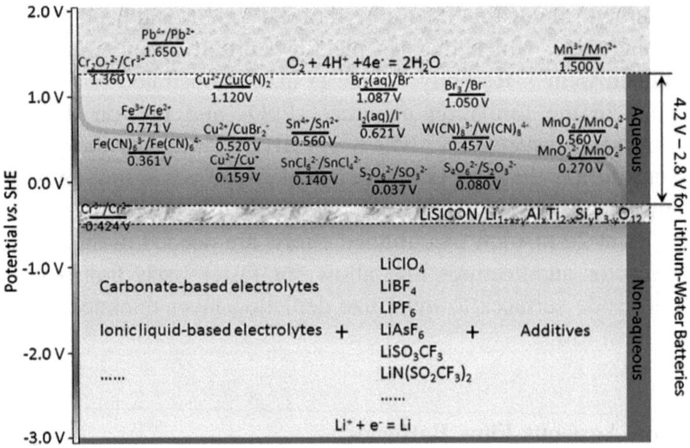

Fig. 11 Systems for lithium-aqueous batteries [125]. Reproduced with permission. Copyright 2011 Journal of the American Chemical Society

$$
\begin{array}{lll}
\text{Positive electrode:} & I_3^- + 2e^- \rightleftharpoons 3I^- & E^0 = +0.536\,\text{V} \\
\text{Negative electrode:} & Li^+ + e^- \rightleftharpoons Li & E^0 = -3.040\,\text{V} \\
\text{Total reaction:} & I_3^- + 2Li \rightleftharpoons 3I^- + 2Li^+ & E\ = 3.576\,\text{V}
\end{array}
$$

Lithium iodide is highly soluble in water, up to 12.4 M (25 °C), which can match the high capacity of the lithium negative electrode. The cycling data of the hybrid lithium-iodine battery are shown in Fig. 12b, where high coulombic efficiency (>99 %) and good capacity retention (>99 %) were obtained for 20 cycles. A key challenge for lithium-aqueous systems is the low ionic conductivity of the solid electrolyte, which is two orders of magnitude lower than that of non-aqueous electrolytes and three orders of magnitude lower than that of aqueous electrolytes. Therefore, greatly improved energy efficiency and higher charge/discharge rates for aqueous-non-aqueous based hybrid redox flow cells can be expected from more conductive solid electrolytes.

4.3.4 Lithium–Polysulfide Flow Batteries

In the pursuit of highly efficient, low cost energy storage systems, the lithium-sulfur (Li–S) hybrid battery is one of the most promising technologies due to its high theoretical energy density [129]. In a Li–S cell, lithium metal is applied as the negative electrode and solid-state sulfur as the positive electrode. The electrode reactions of a Li–S cell are:

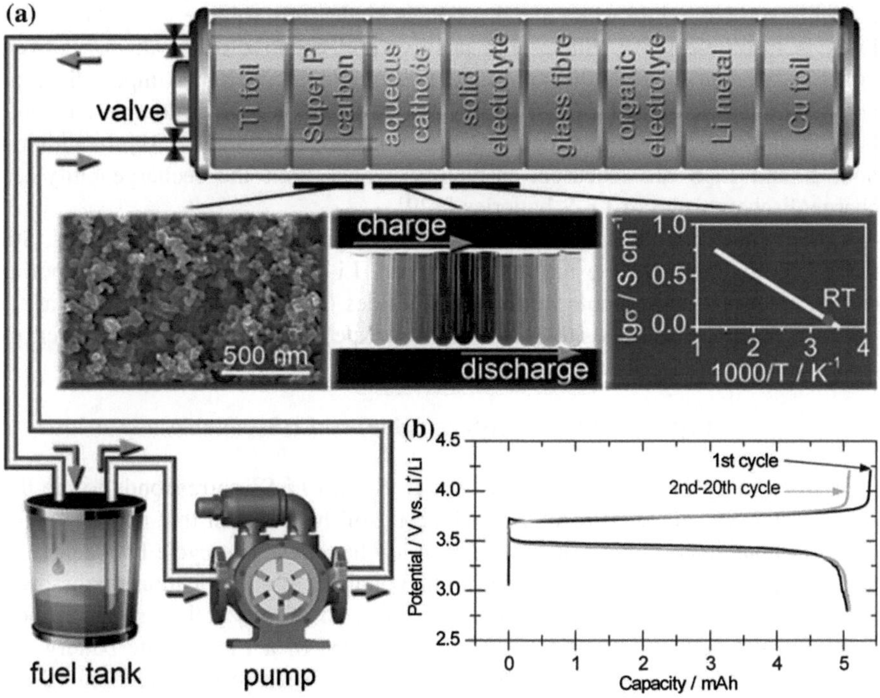

Fig. 12 **a** Schematic illustration of a cathode-flow mode aqueous Li–I battery equipped with an aqueous electrolyte reservoir and pump. The *center insets* from left to right are SEM image of the Super P carbon current collector, an image of color gradients of LiI aqueous electrolytes, and a graph of temperature-dependent ionic conductivity in lithium-ion-conducting LATP solid electrolyte (on the order of 10^{-4} S cm^{-1} at RT). The LiI electrolyte color in the *center-middle inset* corresponds to the depth of charge and discharge (0, 5, 15, 25, 50 and 100 % of charge and 0, 50, 75, 85, 95 and 100 % of discharge from left to right). **b** 20-times cycled charge/discharge profiles of CF/LiIB equipped with an aqueous electrolyte reservoir containing total 0.5 mL of 0.5 m of LiI aqueous electrolyte at a current rate of 2.5 mA/cm^2 and a flow rate of 150 μL/min [128]. Reproduced with permission. Copyright 2014 Advanced Energy Materials

$$\begin{array}{lll} \text{Positive electrode:} & S + 2e^- \rightleftharpoons S^{2-} & E^0 = -0.476\,V \\ \text{Negative electrode:} & Li^+ + e^- \rightleftharpoons Li & E^0 = -3.040\,V \\ \text{Total reaction:} & 2Li + S \rightleftharpoons Li_2S & E = 2.564\,V \end{array}$$

The theoretical capacity of sulfur ($S^0 \rightleftharpoons S^{2-}$) is 1672.0 mAh/g, which is 12.2, 19.7, and 22.6 times higher than that of $LiCoO_2$, $LiFePO_4$, and $LiMn_2O_4$ (assuming 0.5 Li^+ transfer per molecule), respectively. Ether-based electrolytes such as tetrahydrofuran (THF), 1,3-dioxolane (DOL), 1,2-dimethoxyethane (DME), tetraethylene glycol dimethyl ether (TEGDME) are commonly used in Li–S batteries. The evolution of S to Li_2S involves the formation of Li_2S_x intermediates that have varying solubilities in ether-based electrolytes. For example, when $8 \geq x \geq 6$, Li_2S_x

is highly soluble, while when x = 4, Li_2S_x is moderately soluble, and when x < 4, Li_2S_x is insoluble [130]. Since both positive and negative electrodes of conventional Li–S batteries are in the solid state, the soluble intermediates impose the most noticeable technical challenge in a Li–S cell, resulting in low coulombic efficiency (shuttle mechanism) and poor capacity retention (self-discharge) [131]. In addition, both S and Li_2S are electrical insulators, which limit the rechargeability and charge/discharge rate of Li–S batteries [129].

Lithium–polysulfide (Li–PS) flow batteries originated from Li–S batteries. Still using lithium metal as the negative electrode, Li–PS flow cells utilize a porous carbon electrode and soluble lithium polysulfides (Li_2S_x where $8 \geq x \geq 4$) as the positive electrode and positive electrolyte, respectively. The positive electrode reactions can be described as:

$$Li_2S_8(\text{soluble}) \rightleftharpoons Li_2S_6(\text{soluble}) \rightleftharpoons Li_2S_4(\text{soluble})$$

A full charge/discharge cycle between Li_2S_8 and Li_2S_4 corresponds to the theoretical capacity of 209.0 mAh/g S, which is still higher than that of the positive electrode materials in conventional lithium ion batteries. The cycle between Li_2S_8 and Li_2S_4 is realized by controlling cutoff voltages. The application of dissolved Li_2S_x species as the starting positive electrolyte in a Li–S cell was first demonstrated by Rauh et al. in 1979 [132], and the idea of a Li–PS flow battery was proposed by Manthiram et al. in 2013 [133]. It has been recently recognized that $LiNO_3$ can form a stable passivation layer on the surface of the lithium negative electrode, which effectively prevents the reaction between the dissolved Li_2S_x and lithium metal [134, 135]. Therefore, with $LiNO_3$ additive in the electrolyte, ion-selective membranes are not needed in a Li-PS flow cell. Based on this strategy, Yang et al. recently demonstrated the chemistry of Li–PS in a coin-cell configuration. In a constant capacity cycling experiment using 5 M Li_2S_8, the static Li-PS cell displayed superior cyclability at ~200 mAh/g S for more than 1500 cycles [136]. Recently, Fan et al. demonstrated the applicability of semi-solid carbon suspension as the high-surface area (42 m^2/cm^3) current collectors in a Li–PS flow cell [102]. In this proof-of-concept device, potentiostatic cycling experiment where the cell was charged at 2.5 V and discharged at 2.1 V to maintain all polysulfide species solubilized in the solution, the discharge capacity of the cell is 181 mAh/g S with a coulombic efficiency close to 100 %. Moreover, the percolating conductor network with extremely large surface area enabled deep cycling of polysulfide into the precipitation regime of Li_2S [102], which might be due to the mitigated insulating effect of Li_2S on the lower surface area carbon current collector. Therefore, much higher discharge capacity (393 mAh/g S) was achieved by extending the discharge voltage to 1.6 V so that the precipitation regime of Li_2S can be utilized. The coulombic efficiency remained above 99.5 % at the discharge rate of C/22 over 10 cycles.

5 Conclusions and Outlook

Stationary energy storage systems, in particular electrochemical energy storage systems, will play a pivotal role in the widespread integration of renewable, non-dispatchable energy sources (e.g., solar photovoltaic (PV), wind) and in the improvement of energy efficiency of the electric power sector. Redox flow batteries are particularly attractive for these applications due to their favorable combination of performance, cost, and safety. However, present generation technologies are not cost-competitive, which drives research and development efforts towards new redox chemistries, electrolyte formulations, and cell designs. Recent analysis by Darling et al. has indicated that both aqueous and non-aqueous RFBs have pathways to meeting long-term objectives for cost-effective energy storage [137]. As mentioned before, reductions in battery costs ($/kWh) can be achieved via lowering materials cost ($/kg) and increasing materials energy density (kWh/kg). Electroactive organic molecules, both as solo active materials or as components of metal-centered complexes, may offer a pathway forward through both metrics. In principle, organic materials can be synthesized via cost-effective routes and are less dependent on the production and reserves of key elements (e.g., vanadium in VRB). Moreover, the electrochemical and physical properties of electroactive materials can be tailored via modification of the redox moieties and the surrounding molecular structure. This may lead to the development of high capacity and highly soluble redox materials required for energy dense cell operation. To realize this promise, further efforts are needed in the identification and optimization of novel redox active materials, likely leveraging high throughput computations, and in validation of these materials within redox flow cells at high currents and extended cycling. Reductions in battery costs may also be achieved through advances in cell design and engineering. High performance electroreactors, often inspired by more mature electrochemical technologies (e.g., fuel cells), can reduce reactor size requirements and associated costs. Further, new reactor designs are required to efficiently employ non-aqueous chemistries with low ionic conductivity and high voltage, and mixed conductor suspensions that offer higher cell energy densities but frustrate traditional configurations. Together with an evolving regulatory environment, these R&D activities are expected to lead to the commercial success of RFB technologies in the near future and to the deployment of a range of new flowable storage concepts.

Acknowledgments The authors gratefully acknowledge financial support from the Joint Center for Energy Storage Research, an Energy Innovation Hub funded by the U.S. Department of Energy, Office of Science, Basic Energy Sciences, and from the Massachusetts Institute of Technology Energy Initiative's (MITei) Seed Fund Program. In addition, we thank Jarrod Milshtein and Apurba Sakti for stimulating discussions and for assistance in figure development.

References

1. International Energy Agency (2014) Key world energy statistics
2. Hojjati B, Wade SH (2012) US household energy consumption and intensity trends: a decomposition approach. Energy Policy 48:304–314
3. U.S. Energy Information Administration (2011) Annual energy review. http://www.eia.gov/totalenergy/data/annual/index.cfm/
4. Rising above the gathering storm: energizing and employing America for a brighter economic future. In: Committee on prospering in the global economy of the 21st century: an agenda for American science and technology; Committee on Science, Engineering, and Public Policy. National Academies Press, Washington, D.C., 2007
5. Ehleringer JR, Cerling TE, Dearing MD (2005) In: Ehleringer JR, Cerling TE, Dearing MD (eds) A history of atmospheric CO_2 and its effects on plants, animals, and ecosystems. Springer, New York
6. Stocker T, Qin D, Plattner G-K (2013) Climate change 2013: the physical science basis. Working group I contributions to the IPCC fifth assessment report of the intergovernmental panel on climate change
7. International Energy Agency (2012) Energy technology perspectives 2012: pathways to a clean energy system
8. Chu S, Majumdar A (2012) Opportunities and challenges for a sustainable energy future. Nature 488:294–303. doi:10.1038/nature11475
9. U.S. Department of Energy (2010) Smart grid system report
10. U.S. Department of Energy (2013) Grid energy storage
11. Akhil AA, Boyes JD, Butler PC, Doughty DH (2010) Batteries for electrical energy storage applications. In: Reddy T (ed) Linden's handbook of battery, Chap. 30, 4th edn. McGraw-Hill, New York
12. U.S. Department of Energy (2007) Basic research needs for electrical energy storage
13. International Energy Agency (2014) Technology roadmap: energy storage. http://www.iea.org/publications/freepublications/publication/name,36573,en.html. Accessed 10 April 2014
14. Electric Power Research Institute (2003) DOE handbook of energy storage for transmission and distribution applications
15. Akhil AA, Huff G, Currier AB, Kaun BC, Rastler DM, Chen SB, Cotter AL, Bradshaw DT, Gauntlett WD (2013) DOE/EPRI 2013 electricity storage handbook in collaboration with NRECA. Sandia National Laboratories
16. Electric Power Research Institute (2010) Electricity energy storage technology options
17. International Energy Agency (2013) Pumped storage provides grid reliability even with net generation loss. http://www.eia.gov/todayinenergy/detail.cfm?id=11991. Accessed 10 Oct 2014
18. ARPA-E. U.S. Department of Energy. Grid-scale rampable intermittent dispatchable storage. http://www.arpa-e.energy.gov/?q=arpa-e-programs/grids. Accessed 05 Jan 2015
19. U.S. Department of Energy (2014) Joint center for energy storage. http://www.jcesr.org. Accessed 05 Jan 2015
20. Shin S-H, Yun S-H, Moon S-H (2013) A review of current developments in non-aqueous redox flow batteries: characterization of their membranes for design perspective. RSC Adv 3:9095–9116. doi:10.1039/C3RA00115F
21. Schwenzer B, Zhang J, Kim S, Li L, Liu J, Yang Z (2011) Membrane development for vanadium redox flow batteries. ChemSusChem 4:1388–1406. doi:10.1002/cssc.201100068
22. Li X, Zhang H, Mai Z, Zhang H, Vankelecom I (2011) Ion exchange membranes for vanadium redox flow battery (VRB) applications. Energy Environ Sci 4:1147. doi:10.1039/c0ee00770f
23. Chakrabarti MH, Brandon NP, Hajimolana SA, Tariq F, Yufit V, Hashim MA, Hussain MA, Low CTJ, Aravind PV (2014) Application of carbon materials in redox flow batteries. J Power Sources 253:150–166. doi:10.1016/j.jpowsour.2013.12.038

24. Parasuraman A, Lim TM, Menictas C, Skyllas-Kazacos M (2013) Review of material research and development for vanadium redox flow battery applications. Electrochim Acta 101:27–40. doi:10.1016/j.electacta.2012.09.067
25. Li X, Sabir I (2005) Review of bipolar plates in PEM fuel cells: flow-field designs. Int J Hydrogen Energy 30:359–371. doi:10.1016/j.ijhydene.2004.09.019
26. Alotto P, Guarnieri M, Moro F (2014) Redox flow batteries for the storage of renewable energy: a review. Renew Sustain Energy Rev 29:325–335. doi:10.1016/j.rser.2013.08.001
27. Wang W, Luo Q, Li B, Wei X, Li L, Yang Z (2013) Recent progress in redox flow battery research and development. Adv Funct Mater 23:970–986. doi:10.1002/adfm.201200694
28. Leung P, Li X, de León CP, Berlouis L, Low CTJ, Walsh FC (2012) Progress in redox flow batteries, remaining challenges and their applications in energy storage. RSC Adv 2:10125–10156. doi:10.1039/C2RA21342G
29. Yang Z, Zhang J, Kintner-Meyer MCW, Lu X, Choi D, Lemmon JP, Liu J (2011) Electrochemical energy storage for green grid. Chem Rev 111:3577–3613. doi:10.1021/cr100290v
30. Weber AZ, Mench MM, Meyers JP, Ross PN, Gostick JT, Liu Q (2011) Redox flow batteries: a review. J Appl Electrochem 41:1137–1164. doi:10.1007/s10800-011-0348-2
31. Skyllas-Kazacos M, Chakrabarti MH, Hajimolana SA, Mjalli FS, Saleem M (2011) Progress in flow battery research and development. J Electrochem Soc 158:R55–R79. doi:10.1149/1.3599565
32. Viswanathan V, Crawford A, Stephenson D, Kim S, Wang W, Li B, Coffey G, Thomsen E, Graff G, Balducci P, Kintner-Meyer M, Sprenkle V (2014) Cost and performance model for redox flow batteries. J Power Sources 247:1040–1051. doi:10.1016/j.jpowsour.2012.12.023
33. LaMonica M (2013) Enervault novel battery technology. MIT Technol Rev. http://www.technologyreview.com/view/512736/startup-enervault-rethinks-flow-battery-chemistry/. Accessed 10 Sep 2014
34. Vanysek P (2014) Electrochemical series. In: Haynes WM (ed) CRC handbook of chemistry and physics, 95th edn. Taylor & Francis Group, Boca Raton, pp 5-80–5-89
35. Reid MA, Gahn RF (1977) Factors affecting the open-circuit voltage and electrode kinetics of some iron/titanium redox flow cells. NASA-TM X-73669
36. Wang W, Kim S, Chen B, Nie Z, Zhang J, Xia G-G, Li L, Yang Z (2011) A new redox flow battery using Fe/V redox couples in chloride supporting electrolyte. Energy Environ Sci 4:4068–4073. doi:10.1039/C0EE00765J
37. Lopez-Atalaya M, Codina G, Perez JR, Vazquez JL, Aldaz A (1992) Optimization studies on a Fe/Cr redox flow battery. J Power Sources 39:147–154. doi:10.1016/0378-7753(92)80133-V
38. Skyllas-Kazacos M, Grossmith F (1987) Efficient vanadium redox flow cell. J Electrochem Soc 134:2950–2953. doi:10.1149/1.2100321
39. Zhou H, Zhang H, Zhao P, Yi B (2006) A comparative study of carbon felt and activated carbon based electrodes for sodium polysulfide/bromine redox flow battery. Electrochim Acta 51:6304–6312. doi:10.1016/j.electacta.2006.03.106
40. Skyllas-Kazacos M (2003) Novel vanadium chloride/polyhalide redox flow battery. J Power Sources 124:299–302. doi:10.1016/S0378-7753(03)00621-9
41. Xue F-Q, Wang Y-L, Wang W-H, Wang X-D (2008) Investigation on the electrode process of the Mn(II)/Mn(III) couple in redox flow battery. Electrochim Acta 53:6636–6642. doi:10.1016/j.electacta.2008.04.040
42. Fang B, Iwasa S, Wei Y, Arai T, Kumagai M (2002) A study of the Ce(III)/Ce(IV) redox couple for redox flow battery application. Electrochim Acta 47:3971–3976. doi:10.1016/S0013-4686(02)00370-5
43. Bartolozzi M (1989) Development of redox flow batteries. A historical bibliography. J Power Sources 27:219–234. doi:10.1016/0378-7753(89)80037-0
44. Hagedorn N (1984) NASA redox storage system development project. NASA-TM-83677
45. Zhao P, Zhang H, Zhou H, Yi B (2005) Nickel foam and carbon felt applications for sodium polysulfide/bromine redox flow battery electrodes. Electrochim Acta 51:1091–1098. doi:10.1016/j.electacta.2005.06.008

46. Scamman DP, Reade GW, Roberts EPL (2009) Numerical modelling of a bromide–polysulphide redox flow battery: Part 1: modelling approach and validation for a pilot-scale system. J Power Sources 189:1220–1230. doi:10.1016/j.jpowsour.2009.01.071
47. Scamman DP, Reade GW, Roberts EPL (2009) Numerical modelling of a bromide-polysulphide redox flow battery. Part 2: evaluation of a utility-scale system. J Power Sources 189:1231–1239. doi:10.1016/j.jpowsour.2009.01.076
48. Sum E, Rychcik M, Skyllas-kazacos M (1985) Investigation of the V(V)/V(IV) system for use in the positive half-cell of a redox battery. J Power Sources 16:85–95. doi:10.1016/0378-7753(85)80082-3
49. Sum E, Skyllas-Kazacos M (1985) A study of the V(II)/V(III) redox couple for redox flow cell applications. J Power Sources 15:179–190. doi:10.1016/0378-7753(85)80071-9
50. Skyllas-Kazacos M, Rychcik M, Robins RG, Fane AG, Green MA (1986) New all-vanadium redox flow cell. J Electrochem Soc 133:1057–1058. doi:10.1149/1.2108706
51. Li B, Gu M, Nie Z, Shao Y, Luo Q, Wei X, Li X, Xiao J, Wang C, Sprenkle V, Wang W (2013) Bismuth nanoparticle decorating graphite felt as a high-performance electrode for an all-vanadium redox flow battery. Nano Lett 13:1330–1335. doi:10.1021/nl400223v
52. Li B, Gu M, Nie Z, Wei X, Wang C, Sprenkle V, Wang W (2014) Nanorod niobium oxide as powerful catalysts for an all vanadium redox flow battery. Nano Lett 14:158–165. doi:10.1021/nl403674a
53. Kazacos M, Cheng M, Skyllas-Kazacos M (1990) Vanadium redox cell electrolyte optimization studies. J Appl Electrochem 20:463–467. doi:10.1007/BF01076057
54. Li L, Kim S, Wang W, Vijayakumar M, Nie Z, Chen B, Zhang J, Xia G, Hu J, Graff G, Liu J, Yang Z (2011) A stable vanadium redox-flow battery with high energy density for large-scale energy storage. Adv Energy Mater 1:394–400. doi:10.1002/aenm.201100008
55. Kim S, Thomsen E, Xia G, Nie Z, Bao J, Recknagle K, Wang W, Viswanathan V, Luo Q, Wei X, Crawford A, Coffey G, Maupin G, Sprenkle V (2013) 1 kW/1 kWh advanced vanadium redox flow battery utilizing mixed acid electrolytes. J Power Sources 237:300–309. doi:10.1016/j.jpowsour.2013.02.045
56. Poizot P, Dolhem F (2011) Clean energy new deal for a sustainable world: from non-CO_2 generating energy sources to greener electrochemical storage devices. Energy Environ Sci 4:2003–2019. doi:10.1039/C0EE00731E
57. Huskinson B, Marshak MP, Suh C, Er S, Gerhardt MR, Galvin CJ, Chen X, Aspuru-Guzik A, Gordon RG, Aziz MJ (2014) A metal-free organic-inorganic aqueous flow battery. Nature 505:195–198. doi:10.1038/nature12909
58. Huskinson B, Marshak MP, Gerhardt MR, Aziz MJ (2014) Cycling of a quinone-bromide flow battery for large-scale electrochemical energy storage. ECS Trans 61:27–30. doi:10.1149/06137.0027ecst
59. Chambers JQ (1974) Electrochemistry of quinones. In: Patai S (ed) The chemistry of the quinonoid compounds Part 1, Chap. 14. Wiley, New York
60. Bailey SI, Ritchie IM (1985) A cyclic voltammetric study of the aqueous electrochemistry of some quinones. Electrochim Acta 30:3–12. doi:10.1016/0013-4686(85)80051-7
61. Huskinson B, Nawar S, Gerhardt MR, Aziz MJ (2013) Novel quinone-based couples for flow batteries. ECS Trans 53:101–105. doi:10.1149/05307.0101ecst
62. Xu Y, Wen Y-H, Cheng J, Cao G-P, Yang Y-S (2010) A study of tiron in aqueous solutions for redox flow battery application. Electrochim Acta 55:715–720. doi:10.1016/j.electacta.2009.09.031
63. Yang B, Hoober-Burkhardt L, Wang F, Prakash GKS, Narayanan SR (2014) An inexpensive aqueous flow battery for large-scale electrical energy storage based on water-soluble organic redox couples. J Electrochem Soc 161:A1371–A1380. doi:10.1149/2.1001409jes
64. Xu K (2004) Nonaqueous liquid electrolytes for lithium-based rechargeable batteries. Chem Rev 104:4303–4418. doi:10.1021/cr030203g
65. Cappillino PJ, Pratt HD, Hudak NS, Tomson NC, Anderson TM, Anstey MR (2014) Application of redox non-innocent ligands to non-aqueous flow battery electrolytes. Adv Energy Mater 4:n/a–n/a. doi:10.1002/aenm.201300566

66. Sleightholme AES, Shinkle AA, Liu Q, Li Y, Monroe CW, Thompson LT (2011) Non-aqueous manganese acetylacetonate electrolyte for redox flow batteries. J Power Sources 196:5742–5745. doi:10.1016/j.jpowsour.2011.02.020
67. Chakrabarti MH, Dryfe RAW, Roberts EPL (2007) Evaluation of electrolytes for redox flow battery applications. Electrochim Acta 52:2189–2195. doi:10.1016/j.electacta.2006.08.052
68. Zhang D, Lan H, Li Y (2012) The application of a non-aqueous bis(acetylacetone) ethylenediamine cobalt electrolyte in redox flow battery. J Power Sources 217:199–203. doi:10.1016/j.jpowsour.2012.06.038
69. Shinkle AA, Sleightholme AES, Griffith LD, Thompson LT, Monroe CW (2012) Degradation mechanisms in the non-aqueous vanadium acetylacetonate redox flow battery. J Power Sources 206:490–496. doi:10.1016/j.jpowsour.2010.12.096
70. Mun J, Lee M-J, Park J-W, Oh D-J, Lee D-Y, Doo S-G (2012) Non-aqueous redox flow batteries with nickel and iron tris(2,2′-bipyridine) complex electrolyte. Electrochem Solid-State Lett 15:A80–A82. doi:10.1149/2.033206esl
71. Matsuda Y, Tanaka K, Okada M, Takasu Y, Morita M, Matsumura-Inoue T (1988) A rechargeable redox battery utilizing ruthenium complexes with non-aqueous organic electrolyte. J Appl Electrochem 18:909–914. doi:10.1007/BF01016050
72. Chakrabarti MH, Roberts EPL, Bae C, Saleem M (2011) Ruthenium based redox flow battery for solar energy storage. Energy Convers Manage 52:2501–2508. doi:10.1016/j.enconman.2011.01.012
73. Liu Q, Sleightholme AES, Shinkle AA, Li Y, Thompson LT (2009) Non-aqueous vanadium acetylacetonate electrolyte for redox flow batteries. Electrochem Commun 11:2312–2315. doi:10.1016/j.elecom.2009.10.006
74. Liu Q, Shinkle AA, Li Y, Monroe CW, Thompson LT, Sleightholme AES (2010) Non-aqueous chromium acetylacetonate electrolyte for redox flow batteries. Electrochem Commun 12:1634–1637. doi:10.1016/j.elecom.2010.09.013
75. Shinkle AA, Sleightholme AES, Thompson LT, Monroe CW (2011) Electrode kinetics in non-aqueous vanadium acetylacetonate redox flow batteries. J Appl Electrochem 41:1191–1199. doi:10.1007/s10800-011-0314-z
76. Shinkle AA, Pomaville TJ, Sleightholme AES, Thompson LT, Monroe CW (2014) Solvents and supporting electrolytes for vanadium acetylacetonate flow batteries. J Power Sources 248:1299–1305. doi:10.1016/j.jpowsour.2013.10.034
77. Lee D-Y, Lee M-J, Park J-W, Oh D-J, Mun J-Y, Doo S-G (2011) Aromatic ligand coordinated redox couples & their application into redox flow batteries. The International Flow Battery Forum 2011 Edinburgh
78. Gagne RR, Koval CA, Lisensky GC (1980) Ferrocene as an internal standard for electrochemical measurements. Inorg Chem 19:2854–2855. doi:10.1021/ic50211a080
79. Wei X, Cosimbescu L, Xu W, Hu JZ, Vijayakumar M, Feng J, Hu MY, Deng X, Xiao J, Liu J, Sprenkle V, Wang W (2014) Towards high-performance nonaqueous redox flow electrolyte via ionic modification of active species. Adv Energy Mater n/a–n/a. doi:10.1002/aenm.201400678
80. Liang Y, Tao Z, Chen J (2012) Organic electrode materials for rechargeable lithium batteries. Adv Energy Mater 2:742–769. doi:10.1002/aenm.201100795
81. Song Z, Zhou H (2013) Towards sustainable and versatile energy storage devices: an overview of organic electrode materials. Energy Environ Sci 6:2280. doi:10.1039/c3ee40709h
82. Chen Z, Qin Y, Amine K (2009) Redox shuttles for safer lithium-ion batteries. Electrochim Acta 54:5605–5613. doi:10.1016/j.electacta.2009.05.017
83. Brushett FR, Vaughey JT, Jansen AN (2012) An all-organic non-aqueous lithium-ion redox flow battery. Adv Energy Mater 2:1390–1396. doi:10.1002/aenm.201200322
84. Zhang SS (2006) A review on electrolyte additives for lithium-ion batteries. J Power Sources 162:1379–1394. doi:10.1016/j.jpowsour.2006.07.074

85. Zhang L, Zhang Z, Redfern PC, Curtiss LA, Amine K (2012) Molecular engineering towards safer lithium-ion batteries: a highly stable and compatible redox shuttle for overcharge protection. Energy Environ Sci 5:8204–8207. doi:10.1039/C2EE21977H
86. Su L, Ferrandon M, Kowalski JA, Vaughey JT, Brushett FR (2014) Electrolyte development for non-aqueous redox flow batteries using a high-throughput screening platform. J Electrochem Soc 161:A1905–A1914. doi:10.1149/2.0811412jes
87. Wang W, Xu W, Cosimbescu L, Choi D, Li L, Yang Z (2012) Anthraquinone with tailored structure for a nonaqueous metal–organic redox flow battery. Chem Commun 48:6669–6671. doi:10.1039/C2CC32466K
88. Bachman JE, Curtiss LA, Assary RS (2014) Investigation of the redox chemistry of anthraquinone derivatives using density functional theory. J Phys Chem A. doi:10.1021/jp5060777
89. Hernández-Burgos K, Burkhardt SE, Rodríguez-Calero GG, Hennig RG, Abruña HD (2014) Theoretical studies of carbonyl-based organic molecules for energy storage applications: the heteroatom and substituent effect. J Phys Chem C 118:6046–6051. doi:10.1021/jp4117613
90. Hernández-Burgos K, Rodríguez-Calero GG, Zhou W, Burkhardt SE, Abruña HD (2013) Increasing the gravimetric energy density of organic based secondary battery cathodes using small radius cations (Li+ and Mg2+). J Am Chem Soc 135:14532–14535. doi:10.1021/ja407273c
91. Nishide H, Koshika K, Oyaizu K (2009) Environmentally benign batteries based on organic radical polymers. Pure Appl Chem 81:1961–1970. doi:10.1351/PAC-CON-08-12-03
92. Suga T, Nishide H (2011) Rechargeable batteries using robust but redox active organic radicals. In: Hicks R (ed) Stable radicals: fundamentals and applied aspects of odd-electron compounds, Chap. 14, 1st edn. Wiley, New York
93. Nakahara K, Iwasa S, Iriyama J, Morioka Y, Suguro M, Satoh M, Cairns EJ (2006) Electrochemical and spectroscopic measurements for stable nitroxyl radicals. Electrochim Acta 52:921–927. doi:10.1016/j.electacta.2006.06.028
94. Buhrmester C, Moshurchak LM, Wang RL, Dahn JR (2006) The use of 2,2,6,6-tetramethylpiperinyl-oxides and derivatives for redox shuttle additives in Li-ion cells. J Electrochem Soc 153:A1800–A1804. doi:10.1149/1.2221860
95. Li Z, Li S, Liu S, Huang K, Fang D, Wang F, Peng S (2011) Electrochemical properties of an all-organic redox flow battery using 2,2,6,6-tetramethyl-1-piperidinyloxy and N-methyl phthalimide. Electrochem Solid-State Lett 14:A171–A173. doi:10.1149/2.012112esl
96. Wei X, Xu W, Vijayakumar M, Cosimbescu L, Liu T, Sprenkle V, Wang W (2014) TEMPO-based catholyte for high-energy density nonaqueous redox flow batteries. Adv Mater 26:7649–7653. doi:10.1002/adma.201403746
97. Thaller L (1974) Electrically rechargeable redox flow cells. NASA-TM X-71540
98. Aaron DS, Liu Q, Tang Z, Grim GM, Papandrew AB, Turhan A, Zawodzinski TA, Mench MM (2012) Dramatic performance gains in vanadium redox flow batteries through modified cell architecture. J Power Sources 206:450–453. doi:10.1016/j.jpowsour.2011.12.026
99. Darling RM, Perry ML (2014) The influence of electrode and channel configurations on flow battery performance. J Electrochem Soc 161:A1381–A1387. doi:10.1149/2.0941409jes
100. Duduta M, Ho B, Wood VC, Limthongkul P, Brunini VE, Carter WC, Chiang Y-M (2011) Semi-solid lithium rechargeable flow battery. Adv Energy Mater 1:511–516. doi:10.1002/aenm.201100152
101. Li Z, Smith KC, Dong Y, Baram N, Fan FY, Xie J, Limthongkul P, Carter WC, Chiang Y-M (2013) Aqueous semi-solid flow cell: demonstration and analysis. Phys Chem Chem Phys 15:15833–15839. doi:10.1039/C3CP53428F
102. Fan FY, Woodford WH, Li Z, Baram N, Smith KC, Helal A, McKinley GH, Carter WC, Chiang Y-M (2014) Polysulfide flow batteries enabled by percolating nanoscale conductor networks. Nano Lett 14:2210–2218. doi:10.1021/nl500740t
103. Bradley CS (1885) Secondary battery. US Patent 312,802, 24 Feb 1885
104. Barnartt S, Forejt DA (1964) Bromine-zinc secondary cells. J Electrochem Soc 111:1201–1204. doi:10.1149/1.2425960

105. Lim HS, Lackner AM, Knechtli RC (1977) Zinc-bromine secondary battery. J Electrochem Soc 124:1154–1157. doi:10.1149/1.2133517
106. Pavlov D, Papazov G, Gerganska M (1991) Battery energy storage systems. The United Nations Educational, Scientific and Cultural Organization, Regional Office for Science and Technology for Europe, Technical Report No. 7
107. http://www.zbbenergy.com/products/flow-battery/. Accessed 30 Sept 2014
108. Adams GB (1979) Electrically rechargeable battery. US Patent 4,180,623, 25 Dec 1979
109. Magnani NJ, Clark RP, Braithwaite JW, Bush DM, Butler PC, Freese JM, Grothaus KR, Murphy KD, Shoemaker PE(1985) Exploratory battery technology development and testing report for 1985. Sandia National Laboratories
110. Clarke R, Dougherty B, Harrison S, Millington P, Mohanta S (2004) Cerium batteries. US Patent Application Publication 2004/0202925 A1, 14 Oct 2004
111. Clarke R, Dougherty B, Harrison S, Millington J, Mohanta S (2006) Battery with bifunctional electrolyte. US Patent Application Publication 2006/0063065 A1, 23 Mar 2006
112. Hazza A, Pletcher D, Wills R (2004) A novel flow battery: a lead acid battery based on an electrolyte with soluble lead(II) Part I. Preliminary studies. Phys Chem Chem Phys 6:1773. doi:10.1039/b401115e
113. Pletcher D, Wills R (2004) A novel flow battery: a lead acid battery based on an electrolyte with soluble lead(II) Part II. Flow cell studies. Phys Chem Chem Phys 6:1779–1785
114. Pletcher D, Wills R (2005) A novel flow battery: a lead acid battery based on an electrolyte with soluble lead(II) Part III. The influence of conditions on battery performance. J Power Sources 149:96–102. doi:10.1016/j.jpowsour.2005.01.048
115. Hazza A, Pletcher D, Wills R (2005) A novel flow battery: a lead acid battery based on an electrolyte with soluble lead(II) Part IV. The influence of additives. J Power Sources 149:103–111. doi:10.1016/j.jpowsour.2005.01.049
116. Pletcher D, Zhou H, Kear G, Low CTJ, Walsh FC, Wills RGA (2008) A novel flow battery: a lead-acid battery based on an electrolyte with soluble lead(II) Part V. Studies of the lead negative electrode. J Power Sources 180:621–629. doi:10.1016/j.jpowsour.2008.02.024
117. Pletcher D, Zhou H, Kear G, Low CTJ, Walsh FC, Wills RGA (2008) A novel flow battery: a lead-acid battery based on an electrolyte with soluble lead(II) Part VI. Studies of the lead dioxide positive electrode. J Power Sources 180:630–634. doi:10.1016/j.jpowsour.2008.02.025
118. Li X, Pletcher D, Walsh FC (2009) A novel flow battery: a lead acid battery based on an electrolyte with soluble lead(II) Part VII. Further studies of the lead dioxide positive electrode. Electrochim Acta 54:4688–4695. doi:10.1016/j.electacta.2009.03.075
119. Collins J, Kear G, Li X, Low CTJ, Pletcher D, Tangirala R, Stratton-Campbell D, Walsh FC, Zhang C (2010) A novel flow battery: a lead acid battery based on an electrolyte with soluble lead(II) Part VIII. The cycling of a 10 cm × 10 cm flow cell. J Power Sources 195:1731–1738. doi:10.1016/j.jpowsour.2009.09.044
120. Collins J, Li X, Pletcher D, Tangirala R, Stratton-Campbell D, Walsh FC, Zhang C (2010) A novel flow battery: a lead acid battery based on an electrolyte with soluble lead(II) Part IX: Electrode and electrolyte conditioning with hydrogen peroxide. J Power Sources 195:2975–2978. doi:10.1016/j.jpowsour.2009.10.109
121. Verde MG, Carroll KJ, Wang Z, Sathrum A, Meng YS (2013) Achieving high efficiency and cyclability in inexpensive soluble lead flow batteries. Energy Environ Sci 6:1573. doi:10.1039/c3ee40631h
122. Xu W, Wang J, Ding F, Chen X, Nasybulin E, Zhang Y, Zhang J-G (2014) Lithium metal anodes for rechargeable batteries. Energy Environ Sci 7:513–537. doi:10.1039/C3EE40795K
123. Tarascon J-M, Armand M (2001) Issues and challenges facing rechargeable lithium batteries. Nature 414:359–367. doi:10.1038/35104644
124. Christensen J, Albertus P, Sanchez-Carrera RS, Lohmann T, Kozinsky B, Liedtke R, Ahmed J, Kojic A (2011) A critical review of Li/air batteries. J Electrochem Soc 159:R1–R30. doi:10.1149/2.086202jes
125. Lu Y, Goodenough JB, Kim Y (2011) Aqueous cathode for next-generation alkali-ion batteries. J Am Chem Soc 133:5756–5759. doi:10.1021/ja201118f

126. Lu Y, Goodenough JB (2011) Rechargeable alkali-ion cathode-flow battery. J Mater Chem 21:10113–10117. doi:10.1039/C0JM04222F
127. Wang Y, He P, Zhou H (2012) Li-redox flow batteries based on hybrid electrolytes: at the cross road between Li-ion and redox flow batteries. Adv Energy Mater 2:770–779. doi:10.1002/aenm.201200100
128. Zhao Y, Byon HR (2013) High-performance lithium-iodine flow battery. Adv Energy Mater 3:1630–1635. doi:10.1002/aenm.201300627
129. Bruce PG, Freunberger SA, Hardwick LJ, Tarascon J-M (2012) Li–O_2 and Li–S batteries with high energy storage. Nat Mater 11:19–29. doi:10.1038/nmat3191
130. Su Y-S, Fu Y, Cochell T, Manthiram A (2013) A strategic approach to recharging lithium-sulphur batteries for long cycle life. Nat Commun. doi:10.1038/ncomms3985
131. Manthiram A, Fu Y, Chung S-H, Zu C, Su Y-S (2014) Rechargeable lithium–sulfur batteries. Chem Rev doi:10.1021/cr500062v
132. Rauh RD, Abraham KM, Pearson GF, Surprenant JK, Brummer SB (1979) A lithium/dissolved sulfur battery with an organic electrolyte. J Electrochem Soc 126:523–527. doi:10.1149/1.2129079
133. Manthiram A, Fu Y, Su Y-S (2013) Challenges and prospects of lithium–sulfur batteries. Acc Chem Res 46:1125–1134. doi:10.1021/ar300179v
134. Mikhaylik YV (2008) Electrolytes for lithium sulfur cells. US Patent 7,354,680 B2, 8 April 2008
135. Aurbach D, Pollak E, Elazari R, Salitra G, Kelley CS, Affinito J (2009) On the surface chemical aspects of very high energy density, rechargeable Li–sulfur batteries. J Electrochem Soc 156:A694–A702. doi:10.1149/1.3148721
136. Yang Y, Zheng G, Cui Y (2013) A membrane-free lithium/polysulfide semi-liquid battery for large-scale energy storage. Energy Environ Sci 6:1552–1558. doi:10.1039/C3EE00072A
137. Darling RM, Gallagher KG, Kowalski JA, Ha S, Brushett FR (2014) Pathways to low-cost electrochemical energy storage: a comparison of aqueous and nonaqueous flow batteries. Energy Environ Sci doi:10.1039/C4EE02158D

MIX
Papier aus verantwortungsvollen Quellen
Paper from responsible sources
FSC® C105338

If you have any concerns about our products,
you can contact us on
ProductSafety@springernature.com

In case Publisher is established outside the EU,
the EU authorized representative is:
Springer Nature Customer Service Center GmbH
Europaplatz 3, 69115 Heidelberg, Germany

Printed by Libri Plureos GmbH
in Hamburg, Germany